ISBN 978-1-5283-2957-6
PIBN 10902231

Das Tierreich.

Eine Zusammenstellung und Kennzeichnung der rezenten Tierformen.

Begründet von der Deutschen Zoologischen Gesellschaft.

Im Auftrage der

Königl. Preuß. Akademie der Wissenschaften zu Berlin

herausgegeben von

Franz Eilhard Schulze.

21. Lieferung.

Grustacea.
Baron M. Glasebrook.

Amphipoda
I. Gammaridea

by

the Rev. T. R. R. Stebbing, M.A., F.R.S., F.L.S., F.Z.S.,

Fellow of King's College, London;
Formerly Fellow and Tutor of Worcester College, Oxford

With 127 figures.

Berlin.

Verlag von R. Friedländer und Sohn.

Ausgegeben im September 1906.

Das Tierreich.

Eine Zusammenstellung und Kennzeichnung der rezenten Tierformen.

Begründet von der Deutschen Zoologischen Gesellschaft.

Im Auftrage der

Königl. Preuß. Akademie der Wissenschaften zu Berlin

herausgegeben von

Franz Eilhard Schulze.

„Πάντα ῥεῖ.“   „Sine systemate chaos.“

21. Lieferung.

Crustacea.
Beirat: W. Giesbrecht.

Amphipoda
I. Gammaridea

by

the Rev. T. R. R. Stebbing, M. A., F. R. S., F. L. S., F. Z. S.;
Fellow of King's College, London;
formerly Fellow and Tutor of Worcester College, Oxford.

With 127 figures.

Berlin.
Verlag von R. Friedländer und Sohn.
Ausgegeben im September 1906.

Das Tierreich.

Im Auftrage der

Königl. Preuß. Akademie der Wissenschaften zu Berlin

herausgegeben von

Franz Eilhard Schulze.

21. Lieferung.

Crustacea. Beirat: W. Giesbrecht.

Amphipoda

I. Gammaridea

by

the Rev. **T. R. R. Stebbing**, M. A., F. R. S., F. L. S., F. Z. S.;

'Fellow of King's College, London;
formerly Fellow and Tutor of Worcester College, Oxford.

With 127 figures.

Berlin.

Verlag von R. Friedländer und Sohn.

Ausgegeben im September 1906.

11445

A. HOFFER, GUNG.

Contents.

Preface.

It was originally contemplated that this volume would be published within the limits of the nineteenth century. When the date of issue was unavoidably postponed, the internal structure of the work no longer admitted of any extensive remodelling. In the mean time, however, it so happened that our knowledge of the Amphipoda was enlarged by several important contributions. Accordingly the survey of the literature, which had been carried down with some completeness to 1898, was prematurely suffering from that imperfection by which all such surveys are eventually afflicted. To remedy this inconvenience so far as possible, the editor has kindly permitted the addition of many pages which bring the bibliography of the subject well into 1905, with an occasional incursion even into the present year.

In regard to the numerous species recorded as doubtful, it is desirable to explain that in many instances no censure is intended of the original descriptions. The doubtfulness often results only from the fact that recent classification has adopted features of distinction, the employment of which could not easily have been foreseen at an earlier period.

The nature of the present work includes acknowledgment of indebtedness to innumerable authors. To the editorial staff of „Das Tierreich" the thanks of readers should be united with those of contributors, if the most unremitting care in verifying citations is thought worthy of gratitude.

T. R. R. Stebbing,
Ephraim Lodge,
Tunbridge Wells, England.

July 7, 1906.

Abbreviations of the titles of literature.

Abh. Schles. Ges. — Abhandlungen der Schlesischen Gesellschaft für vaterländische Cultur. — Abtheilung für Naturwissenschaften und Medicin. Breslau. 8.

Acta Ac. Petrop. — Acta Academiae Scientiarum Imperialis Petropolitanae. Petropoli. 4.

Acta Univ. Lund. — Acta Universitatis Lundensis. Lunds Universitets Års-Skrift. — Afdelningen för Mathematik och Naturvetenskap ([Tom. 26 & sequ.:] Acta Regiae Societatis physiographicae Lundensis. Kongl. fysiografiska Sällskapets i Lund Handlingar). Lund. 4.

Act. Soc. Helvét. — Actes de la Société Helvétique des Sciences naturelles. Lausanne (Genève, ...). 8.

Agassiz, Nomencl. zool. — Nomenclator zoologicus, continens Nomina systematica Generum Animalium tam viventium quam fossilium. Auctore L. Agassiz. 1 Vol. & Index universalis. Soloduri. 1842—46, 46. 4.

Amer. J. Sci. — The American Journal of Science and Arts. New Haven (New York). 8.

Amer. monthly Mag. — The American monthly Magazine and critical Review. New York. 8.

Amer. Natural. — The American Naturalist. Salem (Philadelphia) (Boston). 8.

Ann. nat. Hist. — The Annals and Magazine of natural History, including Zoology, Botany, and Geology. London. 8.

Ann. N. York Ac. — Annals of the New York Academy of Sciences, late Lyceum of natural History. New York. 8.

Ann. Sci. nat. — Annales des Sciences naturelles. — [Sér. 2—4:] Zoologie ([Sér. 5 & sequ.:] Zoologie et Paléontologie). Paris. 8.

Ann. Soc. ent. Belgique — Annales de la Société entomologique de Belgique. Bruxelles. 8.

Annuaire Mus. St.-Pétersb. — Annuaire du Musée zoologique de l'Académie Impériale des Sciences de St.-Pétersbourg. St.-Pétersbourg. 8.

Annuario Mus. Napoli — Annuario del Museo zoologico della R. Università di Napoli. Napoli. 8.

Ann. Univ. Lyon — Annales de l'Université de Lyon. Paris. 8.

Arb. Inst. Wien — Arbeiten aus dem zoologischen Institute der Universität Wien und der zoologischen Station in Triest. Wien. 8.

Arch. Mus. Lyon — Archives du Muséum d'Histoire naturelle de Lyon. Lyon. 4.

Arch. Mus. Rio Jan. — Archivos do Museu nacional do Rio de Janeiro. Rio de Janeiro. 4.

Arch. Naturg. — Archiv für Naturgeschichte. Berlin. 8.

Arch. Naturv. Kristian. — Archiv for Mathematik og Naturvidenskab. Kristiania. 8.

Atti Acc. Borbon. — Atti della Reale Accademia delle Scienze, Sezione della Società Reale Borbonica. Napoli. 4.

Atti Acc. Napoli — Società Reale di Napoli. Atti della Reale Accademia delle Scienze fisiche e matematiche. Napoli. 4.

Atti Soc. Ital. — Atti della Società Italiana di Scienze naturali. Milano. 8.

Barrois, Cat. Crust. Açores — Catalogue des Crustacés marins recueillis aux Açores durant les Mois d'Août et Septembre 1887. Par Th. Barrois. Lille. 1888. 8.

Barrois, Note Orchesties — Note sur quelques Points de la Morphologie des Orchesties suivie d'une Liste succincte des Amphipodes du Boulonnais. Par Th. Barrois. Lille. 1887. 8.

Bate, Cat. Amphip. Brit. Mus. — Catalogue of the Specimens of Amphipodous Crustacea in the Collection of the British Museum. By C. Spence B a t e. London. 1862 [Preface dat. 1862 XII]. 8.

Bate & Westwood, Brit. sess. Crust. — A History of the British sessile-eyed Crustacea. By C. Spence B a t e and J. O. W e s t w o o d. Vol. 1, 2. London. 1863 (1861—63, 68), 68 (1863, 66—68)*). 8.

Belcher, Last Arct. Voy. — The last of the Arctic Voyages; being a Narrative of the Expedition in H. M. S. Assistance, under the Command of Captain Sir Edward B e l c h e r, in Search of Sir John F r a n k l i n, during the Years 1852—54. With Notes on the natural History, by Sir John R i c h a r d s o n, [Richard] O w e n, Thomas B e l l, J. W. S a l t e r, and Lovell R e e v e. Vol. 1, 2. London. 1855. 8.

Berlin. ent. Z. — Berliner entomologische Zeitschrift. Berlin. 8.

Bih. Svenska Ak. — Bibang till Kongl. Svenska Vetenskaps-Akademiens Handlingar. — Afdelning 4, Zoologi. Stockholm. 8.

Bijdr. Dierk. — Bijdragen tot de Dierkunde. Uitgegeven door het |Koninklijk zoologisch Genootschap Natura Artis Magistra,, te Amsterdam|. Amsterdam. 4.

Biol. Centralbl. — Biologisches Centralblatt. Erlangen (Leipzig). 8.

Boeck, Skand. Arkt. Amphip. — De Skandinaviske og Arktiske Amphipoder, beskrevne af Axel B o e c k. Hefte 1, 2. Christiania. 1872, 76. 4.

Bosc, Crust. — Histoire naturelle des Crustacés, contenant leur Description et leurs Moeurs. Par L. A. G. B o s c. Tom. 1, 2. Paris. X [1802]. 6.

Boston J. nat. Hist. — Boston Journal of natural History. Boston. 8.

Bronn's Kl. Ordn. — Die Klassen und Ordnungen des Thier-Reichs Wissenschaftlich dargestellt in Wort und Bild. Von H. G. B r o n n, fortgesetzt von A. G e r - s t a e c k e r. — 5. Band. Die Klassen und Ordnungen der Arthropoden wissenschaftlich dargestellt in Wort und Bild. Von A. G e r s t a e c k e r. Abtheilung I, II. Crustacea. Leipzig und Heidelberg. 1866—79, 81—95. 8.

Bull. Ac. St.-Pétersb. — Bulletin de l'Académie Impériale des Sciences de St.-Pétersbourg. St.-Pétersbourg. 4 (8).

Bull. Illinois Mus. — Bulletin of the Illinois State Museum of natural History. Springfield. 4.

Bull. Mus. Harvard — Bulletin of the Museum of comparative Zoology at Harvard College, in Cambridge. Cambridge, Mass., U. S. A. 8.

Bull. Mus. Monaco — Bulletin du Musée océanographique de Monaco. Monaco. 8.

Bull. Mus. Paris — Bulletin du Muséum d'Histoire naturelle. Paris. 8.

Bull. N. York Mus. — Bulletin of the New York State Museum. Albany. 8.

Bull. phys.-math. Ac. St.-Pétersb. — Bulletin de la Classe physico-mathématique de l'Académie Impériale des Sciences de St.-Pétersbourg. St.-Pétersbourg. 4.

Bull. sci. France Belgique — Bulletin scientifique de la France et de la Belgique. Paris. 8.

Bull. sci. Nord — Bulletin scientifique, historique et littéraire du Département du Nord et des Pays Voisins ([Tom. 19:] Bulletin scientifique du Nórd de la France et de la Belgique). Lille (Paris). 8.

Bull. Soc. Borda — Bulletin de la Société de Borda. Dax. 8.

Bull. Soc. ent. France — Bulletin des Séances et Bulletin bibliographique de la Société entomologique de France. Paris. 8.

Bull. Soc. ent. Ital. — Bullettino della Società entomologica Italiana. Firenze. 8.

Bull. Soc. Étud. Paris — Bulletin de la Société d'Études scientifiques de Paris. Paris. 8.

Bull. Soc. Moscou — Bulletin de la Société Impériale des Naturalistes de Moscou. Moscou. 8.

Bull. Soc. Nancy — Bulletin de la Société des Sciences de Nancy. Nancy et Paris (Paris). 8.

Bull. Soc. Neuchatel — Bulletin de la Société des Sciences naturelles de Neuchatel. Neuchatel. 8.

*) Cfr.: Thomas R. R. S t e b b i n g in: Rep. Voy. Challenger, v. 29 p. 328, 340, 343, 372.

Bull. Soc. philom. — Bulletin de la Société philomathique de Paris. Paris. 4.

Bull. Soc. Rouen — Bulletin de la Société des Amis des Sciences naturelles de Rouen. Rouen. 8.

Bull. Soc. Vaudoise — Bulletin de la Société Vaudoise des Sciences naturelles. Lausanne. 8.

Bull. Soc. zool. France — Bulletin de la Société zoologique de France. Paris. 8.

Bull. U. S. Bureau Fish. — Department of Commerce and Labor. Bulletin of the Bureau of Fisheries. Washington. 8.

Bull. U. S. Mus. — Bulletin of the United States national Museum. Washington. 8.

Burmeister, Handb. Naturg. — Handbuch der Naturgeschichte. Zum Gebrauch bei Vorlesungen entworfen von Hermann Burmeister. Berlin. 1837. 8.

Carus, Prodr. F. Medit. — Prodromus Faunae Mediterraneae sive Descriptio Animalium Maris Mediterranei Incolarum quam comparata Silva Rerum quatenus innotuit adiectis Locis et Nominibus vulgaribus eorumque Auctoribus in Commodum Zoologorum congessit Julius Victor Carus. Vol. 1, 2. Stuttgart. 1885 (1884, 85), 89—93 (1889, 90, 93). 8.

Costa, Descr. 3 Crost. dal Hope — Achille Costa, Descrizione di tre nuovi Crostacei del Mediterraneo discoperti dal Rev. G. F. Hope. Estratta dal Fascicolo 83⁰ della Fauna del Regno di Napoli. [Napoli.] [1853.] 4.

Costa, Fauna Reg. Napoli — Fauna del Regno di Napoli ossia Enumerazione di tutti gli Animali che abitano le diverse Regioni di questo Regno e le Acque che le bagnano ... di Oronzio-Gabriele Costa. [Continuato da Achille Costa.] — Crostacei ed Aracnidi. Napoli. 1836 (1836—51). 4.

C.-R. Ac. Sci. — Comptes rendus hebdomadaires des Séances de l'Académie des Sciences. Paris. 4.

C.-R. Ass. Franç. — Association Française pour l'Avancement des Sciences. Compte rendu de la ... Session. Paris. 8.

Cuvier, Règne an., ed. 3 — Le Règne animal distribué d'après son Organisation, pour servir de Base à l'Histoire naturelle des Animaux, et d'Introduction à l'Anatomie comparée, par Georges Cuvier. [3.] Édition accompagnée de Planches gravées, ... par une Réunion de Disciples de Cuvier. — Les Crustacés. Avec un Atlas, par [Henri] Milne Edwards. Paris. [1836—49.] 4.

Cuvier, Règne an., n. ed. — Le Règne animal distribué d'après son Organisation, pour servir de Base à l'Histoire naturelle des Animaux et d'Introduction à l' Anatomie comparée. Par [Georges] Cuvier. Nouvelle Édition, revue et augmentée. [Tom. 4 & 5:] Par [Pierre André] Latreille. Tom. 1—5. Paris. 1829, 29, 30, 29, 29. 8.

Danske Selsk. Afh. — Det Kongelige Danske Videnskabernes Selskabs naturviden- skabelige og mathematiske Afhandlinger. Kjöbenhavn. 4.

Danske Selsk. Skr. — Det Kongelige Danske Videnskabernes Selskabs Skrifter. — [Raekke 5 & sequ.:] Naturvidenskabelig og mathematisk Afdeling. Kjöbenhavn. 4.

De Kay, Zool. N.-York — Zoology of New-York, or the New-York Fauna; comprising detailed Descriptions of all the Animals hitherto observed within the State of New-York, with brief Notices of those occasionally found near its Borders, and accompanied by appropriate Illustrations. By James E. De Kay. — Part 6. Crustacea. Albany. 1844. 4.

Denk. Ak. Wien — Denkschriften der Kaiserlichen Akademie der Wissenschaften. — Mathematisch-naturwissenschaftliche Classe. Wien. 4.

Descr. Égypte — Description de l'Égypte, ou Recueil des Observations et des Recherches qui ont été faites en Égypte pendant l'Expédition de l'Armée Française, publié par les Ordres de sa Majesté l'Empereur Napoléon le Grand. — Histoire naturelle. Tom. 1ı—1ıv, 2; Planches. Paris. 1809 [1809—27], 13 [1818, 29]*). 4 & 2.

Desmarest, Consid. gén. Crust. — Considérations générales sur la Classe des Crustacés, et Description des Espèces de ces Animaux, qui vivent dans la Mer, sur les Côtes, ou dans les Eaux douces de la France. Par Anselme-Gaetan Desmarest. Paris, Strasbourg. 1825. 8.

---

*) Cfr.: C. Davies Sherborn in: P. zool. Soc. London, 1897 p. 285.

Dict. Sci. nat. — Dictionnaire des Sciences naturelles, ... Par plusieurs Professeurs du Jardin du Roi, et des principales Écoles de Paris. [Red. par F. Cuvier.] Tom. 1—60; Planches. Strasbourg et Paris (Paris). 1816—30. 8.

Dijmphna Udb. — Dijmphna-Togtets zoologisk-botaniske Udbytte. Udgivet . . . . af Kjøbenhavns Universitets zoologiske Museum ved Chr. Fr. Lütken. Kjøbenhavn. 1887. 8.

Edinb. Enc. — The Edinburgh Encyclopaedia; conducted by David Brewster. With the Assistance of Gentlemen eminent in Science and Literature. Vol. 1—18. Edinburgh. 1830 [1809—31]. 4.

Enc. Brit., ed. 5 — The Encyclopaedia Britannica, a Dictionary of Arts, Sciences, and general Literature. Fifth Edition. Vol. 1—20; Suppl. 1—6. Edinburgh. 1814—17; 16—24. 4.

Enc. méth. — Encyclopédie méthodique, ou par Ordre de Matières; par une Société de Gens de Lettres, de Savans et d'Artistes. — Histoire naturelle. Tom. 4—10: Insectes (Entomologie). Paris, Liège (Paris). 1789, 90, 91, 92, 1811 [& 1812vii], 19 [& 1824vii], 25 [& 1828].*) 4.

Expl. Algérie — Exploration scientifique de l'Algérie pendant les Années 1840, 41, 42. Publiée par Ordre du Gouvernement et avec le Concours d'une Commission académique. — Sciences physiques. Zoologie. I—IV. Histoire naturelle des Animaux articulés par H. Lucas. Partie 1—3; Atl. Paris. 1849 [1845—49]. 4.

Exp. Morée — Expédition scientifique de Morée. Section des Sciences physiques. — Tom. 3. Partie I. Zoologie. Cum Atl. Paris. 1836 (1832, 33); 31—35. 4 & 2.

Expl. Tunisie — Exploration scientifique de la Tunisie, publiée sous les Auspices du Ministère de l'Instruction publique. — Zoologie. Étude sur les Crustacés terrestres et fluviatiles recueillis en Tunisie en 1883, 1884 et 1885 par A. Letourneux, M. Sédillot et Valery Mayet. Par Eugène Simon. Paris. 1886. 8.

Fabricius, Ent. syst. — Joh. Christ. Fabricii Entomologia systematica emendata et aucta. Secundum Classes, Ordines, Genera, Species adjectis Synonimis, Locis, Observationibus, Descriptionibus. Tom. 1—4. [Cum Ind.:] Index alphabeticus in J. C. Fabricii Entomologiam systematicam, emendatam et auctam, Ordines, Genera et Species continens. [Cum Suppl.:] Joh. Christ. Fabricii Supplementum Entomologiae systematicae. [Cum Ind. Suppl.:] Index alphabeticus in J. C. Fabricii Supplementum Entomologiae systematicae, Ordines, Genera et Species continens. Hafniae. 1792, 93, 93/94, 94; 96; 98; 99. 8.

Fabricius, Fauna Groenl. — Fauna Groenlandica, systematice sistens Animalia Groenlandiae occidentalis hactenus indagata, . . . secundum proprias Observationes Othonis Fabricii. Hafniae et Lipsiae. 1780. 8.

Fabricius, Gen. Ins. — Ioh. Christ. Fabricii Genera Insectorum eorumque Characteres naturales secundum Numerum, Figuram, Situm et Proportionem omnium Partium Oris adiecta Mantissa Specierum nuper detectarum. Chilonii. [1777.] 8.

Fabricius, Reise Norweg. — Johann Christian Fabricius Reise nach NorWegen mit Bemerkungen aus der Naturhistorie und Oekonomie. Hamburg. 1779. 8.

Fabricius, Spec. Ins. — Ioh. Christ. Fabricii Species Insectorum exhibentes eorum Differentias specificas, Synonyma Auctorum, Loca natalia, Metamorphosin adiectis Observationibus, Descriptionibus. Tom. 1, 2. Hamburgi et Kilonii. 1781. 8.

Fabricius, Syst. Ent. — Io. Christ. Fabricii Systema Entomologiae, sistens Insectorum Classes, Ordines, Genera, Species, adiectis Synonymis, Locis, Descriptionibus, Observationibus. Flensburgi et Lipsiae. 1775. 8.

Fauna Haw. — Fauna Hawaiiensis, or the Zoology of the Sandwich (Hawaiian) Isles. Being Results of the Explorations instituted by the joint Committee appointed by the Royal Society of London for promoting natural Science and the British Association for the Advancement of Science and carried on with the Assistance of those Bodies and the Trustees of the Bernice Pauahi Bishop Museum at Honolulu. Edited by D. Sharp. Vol. 1—3. Cambridge. 1899—1904. 4.

---

*) Cfr.: C. Davies Sherborn & B. B. Woodward in: P. zool. Soc. London, 1893 p. 583 & 1899 p. 595.

Fedtschenko, Turkestan — Путешествіе въ Туркестанъ члена-основателя Общества
    А. П. Федченко, совершенное отъ Императорскаго Общества Любителей
    Естествознанія по Порученію Туркестанскаго Генералъ-Губернатора К. П.
    Фонъ-Кауфмана. [Die Reise Fedtschenko's in Turkestan, auf Ver-
    anlassung des General-Gouverneurs K. P. v. Kauffmann herausgegeben
    von der Kaiserlichen Gesellschaft der Freunde der Naturwissenschaften zu
    Moskau.] — Том. П. Зоогеографическія Изслѣдованія. [Zoogeographische
    Untersuchungen.] Часть III. Ракообразныя, Crustacea, обработалъ. В. Н.
    Ульянинъ. [W. N. Uljanin.] Тетрадь 1. С.-Петербургъ, Москва. 1875.
    4. [& in: *Izv. Obshch. Moskov.*, *v.* 11 nr. 6.]
Feuille Natural. — Feuille des jeunes Naturalistes. Paris. 8.
F. Fl. Neapel — Fauna und Flora des Golfes von Neapel und der angrenzenden
    Meeres-Abschnitte herausgegeben von der Zoologischen Station zu Neapel.
    — 20. Monographie. Gammarini del Golfo di Napoli. Monografia di Antonio
    Della Valle. Con Atl. Berlin. 1893. 4.
Forbes, Nat. Hist. Sokotra — The natural History of Sokotra and Abd-el-Kuri. Being
    the Report upon the Results of the Conjoint Expedition to these Islands
    in 1898—9, by W. R. Ogilvie-Grant, of the British Museum, and H. O.
    Forbes, of the Liverpool Museums, together with Information from other
    available Sources. Forming a Monograph of the Islands. Edited by Henry
    O. Forbes. Liverpool, London. 1903. 8.
Forh. Selsk. Christian. — Forhandlinger i Videnskabs-Selskabet i Christiania. Chri-
    stiania. 8.
Forh. Skand. Naturf. — Forhandlinger ved de Skandinaviske Naturforskeres. Göthe-
    borg (...). 8.
Frey & Leuckart, Wirbell. Th. — Beiträge zur Kenntnis wirbelloser Thiere mit be-
    sonderer Berücksichtigung der Fauna des Norddeutschen Meeres. Von Heinrich
    Frey und Rudolf Leuckart. Braunschweig. 1847. 4.
Gardiner, Fauna Mald. Laccad. — The Fauna and Geography of the Maldive and
    Laccadive Archipelagoes. Being the Account of the Work carried on and of the
    Collections made by an Expedition during the Years 1899 and 1900. Edited
    by J. Stanley Gardiner. Vol. 1, 2. Cambridge. 1903 (1901, 02, 03), 03—05. 4.
Gay, Hist. Chile — Historia fisica y politica de Chile segun Documentos adquiridos ...
    y publicada bajo los Auspicios del supremo Gobierno por Claudio Gay. —
    Zoologia. Tom. 1—8; Atl. Paris. 1847—54. 8 & 2.
Geer, Mém. Hist. Ins. — Mémoires pour servir à l'Histoire des Insectes. Par Charles
    de Geer. Tom. 1—7. Stockholm. 1752, 71, 73, 74, 75, 76, 78. 4.
Gervais & Beneden, Zool. méd. — Zoologie médicale. Exposé méthodique du Règne
    animal basé sur l'Anatomie, l'Embryogénie et la Paléontologie comprenant
    la Description des Espèces employées en Médecine et de celles qui sont Veni-
    meuses et de celles qui sont Parasites de l'Homme et des Animaux par Paul
    Gervais [et] P.-J. van Beneden. Tom. 1, 2. Paris, Londres, New-York.
    1859. 8.
Gmelin, Gen. Syst. Nat. — A general System of Nature, through the three grand
    Kingdoms of Animals, Vegetables and Minerals. Translated from Gmelin's
    last Edition of the celebrated Systema Naturae by Sir Charles Linné.
    Amended and enlarged by the Improvements and Discoveries of later
    Naturalists and Societies, with appropriate Copper-Plates, by William
    Turton. — Vol. 1—7. London. 1806. 8.
Gmelin, Syst. Nat. — Caroli a Linné Systema Naturae per Regna tria Naturae, secundum
    Classes, Ordines, Genera, Species, cum Characteribus, Differentiis. Synonymis,
    Locis. Editio XIII, aucta, reformata. Cura Jo. Frid. Gmelin. — Tomus I.
    Pars 1—7. Lipsiae. 1788 [Pars 6 & 7: 1791]. 8.
Gosse, Man. mar. Zool. — A Manual of marine Zoology for the British Isles. By
    Philip Henry Gosse. Part 1, 2. London. 1855, 56. 12.
Gosse, Rambles Devonsh. — A Naturalist's Rambles on the Devonshire Coast. By
    Philip Henry Gosse. London. 1853. 8.

Grube, Ausfl. Triest — Ein Ausflug nach Triest und dem Quarnero. Beiträge zur Kenntniss der Thierwelt dieses Gebietes von Adolph Eduard Grube. Berlin. 1861. 8.

Grube, Lussin — Die Insel Lussin und ihre Meeresfauna. Nach einem sechswöchentlichen Aufenthalte geschildert von Adolf Eduard Grube. Breslau. 1864. 8.

Guérin-Méneville, Iconogr. Règne an. — Iconographie du Règne animal de G. Cuvier, ou Représentation d'après Nature de l'une des Espèces les plus remarquables et souvent non encore figurées, de chaque Genre d'Animaux. Avec un Texte descriptif mis au Courant de la Science. Par F. E. Guérin-Méneville. Tom. 1—3. Paris, Londres. 1829—44. 4 (& 8).

Hansen, Choniostom. — The Choniostomatidae. A Family of Copepoda, Parasites on Crustacea Malacostraca. By H. J. Hansen. Copenhagen. 1897. 4.

Harriman Alaska Exp. — Harriman Alaska Expedition. With Cooperation of Washington Academy of Sciences. Alaska. — Vol. 10. Crustaceans. By Mary J. Rathbun, Harriet Richardson, S. J. Holmes, and Leon J. Cole. New York. 1904. 8.

Haswell, Cat. Austral. Crust. — The Australian Museum Sydney. Catalogue of the Australian stalk- and sessile-eyed Crustacea. By William A. Haswell. Sydney. 1882. 8.

Herbst, Naturg. Krabben Krebse — Versuch einer Naturgeschichte der Krabben und Krebse nebst einer systematischen Beschreibung ihrer Verschiedenen Arten von Johann Friedrich Wilhelm Herbst. Band 1—3. Berlin und Stralsund. 1790 (1782—90), 96 (1791—96), 99—1804. 4 & 2.

Herdman, Rep. Ceylon Pearl Fish. — Report to the Government of Ceylon on the Pearl Oyster Fisheries of the Gulf of Manaar, by W. A. Herdman. With supplementary Reports upon the marine Biology of Ceylon, by other Naturalists. Published .... by the Royal Society. Part 1, 2. London. 1903, 04. 4.

Herrmannsen, Ind. Gen. Malac. — Indicis Generum Malacozoorum Primordia. Nomina Subgenerum, Generum, Familiarum, Tribuum, Ordinum. Classium; adjectis Auctoribus, Temporibus, Locis systematicis atque literariis, Etymis, Synonymis. Praetermittuntur Cirripedia, Tunicata et Rhizopoda. Conscripsit A. N. Herrmannsen. Vol. 1, 2. Cassellis. 1846. 47—49. 8.

Hist. An. artic. — Histoire naturelle des Animaux articulés, Annelides, Crustacés, Arachnides, Myriapodes et Insectes. — Histoire naturelle des Crustacés, des Arachnides et des Myriapodes, par [Hippolyte] Lucas. Paris. 1840. 8.

Hope, Cat. Crost. Ital. — Catalogo dei Crostacei Italiani e di molti altri del Mediterraneo per Fr. Gugl. Hope. Napoli. 1851. 8.

Horae Soc. ent. Ross. — Horae Societatis entomologicae Rossicae. Petropoli. 8.

Huxley, Man. Anat. Invert. — A Manual of the Anatomy of invertebrated Animals by Thomas H. Huxley. London. 1877. 8.

J. Ac. Philad. — Journal of the Academy of natural Sciences of Philadelphia. Philadelphia. 8(4).

Jahrb. Hamburg. Anst. — Jahrbuch der Hamburgischen wissenschaftlichen Anstalten. Hamburg. 8.

Jahresber. Comm. D. Meere — Jahresbericht der Commission zur Wissenschaftlichen Untersuchung der Deutschen Meere in Kiel. Berlin. 2.

Jahresber. Ges. Hannover — Jahresbericht der naturhistorischen Gesellschaft zu Hannover. Hannover. 8.

Jahresber. Schles. Ges. — Jahres-Bericht der Schlesischen Gesellschaft für vaterländische Cultur. Breslau. 4, 8.

J. Asiat. Soc. Bengal — Journal of the Asiatic Society of Bengal. — [Vol. 34 & sequ.:] Part II, Natural History etc. Calcutta. 8.

J. Linn. Soc. — The Journal of the Linnean Society. — Zoology. London. 8.

Jurinac, Fauna Kroat. Karst. — Ein Beitrag zur Kenntnis der Fauna des Kroatischen Karstes und seiner unterirdischen Höhlen. Von Adolf E. Jurinac. Jena. 1888. 8.

Koch, C. M. A. — C. L. Koch, Deutschlands Crustaceen, Myriapoden und Arachniden. Ein Beitrag zur Deutschen Fauna. Herausgegeben von [Gottlieb August Wilhelm] Herrich-Schäffer. Heft 1—40. Regensburg. 1835—44. 16.

Kossmann, Reise Roth. Meer. — Zoologische Ergebnisse einer im Auftrage der Königlichen Academie der Wissenschaften zu Berlin ausgeführten Reise in die Küstengebiete des Rothen Meeres. Herausgegeben mit Unterstützung ... von Robby Kossmann. 1. Hälfte. 2. Hälfte, I. Lieferung. Leipzig. 1877, 80. 4.

Lamarck, Hist. An. s. Vert. — Histoire naturelle des Animaux sans Vertèbres, présentant les Caractères généraux et particuliers de ces Animaux, ...; précédée d'une Introduction ... Par [Jean Baptiste] de Lamarck. Tom. 1—5, 61 & II, 7. Paris. 1815 III, 16 III, 16 VIII, 17 III, 18 VII, 19 II—VI, 22 IV, 22 VIII. 8.

Lamarck, Hist. An. s. Vert., ed. 2 — Histoire naturelle des Animaux sans Vertèbres, présentant les Caractères généraux et particuliers de ces Animaux, ...; précédée d'une Introduction .... Par J. B. P. A. de Lamarck. 2. Édition. Revue et augmentée de Notes présentant les Faits nouveaux dont la Science s'est enrichie jusqu'a ce Jour; par G. P. Deshayes et H. Milne Edwards. Tom. 1—11. Paris. 1835—45. 8.

Latreille, Gen. Crust. Ins. — P. A. Latreille Genera Crustaceorum et Insectorum secundum Ordinem naturalem in Familias disposita, Iconibus Exemplisque plurimis explicata. Tom. 1—4. Parisiis et Argentorati. 1806, 07, 07, 09. 8.

Latreille, Hist. Crust. Ins. — Histoire naturelle, générale et particulière, des Crustacés et des Insectes. Ouvrage faisant Suite aux Oeuvres de Leclerc de Buffon, et Partie du Cours complet d'Histoire naturelle rédigé par C. S. Sonnini. Par P. A. Latreille. Tom. 1—14. Paris. X—XIII [1802—1805]. 8.

Leach, Zool. Misc. — The zoological Miscellany; being Descriptions of new, or interesting Animals, by William Elford Leach. Illustrated with coloured Figures, drawn from Nature, by R. P. Nodder. Vol. 1—3. London. 1814, 15, 17. 8.

Linné, Fauna Svec., ed. 2 — Caroli Linnaei Fauna Svecica sistens Animalia Sveciae Regni: Mammalia, Aves, Amphibia, Pisces, Insecta, Vermes. Distributa per Classes & Ordines, Genera & Species, cum Differentiis Specierum, Synonymis Auctorum, Nominibus Incolarum, Locis Natalium, Descriptionibus Insectorum. Editio altera, auctior. Stockholmiae. 1761. 8.

Linné, Syst. Nat., ed. 10 — Caroli Linnaei Systema Naturae per Regna tria Naturae, secundum Classes, Ordines, Genera, Species, cum Characteribus, Differentiis, Synonymis, Locis. Editio X, reformata. — Tomus I. Holmiae. 1758. 8.

Linné, Syst. Nat., ed. 12 — Caroli a Linné Systema Naturae per Regna tria Naturae, secundum Classes, Ordines. Genera, Species, cum Characteribus, Differentiis, Synonymis, Locis. Editio XII, reformata. — Tomus I. Pars 1, 2. Holmiae. 1766, 67. 8.

List Brit. An. Brit. Mus. — List of the Specimens of British Animals in the Collection of the British Museum. — Part 4. Crustacea. [By Adam White.] London. 1850. 12.

Mag. nat. Hist. — The Magazine of natural History, and Journal of Zoology, Botany Mineralogy, Geology, and Meteorology. London. 8.

Mag. Zool. — Magasin de Zoologie |[Sér. 2:], d'Anatomie comparée et de Paléontologie|. Paris. 8.

Mandt, Observ. Groenl. — Observationes in Historiam naturalem et Anatomiam comparatam in Itinere Groenlandico factae. Dissertatio inauguralis, quam ... Die XXII. M. Julii A. 1822 publice defendet Auctor Martinus Guilelmus Mandt. [Berolini.] 1822. 8.

Marschall, Nomencl. zool. — Nomenclator zoologicus continens Nomina systematica Generum Animalium tam viventium quam fossilium, secundum Ordinem alphabeticum disposita. Conscriptus a Comite Augusto de Marschall. Vindobonae. 1873. 8.

Mem. Acc. Napoli — Memorie della Reale Accademia delle Scienze dal 1852 in avanti. Napoli. 4.

Mem. Acc. Verona — Memorie dell' Accademia d'Agricoltura, Commercio ed Arti di Verona. Verona. 8.

Mem. Ac. Washington — Memoirs of the national Academy of Sciences. Washington. 4.

Mem. Boston Soc. — Memoirs read before the Boston Society of natural History, being a new Series of the Boston Journal of natural History. Boston. 4.

Mém. cour. Ac. Belgique — Mémoires couronnés et Mémoires des Savants étrangers, publiés par l'Académie Royale des Sciences, des Lettres et des Beaux-Arts de Belgique. Bruxelles. 4.

Mem. Ist. Veneto — Memorie del Reale Istituto Veneto di Scienze, Lettere ed Arti. Venezia. 4.

Mém. prés. Ac. St.-Pétersb. — Mémoires présentés à l'Académie Impériale des Sciences de Saint-Pétersbourg, par divers Savants, et lus dans ses Assemblées. St.-Pétersbourg. 4.

Mém. Soc. Genève — Mémoires de la Société de Physique et d'Histoire naturelle de Genève. Genève. 4.

Mém. Soc. Kiew — Mémoires de la Société des Naturalistes de Kiew. Записки Кіевскаго Общества Естествонспытателей. Kiew. Кіевъ. 8.

Mém. Soc. zool. France — Mémoires de la Société zoologique de France. Paris. 8.

Middendorff, Reise Sibirien — Reise in den äussersten Norden und Osten Sibiriens während der Jahre 1843 und 1844 mit allerhöchster Genehmigung auf Veranstaltung der Kaiserlichen Akademie der Wissenschaften zu St. Petersburg ausgeführt und in Verbindung mit vielen Gelehrten herausgegeben von A. Th. v. Middendorff. — Band 2. Zoologie. Theil I, II. St. Petersburg. 1851, 53. 4.

Miers, Cat. Crust. N. Zealand — Colonial Museum and geological Survey Department .... Catalogue of the stalk- and sessile-eyed Crustacea of New Zealand. By Edward J. Miers. London. 1876. 8.

Milne Edwards, Hist. nat. Crust. — Histoire naturelle des Crustacés, comprenant l'Anatomie, la Physiologie et la Classification de ces Animaux; par [Henri] Milne Edwards. Tom. 1—3; Planches. Paris. 1834, 37, 40. 8.

Müller, Für Darwin — Für Darwin von Fritz Müller. Leipzig. 1864. 8.

Müller, Zool. Dan., ed. 3 — Zoologia Danica seu Animalium Daniae et Norvegiae rariorum ac minus notorum Descriptiones et Historia. Auctore Othone Friderico Müller. Ad Formam Tabularum denuo edidit Frater Auctoris ([Vol. 3:] Descripsit et Tabulas addidit Petrus Christianus Abildgaard) ([Vol. 4:] Descripserunt et Tabulas dederunt P. C. Abildgaard, M. Vahl, J. S. Holten, J. Rathke). [Editio 3.] Vol. 1—4. Havniae. 1788, 88, 89, 1806. 2.

Nachr. Ges. Götting. — Nachrichten von der Königl. Gesellschaft der Wissenschaften und der Georg-Augusts-Universität zu Göttingen. Göttingen. 8.

N. Acta Ac. Leop. — Nova Acta Academiae Caesareae Leopoldino-Carolinae Germanicae Naturae Curiosorum. Verhandlungen der Kaiserlichen Leopoldinisch-Carolinischen Akademie der Naturforscher. Norimbergae (Erlangen etc.). 4.

N. Acta Soc. Upsal. — Nova Acta Regiae Societatis Scientiarum Upsaliensis. Upsaliae. 4.

Nansen, Norweg. North Polar Exp. — The Norwegian North Polar Expedition 1893—1896. Scientific Results. Edited by Fridtjof Nansen. Vol. 1—6. London, Christiania, New York, Bombay, Leipzig. 1900—05. 4.

Nardo, Prosp. Fauna Venet. — Prospetto della Fauna marina volgare del Veneto Estuario con Cenni sulle principali Specie commestibili dell' Adriatico, ecc. del Gio. Domenico Nardo. Estratto dall' Opera: Venezia e le sue Lagune. Venezia. 1847. 4.

Nardo, Sinon. Spec. Chiereghini — Sinonimia moderna delle Specie registrate nell' Opera intitolato: Descrizione de' Crostacei, de' Testacei e de' Pesci che abitano le Lagune e Golfo Veneto, rappresentati in Figure a Chiaroscuro ed a Colori dall' Abate Stefano Chiereghini. Applicata per Commissione governativa dal Gio. Domenico Nardo. Venezia. 1847. 8.

Nares, Voy. Polar Sea — Narrative of a Voyage to the Polar Sea during 1875—6 in H. M. Ships 'Alert' and 'Discovery' by G. S. Nares. With Notes on the natural History, edited by H. W. Feilden. Vol. 1, 2. London. 1878. 8.

Nat. Hist. Rev. — The natural History Review. A quarterly Journal. London. 8.

Nat. Hist. Tr. Northumb. — Natural History Transactions of Northumberland and Durham ¦and Newcastle-upon-Tyne|, being Papers read at the Meetings of the natural History Society of Northumberland, Durham and Newcastle-upon-Tyne, and the Tyneside Naturalists' Field Club. London (Newcastle). 8.

Nat. Sci. — Natural Science. A monthly Review of scientific Progress. London & New York (London). 8.

Naturaliste — Le Naturaliste. Paris. 4.

Naturh. Tidsskr. — Naturhistorisk Tidsskrift. Kjøbenhavn. 8.

Niederl. Arch. Zool. — Niederländisches Archiv für Zoologie. Haarlem, Leipzig (Leiden, Leipzig). 8.

N. Mém. Soc. Moscou — Nouveaux Mémoires de la Société Impériale des Naturalistes de Moscou. Moscou. 4.

Norman & Scott, Crust. Devon Cornwall — The Crustacea of Devon and Cornwall by Canon A. M. Norman and Thomas Scott. London. 1906. 8.

Norske Nordhavs-Exp. — Den Norske Nordhavs-Expedition 1876—1878. The Norwegian North-Atlantic Expedition 1876—1878. — 6. Bind (Volume). XIV & XV. Crustacea, I & II. Ved (By) G. O. Sars. Christiania. 1885, 86. 4.

Norske Selsk. Skr. — Det Kongelige Norske Videnskabers-Selskabs Skrifter i det 19 de Aarhundrede. Throndhjem. 4 (8).

Nouv. Dict., ed. 2 — Nouveau Dictionnaire d'Histoire naturelle, appliquée aux Arts, à l'Agriculture, à l'Économie rurale et domestique, à la Médecine, etc. Par une Société de Naturalistes et d'Agriculteurs. Nouvelle Édition. Tom. 1—6, 7—18, 19—27, 28—36. Paris. 1816, 17, 18, 19. 8.

Nyt Mag. Naturv. — Nyt Magazin for Naturvidenskaberne. Christiania. 8.

N. Zealand J. Sci. — The New Zealand Journal of Science. Dunedin. 8.

Öfv. Ak. Förh. — Öfversigt af Kongl. Vetenskaps-Akademiens Förhandlingar. Stockholm. 8.

Ov. Danske Selsk. — Oversigt over det Kongelige Danske Videnskabernes Selskabs Forhandlinger og dets Medlemmers Arbeider. Kjöbenhavn. 4 (8).

P. Ac. Philad. — Proceedings of the Academy of natural Sciences of Philadelphia. Philadelphia. 8.

Pallas, Misc. zool. — P. S. Pallas, Miscellanea zoologica quibus novae imprimis atque obscurae Animalium Species describuntur et Observationibus Iconibusque illustrantur. Hagae Comitum. 1766. 4.

Pallas, Reise Ruß. — P. S. Pallas Reise durch verschiedene Provinzen des Rußischen Reichs. Theil 1—3. St. Petersburg. 1771, 73, 76. 4. (2. Auflage: 1. Theil. St. Petersburg. 1801. 4.)

Pallas, Spic. zool. — Spicilegia zoologica quibus novae |imprimis| et obscurae Animalium Species Iconibus, Descriptionibus atque Commentariis illustrantur. Cura P. S. Pallas. Fasciculus 1—14. Berolini. 1767—80. 4.

P. Amer. Ac. — Proceedings of the American Academy of Arts and Sciences. Boston and Cambridge. 8.

Pam. Fizyjogr. — Pamiętnik Fizyjograficzny. Warszawa. 4.

Pap. Boston Soc. — Occasional Papers of the Boston Society of natural History. Boston. 8.

Parry, J. third Voy. — Journal of a third Voyage for the Discovery of a North-West Passage from the Atlantic to the Pacific; performed in the Years 1824—25, in His Majesty's Ships Hecla and Fury, under the Orders of Captain William Edward Parry. London. 1826. 4.

Parry, J. Voy. — Journal of a Voyage for the Discovery of a North-West Passage from the Atlantic to the Pacific; performed in the Years 1819—20, in His Majesty's Ships Hecla and Griper, under the Orders of William Edward Parry. With an Appendix, containing the scientific and other Observations. [Cum Suppl.:] Appendix X, natural History (A Supplement to the Appendix of Captain Parry's Voyage for the Discovery of a North-West Passage, in the Years 1819—20. Containing an Account of the Subjects of natural History). London. 1821, 24. 4.

P. Calif. Ac. — Proceedings of the California Academy of Sciences. -- [Ser. 3:] Zoology. San Francisco. 8.

P. Dublin Univ. zool. bot. Ass. — Proceedings of the Dublin University zoological and botanical Association. Dublin. 8.

Pennant, Brit. Zool., ed. 4 — British Zoology. [Preface:] Thomas Pennant. 4. Edition. Vol. 1—4. Warrington (London). 1776, 77. 8.

Pennant, Brit. Zool., ed. 5 — British Zoology, a new [5.] Edition. Vol. 1—4. London. 1812. 8.

Philippi, Reise Atacama — Reise durch die Wüste Atacama auf Befehl der Chilenischen Regierung im Sommer 1853—54 unternommen und beschrieben von Rudolph Amandus Philippi. Halle. 1860. 4.

Phil. Tr. — Philosophical Transactions of the Royal Society of London. London. 4.

Phipps, Voy. North Pole — A Voyage towards the North Pole undertaken by his Majesty's Command 1773 by Constantine John Phipps. London. 1774. 4.

P. Linn. Soc. N. S. Wales — The Proceedings of the Linnean Society of New SouthWales. Sydney. 8.

P. Liverp. biol. Soc. — Proceedings |[Vol. 4 & sequ.:] and Transactions| of the Liverpool biological Society. Liverpool. 8.

P. nat. Hist. Soc. Glasgow — Proceedings |and Transactions |of the natural History Society of Glasgow. Glasgow. 8.

P. phys. Soc. Edinb. — Proceedings of the Royal physical Society of Edinburgh. Edinburgh. 8.

Pratz, Grundw. Thiere — Über einige im Grundwasser lebende Thiere. Beitrag zur Kenntnis der unterirdischen Crustaceen. Dissert. inaug. E. Pratz. Petersburg. 1866. 8.

P. R. Soc. Edinb. — Proceedings of the Royal Society of Edinburgh. Edinburgh. 8.

P. R. Soc. Tasmania — Papers and Proceedings of the Royal Society of Tasmania. Hobart. 8.

P. R. Soc. Victoria — Proceedings of the Royal Society of Victoria. New Series. Melbourne. 8.

Publ. Expl. Mer — Conseil permanent international pour l'Exploration de la Mer. Publications de Circonstance. Copenhague. 4.

P. U. S. Mus. — Proceedings of the United States national Museum. Washington. 8.

P. zool. Soc. London — Proceedings of the zoological Society of London. London. 8.

Quart. J. geol. Soc. — The quarterly Journal of the geological Society of London. London. 8.

Rad Jugoslav. Ak. — Rad Jugoslavenske Akademije Znanosti i Umjetnosti. Zagreb. 8.

Rafinesque, Anal. Nat. — Analyse de la Nature ou Tableau de l'Univers et des Corps organisés. Par C. S. Rafinesque. Palerme. 1815. 8.

Rafinesque, Ann. Nat. -- Annals of Nature or annual Synopsis of new Genera and Species of Animals, Plants, etc., discovered in North America by C. S. Rafinesque. First annual Number, for 1820. Lexington. 8.

Rafinesque, Précis Découv. somiol. — Précis des Découvertes et Travaux somiologiques de Mr. C. S. Rafinesque-Schmaltz, entre 1800 et 1814. Ou Choix raisonné de ses principales Découvertes en Zoologie et en Botanique, pour servir d'Introduction à ses Ouvrages futurs. Palerme. 1814. 12.

Ray, Hist. Ins. — Historia Insectorum. Autore Joanne Raio. Opus posthumum Jussu Regiae Societatis Londinensis editum. Cui subjungitur Appendix de Scarabaeis Britannicis, Autore M. Lister. Londini. 1710. 4.

Rec. Austral. Mus. — Records of the Australian Museum. Sydney. 8.

Recu. Passage Vénus — Institut de France. Académie des Sciences. Recueil de Mémoires, Rapports et Documents relatifs à l'Observatiou du Passage de Vénus sur le Soleil. — Tom. 3. Partie II. Mission de l'Ile Campbell. Cum Atl. Paris. 1885. 4.

Reise Novara — Reise der Österreichischen Fregatte Novara um die Erde in den Jahren 1857, 1858, 1859 unter den Befehlen des Commodore B. von Wüllerstorf-Urbair. Wissenschaftlicher Theil. — Zoologischer Theil. 2. Band. III. Abtheilung. Crustaceen. Beschrieben von Camil Heller. Wien. 1865. 4.

Rend. Soc. Borbon. — Rendiconto della Società Reale Borbonica. Accàdemia delle Scienze. Napoli. 4.

Rep. Brit. Ass. — Report of the . . . Meeting of the British Association for the Advancement of Science. London. 8.

Rep. Devonsh. Ass. — Reports of the Meetings and Transactions of the Devonshire Association for the Advancement of Science, Literature and Art. Plymouth. 8.

Rep. Fish. Board Scotl. — Annual Report of the Fishery Board for Scotland. — Part III. Scientific Investigations. Edinburgh. 8.

Rep. Peabody Ac. — Annual Report of the Trustees of the Peabody Academy of Sciences. Salem. 8.

Rep. U. S. Fish Comm. — United States Commission of Fish and Fisheries. Report of the Commissioner. Washington. 8.

Rep. U. S. geol. Surv. Terr. — Report of the United States geological |and geographical| Survey of the Territories. Washington. 4.

Rep. Voy. Alert — Report on the zoological Collections made in the Indo-Pacific Ocean during the Voyage of H. M. S. „Alert" 1881 — 2. London. 1884. 8.

Rep. Voy. Challenger — Report on the scientific Results of the Voyage of H. M. S. Challenger during the Years 1873 — 76 under the Command of George S. Nares and Frank Tourle Thomson. Prepared under the Superintendence of |the late| C. Wyville Thomson |and now of John Murray|. — Zoology. Vol. 29. (Part LXVII.) Report on the Amphipoda collected by H. M. S. Challenger during the Years 1873—76. By Thomas R. R. Stebbing. London, Edinburgh, Dublin. 1888. 4.

Résult. Camp. Monaco — Résultats des Campagnes scientifiques accomplies sur son Yacht par Albert Ier Prince souverain de Monaco. Publiés sous sa Direction avec le Concours de Jules de Guerne (Jules Richard). — Fascicule 16. Amphipodes provenant des Campagnes de l'Hirondelle 1885— 1888. Par Ed. Chevreux. Monaco. 1900. 4.

Rev. biol. Nord France — Revue biologique du Nord de la France. Lille. 8.

Rev. Sci. nat. — Revue des Scieces naturelles. Montpellier, Paris. 8.

Risso, Hist. nat. Eur. mérid. — Histoire naturelle des principales Productions de l'Europe méridionale et particulièrement de celles des Environs de Nice et des Alpes maritimes; par A. Risso. Tom. 1—5. Paris et Strasbourg. 1826. 8.

Ritzema Bos, Bijdr. Crust. Hedriophthal. — Bijdrage tot de Kennis van de Crustacea Hedriophthalmata van Nederland en zijne Kusten. Akademisch Proefschrift... door Jan Ritzema Bos. Groningen. 1874. 8.

Rösel, Insecten-Belustig. — Der monathlich=herausgegebenen Insecten=Belustigung . . . Theil, . . . von August Johann Rösel von Rosenhof. Theil 1—4. Nürnberg [1746, 49], 55, 61. 4.

Ross, App. sec. Voy. — Appendix to the Narrative of a second Voyage in Search of a North-West Passage, and of a Residence in the Arctic Regions during the Years 1829, 30, 31, 32, 33. By Sir John Ross. Including the Reports of Commander, now Captain, James Clark Ross, and the Discovery of the northern magnetic Pole. London. 1835. 4.

Ross, Voy. Baffin's Bay — A Voyage of Discovery, made under the Orders of the Admiralty, in His Majesty's Ships Isabella and Alexander, for the Purpose of exploring Baffin's Bay, and inquiring into the Probability of a North-Western Passage. By John Ross. London. 1819. 4.

Sabine, An. north. Exp. — An Account of the Animals seen by the late northern Expedition whilst within the arctic Circle. Being No. X of the Appendix to Capt. Parry's Voyage of Discovery. By Edward Sabine. London. 1821. 4.

Samouelle, Ent. Compend. — The Entomologist's useful Compendium; or an Introduction to the Knowledge of British Insects. By George Samouelle. London. 1819. 8.

Sars, Crust. d'Eau douce Norvège — Histoire naturelle des Crustacés d'Eau douce de Norvège. Par George Ossian Sars. 1. Livraison. Les Malacostracés Christiania. 1867. 4.

Sars, Crust. Norway — An Account of the Crustacea of Norway with short Descriptions and Figures of all the Species by G. O. Sars. Vol. 1—4. Christiania and Copenhagen (Bergen). 1895 (1890—95), 99 (1896—99), 1900 (1899—1900), 03 (1901—03). 4.

Savigny, Mém. An. s. Vert. — Mémoires sur les Animaux sans Vertèbres; par Jules-César Savigny. Partie 1, 2. Paris. 1816 I, 16. 8.

SB. Ak. Wien — Sitzungsberichte der Kaiserlichen Akademie der Wissenschaften. — Mathematisch-naturwissenschaftliche Classe. Wien. 8.

SB. Böhm. Ges. — Sitzungsberichte der Königl. Böhmischen Gesellschaft der Wissenschaften ·in Prag|. — [ann. 1885 & sequ.:] Mathematisch-naturwissenschaftliche Classe. Prag. 8.

Schr. Ges. 'Königsb. — Schriften der |Königlichen| physikalisch-ökonomischen Gesellschaft zu Königsberg. Königsberg. 4.

Scopoli, Ent. Carniol. — Joannis Antonii Scopoli Entomologia Carniolica exhibens Insecta Carnioliae indigena et distributa in Ordines, Genera, Species, Varietates. Methodo Linnaeana. Vindobonae. 1763. 8.

Scoresby, Account arct. Regions — An Account of the arctic Regions, with a History and Description of the Northern Whale-Fishery. By William Scoresby, Jun. Vol. 1, 2. Edinburgh. 1820. 8.

Scudder, Nomencl. zool. — Nomenclator zoologicus. An alphabetical List of all generic Names that have been employed by Naturalists for recent and fossil Animals from the earliest Times to the Close of the Year 1879. I. Supplemental List. II. Universal Index. By Samuel H. Scudder. Washington. 1882. 8.

Slabber, Natuurk. Verlustig. — Natuurkundige Verlustigingen, behelzende microscopise Waarneemingen van in- en uitlandse Water- en Land-Dieren. Door Martinus Slabber. I—XVIII Stukje. Haarlem. 1778 (1769). 4.

Slabber, Phys. Belustig. — Physikalische Belustigungen oder mikroskopische Wahrnehmungen in- und ausländischer Wasser- und Landthierchen. Von Martin Slabber. Aus dem Holländischen übersetzt von P. L. St. Müller. Nürnberg. 1775. 4.

Smithson. Contr. — Smithsonian Contributions to Knowledge. Washington. 4.

Spolia Zeyl. — Spolia Zeylanica. Issued by the Colombo Museum, Ceylon. Colombo. 8.

Svenska Ak. Handl. — Kongliga Svenska Vetenskaps-Akademiens Handlingar. Stockholm. 8 (4).

Syezda Russ. Est. — Труды Съѣзда Русскихъ Естествоиспытателей. [Arbeiten der Versammlung Russischer Naturforscher.] 4.

Tabl. enc. méth. — Tableau encyclopédique et méthodique des trois Règnes de la Nature. — Crustacés, Arachnides et Insectes, par [Pierre André] Latreille. Paris. 1818 [1797, 1818, 18—].*) 4.

Tijdschr. Nederl. dierk. Ver. — Tijdschrift der Nederlandsche dierkundige Vereeniging. s'Gravenhage (Rotterdam) (Leiden). 8.

Tr. Amer. micr. Soc. — Transactions of the American microscopical Society. Buffalo. 8.

Trav. Soc. St.-Pétersb. — Travaux de la Société Impériale des Naturalistes de St.-Pétersbourg. Section de Zoologie et de Physiologie. St.-Pétersbourg. 8.

Tr. Connect. Ac. — Transactions of the Connecticut Academy of Arts and Sciences. New Haven. 8.

Tr. ent. Soc. London — The Transactions of the entomological Society of London. London. 8.

Tr. Irish Ac. — Transactions of the Royal Irish Academy. Dublin. 4.

Tr. Linn. Soc. London — The Transactions of the Linnean Society of London. — [Ser. 2:] Zoology. London. 4.

Tr. N. Zealand Inst. — Transactions and Proceedings of the New Zealand Institute. Wellington (Wellington, London). 8.

*) Cfr.: C. Davies Sherborn & B. B. Woodward in: P. zool. Soc. London, 1893 p. 584.

Tromsø Mus. Aarsh. — Tromsø Museums Aarshefter. Tromsø. 8.

Tr. Tyneside Club — Transactions of the Tyneside Naturalists' Field Club. Newcastle-upon-Tyne. 8.

Trudui Kazan. Univ. — Труды Общества Естествоиспытателей при Имп. Казанскомъ Университетѣ. Казань. [Arbeiten der Gesellschaft der Naturforscher an der Kais. Kasanschen Universität. Kasan.] 4 (8).

Tr. zool. Soc. London — Transactions of the zoological Society of London. London. 4.

Udb. Hauchs — Det videnskabelige Udbytte af Kanonbaaden „Hauchs" Togter i de Danske Have indenfor Skagen i Aarene 1883—86. Udgivet paa Bekostning af Ministeriet . . . ved C. G. Joh. Petersen. Hefte 1—5. Kjöbenhavn. 1889, 89, 90, 91, 93. 4.

U. S. expl. Exp. — United States exploring Expedition. During the Years 1838—42. Under the Command of Charles Wilkes. — Vol. 13. Part I, II. Crustacea. By James D. Dana. Cum Atl. Philadelphia. 1852, 52 [1853]; 55. 4 & 2.

Vega-Exp. — Vega-Expeditionens vetenskapliga Jakttagelser. Bearbetade af Deltagare i Resan och andra Forskare, utgifna af A. E. Nordenskiöld. Bandet 1—5. Stockholm. 1882, 83, 83, 87, 87. 8.

Verh. Ges. Wien — Verhandlungen der kaiserlich-königlichen zoologisch-botanischen Gesellschaft in Wien. Wien. 8.

Verh. Ver. Rheinlande — Verhandlungen des naturhistorischen Vereines der Preußischen Rheinlande. Bonn. 8.

Vetensk. Ak. Handl. — Kongl. Vetenskaps-Akademiens Handlingar. Stockholm. 8.

Vid. Meddel. — Videnskabelige Meddelelser fra den naturhistoriske Forening i Kjøbenhavn for Aarene .... Kjøbenhavn. 8.

Viviani, Phosphor. Maris — Phosphorescentia Maris quatuordecim lucescentium Animalculorum novis Speciebus illustrata a Dominico Viviani. Accedit novi cujusdam Generis, e Molluscorum Familia Descriptio et Anatome. Genuae. 1805. 4.

Voy. Nord — Voyages de la Commission scientifique du Nord, en Scandinavie, en Laponie, au Spitzberg et aux Feröe, pendant les Années 1838, 39 et 40, sur la Corvette la Recherche, commandée par Fabvre; publiés par Ordre du Roi sous la Direction de Paul Gaimard. — Zoologie. Planches. Paris. 2.

Weber, Reise Niederl. O.-Ind. — Zoologische Ergebnisse einer Reise in Niederländisch Ost-Indien. Herausgegeben von Max Weber. Band 1—4. Leiden. 1890—97. 8.

White, Crust. Brit. Mus. — [Adam White] List of the Specimens of Crustacea in the Collection of the British Museum. London. 1847. 12.

White, Hist. Brit. Crust. — A popular History of British Crustacea comprising a familiar Account of their Classification and Habits. By Adam White. London. 1857. 16.

Whymper, Trav. Great Andes — Travels amongst the Great Andes of the Equator by Edward Whymper. 1 Vol. & supplementary Appendix. London. 1892, 91. 8.

Wiss. Meeresunters. — Wissenschaftliche Meeresuntersuchungen. Herausgegeben von der Kommission zur Wissenschaftlichen Untersuchung der Deutschen Meere in Kiel und von der biologischen Anstalt auf Helgoland. Neue Folge. Kiel und Leipzig. 4.

Zaddach, Syn. Crust. Pruss. — Synopseos Crustaceorum Prussicorum Prodromus. Dissertatio . . . Die XI. M. Decembris 1844 publice defendet Ernestus Gustavus Zaddach. Regiomonti. 1844. 4.

Zool. Anz. — Zoologischer Anzeiger. Leipzig. 8.

Zool. J. — The zoological Journal. London. 8.

Zool. Rec. — The zoological Record. London. 8.

Zweite D. Nordpolarf. — Die zweite Deutsche Nordpolarfahrt in den Jahren 1869 und 1870 unter Führung des Kapitän Karl Koldewey. Herausgegeben von dem Verein für die Deutsche Nordpolarfahrt in Bremen. — 2. Band. Wissenschaftliche Ergebnisse. 8. Crustaceen. Bearbeitet von R. Buchholz. Leipzig. 1874. 8.

Z. wiss. Zool. — Zeitschrift für Wissenschaftliche Zoologie. Leipzig. 8.

Systematic Index.

[An obelisk † signifies that the generic or specific name is new.]

Amphipoda

1775 Gen. *Gammarus* (part.) + Gen. *Oniscus* (part.), J. C. Fabricius, Syst. Ent., p. 418, 296 | 1793 Subgen. *Gammarellus* (part.), J. F. W. Herbst, Naturg. Krabben Krebse, v. 2 p. 106 | 1802 Fam. *Gammarinae* (ex Ord. *Branchiogastra*), Latreille, Hist. Crust. Ins., v. 3 p. 38 | 1813/14 Trib. *Gasteruri* (part.), Leach in: Edinb. Enc., v. 7 p. 386 | 1815 Leg. *Edriophthalma* (part.), Leach in: Tr. Linn. Soc. London, v. 11 p. 307, 352 | 1816 & 17 Ord. *Amphipoda*, Latreille in: Nouv. Dict., ed. 2 v. 1 p. 467; v. 8 p. 493 ! 1888 Subord. *A.*, T. Stebbing in: Rep. Voy. Challenger, v. 29 p. 601 | 1890 *A.*, G. O. Sars, Crust. Norway, v. 1 p. 3.

Head carrying 2 pairs of antennae, the mandibles, 2 pairs of maxillae, and the basally coalesced maxillipeds. Cornea of eyes not divided into facets. Peraeon consisting usually of 7 segments, each bearing a pair of legs; first segment sometimes fused with head. Pleon consisting normally of 6 segments and the telson, or rudimentary; if normal, bearing 6 pairs of legs, the first three of which, functioning as swimming organs, differ in structure from the rest. Organs of respiration (branchial vesicles) attached to the peraeon, not to the pleon. Heart (in a pericardiac sinus) never exceeding hinder limit of peraeon, with aorta also at posterior extremity. Antennal gland present, though sometimes small.

The body, often compressed, sometimes depressed, or piriform, or even rodlike, consists of head, peraeon and pleon. The number of its free segments reaches 15 (head 1, peraeon 7, pleon 7). First segment of peraeon sometimes coalesced with head, rarely with 2^d segment of peraeon; fusion of the last 2 segments of peraeon occurs. Pleon often as long as peraeon, or even rather longer; the limits between the hinder segments occasionally disappear; in the Caprellidea the pleon is rudimentary. The last pleon segment is the telson.

Appendages of head: 1) First antenna; peduncle of 3 (rarely fewer) joints, with a many-jointed primary flagellum and usually a shorter accessory flagellum. 2) Second antenna; peduncle of 5 joints, with one flagellum; the first 2 joints of peduncle are short and not always distinct; the 2^d usually with a conical process (gland-cone) for the opening of the antennal gland. The length of antenna 1 and 2, in proportion to each other and to the body varies within wide limits. 3) Mandible (Fig. 7, 8 p. 30, 32); basal joint provided with masticatory structures; palp with 3 or fewer joints, or absent. Between the mandibles the upper lip, attached to the epistome, and the lower lip frame the entrance to the oesophagus. 4) First maxilla (Fig. 78 p. 333) consisting of 5 joints; the palp, that is the 4th and 5th joints, not rarely wanting. 5) Second maxilla (Fig. 88 p. 476) consisting of 3 joints. In maxilla 1 and maxilla 2 the 2^d joint is indistinct,

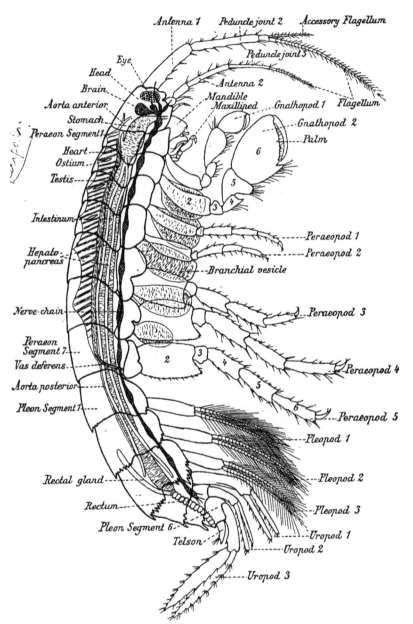

Fig. 1. Scheme of a male of the legion Gammaridea.

and the 1st and 3d are each expanded to a plate, provided with denticles and spines. 6) Maxilliped (Fig. 49 p. 186) 7-jointed; first joint, sometimes also the 2d, of the one maxilliped fused with that of the other maxilliped; 4th—7th joints forming the palp; 1 or 2 or 3 joints of the palp sometimes wanting; articulation still more reduced in the Hyperiidea.

Each segment of peraeon commonly carries one pair of legs. The 7 legs of peraeon are divided into 2 groups: the 1st—4th leg, and the 5th—7th leg; the former forming an angle open to the front, the latter to the rear; the 3d joint in all of them acts as the knee of the angle. The first 2 legs are called gnathopods (not so much on account of their own function, as from their homology with the 2d and 3d maxillipeds of Decapoda); so legs of peraeon are: 2 gnathopods and 5 peraeopods. All are uniramous; the normal number of joints is 7. The first joint, greatly enlarged (or perhaps soldered) to the so-called epimera or side-plates in the Gammaridea, is liable to evanescence in the Hyperiidea, and wanting in the Caprellidea; the last joints of gnathopod 1 and 2 usually, of other peraeopods very rarely (except in some genera of Hyperiidea), form a grasping organ. Close at the base of gnathopod 2 and of the peraeopods there are the branchial vesicles, 2—6 on each side.

The legs of the first 3 segments of pleon are called pleopods, those of the 4th—6th segments uropods. The pleopods consist of a basal joint and almost always 2 many-jointed, often flagelliform rami, beset with swimming-setae; the basal joints of each pair of pleopods are often coupled by hooked spines (Fig. 120 p. 696). The rami of the uropods are sometimes 2-, usually 1-jointed; they usually bear spines, which are sometimes uncinate.

External sexual characters are found especially in antenna 1 and 2, in the grasping hands of gnathopods and peraeopods, and in the marsupial plates of ♀. The ducts of ovaries open in the 5th, those of testes in the 7th segment of peraeon; no proper copulatory organs observed.

Among the anatomical characters especially of value for classification are the number, form and structure of the paired eyes, the number of the long midgut-glands (hepato-pancreatic tubes) flanking the intestinal canal, and of the so-called rectal glands; also the number of the venose ostia of the heart and of the un-fused ganglia of the ventral nerve-chain.

The eggs are laid in the marsupium, formed by the marsupial plates; the young, when hatching, resemble the adult; only in the Hyperiidea sometimes pleon and pleopods are at first scarcely developed.

The great majority of the Amphipoda are marine; only among the Gammaridea are there species living in fresh or brackish Water, or near the sea aboVe highWater-mark, or eVen, though exceptionally, far from Water in damp places. The Gammaridea and Caprellidea are extensiVely littoral, the Hyperiidea pelagic; also in each legion there are species liVing on swimming animals or floating objects. The geographical distribution is cosmopolitan, ascertained to Within less than 600 km of the north pole, to a depth of 5310 m in the ocean, and in fresh Water to a height of 4054 m aboVe sea-leVel.

Their diet is chiefly of animal substances. Some Gammaridea construct dWelling tubes, some liVe in grooVes and holloWs, Which they burrow in ascidians, sponges or wood. The species, liVing on or in coelenterata, sponges, tunicates, mollusca, are perhaps commensals, but at least sometimes feed on the body of their host, sometimes, as Phronima, unite feeding With housebuilding. True parasites are the Cyamidae. Parasites of the Amphipoda are Gregarinida, Trematoda and Copepoda.

Long before J. C. Fabricius' establishment of the Genus Gammarus, Amphipoda Were described under the generic designations of Cancer, Oniscus, Astacus and Squilla.

P. A. Latreille (1802, Hist. Crust. Ins., v. 3 p. 38) was the first who united all species known to him of Amphipoda (as at present understood) in one group, the family of Gammarinae, to the exclusion of all other species; later (1816, in: Nouv. Dict., ed. 2 v. 1 p. 467) he devised for the group the name Amphipoda, separating, however, the genera Cyamus and Caprella as Laemodipoda. The limits of our order, generally accepted to-day, were settled by Krøyer (1843, in: Naturh. Tidsskr., v. 4 p. 490) and J. D. Dana (1852, in: Amer. J. Sci., ser. 2 v. 14 p. 297); from the latter author is derived also the division of the Amphipoda into the 3 tribes of Gammaridea, Hyperiidea and Caprellidea. Only Gerstaecker (1883 & 86, in: Bronn's Kl. Ordn., v. 5 II p. 283, 481) regards the Tanaidea as a tribe of Amphipoda.

Synoptical diagnosis of the three legions:

Head not fused with 1st segment of peraeon. Palp of maxilliped 2- to 4-jointed. Peraeon with 7 pairs of legs; 5 or 6 segments of peraeon with branchial vesicles; 4 segments in ♀ with marsupial plates; 1st joint of gnathopods 1, 2 and of peraeopods 1—5 forming or united to well developed side-plates. Pleon consisting usually of 7 free segments, carrying 3 pairs of pleopods and usually 3 pairs, at least 1 pair, of uropods; uropod 1 always with 2 rami. Eyes varying in size and form, 0—4 in number. Hepato-pancreatic tubes 4, rarely 2; rectal glands 2 or 1, sometimes rudimentary. Heart with 3 pairs, rarely 1 pair, of ostia. Nerve-chain with 4 ganglia in pleon-segments 1—4   I. Leg. **Gammaridea**

Head not fused with 1st segment of peraeon. Maxilliped without palp. Peraeon with 7 pairs of legs; 3—5 (6?) segments of peraeon with branchial vesicles; 4 segments in ♀ with marsupial plates; 1st joint of gnathopods 1. 2 and of peraeopods 1—5 small or wanting. Pleon consisting usually of 7 free segments, with 3 pairs of pleopods and 3 pairs of uropods; rami of uropods often evanescent. Eyes usually of large size. Hepato-pancreatic tubes 2 or none; rectal glands none. Heart with 2, rarely 3, pairs of ostia. Nerve-chain with 3 or 4 ganglia in the anterior segments of pleon . . . .   II. Leg. **Hyperiidea**

Head fused with 1st segment of peraeon. Palp of maxilliped 1- to 4-jointed. Peraeon often with fewer than 7 pairs of legs; 2, rarely 3, segments of peraeon with branchial vesicles; 2 segments in ♀ with marsupial plates; 1st joint of gnathopods and peraeopods wanting. Pleon and its legs rudimentary. Eyes small, 1 pair. Hepato-pancreatic tubes 2; rectal glands none. Heart with 3 pairs of ostia. Posterior ganglia of nerve-chain very small, none situated in pleon . . . .   III. Leg. **Caprellidea**

I. Leg. Gammaridea

1852 Subtrib. *Gammaridea*, J. D. Dana in: Amer. J. Sci., ser. 2 *v.* 14 p. 308 |
1853 Subtrib. *G.*, J. D. Dana in: U. S. expl. Exp., *v.* 13ɪɪ p. 806, 825 | 1861 Trib.
Prostomatae + *Gammaridae*, A. Boeck in: Forb. Skand. Naturf., Møde 8 p. 637, 639 |
1888 Trib. *Amphipoda Gammarina,* T. Stebbing in: Rep. Voy. Challenger, *v.* 29 p. 601 |
1890 Trib. *Gammaridea,* G. O. Sars, Crust. NorWay, *v.* 1 p. 21 | 1893 Subord. *Gammarini,*
A. Della Valle in: F. Fl. Neapel, *v.* 20 p. 297.

Head not fused with 1st peraeon segment.. Eyes usually compound,
even when externally simple, but sometimes reduced to mere pigment patches
or entirely absent; normal number 2, varying to 1, 3, 4 or 6 (p. 104); the
cornea of each eye sometimes forming a lens*). Antenna 1 consisting of 3-jointed
peduncle and flagellum, with or without accessory flagellum. Antenna
2 longer or shorter than antenna 1 or equal to it; peduncle apparently
5-jointed, usually with conical process (gland-cone) opening on 2d joint.
The joints of one or both antennae in ♂, or in ♂ and ♀, sometimes carrying
membranaceous appendages (calceoli, Fig. 82 p. 347). Epistome flat or projecting.
Upper lip with distal margin of outer plate smooth or variously bilobed
(Fig. 41 p. 163). Lower lip divided into 2 principal lobes, prolonged backward
into the so-called mandibular processes, and often having between them 2
smaller inner lobes (Fig. 71 p. 278, Fig. 102 p. 598). Mandible (Fig. 77
p. 333) normally composed of basal joint and 3-jointed palp; the basal joint
produced to a dentate cutting edge with adjacent accessory plate on inner
side and armed with a spine-row between the cutting-plates and the molar,
which is a prominent denticulate tubercle; each one of the features sometimes degraded or absent (Fig. 2 p. 12, Fig. 21 p. 96). Maxilla 1, 1st and 3d joints
with expansion (inner and outer plate), surmounted by spines, which are usually
furcate or serrate; 4th and 5th joints forming the so-called palp (Fig. 78 p. 333,
Fig. 87 p. 476); the parts, especially inner plate and palp. liable to evanescence.
Maxilla 2, 1st and 3d joints with expansion (inner and outer plate); the
plates very variable in relative size, rarely wanting, usually in part
fringed with setae or slender spines (Fig. 88 p. 476). Maxillipeds (Fig. 49
p. 186, Fig. 54 p. 208), 2d and 3d joints usually with expansion (inner
and outer plate); palp 2—4-jointed. Peraeon of 7 distinct segments, rarely
reduced to 6 by coalescence of the last two. Legs of the peraeon overlapped
at base by the side-plates, developed from or soldered to the 1st joints
of the legs; the first 4 side-plates usually larger than the 3 following, the
4th commonly the largest, but with many exceptions. Gnathopods 1 and 2 have
the 4th joint not end to end with the 5th, but more or less underpropping it,
and the 6th joint generally more or less expanded to form the so-called hand.
The branchial vesicles, usually attached to gnathopod 2 and peraeopods 1—4
or 1—5, are simple, twisted, pleated or divided into leaf-like lobes; simple
accessory vesicles rare. In the ♀ peraeon segments 2—5 are furnished at the

---

*) Articulated eye-lobes without visual elements attributed to Ingolfiella Hansen
1903 (see p. 726).

proper age with marsupial plates, fringed with simple setae. The pleon segments 1—3 are always distinct; the pleopods are almost always biramous; the peduncles often connected by coupling-spines (Fig. 14 p. 83); the rami usually tapering, many-jointed, each joint with a pair of plumose setae; the 1st joint of inner ramus often having on its inner margin spines cleft at the apex. Pleon segments 4—6 are usually distinct; outer ramus of the uropods sometimes 2-jointed; uropod 1 never absent, and always biramous; of uropods 2 and 3 rarely one or both obsolete, or without one or both of the rami. Telson (probably) never absent, sometimes divided to the base, sometimes not divided at all, and exhibiting every intermediate gradation.

Distribution see above pag. 3.

41 Families, 304 accepted and 9 doubtful genera, 1076 accepted species and 257 doubtful.

Synopsis of the families:*)

1. Antenna 1 (Fig. 3, 6, 9, 12), 1st joint stout**), with accessory flagellum; mandible (Fig. 7, 8), cutting edge almost smooth***), with palp; gnathopod 2 (Fig. 10, 11), 3^d joint elongate . . . . 1. Fam. **Lysianassidae** . . p.   8
   These characters not combined — **2.**

2. Body (Fig. 18) plump; antenna 1 (Fig. 19) with accessory flagellum; mandible (Fig. 17, 21) without molar, without palp . . . . 2. Fam. **Stegocephalidae** . p. 88
   These characters not combined — **3.**

3. Head (Fig. 25) tapering, truncate; eyes, when present, externally simple, usually 4; antenna 1 without accessory flagellum; telson (Fig. 24) more or less cleft . . . 3. Fam. **Ampeliscidae** . . p. 97
   These characters not combined — **4.**

4. Antenna 1 (Fig. 30—32) with accessory flagellum; mandible with palp normal; peraeopods 3—5 (Fig. 35—38) adapted for burrowing by expansion of joints and armature of spines and setae — **5.**
   These characters not combined — **6.**

5. Peraeopod 4 not greatly longer than peraeopod 5 . . . . 4. Fam. **Haustoriidae** . . p. 118
   Peraeopod 4 greatly longer than peraeopod 5   5. Fam. **Phoxocephalidae** . p. 133

6. Upper lip incised; maxillipeds normal; uropod 3 (Fig. 39) biramous; telson elongate, tapering, entire . . . . 6. Fam. **Amphilochidae** . . p. 148
   These characters not combined — **7.**

7. Antenna 1 without accessory flagellum; maxillipeds (Fig. 44, 49, 51, 54) more or less abnormal; telson entire — **8.**
   These characters not combined — **16.**

8. Gnathopod 1 (Fig. 42) chelate — **9.**
   Gnathopod 1 (Fig. 45) not chelate — **10.**

---

*) For some of the imperfections in this synopsis the reader is invited to believe that a more or less adequate apology could be offered, were space available for its presentment.

**) Except in Amaryllis (Fig. 5 p. 23).

***) Except in Valettia and Podoprion; wanting in Kerguelenia (Fig. 2 p. 12).

9 { Mandible and maxillae 1 and 2 developed 7. Fam. **Leucothoidae** . . p. 161
    { Mandible weak, maxillae 1 and 2 obsolete 8. Fam. **Anamixidae** . . . p. 170

10 { Uropod 3 not biramous — **11.**
     { Uropod 3 biramous — **14.**

11 { Mandible with palp — **12.**
     { Mandible without palp — **13.**

12 { Mandible, 3d joint of palp very short . . 9. Fam. **Metopidae** . . . . p. 171
     { Mandible, 3d joint of palp not short. . . 10. Fam. **Cressidae** . . . . p. 190

13 { Maxillipeds (Fig. 51) with outer plate obsolete 11. Fam. **Stenothoidae** . . . p. 192
     { Maxillipeds with outer plate developed . . 12. Fam. **Phliantidae** . . . p. 200

14 { Mandible without palp . . . . 13. Fam. **Colomastigidae**. . p. 206
     { Mandible with palp — **15.**

15 { Maxillipeds (Fig. 54) with palp 2-jointed . 14. Fam. **Lafystiidae** . . . p. 208
     { Maxillipeds with palp 4-jointed . . . . . 15. Fam. **Laphystiopsidae** . p. 209

16 { Mandible (Fig. 56, 58, 60) with molar weak
     { or wanting; telson more or less divided — **17.**
     { These characters not combined — **19.**

17 { Maxillipeds (Fig. 55), inner plate well developed 16. Fam. **Acanthonotozo-
     {   matidae** . . . . . p. 210
     { Maxillipeds (Fig. 59), inner plate small — **18.**

18 { Gnathopods 1 and 2 (Fig. 57) simple. . . 17. Fam. **Pardaliscidae** . . p. 220
     { Gnathopods 1 and 2 strongly subchelate . 18. Fam. **Liljeborgiidae** . . p. 229

19 { Eyes*), when present, usually dorsal con-
     {   tiguous or confluent (Fig. 61, 62) — **20.**
     { Eyes*), when present, lateral — **22.**

20 { Antenna 1 (Fig. 67) without accessory flagel-
     {   lum; mandible, 3d joint of palp large;
     {   peraeopod 5 (Fig. 64) much longer than
     {   peraeopod 4; telson entire . . . . . . 19. Fam. **Oedicerotidae** . . p. 235
     { Antenna 1 (Fig. 70) with accessory flagellum;
     {   mandible, 3d joint of palp small; peraeo-
     {   pod 5 not exceptionally longer than
     {   peraeopod 4; telson usually cleft**) — **21.**

21 { Peraeopods 1 and 2 with 4th and 5th joints
     {   dilated . . . . 20. Fam. **Synopiidae** . . . . p. 270
     { Peraeopods 1 and 2 with 4th and 5th joints
     {   not dilated . . . . 21. Fam. **Tironidae** . . . . p. 273

22 { Side-plate 4 (Fig. 84) usually excavate behind;
     {   peraeopods 1 (Fig. 91) and 2 not glandular;
     {   telson (Fig. 81, 100) variable; animal
     {   usually not domicolous — **23.**
     { Side-plate 4 usually not excavate behind;
     {   peraeopods 1 and 2 frequently glandular;
     {   telson (Fig. 115, 118) entire; animal
     {   usually domicolous — **31.**

23 { Mandible (Fig. 77) with palp — **24.**
     { Mandible (Fig. 90) without palp — **30.**

24 { Telson entire — **25.**
     { Telson cleft — **27.**
     { Telson variable, usually cleft; antenna 1
     {   usually with accessory flagellum. . . . 30. Fam. **Gammaridae** . . . p. 364

---

*) In the case of blind species or exceptional genera like Ommatogammarus the distinguishing characters under 20 and 21 will give assistance, and for the Gammarids of Lake Baikal the locality will be a guide.

**) Except in Bruzelia (p. 274).

√f

25 { Rostrum Weak . . . . 22. Fam. **Calliopiidae** . . . p. 285
{ Rɔ strum Well marked — **26.**

26 { Side-plates 1—4 rounded; antenna 1 longer
     than antenna 2 . . . . 23. Fam. **Pleustidae** . . . . p. 309
{ Side-plates 1—4 angular; antenna 1 shorter
     than antenna 2 . . . . 24. Fam. **Paramphithoidae** p. 320

27 { Pleon segments 5 and 6 coalesced . . . . 25. Fam. **Atylidae** . . . . p. 327
{ Pleon segments 5 and 6 not coalesced
     (Fig. 80) — **28.**

28 { LoWer lip, inner plates rather large; uropod
     3 greatly elongate . . . . 26. Fam. **Melphidippidae** . p. 334
{ LoWer lip, inner plates small or Wanting;
     uropod 3 (Fig. 81) not greatly elongate —**29.**

29 { Gnathopods 1 and 2, hands poWerful (Fig. 79) 27. Fam. **Eusiridae** . . . . p. 338
{ Gnathopod 1 Without hand . . . . 28. Fam. **Bateidae** . . . . p. 355
{ Gnathopods 1 and 2, hands not poWerful . 29. Fam. **Pontogeneiidae** . p. 356

30 { Uropod 3, both rami Well deVeloped . . . 31. Fam. **Dexaminidae** . . p. 514
{ Uropod 3, one ramus Wanting or Very small 32. Fam. **Talitridae** . . . p. 523

31 { Imperfectly deVeloped eyes coVering the head 41. Fam. **Hyperiopsidae** . p. 714
{ Eyes, when present, not abnormal — **32.**

32 { Uropods 2 and 3 (Fig. 103, 118) deVeloped — **33.**
{ Uropods 2 and 3 (Fig. 121, 127), one or
     other Wanting or rudimentary . . . . 40. Fam. **Podoceridae** . . . p. 694

33 { Pleon (Fig. 110) compressed — **34.**
{ Pleon (Fig. 115, 118), as a rule, depressed —**38.**

34 { Uropod 3 (Fig. 103) not uncinate — **35.**
{ Uropod 3 (Fig. 110) uncinate — **37.**

35 { Gnathopod 1 (Fig. 101) larger than gnatho-
     pod 2 . . . . 33. Fam. **Aoridae** . . . . . p. 585
{ Gnathopod 1 (Fig. 104, 105) not larger than
     gnathopod 2 — **36.**

36 { Peraeopods 1—5 not subchelate . . . . . 34. Fam. **Photidae** . . . . p. 603
{ Peraeopods 1—5 subchelate . . . . 35. Fam. **Isaeidae** . . . . p. 630

37 { LoWer lip With principal lobes notched . . 36. Fam. **Ampithoidae** . . p. 631
{ LoWer lip With principal lobes not notched 37. Fam. **Jassidae** . . . . p. 647

38 { Antenna 2 (Fig. 116, 117), flagellum not
     spatulate . . . . 38. Fam. **Corophiidae** . . . p. 662
{ Antenna 2 (Fig. 119), flagellum spatulate . 39. Fam. **Cheluridae** . . . p. 693

1. Fam. **Lysianassidae**

1849 Subfam. *Lysianassinae*, J. D. Dana in: Amer. J. Sci., ser. 2 *v.* 8 p. 136 | 1856 *Lysianassides*, Bate (& WestWood) in: Rep. Brit. Ass., Meet. 25 p. 21 | 1857 *Lysiassides*, Bate in: Ann. nat. Hist., ser. 2 *v.* 20 p. 524 | 1857 Subfam. *Lisianassini*, A. Costa in: Mem. Acc. Napoli, *v.* 1 p. 173 | 1865 Subfam. *Lysianassina* + Subfam. *Trischizostomatina*, W. Lilljeborg in: N. Acta Soc. Upsal., ser. 3 *v.* 6 nr. 1 p. 18 | 1872 *Prostomatidae* + Subfam. *Lysianassinae*, A. Boeck, Skand. Arkt. Amphip., *v.* 1 p. 94; p. 112 | 1874 *Lysianassidae*, Buchholz in: Zweite D. Nordpolarf., *v.* 2 p. 299 | 1888 *L.* + *Valettidae*, T. Stebbing in: Rep. Voy. Challenger, *v.* 29 p. 606, 723 | 1890 *L.*, G. O. Sars, Crust. NorWay, *v.* 1 p. 28 | 1893 *Lisianassidi*, A. Della Valle in: F. Fl. Neapel, *v.* 20 p. 769.

Of side-plates 1—5 (Fig. 3, 4, 6) one or more deep (except in Valettia). Eyes (Fig. 3, 6) usually large, paired, rarely reduced each to a single lens (Hippomedon holbölli, Acidostoma sp.), sometimes degraded or absent (triple in Hirondellea). Antenna 1 (Fig. 5, 9, 12) not longer than antenna 2. the 1st joint tumid (except in Amaryllis, Fig. 5). Accessory flagellum always present (except in Lepidepecreum longicorne, p. 80). Flagellum of antenna 2 often differentiated in ♂. Lower lip without inner lobes. Mandible (Fig. 2, 7, 8), cutting edge almost simple (dentate in Valettia and Podoprion, wanting in Kerguelenia, Fig. 2); accessory plate generally present on left mandible; molar seldom very robust; palp 3-jointed. Gnathopod 2 (Fig. 10, 11) usually slender and delicate, 3^d joint elongate, 5th and 6th spinulose, 7th minute. Uropod 3 (Fig. 13) biramous (except in Acontiostoma and Stomacontion). Hepato-pancreatic tubes (so far as known) 4.

Marine.

49 accepted and 2 doubtful genera, 136 accepted species and 35 doubtful.

Synopsis of accepted genera:

1	Mandible (Fig. 2) without distinct cutting edge	1. Gen. **Kerguelenia** . . .	p. 11
	Mandible (Fig. 7, 8) with distinct cutting edge — 2.		
2	Mouth-organs greatly projecting below, more or less stiliform — 3.		
	Mouth-organs not stiliform — 6.		
3	Eyes and gnathopod 1 (Fig. 3) powerfully developed	2. Gen. **Trischizostoma** . .	p. 12
	Eyes and gnathopod 1 not powerfully developed — 4.		
4	Uropod 3 biramous	3. Gen. **Acidostoma** . . .	p. 14
	Uropod 3 not biramous — 5.		
5	Uropod 3 without rami	4. Gen. **Acontiostoma** . .	p. 15
	Uropod 3 with a single ramus.	5. Gen. **Stomacontion** . .	p. 16
6	Eyes triple	6. Gen. **Hirondellea** . . .	p. 16
	Eyes, when present, paired — 7.		
7	Gnathopod 1 chelate — 8.		
	Gnathopod 1 not chelate — 12.		
8	Maxilliped, palp 3-jointed; telson not cleft	7. Gen. **Podoprionella** . .	p. 17
	Maxilliped, palp 4-jointed; telson cleft — 9.		
9	Peraeopod 3, 2^d joint deeply indentured .	8. Gen. **Podoprion**	p. 18
	Peraeopod 3, 2^d joint not deeply indentured — 10.		
10	Chela of gnathopod 1 narrow	9. Gen. **Euonyx**	p. 19
	Chela of gnathopod 1 broad — 11.		
11	Telson much longer than broad	10. Gen. **Opisa**	p. 20
	Telson not much longer than broad . . .	11. Gen. **Sophrosyne**. . . .	p. 21
12	Mandible, cutting edge strongly dentate .	12. Gen. **Valettia**	p. 22
	Mandible, cutting edge not strongly dentate — 13.		
13	Maxilla 1 without palp	13. Gen. **Amaryllis**	p. 23
	Maxilla 1 with 2-jointed palp — 14.		
14	Mandible, molar rather strong; maxilliped, palp rather long; gnathopod 1 subchelate; telson entire or not deeply incised *) .	14. Gen. **Onisimus**	p. 25
	These characters not combined — 15.		

*) Onisimus perplexes the arrangement by having the telson sometimes entire, sometimes as deeply cleft as in Paratryphosites (p. 42). From the latter it is separated by the much shorter outer plates of the maxillipeds. In Lysianassa (p. 37) the telson sometimes has a small notch.

15 { Side-plates 1 and 2 (Fig. 6) Very small — **16**.
    { Side-plates 1 and 2 not Very small — **17**.

16 { Peraeopod 3, 2^d joint deeply indentured . 15. Gen. **Cyphocaris** . . . . p. 28
    { Peraeopod 3, 2^d joint not deeply indentured 16. Gen. **Cyclocaris** . . . . p. 30

17 { Telson entire — **18**.
    { Telson cleft — **24.**

18 { Antenna 2, penultimate joint of peduncle
        dilated . . . . 17. Gen. **Lysianella** . . . . p. 31
    { Antenna 2, penultimate joint of peduncle
        not dilated — **19**.

19 { Uropod 3, inner ramus rudimentary . . . 18. Gen. **Onesimoides** . . . p. 32
    { Uropod 3, inner ramus not rudimentary — **20**.

20 { Mandible, molar Well developed . . . . . 19. Gen. **Pseudalibrotus** . . p. 33 l
    { Mandible, molar Weak or obsolete — **21**.

21 { Maxillipeds, palp 4-jointed — **22**.
    { Maxillipeds, palp 3-jointed — **23**.

22 { Maxilla 2, inner plate narroW . . . . . . 20. Gen. **Nannonyx** . . . . p. 34
    { Maxilla 2, inner plate broad. . . . . . . 21. Gen. **Lysianassa** . . . . p. 37

23 { Maxillipeds, inner plate short . . . . . . 22. Gen. **Perrierella** . . . . p. 40
    { Maxillipeds, inner plate long . . . . . . 23. Gen. **Normanion** . . . . p. 41

24 { Telson not deeply cleft — **25**.
    { Telson deeply cleft — **29**.

25 { Mandible, palp attached oVer molar — **26**.
    { Mandible, palp attached behind molar — **27**.

26 { Maxillipeds, 4th joint of palp short . . . 24. Gen. **Paratryphosites** . p. 42
    { Maxillipeds, 4th joint of palp long . . . . 25. Gen. **Paronesimus** . . . p. 43

27 { Gnathopod 1 subchelate . . . . 26. Gen. **Orchomene** . . . . p. 44
    { Gnathopod 1 not subchelate — **28**.

28 { Antenna 1, 2^d and 3^d joints not extremely
        short . . . . 27. Gen. **Socarnoides** . . . p. 47
    { Antenna 1, 2^d and 3^d joints extremely short 28. Gen. **Menigrates** . . . . p. 48

29 { Side-plate 1, lower front angle concealed — **30**.
    { Side-plate 1, loWer front angle not con-
        cealed — **31.**

30 { Maxilla 1 with outer plate, maxilla 2 with
        inner plate, Very broad . . . . 29. Gen. **Aristias** . . . . . p. 49
    { Maxilla 1 With outer plate, maxilla 2 with
        inner plate, not Very broad . . . . . 30. Gen. **Ambasia** . . . . . p. 51

31 { Branchial Vesicles more or less complex — **32**.
    { Branchial Vesicles simple *) — **35**.

32 { Branchial Vesicles (Fig. 10) pleated on both
        sides — **33**.
    { Branchial Vesicles pleated on one side
        only — **34.**

33 { Gnathopod 1 not subchelate . . . . 31. Gen. **Ichnopus** . . . . . p. 52
    { Gnathopod 1 subchelate . . . . 32. Gen. **Anonyx** . . . . . p. 53

34 { Gnathopod 1, finger short. . . . . 33. Gen. **Socarnes** . . . . . p. 56
    { Gnathopod 1, finger long . . . . 34. Gen. **Hippomedon** . . . p. 58

35 { Peraeopod 3, 2^d joint deeply indentured . 35. Gen. **Glycerina** . . . . p. 60
    { Peraeopod 3, 2^d joint not deeply indentured — **36.**

---

*) In Tmetonyx miersi (p. 75) not quite simple.

36 { Mandible, palp attached not behind the
molar — **37**.
Mandible, palp attached behind the molar — **46**.

37 { Gnathopod 1, finger minute, shrouded among
setae . . . . 36. Gen. **Scopelocheirus** . . p. 61
Gnathopod 1, finger not concealed — **38**.

38 { Gnathopod 1 simple . . . . 37. Gen. **Alicella** . . . . . p. 63
Gnathopod 1 **ṇ**ot simple — **39**.

39 { Gnathopod 1 imperfectly subchelate — **40**.
Gnathopod 1 distinctly subchelate — **41**.

40 { Side-plates 1—4 closely packed . . . . . 38. Gen. **Uristes** . . . . . . p. 63
Side-plates 1—4 not closely packed . . . 39. Gen. **Centromedon** . . . p. 65

41 { Gnathopod 1, 6th joint distally widened . 40. Gen. **Cheirimedon** . . . p. 66
Gnathopod 1, 6th joint not distally widened — **42**.

42 { Gnathopod 1, 6th joint much longer than 5th 41. Gen. **Tryphosella** . . . p. 67
Gnathopod 1, 6th joint not much longer
than 5th — **43**.

43 { Maxilla 2, inner plate not much shorter than
outer . . . . 42. Gen. **Tryphosa** . . . . p. 68
Maxilla 2. inner plate much shorter than
outer — **44**.

44 { Epistome not projecting . . . . 43. Gen. **Chironesimus** . . . p. 72
Epistome projecting — **45**.

45 { Uropod 3, rami foliaceous . . . . 44. Gen. **Eurythenes** . . . . p. 72
Uropod 3, rami lanceolate . . . . 45. Gen. **Tmetonyx** . . . . p. 73

46 { Epistome with spiniform projection . . . 46. Gen. **Tryphosites** . . . p. 77
Epistome without spiniform projection — **47**.

47 { Antenna 1 (Fig. 12), 1st joint carinate,
produced . . . . 47. Gen. **Lepidepecreum** . . p. 78
Antenna 1, 1st joint not carinate, not pro-
duced — **48**.

48 { Body robust, epistome sometimes projecting*) 48. Gen. **Orchomenella** . . p. 81
Body slender, epistome not projecting*) . 49. Gen. **Orchomenopsis** . . p. 83

1. Gen. **Kerguelenia** Stebb.

1888 *Kerguelenia* (Sp. un.: *K. compacta*), T. Stebbing in: Rep. Voy. Challenger,
v. 29 p. 1219 | 1891 *K.*, G. O. Sars, Crust. Norway, v. 1 p. 119.

Side-plates, especially 4th, large and deep. Antenna 1 short in both
sexes, with a dense brush only in ♂; antenna 2 more-slender and scarcely
longer than antenna 1. Mandible without cutting edge, spine-row, or molar;
palp slender, apical (Fig. 2). Maxillae 1 and 2 normal, but very small.
Maxillipeds with both plates small; palp elongate. Gnathopods 1 and 2 slender,
elongate; 3d joint long, 7th minute; gnathopod 1 simple, gnathopod 2 chelate.
2d joint broad in peraeopods 4 and 5, not in peraeopod 3; 4th joint broad in
peraeopods 3—5. Uropod 3 very small; rami rudimentary. Telson small,
rounded, entire.

2 species.

---

*) No distinguishing feature between those two genera seems to be absolutely constant.

Synopsis of species:

Gnathopod 1, 7th joint curved; peraeopod 3, 2d joint
    not lobed . . . . 1. **K. compacta** . p. 12
Gnathopod 1, 7th joint straight; peraeopod 3, 2d joint
    lobed below . . . . 2. **K. borealis** . . p. 12

    1. **K. compacta** Stebb. 1888 *K. c.*, T. Stebbing in: Rep. Voy. Challenger,
v. 29 p. 1220 t. 15 A.

    Body compact. Side-plates 1—4 deep, 4th greatly excavate behind
for the 5th. Pleon segment 3 with postero-lateral angles almost quadrate.

Eyes doubtful. Antenna 1 in ♀, 1st joint stout, flagellum
5 joints, accessory flagellum 3 joints. Antenna 2, last
3 joints of peduncle nearly equal, flagellum 4 or 5 joints.
Gnathopod 1, 3d joint longer than 4th, as long as 5th;
6th narrow, tapering; 7th small, with 2 setules, one as
long as the joint and plumose, projecting from its
concave margin. Gnathopod 2 longer than gnathopod 1,
3d joint not quite so long as 5th; 5th much longer than
6th, both of them setose; the chela minute. Peraeopods 1
and 2 with 4th joint a little widened. Peraeopod 3,
2d joint not or scarcely expanded, 4th much wider than
the 3d or 5th, quite overlapping the latter behind; the
6th rather longer, not broader, than the 5th. Feraeopods 4 and 5 with very widely expanded 2d joint,

Fig. 2. **K. compacta.**
Mandible.

otherwise resembling peraeopod 3, but while 2d joint in
peraeopod 5 is much larger than in peraeopod 4, the
rest of the limb is smaller. Uropod 1 longer than uropod 2. Colour in
spirit light brown. L. about 4 mm.

    Cumberland Bay [Kerguelen Island]. Depth 230 m.

    2. **K. borealis** O. Sars 1891 *K. b.*, G. O. Sars, Crust. Norway, *v.* 1 p. 119 t. 40 f. 2.

    Perhaps identical with K. compacta. Eyes triangular, imperfect, reddish
with white border. Antenna 1 in ♀, flagellum 7-jointed, accessory flagellum
4 joints. Antenna 1 in ♂, having last 2 joints of peduncle and base of flagellum
more swollen and provided with dense brush of setae. Antenna 2, flagellum
7 joints. Gnathopod 1, 7th joint continuous with tapering 6th, and provided
with a strong secondary booth and 2 slender setae. Peraeopod 3, 2d joint
expanded below into a lobe. Uropod 3 extremely small, rami much smaller
than peduncle, the outer the smaller. Telson enclosed between the lateral
parts of the preceding segments; its tip obtusely truncated. Colour bright
orange. L. ♀ 5 mm.

    Arctic Ocean (Finmark, Nordland), Hardangerfjord   Depth 190—280 m.

2.. Gen. **Trischizostoma** Boeck

    1853 *Guerinia* (Sp. un.: *G. nicaeensis*) (non Robineau-Desvoidy 1830, Diptera!),
(Hope in MS.) A. Costa, Descr. 3 Crost. dal Hope, p. 3 | 1853 *G.*, A. Costa, Fauna
Reg. Napoli, fasc. Apr. 1853 p. 1 | 1861 *Trischizostoma* (Sp. un.: *T. raschii*). A. Boeck
in: Porh. Skand. Naturf., Møde 8 p. 637 | 1886 *T.*, Bovallius in: N. Acta Soc. Upsal.,
ser. 3 *v.* 13 nr. 9 p. 23 | 1888 *T.*, T. Stebbing in: Rep. Voy. Challenger, *v.* 29 p. xix,
272, 1673, etc. | 1890 *T.*, G. O. Sars, Crust. Norway, *v.* 1 p. 29 | 1895 *T.*, T. Stebbing
in: Nat. Sci., *v.* 6 p. 265 | 1893 *Guerina* + *T.*, A. Della Valle in: F. Fl. Neapel, *v.* 20
p. 775, 779.

Back broadly rounded. Rostrum broad, apically rounded (Fig. 3).
Side-plate 2 the largest. Eyes very large. Antenna 1 in ♀, 1ˢᵗ joint large
in both flagella; in ♂ 1ˢᵗ joint of principal flagellum very large. Antenna 2,
ultimate joint of peduncle in ♀ shorter than penultimate, in ♂ the reverse. Mouth-
organs in ♀ (Sars): Upper lip long, narrow; lower lip with lobes acutely lanceolate.
Mandible, cutting edge very narrow, sharpened and obliquely truncated;
accessory plate inconspicuous; no molar; palp very large and densely setose.
Maxilla 1, inner plate small, unarmed; outer very narrow, apically divided
into small, claw-like teeth (4 strong teeth, Bovallius); palp very minute but
distinctly biarticulate. Maxilla 2, both plates very narrow, stiliform. Maxillipeds,

inner plates very narrow, unarmed,
outer plates linguiform, partly en-
compassing the -other oral parts;
palp geniculate at the middle,
1ˢᵗ joint short, 4ᵗʰ lanceolate. Mouth-
organs in ♂ (Della Valle): Upper
lip long, not narrow, apex not emarginate; lower lip with broad lobes,
and a shorter pointed single lobe
between them. Mandible narrowly
spatulate, without true cutting

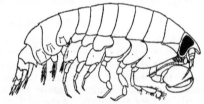

Fig. 3. T. nicaeense.
Lateral view. [After G. O. Sars.]

edges; no molar, 3 spinules; palp very large and setose. Maxilla 1, no
inner plate; outer plate elongate, tipped with 5 large teeth; palp reduced to a
little tubercle. Maxilla 2, inner plate shorter and narrower than outer, apically
pointed, with one setule, outer apically broad, with 2 or 3 little setules.
Maxillipeds, inner plates very narrow, unarmed; outer plates as in ♀, palp
with 1ˢᵗ joint longer than 2ᵈ or 3ᵈ, 4ᵗʰ lanceolate. Gnathopod 1 enormously
developed, subchelate; in the adult of both sexes the hand is so inverted,
that the finger seems to be attached to the wrong end of the palm. Uropods
with rami subequal, broadly lanceolate; outer ramus of 3ᵈ 2-jointed. Telson
small, entire.

1 species.

1. **T. nicaeense** (A. Costa) 1853 *Guerinia nicaeensis*, A. Costa, Descr. 3 Crost.
dal Hope, p. 3 | 1853 *G. n.*, A. Costa, Fauna Reg. Napoli, fasc. Apr. 1853 p. 2 | 1867
G. n., A. Costa in: Annuario Mus. Napoli, *v.* 4 p. 44 t. 3 f. 1 | 1888 *Trischizostoma n.*,
T. Stebbing in: Rep. Voy. Challenger, *v.* 29 p. 272, 1673 | 1861 *T. raschii*, A. Boeck
in: Forb. Skand. Naturf., Møde 8 p. 637 | 1872 *T. r.*, A. Boeck, Skand. Arkt. Amphip.,
v. 1 p. 97 t. 2 f. 1 | 1886 *T. r.*, Bovallius in: N. Acta Soc. Upsal., ser. 3 *v.* 13 nr. 9 p. 24
t. 3 f. 41—60 (♀), 61—67 (♂juv.) | 1890 & 95 *T. r.*, G. O. Sars, Crust. Norway, *v.* 1
p. 31 t. 12 (♀); p. 673 | 1893 *Guerina nicaeensis* + *T. r.*, A. Della Valle in: F. Fl.
Neapel, *v.* 20 p. 776 t. 61 f. 10--22 (♂); p. 780.

Rostrum in ♀ horizontal, in ♂ slightly depressed. Side-plates shallow,
1ˢᵗ small, partly covered by the 2ᵈ. Eyes almost meeting on the top of the
head. Antenna 1, 1ˢᵗ joint large, 2ᵈ and 3ᵈ short; flagellum in ♀ 9-jointed;
accessory flagellum subequal in length to 1ˢᵗ joint of primary, with 1 long
and 2 minute joints; flagellum in ♂ 8-jointed, 1ˢᵗ joint much larger than in ♀,
furnished with rows of hyaline filaments. Antenna 2 in ♀, twice length of
antenna 1; flagellum 24-jointed; flagellum in ♂ 60-jointed, twice length of
peduncle. Gnathopod 1, 6ᵗʰ joint greatly inflated, rounded triangular, the long
palm armed with recurved teeth, those near the prominent prehensile angle
stronger and claw-like; 7ᵗʰ joint strong, curved at the tip. Gnathopod 2 slender;

3[d] joint very long, much more so in ♂ than ♀; the 5th longer in ♀ than ♂; 6th oval, densely setose; 7th minute, inserted on the middle of distal margin of 6th. Peraeopods 1 and 2, 2[d] joint a little dilated, seemingly more in ♂ than in ♀; 4th joint dilated, especially in peraeopod 2, seemingly less in ♂ than in ♀. Peraeopods 3, 4 and 5 rather short, successively longer; 2[d] joint laminar. Telson rounded, almost as broad as long; apex obtusely truncated in ♀, broad, oval in ♂, with a small emargination in young ♂ (Bovallius' figure). Colour whitish with light reddish tinge on sides (Sars), white as ivory; the eyes dark brown (Costa and Bovallius). L. ♀ 22—30, ♂ 12—13 mm.

North-Sea, Skagerrak, North-Atlantic and Arctic Ocean (South-, West- and North-Norway), depth 190--475 m; Mediterranean. Usually on fishes.

3. Gen. **Acidostoma** Lillj.

1865 *Acidostoma* (Sp. un.: *A. obesum*), W. Lilljeborg in: N. Acta Soc. Upsal., ser. 3 v. 6 nr. 1 p. 18, 34 | 1890 *A.*, G. O. Sars, Crust. Norway, v. 1 p. 37.

Body short, robust. Side-plates large, deep. Antenna 1, 2[d] and 3[d] joints short, in ♂ but not in ♀ nearly as stout as the 1st; 1st joint of flagellum small in ♀, in ♂ laminar, densely fringed; accessory flagellum in ♂ and ♀ nearly as long as primary. Antenna 2 rather slender. Upper lip long, narrow. Lower lip with the lobes narrowed. Mandible with minute tooth at each end of cutting edge, narrow, simple; no molar; palp slender, nearly as long as the trunk, armed only at apex. Maxilla 1, inner plate small, narrow; outer plate narrow, tipped with small unguiform teeth, palp a rudiment. Maxilla 2, both plates stiliform. Maxillipeds, inner plates tapering, outer large without marginal teeth, palp scarcely reaching beyond outer plate, its 4th joint rudimentary. Gnathopod 1 rather robust, 5th joint subequal in length to 6th, the latter tapering, without palm. Gnathopod 2, 6th joint narrowly oblong, densely hirsute, 7th wanting. Peraeopods 3, 4 and 5 robust, 2[d] and 4th joints much expanded. Uropod 2 with broad peduncle; uropod 3 very small. Telson short, broad, more or less cleft or emarginate.

2 species.

Synopsis of species:
Telson rather deeply cleft . . . . 1. **A. obesum** . . . p. 14
Telson slightly emarginate . . . . 2. **A. laticorne** . . p. 14

1. **A. obesum** (Bate) 1862 *Anonyx obesus*, Bate, Cat. Amphip. Brit. Mus., p. 74 t. 12 f. 1 | 1865 *Acidostoma obesum*, W. Lilljeborg in: N. Acta Soc. Upsal., ser. 3 v. 6 nr. 1 p. 34 t. 5 | 1890 *A. o.*, G. O. Sars, Crust. Norway, v. 1 p. 38 t. 14 f. 2.

Pleon segment 3 with postero-lateral angles narrowly rounded. Eyes small, rounded, brownish. Antenna 1 in ♀, flagellum 7-jointed, accessory flagellum 5-jointed, in ♂ 1st joint of flagellum large, flattened, but arched below and armed with sensory filaments. Uropod 3 scarcely half as long as uropod 2, the rami lanceolate, unarmed, the outer one the longer. Telson almost as broad as long, rounded, cleft beyond the middle. Colour pale orange with light red bands across the segments. L. 5 mm.

North-Atlantic, North-Sea and Skagerrak (Moray Firth, Shetland Islands, South- and West-Norway, Bohuslän, South-West-England, West-France). Depth, low tide mark to 90 m.

2. **A. laticorne** O. Sars 1879 *A. laticornis*, G. O. Sars in: Arch. Naturv. Kristian., v. 4 p. 440 | 1885 & 86 *A. laticorne*, G. O. Sars in: Norske Nordhavs-Exp., v. 6 Crust. I p. 152 t. 13 f. 3, 3 a (♂) | ?1893· *A. l.*, A. Della Valle in: F. Fl. Neapel, v. 20 p. 782. 287, 769; t. 6 f. 12; t. 28 f. 1—21 (♀) | 1886 *A. obesum* (err., non *Anonyx obesus* Bate 1862!) + *A. l.*, G. O. Sars in: Norske Nordhavs-Exp., v. 6 Crust. II p. 43, 44.

Differs from A. obesum in the following points. Pleon segment 3, postero-lateral angles slightly projecting, but not acute. No eyes (small eyes, with a little biconvex cornea, Della Valle). Uropod 3 with the rami almost tubercular, scarcely reaching beyond peduncle of uropod 2. Telson almost quadrate, slightly emarginate distally. Colour whitish (orange-yellow, Della Valle). L. ♀ 11, ♂ 7·5 mm.

Arctic Ocean (Lofoten Isles), depth 800—1200 m; ? Mediterranean, on Cereactis aurantiaca (Chiaje).

4. Gen. Acontiostoma Stebb.

1888 Acontiostoma (Sp. typ.: A. marionis) (part.), T. Stebbing in: Rep. Voy. Challenger, v. 29 p. 709 | 1893 A., A. Della Valle in: F. Fl. Neapel, v. 20 p. 785.

Body compact (Fig. 4); side-plates large, deep, 1st pair projecting over sides of head. Antenna 1, 2^d joint not extremely short, both flagella with few and very short joints, only in ♂ 1st of primary enlarged and fringed with filaments. Upper lip with pointed apex. Lower lip with the lobes narrowed. Mandible long and narrow, without molar, palp slender, much shorter than the trunk. Maxilla 1, inner plate small, narrow, outer plate narrow with serrate spines, palp small, distinct, 1-jointed. Maxilla 2, both plates narrow. Maxillipeds, the basal part robust, inner plates narrow, apically acute, outer plates apically angular, without marginal teeth, palp geniculate, scarcely reaching beyond outer plate, 1st joint short, 2^d and 3^d long, 4th small. Gnathopod 1 slender, not subchelate, 5th joint shorter than 6th. Gnathopod 2 with 6th and 7th joints forming a feeble chela. Peraeopods 3, 4 and 5 robust, 2^d and 4th joints much expanded. Uropod 3 very short, without rami. Telson short, entire or emarginate.

Fig. 4. A. marionis.
Lateral view.

2 species.

Synopsis of species:

Integument solid, opaque . . . . 1. A. marionis . . . . p. 15
Integument thin, pellucid . . . . 2. A. magellanicum . p. 15

1. A. marionis Stebb. 1888 A. m., T. Stebbing in: Rep. Voy. Challenger, v. 29 p. 709 t. 30 | 1893 A. m., A. Della Valle in: F. Fl. Neapel, v. 20 p. 786.

Pleon segments 1—3 distally carinate on the back, 4th with depression followed by elevated carina produced a little backward. Eyes large, oval. Antenna 1, 1st joint as long as rest of antenna, 2^d as long as the 7-jointed flagellum, accessory flagellum 2-jointed. Antenna 2 little longer than antenna 1, ultimate joint of peduncle much shorter than penultimate, a little shorter than the 7-jointed flagellum. Mandible with spine-row of 3 or 4 spinules, and in place of molar a ciliated tract. Maxilla 1, palp small, acute, 1-jointed. Maxillipeds, inner plates short, 4th joint of palp very short, with small oval spine within the tip. Gnathopod 2, 5th joint a little longer than 6th. Uropods with many spines. Telson not much longer than broad, apex rounded, girt with short spines. L. 7·5, height at centre rather over 5 mm.

Southern Indian Ocean (Marion Island). Depth 90—135 m.

2. A. magellanicum Stebb. 1888 A. m., T. Stebbing in: Rep. Voy. Challenger, v. 29 p. 714 t. 31.

Pleon segments 1—3 not carinate, 4th with depression followed by small hump overhanging the following segments. Eyes small. Antenna 1 and 2 nearly as in preceding species; flagellum 4-jointed, accessory flagellum 2-jointed. Antenna 2 rather shorter than antenna 1, flagellum 4-jointed. Mandible with strap-shaped secondary plate on left mandible, spine-row of 6 spinules, and ciliated tract. Maxilla 1, palp small, acute, 1-jointed. Maxillipeds, inner plates short, 4th joint of palp forming a small but very distinct, straight, tapering finger. Gnathopod 2, 5th joint a little shorter than 6th. Uropods almost entirely devoid of spines. Telson short and broad, emarginate or a little cleft, with 2 spines on each apex. L. 3 mm.

Possibility that this may be the young of A. marionis.

Entrance to the Strait of Magellan (Cape Virgins). Depth 100 m.

5. Gen. **Stomacontion** Stebb.

1888 *Acontiostoma* (part.), T. Stebbing in: Rep. Voy. Challenger, *v.* 29 p. 709 | 1899 *Stomacontion* (Sp. typ.: *Acontiostoma pepinii*), T. Stebbing in: Ann. nat. Hist., ser. 7 *v.* 4 p. 205.

Agreeing in general with Acontiostoma (p. 15), but distinguished by maxilla 1 with palp 2-jointed, maxilliped with 4th joint of palp rudimentary, uropod 3 ending in a tubercular ramus.

1 species.

1. **S. pepinii** (Stebb.) 1888 *Acontiostoma p.* + *A. kergueleni*, T. Stebbing in: Rep. Voy. Challenger, *v.* 29 p. 716 t. 32 (♀ or juv.); p. 720 t. 33 (♂) | 1899 *A. p.*, *Stomacontion sp. typ.*, T. Stebbing in: Ann. nat. Hist., ser. 7 *v.* 4 p. 206.

♀ or juv. Peraeon broadly rounded, afterpart of pleon compressed. Pleon segment 3 dorsally rounded, rising above the 4th, which has a deep dorsal excavation, while the end is strongly upturned, with the process rounded behind. Eyes very small. Antennae about as in Acontiostoma, but peduncle of antenna 1 more robust, flagellum 5-jointed, accessory flagellum 2-jointed. Antenna 2, flagellum 4-jointed. Mandible with 7 spinules and ciliated tract. Maxilla 1, palp minute, 2-jointed. Maxillipeds, inner plates long, 4th joint of palp quite rudimentary. Gnathopod 2, 5th and 6th joints subequal. Peraeopod 5, 2[d] joint at lower hind corner quadrate, the point a little incised. Uropods 1 and 2, each with subequal rami, uropod 3 with a single tubercular ramus. Telson with small apical notch accompanied by spines and cilia. L. 2·5—7·5 mm. — ♂. Pleon segment 3 ending dorsally in a little upturned tip. Eyes large, oval. Antenna 1 distinguished by having 1st joint of 5-jointed flagellum enlarged and fringed with filaments, accessory flagellum 2-jointed. Antenna 2, flagellum 4-jointed. Mandible, with small bifid accessory plate on the left; 2 or 3 spinules seen, but no ciliated tract. Gnathopod 2, 5th joint longer than 6th. Peraeopod 5, 2[d] joint a little rounded at lower hind corner. Uropods 1 and 2, each with the rami unequal. Uropod 3 with one tubercular ramus. Telson with apical notch, and spinules. L. 7·5 mm.

Royal Sound [Kerguelen Island]. Depth 50 m.

6. Gen. **Hirondellea** Chevreux

1889 *Hirondellea* (Sp. un.: *H. trioculata*), Chevreux in: Bull. Soc. zool. France, *t.* 14 p. 285 | 1893 *H.*, A. Della Valle in: F. Fl. Neapel, *v.* 20 p. 836 | 1893 *Hirondella* (laps.), J. V. Carus in: Zool. Anz., Regist. 11—15 p. 142.

Side-plate 1 narrow. 3 eyes. Mouth-organs prominent. Mandible elongate, palp attached behind molar. Maxilla 1 robust, palp broad and long, with teeth few but strong. Maxilla 2, plates broad, outer less so than inner. Maxillipeds, inner plates broad, quadrangular, obliquely truncate, outer plates broad, ovate, not reaching end of 2^d joint of palp. Gnathopod 1 subchelàte, gnathopod 2 minutely chelate. Peraeopods slender. Telson deeply emarginate.

1 species.

1. **H. trioculata** Chevreux 1889 *H. t.*, Chevreux in: Bull. Soc. zool. France. *v.* 14 p. 285 f.

Body thick, peraeon carinate. Head short, lateral lobes broadly rounded. Side-plates 1—3 narrow, 4th with smoothly rounded hind lobe. Pleon segment 3 with postero-lateral angles a little produced and acute, segment 4 with dorsal depression followed by rounded carina. One large oval eye occupying dorsal breadth of the head, two others narrowly crescent-shaped on the lateral lobes. Antenna 1, peduncle short and thick, flagellum with 12 joints, only the 1st long; accessory flagellum with 8 joints, only the 1st long. Antenna 2 not much longer than antenna 1, ultimate joint of peduncle rather shorter than penultimate, flagellum 13-jointed. Gnathopod 1, 6th joint rather longer than 5th, rectangular, palm almost straight, not long. Peraeopods 3—5, expanded 2^d joint much longer than broad, largest in 5th peraeopod, which is shorter than 4th. Uròpods rather long, reaching back equally far. Uropod 2, inner branch a little longer than outer, with spiniferous contraction above the apex. Telson longer than peduncle of uropod 3, divided scarcely to the centre, the lobes strongly divergent. Colour bright amber yellow, sometimes tinted with rose, eyes yellow. L. 13 mm.

North-Atlantic (Azores). Depth 1236 m.

7. Gen. **Podoprionella** O. Sars

1895 *Podoprionella* (Sp. un.: *P. norvegica*). G. O. Sars, Crust. Norway, *v.* 1 p. 687.

Side-plate 1 short, 4th broadly and squarely expanded behind. Upper lip narrowly rounded. Mandibles, cutting edge smooth, spines rudimentary, molar wanting, palp attached far back, its 2^d joint elongate, 3^d short. Maxilla 1, no inner plate observed, outer plate with strong teeth, palp normal. Maxilla 2, plates very short. Maxillipeds, inner plates narrow, tapering, outer plates large, both pairs nearly smooth, palp not reaching end of outer plate, 4th joint obsolete. Gnathopod 1 rather strong, chelate, 6th joint much longer than 5th. Gnathopod 2 minutely chelate. Peraeopods 3—5, expanded 2^d joint with hind margin deeply pectinate. Uropod 1 much longer than uropod 2, which is much longer than the small uropod 3. Telson entire, squamiform.

1 species.

1. **P. norvegica** O. Sars 1895 *P. n.*, G. O. Sars, Crust. Norway, *v.* 1 p. 688 t. v.

Body compact, peraeon thick; pleon small, segment 3 with small upturned tooth at hind corners. Eyes rounded oval, with 10 large and strongly refractive lenses, pigment dark brown. Antenna 1 in ♀, 2^d and 3^d joints not extremely short, flagellum 5-jointed, shorter than peduncle, accessory flagellum 2-jointed. Antenna 2 scarcely as long as antenna 1, ultimate joint of peduncle shorter than penultimate, flagellum 4-jointed, very short. Gnathopod 1, a considerable cavity between the acute thumb and the curved finger forming the chela.

Gnathopod 2 long and slender. 6th joint almost linear, exceeding $^{1}/_{2}$ length
of the long narrow 5th joint. Peraeopods 1—5, joints 3—7 slender, 6th longer
than 5th. Uropod 1 with subequal rami, outer with 2 indents on outer margin,
inner with 1 on each margin. Uropod 2 similar to uropod 1, except in size.
Uropod 3, inner ramus scarcely more than half as long as outer. Telson
rounded, with small indent on each side of rounded apex. Body pellucid,
yellowish, showing a globular dilatation of intestine within hinder part of
peraeon. L. 3 mm.

Trondhjemsfjord. Depth 110—150 m.

8. Gen. **Podoprion** Chevreux

1891 *Podoprion* (Sp. un.: *P. bolivari*), Chevreux in: Mém. Soc. zool. France, *v.* 4
p. 6 | 1893 *P.*, A. Della Valle in: F. Fl. Neapel, *v.* 20 p. 774.

Side-plate 1 much smaller than the following 3. Antenna 1 shorter
than antenna 2; both with well developed flagella. Epistome little prominent,
upper lip rounded. Lower lip with broad lobes, the mandibular processes
turned forward instead of as usual backward. Mandibles broad and short,
cutting edge quadri-dentate, accessory plate having 6 or 7 little conical teeth;
molar little prominent, very near the extremity of the mandible, furnished
with hooked and bearded teeth (perhaps the spine-row, with plumose seta-like
spines); palp with 2^{d} joint much the longest. Maxilla 1, inner plate with
6 plumose setae, outer plate with several simple spines, divergent. of different
sizes; the 2 joints of palp equal in length, 5 little spines on the apex. Maxilla 2,
inner plate the broader with 14 plumose spines, outer plate with 12 simple
spines. Maxillipeds, inner plates apically truncate; outer plates broadly oval not
quite reaching end of long 2^{d} joint of palp, teeth and spines fringing the
inner and distal margins; 4th joint of palp curved, finger-like. Gnathopod 1
chelate, 6th joint much longer than 5th, the thumb short, curved, acute,
leaving only a small space in the chela formed with it by the short hooked
finger. Gnathopod 2 slender, 6th joint much shorter than 5th, forming a
sort of chela with the minute finger, but its produced apex rounded, not
acute. Peraeopods 1—5 slender; all elongate, but especially the 4th and 5th;
6th joint in all longer than 5th. In peraeopod 3 the expanded 2^{d} joint has
the hind margin cut into 6 large teeth; in the following peraeopods it is
only slightly denticulate. Uropod 1 with stiliform equal rami; uropod 2 much
shorter, inner ramus the longer; uropod 3 with long lanceolate rami, the
outer exceeding the other by a spine-like 2^{d} joint. Telson long, deeply
cleft, the apices standing apart, each notched for a spinule.

1 species.

1. **P. bolivari** Chevreux   1891 *P. b.*, Chevreux in: Mém. Soc. zool. France, *v.* 4
p. 6 t. 1.

Side-plate 4 comparatively narrow, shallowly excavate behind. Pleon
segment 3 with corners quadrate, segment 4 with deep dorsal depression.
Eyes large, reniform, red. Antenna 1, 1st joint with a distal tooth, 2^{d} and
3^{d} very short; flagellum with 19 joints, 1st long, with brush; accessory flagellum
5-jointed. Antenna 2 nearly half as long as body; ultimate and penultimate
joints of peduncle subequal; flagellum with 30 short joints. In the telson
the outer arm of each notched apex is longer than the inner. Colour pale
rose. L. 11 mm.

North-Atlantic (Vigo [Spain]). Depth 20 m.

9. Gen. **Euonyx** Norm.

1867 *Euonyx* (Sp. un.: *E. chelatus*), A. M. Norman in: Rep. Brit. Ass., Meet. 36 p. 197, 202 | 1888 *E.*, T. Stebbing in: Rep. Voy. Challenger, *v.* 29 p. 668 | 1891 *E.*, G. O. Sars, Crust. Norway, *v.* 1 p. 116 | 1893 *E.*, A. Della Valle in: F. Fl. Neapel, *v.* 20 p 841 | 1876 *Leptochela* (Sp. un.: *Opis leptochela*) (non Stimpson 1860, Decapoda!), A. Boeck. Skand. Arkt. Amphip., *v.* 2 p. 190.

Side-plate 1 small, partly concealed by 2^d. Mouth-parts differing in the 2 species. Mandibles with cutting edge almost simple, molar weak or obsolete, palp central or subcentral. Maxilla 1, inner plate not large, outer plate broad, obliquely truncate, palp normal. Maxilla 2, inner plate the shorter, obliquely truncate. Maxillipeds, 2^d joint of palp reaching beyond the outer plates. Gnathopods 1 and 2 slender, 3^d joint elongate; gnathopod 1 chelate, gnathopod 2 subchelate. Uropod 3, rami lanceolate, the outer with a small second joint. Telson oblong, deeply cleft.

2 species.

Synopsis of species:

Gnathopod 1, 5th and 6th joints subequal . · . . . . 1. **E. chelatus** . . p. 19
Gnathopod 1, 5th joint much shorter than 6th . . . . . . 2. **E. normani** . . p. 19

1. E. chelatus Norm. 1867 *E. c.*, A. M. Norman in: Rep. Brit. Ass., Meet. 36 p. 197, 202 | 1888 *E. c.*, T. Stebbing in: Rep. Voy. Challenger, *v.* 29 p. 673 | 1891 *E. c.*, G. O. Sars, Crust. Norway, *v.* 1 p. 117 t. 40 f. 1 | 1893 *E. c.*, A. Della Valle in: F. Fl. Neapel, *v.* 20 p. 842 | 1868 *Opis leptochela*, Bate & Westwood. Brit. sess. Crust., *v.* 2 p. 501 ! 1876 *O. l.*, *Leptochela*, A. Boeck. Skand. Arkt. Amphip., *v.* 2 p. 190.

Pleon segment 3, postero-lateral angles produced to a short blunt point; segment 4 with dorsal depression followed by high compressed hump. Eyes a constricted oval, chalky white. Antenna 1 large, 1st joint concave above; flagellum twice length of peduncle, with 10 joints, 1st very large; accessory flagellum 5-jointed. Antenna 2 twice length of antenna 1. ultimate and penultimate joints of peduncle equal, flagellum 22-jointed. Mandible with molar obsolete, palp on distal side of centre. Maxilla 1, inner plate with several apical setae. Maxillipeds, outer plates rather small with slender setae on the margin, but no spine-teeth. Gnathopod 1 with few setules, 3^d joint longer than 5th or 6th, 6th with hind margin straight, the closely adjacent finger and thumb forming a chela of about $1/_3$ length of the hand. Gnathopod 2 longer and stronger than gnathopod 1, 6th joint about $1/_2$ length of 5th, distally widened, palm concave. 7th joint strong and curved. Peraeopods 3—5, 2^d joint not very large, rounded, 4th much expanded. Uropod 3 with the rami unarmed, the inner reaching 2^d joint of outer. Telson without spinules, oblong, cleft much beyond the middle, apices blunt. Colour whitish tinged with yellow. L. ♀ 10 mm.

Arctic Ocean, North-Atlantic and Irish Sea (Finmark. Trondhjemsfjord, Hebrides, Isle of Man). Depth 95—280 m.

2. E. normani Stebb. 1888 *E. n.*, T. Stebbing in: Rep. Voy. Challenger, *v.* 29 p. 669 t. 19.

Pleon segment 3, postero-lateral angles blunt; segment 4 a little raised behind the dorsal depression. Eyes indistinct, narrowly oval. Antenna 1, 1st joint stout, not concave above; flagellum more than twice length of peduncle, with 29 joints, 1st very large; accessory flagellum 9-jointed. Antenna 2,

2*

not much longer than antenna 1, ultimate joint of peduncle shorter than penultimate, flagellum 35-jointed. Mandible with molar distinct but not denticulate,
palp central. Maxilla 1 with 3 plumose setae on inner plate. Maxillipeds, outer
plates rather large, but not reaching end of 2^d joint of palp, spine-teeth of
inner margin minute and numerous. Gnathopod 1 with few setules, 3^d joint
longer than 5th, much shorter than 6th, 4th and 5th joints equal, 6th with
hind margin concave, the closely adjacent finger and thumb forming a chela
of not more than $^1/_4$ length of hand. Gnathopod 2 longer, not stronger than
gnathopod 1, 6th joint slender, not widening distally, about $^1/_2$ length of 5th,
the minute 7th joint not curved, but closely adjusted to the oblique palm.
Peraeopods 3—5, 2^d joint roundly quadrangular, 4th very moderately expanded.
Uropod 3 with many little spines on the short broad rami, of which the
inner reaches the 2^d joint of the outer. Telson with some marginal spinules,
oblong, cleft $^3/_4$ length, apices blunt, notched. L. ♀ 15 mm.

South-Pacific (Kermadec Islands). Depth 1140 m.

10. Gen. Opisa Boeck

1842 Opis (Sp. un.: O. eschrichtii) (non Defrance 1825, Mollusca!), Krøyer in:
Naturh. Tidsskr., v. 4 p. 149 | 1876 Opisa (Sp. typ.: Opis typica), A. Boeck, Skand.
Arkt. Amphip., v. 2 p. 190 | 1890 O., G. O. Sars, Crust. Norway, v. 1 p 36 | 1893 O.
(part.), A. Della Valle in: F. Fl. Neapel, v. 20 p. 806.

Side-plate 1 short, partly concealed by 2^d, 4th deeply excavate behind.
Upper lip rather broadly rounded. Mandible. cutting edge smooth, spine-row
feeble, molar obscure, in any case without triturating surface; palp rather
far back, 2^d joint strongly setiferous. Maxilla 1 normal, 2 setae on narrow
inner plate. Maxilla 2, plates setose only apically. Maxillipeds, inner plates
normal, outer plates angular at apex, inner margin denticulate, palp normal,
rather short. Gnathopod 1 chelate. 5th joint small, 6th greatly widened, the
thumb acute, not very long, leaving a great cavity between itself and the
much curved finger (7th joint). Gnathopod 2 subchelate. Peraeopods 1—5
not elongate, most of the joints slender, 2^d joint in peraeopods 3—5 greatly
expanded. Uropod 3 elongate. Telson elongate, deeply cleft.

1 species.

1. O. eschrichtii (Krøyer) 1842 Opis e., (Holbøll in MS.) Krøyer in:
Naturh. Tidsskr., v. 4 p. 149 | 1876 Opisa eschrichti, A. Boeck, Skand. Arkt. Amphip.,
v. 2 p. 190 | 1890 O. e., G. O. Sars, Crust. Norway, v. 1 p. 36 t. 14 f. 1 | 1893 O. eschrichtii,
A. Della Valle in: F. Fl. Neapel, v. 20 p. 806 | 1846 Opis typica, Krøyer in: Naturh.
Tidsskr., ser. 2 v. 2 p. 46 | 1846 O. t., Krøyer in: Voy. Nord, Crust. t. 17 f 1.

Pleon segment 3, postero-lateral angles rounded. Eyes rather large,
oblong, pigment dark brown. Antenna 1, 1st joint very large, flagellum
8-jointed, about as long as peduncle, 1st joint very large, especially in ♂,
accessory flagellum 5-jointed, rather long. Antenna 2 in ♀ scarcely longer
than antenna 1, in ♂ as long as body, with calceoli. Gnathopod 1, 6th joint
nearly globular. Gnathopod 2, 6th joint about $^1/_2$ length of 5th, slightly
dilated distally, apex truncate. Peraeopods 3—5, 2^d joint large, rounded
oval, nearly as long as rest of limb. Uropod 3, rami much longer than
peduncle, outer ramus the longer, with distinct terminal joint. Telson about
3 times longer than broad, cleft nearly to base, 2 pairs of dorsal and 1 of
terminal spinules. Whitish, with dark brown intestine seen through skin.
L. 7—8 mm.

Arctic Ocean, North-Atlantic, North-Sea and Skagerrak (Iceland, Norway);
Korea Sea, depth 66 m.

11. Gen. **Sophrosyne** Stebb.

1888 *Sophrosyne* (Sp. un.: *S. murrayi*), T. Stebbing in: Rep. Voy. Challenger, *v.* 29 p. 652 | 1891 *S*, T. Stebbing & D. Robertson in: Tr. zool. Soc. London, *v.* 131 p. 31 | 1893 *S.*, A. Della Valle in: F. Fl. Neapel, *v.* 20 p. 795 | 1899 *Paropisa* (Sp. typ.: *Opisa hispana*), T. Stebbing in: Ann. nat. Hist., ser. 7 *v.* 4 p. 206.

Mandible, molar wanting, palp affixed far forward, 2^d and 3^d joints subequal. Maxilla 1, inner plate small, with 1 seta, outer plate feebly armed, palp with few spine-teeth. Maxilla 2, inner plate much shorter than outer, neither strongly setose. Maxillipeds slender, inner and outer plates very small, the palp elongate. Gnathopod 1 robust, 5th joint triangular, somewhat produced, 6th broadly oblong. produced into a sharp tooth, opposed to the rather powerful finger almost chelately. Gnathopod 2 subchelate. Peraeopods, 7th joint rather elongate. 2^d joint broad in peraeopods 3—5. Uropod 3, rami unarmed, the inner very little shorter than the outer. Telson not very elongate, partially cleft.

3 species.

Synopsis of species:

$$1 \begin{cases} \text{Peraeopod 5, hind margin of 2}^d \text{ joint produced} \\ \quad \text{downward} \dots \dots \dots \dots \dots \text{3. S. hispana} \dots \text{p. 22} \\ \text{Peraeopod 5, hind margin of 2}^d \text{ joint not produced} \\ \quad \text{downward} - \textbf{2.} \end{cases}$$

$$2 \begin{cases} \text{Telson cleft beyond the centre} \dots \dots \dots \text{1. S. murrayi} \dots \text{p. 21} \\ \text{Telson not cleft to the centre} \dots \dots \dots \text{2. S. robertsoni.} . \text{p. 21} \end{cases}$$

1. **S. murrayi** Stebb. 1888 *S. m.*, T. Stebbing in: Rep. Voy. Challenger, *v.* 29 p. 652 t. 15 | 1893 *S. m.*, A. Della Valle in: F. Fl. Neapel, *v.* 20 p 795 t. 60 f. 38.

Pleon segment 3, postero-lateral angles acutely upturned, back with a pair of humps and distally squared, segment 4 abruptly narrower. Antenna 1, 1st joint rather elongate, 2^d longer than the 1st joint of the 7-jointed flagellum, accessory flagellum 4-jointed. Antenna 2, ultimate joint of peduncle shorter than penultimate, flagellum 8-jointed. Maxillipeds, inner plates not quite reaching base of palp. Gnathopod 1, 3^d, 4th and 5th joints strongly and densely spined along the hind margin, palm of 6th joint spinulose. Gnathopod 2, 6th joint more than half as long as 5th, distally widened, with concave palm, finger curved. Uropod 3, rami short, the inner a little shorter than the outer. Telson as broad as long, cleft dehiscent, not quite $^2/_3$ of the length, apices broadly rounded. L. ♀ about 13 mm.

Christmas Harbour [Kerguelen Island].

2. **S. robertsoni** Stebb. & D. Roberts. 1891 *S. r.*, T. Stebbing & D. Robertson in: Tr. zool. Soc. London, *v.* 131 p. 31 t. 5A | 1893 *S. r.*, A. Della Valle in: F. Fl. Neapel, *v.* 20 p. 795.

Pleon segment 3, postero-lateral angles acutely upturned, back arched, not distally squared, segment 4 with 3 little dorsal humps. Antenna 1, 1st joint rather tumid, 2^d longer than the short 1st joint of the 6-jointed flagellum, accessory flagellum 3-jointed. Antenna 2, ultimate joint of peduncle shorter than penultimate, flagellum 8-jointed. Maxillipeds, inner plates not nearly reaching base of palp. Gnathopod 1 as in S. murrayi, but more feebly armed. Gnathopod 2, 6th joint narrow just at the base, but thence of uniform width, palm very small, convex. Uropod 3, rami scarcely showing any difference in length. Telson longer than broad, cleft dehiscent, not quite reaching the middle, apices broad, slightly indented. L. ♀ 6 mm.

Firth of Clyde [Scotland].

3. **S. hispana** (Chevreux) 1887 *Opis h.*, Chevreux in: Bull. Soc. zool. France, *v.* 12 p. 567 | 1888 *Opisa h.*, T. Stebbing in: Rep. Voy. Challenger, *v.* 29 p. 1641 | 1893 *O. h.*, A. Della Valle in: F. Fl. Neapel, *v.* 20 p 807 | 1899 *O. h.*, *Paropisa sp. typ.*, T. Stebbing in: Ann. nat. Hist., ser. 7 *v.* 4 p. 206.

Head, lateral angle little prolonged, rounded. Side-plate 1 broader and deeper than the 2 following. Pleon segment 3, postero-lateral corners slightly produced backward, quadrate, 4th slightly gibbous. Eyes wanting. Antenna 1, flagellum 4-jointed, with long setae, accessory flagellum 2-jointed. Antenna 2, penultimate joint rather longer than ultimate of peduncle. Gnathopods 1 and 2 apparently as in S. robertsoni. Peraeopod 5, 2^d joint widely expanded behind and produced downward to middle of 4th, its margin dentate. Uropod 3, rami smooth, a little longer than peduncle. Telson as broad as long, cleft to the centre. L. 2 mm.

North-Atlantic (Cape Finisterre). Depth 510 m.

12. Gen. **Valettia** Stebb.

1888 *Valettia* (Sp. un.: *V. coheres*), T. Stebbing in: Rep. Voy. Challenger, *v.* 29 p. 723 | 1893 *V.*, A. Della Valle in: F. Fl. Neapel, *v.* 20 p. 772.

None of the side-plates especially large. Mandible broad and short, the cutting edge strongly dentate, secondary plate on left mandible also strongly dentate, molar denticulate, palp central. Maxillipeds, inner plates with more than 3 apical teeth, outer plates at apex acutely produced, inner margin without teeth, 1st joint of palp not shorter than 2^d. Gnathopods 1 and 2 subchelate, the 2^d little weaker than the 1st and of similar structure except in having the 3^d joint elongate. Peraeopods 3—5, with the 2^d joint so little expanded, that these limbs are not imbricated as usual, but stand apart. Uropods successively shorter, rami of each pair unequal. Telson short, broad, cleft beyond the middle.

1 species.

1. **V. coheres** Stebb. 1888 *V. c.*, T. Stebbing in: Rep. Voy. Challenger, *v.* 29 p. 724 t. 34.

Side-plate 4 little excavate behind. Pleon segment 3, angles acute, a little upturned, 4th with dorsal depression followed by small distal hump. Eyes not observed. Antenna 1, peduncle short, stout, 1st joint scarcely longer than broad, flagellum with 13 or 14 joints, 1st subequal in length to peduncle, accessory flagellum with 1 long and 3 short joints. Antenna 2 subequal in length to antenna 1, ultimate joint of peduncle shorter than penultimate, flagellum 14-jointed. Maxilla 1, inner plate with 5 short plumose setae. outer with 8 or 9 slender spines, 2^d joint of palp long, with 6 spine-teeth. Maxilla 2, with about 12 spines on each plate. Gnathopod 1, 3^d and 4th joints short, 5th shorter than 6th, which is oblong, not $1^1/_2$ times as long as broad, palm rather deeply concave, 7th joint curved, not massive, impinging on palmar spines. Gnathopod 2 similar to gnathopod 1, but 3^d joint longer than 4th, and 5th longer than 6th, which has the palm not concave, but sloping inwards, a little crenate. Peraeopods 1 and 2, 4th joint broader and much longer than 5th. Peraeopod 3, 2[d] joint a narrow oval. Peraeopod 4, 2[d] joint rather larger than in peraeopod 3. Peraeopod 5, 2[d] joint expanded above, narrowing below. Uropod 3, rami short, broad, the outer the longer by a minute 2[d] joint. Telson with rounded apices. L. 12·5 mm.

Antarctic Ocean (lat. 62° 26′ S.) Depth 3612 m.

13. Gen. **Amaryllis** Hasw.

1880 *Amaryllis*, Haswell in: P. Linn. Soc. N.S.Wales, *v.*4 p.253 | 1888 *A.*, T. Stebbing in: Rep. Voy. Challenger, *v.*29 p.698 | 1893 *A.*, A. Della Valle in: F. Fl. Neapel, *v.*20 p.781.

Head very deep. Peraeon segment 1 shallow, segments successively deeper to 2[d] of pleon. Side-plate 1 very small, 4th abruptly larger than those preceding it. Antenna 1 (Fig. 5), 1st joint not very stout, 2[d] and 3[d] not very short. Mandible, cutting edge simple, secondary plate of left mandible denticulate, spine-row with many short spines, molar feeble, ciliated, not denticulate, palp behind centre of trunk. Maxilla 1, inner plate with 2 plumose setae, palp wanting. Maxillipeds, inner and outer plates large, outer without spine-teeth, · 4th joint of palp small, obtuse, without a nail. Gnathopod 1 simple. Gnathopod 2 subchelate. Telson cleft.

3 species accepted, 1 doubtful.

Synopsis of accepted species:

1 { Antenna 1, 2[d] joint much longer than 3[d] . . 1. **A. haswelli** . . . . . p. 23
  { Antenna 1, 2[d] joint not much longer than 3[d] — 2.

2 { Peraeopod 3, 2[d] joint expanded above. . . . 2. **A. macrophthalma** . . p. 24
  { Peraeopod 3, 2[d] joint not expanded above . . 3. **A. bathycephala** . . . p. 24

1. A. haswelli Stebb. 1888 *A. h.*, T. Stebbing in: Rep. Voy. Challenger, *v.*29 p.703 t.28 | 1893 *A. h.*, A. Della Valle in: F. Fl. Neapel, *v.*20 p.781.

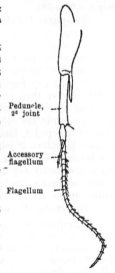

Fig. 5. **A. haswelli.** Antenna 1.

Pleon segment 3, with the angles acute, not upturned, the hind margin bulging above them. Eyes not distinctly observed. Antenna 1 (Fig. 5), 1st joint elongate, and ending in a tooth ¹/₂ length of 2[d] joint; 2[d] thrice length of 3[d]; flagellum with 24 joints. 1st shorter than last of peduncle; accessory flagellum 4-jointed. Antenna 2 subequal to antenna 1, ultimate joint of peduncle about ¹/₂ length of penultimate, flagellum 22-jointed. Maxilla 1, outer plate carrying 11 spines. Gnathopod 1, 2[d] joint as long as the 4 following combined and much wider than any, 3[d] as long as 4th, 5th longer than the long tapering 6th, which has the hind margin pectinate; there are many spines and setae on these joints; 7th joint small, curved. Gnathopod 2, 2[d] joint not as long as 5th and 6th combined, 3[d] longer than 4th, 5th nearly twice as long as 6th, both slender, densely furred; 7th joint small, much curved, overarching the narrow palmar margin. Peraeopods 1 and 2 slender. Peraeopod 3, side-plate much wider than 2[d] joint, with hind lobe the longer, 4th joint longer and much broader than 5th. Peraeopod 4, 2^{d} joint oblong, with rounded lower hind corner. Peraeopod 5, 2^{d} joint larger than in preceding limb, lower hind corner quadrate. Uropod 2 with rami unequal, the longer having a constriction. Uropod 3 with stiliform subequal rami longer than the peduncle. Telson shorter than peduncle of uropod 3, cleft a little beyond centre, narrowing to the divergent apices. L. ♀ about 11 mm.

   North-Atlantic (Azores). Depth 18.9 m.

   2. **A. macrophthalma** Hasw. 1880 *A. macrophthalmus* + *A. brevicornis*, Haswell in: P. Linn. Soc. N.S.Wales, *v.*4 p.253 t. 8 f. 3; p.254 | 1888 *A. m.*, T. Stebbing in: Rep. Voy. Challenger, *v.* 29 p. 707 t. 29 | 1893 *A. m.* (part.), A. Della Valle in: F. Fl. Neapel, *v.* 20 p. 781.

   Pleon segment 3, postero-lateral corners squarely upturned, forming a little pocket above the point, but in the small Patagonian specimen acute, not upturned. Eyes vertically elongated, subcrescentic. Antenna 1, 3 joints of peduncle not very elongate, 2^{d} and 3^{d} nearly equal, flagella variable, primary with 17, 30, 18, 5 joints in different specimens, accessory flagellum with 13, 7, 5, 2 joints, respectively. Antenna 2, ultimate joint of peduncle shorter than penultimate. Gnathopod 1, 3^{d} joint longer than 4th, 5th joint shorter than the tapering strongly pectinate 6th, 7th joint short, slightly curved. Gnathopod 2 not extremely slender, 5th joint only a little longer than 6th, which is elongate, of uniform width, 7th joint small, rather stout at the base. Peraeopods 3, 4 and 5, 2^{d} joint roundly expanded, 4th joint broad, not elongate. Uropod 2 with rami unequal, the longer constricted. Telson reaching a little beyond peduncle of uropod 3, cleft a little beyond the middle, apices broad, not divergent. L. about 6—20 mm.

   South-Pacific (Tasmania; Port Jackson [East-Australia], sublitoral); Strait of Magellan (Cape Virgins, depth 100 m).

   3. **A. bathycephala** Stebb. 1888 *A. bathycephalus*, T. Stebbing in: Rep. Voy. Challenger, *v.*29 p.699 t. 27 | 1893 *A. macrophthalmus* (part.), A. Della Valle in: F. Fl. Neapel, *v.* 20 p. 781.

   Side-plates 4 distinguished from those of 2 preceding species by being broader than deep and having the backward prolongation much more rounded. Pleon segment 3, postero-lateral corners squarely upturned, forming a little pocket above the point. Eyes large, inversely flask-shaped. Antenna 1, 1st joint rather long, 2^{d} longer than 3^{d}, flagellum with 10 or 11 joints, 1st not longer than 2^{d}, accessory flagellum 3-jointed. Antenna 2, ultimate joint of peduncle a little shorter than penultimate, flagellum 9-jointed. Gnathopod 1 as in the preceding species. Gnathopod 2, 5th joint decidedly longer than 6th, 6th long, dilating towards the oblique palm, 7th joint small, curved, not reaching end of palm. Peraeopod 3, hind lobe of side-plate more produced downwards than in A. macrophthalma, 2^{d} joint pear-shaped, narrow above, with a rounded decurrent lobe below. Peraeopods 4 and 5, 2^{d} joint broadly oval, 4th joint as in peraeopod 3 moderately expanded. Uropod 2, rami unequal, the longer constricted. Telson not nearly reaching end of peduncle of uropod 3, cleft scarcely beyond centre, apices rounded, scarcely divergent. L. about 6 mm.

   Port Phillip [Melbourne]. Depth 60 m.

   A. pulchella Bonnier 1896 *A. pulchellus*, J. Bonnier in: Ann. Univ. Lyon, *v.* 26 p. 624 t. 36 f. 3.

   Probably identical with A. haswelli (p. 23). Gnathopod 1, 5th joint shorter than 6th (equal in fig.). Gnathopod 2, 5th joint not very much longer than 6th and rather narrower. L. ♀ 11 mm.

   Bay of Biscay. Depth 950 m.

14. Gen. **Onisimus** Boeck

1871 *Onisimus* (part.), A. Boeck in: Forb. Selsk. Christian., 1870 p. 111 | 1894 *O.*, T. Stebbing in: Bijdr. Dierk., *v.* 17 p. 10 | 1872 & 76 *Onesimus* (part.), A. Boeck, Skand. Arkt. Amphip., *v.* 1 t. 4—6; *v.* 2 p. 161 | 1891 *O.*, G. O. Sars, Crust. NorWay, *v.* 1 p. 104 | 1893 *O.*, J. Bonnier in: Bull. sci. France Belgique, *v.* 24 p. 174 | 1893 *Pseudalibrotus* (Sp. un.: *P. littoralis*) (part.), A. Della Valle in: F. Fl. Neapel, *v.* 20 p. 798.

Body rather robust. Side-plates of moderate size, pleon segment 3 with postero-lateral corners distinctly produced. Antenna 1, peduncle stout, 2^d and 3^d joints very short, both antennae in ♂ longer than in ♀ and provided with distinct calceoli. Epistome very slightly prominent, defined from upper lip by a distinct sinus. Mandible, molar rather strong, palp not large, set well forward. Maxilla 1, inner plate with 2 setae, outer broad, obliquely truncate, with 11 spines, several (5—8) spine-teeth on palp. Maxilla 2, outer plate broader and much longer than inner. Maxillipeds normal, outer plates crenulate, carrying on rounded apex 1—3 spines, not reaching beyond 2^d joint of slender, rather elongate palp. Gnathopod 1 subchelate, not very robust, 6th joint narrowly oblong, little or not at all longer than 5th. Gnathopod 2, 6th joint uniformly narrow or distally widened. Peraeopods 3—5 rather short and robust. Uropod 3 short, outer ramus with small 2^d joint, and sometimes setiferous. Telson short and broad, never deeply incised, sometimes entire.

8 species.

Synopsis of species:

1 { Eyes large; telson incised — **2.**
{ Eyes small; telson not incised — **5.**

2 { Gnathopod 1, 6th joint longer than 5th . , . 1. **O. edwardsii** . . . . . p. 25
{ Gnathopod 1, 6th joint subequal to 5th — **3.**

3 { Peraeopods 3—5, 4th joint Very robust . . . 2. **O. plautus** . . . . . . p. 26
{ Peraeopods 3—5, 4th joint not Very robust — **4.**

4 { Pleon segment 1, antero-lateral angles uncinate 3. **O. normani** . . . . . p. 26
{ Pleon segment 1, antero-lateral angles not
{ uncinate . . . . 4. **O. turgidus** . . . . . p. 26

5 { Telson apically truncate . . . . 5. **O. brevicaudatus** . . p. 27
{ Telson apically indented — **6.**

6 { Gnathopod 2, palm Widened, emarginate . . . 6. **O. caricus** . . . . . . p. 27
{ Gnathopod 2, palm not Widened, nor emargi-
{ nate — **7.**

7 { Eyes imperfect . . . . 7. **O. leucopis** . . . . . p. 28
{ Eyes normal . . . . 8. **O. affinis** . . . . . . p. 28

1. **O. edwardsii** (Krøyer) 1846 *Anonyx e.*, Krøyer in: Naturh. Tidsskr., ser. 2 *v.* 2 p. 1, 41 | 1846 *A. e.*, Krøyer in: Voy. Nord, Crust. t. 16 f. 1 | 1893 *A. e.* (part.), A. Della Valle in: F. Fl. Neapel, *v.* 20 p. 828 | 1871 *Onisimus e.*, A. Boeck in: Forh. Selsk. Christian., 1670 p. 113 | 1894 *O. e.*, T. Stebbing in: Bijdr. Dierk., *v.* 17 p. 10 | 1872 & 76 *Onesimus e.*, A. Boeck, Skand Arkt. Amphip., *v.* 1 t. 6 f. 4; *v.* 2 p. 167 | 1891 *O. e.*, G. O. Sars. Crust. NorWay, *v.* 1 p. 105 t. 36 f. 1.

Head, lateral angles but slightly projecting, evenly rounded. Side-plate 1 rather expanded below, 4th with hinder expansion short, obtusely truncated. Pleon segment 3, postero-lateral angles produced to a narrow slightly upturned lobe. Eyes rather large, oblong oval, slightly expanding below, bright red.

Antenna 1 in ♀, flagellum with 15 joints. 1st equal to 2[d] and 3[d] combined, accessory flagellum 6-jointed. Antenna 2 a little longer than antenna 1, flagellum 18-jointed. Gnathopod 1, 6th joint much longer than 5th, distally a little curved, palm not very oblique, somewhat arcuate, finely denticulate. Gnathopod 2 very slender, 6th joint more than twice as long as broad, more than half as long as 5th, scarcely dilated distally, palm and finger minute. Peraeopods 3—5, 4th joint moderately robust, 2[d] of 5th pair large, as long as rest of limb. Uropod 3 with 2 spinules on inner margin of inner ramus. Telson rounded, scarcely longer than broad, incision very short, apices broad. Colour pale reddish yellow, eggs in pouch bright orange. L. 11—14 mm.

Arctic Ocean; North-Atlantic and North-Sea (West-Norway). Depth 15—94 m.

2. **O. plautus** (Krøyer) 1845 *Anonyx p.*, Krøyer in: Naturh. Tidsskr., ser. 2 *v.* 1 p. 629 | 1846 *A. p.*, Krøyer in: Naturh. Tidsskr., ser. 2 *v.* 2 p. 39 | 1846 *A. p.* (part.), Krøyer in: Voy. Nord. Crust. t. 15 f. 2 | 1893 *A. p.* (part.), A. Della Valle in: F. Fl. Neapel, *v.* 20 p. 828 | 1866 *Lysianassa plauta*, Goës in: Öfv. Ak. Förb., *v.* 22 p 521 | 1871 *Onisimus plautus*, A. Boeck in: Forh. Selsk. Christian., 1870 p. 112 | 1894 *O. p.*, T. Stebbing in: Bijdr. Dierk., *v.* 17 p. 10 | 1872 & 76 *Onesimus p.*, A. Boeck, Skand. Arkt. Amphip., *v.* 1 t. 4 f. 2; *v.* 2 p. 164 | 1891 *O. p.*, G. O. Sars, Crust. Norway, *v.* 1 p. 107 t. 37 f. 1.

Head, lateral angles acute. Side-plate 1 narrow, scarcely expanded below, 4th with hinder expansion obtuse, very slight. Pleon segment 3, postero-lateral angles with upturned point. Eyes much wider below than above, larger in ♂ than in ♀, red. Antenna 1 in ♀, flagellum with 11 joints, 1st equal to next 3 combined, accessory flagellum 4-jointed. Antenna 2 a little longer than antenna 1, flagellum 13-jointed. Both antennae in ♂ considerably longer than in ♀, flagella slender, with more joints. Gnathopod 1, 6th joint subequal to 5th, palm nearly transverse but with corner rounded off. Gnathopod 2, 6th joint more than half as long as 5th, a little dilated distally. Peraeopods 3—5, 4th joint very robust, 2[d] of peraeopod 5 much longer than rest of limb. Uropod 3 with no spinules on either ramus. Telson. rounded oval, incised for ⅓ of length, apices rather narrow, indented. Colour reddish yellow to light brown. L. 8 mm.

Arctic Ocean; North-Atlantic and North-Sea (West-Norway); Skagerrak (Bohuslän).

3. **O. normani** (O. Sars) 1891 & 95 *Onesimus n.*, (Schneider in MS.) G. O. Sars, Crust. Norway, *v.* 1 p. 106 t. 36 f. 2; p. 686 | 1893 *Anonyx n.*, A. Della Valle in: F. Fl. Neapel, *v.* 20 p. 827.

Head, lateral angles acute. Side-plate 1 scarcely expanded below, 4th with hinder expansion short and blunt. Pleon segment 1, antero-lateral angles sharply uncinate, segment 3 with postero-lateral angles acutely uncinate. Eyes oblong oval, red. Antennae 1 and 2 in ♀ short and stout; antenna 1, flagellum with 8 joints, the 1st large, accessory flagellum 3-jointed. Antenna 2, flagellum 8-jointed. Gnathopod 1, 6th joint scarcely or not longer than 5th, palm not curved, very oblique and spinulose, angle indistinct. Gnathopod 2, 6th joint half length of 5th, scarcely widening distally. Peraeopods 3—5, 4th joint not robust, 2[d] of peraeopod 5 as long as rest of limb. Uropod 2, inner ramus constricted. Uropod 3, rami without spinules. Telson a little longer than broad, incised for rather more than ⅓ of the length, apices moderately broad, indented. Colour pure white, eggs in pouch reddish. L. 9 mm.

Arctic Ocean (Tromsö, Finmark), Christianiafjord.

4. **O. turgidus** (O. Sars) 1879 *Anonyx (O.) t.*, G. O. Sars in: Arch. Naturv. Kristian., *v.* 4 p. 437 | 1885 *O. t.*, G. O. Sars in: Norske Nordhavs-Exp., *v.* 6 Crust. 1 p. 147 t. 12 f. 5, 5 a—i | 1893 *Anonyx plautus* (part.), A. Della Valle in: F. Fl. Neapel, *v.* 20 p. 828.

Body unusually tumid. Head, lateral angles acute. Side-plate 1 little expanded below, 4th forming an almost rectangular corner below the hinder emargination. Pleon segment 3, postero-lateral angles acutely uncinate. Eyes oval, the broad end below, rich vermilion in life. Antenna 1, peduncle tumid, flagellum with 10 joints, 1st very large, accessory flagellum with 4 joints, 1st large. Antenna 2, flagellum 7-jointed. Gnathopod 1 rather stout, 6th joint very little longer than 5th, palm rather oblique. Gnathopod 2, 6th joint (as figured) much more than half as long as 5th, with a very small excavate palm. Peraeopods 3—5, 2d joint large, hind margin distinctly serrate, 4th joint only moderately broad. Uropod 3 without spinules. Telson short and broad, incised $^2/_5$ of the length, apices broad. Colour whitish, faintly tinged with red. L. reaching 15 mm.

Arctic Ocean (midway between Jan Mayen and Finmark). Depth 3220—3310 m.

5. O. brevicaudatus H. J. Hansen 1886 *O. b.*, H. J. Hansen in: Dijmphna Udb., p. 216 t. 21 f. 7—7e | 1894 *O. b.* (part.), T. Stebbing in: Bijdr. Dierk., v. 17 p. 12 | 1893 *Pseudalibrotus littoralis* (err., non *Anonyx l.* Krøyer 1845!) (part.), A. Della Valle in: F. Fl. Neapel, v. 20 p. 799.

Head, lateral angles narrowly rounded. Side-plate 1 somewhat expanded below, 4th with a rather small expansion. Pleon segment 3, postero-lateral angles acutely upturned. Eyes on the projecting lobes of the head, narrowed above, almost triangular, not approaching summit of head. Antenna 1, flagellum in ♀ with 11 joints, in ♂ with 17, 1st large, accessory flagellum with 5 joints, 1st long. Antenna 2, flagellum in ♀ 15-, in ♂ 24-jointed. Both antennae have calceoli in ♂. Gnathopod 1, 6th joint longer than 5th, palm oblique, finger projecting a little beyond it. Gnathopod 2, 6th joint in ♂ with margins subparallel, hinder longer than front, finger fixed in middle of apical margin and closing down on oblique palm; in ♀ with margins diverging distally, hinder shorter than front, finger fixed near front angle and when closed leaving a gap between its concave margin and the excavate palm. Peraeopods 3—5, 4th joint moderately broad. Uropod 3 with spinules on the rami. Telson with breadth and length subequal, distal margin straight or faintly rounded or faintly emarginate. L. ♀ 16·5, ♂ 15·5 mm.

Kara Sea (lat. 71° 13′N.). Depth 110 m.

6. O. caricus H. J. Hansen 1886 *O. c.*, H. J. Hansen in: Dijmphna Udb., p. 214 t. 21 f. 6 | 1894 *O. c.*, T. Stebbing in: Bijdr. Dierk., v. 17 p. 11 | 1893 *Anonyx edwardsii* (part.), A. Della Valle in: F. Fl. Neapel, v. 20 p. 828.

Head, lateral angles slightly projecting, broadly rounded. Side-plate 1 considerably expanded below, 4th with hinder expansion rather short, obtusely truncated. Pleon segment 3, postero-lateral angles produced upwards into a short blunt point. Eyes on the projecting lobes of the head, narrowed above, almost triangular, not approaching summit of head. Antenna 1, flagellum in ♀ with 16 joints, 1st equal to 2 or 3 next combined, accessory flagellum 4-jointed; flagellum in ♂ with 30 joints, 1st joint equal to 4 or 5 next combined, accessory flagellum with 5 joints, 1st elongate. Antenna 2, flagellum in ♀ 20-, in ♂ 37-jointed. Maxillipeds with 3 spines on apical margin of outer plates, where other species appear to have only one. Gnathopod 1, 6th joint longer than 5th, palm oblique. spinules fringing it and part of the hind margin. Gnathopod 2, 6th joint in both sexes rather more than half as long as 5th, subtriangular, expanding distally, front margin rather the longer, finger stout, leaving an interval between its concave margin and the excavate palm.

Peraeopods 3—5, 4ᵗʰ joint moderately broad. Uropod 3 with spinules on both rami and setae on outer of ♂. Telson scarcely longer than broad, distally somewhat narrowed, with a minute indent on each side of the shallow central emargination. L. ♀ 29, ♂ 26·5 mm.

Arctic Ocean. Depth 94—143 m.

7. **O. leucopis** (O. Sars) 1879 *Anonyx (O.) l.*, G. O. Sars in: Arch. Naturv. Kristian., v.4 p.438 | 1885 *O. l.*, G. O. Sars in: Norske Nordhavs-Exp., v.6 Crust. I p.149 t.13 f.1, 1a | 1882 *Onesimus l.*, Hoek in: Niederl. Arch. Zool., suppl. 1 nr.7 p 45 | 1893 *Anonyx edwardsii* (part.), A. Della Valle in: F. Fl. Neapel, v.20 p.828.

Head, lateral angles somewhat produced and pointed. Side-plate 1 considerably expanded below, 4ᵗʰ scarcely at all. Pleon segment 3, postero-lateral angles acute, very slightly produced, not upturned. Eyes very small, imperfect, placed near lateral lobes of head. colour light, well-nigh lacteal. Antenna 1, flagellum with 11 joints, 1ˢᵗ rather large, accessory flagellum long, 4-jointed. Antenna 2, flagellum rather long, 16-jointed. Mouth-organs undescribed. Gnathopods 1 and 2, supposed to resemble those of O. turgidus (p. 26). Peraeopods 3—5, 4ᵗʰ joint rather short and thick. Uropod 3 without spinules. Telson short and broad, with central emargination. Colour whitish, semi-translucent. L. 10 mm.

North-Atlantic (between Iceland and Norway). Depth 1520 m.

8. **O. affinis** H. J. Hansen 1886 *O. a.*, H. J. Hansen in: Dijmphna Udb., p.216 t.21 f.9, 9a | ?1894 *O. brevicaudatus* (part.), T. Stebbing in: Bijdr. Dierk., v.17 p.13 | 1893 *Pseudalibrotus littoralis* (err., non *Anonyx l.* Krøyer 1845!) (part.), A. Della Valle in: F. Fl. Neapel, v.20 p.799.

Head, lateral angles slightly projecting, rounded. Side-plate 1 rather expanded below, 4ᵗʰ with hinder expansion short. obtusely truncated. Pleon segment 3, postero-lateral angles produced to an upturned rather blunt process. Eyes short, somewhat triangular as in O. caricus (p.27). Antennae, gnathopod 1, peraeopods much as in O. edwardsii (p. 25). Gnathopod 2 almost as in O. brevicaudatus ♂, 6ᵗʰ joint broader than in O. edwardsii, not twice as long as broad, hind margin rather longer than front, a minute cavity between slightly outdrawn palm and the finger. Telson very little longer than broad, narrowing a little from base to apex, the latter subtruncate. in the middle a little emarginate with an indent on each side. Specimen not adult. L. 13·5 mm.

Kara Sea. Depth 54 m.

15. Gen. **Cyphocaris** Boeck

1871 *Cyphocaris* (Sp. un.: *C. anonyx*), (Lütken in MS.) A. Boeck in: Forh. Selsk. Christian., 1870 p.103 | 1872 *C.*, A. Boeck, Skand. Arkt. Amphip., v.1 p.140 | 1888 *C.*, T. Stebbing in: Rep. Voy. Challenger, v.29 p.656 | 1893 *C.*, J. Bonnier in: Bull. sci. France Belgique. v.24 p.174 | 1893 *C.*, A. Della Valle in: F. Fl. Neapel, v.20 p.846.

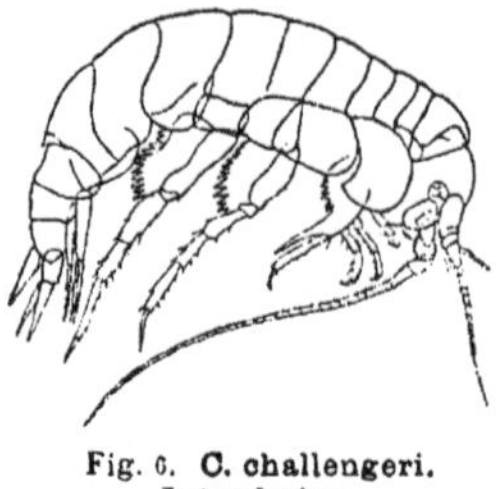

Fig. 6. **C. challengeri.** Lateral view.

Head directed downward beneath the large segment 1 of peraeon (Fig. 6). Side-plates 1, 2 and 3 very small, 4 and 5 large. Pleon segments large. Eyes none or unknown. Epistome not projecting in front of upper lip. Mandible, molar prominent, denticulate, palp affixed above it, large, with joints 2 and 3 subequal. Maxilla 1, inner plate with 7 setae, outer plate with 11 spines, palp with 6 spine-teeth. Maxilla 2, inner plate broader than outer, nearly as long, with setae on inner margin. Maxillipeds, inner plates not very long, with 3 apical spine-teeth, outer plates not large, fringed with spine-teeth, palp elongate. Gnathopod 1, feebly subchelate, 2^d joint longer than rest of limb. Gnathopod 2 elongate. Peraeopods 3—5, 2^d joint deeply serrate on hind margin. Uropod 3, rami subequal, long, lanceolate. Telson long, tapering, deeply cleft.

2 species.

Synopsis of species:

Peraeopod 3, 2^d joint produced into a long spur . . . . . 1. **C. challengeri** . p. 29
Peraeopod 3, 2^d joint not produced into a long spur . . . 2. **C. anonyx** . . . p. 29

1. **C. challengeri** Stebb. 1888 *C. c.*, T. Stebbing in: Rep. Voy. Challenger, *v.* 29 p. 661 t. 17 | 1893 *C. c.*, A. Della Valle in: F. Fl. Neapel, *v.* 20 p. 847.

Top of head facing forward, its proper front margin a little sinuous. Peraeon segment 1 as long as 3^d and 4th combined. Side-plate 4 concealing the 3^d, with long irregularly bowed front and deeply excavate hind margin, 5th much broader than deep. Pleon segment 3, postero-lateral angles squared, minutely produced. Antenna 1, 1st joint equal to 2^d and 3^d combined, flagellum with 15 joints, 1st as long as 1st of peduncle, accessory flagellum 3-jointed. Antenna 2, ultimate joint of peduncle shorter than penultimate, flagellum 40-jointed. Gnathopod 1, 6th joint a little narrower and longer than 5th, narrowing distally, hind margin minutely pectinate, finger denticulate and toothed on inner margin. Gnathopod 2, 5th and 6th joints longer than in gnathopod 1. 6th shorter than 5th, palm small, well defined, finger curved. Peraeopods slender, except in 2^d joint of peraeopods 3—5. Peraeopod 3, 2^d joint, front margin strongly bent, hinder cut into 7 teeth and prolonged sharply almost to end of 6th. Peraeopod 4, 2^d joint almost triangular, with 14 teeth, much overlapping 3^d joint. Peraeopod 5, 2^d joint more oval, with 13 teeth, a little overlapping 3^d joint. Uropod 3, rami long, almost equal, slightly armed. Telson cleft for $^2/_3$ of the length, without lateral, but with a pair of apical spinules. L. about 7 mm.

North-Pacific (north of Sandwich Islands).

2. **C. anonyx** Boeck 1871 *C. a.*, (Lütken in MS.) A. Boeck in: Forh. Selsk. Christian., 1870 p. 104 | 1872 *C. a.*, A. Boeck, Skand. Arkt. Amphip., *v.* 1 p. 141 t. 6 f. 1 | 1887 *C. a.*, H. J. Hansen in: Vid. Meddel., ser. 4 *v.* 9 p. 67 | 1888 *C. micronyx*, T. Stebbing in: Rep. Voy. Challenger, *v.* 29 p. 656 t. 16.

Head partly concealed in overarching, rather sharply projecting, long 1st peraeon segment. Side-plate 4 more angular than in C. challengeri. Pleon segment 3' postero-lateral angles almost quadrate. Antenna 1, 1st joint longer than 2^d and 3^d combined, flagellum with 21—30 joints, 1st very long, accessory flagellum 5-jointed. Antenna 2, ultimate joint of peduncle shorter than penultimate, flagellum 75-jointed. Gnathopod 1 nearly as in C. challengeri, 6th joint not longer than 5th. Gnathopod 2, 6th joint oval, shorter than 5th, both armed on hind margin with graduated sets of pectinate spines and long broad-ended setae, finger minute. Peraeopods 1 and 2 subchelate, the 6th joint being distally widened and having a palm margin set with teeth and spines, finger powerful, curved, acute. Peraeopod 3, 2^d joint with front margin bent, hind margin cut into about 8 teeth, prolonged to end

of 5th joint, inner margin of process cut into about 7 teeth, end of limb
as in peraeopods 1 and 2. Peraeopod 4, 2^d joint little produced behind, the
hind margin divided into 11 teeth, joints 4, 5 and 6 longer than in peraeopod 3,
palm narrower. Peraeopod 5, 2^d joint with hind margin cut into 14 teeth,
5th and 6th joints very elongate, no palm margin, finger nearly straight.
Uropod 3, outer ramus with small 2^d joint, inner not shorter than outer,
both with plumose setae. Telson long, narrow, cleft nearly $^3/_4$ of length,
with 2 or 3 pairs of lateral spinules, and an apical pair. L. 14 mm.

South-Pacific (West of South America, depth 2700 m); South-Atlantic (Tristan
da Cunha, depth 2565 m); North-Atlantic (Cape Farewell, depth 540 m).

16. Gen. Cyclocaris Stebb.

1888 *Cyclocaris*, T. Stebbing in: Rep. Voy. Challenger, *v.* 29 p 664 | 1893 *C.*,
A. Della Valle in: F. Fl. Neapel, *v.* 20 p. 843.

Head deep. Side-plates 1 and 2 very small, 3^d and 4th large. Antenna 1,
peduncle very short. Mandible, cutting edge broad, simple accessory plate
minute, molar weak or wanting, palp central (Fig. 7). Maxilla 1 with
several plumose setae on inner plate, 11 spines on broad apex of outer plate,

2^d joint of palp long, quinque-dentate. Maxilla 2, inner
plate much shorter than outer, inner margin setose. Maxillipeds, inner plates with wide concave distal margin, outer
plates broad and long, fringed with setae, spines and teeth.
Gnathopod 1 simple, 3^d joint much longer than 4th, 5th
longer and slightly wider than the long narrow tapering 6th,
7th slender. Gnathopod 2 slender, 5th joint much longer
than 6th, which is also long and narrow, not tapering;
7th joint small, thick at base, closing over short palm.
Peraeopods 1 and 2 not very slender, side-plate of peraeopod 1 distally dilated, that of peraeopod 2 very little

Fig. 7. C. tahitensis. excavate behind, 4th joint in each rather large. Peraeo-
Mandible.   pods 3—5 pretty strongly spined, 2^d joint broadly expanded,
4th very moderately. Uropod 2 with rami unequal, uropod 3 reaching far beyond
uropod 2, the outer ramus the longer, with a small 2^d joint. Telson long,
narrow, tapering, cleft nearly to the base, a spiniferous notch at each apex
with the inner point the longer.

2 species.

Synopsis of species:

1. **C. tahitensis** Stebb. 1888 *C. t.*, T. Stebbing in: Rep. Voy. Challenger, *v.* 29
p. 664 t. 18.

Side-plate 1 distally narrowed. Pleon segment 3 with blunt angles.
Eyes uncertain. Antenna 1, peduncle tumid, 2^d and 3^d joints very short;
flagellum with 10 joints, 1st stout, as long as peduncle or as rest of flagellum;
accessory flagellum with 6 joints, 1st long. Antenna 2, ultimate joint of
peduncle a little longer than penultimate; flagellum 25-jointed. Mandible with
a tooth at the top of cutting edge, and lower margin behind its lower angle
slightly serrate (Fig. 7), molar obscure. Gnathopod 2, 5th joint much longer
than 3^d. Uropod 1, rami subequal. L. about 10 mm.

Tropical Pacific (Tahiti). Taken in townet.

2. **C. guilelmi** Chevreux 1899 *C. g.*, Chevreux in: Bull. Soc. zool. France, *v.* 24 p. 148 f. 1—5.

Body robust. Head with little recurved rostrum, lateral angles faintly indicated, post-antennal corners more salient, rounded. Side-plate 1 almost rectangular, with spinule on lower margin, 2^d rather larger, rounded below. Pleon segment 3, postero-lateral corners subacute. Eyes wanting. Lower lip with lobes distally obliquely truncate. Mouth-parts in general form agreeing with C. tahitensis (p. 30), but mandible carrying a dentiform molar process, very elongate, curved at the extremity and furnished with a row of small setae. Gnathopod 1, 6th joint rather stouter than in C. tahitensis, regularly tapering. Gnathopod 2, 5th joint little longer than 3^d. Uropod 1, inner ramus $^2/_3$ as long as peduncle, outer notably shorter. Uropod 2, inner ramus a little shorter than peduncle, outer $^2/_3$ as long as inner. L. 11—12 mm.

Arctic Ocean (Lofoten Isles). Depth 1095 m.

17. Gen. **Lysianella** O. Sars

1882 *Lysianella* (Sp. un.: *L. petalocera*), G. O. Sars in: Forh. Selsk. Christian., nr. 18 p. 78 | 1890 *L.*, G. O. Sars, Crust. Norway, *v.* 1 p. 50 | 1893 *L.*, J. Bonnier in: Bull. sci. France Belgique. *v.* 24 p. 174 | 1893 *L.*, A. Della Valle in: F. Fl. Neapel, *v.* 20 p. 797.

Body shorter and stouter in ♂ than in ♀; side-plates deep, 4th strongly excavate behind. Antenna 1 stouter in ♂, 1st joint of flagellum large and setose in ♂. small in ♀, accessory flagellum long in both sexes. Antenna 2 alike in ♂ and ♀, about as long as antenna 1, penultimate joint of peduncle oval, broadly laminar, densely furred on inner surface. Mouth-organs prominent below. Upper lip jutting out much in front of epistome. Mandible strong, molar small and weak, palp short, attached behind molar, its 3^d joint the longest. Maxilla 1, inner plate tapering, with 2 setae, otherwise normal. Maxilla 2, plates narrow, armed at apices. Maxillipeds, inner plates acutely tapering, outer plates narrow oval, palp slender, 2^d joint longer than 1st, 4th rather small. Gnathopod 1 subchelate, 5th and 6th joints subequal, 7th small. Gnathopod 2, tending to minutely chelate. Peraeopods 1—5 slender. Peraeopods 3—5, 2^d joint large, longer than broad. Uropod 3 small, rami subequal, as long as peduncle, outer ramus 2-jointed, the 2^d about $^2/_3$ as long as the 1st. Telson small, oval, entire or partially cleft

2 species.

Synopsis of species:

1. **L. petalocera** O. Sars 1882 *L. p.*, G. O. Sars in: Forh. Selsk. Christian., nr. 18 p. 78 t. 3 f. 3,3a | 1890 *L. p.*, G. O. Sars, Crust. Norway, *v.* 1 p. 51 t. 18 f. 2 | 1893 *L. p.*, A. Della Valle in: F. Fl. Neapel, *v.* 20 p. 797 t. 61 f. 9.

Head, lateral corners subangular, less projecting in ♂ than in ♀. Pleon segment 3. corners obtusely rounded. Eyes oval, dark purplish brown with reddish coating, larger in ♂. Antenna 1, flagellum 8-jointed, accessory flagellum 4-jointed, 1st joint the longest; antenna 2, flagellum 8-jointed. Gnathopod 1, 6th joint of nearly uniform width, palm oblique, not very large, finger short, simple. Gnathopod 2, 6th joint about half length of 5th, produced

at the tip, but not strongly. Telson not twice as long as broad, apex entire,
evenly rounded. Colour whitish, intestine greenish, ova in pouch orange-
coloured. L. ♀ 5, ♂ 3·5 mm.

North-Atlantic, North-Sea and Skagerrak (South- and West-Norway). Down
to 188 m.

2. **L. dellavallei** Stebb.*) 1893 *Anonyx petalocerus* (non *Lysianella petalocera*
G. O. Sars 1882!). A. Della Valle in: F. Fl. Neapel, *v.* 20 p. 816 t. 61 f. 1—9.

Head, lateral angles rounded. Antenna 1 in ♀, flagellum 9-jointed,
accessory flagellum with 5 subequal joints. Antenna 2, flagellum 12-jointed.
Telson cleft a quarter of the length. L. 5 mm.

Gulf of Naples.

18. Gen. **Onesimoides** Stebb.

1888 *Onesimoides* (Sp. un.: *O. carinatus*). T. Stebbing in: Rep. Voy. Challenger,
v. 29 p. 647 | 1893 *O.*, J. Bonnier in: Bull. sci. France Belgique, *v.* 24 p. 174 | 1893 *O.*,
A. Della Valle in: F. Fl. Neapel, *v.* 20 p. 796.

Head with small subacute lateral lobes. Side-plate 1 almost completely clear of the head. Antennae 1 and 2 subequal, short, in antenna 1

first joint of flagellum and of accessory flagellum long,
the latter expanded over the former. Antenna 2, ultimate
and penultimate joints of peduncle subequal. Epistome projecting a little in front of upper lip. Mandible, molar
denticulate, palp set just over it, with 1st joint not very
short, 2d long (Fig. 8). Maxilla 1, inner plate with 2 setae
on the narrow apex, outer plate broad, obliquely truncate,
with 11 spines, palp with 9 spine-teeth on long 2d joint
in right, 12 in left maxilla. Maxilla 2, inner plate little
shorter than outer. Maxillipeds, inner plates with 3 apical
spine-teeth, outer plates reaching end of palp's 2d joint,
margin beset with teeth and spines, palp robust. 4th joint
unguiform, short. Gnathopod 1 subchelate, 5th joint short,

Fig. 8. **O. carinatus.**
Mandible.

6th robust, oblong, palm transverse, finger as long as palm,
4th, 5th and 6th joints strongly setose on hind margin.
Gnathopod 2 weak, minutely chelate, 6th joint rather longer than broad, the
minute finger in middle of apical margin. Peraeopods 1—5, 7th joint short.
Peraeopods 3—5, 2d joint expanded, in 5th greatly, as long as broad.
Uropod 3 short, outer ramus as long as peduncle, with a short 2d joint,
inner ramus rudimentary. Telson short, as broad as long, rounded, apically truncate.

1 species.

1. **O. carinatus** Stebb. 1888 *O. c.*, T. Stebbing in: Rep. Voy. Challenger, *v.* 29
p. 648 t. 14 | 1893 *O. c.*, A. Della Valle in: F. Fl. Neapel, *v.* 20 p. 796 t. 60 f. 39—41.

Peraeon with all but chiefly last 2, and pleon with first 4 segments
carinate, body furred; pleon segment 3, lower corners quadrate. Antenna 1,
flagellum 12-jointed, accessory flagellum 4-jointed. Antenna 2, flagellum
9-jointed. Side-plate 1 broader above than below. Uropod 1, rami unequal;
uropod 2, rami subequal. L. 11—12 mm.

Coral Sea (North-East-Australia). Depth 2560 m.

---

*) Nom. nov. After Antonio Della Valle.

19. Gen. **Pseudalibrotus** Della Valle

1871 *Onisimus* (part.), A. Boeck in: Forh. Selsk. Christian, 1870 p. 112 | 1876
Onesimus (part.), A. Boeck, Skand. Arkt. Amphip., *v.* 2 p. 161 | 1893 *Pseudalibrotus*
(Sp. un.: *P. littoralis*) (part.), A. Della Valle in: F. Fl. Neapel, *v.* 20 p. 798 | 1896 *P.*,
G. O. Sars in: Bull. Ac. St.-Pétersb., ser. 5 *v.* 4 p. 422.

Body rather slender and compressed. Side-plates not very large, 4th
pair deeply but narrowly emarginate behind. Antenna 1 and 2, with many
short joints in the flagella both of ♂ and ♀. Epistome not distinctly
defined from upper lip, both together forming a rounded prominence. Lower
lip with front lobes a little emarginate distally. Mandible, molar rather
strong, palp large, set well forward. Maxilla 1, inner plate small with
2 setae, outer plate broad, obliquely truncate, palp with small apical teeth.
Maxilla 2, inner plate much shorter than outer. Maxillipeds normal, outer
plates crenulate, not reaching end of 2d joint of palp, 1st—3d joints of palp
robust, 4th small, unguiform. Gnathopod 1 subchelate, 6th joint a little
longer than 5th, oblong, palm oblique, 7th joint simple. Gnathopod 2, 6th
joint about half length of 5th, slender, oblong, more or less produced, subchelate or chelate. Peraeopods 3—5, 2d joint large, longer than broad.
Uropod 3, rami longer than peduncle, outer rather the longer. Telson
short, broad, entire.

3 species.

Synopsis of species:

$$\left. 1 \left\{ \begin{array}{l} \text{Gnathopod 1 very robust} \dots\dots\dots\dots\dots \text{1. \textbf{P. litoralis}} \dots \text{p. 33} \\ \text{Gnathopod 1 not very robust — \textbf{2}.} \end{array} \right. \right.$$

$$\left. 2 \left\{ \begin{array}{l} \text{Pleon segment 3. postero-lateral corners considerably} \\ \quad \text{produced} \dots\dots\dots\dots\dots\dots\dots \text{2. \textbf{P. caspius}} \dots \text{p. 34} \\ \text{Pleon segment 3, postero-lateral corners scarcely} \\ \quad \text{produced} \dots\dots\dots\dots\dots\dots\dots \text{3. \textbf{P. platyceras}} \dots \text{p. 34} \end{array} \right. \right.$$

1. P. litoralis (Krøyer) 1845 *Anonyx l.,* Krøyer in: Naturh. Tidsskr., ser. 2
v. 1 p. 621 | 1846 *A. littoralis,* Krøyer in: Voy. Nord, Crust. t. 13 f. 1 a-t | 1862 *Alibrotus l.,*
Bate, Cat. Amphip. Brit. Mus., p. 86 t. 14 f. 7 | 1891 *A. l.,* G. O. Sars, Crust. Norway,
v. 1 p. 102 t. 35 f. 2 | 1894 *A. litoralis,* T. Stebbing in: Bijdr. Dierk., *v.* 17 p. 9 | 1866
Lysianassa l., Goës in: Öfv. Ak. Förh., *v.* 22 p. 521 | 1871 *Onisimus l.,* A. Boeck in:
Forb. Selsk. Christian., 1870 p. 112 | 1875 *O. l.,* Cam. Heller in: Denk. Ak. Wien, *v.* 35
p. 31 t. 2 f. 8—16 | 1876 *Onesimus l.,* A. Boeck, Skand. Arkt. Amphip., *v.* 2 p. 162 t. 5
f. 7 | 1893 *Pseudalibrotus littoralis* (part.), A. Della Valle in: F. Fl. Neapel, *v.* 20 p. 799.

Head, lateral lobes narrowly rounded, little prominent. Side-plate 1
much expanded distally. Pleon segment 3, postero-lateral angles minutely
outdrawn, almost quadrate. Eyes small, rounded oval, red. Antenna 1,
1st joint more than twice as long as 2d and 3d combined, flagellum in ♀ with
26 joints, 1st much the largest, accessory flagellum with 5 joints, 1st as long
as the rest combined. Antenna 2, ultimate joint of peduncle shorter than penultimate, flagellum in ♀ about 30-jointed. Flagella in ♂ rather longer than
in ♀ and with many more joints. Gnathopod 1 strong, 5th joint distally broad,
6th broadly oblong, palm spinulose, with 2 spines at the angle. Gnathopod 2, palm
minutely produced. Peraeopods 1 and 2, 6th joint slightly curved. Peraeopods 3—5, 5th and 6th joints slender. Uropod 2, inner ramus constricted.
Uropod 3, outer ramus with small 2d joint and setose on inner margin.
Telson rounded, scarcely longer than broad, distal border faintly emarginate
in -the centre, carrying two spinules. Colour whitish. L. 13—18 mm.

Arctic Ocean (North-Norway).

2. **P. caspius** O. Sars 1896 *P. c. (Onesimus c. + O. pomposus*, O. Grimm in MS.), G. O. Sars in: Bull. Ac. St.-Pétersb., ser. 5 *v.* 4 p. 422 t. 1 f. 1—20.

Body rather stout, back strongly curved. Head, lateral lobes narrowly rounded. Side-plate 1 distally expanded, much broader than 2^d or 3^d. Pleon segment 3, postero-lateral angles acute, considerably produced. Eyes rather small, rounded oval, probably bright red. Antenna 1, 1st joint large, flattened, twice as long as 2^d and 3^d combined, flagellum about twice as long as peduncle, with 16 joints, 1st much the largest, with 5 fascicles of filaments, accessory flagellum with 4 joints, 1st much the longest. Antenna 2 rather longer, ultimate joint of peduncle not very long, narrower and shorter than penultimate, flagellum longer than in antenna 1, about 20-jointed. Mouth-parts nearly as in P. litoralis, but palp of mandible attached over molar, not in advance of it, palp of maxilliped less robust. Gnathopod 1 not robust, 6th joint scarcely as broad as 5th, somewhat narrowing distally, palm rather short, with single spine at the angle, finger small. Gnathopod 2 very slender, 6th joint produced so as to form a little definite chela. Peraeopods 1—5 shorter and less slender than in P. litoralis. Peraeopods 3—5, 2^d joint with few serrations on the hind margin. Uropods 1 and 2, rami simple, with scattered spines. Uropod 3, peduncle rather short, rami simple, with no 2^d joint to outer ramus. Telson about as long as broad, apex slightly produced between a pair of notches, each of which carries a setule. L. 9 mm.

Caspian Sea, middle part. Depth 150—470 m.

3. **P. platyceras** O. Sars 1896 *P. p. (Onesimus p.* O. Grimm in MS.), G. O. Sars in: Bull. Ac. St.-Pétersb., ser. 5 *v.* 4 p. 426 t. 1 f. 21—23.

Very like P. caspius, but twice as long and more tumid, back broadly rounded, pleon segment 4 with marked dorsal depression. Head, lateral lobes subangular. Side-plate 1 comparatively broader than in P. caspius. Pleon segment 3, postero-lateral corners acute, but very little produced. Eyes narrowed above to an obtuse point. Antenna 1, 1st joint exceedingly large, conspicuously flattened, flagellum rather slender, with 30—40 joints, accessory flagellum with 6 joints. Antenna 2 scarcely longer, flagellum 34-jointed. Gnathopod 1, 6th joint rather shorter in proportion to its breadth, palm less oblique. L. ♀ 20 mm.

Caspian Sea, middle part. Depth 75 m.

20. Gen. **Nannonyx** O. Sars

1871 *Orchomene* (part.), A. Boeck in: Forh. Selsk. Christian., 1870 p. 114 | 1890 & 91 *Nannonyx* (Sp. un.: *N. goësii*), G. O. Sars, Crust. NorWay, *v.* 1 t. 24; p. 71 | 1893 *N.*, J. Bonnier in: Bull. sci. France Belgique, *v.* 24 p. 173 | 1893 *N.*, A. Della Valle in: F. Fl. Neapel, *v.* 20 p. 794.

Body compact, side-plates large. Mandible elongate, molar obsolete, palp affixed far back. Maxilla 1, inner plate with 1 or 2 small setae, plumose or simple. Maxilla 2, both plates long, narrow. Maxillipeds, inner plates generally elongate, narrow, outer plates long and broad, slightly crenulate, reaching beyond the 2^d joint of palp, 4th joint of palp distinct but rudimentary in size. Gnathopod 1 simple. Gnathopod 2 minutely chelate. Uropod 3 very small, the outer ramus the longer, 2-jointed. Telson entire.

5 species.

Synopsis of species:

1 { Maxillipeds, inner plates not elongate or narrow 1. **N. spinimanus** . . p. 35
  { Maxillipeds, inner plates elongate, narrow — **2.**

2 { Uropod 3, outer ramus with 2ᵈ joint conspicuous 2. **N. integricauda** . . p 35
  { Uropod 3, outer ramus with 2ᵈ joint minute — 3.

3 { Side-plate 4 little expanded backward . . . . . 3. **N. goësii** . . . . . p. 36
  { Side-plate 4 greatly expanded backward — 4.

4 { Pleon segment 3, angles acutely upturned . . . 4. **N. thomsoni** . . . p. 36
  { Pleon segment 3, angles rounded . . . . 5. **N. kidderi** . . . . p. 36

1. **N. spinimanus** A. Walker 1895 *N. s.*, A. O. Walker in: P. Liverp. biol. Soc., *v.* 9 p. 292 t. 18 f. 1—11; t. 19 f. 6 a.

Pleon segment 3, postero-lateral angles rather acute with the point blunt, hind margin straight, serrate, 4ᵗʰ segment with prominent rounded hump. Antenna 1, flagellum 7-jointed, accessory flagellum 4-jointed. Antenna 2, flagellum in ♂ and ♀ 5-jointed, shorter than peduncle. Mandible with broad cutting edge. Maxilla 1, inner plate with 2 setae, outer plate broad, apex decurrent. Maxilla 2, inner plate twice as broad as outer. Maxillipeds, inner plates not reaching beyond 1ˢᵗ joint of palp, outer plates with spine-teeth fringing inner margin, 4ᵗʰ joint of palp very small. Gnathopod 1, 5ᵗʰ and 6ᵗʰ joints stout, 6ᵗʰ little longer than 5ᵗʰ, tapering distally more in ♂ than in ♀, leaving a small palm margin. Gnathopod 2 subchelate, 6ᵗʰ joint rather shorter than 5ᵗʰ and nearly as wide. Peraeopods 3—5, 2ᵈ joint expanded, 4ᵗʰ in peraeopods 3 and 4 wider and more produced than in peraeopod 5. Uropods 1 and 2, rami shorter than peduncles. Uropod 3, outer ramus in ♂ rather shorter than peduncle, in ♀ about half as long as peduncle, inner ramus ovate. Telson as broad as long, distally narrowed and rounded, carrying a pair of marginal spinules. Colour brown. L. 4·5 mm.

Menai Strait [North-Wales].

2. **N. integricauda** (Stebb.) 1888 *Ambasia i.*, T. Stebbing in: Rep. Voy. Challenger, *v.* 29 p. 695 t. 26.

Side-plate 1 distally somewhat, 4ᵗʰ very considerably, expanded. Pleon segment 3, postero-lateral angles outdrawn into a little blunt point; 4ᵗʰ without perceptible dorsal depression. Antenna 1, 2ᵈ joint half length of 1ˢᵗ, flagellum 5-jointed, shorter than 1ˢᵗ joint of peduncle, accessory flagellum 2-jointed. Antenna 2 (♀?), last 3 joints of peduncle nearly equal, flagellum small, with 4 joints, the first the longest. Mandible, 1ˢᵗ joint of palp rather long. Maxilla 1, inner plate distally rounded, with one simple seta, outer plate with 11 spines, palp carrying 4 spines on apex of 2ᵈ joint. Maxilla 2, plates subequal, not very narrow. Maxillipeds, inner plates long, narrow, with a few small setae and perhaps 3 tooth-spines on the apex, outer plates reaching halfway along 3ᵈ joint of palp, 4ᵗʰ joint of palp very small, the nail forming half its length. Gnathopod 1, 5ᵗʰ and 6ᵗʰ joints subequal, very scantily armed, not very long, 6ᵗʰ joint tapering without palm. Gnathopod 2, 3ᵈ joint as long as 5ᵗʰ, 6ᵗʰ half length of 5ᵗʰ. Peraeopod 1 slender, 7ᵗʰ joint rather elongate. Peraeopods 3—5, 2ᵈ joint expanded, overlapping 3ᵈ, 4ᵗʰ not very broad. Pleopods, peduncle broad, each ramus only 4-jointed, the outer with an extremely broad 1ˢᵗ joint, 1ˢᵗ joint of inner narrow. Uropod 3, the short outer ramus subequal in length to the peduncle, its 2ᵈ joint nearly as long as the 1ˢᵗ, inner ramus more than half length of outer, each with inner margin serrulate. Telson tapering to a rounded apex. L. 3 mm.

Royal Sound [Kerguelen Island]. Depth 50 m.

3. **N. goësii** (Boeck) 1871 *Orchomene g.*, A'. Boeck in: Forh. Selsk. Christian., 1870 p. 116 | 1872 & 76 *O. g.*, A. Boeck, Skand. Arkt. Amphip., *v.*1 t. 4 f. 5; *v.*2 p. 177 | 1890 & 91 *Nannonyx g.*, G. O. Sars, Crust. NorWay, *v.*1 t. 24 f. 3; p. 72 | 1893 *N. g.*, A. Della Valle in: F. Fl. Neapel, *v.* 20 p. 794.

Back broadly rounded. Head with lateral angles broadly rounded. Side-plates deep, 4th deeply but not widely emarginate behind, 5th rather deeper than broad. Pleon segment 3, postero-lateral angles quadrate, hinder edge faintly crenulate, segment 4 with hump-like dorsal projection. Eyes oval, dark brown. Antenna 1, 1st joint twice length of 2d and 3d combined, flagellum short, of 4 joints, 1st as long as other 3 combined, accessory flagellum 2-jointed. Antenna 2 in ♀, ultimate joint of peduncle shorter than either penultimate or antepenultimate, flagellum short, 4-jointed. Mandible, 1st joint of palp very short. Maxilla 1, inner plate slender, with single simple seta on the acute apex, palp without apical spines. Maxilla 2, plates very narrow. Maxillipeds, with many diverging setae about the base, inner plates elongate, outer large and long, minutely crenulate, 4th joint of palp very small, knoblike. Gnathopod 1 powerful. none of the joints long, 4th, 5th and 6th joints densely setose, 5th and 6th subequal in length, 6th tapering, 7th very small. Gnathopod 2, 3d joint as long as 5th, 6th very narrow, about half length of the 5th and narrower. Peraeopods 1—5 short and stout, 7th joint very small. Peraeopods 3—5, expansion of 2d joint overlapping 3d, 4th joint rather wide. Uropod 3 very small, rami shorter than peduncle, outer nearly twice as long as inner, its second joint minute. Telson rounded quadrangular, apex truncate (notched, Boeck). Colour yellowish, with transverse orange bands. Eggs in pouch dark violet. L. ♀ 4 mm.

Arctic Ocean, North-Atlantic and North-Sea (West-Norway). Shallow Water to 75 m.

4. **N. thomsoni** Stebb.*) 1879 *Lysianassa kröyeri* (err., non *Ephippiphora kröyeri* A. White 1847!), G. M. Thomson in: Tr. N. Zealand Inst., *v.* 11 p. 237.

Side-plate 1 considerably expanded below, 4th very greatly. Pleon segments 3, 4 and 5 slightly carinate, the postero-lateral angles of 3d acutely upturned. Eyes elongate. Antenna 1, 2d and 3d joints not extremely short, flagellum of few joints, all short. Antenna 2 in ♀, ultimate and penultimate joints of peduncle subequal, flagellum 7-jointed; in ♂ flagellum rather long and slender. Mandible, molar linear, 1st joint of palp rather long. Maxilla 1, inner plate with minute seta on narrowly rounded apex, outer plate with 11 spines, 2d joint of palp apparently unarmed. Maxilla 2, inner plate rather the broader, obliquely truncate. Maxillipeds as in N. goësii, but little 4th joint of palp has a minute nail. Gnathopod 1 not especially stout, 6th joint tapering, without palm, rather longer than 5th. Gnathopod 2, 3d joint as long as 5th, 6th not narrower than 5th, more than half as long. Peraeopods 1 and 2, 4th joint long, 7th moderately small. Peraeopods 3—5, expansion of 2d joint overlapping 3d, 4th rather wide. Pleopods narrow. Uropod 3, with inner ramus almost rudimentary, the outer with small 2d joint and not longer than peduncle. Telson very small, rather broader than long, sides slightly converging, apex truncate. L. 8 mm.

South-Pacific (Dunedin [New Zealand]).

5. **N. kidderi** (S. I. Sm.) 1876 *Lysianassa k.*, (S. I. Smith in:) Kidder in: Bull. U. S. Mus., *v.* 3 p. 59.

---

*) Nom. nov. After George Malcolm Thomson.

Head, lateral angles quadrate, slightly rounded. Side-plate 1 very slightly expanded below, 4th greatly, the emargination strongly quadrate. Pleon segment 3, angles rounded. Antenna 1, 2^d and 3^d joints short, flagellum with 6 or 7 joints, accessory flagellum with 4 joints. Antenna 2, ultimate and penultimate joints of peduncle rather short, flagellum in ♀ with 7 or 8 joints, in ♂ stout, with 18 short joints, most with calceoli. Mouth-organs nearly as in N. thomsoni, but 2^d joint of mandible-palp not so long. Maxilla 1, 2 very minute setae on inner plate. Terminal joint of maxillipeds showing no nail. Gnathopods 1 and 2, and peraeopod 3 nearly as in N. thomsoni. Branchial vesicles almost simple. Uropod 3 very short, one branch bluntly conical, the other rudimentary. Telson as broad as long, sides slightly converging, apex slightly concave. L. 3—4 mm.

Southern Indian Ocean (Kerguelen Island). Rocky beaches.

21. Gen. Lysianassa M.-E.

1830 *Lysianassa**) (part.), H. Milne Edwards .in: Ann. Sci. nat., $v.$ 20 p. 364 | 1840 *Callisoma* (nom. nud.), O. G. Costa, Fauna Reg. Napoli, Crost., Cat. p. 5 | 1851 *C.* (non L. Agassiz 1846, Coleoptera!) (part.), A. Costa, Fauna Reg. Napoli, fasc. Marz. 1851 p. 1 | 1855 *Lycianassa*, T. Bell in: Belcher. Last Arct. Voy., $v.$ 2 p. 406 | 1867 Subgen. *Lysianassina*, A. Costa in: Annuario Mus. Napoli, $v.$ 4 p. 43 | 1888 *Lysianax* (part.), T. Stebbing in: Rep. Voy. Challenger, $v.$ 29 p. 681, 1676.

Body compressed. Side-plates large. Antenna 1, 2^d joint longer than usual in this family. Antenna 2 short in ♀, long in ♂. Upper lip defined by a narrow incision and produced forward to a large linguiform plate. Mandible slender, cutting edge simple, molar very small, palp attached behind it, 1st joint not very short. Maxilla 1, inner plate without setae, outer plate with crowded spines, palp cut into small teeth at apex. Maxilla 2, inner plate much broader than outer. Maxillipeds, inner plates normal, outer oval, without spines, minutely crenulate, palp long and slender. Gnathopod 1 simple. Gnathopod 2 subchelate or almost chelate. Peraeopods 1—5 rather slender. Peraeopods 3—5, 2^d joint broadly expanded. Branchial vesicles with accessory folds on one side only. Uropod 2, with one ramus constricted. Uropod 3 not very small, rami narrowly lanceolate one-jointed, setose only in ♂. Telson small, entire or minutely notched.

7 species accepted, 3 doubtful.

Synopsis of accepted species:

1 {	Telson entire — **2.**	
	Telson minutely notched — **6.**	
2 {	Pleon segment 3 uncinate	1. **L. plumosa** . . . p. 38
	Pleon segment 3 not uncinate — **3.**	
3 {	Telson apically truncate — **4.**	
	Telson apically rounded — **5.**	
4 {	Antenna 1, 1st joint bidentate	2. **L. bispinosa** . . . p. 38
	Antenna 1, 1st joint smooth	3. **L. cubensis** . . . p. 38
5 {	Gnathopod 2 tending to chelate	4 **L. longicornis** . . p. 39
	Gnathopod 2 simply subchelate	5. **L. cinghalensis** . p. 39
6 {	Peraeopod 3, 2^d joint rounded	6. **L. variegata** . . . p 39
	Peraeopod 3, 2^d joint narrow proximally, expanded distally	7. **L. punctata** . . . p. 40

*) The name *Lysianassa* is not preoccupied by G. Münster (cfr.: 1846, Herrmannsen, Ind. Gen. Malac., $v.$ 1 p. 637). The Editor.

1. **L. plumosa** Boeck 1861 *L. costae* (err., non H. Milne Edwards 1830!), Bate & Westwood, Brit. sess. Crust., *v.*1 p. 74 f. | 1865 *L. c.*, W. Lilljeborg in: N. Acta Soc. Upsal., ser. 3 *v.*6 nr. 1 p. 21 | 1890 *L. c.*, G. O. Sars, Crust. Norway, *v.* 1 p. 42 t. 16 f. 1 | 1871 *L. plumosa*, A. Boeck in: Forh. Selsk. Christian., 1870 p. 96 | 1872 *L. p.* + *L. costae*, A. Boeck, Skand. Arkt. Amphip., *v.* 1 p. 116 t. 3 f. 5; p. 118 t. 4 f. 1 | 1893 *Lysianax septentrionalis*, A. Della Valle in: F. Fl. Neapel, *v.* 20 p. 788.

Head, lateral angles produced, acuminate. Side-plates close set and deep, 4th deeply and widely emarginate behind. Pleon segment 3, postero-lateral angles sharply uncinate. Eyes broadly reniform, dark purplish. Antenna 1, flagellum with 10—12 joints, about as long as peduncle; accessory flagellum 3- or 4-jointed. Antenna 2 in ♀ scarcely longer than antenna 1, ante-penultimate joint of peduncle rather long, ultimate and penultimate subequal in length, flagellum with 8 joints. Antenna 2 in ♂ longer than the body, ante-penultimate joint of peduncle short, ultimate twice as long as penultimate, flagellum filiform, with 60—70 joints. Gnathopod 1, 1st joint as long as rest of limb, 6th joint narrow, tapering, slightly longer than 5th, 7th small. (The subsigmoid shape attributed .by Boeck to the hand of gnathopod 1 in his single ♂ specimen cannot be relied on as a normal character.) Gnathopod 2, 6th joint short, truncate or slightly excavate, forming a small palm. Peraeo-pods 1 and 2 with very long plumose setae on the 4th joint in ♂. Telson oval, apically rounded, much shorter than peduncle of uropod 3. L. 12 mm.

North-Atlantic and North-Sea (West-Norway, Great Britain, France). Down to 188 m.

2. **L. bispinosa** (Della Valle) 1893 *Lysianax bispinosus*, A. Della Valle in: F. Fl. Neapel. *v.* 20 p. 792 t. 1 f. 5; t. 25 f. 16—21.

Head with lateral lobes little prominent. Side-plates not very deep. Pleon segment 3, postero-lateral angles rounded. Eyes brown. Antenna 1 in ♀, 1st joint distally produced into two unequal teeth. Antenna 2, ultimate and penultimate joints of peduncle tolerably large. Maxillipeds, outer plates reaching a little beyond 2d joint of palp. Peraeopod 3, 2d joint rounded. Telson subrectangular, distal margin straight. The other characters as in L. longi-cornis. Colour white, with brown spots, besides tints of orange and scarlet due to the internal organs, the appendages chalk-white. L. 10 mm. — ♂ unknown.

Gulf of Naples.

3. **L. cubensis** (Stebb.) 1897 *Lysianax c.*, T. Stebbing in: Tr. Linn. Soc. London, ser. 2 *v.* 7 p. 29 t. 7B.

Head, lateral angles broadly rounded. Side-plate 1 expanded below, 5th broader than deep. Pleon segment 3, postero-lateral angles almost quadrate, but with convex hind margin. Eyes large, oval, dark. Antenna 1, 1st joint not produced into apical teeth, flagellum small, of 6 or 7 joints, none long, accessory flagellum 2-jointed. Antenna 2 in ♀ very small, peduncle slender, flagellum imperfect. Gnathopod 1, 5th and 6th joints subequal in length. Gnathopod 2, 6th joint more than twice as long as broad, bulging a little distally, transversely truncate, subchelate, finger minute, closely fitting the palm. Peraeopod 3, 2d joint broader than long. Peraeopod 4, 2d joint with convex front, sinuous hinder margin. Peraeopod 5, 2d joint large, oval. Uropod 2, inner ramus much dilated, and towards the end strongly and abruptly constricted. Uropod 3 short, peduncle stout, rami subequal, slight. Telson boat-shaped, with truncate hind margin. L. 7·5 mm.

Gulf of Mexico or Caribbean Sea (Cuba).

4. **L. longicornis** H. Luc. 1845/46 *L. l.*, H. Lucas in: Expl. Algérie, An. artic., *v.* 1 p. 53 Crust. t. 5 f. 2 | 1866 *L. l.* (part.). E. Grube in: Arch. Naturg., *v.* 821 p. 396 t. 9 f. 8 (♂) | 1866 *L. l.*, Cam. Heller in: Denk. Ak. Wien, *v.* 26 ɪɪ p. 17 t. 2 f. 12—15 | 1892 *L. l.*, A. O. Walker in: Ann. nat. Hist., ser. 6 *v.* 9 p. 135, 136 | 1867 *Lysianassina l.* + *?L. filicornis*, A. Costa in: Annuario Mus. Napoli, *v.* 4 p. 43 | 1893 *Lysianax l.*, A. Della Valle in: F. Fl. Neapel, *v.* 20 p. 790 t. 3 f. 6; t. 25 f. 1—15 | ? 1853 *Lysianassa spinicornis*, A. Costa in: Rend. Soc. Borbon., n. ser. *v.* 2 p. 172 | ? 1857 *L. s.*, A. Costa in: Mem. Acc. Napoli, *v.* 1 p. 185 t. 1 f. 4 | ? 1857 *L. loricata*, A. Costa in: Mem. Acc. Napoli, *v.* 1 p. 186 t. 1 f. 5 | ? 1862 *L. filicornis*, A. Costa in: Annuario Mus. Napoli, *v.* 1 p. 80 t. 2 f. 18—23 | ? 1889 *Lysianax ceratinus*, A. O. Walker in: P. Liverp. biol. Soc., *v.* 3 p. 200 t. 10 f. 1—8 | 1899 *L. c.*, Chevreux in: C.-R. Ass. Franc., Sess. 27 *v.* 2 p. 476.

Head, lateral angles produced, rounded. Side-plates close set, deep, 1ˢᵗ a little widened distally. Pleon segment 3, postero-lateral angles rounded. Eyes broadly reniform, red. Antenna 1, 1ˢᵗ joint with inner distal margin produced into a spiniform process, varying from large to evanescent, 2ᵈ joint about half length of first, or more, 3ᵈ short, flagellum with 10 joints, 1ˢᵗ not long, accessory flagellum with 3—5 joints. Antenna 2, ultimate joint of peduncle much longer than penultimate in ♂, flagellum in ♂ much longer than body. Gnathopod 1, 5ᵗʰ joint a little shorter than 6ᵗʰ, 6ᵗʰ tapering, 7ᵗʰ small. Gnathopod 2, 6ᵗʰ joint with distal hind margin prolonged into a kind of lobe, rough with spinules. Peraeopod 3, 2ᵈ joint elliptical, broader than long. Peraeopod 4, 2ᵈ joint with hind margin rather sinuous. Peraeopod 5, 2ᵈ joint very broad, but longer than broad. Uropods all with awl-shaped rami, more or less armed with spines. Telson elliptical, apically rounded. Colour grey dorsally. Side-plates rusty-yellow and greyish with scattered little white spots. Flagella of antennae crimson. L. 8—10 mm.

Mediterranean; North-Atlantic and North-Sea (Great Britain).

5. **L. cinghalensis** (Stebb.) 1897 *Lysianax c.*, T. Stebbing in: Tr. Linn. Soc. London, ser. 2 *v.* 7 p. 28 t. 7ᴀ.

Head, lateral lobes produced, the apices rounded. Side-plates 1 and 4 distally much widened. Pleon segment 3, postero-lateral angles rounded. Eyes very large and dark, meeting on the top of the head. Antenna 1, 1ˢᵗ joint tumid, 2ᵈ and 3ᵈ not extremely short, flagellum with 6 joints, none long, accessory flagellum with 3 joints. Antenna 2, as usual in ♂, flagellum 35-jointed, slender, reaching about to middle of pleon. Maxillipeds, inner plates long, outer plates large but not reaching beyond 2ᵈ joint of palp. Gnathopod 1, 6ᵗʰ joint slightly longer than 5ᵗʰ. Gnathopod 2, 3ᵈ joint as long as 5ᵗʰ, 6ᵗʰ rather more than half length of 5ᵗʰ, subchelate, minute finger closing down on small transverse palm. Peraeopods 3—5, 2ᵈ joint broadly expanded, produced below the 3ᵈ joint. Uropod 3, peduncle elongate, rami fringed with a few long setae. Telson much longer than broad, apically rounded. L. about 8 mm.

Indian Ocean (Trincomali [Ceylon]). Surface.

6. **L. variegata** (Stimps.) 1855 *Anonyx variegatus*, Stimpson in: P. Ac. Philad., *v.* 7 p. 394 | 1862 *Lysianassa variegata*, Bate, Cat. Amphip. Brit. Mus., p. 67 t. 10 f. 7 | 1888 *Lysianax variegatus*, T. Stebbing in: Rep. Voy. Challenger, *v.* 29 p. 682 t. 23.

Head, lateral lobes produced, rounded. Side-plate 1 distally widened, 4ᵗʰ with lower expansion deep, not broad. Pleon segment 3, postero-lateral angles forming a minute tooth far upturned. Eyes large, reniform, very dark in spirit. Antenna 1, 1ˢᵗ joint tumid, 2ᵈ much longer than 3ᵈ, flagellum

8-jointed, accessory flagellum 4-jointed. Antenna 2 in ♀, ultimate joint of peduncle slightly longer than penultimate, flagellum 8-jointed; in ♂ ultimate joint of peduncle twice length of penultimate, flagellum with 53 joints, calceoli on many of them. Mandible, palp set far back, 1ˢᵗ joint rather long. Maxillipeds, outer plates large, reaching beyond 2ᵈ joint of palp, 4ᵗʰ joint of palp very small. Gnathopod 1, 6ᵗʰ joint much narrower and slightly longer than 5ᵗʰ. Gnathopod 2, 3ᵈ joint rather longer than 5ᵗʰ, 6ᵗʰ oval, with convex palm. Peraeopods 1 and 2, 4ᵗʰ and 5ᵗʰ joints fringed with plumose setae. Peraeopod 3, 2ᵈ joint broad, rounded. Peraeopods 4 and 5, 2ᵈ joint somewhat produced downward. Branchial vesicles with large accessory lobes. Uropod 3, rami shorter than peduncle, broadly lanceolate, setose on opposed margins, the outer rather the longer. Telson little longer than broad, much shorter than peduncle of uropod 3, distally squared, with a small central notch. Colour yellowish mottled with brown, with scattered white dots. L. 14 mm.

Simon's Bay [Cape of Good Hope]. Depth 33 m.

7. **L. punctata** (A. Costa) 1840 *Callisoma p.* (nom: nud.), O. G. Costa, Fauna Reg. Napoli, Crost., Cat. p. 5 | 1851 *C. punctatum*, A. Costa, Fauna Reg. Napoli, fasc. Marz. 1851 p. 4 t. 8 f. 4—7 | 1851 *C. p.*, (A. Costa in:) F. W. Hope, Cat. Crost. Ital., p. 23, 44 | 1893 *Lysianax punctatus*, A. Della Valle in: F. Fl. Neapel, *v.* 20 p. 789 t. 6 f. 6; t. 25 f. 22—32.

Head, lateral angles produced, rounded; Side-plate 4 greatly expanded below, expansion not deep, but very broad. Pleon segment 3, postero-lateral angles rounded. Eyes red-brown. Antenna 1 in ♀, 1ˢᵗ joint tumid, 2ᵈ and 3ᵈ comparatively short, flagellum very short, 6- or 7-jointed, accessory flagellum 3-jointed. Peraeopod 3, 2ᵈ joint attached to a cleft in the side-plate, narrow above, greatly expanded below. Peraeopods 4 and 5, 2ᵈ joint produced downward below the 3ᵈ. Telson elongate, subrectangular, distally with a shallow but rather wide notch. The rest as in L. longicornis (p. 39). Ground colour citron-yellow, with a generally red hue from very numerous scarlet chromatophores; appendages colourless or greyish. L. ♀ 6—7 mm.

Gulf of Naples. Upon the pleon of Paguri laden with eggs, matching the colour of its resting-place.

L. costae M.-E. 1830 *L. c.*, H. Milne Edwards in: Ann. Sci. nat., *v.* 20 p. 365 t. 10 f. 17.

L. 6—7 mm. ♀.
Gulf of Naples.

L. nasuta Dana 1853 & 55 *L. n.*, J. D. Dana in: U. S. expl. Exp., *v.* 13ɪɪ p. 915; t. 62 f. 2 a—m | 1893 *Lysianax n.*, A. Della Valle in: F. Fl. Neapel, *v.* 20 p. 793.

L. 10 mm.
Tropical Atlantic (Rio Janeiro [Brazil]).

L. pilicornis Heller 1866 *L. p.*, Cam. Heller in: Denk. Ak. Wien, *v.* 26ɪɪ p. 17 t. 2 f. 16.

L. 8—9 mm. ♂.
Adriatic (Lesina).

22. Gen. **Perrierella** Chevreux & E. L. Bouv.

1892 *Perrierella* (Sp. un.: *P. crassipes*), Chevreux & E. L. Bouvier in: Bull. Soc. zool. France, *v.* 17 p. 50 | 1893 *P.*, J. Bonnier in: Bull. sci. France Belgique, *v.* 24 p. 181 | 1893 *P.*, A. Della Valle in: F. Fl. Neapel, *v.* 20 p. 840 | 1895 *P.*, G. O. Sars, Crust. Norway, *v.* 1 p. 677 | 1892 *Pararistias* (Sp. un.: *P. audouinianus*), D. Robertson in: P. nat. Hist. Soc. Glasgow, n. ser. *v.* 3 p. 201.

Body compact. Side-plates not large, 1st partly concealed. Both pairs of antennae short, peduncle of antenna 1 rather elongate. A deep groove between epistome and upper lip. Mandible stout, molar obsolete, 1st joint of palp very short. Maxilla 1, inner plate with 3 setae, outer with 7(?) spines, palp with minute teeth and spines. Maxilla 2, plates rather short, inner the broader. Maxillipeds, inner plates minute, carrying 3 spines, outer elongate with spines on inner margin, palp only 3-jointed, not reaching beyond outer plate. Gnathopod 1, 6th joint rather longer than 5th, oval, making an approach to a palm, so as to be weakly subchelate. Gnathopod 2 slender, elongate, subchelate or almost chelate. Peraeopods 1—5 short and stout, 4th joint rather broad, 6th produced to a short tooth-like process. Uropod 3, peduncle short, rami short, outer rather the longer, 2-jointed. Telson entire.

1 species.

1. **P. audouiniana** (Bate) 1857 *Lysianassa a.*, Bate in: Ann. nat. Hist., ser. 2 *v.* 19 p. 138 | 1862 *L. a.*, Bate, Cat. Amphip. Brit. Mus., p. 69 t. 11 f. 1 (excl. 1g) | 1889 *Lysianax audouinianus*, A. O. Walker in: P. Liverp. biol. Soc., *v.* 3 p. 203 t. 10 f. 9, 10 | 1890 *Aristias audouiniana*, Meinert in: Udb. Hauchs. *v.* 3 p. 152 t. 1 f. 1—6 | 1892 *Pararistius audouinianus*, D. Robertson in: P. nat. Hist. Soc. Glasgow, n. ser. *v.* 3 p. 201 | 1893 *Perrierella audouiniana*, J. Bonnier in: Bull. sci. France Belgique, *v.* 24 p. 175 t. 5 | 1895 *P. a.*, G. O. Sars, Crust. Norway, *v.* 1 p. 678 t. II f. 2 | 1875 *Aristias tumidus* (err., non *Anonyx t.* Krøyer 1846!), Cam. Heller in: Denk. Ak. Wien, *v.* 35 p. 30 t. 2 f. 1—7 | 1892 *Perrierella crassipes*, Chevreux & E. L. Bouvier in: Bull. Soc. zool. France, *v.* 17 p. 50 f. | 1893 *P. c.*, A. Della Valle in: F. Fl. Neapel, *v.* 20 p. 841.

Head deep, lateral angles slightly produced. Side-plate 4 little emarginate behind. Pleon segment 3 at postero-lateral angles almost quadrate. Eyes large, irregularly oval, dark with chalky white coating. Antenna 1, 2^{d} and 3^{d} joints not very short, flagellum with 4 joints, 1st the longest, accessory flagellum 2-jointed. Antenna 2, ultimate and penultimate joints of peduncle subequal, flagellum 4-jointed, much shorter than that of antenna 1. Palp of maxilla 1 ends in 4 crenulate teeth (Chevreux & Bouvier), or in 3 spines and a serrulate margin (Robertson, Sars' figure), or in 3 setae (Heller), or in 8 spinules (Bonnier). Maxillipeds, the palp has a tuberculiform 4th joint (Chevreux & Bouvier, contrary to Heller, Bonnier and Sars). Uropod 3 has the rami finely serrulate. Telson nearly twice as long as broad (Sars; not so long in Bonnier's figure), narrows distally to a slightly emarginate apex. Colour yellowish (Sars), whitish slightly tinted with rose on the back (Chevreux), eggs pale green (Bonnier). L. 2—4 mm.

North-Atlantic, North-Sea and Kattegat (Great Britain, France, West-Norway); Mediterranean. Commensal in sponges.

23. Gen. **Normanion** Bonnier

1871 *Normania* (Sp. un.: *N. quadrimana*) (non G. S. Brady 1866, Ostracoda!), A. Boeck in: Forb. Selsk. Christian., 1870 p. 119 | 1890 *N.*, G. O. Sars, Crust. Norway, *v.* 1 p. 32 | 1893 *N.*, A. Della Valle in: F. Fl. Neapel, *v.* 20 p. 796 | 1893 *Normanion*, J. Bonnier in: Bull. sci. France Belgique, *v.* 24 p. 167, 173 | 1895 *N.*, G. O. Sars, Crust. Norway, *v.* 1 p. 674.

Body compact, side-plates not very deep, 4th little excavate behind. Pleon segment 3, postero-lateral angles rounded. Antenna 1 and 2 in ♀ short, 2^{d} and 3^{d} joints of antenna 1 not extremely short. Epistome not defined from upper lip. Mandible, molar weak, palp long, set far back, 1st joint short. Maxilla 1, inner plate with 2 small apical setae, 2^{d} joint of palp broad. Maxilla 2, plates rather narrow, only apically armed. Maxillipeds, inner plates long and narrow, outer long and broad, unarmed, reaching much beyond the 3-jointed palp, of which the 2^d joint is shorter than the 1st. Gnathopod 1 subchelate, 3^d joint longer than 4th, 6th greatly expanded, broader than long, 7th as long as the breadth of 6th. Gnathopod 2 minutely chelate. Peraeopods 1—5 slender. Peraeopods 3—5, 2^d joint longer than broad. Branchial vesicles almost simple. Uropod 3, peduncle long, rami narrowly lanceolate, outer slightly the longer. Telson short, entire, rounded quadrangular.

2 species.

Synopsis of species:

Gnathopod 1, process of 5th joint obtuse . . . . . . 1. **N. quadrimanus** . p. 42
Gnathopod 1, process of 5th joint acute . . . . 2. **N. sarsi** . . . . . p. 42

1. **N. quadrimanus** (Bate & Westw.) 1868 *Opis quadrimana*, Bate & WestWood, Brit. sess. Crust., *v.*2 p. 503 f. | 1895 *Normanion amblyops*, G. O. Sars. Crust. NorWay, *v.*1 p. 674 t. I f. 1.

Eyes large and confluent, visual elements imperfectly developed, light reddish brown, slightly areolated. Antenna 1, 1st joint not twice as long as 2^d and 3^d combined, flagellum with 6 joints, 1st rather long, accessory flagellum 4-jointed. Antenna 2 in ♀, ultimate joint of peduncle shorter than penultimate, flagellum 6-jointed. Gnathopod 1, 3^d joint considerably longer than 4th, 5th with a comparatively short and obtusely rounded expansion, 6th much broader than it is long, palm transverse, defined below by a dentiform projection with a comparatively small spine on each side, 7th joint or finger rather slender and perfectly smooth. Telson, as figured, scarcely longer than broad. Body pellucid, pale yellowish with lateral orange patches. L. ♀ 5·5 (Sars), about 3 mm (Bate & Westwood).

Trondhjemsfjord. Parasitic on fish, from 370—570 m.

2. **N. sarsi** Stebb.[*]) ?1871 *Normania qvadrimana* (err., non *Opis quadrimana* Bate & Westwood 1868!), A. Boeck in: Forh. Selsk. Christian., 1870 p. 120 | ?1872 & 76 *N. q.*, A. Boeck, Skand. Arkt. Amphip., *v.*1 t. 6 f. 3; *v.*2 p. 188 | 1890 *N. q.*, G. O. Sars, Crust. NorWay, *v.*1 p. 83 t. 13 f. 1 | 1895 *Normanion qvadrimanus*, G. O. Sars, Crust. NorWay, *v.*1 p. 674.

Head rather deep, lateral corners angular, only slightly projecting. Body somewhat compressed, back evenly vaulted. Eyes large, oval, visual elements unusually large, brownish, with orange-coloured coating. Antenna 1, flagellum with 5 joints, 1st of no great length, accessory flagellum slender, 3-jointed. Antenna 2 in ♀, flagellum 4-jointed. In other respects antenna 1 and 2 as in N. quadrimanus. Gnathopod 1, 3^d joint a little longer than 4th, 5th forming a narrow projecting lobe, 6th a little broader than long, palm transversely truncated and armed below with 3 strong spines, 7th joint strongly serrate on concave margin. Telson a little longer than broad (Sars), broader than long (Boeck). Body whitish, pellucid, with yellowish intestine and dark bluish ovaries. L. 5 (Sars), 4 mm (Boeck).

North-Atlantic, North-Sea and Skagerrak (South- and West-Norway). Parasitic on fish.

24. Gen. **Paratryphosites** Stebb.

1871 *Hippomedon* (part.), A. Boeck in: Forh. Selsk. Christian., 1870 p. 102|1899 *Paratryphosites* (Sp. typ.: *Lysianassa abyssi*), T. Stebbing in: Ann. nat. Hist., ser. 7 *v.*4 p. 206.

Pleon segment 3, postero-lateral angles acutely upturned. Antenna 1, 1st joint of flagellum not very large, accessory flagellum small. Antenna 2,

---

[*]) Nom. nov. After Georg Ossian Sars.

ultimate and penultimate joints of peduncle subequal. Mandible, molar prominent. palp affixed over it or a little in front, 2^d and 3^d joints subequal. Maxilla 1, inner plate carrying 5 setae, outer with the usual spines, palp broad at apex. Maxilla 2, with setae on inner margin of inner plate. Maxillipeds, outer plates reaching much beyond 2^d joint of palp, with spine-teeth on straight inner margin. Gnathopod 1 subchelate, 6th joint shorter than 5th, of uniform width, palm oblique, well defined, overlapped by finger. Gnathopod 2 subchelate, 5th joint distally widened, 6th a little more than half as long. Peraeopods 3—5, 2^d joint large, serrate on hind margin, 4th expanded in 3^d and 4th pairs, 4th pair the longest. Uropod 3, rami with many spinules, outer the longer. Telson little longer than broad, cleft scarcely reaching the middle, dehiscent, each apex carrying 5 spinules.

1 species.

1. **P. abyssi** (Goës) 1866 *Lysianassa a.*, Goës in: Öfv. Ak. Förb., *v.* 22 p. 519 t. 37 f. 5 | 1871 *Hippomedon a.*, A. Boeck in: Forh. Selsk. Christian., 1870 p. 103 | 1872 *H. a.*, A. Boeck, Skand. Arkt. Amphip., *v.* 1 p 138 | 1890 *H. a.*, G. O. Sars, Crust. Norway, *v.* 1 p. 56 | 1893 *Euryporeia? a.*, A. Della Valle in: F. Fl. Neapel, *v.* 20 p. 848 | 1899 *Lysianassa a.*, *Paratryphosites sp. typ.*, T. Stebbing in: Ann. nat. Hist., ser. 7 *v.* 4 p. 206.

Head, lateral angles acute. Eyes oval, red. Antenna 1, 1st joint long, cylindrical, not produced, flagellum with 10—12 joints, accessory flagellum with 3 subequal joints. Antenna 2, flagellum 22-jointed. L. reaching 17 mm.

Arctic Ocean (north of lat. 68° 36' N.) Depth 38—528 m.

25. Gen. **Paronesimus** Stebb.

1894 *Paronesimus* (Sp. un.: *P. barentsi*), T. Stebbing in: Bijdr. Dierk., *v.* 17 p. 14.

Side-plate 1 not expanded below, 4th only slightly, lower hind margin obliquely truncate. Antenna 1, 1st joint tumid, 2^d and 3^d very short. Antenna 2, ultimate joint of peduncle much shorter than penultimate; flagella of both pairs short in ♀. Mandible, molar not very strong, palp attached over the molar, much longer than trunk, 1st joint short, 3^d subequal to 2^d. Maxilla 1, 2 setae on inner plate, 11 spines on outer, 4 or 5 spine-teeth on long 2^d joint of palp. Maxilla 2, inner plate narrower than, much more than half as long as outer. Maxillipeds, 3 teeth on inner plates, outer plates with minute teeth on inner margin and 1 spine on apical, 3^d and 4th joints of palp rather elongate. Gnathopod 1, 6th joint longer than 5th, its slightly curved oblong narrowed distally, forming a very short palm not in line with insertion of the short finger. Gnathopod 2, 3^d joint as long as 5th, 6th much more than half as long as 5th, widening gradually to a sinuous palm, 7th joint strongly curved. Peraeopods 1 and 2, 2^d joint stout, rather short, 4th rather large. Peraeopods 3—5, 2^d joint much longer than broad, 4th rather broad. Uropods 1—3, rami not elongate. Uropod 3, rami short, smooth, the outer the longer, with a small 2^d joint. Telson $1\frac{1}{2}$ times as long as broad, cleft for nearly half its length.

1 species.

1. **P. barentsi** Stebb. 1894 *P. b.*, T. Stebbing in: Bijdr. Dierk., *v.* 17 p. 14 t. 2.

Pleon segment 3, postero-lateral angles forming small upturned point. Antenna 1 in ♀, flagellum, 1st joint much larger than next, accessory flagellum 4-jointed. Antenna 2 in ♀, flagellum 10-jointed. Telson with a spinule in each of the slightly divergent apices. L.?

Arctic Ocean (lat. 76° N., long. 54° E.) Depth 125 m.

26. Gen. **Orchomene** Boeck

1871 *Orchomene* (part.), A. Boeck in: Forh. Selsk. Christian., 1870 p. 114 | 1876
O., A. Boeck, Skand. Arkt. Amphip., *v.*2 p. 171 | 1890 *O.*, G. O. Sars, Crust. NorWay,
*v.*1 p.59 | 1893 *O.*, J. Bonnier in: Bull. sci. France Belgique, *v.*24 p.174 | 1873 *Orchomena,*
E. v. Martens in: Zool. Rec., *v.*8 p.188.

Side-plates large. Pleon segment 3, postero-lateral angles not produced,
generally serrate. Antenna 1, flagellum in ♂ with calceoli, accessory flagellum
well developed (Fig. 9). Antenna 2 in ♀, antepenultimate joint of peduncle long,
flagellum short, in ♂ flagellum long filiform, with calceoli. Epistome projecting

Flagellum Accessory flagellum

Fig. 9. **O. serrata.**
Antenna 1. [After G. O Sars.]

roundly in front of upper lip. Mandible long,
molar very small, palp slender, set far back.
Maxilla 1, 2 setae on inner plate, outer with
apex scarcely oblique, palp with spine-teeth not
numerous. Maxilla 2, both plates long, narrow,
setiferous only apically. Maxillipeds, inner plates
elongate, outer long oval, weakly crenulate, palp
not very large. Gnathopod 1 short, stout, sub-
chelate, 6th joint longer than distally widened 5th, with distinct transverse
palm, finger short, curved. Gnathopod 2 slender, 6th joint narrow, produced
beneath the minute finger. Peraeopods 3—5, 2^{d} joint very large. Uropod 3
in ♀ scarcely reaching beyond uropod 2, in ♂ much larger and plumose.
Cleft of telson not always reaching the centre.

6 species.

Synopsis of species:

1 { Pleon segment 3, hinder edge coarsely serrate — **2.**
     Pleon segment 3, hinder edge finely serrate or
     smooth — **3.**

2 { Pleon segment 4 With evenly rounded carina . . . 1. **O. serrata** . . . p. 44
     Pleon segment 4 With carina ending acutely . . . . 2. **O. pectinata** . . p. 45

3 { Eyes dark — **4.**
     Eyes not dark — **5.** .

4 { Pleon segment 3, hind margin minutely crenulate . 3. **O. batei** . . . . p. 45
     Pleon segment 3, hind margin perfectly smooth . . 4. **O. hanseni** . . . p. 46

5 { Head, lateral angles acute, greatly projecting . . 5. **O. crispata** . . p. 46
     Head, lateral angles quadrate, slightly projecting . 6. **O. amblyops** . . p. 46

1. O. serrata (Boeck) 1861 *Anonyx serratus,* A. Boeck in: Forh. Skand. Naturf.,
Møde 8 p. 641 | 1893 *A. s.,* A. Della Valle in: F. Fl. Neapel, *v.*20 p. 819 | 1871 *Orcho-
mene s.* (part.), A. Boeck in: Forb. Selsk. Christian., 1870 p. 115 | 1872 & 76 *O. s.,* A. Boeck,
Skand. Arkt. Amphip., *v.*1 t. 5 f. 2; *v.*2 p. 172 | 1890 & 95 *O. s.,* G. O. Sars, Crust.
NorWay, *v.*1 p. 62 t. 23 f. 1; p. 682 t. IV f. 1.

Head, lateral angles considerably projecting and obtuse. Pleon segment 3,
lateral angles quadrate, with their hind margin straight, coarsely serrate, forming
16—20 points, 4th segment with evenly rounded dorsal carina behind the depression.
Eyes narrow oval, tapering above in ♀, broad oval in ♂. Antenna 1 (Fig. 9),
flagellum with 8 joints. 1st very large, especially in ♂, accessory flagellum
6-jointed. Antenna 2, flagellum 6-jointed in ♀, little longer in ♂, but with
8 joints, 3 carrying calceoli. Gnathopod 1, 6th joint of nearly uniform breadth,
nearly twice as long as 5th, which is equal to 4th. Peraeopods 3 – 5, 2^{d} joint
subequal in length to rest of limb. Uropod 3 alike in ♀ and ♂, inner ramus

shorter than basal joint of outer, armed with 3 spinules. Telson broadly
ovate, tapering distally, with 3 pairs of dorsal spinules, cleft narrow, reaching
about to centre. Colour yellow or ochraceous. L. ♀ 10, ♂ about 6 mm.

Arctic Ocean, North-Atlantic and North-Sea (Norway). Depth 56—188 m.

2. **O. pectinata** O. Sars 1882 *O. pectinatus, O. pectinata*, G. O. Sars in: Forh.
Selsk. Christian., nr. 18 p. 80 t. 3 f. 5, 5a; p. 120 | 1890 & 95 *O. pectinatus, O. pectinata*,
G. O. Sars, Crust. Norway, *v.* 1 p. 64 t. 23 f. 3; p. 682 | 1893 *Anonyx pectinatus* (part.),
A. Della Valle in: F. Fl. Neapel, *v.* 20 p. 820.

Head, lateral angles projecting, acute. Side-plate 1 dilated below.
Pleon segment 3, postero-lateral angles scarcely rounded, their hind margin
curved, with coarse and sharp upturned serration of 12—16 points, 4th segment
with high carina ending acutely. Eyes narrow, slightly sigmoid, visual elements
imperfectly developed; almost cream-coloured. Antenna 1 in ♀, flagellum with
13 joints, 1st large, accessory flagellum 7-jointed. Epistome greatly projecting,
evenly rounded. Gnathopod 1, 6th joint not much longer than 5th, a little
tapering. Uropod 3, inner ramus as long as basal joint of outer, carrying
4 spinules. Telson small, with 2 pairs of small dorsal and a pair of apical
spinules, cleft reaching nearly to centre, very dehiscent. Colour pale greyish ·
white. L. ♀ 12 mm.

Arctic Ocean, North-Atlantic and North-Sea (West-Norway). Depth 226 m.

3. **O. batei** O. Sars ?1853 *Lysianassa humilis*, A. Costa in: Rend. Soc. Borbon.,
n. ser. *v.* 2 p. 172 | 1893 *Anonyx h.* + *A. goesii*, A. Della Valle in: F. Fl. Neapel, *v.* 20
p. 817 t. 26 f. 32—37; p. 273, 920 | 1895 *A. h.*, Sowinski in: Mém. Soc. Kiew, *v.* 14
p. 250 t. 4 f. 15—18; t. 5 f. 1—7 | 1857 *A. edwardsii* (err., non Krøyer 1846!), Bate in:
Ann. nat. Hist., ser. 2 *v.* 19 p. 138 | 1861 *Lysianassa longicornis* (part.) (err., non H. Lucas
1845/46!) + *Anonyx edwardsi*, Bate & Westwood, Brit. sess. Crust., *v.* 1 p. 85 (figure of
animal and pleon only); p. 94 f. | 1882 *Orchomene batei,* G. O. Sars in: Forh. Selsk.
Christian., nr. 18 p. 81 | 1890 *O. b.*, G. O. Sars, Crust. Norway, *v.* 1 p. 60 t. 22.

Head, lateral angles projecting, broadly rounded in ♀, narrow linguiform
in ♂. Side-plate 5 broader than deep. Pleon segment 3, postero-lateral angles
almost quadrate, with hinder margin minutely crenulate, dorsal depression of
4th segment stronger in ♂ than in ♀. Eyes oval, slightly reniform, dark
reddish brown, much larger in ♂ than in ♀. Antenna 1, flagellum with 8 joints,
1st large, accessory flagellum with 5 joints, 1st of moderate length. Antenna 2
in ♀, ultimate joint of peduncle shorter than penultimate, scarcely as long as
antepenultimate, flagellum 9-jointed; in ♂, ultimate joint of peduncle stout, full
as long as penultimate and longer than antepenultimate, flagellum reaching the
full length of the animal. Epistome narrow linguiform, projecting much.
Gnathopod 1 short and stout, 6th joint not greatly longer than 5th, distally
tapering. Gnathopod 2, 6th joint very narrow, oblong linear, minutely produced,
the spiny armature unusually coarse. Uropod 3 in ♀, inner ramus shorter than
basal joint of outer, mucroniform, without spinules; in ♂, both rami long,
and densely setose. Telson in ♀ oblong quadrangular, scarcely tapering, about
once and a half as long as broad, with 2 pairs of marginal spinules,
cleft very short and wide, about one fourth of total length; telson in ♂ more
than twice as long as broad, with 3 or 4 pairs of spinules, cleft much
narrower, nearly a third of the length. Body cream-coloured, each segment
of peraeon with a reddish speck at the hind corners. L. ♀ 7, ♂ 8 mm.

North-Atlantic, North-Sea and Skagerrak (South- and West-Norway, Great Britain,
France). Depth 38—76 m.

4. O. hanseni Meinert ?1867 *Anonyx melanophthalmus,* A. M. Norman in: Rep. Brit. Ass., Meet. 86 p. 201 | 1890 *Orchomene hanseni,* Meinert in: Udb. Hauchs, v. 3 p. 154 t. 1 f. 18—24 | 1895 *O. h.,* G. O. Sars, Crust. Norway, v. 1 p. 681 t. III f. 2.

♀. Head, lateral angles somewhat projecting, narrowly rounded. Side-plate 5 rather broader than deep. Pleon segment 3, postero-lateral angles narrowly rounded, with hind margin perfectly smooth. Eyes unusually large, oblong oval, widened below, almost black. Epistome narrowly rounded and less projecting than in O. batei (p. 45). Antennae and limbs not distinguishable from those of O. batei, except that in gnathopod 2 the spines of the 6th joint are less coarse. Uropod 3, inner ramus simple, much shorter than outer. Telson about once and a half as long as broad, with a single pair of marginal spinules, cleft very short and angular, wider than deep. Colour uniformly whitish. L. 7—8 mm. — ♂ unknown.

Christianiafjord, Kattegat, North-Atlantic (Hebrides). Depth 14—90 m.

5. O. crispata (Goës) 1866 *Lysianassa c.,* Goës in: Öfv. Ak. Förh., v. 22 p. 519 t. 37 f. 3 | 1890 *Orchomene crispatus,* G. O. Sars, Crust. Norway, v. 1 p. 63 t. 23 f. 2 | 1893 *Anonyx c.* A. Della Valle in: F. Fl. Neapel, v. 20 p. 819 | 1871 *Orchomene serratus* (part.) (err., non *Anonyx s.* A. Boeck 1861!), A. Boeck in: Forb. Selsk. Christian., 1870 p. 115 | 1882 *O. s.,* G. O. Sars in: Forh. Selsk. Christian., nr. 18 p. 81.

Head, lateral angles greatly produced, acute. Side-plates very deep, 1st much overlapping the head, 4th as deep as broad. Pleon segment 3, postero-lateral angles rounded, with hind margin straight, finely serrated, 4th segment without dorsal projection. Eyes oblong linear, reddish brown with light orange coating. Antenna 1, 1st joint unusually narrow, almost as long as rest of antenna, flagellum with 9 joints, 1st not long, accessory flagellum 6-jointed. Epistome broad, little projecting, obtusely truncate at tip. Gnathopod 1, 3[d] joint conspicuously longer than 4th, 6th much longer than 5th, slightly tapering to the transverse palm. Peraeopods 3—5, 2[d] joint much expanded, peraeopod 5 almost quadrate, Uropod 3, rami unusually broad. Telson short and broad, less than once and a half as long as broad, with 2 pairs of marginal spinules, cleft dehiscent, not quite reaching the centre. Body somewhat flesh-coloured. L. ♀ 12 mm.

Arctic Ocean; North-Atlantic and North-Sea (West-Norway). Depth 188—376 m.

6. O. amblyops O. Sars 1890 & 91 *O. a.,* G. O. Sars, Crust. Norway, v. 1 p. 65; t. 25 f. 1 | 1893 *Anonyx pectinatus* (part.), A. Della Valle in: F. Fl. Neapel, v. 20 p. 820.

Head, lateral angles quadrate, slightly produced. Side-plate 1 expanded below. Pleon segment 3, postero-lateral angles slightly rounded, their hind margin convex, serrate with 14—16 distinct points, 4th segment with triangular dorsal projection ending obtusely. Eyes recurved below, visual elements imperfectly developed, light orange with white reticulation. Antennae as in O. serrata (p. 44). Epistome as in O. crispata, much less prominent than in O. pectinata (p. 45). Gnathopod 1, 6th joint considerably longer than 5th. Uropod 3, inner ramus shorter than basal joint of outer, which carries 3 feathered setae on inner margin. Telson not once and a half as long as broad, with 3 pairs of marginal spinules, cleft gently dehiscent, about reaching the centre. Colour whitish, tinged in front with yellow. L. ♀ 8 mm.

North-Atlantic and North-Sea (West-Norway). Depth 188—376 m.

27. Gen. **Socarnoides** Stebb.

1888 *Socarnoides* (Sp. un.: *S. kergueleni*), T. Stebbing in: Rep. Voy. Challenger, *v.* 29 p. 690 | 1893 *S.*, A. Della Valle in: F. Fl. Neapel, *v.* 20 p. 793 | 1890 *Socarnioides* (laps.), Beddard in: Zool. Rec., *v.* 25 Index p. 15.

·Body compact, side-plates rather large. Antenna 1, peduncle stout, 2^d and 3^d joints not extremely short, flagellum of few joints. Antenna 2 geniculate, ultimate joint of peduncle shorter than penultimate. Epistome projecting rather beyond upper lip, separated by deep incision. Lower lip with narrowed lobes and mandibular processes divergent. Mandible narrowly elongate, cutting edge simple, molar weak, palp of moderate length, attached far back. Maxilla 1, inner plate without setae, outer carrying 11 spines, palp with 1st joint very short, 2^d large, somewhat tapering, without spines or teeth, but slightly serrate. Maxillipeds, inner and outer plates long, scantily armed, the outer extending far along 3^d joint of palp. Gnathopod 1 not subchelate, 7th joint small. Gnathopod 2 minutely chelate. Peraeopods 3—5, 2^d joint greatly expanded, especially in peraeopod 5. Uropod 2, shorter than uropod 1, rami subequal, inner with spiniferous constriction. Uropod 3, shorter than uropod 2, peduncle produced into a dentiform process, rami not long, outer the longer, with small 2^d joint. Telson small, narrowing distally, cleft not reaching the centre, dehiscent.

2 species.

Synopsis of species:

Maxillipeds, outer plates obtusely pointed . . . . 1. **S. kergueleni** . p. 47
Maxillipeds, outer plates apically rounded . . . . 2. **S. stebbingi** . . p. 47

1. S. kergueleni Stebb. 1888 *S. k.*, T. Stebbing in: Rep. Voy. Challenger, *v.* 29 p. 691 t. 25.

Head, lateral angles rounded, projecting. Side-plate 1 not distally produced forward. Pleon segment 3, postero-lateral angles rounded. Eyes large, reniform. Antenna 1, flagellum with 8 joints, 1st not very large, accessory flagellum 4-jointed. Antenna 2, geniculate between penultimate and ante-penultimate joint of peduncle, ultimate of peduncle shorter than penultimate, flagellum 7-jointed. Maxillipeds, outer plates large, with apex obtusely pointed, 1 seta on inner margin. Gnathopod 1, 6th joint tapering, narrower and slightly longer than 5th, 7th joint small. Gnathopod 2, 6th joint slender, much shorter than 5th, the thumb shorter than the small finger, thus making a rather imperfect chela. Peraeopods 1—5, 4th joint rather broad. Peraeopod 3, 2^d joint almost round, scarcely as large as the side-plate. Peraeopod 5, 2^d joint very large, as long as rest of limb. Uropod 3, inner ramus not much shorter than outer. L. about 7 mm.

Southern Indian Ocean (Kerguelen Island). ·Depth 54—230 m.

2. S. stebbingi (G. M. Thoms.) 1893 *Lysianax s.*, G. M. Thomson in: P. R. Soc. Tasmania, 1892 p. 19 t. 3 f. 9—18; t. 5 f. 9, 10.

♂. Closely agreeing with S. kergueleni in most respects, though in a few rather strikingly different. Side-plate 1 large, in figure very large, distally produced forward. Antenna 1, flagellum about 10-jointed, accessory flagellum about 6-jointed. Antenna 2, geniculate between ultimate and penultimate joints of peduncle, flagellum „long and many (30—40) jointed", (number perhaps varying on the two sides). Maxilla 1 (in figure) single-jointed. Maxillipeds, outer plates large, apically rounded, 2 or 3 minute spines on inner margin below the middle. Uropod 3, inner ramus „short and quite rudimentary". Telson doubtful. Colour nearly white. Integument very hard and brittle, without any markings. L. 6 mm. — ♀ unknown.

Pirates Bay [Tasmania]. In a rock-pool.

28. Gen. **Menigrates** Boeck

1871 *Menigrates* (Sp. un.: *M. obtusifrons*), A. Boeck in: Forh. Selsk. Christian., 1870 p. 113 | 1891 *M.*, G. O. Sars, Crust. Norway, *v.* 1 p. 110 | 1894 *M.*, T. Stebbing in: Bijdr. Dierk., *v.* 17 p. 15.

Body tumid, dorsally broad, side-plates rather large. Antennae 1 and 2, ♂ and ♀, short, with few-jointed flagella, longer in ♂ than in ♀. Antenna 1, 2^d and 3^d joints of peduncle very short, 1st joint in each flagellum rather long. Epistome not distinctly defined from upper lip. Lobes of lower lip narrowly rounded, hind lobes little divergent. Mandible elongate, cutting edge simple, molar weak, palp of moderate length, attached far back. Maxilla 1, inner plate with 2 setae, outer obliquely truncate, carrying 11 spines, palp with 6 spine-teeth on the apex. Maxilla 2, inner plate shorter than outer, obliquely truncate. Maxillipeds, inner plates well armed, outer large, reaching beyond the rather short 2^d joint of palp, 4th joint of palp small. Gnathopod 1 not subchelate, 6th joint longer than 5th. Gnathopod 2 elongate, subchelate. Peraeopods 3—5 short and robust, 2^d joint greatly expanded, especially in peraeopod 5. Uropod 3 very small, quite unarmed, outer ramus the longer, with small second joint. Telson short and broad, deeply incised, apices rounded.

1 species.

1. **M. obtusifrons** (Boeck) 1861 *Anonyx o.*, A. Boeck in: Forb. Skand. Naturf., Møde 8 p. 643 | 1871 *Menigrates o.*, A. Boeck in: Forb. Selsk. Christian., 1870 p. 114 | 1872 & 76 *M. o.*, A. Boeck, Skand. Arkt. Amphip., *v.* 1 t. 6 f. 2; *v.* 2 p. 169 | 1891 *M. o.*, G. O. Sars, Crust. Norway, *v.* 1 p. 111 t. 38 f. 1 | 1894 *M. o.*, T. Stebbing in: Bijdr. Dierk., *v.* 17 p. 15 | 1865 *Anonyx brachycercus* + *A. obtusifrons*, W. Lilljeborg in: N. Acta Soc. Upsal., ser. 3 *v.* 6 nr. 1 p. 27 t. 4 f. 42—49; p. 32 | 1893 *Ichnopus nugax* (part.), A. Della Valle in: F. Fl. Neapel, *v.* 20 p. 804.

Head, lateral corners somewhat produced and angular. Side-plate 4 little produced backward. Pleon segment 3, postero-lateral angles forming a short acutely upturned process. Eyes reniform. red. Antenna 1, flagellum 8-jointed in ♀, 13-jointed in ♂, accessory flagellum 4-jointed. Antenna 2, flagellum 8-jointed in ♀, 16-jointed in ♂; in ♂ the joints of the flagella are broad, with large calceoli. Gnathopod 1 rather powerful, 6th joint much longer than 5th, slightly curved, tapering, 7th joint strong, curved. Gnathopod 2, 6th joint half length of 5th, oblong oval, scarcely dilated distally. Peraeopods 3—5 with 2^d and 4th joints much expanded, the 2^d most, the 4th least in the 5th pair. Colour pale yellowish with faint reddish tinge in front. L. 13 mm.

Arctic Ocean, North-Atlantic and North-Sea (West-Norway). Depth 36—183 m.

29. Gen. **Aristias** Boeck

1871 *Aristias* (Sp. un.: *A. tumidus*), A. Boeck in: Forh. Selsk. Christian., 1870 p. 106 | 1890 & 95 *A.*, G. O. Sars, Crust. Norway, *v.* 1 p. 47; p. 675 | 1893 *A.*, A. Della Valle in: F. Fl. Neapel, *v.* 20 p. 843.

Body short and thick. Side-plates rather small, 1st almost concealed, 4th very slightly excavate. Antenna 1 not very turgid, 2d joint not extremely short, 1st joint of flagellum rather large. Antenna 2 in ♀ shorter than antenna 1. Epistome scarcely projecting, defined by a small but distinct sinus. Lower lip, front lobes narrow, hind processes blunt. Mandible strong, cutting edge simple, molar narrow, prominent, acuminate, palp central. Maxilla 1 with 5 or more setae on inner plate, outer plate very broad, palp normal. Maxilla 2 with the plates divergent, inner very broad, outer much narrower. Maxillipeds, inner plates small, outer large, oval, 1st joint of palp the largest. Gnathopod 1 rather robust, 6th joint tumid at base, tapering to an apex rather wider than the base of the finger. Gnathopod 2 slender, minutely chelate. Peraeopods 3—5, 2d joint not greatly dilated. Uropod 3, outer ramus the longer, with small 2d joint. Telson rather short, deeply cleft.

5 species.

Synopsis of species:

1	Uropod 3, inner ramus shorter than 1st joint of outer	1. A. tumidus	p. 49
	Uropod 3, inner ramus not shorter than 1st joint of outer — **2.**		
2	Eyes rudimentary	2. **A. microps**	p. 49
	Eyes not rudimentary — **3.**		
3	Telson cleft to the base	3. **A. commensalis** . .	p. 50
	Telson not cleft to the base — **4.**		
4	Antenna 1, accessory flagellum 5-jointed	4. **A. neglectus** . . .	p. 50
	Antenna 1, accessory flagellum 2-jointed	5. **A. megalops** . . .	p. 51

1. A. tumidus (Krøyer) 1846 *Anonyx t.*, Krøyer in: Naturh. Tidsskr., ser. 2 *v.* 2 p. 16, 40 | 1846 *A. t.*, Krøyer in: Voy. Nord, Crust. t. 16 f. 2 | 1887 *Aristias t.*, H. J. Hansen in: Vid. Meddel., ser. 4 *v.* 9 p. 67 t. 2 f. 3—3b | 1890 *A. t.*, G. O. Sars, Crust. Norway, *v.* 1 p. 49 t. 18 f. 1 | 1893 *A. t.*, A. Della Valle in: F. Fl. Neapel, *v.* 20 p. 846 1884 *Menigrates arcticus, Orchomene? a.*, J. S. Schneider in: Tromsø Mus. Aarsh., *v.* 7 p. 63 t. 1, 2.

Head, lateral corners bluntly rounded, not projecting. Pleon segment 3, lateral angles quadrate, margin faintly serrate. Eyes large, constricted above, black. Antenna 1 in ♀, flagellum 8-jointed, accessory flagellum 4- or 5-jointed. Antenna 2 in ♀, flagellum 7-jointed. Gnathopod 1 rather short and stout, 6th joint with hind margin serrate, apically forming a minute palm. Peraeopods 1—5, 6th joint produced beneath the 7th to a short tooth-like process. Peraeopods 3—5, 4th joint expanded and decurrent. Uropod 3, outer ramus rather large and dilated, 2d joint small, abruptly narrower, inner ramus little more than half as long as outer. Telson about as broad as long, narrowly cleft beyond middle, not to base, with one spinule at each apex. Colour greyish white. L. 8 mm.

Arctic Ocean, Varangerfjord. Depth 110—140 m.

2. A. microps O. Sars 1895 *A. m.*, G. O. Sars, Crust. Norway, *v.* 1 p. 675 t. I f. 2.

Head, lateral corners not produced, front edges nearly straight. Pleon segment 3, angularly produced at lateral corners (Sars, text, not figure). Eyes quite rudimentary, represented each by a small patch of opaque whitish pigment, without visual elements. Antenna 1 in Q, peduncle thick, flagellum not as long as peduncle, 5-jointed, accessory flagellum 2-jointed. Antenna 2, flagellum much shorter than peduncle, 4-jointed. Gnathopod 1, 5th joint stout, 6th a little longer than 5th, hind margin serrulate, a small palm. Gnathopod 2, 6th joint more than half as long as 5th, narrowed distally. Uropod 3, rami rather short, spiniform 2d joint of outer abruptly narrower than 1st joint, inner ramus mucroniform, a little shorter than outer. Telson short and broad, cleft narrowly nearly to the base, with a spinule at each rounded apex. Body semi-pellucid, yellowish. L. Q 3 mm.

Trondhjemsfjord, Arctic Ocean (Nordland, Very deep water; Tromsö).

3. A. commensalis Bonnier 1896 *A. c.*, J. Bonnier in: Ann. Univ. Lyon, *v.* 26 p. 614 t. 35 f. 4.

Body thick, compact. Head, lateral corners slightly produced to obtuse apex. Side-plates not very large; pleon segments 1—3, postero-lateral corners subquadrate. Eyes round, remaining reddish in spirit. Antennae 1 and 2 short, equal. Antenna 1, flagellum with 6 joints, 1st very large, densely clothed, the rest small; accessory flagellum as long as 1st joint of primary, 5-jointed. Antenna 2, ultimate and penultimate joints of peduncle equal, flagellum with 10 joints, 6 calceoli. Epistome carinate. Mandible, palp attached behind the molar. Maxillipeds, inner plates forming a little double crest, outer large, palp small, 1st joint broad, finger very short. Gnathopod 1, 5th joint as long as 6th, 6th tapering, hind margin crenulate, finger small. Gnathopod 2, 5th joint twice 6th, 6th with the minute finger forming a tiny chela. Peraeopods 1—5 tolerably stout. Peraeopods 3—5, 2d joint expanded, hind margin serrate. Uropod 1 spinose; uropod 3, inner ramus broad, with 3 spinules on distal part of outer margin, outer ramus longer, serrate on inner margin, with small 2d joint. Telson cleft to the base, with 1 or 2 pairs of marginal spines and 2 or 3 on the obtuse apices. L. ♂ 10 mm.

Bay of Biscay. Depth 800 and 960 m. On Pheronema grayi S. Kent, and Pteraster personatus P. Sladen.

4. A. neglectus H. J. Hansen 1859 *Anonyx tumidus* (err., non Krøyer 1846!), R. M. Bruzelius in: Svenska Ak. Handl., n. ser. *v.* 3 nr. 1 p. 41 | 1867 *A. t.*, Cam. Heller in: Denk. Ak. Wien, *v.* 26 II p. 25 t. 3 f. 6—12 | 1871 *Aristias t.*, A. Boeck in: Forh. Selsk. Christian., 1870 p. 107 | 1872 *A. t.*, A. Boeck, Skand. Arkt. Amphip, *v.* 1 p. 148 t. 3 f. 4 | ?1861 *Lysianassa ciliata*, E. Grube, Ausfl. Triest, p. 135 | 1887 *Aristias neglectus*, H. J. Hansen in: Vid. Meddel., ser. 4 *v.* 9 p. 67 | 1893 *A. n.*, A. Della Valle in: F. Fl. Neapel, *v.* 20 p. 844 t. 6 f. 9; t. 26 f. 16—31 | 1895 *A. n.*, G. O. Sars, Crust. Norway, *v.* 1 p. 675 | 1890 *A. audouinianus* (err., non *Lysianassa audouiniana* Bate 1856!), G. O. Sars. Crust. Norway, *v.* 1 p. 48 t. 17 f. 2.

Head, with lateral corners almost rectangular and slightly produced. Pleon segment 3, lateral angles narrowly rounded, margin smooth. Eyes rounded oval, black with a whitish coating. Antenna 1 in Q, flagellum 10-jointed, accessory flagellum 5-jointed. Antenna 2 in Q, flagellum 9-jointed. Gnathopod 1, 6th joint fully as long as 5th, hind margin finely serrate, apically scarcely forming a palm. Gnathopod 2 long and slender, 6th joint about half length of 5th. Peraeopods 1—5, 6th joint produced to a short tooth-like process. Peraeopods 3—5, 4th joint very little expanded. Uropod 3, outer ramus little longer or broader than inner. Telson longer than broad, cleft beyond middle, not to base, with rather wide incision, 3 spinules at each apex. Colour corneous yellow. ovaries dark bluish. L. 8 mm.

Arctic Ocean, North-Atlantic, North-Sea, Skagerrak and Kattegat (South- and West-Scandinavia, Shetland Islands); Mediterranean.

5. **A. megalops** O. Sars 1895 *A. m.*, G. O. Sars, Crust. Norway, *v.* 1 p. 676 t. II f. 1.

Head, with lateral corners slightly produced. Pleon segment 3, lateral corners acutely produced. Eyes very large, oblong oval, visual elements imperfectly developed, pigment light reddish brown. Antenna 1 in ♂, peduncle rather long, flagellum 4-jointed, scarcely as long as peduncle, accessory flagellum 2-jointed. Antenna 2 in ♂, flagellum 4-jointed, half as long as peduncle. Gnathopods 1 and 2 and peraeopods 1—5 nearly as in A. microps (p. 49). Uropod 3, outer ramus considerably longer than inner. Telson as in A. microps, but apices more truncate. Colour light yellowish. L. ♂ 3 mm.

Perhaps the ♂ of A. microps (p. 49).

Trondhjemsfjord. Depth 565—753 m.

30. Gen. **Ambasia** Boeck

1871 *Ambasia* (Sp. un.: *A. danielssenii*), A. Boeck in: Forh. Selsk. Christian., 1870 p. 97 | 1890 *A.*, G. O. Sars, Crust. Norway, *v.* 1 p. 45 | 1893 *A.*, A. Della Valle in: F. Fl. Neapel, *v.* 20 p. 805.

Side-plate 1 partly concealed, 3 following large and deep. Antenna 1, 1ˢᵗ joint tumid, 2ᵈ and 3ᵈ very short, flagellum with 1ˢᵗ joint very long, accessory flagellum subequal to peduncle in length. Antenna 2 in ♀ scarcely longer than antenna 1. Epistome projecting in front of upper lip as a large angular plate. Lower lip, mandibular processes narrow, divergent. Mandible strong, cutting edge broad, molar evanescent, palp behind the centre. Maxilla 1 with 2 setae on inner plate, a small number of spines on outer, palp normal. Maxilla 2, plates nearly equal, apically spined. Maxillipeds, inner plates normal, outer large, elliptical, almost reaching end of palp, in which the 4ᵗʰ joint is rudimentary. Gnathopod 1 slender, 6ᵗʰ joint shorter than 5ᵗʰ, tapering, not subchelate, finger small. Gnathopod 2 very slender, 6ᵗʰ joint about half length of 5ᵗʰ, long oval. Peraeopods 1—5, 7ᵗʰ joint very small. Peraeopods 3—5, 2ᵈ joint well expanded. Uropod 3 rather short, outer ramus with small 2ᵈ joint. Telson of moderate size, rather tapering, deeply cleft.

2 species.

Synopsis of species:

Uropod 3 with very unequal rami . . . . 1. **A. danielssenii** . . p. 51
Uropod 3 with equal rami . . . . 2. **A. pulchra** . . . . p. 52

1. **A. danielssenii** Boeck 1871 *A. d.*, A. Boeck in: Forh. Selsk. Christian., 1870 p. 97 | 1872 *A. d.*, A. Boeck, Skand. Arkt. Amphip., *v.* 1 p. 121 t. 3 f. 6 | 1890 *A. danielsseni*, G. O. Sars, Crust. Norway, *v.* 1 p. 46 t. 17 f. 1 | 1893 *A. d.*, A. Della Valle . in: F. Fl. Neapel, *v.* 20 p. 805.

♀. Head, with lateral corners produced, almost quadrate. Pleon segment 3, postero-lateral angles ending in small upturned tooth, above which the margin is broadly rounded; segment 4 with high compressed triangular expansion dorsally. Eyes narrow, deep, subsigmoid, visual elements imperfect, pigment a beautiful red. Antenna 1 in ♀, flagellum 7-jointed, accessory flagellum 5-jointed, scarcely longer than 1ˢᵗ joint of primary. Antenna 2 in ♀, ultimate joint of peduncle shorter than penultimate, flagellum 5-jointed. Gnathopod 1, 2ᵈ joint rather dilated, shorter than rest of limb. Gnathopod 2 subchelate, tending to chelate. Uropod 3, inner ramus much the shorter. Colour dark purplish red in transverse bands of pigment spots. L. 13 mm. — ♂ not known.

Arctic Ocean, North-Atlantic, North-Sea and Skagerrak (Norway).

4*

2. **A. pulchra** (H. J. Hansen) 1887 *Tryphosa p.*, H. J. Hansen in: Vid. Meddel.,
ser. 4 r. 9 p. 78 t. 2 f. 6--6e | 1893 *Ichnopus nugax* (part.), A. Della Valle in: F. Fl. ,
Neapel, *v.* 20 p. 804.

Head, lateral angles little produced, almost quadrate. Pleon segment 3
as in A. danielsseuli, segment 4 not carinate. Eyes subreniform, very deep,
rather narrow, yellowish (red in life). Antenna 1, flagellum with many joints,
accessory flagellum with 8 joints. Antenna 2, ultimate joint of peduncle longer
than penultimate, flagellum with many joints. Both antennae in ♂ have the
flagella more elongate than in ♀: that of antenna 2 is very long. Upper
lip (epistome?) with prominent, upturned, bluntly pointed process. Mouth-
organs not described. Gnathopod 1, 2d joint not much dilated. Gnathopod 2
subchelate. Peraeopods, 4th joint little expanded. Uropod 3, rami equal in
length. L. 15 mm.

Arctic Ocean and North-Atlantic (West-Greenland). Depth 28—188 m.

31. Gen. **Ichnopus** A. Costa

1853 *Ichnopus* (Sp. un.: *I. taurus*), A. Costa in: Rend. Soc. Borbon., n. ser. *v.* 2
p. 169 | 1857 *I.*, A. Costa in: Mem. Acc. Napoli, *v.* 1 p. 188 | 1890 *I.*, G. O. Sars. Crust.
NorWay, *v.* 1 p. 39 | 1893 *I.* (part.), A. Della Valle
in: F. Fl. Neapel. *v.* 20 p. 800.

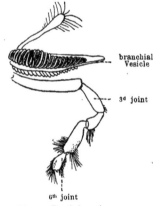

branchial
Vesicle

3d joint

6th joint

Fig. 10. I. spinicornis.
Gnathopod 2. [After G. O. Sars.]

Body rather slender and compressed.
Flagella of antennae 1 and 2 with numerous
short joints, those of antenna 2 elongate in
adult ♂. Upper lip slightly projecting above,
a small incision between it and the epistome.
Lower lip with narrow front lobes, the
mandibular processes slender, very divergent.
Mandible, cutting edge broad, simple, molar
weak, palp large, central. Maxilla 1, inner
plate small, outer broad, obliquely truncate,
spines strong, palp normal. Maxilla 2, plates
rather narrow. Maxillipeds normal, outer plates
large, broadly rounded. Gnathopod 1 rather
thin, not subchelate, 7th joint armed on
concave margin with dense bunch of spinules.
Gnathopod 2 elongate, subchelate (Fig. 10).
Peraeopods 3—5 successively longer. Bran-
chial vesicles large, bipinnate with accessory lobes. Uropod 3, rami rather
large, lanceolate, outer 2-jointed. Telson deeply cleft.

2 species.

Synopsis of species:

Gnathopod 1, finger not expanded at base . . . . 1. **I. spinicornis** . p. 52
Gnathopod 1, finger expanded at base . . . . 2. **I. taurus**. . . . p. 53

1. **I. spinicornis** Boeck 1861 *I. s.*, A. Boeck in: Forh. Skand. Naturf., Møde 8
p. 645 | 1890 *I. s.*, G. O. Sars, Crust. NorWay, *v.* 1 p. 40 t. 15 | 1865 *Lysianassa s.*,
W. Lilljeborg in: N. Acta Soc. Upsal., ser. 3 *v.* 6 nr. 1 p. 20 | 1866 *Ichnopus calceolatus*,
Cam. Heller in: Denk. Ak. Wien, *v.* 26 II p. 20 t. 2 f. 26—28 | 1871 *I. minutus*, A. Boeck
in: Forh. Selsk. Christian., 1870 p. 99 | 1872 *I. spinicornis* + *I. m.*, A. Boeck. Skand.
Arkt. Amphip., *v.* 1 p. 124 t. 2 f. 3; p. 126 t. 3 f. 7 | 1893 *I. taurus* (part.), A. Della
Valle in: F. Fl. Neapel, *v.* 20 p. 801.

Head, lateral corners somewhat projecting and angular. Back rounded.
Side-plates not very deep. Pleon segment 1 produced below in front

into a short tooth, segment 3 with the hind corners toothed and upturned, and the margin above bowed out. Eyes reniform, larger in σ, deep red: Antenna 1 in φ, about $^1/_3$ length of body, longer in σ, 1st joint stout, with short but conspicuous dentiform process, 2[d] and 3[d] joints very short, flagellum in σ with more than 60 joints, none very large, accessory flagellum 10-jointed. Antenna 2, in φ a little, in σ much, longer than antenna 1, ultimate joint of peduncle scarcely shorter than penultimate. Both pairs of antennae in both sexes have numerous calceoli. Gnathopod 1, 6th joint slender, tapering, narrower but not much shorter than 5th, finger evenly curved, with bunch of delicate spinules on inner side of base (Sars), subapically strongly curved, and below armed with 6 or 7 pointed spinules (Heller). Gnathopod 2 nearly twice as long as gnathopod 1, 6th joint oval, half as long as 5th, not apically produced, finger minute on the centre of the hand's apex (Fig. 10). Peraeopod 5 nearly twice as long as peraeopod 3. Uropod 2, the rami unequal, the inner the shorter, with a constriction in which a spine is planted. Uropod 3, rami much longer than peduncle, outer the longer, with distinct terminal joint. Telson twice as long as broad, cleft beyond the middle, a spinule at each apex. Colour light greenish, a crimson tinge at each extremity. L. 12—17 mm.

Arctic Ocean, North-Atlantic and North-Sea (West-Norway); Mediterranean; Java Sea (lat. 3° S., long. 107° E.).

2. **I. taurus** A. Costa 1853 *I. t.*, A. Costa in: Rend. Soc. Borbon., n. ser. c. 2 p. 172 | 1857 *I. t.*, A. Costa in: Mem. Acc. Napoli, v. 1 p. 189 t. 1 f. 3 | 1893 *I. t.* (part.), A. Della Valle in: F. Fl. Neapel, v. 20 p. 801 t. 3 f. 1; t. 27 f. 1—22 | 1866 *Lysianassa longicornis* (part.: φ, non σ!), E. Grube in: Arch. Naturg., v. 321 p. 396 t. 9 f. 8 | 1866 *Ichnopus affinis*, Cam. Heller in: Denk. Ak. Wien, v. 26 II p. 19 t. 2 f. 19—25 | 1895 *I. a.*, *I. taurus?*, Chevreux in: Mém. Soc. zool. France, v. 8 p. 425.

Distinguished from I. spinicornis by not having calceoli (so far as appears) on the antennae of φ. Antenna 1, 1st joint or 1st and 2[d] joints, especially in σ, usually with a little tooth; accessory flagellum 7-jointed. Finger of gnathopod 1 strongly curved (at least at the tip), carrying at the expanded base spines of variable size. Colour grey green, sprinkled with many red spots (Della Valle). L. 10—12 mm.

Gulf of Naples, Adriatic, North-Atlantic (lat. 47° N., long. 8° W., depth 2620 m).

32. Gen. **Anonyx** Kröyer

1838 *Anonyx* (part.), Kröyer in: Danske Selsk. Afh., v. 7 p. 242 | 1871 *A.*, A. Boeck in: Forh. Selsk. Christian., 1870 p. 107 | 1888 *A.*, T. Stebbing in: Rep. Voy. Challenger, v. 29 p. 607 | 1891 *A.*, G. O. Sars, Crust. Norway, v. 1 p. 87 | 1893 *A.* (part.), A. Della Valle in: F. Fl. Neapel, v. 20 p. 810 | 1894 *A.*, T. Stebbing in: Bijdr. Dierk., c. 17 p. 7.

Side-plate 1 expanded below. Pleon segment 3, postero-lateral angles upturned. Antenna 1, accessory flagellum well developed. Antennae 1 and 2 in σ with large calceoli. Epistome not projecting. Upper lip produced in front to a compressed linguiform lobe. Mandible powerful, with one or two teeth at angle of cutting edge, molar prominent, conically produced and ciliated, palp set far forward. Maxilla 1, inner plate with 2 setae, outer broad, obliquely truncate, carrying 11 spines, apex of palp having 5—10 spine-teeth. Maxilla 2, inner plate much smaller than outer. Maxillipeds, outer plates broadly oval, not (or not always?) reaching end of 2[d] joint of palp, which is robust. Gnathopod 1 short and stout, distinctly subchelate, palm nearly transverse, finger short and without prominent secondary

tooth. Gnathopod 2, 6th joint oval, its apex slightly produced beneath the
minute finger. Peraeopods rather long, 2^d joint large in peraeopods 3—5.
Branchial vesicles transversely folded on both sides. Uropod 3 projecting
beyond the preceding pair, rami lanceolate, with spinules and setae on the
margins. Telson deeply cleft, with a pair of apical spinules.

5 species.

Synopsis of species:

1 { Eyes greatly dilated below — **2**.
  { Eyes not greatly dilated below — **3**.

2 { Gnathopod 1, 6th joint not longer than 5th. . . 1. **A. nugax** . . . . . p. 54
  { Gnathopod 1, 6th joint longer than 5th . . . . 2. **A. lagena** . . . . . p. 54

3 { Antenna 1, flagellum and accessory flagellum sub-
  {   equal . . . . 3. **A. affinis** . . . . . p. 55
  { Antenna 1, flagellum and accessory flagellum
  {   unequal — **4**.

4 { Side-plate 1 scarcely widened below . . . . . . 4. **A. lilljeborgii** . . . p. 55
  { Side-plate 1 much widened below . . . . 5. **A. ampulloides** . . p. 55

1. **A. nugax** (Phipps) .1774 *Cancer n.*, Phipps, Voy. North Pole, p. 192 t. 12
f. 2 ; 1781 *Gammarus n.*, J. C. Fabricius, Spec. Ins., *v.* 1 p. 516 | 1826 *Talitrus n.*,
J. C. Ross in: W. E. Parry, J. third Voy., App. p. 119 | 1829 *Gammarus n.*, *Atylus*
(part.)?, Latreille in: G. CuVier, Règne an., n. ed. *v.* 4 p. 120 | 1877 *Anonyx n.*, Miers in:
Ann. nat. Hist., ser. 4 *v.* 19 p. 135 | 1891 & 95 *A. n.* (part.), G. O. Sars, Crust. Norway,
v. 1 p. 88, 686 | 1893 *A. n.* (part.), A. Della Valle in: F. Fl. Neapel, *v.* 20 p. 834 | 1894
A. n., T. Stebbing in: Bijdr. Dierk., *v.* 17 p. 7 | 1845 & 46 *A. ampulla* (part.), Krøyer
in: Naturh. Tidsskr., ser. 2 *v.* 1 p. 578; *v.* 2 p. 43 | 1862 *A. lagena* (err., non Kröyer
1838!), Bate, Cat. Amphip. Brit. Mus., p. 77 t. 12 f. 7 | 1866 *Lysianassa l.*, Goës in:
Öfv. Ak. Förh., *v.* 22 p. 518.

Head, lateral angles little produced, slightly rounded. Side-plate 1
widened below. Pleon segment 3, postero-lateral angles with upturned
process acute, not long, segment 4 slightly carinate. Eyes flask-shaped,
greatly widened below, very dark. Antenna 1, 1st joint not much longer
than broad, flattened on inner side, sometimes also ridged on the outer side,
flagellum in ♀ of 23, in ♂ of 34 joints, 1st not very elongate, accessory flagellum
with 10 joints, 1st the longest. Antenna 2, ultimate joint of peduncle a little
shorter than penultimate, flagellum in ♀ reaching 40, in ♂ 60 joints. Gnathopod 1,
6th joint equal in length to 5th. Gnathopod 2, 6th joint oval, rather more
than half length of 5th. Uropod 2, inner ramus constricted. Uropod 3,
rami broad and flat, inner almost as long as outer, of which the 2^d joint
is small. Telson oblong, breadth two thirds of the length, very deeply
cleft, 2 or 3 pairs of marginal spinules and a pair on the obtuse apices.
Colour light rose-red on the back, nearly white on the sides (Krøyer), light
claret-red (Sars). L. reaching 45 mm or more.

Arctic Ocean, to lat. 83° 19' N. Depth 18—1184 m.

2. **A. lagena** Kröyer 1838 *Lysianassa l.*, *A. l.* (Reinhardt in MS.) + *L. appen-
diculosa*, *A. appendiculosus*, Kröyer in: Danske Selsk. Afh., *v.* 7 p. 237, 244 t. 1 f. 1;
p. 240, 244 t. 1 f. 2 | 1838 *A. l.* + *A. appendiculosus*, Kröyer in: Naturh. Tidsskr., *v.* 2
p. 256, 257 | 1840 *Lysianassa l.* + *L. appendiculata*, H. Milne Edwards, Hist. nat. Crust.,
v. 3 p. 21 | 1855 *Lycianassa l.*, T. Bell in: Belcher, Last arct. Voy., *v.* 2 p. 406 | 1871
Anonyx l., A. Boeck in: Forh. Selsk. Christian., 1870 p. 108 | 1889 *A. kükenthali*,
Vosseler in: Arch. Naturg., *v.* 551 p. 154 t. 8 f. 1—17 | 1891 & 95 *A. nugax* (part.),
.*A. lagena*, G. O. Sars, Crust. Norway, *v.* 1 p. 88 t. 31; p. 686.

In general closely resembling A. nugax. Antenna 1 in ♀, flagellum 15-jointed, accessory flagellum 8-jointed. Antenna 2, flagellum in ♀ 23-jointed. Gnathopod 1, 6th joint longer than 5th. Uropod 3, inner ramus longer than basal joint of outer. Colour whitish, with transverse yellowish band on each segment. L. ♀ 18 mm.

Arctic Ocean; North-Atlantic, North-Sea, Skagerrak and Kattegat (Scandinavia, North Britain). Depth 38—565 m.

3. **A. affinis** Ohlin 1895 *A. a.* (non A. Della Valle 1893), Ohlin in: Acta Univ. Lund., *v.*31 nr.6 p.24 t. f.15—18 | 1895 *A. a., A. kükenthali var.?*, Ohlin in: Zool. Anz., *v.* 18 p.486.

♀. Head, lateral angles a little produced, rounded. Pleon segment 3, postero-lateral angles produced into short sharp process, segment 4 carinate. Eyes very large, oblong, dark. Antennae 1 and 2 short, subequal. Antenna 1, flagellum 6-, accessory flagellum 5- or 6-jointed. Antenna 2, ultimate joint of peduncle shorter than penultimate, flagellum 6- or 7-jointed. Palp of mandible affixed over molar. Gnathopod 1 as in A. lilljeborgii, but palm (in figure) more oblique. Gnathopod 2, 5th joint narrower than 3d and little longer, 6th joint distally truncate. Telson twice as long as broad, cleft nearly to the base. Uropod 3, inner ramus two thirds length of outer, with 3 long setae on inner margin. L. 13 mm. — ♂ unknown.

Baffin Bay (Cape Dudley Digges). Depth 32—47 m.

4. **A. lilljeborgii** Boeck 1871 *A. l.*, A. Boeck in: Forb. Selsk. Christian., 1870 p.109 | 1872 *A. l.*, A. Boeck, Skand. Arkt. Amphip., *v.*1 p.154 t.4 f.3 | 1891 *A. l.*, G. O. Sars, Crust. Norway, *v.*1 p.90 t.32 f.1 | 1894 *A. l.*, T. Stebbing in: Bijdr. Dierk., *v.*17 p.8 | 1893 *A. nugax* (part.), A. Della Valle in: F. Fl. Neapel, *v.*20 p.834.

In general closely resembling A. lagena. Head, lateral angles a little more produced and narrowly rounded. Side-plate 1 of nearly uniform width. Pleon segment 3, postero-lateral angles not very acute. Eyes not greatly dilated below, brownish red. Antenna 1 in ♀, flagellum 14-, accessory flagellum 6-jointed. Antenna 2 in ♀, flagellum 15-jointed. Gnathopod 1 not very robust. 6th joint equal in length to 5th, rather tapering. Gnathopod 2 very slender, 6th joint narrower and more distinctly produced beneath the finger than in A. lagena. Peraeopods 1 and 2, an obtuse spine at end of 6th joint more conspicuous than in the larger species. Uropod 3, rami narrower. Telson rather longer in proportion to breadth. L. ♀ 11 mm.

Arctic Ocean. North-Atlantic, North-Sea and Skagerrak (West-Norway). Depth 113—132 m.

5. **A. ampulloides** Bate 1862 *A. a.*, Bate, Cat. Amphip. Brit. Mus., p.78 t.12 f.8 | 1888 *A. a.*, T. Stebbing in: Rep. Voy. Challenger, *v.*29 p.608 t.3 | 1891 *A. a.*, G. O. Sars, Crust. Norway, *v.*1 p.88 | 1893 *A. nugax* (part.), A. Della Valle in: F. Fl. Neapel, *v.*20 p.834.

Head, lateral angles scarcely produced, rounded. Side-plate 1 considerably widened below. Pleon segments 1—4 subcarinate, segment 3 with postero-lateral angles much upturned but little produced. Eyes large, reniform. nearly meeting at top of head. Antenna 1, accessory flagellum with 7 or 8 joints, 1st elongate. Antenna 2, ultimate joint of peduncle considerably shorter than penultimate. Gnathopod 1, 6th joint narrower than 5th and barely as long, scarcely narrowed distally. Gnathopod 2, 6th joint more than half length of 5th, narrow, palm slightly excavate, finger not extremely minute. Peraeopod 1, 6th joint having on hind margin a minute hooked spine close to hinge of finger. Uropod 2, inner ramus constricted. Uropod 3, rami lanceolate, subequal, outer with small 2^d joint. Telson nearly twice as long as broad, cleft three quarters of length, rather ·dehiscent. L. about 13 mm.

North-Pacific (Japan). Depth 1419 m.

33. Gen. **Socarnes** Boeck

1838 *Anonyx* (part.), Kröyer in: Danske Selsk. Afh., *v.* 7 p. 242 | 1847 *Ephippiphora* (Sp. un.: *E. kroyeri*) (non Duponchel 1834, Lepidoptera!), A. White in: P. zool. Soc. London, *v.* 15 p. 124 | 1871 *Socarnes* (Sp. un.: *S. vahli*), A. Boeck in: Forh. Selsk. Christian., 1870 p. 99 | 1872 *S.,* A. Boeck, Skand. Arkt. Amphip., *v.* 1 p. 128 | 1890 *S.,* G. O. Sars, Crust. Norway, *v.* 1 p. 43 | 1893 *S.,* J. Bonnier in: Bull. sci. France Belgique, *v.* 24 p. 188.

Body dorsally broad, laterally compressed. Side-plates rather large. Antenna 1, joints of flagellum not numerous; antenna 2 short in ♀. Upper lip separated by rather deep incision from epistome and prominent beyond it. Mandibular processes of lower lip not very divergent. Mandible, cutting edge simple, molar very weak, palp attached behind it. Maxilla 1 normal. Maxilla 2 with inner plate obliquely truncate. Maxillipeds normal, inner plates obliquely truncate, outer rounded, 7th joint small normal. Gnathopod 1 not subchelate. Gnathopod 2 elongate, subchelate. Peraeopods 3—5 successively longer. Branchial vesicles with accessory lobes on one side only. Uropod 3, rami of moderate size, outer 2-jointed. Telson deeply cleft.

3 species accepted, 1 obscure.

Synopsis of accepted species:

1 { Pleon segment 3 with central triangular projection of postero-lateral margins . . . . 1. **S. bidenticulatus** . . . p. 56
{ Pleon segment 3 without triangular projection — **2.**

2 { Antenna 1, accessory flagellum with more than 4 joints . . . . 2. **S. vahlii** . . . . p. 57
{ Antenna 1, accessory flagellum with not more than 4 joints . . . . 3. **S. erythrophthalmus** . p. 57

1. S. bidenticulatus (Bate) 1835 *Gammarus nugax* (err., non *Cancer n.* Phipps 1774!), J. C. Ross in: John Ross, App. sec. Voy., nat. Hist. p. 87 | 1862 *Lysianassa n.,* Bate, Cat. Amphip. Brit. Mus., p. 65 t. 10 f. 3 | 1858 *L. bidenticulata,* Bate in: Ann. nat. Hist., ser. 3 *v.* 1 p. 362 | 1877 *Anonyx bidenticulatus,* Miers in: Ann. nat. Hist., ser. 4 *v.* 19 p. 136 | 1885 & 86 *Socarnes b.,* G. O. Sars in: Norske Nordhavs-Exp., *v.* 6 Crust. I p. 139, 276 t. 12 f. 1; Crust. II p. 38 | 1894 *S. b.,* T. Stebbing in: Bijdr. Dierk., *v.* 17 p. 3 | 1893 *Ichnopus b.,* A. Della Valle in: F. Fl. Neapel, *v.* 20 p. 804 | 1866 *Lysianassa vahli* (part.), Goës in: Öfv. Ak. Förh., *v.* 22 p. 518 | 1871 *Socarnes v.* (part.), A. Boeck in: Forb. Selsk. Christian., 1870 p. 100 | 1882 *S. ovalis,* Hoek in: Niederl. Arch. Zool., suppl. 1 nr. 7 p. 42 t. 3 f. 29—29 r.

Body deep and tumid. Head with lateral angles acute. Side-plates deep. Pleon segment 3 with postero-lateral margins showing 2 triangular projections, of which the upper is larger, the lower small and sometimes evanescent. Eyes narrow, reniform or sublinear, darkbrown. Antenna· 1, peduncle short and thick, 1st joint large, 2^d and 3^d very short, flagellum of about 18 joints, the 1st not very large. accessory flagellum 9-jointed, more than half as long as primary. Antenna 2 in ♀ more slender and scarcely longer than antenna 1, flagellum 17-jointed. Gnathopod 1, 6th joint tapering, about as long as the 5th, 7th small. Gnathopod 2, 6th joint shorter than 5th, not forming a chela but produced to angular corner beneath the finger. Peraeopods 3—5, 2^d joint broadly expanded, almost circular. Uropod 3 scarcely reaching beyond uropod 2, rami shortly lanceolate. Telson triangular, cleft to the middle. L. attaining 36 mm.

Arctic Ocean.

2. **S. vahlii** (Kröyer) 1838 *Lysianassa v., Anonyx v.,* (Reinhardt in MS.) Kröyer in: Danske Selsk. Afh., *v.* 7 p. 233, 244 | 1871 *Socarnes vahli* (part.), A. Boeck in: Forh. Selsk. Christian., 1870 p. 100 | 1890 *S. v.,* G. O. Sars, Crust. Norway, *v.*1 p. 44 t. 16 f. 2 ; 1894 *S. vahlii,* T. Stebbing in: Bijdr. Dierk., *v.*17 p. 3 | 1888 *Ephippiphora v., S.* (part.), T. Stebbing in: Rep. Voy. Challenger, *v.* 29 p 177, 1698 | 1893 *Ichnopus nugax* (part.), A. Della Valle in: F. Fl. Neapel, *v.* 20 p. 804.

Back broadly rounded. Head, lateral angles narrowly rounded. Side-plates rather deep. Pleon segment 3 broadly rounded at the postero-lateral corners. Eyes reniform, black. Antenna 1, 2^d joint of peduncle not extremely short, flagellum with 12 joints, the 1st rather small, accessory flagellum 7-jointed, half length of primary. Antenna 2 in ♀ scarcely longer than antenna 1, in ♂ longer than the body. Gnathopod 1, 6th joint narrower than 5th and not longer, tapering, 7th small. Gnathopod 2 elongate, 6th joint half length of 5th, roundly expanded below the finger. Uropod 3, inner ramus considerably shorter than outer. Telson cleft beyond the middle, subequal in length to peduncle of uropod 3, the lobes narrowly rounded apically, a spinule at each⁻apex. Colour whitish, with broad transverse bands of beautiful crimson spots. L. 14 mm.

Arctic Ocean and North-Atlantic (West-Norway, chiefly to the north).

3. **S. erythrophthalmus** D. Roberts. 1892 *S. e.,* D. Robertson in: P. nat. Hist. Soc. Glasgow, n. ser. *v.*3 p. 200 | 1893 *S. e.,* J. Bonnier in: Bull. sci. France Belgique, *v.* 24 p. 183 t. 6.

Head with lateral angles rather sharply advanced, the margin above slightly serrate. Pleon segment 3 with postero-lateral angles rounded. Eyes oval, red, fading in spirit. Antenna 1, peduncle robust, flagellum of 8—13 joints, 1st scarcely larger than 2^d; 2^d, 3^d and 4th in ♂ with calceoli; accessory flagellum with 4 slender joints. Antenna 2, a little longer than antenna 1, ultimate joint of peduncle shorter than penultimate, flagellum 6—10-jointed. Gnathopod 1 as in S. vahlii. Gnathopod 2, 6th joint half length of 5th, rather squared below the 7th joint. Uropod 3 as in S. vahlii. Telson rather longer than peduncle of uropod 3, cleft to the centre, with a spinule at each narrow apex. Colour pellucid with slight greenish tint. The eggs few and large, when ripe golden yellow. L. 3—4 mm.

Firth of Clyde, surface to 26 m; Pas de Calais (Cape Gris-Nez, depth 50 m).

S. kroyeri (A. White) 1847 *Ephippiphora k.,* A. White in: P. zool. Soc. London *v.*15 p. 124 | 1884 *E. kröyeri,* Miers in: Rep. Voy. Alert, p. 312 | 1888 *E. kroyeri, Socarnes* (part.), T. Stebbing in: Rep. Voy. Challenger, *v.* 29 p. 555, 1698 ; 1862 *Lysianassa k., L. kröyeri,* Bate, Cat. Amphip. Brit. Mus., p. 65 t. 10 f. 4 | 1872 *Socarnes? kröyeri,* A. Boeck, Skand. Arkt. Amphip., *v.* 1 p. 128.

L. 25 mm.

Indian Ocean and South-Pacific (Tasmania, Dundas Straits, Prince of Wales Channel. Port Denison).

34. Gen. **Hippomedon** Boeck

1871 *Hippomedon* (part.), A. Boeck in: Forh. Selsk. Christian.. 1870 p. 102|1872 *H.*, A. Boeck, Skand. Arkt. Amphip., *v.* 1 p. 135 | 1890 *H.*, G. O. Sars, Crust. Norway, *v.* 1 p. 55 | 1893 *H.*, J. Bonnier in: Bull. sci. France Belgique, *v.* 24 p. 174 | 1893 *H.*, A. Della Valle in: F. Fl. Neapel, *v.* 20 p. 807 | 1894 *H.*, T. Stebbing in: Bijdr. Dierk., *v.* 17 p. 4 | 1888 *Platamon* (Sp. un.: *P. longimanus*), T. Stebbing in: Rep. Voy. Challenger, *v.* 29 p. 642.

Side-plates rather narrow. Pleon segment 3, postero-lateral angles sharply upturned. Eyes imperfectly developed. Antenna 1, flagellum not large, 1st joint very large, accessory flagellum small. Antenna 2 much longer than antenna 1, flagellum many-jointed; flagellum of antennae 1 and 2 with calceoli in.♂. Epistome not projecting. Mandible short and strong, molar prominent, powerful, palp elongate, affixed in front of molar. Maxilla 1, inner plate short, with 2 setae, outer with the usual spines, palp expanded distally, carrying numerous spine-teeth. Maxilla 2, plates rather short and broad, the inner fringed on inner margin. Maxillipeds, outer plates fringed with spine-teeth on straight inner margin, generally reaching much beyond 2^d joint of the robust palp. Gnathopod 1 slender, 5th joint long, 6th oblong ovate, with ill-defined palm, finger slender. Gnathopod 2 subchelate. Peraeopods slender, except 2^d joint of pairs 3—5. Branchial vesicles of peraeopods 3 and 4 somewhat complex, a small one on peraeopod 5. Uropod 3. rami long, subequal, with marginal spinules, generally without conspicuous setae, the outer 2-jointed. Telson oblong, cleft deep, dehiscent, with 1 pair of apical spinules.

6 species.

Synopsis of species:

1 {	Eyes with one conspicuous lens apiece	1. **H. holbölli** . . . . p. 58
	Eyes without lenses — **2.**	
2 {	Pleon segment 3, lateral processes defined by a notch	2. **H. denticulatus** . . p. 59
	Pleon segment 3, lateral processes without notch — **3.**	
3 {	Gnathopod 1, palm confluent with hind margin — **4.**	
	Gnathopod 1, palm not confluent	6. **H. geelongi** . . . . p. 60
4 {	Pleon segment 3, lateral processes acutely prolonged	3. **H. propinqvus** . . p. 59
	Pleon segment 3, lateral processes rather short and blunt — **5.**	
5 {	Cleft of telson scarcely dehiscent	4. **H. robustus** . . . . p. 59
	Cleft of telson strongly dehiscent	5. **H. longimanus** . . p. 60

1. **H. holbölli** (Krøyer) 1846 *Anonyx h.*, Krøyer in: Naturh. Tidsskr., ser. 2 *v.* 2 p. 8, 38 | 1846 *A. h.*, Krøyer in: Voy. Nord, Crust. t. 15 f. 1 a—s | 1866 *Lysianassa h.*, Goës in: Öfv. Ak. Förh., *v.* 22 p. 520 | 1885 *Hippomedon h. var.*, G. O. Sars in: Norske Nordhavs-Exp., *v.* 6 Crust. I p. 142 t. 12 f. 2 | 1887 *H. h.*, H. J. Hansen in: Vid. Meddel., ser. 4 *v.* 9 p. 63 t. 2 f. 1—1 b | 1890 *H. h.*, *H. holbølli*, G. O. Sars, Crust. Norway, *v.* 1 p. 58 t. 21 f. 2 | 1893 *H. holbölli* (part.), A. Della Valle in: F. Fl. Neapel, *v.* 20 p. 808.

Head, lateral angles not produced. Side-plate 1 scarcely widened below. Pleon segment 3, postero-lateral angles moderately upturned, acute, segment 4 with conspicuous carina. Body above and partly on the side very finely reticulate with longitudinal striae, side-plates and sides of pleon finely granulate. Eyes at lateral angles of head, with a conspicuous watchglass-shaped lens. Antenna 1, joints of peduncle not produced, flagellum 12-jointed, accessory flagellum 4-jointed. Antenna 2, ultimate joint of peduncle little longer than penultimate. Peraeopods, 7th joint elongate. Uropod 3, 2[d] joint of outer ramus rather long. Telson not twice as long as broad, cleft extending ²/₃ of the length, dehiscent. L. 11—16 mm.

Arctic Ocean.

2. **H. denticulatus** (Bate) 1857 Anonyx d., Bate in: Ann. nat. Hist., ser. 2 v. 19 p. 139 | 1861 A. d., Bate & Westwood, Brit. sess. Crust., v. 1 p. 101 f. | 1887 Hippomedon d., H. J. Hansen in: Vid. Meddel., ser. 4 v. 9 p. 65 t. 2 f. 2—2b | 1890 H. d., G. O. Sars, Crust. Norway, v. 1 p. 56 t. 20 | 1893 H. d. (part.), A. Della Valle in: F. Fl. Neapel, v. 20 p. 808 t. 29 f. 33—42 | 1871 H. holbølli (err., non Anonyx holbölli Krøyer 1846!), A. Boeck in: Forh. Selsk. Christian., 1870 p. 102 | 1872 H. h., H. holbölli, A. Boeck, Skand. Arkt. Amphip., v. 1 p. 136 t. 5 f. 6; t. 6 f. 7.

Head, lateral angles slight produced, acute. Side-plate 1 widened below, covering the mouth-organs. Pleon segment 3, postero-lateral angles strongly upturned, unique in having a notch separating the acute process from the hind margin, segment 4 without carina. Body smooth, slightly punctate. Eyes linear, slightly widened below, without lenses, light red, transversely striped with opaque white. Antenna 1, joints 1 and 2 bluntly produced, flagellum in ♀ 11-jointed, in ♂ much longer, accessory flagellum 3-jointed. Antenna 2 in ♀ more than twice as long as antenna 1, in ♂ as long as the body, ultimate joint of peduncle nearly twice penultimate. Gnathopod 1, 6th joint half as long as 5th, palm finely denticulate. Peraeopods, 7th joint scarcely more than half as long as 6th. Uropod 3, 2[d] joint of outer ramus very small. Telson nearly twice as long as broad, cleft extending beyond the middle, distally dehiscent, with one pair of marginal spinules. Body whitish, pellucid, with some transverse orange bands. L. ♀ 14, ♂ 11 mm.

North-Atlantic, North-Sea, Skagerrak and Kattegat (Bohuslän, Denmark, Great Britain, France); Mediterranean.

3. **H. propinqvus** O. Sars 1859 Anonyx holbölli (err., non Krøyer 1846!), R. M. Bruzelius in: Svenska Ak. Handl., n. ser. v. 3 nr. 1 p. 43 | 1890 Hippomedon propinqvus, G. O. Sars, Crust. Norway, v. 1 p. 57 t. 21 f. 1 | 1893 H. propinquus, A. Della Valle in: F. Fl. Neapel, v. 20 p. 810 | 1894 H. squamosus, T. Stebbing in: Bijdr. Dierk., v. 17 p. 4 t. 1.

Head, lateral angles narrowly rounded. Side-plate 1 scarcely widened below. Pleon segment 3, postero-lateral angles strongly upturned, acute process without defining notch, segment 4 slightly produced dorsally. Integument very finely and irregularly reticulated. Eyes as in H. denticulatus. Antenna 1, joints 1 and 2 little produced, 1st joint of flagellum very elongate, accessory flagellum slender, 3-jointed. Antenna 2, ultimate joint of peduncle little longer than penultimate. Gnathopod 1 more slender than in H. denticulatus. Peraeopods 1 and 2, 7th joint nearly as long as 6th, with cap over the apex. Uropod 3, 2[d] joint of outer ramus rather large. Telson more than twice as long as broad, cleft extending much beyond the middle, distally dehiscent, with 2 or 3 pairs of marginal spinules. Colour whitish, pellucid, tinged with crimson at either end. L. ♀ 10 mm.

North-Atlantic and Arctic Ocean (Trondhjemsfjord, Nordland and Finmark to Vadsö). Depth 37—188 m.

4. **H. robustus** O. Sars 1895 H. r., G. O. Sars, Crust. Norway, v. 1 p. 679 t. III f. 1.

Head, lateral angles acute, scarcely produced. Side-plate 1 a little expanded below, 4th much broader than the preceding pairs. Pleon segment 3, postero-lateral angles unusually short and broad, 4th segment with dorsal depression shallow. Integument finely and irregularly reticulated. Eyes narrow oblong, widened a little below, without trace of visual elements, light red, transversely striped with opaque white. Antenna 1, joints 1 and 2 slightly produced, flagellum in ♀ with 11 joints, 1st of moderate length, accessory flagellum 4-jointed. Antenna 2, ultimate joint of peduncle little longer than penultimate; flagellum in ♀ 32-jointed. Telson about twice as long as broad, cleft extending about $^2/_3$ of length, dehiscent only distally, carrying 2 or 3 pairs of marginal spinules. Body whitish, pellucid, without coloured patches, but side-plates and 2^d joint in peraeopods 3—5 mottled with dark greenish spots. L. ♀ 10 mm.

Trondhjemsfjord. Depth 94 m.

5. **H. longimanus** (Stebb.) 1888 *Platamon l.*, T. Stebbing in: Rep. Voy. Challenger, v. 29 p. 643 t. 13 | 1893 *Hippomedon denticulatus* (part.), A. Della Valle in: F. Fl. Neapel, v. 20 p. 808.

Head, lateral angles moderately acute. Side-plate 1 a little widened below, 4th considerably. Pleon segment 3. postero-lateral angles acutely upturned, the process neither very long nor very sharp, segment 4 with slight carina ending acutely. Eyes not observed. Antenna 1, 1st joint carinate, produced over the 2^d, flagellum in ♀ with 7 joints, 1st very long, accessory flagellum small, 3-jointed. Antenna 2, ultimate joint of peduncle scarcely longer than penultimate, flagellum 35-jointed. Maxilla 1 with 19 teeth on very broad apex of palp. Maxillipeds, outer plates not reaching greatly beyond 2^d joint of palp. Gnathopod 1, 6th joint fusiform, palm denticulate, and finger with cap over the apex, as in H. denticulatus, H. propinqvus, and probably other species. Gnathopod 2, 6th joint distally widened, with excavate palm, finger comparatively large. Peraeopods 1 and 2, 7th joint as long as 6th, with cap over apex. Uropod 3, 2^d joint of outer ramus not very large. Telson not twice as long as broad, cleft extending $^3/_5$ of the length, dehiscent throughout, with 3 or 4 pairs of marginal spinules. L. about 17 mm.

North-Atlantic (Cape Finisterre). Depth 2057 m.

6. **H. geelongi** Stebb. 1888 *H. g.*, T. Stebbing in: Rep. Voy. Challenger. v. 29 p. 635 t. 11 | ?1893 *Cheirimedon crenatipalmatus* (err., non T. Stebbing 1888!). A. Della Valle in: F. Fl. Neapel, v. 20 p. 837.

Head, lateral angles acute. Pleon segment 3, postero-lateral angles acutely upturned. the process not very large. No eyes perceived. Antenna 1, 1st joint very obtusely produced, not carinate, accessory flagellum 5-jointed. Antenna 2, ultimate joint of peduncle not longer than penultimate, flagellum 30-jointed. Maxilla 1, inner plate with 7 very unequal setae, outer with 11 spines, palp with 12 spine-teeth. Gnathopod 1, 6th joint as long as 5th. slightly widened distally, the moderately oblique palm overlapped by the finger. Peraeopods, finger long and slender. Uropod 3 with plumose setae as well as spinules on both rami, the outer rather the longer, with a short 2^d joint. Telson with 2 pairs of marginal spinules, cleft extending much beyond the middle, the apices a little dehiscent. L. 12 mm.

The position of this species in the genus is rather doubtful.

Port Phillip [Melbourne]. Depth 60 m.

35. Gen. **Glycerina** Hasw.

1880 *Glycera* (Sp. un.: *G. tenuicornis*) (non Savigny 1822, Vermes!). Haswell in: P. Linn. Soc. N. S. Wales, v. 4 p. 256 | 1882 *Glycerina* (Sp. un.: *G. tenuicornis*), Haswell, Cat. Austral. Crust.. p. 233 | 1888 *G.*, T. Stebbing in: Rep. Voy. Challenger; v. 29 p 648.

Antenna 1 slender, rather long. Mandible, molar prominent. Maxilla 1, outer plate with oblique apical margin, 7 or 8 spine-teeth on apex of palp. Gnathopod 1 imperfectly subchelate. Gnathopod 2 subchelate. Uropod 3, rami broadly lanceolate. Telson deeply cleft.

The genus and its 2 species remain in much obscurity.

2 species.

Synopsis of species:

1. G. tenuicornis (Hasw.) 1880 *Glycera t.*, Haswell in: P. Linn. Soc. N·S.Wales. *v.* 4 p 256 t. 8 f. 6 | 1882 *Glycerina t.*, Haswell, Cat. Austral. Crust., p. 234 t. 4 f. 3.

Head, lateral angles rather acutely produced. Pleon segment 3, postero-lateral angles subquadrate. Eyes long oval, nearly meeting above. Antenna 1, 1st joint stout, flagellum slender, longer than peduncle, accessory flagellum 9-jointed. Antenna 2 rather longer than antenna 1, flagellum 3 times as long as peduncle. Gnathopod 1 long, filiform, 6th joint about $1\frac{1}{3}$ length of 5th, irregularly ovoid, narrowed distally, with curved setae on hind margin. Gnathopod 2 long, slender, stouter than gnathopod 1, 6th joint nearly twice as long as 5th [?], subquadrate, nearly as broad as long, palm concave, defined by acutely produced angle. Peraeopod 3 much shorter than the rest, 2d joint circular with deeply serrate hind margin. Peraeopods 4 and 5, 2d joint oval, not deeply serrate. Apices of telson acute. L. 8 mm.

Coral Sea (Howick Group), Port Jackson [East-Australia].

2. G. affinis Chilton 1885 *G. a.*, Chilton in: P. Linn. Soc. N. S. Wales, *v.* 9 p. 1036 t. 47 f. 1a, b.

In general shape closely resembling G. tenuicornis. Gnathopod 1, 6th joint longer than 5th, tapering, without palm. Gnathopod 2 very long and slender, 6th joint shorter than 5th, of uniform width, nearly 3 times as long as broad, palm rather oblique.

Port Jackson [East-Australia].

36. Gen. **Scopelocheirus** Bate

1851 *Callisoma* (non L. Agassiz 1846, Coleoptera!) (part.), A. Costa, Fauna Reg. Napoli, fasc. Marz. 1851 p. 1 | 1888 *C.*, T. Stebbing in: Rep. Voy. Challenger, *v.* 29 p. 247 | 1890 *C.*, G. O. Sars, Crust. Norway, *v.* 1 p. 52 | 1893 *C.*, A. Della Valle in: F. Fl. Neapel, *v.* 20 p. 838 | 1856 *Scopelocheirus* (nom. nud.), Bate in: Rep. Brit. Ass., Meet. 25 p. 58 | 1857 *S.* (Sp. un.: *S. crenatus*), Bate in: Ann. nat. Hist., ser. 2 *v.* 19 p. 138.

Side-plates large. Antenna 1 much shorter than antenna 2, 1st joint tumid, 1st joint of flagellum in ♂ very large, accessory flagellum rather small. Antenna 2, basal joint swollen, flagellum long and slender, especially in ♂; antennae 1 and 2 in ♂ with calceoli. Epistome projecting, rounded, defined by a sinus from upper lip. Hind lobes of lower lip very divergent. Mandible, molar prominent, tapering, palp affixed above it, large, setose. Maxilla 1, inner plate with several plumose setae along inner margin, outer plate with the usual spines, palp large with bifurcate apical spines. Maxilla 2, both plates short, broad, densely setose, setae of inner plate fringing its inner margin. Maxillipeds, outer plates not reaching end of 2d joint of palp, fringed with spine-teeth, 4th joint of palp long and slender. Gnathopods 1 and 2 subequal in length.

Gnathopod 1, 6th joint long, narrow, with double row of delicate setae at apex, finger rudimentary in dense brush of setae. Gnathopod 2 minutely chelate. Peraeopods 3—5, 2[d] joint expanded; 4th joint of 3[d] pair expanded; 5th pair longest. Uropod 3 prolonged, rami subequal, with spinules and setae. Telson long, narrow, deeply cleft, with a pair of apical spinules.

2 species.

Synopsis of species:

Gnathopod 1, 6th joint much longer than 5th . . . . . . 1. **S. hopei** · . . . p. 62
Gnathopod 1, 6th joint not longer than 5th . . . . 2. **S. crenatus** . . p. 62

1. S. hopei (A. Costa) 1851 *Callisoma h.*, A. Costa, Fauna Reg. Napoli, fasc. Marz. 1851 p. 5 t. 8 II f. 1 | 1851 *C. h.*, (A. Costa in:) F. W. Hope, Cat. Crost. Ital., p. 44 f. 2 | 1893 *C. h.* (part.), A. Della Valle in: F. Fl. Neapel, v. 20 p. 839 t. 6 f. 11; t. 26 f. 1—15; t. 43 f. 19 | 1853 *C. barthelemyi*, A. Costa, Descr. 3 Crost. dal Hope, p. 7 | 1859 *Anonyx kröyeri*, R. M. Bruzelius in: Svenska Ak. Handl., n. ser. v. 3 nr. 1 p. 45 t. 2 f. 7 | 1890 *Callisoma k.*, G. O. Sars, Crust. Norway, v. 1 p. 54 t. 19 f. 2 | 1874 *C. branickii*, Wrześniowski in: Ann. nat. Hist., ser. 4 v. 14 p. 15.

Head, lateral angles little produced. Side-plate 1 widened below, not completely covering ·mouth-organs, expansion of 4th not quadrate. Pleon segment 3, postero-lateral angles almost quadrate, hind margin smooth, 4th with dorsal depression moderately deep, followed by rounded carina. Eyes broad oval, dark brownish (Sars), scarlet (Della Valle). Antenna 1 in ♀, flagellum with 7 or 8 joints, 1st not very large, accessory flagellum with 3 or 4 joints. Mandible, 2[d] joint of palp broad. Gnathopod 1, 6th joint considerably longer than 5th, margins parallel. Gnathopod 2, 5th joint much widened distally. Peraeopods 3—5, 4th joint considerably dilated in the 3[d], but much less so in the other two. Uropod 3, inner ramus a little shorter than outer, but scarcely narrower. Telson about twice as long as broad, cleft almost to the base. Colour yellowish, without distinct pigmentary spots. L. ♀ 5·5 (Sars), 7 (Bruzelius) mm.

Mediterranean; North-Atlantic, North-Sea, Skagerrak, Kattegat and Baltic (Great Britain, Norway, Sweden). Depth 58—78 m.

2. S. crenatus Bate 1856 *S. breviatus* (nom. nud.), Bate in: Rep. Brit. Ass., Meet. 25 p. 58 | 1857 *S. crenatus*, Bate in: Ann. nat. Hist., ser. 2 v. 19 p. 138 | 1861 *Callisoma crenata*, Bate & Westwood, Brit. sess. Crust., v. 1 p. 120 f. | 1890 *C. c.*, G. O. Sars, Crust. Norway, v. 1 p. 53 t. 19 f. 1 | 1890 *Tryphosa serra*, Meinert in: Udb. Hauchs, v. 3 p. 156 t. 1 f. 30—38 | 1893 *Callisoma hopei* (part.), A. Della Valle in: F. Fl. Neapel, v. 20 p. 839.

Head, lateral angles slightly projecting. Side-plate 1 widened below, completely covering mouth-organs, expansion of 4th quadrate. Pleon segment 3 almost quadrate, hind margin crenulate, 4th with deep dorsal depression in front of rounded carina. Eyes broad oval, reddish brown, orange-coated. Antenna 1, flagellum with 10—12 joints, 1st very large only in ♂, accessory flagellum 3-jointed. Antenna 2 in ♀, flagellum 24-jointed. Gnathopod 1, 6th joint subequal to 5th, slender, slightly narrowed at the middle. Gnathopod 2, 6th joint much shorter than the slender 5th. Peraeopod 3, 2[d] joint greatly dilated. Peraeopods 3 and 4, 4th joint dilated. Uropod 3, inner ramus as long as outer. Telson nearly 3 times as long as broad, deeply cleft, with a pair of dorsal spinules. Colour yellowish, with numerous orange spots. L. ♀ 9·5 mm.

North-Atlantic, North-Sea and Skagerrak (Great Britain, South- and West-Norway); Kattegat

37. Gen. **Alicella** Chevreux

1899 *Alicella* (Sp. un.: *A. gigantea*), Chevreux in: Bull. Soc. zool. France, v.24 p.154.

Head short, hollowed in front. Side-plates not so deep as their respective segments. Eyes large, imperfectly developed. Antennae 1 and 2 slender. Antenna 1 with accessory flagellum well developed. Lobes of lower lip apically narrowed. Mandible with broad dentate cutting edge, accessory plate on left mandible.denticulate, molar consisting of a narrow, spinulose, backward-pointing lamina, palp short, 3^d joint dilated, strongly setose. Maxilla 1, inner plate obliquely truncate, fringed with 16 plumose setae, outer plate with 9 spines, 2^d joint of palp long, with 23 apical spines. Maxilla 2, inner plate a little shorter than outer, each carrying long plumose setae. Maxillipeds, inner plates obliquely truncate at apex, outer plates broad, nearly reaching end of palp's 2^d joint, ultimate joint of palp dactyliform. Gnathopod 1 short, simple. Gnathopod 2 long, slender, feebly subchelate. Uropod 3, rami long, equal, lanceolate. Telson elongate, somewhat tapering, cleft nearly to the base.

1 species.

1. **A. gigantea** Chevreux 1899 *A. g.*, Chevreux in: Bull. Soc. zool. France, v. 24 p. 154 f. 1—6.

Pleon segments 3—6 dorsally grooved and slightly bicarinate. Head in front forming a kind of vault, at bottom of which the antennae 1 and 2 are inserted. Side-plate 5 less deep than side-plate 4, much broader than deep. Pleon segment 3, postero-lateral corners quadrate. Eyes opaque white or orange, irregularly reniform, without trace of ocelli. Antenna 1, 2^d joint not half as long as 1st, 3^d very short, flagellum 35-jointed, accessory flagellum 9-jointed. Antenna 2 a little the longer, ultimate joint of peduncle much longer than penultimate, flagellum 60-jointed. Epistome little prominent, upper lip with unsymmetrical apex, feebly bilobed. Gnathopod 1, 6th joint tapering, rather shorter than 5th, finger more than one third as long as 6th. Gnathopod 2 more than twice as long as gnathopod 1, 6th joint very slender, nearly as long as 5th, its front and hind margins parallel, finger feeble and short. Peraeopods 1 and 2 not longer than gnathopod 1, joints 2—4 rather robust, the following slender, very short. Peraeopods 3—5, 2^d joint somewhat expanded, the rest slender. Peraeopods 4 and 5 equal, longer than peraeopod 3, finger very short. Pleopods 1—3, rami very long, many-jointed. Uropods 1 and 2, outer branch nearly smooth, rather shorter and narrower than the spinulose inner branch. Uropod 3, rami fringed with long plumose setae. Telson with numerous little groups of spinules along back of each lobe, apex of each emarginate, carrying 2 spines. Colour white tinted with greenish yellow. L. reaching 140 mm.

North-Atlantic (lat. 31° N., long. 28° W.). Depth 5285 m.

38. Gen. **Uristes** Dana

1849 *Uristes*, J. D. Dana in: Amer. J. Sci., ser. 2 v. 8 p. 136 | 1852 *U.* (Sp. un.: *U. gigas*), J. D. Dana in: P. Amer. Ac., v. 2 p. 209 | 1862 *U.*, Bate, Cat. Amphip. Brit. Mus., p. 89 | 1888 *U.*, T. Stebbing in: Rep. Voy. Challenger, v. 29 p. 263 | 1893 *U.*, A. Della Valle in: F. Fl. Neapel, v. 20 p. 836 | 1899 *U.*, T. Stebbing in: Ann. nat. Hist., ser. 7 v. 4 p. 211 | 1891 *Pseudotryphosa*, G. O. Sars, Crust. Norway, v. 1 p. 83 | 1893 *Tryphosella* (part.), J. Bonnier in: Bull. sci. France Belgique, v. 24 p. 170.

Body stout. Pleon strongly developed. Side-plates 1—4 in position not occupying a greater length than side-plates 5—7 in position. Antenna 1 and 2 calceoliferous, the flagella many-jointed. Antenna 1, flagellum stout, accessory flagellum slender. Antenna 2 not much longer than antenna 1. Epistome little or not at all projecting. Mandible, molar well developed, palp affixed nearly over it, the 3^d joint elongate. Maxilla 1, inner plate with 2 plumose setae, outer with very oblique apex, spine-teeth on apex of palp very small. Maxillipeds, outer plates broad, not quite reaching apex of 2^d joint of palp, palp rather robust, 4th joint unguiform. Gnathopod 1 imperfectly subchelate. Gnathopod 2 subchelate. Peraeopods 3—5, 2^d joint much expanded. Uropod 3 extending much beyond uropod 2, both rami carrying setae on inner margin, inner ramus longer than 1st joint of outer. Telson large, twice as long as broad, deeply cleft.

2 species.

Synopsis of species:

Side-plate 5 smooth . . . . 1. **U. gigas** . . . . p 64
Side-plate 5 with central boss . . . . 2. **U. umbonatus** . p. 64

1. **U. gigas** Dana 1849 *U.*, J. D. Dana in: Amer. J. Sci., ser. 2 *v.*8 p.136 | 1852 *U. gigas*, J.D. Dana in: P. Amer. Ac., *v.*2 p.209 | 1853 & 55 *U. g.*, J. D. Dana in: U. S. expl. Exp., *v.*13 II p.917; t. 62 f. 3 a—g | 1899 *U. g.*, T. Stebbing in: Ann. nat. Hist., ser. 7 *v.*4 p.211 | 1888 *U. g.* + *Tryphosa antennipotens*, T. Stebbing in: Rep. Voy. Challenger, *v.* 29 p. 266; p. 617 t. 6 | 1891 *Pseudotryphosa a.*, G. O. Sars, Crust. NorWay, *v.*1 p 83 | 1893 *Anonyx a.* + *A. ?gigas*. A. Della Valle in: F. Fl. Neapel, *v.*20 p. 827, 836 1893 *Tryphosella a.*, J. Bonnier in: Bull. sci. France Belgique, *v.*24 p. 171.

Head, lateral angles acute. Side-plate 1 narrowed below, concealing the mouth-organs, 2—4 successively deeper, overlapping. Pleon segment 4 dorsally depressed, then carinate, with upturned apex; postero-lateral angles of segment 3 produced, very broadly rounded. Eyes reniform, seemingly large, indistinct in spirit. Antenna 1. 1st joint long, flagellum with 52 joints, 1st not very long, accessory flagellum slight, 4-jointed, shorter than 1st of primary. Antenna 2, ultimate and penultimate joints of peduncle subequal, flagellum 53-jointed, rather longer and thinner than in antenna 1, in both with calceoli in ♀. Epistome a little prominent. Lower lip, mandibular processes not specially divergent. Maxilla 1 with setae on outer margin below palp, and also (according to Dana) at base of its 1st joint. Gnathopod 1, 6th joint a little longer than 5th and nearly as broad, slightly narrowed distally, palm ill-defined, 7th joint less than half length of 6th. Gnathopod 2, 6th joint more than half as long as 5th, palm transverse. Peraeopods 3—5, 2^d joint much expanded, hind margin serrulate. Peraeopod 5 rather shorter than peraeopod 4. Uropod 3, rami long, broadly lanceolate, subequal, carrying spines and plumose setae. Telson long, narrowing a little distally, cleft $^4/_5$ of length, with 5 lateral and 2 apical spines to each lobe. L. 18 mm.

Antarctic Ocean; Southern Indian Ocean (Heard Island, depth 274 m).

2. **U. umbonatus** (O. Sars) 1882 *Ichnopus u.*, G. O. Sars in: Forb. Selsk. Christian., nr. 18 p. 79 t. 3 f. 2 | 1891 & 95 *Pseudotryphosa umbonata*, G. O. Sars, Crust. NorWay, *v.*1 p. 83 t. 29 f. 2; p. 686 | 1893 *Anonyx umbonatus*, A. Della Valle in: F. Fl. Neapel, *v.* 20 p. 825 | 1899 *Uristes u.*, T. Stebbing in: Ann. nat. Hist., ser. 7 *v.*4 p. 211.

Head, lateral angles acute. Side-plate 1 narrowed below, 5th broader than deep, with central outstanding boss or umbo. Pleon segment 3 nearly quadrate, 4th without any dorsal projection. Eyes narrow, sigmoid, light red, evanescent in spirit. Antenna 1, 1st joint rather elongate, flagellum stout, of 25 joints, 1st largest, not very large, accessory flagellum almost filiform, of 6 small joints. Antenna 2 rather longer than antenna 1, ultimate joint of peduncle longer than penultimate, flagellum not very stout, 23-jointed. Epistome not projecting. Gnathopod 1, 6th joint much longer than 5th, oblong, palm very oblique, ill-defined, 7th joint about half as long as 6th. Gnathopod 2, 6th joint rather broader than 5th, and more than half as long. Telson tapering much, carrying 2 pairs of dorsal spinules. Colour whitish. L. 11 mm.

Arctic Ocean, North-Atlantic and North-Sea (West-Norway), Skagerrak. Depth 55—790 m.

39. Gen. Centromedon O. Sars

1891 *Centromedon* (Sp. typ.: *C. pumilus*), G. O. Sars, Crust. Norway. v. 1 p. 99 | 1893 *C.*, J. Bonnier in: Bull. sci. France Belgique, v. 24 p. 173.

Head, lateral angles acutely produced. Side-plates deep. Pleon segment 3, postero-lateral angles acutely upturned. Eyes imperfect or wanting. Antenna 1 and 2 in ♀ stout, subequal. Epistome not projecting. Mandible, molar conical, palp rather large, affixed over molar. Maxilla 1, inner plate with 2 setae, outer with usual spines, apex oblique, palp with few spines on apex. Maxilla 2, inner plate a little shorter than outer. Maxillipeds, outer plates not reaching end of 2^d joint of palp, which has the 4th joint small. Gnathopod 1 stout, imperfectly subchelate. Gnathopod 2 subchelate. Peraeopods 3—5, 2^d joint large; 5th peraeopod much shorter than 4th. Uropod 3 rather small, rami without marginal setae. Telson oblong, deeply cleft, with 2 pairs of apical spinules.

4 species.

Synopsis of species:

1 { Peraeopod 5, 2^d joint with lower hind corner acute 1. **C. calcaratus** . p. 65
{ Peraeopod 5, 2^d joint with lower hind corner rounded — 2.

2 { Head, lateral angles slightly upturned . . . . 2. **C. productus** . . p. 66
{ Head, lateral angles straight pointed — 3.

3 { Pleon segment 4 with rounded carina . . . . 3. **C. pumilus** . . . p. 66
{ Pleon segment 4 with pointed carina . . . . 4. **C. typhlops** . . p. 66

1. C. calcaratus (O. Sars) 1879 *Anonyx c., Hippomedon?*, G. O. Sars in: Arch. Naturv. Kristian., v. 4 p. 440 | 1885 *A. c.*, G. O. Sars in: Norske Nordhavs-Exp., v. 6 Crust. I p. 142 t. 12 f. 3 | 1893 *A. c.*, A. Della Valle in: F. Fl. Neapel, v. 20 p. 829 | 1891 *A. c., Centromedon?*, G. O. Sars, Crust. Norway, v. 1 p. 100.

Head, lateral angles produced to a straight spine-like point. Side-plate 1 not narrowed below. Pleon segment 3, postero-lateral angles acutely upturned to a greater extent than in any other species of the genus, 4th without carina. Eyes wanting, unless represented by a diffuse whitish-yellow pigment. Antenna 1, peduncle rather stout, flagellum with 9 joints, 1st not very long, accessory flagellum 3-jointed. Antenna 2 about as long as antenna 1, ultimate joint of peduncle much shorter than penultimate, flagellum 9-jointed. Gnathopod 1, 6th joint longer than 5th, rather narrow, tapering, palm undefined. Gnathopod 2, 6th joint comparatively broad, palm not produced. Peraeopods 1 and 2, 7th joint almost straight. Peraeopods 3 and 4, 2^d joint oval. Peraeopod 5, 2^d joint rather quadrate, the lower hind corner outdrawn to a spine-like point. Uropod 3 nearly as in C. pumilus, but 2^d joint of outer ramus less elongate. Telson very deeply cleft, much tapering. Colour whitish. L. 8 mm.

Arctic Ocean (between Iceland and Jan Mayen, North West of Bear Island). Depth 1240—2260 m.

2. **C. productus** (Goës) 1866 *Lysianassa producta*, Goës in: Öfv. Ak. Förh., *v.* 22 p. 519 t. 37 f. 4 | 1891 *Centromedon affinis*, G. O. Sars, Crust. Norway, *v.* 1 p. 101 | 1893 *C. a., Anonyx a.*, A. Della Valle in: F. Fl. Neapel, *v.* 20 p. 831, 920.

Nearly allied to C. pumilus, but differs by its much larger size, by the lateral corners of the cephalon being not straight but slightly upturned at the tip, and by having the postero-lateral angles of pleon segment 3 considerably narrower and more produced (Sars). L. 8 mm (Goës).

Arctic Ocean and North-Atlantic (Spitzbergen, Norway).

3. **C. pumilus** (Lillj.) 1865 *Anonyx p.*, W. Lilljeborg in: N. Acta Soc. Upsal., ser. 3 *v.* 6 nr. 1 p. 26 t. 4 f. 35—41 | 1893 *A. p.*, A. Della Valle in: F. Fl. Neapel, *v.* 20 p. 831 | 1891 *Centromedon p.*, G. O. Sars, Crust. Norway, *v.* 1 p. 100 t. 34 f. 2.

Head, lateral angles drawn out into a straight spine-like point. Side-plate 1 narrowed below. Pleon segment 4 with small rounded carina. Eyes wanting. Antenna 1, peduncle not very stout, flagellum with 8 joints, 1st not elongate, accessory flagellum 3-jointed. Antenna 2 scarcely longer than antenna 1, ultimate joint of peduncle a little shorter than penultimate, flagellum 9-jointed. Gnathopod 1, 6th joint as long as 5th, tapering distally, palm oblique, ill-defined, 7th half length of 6th. Gnathopod 2, 6th joint oblong oval, more than half length of 5th, palm transverse. Peraeopods 1 and 2, 6th joint slender, 7th long. Uropod 3, inner ramus longer than basal joint of outer, which has the 2^d joint nearly as long as the basal. Telson with 1 or 2 pairs of dorsal spinules. Colour whitish, tinged with red at ends of segments and joints. L. ♀ 5 mm.

Arctic Ocean, North-Atlantic (North-America, Norway), Skagerrak (Bohuslän). Depth 94—188 m.

4. **C. typhlops** (O. Sars) 1879 *Anonyx t.*, G. O. Sars in: Arch. Naturv. Kristian., *v.* 4 p. 436 | 1885 *A. t.*, G. O. Sars in: Norske Nordhavs-Exp., *v.* 6 Crust. I p. 145 t. 12 f. 4 a—k | 1891 *A. t., Centromedon?*, G. O. Sars, Crust. Norway, *v.* 1 p. 100 | ? 1889 *A. caecus*, Vosseler in: Arch. Naturg., *v.* 551 p. 155 t. 8 f. 8—14 | 1893 *A. nugax* (part.), A. Della Valle in: F. Fl. Neapel, *v.* 20 p. 834.

Head, lateral angles produced, somewhat acutely. Side-plate 1 narrowed below. Pleon segment 3, postero-lateral angles shortly but acutely upturned, 4th with an acutely ending carina. Eyes wanting. Antenna 1, peduncle short and thick, flagellum of 11 joints, 1st large, accessory flagellum slender, 4-jointed. Antenna 2 rather longer than antenna 1, ultimate and penultimate joints of peduncle subequal, flagellum 18-jointed. Gnathopod 1, rather stout, 6th joint a little longer than 5th, with oblique but distinct palm. Peraeopods 3—5, 2^d joint oval. Uropod 3 as in C. calcaratus (p. 65), telson with cleft not quite so deep. Colour whitish, tinged with red. L. 15 mm.

Arctic Ocean (between Jan Mayen and Finmark). Depth 3200 m.

40. Gen. **Cheirimedon** Stebb.

1888 *Cheirimedon* (Sp. un.: *C. crenatipalmatus*), T. Stebbing in: Rep. Voy. Challenger, *v.* 29 p. 638 | 1890 *C.*, G. O. Sars, Crust. Norway, *v.* 1 p. 34 | 1893 *C.*, J. Bonnier in: Bull. sci. France Belgique, *v.* 24 p. 174 | 1893 *C.*, A. Della Valle in: F. Fl. Neapel, *v.* 20 p. 837.

Head, lateral corners angularly projecting. Body compressed; side-plates deep, 4th deeply not broadly excavate. Pleon segment 4 with high compressed projection. Antenna 1, 2[d] and 3[d] joints short, 1st of flagellum long. Epistome slightly projecting, defined by short incision from upper lip. Mandible with denticulate molar, palp elongate, attached above the molar, its 1st joint short. Maxilla 1, inner plate with 2 plumose setae, outer with 11 spines, palp with many apical spine-teeth. Maxilla 2, plates fringed at the apices. Maxillipeds, plates of moderate size, 2[d] joint of palp not longer than 1st. Gnathopod 1 subchelate, 5th joint very short, triangular, 6th large, expanding gradually to the transverse palm, 7th joint as long as breadth of palm. Gnathopod 2 subchelate. Peraeopods 1 and 2, 4th joint rather long. Peraeopods 3—5, 2[d] joint longer than broad. Branchial vesicles simple. Uropod 3, peduncle not very long, outer ramus 2-jointed, much longer than the inner. Telson narrowing distally, deeply cleft.

2 species.

Synopsis of species:

Pleon segment 3, postero-lateral angles uncinate . . 1. **C. crenatipalmatus**. . p. 67
Pleon segment 3, postero-lateral angles quadrate . . 2. **C. latimanus** . . . . . p. 67

1. C. crenatipalmatus Stebb. 1888 *C. c.*, T. Stebbing in: Rep. Voy. Challenger, *v.* 29 p. 638 t. 12 | 1893 *C. c.*, A. Della Valle in: F. Fl. Neapel, *v.* 20 p. 837.

Head, lateral lobes sharply produced. Pleon segment 3, postero-lateral angles acutely and strongly upturned. Antenna 1, flagellum with 12 joints, of which 1st very long, longer than 3-jointed accessory flagellum; antenna 2, ultimate and penultimate joints of peduncle subequal, flagellum 7-jointed. Maxilla 1, palp with 9 teeth on right, 12 on left maxilla. Gnathopod 1, palm margin crenulate. Gnathopod 2, 3[d] joint shorter than 5th, which is rather robust, 6th scarcely half length of 5th, palm not at all produced, finger shutting closely upon it. Telson nearly once and a half as long as broad. L. ♀ 7·5 mm.

Cumberland Bay [Kerguelen Island]. Depth 230 m.

2. C. latimanus (O. Sars) 1882 *Normania latimana*, G. O. Sars in: Forh. Selsk. Christian., nr. 18 p. 83 t. 3 f. 6, 6 a | 1890 *Cheirimedon latimanus*, G. O. Sars, Crust. Norway, *v.* 1 p. 35 t. 13 f. 2 | 1893 *C. l.*, A. Della Valle in: F. Fl. Neapel, *v.* 20 p. 838 t. 60 f. 50.

Head, with lateral lobes produced almost acutely. Pleon segment 3, postero-lateral angles quadrate. Eyes narrow oval. Antenna 1, flagellum with 7 joints, of which the 1st moderately long, about half as long as the slender 3-jointed accessory flagellum; antenna 2, ultimate joint of peduncle much shorter than penultimate, flagellum 6-jointed. Gnathopod 1, 6th joint with palm margin not crenulate. Gnathopod 2, 6th joint fully half length of 5th. Telson almost twice as long as broad. L. ♀ 6 mm.

North-Sea (Bukken [West-Norway]).

41. Gen. **Tryphosella** Bonnier

1890 *Orchomenella?* (part.), G. O. Sars, Crust. Norway, *v.* 1 p. 66 | 1893 *Tryphosella* (part.), J. Bonnier in: Bull. sci. France Belgique, *v.* 24 p. 170.

Antenna 2 in ♀ little longer than antenna 1. Mandible with denticulate molar, which is prominent, pointing backward, palp with 1st joint short,

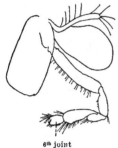

6th joint

Fig. 11. **T. barbatipes.**
Gnathopod 2.

1 species.

attached just over molar. Maxilla 1, inner plate with 2 setae, outer with 11 spines on slightly decurrent apex. Maxilla 2, outer plate rather longer than inner. Maxillipeds, inner plates normal, outer fully reaching end of 2^d joint of palp, fringed on inner margin with stout spine-teeth, 2^d joint of palp not longer than 1st. Gnathopod 1 setose, subchelate, 3^d joint not longer than 4th, 6th much longer than 5th, oblong, palm transverse. Gnathopod 2 (Fig. 11) minutely chelate, 6th joint less than half as long as 5th. Peraeopods 3—5, 2^d joint large. Peraeopod 4 longer than peraeopod 5. Uropod 3, inner ramus as long as basal joint of outer. Telson cleft beyond the centre.

1. **T. barbatipes** (Stebb.) 1888 *Tryphosa b.*, T. Stebbing in: Rep. Voy. Challenger, *v.* 29 p. 621 t. 7 | 1890 *T. b., Orchomenella?*, G. O. Sars, Crust. Norway, *v.* 1 p. 66 | 1893 *Anonyx b.*, A. Della Valle in: F. Fl. Neapel, *v.* 20 p. 814 | 1893 *Tryphosella b.*, J. Bonnier in: Bull. sci. France Belgique, *v.* 24 p. 171, 195.

Head, lateral corners much produced, acute. Side-plate 1 narrowed below, 5th as broad as deep. Pleon segment 3, postero-lateral angles slightly rounded. Antenna 1 in ♀, flagellum with 8 joints, 1st large and equal to 4-jointed accessory flagellum. Antenna 2 in ♀, ultimate and penultimate joints of peduncle subequal, flagellum about 7-jointed. Peraeopods, 4th joint only little widened. Telson oblong oval, cleft for $^2/_3$ of its length. L. ♀ 9 mm.

Southern Indian Ocean (Kerguelen Island). Depth 228 m.

42. Gen. **Tryphosa** Boeck

?1849 *Stenia*, J. D. Dana in: Amer. J. Sci., ser. 2 *v.* 8 p. 136 | 1871 *Tryphosa* (part.), A. Boeck in: Forh. Selsk. Christian., 1870 p. 117 | 1876 *T.* (part.), A. Boeck, Skand. Arkt. Amphip., *v.* 2 p. 180 | 1891 & 95 *T.*, G. O. Sars, Crust. Norway, *v.* 1 p. 75; p. 683, 684 | 1893 *Tryphosella* (part.), *Typhosa* (laps.), J. Bonnier in: Bull. sci. France Belgique, *v.* 24 p. 170.

Side-plate 1 usually narrowed below. Accessory flagellum of antenna 1 well developed. Antenna 2 in ♂ scarcely half the length of the body. Epistome usually projecting in front of upper lip, rounded. Mandible, molar denticulate, prominent, palp attached just over it, with 1st joint short, 2^d elongate. Maxilla 1, inner plate short with 2 setae, outer obliquely truncate with crowded spines. Maxilla 2, outer plate a little longer than inner. Maxillipeds normal, outer plates reaching beyond 2^d joint of palp, the margin crenulate. Gnathopod 1 slender, subchelate, 6th joint scarcely as long as 5th, both narrow. Gnathopod 2 subchelate, 6th joint rather stout, more than half as long as 5th. Peraeopods 3—5, 2^d joint large, rest of limb slender. Uropod 3, inner ramus at least as long as basal joint of outer. Telson oblong oval, deeply cleft.

9 species.

Synopsis of species:

1 { Eyes Wanting — **2.**
  { Eyes present — **5.**

2 { Pleon segment 4 not dorsally eleVated — **3.**
  { Pleon segment 4 dorsally eleVated — **4.**

1. T. insignis Bonnier 1896 *T. i.*, J. Bonnier in: Ann. Univ. Lyon, v.26 p.619 t. 36 f. 1.

Head, lateral corners produced to a right-angled lobe with straight sides. Side-plate 1 scarcely narrowed at rounded distal end, 4th large, much produced backward. Pleon segment 3, postero-lateral corners produced backward in narrowly rounded lobe, segment 4 not dorsally elevated. Eyes wanting. Antenna 1 in young ♂, flagellum of 7 joints, 1st as long as the peduncle's short 2d and 3d joints combined, the other 6 joints small, without calceoli, accessory flagellum 3-jointed. Antenna 2 scarcely longer than antenna 1, flagellum with 7 small joints. Epistome not projecting in front of upper lip. Gnathopod 1 slender, 5th and 6th joints equal, palm small, defined by 2 spinules, finger bifid. Gnathopod 2 scarcely longer, 5th joint considerably longer than 6th. Peraeopod 3, 2d joint nearly as broad as long, smaller than the side-plate. Peraeopod 4, 2d joint oblong oval, rest of limb more elongate than in peraeopods 3 and 5. Peraeopod 5, 2d joint very large, rest of limb small. Uropod 3, 2-jointed outer ramus with 1 seta on outer margin, inner ramus quite unarmed. Telson deeply cleft, with 2 pairs of marginal and 1 of apical spines. L. 6 mm.

Bay of Biscay. Depth 950 m.

2. T. kergueleni (Miers) 1875 *Lysianassa k.*, Miers in: Ann. nat. Hist., ser. 4 v. 16 p. 74 | 1879 *Anonyx k.*, Miers in: Phil. Tr.. v. 168 p. 207 t. 11 f. 4 | 1888 *Hippomedon k.*, T. Stebbing in: Rep. Voy. Challenger, v. 29 p. 623 t. 8 | 1893 *H. holbölli* (part.), A. Della Valle in: F. Fl. Neapel, v. 20 p. 808.

Head, lateral angles acutely produced. Side-plate 1 not narrowed below. Pleon segment 3, postero-lateral angles upturned, the process long, narrow, acute, segment 4 not apically elevated. Eyes doubtful. Antenna 1, 1st joint broad and long, flagellum with 14 joints, 1st not very long, accessory flagellum 5-jointed. Antenna 2, ultimate joint of peduncle a little shorter than penultimate, flagellum 16-jointed. Maxilla 1, with 12 spine-teeth on apex of palp. Maxillipeds, with spine-teeth on inner margin of outer plates. Gnathopod 1, 6th joint very little shorter than 5th, rather narrow, of uniform width, palm oblique. Gnathopod 2, 6th joint long ovate. Uropod 3, inner ramus longer than basal joint of outer. Telson nearly twice as long as broad, cleft extending more than three fourths of length, not dehiscent, with 3 pairs of marginal spinules and a pair in the emarginate apices. L. 10-mm.

Southern Indian Ocean (Kerguelen Island). Depth 37 m.

3. **T. trigonica** (Stebb.) 1888 *Hippomedon trigonicus*, T. Stebbing in: Rep.
Voy. Challenger, v. 29 p. 630 t. 9 | 1893 *Anonyx miersi?*, A. Della Valle in: F. Fl. Neapel,
v. 20 p. 814, 932.

Head, lateral angles moderately acute. Pleon segment 3, postero-lateral
angles upturned, the process short, only subacute, segment 4 with raised
angular apex. Antenna 1, 1st joint long and narrow, flagellum with 11 joints,
1st not very long, accessory flagellum 3-jointed. Antenna 2, ultimate and
antepenultimate joints of peduncle equal, shorter than penultimate, flagellum
9-jointed. Epistome a little prominent, rounded. Maxillipeds and gnathopod 1
as in T. kergueleni (p. 69). Gnathopod 2, 6th joint rather stouter. Uropod 3,
inner branch not longer than basal joint of outer. Telson not greatly longer
than broad, the cleft not dehiscent, $^2/_3$ of the length, with 3 pairs of marginal
spinules and a pair on the truncate apices. L. 6 mm.

Southern Indian Ocean (Kerguelen Island).

4. **T. pusilla** (O. Sars) 1879 *Anonyx (T.) pusillus*, G. O. Sars in: Arch. Naturv.
Kristian., v. 4 p. 439 | 1885 *T. pusilla*, G. O. Sars in: Norske Nordhavs-Exp., v. 6 Crust. I
p. 151 t. 13 f. 2, 2 a.

Head, lateral angles prominent, obtuse. Pleon segment 3, postero-lateral
angles acute, scarcely produced, segment 4 hunched above. Eyes wanting.
Antenna 1, flagellum with 11 joints, 1st rather large, accessory flagellum very
slender, 3-jointed. Antenna 2 about equal to antenna 1, ultimate joint of
peduncle very short, flagellum 12-jointed. Epistome not described. Gnathopod 1, 6th joint subequal to 5th, slender, palm rather oblique. Telson tapering
much. Colour whitish. L. 5·5 mm.

Arctic Ocean (between Jan Mayen and Iceland). Depth 1890 m.

5. **T. sarsi** (Bonnier) 1891 *T. nana* (err., non *Anonyx nanus* Krøyer 1846!),
G. O. Sars, Crust. Norway, v. 1 p. 76 t. 27 f. 1 | 1893 *Tryphosella sarsi*, J. Bonnier in:
Bull. sci. France Belgique, v. 24 p. 171 | 1895 *Tryphosa s.*, G. O. Sars, Crust. Norway.
v. 1 p. 684.

Head, lateral angles little projecting, rounded. Pleon segment 3, postero-
lateral angles quadrate, segment 4 without dorsal projection. Eyes oblong
oval, light red. Antenna 1, flagellum with 9 joints, 1st rather short, accessory
flagellum slender, of 5 joints, 1st short. Antenna 2 in ♀ little longer than
antenna 1, flagellum 9-jointed; antenna 2 in ♂ twice as long as antenna 1.
Epistome obtusely angular in front. Gnathopod 1, 6th joint about as long
as 5th, palm nearly transverse. Gnathopod 2, 6th joint oval, scarcely narrower
than 5th. Uropod 3, inner ramus equal to basal joint of outer. Telson
with 2 pairs of marginal spinules. Body whitish, pellucid, with a few reddish
spots. L. ♀ 4 mm.

North-Atlantic (Norway). Depth 11—38 m.

6. **T. angulata** O. Sars 1891 *T. a.*, G. O. Sars, Crust. Norway, v. 1 p. 78 t. 28
f. 1 | 1893 *Tryphosella a.*, J. Bonnier in: Bull. sci. France Belgique, v. 24 p. 171 | 1893
Anonyx angulatus, A. Della Valle in: F. Fl. Neapel, v. 20 p. 825.

Head, lateral angles drawn out into a little tooth. Pleon segment 3,
postero-lateral angles quadrate, segment 4 with high carina, the apex vertically
two-angled. Eyes very large, narrow oblong, light red. Antenna 1, peduncle
narrower than usual, flagellum setose, of 14 joints, 1st not long; accessory
flagellum of 5 joints, none long. Antenna 2 in ♀ a little longer than antenna 1,
flagellum 15-jointed. Epistome prominent, evenly rounded. Gnathopod 1,

6th joint scarcely as long as 5th, palm oblique, finger with secondary tooth conspicuous. Gnathopod 2, 6th joint as broad as 5th. Uropod 3, inner ramus equal to basal joint of outer. Telson with 3 or 4 pairs of marginal spinules. Colour pale reddish. L. ♀ 7 mm.

Arctic Ocean (Nordland, Finmark), Trondhjemsfjord. Depth 188—282 m.

7. T. compressa O. Sars 1891 & 1895 *T. c.*, G. O. Sars, Crust. Norway, *v.* 1 p. 76; p. 685 t. IV f. 2 | 1893 *Tryphosella c.*, J. Bonnier in: Bull. sci. France Belgique, *v.* 24 p. 171.

Head, lateral angles much projecting, narrowly rounded. Pleon segment 3, postero-lateral angles less than a right angle, the point not sharpened, segment 4 as in T. angulata. Eyes long, narrow, slightly sigmoid, light red. Antenna 1, 1st joint of peduncle rather long, flagellum 12-jointed, accessory flagellum with 6 joints, 1st short. Antenna 2 in ♀ subequal to antenna 1, flagellum 14-jointed. Epistome large, evenly rounded. Gnathopod 1, 6th joint shorter than 5th, oblong linear, the palm small, a little oblique, finger with secondary tooth conspicuous. Gnathopod 2, 6th joint large, hirsute, broader than 5th, distally widened, palm transverse, defined by a projecting angle, finger rather strong and curved. Uropod 3, inner ramus scarcely longer than basal joint of outer. Telson with 3 pairs of marginal spinules. L. ♀ 8 mm.

Arctic Ocean (West of Spitzbergen, Nordland).

8. T. nanoides (Lillj.) 1865 *Anonyx n.*, W. Lilljeborg in: N. Acta Soc. Upsal., ser. 3 *v.* 6 nr. 1 p. 25 t. 3 f. 32—34 | 1893 *A. n.*, A. Della Valle in: F. Fl. Neapel, *v.* 20 p. 832 | 1871 *Tryphosa n.*, A. Boeck in: Forh. Selsk. Christian., 1870 p. 118 | 1876 *T. n.*, A. Boeck, Skand. Arkt. Amphip., *v.* 2 p. 186 | 1891 & 95 *T. n.*, G. O. Sars, Crust. Norway, *v.* 1 p. 79 t. 28 f. 2; p. 684 | 1893 *Tryphosella n.*, J. Bonnier in: Bull. sci. France Belgique, *v.* 24 p. 171.

Head, lateral angles little projecting, rounded. Pleon segment 3, postero-lateral angles a little obtusely upturned, segment 4 with rounded carina overlapping segment 5. Eyes large, reniform, pale red. Antenna 1, peduncle massive, flagellum with 11 joints, 1st very large, accessory flagellum with 10 joints, 1st comparatively large. Antenna 1 in ♀ rather longer than antenna 2, flagellum 22-jointed. Epistome with narrow rounded lobe overhanging upper lip. Gnathopod 1, 6th joint a little shorter than 5th, palm transverse. Gnathopod 2, 6th joint about half as long as 5th, scarcely as broad. Uropod 3 larger than usual, inner ramus subequal to basal joint of outer, both with small setae on inner margin. Telson very slightly tapering, with 3 or 4 pairs of marginal spinules. Body whitish pellucid, with diffuse orange patches. L. ♀ 8 mm.

Arctic Ocean; North-Atlantic, North-Sea, Skagerrak, Kattegat and Baltic (Norway, Sweden, Denmark, Scotland). Depth 94—188 m,

9. T. høringii Boeck 1871 *T. h.*, A. Boeck in: Forh. Selsk. Christian., 1870 p. 118 | 1876 *T. höringii*, A. Boeck, Skand. Arkt. Amphip., *v.* 2 p. 182 | 1891 *T. höringii*, . *T. höringii*, G. O. Sars, Crust. Norway, *v.* 1 p. 77 t. 27 f. 2 | 1893 *Tryphosella höringii*, J. Bonnier in: Bull. sci. France Belgique, *v.* 24 p. 171 | 1893 *Anonyx nanus* (part.), A. Della Valle in: F. Fl. Neapel, *v.* 20 p. 820.

Head, lateral angles projecting, rounded. Pleon segment 3, postero-lateral angles almost quadrate, segment 4 with high carina angularly produced at apex. Eyes large, oblong oval, widened below, light red. Antenna 1, peduncle stout, flagellum with 11 joints, 1st not very large, accessory flagellum with 6 joints,

1st the longest. Epistome rather prominent, rounded. Gnathopod 1 slender, 6th joint a little shorter than 5th, palm nearly transverse. Gnathopod 2, 6th joint scarcely as broad as 5th. Uropod 3, inner ramus rather longer than basal joint of outer. Telson with 3 pairs of marginal spinules. Colour pale reddish. L. ♀ 7, ♂ 6 mm.

Arctic Ocean and North-Atlantic (North America, Scandinavia). Depth 94—282 m.

43. Gen. **Chironesimus** O. Sars

1891 Chironesimus (Sp. un.: C. debruynii), G. O. Sars, Crust. Norway, v. 1 p. 108 | 1894 C., T. Stebbing in: Bijdr. Dierk., v. 17 p. 13.

Head, lateral lobes slightly produced, subacute. Side-plate 1 scarcely expanded below, 4th rather considerably. Pleon segment 3, postero-lateral angles acutely upturned. Epistome not projecting, upper lip forming in front a rather large compressed linguiform lobe. Mandible, molar tapering to a blunt apex, palp rather large, set well forward. Maxilla 1, 2 setae on inner plate, 11 spines on outer, 9 spine-teeth on palp. Maxilla 2, inner plate not very narrow, more than half as long as outer. Maxillipeds, broad outer plates set with minute teeth on inner and apical margin, 4th joint of palp not elongate. Gnathopod 1 subchelate, 6th joint shorter and much narrower than 5th, oblong, slightly narrowed towards the short almost transverse palm. Gnathopod 2, 2d joint rather shorter than 5th, 6th more than half as long as 5th, widening distally, front margin longer than hind one, palm concave, the curved finger not reaching its outer angle. Peraeopods comparatively slender. Peraeopods 3—5, 2d joint large, 4th not very robust. Uropod 2, inner branch constricted. Uropod 3 reaching beyond uropod 2, rami longer than peduncle, subequal, lanceolate, outer with small second joint, both carrying spinules and setae. Telson nearly twice as long as broad, cleft more than 2/3 length, a spinule in each of the blunt rather divergent apices.

1 species.

1. **C. debruynii** (Hoek) 1882 Anonyx d., Hoek in: Niederl. Arch. Zool., suppl. 1 nr. 7 p. 44 t. 3 f. 30—30x | 1893 A. d., A. Della Valle in: F. Fl. Neapel, v. 20 p. 830 ' 1891 & 95 Chironesimus d., G. O. Sars, Crust. Norway, v. 1 p. 109 t. 37 f. 2; p. 687 | 1894 C. d., T. Stebbing in: Bijdr. Dierk., v. 17 p. 13.

Eyes oblong oval, a little broader below, red. Antenna 1, 1st joint tumid, ridged on inner margin, 2d and 3d very short, flagellum in ♀ with 11 joints, in ♂ with 18 (or more), 1st largest, accessory flagellum with 6 joints, 1st large and somewhat expanded. Antenna 2, ultimate and penultimate joints of peduncle subequal, flagellum in ♀ with 15 joints, in ♂ with 20 (or more). Gnathopod 1, the 6th joint is as long as the 5th and tapers too much to form a palm (Hoek, not Sars or Stebbing). Peraeopod 5, 2d joint broadly oval, about as long as rest of limb. Colour yellowish white, intestine dark brown, ovaries reddish. L. (♂juv.?) 20 (Hoek), ♀ 14 (Sars) mm.

Arctic Ocean; North-Atlantic and North-Sea (West-Norway). Depth 94—240 m.

44. Gen. **Eurythenes** S. I. Sm.

1865 Eurytenes (Sp. un.: E. magellanicus) (non Arn. Foerster 1862, Hymenoptera!). W. Lilljeborg in: N. Acta Soc. Upsal., ser. 3 v. 6 nr. 1 p. 11 | 1882 Eurythenes, (S. I. Smith in:) Scudder, Nomencl. zool., suppl. L. p. 135 | 1884 E., S. I. Smith in: Amer. J. Sci., ser. 3 v. 28 p. 54 | 1891 Euryporeia, G. O. Sars, Crust. Norway, v. 1 p. 85 | 1893 E., A. Della Valle in: F. Fl. Neapel, v. 20 p. 847.

Body massive. Head, lateral angles little produced. Side-plates not very deep, 1st very small, rounded, not covering the head or the protruding mouth-organs, 5th much broader than deep. Antenna 1, peduncle not very stout, flagellum with many short joints, accessory flagellum well developed. Antenna 2 in ♀ much longer than antenna 1, ultimate joint of peduncle longer than penultimate, flagellum many-jointed. Epistome not defined by an incision from upper lip, projecting broadly rounded in front of mouth-organs. Lower lip with front lobes apically emarginate. Mandible broad, molar large and prominent, palp not very long, affixed a little in front of molar. Maxilla 1, inner plate carrying many setae, outer with the usual 11 spines on oblique apex, palp very narrow, apical spines few. Maxilla 2, inner plate much shorter than outer, both with many setae along inner margin. Maxillipeds normal, the large outer plates not reaching beyond 2^d joint of the large setose palp, their inner margin furnished with small nodular teeth. Gnathopod 1 strong, subchelate, 6th joint rather longer than 5th, almost oblong, a little tapering, palm small, slightly excavate. Gnathopod 2 very slender, subchelate, 6th joint much more than half length of 5th, nearly linear, not produced to form a chela with the minute finger. Peraeopods 3—5 rather short and stout, 2^d joint moderately large, 4th somewhat expanded and produced. Uropod 3 projecting beyond uropods 1 and 2, rami foliaceous, setose, inner ramus as long as basal joint of outer. Telson long, free from spinules, sharply tapering, cleft deep, not dehiscent.

1 species.

1. E. gryllus (Lcht.) 1822 *Gammarus g.*, (H. Lichtenstein in:) Mandt, Observ. Groenl., p. 34 | 1866 *Lysianassa g.*, Goës in: Öfv. Ak. Förh., *v.* 22 p. 517 t. 36 f. 1 | 1884 *Eurythenes g.*, S. I. Smith in: Amer. J. Sci., ser. 3 *v.* 28 p. 54 | 1891 *Euryporeia g.*, G. O. Sars, Crust. Norway, *v.* 1 p. 86 t. 30 | 1893 *E. g.*, A. Della Valle in: F. Fl. Neapel, *v.* 20 p. 848 t. 60 f. 58 | 1848 *Lysianassa magellanica*, H. Milne Edwards in: Ann. Sci. nat., ser. 3 *v.* 9 p. 398 | 1865 *Eurytenes magellanicus*, W. Lilljeborg in: N. Acta Soc. Upsal., ser. 3 *v.* 6 nr. 1 p. 11 t. 1—3 f. 1—22.

Back broadly rounded (Sars), last segment of peraeon and 1st—5th of pleon with a low longitudinal ridge (Lilljeborg), segments 4—7 of peraeon and 1—4 of pleon carinate. Pleon segment 3, postero-lateral angles rounded, segments 3 and 4 with dorsal depression. Eyes very large, irregularly flask-shaped, light orange. Antenna 1, 1st joint not very long, flagellum with 30 joints, 1st large, accessory flagellum with 10 joints,' 1st the largest. Antenna 2 twice as long as antenna 1, flagellum slender, setose, 50-jointed. Gnathopod 2 twice as long as gnathopod 1. Telson nearly reaching apex of uropod 3. Colour rosy with yellowish tinge, margin of legs vermilion. L. reaching 90 mm.

● Arctic Ocean and North-Atlantic (Finmark, North East of America, Bay of Biscay, Azores), tropical Atlantic (lat. 4° S., long. 18° W.). South-Atlantic (Cape Horn). From Procellaria glacialis L., from stomachs of sharks, and fished up from great depths.

45. Gen. Tmetonyx Stebb.*)

1891 *Hoplonyx* (non Jam. Thomson 1858, Coleoptera!), G. O. Sars, Crust. Norway, *v.* 1 p. 91 | 1894 *H.*, T. Stebbing in: Bijdr. Dierk., *v.* 17 p. 9.

Epistome more or less projecting and rounded in front. Antennae 1 and 2 in ♂ with small calceoli. Mandible, molar large, obliquely truncate, palp affixed over it, 2^d and 3^d joints subequal. Maxilla 1, 2 setae on inner plate.

---

*) Nom. nov.; τμητός, shaped by cutting, ὄνυξ, nail. — The name *Hoplonyx* is preoccupied (1858, James Thomson in: Arch. ent., *v.* 2 p. 98).

11 spines on the broad obliquely truncate outer, several spine-teeth on palp.
Maxilla 2, plates rather longer than in Anonyx (p. 53). Maxillipeds, outer plates
large, oblong oval, reaching end of 2^d joint of palp, which is not specially
robust. Gnathopod 1 slender, 3^d joint generally longer than 4th, 6th oblong,
palm very oblique, rather indistinctly defined, 7th rather long, minutely
denticulate, having a secondary tooth on inner margin and a cap over the
apex. Gnathopod 2, 6th joint usually not at all produced beneath the
minute finger. Peraeopods 1, rather elongate, 2^d joint large. Branchial
vesicles usually simple. Uropod 3 projecting beyond uropod 2, rami minutely
denticulate. Telson oblong, deeply cleft.

8 species.

Synopsis of species:

1 { Pleon segment 3 little or not at all produced — **2.**
   { Pleon segment 3 acutely produced or upturned — **5.**

2 { Head, lateral corners rounded — **3.**
   { Head, lateral corners not rounded — **4.**

3 { Antenna 1, 1st joint of flagellum not very large 1. **T. cicada** . . . . . p. 74
   { Antenna 1, 1st joint of flagellum very large . . 2. **T. miersi** . . . . . p. 75

4 { Head, lateral corners acute, much produced; eyes
       bright red, angled . . . . 3. **T. acutus** . . . . . p. 75
   { Head, lateral corners quadrate; eyes light red,
       scarcely angled . . . . 4. **T. albidus** . . . . p. 75

5 { Side-plate 1 expanded below . . . . 5. **T. cicadoides** . . . p. 75
   { Side-plate 1 not expanded below — **6.**

6 { Eyes bright red . . . . 6. **T. similis** . . . . . p. 76
   { Eyes white . . . . 7. **T. leucophthalmus** p. 76
   { Eyes wanting . . . . 8. **T. caeculus** . . . . p. 76

1. T. cicada (O. Fahr.) 1780 *Oniscus c.*, O. Fabricius, Fauna Groenl., p. 258 |
1888 *Anonyx c.*, T. Stebbing in: Rep. Voy. Challenger, *v.* 29 p. 47, 617 | 1893 *A. c.* (part.),
A. Della Valle in: F. Fl. Neapel, *v.* 20 p. 833 | 1891 *Hoplonyx c.*, G. O. Sars, Crust.
Norway, *v.* 1 p. 92 t. 32 f. 2 | 1894 *H. c.*, T. Stebbing in: Bijdr. Dierk., *v.* 17 p. 9 | 1845
Anonyx gulosus, Krøyer in: Naturh. Tidsskr., ser. 2 *v.* 1 p. 611 | 1846 *A. g.*, Krøyer in:
Voy. Nord, Crust. t. 14 f. 2 a—t | 1866 *Lysianassa gulosa*, Goës in: Öfv. Ak. Förh.,
v. 22 p. 520 | 1851 *Anonyx norvegicus*, W. Liljeborg in: Öfv. Ak. Förb., *v.* 8 p. 22 | 1861
A. bruzelii, A. Boeck in: Forh. Skand. Naturf., Møde 8 p. 643.

Head, lateral angles very slightly projecting, slightly rounded. Side-
plate 4 expanded in a narrow obtuse lobe. Pleon segment 3, postero-
lateral angles almost quadrate, with very short produced point. Eyes angled,
narrow linear above, lower arm oval, bright red. Antenna 1 in ♀, flagellum with
16 joints, 1st not very large, accessory flagellum 7-jointed. Antenna 2 in ♀,
flagellum 28-jointed. Gnathopod 1, 5th and 6th joints equal in length, palm
arcuate, finely serrate, not defined by any trace of angle. Gnathopod 2,
6th joint half as long as 5th. Uropod 2 with neither ramus constricted.
Uropod 3, inner ramus scarcely longer than basal joint of outer. Telson
not quite twice as long as broad, cleft nearly to base, with 2 pairs of
dorsal spinules and an apical pair. Colour creamy with tinge of rose and
on the back bright apple red; eggs in pouch dark violet. L. ♀ 18—24 mm.

Arctic Ocean; North-Atlantic, North-Sea and Skagerrak (South- and West-
Norway, Bohuslän, North Britain, France).

2. **T. miersi** (Stebb.) 1888 *Hippomedon m.*, T. Stebbing in: Rep. Voy. Challenger, v. 29 p. 631 t. 10 | 1893 *Anonyx miersii;* A. Della Valle in: F. Fl. Neapel, v. 20 p. 813.

Head, lateral angles rounded. Side-plate 1 narrowed below. Pleon segment 3, postero-lateral angles rounded, almost quadrate. Eyes doubtful. Antenna 1, flagellum with 11 joints, 1st very large, accessory flagellum 4-jointed. Antenna 2 in ♂, ultimate joint of peduncle longer than penultimate, flagellum 38-jointed. Epistome? Gnathopod 1, 3^d joint not longer than 4th, 6th rather longer than 5th. Gnathopod 2, 6th joint not half as long as 5th, apically a little produced. Branchial vesicles, with approximation to Anonyx (p. 53), showing some irregular folds. Uropod 3, inner ramus shorter than basal joint of outer, each carrying a fringe of very long plumose setae. Telson much longer than broad, cleft about $^4/_5$ of the length, with 2 pairs of marginal spinules and 1 pair in the broad emarginate apices. L. about 12 mm.

Bass Strait (East Moncoeur Island). Depth 68 m.

3. **T. acutus** (O. Sars) 1891 *Hoplonyx a.*, G. O. Sars, Crust. Norway, v. 1 p. 95 t. 33 f. 2 | 1893 *Anonyx cicada* (part.), A. Della Valle in: F. Fl. Neapel, v. 20 p. 833.

Head, lateral angles much and very acutely produced. Side-plate 1 neither narrowed nor expanded below, 4th much expanded into a narrow obtuse lobe. Pleon segment 3, postero-lateral angles scarcely produced, quadrate. Eyes as in T. similis (p. 76). Gnathopod 1, palm straight. Gnathopod 2, 6th joint fully half as long as 5th, palm obliquely truncate, finger stronger than usual. Uropod 3, rami nearly equal in length. Telson with 2 pairs of dorsal spinules. Eggs in pouch orange-coloured. L. ♀ 13 mm. — Other points in practical agreement with the account given of T. similis (p. 76).

North-Atlantic (West-Norway).

4. **T. albidus** (O. Sars) 1891 *Hoplonyx a.*, G. O. Sars, Crust. Norway, v. 1 p. 96 t. 33 f. 3 | 1893 *Anonyx a.*, A. Della Valle in: F. Fl. Neapel, v. 20 p. 826.

Head, lateral angles nearly quadrate. Side-plate 1 narrowed below, 4th expanded into a short broad quadrate lobe. Pleon segment 3, postero-lateral angles not at all produced, quadrate, 4th deeply depressed dorsally. Eyes large, scarcely prolonged backward as in other species, visual elements indistinct, pigment light red. Antenna 1 in ♀, flagellum 15-jointed, accessory flagellum 8-jointed. Antenna 2 considerably longer than antenna 1, flagellum 28-jointed. Epistome much projecting, obtusely angled. Gnathopod 1, 6th joint subequal to 5th, palm indistinctly defined below. Gnathopod 2, 6th joint more than half length of 5th, narrow. Peraeopods 3—5 shorter than usual, 2^d joint very large. Uropod 3, rami rather broad, inner shorter than outer. Telson not twice as long as broad, with 2 pairs of dorsal spinules. Colour whitish, not tinged with red, eggs in pouch bright red. L. ♀ 12 mm.

Arctic Ocean, North-Atlantic and North-Sea (North- and West-Norway). Depth 150—282 m.

5. **T. cicadoides** (Stebb.) 1888 *Anonyx c.*, T. Stebbing in: Rep. Voy. Challenger, v. 29 p. 612 t. 4, 5 | 1891 *A. c., Hoplonyx* (part.), G. O. Sars, Crust. Norway, v. 1 p. 92 | 1893 *Ichnopus sp.?* + *A. nugax* (part.)?, A. Della Valle in: F. Fl. Neapel, v. 20 p. 835.

Head, lateral angles a little produced, rounded. Side-plate 1 expanded below, 4th expanded into a rather narrow rounded lobe. Pleon segment 3, postero-lateral angles much produced, acutely upturned. Character of eyes uncertain. Antenna 1 in ♀, flagellum with 20 joints, 1st not very large, accessory flagellum with 9 joints, 1st not very long. Antenna 2 in ♀, ultimate and penultimate joints of peduncle subequal, flagellum 30-jointed. Antenna 2 in ♂, ultimate joint of peduncle longer than penultimate, flagellum with 50 joints, most with calceoli as in flagellum of antenna 1. Maxilla 1, 10 spine-teeth on apex of palp. Maxillipeds, 5 small spines on apical border of outer plates. Gnathopod 1, 6th joint longer than 5th, palm very oblique and very ill-defined. Gnathopod 2, 6th joint not half as long as 5th, narrow, with a little convex palm. Uropod 2, inner ramus strongly constricted. Uropod 3, rami narrowly lanceolate, carrying spinules and setae, inner scarcely longer than basal joint of outer. Telson not quite twice as long as broad, cleft nearly to base, with 3 pairs of dorsal spinules, and outer angle of each apex produced into a small tooth. Some specimens in spirit, of a deep brown colour, others cream-coloured. L. 18 mm.

Southern Indian Ocean (Kerguelen Island). Depth 36—228 m.

6. **T. similis** (O. Sars) 1891 *Hoplonyx s.*, G. O. Sars, Crust. NorWay, *v.* 1 p. 93 t. 33 f. 1 | 1893 *Anonyx cicada* (part.), A. Della Valle in: F. Fl. Neapel, *v.* 20 p. 833.

Head, lateral angles not rounded. Side-plate 1 narrowed below, 4th expanded into a short, broad, rather squared lobe. Pleon segment 3, postero-lateral angles acutely produced. Eyes angled, upper arm very narrow, lower bulb-like, bright red. Antenna 1 in ♀ rather more slender than in T. cicada (p. 74), flagellum with 20 joints, 1st rather large, accessory flagellum 6-jointed. Antenna 2 in ♀ little longer than antenna 1, flagellum 24-jointed. Epistome somewhat more projecting than in T. cicada. Gnathopod 1, 6th joint a little longer than 5th, palm slightly flexuous, defined by an obtuse angle. Gnathopod 2, 6th joint scarce half as long as 5th. Uropod 3, inner ramus distinctly longer than basal joint of outer. Telson about twice as long as broad, deeply cleft, with only 1 pair of dorsal spinules. Pellucid, with faint tinge of pale reddish yellow. L. 14 mm.

Arctic Ocean, North-Atlantic and North-Sea (West-NorWay, Scotland). Depth 38—282 m.

7. **T. leucophthalmus** (O. Sars) 1891 *Hoplonyx l.*, G. O. Sars. Crust. NorWay, *v.* 1 p. 97 t. 34 f. 1 | 1893 *Anonyx cicada* (part.), A. Della Valle in: F. Fl. Neapel, *v.* 20 p. 833.

Head, lateral angles obtusely projecting. Side-plate 1 slightly narrowed below, 4th rather broadly and squarely expanded. Pleon segment 3, postero-lateral angles rather sharply produced. Eyes angled, narrow, without distinct visual elements, pigment whitish. Antenna 1 in ♀, flagellum with 18 joints, 1st rather long, accessory flagellum 6-jointed. Antenna 2 in ♀ little longer than antenna 1. Epistome much projecting, evenly rounded. Gnathopod 1, 6th joint as long as 5th, palm flexuous, defined by an indistinct obtuse angle. Gnathopod 2, 6th joint short and broad, not half length of 5th, palm transverse. Peraeopods rather long and slender. Uropod 3, inner ramus scarcely longer than basal joint of outer. Telson not nearly twice as long as broad, with only one pair of dorsal spinules. Colour pale reddish yellow, pinkish in front. L. ♀ 15 mm.

Hardangerfjord and Trondhjemsfjord. Depth to 282 m.

8. **T. caeculus** (O. Sars) 1891 *Hoplonyx c.*, G. O. Sars, Crust. NorWay, *v.* 1 p. 98 t. 35 f. 1 | 1893 *Anonyx cicada* (part.), A. Della Valle in: F. Fl. Neapel, *v.* 20 p. 833.

Head, lateral angles acutely projecting. Side-plate 1 scarcely narrowed below, 4th with expansion short, broad, obtuse. Pleon segment 3, postero-lateral angles acutely upturned, 4th with rounded carina behind the dorsal depression. Eyes wholly wanting. Antenna 1, flagellum with 11 joints, 1st very large, accessory flagellum with 4 joints, 1st long. Epistome scarcely projecting in front of upper lip, but defined from it by distinct incision. Gnathopod 1, 3d joint less elongate than in other species, 6th joint as long as 5th, palm very oblique, finely denticulate, defined by obtuse angle. Gnathopod 2, 6th joint oblong oval, more than half as long as 5th, palm transverse. Peraeopods rather slender, 7th joint long and narrow. Peraeopod 5, 2d joint very large, as long as rest of limb. Uropod 3, rami narrow, mucroniform, without the marginal setae found in the other species, the outer the longer, with spiniform 2d joint. Telson not nearly twice as long as broad, deeply cleft, with 1 or 2 pairs of dorsal spinules. Body pellucid whitish, unpigmented. L. ♀ 5 mm.

Trondhjemsfjord. Depth about 280 m.

46. Gen. **Tryphosites** O. Sars

1871 *Tryphosa* (part.), A. Boeck in: Forh. Selsk. Christian., 1870 p. 117 | 1891 *Tryphosites* (Sp. typ.: *Anonyx longipes*), G. O. Sars, Crust. Norway, v. 1 p. 81.

Side-plate 1 of uniform width. Pleon segment 3, postero-lateral angles acutely upturned. Antenna 1, accessory flagellum rather small, antennae 1 and 2 subequal in ♀, antenna 1 in ♂ long, antenna 2 very long. Epistome produced into an acutely lanceolate process. Mandible with prominent molar, palp slender, affixed scarcely behind it. Maxilla 1, inner plate not very short, otherwise as in Tryphosa (p. 68). Maxilla 2, plates moderately wide, inner plate a little the shorter. Maxillipeds, outer plates large, reaching much beyond 2d joint of palp, fringed with small spine-teeth. Gnathopod 1 slender, subchelate, 6th joint shorter than 5th. Gnathopod 2 slender, subchelate, 6th joint nearly fusiform, much more than half length of linear 5th. Peraeopods slender, long. Peraeopods 3—5, 2d joint much expanded. Uropod 2, inner branch constricted. Uropod 3 rather large, inner ramus a little longer than basal joint of outer, both carrying spinules and setae in both sexes. Telson deeply cleft.

1 species.

1. **T. longipes** (Bate & Westw.) 1861 *Anonyx l.* + *A. ampulla* (err., non *Cancer a.* Phipps 1774!), Bate & Westwood, Brit. sess. Crust., v. 1 p. 113 f.; p. 116 f. | 1862 *A. l.* + *A. a.*, Bate, Cat. Amphip. Brit. Mus., p. 79 t. 13 f. 4, 5 | 1871 *Tryphosa l.*, A. Boeck in: Forb. Selsk. Christian., 1870 p. 118 | 1891 *Tryphosites l.*, G. O. Sars, Crust. Norway, v. 1 p. 81 t. 28 f. 3; t. 29 f. 1 | 1893 *Anonyx l.*, A. Della Valle in: F. Fl. Neapel, v. 20 p. 830.

Head, lateral angles a little produced, acute. Side-plate 5 much broader than deep. Pleon segment 4 with no dorsal projection. Eyes oval, not very large, light red. Antenna 1 in ♀, flagellum with 18 joints, 1st very large, accessory flagellum 5-jointed; flagellum in ♂ 30-jointed. Antenna 2, penultimate joint of peduncle rather expanded, flagellum in ♀ 15-jointed, in ♂ filiform, longer than the body. Gnathopod 1, 6th joint of uniform width, palm slightly oblique. Peraeopods 1 and 2 very setose on hind margin. Telson only slightly tapering, carrying 3 pairs of marginal spinules and 3 pairs on the truncate apices. Body whitish pellucid. L. 12 mm.

Arctic Ocean, North-Atlantic, North-Sea, Skagerrak, Kattegat and Baltic (Norway, Denmark, Great Britain, West-France); Mediterranean.

47. Gen. **Lepidepecreum** Bate & Westw.

1868 *Lepidepecreum,* Bate & Westwood, Brit. sess. Crust., *v.* 2 p. 509 | 1888 *L.,*
T. Stebbing in: Rep. Voy. Challenger, *v.* 29 p. 686 | 1891 *L.,* G. O. Sars, Crust. Norway,
v. 1 p. 112 | 1871 *Orchomene* (part.), A. Boeck in: Forh. Selsk. Christian., 1870 p. 114.

Body carinate, as a rule extensively, integument firm, calcareous. Antenna 1,
1st joint usually carinate, produced (Fig. 12). Antenna 2 in ♀ not longer than
antenna 1, antepenultimate joint of peduncle rather
long, in ♂ flagellum slender, very elongate. Epistome
forming a broad compressed plate projecting in front
of upper lip. Mandible, molar
slight, palp slender, affixed far
back. Maxilla 1, inner plate with
2 setae, outer with almost transverse apex. Maxilla 2, plates
rather elongate, inner narrower
and a little shorter than outer.
Maxillipeds, outer plates reaching
beyond 2d joint of palp. Gnathopod 1 subchelate, 6th joint narrow.
Gnathopod 2 minutely chelate.
Peraeopods 1 and 2 slender.

Fig. 12.
L. longicorne, ♂.
Antenna 1.
[After G. O. Sars.]

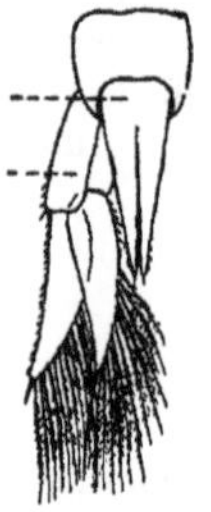

Fig. 13.
L. longicorne, ♂.
Uropod 3 and telson.
[After G. O. Sars.]

Peraeopods 3—5, 2d joint large. Uropod 3, rami in ♀ scarcely reaching beyond
uropod 2, with few setae or none, in ♂ much larger, densely setose on
inner margin (Fig. 13). Telson conically attenuated, cleft (Fig. 13).

5 species.

Synopsis of species:

1 { Carina of small extent . . . . 1. **L. typhlops** . . . . p. 78
  { Carina of great extent — **2.**

2 { Telson not cleft beyond the centre . . . . . . 2. **L. foraminiferum** . p. 79
  { Telson cleft beyond the centre — **3.**

3 { Peraeopod 5, 2d joint greatly produced down-
  {   ward . . . . 3. **L. clypeatum** . . p. 79
  { Peraeopod 5, 2d joint not greatly produced down-
  {   ward — **4.**

4 { Side-plate 5 not deeper than broad . . . . . . 4. **L. longicorne** . . . p. 80
  { Side-plate 5 deeper than broad . . . . 5. **L. umbo** . . . . . p. 80

1. **L. typhlops** Bonnier 1896 *L. t.,* J. Bonnier in: Ann. Univ. Lyon, *v.* 26
p. 621 t. 36 f. 2.

Back somewhat rounded, not carinate, apparently with single exception
of pleon segment 4, which is raised into a sharp, apically upward pointing
keel, and in ♂, but not in ♀, has a deep transverse sinus. Head very narrow,
rostrum small, lateral corners greatly produced into a narrow, somewhat
downward directed, apically rounded lobe. Side-plates 1—4 very large; pleon
segment 3, postero-lateral corners quadrate. Eyes entirely wanting. Antenna 1,
1st joint large, scarcely produced apically, rest disposed at right angles to 1st,
flagellum in ♂ with 6 joints, with some calceoli, 1st joint longer than other
5 combined, in ♀ with 5 joints, 1st not very long; accessory flagellum very small,
2-jointed. Antenna 2 in ♂ as long as body, ultimate and penultimate joints of

peduncle equal, each narrower and a little shorter than antepenultimate; flagellum with many short joints, each with a calceolus. Epistome carinate with 2 successive apices above, below dividing into two keels. Maxilla 1, outer plate apically carrying 12 spines. Gnathopod 1, side-plate said to be almost as long as the remainder of the appendage (but?), 5th and 6th joints equal, margins parallel, palm small, well defined. Gnathopod 2 longer, very slender, 6th joint inflated at the distal end, which prolongs itself in a rounded lobe beyond the finger. Peraeopods 1 and 2, 5th joint shorter than 6th, finger long. Peraeopod 5, 2^d joint broader and longer than in preceding peraeopods. Uropods 1 and 2 with few spines on the rami; uropod 3, outer ramus 2-jointed, fringed in ♂, not in ♀, with long setae on inner margin, inner ramus carrying 3 spines and 2 setae. Telson long, deeply cleft, with a pair of apical spinules. L. about 6 mm.

Bay of Biscay. Depth 650—950 m.

2. **L. foraminiferum** Stebb. 1888 *L. f.*, T. Stebbing in: Rep. Voy. Challenger, *v.* 29 p. 686 t. 24 | 1893 *Anonyx longicornis* (part.), A. Della Valle in: F. Fl. Neapel, *v.* 20 p. 814.

Carina extending from 2^d joint of antenna 1 to pleon segment 4, on peraeon segments 6 and 7 and pleon segments 1—3 the ridge forming a small distal tooth. Head slightly rostrate, lateral angles outdrawn into long narrow lobes ending obtusely. Side-plate 5 with breadth and depth subequal. Pleon segment 3, postero-lateral angles acute, slightly upturned, segment 4 forming a strong upturned carinal tooth. Eyes not perceived. Antenna 1 in ♀, flagellum with 5 joints, 1st not very large, accessory flagellum 2-jointed; in ♂, flagellum with 6 joints, 1st large, accessory flagellum 3-jointed, slender. Antenna 2 in ♀, ultimate joint of peduncle shorter than either of the preceding, in ♂ rather longer than either; flagellum in ♀ with 5 joints, in ♂ with many. Epistome projecting. Gnathopod 1, 6th joint nearly as long as 5th, of uniform width, palm slightly concave and oblique. Gnathopod 2, 6th joint more than half length of 5th, both distally widened. Peraeopods 3—5, 2^d joint large, overlapping 3^d, in peraeopod 5 very large, much longer than rest of limb, in peraeopods 3—5 4th joint expanded. Uropod 3, rami rather broad, outer the longer, with setae also in ♀. Telson cleft not quite to the centre, with 2 pairs of marginal spinules. L. 5 mm.

Southern Indian Ocean (Kerguelen Island). Depth 230 m.

3. **L. clypeatum** Chevreux 1888 *L. c.*, Chevreux in: Bull. Soc. zool. France, *v.* 13 p. 40 | 1893 *Anonyx longicornis* (part.), A. Della Valle in: F. Fl. Neapel, *v.* 20 p. 814.

♀. Carina as in L. umbo (p. 80). Head, lateral angles little produced, acute. Side-plate 5 of equal breadth and depth. Pleon segment 3, postero-lateral angles acute, carina of this and segment 4 produced into a sharp strong tooth. Eyes inconspicuous. Antenna 1, 1st joint very stout, produced over half length of 2^d. Antenna 2 subequal to antenna 1, flagellum in both short, with long cilia. Gnathopod 1, 6th joint a little longer than 5th, palm oblique. Gnathopod 2, 6th joint half length of 5th, distally dilated, angle of palm produced. Peraeopods 3 and 4, 2^d joint expanded, serrate on front margin, produced below 3^d joint, 4th much dilated and produced. Peraeopod 5, 2^d joint very large, produced almost to the end of the 6th joint. Telson very elongate, cleft to the base. L. 5 mm. — ♂ unknown.

North-Atlantic (lat. 46° N.). Depth 180 m.

4. **L. longicorne** (Bate & Westw.) 1861 *Anonyx longicornis*, Bate & Westwood, Brit. sess. Crust., *v.* 1 p. 91 f. | 1862 *A. l.*, Bate, Cat. Amphip. Brit. Mus., p. 72 t. 11 f. 4 | 1893 *A. l.* (part.), A. Della Valle in: F. Fl. Neapel, *v.* 20 p. 814 t. 60 f. 47—49 | 1888 *Lepidepecreum longicorne*, T. Stebbing in: Rep. Voy. Challenger, *v.* 29 p. 373 | 1868 *L. carinatum* + *Anonyx longicornis*, Bate & Westwood, Brit. sess. Crust., *v.* 2 p. 509, 510 f. | 1891 & 95 *L. c.*, G. O. Sars, Crust. Norway, *v.* 1 p. 113 t. 38 f. 2; t. 39 f. 1; p. 687 | 1890 *L. mirabile*, Meinert in: Udb. Hauchs, *v.* 3 p. 153 t. 1 f. 7—12.

Dorsal carina extending from 2^d joint of antenna 1 back to 4th segment of pleon. Head slightly rostrate, lateral angles produced to large slightly deflexed linguiform lobes. Side-plates 4—6 carinate near the base, 5th quadrate, not deeper than broad. Pleon segment 3, postero-lateral angles quadrate, dorsal carina of this and of segment 4 acutely produced. Eyes oval, larger in ♂ than in ♀, reddish with opaque white coating. Antenna 1 in ♀, 1st joint large, its dorsal carina .jutting out over 2^d joint, which on a smaller scale is similar, flagellum short, with 7 joints, 1st not large, accessory flagellum wanting. Antenna 1 in ♂ a little larger, 2^d joint scarcely produced, flagellum with 9 joints, 1st very large (Fig. 12). Antenna 2 in ♀ shorter than antenna 1, ultimate joint of peduncle shorter than penultimate or antepenultimate, flagellum 4-jointed. Antenna 2 in ♂, ultimate joint of peduncle as long as either of 2 preceding joints, flagellum as long as body. Epistome large, nearly rectangular. Gnathopod 1, 6th joint shorter than 5th, of nearly uniform width. Gnathopod 2 very slender, 6th joint more than half length of 5th, oblong triangular. Peraeopod 3, 2^d joint rounded, 4th expanded. Peraeopods 4 and 5, 2^d joint oval, narrowed below, 4th expanded. Uropod 3, rami very nearly equal (Fig. 13), in ♀ without setae. Telson narrow and long, nearly 3 times as long as broad, deeply cleft, apices closely adjacent, acute (Fig. 13). Colour chalky white, with dorsal orange patch on each segment. L. ♀ 7, ♂ 8 mm.

North-Atlantic, North-Sea, Skagerrak, Kattegat and Baltic (South-Norway, Denmark, France, Great Britain); Mediterranean.

5. **L. umbo** (Goës) 1866 *Lysianassa u.*, Goës in: Öfv. Ak. Förh., *v.* 22 p. 520 t. 37 f. 6 | 1871 *Orchomene u.*, A. Boeck in: Forh. Selsk. Christian., 1870 p. 117 | 1882 *Lepidepecreum u.*, G. O. Sars in: Forb. Selsk. Christian., nr. 18 p. 81 | 1891 *L. u.*, G. O. Sars, Crust. Norway, *v.* 1 p. 115 t. 39 f. 2 | 1893 *Anonyx u.*, A. Della Valle in: F. Fl. Neapel, *v.* 20 p. 815.

Dorsal carina extending from 1st joint of antenna 1 to 4th pleon segment. Head, lateral angles acutely produced. Side-plates large and deep, 5th oval, much deeper than broad, lower hind corner produced downward, the centre forming a conspicuous boss. Pleon segment 3, postero-lateral angles quadrate, minutely produced, carinate process of this and segment 4 acutely upturned. Eyes narrow oblong, red. Antenna 1, only 1st joint produced at apex, flagellum in ♀ with 8 joints, 1st of moderate size, accessory flagellum 3-jointed, very small. Antenna 2 in ♀ nearly as long as antenna 1, flagellum 8-jointed. Antenna of ♂ modified as in L. longicorne. Epistome obtusely rounded in front. Gnathopod 1, 6th joint about as long as 5th, slightly widened distally. Gnathopod 2, 5th joint considerably expanded distally, and 6th rather stouter than in L. longicorne. Peraeopods 3—5, 2^d joint not narrowed below, 4th not expanded. Uropod 3, inner ramus notably shorter than outer; outer in ♀ carrying 2 setae. Telson oblong triangular, scarcely twice as long as broad, deeply cleft, apices acute. Colour bright carneous red. L. ♀ 11 mm.

Arctic Ocean, North-Atlantic (Southernmost range, Brönösund [Nordland]). Depth 58—188 m.

48. Gen. **Orchomenella** O. Sars

1871 *Orchomene* (part.) + *Tryphosa* (part.), A. Boeck in: Forh. Selsk. Christian., 1870 p. 114, 117 | 1890 & 95 *Orchomenella* (part.), G. O. Sars, Crust. Norway, v. 1 p. 66; p. 683 | 1894 *O.*, T. Stebbing in: Bijdr. Dierk., v. 17 p. 6.

Side-plates large. Antenna 1, accessory flagellum moderately developed. Antenna 2 in ♀ little longer than antenna 1, in ♂ not greatly elongated. Epistome little or not projecting beyond upper lip. Mouth-organs nearly as in Orchomene (p. 44), molar of mandible rather stronger, outer plate of maxilla 1 more obliquely truncate, inner and outer plates of maxillipeds shorter and palp longer. Gnathopods, peraeopods and uropod 3 in ♀ as in Orchomene, uropod 3 in ♂ nearly the same as in ♀, telson larger than in Orchomene, oblong triangular, cleft extending beyond the centre, scarcely dehiscent.

5 species.

Synopsis of species:

1 ⎰ Epistome not at all prominent — **2.**
  ⎱ Epistome rather prominent — **4.**

2 ⎰ Eyes Wanting . . . . 1. **O. laevis.** . . . . p. 81
  ⎱ Eyes present — **3.**

3 ⎰ Side-plate 5 scarcely produced doWnWard behind . . 2. **O. nanus** . . . p. 81
  ⎱ Side-plate 5 produced doWnWard behind . . . . . 3. **O. pinguis** . . . p. 82

4 ⎰ Side-plate 5 deeper than broad, produced doWnWard 4. **O. minuta** . . . p. 82
  ⎱ Side-plate 5 broader than deep. not produced . . . 5. **O. groenlandica** p. 83

1. **O. laevis** Bonnier 1896 *O. l.*, J. Bonnier in: Ann. Univ. Lyon, v. 26 p. 617 t. 35 f. 5.

Pleon segment 4 with deep dorsal depression. Head, lateral corners sharply produced. Pleon segment 3, postero-lateral corners strongly produced. apparently subacute. Eyes entirely wanting. Antenna 1 in ♂ short, 2^d and 3^d joints very short, flagellum without calceoli, with 11 joints, 1st not very long, shorter than accessory flagellum. Antenna 2 nearly as long as body, with a calceolus on each joint. Epistome with crest not projecting in front of upper lip. Lower lip, mandibular processes acute. Maxilla 1, 2^d joint of palp with 6 spines on apical margin. Maxilla 2 (in fig.), inner plate shorter than outer. Maxillipeds, outer plates fully reaching end of palp's 2^d joint, inner margin crenulate, apical carrying a spine and 4 setae. Gnathopod 1, 5th joint subequal to 6th, 6th oblong, the short oblique palm defined by a spine at the angle. Gnathopod 2, 5th joint not nearly twice as long as the subchelate 6th. Peraeopods 1 and 2, finger longer than 6th joint. Peraeopods 3—5 successively shorter (text. but in fig. 4th the longest), 2^d joint oblong oval in peraeopods 3 and 4, much more expanded in peraeopod 5, very broad and longer than the remaining joints combined. Uropods 1 and 2 short, rami as long as peduncle; uropod 3 much longer, inner ramus with 3 spines on inner margin, outer much longer, 2-jointed, with 3 spines on outer margin. Telson. deeply cleft, with a pair of lateral and a pair of apical spines. L. 5 mm.

Bay of Biscay. Depth 950 m.

2. **O. nanus** (Krøyer) 1846 *Anonyx n.*, Krøyer in: Naturh. Tidsskr., ser. 2 v. 2 p. 30 | 1846 *A. n.*, Krøyer in: Voy. Nord, Crust. t. 17 f. 2 a—t | ? 1893 *A. n.* (part.), A. Della Valle in: F. Fl. Neapel, v. 20 p. 820 | 1871 *Tryphosa n.*, A. Boeck in: Forh. Selsk. Christian., 1870 p. 117 | 1893 *T. nana*, J. Bonnier in: Bull. sci. France Belgique, v. 24 p. 170, 191 t. 7 | 1895 *Orchomenella n.* (*Anonyx namus*, laps.!), G. O. Sars, Crust. Norway, v. 1 p. 683 | 1882 *Tryphosa ciliata*, G. O. Sars in: Forh. Selsk. Christian., nr. 18 p. 81 t. 3 f. 4 | 1891 *Orchomenella c.*, G. O. Sars, Crust. Norway, v. 1 p. 69 t. 25 f. 2.

Head, lateral angles broadly rounded in ♀, narrower in ♂. Side-plate 1 of uniform width, 5th of equal breadth and depth, lower hinder angle scarcely produced. Pleon segment 3, postero-lateral angles rounded, hind margin smooth, 4th segment with deep dorsal depression. Eyes large, oval, widened below, light red. Antenna 1, flagellum in ♀ with 8 joints, in ♂ with 12, 1st large and hirsute, accessory flagellum with 3 joints, 1st long, slightly dilated, hirsute. Antenna 2 in ♀ little longer than antenna 1, in ♂ about once and a half as long. Epistome flattened in front, not projecting beyond upper lip. Gnathopod 1 rather robust, 6th joint a little longer than 5th, slightly tapering, palm transverse. Gnathopod 2, 6th joint oval, fully half as long as 5th, obtusely produced beneath the minute finger. Peraeopods 3—5, 2^{d} joint nearly as long as rest of leg. Uropod 3 in ♀, inner ramus not longer than basal joint of outer. Telson rather broad, triangular, with 1 pair of marginal spinules, cleft narrow, extending beyond the centre. Colour greyish white. L. ♀ over 5 mm.

North-Atlantic, North-Sea and Skagerrak (South-Norway, France, Holland, Great Britain).

3. **O. pinguis** (Boeck) 1861 *Anonyx p.*, A. Boeck in: Forb. Skand. Naturf., Møde 8 p. 642 | 1893 *A. p.*, A. Della Valle in: F. Fl. Neapel, *v.* 20 p. 821 t. 28 f. 22—35 | 1871 *Orchomene pingvis*, A. Boeck in: Forh. Selsk. Christian., 1870 p. 115 | 1872 & 76 *O. p.*, A. Boeck, Skand. Arkt. Amphip., *v.* 1 t. 5 f. 1; *v.* 2 p. 176 | 1890 & 95 *Orchomenella p.*, G. O. Sars, Crust. Norway, *v.* 1 p. 67 t. 24 f. 2; p. 683 | 1893 *Tryphosa pinguis*, J. Bonnier in: Bull. sci. France Belgique, *v.* 24 p. 196.

Head, lateral angles narrowly rounded, produced. Side-plate 1 a little expanded or narrowed (Della Valle) below, 5th deeper than wide, produced downward behind. Pleon segment 3, postero-lateral angles scarcely rounded, hind margin minutely crenulate or smooth (Della Valle). Eyes narrow, reniform, light red. Antenna 1, flagellum with 8—11 joints, 1st large, hirsute, accessory flagellum with 4 joints. Antenna 2 in ♀ decidedly longer than antenna 1, flagellum with 15 joints, in ♂ with 20 joints. Epistome scarcely projecting in front of upper lip, defined from it by distinct incision. Gnathopod 1 rather short and stout, 6th joint much longer than 5th, tapering to the transverse palm. Peraeopods 3—5, 2^{d} joint shorter than the rest of the limb. Uropod 3, inner ramus scarcely as long as basal joint of outer, which in ♀ has 3 setae on inner margin. Telson tapering, with 2 pairs of marginal spinules, cleft usually extending beyond the centre, scarcely dehiscent. Colour whitish. L. ♀ 4—7·5 mm.

Arctic Ocean, North-Atlantic, North-Sea and Skagerrak (Siberia, South- and West-Norway, Malangenfjord [Finmark]); Mediterranean.

4. **O. minuta** (Krøyer) 1846 *Anonyx minutus*, Krøyer in: Naturh. Tidsskr., ser. 2 *v.* 2 p. 23 | 1846 *A. m.*, Krøyer in: Voy. Nord, Crust. t. 18 f. 2 a—t | 1893 *A. m.*, A. Della Valle in: F. Fl. Neapel, *v.* 20 p. 826 | 1866 *Lysianassa minuta*, Goës in: Öfv. Ak. Förh., *v.* 22 p 520 | 1871 *Orchomene minutus*, A. Boeck in: Forh. Selsk. Christian., 1870 p. 116 | 1893 *O. m.*, J. Bonnier in: Bull. sci. France Belgique, *v.* 24 p. 194 | 1890 & 95 *Orchomenella minuta*, G. O. Sars, Crust. Norway, *v.* 1 p. 66 t. 24 f. 1; p. 683 | 1895 *O. m.*, Ohlin in: Acta Univ. Lund., *v.* 31 nr. 6 p. 22 | 1894 *O. minutus*, T. Stebbing in: Bijdr. Dierk., *v.* 17 p. 6.

Head, lateral angles in ♀ almost quadrate, in ♂ rather narrow and acute. Side-plate 1 of uniform width, 5th deeper than broad, produced downward behind in triangular lobe. Pleon segment 3, postero-lateral angles a very little produced, hind margin smooth, 4th segment with shallow dorsal depression. Eyes oval, slightly widened below, light red. Antenna 1, flagellum in ♀ with 10 joints, accessory flagellum with 4 or 5. Antenna 2 in ♀ little longer than antenna 1, flagellum with 10—12 joints, much longer in ♂. Epistome slightly projecting in front of upper lip, evenly rounded. Gnathopod 1 not robust, 6th joint scarcely longer than 5th, palm very slightly oblique. Gnathopod 2, palm produced beneath the minute finger. Peraeopods 3—5 unusually short, 2^{d} joint large, as long as rest of limb, 7th joint very small. Uropod 3, inner ramus as long as basal joint of outer, which in ♀ has 1 seta on inner margin. Telson nearly twice as long as broad, carrying 2 pairs of marginal spinules, cleft extending beyond the centre, apically dehiscent. Colour pale yellowish red, with orange speck at corners of each segment. L. ♀ 6 (Sars), reaching 11 mm (Ohlin).

Arctic Ocean, North-Atlantic, North-Sea and Skagerrak (whole Norway). British and Mediterranean localities remain doubtful. Depth 2—112 m.

5. **O. groenlandica** (H. J. Hansen) 1887 *Anonyx groenlandicus*, H. J. Hansen in: Vid. Meddel, ser. 4 v. 9 p. 72 t. 2 f. 5—5g | 1893 *A. g.*, A. Della Valle in: F. Fl. Neapel, v. 20 p. 832 | 1891 & 95 *Orchomenella groenlandica*, G. O. Sars, Crust. Norway, v. 1 p. 70 t. 26 f. 1; p. 684 | 1893 *Orchomene g.*, J. Bonnier in: Bull. sci. France Belgique, v. 24 p. 194.

Head, lateral corners slightly projecting, subangular (Sars), rounded (Hansen). Side-plate 1 narrowed below, 5th broader than deep, not produced behind. Pleon segment 3, angles upturned, acute, 4th with deep depression and low carina ending acutely. Eyes oval, widened below, visual elements imperfectly developed, pigment dark red with opaque white coating. Antenna 1, flagellum in ♀ with 8, in ♂ with 14 joints, 1st rather large, accessory flagellum with 4 joints, 1st long. Antenna 2 in ♀ scarcely longer than antenna 1, not nearly twice as long in ♂. Epistome considerably projecting in front of upper lip (as in Orchomene, p. 44), evenly rounded. Mandible rather stout. Gnathopod 1, 6th joint longer than 5th, scarcely tapering, palm transverse. Gnathopod 2 subchelate, 6th joint half length of 5th, slightly produced at angle of palm. Peraeopods 3—5 of moderate length, 2^{d} joint large, 4th little widened. Uropod 2, inner ramus constricted. Uropod 3, inner ramus in ♀ nearly as long as outer, in ♂ somewhat larger, with both rami setose on inner margin. Telson oblong oval, nearly twice as long as broad, with 2 pairs of marginal spinules, cleft narrow, extending beyond the centre. Colour whitish, pellucid. L. 7 mm.

Arctic Ocean. Depth 19—94 m.

49. Gen. **Orchomenopsis** O. Sars

1891 *Orchomenopsis*, G. O. Sars, Crust. Norway, v. 1 p. 73 | 1893 *O.*, J. Bonnier in: Bull. sci. France Belgique, v. 24 p. 174.

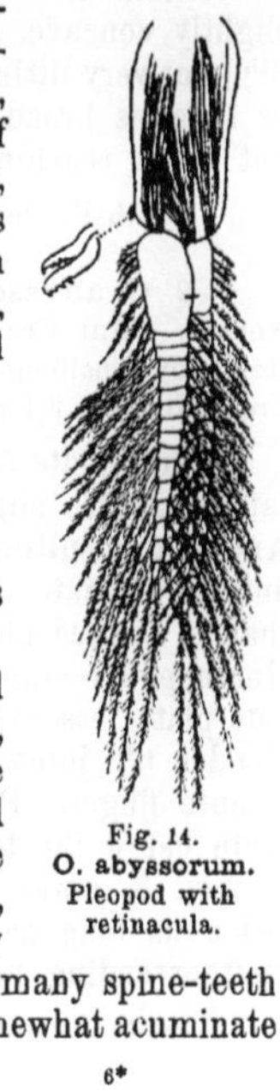

Head, lateral angles rounded. Side-plate 1 expanded below, 4th with rather short obtuse expansion. Pleon segment 3, postero-lateral angles not produced. Antenna 1, peduncle stout, 2^{d} and 3^{d} joints very short, accessory flagellum well developed. Antenna 2 notably longer than antenna 1. Epistome not projecting. Mandible strong, molar weak, much ciliated, palp set far back, its 1st joint short. Maxilla 1, 2 setae on narrow inner plate, 11 spines on broad obliquely truncate outer plate, many spine-teeth on apex of palp. Maxilla 2, plates rather long and narrow, somewhat acuminate

Fig. 14.
O. abyssorum.
Pleopod with
retinacula.

6*

apically. Maxillipeds, outer plates oval, with little nodulous teeth on inner margin, length of plates compared with joints of palp a little variable. Gnathopod 1 powerful, subchelate, 6th joint considerably longer than 5th, palm transverse, well defined. Gnathopod 2, 5th joint rather expanded, 6th narrower, about half length of 5th. Peraeopods strong, 2^d joint of peraeopods 3—5 moderately expanded, narrowed below. Uropod 3, rami projecting beyond uropod 2, setiferous. Telson more or less tapering and more or less deeply cleft.

4 species.

Synopsis of species:

1 $\begin{cases}$ Cleft of telson not quite reaching the centre. . . . 1. **O. musculosa** . p. 84 \\ Cleft of telson extending beyond the centre — **2.** $\end{cases}$

2 $\begin{cases}$ Gnathopod 2 minutely chelate . . . . 2. **O. abyssorum** . p. 84 \\ Gnathopod 2 subchelate — **3.** $\end{cases}$

3 $\begin{cases}$ Cleft of telson not dehiscent . . . . 3. **O. obtusa** . . . p. 85 \\ Cleft of telson dehiscent . . . . 4. **O. zschauii** . . p. 85 $\end{cases}$

1. **O. musculosa** (Stebb.) 1888 *Orchomene musculosus*, T. Stebbing in: Rep. Voy. Challenger, v. 29 p. 673 t. 20 | 1891 *O. m.. Orchomenopsis* (part.), G. O. Sars. Crust. NorWay, v. 1 p. 74 | 1893 *Anonyx m.*, A. Della Valle in: F. Fl. Neapel, v. 20 p. 823.

Side-plate 5 deeper behind than in front. Pleon segment 3, postero-lateral angles much rounded. Antenna 1, flagellum with 11 joints, 1st large, accessory flagellum with 4 joints, 1st long. Antenna 2, ultimate and penultimate joints of peduncle subequal, flagellum with 13 joints. Maxilla 1, one palp with 11, the other with 8 spine-teeth. Maxillipeds, inner plates with apical margin excavate, outer plates reaching beyond 2^d joint of palp, with 2 spine-teeth on apical margin. Gnathopod 1, 2^d joint short and massive, not longer than 5th and 6th combined, 5th short, cup-shaped, 6th oblong, thickest near base, hind margin slightly concave, palm transverse, with accurately fitting finger. Gnathopod 2, 6th joint very little apically produced. Uropod 3, rami nearly equal. Telson twice as long as broad, with a pair of dorsal and a pair of apical spinules, cleft not quite reaching centre, slightly dehiscent throughout. L. about 12 mm.

North-Pacific (South of Japan). Surface.

2. **O. abyssorum** (Stebb.) ?1865 *Anonyx chilensis*, Cam. Heller in: Reise NoVara, v. 2 III Crust. p. 129 t. 11 f. 5 | 1888 *Orchomene abyssorum*, T. Stebbing in: Rep. Voy. Challenger, v. 29 p. 676 t. 21 | 1891 *O. a., Orchomenopsis* (part.). G. O. Sars, Crust. NorWay, v. 1 p. 74 | 1893 *Anonyx a.*, A. Della Valle in: F. Fl. Neapel, v. 20 p. 824.

Side-plate 5 not deeper behind than in front. Pleon segment 3, postero-lateral angles slightly rounded. Antenna 1 similar to that of O. musculosa. Antenna 2, ultimate joint of peduncle shorter than penultimate, equal to antepenultimate, flagellum with 15 joints. Mandible, 2^d joint of palp longer than in O. musculosa, but mouth-organs of the two species in close agreement. Gnathopod 1 similar to that of O. musculosa, but of slighter construction, the side-plate less expanded below, and the joints more slender. Gnathopod 2, the slender 6th joint strongly produced at apex to form a little chela with the minute finger. Pleopods, coupling spines very small, yet with 3 retroverted teeth below the terminal hook; 6 or 7 cleft spines on inner ramus (Fig. 14). Uropod 3, outer ramus pretty distinctly longer than inner. Telson nearly twice as long as broad, with 3 pairs of marginal and 1 of apical spinules, cleft extending beyond the middle, dehiscent. L. about 9 mm.

South-Atlantic (Buenos Ayres), depth 3578 m; South-Pacific (Chili)?.

3. O. obtusa O. Sars 1891 & 95 *O. o.*, G. O. Sars, Crust. Norway, *v.* 1 p. 74 t. 26 f. 2; p. 684 | 1893 *Anonyx obtusus*, A. Della Valle in: F. Fl. Neapel, *v.* 20 p. 824.

Side-plate 5 not deeper behind than in front. Pleon segment 3, postero-lateral angles smoothly rounded, 4th forming behind the dorsal depression a rounded carina. Eyes rather large, irregularly oval, light red. Antenna 1 nearly as in preceding species, flagellum with 10 joints, accessory flagellum with 5 joints. Antenna 2, ultimate joint of peduncle shorter than penultimate, flagellum with 18 joints. Maxillipeds, inner plates with apical margin not excavate. Gnathopod 1, intermediate in robustness between the 2 preceding species, structure similar, except 3[d] joint slightly longer. Gnathopod 2, 6th joint not sufficiently produced to be considered chelate. Uropod 3, inner ramus very little shorter than outer. Telson about twice as long as broad, regularly tapering, with 6 or 7 pairs of dorsal and marginal and 1 of apical spinules, cleft extending beyond the middle, at no part dehiscent. Colour uniformly whitish. L. ♀ 12 mm.

Trondhjemsfjord, depth at about 188 m; Stavangerfjord, depth 753 m.

4. O. zschauii (Pfeff.) 1888 *Anonyx z.*, Pfeffer in: Jahrb. Hamburg. Anst., *v.* 5 p. 87 t. 2 f. 1 | 1893 *A. z.*, A. Della Valle in: F. Fl. Neapel, *v.* 20 p. 823 | 1888 *Orchomene cavimanus*, T. Stebbing in: Rep. Voy. Challenger, *v.* 29 p. 679 t. 22 | 1891 *O. excavatus* (err., pro: *cavimanus*) + *Anonyx zschauii*, G. O. Sars, Crust. Norway, *v.* 1 p. 74, 88.

Side-plate 5 not deeper behind than in front. Pleon segment 3, postero-lateral angles rounded, 4th forming behind the dorsal depression a carinate point lifted above the 5th segment. . Eyes large, wider below than above. Antenna 1, flagellum with 13—20 joints, accessory flagellum with 5—7 joints. Antenna 2, ultimate joint of peduncle shorter than penultimate, equal to antepenultimate, flagellum with 15—18 joints. Epistome appears to be somewhat projecting. Maxilla 1 with inner plate elongate, and having (Pfeffer) about 14 spine-teeth on apex of palp. Maxilla 2, inner plate very little shorter than outer. Maxillipeds, inner plates not apically excavate, outer plates scarcely reaching end of 2[d] joint of palp. Gnathopod 1 nearly as in O. musculosa. Gnathopod 2, 5th joint wide, 6th joint with the little palm excavate, leaving a cavity when the small finger is closed upon it. Uropod 3, inner ramus considerably shorter than outer. Telson twice as long as broad, with 2 pairs of marginal and 1 pair of apical spinules, cleft extending for ³/₄ of the length, dehiscent. L. 10—15·5 mm.

South-Atlantic (South Georgia); southern Indian Ocean (Kerguelen Island, depth 228 m).

Lysianassidarum genera dubia et species dubiae.

Alibrotus M.-E. 1830 *Lysianassa* (part.), H. Milne Edwards in: Ann. Sci. nat., *v.* 20 p. 364 | 1840 *Alibrotus* (Sp. un.: *A. chauseicus*), H. Milne Edwards, Hist. nat. Crust., *v.* 3 p. 23 | 1846 *Halibrotus* (nom. emend.), L. Agassiz, Nomencl. zool., Index p. 14, 171.

Anonyx albus Gosse 1850 *A. a.* (nom. nud.), A. White in: List Brit. An. Brit. Mus., *v.* 4 p. 50 | 1855 *A. a.*, Gosse, Man. mar. Zool., *v.* 1 p. 139 f. 261; p. 142 | 1857 *A. a.*, A. White, Hist. Brit. Crust., p. 169.

North-Atlantic (Great Britain).

A. amaurus Giles 1888 *A. a.*, G. M. Giles in: J. Asiat. Soc. Bengal, *v.* 57 p. 220 t. 6 f. 1.

Blind. Side-plates subequal in depth. Gnathopod 1 strongly subchelate. All limbs short. Colour ivory white. L. 12 mm.

Bay of Bengal (Burmah). Depth 2478 m, in water-logged seeds.

A. annulatus Bate 1862 *A. a.*, (Stimpson in MS.) Bate, Cat. Amphip. Brit. Mus., p. 79 t. 13 f. 3.

Perhaps belonging to Tryphosa (p. 68). L. 6 mm.

North-Pacific (Japan).

A. brocchii Catta 1875 *A. b.*, Catta in: Rev. Sci. nat., *v.* 4 p. 164.

Mediterranean (Marseilles). Depth 25—30 m.

A. elegans W. Thomps. 1847 *A. e.*, W. Thompson in: Ann. nat. Hist., *v.* 20 p. 243.

Perhaps identical with Orchomene batei (p. 45). L. 12 mm.

North-Atlantic (Ireland).

A. exiguus Stimps. 1853 *A. e.*, Stimpson in: Smithson. Contr., *v.* 6 nr. 5 p. 51 | 1862 *A. e.*, Bate, Cat. Amphip. Brit. Mus., p. 75 t. 12 f. 3.

Fundy Bay (Grand Manan). Depth 14—27 m.

A. femoratus Pfeff. 1888 *A. f.*, Pfeffer in: Jahrb. Hamburg. Anst., *v.* 5 p. 93 t. 2 f. 2.

Possibly belonging to Onisimus (p. 25). L. about 12 mm.

South-Atlantic (South Georgia).

A. filiger Stimps. 1864 *A. f.*, Stimpson in: P. Ac. Philad., p. 157 | 1888 *A. f.*, T. Stebbing in: Rep. Voy. Challenger, *v.* 29 p. 351.

To Hippomedon (p. 58)?, or Lepidepecreum (p. 78)? L. 8 mm.

Puget Sound.

A. indicus Giles 1890 *A. i.*, G. M. Giles in: J. Asiat. Soc. Bengal, *v.* 59 p. 69 t. 2 f. 5, 5a.

If the palp of mandible is really only 2-jointed, the species requires a new genus. L. about 5 mm.

Bay of Bengal (Seven Pagodas near Madras). Depth 9—18 m.

A. nardonis Heller 1866 *A. n.*, Cam. Heller in: Denk. Ak. Wien, *v.* 26 II p. 26 t. 2 f. 17, 18 | 1876 *A. n.*, *Tryphosa* (part.)?. A. Boeck, Skand. Arkt. Amphip., *v.* 2 p. 180 | 1893 *A. n.*, *Ichnopus* (part.)?, A. Della Valle in: F. Fl. Neapel, *v.* 20 p. 836.

Excluded from Anonyx and Tryphosa by the gnathopod 1, from Ichnopus by the antennae; perhaps a Socarnes (p. 56). L. 4—5 mm.

Adriatic.

A. nobilis Stimps. 1853 *A. n.*, Stimpson in: Smithson. Contr., *v.* 6 nr. 5 p. 50 | 1862 *A. n.*, Bate, Cat. Amphip. Brit. Mus., p. 76 t. 12 f. 5.

Fundy Bay (Grand Manan). Low water mark.

A. pallidus Stimps. 1853 *A. p.*, Stimpson in: Smithson. Contr., *v.* 6 nr. 5 p 50 | 1862 *A. p.*, Bate, Cat. Amphip. Brit. Mus., p. 81.

Fundy Bay (Grand Manan). Depth 7—36 m.

A. politus Stimps. 1853 *A. p.*, Stimpson in: Smithson. Contr., *v.* 6 nr. 5 p. 50 | 1862 *A. p.*, Bate, Cat. Amphip. Brit. Mus., p. 80.

Fundy Bay (Grand Manan). Depth 72 m.

A. punctatus Bate 1862 *A. p.*, (Stimpson in MS.) Bate, Cat. Amphip. Brit. Mus., p. 78 t. 13 f. 2.

L. 13 mm.

Behring Strait.

A. schmardae Heller 1866 *A. s., Ichnopus s.* + *A. filicornis* (♂), Cam. Heller in: Denk. Ak. Wien, *v.* 26 II p. 21 t. 2 f. 29—33; p. 23 t. 3 f. 13—16 | ? 1893 *I. s.*, A. Della Valle in: F. Fl. Neapel, *v.* 20 p. 803 t. 5 f. 4; t. 27 f. 23—32.

Della Valle's species differs from that of Heller in gnathopod 1, uropod 3 and telson, and is excluded from Ichnopus by the branchial vesicles with accessory lobes on one side only, by uropod 3 with 1-jointed outer ramus and the large 1st joint of flagellum in antenna 1.

Adriatic, Gulf of Naples.

A. sp., Bate 1862 *A. plautus* (err., non Krøyer 1845!), Bate, Cat. Amphip. Brit. Mus., p. 78 t. 13 f. 1.

Approaching to Nannonyx spinimanus (p. 35). L. 5 mm.

North-Sea (Scotland).

A. sp., Della Valle 1893 *A. nanus* (err., non Krøyer 1846!), A. Della Valle in: F. Fl. Neapel, *v.* 20 p. 820 t. 28 f. 36—42.

Mediterranean.

A. sp., G. M. Thoms. 1882 *A. exiguus* (err., non Stimpson 1853!), G. M. Thomson in: Tr. N. Zealand Inst., *v.* 14 p. 232 t. 18 f. 2.

Paterson Inlet [Stewart Island near New Zealand]. Depth 14 m.

Lysianassa affinis Hasw. 1880 *L. a.*, Haswell in: P. Linn. Soc. N.S. Wales, *v.* 4 p. 255 | 1885 *L. a.*, Haswell in: P. Linn. Soc. N.S. Wales, *v.* 10 p. 99 t. 12 f. 5, 6.

Perhaps ♂ of Lysianassa nitens (p. 88).

Port Jackson [East-Australia].

L. australiensis Hasw. 1880 *L. a.*, Haswell in: P. Linn. Soc. N.S. Wales, *v.* 4 p. 323 t. 18 f. 3 | 1885 *L. a.*, Haswell in: P. Linn. Soc. N.S. Wales, *v.* 10 p. 99 t. 12 f. 3, 4.

Port Jackson [East-Australia].

L. brasiliensis Dana 1852 *L. b.*, J. D. Dana in: P. Amer. Ac., *v.* 2 p. 208 | 1853 & 55 *L.? b.*, J. D. Dana in: U. S. expl. Exp., *v.* 13 II p. 914; t. 62 f. 1 a—f.

Tropical Atlantic (Rio Janeiro).

L. chauseica M.-E. 1830 *L. c.*, H. Milne Edwards in: Ann. Sci. nat., *v.* 20 p. 365 | 1840 *Alibrotus chauseicus*, H. Milne Edwards, Hist. nat. Crust., *v.* 3 p. 23.

Golfe de St. Malo (Iles Chausay).

L. cymba Goës 1866 *L.? c.*, Goës in: Öfv. Ak. Förh., *v.* 22 p. 521 t. 38 f. 7 | 1872 *L. c.*, A. Boeck, Skand. Arkt. Amphip., *v.* 1 p 118 | 1893 *L. c.*, A. Della Valle in: F. Fl. Neapel, *v.* 20 p. 849.

Arctic Ocean (Spitzbergen). Depth 9 m.

L. fisheri Lockington 1877 *L. f.*, Lockington in: P. Calif Ac., *v.* 7 p. 48.

Behring Sea (Alaska).

L. marina Bate 1857 *L. m.*, Bate in: Ann. nat. Hist., ser. 2 *v.* 19 p. 138 | 1862 *L. atlantica* (part.), Bate, Cat. Amphip. Brit. Mus., p. 68 t. 10 f. 10.

Plymouth Sound; North-Sea (Banff).

L. martensi Goës 1866 *L. m.*, Goës in: Öfv. Ak. Förh., *v.* 22 p. 519 t. 37 f. 2 | 1871 *Anonyx m.*, A. Boeck in: Forh. Selsk. Christian., 1870 p. 109 | 1872 *A. m.*, A. Boeck, Skand. Arkt. Amphip., *v.* 1 p. 156 | 1893 *A. m.*, A. Della Valle in: F. Fl. Neapel, *v.* 20 p. 836.

Nearer to Anonyx lilljeborgii (p. 55) than to A. nugax (p. 54). L. 7·5 mm.

Arctic Ocean (Spitzbergen). Depth 38 m.

L. nitens Hasw. 1880 *L. n.*, Haswell in: P. Linn. Soc. N.S. Wales, *v.*4 p.255
t.8 f.5|1885 *Anonyx n.*, Haswell in: P. Linn. Soc. N.S. Wales, *v.*10 p.98 t.12 f.1, 2|
1893 *A. n.*, A. Della Valle in: F. Fl. Neapel, *v.*20 p.836.

L. 6 mm.

Port Jackson [East-Australia].

L. woodmasoni Giles 1890 *L. w.*, G. M. Giles in: J. Asiat. Soc. Bengal, *v.*59
p.68 t.2 f.4.

Agreeing in several points with Podoprion bolivari (p.18). L. 8 mm.

Macpherson's Strait (Andaman Islands). Coral sand, depth 30 m.

Orchomene grimaldii Chevreux 1890 *O. g.*, Chevreux in: Bull. Soc. zool.
France, *v.* 15 p.164.

Separated from Orchomene by the entire telson, from Nanuonyx by the subchelate gnathopod 1, from Onisimus by having the rami of uropod 3 shorter than
the peduncle. Mouth-organs not described.

Mediterranean (Monaco). Depth 475 m.

Stenia Dana 1849 *S.* (non Guenée 1846, Lepidoptera, fide: A. Marschall,
Nomencl. zool., p.313), J. D. Dana in: Amer. J. Sci., ser.2 *v.*8 p.136 | 1852 *S.* (Sp.
un.: *S. magellanica*), J. D. Dana in: P. Amer. Ac., *v.*2 p.209.

S. magellanica Dana 1852 *S. m.*, J. D. Dana in: P. Amer. Ac., *v.*2 p.209 |
1888 *Anonyx magellanicus*, T. Stebbing in: Rep. Voy. Challenger, *v.*29 p.266 | 1853
& 55 *A. fuegiensis*, J. D. Dana in: U. S. expl. Exp., *v.*13 II p.919; t.62 f.4 a—p | 1893
A. f., A. Della Valle in: F. Fl. Neapel, *v.*20 p.836.

Epistome and telson agree with Onisimus (p.25), but head and mouth-organs
in general and gnathopod 1 with Tryphosa (p.68).

Good Success Bay [Tierra del Fuego].

---

2. Fam. **Stegocephalidae**

1852 Subfam. *Stegocephalinae*, J. D. Dana in: Amer. J. Sci., ser.2 *v.*14 p.310 |
1853 Subfam. *S.*; J. D. Dana in: U. S. expl. Exp., *v.*13 II p.907 | 1882 *Stegocephalidae*,
G. O. Sars in: Forh. Selsk. Christian., nr. 18 p.23 | 1886 Subfam. *Lysianassina* (part.),
Gerstaecker in: Bronn's Kl. Ordn., *v.*5 II p.499 | 1888 *Stegocephalidae*, T. Stebbing in:
Rep. Voy. Challenger, *v.*29 p.727 | 1891 *S.*, G. O. Sars, Crust. Norway, *v.*1 p.196 | 1893
Gammaridi (part.), A. Della Valle in: F. Fl. Neapel, *v.*20 p.620.

Head short, retiring. Peraeon broad, deflexed in front. Side-plate 4
large (Fig. 18). Antenna 1, accessory flagellum 1- or 2-jointed (Fig. 19).
Antenna 2 seldom much longer than antenna 1. Upper lip bilobed. Lower
lip without inner lobes. Mandible (Fig. 17, 21) without molar or palp,
accessory plate on one of the pair. Maxilla 1 (Fig. 15, 20), inner plate with
numerous setae, outer usually with 9 principal spines. Maxilla 2 (Fig. 16),
inner plate very broad and setose, outer narrow. Maxillipeds, plates broad rather
than long, last 2 joints of palp narrow. Gnathopods 1 and 2 not subchelate.
Peraeopod 3, 2^d joint not expanded. Uropod 3 biramous. Telson small.

Marine.

9 genera, 12 accepted species. 2 obscure.

Synopsis of the genera:

1 { Mandible (Fig. 17, 21) denticulate — **2**.
  { Mandible not denticulate — **6**.

2 { Telson cleft — **3**.
  { Telson not cleft — **5**.

3 { Maxilla 1 (Fig. 15), palp 2-jointed . . . . 1. Gen. **Phippsia** . . . . . . p. 89
  { Maxilla 1, palp 1-jointed — **4**.

4 { Peraeopod 4, 2^d joint expanded . . . . . 2. Gen. **Stegocephalus** . . . p. 90
  { Peraeopod 4, 2^d joint not expanded . . . 3. Gen. **Stegocephaloides** . . p. 91

5 { Maxillipeds, inner plates of normal size . 4. Gen. **Andaniopsis** . . . . p. 92
  { Maxillipeds, inner plates very short . . . 5. Gen. **Andaniella** . . . . p. 93

6 { Telson not cleft — **7**.
  { Telson cleft — **8**.

7 { Maxilla 1, palp 2-jointed . . . . 6. Gen. **Andaniexis** . . . . p. 94
  { Maxilla 1 (Fig. 20), palp 1-jointed . . . . 7. Gen. **Parandania** . . . . p. 95

8 { Pleon segment 6 very elongate . . . . . 8. Gen. **Andaniotes** . . . . p. 96
  { Pleon segment 6 not very elongate . . . 9. Gen. **Euandania** . . . . . p. 97

1. Gen. **Phippsia** Stebb.*)

1793 Subgen. *Gammarellus* (part.), J. F. W. Herbst, Naturg. Krabben Krebse, v. 2 p. 106 | 1891 *Aspidopleurus* (non F. J. Pictet & A. Humbert 1866, Pisces!), G. O. Sars, Crust. Norway, v. 1 p. 203 | 1893 *A.*, A. Della Valle in: F. Fl. Neapel, v. 20 p. 632.

Side-plate 4 immensely expanded, overlapping side-plates 5 and 6. Eyes wanting. Epistome projecting as rounded compressed lobe. Upper lip slightly and centrally notched. Lower lip, lobes narrow, not strongly dehiscent. Mandible, cutting edge narrow, finely dentate. Maxilla 1 (Fig. 15), inner plate setose on inner margin of narrowed distal half, palp very small, but 2-jointed. Maxilla 2 (Fig. 16), inner plate long, distally rather widened, outer plate wide apart, linear, tipped with long hooked spines.

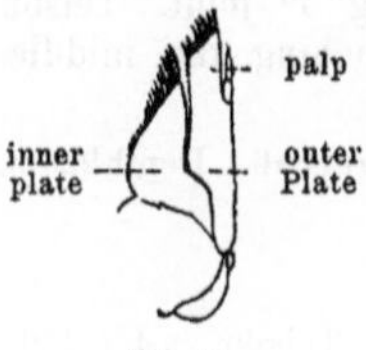

Fig. 15.
P. gibbosa.
Maxilla 1.
[After
G. O. Sars.]

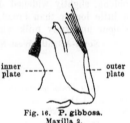

Fig. 16. P. gibbosa.
Maxilla 2.
[After G. O. Sars.]

Maxillipeds, inner plates narrow, obliquely truncate, outer reaching beyond acutely produced 2^d joint of palp. Gnathopod 1 rather stouter than gnathopod 2. Peraeopod 4, 2^d joint linear. Peraeopod 5, 2^d joint expanded. Uropod 3, rami lanceolate, unarmed. Telson oval, more or less cleft.

2 species.

Synopsis of species:

Head with acute rostrum . . . . 1. **P. ampulla** . . . p. 89
Head without distinct rostrum . . . . 2. **P. gibbosa** . . . p. 90

1. **P. ampulla** (Phipps) 1774 *Cancer a.*, Phipps, Voy. North Pole, p. 191* t. 12 f. 3 | 1781 *Gammarus a.*, J. C. Fabricius, Spec. Ins., v. 1 p. 515 | 1793 *Cancer (Gammarellus) a.*, J. F. W. Herbst, Naturg. Krabben Krebse, v. 2 p. 116 t. 35 f. 1 | 1840 *Lysianassa? a.*, H. Milne Edwards, Hist. nat. Crust., v. 3 p. 22 | 1866 *Stegocephalus a.* (part.), Goës

---

*) Nom. nov. After Constantine John Phipps. — The name *Aspidopleurus* is preoccupied (1866, F. J. Pictet & A. Humbert, Poiss. foss. Liban, p. 107).

in: Öfv. Ak. Förb., *v.*22 p.521 t.38 f.9 (not f.8) | 1893 *Aspidopleurus a.*, A. Della Valle in: F. Fl. Neapel, *v.* 20 p. 633 t. 59 f. 45 | 1880 *Stegocephalus kessleri* (nom. nud.), Stuxberg in: Bih. Svenska Ak., *v.*5 nr.22 p. 65, 72 | 1882 *S. k.*, Stuxberg in: Vega-Exp., *v.*1 p.713 f. | 1891 *S.k.*, *Aspidopleurus* (part.)?, G. O. Sars, Crust. Norway, *v.*1 p.204.

Head with conical acute down-bent rostrum. Side-plate 4 with the great backward expansion moderately deep. Pleon segment 3 not dorsally produced, postero-lateral corners much curved upward to a subacute point. Mouth-organs undescribed. Peraeopod 4, the narrow 2^d joint with lower hind corner ending in a little rounded lobe (Goës). Telson oval, a little longer than broad, cleft dehiscent, not quite reaching the middle. Colour almost white (J. C. Fabricius). — L. attaining 56·5 mm.

Arctic Ocean.

2. **P. gibbosa** (O. Sars) 1882 *Stegocephalus gibbosus*, G. O. Sars in: Forh. Selsk. Christian., nr. 18 p. 85 t. 3 f. 7 | 1891 *Aspidopleurus g.*, G. O. Sars, Crust. Norway, *v.* 1 p.204 t.71 f.1 | 1893 *A. g.*, A. Della Valle in: F. Fl. Neapel, *v.* 20 p. 634 t. 59 f. 46, 47.

Head without rostrum. Side-plate 4 with lower margin horizontal, backward expansion very deep, rounded behind. Pleon segment 3 produced at the end dorsally to a gibbous projection, pointed at the tip, postero-lateral corners quadrate below, acutely upturned above. Antenna 1, 1st joint of peduncle scarcely longer than 2^d and 3^d combined, flagellum with 6 joints, 1st not very elongate, accessory flagellum small. Antenna 2 subequal to antenna 1, ultimate joint of peduncle rather longer than penultimate. Gnathopods 1 and 2, 6th joint narrow, tapering, scarcely longer than 5th. Peraeopod 4, 2^d joint with lower hind angle rounded, not produced. Peraeopod 5, 2^d joint oblong, slightly widened distally, rounded end overlapping 3^d joint. Telson a little longer than broad, cleft dehiscent, not nearly reaching the middle. Colour uniformly milk white. L. ♀ 8 mm.

North-Atlantic (West-Norway). Among living Lophelia prolifera (Pall.). Depth 226 m.

2. Gen. **Stegocephalus** Krøyer

1842 *Stegocephalus* (Sp. un.: *S. inflatus*), Krøyer in: Naturh. Tidsskr., *v.*4 p.150 | 1891 *S.*, G. O. Sars, Crust. Norway, *v.*1 p.197 | 1893 *S.* (part.), A. Della Valle in: F. Fl. Neapel, *v.* 20 p. 626.

Head more or less rostrate. Side-plate 4 deep, completely overlapping the 5th, lower margin evenly curved. Eyes wanting. Antenna 1 with bundles

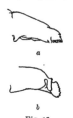

of sensory setae on flagellum, accessory flagellum with 2^d joint minute. Epistome flattened in front. Upper lip deeply and unsymmetrically bilobed. Lower lip, lobes dehiscent with incurved apical tooth. Mandible (Fig. 17), cutting edge coarsely dentate, accessory plate of left mandible finely serrate. Maxilla 1, inner plate fringed with setae all along oblique upper margin, outer carrying about a score of spines, palp 1-jointed, tipped with a few spines. Maxilla 2, inner plate broad, with numerous spines, outer linear, standing apart, tipped with short hooked spines. Maxillipeds, plates apically rounded, outer large, broad, slightly serrate on inner margin, not reaching beyond 2^d joint of palp. Gnathopod 1 rather stouter than gnathopod 2, 6th joint narrow in both, tapering only in gnathopod 2. Peraeopods 4

Fig. 17.
S. inflatus.
Mandible *a* right, *b* left hand.

and 5, 2^d joint expanded. Uropod 3, rami lanceolate, minutely denticulate. Telson triangular, cleft, unarmed.

2 species.

Synopsis of species:
Pleon segment 3, lower margin serrate, hinder smooth . . . . 1. **S. inflatus** . . p. 91
Pleon segment 3, lower margin smooth, hinder serrate . . . . 2. **S. similis** . . p. 91

1. **S. inflatus** Krøyer 1842 *S. i.*, Krøyer in: Naturh. Tidsskr., *v.* 4 p. 150 | 1845 *S. i.*, Krøyer in: Naturh. Tidsskr., ser. 2 *v.* 1 p. 522 t. 7 f. 3 ; 1887 *S. i.*, H. J. Hansen in: Dijmphna Udb., p. 218 t. 21 f. 10—10 c | 1888 *S. i.*, T. Stebbing in: Rep. Voy. Challenger, *v.* 29 p. 728, 1721 t. 137 A | 1891 *S. i.*, G. O. Sars, Crust. Norway, *v.* 1 p. 198 t. 69 | 1893 *S. i.*, A. Della Valle in: F. Fl. Neapel, *v.* 20 p. 627 t. 59 f. 32—34 | 1855 *S. ampulla* (err., non *Cancer a.* Phipps 1774!), T. Bell in: Belcher, Last arct. Voy., *v.* 2 p. 406 t. 35 f. 1 | 1866 *S. a.* (part.), Goës in: Öfv. Ak. Förh.. *v.* 22 p. 521 t. 38 f. 8 (not f. 9).

Head with deflexed rostrum reaching end of 1st joint of antenna 1, lateral corners rounded, somewhat projecting. Peraeon segment 1 not so long as 2^d and 3^d combined. Side-plate 4 twice as deep as segment, as broad as side-plates 1—3 combined. Pleon segment 3, lower margin serrate, hinder smooth, angles acute. Antenna 1, 1st joint nearly twice as long as 2^d and 3^d combined, with slight setulose expansion; flagellum with 12 joints, 1st not elongate, shorter than accessory flagellum. Antenna 2 rather longer than antenna 1, ultimate joint of peduncle longer than penultimate, flagellum slender, subequal to peduncle. Gnathopod 1, 6th joint as long as the slightly dilated 5th. Gnathopod 2, 6th joint sublinear, longer than 5th. Peraeopod 4, 2^d joint oblong quadrangular, lower hind corner quadrate. Peraeopod 5, 2^d joint much expanded, hind margin very convex, serrulate, lower corner acute. Uropod 3, rami nearly twice as long as peduncle. Telson nearly twice as long as broad, ending acutely, cleft beyond centre. Colour yellowish with brown patches, legs and antennae banded. L., fully extended, 47 mm in rare instances.

Arctic Ocean; North-Atlantic (West-Norway, Shetland Isles, Nova Scotia). Depth 188—376 m.

2. **S. similis** O. Sars 1891 *S. s.*, G. O. Sars, Crust. Norway, *v.* 1 p. 200 t. 70 f. 1 ; 1893 *S. s.*, A. Della Valle in: F. Fl. Neapel, *v.* 20 p. 627 t. 50 f. 35.

Head with very short obtuse rostrum, lateral corners rounded, very slightly projecting. Peraeon segment 1 longer than 2^d and 3^d combined. Side-plate 4 not nearly twice as deep as segment, nor as broad as side-plates 1—3 combined. Pleon segment 3, lower margin smooth, hinder serrate, angle not produced. Antenna 1 as in S. inflatus, but flagellum with 9 joints, 1st longer than accessory flagellum. Antenna 2 scarcely longer than antenna 1, ultimate and penultimate joints of peduncle subequal, flagellum shorter than peduncle. Gnathopod 1, 5th joint much dilated, shorter than 6th. Peraeopods 4 and 5, 2^d joint oval, longer in 5th pair than in 4th. Uropod 3 as in S. inflatus. Telson not nearly twice as long as broad, ending subacutely, cleft beyond centre. Straw-coloured, mottled with brown and reddish spots. L. ♀ 12 mm.

Arctic Ocean and North-Atlantic (West-Norway). Often but not always with S. inflatus.

3. Gen. **Stegocephaloides** O. Sars

1891 *Stegocephaloides*, G. O. Sars, Crust. Norway, *v.* 1 p. 201 | 1893 *S.* (part.), A. Della Valle in: F. Fl. Neapel, *v.* 20 p. 629.

Like Stegocephalus (p. 90) in mouth-organs and general appearance (Fig. 18), but side-plate 4 with lower and hinder margins meeting in an angle; 6th smaller

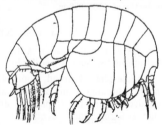

Fig. 18. **S. christianiensis**, ♀. Lateral view. [After G. O. Sars.]

· than 7th, greatly narrowed below. Antenna 1, peduncle rather flattened, flagellum with 4 joints, 1st very large, last spiniform. Gnathopods 1 and 2 subequal and similar, 6th joint oblong, tapering distally, rather longer than 5th, finger very short. Peraeopod 4, 2[d] joint linear; peraeopod 5, 2[d] joint large, produced downward in an acute lobe. Uropod 3, rami unarmed. Telson cleft. ·

2 species.

Synopsis of species:

Peraeopod 5, apex of 2[d] joint not very acute nor
produced below 4th . . . . 1. S. christianiensis. . . p. 92
Peraeopod 5, apex of 2[d] joint very acute, produced
much below 4th . . . . 2. S. auratus . . . . . . p. 92

1. S. christianiensis (Boeck) 1871 Stegocephalus c., A. Boeck in: Forh. Selsk. Christian., 1870 p.128 | 1876 S. c., A. Boeck, Skand. Arkt. Amphip., v.2 p.424 t.8 f.4; t.9 f.1 | 1891 Stegocephaloides c., G. O. Sars, Crust. Norway, v.1 p.202 t.70 f.2 | 1893 S. c., A. Della Valle in: F. Fl. Neapel, v.20 p.631 t.59 f.41.

Head with very short rostrum, lateral angles acute (Fig. 18). Side-plate 4 nearly as broad as deep. Pleon segment 3, postero-lateral angles produced, minutely bidentate. Antenna 1, accessory flagellum scarce half as long as laminar 1st joint of flagellum. Antenna 2, ultimate and penultimate joints of peduncle subequal, flagellum short, 8- or 9-jointed. Peraeopod 5, 2[d] joint as long as rest of limb, overlapping 4th joint, hind margin serrate, angularly curved, obtusely pointed below. Telson oblong oval, moderately tapering, cleft nearly to the middle. Colour tessellated with angular patches of pigment spots, dark greenish brown. L. ♀ normally 7 mm; North-Atlantic specimens more than twice this size, perhaps therefore a different species (Sars).

Arctic Ocean, North-Atlantic, North-Sea and Skagerrak (Bohuslän, South- and West-Norway). Depth 38—188 m.

2. S. auratus (O. Sars) 1882 Stegocephalus a., G. O. Sars in: Forh. Selsk. Christian., nr.18 p.86 t.3 f.8 | 1891 Stegocephaloides a., G. O. Sars, Crust. Norway, v.1 p.203 t.70 f.3 | 1893 S. a., A. Della Valle in: F. Fl. Neapel, v.20 p.631 t.59 f.42.

Head with minute rostrum, lateral angles narrowly rounded. Side-plate 4 much deeper than broad. Pleon segment 3, postero-lateral angles produced, minutely serrate. Antenna 1, accessory flagellum more than half as long as 1st joint of peduncle. Peraeopod 5 short, 2[d] joint much longer than rest of limb, overlapping 5th joint, hind margin serrate, smoothly curved, acutely pointed below. Telson subtriangular, sharply tapering, cleft beyond the middle. Colour semipellucid, with broad orange band round the middle. L. ♀ 5 mm.

Arctic Ocean, North-Atlantic and North-Sea (North- and West-Norway). Solitary, in depths of 150—376 m.

4. Gen. Andaniopsis O. Sars

1871 Andania (part.), A. Boeck in: Forb. Selsk. Christian., 1870 p.128 | 1891 Andaniopsis (Sp. un.: A. nordlandica), G. O. Sars, Crust. Norway, v.1 p.208.

Side-plate 4 broad, completely overlapping side-plate 5. Eyes distinct, small. Antennae 1 and 2 subequal; antenna 1, flagellum 4-jointed. Epistome rounded in front. Upper lip unsymmetrically bilobed. Mandible, cutting edge straight, finely denticulate. Maxilla 1, inner plate with many setae, outer rather narrow, palp small, 1-jointed. Maxilla 2, inner plate short, broad, with many setae or spines, outer short, sublinear, tipped with spines.

Maxillipeds, inner plates transversely truncate, reaching end of palp's 1st joint, outer rather large, finely spinulose inside, palp very slender. Gnathopods 1 and 2 similar, 6th joint narrow, tapering, rather longer than 5th, finger slight, smooth. Peraeopod 4, 2d joint a little dilated below. Peraeopod 5 much shorter than preceding pairs, 2d joint greatly expanded, decurrent lobe obtusely pointed, 4th joint expanded, decurrent. Uropod 3, rami narrow, unarmed, outer 2-jointed. Telson very small, triangular, entire.

1 species.

1. **A. nordlandica** (Boeck) 1871 *Andania n.*, A. Boeck in: Forb. Selsk. Christian., 1870 p. 129 | 1876 *A. n.*, A. Boeck. Skand. Arkt. Amphip., *v.* 2 p. 428 t. 9 f. 3 | 1891 *Andaniopsis n.*, G. O. Sars. Crust. Norway. *v.* 1 p. 209 t. 72 f. 2 | 1893 *Stegocephaloides nordlandicus*, A. Della Valle in: F. Fl. Neapel, *v.* 20 p. 630 t. 59 f. 40.

Head not rostrate, lateral corners quadrate. Peraeon segment 1 scarcely as long as 2d and 3d combined. Side-plate 4 nearly as broad as deep. Pleon segment 3, postero-lateral angles quadrate. Eyes narrow, oblong, white, tinged with red below. Antenna 1, 1st joint of flagellum very large, accessory flagellum scarcely more than half its length. Antenna 2, ultimate joint of peduncle shorter than penultimate; flagellum as long as peduncle. Uropod 3, rami longer than peduncle, 2d joint of outer ramus subequal to 1st. Telson rather broader than long, apex obtusely pointed. Colour yellowish, transversely banded with mottling of dark brown and reddish spots. L. ♀ 5 mm.

Arctic Ocean, North-Atlantic and North-Sea (West-Norway). Depth 38—188 m.

5. Gen. **Andaniella** O. Sars

1891 *Andaniella* (Sp. un.: *A. pectinata*), G. O. Sars, Crust. Norway, *v.* 1 p. 210.

Side-plate 4 broad, completely overlapping 5th. Eyes wanting. Antenna 1 longer than antenna 2, flagellum 4-jointed. Epistome produced in front to an acuminate lappet. Upper lip rather wide, unsymmetrically bilobed. Lower lip with the lobes rather narrow (with a bidentate process, Aurivillius), apically rounded. Mandible strong, cutting edge not straight, coarsely dentate. Maxilla 1. inner plate rather small, with several setae, outer large, with 9 strong spines, palp very small, 1-jointed (with an obscure suture near the base, Aurivillius). Maxilla 2, inner plate short, broad, outer not extremely narrow, with 4 strong apical spines. Maxillipeds, inner plates very short, only reaching base of palp, outer large, reaching much beyond 2d joint of slender, slightly armed palp. Gnathopod 1 shorter than gnathopod 2, 6th joint longer than 5th, finger strong, curved, armed with spines. Peraeopod 4, 2d joint linear. Peraeopod 5 much shorter than preceding pairs, 2d joint oval, decurrent part rounded, 4th joint not much expanded or decurrent. Uropod 3, rami shorter than peduncle, outer 2-jointed, longer than inner. Telson small, triangular, entire.

1 species.

1. **A. pectinata** (O. Sars) 1882 *Andania p.*, G. O. Sars in: Forh. Selsk. Christian., nr. 18 p. 86 t. 3 f. 9 a, b | 1885 *A. p.*, C. W. S. Aurivillius in: Vega-Exp., *v.* 4 p. 226 t. 7 f. 1—12 | 1888 *Stegocephalus pectinatus*, T. Stebbing in: Rep. Voy. Challenger, *v.* 29 p. 557 | 1891 *Andaniella pectinata*, G. O. Sars, Crust. Norway, *v.* 1 p. 211 t. 72 f. 3 | 1893 *Stegocephaloides pectinatus*, A. Della Valle in: F. Fl. Neapel, *v.* 20 p. 630 t. 59 f 39.

Body capable of rolling up into a ball. Head not rostrate, lateral corners quadrate. Peraeon segment 1 little longer than 2d. Pleon segment 3, postero-lateral corners slightly rounded. Antenna 1, peduncle compressed, flagellum, 1st joint longer than other 3 combined, twice as long as accessory flagellum. Antenna 2, ultimate joint of peduncle much shorter than the stout penultimate, flagellum very short. Gnathopod 1, 6th joint slightly dilated in the middle, finger with 4 spines. Gnathopod 2, 6th joint sublinear, finger with 2 spines. Peraeopod 5, 2d joint oval, scarcely longer than rest of limb, hind margin smooth reaching middle of slightly decurrent 4th joint. Telson much broader than long. Colour yellowish, mottled with light brown spots. L. ♀ 4 mm.

Arctic Ocean and North-Atlantic (Norway, Greenland, Spitzbergen). Sometimes in branchial cavity of Ascidians.

6. Gen. **Andaniexis** Stebh.*)

1871 *Andania* (non F. Walker 1860, Lepidoptera!) (part.), A. Boeck in: Forh. Selsk. Christian., 1870 p. 128 | 1876 *A.*, A. Boeck. Skand. Arkt. Amphip., *v.* 2 p. 426 | 1888 *A.* (part.), T. Stebbing in: Rep. Voy. Challenger, *v.* 29 p 30 | 1891 *A.*, G. O. Sars, Crust. Norway, *v.* 1 p. 206 | 1893 *A.*, A. Della Valle in: F. Fl. Neapel, *v.* 20 p. 632.

Side-plate 4 not completely overlapping 5th. Eyes distinct, but without any trace of visual elements. Antennae 1 and 2 subequal; flagellum of antenna 1 slender, 5-jointed, accessory flagellum as long as 1st joint of primary. Epistome rounded in front. Upper lip small, slightly emarginate. Lower lip, lobes dehiscent, narrowly rounded with small dentiform process. Mandible broad, cutting edge straight, simple. Maxilla 1, inner plate broad, densely setose, outer with strong denticulate spines, palp well developed, 2-jointed. Maxilla 2, inner plate short, broad, with many plumose setae, outer short, narrow, with simple setiform spines. Maxillipeds, inner plates reaching little beyond base of palp, obliquely truncate, outer short, rounded, with curved spinules on inner margin, palp robust, rapidly tapering. Gnathopod 1 rather robust, 6th joint oblong oval, longer than 5th. Gnathopod 2, 6th joint linear, much longer than 5th. Peraeopod 4, 2d joint moderately expanded, oblong oval, 6th joint long. Peraeopod 5, 2d joint much expanded, partly overlapping the rather small 4th joint. Uropod 3, rami narrow, unarmed, shorter than peduncle, 2d joint of outer ramus much shorter than 1st. Telson small, triangular, as long as broad, entire.

1 accepted species, 1 incompletely known.

1. **A. abyssi** (Boeck) 1871 *Andania a.*, A. Boeck in: Forh. Selsk. Christian., 1870 p. 129.| 1876 *A. a.*, A. Boeck, Skand. Arkt. Amphip., *v.* 2 p. 426 t. 9 f. 2 | 1891 *A. a.*, G. O. Sars, Crust. Norway, *v.* 1 p 207 t. 71 f. 2; t. 72 f. 1 | 1893 *A. a.*, A. Della Valle in: F. Fl. Neapel, *v.* 20 p. 632 t. 59 f. 43—44.

Head not rostrate, lateral angles rounded, projecting. Peraeon segment 1 nearly as long as 2d and 3d combined. Side-plate 4 obliquely quadrangular, deeper than broad. Pleon segment 3 smooth, postero-lateral angles produced, not acute. Eyes conspicuous, subquadrate, milky-white. Antenna 1, 1st joint large, somewhat flattened; flagellum, 1st joint rather long, 5th spiniform. Antenna 2, ultimate joint of peduncle much longer than penultimate, flagellum 9-jointed. Peraeopod 5, 2d joint much longer than rest of limb, hind margin evenly curved, serrate. Colour brownish grey, with transverse bands of spots, partly dark greenish. L. ♀ 7 mm.

Arctic Ocean and North-Atlantic (West-Norway); Christiániafjord. Depth 376—752 m.

*) Nom. nov. — The name *Andania* is preoccupied (1860, Francis Walker, List Lep. Brit. Mus., *v.* 20 p. 57, 223).

A. spinescens (Alcock) 1894 *Andania s.*, A. Alcock in: Ann. nat. Hist., ser. 6 *v.* 13 p. 411 f.

Head concealed under peraeon segment 1; pleon segments 1—4 carinate, apically dentate. L. nearly 40 mm.

Bay of Bengal (lat. 10⁰ N., long. 86⁰ E.) Depth 3646 m.

7. Gen. **Parandania** Stebb.

1899 *Parandania* (Sp. typ.: *Andania boecki*), T. Stebbing in: Ann. nat. Hist., ser. 7 *v.* 4 p. 206.

Side-plate 4 as broad as deep, not deeper than the segment, only in part overlapping the 5ᵗʰ. Peraeon segment 1 as long as 2ᵈ and 3ᵈ combined. Pleon rather large. Pleon segment 3, postero-lateral angles produced, not acute. Eyes wanting. Antennae very unequal, flagella many-jointed. Antenna 1, 1ˢᵗ joint of flagellum and accessory flagellum long (Fig. 19); antenna 2, ultimate joint of peduncle longer than penultimate. Epistome carinate. Upper lip short, emarginate. Lower lip, lobes very broad, distally truncate, with denticle at outer corner. Mandible short and broad, cutting edge very broad, straight, smooth. Maxilla 1, inner plate large with a score of plumose setae, outer plate with 9 spines, palp 1-jointed (Fig. 20). Maxilla 2, inner plate broad, with many spines and setae, outer narrow, apically fringed, and with 2 setules on outer margin. Maxillipeds, inner plates broad, inflated, setose, not reaching apex of 1ˢᵗ joint of palp, outer not nearly reaching apex of 2ᵈ joint of palp, fringed with slender spines; palp's first 2 joints robust. Gnathopod 1 robust, 6ᵗʰ joint oval, tapering distally, rather longer than the stouter 5ᵗʰ, 7ᵗʰ linear, curved. Gnathopod 2

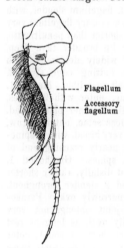

Fig. 19. P. boecki. Antenna 1.

inner plate

Fig. 20.
P. boecki.
Maxilla 1.

like gnathopod 1, but longer and less robust, 3ᵈ joint in both pairs short. Peraeopods 1—5, 4ᵗʰ joint little expanded, long, except in peraeopod 5. Peraeopod 4, 2ᵈ joint oblong oval, slightly decurrent. Peraeopod 5, 2ᵈ joint more expanded and decurrent than in peraeopod 4. Uropods 1—3, all the rami subequal, except in the 1ˢᵗ pair slightly armed, in each pair much shorter than peduncle. Telson a little longer than broad, oval, entire.

1 species.

1. **P. boecki** (Stebb.) 1888 *Andania b.*, T. Stebbing in: Rep. Voy. Challenger, *v.* 29 p. 735 t. 36 | 1893 *Stegocephalus boeckii*, A. Della Valle in: F. Fl. Neapel, *v.* 20 p. 628 t. 59 f. 36 | 1899 *Andania boecki, Parandania* sp. typ., T. Stebbing in: Ann. nat. Hist., ser. 7 *v.* 4 p. 206.

Head short. Antenna 1 (Fig. 19), peduncle short and thick, flagellum with 14 joints, 1ˢᵗ longer than rest combined, much longer than peduncle, channelled to receive accessory flagellum, which is 1-(2-?)jointed and nearly as long as 1ˢᵗ joint of primary. Antenna 2 much longer than antenna 1, ultimate joint of peduncle more than twice as long as penultimate, flagellum 25-jointed, longer than peduncle. Branchial vesicles large, even on peraeopod 5 well developed. Uropods 1 and 2 with numerous spines and setae on peduncles, rami ·carinate and spinulose. Uropod 3, peduncle unarmed, rami lanceolate, not carinate. L. 22 mm.

Tropical-Atlantic (Pernambuco). Depth 1234 m.

8. Gen. **Andaniotes** Stebb.

1897 *Andaniotes* (Sp. un.: *A. corpulentus*), T. Stebbing in: Tr. Linn. Soc. London, ser. 2 *v.* 7 p. 30 | 1897 *Andaniodes*, J. V. Carus in: Zool. Anz., Bibliogr. *v.* 2 p. 622.

Side-plate 4 slightly deeper than broad, overlapping 5th and part of 6th. Peraeon segment 1 subequal to 2d and 3d combined. Pleon segment 6 longer than 4th or 5th. Eyes wanting. Antennae subequal. Antenna 1, 1st joint of peduncle very thick, longer than 2d and 3d combined, flagellum slender, with 4 joints, 1st longer than 1st of peduncle, 4th spiniform; accessory flagellum very ·small, 1-jointed. Antenna 2, ultimate joint of peduncle shorter than penultimate, flagellum short. Epistome slightly carinate. Upper lip broader than deep, faintly emarginate. Lower lip, lobes narrow, very widely dehiscent, with minute upright spine at apex. Mandible (Fig. 21), cutting edge straight, smooth, rather broad, tooth-like secondary plate on left mandible. Maxilla 1,

Fig. 21.
A. corpulentus.
Left mandible.

inner plate with 7—11 plumose setae, outer with 9 spines, palp 1-jointed, with 7 spines. Maxilla 2, inner plate broad, densely fringed with spines and plumose setae, outer narrow, with 9 setae. Maxillipeds, inner plates very broad, with 3 spine-teeth but not many setae, outer not nearly reaching end of palp's 2d joint, fringed with short spines. Gnathopod 1, 2d joint broad, especially in ♂, 6th abruptly narrowed distally, rather shorter and a good deal narrower than the 5th. Gnathopod 2 slender throughout, 3d joint elongate, 5th and 6th joints subequal, 6th narrowly oval. Peraeopods 1—5, 4th joint decurrent. Peraeopod 4, 2d joint oblong, not very widely expanded. Peraeopod 5, 2d joint very broadly oval, as long as rest of limb. Uropod 1—3 in ♂, peduncle robust, rami very short; uropod 1, outer ramus· thick, inner thin; uropod 3, rami minute, outer 2-jointed, nearly twice as long as inner. In the ♀ uropods 1—3, peduncle less robust, rami all slender, in uropod 3 subequal, nearly as long as peduncle. Telson oval, cleft nearly to middle.

1 species.

1. **A. corpulentus** (G. M. Thoms.) 1882 *Anonyx c.*, G. M. Thomson in: Tr. N. Zealand Inst., *v.* 14 p. 231 t. 17 f. 1a—f | 1897 *Andaniotes c.*, T. Stebbing in: Tr. Linn. Soc. London, ser. 2 *v.* 7 p. 31 t. 8 | 1888 *Andania abyssorum*, T.·Stebbing in: Rep. Voy. Challenger, *v.* 29 p. 739 t. 37 | 1893 *Stegocephalus a.*, A. Della Valle in: F. Fl. Neapel, *v.* 20 p. 629.

Head with small rostrum, lateral corners rounded, sides produced downward. Pleon segment 3, lateral angles narrowly rounded, dorsum bent, in ♂ distally furnished with 2 ·small humps; segment 6 doubly ridged. Antenna 2, flagellum with 8 joints. Apical narrowing of 6th joint of gnathopod 1 more abrupt in ♀ than in ♂; gnathopods 1 and 2 are furnished with various spines and setae, many serrate or plumose. Peraeopod 3, 2d joint scarcely perceptibly dilated. The sexual variation in the uropods is very marked. L. 7 mm.

South-Pacific (New Zealand). Depths 2012 (Challenger Exp.), 15 m (G. M. Thomson).

9. Gen. **Euandania** Stebb.

1899 *Euandania* (Sp. typ.: *Andania gigantea*), T. Stebbing in: Ann. nat. Hist., ser. 7 *v.* 4 p. 206.

Side-plates much shallower than the segments, 4th rather broader than deep, only in part overlapping 5th. Peraeon segment 1 as long as 2[d] and 3[d] combined. Pleon much smaller than peraeon. Pleon segment 3, postero-lateral angles produced, scarcely acute. Eyes wanting. Antennae not very unequal, flagella many-jointed. Antenna 1, 1st joint of flagellum and accessory flagellum long. Antenna 2, ultimate joint of peduncle longer than penultimate, flagellum shorter than in Parandania (p. 95). Mandible widened at the middle, cutting edge broad, straight, smooth, accessory plate of right mandible small. Lower lip, lobes widely, not deeply, dehiscent, broadly rounded, without mandibular processes. Other parts of the mouth as in Parandania, except that setae and spines on inner plate of maxillae 1 and 2 are more numerous, and there are no setules on outer margin of outer plate of maxilla 2. Gnathopods 1 and 2, peraeopods 1—5 and uropods 1—3 nearly as in Parandania. Telson small, oval, cleft less than half its length.

· 1 species.

1. **E. gigantea** (Stebb.) 1888 *Andania g.*, T. Stebbing in: Rep.Voy.Cballenger, *v.* 29 p. 730 t. 35 | 1893 *Stegocephalus giganteus*, A. DellaValle in: F. Fl. Neapel, *v.* 20 p. 628 t. 59 f. 37 | 1899 *Andania gigantea*, *Euandania sp. typ.*, T. Stebbing in: Ann. nat. Hist., ser. 7 *v.* 4 p. 206.

Head short. Peraeon segments greatly dilated. Antenna 1, peduncle short and thick, flagellum with 14 joints, 1st much longer than peduncle or rest of flagellum, grooved to receive accessory flagellum, which consists of 1 joint nearly as long as 1st of primary. Antenna 2, ultimate joint of peduncle much thinner than penultimate, twice as long, flagellum with 25 joints, shorter than peduncle. L. 38—50, height 18—38 mm.

South-Pacific (between lat. 46[o] and 47[o] S.). Depths 2926 and 3430 m.

Stegocephalidarum species dubia.

Stegocephalus latus Hasw. 1880 *S. l.*, Haswell in: P. Linn. Soc. N.S. Wales, *v.* 4 p. 252 t. 8 f. 2 | 1885 *S. l.*, Haswell in: P. Linn. Soc. N. S. Wales, *v.* 10 p. 97 t. 11 f. 7—12.

Resembling Stegocephaloides (p. 91). L. about 10 mm.

South-Pacific (Tasmania, New South Wales).

---

3. Fam. **Ampeliscidae**

1857 *Tetromatides*, Bate in: Ann. nat. Hist., ser. 2 *v.* 19 p. 139 | 1857 *Ampeliscades*, Bate in: Ann. nat. Hist., ser. 2 *v.* 20 p. 525 | 1857 Subfam. *Ampeliscini*, A. Costa in: Mem. Acc. Napoli, *v.* 1 p. 173 | 1861 Subfam. *Ampeliscides*, Bate & Westwood, Brit. sess. Crust., *v.* 1 p. 124 | 1865 Subfam. *Ampeliscina*, W. Lilljeborg in: N. Acta Soc. Upsal., ser. 3 *v.* 6 nr. 1 p. 18 | 1871 Subfam. *Ampeliscinae*, A. Boeck in: Forh. Selsk. Christian., 1870 p. 220 | 1876 *Ampeliscaidae*, A. Boeck, Skand. Arkt. Amphip., *v.* 2 p. 516 | 1882 *Ampeliscidae*, G. O. Sars in: Forh. Selsk. Christian., nr. 18 p. 29 | 1891 *A.*, G. O. Sars, Crust. Norway, *v.* 1 p. 162.

Front of head narrowly truncate, without rostrum. Side-plate 1 setose on lower margin. Pleon segments 5 and 6 coalesced (Fig. 25 p. 108). Eyes frequently 4, exceptionally 6, sometimes 2, or wanting; each eye with or without a lens-shaped thickening of its cornea; sometimes ocular pigment present without lenses. Antenna 1 attached at apex of head, without accessory appendage; antenna 2 far back below. Upper lip helmet-shaped, minutely incised; lower lip 4-lobed. Mandible (Fig. 28), maxillae 1 and 2, maxillipeds with all the normal parts well developed. Inner plate of maxilla 1 sometimes without setae, sometimes with 1 or 2. Base of maxillipeds fringed with feathered setae and the outer plates armed with strong spine-teeth on inner margin. Gnathopods 1 and 2 imperfectly subchelate, the 5th joint never shorter than the 6th; gnathopod 2 longer and more slender than gnathopod 1 (unless Ampelisca australis be an exception). Peraeopods 1 and 2 (Fig. 22) with 4th joint very large, usually longer and more setose in peraeopod 1 than in peraeopod 2, 7th joint slender, elongate. Peraeopods 3 and 4, 2^{d} joint very broad, 5th long and strongly spined, 6th and the minute 7th retroverted. Peraeopod 5 (Fig. 23, 26, 27, 29) unlike the preceding, varying in the different genera. Branchial vesicles transversely pleated. Cleft of telson (Fig. 24) generally very deep, sometimes minute. Antennae 1 and 2, uropod 3 generally, and occasionally peraeopod 5 and telson vary in the two sexes. Glutiniferous apparatus extensive.

Marine.

3 genera, 40 accepted and 7 doubtful species.

Synopsis of the genera:

$1\Big\{$ Peraeopod 5 (Fig. 23, 26), 2^{d} joint without setae between its expansion and the 3^{d} joint, 6th foliaceous, 7th lanceolate . . . . 1. Gen. **Ampelisca** . p. 98
Peraeopod 5 (Fig. 27, 29), 2^{d} joint with setae between its expansion and the 3^{d} joint, 6th narrow, 7th spiniform — 2.

$2\Big\{$ Peraeopod 5 (Fig. 27), 2^{d} joint much widened distally . . . . 2. Gen. **Byblis** . . . p. 111
Peraeopod 5 (Fig. 29), 2^{d} joint not widened distally   3. Gen. **Haploops** . . p. 116

1. Gen. **Ampelisca** Krøyer

1842 *Ampelisca* (Sp. un.: *A. eschrichtii*), Krøyer in: Naturh. Tidsskr., v. 4 p. 154 | 1888 *A.*, T. Stebbing in: Rep. Voy. Challenger, v. 29 p. 1035 | 1891 *A.*, G. O. Sars, Crust. Norway, v. 1 p. 164 | 1893 *A.* (part.), A. Della Valle in: F. Fl. Neapel, v. 20 p. 469 | 1846 *Ampelisia* (laps.), Krøyer in: Voy. Nord, Crust. t. 23 f. 1 | 1853 *Pseudophthalmus*, Stimpson in: Smithson. Contr., v. 6 nr. 5 p. 57 | 1853 *Araneops*, A. Costa in: Rend. Soc. Borbon., n. ser. v. 2 p. 169 | 1856 *Tetromatus*, Bate in: Rep. Brit. Ass., Meet. 25 p. 58 | 1857 *T.*, Bate in: Ann. nat. Hist., ser. 2 v. 19 p. 139 | 1887 *Amplisca* (laps.), Chevreux in: Bull. Soc. zool. France, v. 12 p. 574.

Head with postero-antennal corners obsolete. Side-plate 1 scarcely deeper than 2^{d}, often concealing base of antenna 2; side-plate 4 with rare exceptions (as A. odontoplax), obliquely truncated below the posterior angle (Fig. 25). Corneal lenses 4 or none. Mandibular palp with 2^{d} joint generally laminar, 3^{d} linear and rather short. Peraeopods 3 and 4 with 2^{d} joint very broad, its front edge strongly curved in the middle, 5th joint carrying a simple series of spines within the hinder margin. Peraeopod 5 (Fig. 23, 26), 2^{d} joint variously expanded so that the greatest breadth is sometimes lowest (transversely truncate), sometimes more or less median (obliquely truncate), the margin forming the lower side of the triangle being fringed with plumose setae; the following joints varying much in relative size, the 6th foliaceous, the 7th lanceolate. Uropod 3 (Fig. 24) reaching considerably beyond the others, rami foliaceous,

inner broader than outer. Telson (Fig. 24) oblong, cleft nearly to the base.
Two pairs of hepato-pancreatic caeca.

25 species accepted, 7 insufficiently described.

Synopsis of the accepted species:

1 { Peraeopod 5, 3^d joint shorter than 4th — **2.**
{ Peraeopod 5, 3^d joint longer than 4th — **10.**

2 { Eyes present — **3.**
{ Eyes Wanting — **8.**

3 { Pleon segment 3, postero-lateral margin bisi-
{ nuate — **4.**
{ Pleon segment 3, postero-lateral margin not
{ bisinuate — **7.**

4 { Telson without dorsal setules . . . .   1. **A. eschrichtii** . . . p. 100
{ Telson with dorsal setules — **5.**

5 { Peraeopod 1, 4th joint With produced distal lobe   2. **A. brevicornis** . . . p. 100
{ Peraeopod 1, 4th joint without produced distal
{ lobe — **6.**

6 { Pleon segment 4, dorsal carina little eleVated   3. **A. macrocephala** . . p. 101
{ Pleon segment 4, dorsal carina much eleVated   4. **A. gibba** . . . . . . p. 101

7 { Pleon segment 3, postero-lateral angles qua-
{ drate, antenna 1 much shorter than antenna 2   5. **A. chiltoni** . . . . . p. 102
{ Pleon segment 3, postero-lateral angles rounded,
{ antenna 1 not shorter than antenna 2 . .   6. **A. fusca** . . . . . . p. 102

8 { Antenna 1, flagellum many-jointed . . . . .   7. **A. odontoplax** . . . p. 103
{ Antenna 1, flagellum few-jointed — **9.**

9 { Peraeopod 5, 6th joint longer than 5th . . .   8. **A. uncinata** . . . . p. 103
{ Peraeopod 5, 6th joint shorter than 5th . . .   9. **A. abyssicola** . . . p. 104

10 { Gnathopod 2 shorter and stouter than gnatho-
{ pod 1 . . . .   10. **A. australis** . . . . p. 104
{ Gnathopod 2 not shorter or stouter than
{ gnathopod 1 — **11.**

11 { Antenna 2 not Very much longer than antenna 1 —**12.**
{ Antenna 2 Very much longer than antenna 1 —**18.**

12 { Cornea not thickened to a lens — **13.**
{ Cornea thickened to a lens — **15.**

13 { Antenna 2, ultimate joint of peduncle not longer
{ than penultimate . . . .   11. **A. rubella** . . . . . p. 104
{ Antenna 2, ultimate joint of peduncle longer
{ than penultimate — **14.**

14 { Pleon segment 3, angles slightly produced . .   12. **A. amblyops** . . . . p. 105
{ Pleon segment 3, angles not produced . . .   13. **A. pusilla** . . . . . p. 105

15 { Corneal lenses large . . . .   14. **A. anomala** . . . . p. 106
{ Corneal lenses small — **16.**

16 { Pleon segment 4 with eleVated carina . . . .   15. **A. spinipes** . . . . p. 106
{ Pleon segment 4, carina not eleVated — **17.**

17 { Peraeopods 1 and 2, 7th joint not longer than
{ 5th and 6th . . . .   16. **A. aequicornis** . . . p. 106
{ Peraeopods 1 and 2, 7th joint longer than 5th
{ and 6th . . . .   17. **A. serraticaudata** . p. 107

18 { Peraeopods 1 and 2, 7th joint much longer than
{ 5th and 6th — **19.**
{ Peraeopods 1 and 2, 7th joint not much longer
{ than 5th and 6th — **20.**

19 { Head obliquely truncate . . . . 18. **A. diadema** . . . . p. 107
{ Head transversely truncate . . . . 19. **A. acinaces** . . . . p. 108

20 { Antenna 2 very short . . . . 20. **A. spinimana** . . . p. 109
{ Antenna 2 not very short — 21.

21 { Cornea not forming a lens . . . . 21. **A. compacta** . . . . p. 109
{ Cornea forming a lens — 22.

22 { Pleon segment 4, dorsal apex raised above
segment 5 — 23.
{ Pleon segment 4, dorsal apex not raised above
segment 5 — 24.

23 { Peraeopod 5, finger not longer than 6th joint 22. **A. typica** . . . . . p. 109
{ Peraeopod 5, finger longer than 6th joint . . 23. **A. zamboangae** . . p. 110

24 { Head obliquely truncate . . . . 24. **A. tenuicornis** . . . p. 110
{ Head transversely truncate . . . . 25. **A. sarsi** . . . . . p. 111

1. **A. eschrichtii** Krøyer 1842 *A. e.*, Krøyer in: Naturh. Tidsskr., *v.* 4 p. 155 | 1891
A. e., *A. eschrichti*, G. O. Sars, Crust. Norway, *v.* 1 p. 174 t. 61 f. 1 | 1893 *A. eschrichtii*
(part.), A. Della Valle in: F. Fl. Neapel, *v.* 20 p. 475 | 1894 *A. e.*, T. Stebbing in: Bijdr.
Dierk., *v.* 17 p. 17 | 1862 *A. ingens* (*Pseudophthalmus* i. Stimpson in MS.), Bate, Cat.
Amphip. Brit. Mus., p. 92 t. 15 f. 2 | 1871 *A. dubia* + *A. eschrichti* + *A. propinqva*,
A. Boeck in: Forh. Selsk. Christian., p. 224, 225.

Peraeon segments 6 and 7 and pleon segments 1—4 dorsally carinate,
composite 5th and 6th of pleon with a pair of tubercles flanking an excavation.
Pleon segment 3 with margin moderately sinuous over acutely prominent
postero-lateral angles. Corneal lenses distinct, lower pair a little removed
from lower front corners of head, pigment well defined, bright red. Antenna 1 in
♀ reaching beyond peduncle of antenna 2, flagellum $2^1/_2$ times as long as
peduncle, with 30 joints. Antenna 2 twice length of antenna 1, ultimate and
penultimate joints of peduncle subequal. Peraeopods 1 and 2, 7th joint considerably longer than 5th and 6th combined. Peraeopod 5, 2d joint longer than
all the rest of the leg, produced below the 3d, expansion obliquely truncated, and
lower corner broadly rounded, 5th joint projecting anteriorly, 6th narrow, shorter
than 4th and 5th combined, 7th more than half length of 6th. Uropod 3
with rami lanceolate, fringed with plumose setae and a few spinules. Telson
oblong, with a small denticle at the side of each acute apex, without any
dorsal spinules. Body semipellucid, tinged with yellowish, mottled with
orange and pinkish pigment. L. reaching 30 mm.

Arctic Ocean, widely distributed; Skagerrak (Bohuslän). Depths to 254 m.

2. **A. brevicornis** (A. Costa) 1853 *Araneops b.*, A. Costa in: Rend. Soc. Borbon.,
n. ser. *v.* 2 p 171 | 1893 *Ampelisca b.* (part.), A. Della Valle in: F. Fl. Neapel, *v.* 20 p. 473
t. 4 f. 4; t. 37 f. 29; t. 38 f. 3. 5, 6, 9, 13; t. 43 f. 20; t. 44 f. 26—28; t. 45 f. 5—10 (but
see p. 109); t. 47 f. 5--16 | 1855 *A. laevigata*, W. Liljeborg in: Öfv. Ak. Förh., *v.* 12
p. 123 | 1891 *A. l.*, G. O. Sars, Crust. Norway, *v.* 1 p. 169 t. 59 f. 1 | 1856 *Tetromatus
bellianus*, Bate in: Rep. Brit. Ass., Meet. 25 p. 58 t. 17 f. D5 | 1857 *T. b.*, Bate in: Ann.
nat. Hist., ser. 2 *v.* 19 p. 139.

Pleon segment 3 deeply bisinuated over acutely prominent postero-
lateral angles, segment 4 with small dorsal hump at the end. Corneal lenses
small, distinct, lower pair at lower corners of head, pigment dark brownish
(Sars) or bright red (Della Valle). Antenna 1 in ♀ not (in ♂ scarcely) reaching
beyond penultimate joint of peduncle of antenna 2, flagellum with 10 joints, not
twice length of peduncle; antenna 2 in ♀ scarcely more than half length of body,
ultimate joint of peduncle much shorter than penultimate. Peraeopod 1, 4th joint

with produced lobe. Peraeopods 1 and 2, 7th joint considerably longer than 5th and 6th combined. Peraeopod 5, 2[d] joint shorter than rest of leg, the expansion transversely truncate, reaching end of 3[d] joint, 4th completely. overlapping the triangular 5th with a setose lobe behind, 6th oval, broad, as long nearly as the 3 preceding combined, 7th narrowly lanceolate about half length of 6th. Uropod 3 with rather broad foliaceous rami, carrying plumose setae on confronted edges. Telson about twice as long as broad, slightly constricted near base, triangularly tapering distally, here carrying within each outer margin a row of setules; cleft for $^2/_3$ length. Body whitish, pellucid, mottled with dark brown stars, head and front legs speckled with light yellow. L. 7—9 (Della Valle), 12 mm (Sars).

Arctic Ocean, North-Atlantic, North-Sea and Skagerrak (South- and West-Norway, northwards to Lofoten Isles, British Isles, France); Kattegat; Mediterranean.

3. **A. macrocephala** Lilj. 1852 *A. m.*, W. Liljeborg in: Öfv. Ak. Förh., *v.* 9 p. 7 | 1891 *A. m.*, G. O. Sars, Crust. Norway, *v.* 1 p. 172 t. 60 f. 1 | 1894 *A. m.*, T. Stebbing in: Bijdr. Dierk., *v.* 17 p. 17 | 1874 *A. eschrichtii* (err., non Krøyer 1842!), Buchholz in: Zweite D. Nordpolarf., *v.* 2 p. 375 Crust. t. 13 f. 1 | 1893 *A. e.* (part.), A. Della Valle in: F. Fl. Neapel, *v.* 20 p. 475.

Pleon segment 3 sinuous above acutely prominent postero-lateral angles, segment 4 with dorsal carina little elevated. Corneal lenses very small, distinct, lower pair at lower corners of head, pigment somewhat irregular, bright red. Antenna 1 in ♀ reaching end of peduncle of antenna 2, flagellum twice length of peduncle, 12-jointed. Antenna 2 in ♀ scarcely more than half length of body, ultimate joint of peduncle shorter than penultimate. Peraeopods 1 and 2, 7th joint subequal to 5th and 6th combined. Peraeopod 5, 2[d] joint subequal in length to rest of leg, expansion transversely truncate, descending below 3[d] joint with its full breadth instead of in a lobe, 4th joint forming a small setose lobe behind, 5th carrying 3 long plumose setae on the hinder angle, 6th equal in length to 4 and 5th combined. Uropod 2, outer ramus with long spine near apex. Uropod 3 with rami broadly lanceolate. Telson oblong oval, with 4 pairs of dorsal spinules and 1 pair at the blunted tip. Body whitish, pellucid, sides mottled with pinkish and yellowish specks. L. ♀ 14 mm, ♂ rather less.

Arctic Ocean; North-Atlantic, North-Sea and Skagerrak (British Isles, Whole Norway); Kattegat.

4. **A. gibba** O. Sars 1882 *A. g.*, G. O. Sars in: Forh. Selsk. Christian., nr. 18 p. 107 t. 6 f. 1, 1 a | 1891 *A. g.*, G. O. Sars, Crust. Norway, *v.* 1 p. 171 t. 59 f. 2 | 1893 *A. brevicornis* (part.), A. Della Valle in: F. Fl. Neapel, *v.* 20 p. 473.

Pleon segment 3 bisinuate not very deeply over acute, moderately prominent postero-lateral angles, segment 4 with pronounced dorsal hump at the end. Corneal lenses distinct, lower pair a little behind lower corners of head, pigment well defined, bright red. Antenna 1 in ♀ scarcely reaching beyond penultimate joint of peduncle of antenna 2, 2[d] joint more elongated than in A. brevicornis (p. 100), flagellum with 6 joints, about as long as peduncle; antenna 1 in ♂ not reaching end of peduncle of antenna 2. Antenna 2 in ♀ not quite as long as body, very slender, ultimate joint of peduncle a little shorter than penultimate. Peraeopods 1 and 2, 4th joint narrower than in A. brevicornis, without produced distal lobe, 7th longer than 5th and 6th combined. Peraeopod 5, 2[d] joint subequal to rest of leg, expansion transversely truncate, characters of limb as in A. brevicornis, but lobe of 4th joint much smaller with only

3 setae at the tip, the 7th joint ending in a little curved nail. Uropod 3
with foliaceous rami, not very broad and nearly naked. Telson in ♀ shorter
and broader than in A. brevicornis, otherwise similar, narrow in ♂. Pellucid,
except for a few yellowish specks in front, ova in pouch orange-coloured.
L. ♀ 8, ♂ 7 mm.

North-Atlantic, North-Sea and Skagerrak (South- and West-Norway, northwards
to Trondhjemsfjord). Depth 94—282 m, muddy ground.

5. **A. chiltoni** Stebb. 1888 *A. c.*, T. Stebbing in: Rep. Voy. Challenger, *v.* 29
p. 1042 t. 103 | 1893 *A. propinqua* (part.), A. Della Valle in: F. Fl. Neapel, *v.* 20 p. 484.

Head transversely truncate. Pleon segment 3 with postero-lateral
corners almost quadrate, minutely produced, segment 4 dorsally ending
angularly. Corneal lenses 4, lower pair occupying lower corners of head.
Antenna 1 in ♀ not reaching end of peduncle of antenna 2, 2^d joint scarcely
twice length of 1st, flagellum with 9 joints, somewhat longer than peduncle;
antenna 2 in ♀ about half length of body, ultimate joint of peduncle a little
shorter than penultimate, flagellum with about 25 joints, not much longer than
peduncle. Mandibular palp slender, 3^d joint shorter than 2^d. Maxilla 1 with
2 small setae on inner plate. Peraeopods 1 and 2, 7th joint longer than 5th and
6th combined. Peraeopod 5 as in A. eschrichtii (p. 100). Uropod 3 as in
A. eschrichtii. Telson oblong, with 3 pairs of dorsal spinules and 1 pair in
the blunt emarginate apices. L. 15 mm.

South-Pacific (New Zealand). Depth 282 m.

6. **A. fusca** Stebb. 1888 *A. f.*, T. Stebbing in: Rep. Voy. Challenger, *v.* 29
p. 1052 t. 105 | 1893 *A. f.*, A. Della Valle in: F. Fl. Neapel, *v.* 20 p. 483.

Head transversely truncate. Side-plates 1—3 with a tooth at lower hind
corner, 4th with lower and hind margins at right angles. Pleon segments 3

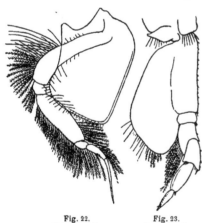

and 4 slightly carinate, segment 3
with postero-lateral angles gently
rounded. Corneal lenses 4, lower
pair removed from lower corners of
the head, projecting from lower
margin. Antenna 1 in ♀ longer than
antenna 2, 2^d joint more than twice
as long as 1st, flagellum with 34
joints, much longer than peduncle;
antenna 2 nearly as long as the
body, ultimate joint of peduncle
shorter than penultimate, flagellum
with 18 slender joints, shorter than
peduncle of antenna 2 and shorter
than flagellum of antenna 1. Mandibular palp with 2^d joint rather
broad, 3^d narrower and shorter.
Maxilla 1 with 2 small setae on
inner plate. Peraeopods 1 and 2
(Fig. 22), 7th joint much longer
than 5th and 6th combined. Peraeo-

Fig. 22.   Fig. 23.
Peraeopod 2.   Peraeopod 5.
Fig. 22 and 23. **A. fusca.**
Peraeopods 2 and 5.

pod 5 (Fig. 23), 2^d joint with expansion obliquely and evenly rounded, reaching
end of 4th joint, which with its strongly plumose hind lobe is much longer

than the 3ᵈ; the 5ᵗʰ produced in front is as long as the 4ᵗʰ; the 6ᵗʰ is a little longer than the 5ᵗʰ; the lanceolate 7ᵗʰ ³/₄ length of 6ᵗʰ. Uropod 3 (Fig. 24) with broad foliaceous rami, plumose on the confronted margins. Telson (Fig. 24) nearly twice as long as broad, cleft ⁴/₅ length, the distal half triangularly tapering to the acute apices, each carrying a spinule; on the surface within the margins are 6 or more pairs of setules. The specimens in spirit dark coloured, the branchial vesicles in particular being port-wine coloured. L. about 13 mm.

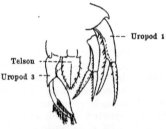

Uropod 1

Telson

Uropod 3

Southern Indian Ocean (Cape Agulhas). Depth 282 m.

Fig. 24. A. fusca.
Uropod 1—3 and telson.

7. **A. odontoplax** O. Sars 1879 *A. o.*, G. O. Sars in: Arch. Naturv. Kristian., *v.* 4 p. 454 | 1885 *A. o.*, G. O. Sars in: Norske Nordhavs-Exp., *v.* 6 Crust. I p. 196 t. 16 f. 4 | 1891 *A. o.*, G. O. Sars, Crust. Norway, *v.* 1 p. 176 t. 61 f. 2 | 1893 *A. o.*, A. Della Valle in: F. Fl. Neapel, *v.* 20 p. 485.

Head carinate dorsally. Side-plates 1—3 with distinct tooth at lower hind corner, 4ᵗʰ with hind and lower margins in a continuous curve. Pleon segments 1—4 with carina, that of 4ᵗʰ ending angularly, postero-lateral angles of 3ᵈ segment slightly produced and acute. Eyes none. Antenna 1 in ♀ about half length of body, rather more than half of antenna 2, 2ᵈ joint of peduncle twice length of 1ˢᵗ, flagellum with 32 joints, scarcely twice length of peduncle. Antenna 2, ultimate joint of peduncle shorter than penultimate. Gnathopod 1, 6ᵗʰ joint a little shorter than 5ᵗʰ. Peraeopods 1 and 2, 7ᵗʰ joint considerably longer than 5ᵗʰ and 6ᵗʰ combined. Peraeopod 5, 2ᵈ joint broad, scarcely as long as rest of leg, expansion nearly transversely truncate, reaching a little below the short 3ᵈ joint, 6ᵗʰ longer than 4ᵗʰ and 5ᵗʰ combined, 7ᵗʰ more than half length of 6ᵗʰ, narrowly lanceolate. Uropod 2 with rami very spinose on inner margin, and carrying a long terminal spine; uropod 3, rami lanceolate, fringed with setae and spines. Telson oblong oval, with 2 pairs of dorsal spinules and a spinule in the emargination of each apex. Whitish, pellucid, with no distinct pigmentation. L. ♀ 18 mm.

North-Atlantic and Arctic Ocean (Trondhjemsfjord, Helgeland, West-Finmark). Depth 267 m.

8. **A. uncinata** Chevreux 1887 *A. u.*, Chevreux in: Bull. Soc. zool. France, *v.* 12 p. 573 | 1893 *A. brevicornis* (part.), A. Della Valle in: F. Fl. Neapel, *v.* 20 p. 473.

Resembling A. gibba and A. brevicornis (p. 101, 100). Head obliquely truncate. Side-plates 1—3 with small tooth at lower hind corner. Pleon segment 3, postero-lateral angles acute and recurved, segment 4 with high compressed carina, ending angularly, following segments slightly carinate. Eyes none. Antenna 1 not reaching end of penultimate joint of peduncle of antenna 2, 2ᵈ joint ¹/₃ longer than 1ˢᵗ, flagellum 10-jointed. Antenna 2 a little longer than the body, peduncle ²/₃ of whole length, ultimate joint of peduncle a little shorter than penultimate. Peraeopods 1 and 2, 7ᵗʰ joint longer than 5ᵗʰ and 6ᵗʰ combined. Peraeopod 5, 2ᵈ joint reaching nearly to middle of 4ᵗʰ; the 4ᵗʰ with curved and angular lobe descending below the 5ᵗʰ; the 6ᵗʰ broad and oval; the 7ᵗʰ nearly as long as the 6ᵗʰ, very strongly curved in form of a hook at its extremity. Uropod 3, rami lanceolate, dilated at the base, almost entirely smooth.

Telson with 6 or 7 stiff setae on each side, its shape as in A. gibba ♀.
Colour after long preservation in spirit still keeping a green tint. L. 7 mm.

North-Atlantic (Cape Finisterre). Depth 510 m.

9. **A. abyssicola** Stebb. 1888 *A. a.*, T. Stebbing in: Rep. Voy. Challenger,
v. 29 p. 1047 t. 104 | 1893 *A. a.*, A. Della Valle in: F. Fl. Neapel, *v.* 20 p. 477.

Head transversely truncate. Side-plates 1—3 with small tooth at lower
hind corner, 4th with lower and hind margins almost continuous. Pleon segment 3
with postero-lateral corners quadrate, segment 4 impressed and with carina
dorsally ending angularly. Eyes wanting. Antenna 1 in ♀ not nearly reaching
end of peduncle of antenna 2, 2^d joint nearly twice length of 1st, flagellum
10- or 11-jointed, scarcely as long as peduncle; antenna 2 nearly as long as body,
ultimate joint of peduncle slightly longer than penultimate. Mandibular palp
with 2^d joint long and rather broad, 3^d much shorter and narrower, but not
very narrow. Maxilla 1 with two setae on inner plate. Gnathopod 1, 5th
and 6th joints stout. Peraeopods 1 and 2, 7th joint a little longer than 5th and
6th combined. Peraeopod 5, expansion obliquely rounded, reaching nearly
to end of 4th joint, 3^d and 4th joints both short, 5th longer than both combined,
produced downwards in front, 6th oval, considerably shorter than 5th and
scarcely longer than the lanceolate 7th. Uropod 3 with lanceolate rami,
serrate on both margins and furnished with spines and plumose setae. Telson
not twice as long as broad, deeply cleft, with 4 pairs of spinules within the
margins, the apices broad, each with a spinule. L. 16 mm.

Tropical Atlantic (Virgin Islands Culebra and St. Thomas). Depth 714 m.

10. **A. australis** HasW. 1880 *A. a.*, Haswell in: P. Linn. Soc. N.S.Wales, *v.* 4
p. 257 t. 9 f. 1 | 1882 *A. a.*, Haswell, Cat. Austral. Crust., p 235 | 1885 *A. a.* (part.?),
HasWell in: P. Linn. Soc. N.S.Wales, *v.* 10 p 97 t. 12 f. 7—16; t. 13 f. 1—4 | 1893 *A. a.*,
A. Della Valle in: F. Fl. Neapel, *v.* 20 p. 471.

Eyes not mentioned or figured, perhaps but not certainly to be presumed
from generic account adopted in Haswell's Catalogue. Antenna 1 not nearly
reaching end of peduncle of antenna 2; antenna 2 about twice as long as
antenna 1, ultimate joint of peduncle shorter than penultimate, flagellum with
about 10 slender joints. Gnathopod 1, 6th joint about as long as 5th. Gnathopod 2
similar to gnathopod 1, but shorter and slightly stouter [?]. Peraeopods 1
and 2, 7th joint as long as 5th and 6th combined. Peraeopod 5 with
4th joint very short and broad, 5th longer than 4th or 6th, 7th long, slender
and slightly curved (measurements agree nearly with those of peraeopod 5
in A. abyssicola). Uropod 3, rami broad lanceolate, the outer armed on one
border, the inner on both with slender setae. Telson squamiform, cleft,
rounded posteriorly. L. about 9 mm. (Figures of peraeopod 5 given by
Haswell 1885 are at variance with the descriptions of 1880, 1882.)

Port Jackson, Port Denison and Port Stephens [East-Australia]. Depth 9—11 m.

11. **A. rubella** A. Costa 1864 *A. r.*, A. Costa in: Annuario Mus. Napoli, *v.* 2
p. 153 t. 2 f. 7 | 1888 *A. r.*, T. Stebbing in: Rep. Voy. Challenger, *v.* 29 p. 346 | 1893
A. r. (part.), A. Della Valle in: F. Fl. Neapel, *v.* 20 p. 109, 482 t. 2 f. 4; t. 37 f. 21; t. 38
f. 1, 4, 10, 16; t. 45 f. 6.

Head truncate nearly transversely, but with upper corner a little
prominent. Pleon segment 3 moderately sinuous above bluntly projecting
postero-lateral angles. Corneal lenses wanting; crimson pigment not very
brilliant; rudiments of a 3^d pair of eyes conspicuous. Antenna 1 about as
long as antenna 2, peduncle reaching beyond penultimate joint of peduncle of

antenna 2, 2^d joint not quite twice as long as 1st, flagellum with 20 joints, more than twice as long as peduncle; ultimate and penultimate joints of antenna 2 equal. Peraeopods 1 and 2, 7th joint shorter than 5th and 6th combined. Feraeopod 5, 2^d joint longer than rest of leg, expansion broadly rounded, reaching end of 3^d joint, which is longer than 4th; 6th fully as long as 4th and 5th combined, itself short and broad; 7th not much shorter than 6th. Uropod 3 with broad rami. Telson having only a spinule at each apex. Colour deep red (Costa) or pearl-grey, pellucid with a slight tendency on the appendages to violet (Della Valle). L. 6 mm.

Mediterranean. Among algae.

12. **A. amblyops** O. Sars 1891 *A. a.*, G. O. Sars, Crust. Norway, *v.* 1 p. 180 t. 63 f. 1 | 1893 *A. aequicornis* (part.), A. Della Valle in: F. Fl. Neapel, *v.* 20 p. 907, 917.

Side-plate 1 not concealing base of antenna 2, 4th having hind and lower margins defined by an obtuse angle. Pleon segment 3 with posterolateral angles slightly produced, 4th with carina ending in a prominent blunted angle. Corneal lenses absent, but patches of reddish pigment present at the ocular positions. Antenna 1 in ♀ not much shorter than antenna 2, peduncle reaching beyond penultimate joint of peduncle of antenna 2, 2^d joint more than twice as long as 1st, flagellum with 25 joints, twice length of peduncle; antenna 2 not as long as body, ultimate joint of peduncle longer than penultimate. Gnathopod 1, 6th joint a little shorter than 5th. Peraeopods 1 and 2, 7th joint subequal to 5th and 6th combined. Peraeopods 3 and 4 comparatively elongate, slender. Peraeopod 5, 2^d joint considerably longer than rest of leg, expansion obliquely truncate, reaching to end of 3^d joint, which is longer than the 4th, 6th scarcely as long as 4th and 5th combined, 7th lanceolate, about half length of 6th. Uropod 3 broad, foliaceous, with a few scattered setae. Telson oblong oval, with 2 pairs of dorsal spinules and a pair at the unemarginate apices. Highly pellucid, nearly colourless. L ♀ 8 mm.

Christianiafjord and Trondhjemsfjord. Depth 188—282 m.

13. **A. pusilla** O. Sars 1891 *A. p.*, G. O. Sars, Crust. Norway, *v.* 1 p. 181 t. 63 f. 2 | 1893 *A. aequicornis* (part.), A. Della Valle in: F. Fl. Neapel, *v.* 20 p. 907, 917.

Closely approaching A. rubella (p. 104). Front of head obliquely emarginate, lower corner little projecting. Side-plate 1 not concealing base of antenna 2. Pleon segment 3, lower hind corners quadrate, 4th scarcely carinate in ♂, more distinctly and with deep anterior saddle-shaped depression in ♂. Corneal lenses absent, but patches of pigment representing the eyes. Antennae 1 and 2 in ♀ subequal, little more than half length of body; peduncle of antenna 1 reaching beyond penultimate joint of peduncle of antenna 2, 2^d joint twice as long as 1st; antenna 2, ultimate joint of peduncle longer than penultimate. Gnathopod 1, 6th joint fully as long as 5th. Peraeopods 1—4 as in A. amblyops. Peraeopod 5, 2^d joint little longer than rest of leg, expansion broad, obtusely truncated in ♀, narrower and more obliquely truncated in ♂, reaching below the 3^d joint, which is longer than the 4th; 6th longer than 4th and 5th combined, 7th more than half the length of the 6th. Uropod 3, rami not very broad, slightly armed. Telson without dorsal spinules. Body whitish, pellucid, with a faint rosy tinge on front of body and side-plate 1. L. ♀ scarcely more than 5 mm.

North-Sea, Skagerrak, North-Atlantic and Arctic Ocean (South- and West-Norway, and northwards to Selsövig exactly at the polar circle). Depth 188—376 m, muddy ground.

14. A. anomala O. Sars 1882 *A. a.*, G. O. Sars in: Forh. Selsk. Christian., nr. 18 p. 108 t. 6 f. 2 | 1891 *A. a.*, G. O. Sars, Crust. Norway, *v.* 1 p. 178 t. 62 f. 2 | 1893 *A. aequicornis* (part.), A. Della Valle in: F. Fl. Neapel, *v.* 20 p. 478.

Head transversely truncate, emarginate, lower corner acutely produced. Side-plate 1 not concealing base of antenna 2. Pleon segment 3, lower hind corners quadrate; 4th with distinct carina ending angularly. Corneal lenses larger but less refractive than usual, lower pair removed from lower corners and lower margin of head, pigment well defined, reddish. Antenna 1 in ♀ considerably more than half as long as the body, peduncle reaching beyond penultimate joint of peduncle of antenna 2, 2d joint twice as long as 1st, flagellum 30-jointed, 2$\frac{1}{2}$ times as long as peduncle; antenna 2 nearly as long as body, 1$\frac{1}{2}$ length of antenna 1, ultimate joint of peduncle longer than penultimate. Gnathopod 1, 6th joint nearly as long as 5th; gnathopod 2 differing from gnathopod 1 less than usual. Peraeopods 1 and 2, 7th joint considerably longer than 5th and 6th combined. Peraeopods 3 and 4 with 5th joint rather elongate. Peraeopod 5. 2d joint longer than rest of leg, expansion obliquely rounded, not reaching below 3d joint, which is longer than the 4th, 6th fully as long as 4th and 5th combined, 7th more than half length of 6th. Uropod 3, rami foliaceous, fringed with a few setules. Telson oval, scarcely twice as long as broad, without dorsal spinules. Body highly pellucid, nearly colourless. L. ♀ 7 mm.

North-Sea (Korshavn [West-Norway]), Hardangerfjord, Christianiafjord, North-Atlantic (Cape Finisterre). Depth 188—376 m.

15. A. spinipes Boeck 1861 *A. s.*, A. Boeck in: Forh. Skand. Naturf., Møde 8 p. 653 | 1876 *A. s.*, A. Boeck, Skand. Arkt. Amphip., *v.* 2 p. 526 t. 31 f. 5 | 1891 *A. s.*, G. O. Sars, Crust. Norway, *v.* 1 p. 173 t. 60 f. 2 | 1893 *A. aequicornis* (part.), A. Della Valle in: F. Fl. Neapel, *v.* 20 p. 478.

Head truncate nearly transversely, but with upper corner rather more projecting than lower. Pleon segments 1—3 slightly carinate, 3d with lower hind corners nearly quadrate; 4th deeply impressed in front and with high rounded carina behind. Corneal lenses small, distinct, lower pair removed from lower corners of head; pigment irregular, reddish, partly coated with chalky white. Antenna 1 in ♀ nearly half as long as body, peduncle just reaching beyond penultimate joint of peduncle of antenna 2, 2d joint not twice length of 1st, flagellum 32-jointed, about 3 times as long as peduncle; antenna 2 in ♀ 1$\frac{1}{2}$ length of antenna 1, ultimate and penultimate joints of peduncle equal. Gnathopod 1, 6th joint decidedly shorter than 5th; gnathopod 2 much more slender than gnathopod 1. Peraeopods 1 and 2, 7th joint equal to 5th and 6th combined. Peraeopods 3 and 4 as usual very spinous. Peraeopod 5, 2d joint as long as rest of leg, expansion obliquely truncated. reaching to end of 3d joint, which is much longer than the 4th; 6th longer than 4th and 5th combined; 7th $\frac{3}{4}$ length of 6th. Uropod 3 foliaceous, with short setae on the confronted margins in ♀. Telson very narrow, without dorsal spinules, but with 3 marginal spinules near each apex. Semipellucid, light yellowish, with orange shadows and specks, peduncles of antennae with reddish tips. L. ♀ 15 mm, ♂ rather less.

North-Sea, Skagerrak, North-Atlantic and Arctic Ocean (South- and West-Norway, northwards to Lofoten Isles). Depth 56—188 m.

16. A. aequicornis Bruz. 1859 *A. a.*, R. M. Bruzelius in: Svenska Ak. Handl., n. ser. *v.* 3 nr. 1 p. 82 t. 4 f. 15 | 1891 *A. aequicornis*, G. O. Sars. Crust. Norway, *v.* 1 p. 177 t. 62 f. 1 | 1893 *A. aequicornis* (part.), A. Della Valle in: F. Fl. Neapel, *v.* 20 p. 478.

Back not carinate, except slightly at pleon segment 4. Head truncate nearly transversely, but with upper corner rather more projecting than lower. Side-plate 4 with hind and lower margins defined by an obtuse angle. Pleon segment 3, lower hind corners quadrate. Corneal lenses small, distinct, lower pair removed from lower corners of head, pigment well defined, reddish. Antenna 1 in ♀ very little shorter than antenna 2, peduncle reaching beyond penultimate joint of peduncle of antenna 2, 2^d joint twice as long as 1st, flagellum 24-jointed, more than twice as long as peduncle; antenna 2 a little over half length of body, ultimate joint of peduncle longer than penultimate. Gnathopod 1, 6th joint shorter than 5th; gnathopod 2 much more slender than gnathopod 1. Peraeopods 1 and 2, 7th joint scarcely equal to 5th and 6th combined. Peraeopods 3 and 4 unusually short and robust. Peraeopod 5, 2^d joint longer than rest of leg, expansion broad, obtusely truncate, not reaching end of 3^d joint, which is much longer than the 4th, 6th shorter than 4th and 5th combined, 7th broadly lanceolate, more than half length of 6th. Uropod 3, rami foliaceous, fringed with spinules. Telson oblong, a little tapering distally, with 2 pairs of dorsal spinules and 1 pair at the blunted apices. L. ♀ 11 mm.

North-Sea, Skagerrak and North-Atlantic (South- and West-Norway, northwards to Lofoten Isles, Bohusläu, British Isles).

17. **A. serraticaudata** Chevreux 1888 *A. s.*, *A. serraticandata* (laps.), Chevreux in: C.-R. Ass. Franç., Sess. 17 *v.* 2 p. 349 t. 6 f. 3—9 | 1893 *A. rubella* (part.), A. Della Valle in: F. Fl. Neapel, *v.* 20 p. 482.

Head transversely truncate. Pleon segments 1—3 with postero-lateral angles acute, slightly produced, segment 4 without any of the usual gibbosity. Eyes figured as having small corneal lenses. Antenna 1 in ♀ little shorter than antenna 2, about half length of the body, peduncle reaching beyond penultimate joint of peduncle of antenna 2, 2^d joint figured as not longer than 1st, flagellum more than twice as long as peduncle; ultimate joint of peduncle of antenna 2 figured as longer than penultimate. Gnathopod 1, 6th joint sub-equal in length to 5th. Peraeopods 1 and 2, 7th joint much longer than 5th and 6th combined. Peraeopods 3 and 4 short and robust. Peraeopod 5, 2^d joint longer than rest of leg, expansion very obliquely truncate, reaching end of 3^d joint, which is longer than the 4th; 6th joint not as long as 4th and 5th combined; 7th very short and broad. Uropod 3, rami broad, with plumose setae on the confronted margins, outer ramus curved at the apex, serrate in the upper part. Telson long, very deeply cleft, with a long spinule at each apex. L. ♀ 5·5 mm.

Mediterranean (Cherchell [Algeria]).

18. **A. diadema** (A. Costa) 1853 *Araneops d.*, A. Costa in: Rend. Soc. Borbon, n. ser. *v.* 2 p. 171 | 1867 *Ampelisca d.*, A. Costa in: Annuario-Mus. Napoli, *v.* 4 p. 45 | 1893 *A. d.* (part.), A. Della Valle in: F. Fl. Neapel, *v.* 20 p. 479 t. 4 f. 2; t. 37 f. 19, 20, 22—28, 30—38; t. 38 f. 2, 7, 8, 11, 12, 14, 15; t. 40 f. 39, 40; t. 41 f. 23; t. 44 f. 4, 8, 9, 11; t. 45 f. 17. 18; t. 46 f. 4—6; t. 47 f. 29; t. 48 f. 19 | 1862 *A. gaimardii* (part.), Bate, Cat. Amphip. Brit. Mus., p. 91 | 1871 *A. assimilis*, A. Boeck in: Forh Selsk. Christian., 1870 p. 222 | 1891 *A. a.*, G. O. Sars, Crust. Norway, *v.* 1 p. 168 t. 58 f. 2.

Head obliquely truncated. Ventral surface of peraeon segments 5 and 6 each carrying a backward directed hook, segment 7 with 2 unequal hooks directed forward. Pleon segment 3 with postero-lateral angles rounded, segment 4 with slight anterior depression and carina barely indicated (rather high and evenly rounded, Boeck, Sars). Corneal lenses 4, lower pair at lower corners of head. Antenna 1 in ♀ a little longer than peduncle of antenna 2, 2ᵈ joint of peduncle not much longer than 1ˢᵗ, flagellum 10-jointed, about twice as long as peduncle; ultimate joint of peduncle of antenna 2 in ♂ much, in ♀ a little longer than penultimate, flagellum in ♀ 20-jointed, 1½ length of peduncle. Peraeopods 1 and 2, 7ᵗʰ joint much longer (shorter, Boeck) than 5ᵗʰ and 6ᵗʰ combined. Peraeopod 5, 2ᵈ joint longer than rest of leg, not reaching end of 3ᵈ, expansion obliquely rounded, 3ᵈ joint longer than 4ᵗʰ, 6ᵗʰ longer than 4ᵗʰ and 5ᵗʰ combined, 7ᵗʰ almost as long as 6ᵗʰ. Uropod 3 with rami broad, carrying some feathered setae on the confronted edges. Telson long oval, cleft almost to the base, without dorsal spinules, but with some setules on the margins. Colour greyish, passing into yellow-brown on the sides, with red or violet on sides of pleon and legs. L. 7—12 mm.

Mediterranean; North-Atlantic, North-Sea and Skagerrak (West- and South-Norway, Bohuslän).

19. **A. acinaces** Stebb. 1888 *A. a.*, T. Stebbing in: Rep. Voy. Challenger, *v.*29 p.1036 t.101, 102 | 1893 *A. a.*, A. Della Valle in: F. Fl. Neapel, *v.*20 p.476.

Head transversely truncate. Side-plates 1 and 2 minutely toothed at lower hind corner, 1ˢᵗ distally much widened, 4ᵗʰ with lower and hind margins forming a very obtuse rounded angle (Fig. 25). Corneal lenses 4, the lower pair behind the obtusely projecting lower corners of the head. Antenna 1 in ♀ not nearly reaching end of peduncle of antenna 2, 2ᵈ joint about twice length of 1ˢᵗ, flagellum 10-jointed, longer than peduncle; antenna 2 in ♀ as long as the body, ultimate joint of peduncle rather shorter than penultimate, which is longer than whole peduncle of antenna 1, flagellum 34-jointed, longer than peduncle. Mandibular palp with 2ᵈ joint moderately broad, 3ᵈ much shorter and

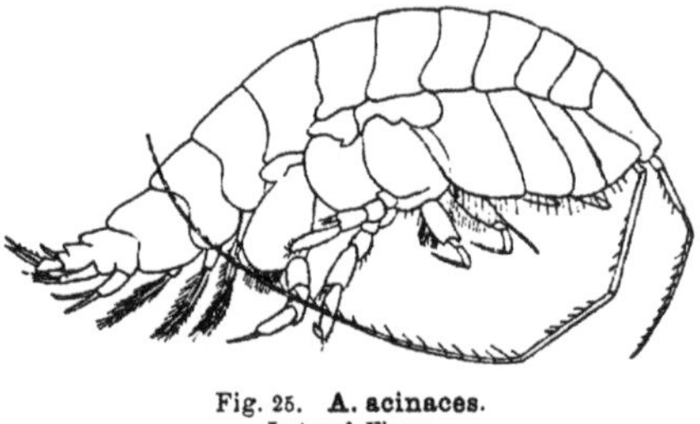

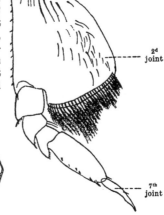

2ᵈ joint

7ᵗʰ joint

Fig. 25. **A. acinaces.**
Lateral View.

Fig. 26. **A. acinaces.**
Peraeopod 5.

narrower. Maxilla 1 with 2 small setae on inner plate. Peraeopods 1 and 2, 7ᵗʰ joint much longer than 5ᵗʰ and 6ᵗʰ combined. Peraeopod 5 (Fig. 26), 2ᵈ joint with the expansion broad, almost transversely truncate, not reaching end of 3ᵈ joint, which is longer than 4ᵗʰ; 5ᵗʰ joint a little longer than the 4ᵗʰ, each of them being lobed in front but not deeply; 6ᵗʰ nearly as long as 3 preceding joints combined; 7ᵗʰ narrowly lanceolate, about ³/₄ length of 6ᵗʰ. Uropod 3, rami lanceolate, furnished with spines and setules. Telson distally tapering, with evenly curved margins, within which are planted 5 pairs of setules; a pair of setules adjoin the moderately acute apices. L. 21 mm.

Port Jackson [East-Australia]. Depth 63 m.

20. **A. spinimana** Chevreux 1887 *Amplisca* (laps.) *s.*, Chevreux in: Bull. Soc. zool. France, *v.* 12 p. 574 | 1893 *Ampelisca diadema* (part.), A. Della Valle in: F. Fl. Neapel, *v.* 20 p. 479.

Head obliquely truncate. Pleon segment 3 with postero-lateral angles slightly produced, almost quadrate; segment 4 slightly carinate in ♀, deeply impressed and more strongly carinate in ♂. Eyes normal. Antenna 1 in ♀ extremely short, 1st joint scarcely longer than broad, 2^d 1$^1/_3$ length of 1st, flagellum 6-jointed. Antenna 2 in ♀ scarcely $^1/_3$ length of body, ultimate and penultimate joints of peduncle equal, flagellum 11-jointed. Antenna 2 in ♂ much longer than the body. Gnathopod 1, 6th joint as long as 5th, armed with long spines and setae on the concave palm. Peraeopods 1 and 2, 7th joint longer than 5th and 6th combined. Peraeopod 5, 2^d joint not quite reaching end of 3^d, which is armed with 4 spines on the front; 7th joint as long as 6th. Uropod 3 in ♀ with long smooth rami ending in a little spine, the rami in ♂ setose on the confronted edges. Telson cleft almost to the base, with 4 spines at the extremity. Colour uniform pale rose. L. 7 mm.

North-Atlantic (Cape Finisterre, depth 510 m; Croisic, depth 15—20 m).

21. **A. compacta** Norm. 1882 *A. c.*, A. M. Norman in: P. R. Soc. Edinb., *v.* 11 p 688.

Body rounded, compact. Pleon not carinate. segment 3 with postero-lateral angles acute, not much produced nor upturned, segment 4 with slight saddle-shaped depression. Eyes apparently wanting. Antenna 1 longer than peduncle of antenna 2, flagellum 10-jointed; antenna 2 much longer than antenna 1, ultimate and penultimate joints of peduncle subequal. Peraeopods 1 and 2, 7th joint subequal to 5th and 6th combined. Peraeopod 5, 2^d joint reaching end of 3^d, expansion transversely truncate. lower margin slightly concave(?), 3^d joint much longer than 4th, 6th not quite equal to 4th and 5th combined, 7th half length of 6th. Uropod 3 with the rami sparingly ciliated. Telson deeply cleft. L. about 8 mm.

North-Atlantic. Depth 968 m.

22. **A. typica** (Bate) 1856 *Tetromatus typicus*, Bate in: Rep. Brit. Ass., Meet. 25 p. 58 t. 17 f. 8, D 4 | 1857 *T. t.*, Bate in: Ann. nat. Hist., ser. 2 *v.* 19 p. 139 | 1871 *Ampelisca typica*, A. Boeck in: Forb. Selsk. Christian., 1870 p. 222 | 1891 *A. t.*, G. O. Sars, Crust. Norway, *v.* 1 p. 165 t. 57 | ?1859 *A. carinata*, R. M. Bruzelius in: Svenska Ak. Handl., n. ser. *v.* 3 nr. 1 p. 87 t. 4 f. 16 | 1862 *A. gaimardii* (part.), Bate, Cat. Amphip. Brit. Mus., p. 91 t. 15 f. 1 | 1893 *A. diadema* (part.), A. Della Valle in: F. Fl. Neapel, *v.* 20 p. 479.

Head truncate nearly transversely. Pleon segment 3 with postero-lateral corners quadrate, segment 4 deeply impressed, especially in ♂, with rather high carina projecting angularly. Corneal lenses distinct, lower pair a little removed from lower corners of head, pigment well defined, red with chalky white coating. Antenna 1 in ♀ very small, much shorter than peduncle of antenna 2, 2^d joint longer than 1st, flagellum about 7-jointed, a little longer than peduncle; antenna 1 in ♂ reaching beyond peduncle of antenna 2. Antenna 2 in ♀ scarcely more than half length of body, in ♂ as long as body, ultimate joint of peduncle equal to penultimate in ♀, longer than it in ♂. Gnathopod 1, 6th joint a little shorter than 5th. Peraeopods 1 and 2, 7th joint

longer than 5th and 6th combined. Peraeopod 5, 2[d] joint longer than rest
of leg, expansion obliquely truncated. reaching beyond 3[d] joint, which is
longer than 4th, 6th scarcely longer than 4th and 5th combined, 7th about
as long as 6th. Uropod 3 in ♀ only fringed with simple setules, in ♂ with
plumose setae on the confronted edges. Telson about twice as long as broad,
apically blunted, carrying several pairs of dorsal spinules, and in ♀ 5 pairs
of marginal setules. Whitish, pellucid, mottled with light yellowish and
a few reddish patches, intestine orange-coloured, ova in pouch rose-coloured.
L. 10 mm.

North-Atlantic, North-Sea and Skagerrak (British Isles, South- and West-Norway
up to Trondhjemsfjord); Kattegat. Depth 37—112 m.

23. **A. zamboangae** Stebb. 1888 *A. z.*, T. Stebbing in: Rep. Voy. Challenger,
v. 29 p. 1057 t. 106 | 1893 *A. diadema* (part.), A. Della Valle in: F. Fl. Neapel, *v.* 20 p. 479.

♀ unknown. — ♂. Closely allied to A. typica (p. 109). Pleon segment 3
with postero-lateral corners almost quadrate, slightly rounded. segment 4
impressed and with a carina raised at the apex above the next segment.
Corneal lenses 4, lower pair at the lower corners of the head. Antenna 1
reaching a little beyond peduncle of antenna 2, 2[d] joint not longer than
1st, flagellum about 24-jointed, much longer than peduncle, antenna 2 as
long as body, ultimate joint of peduncle much longer than penultimate, flagellum
38-jointed, much longer than peduncle. Palp of mandible very large, 2[d] joint
very broad, 3[d] much narrower but subequal in length. Gnathopod 1, 6th
joint not much shorter than 5th. Peraeopods 1 and 2, 7th joint little longer
than 5th and 6th combined. Peraeopod 5, 2[d] joint with expansion obliquely,
but evenly rounded, reaching end of 3[d] joint. which is much longer than
the 4th, 6th not longer than the 4th and 5th combined, 7th longer than
the 6th. Uropod 3 with broad lanceolate rami serrate on the confronted margins
and furnished with feathered setae and spines. Telson not twice as long
as broad, with one pair of dorsal spinules, cleft nearly ⁴/₅ length, the distal
half triangularly tapering to the acute apices, each of which is notched for
a setule. L. (in the position of usual curvature) 6 mm.

Basilan Strait (Samboanga [Philippine Islands]). Surface.

24. **A. tenuicornis** Lilj. 1855 *A. t.*, W. Liljeborg in: Öfv. Ak. Förh., *v.* 12
p. 123 | 1891 *A. t.*, G. O. Sars, Crust. Norway, *v.* 1 p. 167 t. 58 f. 1 | 1862 *A. laevigata*
(err., non W. Liljeborg 1855!), Bate, Cat. Amphip. Brit. Mus., p. 96 | 1893 *A. diadema*
(part.), A. Della Valle in: F. Fl. Neapel. *v.* 20 p. 479.

Head strongly curved anteriorly and obliquely truncate. Pleon segment 3
with postero-lateral corners quadrate, sharpened in ♀, in ♂ more obtuse;
segment 4 slightly impressed in ♀, with very low and scarcely projecting
carina, in ♂ deeply impressed, with distinct though rounded carina. Corneal
lenses distinct, lower pair a little removed from lower corners of head,
pigment irregular. bright red. Antenna 1 in ♀ not reaching end of peduncle of
antenna 2, 2[d] joint about twice as long as 1st, flagellum about 11-jointed,
nearly twice as long as peduncle; antenna 2 in ♀ nearly as long as body, ultimate
and penultimate joints of peduncle subequal. Antennae in ♂ as in A. typica
(p. 109). Peraeopods 1 and 2, 7th joint subequal to 5th and 6th combined.
Peraeopod 5, 2[d] joint scarcely longer than rest of leg, expansion nearly
transversely truncate, not reaching beyond 3[d] joint, which is longer than
4th, 6th joint longer than 4th and 5th combined, 7th much shorter than 6th.
Uropod 3 in ♀ with fewer marginal setules than in A. typica. Telson slightly
constricted near the base, without dorsal spinules, but with 3 pairs of marginal

setules. Whitish, pellucid, with a conspicuous carmine patch on side-plate 1. L. scarcely more than 8 mm.

North-Atlantic, North-Sea and Skagerrak (South- and West-Norway up to Trondhjemsfjord, British Isles); Kattegat. Depth 19—188 m.

25. **A. sarsi** Chevreux 1888 *A. s.*, Chevreux in: C.-R. Ass. Franç., Sess. 16 *v.* 2 p. 666.

Closely allied to A. tenuicornis. Head transversely truncate, similar to A. typica (p. 109). Pleon segment 3 with lateral margin slightly prolonged backward, rounded, segment 4 scarcely carinate in ♀, in ♂ impressed and with a high gibbous carina. Eyes not mentioned, but perhaps to be presumed, as the head is compared to that of *A. laevigata* Liljeborg (p. 100). Antenna 1 in ♀ much shorter than peduncle of antenna 2. Antenna 2 in ♀ $\frac{2}{3}$ length of body, in ♂ longer than the body. Peraeopods 1 and 2, 7th joint as long as 5th and 6th combined. Peraeopod 5, 2d joint not reaching end of 3d, 4th joint not overlapping 5th. L. 7 mm.

North-Atlantic (Croisic). Depth 15—20 m.

A. agassizi (Judd) 1896 *Byblis a.*, S. D. Judd in: P. U. S. Mus., *v.* 18 p. 599 f. 9—11.

Near to A. diadema (p. 107).

Narragansett Bay (Newport [Rhode Island]).

A. daleyi Giles 1890 *A. d.*, G. M. Giles in: J. Asiat. Soc. Bengal, *v.* 59 p. 66 t. 2 f. 3.

Side-plate 5 having hind border of hind lobe notched. Antenna 2, last 3 joints of peduncle very long and slender, antepenultimate longest, ultimate shortest. Peraeopods 3 and 4 with 7th joint (finger or spine?) remarkably long and slender. L. 11 mm.

Bay of Bengal (Seven Pagodas near Madras). Depth 13 m.

A. japonica Bate 1862 *A. j.*, (Stimpson in MS.) Bate, Cat. Amphip. Brit. Mus., p. 94 t. 15 f. 5.

L. 9 mm.

North-Pacific (Japan).

A. limicola (Stimps.) 1853 *Pseudophthalmus l.*, *P. limicolus*, Stimpson in: Smithson. Contr., *v.* 6 nr. 5 p. 57, 65 | 1862 *Ampelisca limicola*, Bate, Cat. Amphip. Brit. Mus., p. 93 t. 15 f. 4.

But 1 clear eye-spot on each side of head. Peraeopod 5, 7th joint not lanceolate, but stiliform and curved. L. 15 mm.

North-Atlantic (Charleston). Low water, in mud-holes.

A. nordmannii (M.-E.) 1840 *Acanthonotus n.*. H. Milne Edwards, Hist. nat. Crust., *v.* 3 p. 24 | 1893 *Ampelisca n.*, A. Della Valle in: F. Fl. Neapel, *v.* 20 p. 469.

Black Sea (Crimea).

A. pelagica (Stimps.) 1853 *Pseudophthalmus pelagicus*, Stimpson in: Smithson. Contr., *v.* 6 nr. 5 p. 57 | 1862 *Ampelisca pelagica*, Bate. Cat. Amphip. Brit. Mus., p. 94.

L. 10 mm.

Long Island Sound and Hake Bay [North-America]. Depth 55—91 m.

A. pugetica Stimps. 1864 *A. pugeticus*, Stimpson in: P. Ac. Philad., p. 158.

Puget Sound [Western North-America]. Depth 18 m.

2. Gen. **Byblis** Boeck

1871 *Byblis* (Sp. un.: *B. gaimardi*), A. Boeck in: Forh. Selsk. Christian., 1870 p. 228 | 1891 *B.*, G. O. Sars, Crust. Norway, *v.* 1 p. 182.

Head with postero-antennal corners distinct. Side-plate 1 scarcely deeper than 2, not concealing base of antenna 2; side-plate 4 evenly curved below the prominent posterior angle. Pleon segment 3, postero-lateral angles obtusely rounded. Corneal lenses **4** or none. Mandibular palp with 2^d joint narrow, 3^d narrow and rather short. Peraeopods 3 and 4, 2^d joint rather broad, its front edge moderately curved, 5th joint carrying several transverse rows of spines. Peraeopod 5 (Fig. 27), 2^d joint strongly produced, the long hind margin nearly straight, inner and lower margins of the squarely descending lobe carrying a dense fringe of plumose setae; 3^d joint shorter than 4th, 4th and 5th rather wide. 6th sublinear, 7th spiniform. Uropod 3 scarcely reaching beyond the others, rami narrowly lanceolate, not setose in ♀, confronted edges more or less serrate. Telson short and broad, not very deeply cleft. Two pairs of hepato-pancreatic caeca.

10 species.

Synopsis of species:

1 { Cleft of telson minute — **2.**
  { Cleft of telson not minute — **5.**

2 { Antenna 2, ultimate joint of peduncle longer than
  {   penultimate . . . .   1. **B. erythrops** . . p. 112
  { Antenna 2, ultimate joint of peduncle not longer
  {   than penultimate — **3.**

3 { Corneal lenses Wanting . . . .   2. **B. abyssi** . . . . p. 113
  { Corneal lenses not Wanting — **4.**

4 { Antenna 2, ultimate joint of peduncle shorter than
  {   penultimate; peraeopods 1 and 2, 6th joint much
  {   longer than 5th . . . .   3. **B. gaimardii** . . p. 113
  { Antenna 2, ultimate joint of peduncle equal to
  {   penultimate; peraeopods 1 and 2, 6th joint not
  {   much longer than 5th . . . .   4. **B. longicornis** . . p. 113

5 { Antenna 1 reaching Well beyond peduncle of
  {   antenna 2 — **6.**
  { Antenna 1 reaching scarcely or not beyond
  {   peduncle of antenna 2 — **8.**

6 { Antenna 1 not Very long . . . .   5. **B. affinis** . . . . p. 114
  { Antenna 1 Very long — **7.**

7 { Corneal lenses absent . . . .   6. **B. crassicornis** . p. 114
  { Corneal lenses present . . . .   7. **B. guernei** . . . p. 114

8 { Corneal lenses absent . . . .   8. **B. serrata** . . . p. 114
  { Corneal lenses present — **9.**

9 { Branchial vesicles and peraeopod 2 of normal length   9. **B. kallarthra** . . p. 115
  { Branchial Vesicles and peraeopod 2 of abnormal
  {   length . . . .   10. **B. lepta** . . . . . p. 115

1. **B. erythrops** O. Sars 1882 *B. e.*, G. O. Sars in: Forh. Selsk. Christian., nr. 18 p. 109 t. 6 f. 3, 3a | 1891 *B. e.*, G. O. Sars, Crust. Norway, *v.* 1 p. 187 t. 65 f. 3 | 1893 *Ampelisca gaimardii* (part.), A. Della Valle in: F. Fl. Neapel, *v.* 20 p. 472.

Corneal lenses very small, lower pair removed from lower corners of front, pigment bright red. Antenna 1 little shorter than antenna 2, peduncle reaching considerably beyond end of penultimate joint of peduncle of antenna 2, flagellum in ♀ 35-jointed, more than twice as long as peduncle; ultimate joint of peduncle of antenna 2 notably longer than penultimate. Peraeopod 5, 2^d joint produced below the 4th, its front and lower edges in a continuous curve, 5th joint as long as 3^d and 4th combined, a little longer than the 6th. Uropod 3 with rami very slightly serrate on the confronted edges. Telson almost square, distal angles rounded, cleft short. Body pellucid, with dark violet intestine shining through the integument, sides mottled with white. L. ♀ 8 mm, or slightly more.

Arctic Ocean and North-Atlantic (Magerö and Bejan [Norway]). Depth 150—188 m, muddy bottom.

2. **B. abyssi** O. Sars 1879 *B. a.*, G. O. Sars in: Arch. Naturv. Kristian., *v.* 4 p. 456 | 1885 *B. a.*, G. O. Sars in: Norske Nordhavs-Exp., *v.* 6 Crust. 1 p. 201 t. 16 f. 6 | 1891 *B. a.*, G. O. Sars, Crust. Norway, *v.* 1 p. 189 t. 66 f. 2 | 1893 *Ampelisca gaimardii* (part.), A. Della Valle in: F. Fl. Neapel, *v.* 20 p. 472.

Corneal lenses wanting. Antenna 1, peduncle not nearly reaching end of penultimate joint of peduncle of antenna 2, flagellum in ♀ 28-jointed, twice as long as peduncle; ultimate joint of peduncle of antenna 2 notably shorter than penultimate, antenna 2 nearly twice as long as antenna 1. Peraeopod 5, 2^d joint broad, reaching a little below 4th, front and lower edges in a continuous curve, but almost quadrate, 5th joint as long as 3^d and 4th combined, slightly longer than 6th, 7th joint very short. Uropod 3, inner margin of inner ramus serrate, that of outer ramus almost smooth. Telson almost quadrate, but broader than long, slightly narrowed distally, distal angles rounded, cleft nearly obsolete. Colour whitish, pellucid. L. ♀ 12 mm.

Arctic Ocean and North-Atlantic (West- and Northwest-Norway). Depth 659—1186 m.

3. **B. gaimardii** (Krøyer) 1846 *Ampelisia* (laps., corr.: *Ampelisca*) *g.*, Krøyer in: Voy. Nord., Crust. t. 23 f. 1 a—y | 1871 *Byblis gaimardi*, A. Boeck in: Porh. Selsk. Christian., 1870 p. 228 | 1891 *B. gaimardii*, G. O. Sars, Crust. Norway, *v.* 1 p. 183 t. 64 | 1893 *Ampelisca g.* (part.), A. Della Valle in: F. Fl. Neapel, *v.* 20 p. 472 t. 57 f. 39—41 | 1896 *Byblis serrata* (err.?, non S. I. Smith 1873!), S. D. Judd in: P. U. S. Mus., *v.* 18 p. 596 f. 4—8.

Corneal lenses large, lower pair occupying lower corners of front, pigment brownish. Antenna 1, peduncle not reaching end of penultimate joint of peduncle of antenna 2, flagellum in ♀ 20-jointed, twice as long as peduncle, ultimate joint of peduncle of antenna 2 shorter than penultimate. Peraeopods 1 and 2 having 6th and 7th joints each much longer than the 5th; peraeopod 5 with 2^d joint produced below the 4th, its front and lower edges defined by an obtuse angle, 5th joint as long as 3^d and 4th combined and rather longer than the 6th. Uropod 3 having the rami acute, with confronted edges serrate and a projecting corner near the base. Telson with nearly equal length and breadth, cleft very short. Colour whitish with faint orange markings, front of head mottled with dark violet. L. ♀ 15, ♂ rather less (Sars), reaching 23 mm (Ohlin).

Arctic Ocean, North-Atlantic and North-Sea (whole West-Norway); Christiania-fjord; Kattegat. Depth 37—150 m.

4. **B. longicornis** O. Sars 1891 *B. l.*, G. O. Sars, Crust. Norway, *v.* 1 p. 185 t. 65 f. 1 | 1893 *Ampelisca gaimardii* (part.), A. Della Valle in: F. Fl Neapel, *v.* 20 p. 922 | 1894 *Byblis intermedius*, T. Stebbing in: Bijdr. Dierk., *v.* 17 p. 18.

Corneal lenses smaller than in B. gaimardii, lower pair removed from lower corners of front, pigment light brown. Antenna 1, peduncle usually reaching end of penultimate joint of peduncle of antenna 2, flagellum in ♀ 27—35-jointed, twice as long as peduncle; ultimate and penultimate joints of peduncle of antenna 2 subequal; both pairs longer than in B. gaimardii. Peraeopods 1 and 2 having 6th joint not much longer than 5th; peraeopod 5 with 2^d joint not produced below the 4th, its front and lower edges in a continuous curve, 5th joint scarcely as long as 3^d and 4th combined and shorter than the 6th. Uropod 3 with rami acute and confronted edges serrate, but without projecting corner near the base. Telson broader than long, slightly tapering, cleft very short. Head not mottled. L. ♀ 12 mm.

Arctic Ocean (Lofoten Isles, Finmark, South-Spitzbergen, Barents Sea).

5. B. affinis O. Sars 1891 *B. a.*, G. O. Sars, Crust. Norway, *v.* 1 p. 186 t. 65 f. 2 | 1893 *Ampelisca gaimardii* (part.), A. Della Valle in: F. Fl. Neapel, *v.* 20 p. 907, 922.

Corneal lenses smaller than in B. longicornis (p. 113), lower pair remote from lower corners of front, pigment dark brownish. Antenna 1, peduncle reaching end of penultimate joint of peduncle of antenna 2, flagellum in ♀ 24-jointed, twice as long as peduncle; ultimate and penultimate joints of peduncle of antenna 2 subequal; both pairs shorter than in B. longicornis. Gnathopod 1 with 6th joint little shorter than 5th, a character which this species has in common only with B. serrata and B. guernei. Peraeopods 1 and 2 having 6th joint considerably longer than 5th; peraeopod 5 with 2[d] joint not produced below the 4th, front and lower edges in a continuous curve, but almost quadrate, 5th joint as long as 3[d] and 4th combined, and rather longer than the 6th. Uropod 3 lanceolate, with confronted edges slightly serrate. Telson very little broader than long, very slightly tapering to the truncate distal margin, cleft extending almost to the middle. Body pellucid with reddish intestine, head dotted with whitish and brownish pigment not ramified, orange patches on 2[d] joint in peraeopods 3—5 and on pleon. L. ♀ 9 mm.

Trondhjemsfjord. Depth 75—113 m.

6. B. crassicornis Metzg. 1875 *B. c.*, Aug. Metzger in: Jahresber. Comm. D. Meere, *v.* 2/3 p. 297 t. 6 f. 9, 9 a, b | 1891 *B. c.*, G. O. Sars, Crust. Norway, *v.* 1 p. 188 t. 66 f. 1 | 1893 *Ampelisca gaimardii* (part.), A. Della Valle in: F. Fl. Neapel, *v.* 20 p 472.

Corneal lenses wanting. Antenna 1, peduncle reaching considerably beyond penultimate joint of peduncle of antenna 2, flagellum in ♀ 20-jointed, scarcely twice as long as peduncle; ultimate joint of peduncle of antenna 2 a little shorter than penultimate, antenna 2 little longer than antenna 1. Peraeopod 5 with 2[d] joint less expanded than usual, scarcely reaching beyond the 4th joint, front and lower margins in a continuous curve, 5th joint little longer than 4th, not longer than 6th. Uropod 3, rami not serrate on the confronted edges. Telson fully as long as broad, much tapering, cleft nearly reaching the middle. Body pellucid, nearly colourless. L. ♀ 7 mm.

Bukkenfjord (Jaederen, Hvitingsö), Christianiafjord. Depth 188—282 m.

7. B. guernei Chevreux 1887 *B. g.*, Chevreux in: Bull. Soc. zool. France, *v.* 12 p. 576.

Corneal lenses very distinct. Antenna 1 more than $^3/_4$ length of the body; antenna 2 as long as the body, ultimate joint of peduncle a little shorter than penultimate. Gnathopod 1, 5th joint a little longer than 6th. Peraeopod 5 having the 2[d] joint much expanded and reaching to the end of the 5th joint. Telson a little broader than long, distally truncate, cleft extending to the middle. L. 5 mm.

North-Atlantic (Cape Finisterre). Depth 510 m.

8. B. serrata S. I. Sm. 1873 *B. s.*, (S. I. Smith in:) A. E. Verrill in: Rep. U. S. Fish Comm., *v.* 1 p. 561 | 1879 *Ampelisca minuticornis*, G. O. Sars in: Arch. Naturv. Kristian, *v.* 4 p. 455 | 1885 *A. m.*, G. O. Sars in: Norske Nordhavs-Exp., *v.* 6 Crust. I p. 198 t. 16 f. 5 a—o | 1893 *A. m.*, A. Della Valle in: F. Fl. Neapel, *v.* 20 p. 477 | 1891 *Byblis m.*, G. O. Sars, Crust. Norway, *v.* 1 p. 190 t. 66 f. 3.

Lower margin of side-plate 1 (Sars), or side-plate 1 and 2 (Smith) serrate between marginal setae. Corneal lenses wanting (Sars, not mentioned by Smith). Antenna 1 scarcely reaching beyond penultimate joint of peduncle of antenna 2, flagellum in ♀ 6-jointed, shorter than peduncle; antenna 2 little more than half as long as the body, ultimate joint of peduncle shorter than penultimate. Gnathopod 1 with 5th joint scarcely if at all longer than 6th (Smith and Sars 1885, but considerably longer in Sars's figure). Peraeopod 5, 2^d joint much expanded and reaching to the end of the 5th joint, anterior corner rounded off (Sars), anterior and inferior margins evenly arcuated (Smith), 5th joint about as long as 3^d and 4th combined and about as long as the 6th. Uropod 3 with confronted edges not having any distinct serration. Telson subquadrate, about as long as broad, converging (rapidly, Smith) to distal margin, cleft extending about to the middle (Sars), rather more than half its length (Smith and Sars 1885). Body highly pellucid, and nearly colourless (Sars), specked and mottled on side-plates; 2^d joint of peraeopods 3—5. and sides of pleon with crowded points of dark pigment (Smith). L. ♀ 8—12 mm.

Arctic Ocean and North-Atlantic (Spitzbergen, West- and North-Norway, depth 659—1193 m; off Vineyard Sound and Buzzard Bay, deep water).

9. **B. kallarthra** Stebb. 1886 *B. kallarthrus.* T. Stebbing in: P. zool. Soc. London, p. 4 | 1887 *B. k.,* T. Stebbing in: Tr. zool. Soc. London, *v.* 12 VI p. 199 t. 38 | 1893 *Ampelisca k.,* A. Della Valle in: F. Fl. Neapel, *v.* 20 p. 476.

Lower margins of side-plates 1 and 2 serrate, side-plate 4 produced backwards with sinuate lower margin, not observed in any other species. The 4 corneal lenses surrounded with dark pigment. Antenna 1 scarcely more than reaching end of penultimate joint of peduncle of antenna 2, flagellum in ♀ 11-jointed, a little longer than peduncle; ultimate joint of peduncle of antenna 2 rather shorter than penultimate, antenna 2 not quite $^3/_5$ length of body. Gnathopod 1, 5th joint much longer than 6th. Peraeopods 1 and 2, 6th joint considerably longer than 5th; peraeopod 5 (Fig. 27) with 2^d joint produced fully to the end of the 5th, the front and lower margins forming a gently continuous curve, 5th joint considerably longer than 3^d and 4th combined and much longer than 6th. Uropod 3 with confronted margins of rami serrate and angled near the base. Telson slightly longer than broad, a little rounded, cleft extending a little beyond the middle. Branchial vesicles with secondary lamellae on both sides. L. ♀ 10 mm.

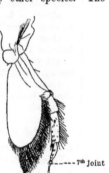

- - - - 7th Joint

Fig. 27.
B. kallarthra.
Peraeopod 5.

Singapore Strait.

10. **B. lepta** (Giles) 1888 *Ampelisca l.,* G. M. Giles in: J. Asiat. Soc. Bengal, *v.* 57 p. 223 t. 8, 9 | 1893 *A. l.,* A. Della Valle in: F. Fl. Neapel, *v.* 20 p. 894.

The 4 corneal lenses surrounded with dark brown pigment. Antenna 1 reaching a little beyond peduncle of antenna 2, flagellum with 10 very slender joints; antenna 2, $^2/_3$ length of body, ultimate joint of peduncle a little shorter than penultimate, flagellum 14- or 15-jointed. Peraeopod 2 unusually elongate. Peraeopod 5, 2^d joint descending below the 4th. Uropod 3, rami apparently serrate on confronted margins and angled near the base. Telson deeply cleft, semilunar. Colour ivory white. Branchial vesicles of great length, with conspicuous accessory laminae on each surface. L. about 6 mm.

Head of Bay of Bengal. Depth 195 m.

8*

3. Gen. **Haploops** Lilj.

1855 *Haploops*, W. Liljeborg in: Öfv. Ak. Förh., *v.* 12 p. 135 | 1891 *H.*, G. O. Sars, Crust. Norway, *v.* 1 p. 191 | 1893 *H.*, A. Della Valle in: F. Fl. Neapel, *v.* 20 p. 111, 126, 485 | 1894 *H.*, T. Stebbing in: Bijdr. Dierk., *v.* 17 p. 18.

Head with postero-antennal corners rounded off. Side-plate 1 deeper than next pair, partly concealing base of antenna 2; 4th evenly curved below the prominent posterior angle. Eyes less perfect than in the other genera, corneal lenses 4 or 2 or none. Mandibular palp (Fig. 28) setose, with elongate 3[d] joint. Peraeopods 3 and 4, 2[d] joint rather broad, its front edge moderately curved, 5th carrying several transverse rows of spines. Peraeopod 5 (Fig. 29), 2[d] joint squarely dilated above, so that the front and hind margin are subparallel, while the descending lobe is narrowly or broadly rounded, its margin fringed with setae, of which there are many also on the hind margin and surface of the expansion; 3[d] joint shorter than 4th, 4th and 5th broad, 6th very narrow, 7th spiniform. Uropod 3 reaching considerably beyond the others, rami foliaceous, setose at the edges, inner margin of inner ramus armed with strong spines. Telson of moderate size, usually deeply cleft. One pair of hepato-pancreatic caeca (Della Valle).

5 species.

Synopsis of species:

1 { Body smooth or nearly so, corneal lenses present — **2.**
 Body with dorsal fascicles of setae, corneal lenses absent — **4.**

2 { Antenna 1 very long; uropod 1, inner ramus very short . . . . 1. **H. dellavallei** . p. 116
 Antenna 1 not very long; uropod 1, inner ramus not very short — **3.**

3 { Corneal lenses 2; uropod 3 in ♀, rami subequal, setose . . . . 2. **H. tubicola.** . . p. 117
 Corneal lenses 4; uropod 3 in ♀, rami unequal, smooth . . . . 3. **H. laevis.** . . . . p. 117

4 { Telson cleft a little beyond the middle . . . . . 4. **H. setosa** . . . p. 117
 Telson cleft nearly to the base . . . . 5. **H. robusta** . . . p. 118

1. **H. dellavallei** Stebb.*) 1893 *H. tubicola* (err., non W. Liljeborg 1855!), A. Della Valle in: F. Fl. Neapel, *v.* 20 p. 486 t. 3 f. 2; t. 37 f. 1—18.

Corneal lenses 4, colour of eyes crimson, standing out on a ground of bright yellow. Antenna 1 much longer than the body, flagellum 7 times as long as peduncle, of about 30 joints with very long setae. Antenna 2 subequal to antenna 1, ultimate joint of peduncle a little longer than penultimate, flagellum as in antenna 1. Peraeopod 5, 2[d] joint rather narrow, broadest at base, hind margin concave, lower lobe narrow, scarcely reaching beyond the short 3[d], 4th and 5th equal, 5th with both margins crenate, 6th and 7th very small. Uropod 1 perfectly smooth, inner ramus spine-like, $\frac{1}{3}$ length of outer; uropod 2 furnished with spines; uropod 3 with the rami subequal and foliaceous, armed with spinules and setules. Telson heart-shaped, cleft almost to the base, with a setule on each side of the almost acute apex. Colour pearl-grey, changing into yellow-brown on the sides; a large crimson blotch occupies part of the head and peraeon segment 1.' L. 8 mm.

Gulf of Naples. Depth 20—40 m.

---

*) Spec. nov. After Antonio Della Valle.

2. H. tubicola Lilj. 1852 *Ampelisca eschrichti?* (err., non *A. eschrichtii* Krøyer 1842!), W. Liljeborg in: Öfv. Ak. Förh., *v.*9 p.6 | 1855 *Haploops tubicola* + *H. carinata*, W. Liljeborg in: Öfv. Ak. Förh., *v.*12 p.135, 136 | 1880 *H. t.*, Stuxberg in: Bih. Svenska Ak., *v.*5 nr.22 p.65 | 1891 *H. t.*, G. O. Sars, Crust. Norway, *v.*1 p.192 t.67 | 1894 *H. t.*, T. Stebbing in: Bijdr. Dierk., *v.*17 p.19.

Pleon in ♂ carinate. Corneal lenses 2, pigment bright reddish. Antenna 1 about half length of body, 2^d joint much longer than 1st, flagellum $2\frac{1}{2}$ times as long as peduncle, with about 23 joints carrying setae. Antenna 2 subequal to antenna 1, ultimate and penultimate joints of peduncle subequal. Peraeopod 5, 2^d joint rather narrow, broadest at base, hind margin concave, lower lobe narrow, scarcely reaching beyond 3^d joint, 4th longer than 5th, 6th and 7th very small. Uropod 1, inner ramus $\frac{3}{4}$ length of outer, carrying spinules; uropod 3, rami subequal and foliaceous, fringed in ♂ with long ciliated setae, in ♀ with shorter setae, the outer ramus having also 4 spinules. Telson a little longer than broad, cleft extending far beyond the middle, apices somewhat rounded. Body whitish, pellucid, with dark violet intestine. L. ♂ 11, ♀ 10—22 mm.

Arctic Ocean, North-Atlantic, North-Sea, Skagerrak, Kattegat and Baltic (Norwegian, Danish, Dutch, French and British coasts).

3. H. laevis Hoek ?1866 *H. tubicola* (part.), Goës in: Öfv. Ak. Förh., *v.*22 p 528 | 1893 *H. t.* (part.), A. Della Valle in: F. Fl. Neapel. *v.*20 p.486 | 1882 *H. laevis*, Hoek in: Niederl. Arch. Zool., suppl. 1 nr.7 p.61 t.3 f.31 | 1894 *H. l.*, T. Stebbing in: Bijdr. Dierk., *v.*17 p.19 t.3.

Several segments of peraeon with a patch of fine down on the back. Corneal lenses 4. Antenna 1 about half length of body, 1st joint nearly as long as 2^d, flagellum in ♀ 30-jointed, scarcely twice as long as peduncle; antenna 2 rather longer than antenna 1, ultimate joint of peduncle a little longer than penultimate. Gnathopod 1, 6th joint distally rather widened instead of narrowed as is more usual. Peraeopod 5 (Fig. 29) nearly as in H. tubicola, but the little 6th and 7th joints more elongate. Uropod 1, inner ramus $\frac{2}{3}$ length of outer, both smooth; uropod 2 furnished with spines; uropod 3 in ♀ with rami not longer or broader than those of uropod 2, setose, the inner and shorter branch carrying also 2 spines.

Fig. 28. **H. laevis.** Mandible.

Fig. 29. **H. laevis.** Peraeopod 5.

Telson scarcely longer than broad, cleft fully $\frac{2}{3}$ length, sides slightly converging, apices broadly rounded. L. reaching 19 mm.

Arctic Ocean (between lat. 69° and 76° N., long. 36° and 65° E.). Depth 45—276 m.

4. H. setosa Boeck 1871 *H. s.*, A. Boeck in: Forh. Selsk. Christian., 1870 p.228 | 1876 *H. s.*, A. Boeck, Skand. Arkt. Amphip., *v.*2 p.541 t.30 f.7 | 1891 *H. s.*, G. O. Sars, Crust. Norway, *v.*1 p.194 t.68 f.1 | 1893 *H. s.*, A. Della Valle in: F. Fl. Neapel, *v.*20 p.489.

Fascicles of long slender setae rising from dorsal end of peraeon segments 5—7 and pleon segments 1—3. Head with lower corners distinctly angular. Postero-lateral angles of 3^d pleon segment very slightly produced. Patches of reddish pigment in the position of the absent corneal lenses. Antenna 1 more than half as long as body, 2^d joint of peduncle much longer than 1st, flagellum 24-jointed, twice as long as peduncle; antenna 2 subequal to antenna 1, ultimate and penultimate joints of peduncle subequal. Gnathopod 1, 6th joint distally narrow. Peraeopod 5, 2^d joint broad, hind margin straight, lower lobe broad, descending much below the 3^d, the expansion on surface and margin being densely setose, 4th and 5th joints subequal, 5th not equalling in length the narrow 6th and 7th joints combined. Uropod 1 slightly armed, inner ramus $^3/_4$ length of outer; uropod 2 with rami much shorter than those of uropod 3, which are foliaceous and setose, the inner and shorter carrying 2 spines. Telson about as long as broad, cleft scarcely beyond the middle, tapering to an almost acute apex. Colour greyish white, with brownish intestine. L. ♀ 13 mm.

Arctic Ocean; North-Atlantic, North-Sea and Skagerrak (South- and West-Norway).

5. **H. robusta** O. Sars 1891 *H. r.*, G. O. Sars, Crust. Norway, *v.* 1 p. 195 t. 68 f. 2 | 1893 *H. r., H. setosa?*, A. Della Valle in: F. Fl. Neapel, *v.* 20 p. 907, 932 | 1894 *H. r.*, T. Stebbing in: Bijdr. Dierk., *v.* 17 p. 18.

Perhaps not distinct from H. setosa. Fascicles of setae as in H. setosa. Head with lower angles more or less rounded. Postero-lateral angles of 3^d pleon segment not produced. Corneal lenses wanting. Antenna 1 in the type specimen less than half length of body, with 2^d joint of peduncle little longer than 1st, but sometimes little differing from the proportions in H. setosa; flagellum sometimes consisting of 33 joints; antenna 2, flagellum sometimes of 30 joints. Peraeopod 5 as in H. setosa, but 4th and 5th joints rather longer in proportion to breadth. Uropod 3 with 3 spines on the inner ramus. Telson rather longer than broad, cleft almost to the base. L. reaching 22·5 mm.

Arctic Ocean (Finmark, and between lat. 72⁰ and 76⁰ N., long. 16⁰ and 54⁰ E.).

---

4. Fam. **Haustoriidae**

1882 *Pontoporeiidae*, G. O. Sars in: Forh. Selsk. Christian., nr. 18 p. 22 | 1888 *P.*, T. Stebbing in: Rep. Voy. Challenger, *v.* 29 p. 804 | 1891 *P.*, G. O. Sars, Crust. Norway. *v.* 1 p. 121.

Head seldom rostrate. Side-plates of moderate size, generally fringed with setae, side-plate 5 bilobed. Antenna 1 (Fig. 30, 32) usually shorter than antenna 2 (Fig. 31, 33), with accessory flagellum, joints of peduncle sharply defined. Epistome not projecting. Upper lip with rounded margin. lower lip quadrilobate. Mandible with dentate cutting edge, accessory plate, prominent and large molar, palp 3-jointed. Maxillae 1 and 2 (Fig. 34) and maxillipeds usually but not always normal. Gnathopods 1 and 2 seldom powerful, weakly subchelate or chelate. Peraeopods 3—5 (Fig. 35, 36) varied, often adapted for burrowing. Branchial vesicles simple. Pleopods as a rule well developed, especially in ♂. Uropods 1—3 biramous, uropod 3 unlike uropods 1 and 2. Telson flattened, more or less deeply cleft.

Marine; also in brackish and fresh water.

8 genera, 21 accepted, 5 doubtful species.

Synopsis of the genera:

1. Gen. **Bathyporeia** Lindstr.

1855 *Bathyporeia* (Sp. un.: *B. pilosa*). Lindström in: Öfv. Ak. Förh.; *v.* 12 p. 59 |
1857 *Bathyporea* (laps.), Bate in: Ann. nat. Hist., ser. 2 *v.* 19 p. 271 | 1891 *Bathyporeia*,
G. O. Sars, Crust. Norway, *v.* 1 p. 127 | 1893 *B.*, A. Della Valle in: F. Fl. Neapel, *v.* 20
p. 751 | 1856 *Thersites* (nom. nud.), Bate in: Rep. Brit. Ass., Meet. 25 p. 59 | 1857 *T.*
(non L. Pfeiffer 1855, Mollusca!), Bate in: Ann. nat. Hist., ser. 2 *v.* 19 p. 146.

Body compressed. Head without rostrum. Side-plates not very large, 1st
narrow, bent forward. Pleon segment 4 with 2 dorsal setules curving forward.
Antenna 1 (Fig. 30) shorter than antenna 2, geniculate between the 1st and
2d joint, 1st about twice as long as 2d and 3d combined, accessory flagellum small,
2-jointed, flagellum with calceoli on upper margin in ♂. Antenna 2, ultimate
joint of peduncle shorter than penultimate, flagellum
short in ♀, long, with calceoli, in ♂. Mandible,
cutting plate narrowly produced, simple, few spines
in spine-row, 3d joint of palp narrow, curved. Maxilla 1,
inner plate with many setae, 2d joint of palp with
apex incurved, hairy. Maxillipeds, outer plates rather
short, with strong spines on inner margin, 2d joint
of palp large, hairy, strongly produced at inner apex,
3d narrow, incurved, 4th small. Gnathopod 1, 6th joint
oval, shorter than 5th, scarcely subchelate, finger small.

Peduncle, 1st joint

Accessory flagellum

Fig. 30.
B. guilliamsoniana.
Antenna 1.

Gnathopod 2, fringed with long setae, the 6th joint slenderly subfusiform,
finger wanting. Peraeopods 1 and 2 short, with 4th joint rather stout. Peraeopod 3 doubly geniculate in posture, 2d joint narrowest at the base, 4th
expanded, setose, 5th and 6th narrow, 7th wanting. Peraeopod 4, 2d and
4th joints more expanded than in peraeopod 5, both pairs with finger rudimentary, and having terminal joints fringed with setae and groups of spines.
Uropod 3, outer ramus long, 2-jointed, inner small, laminar. Telson cleft
to base, carrying spines on truncate apices and outer lateral margins.

5 species accepted, 1 doubtful.

Synopsis of accepted species:

1 ⎰ Pleon segment 3, postero-lateral angles produced
  ⎱   to a tooth . . . . 1. **B. guilliamsoniana** . p. 120
     Pleon segment 3, postero-lateral angles not
        produced to a tooth — **2.**

2 ⎰ Pleon segment 4 with dorsal spinules — **3.**
  ⎱ Pleon segment 4 without dorsal spinules — **4.**

3 ⎰ Pleon segment 4 with 1 pair of dorsal spinules 2. **B. pelagica** . . . . . p. 120
  ⎱ Pleon segment 4 with 2 pairs of dorsal spinules 3. **B. gracilis** . . . . . p. 121

4 ⎰ Peraeopods 4 and 5, 2ᵈ joint less than half
  ⎱   as long as rest of limb . . . . 4. **B. robertsoni** . . . . p. 121
     Peraeopods 4 and 5, 2ᵈ joint more than half
        as long as rest of limb . . . . 5. **B. pilosa** . . . . . . p. 121

1. **B. guilliamsoniana** (Bate) 1856 *Thersites guilliamsonia* (nom. nud.). Bate in: Rep. Brit. Ass., Meet. 25 p. 59 | 1857 *T. guilliamsoniana*, Bate in: Ann. nat. Hist., ser. 2 *v.* 19 p. 146 | 1857 *Bathyporea guilliamsonia, B. pilosa?*, Bate in: Ann. nat. Hist., ser. 2 *v.* 19 p. 271 | 1862 *Bathyporeia p.* (err., non Lindström 1855!), Bate, Cat. Amphip. Brit. Mus., p. 172 t. 31 f. 4 | 1891 *B. norvegica*, G. O. Sars, Crust. Norway, *v.* 1 p. 128 t. 43 | 1893 *B. n.*, A. Della Valle in: F. Fl. Neapel, *v.* 20 p. 754 | 1893 *B. n.*, T. Scott in: Rep. Fish. Board Scotl., *v.* 11 p. 213 t. 5 f. 22.

Back comparatively broad. Head scarcely as long as peraeon segments 1 and 2 combined. Side-plate 1 acutely pointed, 2ᵈ and 3ᵈ with small tooth at hinder angle. Pleon segment 3 with postero-lateral angles forming a short but distinct tooth; segment 4 dorsally depressed deeply at base, with a pair of spinules behind the bent setules. Eyes in ♀ reniform, in ♂ much longer, widening upward, dark red. Antenna 1 (Fig. 30), flagellum in ♀ with 8, in ♂ with 13 joints, 2ᵈ joint of accessory flagellum ¹/₃ as long as the 1ˢᵗ, which has spinules on both edges. Antenna 2, flagellum in ♀ with 8 joints, shorter than last 2 of peduncle, in ♂ slender, considerably longer than the body. Gnathopod 1, 6ᵗʰ joint broadly oval, much shorter than 5ᵗʰ. Peraeopod 3, 4ᵗʰ joint much expanded, setose, as long as 5ᵗʰ and 6ᵗʰ combined. Peraeopods 4 and 5, 2ᵈ joint about half as long as rest of limb. Uropod 3, 1ˢᵗ joint of outer ramus a little dilated distally, 2ᵈ about ¹/₃ as long, setose on both margins. Each half of telson sublinear, with about 9 apical spines and 4 lateral. Body pellucid, nearly colourless. L. ♀ 7, ♂ 8 mm.

North-Atlantic and North-Sea (England, between tide-marks; Scotland; South-Norway, depth 4—11 m).

2. **B. pelagica** (Bate) 1856 *Thersites p.* (nom. nud.), Bate in: Rep. Brit. Ass., Meet. 25 p. 59 | 1857 *T. p.*, Bate in: Ann. nat. Hist., ser. 2 *v.* 19 p. 146 | 1862 *Bathyporeia p.*, Bate, Cat. Amphip. Brit. Mus., p. 174 t. 31 f. 6 | 1891 *B. p.*, G. O. Sars. Crust. Norway, *v.* 1 p. 129 t. 44 f. 1 | 1893 *B. p.*, T. Scott in: Rep. Fish. Board Scotl., *v.* 11 p. 213 t. 5 f. 23—25 | 1875 *B. pilosa* (part.), T. Stebbing in: Ann. nat. Hist., ser. 4 *v.* 15 p 74 t. 3 | 1893 *B. p.* (part.), A. Della Valle in: F. Fl. Neapel, *v.* 20 p. 752 | 1877 *B. tenuipes*, Meinert in: Naturh. Tidsskr., ser. 3 *v.* 11 p. 201.

More slender than the preceding species; head fully as long as peraeon segments 1 and 2. Side-plate 1 less narrowed in front, obtusely pointed, in 2ᵈ and 3ᵈ tooth obsolete. Pleon segment 3, lateral angles rounded, without tooth; segment 4 with a like pair of spinules. Eyes in ♀ small, rounded oval, larger in ♂, bright red. Antenna 1, flagellum in ♀ 6-, in ♂ 9-jointed; accessory flagellum, 2ᵈ joint nearly half as long as 1ˢᵗ, which has spinules on outer edge only. Antenna 2, flagellum in ♀ with 7 joints, much shorter than last 2 of

peduncle, in ♂ about as long as the body. Gnathopod 1, 6ᵗʰ joint oblong
oval, somewhat shorter than 5ᵗʰ. Peraeopods less setose. Uropod 3, 1ˢᵗ joint
of outer ramus scarcely dilated distally, 2ᵈ nearly half as long, and in ♀
without lateral setae. Telson with about 6 apical spines and 2 lateral on
each half. Body pellucid, nearly colourless. L. scarcely more than 5 mm.

Arctic Ocean, North-Atlantic, North-Sea, Skagerrak and Kattegat (Scandinavia,
France, Great Britain).

3. **B. gracilis** O. Sars 1891 *B. g.*, G. O. Sars, Crust. Norway, *v.* 1 p. 132 t. 45
f. 1 | 1893 *B. pilosa* (part.), A. Della Valle in: F. Fl. Neapel, *v.* 20 p. 752.

Rather slender though not much compressed; head about as long as
peraeon segments 1 and 2. Side-plate 1 rather narrow, obtusely pointed, 2ᵈ and
(?) 3ᵈ with tooth. Pleon segment 3, angles rounded, without tooth; segment 4
not deeply depressed and carrying 2 pairs of spinules. Eyes imperfectly
developed, and not at all visible in spirit specimens. Antenna 1, flagellum
in ♀ with 6, in ♂ with 8 joints; accessory flagellum, 2ᵈ joint scarcely more than
¹/₅ length of 1ˢᵗ, which has spinules only on outer edge. Antenna 2 in ♀ twice as
long as antenna 1, flagellum slender, subequal to last 2 joints of peduncle.
8-jointed; in ♂ scarcely longer than in ♀, flagellum with 13 joints, distinct calceoli
on the first 5. Gnathopod 1, 6ᵗʰ joint regularly oval, much shorter than 5ᵗʰ.
Peraeopods 3—5 much more slender than in any of the other species known.
4ᵗʰ joint little expanded in peraeopod 3 and scarcely as long as 5ᵗʰ and
6ᵗʰ combined; 5ᵗʰ and 6ᵗʰ very elongate and narrow in peraeopods 4 and 5.
Uropod ᵥ3, 2ᵈ joint of outer ramus about ¹/₃ length of 1ˢᵗ, both margins
smooth. Each half of telson rather narrow, with 4 apical spines and
2 lateral. L. about 6 mm.

North-Atlantic (West-Norway). From rather deep water.

4. **B. robertsoni** Bate 1862 *B. r.*, Bate, Cat. Amphip. Brit. Mus., p. 173 t. 31 f. 5 |
1891 *B. robertsonii*, G. O. Sars, Crust. Norway, *v.* 1 p. 131 t. 44 f. 2 | 1893 *B. robertsoni*,
T. Scott in: Rep. Fish. Board Scotl., *v.* 11 p. 213 t. 5 f. 26—29 | 1875 *B. pilosa* (part.),
T. Stebbing in: Ann. nat. Hist., ser. 4 *v.* 15 p. 74 | 1893 *B. p.* (part.), A. Della Valle
in: F. Fl. Neapel, *v.* 20 p. 752.

In general agreement with B. pelagica. Pleon segment 4 without dorsal
spinules. Eyes rather large, oblong oval or reniform, very dark. Antenna 1
in ♂, flagellum 9-jointed; accessory flagellum, 2ᵈ joint extremely small.
1ˢᵗ with group of spinules in middle of inner margin. Antenna 2 in adult ♂
scarcely more than half length of body, flagellum about twice as long as
peduncle, with 17 joints, the first 15 each with rather large calceolus. Gnathopod 1, 6ᵗʰ joint ovate, about as long as 5ᵗʰ, finger hook-like. Uropod 3.
2ᵈ joint of outer ramus scarcely more than ¹/₅ length of 1ˢᵗ, which is rather
expanded. L. 6 mm.

Arctic Ocean, North-Atlantic and North-Sea (Finmark, Great Britain, France).

5. **B. pilosa** Lindstr. 1855 *B. p.*, Lindström in: Öfv. Ak. Förh., *v.* 12 p. 59 t. 2
f. 1—11 | 1891 *B. p.*, G. O. Sars, Crust. Norway, *v.* 1 p. 133 t. 45 f. 2 | 1893 *B. p.* (part.),
A. Della Valle in: F. Fl. Neapel, *v.* 20 p. 752.

In general agreement with B. pelagica. Pleon segment 4 slightly impressed
and with no dorsal spines behind the forward curving setules. Eyes in ♀ small,
rounded oval, in ♂ somewhat larger, blackish. Antenna 1, 1ˢᵗ joint scarcely
twice as long as 2ᵈ and 3ᵈ combined, flagellum in ♀ 6-, in ♂ 12-jointed;

accessory flagellum, 1st joint much coarser than in B. pelagica, 2d little more than ¹/₃ as long. Antenna 2 in ♀ little longer than antenna 1, flagellum shorter than ultimate and penultimate joints of peduncle combined, 8-jointed, in ♂ nearly as long as the body. Gnathopod 1, 6th joint oval, about as long as 5th. Peraeopod 3, 4th joint considerably expanded, densely fringed, much longer than the short 5th and 6th combined. Peraeopods 4 and 5 much shorter and coarser in structure than in the other 4 species. Uropod 3, 1st joint of outer ramus rather expanded, 2d scarcely more than ¹/₄ as long. Lobes of telson comparatively short, each with 6 apical and 3 lateral spines. L. about 5 mm.

Baltic, Kattegat. Depth 4—75 m.

B. lindströmi Stebb.*) 1893 B. pilosa (part.) (err., non Lindström 1855!), A. Della Valle in: F. Fl. Neapel, v. 20 p. 752 t. 5 f. 1; t. 36 f. 19—32.

General aspect robust. Side-plate 1 ending obtusely, 2d with tooth at hinder angle (front in text). Pleon segment 3, angles rounded, segment 4 carrying a single forward bending setule and a pair of spinules. Eyes oval, scarlet. Antenna 1 in ♀, flagellum 5- or 6-jointed. Antenna 2 in ♀, flagellum 10-jointed, nearly equal ultimate and penultimate joints of peduncle combined. Gnathopod 1, 6th joint broadly oval, much shorter than 5th. Peraeopods 2, 3 and 4 as in B. pelagica (p. 120), peraeopod 5 elongate as in B. gracilis (p. 121), but the finger said to be wanting. Uropod 3, outer ramus, 2d joint about ¹/₃ length of 1st. Apices of telson said to be rounded. No apical spines mentioned or figured. Body pellucid, colourless, or with a yellowish tinge. L. 5 mm.

Bay of Naples. Depth 10—20 m, in sand.

2. Gen. **Platyischnopus** Stebb.

1888 Platyischnopus (Sp. un.: P. mirabilis), T. Stebbing in: Rep. Voy. Challenger, v. 29 p. 830 | 1893 Platyschnopus, A. Della Valle in: F. Fl. Neapel, v. 20 p. 784.

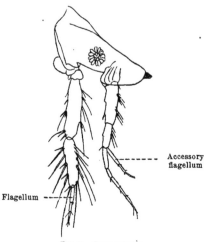

Fig. 31. **P. mirabilis.**
Head and antennae.

Head (Fig. 31) long, rostrate. Side-plates 1—3 small, 4th quadrate. Antenna 2 attached in front of eyes to lower surface of head, 2d joint of peduncle longer than 1st or 3d, flagellum in ♀ few-jointed, not longer than peduncle, accessory flagellum half as long, 3-jointed. Antenna 2 (Fig. 31) very little longer than antenna 1, attached behind the eyes, ultimate joint of peduncle shorter than penultimate, longer than the 3-jointed flagellum. Upper lip evenly rounded. Lower lip without inner lobes. Mandible, cutting edge narrowly produced, almost simple, accessory plate perhaps wanting, no spine-row, molar prominent, denticulate, palp with 3d joint shorter than 2d, both long and slender. Maxilla 1, inner plate small, outer with 8 or 9 nearly simple spines, palp 1-jointed with 3 setae. Maxilla 2, outer plate the broader. Maxillipeds, inner plates not large, outer narrow, with strong

*) Spec. nov. After Gustaf Lindström.

spine-teeth on inner margin, reaching beyond long 2^d joint of palp. Gnathopods 1 and 2 long and slender, distinctly chelate, or gnathopod 1 nearly simple, 2^d and 5th joints elongate. Peraeopods 1 and 2 small, 4th joint longer than 5th or 6th. Peraeopod 3, 2^d, 4th and 5th joints more or less widened. Peraeopod 4, 2^d joint oval, 4th and 5th much expanded. Peraeopod 5, 2^d joint much broader than long, 4th and 5th well expanded. Uropod 1, outer ramus a little, uropod 2 (and probably uropod 3), outer ramus much longer than inner. Telson rather longer than broad, widely emarginate at apex or deeply cleft.

The hood-like shape of head and the 1-jointed palp of maxilla 1 approximate this anomalous genus to the Phoxocephalidae (p. 133), the long, slender, chelate gnathopods 1 and 2 and the emarginate telson to some Acanthonotozomatidae (p. 210).

2 species.

Synopsis of species:

Gnathopods 1 and 2, 3^d joint elongate . . . . 1. **P. mirabilis** . . p. 123
Gnathopods 1 and 2, 3^d joint very short . . . . 2. **P. neozelanicus** . p. 123

1. P. mirabilis Stebb. 1888 *P. m.*, T. Stebbing in: Rep. Voy. Challenger, v. 29 p. 830 t. 58 | 1893 *Platyschnopus m.*, A. Della Valle in: F. Fl. Neapel, v. 20 p. 785 t. 60 f. 36.

Head longer than peraeon segments 1—3 combined, shallower where the small, round, dark eyes are placed than before or behind them (Fig. 31). Rostrum pointed. Pleon segment 3, postero-lateral angles produced into a small, acutely upturned point. In antenna 1 and 2 all the flagella slender, 1st joint longest. Gnathopod 1, 3^d joint elongate, 6th joint the widest of all, rather shorter than the 5th, its front margin much shorter than the hinder, being thus produced into an acute triangle for antagonism with the small nail-tipped finger. Gnathopod 2, all the joints except the finger longer than in gnathopod 1, 6th joint rather more than half as long as 5th, the chela-forming process shorter and sharper than in gnathopod 1. Peraeopods 1 and 2, 4th joint rather long and broad, finger slender, tapering. Peraeopods 3—5, 4th and 5th joints with many groups of spines; in peraeopod 3 2^d joint widened below, in peraeopod 4 regularly and rather broadly oval, in peraeopod 5 partially fused with the side-plate, twice as broad as long, with acute hinder angle. Peraeopod 5, 4th and 5th joints greatly expanded. Uropod 3, peduncle rather stout, much longer than one of the rami, the other ramus unknown. L. about 8 mm.

Port Jackson [East-Australia], depth 4—19 m; tropical Atlantic (harbour of Bahia).

2. P. neozelanicus Chilton 1897 *P. n.*, Chilton in: Ann. nat. Hist., ser. 6 v. 19 p. 1 t. 5.

Body rather broad, chiefly at peraeon. Head and position of antennae as in P. mirabilis. Antenna 1 rather shorter than antenna 2, peduncle longer than flagellum, 2^d joint a little longer than 1st, oblong, with 4 or 5 long plumose setae, flagellum 5-jointed, accessory flagellum 3-jointed. Antenna 2, peduncle stout, flagellum apparently about as long as ultimate joint of peduncle. Mouth-parts not precisely known. Gnathopod 1, 2^d joint ending in 2 long setae, 3^d short, 5th longer and wider than 6th, setose on sinuous hind margin, 6th slightly widened distally, with long setae on distal half of hind margin, no distinct palm, finger rather small. Gnathopod 2 similar in general shape but rather longer, 5th joint twice as long as 6th, and widened near the base, 6th produced distally to a small tooth, forming a little chela with the minute finger. Peraeopod 1, 6th joint rather longer than 5th, apex oblique with about 6 spiniform setae, a little longer and more slender than the finger. Feraeopod 2 rather longer than peraeopod 1, 5th joint with convex hind margin, carrying about 6 plumose setae, which reach beyond the finger. Peraeopod 3, 2[d] joint as broad as long, remainder not very different from P. mirabilis. Peraeopod 4 similar to peraeopod 3. Peraeopod 5, 2[d] joint broader than long, 4th and 5th moderately expanded. Uropod 3, peduncle much shorter than rami, outer slightly longer, 2-jointed, each with 2 or 3 long plumose setae. Telson double or deeply cleft, each lobe with 2 lateral setules and a stout apical spine. L. about 4 mm.

Otago Harbour [New Zealand].

3. Gen. **Haustorius** St. Müll.

1775 *Haustorius* (Sp. un.: *H. arenarius*), St. Müller in: Slabber, Phys. Belustig., p. 48 | 1888 *H.*, T. Stebbing in: Rep. Voy. Challenger, *v.* 29 p. 39 | 1891 *H.*, G. O. Sars, Crust. Norway, *v.* 1 p. 134 | 1893 *H.*, A. Della Valle in: F. Fl. Neapel, *v.* 20 p. 750 | 1818 *Lepidactylis* (Sp. un.: *L. dytiscus*), Say in: J. Ac. Philad., *v.* 1 II p. 379 | 1825 *Pterygocerus* (Sp. un.: *Oniscus arenarius*), Latreille in: Enc. méth., *v.* 10 p. 236 | 1829 *Pterygocera*, Latreille in: G. Cuvier, Règne an., n. ed. *v.* 4 p. 124 | 1851 *Bellia* (Sp. un.: *B. arenaria*) (non H. Milne Edwards 1848, Decapoda!), Bate in: Ann. nat. Hist., ser. 2 *v.* 7 p. 318 | 1854 *Sulcator*, Bate in: Ann. nat. Hist., ser. 2 *v.* 13 p. 504.

Back broad, except in pleon segments 4—6, which are usually folded in. Head with small rostrum. Side-plates 1—3 obtusely pointed, 4th little

Accessory flagellum

Inner plate

Fig. 32. Antenna 1.   Fig. 33. Antenna 2.   Fig. 34. Maxilla 2.

Fig. 32—34. H. arenarius. [After G. O. Sars.]

excavate. Antennae 1 and 2 (Fig. 32, 33), joints of peduncle very distinct, with fan-like armature of plumose setae chiefly on penultimate, flagellum

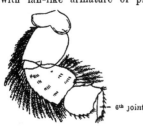

shorter than peduncle; accessory flagellum of antenna 1 well developed; penultimate joint of peduncle of antenna 2 longer and greatly wider than ultimate. Mandible normal, palp large, 3[d] joint not shorter than 2[d]. Maxilla 1, inner plate small, with several setae, outer apically narrowed, palp with 2 sets of setae on 2[d] joint. Below the base is a setose lamina, of which 6th joint the homology and function have not hitherto been explained. Maxilla 2 (Fig. 34), inner plate with a sinuous setose ridge on the surface, outer much larger, semilunar, densely

Fig. 35. H. arenarius.
Peraeopod 4. [After G. O. Sars.]

setose on inner and ciliated on outer margin. Maxillipeds, inner and outer plates subequal, narrow, fringed on inner margin, 2[d] joint of palp large, much produced, 3^d sharply bent, 4th wanting. Gnathopod 1 simple, 5th joint much broader and longer than 6th, finger small. Gnathopod 2, forming a minute chela. Peraeopods 1 and 2, 4th and 5th joints rather broad, 5th and 6th short, 5th widened distally. Peraeopods 3—5 (Fig. 35) with 2^d, 4th and 5th joints much expanded and carrying many spines and setae. Finger wanting in all peraeopods. Pleopods weak. Uropod 1 strong, spinose; uropod 2 smaller, setose; uropod 3, outer ramus 2-jointed, longer than inner, setose at tip. Telson short, much broader than long, not tapering, a little incised.

1 species.

1. H. arenarius (Slabber) 1769 *Oniscus a., O. arenatius,* Slabber, Natuurk. Verlustig., p. 92 t. 11 f. 3, 4 | 1775 *Haustorius arenarius,* St. Müller in: Slabber, Phys. Belustig., p. 48 | 1891 *H. a.,* G. O. Sars, Crust. Norway, v. 1 p. 135 t. 46 | 1893 *H. a.,* A. Della Valle in: F. Fl. Neapel, v. 20 p. 750 t. 60 f. 22, 23 | 1825 *Oniscus a., Pterygocerus,* Latreille in: Enc. méth., v. 10 p. 236 | 1829 *O. a., Pterygocera,* Latreille in: G. Cuvier, Règne an., n. ed. v. 4 p. 124 | 1878 *P. arenaria,* Bovallius in: Bih. Svenska Ak., v. 4 nr. 8 p. 1 t. 1—4 | 1851 *Bellia a.,* Bate in: Ann. nat. Hist., ser. 2 v. 7 p. 318 t. 11 f. 1—8 | 1854 *Sulcator arenarius,* Bate in: Ann. nat. Hist., ser. 2 v. 13 p. 504 | 1871 *S. arenatius,* A. Boeck in: Forh. Selsk. Christian., 1870 p. 137 | 1818 *Lepidactylis dytiscus,* Say in: J. Ac. Philad., v. 1 п p. 380.

Pleon segment 3 much larger than 1st or 2^d, angles rounded, setose. Eyes small, imperfectly developed, light-coloured. Antenna 1 (Fig. 32), flagellum 9-jointed, accessory flagellum 5-jointed. Antenna 2 (Fig. 33) little longer than antenna 1, flagellum 10-jointed. Gnathopod 2 more slender than gnathopod 1, 6th joint much shorter than 5th. In peraeopod 4 (Fig. 35), 4th joint much larger than in the other pairs, longer than broad. In peraeopod 5 4th joint much broader than long. Telson twice as broad as long, cleft not reaching the middle, each lobe with an apical tuft of spinules and 2 lateral spinules. Colour paler than sand. L. ♀ 11 mm.

Burrows with great dexterity in wet sand. Will live long in suitable confinement.

North-Atlantic (Holland, France and Britain, North-America); Kattegat.

4. Gen. **Cardenio** Stebb.

1888 *Cardenio* (Sp. un.: *C. paurodactylus*). T. Stebbing in: Rep. Voy. Challenger, v. 29 p. 806 | 1893 *C.,* A. Della Valle in: F. Fl. Neapel, v. 20 p. 749 | 1890 *Cardenis,* Warburton in: Zool. Rec., v. 25 Crust. p. 18.

Head somewhat produced. Side-plate 1 very small, 4th broader than deep, scarcely excavate. Antenna 1 shorter than antenna 2, accessory flagellum small, with 2 joints, 2^d minute. Mandible short, strong, cutting edge quadri-dentate, accessory plate rather stronger on left than on right mandible, left with 3, right with 2 spines in spine-row, molar strong, palp broad, 3^d joint much shorter than 2^d. Maxilla 1, 12 setae on inner plate, 9 spines on outer, several spinules and setules on 2^d joint of palp. Maxilla 2, both plates broad, inner with lateral as well as apical armature. Maxillipeds, inner plates moderate, outer fringed with spinules, nearly reaching apex of palp's long 2^d joint; 1st of palp very short, 4th rudimentary or wanting. Gnathopods resembling those of Synopia. Gnathopod 1, 5th joint broad and long, 6th narrower, only half as long, both fringed with spines and setae, finger rudimentary. Gnathopod 2, 2^d, 5th and 6th joints long and slender, 6th more than half as long as 5th, capable of closing against it; both carrying long setae, finger wanting. Peraeopods 1—5, 4th joint widened, little in 2^d, greatly in 3^d pair; longer than 5th in all but the 1st pair; 5th joint rather wide; finger wanting in 1st, 2^d and 5th pairs; short and blunt, tipped with long spine in 3^d and 4th pairs. Uropod 1, peduncle longer than subequal rami; uropod 2, peduncle intermediate in length between rami; uropod 3, peduncle much shorter than lanceolate rami. Telson long, tapering, cleft nearly to base.

1 species.

1. **C. paurodactylus** Stebb. 1888 *C. p.*, T. Stebbing in: Rep. Voy. Challenger, *v.*29 p.806 t.53 | 1893 *C. p.*, A. Della Valle in: F. Fl. Neapel, *v.*20 p.750 | 1890 *Cardenis p.*, Warburton in: Zool. Rec., *v.* 25 Crust. p. 18.

Back rounded, head as long as peraeon segments 1 and 2 combined. Hind margin of pleon segments dorsally denticulate. Pleon segment 3, postero-lateral angles acute, segments 4—6 comparatively long. Eyes reniform, small but with numerous components, set near front of head, dark in spirit. Antenna 1, joints of peduncle nearly equal, not long, flagellum in ♀ 4-jointed. Antenna 2, ultimate joint of peduncle rather longer than penultimate, flagellum in ♀ 5—7-jointed. Gnathopod 2, 6th joint with an apical projecting point. Peraeopod 3, 2^d joint almost circular, but wider than deep. Peraeopod 4, 2^d joint with sinuous hind margin, little wider than the 4th and not so long. Peraeopod 5, 2^d joint more expanded, 4th shorter, 6th longer than in preceding pair. L. 5 mm.

Southern Indian Ocean (Kerguelen Islands).

5. Gen. **Priscillina** Stebb.

1871 *Priscilla* (Sp. un.: *P. armata*) (non Jam. Thomson 1864, Coleoptera!), A. Boeck in: Forh. Selsk. Christian., 1870 p.124 | 1891 *P.*, G. O. Sars, Crust. Norway, *v.* 1 p.125 | 1888 *Priscillina*, T. Stebbing in: Rep. Voy. Challenger, *v.* 29 p. 1680 | 1893 *P.*, A. Della Valle in: F. Fl. Neapel, *v.* 20 p. 754.

Side-plates 1—4 strong, ending below in an obtuse, setose point, sideplate 5 small. Antenna 1 shorter than antenna 2, accessory flagellum rather large; peduncle setose in both pairs. Mandible normal, palp robust. Maxilla 1, inner plate large with many setae; maxilla 2, plates subequal. Maxillipeds, outer plates scarcely reaching end of 2^d joint of palp. Gnathopods 1 and 2 subchelate, 6th joint as broad as 5th, finger rather short. Peraeopods 1—5 with many fascicles of spines, finger minute. Peraeopods 3 and 4, 2^d joint little expanded, upper hind angle (at least in ♀) produced to a hook-like process. Peraeopod 5, 2^d joint greatly expanded, densely setose. Uropods 1—3 rather robust. Uropod 3, outer ramus spinose, with minute 2^d joint, inner very small. Telson laminar, deeply notched.

1 species.

1. **P. armata** (Boeck) 1861 *Pontoporeia a.*, A. Boeck in: Forh. Skand. Naturf., Møde 8 p. 648 | 1871 *Priscilla a.*, A. Boeck in: Forh. Selsk. Christian., 1870 p.124 | 1891 *P. a.*, G. O. Sars, Crust. Norway, *v.*1 p.126 t. 42 | 1888 *Priscillina a.*, T. Stebbing in: Rep. Voy. Challenger, *v.*29 p. 1719 | 1893 *P. a.*, A. Della Valle in: F. Fl. Neapel, *v.*20 p. 754 t. 60 f. 24, 25.

Head with sharply produced postero-antennal angles. Body rather robust, back broad. Pleon segments 2 and 3 dorsally ending in upturned acute process, postero-lateral corners rounded. Eyes inconspicuous (in spirit). Antenna 1, flagellum much shorter than peduncle, 10-jointed, accessory flagellum more than half length of primary, 5-jointed. Antenna 2, ultimate joint of peduncle much smaller than large penultimate, flagellum in ♀ about half length of peduncle. Gnathopod 1, 6th joint as long as 5th, rather broader, palm oblique. Gnathopod 2, palm transverse. Peraeopod 3, hook of 2^d joint larger than in peraeopod 4. Peraeopod 5, 2^d joint nearly orbicular, 5th gradually expanded distally, with series of short spines. Telson little longer than broad, notched widely a fourth of length, each truncate apex with 4 spinules. L. 11 mm.

Arctic Ocean and North-Atlantic (Greenland; West-Norway?).

6. Gen. **Pontoporeia** Krøyer

1842 *Pontoporeia* (Sp. un.: *P. femorata*), Krøyer in: Naturh. Tidsskr., *v.*4 p.152 | 1891 *P.*, G. O. Sars, Crust. Norway, *v.*1 p.122 | 1893 *P.*, A. Della Valle in: F. Fl. Neapel, *v.*20 p.716 | 1846 *Pontoporia*, L. Agassiz, Nomencl. zool., Index p.305 | 1853 *Pontiporeia*, J. D. Dana in: U. S. expl. Exp., *v.*13 π p. 912.

Side-plate 5, front lobe the larger. Antennae 1 and 2 in ♀ subequal, with short flagella, in ♂ flagella long. Mandible normal, palp slender, setose. Maxilla 1, inner plate with a few setae. Maxilla 2, outer plate the broader. Maxillipeds, outer plates not quite reaching end of broad 2^d joint of palp, palp's 4th joint small. Gnathopod 1 setose, 5th joint laminarly expanded, 6th smaller than 5th, palm ill-defined, finger feeble. Gnathopod 2 slender, subchelate or almost chelate, 6th joint narrow, finger small. Peraeopod 4 the longest. Peraeopods 1 and 2, 4th joint rather long. Peraeopods 3 and 4, 2^d joint slightly expanded, narrowed below. Peraeopod 5 (Fig. 36), 2^d joint greatly expanded, setose. Peraeopods 1—5, finger small. Uropod 3 short, rami 1-jointed, outer the larger. Telson squamiform, deeply cleft.

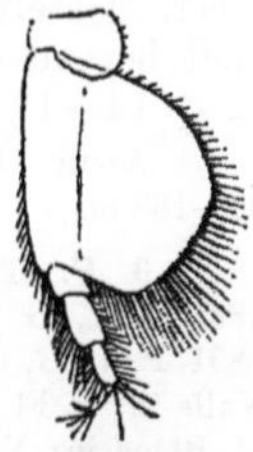

Fig. 36.
P. femorata.
Peraeopod 5.
[After G. O. Sars.]

3 species accepted, 1 doubtfully distinct.

Synopsis of accepted species:

1 { Gnathopod 1, palm short . . . . 1. **P. microphthalma** . . p. 127
  { Gnathopod 1, palm not short — **2.**

2 { Pleon segment 4 with furcate dorsal process . 2. **P. femorata** . . . . . p. 128
  { Pleon segment 4 without dorsal process or
    apical spines . . . . 3. **P. affinis** . . . . . . p. 128

1. **P. microphthalma** O. Sars 1896 *P. m.*, (O. Grimm in MS.) G. O. Sars in: Bull. Ac. St.-Pétersb., ser. 5 *v.*4 p. 428 t. 2 f. 1—7.

Back not broad. Pleon segment 4 dorsally raised at apex with 2 denticles (spines?) at top and a spinule on each side. Head, lateral corners narrowly rounded. Side-plates and pleon segment 3 as in P. affinis (p. 128). Eyes small, irregularly oval, pigment light. Antenna 1, flagellum subequal to peduncle, 9-jointed, accessory flagellum 3-jointed. Antenna 2, last 3 joints of peduncle thick, hirsute, flagellum 9-jointed. Gnathopod 1 with much shorter palm than in the other 2 species. Gnathopod 2, palm transverse, not as in the other 2 species produced to a thumb-like prominence. Peraeopod 5, 2^d joint longer than rest of limb, more regularly rounded oval than in the other 2 species, with which the uropods and telson agree. L. (♀?) 6 mm.

Caspian Sea. Depth 144—162 m.

2. **P. femorata** Krøyer 1842 *P. f.*, Krøyer in: Naturh. Tidsskr., *v.* 4 p. 153 |
1846 *P. f.*, Krøyer in: Voy. Nord, Crust. t. 23 f. 2a—y | 1891 *P. f.*, G. O. Sars, Crust.
NorWay, *v.* 1 p. 123 t. 41 f. 1 | 1893 *P. f.*, A. DellaValle in: F. Fl. Neapel, *v.* 20 p. 717
t. 60 f. 7 | 1859 *P. furcigera*, R. M. Bruzelius in: Svenska Ak. Handl., n. ser. *v.* 3 nr. 1
p. 49 t. 2 f. 8 | 1884 *P. f.*, H. Blanc in: N. Acta Ac. Leop., *v.* 47 p. 60 t. 7 f. 40—44.

Back broad. Head, lateral corners angular. Side-plates 1—3 with
tooth at lower hind corner, 1st broadly expanded, 5th with front lobe deep.
Pleon segment 3, postero-lateral angles narrowly rounded; segment 4 with
bifurcate process on the back. Eyes reuiform, bright red. Antenna 1 in
♀, flagellum shorter than peduncle, 9-jointed, accessory flagellum 2-jointed.
Antenna 2 in ♀, ultimate and penultimate joints of peduncle subequal, together
equal to 12-jointed flagellum. Gnathopod 1, 5th joint very broad, 6th obliquely
and broadly oval, finger very slender and curved. Gnathopod 2, 6th joint
a little shorter than 5th. Peraeopod 5 (Fig. 36), 2d joint very broad, much
longer than rest of limb. Uropod 3, outer ramus considerably larger than
inner, densely spinose on outer edge. Telson rather longer than broad,
cleft beyond centre, 3 setules on each rounded apex. Colour paleyellowish.
L. ♀ 14—17 mm.

Arctic Ocean; North-Atlantic (Labrador, NorWay); Kattegat; Baltic. Depth
38—188 m.

3. **P. affinis** Lindstr. 1855 *P. a.*, Lindström in: Öfv. Ak. Förh., *v.* 12 p. 63 |
1867 *P. a.*, G. O. Sars, Crust. d'Eau douce Norvége, p. 82 t. 7 f. 10—25; t. 8 f. 1—5 |
1891 *P. a.*, G. O. Sars, Crust. NorWay, *v.* 1 p. 124 t. 41 f. 2 | 1893 *P. a.* (part.), A. Della
Valle in: F. Fl. Neapel, *v.* 20 p. 717 t. 54 f. 1 | 1885 *P. femorata* (err., non Krøyer 1842!),
H. Blanc in: N. Acta Ac. Leop., *v.* 47 p. 58 t. 6 f. 33; t. 7 f. 34—39.

Back less broad than in P. femorata. Pleon dorsally hirsute, segment 4
without process. Head, lateral corners narrowly rounded. Side-plates 1—3 without
tooth, 1st little expanded, 5th with front lobe not very deep. Pleon segment 3,
corners subquadrate. Eyes small, oval, black. Antenna 1 in ♀, flagellum as
long as peduncle, accessory flagellum 3-jointed. Antenna 1 in ♂ (*P. filicornis*
Smith), flagellum 35-jointed; antenna 2 in ♂, flagellum 50-jointed. Gnathopod 1,
5th joint of moderate breadth, 6th narrow oval. Gnathopod 2, 6th joint
about as long as 5th. Peraeopod 5, 2d joint scarcely longer than rest of
limb. Uropod 3 in ♀ very small. Telson small, broader than long, cleft
about to centre (Sars) or nearly to base (Blanc). Colour yellowish, orange-
tinged, bluish-green borders. L. ♀ 8 mm.

Fresh Water lakes (NorWay, SWeden, Russia, North-America); Baltic, Kattegat,
Kara Sea, North-Atlantic (France).

P. hoyi S. I. Sm. 1871 *P. affinis* (err.?, non Lindström 1855!), S. I. Smith (&
A. E. Verrill) in: Amer. J. Sci., ser. 3 *v.* 2 p. 452 | 1893 *P. a.* (part.), A. DellaValle in:
F. Fl. Neapel, *v.* 20 p. 717 | 1874 *P. hoyi* + *Pontiporeia filicornis*, S. I. Smith in: Rep.
U. S. Fish. Comm., *v.* 2 p. 647 t. 2 f. 5; p. 649.

Neither the „papilliform appendages" (mentioned also by Sars 1867 in P. affinis)
on the sternal portion of the thoracic segments in P. hoyi, nor the long flagella with
calceoli in *P. filicornis* ♂ are of specific Value.

North-America (Lake Superior, Lake Michigan).

7. Gen. **Urothoe** Dana

1852 *Urothoe*, J. D. Dana in: Amer. J. Sci., ser. 2 *v.* 14 p. 311 | 1853 *U.* (part.), J. D.
Dana in: U. S. expl. Exp., *v.* 13 II p. 920 | 1857 *Urothoë*, Bate in: Ann. nat. Hist., ser. 2 *v.* 19
p. 145 | 1891 *Urothoe*, T. Stebbing in: Tr. zool. Soc. London, *v.* 13 I p. 1 | 1891 *Urothoë*, G. O.
Sars, Crust. NorWay, *v.* 1 p 137 | 1893 *Urothoe*, A. DellaValle in: F. Fl. Neapel. *v.* 20 p. 663 |
1853 *Egidia* (Sp. un.: *E. pulchella*). A. Costa in: Rend. Soc. Borbon., n. ser. *v.* 2 p. 170.

Body robust. Head slightly produced, sides triangular downward below the antennae. Side-plates not large. Gland-cells numerous. Antenna 1, joints of peduncle not very unequal, flagellum and accessory flagellum very short. Antenna 2, peduncle spinose, ultimate joint in $\male$ with calceoli, flagellum in $\female$ 2- or 3-jointed, in $\male$ long, many-jointed, with calceoli. Upper lip rounded. Mandible, cutting edge scarcely denticulate, accessory plate small, spine-row wanting, molar strong, palp slender, 1st joint not short, 3^d as long as or longer than 2^d. Maxilla 1, inner plate slight, number of setae small, variable, outer plate with 11 spines, 2^d joint of palp equal to or longer or shorter than 1st, with 3 apical setae. Maxilla 2, inner plate with armature on inner margin, outer rather the broader. Maxillipeds, plates of moderate size, outer fringed with spines, 2^d joint of palp produced, 3^d club-shaped, 4th slender. Gnathopods 1 and 2 similar, weakly subchelate, 6th joint shorter than 5th; finger longer in gnathopod 1 than in gnathopod 2. Peraeopods 1 and 2, 4th joint longer than either 5th or 6th, both of which are strongly spined. Peraeopod 3, joints 2—5 dilated, 2^d quadrate, joints 4—6 strongly spinose, joint 5 as wide as or wider than 4th; peraeopod 4 the longest; peraeopods 4 and 5, 2^d joint oval, others not expanded. All the legs have a cap over finger-tip. Peraeopods 1—5, finger nodulous or serrulate. Pleopods, inner ramus the shorter. Uropods 1 and 2, rami narrow. Uropod 3, rami foliaceous, not very unequal, setose especially in $\male$. Telson cleft nearly to base. As a rule peraeopods 1 and 2 have a row of setae on outer surface of joints 4 and 5; there are long plumose setae in peraeopod 3 on inner distal margin of joints 4 and 5 and on inner surface of joint 6, in peraeopod 4 on inner surface of 2^d joint and hind margin of 4th, in pleon segment 2 on lower inner surface.

7 species accepted, 3 obscure.

Synopsis of the accepted species:

1 { Without eyes . . . . 1. **U. abbreviata** . p. 129
  { With eyes — **2.**

2 { Uropod 1, rami strongly curved . . . . 2. **U. marina** . . . p. 130
  { Uropod 1, rami not strongly curved — **3.**

3 { Peraeopod 3, 5th joint extremely broad — **4.**
  { Peraeopod 3, 5th joint not extremely broad — **5.**

4 { Uropod 3, inner ramus longer than outer . . . . 3. **U. grimaldii** . . p. 130
  { Uropod 3, inner ramus shorter than outer . . . . 4. **U. pulchella** . . p. 130

5 { Uropods 1 and 2 strongly armed . . . . 5. **U. poucheti** . . p. 131
  { Uropods 1 and 2 weakly armed — **6.**

6 { Antenna 1, accessory flagellum long . . . . 6. **U. brevicornis** . p. 131
  { Antenna 1, accessory flagellum short . . . . . . 7. **U. elegans** . . . p. 131

1. **U. abbreviata** O. Sars 1879 *U. a.*, G. O. Sars in: Arch. Naturv. Kristian., *v.* 4 p. 446 | 1885 *U. a.*, G. O. Sars in: Norske Nordhavs-Exp., *v.* 6 Crust. I p. 164 t. 14 f. 1 | 1891 *U. a.*, T. Stebbing in: Tr. zool. Soc. London, *v.* 13i p. 3, 9 | 1893 *U. irrostrata* (part.), A. Della Valle in: F. Fl. Neapel, *v.* 20 p. 664.

Body remarkably short and thick-set. Pleon segment 3, postero-lateral corners quadrate. Eyes wanting. Antenna 1 much longer than antenna 2, 3 joints of peduncle uniform in length, flagellum 4-jointed, accessory flagellum 1-jointed. Peraeopod 3, 2^d joint short and broad, the others greatly (moderately in figure) dilated at the extremity, finger almost straight. Uropod 3 very small. Telson small. Colour yellowish white. L. 4 mm.

Perhaps the young of U. elegans (p. 131).

Arctic Ocean (Finmark). Depth 1167 m.

2. **U. marina** (Bate) 1857 *Sulcator marinus*, Bate in: Ann. nat. Hist., ser. 2 *v.* 19 p. 140 | 1862 *Urothoë m.*, Bate, Cat. Amphip. Brit. Mus., p. 115 t. 19 f. 2 | 1869 *U. m.? var. pectinatus*, E. Grube in: Abh. Schles. Ges., 1868/69 p 119 | 1891 *U. m.*, T. Stebbing in: Tr. zool. Soc. London, *v.* 13ı p. 16 t. 2 | 1893 *U. irrostrata* (part.), A. Della Valle in: F. Fl. Neapel, *v.* 20 p. 664.

Eyes very large in adult ♂, nearly meeting on top of head. Antenna 1, 3^d joint $^2/_3$ length of 2^d, flagellum 9-jointed, accessory flagellum 5-jointed. Antenna 2, flagellum in ♀ as long as ultimate joint of peduncle, flagellum in ♂ 50-jointed. Maxilla 1, inner plate with 4 or 5 setae, joints of palp equal. Gnathopod 1, 5th joint having at the distal margin a row of 12 pectinate spines and a row of microscopic spinules; gnathopod 2 without these rows, palm less oblique, finger shorter than in gnathopod 1. Peraeopods 1 and 2, finger with 5 tubercles on hind margin. Peraeopod 3, 2^d joint very broad, hind corners obtuse, 3^d broader than long, 4th rather longer, scarcely wider, inner apical margin fringed with 16 very long plumose setae, 5th nearly as long as 3^d and 4th combined and slightly wider, 6th longer, much narrower than 5th, 7th nearly as long as 6th, rather broad, except at tip, with about 13 nodules on front margin. Uropod 1, peduncle equal to outer ramus, rami smooth, strongly curved, inner much less than outer; uropod 2 smaller, rami smooth, slightly curved, outer scarcely longer than inner; uropod 3, rami subequal, rather long, moderately broad, in ♂ densely plumose, outer with minute 2^d joint. Telson, length and breadth equal, cleft nearly to base, each apex with feathered cilium, spine and 3 setae. L. 8 mm.

North-Atlantic (Shetland Islands, Moray Firth, Firth of Clyde, Liverpool Bay, France).

3. **U. grimaldii** Chevreux 1895 *U. g.*, Chevreux in: Mém. Soc. zool. France, *v.* 8 p. 428 f. 1—4.

♀ unknown. — ♂. Eyes very large, oval, meeting on top of head. Antenna 1, 2^d joint of peduncle rather longer than 1st, 1st than 3^d, flagellum shorter than 3^d, 4-jointed. Antenna 2, flagellum 40-jointed. Mouth-organs more robust than usual. Maxilla 1, inner plate small, with 3 setae, outer plate very broad, 1st joint of palp longer than 2^d. Gnathopods 1 and 2 rather slender. Peraeopod 3, 2^d joint with hinder angles rounded, 3^d and 4th subequal, 5th $2^1/_2$ times as broad as long, with very numerous spines, finger having 5 or 6 long spines on front margin. Uropods 1 and 2 with slender subequal rami, inner branch in 1st slightly armed. Uropod 3 uniquely having the inner ramus rather longer than the outer; 2^d joint of outer minute. Telson cleft nearly to the base, a little broader than long, with 2 spinules on each apex. L. 3·5 mm.

Mediterranean (Morocco). Depth 2 m.

4. **U. pulchella** (A. Costa) 1853 *Egidia p.*, A. Costa in: Rend. Soc. Borbon., n. ser. *v.* 2 p. 172 | 1876 *Urothoë p.*, A. Boeck, Skand. Arkt. Amphip., .*v.* 2 p. 225 | 1891 *U. p.*, T. Stebbing in: Tr. zool. Soc. London, *v.* 13ı p. 11 t. 4 ᴀ | 1893 *U. irrostrata* (part.), A. Della Valle in: F. Fl. Neapel, *v.* 20 p. 664 t. 5 f. 3, 8; t. 36 f. 1—18; t. 60 f. 11, 12.

Eyes in ♂ reniform, except in smaller stages, in ♀ circular. Antenna 1, 1st and 2^d joints equal, 3^d rather shorter, flagellum 5- or 6-jointed, little longer than ultimate joint of peduncle, accessory flagellum 3- or 4-jointed, longer than half primary. Antenna 2, flagellum in ♀ scarcely as long as ultimate joint of peduncle, flagellum in ♂ 34-jointed. Maxilla 1, inner plate slender, with 6 setae, 1st joint of palp rather broader and longer than 2^d. Gnathopod 1, 7 spines in the row on distal margin of 5th joint. Gnathopod 2, 6th joint about $^4/_5$ length of 5th, with the small convex palm more prominent than usual. Peraeopod 3, 4th joint scarcely longer or broader than 3^d, 5th more than 1$^1/_2$ times as broad as long, with many spines, finger narrowing abruptly to a sharp apex, its front margin not nodulous, but finely serrulate. Uropod 1, rami curved, armed with 2 spines; uropod 2 smaller, rami smooth; uropod 3, rami plumose but not densely, outer exceeding length of inner by its small 2^d joint. Telson short, a little broader than long, sides convex, each apex armed with a spinule. Colour of ♀ grey, tinted on the back with violet, of ♂ crimson. L. 5 mm.

Gulf of Naples, North-Atlantic (West-France).

5. **U. poucheti** Chevreux 1888 *U. p.*, Chevreux in: Bull. Soc. zool. France, *v.* 13 p. 34 | 1891 *U. p.*, T. Stebbing in: Tr. zool. Soc. London, *v.* 13i p. 9, 25 | 1893 *U. irrostrata* (part.), A. Della Valle in: F. Fl. Neapel, *v.* 20 p. 664.

♀ unknown. — ♂. Body comparatively slender. Eyes very large and dark, meeting on the top of the head. Peraeopod 3, 2^d joint with hinder angles well rounded, 4th not broader than 3^d, not strongly spined, 5th not broader than 4th, finger long, slender, distal half serrate. Uropod 1, rami equal, slender, not quite as long as peduncle, which is more spinose than in other species, outer ramus with 4 spines; uropod 2, rami equal, slender, shorter than peduncle, outer with 3 spines; uropod 3, rami long. L. 5 mm.

North-Atlantic (Azores).

6. **U. brevicornis** Bate 1862 *U. b.*, Bate, Cat. Amphip. Brit. Mus., p. 116 t. 20 f. 1 | 1891 *U. b.*, T. Stebbing in: Tr. zool. Soc. London, *v.* 13i p. 23 t. 3, 4c | 1893 *U. irrostrata* (part.), A. Della Valle in: F. Fl. Neapel, *v.* 20 p. 664.

Eyes in ♂ conspicuous, almost contiguous at the top. Antenna 1 in ♂, 1st joint longer than 2^d, flagellum 7-jointed, accessory flagellum 6-jointed, more than half length of primary. Antenna 2, flagellum in ♀ shorter than ultimate joint of peduncle, flagellum in ♂ 23- or 24-jointed. Maxilla 1, inner plate with 3 setae, 1st joint of palp rather longer than 2^d. Gnathopod 1 as in U. marina (p. 130), spines on distal margin of 5th joint varying in number. Peraeopod 3, 2^d joint with angles more acute than usual, other joints as in U. marina, except finger, which tapers gradually and has finer nodules. Uropod 1, peduncle longer than in U. marina, equal to the rami, which are slender, subequal, nearly straight, reaching beyond the 2^d pair, outer with 3 spinules, inner with 1—3 setules; uropod 2, peduncle shorter than rami, which are subequal, straight, with 2 or 3 spinules on outer and a spinule on the inner. Uropod 3 and telson as in U. marina. L. about 8 mm.

St. George's Channel, Bristol Channel and English Channel (Wales and Devonshire).

7. **U. elegans** Bate 1857 *U. e.*, Bate in: Ann. nat. Hist., ser. 2 *v.* 19 p. 145 | 1861 *U. norvegica*, A. Boeck in: Forh. Skand. Naturf., Møde 8 p. 647 | 1891 *U. n.*, G. O. Sars, Crust. Norway, *v.* 1 p. 138 t. 47 ! 1891 *U. elegans* + *U. n.*, T. Stebbing in: Tr. zool. Soc. London. *v.* 13i p. 13 t. 4; p. 21 t. 4в | 1893 *U. irrostrata* (part.), A. Della Valle in: F. Fl. Neapel, *v.* 20 p. 664.

Eyes moderately large in adult ♂. Antenna 1, 3^d joint $^2/_3$ length of 2^d, flagellum 6-jointed, accessory flagellum 3-jointed. Antenna 2, flagellum in ♀ equal ultimate joint of peduncle, 2- or 3-jointed, flagellum in ♂ 40-jointed. Maxilla 1, inner plate with 2 setae, joints of palp equal. Peraeopods 1 and 2, finger with 3 or 4 distant pointed tubercles. Peraeopod 3, 2^d joint rather variable

9*

in width, 4th joint longer and wider than 3d, 5th longer than broad, not broader than 4th, finger rather narrow, distal half slenderly tapering, with 7 small tubercles on front margin. Uropod 1, rami smooth or almost smooth, nearly straight, subequal; uropod 2 much smaller, otherwise similar; uropod 3, rami rather broad, plumose, outer rather the longer. Telson longer than broad, cleft nearly to the base, with spinule and setule on each apex. Colour in life, adorned with rose-coloured markings (Bate), yellowish, changing in the ♀ to light orange (Sars). L. 4—6 mm.

North-Atlantic (England, Scotland, West-France, Azores, Norway).

U. bairdii Bate 1862 *U. b.*, Bate, Cat. Amphip. Brit. Mus., p. 114 t. 19 f. 1 | 1871 *U. norvegica* (err., non A. Boeck 1861!), A. Boeck in: Forb. Selsk. Christian., 1870 p. 138 | 1891 *U. marinus* (part.), T. Stebbing in: Tr. zool. Soc. London, v. 13 i p. 7 | 1893 *U. irrostrata* (part.), A. Della Valle in: F. Fl. Neapel, v. 20 p. 664.

L. 5 mm.

Moray Firth.

U. irrostrata Dana 1853 & 55 *U. irrostratus*, J. D. Dana in: U. S. expl. Exp., v. 13 ii p. 922 t. 62 f. 6 a—f | 1891 *U. i.*, T. Stebbing in: Tr. zool. Soc. London, v. 13 i p. 10 | 1893 *U. irrostrata* (part.), A. Della Valle in: F. Fl. Neapel, v. 20 p. 664.

Antenna 1, flagellum 6- or 7-jointed, accessory flagellum 2- or 3-jointed. Peraeopods 4 and 5, finger nodulose on front margin.

Sooloo Sea.

U. rubra Giles 1888 *U. ruber*, G. M. Giles in: J. Asiat. Soc. Bengal, v. 57 p. 246 t. 11 | 1893 *U. r.*, A. Della Valle in: F. Fl. Neapel, v. 20 p. 895.

Colour bright brick red. L. about 3 mm.

Bay of Bengal (Banks of Chittagong). Surface.

8. Gen. **Urothoides** Stebb.

1891 *Urothoides* (Sp. un.: *U. lâchneëssa*), T. Stebbing in: Tr. zool. Soc. London, v. 13 i p. 26.

♀. Head with obtusely produced and deflexed rostrum. Antennae 1 and 2, mouth-organs, gnathopods 1 and 2, peraeopods 1 and 2 nearly as in Urothoe (p. 128). The anterior gastric lobes (triturating organs) scantily armed, not profusely as in Urothoe. Peraeopods 1—5, finger slender, not nodulose. Peraeopod 3, 2d joint expanded but not quadrate, 4th wider than 5th. Peraeopod 4 with 2d, 4th and 5th joints much expanded, 4th longer and wider than 5th. Peraeopod 5 much shorter than peraeopod 4, 2d joint greatly expanded and produced, as long as rest of limb, which is not expanded. Uropod 1, rami stiliform, subequal. almost smooth; uropod 2 much smaller, rami slender, smooth, subequal; uropod 3, outer ramus of 2 equal joints, inner spiniform, not half the length of the outer. Telson little longer than broad, cleft nearly to the base, apices acute, not dehiscent. — ♂ unknown.

1 species.

1. U. lachneëssa (Stebb.) 1888 *Urothoë l.*, T. Stebbing in: Rep. Voy. Challenger, v. 29 p. 825 t. 57 | 1893 *Urothoe lachneessa*, A. Della Valle in: F. Fl. Neapel, v. 20 p. 667 t. 60 f. 13 | 1891 *Urothoides lachneëssa*, T. Stebbing in: Tr. zool. Soc. London, v. 13 i p. 26.

Back covered with a sort of bristly down. Head at the base and peraeon dorsally very broad. Pleon segment 2 without plumose setae; segment 3, postero-lateral angles rounded. Eyes wanting or not perceived.

Antenna 1 in ♀, flagellum shorter than peduncle, 5-jointed, accessory flagellum 2-jointed. Antenna 2 in ♀, ultimate joint of peduncle shorter than penultimate, each with a few strong spines, flagellum 2- or 3-jointed. Gnathopod 1, finger nearly as long as the oval hand; no row of pectinate spinules on distal margin of 5th joint was noticed. Peraeopod 4, the large 2d joint with front margin setose, hind margin sinuous, 4th widening and 5th narrowing distally. Peraeopod 5, 2d joint strongly serrate behind, produced below the 4th joint. Telson broad at base, narrowing gradually to the acute apices. L. about 4 mm.

Cumberland Bay [Kerguelen Island]. Depth 236 m.

5. Fam. Phoxocephalidae

1857 *Phoxides*, Bate in: Ann. nat. Hist., ser. 2 *v.* 20 p. 525 | 1865 Subfam. *Phoxina*, W. Lilljeborg in: N. Acta Soc. Upsal., ser. 3 *v.* 6 nr. 1 p. 18 | 1871 Subfam. *Phoxinae*, A. Boeck in: Forh. Selsk. Christian., 1870 p. 133 | 1885 *Phoxidae*, G. O. Sars in: Norske Nordhavs-Exp., *v.* 6 Crust. I p. 154 | 1891 *Phoxocephalidae*, G. O. Sars, Crust. Norway, *v.* 1 p. 142.

Body fusiform. Hooded rostrum covering base of antenna, lateral corners of head obsolete, postero-antennal angles distinct. Side-plates rather large, 1st—4th obtusely truncate, hinder lobe of side-plate 5 the deeper. Pleon segments 4—6 in ♂ narrower than in ♀. Eyes dorsal, lateral or wanting. Antennae more or less modified in ♂, short in ♀, accessory flagella well developed. Epistome not projecting. Upper lip rounded. Inner lobes of lower lip usually small. Mandible short, cutting edges distinctly developed, molar variable, seldom large, palp rather large. Maxilla 1, inner plate small, palp 1- or 2-jointed. Maxilla 2, plates short, rounded at apex. Maxillipeds, plates small, palp large. Gnathopods 1 and 2 generally similar in form, hands subchelate, large, with palmar tooth fortified by a spine. Peraeopod 4 the longest. Peraeopod 5 (Fig. 37, 38) short, 2d joint greatly expanded. Branchial vesicles simple. Pleopods stronger in ♂ than in ♀. Uropod 3 often varying sexually, outer ramus the longer, 2-jointed. Telson cleft to or nearly to the base.

Marine.

7 genera, 23 accepted, 7 doubtful species.

Synopsis of genera:

1 { Maxilla 1, palp 1-jointed — 2.
{ Maxilla 1. palp 2-jointed — 5.

2 { Mandible, molar well developed . . . . . . 1. Gen. **Phoxocephalus** . p. 134
{ Mandible, molar feebly developed — 3.

3 { Maxillipeds, 3d joint of palp with strongly
{ produced apex . . . . 2. Gen. **Leptophoxus** . . p. 136
{ Maxillipeds, 3d joint of palp with apex not
{ strongly produced — 4.

4 { Gnathopods 1 and 2 alike in size and shape 3. Gen. **Paraphoxus** . . . p. 137
{ Gnathopods 1 and 2 differing in size and shape 4. Gen. **Metaphoxus** . . . p. 138

5 { Eyes wanting . . . . 5. Gen. **Harpinia** . . . . p. 140
{ Eyes present — 6.

6 { Peraeopods 3 and 4, 4th joint strongly expanded 6. Gen. **Pontharpinia** . . p. 145
{ Peraeopods 3 and 4. 4th joint not strongly
{ expanded . . . . 7. Gen. **Parharpinia** . . . p. 147

1. Gen. **Phoxocephalus** Stebb.

1842 *Phoxus* (non Billberg 1820, Coleoptera!) (part.), Krøyer in: Naturh. Tidsskr.,
v. 4 p. 150 | 1842 *Spinifer* (non Rafinesque 1831, Mollusca! fide: L. Agassiz, Nomencl.
zool., Moll. p. 84) (part.), (Holbøll in MS.) Krøyer in: Naturh. Tidsskr., *v.* 4 p. 151 |
1888 *Phoxocephalus* (nom. nov.), T. Stebbing in: Rep. Voy. Challenger, *v.* 29 p. 810 |
1891 *P.*, G. O. Sars, Crust. Norway, *v.* 1 p. 143 | 1893 *P.* (part.), A. Della Valle in:
F. Fl. Neapel, *v.* 20 p. 738.

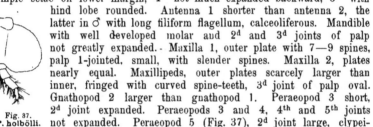

Fig. 37.
P. holbölli.
Peraeopod 5.
[After G. O.
Sars]

Body compressed. Hood more or less acute. Side-plates with a few
simple setae on lower margin, 4th not much expanded backward, 5th with
hind lobe rounded. Antenna 1 shorter than antenna 2, the
latter in ♂ with long filiform flagellum, calceoliferous. Mandible
with well developed molar and 2d and 3d joints of palp
not greatly expanded. - Maxilla 1, outer plate with 7—9 spines,
palp 1-jointed, small, with slender spines. Maxilla 2, plates
nearly equal. Maxillipeds, outer plates scarcely larger than
inner, fringed with curved spine-teeth, 3d joint of palp oval.
Gnathopod 2 larger than gnathopod 1. Peraeopod 3 short,
2d joint expanded. Peraeopods 3 and 4, 4th and 5th joints
not expanded. Peraeopod 5 (Fig. 37), 2d joint large, clypeiform. Uropod 3 in ♀, inner ramus unarmed, much shorter than
outer, in ♂ much larger than in ♀, both rami well developed,
lanceolate, fringed with plumose setae. Telson. lobes rather
narrow, especially in ♂.

3 species accepted, 4 doubtful.

Synopsis of the accepted species:

1 ⎰ Eyes ill developed . . . . 1. **P. holbölli** . . p. 134
  ⎱ Eyes well developed — **2.**

2 ⎰ Gnathopod 2, 6th joint not longer than broad . . . 2. **P. bassi** . . . p. 135
  ⎱ Gnathopod 2, 6th joint longer than broad . . . . . 3. **P. kergueleni** . p. 135

1. P. holbölli (Krøyer) 1842 *Phoxus h.* (*Spinifer spinosissimus + S. flagelli-
formis* Holbøll in MS.), Krøyer in: Naturh. Tidsskr., *v.* 4 p. 151 | 1845 *P. h.*, Krøyer
in: Naturh. Tidsskr., ser. 2 *v.* 1 p. 551 | 1876 *P. holbollii*, S. I. Smith (& Harger) in: Tr.
Connect. Ac., *v.* 3 p 29 | 1888 *Phoxocephalus holbölli*, T. Stebbing in: Rep. Voy. Challenger,
v. 29 p. 1717 | 1891 *P. h.*, G. O. Sars, Crust. Norway, *v.* 1 p. 144 t. 49 | 1893 *P. h.*, A. Della
Valle in: F. Fl. Neapel, *v.* 20 p. 740 | 1853 *Phoxus kroyeri*, Stimpson in: Smithson. Contr.,
v. 6 nr. 5 p 58.

Body glabrous. Hood acute, triangular, postero-antennal corners of head
nearly quadrate. Side-plate 1 little expanded below. 1st—3d with about 10 setae
on hinder part of lower margin, 4th angled at the emargination. Pleon segment 3,
postero-lateral angles slightly rounded. Eyes small, nearly round. an irregular
mass of whitish pigment without any trace of visual elements (Sars). Antenna 1,
1st joint rather longer than 2d and 3d combined. flagellum 6-, accessory flagellum
4-jointed. Antenna 2, penultimate joint of peduncle rather large, with 3 rows of
spines, flagellum in ♀ 6-jointed, in ♂ with numerous joints, nearly reaching end of
body. Gnathopods 1 and 2, 6th joint oblong oval, a little dilated distally, palm
shorter than hind margin; gnathopod 2 only a little larger than gnathopod 1.
Peraeopod 3. 2d joint broader above than below. Peraeopod 4 about half the
length of the body. Peraeopod 5 (Fig. 37), 2d joint as long as rest of limb,
as broad as long, hind margin strongly curved, finely serrate. Uropod 3 in ♀,
inner ramus scarcely half as long as outer, 2d joint of outer less than half
length of 1st; in ♂ rami nearly equal, narrowly lanceolate, plumose, outer
with 3 spines on outer margin. Telson much longer than broad, cleft to

the base, narrowing distally, with 2 spinules and a setule on each rounded apex. Colour (rather variable) light buff changing to orange, with opaque whitish shadows. L. 7 mm.

Arctic Ocean; North-Atlantic, North-Sea, Skagerrak, Kattegat and Baltic (Scandinavia, Great Britain, France, North-America). Sandy bottom, depth 38—94 m.

2. **P. bassi** Stebb. 1888 *P. b.*, *Phoxus b.*, T. Stebbing in: Rep. Voy. Challenger, *v.* 29 p. 811 t. 54 | 1893 *Phoxocephalus b.*, A. Della Valle in: F. Fl. Neapel, *v.* 20 p. 743.

♀ unknown. — ♂. Hood with narrowly rounded apex, postero-antennal corners subquadrate. Side-plate 1 well expanded below, with 14 setae, 2^d—4th with about 10 setae, 4th pair nearly quadrate, but a little deeper than broad, 5th with hind lobe little deeper than front. Pleon segment 3, postero-lateral angles rounded, segment 4 dorsally impressed. Eyes large, irregularly quadrate, dark in spirit. Antenna 1, flagellum 8-jointed, with calceoli, accessory flagellum 5-jointed. Antenna 2, calceoli on ultimate joint of peduncle, flagellum 37-jointed, nearly reaching end of body. Mandible, 3 spines in spine-row, molar small, but strongly dentate. Maxilla 1, outer plate with 9 spines. Maxillipeds, 6 spine-teeth on margin of outer plates. Gnathopod 1, 5th joint triangular, nearly as long as 6th, 6th oblong, twice as long as broad, palm convex, only slightly oblique, the finger with a membranous cap over the tip. Gnathopod 2 much larger, 5th joint small, with the 4th forming a cup for the hand or 6th joint. which is as broad as long, slightly widened at the palm, the palmar tooth strong. Peraeopods 1 and 2, spines at apex of 5th joint nearly as long as 6th joint, finger small. Peraeopods 3—5 and uropods 1—3 nearly as in P. holbölli. Telson much longer than broad, scarcely tapering, cleft almost to the base, 1 spinule and 1 setule on each apex. L. 10 mm.

Bass Strait. Surface.

3. **P. kergueleni** Stebb. 1888 *P. k.*, *Phoxus k.*, T. Stebbing in: Rep. Voy. Challenger, *v.* 29 .p 816 t. 55 | 1893 *Phoxocephalus k.*, A. Della Valle in: F. Fl. Neapel, *v.* 20 p. 742.

Hood with tolerably acute apex. Side-plate 1 a little expanded below, with 5 setae on lower margin. 4th little deeper than broad, hind margin convex, oblique, 5th with hind lobe much deeper than front. Pleon segment 3, postero-lateral angles rounded. Eyes small, distant, not dark in spirit. Antenna 1, flagellum 5-, accessory flagellum 3-jointed. Antenna 2, ultimate joint of peduncle half length of penultimate in ♂, more than half in ♀, flagellum in ♀ with 5 joints, in young ♂ without calceoli with 15 rather stout joints. Mandible, 3 spines in spine-row, molar strongly dentate, 3^d joint of palp shorter than 2^d in ♀, equal to it in ♂. Maxilla 1, outer plate apparently with only 7 spines. Maxillipeds, outer plates with 3 spine-teeth. Gnathopod 1, 5th joint much shorter than 6th, which is oblong, much broader than half the length, palm convex, slightly oblique. Gnathopod 2, 4th and 5th joints short, side by side, 6th much as in gnathopod 1, but much larger both in ♀ and ♂, the palm more oblique and palmar tooth stronger. Peraeopods 1 and 2, apical spine on 5th joint longer than 6th joint. Peraeopod 3, 2^d joint as broad below as above. Peraeopod 5. 2^d joint very broad, longer than rest of limb, hind margin slightly serrate, descending below the 4th joint. Uropod 3 in young ♂, inner ramus longer than 1st joint of outer, 2^d joint of outer half length of 1st. Telson not greatly longer than broad, not cleft to the base but beyond the middle, with a long spine at each broad apex. L. 5 mm.

Cumberland Bay [Kerguelen Island]. Depth 226 m.

P. erythrophthalmus (Catta) 1875 *Phoxus e.*, Catta in: Rev. Sci. nat., *v.* 4 p. 163 | 1888 *Phoxocephalus e.*, T. Stebbing in: Rep. Voy. Challenger, *v.* 29 p. 1717.

Differs from P. holbölli by the very perfect eyes.

Mediterranean.

P. geniculatus (Stimps.) 1855 *Phoxus g.*, Stimpson in: P. Ac. Philad., *v.* 7 p. 382 | 1888 *Phoxocephalus g.*, T. Stebbing in: Rep. Voy. Challenger, *v.* 29 p. 1717 | 1893. *P. g.*, A. Della Valle in: F. Fl. Neapel, *v.* 20 p. 743.

Rostrum long, acute. Eyes white. Antenna 1, flagellum and accessory flagellum equal, 10-jointed. Peraeopods 1 and 2, 4th and 5th joints dilated. Uropod 3, rami unequal, outer long, 3-jointed. Colour white. L. 6 mm.

North-Pacific (Japan).

P. obtusus (Stimps.) 1855 *Phoxus o.*, Stimpson in: P. Ac. Philad., *v.* 7 p. 382 | 1888 *Phoxocephalus o.*, T. Stebbing in: Rep. Voy. Challenger, *v.* 29 p. 1717.

L. 6 mm.

North-Pacific (Japan).

P. simplex (Bate) 1857 *Phoxus kröyerii* (non *P. kroyeri* Stimpson 1853!), Bate in: Ann. nat. Hist., ser. 2 *v.* 19 p. 140 | 1857 *P. simplex*, Bate in: Ann. nat. Hist., ser. 2 *v.* 20 p. 525 | 1888 *Phoxocephalus s.*, T. Stebbing in: Rep. Voy. Challenger, *v.* 29 p. 1717 | 1893 *P. s.*, A. Della Valle in: F. Fl. Neapel, *v.* 20 p. 743 | 1896 *P. s.*, A. O. Walker in: Ann. nat. Hist., ser. 6 *v.* 18 p. 157.

Plymouth Sound.

2. Gen. **Leptophoxus** O. Sars

1891 *Leptophoxus* (Sp. un.: *L. falcatus*), G. O. Sars, Crust. Norway, *v.* 1 p. 146.

Body much compressed. Hood slightly carinate and deflexed at apex. Side-plates with a few simple setae on lower margin, 4th broad, 5th with hinder lobe subtruncate. Antenna 1 much smaller than antenna 2. Antenna 2, peduncle stout, flagellum in ♀ short, in ♂ long and slender. Mandible, cutting-plates, spine-row and molar feebly developed, 2^d and 3^d joints of palp expanded. Maxillae 1 and 2 about as in Phoxocephalus (p. 134). Maxillipeds, inner plates very small, acuminate, 3^d joint of palp strongly produced at outer apex. Gnathopods 1 and 2, 6th joint nearly quadrangular, that of gnatho-pod 2 much the larger. Peraeopod 3, 2^d joint much expanded. Peraeopod 4, except 2^d joint, very slender. Peraeopod 5, 2^d joint large and clypeiform. Uropod 3 alike in ♂ and ♀, inner ramus very small, spiniform, outer slender with spines but no setae. Telson as in Phoxocephalus.

1 species.

1. **L. falcatus** (O. Sars) 1871 *Phoxus simplex* (err., non Bate 1857!), A. Boeck in: Forh. Selsk. Christian., 1870 p. 135 | 1876 *P. s.*, A. Boeck, Skand. Arkt. Amphip., *v.* 2 p. 217 t. 8 f. 3 | 1882 *P. falcatus*, G. O. Sars in: Forh. Selsk. Christian., nr. 18 p. 84 | 1888 *Phoxocephalus f.*, T. Stebbing in: Rep. Voy. Challenger, *v.* 29 p. 1717 | 1893 *P. f.*, A. Della Valle in: F. Fl. Neapel, *v.* 20 p. 739 t. 60 f. 15, 16 | 1891 *Leptophoxus f.*, G. O. Sars, Crust. Norway, *v.* 1 p. 147 t. 50.

Body very slender, especially in ♂. Apex of hood hook-like. Side-plates 1—3 each with 3 small setae at lower hind corner, 1st expanded below, 4th as broad as deep, almost rectangular below the emargination. Pleon segment 3, postero-lateral angles obtuse. Eyes wanting. Antenna 1, 1st joint thick, much longer than 2^d and 3^d combined, flagellum 5-, accessory flagellum 3-jointed. Antenna 2 in ♀ nearly twice as long as antenna 1, ·bent, flagellum 5-jointed, scarcely half length of peduncle, in ♂ nearly as long as body. Maxillipeds, 3ᵈ joint of palp produced to a narrow conical process tipped with 4 plumose setae. Gnathopod 1, 6ᵗʰ joint narrow quadrangular; gnathopod 2, 6ᵗʰ joint nearly twice as large, subquadrate; in both palmar tooth stout. Peraeopod 3, 2ᵈ joint as broad as long, hind margin evenly rounded. Peraeopod 4 nearly half length of the body. Peraeopod 5, 2ᵈ joint shorter than rest of limb, narrower above than below, slightly serrate, as broad as long. Uropods 1 and 2, rami slender, unarmed. Uropod 3, inner ramus extremely small, outer slender, elongate, 2ᵈ joint spiniform, about half length of 1ˢᵗ, Telson cleft to the base, slightly dilated in the middle, each apex narrow. tipped with a spinule and setule. Colour whitish, pellucid. L. 4 mm.

Arctic Ocean, North-Atlantic, North-Sea and Skagerrak (Norway, Bohuslän). Depth 56--376 m.

3. Gen. **Paraphoxus** O. Sars

1891 *Paraphoxus* (Sp. un.: *P. oculatus*), G. O. Sars, Crust. Norway, *v*.1 p. 148.

Body rather stout. Hood not carinate or deflexed. Side-plates 1—4 with several simple setae on lower margin, 4ᵗʰ rather broad, 5ᵗʰ with hind lobe rounded. Eyes well developed. Antennae 1 and 2 in ♀ nearly equal. antenna 2 in ♂, flagellum of moderate length, slender, calceoliferous. Mandible, cutting edges and spine-row well developed, molar feeble, tipped with spinules, palp narrow with few setae. Maxilla 1, palp 1-jointed, well developed. Maxilla 2, inner plate smaller than outer. Maxillipeds, inner plates obtusely rounded at apex. 3ᵈ joint of palp not produced, 4ᵗʰ slender, curved. Gnathopods 1 and 2 alike in structure. Peraeopods 1—5 about as in Phoxocephalus (p. 134). Uropod 3 in ♀, inner ramus much smaller than outer. Uropod 3 in ♂ much larger than in ♀, inner ramus nearly as long as outer, both plumose. Telson with rather narrow lobes.

1 species accepted, 1 incompletely described.

1. **P. oculatus** (O. Sars) 1879 *Phoxus o.*, G. O. Sars in: Arch. Naturv. Kristian, *v*.4 p. 441 | 1885 *P. o.*, G. O. Sars in: Norske Nordhavs-Exp., *v*.6 Crust. I p. 154 t. 13 f. 4a—e | 1888 *Phoxocephalus o.*, T. Stebbing in: Rep. Voy. Challenger, *v*.29 p. 1717 | 1893 *P. o.*, A. Della Valle in: F. Fl. Neapel, *v*.20 p. 740 t. 5 f. 5; t. 35 f. 19—28 | 1891 *Paraphoxus o.*, G. O. Sars, Crust. Norway, *v*.1 p. 149 t. 51.

Body more tumid in ♀ than in ♂. Hood with rounded apex. Side-plate 1 a little expanded distally, with about 14 setae, 2ᵈ with about 8 (Sars; 4 and 6 in Della Valle's figures), 4ᵗʰ as broad as long, expansion evenly rounded. Pleon segment 3, postero-lateral angles produced in a broad linguiform lobe. Eyes in ♀ small, rounded, brownish, in ♂ very large, oblong oval or reniform, dark. Antenna 1 in ♀, 1ˢᵗ joint as long as 2 and 3 combined, flagellum rather shorter than peduncle, 7-jointed, accessory flagellum 4-jointed; in ♂ with a dense brush on inner side of 1ˢᵗ joint, flagellum subequal to peduncle, calceoliferous. Antenna 2 in ♀, flagellum 8-jointed, in ♂ scarcely longer than half the body, penultimate and antepenultimate joints of peduncle densely hairy, ultimate with 2 calceoli, flagellum 20-jointed. Gnathopods 1 and 2, 6ᵗʰ joint oval, a little widened distally, palm shorter than the hind margin, palmar tooth rather short; this joint a little smaller in gnathopod 2 than in gnathopod 1. Peraeopods 1 and 2, 7ᵗʰ joint fully as long as 6ᵗʰ. Peraeopod 3, 2ᵈ joint oblong, much longer than broad. Peraeopod 4 very slender, much longer than half the body. Peraeopod 5, 2ᵈ joint obliquely expanded, as long as rest of limb, not so broad as long, hind margin distinctly serrate, over-lapping the 4th joint. Uropod 3 in ♀, inner ramus scarcely half length of outer, conical, in ♂ rami subequal, lanceolate, plumose. Telson, apices obliquely truncate, with 3 apical setules. Colour greyish white, semi-pellucid, or tending in parts to yellowish. L. ♀ 5, ♂ 4 mm.

Arctic Ocean, North-Atlantic and North-Sea (Norway, depth 38—188 m; Jan Mayen; Greenland, depth 377 m; France); Mediterranean.

P. maculatus (Chevreux) 1888 Phoxus m., Chevreux in: Bull. Soc. zool. France, v. 13 p. 40 | 1893 P. m., A. Della Valle in: F. Fl. Neapel, v. 20 p. 743 | 1888 Phoxocephalus m., T. Stebbing in: Rep. Voy. Challenger, v. 29 p. 1717 | 1899 Paraphoxus m., Chevreux in: C.-R. Ass. Pranç., Sess. 27 v. 2 p. 477.

Perhaps identical with P. oculatus. L. 3·5 mm.

North-Atlantic (lat. 46° N., long. 7° W.).

4. Gen. Metaphoxus Bonnier

1896 Metaphoxus (Sp. un.: M. typicus), J. Bonnier in: Ann. Univ. Lyon, v. 26 p. 630.

Hood not carinate or deflexed. Side-plates 1—4 with few setae on lower margin, 5th with hind lobe rounded. Eyes well developed. Antennae 1 and 2 in ♀ short.ꞏ Antenna 2 in ♂, flagellum long, slender, calceoliferous. Mandible, cutting edges and spine-row well developed, molar feeble, 3^d joint of palp distally rather expanded, carrying setae. Maxilla 1, inner plate without setae, palp 1-jointed, linear, not very elongate. Maxilla 2, inner and outer plate subequal. Maxillipeds, outer plates reaching beyond inner, but not greatly, 3^d joint of palp not apically produced. Gnathopods 1 and 2 unequal, differing somewhat in shape. Peraeopods 1—5 about as in Phoxocephalus (p. 134). Uropod 3 as in Paraphoxus (p. 137). Telson with lobes not very narrow.

3 species.

Synopsis of species:

1. M. typicus Bonnier 1896 M. t., J. Bonnier in: Ann. Univ. Lyon, v. 26 p. 630 t. 37 f. 1.

Extremely near to M. pectinatus (p. 139). Head, postero-antennal angle distinct. Eyes round, small, well developed, remaining black in spirit. Antenna 1 in ♀ short, flagellum 4-jointed, accessory flagellum nearly as long, 3-jointed. Antenna 2, ultimate joint of peduncle shorter than the stout penultimate, flagellum 4-jointed. Gnathopod 1, 6th joint oblong, palm slightly oblique, defined by a small tooth. Gnathopod 2 larger, distally widened, palm more convex and more oblique, defined by a considerable tooth. Peraeopods 1 and 2, finger almost as long as 6th joint. Peraeopod 4, 2^d joint very convex in front, 4th much shorter than 5th. Peraeopod 5, 2^d joint nearly as wide as length of rest of limb. Uropod 3 in ♀, inner ramus very short, outer long, 2-jointed. Telson cleft nearly to base, lobes rather dehiscent near the apices, each of which has 3 spines. L. ♀ a little over 5 mm.

Bay of Biscay. Depth 950 m.

2. **M. fultoni** (T. Scott) 1890 *Phoxocephalus f.*, T. Scott in: Rep. Fish. Board Scotl., *v.* 8 p. 327 t. 12 f. 10—12; t. 13 f. 13—19 | 1892 *P. f.*, D. Robertson in: P. nat. Hist. Soc. Glasgow, n. ser. *v.* 3 p. 207 | 1896 *P. f.*, Calman in: Tr. Irish Ac., *v.* 30 p. 743 t. 31 f. 1, 2 | 1893 *P. chelatus*, A. Della Valle in: F. Fl. Neapel, *v.* 20 p. 742 t. 5 f. 10; t. 35 f. 29—35.

Body compressed, especially in ♂. Hood nearly straight, not very long. Side-plate 1 expanded distally, 1st—3d with 3 or 5 setae, 4th as broad as long, with pretty evenly rounded hind margin. Pleon segment 3, postero-lateral angles squarely rounded. Eyes well developed, dark, larger in ♂. Antenna 1 in ♀, 1st joint as long as 2d and 3d combined, flagellum shorter than peduncle, 4-jointed, accessory flagellum 3-jointed; in ♂ longer, 1st joint with dense brush, flagellum longer than peduncle, 5- or 6-jointed with 1 or 2 calceoli, terminal setae long, accessory flagellum 3-jointed. Antenna 2 in ♀ about $^1/_3$ longer, flagellum very short, 3-jointed; in ♂ about as long as body, penultimate joint of peduncle and 2 preceding joints densely hirsute, ultimate joint with 1 calceolus, flagellum about 22-jointed, calceoliferous. Palp of mandible not very slender. Maxilla 1, the 1-jointed palp narrow but rather long. Gnathopods 1 and 2, 5th joint rather longer and more triangular in gnathopod 1, the subquadrate 6th joint narrower, but in both the front margin of this joint is shorter than the hind one, with which the oblique convex palm forms an acute angle, thus tending to a chelate form. Peraeopods 1 and 2, apical spine of 5th joint sometimes shorter than 6th joint. Peraeopod 3 as in Paraphoxus oculatus (p. 137). Peraeopod 4 not very slender, about equal $^1/_3$ length of body, hind margin of 2d joint a little produced, 4th joint as long as 5th. Peraeopod 5 rather longer in ♂ than in ♀, 2d joint as long as rest of limb, nearly as broad as long, overlapping 3d joint, hind margin crenate, with setules. Uropod 3 in ♀ short and stout, inner ramus half length of 1st joint of outer, in ♂ more elongate, inner ramus equal to 1st joint of outer, both with long plumose setae. Telson deeply cleft, lobes truncate, with 1 spinule and some setules. Colour greyish pellucid, with violet tinting on middle of back. L. 2·5 mm.

Firth of Clyde, Firth of Forth, Menai Strait, Galway Bay; Gulf of Naples. From low water to 40 m.

3 **M. pectinatus** (A. Walker) 1896 *Phoxocephalus simplex* (err.?, non *Phoxus s.* Bate 1857!). Calman in: Tr. Irish Ac., *v.* 30 p. 748 t. 32 f. 3 | 1896 *P. pectinatus*, A. O. Walker in: Ann nat. Hist., ser. 6 *v.* 17 p. 343 t. 16 f. 1—6; *v.* 18 p. 156 | 1899 *Metaphoxus p.*, Chevreux in: C.-R. Ass. Franç., Sess. 27 *v.* 2 p 477.

Hood slightly deflexed at the apex, postero-antennal corners slightly rounded. Side-plate 1 well expanded distally, 1st—3d with 3 or 4 simple setae on lower margin, 4th little deeper than broad, hind margin convex, oblique, 5th with hind lobe much deeper than front. Pleon segment 3, postero-lateral angles slightly rounded, segment 4 with 1 or 2 spinules on dorsal tubercle (♂). Eyes round, dark, large, especially in ♂. Antenna 1, 1st joint longer than 2d and 3d combined, upper margin produced in ♂, flagellum 4-jointed, in ♂ with 2 calceoli and ending in a long seta, accessory flagellum 3-jointed. Antenna 2, flagellum in ♀ 4-jointed. as long as the stout penultimate joint of peduncle, in ♂ about $^2/_3$ length of body, calceolated along with ultimate joint of peduncle. Mandible with spine-row, molar rudimentary (at least in ♂), 2d joint of palp longer than 3d, 3d distally expanded, carrying many setae. Gnathopod 1 smaller than gnathopod 2, with less oblique palm and less powerful palmar tooth, in both gnathopods joints 4 and 5. combining to form a small cup for the large 6th joint, which in gnathopod 1 is much

longer than broad, a little tapering distally, while in gnathopod 2 it widens
a little distally, with a length not much exceeding its considerable breadth.
Peraeopods 1 and 2, 7th joint about $^3/_4$ length of 6th. Peraeopod 4 nearly $^2/_8$
length of the body, 4th joint as long as 5th. Peraeopod 5, 2^d joint as long
as rest of limb, not quite so broad as long, hind margin with about 3
shallow teeth, not produced below the 4th joint, joints 3 and 4 in ♀ with plumose
setae, joint 5 in ♂ with 3 or 4 curved blunt spines forming comb on
distal half of front margin. Uropod 3 in ♀ as in Phoxocephalus holbölli
(p. 134), in ♂ inner ramus as long as 1st joint of outer, the latter without
spines on outer margin. Telson as in P. bolbölli. L. 3 mm.

Firth of Clyde; English Channel (Guernsey, depth 13 m; France).

5. Gen. **Harpinia** Boeck

1842 *Phoxus* (part.) (*Spinifer* Holbøll in MS., part.), Krøyer in: Naturh. Tidsskr.,
v. 4 p. 150 | 1871 *Harpina*. (non Burmeister 1844, Coleoptera!). A. Boeck in: Forh.
Selsk. Christian., 1870 p. 135 | 1876 *Harpinia*, A. Boeck, Skand. Arkt. Amphip., *v.*2
p.218 | 1888 *H.*, T. Stebbing in: Rep. Voy. Challenger, *v.* 29 p. 819 | 1891 *H.*, G. O. Sars,
Crust. Norway, *v.* 1 p. 150 | 1893 *H.*, A. Della Valle in: F. Fl. Neapel, *v.* 20 p. 744.

Body rather broad. Head with evenly vaulted hood, not carinate or
deflexed. Side-plates plumose, 4th broadly produced backward, 5th with

oblique hind lobe. Eyes wanting. Antenna 1, 1st joint very
large, carrying sensory bristles and in ♂ densely setose, accessory
flagellum long. Antenna 2 with many spines and setae on
peduncle, of which the penultimate joint is broad. Primary
flagellum not long in either antenna in either sex. No calceoli.
Mandible with short cutting edge, denticulate accessory plate
on left mandible, spine-row, feeble molar tipped with slender
spines, palp slender, armed chiefly on truncate tip of long
3^d joint. Maxilla 1, 9 spines on outer plate, palp 2-jointed.

Fig. 38.
H. mucronata.
Peraeopod 5.
[After G. O.
Sars.]

Maxilla 2, plates subequal. Maxillipeds, plates slender, palp
elongate. In ♂ the mouth-organs are subject to degradation.
Gnathopods 1 and 2 similar, 6th joint oval, palm more or
less oblique. Peraeopod 3, 2^d joint linear. Peraeopod 4 much
longer than the rest, 2^d joint narrowed distally. Peraeopod 5 (Fig 38) rather
small, 2^d joint much expanded. Uropod 3 in ♀ rather short, outer ramus
the longer, with spines; in ♂ longer, both rami knife-shaped. and as a
rule unarmed. Telson with rounded apices.

12 species.

Synopsis of species:

1. **H. mucronata** O. Sars 1879 *H. m.*, G. O. Sars in: Arch. Naturv. Kristian., v. 4 p. 446 | 1885 *H. m.*, G. O. Sars in: Norske Nordhavs-Exp., v. 6 Crust. I p. 161 t. 13 f. 7, 7 a—g | 1891 *H. m.*, G. O. Sars. Crust. Norway, v. 1 p. 157 t. 54 f. 3 | 1893 *H. m.*, A. Della Valle in: F. Fl. Neapel. v. 20 p. 746 t. 60 f. 17.

Body rather compressed, glabrous above. Hood prominent, slightly convex. Side-plates 1—3 with small tooth at lower hind corner, setae few. Pleon segment 3, postero-lateral angles slightly upturned in long spiniform process. Antennae 1 and 2 comparatively slender, accessory flagellum nearly as long as principal. Gnathopods 1 and 2, palm slightly longer than hind margin in 6ᵗʰ joint. Peraeopod 4 not very slender, scarcely more than half the length of the body. Peraeopod 5 (Fig. 38) very small, 2ᵈ joint less expanded than usual, lower hind corner carrying a strong backward-pointing acute spur, with a slight excavation above it and a few points of serration below. Uropod 3, inner ramus as long as 1ˢᵗ joint of outer, 2ᵈ joint of outer not nearly half as long as 1ˢᵗ. Telson broader than long, apices very obtuse. Colour greyish white. L. ♀ 5 mm.

Arctic Ocean and North-Atlantic (Finmark. Greenland), North-Sea.

2. **H. crenulata** (Boeck) 1871 *Harpina c.*, A. Boeck in: Forh. Selsk. Christian., 1870 p. 136 | 1876 *Harpinia c.*, A. Boeck, Skand. Arkt. Amphip., v. 2 p. 221 t. 8 f. 2 | 1891 *H. c.*, G. O. Sars. Crust. NorWay, v. 1 p. 158 t. 55 f. 2 | 1893 *H. c.* (part.), A. Della Valle in: F. Fl. Neapel, v. 20 p. 745 | 1896 *H. nana*, J. Bonnier in: Ann. Univ. Lyon, v. 26 p. 633 t. 37 f. 2.

Body rather short and stout. Hood prominent. Side-plates, setae not numerous. Pleon segment 3, postero-lateral angles rounded, with a little upturned, scarcely projecting denticle above, the margin below it sometimes crenulate; segments 1—3 dorsally densely hirsute; segment 4 with dorsal depression in ♂. Antenna 1, flagellum 6-jointed, accessory flagellum 5-jointed. Antenna 2, flagellum 5-jointed. Gnathopods 1 and 2, palm longer than hind margin of 6ᵗʰ joint, especially in the more robust gnathopod 2. Peraeopod 4 slender, ⁸/₄ length of body. Peraeopod 5, 2ᵈ joint with expansion not entirely overlapping 4th joint, hind margin in ♀ carrying long setae between about 12 points of serration, the lower ones bipartite; in ♂ a few setules between 5 or 6 obscure points. Uropod 3 in ♀, outer ramus with 3 spines on outer margin of 1st joint, 2^d joint not half length of 1st, inner ramus little more than half length of same; in ♂ 2^d joint more than half length of 1st, and inner ramus longer than 1st joint of outer. Telson scarcely broader than long, apices divergent, narrowly rounded. Colour greyish white, pellucid. L. about 4 mm.

Arctic Ocean, North-Atlantic, North-Sea and Skagerrak (Norway, depth 56—188 m; Bohuslän); Kattegat; Bay of Biscay, depth 950 m.

3. **H. pectinata** O. Sars 1891 *H. p.*, G. O. Sars, Crust. Norway, *v.* 1 p. 154 t. 53 f. 2 | 1893 *H. p.*, A. Della Valle in: F. Fl. Neapel, *v.* 20 p. 746.

Body rather compressed, glabrous above. Hood prominent, postero-antennal corners not produced. Side-plates 1—3 with small tooth at lower hind corner, setae few. Pleon segment 3, postero-lateral angles slightly upturned in rather short spiniform process. Antennae 1 and 2 comparatively slender. Gnathopods 1 and 2, palm about equal to hind margin in 6th joint. Peraeopod 4 rather slender, about half length of body. Peraeopod 5, 2^d joint obliquely expanded, overlapping 4th joint, hind margin coarsely serrate with 5 or 6 unequal downward directed points. Uropod 3 in ♀ small, inner ramus longer than 1st joint of outer, outer with 2 spines on outer margin, 2^d joint half as long as 1st. Telson small, broader than long, apices very obtuse and divergent. Colour greyish white, pellucid. L. ♀ 4 mm.

Arctic Ocean, North-Atlantic, North-Sea and Skagerrak (Norway, Bohuslän).

4. **H. serrata** O. Sars 1879 *H. s.*, G. O. Sars in: Arch. Naturv. Kristian., *v.* 4 p. 445 | 1885 *H. s.*, G. O. Sars in: Norske Nordhavs-Exp., *v.* 6 Crust. I p. 162 t. 13 f. 8, 8a—d | 1891 *H. s.*, G. O. Sars, Crust. Norway, *v.* 1 p. 155 t. 54 f. 1 | 1893 *H. s.*, A. Della Valle in: F. Fl. Neapel, *v.* 20 p. 747 t. 60 f. 18.

Body robust, broadly vaulted. Hood rather convex with blunted tip. Side-plates without tooth at lower hind corner, setae many; 1st much widened distally, 5th with hind lobe unusually deep. Pleon segments 1—3 dorsally hirsute, postero-lateral angles of 3^d with spiniform process slightly upturned. Antenna 1, 1st joint of peduncle unusually large, flagellum rather short, 5-jointed, accessory flagellum 4-jointed. Antenna 2 strong, flagellum short, 5-jointed. Gnathopods 1 and 2, palm of 6th joint about equal to hind margin. Peraeopod 4 very slender, and much more than half length of body. Peraeopod 5, 2^d joint obliquely oval, overlapping 4th, hind margin cut into 5 very large backward directed points. Uropod 3 in ♀, inner ramus much shorter than 1st joint of outer, which has 4 spines on outer margin, the 2^d joint minute with long apical setae. Telson broader than long, apices obtuse, not very divergent. Colour greyish white. L. ♀ 6 mm.

Arctic Ocean (Jan Mayen; Norway?).

5. **H. excavata** Chevreux 1887 *H. e.*, Chevreux in: Bull. Soc. zool. France, *v.* 12 p. 568 | 1893 *H. neglecta* (part.), A. Della Valle in: F. Fl. Neapel, *v.* 20 p. 747.

Body elongate. Hood almost straight. Pleon segment 3, postero-lateral angles strongly produced, acutely curved upward. Antenna 2, penultimate joint of peduncle with 2 rows of strong spines. Gnathopod 1, 6th joint long oval, palm crenulate, defined by a spine. Gnathopod 2, 6th joint more broadly oval, palm defined by a long sharp tooth, with another stout and short at its middle. Peraeoped 4 long, robust, 4th joint with 2 very long curved apical spines, 6th much narrower than 5th, more than twice as long, finger short, straight, scarcely $^1/_3$ as long as 6th. Peraeopod 5, 2^d joint finely crenulate on regularly rounded hind margin, produced scarcely half-way down the 3^d, which has very long plumose setae in front. Uropod 3 a little shorter than the preceding. Telson cleft to the base, each half with 2 long plumose cilia, planted vertically near the apex. L. 6 mm.

Bay of Biscay (about lat. 43° N.). Depth 5110 m.

6. **H. antennaria** Meinert 1857 *Phoxus plumosus* (err., non Krøyer 1842!), Bate in: Ann. nat. Hist., ser. 2 v.19 p.140 | 1862 *P. p.*, Bate. Cat. Amphip. Brit. Mus., p. 99 t. 16 f. 3 | 1876 *Harpinia plumosa* (part.), A. Boeck. Skand. Arkt. Amphip., v.2 p.219 t. 8 f. 1 | 1890 *H. antennaria*, Meinert in: Udb. Hauchs, v.3 p. 160 t. 1 f. 39—41 | 1891 *H. neglecta*, G. O. Sars, Crust. Norway, v.1 p.153 t. 53 f. 1 | 1893 *H. n.* (part.), A. Della Valle in: F. Fl. Neapel, v.20 p. 747 t. 5 f. 6; t. 35 f. 1—18; t. 60 f. 19.

Body rather compressed. Hood prominent, postero-antennal corners drawn out to a forward pointing process. Pleon segment 3, postero-lateral angles produced, more in ♀ than in ♂, acute, slightly upturned; segments 1—3 in ♀ dorsally densely hirsute. Side-plates and antennae as in H.plumosa (p.144). Gnathopods 1 and 2, palm longer than hind margin of 6th joint, defined by a process in ♀, not in ♂. Peraeopod 4 longer than half of body. Peraeopod 5 short, 2^d joint not overlapping whole of 4th, serrate points of hind margin small but sharp, about 9, with setules between. Uropod 3 slender, outer ramus with 4 spines on outer margin, 2^d joint much less than half as long as inner ramus and shorter than 1st joint of outer. Telson little broader than long, apices narrowly rounded. Colour greyish white. L. ♀ 5 mm, ♂ rather less.

Arctic Ocean, North-Atlantic, North-Sea and Skagerrak (Norway, depth 56—282 m; Bohuslän; Great Britain; France); Kattegat.

7. **H. propinqua** O. Sars 1891 *H. propinqva*, G. O. Sars, Crust. Norway, v.1 p. 156 t. 54 f. 2 | 1893 *H. propinqua*, A. Della Valle in: F. Fl. Neapel, v.20 p. 746.

Body rather slender. Hood prominent, postero-antennal corners apparently quadrate. Side-plates not densely setose. Pleon segment 3, postero-lateral angles acutely but minutely produced; segments 1—3 dorsally finely hirsute. Gnathopods 1 and 2, palm very oblique, but scarcely longer than hind margin of 6th joint. Peraeopod 4 about half length of the body. Peraeopod 5, 2^d joint not completely overlapping 4th, hind. margin very obscurely serrate. Uropod 3, outer ramus with 3 spines on outer margin, 2^d joint more than half as long, inner ramus about equal 1st joint of outer. Telson rather short, broader than long, apices broadly rounded. L. ♀ scarce 5 mm.

Arctic Ocean (Jan Mayen).

8. **H. obtusifrons** Stebb. 1888 *H. o.*, T. Stebbing in: Rep. Voy. Challenger, v. 29 p. 820 t. 56 | 1893 *H. neglecta* (part.), A. Della Valle in: F. Fl. Neapel, v.20 p.747.

Body robust. Hood prominent, very broadly rounded at the apex. Side-plates with feathered setae very numerous. Pleon segment 3, postero-lateral angles pretty strongly upturned in a rather long acute tooth; the segment covered with a fine down. Antenna 1, 1st joint about twice as long as 2^d and 3^d combined, flagellum 7-jointed, accessory flagellum 5-jointed. Antenna 2, flagellum 8-jointed. Lower lip with small conical tooth at inner side of each principal lobe. Mandible, spine-row of 7—9 spines, molar unarmed. Maxillipeds, outer plates with about 14 spine-teeth on inner margin

and apex, and about 7 plumose setae on outer margin; 4th joint of palp
with nail shorter than the base. Gnathopods 1 and 2, palm very oblique, longer
than hind margin of 6th joint; gnathopod 2 more robust than gnathopod 1.
Peraeopod 4 much longer than half the body, 5th joint longer than 4th,
6th longer than 5th. Peraeopod 5, 2^{d} joint much expanded, but longer than
broad, only partly overlapping 4th joint, the hind margin weakly and not
acutely serrate. Uropod 3, outer ramus with 4 spines on outer and 2 on
inner margin, 2^{d} joint rather less than half length of 1st, inner ramus
unarmed, equal 1st joint of outer. Telson about as long as broad, tapering
to the broadly rounded scarcely dehiscent apices. L. ♀ about 8 mm.

Southern Indian Ocean (Kerguelen Islands). Depth 55—220 m.

9. **H. plumosa** (Krøyer) 1842 *Phoxus plumosus*, (Holbøll in MS.) Krøyer in:
Naturh. Tidsskr., *v.*4 p.152 | 1871 *Harpina plumosa*, A. Boeck in: Forh. Selsk. Christian.,
1870 p.135 | 1876 *Harpinia p.* (part.), A. Boeck, Skand. Arkt. Amphip., *v.*2 p.219 t.8
f.5 | 1891 *H.p.*, G. O. Sars, Crust. Norway, *v.*1 p.151 t.52 | 1893 *H.p.*, A. DellaValle
in: F. Fl. Neapel, *v.*20 p.749 | 1853 *Phoxus fusiformis*, Stimpson in: Smithson. Contr.,
*v.*6 nr.5 p.57 | 1876 *Harpina f.*, S. I. Smith (& Harger) in: Tr. Connect. Ac., *v.*3 p.29.

Body moderately compressed, glabrous above. Hood slightly projecting,
apically narrow; postero-antennal corners nearly right-angled. Side-plates 1—5
fringed with about 7, 6, 6, 10, 4 strong plumose setae. Pleon segment 3,
postero-lateral angles forming long, acute, slightly upturned process. Antenna 1,
1st joint of peduncle very large; flagellum 6-jointed, about half length of peduncle,
accessory flagellum 5-jointed. Antenna 2, penultimate joint of peduncle
heart-shaped, flagellum 7-jointed. Maxillipeds (in attachment with peraeon
instead of with head, Krøyer), outer plates armed on inner margin and apex
with 7—9 serrate spine-teeth. Gnathopods 1 and 2 nearly equal, in 6th joint
palm shorter than hind margin. Peraeopod 4 scarcely longer than half
the body. Peraeopod 5, 2^{d} joint subquadrate overlapping most of 4th,
hind margin with 3 or 4 feeble points over a well marked excavation.
Uropod 3 very short, outer ramus with 3 spines on outer margin, 2^{d} joint
more than half length of 1st, inner ramus unarmed, longer than 1st joint
of outer. Telson broader than long, apices obtuse. Colour pale yellow. L. ♀ 7 mm.

Arctic Ocean and North-Atlantic (Greenland, Spitzbergen, Nova Scotia, Kara
Sea, Siberia, America; Norway?).

10. **H. truncata** O. Sars 1891 *H. t.*, G. O. Sars. Crust. Norway, *v.*1 p.157 t.55
f.1 | 1893 *H. crenulata* (part.), A. Della Valle in: F. Fl. Neapel, *v.*20 p.745.

♀. Body rather robust. Hood prominent. Side-plates, setae not
numerous. Pleon segment 3, postero-lateral angles rounded, without denticle;
segments 1—3 densely hirsute. Antenna 1, flagellum 7-jointed, accessory
flagellum 6-jointed. Antenna 2, flagellum 6-jointed. Gnathopods 1 and 2, palm
longer than hind margin of 6th joint, especially in the more robust gnathopod 2. Peraeopod 4 more than half length of body, 6th joint elongated,
with 4 recurved feathered setae on hind margin. Peraeopod 5, 2^{d} joint
much expanded, hind margin little serrate and transversely truncate below.
Uropod 3, outer ramus with 5 spines on outer margin of 1st joint, 2^{d} very
small, inner ramus $^{3}/_{4}$ length of 1st joint of outer. Telson rather longer
than broad, apices divergent. Colour greyish white. L. 6 mm. — ♂ unknown.

Trondhjemsfjord, depth 188—282 m; Skagerrak (Bohuslän).

11. **H. abyssi** O. Sars 1879 *H. a.* + *H. carinata* (♂), G. O. Sars in: Arch. Naturv.
Kristian., *v.*4 p.443, 444 | 1885 *H. a.* + *H. c.*, G. O. Sars in: Norske Nordhavs-Exp., *v.*6
Crust. 1 p.157 t.13 f.5, 5a—m; p.159 t.13 f.6, 6a—e | 1891 *H. a.*, G. O. Sars, Crust.
Norway, *v.*1 p.160 t.56 f.1 | 1893 *H. a.*, A. Della Valle in: F. Fl. Neapel, *v.* 20 p.745.

♀. Body robust. Hood moderately prominent. Side-plates, setae numerous; 1st much expanded distally, 5th with deep transverse hind lobe. Pleon segment 3, postero-lateral angles rounded, lower edge setose; segments 1—3 very finely hirsute; segment 4 with dorsal hump in both sexes. Antenna 1, 1st joint twice as long as 2d and 3d combined, flagellum short, 7-jointed, accessory flagellum 6-jointed. Antenna 2 scarcely as long as antenna 1, flagellum 6-jointed. Penultimate joint of peduncle in both antennae 1 and 2 with dense set of plumose setae. Gnathopods 1 and 2 not very unequal, palm very oblique, much longer than hind margin of 6th joint, palmar tooth almost obsolete. Peraeopod 4 more than half length of body, 6th joint long and slender. Peraeopod 5 very small, front margin angled, setose below, hind margin broadly rounded. little produced downward, finely crenulate below. Uropod 3 not very large, 1st joint of outer ramus with 6 spines on outer margin, 2d joint very small, inner much shorter than 1st joint of outer. Telson short, broader than long. apices divergent. Colour greyish white to rusty yellow. — ♂. More slender and compressed, pleon segment, 6th segment dorsally humped as well as 4th. Antenna 1, the expanded 1st joint of flagellum densely setose. Peraeopod 5, 2d joint, front margin almost straight, hind margin nearly smooth. Uropod 3, rami rather broad. outer with 6 setules, 2d joint minute, inner ramus nearly as long as outer. — L. ♀ & ♂ 12 mm.

North-Atlantic (lat. 63°—75° N.). Depth 660—2288 m.

12. **H. laevis** O. Sars 1891 *H. l.*. G. O. Sars, Crust. Norway, *v.* 1 p. 161 t. 56 f. 2 | 1893 *H. l.*, A. Della Valle in: F. Fl. Neapel, *r.* 20 p. 745.

♀. Body stout, glabrous. Hood very prominent. Side-plates, setae very few (3). Pleon segment 3, postero-lateral angles narrowly rounded, lower margin quite smooth. Antennae 1 and 2 unusually slender, with few setae on penultimate joint of peduncle, antenna 1, flagellum 6-jointed, accessory flagellum 5-jointed. Antenna 2, flagellum 5-jointed. Gnathopods 1 and 2, palm rather oblique, subequal to hind margin of 6th joint, palmar tooth pretty strong. Gnathopod 2 more robust than gnathopod 1. Peraeopod 4 less than half length of body. Peraeopod 5, 2d joint large, fully overlapping 4th joint, hind margin feebly crenulate. Uropod 3, 1st joint of outer ramus with 1 spine on outer margin, 2d joint more than half as long, inner ramus longer than 1st joint of outer. Telson longer than broad, apices narrowly rounded. Colour greyish white, pellucid. L. 4 mm. — ♂ undescribed.

Hardangerfjord and Trondhjemsfjord. Depth 94—188 m.

6. Gen. **Pontharpinia** Stebb.

1853 *Urothoe* (part.), J. D. Dana in: U. S. expl. Exp., *v.* 13 II p. 920 | 1888 *Phoxocephalus* (part.), T. Stebbing in: Rep. Voy. Challenger, *v.* 29 p. 1717 | 1897 *Pontharpinia* (Sp. un.: *P. pinguis*), T. Stebbing in: Tr. Linn. Soc. London‿ ser. 2 *v.* 7 p. 32.

Back broad. Hood obtuse. Eyes distinct. Antenna 1, 3d joint short, accessory flagellum many-jointed. Antenna 2, penultimate joint of peduncle long, broad, carrying setae and spines. Mandible, molar small or obsolete, palp long and slender, 3d joint shorter than 2d. Maxilla 1, palp 2-jointed. Maxillipeds, outer plates short, 4th joint of palp long and slender. Gnathopods 1 and 2 similar. Peraeopods 1 and 2, 4th joint longer than 5th or 6th. Peraeopods 3 and 4 with 2d, 4th and 5th joints expanded. Peraeopod 5, expanded 2d joint greatly produced. Uropod 3, rami lanceolate. Telson deeply cleft.

2 accepted species, 2 uncertain.

1. **P. pinguis** (HasW.) 1880 *Urothoë p.*, HasWell in: P. Linn. Soc. N.S.Wales, *v.*4 p. 325 t. 19 f. 2 | 1891 *Harpinia? p.*, T. Stebbing in: Tr. zool. Soc. London, *v.*131 p.4 | 1897 *Pontharpinia p.*, T. Stebbing in: Tr. Linn. Soc. London, ser.2 *v.*7 p.33 t. 9 B | ?1893 *Urothoe irrostrata* (part.), A. Della Valle in: F. Fl. Neapel, *v.*20 p. 667.

Lower margin of side-plates and of pleon segments 1—3 more or less fringed with setae, pleon segment 3 serrulate above small produced point of postero-lateral angles. Antenna 1, flagellum in ♀ 9-jointed, accessory flagellum 7-, flagellum in ♂ 15-, accessory flagellum 10-jointed; the joints apically oblique. Antenna 2, flagellum in ♂ 21-, in ♀ 10-jointed, the joints apically oblique. Maxilla 1, outer plate with 11 spines. Gnathopods 1 and 2 in ♀, 5th joint broadly fusiform, 6th widening from base, then of uniform width to the transverse palm, ²/₃ length of front margin strongly setose, finger in gnathopod 1 slightly overlapping palm, but not in gnathopod 2. Gnathopod 1 in ♂, 6th joint a little longer than 5th, palm not defined, finger half as long as 6th joint; gnathopod 2, 5th joint short, 6th ovate, palm defined by a prominent angle, finger more than half as long as 6th joint. Peraeopods 3 and 4, 4th joint squarish, hind margin serrate, in peraeopod 4 deeply incised. Peraeopod 5, 2^d joint membranous-looking. Uropod 1, rami slightly curved; uropod 2, rami shorter, straight; uropod 3, rami (at least in ♂) setose. Telson as broad as long, apices divergent, sides (in ♂) setose. L. ♀ 7·5, ♂ 10 mm.

South-Pacific (New South Wales). Cast on beach.

2. **P. rostrata** (Dana) 1853 & 55 *Urothoe rostratus*, J. D. Dana in: U. S. expl. Exp., *v.*13 II p. 921; t. 62 f. 5 a—p | 1876 *Phoxus r.*, A. Boeck, Skand. Arkt. Amphip., *v.*2 p. 214 | 1880 *P. batei*, HasWell in: P. Linn. Soc. N.S. Wales, *v.* 4 p. 259 t. 9 f. 3 | 1888 *Phoxocephalus b.* + *P. rostratus*, T. Stebbing in: Rep. Voy. Challenger, *v.* 29 p. 1717 | 1893 *P.? b.* + *P.? r.*, A. Della Valle in: F. Fl. Neapel, *v.* 20 p. 743, 744.

Hood long, straight, obtuse. Side-plate 1 distally expanded, with a few setae (not mentioned or figured by Dana) on lower hind corner of 1st to 4th; 4th broad, a little narrowed distally, hind excavation shallow. Pleon segment 3, postero-lateral angles squarely rounded, with setae on hind margin above the angle. Eyes round (Dana), long-oval (Haswell), subrotund, large, conspicuous. Antenna 1, 1st joint about as long as 2^d and 3^d combined, flagellum 10—13 joints, accessory flagellum ²/₃ as long as primary, 6—10-jointed. Antenna 2, in ♀ rather, in ♂ much longer than antenna 1, penultimate joint of peduncle in ♀ with numerous stout spines, in ♂ (Dana) towards apex furnished with stout reversed setae, flagellum in ♀ scarcely longer than peduncle, 13-jointed, ♂ (Dana) with 15 long slender joints carrying calceoli. Upper lip broadly rounded. Mandible, cutting edge sharply tridentate, spine-row of about 6 spines, molar very small, palp long and slender, 2^d joint longest, a little curved, 3^d with setae on obliquely truncate apex. Maxilla 1, inner plate oval, with a few apical setae, outer plate with 9 spines, palp long, 2-jointed, the short 1st joint faintly distinct. Maxilla 2, plates broad, subequal, inner with 4 spines on distal half of inner margin. Maxillipeds, outer plates scarcely reaching beyond 1st joint of palp, with 5 slender spines on inner margin, 2^d joint of palp elongate, 3^d with apex reaching a little beyond insertion of finger, finger not longer than 3^d joint unless by inclusion of its apical spine. Gnathopods 1 and 2 similar, 2^d rather larger than 1st, 6th joint broader than 5th, but scarcely if at all longer, oblong, palm setulose, curved, somewhat oblique, defined by a spiniferous process, against which the finger impinges. Peraeopods 1 and 2, 5th joint short, with long, stout, apical spine, 6th with strong spines on lower half of hind margin, finger small. Peraeopod 3, 2d joint broadly expanded, though in one specimen much longer than broad; 4th joint distally broader than the length. subtriangular, with many spines on lower margin, 5th narrower than 4th, subquadrate, almost surrounded by spines, 6th narrow, spinose, finger smooth, slender. Peraeopod 4, 2d joint broadly oval, with long setae on front margin, 4th longer and broader than in peraeopod 3, spinose, remaining joints nearly as in peraeopod 3. Peraeopod 5 much shorter than peraeopod 4, 2d joint large, the wide expansion produced to overlap 3d and 4th joints, 4th and 5th shorter but wider than 6th, all three spinose, especially 5th, finger slender. Uropod 1, rami stout, 5 spines on one, 2 on the other. Uropod 2, 12 spines along margin of peduncle, rami short and stout. Uropod 3, rami in ♂ long, subequal, plumose (Dana), in ♀ outer lanceolate, plumose, with minute terminal joint, inner about $^2/_3$ as long, feebly armed and narrower. Telson cleft to the base or nearly so, apices rounded, with spinule and setules on outer point. L. 6—9 mm.

Sooloo Sea, Port Jackson [East-Australia].

P. grandis (Stimps.) 1857 *Phoxus g.*, Stimpson in: Boston J. nat. Hist., *v.* 6 p. 521 | 1888 *Phoxocephalus g.*, T. Stebbing in: Rep. Voy. Challenger, *v.* 29 p. 1717 | 1893 *P. g.*, A. Della Valle in: F. Fl. Neapel, *v.* 20 p. 743.

L. 12·5 mm.

Entrance of San Francisco Bay. Sandy bottom, depth 19 m.

P. uncirostrata (Giles) 1890 *Phoxus uncirostratus*, G. M. Giles in: J. Asiat. Soc. Bengal, *v.* 59 p. 65 t. 2 f. 2 | 1893 *Phoxocephalus u.*, A. Della Valle in: F. Fl. Neapel, *v.* 20 p. 895.

L. about 5 mm.

Bay of Bengal (Seven Pagodas near Madras). Sandy bottom, depth 9—19 m.

7. Gen. **Parharpinia** Stebb.

1888 *Phoxocephalus* (part.), T. Stebbing in: Rep. Voy. Challenger, *v.* 29 p. 1717 | 1899 *Parharpinia* (Sp. typ.: *Phoxus villosus*), T. Stebbing in: Ann. nat. Hist., ser. 7 *v.* 4 p. 207.

Hood obtuse. Eyes distinct. Antenna 1, 3d joint short, accessory flagellum many-jointed. Antenna 2, penultimate joint of peduncle broad, carrying setae and spines. Mandible, molar small or obsolete, palp long and slender, 3d joint longer than 2d. Maxilla 1, palp 2-jointed. Maxillipeds, outer plates elongate, fringed with serrate spines on inner margin, 4th joint of palp long and slender. Gnathopods 1 and 2 similar. Peraeopods 1 and 2, 4th joint longer than 5th or 6th. Peraeopods 3 and 4 with 4th and 5th joints not expanded. Peraeopod 5, expanded 2d joint moderately produced. Uropod 3, rami lanceolate. Telson deeply cleft.

1 species.

1. P. villosa (Hasw.) 1880 *Phoxus villosus*, Haswell in: P. Linn. Soc. N. S. Wales, *v.* 4 p. 258 t. 9 f. 2 a, b | 1888 *Phoxocephalus v.*, T. Stebbing in: Rep. Voy. Challenger, *v.* 29 p. 1717 | 1893 *P. v.*, A. Della Valle in: F. Fl. Neapel, *v.* 20 p. 743, 744 | 1899 *Phoxus v.*, *Parharpinia* sp. typ., T. Stebbing in: Ann. nat. Hist., ser. 7 *v.* 4 p. 207 | 1882 *Phoxus batei* (err., non Haswell 1880!), G. M. Thomson in: Tr. N. Zealand Inst., *v.* 14 p. 232 t. 17 f. 2 a—e.

10*

Hood long, reaching almost end of peduncle of antenna 1. Side-plates 1—4, hinder half of lower margin well fringed with setae, 4th as broad as deep, excavation shallow. Pleon segment 3. postero-lateral corners subquadrate. Eyes ovate, conspicuous. Antenna 1, 1st joint as long as 2d and 3d combined, 2d with fascicle of setae, flagellum subequal to peduncle or a little longer, 13-jointed, accessory flagellum not much shorter, 10-jointed. Antenna 2, penultimate joint of peduncle not greatly longer than ultimate, with long setae, flagellum rather longer than peduncle. 17-jointed. The long straight 3d joint of the mandibular palp is obliquely truncate at the apex. Palp of maxilla 1 comparatively large. Outer plates of maxillipeds reach more than half-way along the long 2d joint of the palp and are fringed with a dozen graduated spines serrate on both edges. As in Pontharpinia pinguis (p. 146) the 3d joint of the palp is produced a little beyond the insertion of the finger, which is nearly as long as the 3d joint, not tipped with a spine. Gnathopod 1, 5th joint much shorter than the oblong oval 6th; palm very oblique, defined by a tooth with palmar spines. Gnathopod 2, 5th joint about as long as 6th, which is similar to that of gnathopod 1, not broader (rather broader, Haswell). Peraeopods 1 and 2, the 5th joint has the apical spine nearly as long as the 6th joint, finger rather short. Peraeopod 3, 2d joint not greatly expanded, more so above than below, 4th to 6th with spines and long setae. 4th rather shorter but distally wider than 5th, 5th wider, scarcely longer. than 6th. Peraeopod 4, 2d joint strongly curved and setose in front, hind margin nearly straight. Peraeopod 5, 2d joint as broad as long, not produced to end of 4th. Branchial vesicles on gnathopod 2 and peraeopods 1—5 large. Uropod 1, peduncle little longer than the equal rami, of which one has 4 lateral spines, the other only 1. Uropod 2, peduncle fringed with 10 outstanding spines, rami a little unequal. Uropod 3, rami longer than peduncle. not narrow, inner smooth. only a little shorter than the 2-jointed outer, which carries spines and setae. Telson as in Pontharpinia pinguis, the apices rounded, with setae at extremity of outer margin. Colour yellowish (Thomson). L. about 9—14 mm.

Port Jackson [East-Australia]; Paterson Inlet [Stewart Island by New Zealand], depth 13 m.

---

6. Fam. **Amphilochidae**

1871 Subfam. *Amphilochinae*, A. Boeck in: Forh. Selsk. Christian., 1870 p. 129 |
1876 Subfam. *A.*, A. Boeck. Skand. Arkt. Amphip., v. 2 p. 430 | 1882 *Amphilochidae*,
G. O. Sars in: Forb. Selsk. Christian., nr. 18 p. 23 | 1888 *A.*, T. Stebbing in: Rep. Voy.
Challenger, v. 29 p. 743 | 1891 *A.*, G. O. Sars. Crust. NorWay, v. 1 p. 212.

Body rather short and stout. pleon narrow and strongly flexed. Of the side-plates some and always the 4th rather or very large (Fig. 40). Antennae 1 and 2 rather short, seldom differing much in length, accessory flagellum of antenna 1 wanting or small. Upper lip apically incised. Lower lip, inner lobes wanting or rudimentary. Mandible normal. except as to molar, which varies from feeble or obsolete to strong. Maxilla 1, inner plate small, palp 1- or 2-jointed. Maxilla 2 normal or degraded. Maxillipeds, outer plates not very large, palp long. Gnathopods 1 and 2 not very powerful. Peraeopods 1—5 slender. Branchial vesicles usually simple. Marsupial plates often large with long setae. Uropods 1—3, rami slender. Telson entire, unarmed, (Fig. 39 p. 154).

Marine.

9 genera, 19 accepted species and 7 doubtful.

Synopsis of genera:

1 { Side-plates 3 and 4, contiguous margins not
       exactly fitting — **2.**
     Side-plates 3 and 4 (Fig. 40), contiguous
       margins exactly fitting — **5.**

2 { Mandible, molar small or obsolete — **3.**
     Mandible, molar large and prominent — **4.**

3 { Maxilla 2 normal . . . . 1. Gen. **Amphilochus** . . . p. 149
     Maxilla 2 degraded . . . . 2. Gen. **Amphilochoides** . p. 152

4 { Maxilla 1, palp 2-jointed . . . . 3. Gen. **Gitanopsis** . . . . p. 153
     Maxilla 1, palp 1-jointed . . . . 4. Gen. **Gitana** . . . . . . p. 155

5 { Side-plates 1—4 combined in forming lateral
       shields . . . . 5. Gen. **Tetradeion** . . . . p. 157
     Side-plates 1 and 2 (Fig. 40) not forming
       constituents of lateral shields — **6.**

6 { Mandible, molar Wanting . . . . 6. Gen. **Cyproidea** . . . . p. 157
     Mandible, molar developed — **7.**

7 { Uropod 2 not reaching so far as uropod 3 7. Gen. **Stegoplax** . . . . p. 158
     Uropod 2 reaching at least as far as uropod 3 — **8.**

8 { Gnathopod 1, 4th joint not greatly or acutely
       produced . . . . 8. Gen. **Peltocoxa** . . . . . p. 159
     Gnathopod 1, 4th joint greatly and acutely
       produced . . . . 9. Gen. **Paracyproidea** . . p. 160

1. Gen. **Amphilochus** Bate

1862 *Amphilochus* (Sp. un.: *A. manudens*), Bate, Cat. Amphip. Brit. Mus., p. 107 |
1888 *A.*, T. Stebbing in: Rep. Voy. Challenger, v. 29 p. 743 | 1892 *A.*, G. O. Sars, Crust.
Norway, v. 1 p. 215 | 1893 *A.*, A. Della Valle in: F. Fl. Neapel, v. 20 p. 593 | 1876 *Calli-
merus* (Sp. un.: *C. acudigitata*), T. Stebbing in: Ann. nat. Hist., ser. 4 v. 18 p. 445.

Body rather stout. Rostrum curved acuminate, lateral angles of head
projecting, postero-antennal corners obsolete. Side-plate 1 small, obscured,
2^d—4th large and deep, 4th rather strongly emarginate behind, hind lobe of
5th the deeper. Antennae small, peduncles rather long, flagella short.
Antenna 1 larger in ♂ than in ♀, without accessory flagellum. Epistome rounded
in front. Upper lip distally incised, lower lip without inner lobes. Mandible,
cutting edge narrow, denticulate, accessory plate only on left mandible,
molar feeble, 3^d joint of palp rather long. Maxilla 1, inner plate very
small, outer probably with 7 spines, palp 2-jointed. Maxilla 2, inner plate
broader than outer. Maxillipeds, inner plates elongate, outer of moderate size,
palp long, 4th joint short. Gnathopods 1 and 2 subchelate, palm distinct, 5th
joint more or less produced, finger long; gnathopod 2 the larger. Feraeopods 1—5 slender, subequal; peraeopods 3—5, 2^d joint expanded. Uropod 2,
rami unequal, uropod 3, peduncle longer than the subequal rami. Telson
conically tapering, entire.

5 species accepted, 3 doubtful.

Synopsis of accepted species:

1 { Gnathopod 1, front margin of 6th joint acutely
       produced . . . . 1. **A. manudens** . . p. 150
     Gnathopod 1, front margin of 6th joint not acutely
       produced — **2.**

$$2 \begin{cases} \text{Gnathopod 2, process of 5}^{\text{th}} \text{ joint overlapping} \\ \qquad \text{palm of 6}^{\text{th}} \ldots \ldots \ldots \ldots \ldots \text{ 2. A. neapolitanus } . \text{ p. 150} \\ \text{Gnathopod 2, process of 5}^{\text{th}} \text{ joint not overlapping} \\ \qquad \text{palm of 6}^{\text{th}} — 3. \end{cases}$$

$$3 \begin{cases} \text{Telson longer than 3 times the breadth } \ldots . \text{ 3. A. tenuimanus } . \text{ . p. 150} \\ \text{Telson shorter than 3 times the breadth } — 4. \end{cases}$$

$$4 \begin{cases} \text{Mandibular palp, 3}^{\text{d}} \text{ joint longer than 2}^{\text{d}} \ldots . \text{ 4. A. marionis } \ldots \text{ p. 151} \\ \text{Mandibular palp, 3}^{\text{d}} \text{ joint shorter than 2}^{\text{d}} \ldots . \text{ 5. A. brunneus} \ldots \text{ p. 151} \end{cases}$$

1. **A. manudens** Bate 1862 *A. m.*, Bate, Cat. Amphip. Brit. Mus., p. 107 t. 17 f. 6 | 1892 *A. m.*, G. O. Sars, Crust. Norway, *v.* 1 p. 217 t. 74 | 1893 *A. m.*, A. Della Valle in: F. Fl. Neapel, *v.* 20 p. 594 t. 59 f. 4 | 1871 *A. manudens*, A. Boeck in: Forh. Selsk. Christian., 1870 p. 130 | 1876 *A. m.*, A. Boeck, Skand. Arkt. Amphip., *v.* 2 p. 432 t. 11 f. 1 | 1876 *A. concinna* + *Callimerus acudigitata*, T. Stebbing in: Ann. nat. Hist., ser. 4 *v.* 18 p. 443 t. 19 f. 1—1 b; p. 445 t. 20 f. 3 | 1890 *A. boeckii*, Meinert in: Udb. Hauchs, *v.* 3 p. 160.

Head, rostrum longer in ♂ than in ♀, lateral corners ending in a sharp deflexed point. Side-plates deeper in ♀ than in ♂, 1ˢᵗ subquadrate, 2ᵈ and 3ᵈ with lower margin coarsely dentate, 4ᵗʰ with lower and hind margin denticulate. Pleon segment 3, postero-lateral angles produced, ending in small acute point. Eyes round, dark red. Antenna 1, 1ˢᵗ joint apically indented, a little longer than 2ᵈ, 2ᵈ than 3ᵈ, flagellum in ♀ about 6-jointed with slender sensory setae, in ♂ longer, with band-like sensory filaments. Antenna 2 in ♀ subequal to antenna 1, in ♂ much shorter, ultimate joint of peduncle shorter than penultimate, flagellum about half length of peduncle. Mandible, 3ᵈ joint of palp as long as 1ˢᵗ and 2ᵈ combined. Gnathopods 1 and 2, 6ᵗʰ joint widening to convex, rather oblique, denticulate palm, front margin apically produced to an acute tooth, finger slender, curved, reaching beyond the palm, proximal half of inner margin denticulate. In gnathopod 1, 5ᵗʰ joint only moderately produced, in the much larger gnathopod 2, narrowly produced as far as palm margin, which forms an angle with the hind margin. Peraeopods 3—5, 2ᵈ joint oval, serrulate on hind margin, largest in peraeopod 5. Uropod 3 projecting beyond uropods 1 and 2, rami very narrow, with scattered spinules. Telson scarcely more than half length of peduncle of uropod 3, quite simple, acute, length nearly thrice breadth. Colour brownish to reddish. L. ♀ 5, ♂ 4 mm.

Arctic Ocean, North-Atlantic, North-Sea and Skagerrak (Norway, Greenland, France, Great Britain); Kattegat. Depth 188 m.

2. **A. neapolitanus** Della Valle 1893 *A. n.*, A. Della Valle in: F. Fl. Neapel, *v.* 20 p. 595 t. 29 f. 16, 17.

Eyes maroon, reticulated with white. Gnathopods 1 and 2, front margin of 6ᵗʰ joint not produced into an apical tooth; in gnathopod 1 process of 5ᵗʰ joint not reaching palm of the broad 6ᵗʰ; in gnathopod 2, 6ᵗʰ joint larger than in gnathopod 1, process of 5ᵗʰ joint overlapping the palm with a bent point. Other appendages in agreement with A. brunneus (p. 151). Colour greenish-brown in part, in part pearly white lightly flecked with green. L. 3—4 mm.

Gulf of Naples. Among algae.

3. **A. tenuimanus** Boeck 1871 *A. t.*, A. Boeck in: Forh. Selsk. Christian., 1870 p. 131 | 1876 *A. t.*, A. Boeck, Skand. Arkt. Amphip., *v.* 2 p. 437 t. 9 f. 6, 7? | 1892 *A. t.*, G. O. Sars, Crust. Norway, *v.* 1 p. 218 t. 75 f. 1 | 1893 *A. t.* (part.), A. Della Valle in: F. Fl. Neapel, *v.* 20 p. 595 t. 59 f. 5.

Head, rostrum long, lateral corners projecting bluntly. Side-plate 1 a little produced obtusely, 2^d—4th scarcely denticulated. Pleon segment 3, postero-lateral angles without any acute point. Eyes small, rounded; visual elements imperfectly developed, light red with whitish coating. Antenna 1 nearly as in A. manudens (p. 150), antenna 2 comparatively shorter and stouter, not so long as antenna 1. Mandible, 3^d joint of palp rather longer than 2^d. Gnathopods 1 and 2 smaller and less unequal than in A. manudens, without apical tooth, palm nearly straight, irregularly denticulate, finger scarcely overlapping it, process of 5th joint not quite reaching the palm even in gnathopod 2. Peraeopods as in A. manudens, but uropod 3 much shorter, scarcely reaching beyond uropod 1, peduncle little longer than rami, which are without spinules. Telson overlapping peduncle of uropod 3, more than thrice as long as broad, acutely tapering. Colour pale yellowish, with light red patches. L. ♀ 4 mm.

Arctic Ocean, North-Atlantic and North-Sea (Norway, depth 188—376 m; Firth of Clyde).

4. **A. marionis** Stebb. 1888 *A. m.*, T. Stebbing in: Rep. Voy. Challenger, v.29 p. 743 t. 38 | 1893 *A. tenuimanus* (part.), A. Della Valle in: F. Fl. Neapel, v. 20 p. 595.

♀. Side-plate 1 with front margin semicircular, 2^d—4th with lower margin only slightly crenate. Pleon segment 3, postero-lateral angles forming a scarcely perceptible tooth; segment 6 outdrawn on either side of the telson. Eyes small, oval. Mandibular palp, 3^d joint longer than 1st and 2^d combined. Maxillipeds, outer plates broader than in European species, distal margin finely pectinate on inner and ciliated on outer part, with a simple spine between. Gnathopods 1 and 2 without apical tooth to 6th joint, otherwise nearly as in A. manudens (p. 150); gnathopod 2 very much larger than gnathopod 1. Peraeopods 1—5 about as in A. manudens. Telson triangular-oval, not twice as long as broad L. about 3 mm. — ♂ unknown.

Southern Indian Ocean (Marion Island). Depth 188 m.

5. **A. brunneus** Della Valle 1893 *A. b.*, A. Della Valle in: F. Fl. Neapel, v. 20 p. 596 t. 4 f. 5; t. 29 f. 1—15.

Head, rostrum short, lateral corners of head obtuse. Side-plate 1 rectangular, 2^d—4th (as figured) quite smooth. Eyes oval, brown. Antenna 1, 1st joint rather shorter than 2^d, 3^d very short, flagellum 6-jointed, shorter than peduncle. Antenna 2 subequal in length to antenna 1, ultimate and penultimate joints of peduncle equal in length, flagellum 5-jointed. Mandible, 3^d joint of palp shorter than 2^d. Gnathopods 1 and 2 without apical tooth to the 6th joint, which widens distally rather less than in other species; process of the 5th joint does not nearly reach palm; finger does not reach beyond palm. In gnathopod 2 the hand is moderately larger than in gnathopod 1, and has a straighter front margin. Peraeopods normal. Uropod 3 the longest. Telson triangular, about $^2/_3$ length of peduncle of uropod 3. Colour yellow brown to red. L. 4—5 mm.

Gulf of Naples. Depth to 10 m.

A. longimanus Chevreux 1888 *A. l.*, Chevreux in: Bull. Soc. zool. France, v. 13 p. 41.

L. ♀ 3 mm.

North-Atlantic (lat. 46° N., long. 7° W.). Depth 180 m.

A. melanops A. Walker 1894 *A. m.*, A. O. Walker in: Rep. Brit. Ass., Meet. 63 p. 535 | 1895 *A. m.*, A. O. Walker in: P. Liverp. biol. Soc., *v.* 9 p. 298 t. 18 f. 12; t. 19 f. 13—15.

Probably identical with A. brunneus (p. 151). L. 2·5 mm.

Menai Strait, Liverpool Bay. Depth 9—19 m.

A. oculatus H. J. Hansen 1887 *A. o.*, H. J. Hansen in: Vid. Meddel., ser. 4 *v.* 9 p. 89 t. 3 f. 2—2c | 1892 *A. o.*, G. O. Sars, Crust. Norway, *v.* 1 p. 226 | 1893 *A. tenuimanus* (part.), A. Della Valle in: F. Fl. Neapel, *y.* 20 p. 595.

L. 4 mm.

Davis Strait (Godthaab, depth 47 m; Sukkertoppen, depth 9—19 m).

2. Gen. **Amphilochoides.** O. Sars

1892 *Amphilochoides* (Sp. typ.: *Amphilochus odontonyx*). G. O. Sars, Crust. Norway, *v.* 1 p. 220 | 1893 *A.*, A. Della Valle in: F. Fl. Neapel, *v.* 20 p. 593.

General form like Amphilochus (p. 149). Upper lip distally wider, less deeply incised. Lower lip with lobes more strongly inflexed. Mandible, molar obsolete, palp longer, 3^d joint elongate. Maxilla 1, inner plate very small, outer with dentate expansion below the apical spines, palp broad, 2-jointed. Maxilla 2, plates very small, almost unarmed. Maxillipeds, plates moderately broad, not elongate, palp long. Gnathopods 1 and 2 similar, 2^d much the larger, with setiferous process of 5th joint more produced, 6th joint almost fusiform, palm being very oblique, finger long with nodiform denticle in one or both of the pairs on inner margin near hinge.

3 species.

Synopsis of species:

1 { Gnathopods 1 and 2 with nodiform denticle on
      finger . . . . 1. **A. boeckii** . . . p. 152
   { Gnathopod 2, not gnathopod 1, with nodiform denticle
      on finger — **2.**

2 { Pleon segment 3, postero-lateral angles not produced
      into a tooth . . . . 2. **A. odontonyx** . p. 153
   { Pleon segment 3, postero-lateral angles produced
      into a tooth . . . . 3. **A. intermedius** . p. 153

1. A. boeckii O. Sars 1892 *A. odontonyx* (err., non *Amphilochus o.* A. Boeck 1871!), G. O. Sars, Crust. Norway, *v.* 1 p. 221 t. 75 f. 2 | 1893 *A. o.*, A. Della Valle in: F. Fl. Neapel, *v.* 20 p. 593 t. 59 f. 2, 3 | 1895 *A. boeckii*, G. O. Sars, Crust. Norway, *v.* 1 p. 690.

Head, rostrum large, much deflexed, lateral corners nearly rectangular. Side-plate 1 obliquely quadrate, lower front angle acute, 2^d slightly denticulate. Pleon segment 3, postero-lateral angles produced to a distinct acute tooth. Eyes large, rounded oval, dark red. Antenna 1, 2^d joint produced on inner side to acute apical process, flagellum much shorter than peduncle, about 7-jointed, carrying setae. Antenna 2 rather longer than antenna 1, ultimate joint of peduncle much shorter than penultimate, flagellum much shorter than peduncle, 7-jointed. Gnathopod 1, process of 5th joint not reaching palm, 6th joint twice as long as broad, palm denticulate, as long as hind margin, finger with nodiform denticle, tip much overlapping the palm. Gnathopod 2, process of 5th joint reaching palm, 6th joint more than twice as long as broad, palm denticulate, much longer than hind margin, finger with nodiform denticle and slightly overlapping palm. Peraeopods 3—5, 2^d joint broadly expanded, hind margin serrate. Uropod 3 scarcely reaching beyond uropod 1, peduncle rather longer than rami. Telson elongate, reaching beyond peduncle of uropod 3, with a secondary denticle on left of acute apex. Integument firm, squamose. Colour yellowish, mottled with red brown spots, limbs and distal half of pleon tinged with dark crimson. L. ♀ 6 mm.

North-Atlantic and North-Sea (West-Norway). Depth 94—282 m.

2. **A. odontonyx** (Boeck) 1871 *Amphilochus o.*, A. Boeck in: Forh. Selsk. Christian., 1870 p. 131 | 1876 *A. o.*, A. Boeck, Skand. Arkt. Amphip., v. 2 p. 484 t. 11 f. 3 | 1895 *Amphilochoides o.*, G. O. Sars, Crust. Norway, v. 1 p. 690 | 1892 *A. pusillus*, G. O. Sars, Crust. Norway, v. 1 p. 222 t. 76 f. 1.

Head, rostrum not very large, lateral corners somewhat produced. Side-plate 1 bluntly rounded at front corner, 2^d not expanded as in the other species. Pleon segment 3, postero-lateral angles not produced into a tooth, nearly quadrate. Eyes small, rounded. Antennae 1 and 2, flagellum 5-jointed, otherwise as in A. boeckii, but smaller lobe to joint 2 of antenna 1. Gnathopod 1, 6th joint scarcely twice as long as broad, palm denticulate, straight, forming a distinct angle with hind margin, finger without nodiform denticle, finely ciliated on inner margin. Gnathopod 2, process of 5th joint reaching a little beyond palm, which is very oblique, forming an obtuse angle with the short hind margin, and is not denticulate near the hinge, finger with nodiform denticle. Uropod 3 reaching a little beyond uropod 1, peduncle much longer than rami. Telson very long and tapering, apex tridentate. L. ovigerous ♀ under 3 mm.

Arctic Ocean (Vadsö, depth 56—113 m); North-Atlantic and North-Sea (Scotland); Skagerrak, Christianiafjord and Kattegat; Liverpool Bay.

3. **A. intermedius** T. Scott 1896 *A. i.*, T. Scott in: Rep. Fish. Board Scotl., v. 14 p. 159 t. 4 f. 1—3.

Close to A. boeckii, but side-plate 1 notched at lower front angle. Gnathopod 1, palm not denticulate, but fringed with minute setae, finger without nodiform denticle, gnathopod 2, 4th joint with produced bifid apex very conspicuous, palm quite smooth. Pleon segment 3 agrees with A. boeckii, not with A. odontonyx.

Firth of Forth.

3. Gen. **Gitanopsis** O. Sars

1892 *Gitanopsis*, G. O. Sars, Crust. Norway, v. 1 p. 223 | 1893 *G.*, A. Della Valle in: F. Fl. Neapel, v. 20 p. 598.

Rostrum curved. Side-plate 1 not always exceedingly small. Upper lip incised at narrowed apex. Lower lip with lobes narrowed in front, deeply incised on inner margin. Mandible, cutting plate dentate, accessory plate on left mandible, spine-row well developed, molar powerful, palp with 3^d joint the longest. Maxilla 1, inner plate very small, outer with (7?) spines, palp 2-jointed. Maxilla 2, well developed, inner plate broader than outer. Maxillipeds, inner plates narrow, long, outer reaching beyond 1st joint of rather robust and setose palp. Gnathopods 1 and 2 subchelate. Peraeopods 1—5, uropods 1—3 and telson (Fig. 39 p. 154) about as in Amphilochus (p. 149).

3 species.

Synopsis of species:

1 { Pleon segments 1 and 2 dorsally produced to a tooth 1. **G. bispinosa** . p. 154
  { Pleon segments 1 and 2 not dorsally produced — 2.

2 { Gnathopods 1 and 2 strong, Very unequal . . . . . 2. **G. inermis** . . p. 154
  { Gnathopods 1 and 2 weak, not very unequal . . . 3. **G. arctica**. . . p. 155

1. G. bispinosa (Boeck) 1871 *Amphilochus b.*, A. Boeck in: Forh. Selsk. Christian., 1870 p.131 | 1876 *A. bispinosus*, A. Boeck, Skand. Arkt. Amphip., v.2 p.435 t· 10 f. 1 | 1892 *Gitanopsis bispinosa*, G. O. Sars, Crust. Norway, v. 1 p.224 t. 76 f. 2 | 1893 *G. b.*, A. Della Valle in: F. Fl. Neapel, v. 20 p. 598 t. 59 f. 6, 7.

Head, rostrum rather short, lateral corners produced, scarcely obtuse. Side-plates not very large, 1st very partially concealed, tapering to an obtuse denticulated point, 2d finely serrate on truncate distal margin. Pleon segments 1 and 2, produced dorsally to a spiniform recurved process, segment 3, postero-lateral corners quadrate. Eyes rounded oval, light red. Antennae 1 and 2 rather long. Antenna 1, 1st joint scarcely longer than 2d, flagellum nearly twice as long as peduncle, with 15 joints, each with fascicle of delicate sensory setae. Antenna 2 rather longer, slender, ultimate and penultimate joints of peduncle subequal, flagellum 15- or 16-jointed. Gnathopods 1 and 2 rather feeble, not very unequal, 5th joint scarcely narrower than 6th, the lamellar setose lobe not quite reaching the palm even in gnathopod 2, 6th joint oval, about twice as long as broad, palm convex, very oblique, not defined by any distinct angle, with small denticles and bristles on either side of it, finger hirsute. Peraeopods 1—5 slender, spinulose, in 3d—5th pairs 2d joint well expanded, especially in peraeopod 5. Uropods 1—3, rami densely spinulose, uropod 3 reaching as far as uropod 1, peduncle scarcely longer than rami. Telson conically tapering, not reaching end of peduncle of uropod 3. Colour whitish, banded with light red. L. ♀ 5·5 mm.

Arctic Ocean, North-Atlantic, North-Sea and Skagerrak (Norway, depth 94—188 m; Greenland; France; Great Britain).

2. G. inermis (O. Sars) 1882 *Amphilochus i.*, G. O. Sars in: Forb. Selsk. Christian., nr. 18 p.87 t. 3 f. 10 | 1892 *Gitanopsis i.*, G. O. Sars, Crust. Norway, v. 1 p. 225 t. 77 f. 1 | 1893 *G. i.*, A. Della Valle in: F. Fl. Neapel, v. 20 p. 598 t. 59 f. 8.

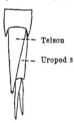

Fig. 39.
G. inermis.
Telson and
uropod 3.
[After G. O. Sars.]

Head, rostrum strongly curved, lateral corners nearly quadrate. Side-plate 1 very small, obliquely quadrangular, 2d slightly expanded, distal margin rounded, coarsely dentate. Pleon segment 3, postero-lateral corners subquadrate. Eyes rather large, rounded oval. Antennae 1 and 2 rather short. subequal. Antenna 1, 1st and 2d joints equal, flagellum 6-jointed, much shorter than peduncle. Antenna 2, ultimate and penultimate joints of peduncle equal, flagellum scarcely half length of peduncle. Gnathopods 1 and 2, resembling those of Amphilochus (p. 149), rather powerful, 6th joint widening to a convex, slightly oblique, denticulate palm, which is a little overlapped by the finger. In gnathopod 1, process of 5th joint not very long, palm defined by 2 slender spines on each side. Gnathopod 2 much larger than gnathopod 1, process of 5th joint long, slender, nearly reaching palm, which has a dense series of small hairs. Uropod 3 scarcely reaching so far back as uropod 1, peduncle rather longer than the smooth rami. Telson longer than peduncle of uropod 3 (Fig. 39), tapering uniformly to acute apex. L. ♀ under 4 mm.

Varangerfjord (Vadsö [Finmark]). Depth 38—94 m.

3. **G. arctica** O. Sars 1892 *G. a.*, G. O. Sars, Crust.Norway, *v.*1 p.227 t.77 f.2 |
1893 *G. a.*, A. Della Valle in: F. Fl. Neapel, *v.*20 p.599 t.59 f.9.

Head, rostrum strongly curved, lateral corners not projecting, evenly rounded. Side-plate 1 very partially concealed, linguiform, slightly indented at apex, 2^d distinctly dentate on rounded distal margin. Pleon segment 3, postero-lateral angles produced, but without tooth. Eyes large, oval. Antenna 1, 2^d joint shorter than 1st, 3^d shorter than 2^d, flagellum 10-jointed, much longer than peduncle. Antenna 2 longer than antenna 1, ultimate and penultimate joints of peduncle subequal, flagellum as long as peduncle, 14-jointed. Gnathopods 1 and 2 rather feeble, not very unequal, 5th joint with process densely setose, not nearly reaching palm even in gnathopod 2, 6th joint scarcely wider than 5th, widening to well defined, denticulate, nearly transverse palm, finger minutely spinulose on inner margin, overlapping palm. Peraeopods 1—5 as in G. bispinosa. Uropod 3 reaching as far back as uropod 1, peduncle longer than unarmed rami. Telson scarcely reaching beyond peduncle of uropod 3, conically tapering, minutely tridentate. L. ♀ 5 mm.

Varangerfjord (Vadsö [Finmark]).

4. Gen. **Gitana** Boeck

1871 *Gitana*, A. Boeck in: Porh. Selsk. Christian., 1870 p.132 | 1892 *G.*, G. O. Sars, Crust. NorWay, *v.*1 p.228 | 1893 *G.*, A. Della Valle in: F. Fl. Neapel, *v.*20 p.589.

Rostrum broad at base. Side-plate 1 very small, almost concealed. Antennae 1 and 2 rather slender. Mandible well developed, molar strong, 3^d joint of palp shorter than 2^d. Maxilla 1, palp 1-jointed. Maxilla 2, inner plate shorter and broader than outer. Maxillipeds as in Gitanopsis (p.153), but palp more elongate. Gnathopods 1 and 2 rather feeble, not very unequal, scarcely subchelate. Peraeopods 1—5, uropods 1—3 and telson about as in Amphilochus (p.149).

3 species.

Synopsis of species:

1 { Gnathopod 2, 6th joint longer than front margin of 5th 1. **G. sarsi**. . . . p.155
   { Gnathopod 2, 6th joint shorter than front margin
     of 5th — 2.

2 { Gnathopods 1 and 2, 5th joint distinctly produced . 2. **G. abyssicola**. p.156
   { Gnathopods 1 and 2, 5th joint not distinctly produced 3. **G. rostrata** . . p.156

1. **G. sarsi** Boeck 1871 *G. s.*, A. Boeck in: Forb. Selsk. Christian., 1870 p.132 | 1876 *G. s.*, A. Boeck, Skand. Arkt. Amphip., *v.*2 p 439 t.11 f.2 j 1893 *G. s.*, Chevreux & E. L. Bouvier in: Ann. Sci. nat., ser. 7 *v.*15 p.122 | 1892 *G. sarsii*, G. O. Sars, Crust. NorWay, *v.*1 p.228 t.78 f.1 | ?1893 *G. s.* (part.), A. Della Valle in: F. Fl. Neapel, *v.*20 p.590 t.29 f.18—32 | 1878 *Amphilochus sabrinae*, T. Stebbing in: Ann. nat. Hist., ser. 5 *v.*2 p.364 t.15 f.1.

Body short and stout. Rostrum strongly curved, rather obtuse, lateral corners of head nearly quadrate. Side-plate 2 tapering to a rounded distal margin with 3 serrations in the middle, 4th pair finely serrate continuously round lower and much of hind margin. Pleon segment 3, postero-lateral angles rather produced, not dentiform in this or the other species of the genus. Eyes of moderate size, rounded oval, dark brown. Antenna 1, 2^d joint slightly shorter than 1st, 3^d a good deal shorter than 2^d, flagellum scarcely longer than peduncle, 7-jointed. Antenna 2 rather longer than antenna 1, ultimate joint of peduncle rather longer than penultimate, flagellum nearly as long as peduncle. Gnathopods 1 and 2 less feeble than in the other 2 species; 5th joint in gnathopod 1 with front margin shorter than 6th, produced setiferous hind margin equal to it; 5th joint in gnathopod 2 with front margin shorter, but produced setiferous hind margin much longer than 6th, which in both is narrowly oval, with fascicles of spinules on hind margin, finger rather small, hirsute on inner margin. Peraeopods 1—5 not quite so slender as in the other species, 2^d joint in 3^d—5th pairs oval. Uropod 3 reaching beyond uropod 1, peduncle much longer than unarmed rami. Telson not reaching apex of peduncle of uropod 3, tapering to tridentate apex. Colour dark brown or blackish violet, the crowded spots usually in transverse bands. L. ♀ about 3 mm.

Arctic Ocean, North-Atlantic, North-Sea and Skagerrak (Norway, Spitzbergen, France, Great Britain); Kattegat. Sublitoral.

2. G. abyssicola O. Sars 1892 *G. a.*, G. O. Sars, Crust. Norway, *v.* 1 p. 229 t. 78 f. 2 | ? 1893 *G. sarsii* (part.), A. Della Valle in: F. Fl. Neapel, *v.* 20 p. 590.

Body short and stout. Head, rostrum shorter and less strongly curved than in G. sarsi (p. 155), lateral corners rather more produced. Side-plate 2 with only 2 serrations on the obtusely pointed tip, 4th quite smooth. Pleon segment 3, postero-lateral angles less produced. Eyes larger, rounder, less refractive, light red. Antennae 1 and 2 about as in G. sarsi, but flagellum of antenna 2 comparatively shorter, 6-jointed. Gnathopod 1, 5th joint little produced, little longer than 6th; gnathopod 2 longer and more slender than gnathopod 1, 5th joint more but still not greatly produced, on both margins much longer than 6th joint. Peraeopods 1—5 more slender, 2^d joint of 3^d—5th pairs broader than in G. sarsi. Uropods 1—3 and telson nearly as in G. sarsi. Colour whitish, pellucid, with patches of light red. L. ♀ nearly 5 mm.

Arctic Ocean (Selsövik [Nordland]. exactly within the polar circle). Depth 188—282 m.

3. G. rostrata Boeck 1871 *G. r.*, A. Boeck in: Forh. Selsk. Christian., 1870 p. 132 | 1876 *G. r.*, A. Boeck, Skand. Arkt. Amphip., *v.* 2 p. 441 t. 11 f. 4 | 1892 *G. r.*, G. O. Sars, Crust. Norway, *v.* 1 p. 230 t. 79 f. 1 | 1893 *G. r.*, A. Della Valle in: F. Fl. Neapel, *v.* 20 p. 592 t. 59 f. 1.

Body less stout. Head, rostrum slightly curved, more produced, acuminate, lateral corners rounded, little produced. Side-plate 2 comparatively much larger, all the narrowly rounded distal margin serrate, 4th seemingly a little serrate. Pleon segment 3, postero-lateral angles slightly produced. Eyes small, rounded, feebly developed, light red with whitish coating. Antenna 1, 2^d joint of peduncle longer than 1st, flagellum shorter than peduncle, 7-jointed. Antenna 2 much longer than antenna 1, ultimate and penultimate joints of peduncle subequal, flagellum about half length of peduncle. Gnathopods 1 and 2 very slender and feeble, 5th joint scarcely at all produced on setose hind margin, much longer than almost linear 6th; gnathopod 2 rather longer than gnathopod 1. Peraeopods 1—5 very long and slender, 2^d joint of 3^d—5th pairs large and laminar. Uropod 3 rather long, but scarcely reaching beyond uropod 1, peduncle much longer than the unarmed rami. Telson not reaching apex of peduncle of uropod 3, tapering to a simple apex. Colour whitish, without spots, sometimes faintly tinged with red. L. ♀ 7 mm.

Arctic Ocean, North-Atlantic and North-Sea (West-Norway). Depth 188—377 m.

5. Gen. **Tetradeion** Stebb.

1899 *Tetradeion* (Sp. typ.: *Cyproidia crassa*) (non *Tetradium* J. D. Dana 1846, Cnidaria), T. Stebbing in: Ann. nat. Hist., ser. 7 v. 4 p. 207.

Body short and stout. pleon shorter than peraeon. Head small, rostrum obsolete. Side-plates 1—4 together forming a continuous shield, the confronted margins of the contiguous side-plates neatly fitting, 4th much broader than 1st to 3d combined, 5th much broader than deep, fitting hind emargination of 4th. 6th and 7th concealed. Eyes well developed. Antennae 1 and 2 small. Antenna 1 the stouter, without accessory flagellum. Antenna 2, penultimate joint of peduncle shorter than antepenultimate. Mouth-parts unknown. Gnathopods 1 and 2 equal, similar, imperfectly subchelate, 4th and 5th joints slightly produced. Peraeopods 1—5 slender, character of 2d joint unknown, but expansion rendered needless by the great extent of side-plate 4. Uropod 1, rami shorter than peduncle, subequal. Uropod 2 reaching as far back as uropod 1, rami a little unequal. Uropod 3 not reaching so far back as the other pairs, stouter, rami decidedly unequal. Telson entire, oval, short.

1 species.

1. **T. crassum** (Chilton) 1883 *Cyproidia? crassa*, Chilton in: Tr. N. Zealand Inst., v. 15 p. 80 t 3 f. 1 | 1899 *C. c.*, *Tetradeion sp. typ*, T. Stebbing in: Ann. nat. Hist., ser. 7 v. 4 p. 207.

Side-plate 1 broader than 2d or 3d, but not so deep, 4th deeper than the rest, much broader than deep. Pleon segment 3, postero-lateral angles not produced. Eyes large. Antenna 1, 1st and 2d joints stout, subequal, 2d produced into a tooth over the small stout 3d, flagellum 6- or 7-jointed, tapering, with long sensory filaments. Antenna 2, antepenultimate joint much longer, ultimate rather longer than penultimate joint of peduncle, flagellum subequal ultimate joint of peduncle, with 4 joints, 1st the longest. Gnathopods 1 and 2, 5th joint bluntly produced, 6th rather narrow, slightly tapering, scarcely forming a palm for the small finger. Peraeopods 1—5 subequal, finger small. Uropods 1 and 2, the longer ramus with 2 spinules. Uropod 3, rami unarmed, one shorter, the other longer than the peduncle. Telson apically a little narrower than at base. Colour brown. L. about 3·5 mm.

Lyttelton Harbour [New Zealand].

6. Gen. **Cyproidea** Hasw.

1880 *Cyproidea*, Haswell in: Ann. nat. Hist., ser. 5 v. 5 p 31 | 1880 *Cyproidia* (part.), Haswell in: P. Linn. Soc. N.S. Wales, v. 4 p. 320 | 1885 *C.* (part.), T. Stebbing in: Ann. nat. Hist., ser. 5 v. 15 p. 59 | 1888 *C.*, *Peltocoxa?*, T. Stebbing in: Rep. Voy. Challenger, v. 29 p. 513 | 1893 *P.* (part.), A. Della Valle in: F. Fl. Neapel. v. 20 p. 647 | 1881 *Cypridoidea*, E. C. Rye in: Zool. Rec., v. 16 Index p. 4.

Body short and stout. Side-plates 1 and 2 rudimentary, 3d and 4th very large, confronted margins closely fitting. 6th and 7th concealed. Eyes well developed. Antennae 1 and 2 small. Antenna 1, peduncle rather stout, accessory flagellum wanting. Antenna 2 slender. Epistome strongly projecting. Upper lip unsymmetrically bilobed. Lower lip apparently with inner lobes obsolete. Mandible, cutting edge with numerous teeth, accessory plate on left mandible also dentate, spine-row well developed, molar wanting, palp slender, 3-jointed, 1st joint not very short. Maxilla 1, inner plate with 1 spinule on rounded apex, outer with 7 or 8 spines, palp large, 1-jointed. Maxilla 2, inner plate the broader, both short, with 2 or 3 apical spines. Maxillipeds, inner plates with apical margin produced at inner angle and having a denticle at the outer, 2 spinules on inner surface below the apex, outer plates short, broader than inner, not reaching end of 1st joint of palp, with 2 apical spinules, palp long, 2^d joint scarcely as long as 1st or 3^d, finger slender, curved. Gnathopod 1 subchelate, 5th joint with produced process. Gnathopod 2 complexly subchelate, 5th joint forming a chela with 6th, but not with 7th. Peraeopods 1—5 slender, 2^d joint not expanded. Uropods 1—3, rami slender, not very unequal, shorter than peduncle. Uropod 2 extending beyond uropod 3. Telson entire.

1 species.

1. **C. ornata** (HasW.) 1880 *C. sp.* (part.), Haswell in: Ann. nat. Hist., ser. 5 *v.* 5 p. 31 | 1880 *Cyproidia ornata*, Haswell in: P. Linn. Soc. N.S. Wales, *v.* 4 p. 320 t. 18 f. 1.

Head small, with small rostrum. Peraeon broad and deep. Pleon folding under, narrow, laterally angled, segment 5 very short, segment 6 produced to a point overlapping but raised above the telson. In lateral view the raised point looks obtuse and the telson pointed, the true character as seen in dorsal view being just the reverse. Side-plate 4 much the largest, with rather broad but shallow excavation behind, into which side-plate 5 neatly fits, concealing the 6th and 7th. Pleon segment 3, postero-lateral corners subquadrate. Eyes rather small, rounded. Antenna 1, 2^d joint longer than 1st, produced in triangular process over half the short narrow 3^d, flagellum slender, with 9 joints carrying hyaline sensory appendages, the last 3 joints extremely thin. Antenna 2 rather longer, gland-cone short, obtuse, antepenultimate joint of peduncle comparatively long, with one margin sinuous, penultimate joint narrower, much longer, ultimate joint a little shorter than penultimate, but longer than the very thin 3-jointed flagellum including its apical spine. Projecting edge of epistome rounded, sharp. Gnathopod 1, 2^d joint narrowly oblong, rather abruptly widened some distance down, 3^d at least as long as 4th, 5th triangular, hind margin setulose, inner margin and apex of process carrying spines, 6th tending to oval but front nearly straight, palm long, scarcely defined, matched by the long finger, which has much of inner margin finely serrulate. Gnathopod 2, 2^d joint a little widened downward, 3^d very short, but hind apex produced to a long curved narrow process, tipped with a spine, 4th with short process, tipped with 2 spines, 5th with broad base nearly as long as 6th, and a tapering process which reaches the truncate apex of oblong 6th, the short finger overlapping the serrulate rounded angle of the 6th and tip of 5th. Peraeopods 1—5, 2^d joint long, and narrow, finger short; in peraeopods 1 and 2, 4th joint rather longer than 5th or 6th, but not so in following pairs. Uropod 1, peduncle very long, slender, rami much shorter, equal. Uropod 2 similar to uropod 1, but shorter. Uropod 3, peduncle stouter than in uropods 1 and 2, outer ramus rather shorter than inner. Telson oblong, apex roundly truncate. Colour light pink, with minute brown and red dots forming a lobed pattern on the side-plates. L. reaching 5 mm.

Port Jackson, Port Stephens and Watson's Bay [East-Australia].

7. Gen. **Stegoplax** O. Sars

1882 *Stegoplax* (Sp. un.: *S. longirostris*), G. O. Sars in: Forh. Selsk. Christian., nr. 18 p. 88 | 1892 *S.*, G. O. Sars, Crust. Norway, *v.* 1 p. 232.

Body short and stout. Head boldly rostrate. Side-plates 1 and 2 rudimentary, 3^d and 4th very large, confronted margins closely fitting, 6th and 7th not concealed (Fig. 40). Eyes feebly developed. Antennae 1 and 2 and mouth-parts as in Cyproidea (p. 157), except that the mandible has a well developed molar, and the maxillipeds have the apex of inner plates transverse, not oblique, and outer reaching end of 1st joint of palp. Gnathopods 1 and 2 subequal, similar, scarcely subchelate, slender, 5th joint a little produced. Peraeopods 1—5 slender. but 2^d joint expanded in 4th and 5th pairs. Uropod 3 larger than uropod 2, and reaching much beyond it. Telson entire.

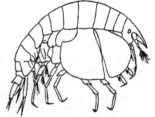

Fig. 40. **S. longirostris**, ♀.
Lateral view.
[After G. O. Sars.]

1 species.

1. S. longirostris O. Sars 1882 *S. l.*, G. O. Sars in: Forh. Selsk. Christian., nr. 18 p. 88 t. 3 f. 11 | 1892 *S. l.*, G. O. Sars, Crust. Norway, *v.* 1 p. 232 t. 79 f. 2 | 1893 *Peltocoxa l.*, A. Della Valle in: F. Fl. Neapel, *v.* 20 p. 650 t. 59 f. 64, 65.

Head, rostrum very long, slightly curved, acuminate. Side-plates 1 and 2 concealed, 3 and 4 with the margins in a uniform curve to the emargination of the 4th, which almost completely receives the 5th; 5th deeper behind than in front. Pleon segment 3, postero-lateral angles rather produced backward, not acutely; segment 6 with 2 longitudinal ridges apically acute (Fig. 40). Eyes small, rounded, 3 components, pigment red with whitish coating. Antenna 1, 1st joint as long as 2^d and 3^d combined, flagellum with 5 joints, 1st much the largest, all with long band-like sensory filaments. Antenna 2, ultimate joint of peduncle twice length of penultimate. flagellum with 4 joints, 1st much the longest. Gnathopods 1 and 2 similar, 6th joint narrow, scarcely longer than 4th and 5th combined, palm not distinctly defined, minutely ciliated, with 2 spinules in gnathopod 1, and one in the narrower gnathopod 2, 5th joint produced behind to a setiferous lobe, finger slender, finely spinulose on inner margin. Peraeopods 1 and 2 very slender, longer than the others. Peraeopods 4 and 5, 2^d joint with sinuosity in hind margin of expansion. Uropod 3 much longer than uropod 2, peduncle longer than the slender unarmed rami. Telson conically triangular and flattened, reaching end of peduncle of uropod 3. Colour greyish white. L. ♀ about 2 mm.

Hardangerfjord, Trondhjemsfjord, Arctic Ocean (Lofoten Isles). Depth 282—565 m.

8. Gen. **Peltocoxa** Catta

1875 *Peltocoxa* (Sp. un.: *P. marioni*), Catta in: Rev. Sci. nat., *v.* 4 p. 161 | 1893 *P.* (part.), A. Della Valle in: F. Fl. Neapel, *v.* 20 p. 647.

Body short and stout. Head with very small rostrum. Side-plates and peraeopods 1—5 nearly as in Stegoplax. Eyes well developed. Antennae 1 and 2 small. Antenna 1 much the stouter, with small, 1-jointed accessory flagellum, the principal flagellum having a stout 1st joint fringed with very long sensory filaments. Mouth-parts apparently in close agreement with those of Stegoplax, except that both plates of maxilla 2 are slender, and the outer plates of maxillipeds reach beyond 1st joint of palp. Gnathopod 1 scarcely subchelate, 5th joint a little produced. Gnathopod 2 strongly subchelate, 5th joint very little produced. Uropod 1 longer than uropod 2, uropod 2 longer than uropod 3 and reaching beyond it. Telson entire, large, boat-shaped.

1 species accepted, 1 doubtful.

1. **P. marioni** Catta   1875   *P. m.*, Catta in: Rev. Sci. nat., *v.* 4   p. 161 | 1885 *Cyproidia damnoniensis*, T. Stebbing in: Ann. nat. Hist., ser. 5 *v.* 15   p. 59   t. 2 | 1893 *Peltocoxa d.* + *P. marionis*, A. Della Valle in: F. Fl. Neapel, *v.* 20   p. 648   t. 30   f. 19—32; t. 60   f. 11, 12;   p. 648.

Head, rostrum small. Side-plates and angles of pleon segment 3 as in Stegoplax longirostris (p. 159). Eyes small, round, about 20 components, pigment red. Antenna 1, 1st joint as long as 2d and 3d combined, flagellum with 4 joints, 1st much the longest, with long band-like filaments, accessory flagellum minute, 1-jointed. Antenna 2, ultimate and penultimate joints of peduncle subequal, flagellum 4-jointed, tapering. Mandible, palp of very delicate structure or wanting. Maxilla 1, inner plate weak, unarmed or wanting; outer plate probably armed with 7 spines, palp rather long, though but 1-jointed. Gnathopod 1, 5th joint slightly produced, 6th narrow, without distinct palm, with spinules on the hinder margin, finger denticulate on inner margin. Gnathopod 2 similar to gnathopod 1, except that the 6th joint widens distally to a distinct convex palm defined by a tooth and spinules. Peraeopods 1—3 slender, peraeopod 3 with the 4 terminal joints shorter than in peraeopods 1 and 2. Peraeopod 4, 2d joint with hind margin of expansion uniformly curved. Peraeopod 5, expansion of 2d joint subquadrate, produced broadly below the 3d. Uropod 1, rami equal, uropods 2 and 3, rami a little unequal. Uropod 3 smaller than uropod 2, peduncle shorter than the serrate rami. Telson triangular, boat-shaped, reaching apex of shorter ramus of uropod 3. Colour greyish with many red spots or red and purple. L. 2·5 mm.

English Channel (South-Devon), Firth of Clyde, Firth of Forth, Liverpool Bay, Gulf of Naples.

P. brevirostris (T. & A. Scott)   1893   *Cyproidia? b.*, T. & A. Scott in: Ann. nat. Hist., ser. 6 *v.* 12   p. 244   t. 13 | 1895   *C. b.*, A. O. Walker in: P. Liverp. biol. Soc., *v.* 9   p. 300.

In agreement with P. marioni, yet without accessory flagellum to antenna 1, palp of mandible rather longer, finger of gnathopods 1 and 2 not denticulate, side-plates 6 and 7 almost entirely concealed, and 2d joint of peraeopod 5 with acutely produced lower hind corner. L. about 1·5 mm.

Moray Firth.   Depth 73 m, from Filograna filograna (L.) [*F. implexa* Berkeley].

9. Gen. **Paracyproidea** Stebb.

1880 *Cyproidia* (part.). Haswell in: P. Linn. Soc. N. S. Wales, *v.* 4   p. 320 | 1899 *Paracyproidea* (Sp. typ.: *Cyproidea lineata*), T. Stebbing in: Ann. nat. Hist., ser. 7 *v.* 4   p. 207.

Like Cyproidea (p. 157) in general, but mandible with well developed molar, maxillipeds with apex of inner plates transverse not oblique; gnathopods 1 and 2 much more slender, rather feebly subchelate, gnathopod 1 having the 4th joint produced along the 5th as in Aora, gnathopod 2 with the 5th joint produced. Uropods 1—3 with rami subequal, much shorter than peduncle, 1st and 2d reaching only a little beyond 3d, telson entire, very large, strongly compressed, extending back almost to extremity of uropods.

1 species.

1. **P. lineata** (Hasw.)   1880   *Cyproidia l.*, Haswell in: P. Linn. Soc. N. S. Wales, *v.* 4   p. 321   t. 18   f. 2 | 1899   *Cyproidea l.*, *Paracyproidea sp. typ.*, T. Stebbing in: Ann. nat. Hist., ser. 7 *v.* 4   p. 207.

Head, rostrum very small. Pleon segments 5 and 6 very short, 6th not dorsally produced. Side-plates as in Cyproidea ornata (p. 158), but 7th perhaps a little exposed. Eyes very large, round, red. Antenna 1, 1st and 2^d joints stout, 2^d a little longer than 1st, produced in a tooth over the whole of the 3^d, flagellum tapering, its 7 joints smooth. Antenna 2 not longer, antepenultimate joint short, ultimate and penultimate joints of peduncle not long, subequal, flagellum 7—9-jointed, subequal to peduncle. Mandibular palp slender, 1st joint little shorter than 2^d or 3^d. Gnathopod 1, 2^d joint distally widened, hind margin very sinuous, 4th joint as long as 5th, tapering, nearly reaching apex of 5th, which is as long as 6th, 6th narrowly oblong, slightly widened at short, curved, sloping palm, which is overlapped by the small finger. Gnathopod 2, 2^d joint straight, 4th with transversely truncate apex, narrower but not shorter than 3^d, 5th produced along 6th, but not reaching the palm, obliquely truncate and inner margin carrying spines, 6th widened to the oblique, curved, spinulose palm, which is shorter than the hind margin and matches the smooth finger. Peraeopods 1—5, 2^d joint slender. Uropods 1—3, peduncle long, rami short. Telson with a very narrow triangular flattened top, thence sloping to meet the straight lower margin, the sides being flat, deep and long. Colour, numerous brown dots disposed in lines on the lateral shields and the body. L. about 2·5 mm.

Port Jackson [East-Australia].

Amphilochidarum species dubiae.

Amphilochus squamosus G. M. Thoms.   1880 *A. s.*, G. M. Thomson in: Ann. nat. Hist., ser.5 *v.*6 p.4 t.1 f.4, 4a | 1881 *A. s.*, G. M. Thomson in: Tr. N. Zealand Inst., *v.*13 p.214 t.7 f.5a, b | 1893 *A.? s.*, A. Della Valle in: F. Fl. Neapel, *v.*20 p.597.

Terminal joints of maxillipeds spinous, not clawed. L. about 4 mm.

Dunedin Harbour [New Zealand].   Depth 7—9 m.

Probolium serratipes (Norm.)   1869 *P. s.*, A. M. Norman in: Rep. Brit. Ass., Meet. 38 p. 273.

L. about 2 mm.

North-Atlantic (Shetland Islands).

P. spence-batei Stebb.   1876 *P. s.-b.*, T. Stebbing in: Ann. nat. Hist., ser. 4 *v.* 17 p. 344 t. 19 f. 4, 4a—c | 1888 *Amphilochus? s.-b.*, T. Stebbing in: Rep. Voy. Challenger, *v.* 29 p. 460.

Probably young of Gitanopsis inermis (p. 154). L. about 3 mm.

Torbay [South-Devon].

---

7. Fam. Leucothoidae

1852 Subfam. *Leucothoinae*, J. D. Dana in: Amer. J. Sci., ser. 2 *v.*14 p. 311 | 1856 *Leucothoides*, Bate (& Westwood) in: Rep. Brit. Ass., Meet.25 p.21 | 1865 Subfam. *Leucothoina*, W. Lilljeborg in: N. Acta Soc. Upsal., ser. 3 *v.*6 nr. 1 p. 18 | 1882 *Leucothoidae*, G. O. Sars in: Forh. Selsk. Christian., nr. 18 p. 27 | 1888 *L.*, T. Stebbing in: Rep. Voy. Challenger, *v.* 29 p. 771 | 1892 *L.*, G. O. Sars, Crust. Norway, *v.* 1 p. 281.

Antennae 1 and 2 small, not very unequal. Antenna 1 with rudimentary accessory flagellum. Antenna 2 the more slender. Epistome (Fig. 41 p. 163) produced in front. Lower lip without inner lobes. Mandible without molar, palp slender with short 3^d joint. Maxilla 1, inner plate small, outer with not more than 7 spines, palp 1- or 2-jointed. Maxillipeds with outer plates small or rudimentary, palp well developed, 4th joint long. Of the gnathopods one or both pairs chelate. Uropods 1—3 almost or entirely unarmed. Telson entire.

Marine.

8 genera, 14 accepted species and 7 doubtful.

Synopsis of genera:

1 $\begin{cases}$ Gnathopod 1, chela formed between 6th and 7th joints; uropod 3 uniramous . . . . . 1. Gen. **Seba** . . . . . p. 162 Gnathopod 1, chela formed between 5th and 6th joints; uropod 3 biramous — 2. $\end{cases}$

2 $\begin{cases}$ Maxillipeds, outer plates rudimentary . . . . 2. Gen. **Leucothoe** . . . p. 163 Maxillipeds, outer plates small, but well developed 3. Gen. **Paraleucothoe** . p. 169 $\end{cases}$

1. Gen. Seba Bate*)

1862 *Seba* (Sp. un.: *S. innominata*), (A. Costa in MS.) Bate, Cat. Amphip. Brit. Mus., p. 159 | 1888 *S.*, T. Stebbing in: Rep. Voy. Challenger, v. 29 p. 782 | 1893 *S.*, A. Della Valle in: F. Fl. Neapel, v. 20 p. 773 | 1884 *Teraticum* (Sp. un.: *T. typicum*), Chilton in: Tr. N. Zealand Inst., v. 16 p. 257 | 1889 *Grimaldia* (Sp. un.: *G. armata*), Chevreux in: Bull. Soc. zool. France, v. 14 p. 283.

Antenna 1 stouter than antenna 2, both with peduncle long, flagellum short, accessory flagellum very small, 2-jointed. Upper lip with symmetrical slightly sinuous margin, epistome bluntly conical. Lower lip without inner lobes. Mandible, cutting plate strongly dentate, accessory plate denticulate, smaller on right than on left, spine-row of 4 spines, molar obsolete, palp with 3d joint much shorter and narrower than 2d. Maxilla 1, inner plate small, outer with 7 spines, palp 1-jointed. Maxilla 2, plates short, inner shorter and broader than outer. Maxillipeds, plates rather small, feebly armed, not rudimentary, palp well developed, 4th joint elongate. Gnathopod 1 chelate or subchelate; gnathopod 2 chelate, 3d joint elongate; in both the basal part of the 6th joint much larger than the thumb. Uropod 2 not shorter than the others. Uropod 3 uniramous. Telson triangular, entire.

2 species.

Synopsis of species:

Antenna 1, flagellum longer than peduncle; peraeopods 3—5,
2d joint narrow . . . . 1. **S. innominata** . p. 162
Antenna 1, flagellum shorter than peduncle; peraeopods 3—5,
2d joint broad . . . . 2. **S. saundersii** . . p. 163

1. **S. innominata** Bate 1862 *S. i.*, (A. Costa in MS.?) Bate, Cat. Amphip. Brit. Mus., p. 159 t. 29 f. 5.

Peraeon segments subequal. Pleon segments 1—3 with sinus above the postero-lateral angle, well marked in segment 3. Eyes small. Antenna 1 nearly half as long as the body, flagellum a little longer than peduncle. Antenna 2 rather shorter than antenna 1, but with peduncle equally long. Gnathopod 1, 5th joint as broad as 6th but only half as long, 6th with chela-forming process as long as finger. Gnathopod 2 similar, but rather larger, with 5th joint rather narrower and very much shorter than 6th. Peraeopods 3—5 (in fig.) with narrowly oval 2d joint.

Gulf of Naples.

---

*) Bate, 1862, attributes the authorship of the genus (and of the species *innominata*) probably by error to A. Costa.

2. **S. saundersii** Stebb. 1875 *S. s.*, T. Stebbing in: Ann. nat. Hist., ser. 4 *v.* 15
p. 186 t. 15 f. 2, 2a—c | 1888 *S. s.*, T. Stebbing in: Rep. Voy. Challenger, *v.* 29 p. 783
t. 49 | 1893 *S. s.*, A. Della Valle in: F. Fl. Neapel, *v.* 20 p. 774 t. 60 f. 32—34 | 1884
S. typica, Chilton in: N. Zealand J. Sci., *v.* 2 p. 230 | 1884 *Teraticum typicum*, Chilton
in: Tr. N. Zealand Inst., *v.* 16 p. 257 t. 18 f. 1a—g | 1889 *Grimaldia armata*, Chevreux
in: Bull. Soc. zool. France, *v.* 14 p. 284 f. | 1899 *Seba a.*, Chevreux in: C.-R. Ass. Franç.,
Sess. 27 *v.* 2 p. 483.

Side-plate 1 directed forwards, 4th not specially larger than any of the
preceding (smaller than any in Chevreux's fig.). Pleon segment 3, postero-
lateral angles rounded, produced backward; segments 5 and 6 coalescent
(Chevreux). Antenna 1, 1st joint thicker but rather shorter than 2^d, 3^d about
$^1/_3$ length of 2^d, flagellum 5-jointed, accessory flagellum with 2 joints, of which
the 2^d is minute. Antenna 2, ultimate joint of peduncle shorter than penultimate,
flagellum 3-jointed. Gnathopods 1 and 2, 5th joint triangular, not produced,
6th with a base which in gnathopod 1 is broad and setose on the hind
margin, but longer, narrower, and smooth in gnathopod 2; in both produced
so as to form with the finger a perfect chela; gnathopod 1 sometimes (in
♂?) subchelate, 6th joint being subquadrate, with the palm transverse, crenate
near the finger hinge, hollowed in the centre, with a tooth in the hollow
(Chilton). Peraeopods 1 and 2 slender. Peraeopods 3—5, 2^d joint oval, hind
margin most serrate in the 5th pair; 3^d pair smaller than 4th or 5th. Uropod 1,
rami longer than peduncle, inner rather the longer. Uropod 2 stouter, peduncle
more nearly equal to the long, subequal rami. Uropod 3, the simple ramus longer
than peduncle. Telson triangular oval, end blunt. L. ♀ 4 mm.

Strait of Magellan (Cape Virgins, depth 100 m); Algoa Bay?, from sponges;
South-Pacific (New Zealand); North-Atlantic (Azores, depth 1287 m; Belle-Ile-en Mer,
depth 180 m).

2. Gen. **Leucothoe** Leach

1793 Subgen. *Gammarellus* (part.), J. F. W. Herbst, Naturg. Krabben Krebse,
v. 2 p. 106 | 1813/14 *Leucothöe* (Sp. un.: *L. articulosa*), Leach in: Edinb. Enc., *v.* 7
p. 403, 432 | 1826 *Leucothoë*, Audouin in: Descr. Égypte, *v.* 1 iv p. 92 | 1840 *Leucothoe*,
O. G. Costa, Fauna Reg. Napoli, Crost., Cat. p. 5 | 1888 *Leucothoë*, T. Stebbing in:
Rep. Voy. Challenger, *v.* 29 p. 771 | 1892 *L.*, G. O. Sars, Crust. Norway, *v.* 1 p. 282 | 1893
Leucothoe, A. Della Valle in: F. Fl. Neapel, *v.* 20 p. 651 | 1816 *Lycesta* (Sp. un.: *L. furina*),
Savigny, Mém. An. s. Vert., *v.* 1 p. 109.

Side-plates 1—4 broad compared with their depth. Antenna 1 stouter
than antenna 2, both with peduncle long, flagellum short, accessory flagellum
of antenna 1 very small, 1-jointed. Epistome conically projecting.
Upper lip unsymmetrically bilobed (Fig. 41). Lower lip without
inner lobes. Mandible, cutting edge strongly dentate, accessory
plate well developed on left, very small on right, molar obsolete,
3^d joint of palp much smaller than 2^d. Maxilla 1, inner plate
very small, with 1 seta, outer with 7 spines, palp 2-jointed.
Maxilla 2, inner plate broader than outer. Maxillipeds, inner
plates short and broad, partly coalesced, outer very small or
rudimentary, the inner edge of the joint carrying it being
sharpened, palp large. Gnathopod 1 chelate between 5th and
6th joints, 5th bulbous at base and produced into a slender
thumb, parallel to the 6th, 7th small, folding over apex of 5th.
Gnathopod 2 powerful, subchelate, 5th joint setose, produced
along hind margin of the large oval 6th, finger strong,
curving over the long oblique palm nearly to the process of 5th joint.

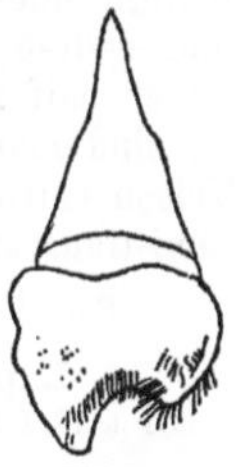

Fig. 41.
L. miersi.
Epistome and
upper lip.

11*

Peraeopods 3—5 subequal and similar. Uropod 2 much the shortest. Uropod 3, peduncle long, rami lanceolate, minutely spinulose.. Telson triangular, long, unarmed.

Frequently lodged in tunicates or sponges.

11 species accepted, 7 doubtful.

Synopsis of accepted species:

1 { Gnathopod 1, finger serrate . . . . 1. **L. traillii** . . . . . p. 164
    Gnathopod 1, finger smooth — **2.**

2 { Antenna 1 in ♂, basal joint very tumid . . . 2. **L. pachycera** . . . p. 164
    Antenna 1 in ♂, basal joint not very tumid — **3.**

3 { Mandibular palp extremely short . . . . 3. **L. miersi** . . . . . p. 165
    Mandibular palp moderately short — **4.**

4 { Pleon segment 3, postero-lateral -angles not
      incised — **5.**
    Pleon segment 3, postero-lateral angles incised — **9.**

5 { Antenna 1, 3d joint not half length of 2d — **6.**
    Antenna 1, 3d joint half length of 2d — **8.**

6 { Gnathopod 2, 6th joint irregularly denticulate . 4. **L. furina** . . . . . p. 165
    Gnathopod 2, 6th joint regularly denticulate — **7.**

7 { Eyes red . . . . 5. **L. spinicarpa** . . p. 165
    Eyes black . . . . 6. **L. commensalis** . p. 166

8 { Gnathopod 1, 6th joint widest at base . . . . 7. **L. tridens** . . . . p. 166
    Gnathopod 1, 6th joint narrowest at base . . . 8. **L. brevidigitata** . p. 167

9 { Gnathopod 1, 5th joint, inner margin smooth . 9. **L. richiardii** . . . p. 167
    Gnathopod 1, 5th joint, inner margin serrate — **10.**

10 { Telson twice as long as broad . . . . 10. **L. incisa** . . . . . p. 167
     Telson not twice as long as broad . . . . . . 11. **L. lilljeborgii** . . . p. 167

1. **L. traillii** G. M. Thoms. 1882 *L. t.*, G. M. Thomson in: Tr. N. Zealand Inst., v. 14 p. 234 t. 18 f. 1 a—d | 1893 *L. spinicarpa* (part.), A. Della Valle in: F. Fl. Neapel, v. 20 p. 652.

Body rather slender. Eyes rounded, large. Antenna 1, 1st joint stout, 2d subequal, slender, flagellum 4-jointed(?). Antenna 2, peduncle rather longer than that of antenna 1, flagellum 5-jointed, shorter than ultimate joint of peduncle. Mandibular palp very slender. Gnathopod 1, process of 5th joint about $^2/_3$ length of 6th joint, apex acute, finger curved, about $^1/_3$ length of 6th joint, inner margin finely serrate. Gnathopod 2, 5th joint produced to $^2/_3$ length of 6th and ending in a curved spine, 6th large, oval, with numerous dentations, finger half as long. Peraeopods slender, 3—5 with 2d joint wide, crenate on hind margin. Uropods 1—3, rami narrowly lanceolate; nearly smooth. Telson narrow, tapering to a subacute, entire apex. Integument rather thin, semi-transparent. L. 9 mm.

Port Pegasus and Paterson Inlet [New Zealand]. Depth 9—18 m.

2. **L. pachycera** Della Valle 1893 *L. p.*, A. Della Valle in: F. Fl. Neapel, v. 20 p. 651 t. 19 f. 22, 23, 29—34.

Pleon segment 2, postero-lateral angles not acute, in segment 3 incised, with upturned acute angle. Antenna 1 in ♂, 1st joint very tumid, ellipsoidal, 2d widening from narrow base, distally truncate, flagellum 7-jointed. Antenna 2 slender, flagellum 5-jointed. Gnathopod 1, process of 5th joint irregularly serrate, uncinate at apex. Gnathopod 2, 2d joint narrow at base, 6th elongate, with irregular outline, finger large. Telson very short, not twice as long as broad, ending only subacutely. Colour pearl-grey, a little interspersed with greenish-yellow, with flecks of crimson on the sides and smaller ones on the back. L. 3—4 mm.

, Gulf of Naples. Depth 10—12 m, in sand.

3. **L. miersi** Stebb. 1888 *L. m.*, T. Stebbing in: Rep. Voy. Challenger, *v.* 29 p. 772 t. 46 | 1893 *L. spinicarpa* (part.), A. Della Valle in: F. Fl. Neapel, *v.* 20 p. 652.

Nearly allied to L. spinicarpa. Side-plate 3 tending to oblong. Antenna 1, flagellum longer than 2^d and 3^d joints of peduncle combined, 17—21-jointed, accessory flagellum rudimentary, 1-jointed. Antenna 2, ultimate joint of peduncle much more than half length of penultimate, flagellum 12-jointed. Mandible with accessory plate on both mandibles multidenticulate, that on the right much the smaller, spine-row of about 30 spines, 3^d joint of palp only about $^1/_4$ length of 2^d. Gnathopod 1, inner margin of process of 5th joint minutely tuberculate. Gnathopod 2, palmar margin serrate more and more deeply as it approaches the finger hinge, close to which the serration is minute. Peraeopods 3—5, oval 2^d joint almost imperceptibly serrate on hind margin. Uropod 3, rami little more than half as long as peduncle. Telson thrice as long as broad, tapering gradually till near the acute apex, then more rapidly. L. 12 mm.

Southern Indian Ocean (Cape Agulhas). Depth 274 m.

4. **L. furina** (Sav.) 1816 *Lycesta f.*, Savigny, Mém. An. s. Vert., *v.* 1 p. 109 t. 4 f. 2 | 1826 *L. f.*, Audouin in: Descr. Égypte, *v.* 1 iv p. 92; Crust. t. 11 f. 2 | 1855 *Leucothoe f.*, W. Liljeborg in: Öfv. Ak. Förb., *v.* 12 p. 128 | 1857 *L. procera*, Bate in: Ann. nat. Hist., ser. 2 *v.* 19 p. 146 | 1893 *L. spinicarpa* (part.), A. Della Valle in: F. Fl. Neapel, *v.* 20 p. 652.

Pleon segment 3, postero-lateral angles slightly rounded, almost quadrate. Eyes irregularly oval. Antenna 1, 1st and 2^d joints long, subequal, 3^d fully $^1/_3$ length of 2^d, flagellum 8-jointed. Antenna 2, ultimate joint of peduncle not much shorter than penultimate, flagellum 4-jointed. Gnathopod 2, 5th joint apically much widened and emarginate, 6th strikingly narrowed to the finger-hinge, near to which it has a small sharp notch and a deeper one in the almost straight palm margin. L. 8 mm.

Mediterranean (Egypt), North-Sea (Banff).

5. **L. spinicarpa** (Abildg.) 1789 *Gammarus spinicarpus*, Abildgaard in: O. F. Müller, Zool. Dan., ed. 3 *v.* 3 p. 66 t. 119 f. 1—4 | 1793 *Cancer (Gammarellus) s.*, J. F. W. Herbst. Naturg. Krabben Krebse, *v.* 2 p. 135 t. 36 f. 6, 7 | 1861 *Gammarus s.*, *Leucothoë articulosa*, A. Boeck in: Forh. Skand. Naturf., Møde 8 p. 654 | 1892 *L. spinicarpa*, *L. a.* (err.), G. O. Sars, Crust. Norway, *v.* 1 p. 283 t. 100; t. 101 f. 1 | ? 1893 *L. s.* (part.), A. Della Valle in: F. Fl. Neapel, *v.* 20 p. 652 t. 6 f. 4; t. 19 f. 1—20 | 1804 *Cancer articulosus*, Montagu in: Tr. Linn. Soc. London, *v.* 7 p. 70 t. 6 f. 6 | 1812 *Astacus a.*, Pennant, Brit. Zool., ed. 5 *v.* 4 p. 36 | 1813/14 *Leucothöe articulosa*, Leach in: Edinb. Enc., *v.* 7 p. 403 | 1818 *Gammarus articulosus*, Lamarck, Hist. An. s. Vert., *v.* 5 p. 181 | 1838 *G. a.*, „Leucothoë" (part.)?, H. Milne Edwards in: Lamarck, Hist. An. s. Vert., ed. 2 *v.* 5 p. 311 | 1826 *Leucothoë articulata* (err.), Audouin in: Descr. Égypte, *v.* 1 iv p. 92 | ? 1851 *L. denticulata*, A. Costa, Fauna Reg. Napoli, fasc. April 1851 t. 9 f. 3.

Body slender, back broadly rounded. Side-plate 1 somewhat expanded distally, front corner transversely truncate, 2^d broader than deep, 3^d tending to semicircular, 4th with lower front angle rounded and upper hind one a little blunted. Pleon segment 3, postero-lateral corners quadrate, with minutely produced point. Eyes oval, bright red. Antenna 1 scarcely $^1/_3$ length

of body, 1ˢᵗ joint with acute apical process, 2ᵈ narrower, subequal, flagellum
about 16-jointed. Antenna 2 more slender, little shorter, ultimate joint of peduncle
about or a little more than half length of penultimate, flagellum 9-jointed.
Mandible, 3ᵈ joint of palp more than ¹/₃ length of 2ᵈ. Gnathopod 1. process
of 5ᵗʰ joint spiniform, inner margin quite smooth, apex slightly curved,
6ᵗʰ joint parallel-sided, inner margin finely serrate, and with a row of curved
setules, finger about half its length, slender, curved. Gnathopod 2, process
of 5ᵗʰ joint densely setose, apex serrulate, 6ᵗʰ joint massive, oval, with acute
setose apex in front, palm convex, minutely serrate, rather more strongly
near the hinge of finger. Peraeopods 1—5 subequal, marginal spinules short,
finger small. Peraeopod 5, the oval 2ᵈ joint more distinctly serrate than
in 3ᵈ and 4ᵗʰ pairs. Uropod 3, rami much more than half length of peduncle.
Telson thrice as long as broad, uniformly tapering to acute apex. Colour
pale flesh-colour with darker transverse bands; ova in pouch grass-green.
L. ♀ 14, ♂ 18 mm.

    Arctic Ocean, North-Atlantic. North-Sea and Skagerrak (Greenland; Norway,
depth 56—282 m; British Isles; Azores; France); Kattegat; Mediterranean? Sometimes
in Ascidians.

    6. **L. commensalis** HasW. 1880 *L. c.*, HasWell in: P. Linn. Soc. N.S.Wales,
v. 4 p. 261 t. 10 f. 3 | 1884 *L. spinicarpa var.*, Miers in: Rep. Voy. Alert, p. 312 | 1885
L. s. var., HasWell in: P. Linn. Soc. N.S.Wales, *v.* 10 p. 101 | 1893 *L. s.* (part.), A. Della
Valle in: F. Fl. Neapel, *v.* 20 p. 653.

    Close to L. spinicarpa (p. 165). Eyes black. Antenna 1 ¹/₄ longer than
antenna 2. Maxillipeds, inner plates less broadly truncate, rudimentary outer
plates smaller, 1ˢᵗ joint of palp more produced on the outer side. Peraeopods of
varying slenderness; peraeopod 5 has the 2ᵈ joint very broadly oval, the hind
margin very convex to the middle or a little below it and then obliquely truncate,
the same form of margin being indicated also in peraeopods 3 and 4.
Apex of telson moderately acute. Colour varying, brick-red or greenish,
sometimes light pink with innumerable minute crimson dots. L. reaching
to 12—14 mm.

    South-Pacific (New South Wales). In Sponges, Ascidians and other positions.

    7. **L. tridens** Stebb. 1888 *L. t.*, T. Stebbing in: Rep. Voy. Challenger, *v.* 29
p. 777 t. 47 | 1893 *L. spinicarpa* (part.), A. Della Valle in: F. Fl. Neapel, *v.* 20 p. 653.

    Side-plate 1 produced forward below, 2ᵈ much broader than deep,
4ᵗʰ foursided, broader below than above. Pleon segment 3, postero-lateral
corners quadrate. Eyes round-oval, dark. Antenna 1, 1ˢᵗ joint with small
apical tooth, 3ᵈ half length of 2ᵈ, flagellum very short, accessory flagellum
shorter than the very short 1ˢᵗ joint of flagellum. Antenna 2, ultimate joint of
peduncle nearly ³/₄ length of penultimate, flagellum of 6 slender joints.
Mandible, cutting plates as in L. miersi (p. 165), spines of spine-row fewer, 3ᵈ joint
of palp rather more than half length of 2ᵈ. Maxillipeds, outer plates minutely
rudimentary. Gnathopod 1, process of 5ᵗʰ joint smooth on inner margin,
apex curving round quite to hinge of finger, 6ᵗʰ joint rather broader at base
than apex, inner margin finely serrate, finger about ¹/₃ length of 6ᵗʰ joint.
Gnathopod 2, apex of 5ᵗʰ joint cut into 5 denticles, 6ᵗʰ joint broadly oval,
with no outer apical process. palm with 3 little teeth near hinge of finger,
rest of serration microscopic. Peraeopods 1—5, armature slight, finger not
very small. Telson scarcely twice as long as broad, apex a little obtuse.
L. 5 mm.

    South-Pacific (New Zealand). Depth 2000 m.

8. **L. brevidigitata** Miers 1884 *L. b.*, Miers in: Rep. Voy. Alert, p. 313 t. 34
f. A | 1888 *L. flindersi*, T. Stebbing in: Rep. Voy. Challenger, *v.* 29 p. 779 t. 48.

Side-plate 1 expanded below, scarcely produced forward, 2^d not broader than long, 4th rotundo-quadrate. Pleon segment 3, postero-lateral angles not acute or incised. Eyes oval. Antenna 1, 1st joint longer than 2^d, 3^d more than half length of 2^d, accessory flagellum not longer than broad. Antenna 2, ultimate joint of peduncle a little shorter than penultimate, flagellum 4-jointed. Upper lip, the narrow lobe very short. Mandible, 10 spines in spine-row, 2^d joint of palp robust, 3^d narrow, half as long as 2^d. Maxillipeds, inner plates longer than in other species, outer very small, 4th joint of palp with a long nail. Gnathopod 1, 2^d joint moderately dilated, 5th scarcely bulbous at base, broadly tapering to a slightly curved apex, with a row of setae on the hind margin, 6th joint extremely narrow at base, then narrowly oval, finger extremely small. Gnathopod 2, 5th joint very short and produced along the hind margin of 6th for less than half its length; 6th joint widest distally, without apical process, palm flat, tuberculate, very long, but convex and less elongate than usual in small specimens. Peraeopod 3, 2^d joint with hind margin smooth, whereas in peraeopod 4 it is serrate. Telson not twice as long as broad, apex a little obtuse. L. 4—16 mm.

Torres Strait (Thursday Island, Flinders Passage). Depth 13—15 m, coral mud.

9. **L. richiardii** Mich. Lessona 1865 *L. r.*, Mich. Lessona in: Atti Soc. Ital., *v.* 8 p. 426 | 1893 *L. r.*, A. Della Valle in: F. Fl. Neapel, *v.* 20 ⃘ p. 654 t. 3 f. 4; t. 19 f. 21.

Agreeing almost entirely with L. spinicarpa (p. 165); differs in having a small incision at the postero-lateral corners of pleon segment 3, and in colour; on peraeon brilliant alternating bands of red and orange, on pleon segments 1—3 large red spots in rows; the small, circular eyes pale red; antennae are red, with a white spot at apex of 2^d joint of antenna 1 and of ultimate and penultimate joints of peduncle of antenna 2; side-plate 4 bright red; gnathopod 2 particoloured, red and yellow; uropod reddish. L. 6—7 mm.

Mediterranean (Genoa, Naples).

10. **L. incisa** D. Roberts. 1888 *L. furina* (err., non *Lycesta f.* Savigny 1816!), Chevreux in: Bull. Soc. Étud. Paris, *v.* 11 p. (9) | 1892 *L. incisa*, D. Robertson in: P. nat. Hist. Soc. Glasgow, n. ser. *v.* 3 p. 217.

Near to L. lilljeborgii, but side-plate 4 with front angle rounded not acute; gnathopod 1, 5th joint with tip of process strongly hooked, inner margin of 6th joint faintly crenulate, finger not very small; gnathopod 2 palm convex, faintly but broadly crenulate, finger not abruptly bent at the base; telson fully twice as long as broad, apex almost acute. As in L. lilljeborgii, postero-lateral angles of pleon segment 3 are sharply upturned forming a sinus, and pleon segment 2 shows a similar tendency. Also in gnathopod 1, inner margin of 5th joint is serrate. Mandible, 3^d joint of palp not much shorter than 2^d. L. 7 mm.

North-Atlantic (France; Firth of Clyde, low water and at 38 m).

11. **L. lilljeborgii** Boeck 1855 *L. articulosa* (err., non Leach 1813/14!), W. Lilljeborg in: Öfv. Ak. Förh, *v.* 12 p. 126 | 1861 *L. lilljeborgii*, A. Boeck in: Forh. Skand. Naturf., Møde 8 p. 654 | 1892 *L. l.*, G. O. Sars, Crust. Norway, *v.* 1 p. 284 t. 101 f. 2 | 1889 *L. imparicornis*, A. M. Norman in: Ann. nat. Hist., ser. 6 *v.* 4 p. 114 t. 10 f. 1—4 | 1893 *L. serraticarpa*, *L. lilljeborgii*, A. Della Valle in: F. Fl. Neapel, *v.* 20 p. 656 t. 19 f. 24—28; p. 908.

Side-plate 1 slightly widened below, 2^d broader than long, 4th pentagonal, with lower front angle acúte. Pleon segment 3, postero-lateral angles acutely upturned forming a sinus. Eyes rounded triangular, dark brownish. Antenna 1 much larger than antenna 2, 1st joint without apical process, as long as 2^d, flagella 9-jointed. Antenna 2 slender, ultimate joint of peduncle not much shorter than penultimate, flagellum 7- or 8-jointed. Gnathopod 1, 2^d joint densely setose on both margins, 5th dilated at base, process not very slender, inner margin serrate, apex a little bent, 6th narrow, inner margin smooth, finger $1/4$ length of 6th joint. Gnathopod 2, 5th joint tridentate at apex, 6th tapering distally, palm very oblique, its edge quite smooth and nearly straight, finger abruptly bent at base. Peraeopods 1—5 very slender, 6th joint elongate. Uropod 3, rami nearly as long as peduncle. Telson triangular oval, little longer than broad, apex well rounded. Colour pellucid with yellowish tinge and irregular orange and pinkish specks. L. ♀ 6 mm.

North-Atlantic, North-Sea and Kattegat (Kullaberg [Sweden]; Kopervik [South-West-Norway], depth 75—113 m; Shetland Isles).

L. affinis Stimps. 1855 *L. a.*, Stimpson in: P. Ac. Philad., *v.* 7 p. 394 | 1862 *L. a.*, Bate, Cat. Amphip. Brit. Mus., p. 378 | 1893 *L. richiardii?* (err., non Mich. Lessona 1865!), A. Della Valle in: F. Fl. Neapel, *v.* 20 p. 656.

L. 12·5 mm.

False Bay [Cape of Good Hope]. On gravelly bottom in the coralline zone.

L. antarctica Pfeff. 1888 *L. a.*, Pfeffer in: Jahrb. Hamburg. Anst., *v.* 5 p. 128 t. 2 f. 4 | 1893 *L. spinicarpa* (part.), A. Della Valle in: F. Fl. Neapel, *v.* 20 p. 653.

Agrees generally with L. spinicarpa (p. 165). L. about 5 mm.

South-Atlantic (South Georgia).

L. crassimana Kossm. 1880 *L. c.*, Kossmann, Reise Roth. Meer., *v.* 21 Malacost. p. 131 t. 13 f. 9, 10 | 1898 *L. c.*, Sowinski in: Mém. Soc. Kiew, *v.* 15 p. 492 t. 11 f. 20—22; t. 12 f. 7, 8 | 1884 *L. spinicarpa* (err., non *Gammarus spinicarpus* Abildgaard 1789!), Miers in: Rep. Voy. Alert, p. 313 | 1893 *L. s.* (part.), A. Della Valle in: F. Fl. Neapel, *v.* 20 p. 653.

L. 7—7·5 mm.

Red Sea, Bosphorus.

L. diemenensis Hasw. 1880 *L. d.*, Haswell in: P. Linn. Soc. N.S.Wales, *v.* 4 p. 262 t. 9 f. 5 | 1893 *L. spinicarpa* (part.), A. Della Valle in: F. Fl. Neapel, *v.* 20 p. 653.

Scarcely distinguishable from L. commensalis (p. 166). But eyes round, large; gnathopod 2, the process of the 5th joint bifurcate and the finger relatively longer.

South Pacific (Tasmania).

L. gracilis Hasw. 1880 *L. g.*, Haswell in: P. Linn. Soc. N.S.Wales, *v.* 4 p. 263 t. 10 f. 2 | 1893 *L. spinicarpa* (part.), A. Della Valle in: F. Fl. Neapel, *v.* 20 p. 653.

Not distinguished from L. commensalis or L. diemenensis (p. 166, 168). L. 10 mm.

South-Pacific (Tasmania).

L. grandimana Stimps. 1853 *L. grandimanus*, Stimpson in: Smithson. Contr., *v.* 6 nr. 5 p. 51 t. 3 f. 37 | 1893 *L. g.*, A. Della Valle in: F. Fl. Neapel, *v.* 20 p. 656.

Probably identical with L. spinicarpa (p. 165). L. 11 mm.

Fundy Bay (Grand Manan). On shelly bottom, depth 55 m.

L. stylifera Stimps. 1855 *L. s.*, Stimpson in: P. Ac. Philad., *v.* 7 p. 383 | 1893 *L. s.*, A. Della Valle in: F. Fl. Neapel, *v.* 20 p. 656.

L. 8 mm.

North-Pacific (Japan).

3. Gen. **Paraleucothoe** Stebb.

1899 *Paraleucothoe* (Sp. typ.: *Leucothoe novae-hollandiae*), T. Stebbing in: Ann. nat. Hist., ser. 7 *v.* 4 p. 208.

Body stout; peraeon segment 1 longer than 2^d. Antenna 1 larger than antenna 2, with rudimentary accessory flagellum. Antenna 2, flagellum shorter than peduncle. Mandible, cutting edge with few teeth, accessory plate larger on left than right mandible, molar wanting, 3^d joint of palp slight, but longer than half 2^d. Maxilla 1, outer plate with 5 seta-like spines, palp very large, 1-jointed, with expanded base. Maxilla 2, inner plate the longer, both plates apparently almost devoid of spines or setae. Maxillipeds, inner plates broad and long, narrow at base, coalesced, but with median line of separation, apical margin slightly concave, the median part produced inward so as to give the effect of an additional pair of plates, also coalesced, not quite reaching the apical margin of the others, but surmounted by the 3 pairs of spine-teeth usually found on the true apex, outer plates not rudimentary but small, scarcely reaching apex of 1st joint of palp, armed with a single spinule at inner apex, 2^d joint of palp the shortest. Gnathopod 1 chelate between 5th and 6th joints, apex of 6th joint grooved to received the finger. Gnathopod 2 subchelate, 5th joint slightly produced, finger strong, but not nearly reaching process of 5th joint. Peraeopods 3—5 subequal, 2^d joint expanded. Uropod 3 reaching beyond uropod 2, peduncle longer than the smooth equal rami. Telson triangular oval.

1 species.

1. **P. novaehollandiae** (Hasw.) 1880 *Leucothoë n.-h.*, Haswell in: P. Linn. Soc. N. S. Wales, *v.* 4 p. 329 t. 20 f. 2 | 1884 *L. n.-h.*, Miers in: Rep. Voy. Alert, p. 314 | 1893 *L. n. h.*, A. Della Valle in: F. Fl. Neapel, *v.* 20 p. 656 | 1899 *L. n.-h.*, *Paraleucothoe sp. typ.*, T. Stebbing in: Ann. nat. Hist., ser. 7 *v.* 4 p. 208.

Head rather small. Peraeon segment 1 rather tumid, longer than any succeeding peraeon segment. Side-plate 1 broader than 2^d or 3^d, somewhat produced forward. Pleon segment 3, postero-lateral corners subquadrate, not produced. Eyes large, oval. Antenna 1, 2^d joint slightly longer than 1st, or shorter, and in the latter case the 3^d, though much narrower is more than $^1/_3$ as long as 2^d; flagellum 13-jointed, much shorter than peduncle, accessory flagellum like a tubercle, yet seemingly 2-jointed. Antenna 2 subequal in length to antenna 1, ultimate joint of peduncle shorter than penultimate, flagellum very short, 9-jointed. Lower lip, mandibular processes quadrate. Mandible, 8 spines in the spine-row, tapering 3^d joint of palp tipped with a long spine. Maxillipeds very narrow, outer plates a little longer than broad, one overlapping the other, the peculiar structure of the inner plates making it impossible to flatten down the outer plates and elongate palps. Gnathopod 1 large in correspondence with the size of the peraeon segment 1 and its side-plate, 3^d joint larger than 4th; 5th with base massive, broad and long, produced to a process shorter than base, but tapering to end of hind margin of the 6th, which from a narrow base becomes subquadrate and has a grooved truncate apex, within which is almost concealed the finger, its short triangular naïl and its slightly curved upper margin alone protruding. Gnathopod 2 not much larger than gnathopod 1, 3^d joint larger than 4th, 5th with strongly setose, grooved, shortly produced hind part, 6th powerful, elongate, basal part broad and longer than broad, palm oblique, sharply defined; the strong finger closing over several teeth into a rather large concave portion with which the palm ends. Peraeopods 1—5 smooth. Peraeopods 3—5, 2^d joint oval, 4th a little produced downward behind. Uropods 1—3 smooth. Uropod 1 the longest, peduncle equal to the equal rami. Uropod 2, peduncle as long as the longer ramus. Uropod 3 as long as uropod 2, peduncle longer than the equal rami. Telson about as long as peduncle of uropod 3, its apex narrowly rounded. Colour light pink, nearly white. L. 18 mm.

Port Jackson [East-Australia].

---

8. Fam. **Anamixidae**

1897 *Anamixidae*, T. Stebbing in: Tr. Linn. Soc. London, ser. 2 *v.* 7 p. 36 | 1899 *A.*, T. Stebbing in: Ann. nat. Hist., ser. 7 *v.* 4 p. 210.

Side-plates 2—4 shield-like. Mandible weak. Maxillae 1 and 2 obsolete. Maxillipeds without plates. Gnathopod 1 (Fig. 42) chelate. Telson simple.

Marine.

1 genus, 1 species.

1. Gen. **Anamixis** Stebb.

1897 *Anamixis* (Sp. un.: *A. hanseni*), T. Stebbing in: Tr. Linn. Soc. London, ser. 2 *v.* 7 p. 35.

Head hood-like. Side-plate 1 very small, 2^d—4th very large. Antenna 1, peduncle long, no accessory flagellum. Antenna 2 remote from antenna 1, more slender, flagellum small. Mouth-organs degraded and abnormal. Gnathopod 1 (Fig. 42) delicately chelate. Gnathopod 2 (Fig. 43) massive, complexly subchelate. Peraeopods 1—5 slender. Pleopods small.

1 species.

1. **A. hanseni** Stebb. 1897 *A. h.*, T. Stebbing in: Tr. Linn. Soc. London, ser. 2 *v.* 7 p. 35 t. 11.

Head narrowed distally, apex rounded, lateral angles indistinct. Side-plate 1 triangular, 2^d the largest. Pleon segment 3, postero-lateral angles a little blunted, segment 5 very short, segment 6 projecting on either side of the telson. Eyes round, lateral. Antenna 1 inserted below apex of head, peduncle rather long, 2^d joint nearly $^3/_4$ of 1st, much more slender, 3^d $^2/_5$ of 2^d, flagellum of 11 joints, some with sensory filaments. Antenna 2 attached near the base of head, penultimate joint of peduncle longest, ultimate joint rather longer than antepenultimate, flagellum with 4 small joints. From the slightly keeled underside of the head a vertical plate projects between the 2^d pair of antennae, having on its truncate front edge some microscopic teeth, and suggesting a coalescence of the mandibles. Maxillipeds, the coalesced 2^d joint showing a slight division between 2 rounded apices on the outer surface, the 3^d joint like one of the joints of the palp, of which the 3^d (6th of the maxillipeds) is the longest, terminal joint or finger long, slender, curved. Gnathopod 1 (Fig. 42) very slight, 5th joint wider and longer than the 2^d, its long slender curved process tipped with a needle-like spine, 6th joint slender, tapering, together with a needle-like finger opposed to the process of the 5th joint to form a long delicate chela. Gnathopod 2 (Fig. 43), 2^{d} joint narrow, wider distally, 3^{d} in appearance if not in reality articulating not only with 2^{d} and 4th, but also with the 5th and 6th joints, 5th subequal to 2^{d}, broadest at base, apically acute, not fully reaching the end of the broad 6th joint, which is 3 or 4 times as long as broad, the palm short, oblique, tridentate, finger more than half as long as 6th joint, strong, with curved acute tip much overlapping apex of

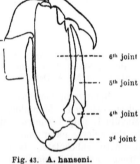

5th joint. Peraeopods 3—5, 2^{d} joint expanded, oval. Pleopods with 5- or 6-jointed rami, shorter than peduncle. Uropod 1 longer than uropod 2;

5th joint

Fig. 42. **A. hanseni.**
Gnathopod 1.

Fig. 43. **A. hanseni.**
Gnathopod 2.

6th joint

5th joint

4th joint

3^{d} joint

in both the inner ramus is a little longer, the outer much shorter than the peduncle; uropod 3 unknown. Telson a little longer than broad, apex broadly rounded. L. about 3 mm.

Tropical Atlantic (West Indies). From Goniastraea varia Dana.

---

9. Fam. **Metopidae**

1899 *Metopidae*, T. Stebbing in: Ann. nat. Hist., ser. 7 v. 4 p. 210.

Side-plate 1 rudimentary, side-plates 2—4 together forming large lateral shield (Fig. 47). Antenna 1 sometimes with rudimentary accessory flagellum. Upper lip bilobed. Lower lip with inner lobes coalesced. Mandible, cutting edge dentate, molar weak or wanting, palp small, 3-jointed, 3^{d} joint very small. Maxilla 1, inner plate very small, outer with 6 spines, palp 1- or 2-jointed. Maxilla 2, inner plate the smaller. Maxillipeds (Fig. 44, 49), inner plates coalesced or separate, outer small or wanting, palp elongate. Gnathopod 1 (Fig. 45) simple or sometimes subchelate, generally feeble. Gnathopod 2 (Fig. 46, 50) generally robust, subchelate. Peraeopod 1 usually more slender than peraeopod 2 (Fig. 48), 2^{d} joint little or not at all expanded in peraeopod 3, usually much but sometimes not at all in peraeopods 4 and 5. Uropods 1 and 2 biramous, uropod 3 with a single 2-jointed ramus. Telson oval, entire.

The adjustment of this family is involved in much difficulty, because in regard to many species it remains uncertain whether they have or have not an accessory flagellum to antenna 1, whether the inner lobes of the lower lip and the inner plates of the maxillipeds be or be not coalesced, and whether the maxilla 1 have the palp 1- or 2-jointed.

Marine.

4 genera, 37 accepted species and 2 doubtful.

Synopsis of genera:

1 { Maxilla 1, palp 1-jointed . . . . 1. Gen. **Metopa** . . . p. 172
{ Maxilla 1, palp 2-jointed — **2.**

1. Gen. **Metopa** Boeck

1871 *Metopa* (part.), A. Boeck in: Forh. Selsk. Christian., 1870 p. 140 | 1876 *M.*, A. Boeck, Skand. Arkt. Amphip., *v.* 2 p. 451 | 1892 *M.*, G. O. Sars, Crust. Norway, *v.* 1 p. 248 | 1893 *M.*, A. Della Valle in: F. Fl. Neapel, *v.* 20 p. 634.

Relative proportions of antennae 1 and 2 varying. Lower lip with inner lobes coalesced. Mandible, molar wanting, palp small, 3ᵈ joint very short. Maxilla 1, palp 1-jointed. Maxillipeds (Fig. 44), inner plates coalesced almost to the apices, outer plates wanting. Gnathopod 1 (Fig. 45) small, seldom distinctly subchelate. Gnathopod 2 (Fig. 46) distinctly subchelate, usually strong, often differing in the two sexes. Peraeopod 3, 2ᵈ joint not expanded. Peraeopod 4 usually, peraeopod 5 always, with 2ᵈ joint expanded. Uropod 3, peduncle longer or shorter than the 2-jointed ramus.

21 species.

Synopsis of species:

13 { Gnathopod 1 distinctly subchelate . . . . . . 13. **M. leptocarpa** . . p. 178
   { Gnathopod 1 not distinctly subchelate — **14.**

14 { Antennae 1 and 2 long — **15.**
   { Antennae 1 and 2 not long — **17.**

15 { Peraeopods 4 and 5, 4ᵗʰ joint narrow, scarcely
       decurrent . . . . 14. **M. longicornis** . . p. 179
   { Peraeopods 4 and 5, 4ᵗʰ joint decurrent, not
       narrow — **16.**

16 { Side-plate 4 nearly as deep as broad . . . . . 15. **M. pusilla** . . . . p. 179
   { Side-plate 4 not nearly as deep as broad . . . 16. **M. aequicornis** . . p. 180

17 { Gnathopod 2, palm serrate — **18.**
   { Gnathopod 2, palm smooth — **19.**

18 { Gnathopod 2, palm irregularly serrate . . . . 17. **M. borealis** . . . . p. 180
   { Gnathopod 2, palm regularly serrate . . . . . 18. **M. rubrovittata** . p. 180

19 { Gnathopod 2 strong . . . . 19. **M. sölsbergi** . . . p. 181
   { Gnathopod 2 feeble — **20.**

20 { Body slender; gnathopod 2, 6ᵗʰ joint oblong
       triangular . . . . 20. **M. tenuimana** . . p. 181
   { Body stout; gnathopod 2, 6ᵗʰ joint oblong
       quadrangular . . . . 21. **M. invalida** . . . p. 182

1. **M. esmarki** Boeck 1872 *M. e.*, A. Boeck in: Forb. Selsk. Christian., 1871 p. 47 | 1893 *M. esmarkii*, A. Della Valle in: F. Fl. Neapel, *v.* 20 p. 644.

Back round. Side-plate 4 very large. Antenna 1 in ♂ much, in ♀ little shorter than antenna 2, 1ˢᵗ joint much longer than 2ᵈ, flagellum 10- or 11-jointed. Antenna 2, ultimate and penultimate joints of peduncle subequal, flagellum very short. Gnathopod 1 in ♂ slender, 4ᵗʰ joint produced, 5ᵗʰ wider and considerably longer than the narrowly oval 6ᵗʰ, which has no definite palm. Gnathopod 2 in ♂ robust, 5ᵗʰ joint cup-shaped, 6ᵗʰ powerful, with a deep and wide cavity between a stout process of the hind margin and a small tooth, over which the finger curves to meet the process, leaving a small cavity between its hinge and the tooth. Gnathopod 2 in ♀ with much smaller process and cavity, the margin serrate above and below the small tooth. Peraeopods 1—5, joints slender, except the 2ᵈ of peraeopod 5, which is almost broader than deep, the 4ᵗʰ of peraeopod 2, which is not much expanded, and the 4ᵗʰ of peraeopods 4 and 5, which is very much widened and overlaps all the next joint. Uropod 3, peduncle with 6 spines, longer than ramus. L. ♀ 4, ♂ 5 mm.

North-Pacific (California).

2. **M. robusta** O. Sars 1892 *M. r.*, G. O. Sars, Crust. Norway, *v.* 1 p. 270 t. 96 f. 1.

Body stout, integument very firm. Head, lateral corners angular. Side-plate 4 nearly twice as large as 2ᵈ and 3ᵈ combined, lower margin very sinuous. Pleon segment 3, postero-lateral angles rather produced. Eyes round, bright red. Antennae 1 and 2 subequal. Antenna 1, flagellum longer than peduncle, 14-jointed. Antenna 2, ultimate and penultimate joints of peduncle subequal, flagellum 9-jointed. Gnathopod 1, 2ᵈ joint rather broad and lamellar, 4ᵗʰ forming a broadly rounded setiferous lobe, 5ᵗʰ very narrow and elongated, rather tapering, 6ᵗʰ about half as long, linear, finger short, finely ciliated within. Gnathopod 2 powerful, as in M. norvegica (p. 177), the 4ᵗʰ and 5ᵗʰ joints both cup-shaped; 6ᵗʰ subquadrate, but gradually expanded towards the slightly oblique, convex, in part coarsely serrate, palm, which is defined by a strong tooth. Peraeopod 1 not like peraeopod 2, but slender, while peraeopods 2—5 are powerful, with the 4th joint expanded and produced, the finger strong, minutely serrate. Peraeopods 4 and 5 have the 2^d joint expanded, in peraeopod 5 rounded quadrangular, and the 4th joint produced below the 5th. Uropod 3, peduncle with 1 spine, little longer than 1st joint of ramus, 2^d joint of ramus shorter than 1st. Telson oval, obtusely pointed, with 2 pairs of strong spines. Colour whitish pellucid, with orange patches. L. ♀ 6 mm.

Arctic Ocean and North-Atlantic (Norway). Depth 55—94 m.

3. **M. sinuata** O. Sars 1887 *M. bruzelii* (part.), H. J. Hansen in: Vid. Meddel., ser. 4 *v.* 9 p. 97 t. 4 f. 2, 2a, b | 1892 *M. sinuata*, G. O. Sars, Crust. Norway, *v.* 1 p. 263 t. 92 f. 2.

Body rather compressed, not very slender. Head, lateral corners angular. Side-plate 4 broader than deep, much larger than 2^d and 3^d combined, lower margin sinuous. Pleon segment 3, postero-lateral corners a little produced. Eyes rather large, irregularly rounded. Antennae 1 and 2 subequal, short and stout. Antenna 1, 1st joint longer than 2^d and 3^d combined, flagellum little longer than peduncle, 12-jointed. Antenna 2, ultimate and penultimate joints of peduncle subequal, together longer than flagellum. Gnathopod 1 not weak, 6th joint longer than 5th, very narrowly oval, with 5 spinules on hind margin. Gnathopod 2, 6th joint oblong oval, palm minutely serrate, rather oblique in ♀, less so in ♂, defined by a slight angular projection. which like the joint itself is more strongly developed in ♂ than in ♀. Peraeopods 2—5 rather stout, 2^d joint in peraeopod 5 broadly oval, and 4th joint produced a little beyond 5th. Uropod 3, peduncle with 2 spines, nearly as long as ramus, of which 2^d joint is shorter than 1st. Telson with 2 pairs of lateral spines, oval, apex obtuse. L. ♀ 4 mm.

Arctic Ocean and North-Atlantic (Nordland, depth 56—75 m; Greenland).

4. **M. propinqua** O. Sars 1892 *M. propinqva*, G. O. Sars, Crust. Norway, *v.* 1 p. 264 t. 93 f. 1.

Very near to M. sinuata; but antennae 1 and 2 longer and more slender, antenna 1 longer than antenna 2, 1st joint as long as 2^d and 3^d combined, flagellum decidedly longer than peduncle. Antenna 2, flagellum as long as ultimate and penultimate joints of peduncle combined. Gnathopod 1, 6th joint shorter than 5th and much narrower, without spinules. Gnathopod 2 rather powerful, defining tooth of palm well developed. Peraeopod 5, 2^d joint irregularly rounded, hind margin bulging in the middle, 4th joint produced in a curved lobe beyond 5th. Uropod 3, peduncle with 3 strong blunt spines, decidedly shorter than ramus, of which 2^d joint is quite as long as 1st. Telson with 3 pairs of lateral spines. Colour whitish, with dark brown stellate spots on back, brown shadows on sides. L. about 3 mm.

Trondhjemsfjord. Depth 75 m.

5. **M. palmata** O. Sars 1892 *M. p.*, G. O. Sars, Crust. Norway, *v.* 1 p. 272 t. 96 f. 2.

Body stout. Head, lateral corners subangular. Side-plate 4 larger than 2^d and 3^d combined, lower margin in a single curve. Pleon segment 3, postero-lateral angles little produced. Eyes round, rather large. Antennae 1 and 2 subequal, short. Antenna 1, 1st joint much longer than 2^d and 3^d combined, flagellum a little longer than peduncle, 12-jointed. Antenna 2, ultimate joint of peduncle rather longer than penultimate, flagellum 7-jointed. Gnathopod 1 slender, 2[d] joint narrow, 4th rather long, narrowly and triangularly produced, 5th long, linear, 6th narrower and rather shorter, finger short, scale-like, fringed closely within with long curved setae. Gnathopod 2 powerful, 5th but not 4th joint cup-shaped, 6th oblong quadrangular, palm transverse, gently concave, obscurely crenulate, defined by a broad setose lobe slightly emarginate at tip, finger ending with a small hook. Peraeopods 1—5 nearly as in M. robusta (p. 173), but rather less strong. Uropod 3, peduncle with 3 spines, nearly as long as ramus, of which the 2[d] joint is nearly as long as the 1st. Telson oval, obtusely pointed, with 2 or 3 pairs of lateral spines. L. 5 mm.

Arctic Ocean (Hammerfest [Finmark]).

6. **M. clypeata** (Kröyer) 1842 *Leucothoe c.*, Kröyer in: Naturh. Tidsskr., *v.* 4 p. 157 | 1845 *L. c.*, Kröyer in: Naturh. Tidsskr., ser. 2 *v.* 1 p. 545 t. 6 f. 2 a—f | 1846 *L. c.*, Kröyer in: Voy. Nord., Crust. t. 22 f. 2 a—o | 1862 *Montagua c.*, Bate, Cat. Amphip. Brit. Mus., p. 58 t. 9 f. 4 | 1871 *Metopa c.*, A. Boeck in: Forb. Selsk. Christian., 1870 p. 140 | 1876 *M. c.* (part.), A. Boeck, Skand. Arkt. Amphip., *v.* 2 p. 451 t. 18 f. 4; t. 19 f. 3 | 1887 *M. c.*, H. J. Hansen in: Vid. Meddel., ser. 4 *v.* 9 p. 90 t. 3 f. 3—3 b | 1893 *M. c.*, A. Della Valle in:·F. Fl. Neapel, *v.* 20 p. 638 t. 59 f. 50, 51.

Body somewhat compressed, strong. Side-plate 4 much broader than deep, much larger than 2[d] and 3[d] combined. Pleon segment 3, postero-lateral corners subquadrate. Eyes very small, round. Antenna 1 very long, 1st joint long, a little longer than 2[d], flagellum slender, twice as long as peduncle, 25-jointed. Antenna 2 rather shorter than antenna 1, ultimate and penultimate joints of peduncle subequal, flagellum less than half length of peduncle, 12-jointed. Maxilla 1 with large 1-jointed palp. Maxillipeds, inner plates separate (?). Gnathopod 1, 4th joint broadly produced and setose, 5th long, tapering, fringed with long setae on the hind margin, 6th much shorter, linear, little curved, with setae on front and apical part of hind margin, finger rudimentary, with a comb of graduated spines. Gnathopod 2 stoutly built, 6th joint very large, longer than the 1st, oblong oval widening distally, palm finely serrate, slightly convex and oblique, defined by a strong tooth. Peraeopod 1 more slender than peraeopod 2, which resembles peraeopod 3. Peraeopod 5, 2[d] joint shorter and broader, also 4th with the laminar part shorter and broader, but the produced part longer than in peraeopod 4 (according to Kröyer's text and to the fig. 3 b of Hansen, but not according to Kröyer's fig. and Boeck's fig. 4n). Uropod 3 stout, peduncle with some spinules, not quite so long as the 2 subequal joints of the ramus. Telson oval, obtuse at apex, figured by Boeck with 2 pairs of lateral spines. L. 6—8 mm.

Arctic Ocean and North-Atlantic (Greenland; Bell Sound [Spitzbergen]?).

7. **M. alderii** (Bate) 1857 *Montagua a.*, Bate in: Ann. nat. Hist., ser. 2 *v.* 19 p. 137 | 1862 *M. a.*, Bate, Cat. Amphip. Brit. Mus, p 57 t. 8 f. 6 | 1871 *Metopa a.*, A. Boeck in: Forb. Selsk. Christian., 1870 p. 141 | 1892 *M. alderi*, G. O. Sars, Crust. Norway, *v.* 1 p. 250 t. 86 | 1893 *M. alderii*, A. Della Valle in: F. Fl. Neapel, *v.* 20 p. 638 t. 59 f. 52 | 1874 *Stenothoe alderi*, M'Intosh in: Ann. nat. Hist., ser. 4 *v.* 14 p. 265.

Body rather compressed, but back broadly rounded. Head, lateral corners broadly truncate. Side-plate 4 much larger than 2[d] and 3[d] combined. Pleon segment 3, postero-lateral corners quadrate. Eyes large, round oval, dark red. Antenna 1 much shorter than antenna 2, 1st joint as long as 2[d] and 3[d] combined, flagellum in ♀, but not in ♂, longer than peduncle. Antenna 2 strong, especially in ♂, ultimate joint of peduncle almost or quite as long as penultimate and longer than flagellum. Gnathopod 1 small, 4th joint not much produced, 5th as long and broad as 6th, which is imperfectly subchelate. Gnathopod 2 strong, especially in ♂, 4th joint (as in M. norvegica, M. spectabilis and M. robusta) produced on the inner side in a thin triangular lamella, so as to have a cup-shaped appearance like the 5th, 6th in ♀ oblong oval, palm oblique, armed with 8 teeth beginning at the hinge of the finger and followed by a deeply sinuous part which reaches the defining tooth; 6th in ♂ much larger, palm with 5 or 6 teeth followed by a deep excavation and a large defining tooth. Peraeopods 3—5, with 4th joint decurrent and rather wide, 2[d] joint well expanded in peraeopods 4 and 5, in the latter rather widened below. Uropod 3, peduncle with 5 minute spines, nearly as long as ramus, of which the 2 joints are equal. Telson large, oblong oval, broadly rounded at apex. Colour whitish, obliquely banded with reddish patches. L. 7 mm.

Arctic Ocean, North-Atlantic, North-Sea and Skagerrak (South- and West-Norway, depth 38—118 m; Spitzbergen; Murman coast; Iceland; British Isles); Kattegat.

8. **M. spectabilis** O. Sars 1876 *M. clypeata var.*, A. Boeck, Skand. Arkt. Amphip., v. 2 t. 18 f. 5 | 1876 *M. alderi* (err., non *Montagua alderii* Bate 1857!), G. O. Sars in: Arch. Naturv. Kristian., v. 2 p. 255 | 1879 *M. spectabilis,* G. O. Sars in: Arch. Naturv. Kristian., v. 4 p. 451 | 1885 *M. s.,* G. O. Sars in: Norske Nordhavs-Exp., v. 6 Crust. I p. 185 t. 15 f. 4 a—n | 1892 *M. s.,* G. O. Sars, Crust. Norway, v. 1 p. 251 t. 87 | 1893 *M. s.,* A. Della Valle in: F. Fl. Neapel, v. 20 p. 641 t. 59 f. 58.

Body rather slender. Head, lateral corners obtusely rounded. The much greater size is the chief distinction between this species and M. alderii (p. 175) but also, eyes comparatively much smaller; antenna 1 in both ♂ and ♀ much shorter than antenna 2, which have ultimate and penultimate joints of peduncle long and strong and flagellum very short; gnathopod 1, 6th joint rather shorter than 5th; gnathopod 2, 6th joint in ♀ with palm more oblique and its sinus deeper, in ♂ very large, the defining tooth much larger, the excavation deeper, the dentate portion of the palm having the hindmost tooth much larger than the others; uropod 3, peduncle as long as ramus, 2[d] joint of ramus shorter than 1st; telson rather narrower. L. ♀ 14 mm.

Arctic Ocean (Hammerfest [Finmark], depth 94—149 m); North-Atlantic.

9. **M. affinis** Boeck 1871 *M. a.,* A. Boeck in: Forh. Selsk. Christian., 1870 p. 142 | 1876 *M. a.,* A. Boeck, Skand. Arkt. Amphip., v. 2 p. 459 t. 19 f. 2 | 1892 *M. a.,* G. O. Sars, Crust. Norway, v. 1 p. 260 t. 91 f. 2 | 1893 *M. a.,* A. Della Valle in: F. Fl. Neapel, v. 20 p. 644.

Body in ♂ slender, but not much compressed. Head, lateral corners rounded. Side-plates smaller than usual. Pleon segment 3, postero-lateral corners nearly quadrate. Eyes small, round. Antenna 1 longer than antenna 2, 1st joint longer than 2[d], flagellum nearly twice as long as peduncle, not very slender. Antenna 2, ultimate and penultimate joints of peduncle subequal, together as long as flagellum. Gnathopod 1, 4th joint broadly produced, setose, 5th rather long, subfusiform, 6th a little shorter, much narrower, nearly linear, finger very short, stout, setose on inner margin. Gnathopod 2 powerful, 5th joint wide, cup-shaped, 6th very large and broad, subquadrate, though somewhat widening distally, palm nearly transverse, smooth, little curved, defined by a small tooth. Peraeopods 1—5 powerful, 6th joint somewhat curved, finger strong and curved. In peraeopod 5, 2[d] joint comparatively small, oval quadrangular, 4th strongly dilated, but only little decurrent.

Uropod 3, peduncle short and broad, little longer than one of the subequal joints of the ramus; its one spine is apical. Telson scarcely twice as long as broad, devoid of spines, apex obtuse. L. 3—4 mm.

Arctic Ocean, North-Sea and Christianiafjord (Norway).

10. **M. latimana** H. J. Hansen 1887 *M. l.*, H. J. Hansen in: Vid. Meddel., ser. 4 *v.* 9 p. 92 | 1893 *M. l.*, A. Della Valle in: F. Fl. Neapel, *v.* 20 p. 644 | 1892 *M. affinis?*, G. O. Sars, Crust. Norway, *v.* 1 p. 260.

Close to M. affinis. Head, lateral corners broadly rounded. Side-plates not very deep. Antenna 1 little shorter than the whole body, peduncle about as long as half flagellum, 1st joint a little longer than 2d. Antenna 2 about ²/₃ length of antenna 1, ultimate joint of peduncle rather longer than penultimate, flagellum 6-jointed. Gnathopod 1 a little shorter and thicker than in M. norvegica, especially as to 6th joint, in which oblique palm occupies whole breadth of hand. Gnathopod 2 rather short, powerful, 6th joint widening distally, rather longer than broad, palm long, smooth, nearly straight, defined by a small tooth. Peraeopods 1—5 long, the fingers very long; peraeopod 5, 2d joint pretty well expanded, only a little longer than broad. L. 3 mm.

Arctic Ocean and North-Atlantic (Greenland).

11. **M. norvegica** (Lilj) 1850 *Leucothoë n.*, W. Liljeborg in: Öfv. Ak. Förb., *v.* 7 p. 82 | 1851 *L. n.*, W. Liljeborg in: Vetensk. Ak. Handl., 1850 p. 335 t. 20 f. 4 | 1862 *Montagua n.*, Bate, Cat. Amphip. Brit. Mus., p. 370 | 1888 *Metopa n.*, T. Stebbing in: Rep. Voy. Challenger, *v.* 29 p. 236 | 1856 *Montagua pollexianus* (nom. nud.), Bate in: Rep. Brit. Ass., Meet. 25 p. 57 | 1857 *M. pollexiana*, Bate in: Ann. nat. Hist., ser. 2 *v.* 19 p. 137 | 1869 *Probolium pollexianum*, A. M. Norman in: Rep. Brit. Ass., Meet. 38 p. 274 | 1874

6th joint

Stenothoe pollexiana, M'Intosh in: Ann. nat. Hist., ser. 4 *v.* 14 p. 265 | 1875 *Metopa p.*, Ad. Metzger in: Jahresber. Comm. D. Meere, *v.* 2/3 p. 299 | 1887 *M. p.*, H. J. Hansen in: Vid. Meddel., ser. 4 *v.* 9 p. 92 t. 3 f. 5, 5a | 1892 *M. p.*, G. O. Sars, Crust. Norway, *v.* 1 p. 269 t. 95 | 1876 *M. clypeata* (part.), A. Boeck, Skand. Arkt. Amphip., *v.* 2 p. 454 | 1893 *M. c.*, *M. pollexiana*, A. Della Valle in: F. Fl. Neapel, *v.* 20 p. 643, 645, 569.

Fig. 45. Gnathopod 1.

Inner plates

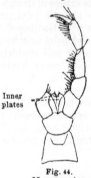

Fig. 44.
M. norvegica.
Maxillipeds.
[After G. O. Sars.]

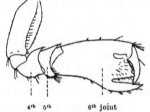

4th 5th   6th joint

Fig. 46. Gnathopod 2, ♂.
Fig. 45, 46. **M. norvegica.**
[After G. O. Sars.]

Body rather compressed, back rounded. Head, lateral corners broadly rounded. Side-plate 4 fully as large as 2d and 3d combined. Pleon segment 3, postero-lateral corners rather acute than quadrate. Eyes round, dark red. Antenna 1 longer than antenna 2, 1st joint longer than 2d and 3d combined, flagellum slender, twice as long as peduncle, 24-jointed. Antenna 2, ultimate and penultimate joints of peduncle subequal, flagellum scarce half length of peduncle. Maxillipeds (Fig. 44), inner plates cleft only at apex, nearly as long as the next joint, which displays no outer plate. Gnathopod 1 (Fig. 45), 4th joint broadly round at produced setose apex, 5th long, densely setose, nearly twice as long as the 6th, which is narrow but has a distinct nearly transverse palm.

Gnathopod 2 (see Lilljeborg's figure) very large, especially in ♂ (Fig. 46), 4th and
5th joints cup-shaped as in M. alderii (p. 175), 6th long, gradually widening distally,
palm angularly produced in the middle, minutely crenulate thence to the hinge,
this part separated by a deep narrow incision from a defining tooth which extends
beyond the rest of the joint and beyond the finger when closed.   Peraeopod 1
more slender than peraeopod 2, but with the 2^d joint rather broad and laminar.
Peraeopods 2—5, finger strong, minutely serrulate within, 4th joint produced,
especially in peraeopod 5, 2^d joint oval, not very long, in peraeopods 4 and 5.
Uropod 3, peduncle with 1 spine on nodiform projection of apex, 1st joint
of ramus longer than 2^d, nearly as long as peduncle.   Telson triangular
oval, tapering to a narrow apex, 2 pairs of lateral spinules.   Colour whitish-
yellowish, with narrow orange band to each segment; ova bright bluish. L. ♀
8 mm, ♂ rather larger.

Arctic Ocean, North-Atlantic and North-Sea (Norway, depth 55—94 m; Greenland;
Murman coast; British Isles).

12. **M. boeckii** O. Sars 1868 *Montagua norvegica* (err., non *Leucothoë n.*
Liljeborg 1850!), Bate & Westwood, Brit. sess. Crust., *v.* 2 p. 500 f. | 1871 *Metopa bruzelii*
(err., non *Montagua b.* Goës 1866!), A. Boeck in: Forh. Selsk. Christian., 1870 p. 142 |
1876 *M. b.*, A. Boeck. Skand. Arkt. Amphip., *v.* 2 p. 458 t. 18 f. 2 ♀ | 1882 *M. borealis*
(part.), G. O. Sars in: Forb. Selsk. Christian., nr. 18 p. 91 t. 4 f. 4a | 1892 *M. boeckii,*
G. O. Sars, Crust. Norway, *v.* 1 p. 252 t. 88.

Body slender and compressed. Head, lateral corners angular. Side-plate 4
larger than 2^d and 3^d combined. Pleon segment 3, postero-lateral corners rather
acute. Eyes round, dark red. Antenna 1 in ♀ rather longer than antenna 2,
scarcely so in ♂, 1st joint not so long as 2^d, flagellum slender, longer than peduncle. Antenna 2, ultimate and penultimate joints of peduncle subequal, stronger
in ♂ than in ♀, with flagellum shorter. Gnathopod 1 feeble, 4th joint not
broadly produced, 5th setose, little longer than the fusiform 6th. Gnathopod 2
rather powerful, 4th joint not cup-shaped, 6th in ♀ triangular oval, palm somewhat
oblique, near the hinge of finger scarcely dentate, in the other half minutely,
being then defined by a strong tooth and 2 spines; in the ♂ larger, having
the minute dentation replaced by a deep and rather wide excavation bounded
by a sharp tooth.   Peraeopods 1—5 rather slender, 4th joint not greatly
expanded or much decurrent even in peraeopods 4 and 5, which have the 2^d joint
oblong oval. Uropod 3 rather slender, peduncle with 4 spines, subequal to
ramus, in which 1st joint is longer than the 2^d. Telson oblong oval, distally
tapering to acute apex, 3 pairs of lateral spines. Colour whitish with scattered
red-brown patches; ova bluish. L. 6 mm.

Arctic Ocean, North-Atlantic and North-Sea (Norway).

13. **M. leptocarpa** O. Sars 1882 *M. l.*, G. O. Sars in: Forb. Selsk. Christian.,
nr. 18 p. 91 f. 3, 3a | 1892 *M. l.*, G. O. Sars, Crust. Norway, *v.* 1 p. 265 t. 93 f. 2 |
1893 *M. l.*, A. Della Valle in: F. Fl. Neapel, *v.* 20 p. 639 t. 59 f. 53.

Body very slender, not much compressed. Head, lateral corners angular.
Side-plate 4 scarcely larger than 2^d and 3^d combined. Pleon segment 3, postero-
lateral corners little produced, rather obtuse. Eyes round. Antennae 1 and 2
subequal, rather short. Antenna 1, 1st joint scarcely as long as 2^d and 3^d combined,
flagellum shorter than peduncle, 9-jointed. Antenna 2, ultimate joint of peduncle
shorter than penultimate, flagellum about half length of peduncle. Gnathopod 1
slender, longer than gnathopod 2, 5th joint extremely narrow and elongated,
6th much shorter, narrowly oblong but widening distally, the small palm nearly
transverse, defined by a small tooth. Gnathopod 2 powerful, 6th joint broadly

oval, palm smooth, convex, rather oblique, defined by a sharp tooth. Feraeopods 1 and 2 little unequal, peraeopods 3—5, 4th joint not greatly expanded, in peraeopod 5 produced below middle of 5th, 2d joint of peraeopod 5 oblong oval. Uropod 3, peduncle with 2 spinules, longer than 1st joint of ramus, 2d as long as 1st. Telson unarmed, rather broadly oval, apex obtuse. L. ♀ 4 mm.

North-Atlantic (Christiansund [West-Norway]). Depth 110—146 m.

14. M. longicornis Boeck 1871 M. l., A. Boeck in: Forh. Selsk. Christian., 1870 p. 148 | 1876 M. l., A. Boeck, Skand. Arkt. Amphip., v. 2 p. 460 t. 19 f. 1 | 1887 M. l., H. J. Hansen in: Vid. Meddel., ser. 4 v. 9 p. 97 t. 4 f. 1, 1 a | 1892 M. l., G. O. Sars, Crust. Norway, v. 1 p. 258 t. 90 f. 2 | 1893 M. l., A. Della Valle in: F. Fl. Neapel, v. 20 p. 644.

Body slender. Head, lateral corners bluntly produced. Side-plate 4 as deep as broad, rather larger than 2d and 3d combined. Pleon segment 3, postero-lateral corners produced but not very acutely. Eyes round, rather large. Antennae 1 and 2 subequal, rather long. Antenna 1, 1st and 2d joints long, sub-equal, flagellum scarcely longer than peduncle, with 10 joints (20, Sars in text). Antenna 2, ultimate joint of peduncle as long as flagellum, rather shorter than penultimate. Gnathopod 1, 6th joint as long as 5th, subfusiform, with a row of seven minute spinules on what may be considered the palmar part of the hind margin. Gnathopod 2 comparatively feeble, scarcely longer than gnathopod 1, 6th joint oblong triangular, palm oblique, smooth, defined by a small tooth. Peraeopods 1—5 rather slender, 4th joint little decurrent, sublinear; 2d rather narrowly oval in peraeopod 4, more broadly so in peraeopod 5, with hind margin rather abruptly curved in the middle. Uropod 3, peduncle with 3 spines, nearly as long as ramus, the joints of which are subequal. Telson with 3 pairs of lateral spines, narrowly oval, blunt at apex. L. ♀ 4 mm.

Arctic Ocean. North-Atlantic and Christianiafjord (Norway; Greenland, depth 56—112 m).

15. M. pusilla O. Sars 1892 M. p., G. O. Sars, Crust. Norway, v. 1 p 256 t. 90 f. 1.

Body short, compressed. Head, lateral corners sharply angular. Side-plates very large, 4th nearly as deep as broad, rather larger than 2d and 3d combined. Pleon small. Pleon segment 3, postero-lateral corners little produced but acute. Eyes minute, round, dark red. Antennae 1 and 2 rather slender and tolerably long, though shorter than in M. longicornis. Antenna 1, 2d joint scarcely shorter than 1st, flagellum longer than peduncle, 11-jointed. Antenna 2 scarcely shorter than antenna 1, ultimate and penultimate joints of peduncle subequal, slender, flagellum shorter than the 2 together, 9-jointed. Gnathopod 1 rather feeble, 6th joint decidedly shorter than 5th, sublinear. Gnathopod 2 moderately strong, 6th joint oblong triangular, widening distally, palm somewhat ˙oblique, nearly straight, slightly serrated, defined by an acute tooth; in ♂ rather longer than in ♀. Peraeopod 1 rather slender and elongated, the others stouter, 3—5 successively shorter, the finger strong and curved. Peraeopod 5, 2d joint rather a short broad oval, 4th produced beyond middle of 5th. Uropod 3, peduncle with 2 spines, scarcely longer than 1st joint of ramus, and subequal to the 2d. Telson with 2 pairs of lateral spines, oblong oval. Colour whitish with stellate brownish grey patches, side-plates also tending to greenish. L. ♀ 3 mm.

North-Atlantic, North-Sea and Skagerrak (West- and South-Norway). In shallow water among Algae and Hydroids.

16. **M. aequicornis** O. Sars 1879 *M. aeqvicornis*, G. O. Sars in: Arch. Naturv. Kristian., *v.* 4 p. 453 | 1885 *M. a.*, G. O. Sars in: Norske Nordhavs-Exp., *v.* 6 Crust. I p. 188 t. 15 f. 5 | 1893 *M. aequicornis*, A. Della Valle in: F. Fl. Neapel, *v.* 20 p. 644.

Body rather stout, back round. Head, lateral corners little projecting. Side-plate 4 elliptic, broader than deep, larger than 2^d and 3^d combined, concealing 5th and 6th. Pleon segment 3, postero-lateral corners slightly produced but obtuse. Eyes small, round, red. Antennae 1 and 2 nearly as long as the body, subequal. Antenna 1, 1st joint a little shorter than 2^d, flagellum slender, a little longer than peduncle, with many joints. Antenna 2, ultimate joint of peduncle longer than penultimate and as long as flagellum. Gnathopod 1 as in M. spectabilis (p. 176). Gnathopod 2, 6th joint very large, distally dilated, palm somewhat oblique, almost straight, coarsely dentate throughout, and defined by a short triangular projection. Peraeopods 1—5, uropods 1—3 and telson resembling those of M. spectabilis. Colour whitish, pellucid, with some very faint pigmentation. L. 7·5 mm.

Arctic Ocean (South-West of Spitzbergen). Depth 1400 m.

17. **M. borealis** O. Sars 1882 *M. b.* (part.), G. O. Sars in: Forh. Selsk. Christian., nr. 18 p. 91 t. 4 f. 4 | 1887 *M. b.?*, H. J. Hansen in: Vid. Meddel., ser. 4 *v.* 9 p. 91 t. 3 f. 4, 4a | 1892 *M. b.*, G. O. Sars, Crust. Norway, *v.* 1 p. 254 t. 89 f. 1 | 1893 *M. b.*, A. Della Valle in: F. Fl. Neapel, *v.* 20 p. 644.

Body rather stout. Head, lateral corners rounded. Side-plates large, 4th much larger than 2^d and 3^d combined. Pleon segment 3, postero-lateral corners acutely produced. Eyes round, dark red. Antennae 1 and 2 short, antenna 1 slightly shorter than antenna 2, 1st joint as long as 2^d and 3^d combined, flagellum longer than peduncle, 12-jointed. Antenna 2, ultimate joint of peduncle rather longer than penultimate, and equal to flagellum. Gnathopod 1 small, 6th joint about as long as 5th, narrowly oval, with no distinct palm. Gnathopod 2, 6th joint oblong, slightly expanded distally, palm slightly oblique, little curved, coarsely and irregularly denticulate throughout, defined by a tooth. Peraeo-pods 2—5 stout; in peraeopod 4, 2^d joint broadly oval, 4th broad and decurrent almost to end of 5th. Uropod 3, peduncle with 2 spinules, much shorter than ramus, of which 2^d joint is shorter than 1st. Telson without spines, oblong oval, apex obtuse. L. ♀ 5—7 mm.

Arctic Ocean and North-Atlantic (Norway, Greenland).

18. **M. rubrovittata** O. Sars ?1869 *Probolium alderi* (err., non *Montagua alderii* Bate 1857!), A. M. Norman in: Rep. Brit. Ass., Meet. 38 p. 273 | 1882 *Metopa rubrovittata*, G. O. Sars in: Forh. Selsk. Christian., nr. 18 p. 90 t. 4 f. 2, 2a | 1889 *M. r.*, Hoek in: Tijdschr. Nederl. dierk. Ver., ser. 2 *v.* 2 p. 188 t. 7 f. 4, 4' | 1892 *M. r.*, G. O. Sars, Crust. Norway, *v.* 1 p. 255 t. 89 f. 2 | 1893 *M. r.*, A. Della Valle in: F. Fl. Neapel, *v.* 20 p. 645.

Body stout and compact. Head, lateral corners angular. Side-plate 4 much broader than deep, nearly twice as large as 2^d and 3^d combined. Pleon segment 3, postero-lateral corners produced but obtuse at apex. Eyes rather small, round, dark red. Antenna 1 short, a little longer than antenna 2, 1st joint rather longer than 2^d and 3^d combined, flagellum longer than peduncle, 10-jointed. Antenna 2, ultimate and penultimate joints of peduncle subequal, flagellum longer than both combined. Gnathopod 1, 6th joint narrower than 5th, subequal to it in length, palm imperfectly defined by a spine about in the middle of the hind margin. Gnathopod 2 rather powerful, 6th joint oblong quadrangular, but widening a little distally, palm nearly transverse, convex, regularly serrate, defined by a strong acute tooth. Peraeopods 2—5 moderately stout, expansion of 2^d joint in peraeopod 5 oblong oval, and the 4th produced nearly to end of 5th. Uropod 3, peduncle with one spine, not longer than 1st joint of ramus, which is equal to 2^d. Telson without spines, more than twice as long as broad, oval, apex obtuse. Colour whitish with bands of crimson, which on the side-plates are oblique and undulating; ova bluish. L. 2·5—4 mm.

Arctic Ocean, North-Atlantic and North-Sea (Christiansund, depth 56 m, and Vadsö [Norway]; Holland; France; Firth of Clyde; Liverpool Bay); Kattegat.

19. **M. sölsbergi** J. S. Schn. 1884 *M. s.*, J. S. Schneider in: Tromsø Mus. Aarsb., *v.* 7 p. 71 t. 3, 4 | 1892 *M. s.*, G. O. Sars, Crust. Norway, *v.* 1 p. 266 t. 94 f. 1 | 1893 *M. s.*, A. Della Valle in: F. Fl. Neapel, *v.* 20 p. 645.

Body robust, little compressed, integument unusually soft and thin. Head, lateral corners rounded. Side-plate 4 as deep as broad, rather larger than 2^d and 3^d combined. Pleon segment 3, postero-lateral corners subquadrate. Eyes round, yellowish red. Antenna 1 rather shorter than antenna 2, 1st joint as long as 2^d and 3^d combined, flagellum much longer than peduncle, 14-jointed. Antenna 2, ultimate joint of peduncle equal to penultimate (Sars) or longer than it (Schneider), flagellum half length of peduncle, 10-jointed. Maxilla 1, palp 1-jointed. Maxillipeds with inner plates apparently separate (see Schneider t. 3 f. 6). Gnathopod 1 rather strong, 4th joint rectangular, little produced, 5th and 6th subequal, 6th oblong oval, with 2 spinules on the hind margin, but no definite palm. Gnathopod 2 strong, 6th joint oblong quadrangular, but slightly expanded distally, palm smooth, slightly oblique, forming an obtuse angle with the hind margin, defined by 2 spinules, and (according to Schneider, but not Sars) by a small process. Peraeopod 1 not quite so stout as peraeopods 2 and 3. Peraeopods 4 and 5, 2^d joint oblong oval, unusually narrow, 4th moderately expanded, and decurrent to middle of 5th. Uropod 3, peduncle with 4 spinules, nearly as long as ramus, of which the 2 joints are equal. Telson unarmed, oval, obtuse at apex. Colour pale carneous or yellowish. L. ♀ nearly 6 mm.

Hardangerfjord and Malangenfjord [Norway]. Depth 18 m.

20. **M. tenuimana** O. Sars 1892 *M. t.*, G. O. Sars, Crust. Norway. *v.* 1 p. 259 t. 91 f. 1.

Body slender. Head, lateral corners angular. Side-plate 4 broader than deep, much larger than 2^d and 3^d combined. Pleon segment 3, postero-lateral corners much produced. Eyes round. Antennae 1 and 2 rather short. Antenna 1 rather the longer, 1st joint much longer than 2^d, flagellum longer than peduncle, 9-jointed. Antenna 2, ultimate and penultimate joints of peduncle subequal, together as long as flagellum. Gnathopod 1, 6th joint rather shorter than 5th, tapering, without any armature to indicate a palm. Gnathopod 2 rather feeble, 6th joint slenderly oblong triangular, widening distally, palm quite smooth, slightly oblique, defined by a tooth and spines. Peraeopods 1—5 nearly as in M. pusilla (p. 179), but rather less stout, though with 2^d and 4th joints of peraeopod 5 more strongly developed. Uropod 3, peduncle with only an apical spinule, not longer than 1st joint of ramus, which has the 2^d joint shorter than 1st. Telson with 1 or 2 pairs of lateral spines, oval, obtuse at apex. L. ♀ 3 mm.

North-Atlantic (West-Norway).

21. **M. invalida** O. Sars 1876 *M. alderii* (part.), A. Boeck, Skand. Arkt. Amphip.,
v. 2 p. 456 t. 17 f. 4, 4 k | 1892 *M. invalida*, G. O. Sars, Crust. Norway, *v.* 1 p. 267 t. 94 f. 2.

Body stout and compact. Head, lateral corners obtusely projecting. Side-plate 4 broader than deep, much larger than 2^d and 3^d combined. Pleon segment 3, lateral corners rather produced. Eyes round, bright red. Antennae 1 and 2 subequal, short. Antenna 1, 1st joint shorter than 2^d and 3^d combined, flagellum scarcely as long as peduncle, 9-jointed. Antenna 2, ultimate and penultimate joints of peduncle subequal, flagellum scarcely longer than ultimate. Gnathopod 1, 6th joint as long as 5th, narrow, tapering, with a single spine about in middle of hind margin. Gnathopod 2 rather feeble, 5th joint widely cup-like, setose at rather produced apex, 6th oblong quadrangular, slightly dilated distally, palm nearly straight, quite smooth, slightly oblique, forming obtuse angle with hind margin. Peraeopods 1—5 rather slender, but 2^d joint of peraeopod 5 rather broad, oval quadrangular, and 4th joint produced beyond middle of 5th. Uropod 3, peduncle with 2 spinules, shorter than ramus, of which the joints are subequal. Telson unarmed, 3 times as long as broad, a little constricted above, apex obtusely pointed. Colour pale greyish white, with transverse yellowish bands; ova bluish. L. ♀ 4 mm.

Arctic Ocean (Hammerfest and Selsövig [North-Norway]). Depth 74—92 m.

2. Gen. **Metopella** O. Sars

1871 *Metopa* (part.), A. Boeck in: Forh. Selsk. Christian., 1870 p. 140 | 1892 *Metopella*, G. O. Sars, Crust. Norway. *v.* 1 p 274 | 1893 *Proboloides* (part.) + *Metopoides* (part.), A. Della Valle in: F. Fl. Neapel, *v.* 20 p. 907.

Antennae 1 and 2 not very long nor very unequal, with or without accessory flagellum to antenna 1. Mandibular palp, 3^d joint small. Maxilla 1, palp (at least sometimes) 2-jointed. Maxillipeds, inner plates not very small, separate (at least sometimes), outer plates very small (mouth-organs ascertained for M. nasuta and M. nasutigenes). Gnathopod 1 simple or subchelate. Gnathopod 2 subchelate. Peraeopods 3—5, 2^d joint little or not at all expanded. Uropod 3, peduncle shorter than ramus.

6 species.

Synopsis of species:

1. **M. nasuta** (Boeck) ˙1871 *Metopa n.*, A. Boeck in: Forh. Selsk. Christian., 1870 p. 144 | 1876 *M. n.*, A. Boeck, Skand. Arkt. Amphip.. *v.* 2 p. 465 t. 18 f. 6 | 1887 *M. n.?*, H. J. Hansen in: Vid. Meddel., ser. 4 *v.* 9 p. 102 | 1893 *M. n.* (part.), A. Della Valle in: F. Fl. Neapel, *v.* 20 p. 637 | 1892 *M. n.*, *Metopella* (part.). G. O. Sars, Crust. Norway, *v.* 1 p. 276 t. 98 f. 1.

Body short, compact and firm. Peraeon segment 4 much the largest, with rounded dorsal carina. Head, lateral corners truncate.. Side-plate 2 with denticle at lower hind corner, 4th much larger than 2^d and 3^d combined. Pleon segment 3, postero-lateral angles subacute, segment 4 carinate, overarching segment 5. Eyes small, round. Antenna 1, 1st joint completely overlapping 2^d, flagellum 8-jointed. Antenna 2 rather shorter than antenna 1, ultimate and penultimate joints of peduncle subequal, flagellum 7-jointed. Maxilla 1, palp 2-jointed. Maxillipeds, inner plates separate. Gnathopod 1 feeble, 4th joint shorter than 5th, 6th joint as long as 5th, almost linear. Gnathopod 2 of moderate size, 5th joint cup-like, 6th oblong triangular. widening distally, palm a little oblique, coarsely serrate, defined by a tooth and 2 spines. Peraeopods 1—5 very slender, none of the joints expanded. Uropod 3, peduncle armed with 1 spine, 2 joints of ramus subequal. Telson narrow, linguiform, obtuse at apex, quite unarmed (Sars) or with 2 lateral spines. L. ♀ 3—4 mm.

Arctic Ocean, North-Atlantic and North-Sea (Norway, depth 94—188 m; ? Greenland, depth 72 m; Firth of Clyde; Firth of Forth; Moray Firth).

2. **M. nasutigenes** (Stebb.) 1888 Metopa n., T. Stebbing in: Rep. Voy. Challenger v. 29 p. 753 t. 40 | 1893 M. nasuta (part.), Proboloides nasutigenes, A. Della Valle in: F. Fl. Neapel. v. 20 p. 637, 907, 945.

Body compact and firm. Peraeon segment 4 much the largest, slightly angled dorsally, not carinate. Head, lateral corners little produced. Side-plate 2 with serration of 4 points at bottom of hind margin, 4th much larger than 2^d and 3^d combined. Pleon segment 3 almost quadrate, segment 4 with slight dorsal depression. Eyes round, not large. Antenna 1, 1st joint acutely produced beyond 2^d, 2^d longer than 3^d, flagellum 10-jointed. Antenna 2 rather shorter than antenna 1, ultimate joint of peduncle rather longer than penultimate, flagellum 8-jointed. Maxilla 1, palp 2-jointed. Maxillipeds, inner plates separate. Gnathopod 1, 4th joint as long as 5th, 5th much shorter than 6th, almost cup-like, 6th almost oblong, subchelate, with oblique, finely pectinate palm, defined by an obtuse angle and spines. Gnathopod 2 not very unlike gnathopod 1, but larger, 4th joint shorter than the cup-like 5th, 6th 3 times as long as broad, almost parallel-sided, palm almost smooth, defined as in gnathopod 1. Peraeopods 1—5 very slender, none of the joints expanded. Uropod 3, peduncle armed with 1 spine, 1st joint of ramus armed with 3 spines, rather longer than 2^d joint. Telson long, with upturned sides, apex acute. L. ♀ 4·5 mm.

Cumberland Bay [Kerguelen Island]. Depth 222 m.

3. **M. ovata** (Stebb.) 1888 Metopa o., T. Stebbing in: Rep. Voy. Challenger, v. 29 p. 764 t. 44 | 1893 M. o., Metopoides ovatus, A. Della Valle in: F. Fl. Neapel, v. 20 p. 645, 907, 938.

Fig. 47. **M. ovata**.
Lateral View, ♀.

Fig 48. **M. ovata**
Peraeopod 2.

Head. lateral corners little produced. Pleon segment 3, postero-lateral corners sub-quadrate. Side-plate 2 completely covering the 1st, 4th much larger than 2^d and 3^d combined, completely covering 5th. 6th and 7th (Fig. 47). Eyes round. Antenna 1, flagellum longer than peduncle, 10-jointed; a rudimentary 2-jointed accessory flagellum, as in Metopoides (p. 185). Antenna 2 nearly as long as antenna 1, ultimate and penultimate joints of peduncle subequal, flagellum longer than peduncle, 10- or 11-jointed. Maxilla 1, palp 2-jointed. Maxillipeds, inner plates separate. Gnathopod 1 subchelate, 4th joint produced along the short cup-like 5th, 6th much longer than 4th or 5th, almost oblong, palm convex, not very oblique, very minutely pectinate, defined by a denticle and 2 spines. Gnathopod 2 very similar to gnathopod 1, but larger, 6th joint longer, a little widened distally, palm smooth, convex, scarcely oblique. Peraeopods 1—5 (Fig. 48) slender, all joints linear, but 2^{d} joint in peraeopods 3 and 4 (not in 5) a little wider above than below. Uropod 3, peduncle scarcely longer than 1st joint of ramus, 2^{d} of ramus rather shorter than 1st. Telson not twice as long as broad, apex narrow. L. about 3 mm.

Strait of Magellan (Cape Virgins). Depth 100 m.

4. **M. neglecta** (H. J. Hansen) 1876 *Metopa longimana* (part.), A. Boeck, Skand. Arkt. Amphip., v.2 t.17 f.5—5n | 1887 *M. neglecta*, H. J. Hansen in: Vid. Meddel., ser.4 v.9 p.96 t.3 f.9—9b | 1893 *M. n.*, A. Della Valle in: F. Fl. Neapel, v.20 p.640 t.59 f.56 | 1892 *M. n.*, *Metopella* (part.), G. O. Sars. Crust. Norway, v.1 p.274 t.97 f.2.

Body rather compressed, integument thin. Head, lateral corners sub-quadrate. Side-plate 4 larger than 2^{d} and 3^{d} combined. Pleon segment 3, postero-lateral angles somewhat produced. Eyes small, round. Antenna 1, flagellum scarcely as long as peduncle, about 8-jointed. Antenna 2 in ♀ subequal to antenna 1, in ♂ rather longer, ultimate joint of peduncle shorter than penultimate, flagellum subequal to the latter. Maxilla 1, palp very long, 1-jointed (Boeck perhaps overlooked the division into 2 joints). Maxillipeds, inner plates separate (Boeck's figure). Gnathopod 1, 4th joint not much produced, 6th almost linear, not much shorter than 5th. Gnathopod 2, 5th joint narrowly cup-like, 6th long and narrow, widening distally, palm rather oblique, curved, smooth, defined by a denticle and 2 spines. Peraeopods 1—5 extremely slender, only 2^{d} joint in last pair at all expanded, and that only in the upper part. Uropod 3, peduncle armed with 2 spinules, joints of ramus subequal. Telson long and narrow, apex obtuse, 2 pairs of lateral spines. L. ♀ 3 mm.

Arctic Ocean, North-Atlantic and North-Sea (West-Norway; Greenland. depth 19—113 m).

5. **M. carinata** (H. J. Hansen) 1887 *Metopa c.*, H. J. Hansen in: Vid. Meddel., ser.4 v.9 p.99 t.4 f.3—3e | 1893 *M. c.*, A. Della Valle in: F. Fl. Neapel, v.20 p.637 t.59 f.49.

Body deep, little compressed in ♀, more so in ♂. Peraeon segment 4 much the largest. with rounded dorsal carina, more elevated in ♀ than in ♂. Side-plate 2 with subacute apex, 4th much larger than 2^{d} and 3^{d} combined. Eyes small, round. Antennae 1 and 2 subequal, very short and slender in ♀, thicker and longer in ♂. Antenna 2, ultimate and penultimate joints of peduncle subequal. Gnathopod 1 feeble, 6th joint twice as long as 5th, more slender, tapering. Gnathopod 2, 6th joint in ♀ large, widening distally, palm curved, a little oblique, defined by a long narrow projecting process; 6th joint in ♂ longer and narrower, not dilated, palm very oblique, defined by a minute tubercle. Peraeopods 1—5, all joints slender, 2^{d} in peraeopod 5 with slight linear expansion. L. ♀ 3, ♂ 2·1 mm.

Arctic Ocean and North-Atlantic (Greenland). Depth 4—94 m.

6. **M. longimana** (Boeck) 1871 *Metopa l.*, A. Boeck in: Forh. Selsk. Christian.,. 1870 p. 144 | 1876 *M. l.* (part.), A. Boeck, Skand. Arkt. Amphip., *v.* 2 p. 464 t. 17 f. 6 | 1887 *M. l.*, H. J. Hansen in: Vid. Meddel., ser. 4 *v.* 9 p. 95 t. 3 f: 8—8b | 1893 *M. l.*, A. Della Valle in: F. Fl. Neapel, *v.* 20 p. 643 t. 59 f. 61 | 1892 *M. l.*, *Metopella* (part.), G. O. Sars, Crust. Norway, *v.* 1 p. 273 t. 97 f. 1.

Body strongly built, integument firm. Head, lateral angles acute. Side-plate 4 unusually large, completely covering 5th, 6th and 7th. Pleon segment 3 slightly produced, subquadrate. Eyes round. Antennae 1 and 2 subequal in ♀, antenna 1 the shorter in ♂. Antenna 1, flagellum 8-jointed. Antenna 2,. ultimate joint of peduncle much shorter than penultimate. Gnathopod 1, 4th joint moderately produced, 6th decidedly shorter than 5th, almost linear. Gnathopod 2, 6th joint in ♀ nearly as in M. neglecta, but palm almost straight, slightly serrate,. defined by a decided tooth; in ♂ larger, with 2 excavations in the palm near the tooth. Peraeopods 1—5 slender, throughout, only 2^{d} joint of 5th with slight linear widening. Uropod 3, peduncle with 2 spines, joints of ramus. subequal. Telson oblong oval, obtuse at apex, with 2 pairs of lateral spines. L. about 3 mm.

Arctic Ocean and North-Atlantic (Greenland); North-Sea (Norway) and Christiania-fjord. Depth 72—110 m.

3. Gen. **Metopoides** Della Valle

1893 *Metopoides* (part.), A. Della Valle in: F. Fl. Neapel, *v.* 20 p. 907.

Antenna 1 shorter than antenna 2, with very small 2-jointed accessory flagellum. Mandibular palp, 3^{d} joint small, with 2 apical setae. Maxilla 1, palp large, 2-jointed. Maxillipeds (Fig. 49), inner plates not very small, doubtfully separate, outer plates very small but apparent. Gnathopods 1 and 2 (Fig. 50) subchelate. Peraeopods 4 and 5, 2^{d} joint expanded. Uropod 3, peduncle shorter than ramus.

3 species.

Synopsis of species:

1 { Peraeopods 3—5, 4th joint narrow . . . . 1. **M. magellanicus** . . p. 185
  { Peraeopods 3—5, 4th joint not narrow — **2.**

2 { Gnathopod 2, 6th joint elongate . . . . 2. **M. parallelocheir** . p. 186
  { Gnathopod· 2, 6th joint short . . . . 3. **M. compactus** . . . p. 186

1. **M. magellanicus** (Stebb.) 1888 *Metopa magellanica*, T. Stebbing in: Rep. Voy. Challenger, *v.* 29 p. 756 t. 41 | 1893 *M. m.*, *Metopoides magellanicus*, A. Della Valle in: F. Fl. Neapel, *v.* 20 p. 644, 907, 938.

Head, lateral angles little produced, obtuse. Side-plate 3 a little narrowed downward, 4th of equal breadth and depth, larger than 2^{d} and 3^{d} combined. Pleon segment 3, postero-lateral corners subquadrate. Eyes round. Antenna 1,. 3^{d} joint longer than half 2^{d}, flagellum longer than peduncle, 13-jointed. Antenna 2, penultimate joint of peduncle long and slender, rather shorter than ultimate, flagellum 8- or 9-jointed. Maxillipeds, inner plates probably not separate except at apex, outer plates consisting of small apical processes on joints of considerable length and width. Gnathopod 1, 5th joint rather· shorter than the oval 6th, palm very oblique, finely pectinate, defined by 2 spines, to which the long finger reaches. Gnathopod 2, 5th joint much shorter than 6th, cup-like, 6th widest where the long oblique palm with close row of spines meets the much shorter hind margin; finger long, curved.. Marsupial plates almost circular. Peraeopods 1—5, 4th joint long, scarcely decurrent, 2^d joint more expanded in peraeopod 5 than in **4**. Uropod **3**, 2^d joint of ramus shorter than 1st. Telson twice as long as broad, distally tapering to acute apex, 3 spines on each margin. L. ♀ 5 mm.

Strait of Magellan (Cape Virgins). Depth 100 m.

2. M. parallelocheir (Stebb.) 1888 *Metopa p.*, T. Stebbing in: Rep. Voy. Challenger, *v.*29 p.762 t.43 | 1893 *M. p.*, *Metopoides p.*, A. Della Valle in: F. Fl. Neapel, *v.*20 p.642, 907, 938; t. 59 f. 59.

Head, lateral corners little produced. Side-plate 3 of uniform breadth. Pleon segment 3, postero-lateral corners quadrate. Eyes round. Antenna 1, 3^d joint half length of 2^d, flagellum shorter than peduncle, about 8-jointed. Antenna 2 as in preceding species. Maxillipeds, inner plates not very large, probably not separate except at apex, outer plates triangular, on joints which are broad but not elongate. Gnathopod 1, 6th joint rather long and narrow, palm almost smooth, extremely oblique and ill-defined except by ending of spine-row, finger finely pectinate, a denticle at base of the nail. Gnathopod 2, 5th joint cup-like, much shorter than the long hand, which has front and hind margins almost parallel, palm short but oblique, defined by a tooth and group of spines, finger short, curved. Peraeopods 1—5, 4th joint rather long, wider and more decurrent in peraeopods 3 and 4 than in 1 and 2. Uropod 3, 2^d joint of ramus a little shorter than 1st. Telson long oval, apex not very acute, 2 spines on each margin. L. 3·5 mm.

Strait of Magellan (Cape Virgins). Depth 100 m.

3. M. compactus (Stebb.) 1888 *Metopa compacta*, T. Stebbing in: Rep. Voy. Challenger, *v.*29 p.767 t.45 | 1893 *M. c.*, *Metopoides compactus*, A. Della Valle in: F. Fl. Neapel, *v.*20 p.644, 907, 938.

Head, lateral corners a little prominent. Side-plate 2 broad, semi-circular, 3^d oblong, 4th broader than deep. Pleon segment 3, postero-lateral corners obtusely quadrate. Eyes round. Antenna 1, 3^d joint longer than half 2^d, flagellum shorter than peduncle, 10-jointed. Antenna 2 little longer than antenna 1, penultimate joint of peduncle rather longer than ultimate, flagellum 8-jointed. Mandibular palp with 2^d joint broader and 3^d a little more developed than usual in this family. Maxillipeds (Fig. 49), inner plates rather large, separation doubtful, outer plates apically rounded, the supporting joint almost as broad as long. Gnathopod 1, 4th and 5th joints strongly armed with pectinate spines, 5th as long as 6th, which widens to the convex, rather oblique, finely pectinate and denticulate palm, forming obtuse angle with hind margin and defined by a row of spines; finger pectinate. Gnathopod 2 (Fig. 50), 5th joint cup-like, rather shorter than the stout oblong 6th, palm very little oblique, sinuous, denticulate, defined

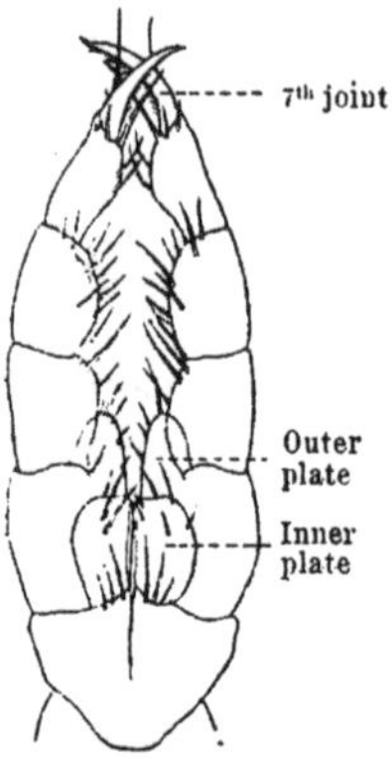

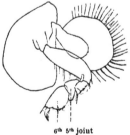

---- 7th joint

Outer plate

Inner plate

Fig. 49.
M. compactus.
Maxillipeds.

6th 5th joint
Fig. 50. M. compactus.
Gnathopod 2.

by a small tooth. Marsupial plates almost circular. Peraeopods 1—5 nearly as in M. parallelocheir. Uropod 2, rami less unequal than in 2 preceding species. Uropod 3, 2^d joint of ramus much shorter than 1st. Telson very broad, though longer than broad, distally converging to an obtuse apex; 3 small spines on each margin. L. ♀ 3·5 mm.

Strait of Magellan (Cape Virgins). Depth 100 m.

4. Gen. **Proboloides** Della Valle

1871 *Metopa* (part.), A. Boeck in: Forh. Selsk. Christian., 1870 p. 140 | 1893 *Proboloides* (part.), A. Della Valle in: F. Fl. Neapel, *v.* 20 p. 907.

Antennae 1 and 2 subequal, not large; antenna 1 without accessory flagellum or by exception perhaps with rudiment. Mandible, molar wanting, palp slender, 3-jointed, apical joint small. Maxilla 1, palp large, 2-jointed. Maxillipeds, inner plates separate, usually not very small, outer plates very small or obsolete. Gnathopod 1 more or less subchelate. Gnathopod 2 distinctly subchelate in ♀, in ♂ powerful, with palm defined or undefined. Peraeopods 3—5, 4th joint usually expanded and produced, 2^d joint expanded in peraeopods 4 and 5, not or scarcely in peraeopod 3.

7 species accepted, 1 doubtful.

Synopsis of accepted species:

1 { Peraeopod 3, 4th joint rather narrow . . . . 1. **P. grandimanus** . . . p. 187
    { Peraeopod 3, 4th joint rather broad — **2.**

2 { Gnathopod 2 in ♂, palm undefined or imperfectly
        defined — **3.**
    { Gnathopod 2 in ♂, palm well defined — **6.**

3 { Peraeopod 3, 2^d joint distally widened . . . 2. **P. crenatipalmatus** . p. 188
    { Peraeopod 3, 2^d joint not widened — **4.**

4 { Gnathopod 2, palm quite smooth . . . . . . 3. **P. bruzelii** . . . . . p. 188
    { Gnathopod 2, palm not quite smooth — **5.**

5 { Gnathopod 2, palm conspicuously dentate . . 4. **P. gregarius** . . . . p. 189
    { Gnathopod 2, palm inconspicuously dentate . 5. **P. calcaratus** . . . . p. 189

6 { Gnathopod 2, 2^d joint broad . . . . 6. **P. glacialis** . . . . . p. 189
    { Gnathopod 2, 2^d joint narrow . . . . 7. **P. groenlandicus** . . p. 190

1. P. grandimanus (Bonnier) 1896 *Probolium grandimanum*, J. Bonnier in: Ann. Univ. Lyon, *v.* 26 p. 638 t. 37 f. 4.

♀ unknown. — ♂. Head without rostrum. Side-plate 1 much more developed than usually in this family, subtriangular, 2^d considerably smaller, with acute apex directed forward. 3^d and 4th large. Pleon segment 3, postero-lateral corners rounded. Eyes small, round. Antenna 1, 1st and 2^d joints long, equal, 3^d very short, flagellum with more than 12 joints. Antenna 2, ultimate joint of peduncle rather shorter than the long penultimate, flagellum with more than 10 joints. Epistome very acute. Upper lip bilobed. Mandible small, but these and other mouth-parts in accord with generic character. Maxillipeds, outer plates quite wanting. Gnathopod 1, 4th joint well produced, obtuse at apex, 5th longer than 6th, 6th distally widened, subchelate. Gnathopod 2, 2^d joint channelled in front, 4th joint apically acute, 5th short, cup-shaped, 6th large, elongate, having near the hinge of finger 4 small teeth, only that most remote from hinge acute, the rest of the margin nearly straight, fringed with setules, finger as long as 6th joint, with setules on inner margin. Peraeopod 3, 2^d joint narrow, 4th scarcely widened distally. Peraeopods 4 and 5, 2^d joint widely expanded, 4th moderately produced. Uropod 1, rami equal, one with 2, the other with 4 spines. Uropod 2, outer ramus shorter than inner, with 1 spine. Uropod 3, ramus with 2 lateral spines, 2^d joint shorter than 1st. Telson with 1 pair of minute and 2 pairs of stout lateral spines. L. less than 5 mm.

Bay of Biscay. Depth 950 m.

2. **P. crenatipalmatus** (Stebb.) 1888 *Metopa crenatipalmata*, T. Stebbing in: Rep. Voy. Challenger, *v.* 29 p. 759 t. 42 | 1893 *M. c., Proboloides crenatipalmatus*, A. Della Valle in: F. Fl. Neapel, *v.* 20 p. 907, 945.

Head, lateral corners very obtuse. Side-plate 4 scarcely equal to 2^d and 3^d combined. Pleon segment 3, postero-lateral corners subquadrate. Eyes round. Antenna 1, 1st joint not longer than 2^d, flagellum longer than peduncle, about 12-jointed; perhaps a rudimentary accessory flagellum present. Antenna 2, ultimate joint of peduncle slightly longer than penultimate, flagellum 8-jointed. Mandibular palp with 2^d joint long, 3^d scarcely longer than 1st, carrying 1 or 2 apical setae. Gnathopod 1, 4th joint not much produced, apically spinose, 5th nearly as long as 6th and rather broader, palm of 6th oblique, finely crenulate, finger finely pectinate on inner margin. Gnathopod 2, 5th joint cup-like, 6th oblong oval, palm crenulate in 2 divisions, rather oblique, defined by a sharp tooth, finger broad, smooth on inner margin. Peraeopod 3, 2^d joint distally lobed behind, 4th broad, a little decurrent. Peraeopods 4 and 5, 2^d and 4th joints expanded, 4th moderately decurrent. Uropod 3, peduncle a little shorter than ramus, with 4 spines; 2^d joint of ramus rather longer than 1st. Telson long oval, obtusely pointed, 3 spines on each margin. L. ♀ 6 mm.

Strait of Magellan (Cape Virgins). Depth 100—270 m.

3. **P. bruzelii** (Goës) 1866 *Montagua b.*, Goës in: Öfv. Ak. Förh., *v.* 22 p. 522 t. 38 f. 10 | 1886 *Metopa b.*, G. O. Sars in: Norske Nordhavs-Exp., *v.* 6 Crust. II p. 48 | 1887 *M. b.* (part.), H. J. Hansen in: Vid. Meddel., ser. 4 *v.* 9 p. 97 t. 4 f. 2c, d | 1892 *M. b.*, G. O. Sars, Crust. Norway, *v.* 1 p. 261 t. 92 f. 1.

Body slender and rather compressed. Head, lateral corners angular. Side-plate 4 much larger than 2^d and 3^d combined. Pleon segment 3, postero-lateral angles rather acute than quadrate. Eyes round, dark red. Antennae 1 and 2 subequal. Antenna 1, 1st joint longer than 2^d and 3^d combined, flagellum longer than peduncle, about 12-jointed. Antenna 2, ultimate and penultimate joints of peduncle subequal, and together subequal to flagellum. Maxilla 1 (Goës) with 2-jointed palp, 1st joint large. Maxillipeds (Goës), inner plates rather small, separate, outer quite obsolete. Gnathopod 1, 4th joint rather broadly produced, 6th narrower than 5th, equal to it in length, tapering distally, smooth. Gnathopod 2, 6th joint oval, palm quite smooth, rather oblique, defined by a slight angular projection and 2 spines; alike in ♂ and ♀, but larger in ♂. Peraeopods 2—5 rather stout, 2^d joint broadly expanded in peraeopod 5, and 4th produced below the middle of the 5th. Uropod 3, peduncle nearly as long as ramus, with 4 spines, 2^d joint of ramus nearly as long as 1st. Telson long oval, tapering to acute apex, 2 pairs of lateral spines (Sars), 3 pairs and a blunt apex (Goës). Colour whitish, with a few orange patches; ova bluish. L. 4 mm.

Arctic Ocean and North-Atlantic (Spitzbergen, Greenland, Norway). Depth 56—118 m.

4. **P. gregarius** (O. Sars) 1882 *Metopa gregaria*, G. O. Sars in: Forh. Selsk.
Christian., nr. 18 p. 93 t. 4 f. 6, 6 a | 1892 *Probolium gregarium*, G. O. Sars, Crust. Norway,
v. 1 p. 245 t. 84 | 1893 *Metopa gregaria, Proboloides gregarius*, A. Della Valle in: F. Fl.
Neapel. *v.* 20 p. 643, 907, 945; t. 59 f. 62. 63.

Head, lateral corners nearly rectangular. Side-plate 4 subequal to 2^d and
3^d combined. Pleon segment 3, postero-lateral angles somewhat produced. Eyes
rounded oval, well developed, dark red. Antenna 1, 1st and 2^d joints subequal
in length, flagellum longer than peduncle. Antenna 2, ultimate joint of peduncle
rather shorter than penultimate. Lower lip, inner plates coalesced. Mandibular
palp very slight, 3^d joint minute. Gnathopod 1, 4th joint considerably
produced, 5th and 6th joints subequal, palm convex, undefined. Gnathopod 2 in ♀,
6th joint oblong oval, palm crenulate in 2 divisions, oblique, defined by a
distinct angle armed with a spinule; in ♂ much larger, much narrower in
projection to length, the hind margin or palm pubescent, nearly straight,
interrupted by 2 denticles and ending in a finely serrate apical expansion,
finger very long, ciliated on inner margin. Peraeopods 3—5, 4th joint broad,
decurrent, in the 5th peraeopod greatly; 2^d joint linear in 3^d, broadly oval
in 4th, obliquely rounded in 5th. Uropod 3 rather short and stout, peduncle
a little longer than ramus, with 6 spines, 2^d joint of ramus shorter than 1st.
Telson twice as long as broad, apex acute, 4 spines on each margin. Colour
whitish, pellucid, with orange patches. L. ♀ 5, ♂ 6 mm.

North-Atlantic (Norway). Depth 75—188 m.

5. **P. calcaratus** (O. Sars) 1882 *Metopa calcarata*, G. O. Sars in: Forh. Selsk.
Christian., nr. 18 p. 92 t. 4 f. 5, 5 a | 1892 *Probolium calcaratum*, G. O. Sars, Crust.
Norway, *v.* 1 p. 247 t. 85 | 1893 *Metopa bruzelii* (part.), *M. calcarata, Proboloides calcaratus*,
A. Della Valle in: F. Fl. Neapel, *v.* 20 p. 641, 907, 945.

Head, lateral corners angular in ♀, more rounded in ♂. Side-plate 4 scarcely
as broad as 2^d and 3^d combined. Pleon segment 3, postero-lateral angles rather
produced. Eyes very large, oval quadrangular, imperfectly developed, bright red.
Antenna 1, 1st joint longer than 2^d, flagellum little longer than peduncle.
Antenna 2, ultimate joint of peduncle rather shorter than penultimate. Gnathopod 1
as in P. gregarius. Gnathopod 2 in ♀, 6th joint triangular oval, palm oblique, nearly
straight, slightly serrated, defined by an obtuse angle with 3 spinules, finger
of moderate size; in ♂, 6th joint very long, palm nearly as in P. gregarius,
but rather concave and without the 2 denticles, finger long and ciliated.
Peraeopods 3—5 more robust than in P. gregarius, the 4th joint more strongly
decurrent. Uropod 3, peduncle as long as ramus, 5 spines. Telson comparatively
shorter and broader, 3 spines on each margin. Colour uniformly whitish, ova
reddish. L. ♀ 5. ♂ 6 mm.

North-Atlantic (Norway). Depth 150—282 m.

6. **P. glacialis** (Krøyer) 1842 *Leucothoe g.*, Krøyer in: Naturh. Tidsskr., *v.* 4
p. 159 | 1846. *L. g.*, Krøyer in: Voy. Nord, Crust. t. 22 f. 3 a—p | 1862 *Montagua g.*,
Bate, Cat. Amphip. Brit. Mus., p. 58 t. 9 f. 3 (inaccurate) | 1871 *Metopa g.*, A. Boeck in:
Forh. Selsk. Christian., 1870 p. 141 | 1887 *M. g.*, H. J. Hansen in: Vid. Meddel., ser. 4
v. 9 p. 93 t. 3 f. 6, 6 a | 1893 *M. g.*, A. Della Valle in: F. Fl. Neapel, *v.* 20 p. 639 t. 59 f. 54.

Side-plate 4 not as broad as 2^d and 3^d combined. Pleon segment 3, postero-
lateral corners bluntly subquadrate. Eyes small, round, dark. Antenna 1, 1st joint
slightly longer than 2^d, flagellum shorter than peduncle, 10-jointed. Antenna 2
scarcely shorter than antenna 1, ultimate joint of peduncle a little shorter than penultimate, longer than 6-jointed flagellum. Maxilla 1, 1st joint of palp unusually

large. Maxillipeds, inner plates distinctly separate and outer altogether wanting. Gnathopod 1 small but rather strong, 5th joint broadly and stoutly oval, 6th shorter and much narrower, widening distally to a transverse palm, which is slightly overlapped by the finger. Gnathopod 2 large, 2^d joint rather short but unusually broad, 6th longer than 2^d, oblong, palm defined by a slender outstanding tooth, within which it is crenulate, then forming a denticle and a larger tooth over which the strong finger curves. Peraeopods 1—5 agreeing with the generic character. Uropod 3, peduncle shorter than the 2-jointed ramus. Telson long oval, without spines. Colour whitish-yellow. L. 7—8 mm.

Arctic Ocean and North-Atlantic (Spitzbergen, Greenland, Iceland).

7. P. groenlandicus (H. J. Hansen) 1887 *Metopa groenlandica*, H. J. Hansen in: Vid. Meddel., ser. 4 *v.* 9 p. 94 t. 3 f. 7—7 e | 1893 *M. g.*, A. Della Valle in: F. Fl. Neapel, *v.* 20 p. 640 t. 59 f. 55.

Very near to P. glacialis (p. 189), but side-plates deeper, 4th larger than 2^d and 3^d combined. Antenna 1 in ♂, 2^d joint longer than 1st, a little shorter than flagellum. Mouth-organs not described. Gnathopod 2, 2^d joint not unusually wide, 6th with a deep and broad sinus in the palm between the defining process and a smaller tooth forming a subobtuse angle with the serrate portion near the hinge of the finger. Gnathopod 2 in ♀, 6th joint oval, a convex serrate palm sloping to a small tooth. Peraeopod 5, 2^d joint much dilated, 4th little. L. ♂ 8·4, ♀ 7·5 mm.

Arctic Ocean and North-Atlantic (Greenland).

P. sarsii (Pfeff.) 1888 *Metopa s.*, Pfeffer in: Jahrb. Hamburg. Anst., *v.* 5 p. 84 t. 2 f. 3, 8; t. 3 f. 2 | 1893 *M. s.*, A. Della Valle in: F. Fl. Neapel, *v.* 20 p. 645.

L. 3 mm.

South-Atlantic (South Georgia). Low tide.

Metopidarum species dubia.

Metopa normani Hoek 1889 *M. n.*, Hoek in: Tijdschr. Nederl. dierk. Ver., ser. 2 *v.* 2 p. 190 t. 7 f. 5, 5'.

Said to be near to Metopella longimana (p. 185), but being nearer to Metopoides parallelocheir (p. 186). L. 4·8 mm.

North-Sea (Dutch coast). Depth 24 m.

---

10. Fam. Cressidae

1899 *Cressidae*, T. Stebbing in: Ann. nat. Hist., ser. 7 *v.* 4 p. 210.

Side-plate 1 rudimentary, 2^d to 4th large but normal in shape. Upper lip bilobed. Lower lip with inner lobes coalesced. Mandible, cutting edge dentate, molar weak, palp elongate, 3^d joint long. Maxilla 1, palp 1-jointed. Maxilla 2, inner plate very small. Maxillipeds, inner plates separate, outer small but distinct. Gnathopod 1 simple. Gnathopod 2 subchelate. Peraeopods 3—5, 2^d joint expanded. Uropods 1 and 2, rami very unequal. Uropod 3 with a single 2-jointed ramus. Telson coalesced with pleon segment 6.

Marine.

1 genus, 3 species.

1. Gen. **Cressa** Boeck

1857 *Danaia* (Sp. un.: *D. dubia*) (non H. Milne Edwards & Haime 1855, Authozoa!),
Bate in: Ann. nat. Hist., ser. 2 *v.* 19 p. 137 | 1871 *Cressa*, A. Boeck in: Forb. Selsk.
Christian., 1870 p. 145 | 1890 *C.*, J. Bonnier in: Bull. sci. France Belgique, *v.* 22 p. 186, 191 |
1892 *C.*, G. O. Sars, Crust. NorWay, *v.* 1 p. 277 | 1893 *C.*, A. Della Valle in: F. Fl.
Neapel, *v.* 20 p. 580.

Antenna 1 much longer than antenna 2. Gnathopod 1 slender and
feeble. Peraeopods 1—5 slender. Telson acute at apex.

3 species.

Synopsis of species:

1 { Some of the segments dorsally dentate . . . . . . 1. **C. dubia** . . . p. 191
  { None of the segments dorsally dentate — **2.**

2 { Eyes present . . . . 2. **C. minuta** . . p. 192
  { Eyes Wanting . . . . 3. **C. abyssicola** . p. 192

1. C. dubia (Bate) 1856 *Montagua dubius* (nom. nud.), Bate in: Rep. Brit.
Ass., Meet. 25 p. 57 | 1857 *Danaia dubia*, Bate in: Ann. nat. Hist., ser. 2 *v.* 19 p. 137 |
1876 *D. d.*, T. Stebbing in: Ann. nat. Hist., ser. 4 *v.* 18 p. 444 t. 19 f. 2 | 1888 *Cressa d.*,
T. Stebbing in: Rep. Voy. Challenger, *v.* 29 p. 293 | 1890 *C. d.* (part.), J. Bonnier in:
Bull. sci. France Belgique, *v.* 22 p. 186 t. 10 | 1892 *C. d.*, G. O. Sars, Crust. NorWay, *v.* 1
p. 278 t. 98 f. 2; t. 99 f. 1 | 1893 *C. d.* (part.), A. Della Valle in: F. Fl. Neapel, *v.* 20
p. 581 t. 58 f. 85 | 1871 *C. schiødtei*, A. Boeck in: Forb. Selsk. Christian., 1870 p 145.

Head, lateral corners acutely produced, with a small tooth below and
to the rear of the principal projection. Peraeon segments 6 and 7 and pleon
segments 1 and 2 produced backward in a curved tooth. Side-plates 2 and 3
with 4 or 5 coarse serrations curving forward at lower hind corner; 4th
emarginate behind, angle of emargination acute. Pleon segment 3, postero-
lateral corners much produced. Eyes large, oval, dark red. Antenna 1, 1st joint
long, rather longer than 2^d, both triangularly produced at apex, flagellum
slender, twice as long as peduncle, 20-jointed; in ♂ longer than in ♀, with
long sensory setae on flagellum. Antenna 2 much shorter and more slender
than antenna 1, ultimate joint of peduncle shorter than penultimate; flagellum
rather longer than peduncle. Lower lip not easily dissected ont (see the greatly
differing figures by Bonnier and Sars). Mandible with accessory plate, a
minute fringe representing spine-row, and a smooth projection answering to
molar, long 3^d joint of palp finely furred. Maxilla 1, inner plate without
setae, outer probably with 6 spines, palp elongate, 1-jointed (in Boeck's figure
2-jointed). Gnathopod 1, 6th joint nearly linear, shorter than the long,
narrow 5th. Gnathopod 2, 5th joint narrowly cup-shaped, 6th subtriangular,
constricted at base, much widened distally, palm little oblique, fringed with
spinules, defined by a tooth. Peraeopods 1—5 very slender, 2^d joint in peraeopods 3—5 oblong oval, obliquely truncated at lower hind corner. Uropods 1 and 2,
rami narrow, acute, constricted towards the end and furnished with 2 spines
at the constriction. Uropod 3, peduncle broad at base, produced at apex,
longer than ramus, of which the 1st joint is twice as long as 2^d and carries
2 apical spines. Telson ending in an acute piece flanked on each side of
its base by an acute little tooth (not shown in Bonnier's figure). Colour
dark brown, spotted. L. ♀ 3 to about 6 mm, ♂ much less (Sars).

North-Atlantic and North-Sea (NorWay, depth 36—146 m; France; Great Britain).

2 **C. minuta** Boeck 1871 *C. m.*, A. Boeck in: Forh. Selsk. Christian., 1870
p. 146 | 1876 *C. m.*, A. Boeck, Skand. Arkt. Amphip., *v.* 2 p. 469 t. 18 f. 7 | 1892 *C. m.*,
G. O. Sars, Crust. Norway, *v.* 1 p. 280 t. 99 f. 2 | 1885 *Danaia m.*, G. O. Sars in: Norske
Nordhavs-Exp., *v.* 6 Crust. I p. 190 | 1890 *Cressa dubia* (part.), J. Bonnier in: Bull.
sci. France Belgique, *v.* 22 p. 186 | 1893 *C. d.* (part.), A. Della Valle in: F. Fl. Neapel,
v. 20 p. 581.

In general resembling C. dubia (p. 191). Back not dentate, though the
imbrication of the segments may produce the appearance of dentation. Head,
lateral corners acutely produced, with no inferior denticle. Side-plates 2 and 3
with a single tooth curving forward at lower hind corner, 4th with angle of
emargination not very acute. Pleon segment 3. postero-lateral angles
moderately produced. Eyes rather small, round. Antenna 1, flagellum 12-jointed.
Gnathopod 1, 6th joint little more than half length of 5th. Peraeopods 3—5,
2^d joint regularly oval, not truncate at lower hind corner. L. ♀ 3—5 mm.

North-Atlantic and North-Sea (Norway, Firth of Forth).

3. **C. abyssicola** O. Sars 1879 *C. a.*, G. O. Sars in: Arch. Naturv. Kristian.,
v. 4 p. 453 | 1892 *C. a.*, G. O. Sars, Crust. Norway, *v.* 1 p. 278 | 1885 *Danaia a.*, G. O.
Sars in: Norske Nordhavs-Exp., *v.* 6 Crust. I p. 190 t. 16 f. 1, 1 a | 1890 *Cressa dubia* (part.),
J. Bonnier in: Bull. sci. France Belgique, *v.* 22 p. 186 | 1893 *C. d.* (part.), A. Della Valle
in: F. Fl. Neapel, *v.* 20 p. 581.

In several respects agreeing with C. dubia (p. 191). Body compact, not dorsally
dentate. Head, lateral corners acutely produced. No eyes. Antenna 1 longer
than the body, 1st and 2^d joints not apically produced. Antenna 2 scarcely
half length of antenna 1. Gnathopod 2, 5th joint narrowly produced, 6th large,
palm almost transverse and straight, with a regular series of 8 spines on
either side. Peraeopods 3—5, 2^d joint regularly oval. Colour whitish.
L. 6 mm.

Arctic Ocean (between Finmark and Beeren Island, depth 841 m; Greenland,
depth 376 m).

---

11. Fam. **Stenothoidae**

1871 Subfam. *Stenothoinae*, A. Boeck in: Forh. Selsk. Christian., 1870 p. 138 |
1888 *Stenothoidae*, T. Stebbing in: Rep. Voy. Challenger, *v.* 29 p. 747 | 1892 *S.*, G. O.
Sars, Crust. Norway, *v.* 1 p. 234.

Agreeing with the Metopidae (p. 171), except in some points of the
mouth-organs. Lower lip with the inner lobes rudimentary, separate. Mandible
without molar or palp. Maxilla 1, palp 2-jointed. Maxillipeds (Fig. 51 p. 193),
inner plates small, separate, outer obsolete. Peraeopod 3, 2^d joint perhaps
sometimes, but generally not, expanded; peraeopods 4 and 5 with 2^d joint expanded.

Marine.

1 genus, 15 accepted species and 3 doubtful.

1. Gen. **Stenothoe** Dana

1852 *Stenothoe*, J. D. Dana in: Amer. J. Sci., ser. 2 *v.* 14 p. 311 | 1853 *S.* (Sp. un.:
S. validus), J. D. Dana in: U. S. expl. Exp., *v.* 13 II p. 923 | 1888 *S.*, T. Stebbing in:
Rep. Voy. Challenger, *v.* 29 p. 748 | 1892 *S.*, G. O. Sars, Crust. Norway, *v.* 1 p. 235 | 1893
S., A. Della Valle in: F. Fl. Neapel, *v.* 20 p. 564 | 1853 *Probolium* (Sp. un.: *P. polyprion*),

A. Costa in: Rend. Soc. Borbon., n. ser. $v.2$ p.170 | 1856 *Montagua* (non Leach 1813/14, Decapoda!), Bate in: Rep. Brit. Ass., Meet. 25 p. 57 | 1857 *M.* (part.), Bate in: Ann. nat. Hist., ser. 2 $v.19$ p.137 | 1871 *Metopa* (part.), A. Boeck in: Forh. Selsk. Christian., 1870 p.140 | 1883 *Montaguana,* Chilton in: Tr. N. Zealand Inst., $v.15$ p.78.

Antennae 1 and 2 usually subequal, flagellum long in antenna 1, peduncle in antenna 2. Mandible, molar evanescent. Maxilla 1, palp large. Maxillipeds (Fig. 51), inner plates very small, outer evanescent or wholly absent, palp elongate, 4th joint ciliated on concave margin. Gnathopods 1 and 2 more or less distinctly subchelate. Gnathopod 1, 4th joint produced into a lobe along side of the 5th (except in S. peltata, p. 194). Gnathopod 2, 6th joint powerful, usually differing to some extent in the 2 sexes. Peraeopod 3, 2^{d} joint linear (except in S. valida, p. 194).

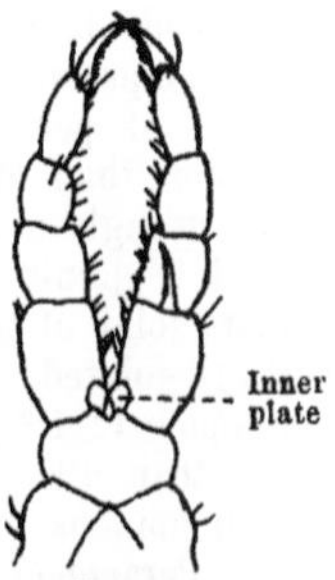

Inner plate

Fig. 51.
S. adhaerens.
Maxillipeds.

15 species accepted, 3 obscure.

Synopsis of accepted species:

1	Back carinate	1. S. richardi	p. 194
	Back not carinate — 2.		
2	Peraeopod 3, 2^{d} joint expanded	2. S. valida	p. 194
	Peraeopod 3, 2^{d} joint not expanded — 3.		
3	Gnathopod 1, 4th joint not-distinctly produced	3. S. peltata	p. 194
	Gnathopod 1, 4th joint distinctly produced — 4.		
4	Gnathopod 1, 2^{d} joint strongly bent at base .	4. S. antennulariae .	p. 195
	Gnathopod 1, 2^{d} joint straight — 5.		
5	Uropod 3, 2^{d} joint of ramus geniculate	5. S. cattai	p. 195
	Uropod 3, 2^{d} joint of ramus straight — 6.		
6	Gnathopod 2, palm with a single deep excavation	6. S. clypeata	p. 195
	Gnathopod 2, palm with more than one or with no excavation — 7.		
7	Uropod 3, peduncle much shorter than ramus — 8.		
	Uropod 3, peduncle longer or scarcely shorter than ramus — 9.		
8	Gnathopod 2, 6th joint subquadrate, palm slightly oblique	7. S. monoculoides .	p. 196
	Gnathopod 2, 6th joint elongate, palm extremely oblique	8. S. dollfusi	p. 196
9	Peraeopods 3—5, 4th joint narrow and little produced -- 10.		
	Peraeopods 3—5, 4th joint widened and produced — 11.		
10	Antennae 1 and 2 of considerable length, palm of gnathopod 2 in ♂ not deeply incised . .	9. S. tenella	p. 196
	Antennae 1 and 2 extremely long, palm of gnathopod 2 in ♂ deeply incised	10. S. megacheir . . .	p. 197
11	Eyes ill developed	11. S. microps	p. 197
	Eyes well developed — 12.		
12	Gnathopod 2 in ♂, palm denticulate throughout	12. S. marina	p. 198
	Gnathopod 2 in ♂, palm denticulate only at apex	13. S. bosphorana*) .	p. 198
13	Gnathopod 2 in ♀, palm defined	14. S. brevicornis . .	p. 199
	Gnathopod 2 in ♀, palm undefined	15. S. adhaerens . . .	p. 199

*) S. bosphorana ♀ is distinguished from S. brevicornis ♀ by the gnathopod 2 with palm ill-defined, and from S. adhaerens ♀ by the gnathopod 1 with 4th joint not produced to apex of 5th. In S. marina the telson has 2 spines on each side, in S. adhaerens 4, in S. brevicornis none.

1. **S. richardi** Chevreux 1895 *S. r.*, Chevreux in: Mém. Soc. zool. France, *v.* 8 p. 432 f.

Head, lateral angles broadly rounded, rostrum little curved. Body much compressed, dorsal carina produced into a tooth on each segment from 3^d of peraeon to 3^d of pleon. Side-plate 2 irregularly pentagonal, 4th wider than deep, rounded behind. Pleon segment 3, postero-lateral angles strongly produced. sharply uncinate. Eyes very large, round. Antenna 1, 1st and 2^d joints subequal, 3^d very short, flagellum 18-jointed. Antenna 2, ultimate joint of peduncle rather longer than long penultimate, flagellum very short, 10-jointed. Maxillipeds, ultimate joint of palp longer than penultimate. Gnathopod 1, 4th joint not quite reaching apex of 5th, 6th long oval, rather longer than 5th. Gnathopod 2, 6th joint broadest at base, where there is a small spinous protuberance, reached by long finger curving over crenulate palm. Peraeopod 5, 2^d joint exceptionally broad and expanded. Uropod 3, peduncle much longer than ramus. of which 2^d joint is rather longer than 1st. Telson broadly oval, 4 spines on each side, one median, apex acute. L. ♂ 5 mm.

Bay of Biscay (lat. 47⁰ N., long 8⁰ W.). Depth 748—1262 m.

2. **S. valida** Dana 1852 *S.*, J. D. Dana in: Amer. J. Sci.. ser. 2 *v.* 14 p. 311 1853 & 55 *S. validus*, J. D. Dana in: U. S. expl. Exp., *v.* 13 II p. 924; t. 63 f. 1 a— o | 1893 *S. valida* (part.), A. Della Valle in: F. Fl. Neapel, *v.* 20 p. 566 t. 58 f. 74—78 | ?1853 *Probolium polyprion*, A. Costa in: Rend. Soc. Borbon., n. ser. *v.* 2 p. 173 | 1876 *Stenothoe p.*, A. Boeck, Skand. Arkt. Amphip.. *v.* 2 p. 446 | 1866 *Probolium megacheles*, Cam. Heller in: Denk. Ak. Wien, *v.* 26 II p. 13 t. 2 f. 1. 2.

Eyes small, round. Antennae 1 and 2 nearly equal; antenna 1 as figured is the shorter in ♂, the longer in ♀. Antenna 2, ultimate joint of peduncle shorter than penultimate. Gnathopod 1, 4th joint longer than 5th, its base in ♂ longer, in ♀ shorter, than the produced part, 6th narrow oval, finger (as figured) much larger in ♂ than in ♀. Gnathopod 2, 4th and 5th joints short, 6th in ♂ very large, oblong, margins nearly parallel, an obtuse tooth at lower apex, finger long and stout, hind margin of hand partly pubescent (Costa: minutely serrate); in the ♀ this joint is rather smaller. more oval, with a small tubercle in place of the tooth, finger shorter and less powerful. Peraeopods 3—5, 2^d joint well expanded, 4th joint moderately so. Uropod 1 extending beyond uropod 2, and uropod 2 beyond uropod 3. Uropod 3 (as figured) has peduncle stout, rather shorter than ramus, of which 2^d joint is longer than 1st. Telson apparently carrying 3 lateral spines. L. 6—8 mm.

Tropical Atlantic (Rio Janeiro); Mediterranean?

3. **S. peltata** S. I. Sm. 1876 *S. p.*, S. I. Smith (& Harger) in: Tr. Connect. Ac., *v.* 3 p. 29 t. 3 f. 5—8 | 1893 *S. p.*, A. Della Valle in: F. Fl. Neapel, *v.* 20 p. 570.

Side-plate 3 deeper, not wider than 2^d, 4th as wide (or long) as length of peraeon segments 1—5. Eyes round, nearly white in spirit. Antenna 1 in ♀, flagellum 8-jointed, scarcely longer than peduncle. Antenna 2 rather longer than antenna 1, ultimate and penultimate joints of peduncle subequal, flagellum subequal to that of antenna 1. Gnathopod 1, 4th joint triangular, distally broader than 5th, 5th parallel-sided, 6th narrower but slightly longer than 5th, distally narrowed, finger about half as long. Gnathopod 2, 6th joint nearly twice as long as broad, palm convex, slightly oblique, defined by an acute lobe within which tip of finger closes. Peraeopods 1 and 2 slender. Uropod 3, ramus slightly longer than peduncle. L. ♀ about 6 mm.

North-Atlantic (St. George's Banks [United States of America]). Depth 55 m.

4. S. antennulariae Della Valle 1893 *S. a.*, A. Della Valle in: F. Fl. Neapel, *v.* 20 p. 565 t. 30 f. 1—18 | 1896 *S. crassicornis* (part.), A. O. Walker in: Rep. Brit. Ass., Meet. 66 p. 420 | 1897 *S. c.*, A. O. Walker in: J. Linn. Soc., *v.* 26 p. 229 t. 18 f. 3—8 e.

Body robust. Side-plate 1 very small, 4th large, subtriangular; depth and width subequal. Eyes small, circular, rosy. Antenna 1, 1st and 2^{d} joints equal in length, flagellum with 7—11 slender joints. Antenna 2, ultimate and penultimate joints of peduncle equal, flagellum short, 6- or 7-jointed. Maxillipeds, inner plates exceedingly small, outer quite wanting, palp slender. Gnathopod 1. 2^{d} joint notably bent at the base, 4th not nearly reaching apex of 5th. 6th oval, rather longer than 5th. Gnathopod 2, 4th and 5th joints short. 5th wide distally, 6th much larger in ♂ than in ♀, in ♂ having the palm defined at some distance from the base by a tooth to which the long finger reaches curving over another tooth or projection much nearer the hinge, (this projection being subdivided in ♂ as well as in ♀: Walker); 6th joint in ♀ almond-shaped, with some slight teeth at intervals over which the moderately sized finger curves. Peraeopods 3—5, 4th joint decurrent, 2^{d} and 4th expanded in peraeopods 4 and 5. Uropod 3 with very short peduncle, 1st joint of the ramus with several spines, 2^{d} rather shorter, with 1 spine (unless supposed 1st joint be in reality the peduncle, and supposed 2^{d} joint an indistinctly 2-jointed ramus). Telson oval, with 2 (Della Valle; 3, Walker) spines on each side; apex acute. Colour pale greyish with some yellow spots, ova bluish-green. L. 1·5—2 mm.

Gulf of Naples (depth 50—80 m, on Hydroids, especially Antennularia and Aglaophenia myriophyllum (L.)); Irish Sea (W.-S.-W. of Calf of Man, depth 42 m).

5. S. cattai Stebb.*) 1876 *Probolium polyprion* (err., non A. Costa 1853!), Catta in: Ann. Sci. nat., ser. 6 *v.* 3 nr. 1 p. 15 t. 2 f. 1 | 1893 *Stenothoe valida* (part.), A. Della Valle in: F. Fl. Neapel, *v.* 20 p. 566 | 1897 *S. crassicornis* (part.), A. O. Walker in: J. Linn. Soc., *v.* 26 p. 229.

Eyes small, round, black. Antenna 1 in ♂ rather shorter than antenna 2. Antenna 1, flagellum 20—25-jointed; antenna 2, ultimate and penultimate joints of peduncle subequal, flagellum 15—24-jointed, thicker than flagellum of antenna 1. Gnathopod 1, 4th and 5th joints side by side, much shorter than the 6th joint, which is trapezoidal, twice as long as broad, back and front margins parallel, palm oblique, finely crenulate, well defined. Gnathopod 2 nearly as in S. valida and S. dollfusi (p. 194, 196), 6th joint with hind margin pubescent and ending in 2 apical teeth, finger cleft at the apex. Peraeopod 3, 2^{d} joint expressly stated to be not expanded. Uropod 2 shorter than the others. Uropod 3; peduncle a little longer than the ramus,—of which the 2^{d} joint is geniculate. Telson twice as long as broad, produced to a little apical point, 3 spines on each side, scattered hairs on the surface. L. ♀ 3 mm.

Mediterranean.

6. S. clypeata Stimps. 1853 *S. c.*, Stimpson in: Smithson. Contr., *v.* 6 nr. 5 p. 51 | 1893 *S. c.*, A. Della Valle in: F. Fl. Neapel, *v.* 20 p. 569 | ?1862 *S. clypeatus*, Bate, Cat. Amphip. Brit. Mus., p. 61 t. 9 f. 7.

Eyes conspicuous, red. Gnathopod 1 slender, 6th joint small. Gnathopod 2, 6th joint very large, palm excavate between 2 teeth, of which the proximal

---

*) Spec. nov. After J. D. Catta.

is the larger. Peraeopod 3, 2^d joint not expanded. Colour bright yellow; in the young pale bluish. L. 12·5 mm.

Fundy Bay (Grand Manan). Depth 55 m.

7. S. monoculoides (Mont.) 1815 *Cancer (Gammarus) m.*, Montagu in: Tr. Linn. Soc. London, *v.*11 p.5 t.2 f.3 | 1828 *G. m.*, G. Johnston in: Zool. J., *v.*3 p.179 | 1856 *Montagua m.*, Bate in: Rep. Brit. Ass., Meet. 25 p.57 | 1869 *Probolium m.*, A. M. Norman in: Rep. Brit. Ass., Meet. 38 p.273 | 1871 *Stenothoe m.*, A. Boeck in: Forb. Selsk. Christian., 1870 p.140 | 1892 *S. m.*, G. O. Sars, Crust. Norway, *v.*1 p.240 t.82 f.1 | 1893 *S. m.* (part.), A. DellaValle in: F. Fl. Neapel, *v.*20 p.568 t.58 f.79 | 1881 *Probolium tergestinum*, Nebeski in: Arb. Inst. Wien, *v.*3 p.143 t.13 f.39.

Back broad. Head scarcely produced in front, lateral corners obtusely angular. Side-plate 4 larger than 2^d and 3^d combined. Pleon segment 3, postero-lateral angles somewhat produced. Eyes small, round, dark red. Antennae 1 and 2 subequal, rather short and stout. Antenna 1, flagellum of 12 very distinct joints, longer than peduncle. Antenna 2, ultimate joint of peduncle longer than penultimate, flagellum 10-jointed, longer than peduncle. Maxillipeds, joint supporting the palp large and laminar, palp rather short and stout. Gnathopods 1 and 2 similar in the 2 sexes, also similar to one another in structure, not in size, 4th and 5th joints short, 5th distally wide, 6th oblong quadrangular, palm well defined, nearly transverse in gnathopod 1, more oblique in the larger gnathopod 2. Peraeopods 1—5 slender throughout, except for expanded 2^d joint in 4th and 5th. Uropod 3, peduncle very short and broad, with 3 strong spines, each joint of ramus about equal to it in length. Telson oblong oval, without spines, apex evenly rounded. Colour whitish with orange-red blotches about middle; ova in pouch dark bluish. L. ♀ 3 mm.

North-Atlantic. North-Sea, Skagerrak and Kattegat (Norway, Denmark, France, Great Britain); Black Sea. Littoral, among Algae.

8. S. dollfusi Chevreux 1887 *S. d.*, Chevreux in: Bull. Soc. zool. France, *v.*12 p.327, 297 f. | 1891 *S. d.*, Chevreux in: Bull. Soc. zool. France. *v.*16 p.260 f.6—10 | 1893 *S. d.*, A. DellaValle in: F. Fl. Neapel, *v.*20 p.570.

Head, lateral angles rounded, little produced. Body compressed. Side-plate 4 as wide as 2^d and 3^d combined, hind margin truncate. Eyes round. Antenna 1 long, 1st joint rather longer than 2^d, flagellum with 15 long slender joints. Antenna 2 rather shorter, ultimate joint of peduncle much shorter than penultimate, flagellum 10-jointed, nearly as long as peduncle. Gnathopod 1, 4th and 5th joints rather long, side by side, 6th oval, subequal to 4th or a little longer. Gnathopod 2, 4th and 5th joints subequal, not elongate, 6th in ♂ very long and narrow, hind margin densely hirsute, ending in 3 coarse teeth or tubercles; the finger almost as long as the 6th joint, with cilia on the concave margin, the apex cleft; in ♀ the 6th joint is broader and less long, the hind margin not very hirsute, with 2 tubercles following an excavation; the finger shorter than in ♂, smooth and apically pointed. Peraeopods 1—5 long and very slender, 2^d joint of peraeopods 4 and 5 less expanded than usual and with the hind margin almost straight. Uropod 3, peduncle much shorter than the spinose ramus. Telson long, with 3 spines on each side. L. 3 mm.

Mediterranean (Cannes); North-Atlantic (Azores). Depth 130 m.

9. S. tenella O. Sars 1882 *S. t.*, G. O. Sars in: Forh. Selsk. Christian., nr. 18 p.88 t.3 f.12 | 1892 *S. t.*, G. O. Sars, Crust. Norway, *v.*1 p.238 t.81 f.2 | 1893 *S. t.*, A. DellaValle in: F. Fl. Neapel, *v.*20 p.570.

Body slender. Head, lateral corners angular. Side-plate 3 distally widened, 4th scarcely as wide as 2[d] and 3[d] combined. Pleon segment 3, postero-lateral corners triangularly produced. Eyes large, round, imperfectly developed, light red. Antennae 1 and 2 long, slender, subequal, more than half length of body. Antenna 1, 2[d] joint longer than 1st, flagellum about twice as long as peduncle, 21-jointed. Antenna 2, ultimate and penultimate joints of peduncle subequal, flagellum about 11-jointed. Gnathopod 1 small, 4th joint produced about to apex of 5th, 6th scarcely broader than 5th. Gnathopod 2 in ♂, 6th joint narrow oblong, palm undefined, concave, pubescent, distally dentate and incised almost at right angles to the length, finger long, powerful, concave margin hairy; in ♀ 6th joint somewhat tapering, palm irregularly serrated, finger and hand less powerful than in ♂. Peraeopods 1—5 long and slender, none of the joints expanded, except 2[d] in peraeopod 4 narrow oval and in peraeopod 5 broad qval. Uropod 3, peduncle much longer than ramus; armed with 8 spines, joints of ramus equal. Telson oblong oval, nearly twice as long as broad, obtusely pointed, 3 spines on each side. Colour whitish, banded irregularly with patches of light red; ova violet. L. ♀ 5·5 mm.

North-Atlantic and North-Sea (West-Norway). Depth 150—282 m.

10. **S. megacheir** (Boeck) 1871 *Metopa m.*, A. Boeck in: Forb. Selsk. Christian., 1870 p. 143 | 1876 *M. m.*, A. Boeck, Skand. Arkt. Amphip., v. 2 p. 462 t. 18 f. 1 | 1892 *Stenothoe m.*, G. O. Sars, Crust. Norway, v. 1 p. 242 t. 83.

Body slender. Head, lateral corners bluntly angular. Side-plate 3 distally widened, 4th scarcely as wide as 2[d] and 3[d] combined. Pleon segment 3, postero-lateral corners triangularly produced. Eyes large, round, indistinctly developed, light yellowish red. Antennae 1 and 2 extremely long and slender. Antenna 1 as long as body, 2[d] joint of peduncle longer than 1st, flagellum filiform, twice as long as peduncle. Antenna 2 in ♀ a little shorter, in ♂ a little longer than antenna 1, ultimate and penultimate joints of peduncle subequal, very long, flagellum much shorter than either. Maxillipeds very slender and elongated. Gnathopod 1 small and feeble, 4th joint not quite produced to apex of 5th, 6th not longer and scarcely broader than 5th. Gnathopod 2, large and powerful, especially in ♂, 6th joint oblong oval, palm long, defined, having in ♂ 3 deep incisions with denticulation above and below them, in ♀ slightly concave, irregularly serrate; finger curved, strong, inner margin in ♂ ciliated. Peraeopods 1—5 long and slender, none of the joints expanded, except 2[d] in peraeopods 4 and 5 not very broadly oval. Uropod 3, peduncle with 10 spines, nearly twice as long as ramus, of which the 2 joints are subequal, very slender. Telson oblong oval, apically narrowed, 3 spines on each side. Colour whitish, pellucid, with scattered orange patches; ova violet. L. 8 mm.

North-Atlantic and North-Sea (West-Norway). Depth 150—282 m.

11. **S. microps** O. Sars 1892 *S. m.*, G. O. Sars, Crust. Norway, v. 1 p. 237 t. 81 f. 1 | 1893 *S. monoculoides* (part.), A. Della Valle in: F. Fl. Neapel, c. 20 p. 907.

Not very sharply distinguished from S. marina (p. 198). Body rather robust. Head, lateral corners broadly rounded. Side-plates and pleon segment 3 nearly as in S. marina. Eyes very small, round, imperfectly developed, light red. Antennae 1 and 2 comparatively shorter than in S. marina, nearly equal. Gnathopod 2 in ♀, 6th joint elongate, palm coarsely dentated, reaching nearly to base. Uropod 3, peduncle with 8 spines, longer than ramus, of which 2[d] joint is shorter than 1st. Telson rather broadly oval, 4 spines on each side. Colour highly pellucid, whitish, with reddish patches; ova small, bluish grey. L. ♀ 8·5 mm.

North-Atlantic (Norway); North-Sea, outside the great fishing banks. Depth 188 m.

12. **S. marina** (Bate) 1856 *Montagua marinus* (nom. nud.), Bate in: Rep. Brit. Ass., Meet. 25 p. 57 | 1857 *M. m.*, Bate in: Ann. nat. Hist., ser. 2 v. 19 p. 137 | 1866 *Probolium marinum*, Cam. Heller in: Denk. Ak. Wien, c. 26 II p. 14 | 1871 *Stenothoe marina*, A. Boeck in: Forh. Selsk. Christian., 1870 p 139 | 1892 *S. m.*, G. O. Sars, Crust. Norway, v. 1 p. 236 t 80 | 1861 *S. danai*, A. Boeck in: Forh. Skand. Naturf., Møde 8 p 655 | 1893 *S. monoculoides* (part.), A. Della Valle in: F. Fl. Neapel, c. 20 p. 568.

Body moderately slender. Head, lateral corners rounded. Side-plate 2 rather small, with little tooth at lower hind corner, 3^d distally expanded, 4th nearly as wide as 2^d and 3^d combined. Pleon segment 3, postero-lateral corners not strongly produced. Eyes small, round, well developed, dark red. Antenna 1 in ♀ nearly half as long as body, 2^d joint of peduncle shorter than 1st, flagellum longer than peduncle, 14—18-jointed. Antenna 2 rather shorter than antenna 1, ultimate and penultimate joints of peduncle subequal, flagellum 10—12-jointed. Antennae 1 and 2 in ♂ rather longer than in ♀. Gnathopod 1, 4th joint produced nearly to apex of 5th, which is distally widened. 6th rather longer than 5th, distally widened, palm obliquely convex. Gnathopod 2 powerful, 6th joint broad at base and tapering in ♀, longer and narrower at base in ♂, palm well defined, almost all the length of the hand, denticulate throughout and in ♂ coarsely so at the apex, finger very long, in ♂ ciliate on inner margin. Peraeopods 3—5, 4th joint rather expanded and produced, 2^d joint long oval in peraeopod 4, broadly oval in peraeopod 5. Uropod 3, peduncle with 6 spines, subequal to ramus, of which 2^d joint is little shorter than 1st. Telson oblong oval, obtusely pointed. 2 spines on each side. Colour pellucid whitish, mottled with yellow and with pinkish patches. L. ♀ 5, ♂ 6 mm.

North-Atlantic and North-Sea (Norway, France, Great Britain); Adriatic. Depth 37—94 m.

13. **S. bosphorana** Sowinski 1898 *S. b.*, Sowinski in: Mém. Soc. Kiew, v. 15 p. 493 t. 11 f. 23—25 (24?); t. 12 f. 9—19.

Closely resembling Proboloides grandimanus (p. 187), except in regard to mandible. Head, rostrum feeble. Eyes rather large, round. Antenna 1 in ♂ shorter than antenna 2, first 2 joints equal, 3^d about $\frac{1}{8}$ as long as 2^d, all 3 almost bare, flagellum with 11 or 12 joints. 1st the longest; accessory flagellum rudimentary, 1-jointed. Antenna 2 in ♂, ultimate joint of peduncle considerably longer than penultimate, flagellum rather shorter than ultimate joint of peduncle, with 8 joints. 1st the longest. Antennae 1 and 2 in ♀ differing from those of ♂, antenna 1, flagellum with 15 joints, 1st not longer than the following; antenna 2, ultimate and penultimate joints of peduncle much shorter than in ♂, flagellum thrice as long as ultimate joint of peduncle. 11- or 12-jointed. Mouthparts as usual in the genus. Gnathopod 1 in ♂, 4th joint not nearly reaching apex of 5th. 5th rather longer than the narrowly oblong oval 6th, palm oblique, ill-defined, finger not large. Gnathopod 1 in ♀, 5th joint much shorter than 6th. Gnathopod 2 in ♂. 4th and 5th joints very short, 4th apically acute, 5th cup-shaped, 6th rather longer and much stouter than 2^d, oblong oval, but with transverse apex serrate in 4 denticles and projecting outward in a small tooth, over which the great finger bends, reaching nearly to end of hind (or palmar) margin, this being bordered with setules interspersed with 4 or 5 minute spines; inner margin of finger setulose, apex not cleft.

Gnathopod 2 in ♀, 6th joint almond-shaped with almost rectilinear front margin and convex hind margin. Peraeopod 1 longer than peraeopod 2; peraeopod 3 similar to peraeopods 1 and 2; peraeopods 4 and 5, 2^d joint oval, wider and more produced in peraeopod 5 than in 4, 4th joint produced downward behind. Uropod 1, rami equal, each figured with 2 (outer with 3 in text), peduncle with 6 spines. Uropod 2 little more than half as long as uropod 1. Uropod 3, peduncle with 3 spines, basal joint of ramus with 3; fig. 24 represents a biramous pair, neither branch 2-jointed. Telson oval, with 2 spines above, 2 setules below on each lateral margin. L. about 3·5 mm.

Bosphorus. Depth 32 m.

14. **S. brevicornis** O. Sars 1882 *S. b.*, G. O. Sars in: Forb. Selsk. Christian., nr. 18 p. 89 t. 4 f. 1 | 1892 *S. b.*, G. O. Sars, Crust. Norway, v. 1 p. 241 t. 82 f. 2 | 1893 *S. b.*, A. Della Valle in: F. Fl. Neapel, v. 20 p. 569.

Body rather robust. Head, lateral corners obtusely angular. Side-plate 3 distally widened, 4th about as wide as 2^d and 3^d combined. Pleon segment 3, postero-lateral corners almost quadrate. Eyes round, well developed, dark red. Antennae 1 and 2 very short. Antenna 1 about $^1/_3$ length of the body, 2^d joint shorter than 1st, flagellum scarcely longer than peduncle, 12-jointed. Antenna 2 subequal to antenna 1, ultimate and penultimate joints of peduncle subequal, flagellum as long as either. Gnathopod 1, 4th joint not reaching apex of 5th, 6th as long as 5th, much wider, widening distally, palm well defined, straight, moderately oblique. Gnathopod 2 in ♀, 6th joint much larger than in gnathopod 1, but similar in shape, palm defined by a projecting angle, sinuous, denticulate near the finger hinge, finger not very strong. Peraeopods 1—5 rather short and stout, 4th joint in peraeopods 3—5 well expanded and produced, 2^d joint in peraeopod 5 more broadly oval than in peraeopod 4. Uropod 3, peduncle rather broad, longer than ramus, with 5 minute spines; 2^d joint of ramus smaller than 1st. Telson oblong oval, obtusely pointed, devoid of spines. Colour whitish, ova bluish green. L. ♀ 8 mm.

Arctic Ocean, North-Atlantic and North-Sea (Norway). Depth 94 m.

15. **S. adhaerens** Stebb. 1888 *S. a.*, T. Stebbing in: Rep. Voy. Challenger, v. 29 p. 748 t. 39 | 1893 *S. a.*, A. Della Valle in: F. Fl. Neapel, v. 20 p. 569.

Head, lateral corners rounded. Side-plate 4 scarcely as wide as 2^d and 3^d combined. Pleon segment 3, postero-lateral angles little produced, obtuse. Eyes rounded oval, well developed. Antennae 1 and 2 subequal, not elongate. Antenna 1, flagellum 17-jointed. Antenna 2, ultimate joint of peduncle shorter than penultimate, flagellum 15-jointed, nearly as long as peduncle. Maxillipeds (Fig. 51 p. 193), inner plates minute, outer wanting. Gnathopod 1, 4th joint produced quite to apex of 5th, 6th longer than 5th, palm well defined, somewhat oblique, finely pectinate. Gnathopod 2 in ♀, 6th joint not twice as long as broad, palm undefined, irregularly but finely denticulate on the part near to the hinge of the long broad finger. Peraeopods 3—5, 4th joint expanded and produced strongly, 2^d joint more broadly oval in peraeopod 5 than in peraeopod 4. Uropod 3, peduncle subequal to ramus, with several spines, 2^d joint of ramus shorter than 1st. Telson long oval, apex subacute, 4 pairs of lateral spines. L. about 6 mm.

Southern Indian Ocean (Cape Agulhas). On screw of H. M. S. Challenger.

S. guerinii (Bate) 1862 *Montagua g.*, Bate, Cat. Amphip. Brit. Mus., p. 59 t. 9 f. 5 | 1893 *Stenothoe valida?* (part.), A. Della Valle in: F. Fl. Neapel, v. 20 p. 569.

Indian Ocean (Madagascar).

S. longimana (Bate) 1862 *Montagua l.*, Bate, Cat. Amphip. Brit. Mus., p. 57 t. 9 f. 1 | 1885 *Probolium longimanum*, J. V. Carus, Prodr. F. Medit., *v.* 1 p. 407 | 1893 *Stenothoe?*, A. Della Valle in: F. Fl. Neapel, *v.* 20 p. 569.

„Piedmont".

S. miersii (Hasw.) 1880 *Montagua m.* + *M. longicornis* (♂?), Haswell in: P. Linn. Soc. N. S. Wales, *v.* 4 p. 323 t. 24 f. 4; p. 323 t. 24 f. 5 | 1883 *Montaguana m.*, Chilton in: Tr. N. Zealand Inst., *v.* 15 p. 79 | 1885 *Probolium m.*, Chilton in: P. Linn. Soc. N. S. Wales, *v.* 9 p. 1043 | 1893 *Stenothoe monoculoides?* + *Montagua miersii*, A. Della Valle in: F. Fl. Neapel, *v.* 20 p. 569.

L. about 4 mm.

South-Pacific (Port Jackson [East-Australia]; Timaru and Lyttelton Harbour [New Zealand]).

---

12. Fam. **Phliantidae**

1899 *Phliadidae*, T. Stebbing in: Tr. Linn. Soc. London, ser. 2 *v.* 7 p. 414.

Peraeon strongly developed. Pleon segments 5 and 6 subject to degradation (Fig. 52). Antennae 1 and 2 short. Antenna 1 without accessory flagellum, flagellum with sensory filaments. Upper lip with distal margin usually undivided. Lower lip without inner lobes. Mandible without palp. Maxilla 1 with palp obsolete. Maxillipeds variable. Gnathopods 1 and 2 simple or only feebly subchelate. Peduncle laterally produced in one or more of the pleopods (Fig. 53). Uropod 3 usually not biramous. Telson short, entire.

Marine.

6 genera, 6 species.

Synopsis of genera:

1 { Maxillipeds, palp 3-jointed — **2.**
  { Maxillipeds, palp 4-jointed — **4.**

2 { Uropod 3 biramous . . . . 1. Gen. **Phlias** . . . . p. 200
  { Uropod 3 not biramous — **3.**

3 { Uropod 3, peduncle and ramus distinct . . . . . 2. Gen. **Pereionotus** . p. 201
  { Uropod 3, ramus not distinct from peduncle . . 3. Gen. **Palinnotus** . . p. 202

4 { Pleopod 3 (Fig. 53), inner ramus rudimentary . 4. Gen. **Iphiplateia** . . p. 203
  { Pleopod 3, inner ramus well developed — **5.**

5 { Side-plates 1—4 very deep . . . . 5. Gen. **Iphinotus** . . p. 204
  { Side-plates 1—4 very shallow . . . . 6. Gen. **Bircenna** . . . p. 205

1. Gen. **Phlias** Guér.

1836 *Phlias*, Guérin (-Méneville) in: Mag. Zool., Cl. 7 t. 19 | 1840 *P.*, H. Milne Edwards, Hist. nat. Crust., *v.* 3 p. 23 | 1853 *P.*, J. D. Dana in: U. S. expl. Exp.. *v.* 13 II p. 908 | 1862 *P.*, Bate, Cat. Amphip. Brit. Mus., p. 83 | 1888 *P.*, T. Stebbing in: Rep. Voy. Challenger, *v.* 29 p. 165 | 1899 *P.*, T. Stebbing in: Tr. Linn. Soc. London, ser. 2 *v.* 7 p. 414.

Body short, laterally compressed. Peraeon segments 1—7 tuberculately carinate. Pleon segments 5 and 6 coalesced, or 5th missing. Head small, partially immersed in peraeon segment 1. Side-plates 1—4 large. Eyes prominent. Antennae 1 and 2 short, antenna 1 the longer and much the stouter, both closely similar to those of Pereionotus, as likewise the limbs, so far as is known. Maxillipeds, palp 3-jointed. Pleopod 3 (uropod 1 according to

Guérin) with peduncle short, and 2 much longer, equal, ciliated rami. Uropod 1 with blunt equal rami, as long as peduncle. Uropod 2 wanting as in Pereionotus ($\mathcal{Q}$). Uropod 3 very short, with 2 conical rami as long as peduncle. Telson small, transverse, a little rounded (text) or a little emarginate.

1 species.

1. **P. serratus** Guér. 1836 *P. s.*, Guérin (-Méneville) in: Mag. Zool., Cl. 7 t. 19 f. 1—4 | 1862 *P. s.*, Bate, Cat. Amphip. Brit. Mus., p. 88 t. 14 a f. 2 (on t. 21) | 1888 *P. s.*, T. Stebbing in: Rep. Voy. Challenger, *v.* 29 p. 165 etc. | 1893 *P. s.*, A. Della Valle in: F. Fl. Neapel, *v.* 20 p. 561 | 1899 *P. s.*, T. Stebbing in: Tr. Linn. Soc. London, ser. 2 *v.* 7 p. 417.

Colour opaque yellow brown. L. 5—6 mm.

South-Pacific (between the Malouines (Falkland) Isles and Port Jackson [East-Australia]); Mediterranean.

2. Gen. **Pereionotus** Bate & Westw.

1862 *Pereionotus* (Sp. un.: *P. testudo*), Bate & Westwood, Brit. sess. Crust., *v.* 1 p. 226 | 1888 *P.*, T. Stebbing in: Rep. Voy. Challenger, *v.* 29 p. 340 | 1893 *P.*, A. Della Valle in: F. Fl. Neapel, *v.* 20 p. 559 | 1899 *P.* (part.), T. Stebbing in: Tr. Linn. Soc. London, ser. 2 *v.* 7 p. 416 | 1864 *Icridium* (Sp. un.: *I. fuscum*), E. Grube in: Jahresber. Schles. Ges., *v.* 41 p. 58.

Body depressed, ridged, hind part of pleon ventrally flexed. Head, rostrum short, obtuse, lateral corners carrying prominent eyes. Antennae 1 and 2 small, antenna 1 the larger, flagellum in both small. Upper lip bilobed. Lower lip without inner lobes. Mandible, cutting edge dentate, spine-row of very few spines and a seta, molar and palp wanting. Maxilla 1, inner plate wanting, outer plate with 5 spines, palp wanting, its place indicated by a bulge of the margin. Maxilla 2, inner and outer plate coalesced except at apex. Maxillipeds, inner plates with 3 spine-teeth on truncate apex, outer plates almost unarmed, reaching end of 2^d joint of palp or further, palp small, of 3 joints, 1st largest. Gnathopods 1 and 2 simple. Peraeopods 1—5 as in gnathopods 1 and 2 with hooked finger. Peraeopods 3 and 4, 2^d joint roundly expanded, 4th joint also wide. Peraeopod 5, 2^d and 4th joints smaller than in peraeopods 3 and 4. Pleopods 2 and 3 sending out an arm from peduncle (as in Iphiplateia, p. 203, but much shorter), tipped with 3 coupling spines. Uropod 1 reaching much beyond the others, peduncle as long as or longer than rami, outer ramus rather the shorter, slightly curved, both rami obtuse, tipped with short, blunt spine. In $\mathcal{Q}$ (?) uropod 2 not developed, uropod 3, peduncle subequal to single spine-tipped ramus. In $\male$ (?) uropod 2 much shorter than uropod 1, but otherwise similar, especially as to the obtuse rami. the pair of peduncles distally convergent, uropod 3 very short. with tubercle on inner margin of peduncle. single ramus not longer than broad, much shorter than peduncle. Telson small, triangular, depressed.

1 species.

1. **P. testudo** (Mont.) 1808 *Oniscus t.*, Montagu in: Tr. Linn. Soc. London, *v.* 9 p. 102 | 1862 *Pereionotus t.*, Bate & Westwood, Brit. sess. Crust., *v.* 1 p. 228 f. | 1893 *P. t.*, A. Della Valle in: F. Fl. Neapel, *v.* 20 p. 559 t. 3 f. 7; t. 31 f. 1—19 | 1862 *Phlias rissoanus*, Bate, Cat. Amphip. Brit. Mus., p. 88 t. 14 a f. 3 (on t. 21) | 1875 *Icridium rissoanum*, Catta in: Rev. Sci. nat., *v.* 4 p. 161 | 1864 *I. fuscum*, E. Grube in: Jahresber. Schles. Ges., *v.* 41 p. 58 | 1864 *I. f.*, E. Grube in: Arch. Naturg., *v.* 30 i p. 209 t. 5 f. 3, 3 a—f.

Peraeon segments 1—7, pleon segments 1 and 2 tuberculately carinate and setulose, lateral tubercles on peraeon segments 2—7. Pleon segment 5 apparently missing in ♀, short in ♂. Head flat, widening distally, a small point projecting below the eye. Side-plates 1—4 wide, outspread, fringed, 4th a little emarginate. Eyes subglobose, red (Montagu: black). Antenna 1, 1st joint very broad, with short obtuse apical process, 2d shorter and narrower with small process. 3d scarcely shorter, this and the little 1-jointed flagellum carrying long setae. Antenna 2 inserted some way behind antenna 1, much shorter, only 3 free joints of peduncle, ultimate and penultimate joints subequal, flagellum 2- or 3-jointed, shorter than antepenultimate joint of peduncle, tipped with long setae. Gnathopods 1 and 2, 3d joint longer than broad, 4th triangular, underriding short 5th, 6th simple as in peraeopods, tipped with spine. Peraeopods 1 and 2, 2d joint rather stout, not long, 4th and 5th short, and broad in comparison with the length, 6th joint and finger similar in all the limbs. Colour dull red, with a white spot on the anterior part of the back, but as the insect dies this mark is lost (Montagu); uniform yellow brown on segments of peraeon and pleon, with large white spots on 2d peraeon segment (Della Valle). L. 3—4 mm.

English Channel (South of Great Britain). Mediterranean.

3. Gen. **Palinnotus** Stebb.

1900 *Palinnotus* (Sp. typ.: *P. thomsoni*), T. Stebbing in: Ann. nat. Hist., ser. 7 v. 5 p. 16.

♀. In general agreement with Pereionotus, but distinguished as follows. Upper lip not bilobed. Maxilla 1 with a small spinule representing the palp. Maxillipeds, outer plates reaching slightly beyond the 3-jointed palp, and minutely fringed on distal half of inner margin. Pleopod 3, but not pleopod 2, with inner side of peduncle produced. Uropod 2 developed, short, uniramous. Uropod 3 without distinction of peduncle and ramus. — ♂ unknown.

1 species.

1. **P. thomsoni** (Stebb.) 1899 *Pereionotus t.*, T. Stebbing in: Tr. Linn. Soc. London, ser. 2 v. 7 p. 417 t. 35A | 1900 *Palinnotus t.*, T. Stebbing in: Ann. nat. Hist., ser. 7 v. 5 p. 16.

Body broad oval. Peraeon segments 1—7 and pleon segments 1 and 2 carinate, the medio-dorsal processes not quite as long as their segments, that on peraeon segment 1 preceded by an acute point directed forward. Head, rostrum small, but distinct, lateral angles a little in advance. Side-plates 1—4 without conspicuous setules on distal margin. Eyes rounded oval, dark, on lateral lobes of head. Antenna 1, 1st joint broad, 2d cylindrical, 3d conical, flagellum small, 2-jointed. Antenna 2 shorter, much more slender, attached to the rear of antenna 1 on under surface of head, ultimate joint of peduncle rather longer than penultimate, flagellum with long setae, 1- or 2-jointed. Gnathopods 1 and 2 and peraeopods 1 and 2 as in Iphiplateia whiteleggei (p. 203). All the limbs have a rather strong spine at apex and at middle of inner margin of 6th joint. Peraeopod 3, 2d joint broader than long, 4th not longer than 3d, very broadly lobed. Peraeopods 4 and 5 with 2d and 4th joints successively much smaller than in peraeopod 3, 2d in peraeopod 5 with hind lobe not reaching instead of overlapping the 3d joint. Pleopod 2, peduncle broader than long, inner margin convex, but not produced. Pleopod 3 with the coupling spines on a short, but very distinctly produced process. Uropod 1, peduncle rather longer than the longer of the two rami, both rami narrow. Uropod 2, peduncle not reaching end of telson, the single ramus narrowly oval. Telson triangular, with rounded apex. L. rather under 5 mm.

Watson's Bay [East-Australia]. Low-tide line.

4. Gen. **Iphiplateia** Stebb.

1899 *Iphiplateia* (Sp. un.: *I. whiteleggei*), T. Stebbing in: Tr. Linn. Soc. London, ser. 2 *v.* 7 p. 414.

Body much depressed, pleon strongly flexed. Head immersed between projecting 1st pair of side-plates, square, feebly rostrate, front corners prominent, carrying the eyes. Side-plates 1—4 large, outspread. Antennae 1 and 2 short, subequal in length, antenna 1 the broader, antenna 2 attached to the rear of antenna 1, on under side of head. Upper lip with convex distal margin. Maxilla 1, inner plate wanting, outer with 5 spines, palp wanting, but position for it indicated. Maxilla 2, inner plate short and broad, with apical spine-teeth, outer narrow, continuous with the base. Maxillipeds rather broad, outer plates feebly armed, reaching apex of 2d joint of palp, finger of palp small, cylindrical, tipped with a long seta. Gnathopods 1 and 2 simple. Peraeopods 3—5, 2d joint very large, 4th broad. Pleopods 2 and 3, peduncle with produced process of inner margin. Pleopod 3, inner ramus rudimentary. Uropods 1 and 2 biramous, 2d the shorter. Uropod 3 very small, 1-jointed. Telson entire.

1 species.

1. I. whiteleggei Stebb. 1899 *I. w.*, T. Stebbing in: Tr. Linn. Soc. London, ser. 2 *v.* 7 p. 415 t. 34.

Body (Fig. 52) forming a broad oval, by help of the 1st joint of antenna 1 and of 2d joint of peraeopods 3—5; dorsal line feebly angular or smooth, peraeon segment 7 slightly upraised, pleon segment 1 projecting backward in a prominent tubercle, pleon small, segment 6 dorsally undeveloped.

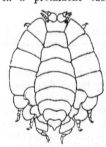

Fig. 52. **I. whiteleggei.**
Dorsal view.

Fig. 53. **I. whiteleggei.**
Pleopod 3
with hooked spines.

Side-plate 1 subtriangular, 2d and 3d oblong, 4th very broad, excavate behind, 5th to 7th small, bilobed, front lobe the larger. Eyes small, oval, dark. Antenna 1, 1st joint nearly as broad as long, with large inner lobe, 2d similar, but smaller, 3d narrow, flagellum very small, 2-jointed. Antenna 2, basal joints apparently soldered to underside of head, ultimate joint of peduncle rather smaller than penultimate, slightly longer than the small 2-jointed flagellum. Epistome rounded above. Mandible, cutting edge quadri-dentate, spine-row of 3 minute spinules; a broad pellucid spine tipped with a setule perhaps represents the molar. Gnathopod 1, 2d joint not nearly reaching distal border of side-plate, 3d as long as 4th, 5th a little longer than the tapering 6th, which forms no palm, finger small, curved. Gnathopod 2 like gnathopod 1, but

5th and 6th joints more nearly equal. Peraeopods 1 and 2, 4th joint a little wider, not longer than 3^d, 5th as broad as long, much shorter than 6th, 6th and 7th nearly agreeing in all the limbs of peraeon. Peraeopod 3, 2^d joint with the broad expansion produced below the 3^d joint, 4th as broad as long, lobed behind. Peraeopod 4, 2^d and 4th joints larger than in peraeopod 3. Peraeopod 5 with 2^d joint shorter than in peraeopods 3 and 4, but even wider, otherwise nearly as peraeopod 3. Pleopod 1, peduncle without process, but carrying 5 or 6 coupling spines. Pleopod 2, peduncle much shorter than in pleopod 1, with short, broad process carrying 5 or 6 coupling spines. Pleopod 3 (Fig. 53), peduncle very short, with long narrow process carrying 3 coupling spines, inner ramus minute, oval, unjointed, without setae, outer ramus normal. Uropod 1, peduncle shorter than the curved outer ramus, longer than the straight inner one. Uropod 2 much smaller, otherwise like the 1st. Uropod 3, each consisting of a small oval lamella, nearly concealed by the telson. Telson semi-oval, with subacute apex. L. about 5 mm.

Watson's Bay [East-Australia].

5. Gen. **Iphinotus** Stebb.

1882 *Iphigenia* (Sp. un.: *I. typica*) (non C. F. Schumacher 1817, Mollusca!), G. M. Thomson in: Tr. N. Zealand Inst.. *v.* 14 p. 237 | 1899 *Iphinotus* (Sp. un.: *I. chiltoni*), T. Stebbing in: Tr. Linn. Soc. London, ser. 2 *v.* 7 p. 419.

Body much depressed, pleon strongly flexed. Head immersed between the projecting 1st pair of side-plates, square, feebly rostrate, front corners prominent, carrying the eyes. Side-plates 1—4 large, outspread. Antennae 1 and 2 nearly as in Iphiplateia (p. 203), as also the mouth-parts, except that the maxillipeds have the 4th joint well developed, unguiform. Limbs of the peraeon nearly as in Iphiplateia, except that the 2^d and 4th joints are very much smaller in peraeopod 5 than in peraeopods 3 and 4. Pleopods 1—3 with both rami well developed. Pleopods 2 and 3, peduncle laterally produced into a long and strong process. Uropods 1 and 2 biramous, peduncle much longer than the rami. Uropod 1 slender, uropod 2 stout. Uropod 3 membranous, not biramous, small. Telson entire.

1 species.

1. I. typicus (G. M. Thoms.) 1882 *Iphigenia typica*, G. M. Thomson in: Tr. N. Zealand Inst., *v.* 14 p. 237 t. 18 f. 4 | 1899 *Iphinotus chiltoni* + *Iphigenia typica*, T. Stebbing in: Tr. Linn. Soc. London, ser. 2 *v.* 7 p. 419 t. 35 B; p. 420.

Body broad oval, with dorsal carina. Eyes rounded oval, dark. Antenna 1, 1st joint large, distally widened, 2^d cylindrical, 3^d scarcely longer than broad, flagellum with 3 small joints. Antenna 2, moderately stout, ultimate joint of peduncle longer than penultimate, flagellum with 5 joints, last 4 very small, setose. Upper lip broader than deep, distal margin almost straight. Gnathopods 1 and 2 and peraeopods 1 and 2 in general as in Palinnotus thomsoni (p. 202), except that the finger is abruptly narrowed at base of the sharp hooked nail, and has there a strong setule. Peraeopods 3 and 4, 2^d joint very large, a little longer in peraeopod 3 than in peraeopod 4, 4th joint greatly expanded, hind lobe nearly double the length of the front margin. Peraeopod 5, 2^d and 4th joints very much smaller than in peraeopods 3 and 4. Pleopod 1, peduncle not very long, not expanded. Pleopod 2, peduncle short, produced on inner side to a long, powerful process, carrying 4 coupling spines. Pleopod 3 like pleopod 2.

but the process a little less massive. Uropod 1, peduncle slender, more than twice as long as the slender, subequal, finely ciliated rami. Uropod 2 shorter but much stouter than uropod 1, peduncle about twice as long as the stumpy rami, fringed near outer margin with about 11 short spines. Uropod 3 broad above, the pointed apex projecting just beyond the telson, the pair together not so broad as the telson. Telson much wider than long, obtuse-angled, a few slight setules at the sides. L. about 5 mm.

Lyttelton Harbour and Otago Harbour [New Zealand].

6. Gen. **Bircenna** Chilton

1884 *Bircenna* (Sp. un.: *B. fulvus*), Chilton in: Tr. N. Zealand Inst., *v.* 16 p. 264 | 1893 *B.*, A. Della Valle in: F. Fl. Neapel, *v.* 20 p. 561 | 1899 *B.*, T. Stebbing in: Tr. . Linn. Soc. London, ser. 2 *v.* 7 p. 421.

Body broad. Pleon segment 5 very short, 6th indistinct. Head broad, depressed in front. Side-plates all very shallow. Pleon segments 1—3, postero-lateral corners rounded. Antennae 1 and 2 short, antenna 1 rather the longer. Mandible without palp. Maxilla 1, inner plate with apical setae, outer with 8 apical spines, palp wanting, but margin of maxilla interrupted at its point of disappearance. Maxillipeds, inner plates distally truncate, with 3 spine-teeth, outer slightly armed, both reaching apex of 2[d] joint of palp, palp short, 1st joint largest, 3 following successively both shorter and narrower, 4th very small. Gnathopods 1 and 2 almost simple. Peraeopods 1 and 2, 4th joint a little widened. Peraeopods 3—5, 2[d] joint expanded, larger in 4th than in 3[d], and much larger in 5th peraeopod; 4th joint rather expanded and decurrent. Pleopods 1—3 biramous, with peduncle broadly produced on inner side. Uropod 1, peduncle shorter than acute, curved, very unequal rami; uropod 2 similar, but rather stouter and shorter; uropod 3, a single joint bifid, with outer apex rounded, inner (perhaps representing ramus) acute, setiferous. Telson short, entire.

1 species.

1. **B. fulva** Chilton 1884 *B. fulvus*, *B. fulva*, Chilton in: Tr. N. Zealand Inst., *v.* 16 p. 264, 265; t. 21 f. 1, 1 a—e | 1893 *B. fulva*, A. Della Valle in: F. Fl. Neapel, *v.* 20 p. 562 t. 58 f. 73, 73* | 1899 *B. f.*, T. Stebbing in: Tr. Linn. Soc. London, ser. 2 *v.* 7 p. 421.

Peraeon much larger than pleon. Head very prominent, protruding a little below the antennae. Eyes minute. Antenna 1, 1st joint as long as 2[d], flagellum as long as peduncle, 6 joints with sensory filaments. Antenna 2 with only 3 free joints of peduncle, ultimate as long as the 2 preceding combined, flagellum about as long as free part of peduncle, 4-jointed. Gnathopods 1 and 2, 3[d] joint longer than broad, as long as 4th, 5th a little shorter than 6th, which is not expanded, but a little produced at apex, yet not enough to make a chela with the short apically toothed finger. Peraeopods 1 and 2 very like the gnathopods 1 and 2, except as regards the 4th joint. Peraeopod 5 much the largest. Telson nearly an equilateral triangle, with a setule on either side of subacute apex. Colour yellow. L. 3 mm.

Lyttelton Harbour [New Zealand].

13. Fam. **Colomastigidae**

1893 Subord. *Subiperini*, A. Della Valle in: F. Fl. Neapel, *v.* 20 p 853 | 1899
Colomastidae, Chevreux in: C.-R. Ass. Pranç., Sess. 27 *v.* 2 p. 483 | 1899 *Colomastigidae.*
T. Stebbing in: Ann. nat. Hist., ser. 7 *v.* 4 p 211.

Body cylindric, subdepressed. Head, rostrum small, acute. Side-plates all shallow, none bilobed. Eyes distinct. Antennae 1 and 2, peduncle well developed, flagellum minute, antenna 1 (always?) the longer, without accessory flagellum. Epistome conical. Upper lip bilobed. Lower lip probably without inner lobes. Mandible, cutting edge divided into spine-like teeth, no accessory plate, spine-row, or palp; molar small. Maxilla 1, inner plate wanting, outer feebly armed, palp 1-jointed. Maxilla 2, inner plate the broader, only slightly separated from outer. Maxillipeds, inner plates completely coalesced, outer broad, reaching to middle of palp's 2^d joint, palp 4-jointed, 3^d joint longest. Gnathopods 1 and 2 simple in ♀, in ♂ gnathopod 2 subchelate. Peraeopods 1—5 subequal, 2^d joint but little expanded. Pleopods, rami few-jointed. Uropod 3, rami lanceolate. Telson entire.

Marine.

1 genus, 3 species.

1. Gen. **Colomastix** Grube

1861 *Colomastix* (Sp. un.: *C. pusilla*), E. Grube, Ausfl. Triest, p. 137 | 1888 *C.*,
T. Stebbing in: Rep. Voy. Challenger, *v.* 24 p. 329 | 1893 *C., J.* Bonnier in: Bull. sci.
France Belgique, *v.* 29 p. 202 | 1893 *C.*, A. Della Valle in: F. Fl. Neapel, *v.* 20 p. 854
1862 *Cratippus* (Sp. un.: *C. tenuipes*), Bate, Cat. Amphip. Brit. Mus., p. 275 | 1869
Exunguia (Sp. un.: *E. stilipes*), (A. M. Norman in:) G. S. Brady & D. Robertson in:
Ann. nat. Hist., ser. 4 *v.* 3 p. 359.

With the characters of the family.

3 species.

Synopsis of species:

1 { Uropod 3, rami very unequal . . . . 1. **C. brazieri** . p. 206
  { Uropod 3, rami subequal — **2.**

2 { Gnathopod 1, 3^d joint rather elongate . . . . . . 2. **C. pusilla** . . p 207
  { Gnathopod 1, 3^d joint not elongate . . . . 3. **C. hamifera** . p. 207

1. C. brazieri Hasw. 1880 *C. b.*, Haswell in: P. Linn. Soc. N. S. Wales, *v.* 4
p. 341 t. 22 f. 4 | 1893 *C. pusilla* (part.), A. Della Valle in: F. Fl. Neapel, *v.* 20 p. 854.

Eyes round, rather prominent. Antennae 1 and 2 subequal, spinose, peduncle slightly compressed; antenna 1, 1st joint stouter and rather longer than 2^d, not as long as 2^d and 3^d combined, flagellum shorter than 3^d joint of peduncle, 4- (Haswell) or (?)3-jointed. Antenna 2, 3^d joint stout, as long as ultimate joint of peduncle, penultimate rather longer, flagellum with 1 distinct joint, perhaps followed by 1 or 2 minute terminal joints. Epistome very acute, maxillipeds and probably other mouth-parts as in C. pusilla. Gnathopod 1, 2^d joint membranous, slightly widened, 3^d rather long, 4th a little longer, 5th and 6th linear, 5th much longer than 6th, the latter tipped with fascicle of spines, one of which may be the finger. Gnathopod 2 more robust, 3^d and 4th joints not elongate, 5th cup-like, rather broad, much shorter than 6th, very spinose. 6th in ♀ elongate oval, apically narrowed, hind margin spinose, palm short,

ill-defined, overlapped by small slender finger, in ♂ broadly oval, narrowed
at the oblique obscurely bidentate palm, finger rather stout, not overlapping
palm. Peraeopods 1—5 moderately robust, 4th joint a little widened.
Branchial vesicles narrow, marsupial plates very long and broad. Uropods 1
and 2, peduncle subequal to rami, rami nearly equal, longer in uropod 1 than in
uropod 2; uropod 3, peduncle shorter than telson, inner ramus long narrowly
lanceolate, reaching much beyond those of uropods 1 and 2, outer very short,
spine-like. Telson conical, compressed, almost acute at apex. Colour light
green. L. about 10 mm.

Port Jackson [East-Australia]. Depth 4—18 m.

2. **C. pusilla** Grube 1861 *C. p.*, E. Grube. Ausfl. Triest, p. 137 | 1864 *C. p.*,
E. Grube in: Arch. Naturg., *v.* 30 i p. 206 t. 5 f. 2, 2 a—b | 1888 *C. p.*, T. Stebbing in:
Rep. Voy. Challenger, *v.* 29 p. 329 | 1893 *C. p.*, J. Bonnier in: Bull. sci. France Belgique,
v. 24 p. 203 t. 8 | 1893 *C. p.* (part.), A. Della Valle in: F. Fl. Neapel, *v.* 20 p. 854 t. 6
f. 2; t. 61 f. 23—37 | 1862 *Cratippus tenuipes*, Bate, Cat. Amphip. Brit. Mus., p 276
t. 46 f. 10 | 1876 *C. t.*, T. Stebbing in: Ann. nat. Hist., ser. 4 *v.* 18 p. 447 t. 20 f. 4. 4a— c !
1866 *C. crassimanus* + *C. pusillus*, Cam. Heller in: Denk. Ak. Wien, *v.* 26 ii p. 50;
t. 4 f. 12, 13 | 1869 *Exunguia stilipes*, (A. M. Norman in:) G. S. Brady & D. Robertson
in: Ann. nat. Hist., ser. 4 *v.* 3 p. 359 t. 22 f. 7—12.

Head, lateral corners rounded. Pleon segments 1—3, lateral angles rounded.
Eyes small, round, red, the lenses outlined with white. Antenna 1 longer and
much stouter than antenna 2, 1st joint not as long as 2[d] and 3[d] combined,
flagellum of 1 short and 2 minute joints. Antenna 2, penultimate joint rather
longer than the stout antepenultimate or narrower ultimate, all serrulate, flagellum
as in antenna 1. Epistome subacute. Lower lip apically broad, prominent
at centre. Maxilla 1, outer plate small on large base, armed with 2 or
3 spinules, the short 1-jointed palp carrying 4 slender spines. Maxillipeds,
inner plates forming one piece, simply conical or blunt, outer plates feebly armed,
reaching middle of 2[d] joint of palp, of which 3[d] is rather longer than 2[d], finger
obtuse, setulose. Gnathopod 1, 2[d] joint widened distally, 4th rather longer than 3[d],
5th and 6th subequal, finger spine-like, distinguished from the group of
spines about it by the attached tendon; gnathopod 1 atrophied in ♂ (Della
Valle). Gnathopod 2, 3[d] and 4th joints short, 5th and 6th in ♀ subequal in
length, spinose, 6th rather tapering, palm ill-defined, finger moderately
strong, in ♂ 5th short, cup-like, 6th greatly enlarged, palm divided into
3 large blunt processes, finger robust (Heller's *C. crassimanus*). Peraeopods 1—5 subequal, not very robust, 2[d] joint little dilated. Branchial
vesicles narrow. Uropods 1—3, rami broad, lanceolate, slightly serrulate,
3[d] shortest, in all rami subequal. Telson suboval, entire, longer than broad
(Della Valle), or distally widened with apical border slightly concave, broader
than long (Bonnier). Colour ivory white with pale dorsal markings, or
greyish, dorsally tinted with ochre. L. 4—5 mm.

Mediterranean, North-Atlantic (France, Great Britain).

3. **C. hamifera** Kossm. 1880 *C. hamifer*, Kossmann, Reise Roth. Meer., *v.* 21
Malacost. p. 136 t. 15 f. 1—10 | 1893 *C. pusilla* (part.), A. Della Valle in: F. Fl. Neapel,
v. 20 p. 854.

Like C. pusilla (p. 207) (of which it is perhaps a young ♂) except in
the following respects. Antenna 1 rather more slender than the equally long
antenna 2. Gnathopod 1, 3[d] joint short. Gnathopod 2, 5th joint as broad as

long, shorter than the large 6th, which is oblong oval, with obscurely defined
undivided palm. Uropod 1, one ramus slender, the other a little longer
and very much broader, except at the rather abruptly acute apex. Telson
(in fig.) seemingly oval. L. 2·5 mm.

Red Sea.

---

14. Fam. Lafystiidae

1893 *Laphystiidae*, G. O. Sars, Crust. NorWay, v. 1 p. 382.

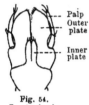

Fig. 54.
L. sturionis.
Maxillipeds.

Body depressed. Head, rostrum broad, tapering, not acute. Side-
plate 4 the deepest. Eyes distinct. Antenna 1 stouter and longer than
antenna 2, no accessory flagellum. Upper lip entire,
distally narrowed. Lower lip without inner lobes. Mandible,
cutting edge dentate, accessory plates on left and right, no
spine-row or molar, palp large, 3[d] joint longest. Maxilla 1,
inner plate with 3 setae, outer with 8 spines, palp rudimentary, nodiform. Maxilla 2, outer plate longer than
inner, inner setiferous on inner margin. Maxillipeds
(Fig. 54), inner plates narrow, not short, outer large,
overlapping the 2-jointed palp. Gnathopod 1 simple,
gnathopod 2 weakly subchelate. Peraeopods 1—5 stout,
subequal, finger uncinate. Uropod 3, rami obtuse, inner
the longer. Telson small, entire.

Marine.

1 genus, 1 species.

---

1. Gen. Lafystius Krøyer

1842 *Lafystius* (Sp. un.: *L. sturionis*), Krøyer in: Naturh. Tidsskr., v. 4 p. 156 |
1888 *L.*, T. Stebbing in: Rep. Voy. Challenger, v. 29 p. 898 | 1893 *L.*, A. Della Valle
in: F. Fl. Neapel, v. 20 p. 587 | 1846 *Laphystius* (nom. em.), L. Agassiz, Nomencl. zool.,
Index p. 200, 202 | 1876 *L.*, *Lafystius*, A. Boeck, Skand. Arkt. Amphip., v. 2 p. 250. 712 |
1893 *Laphystius*, G. O. Sars, Crust. NorWay, v. 1 p. 383 | 1856 *Darwinea* (nom. nud.),
Bate in: Rep. Brit. Ass., Meet. 25 p. 58 | 1857 *Darwinia* (Sp. un.: *D. compressa*), Bate
in: Ann. nat. Hist., ser. 2 v. 19 p. 141 | ? 1870 *Dermophilus* (Sp. un.: *D. lophii*) (non
Dermatophilus Guérin-Méueville, 1829—38, Diptera), E. Beneden & Bessels in: Mém.
cour. Ac. Belgique, v. 34 nr. 4 p. 26 | ? 1873 *Ichthyomyzocus* (part.), E. Hesse in: Ann.
Sci. nat., ser. 5 v. 17 nr. 7 p. 5 | ? 1877 *Desmophilus*, Huxley, Man. Anat. Invert., p. 367.

With the characters of the family.

1 species.

---

1. L. sturionis Krøyer 1842 *L. s.*, Krøyer in: Naturh. Tidsskr., v. 4 p. 157 |
1875 *L. s.*, Schiødte in: Naturh. Tidsskr., ser. 3 v. 10 p. 237 t. 5 f. 9 ···18 | 1876 *L. s.*,
A. Boeck, Skand. Arkt. Amphip., v. 2 p. 252 t. 19 f. 6 | 1888 *L. s.*, T. Stebbing in: Rep.
Voy. Challenger, v. 29 p. 899 t. 187D | 1893 *L. s.*, A. Della Valle in: F. Fl. Neapel, v. 20
p. 588 t. 6 f. 8; t. 32 f. 20—37 | 1893 *Laphystius s.*, G. O. Sars, Crust. NorWay, v. 1
p. 384 t. 134 | 1856 *Darwinea compressus* (nom. nud.), Bate in: Rep. Brit. Ass., Meet. 25
p. 58 | 1857 *Darwinia compressa*, Bate in: Ann. nat. Hist., ser. 2 v. 19 p. 141 | ? 1870
Dermophilus lophii, E. Beneden & Bessels in: Mém. cour. Ac. Belgique, v. 34 nr. 4 p. 26 |
? 1873 *Ichthyomyzocus morrhuae*, E. Hesse in: Ann. Sci. nat., ser. 5 v. 17 nr. 7 p. 7
t. 4 f. 3—7.

Back broad, smooth. Head depressed, rostrum tapering to truncate apex, reaching end of 1st joint of antenna 1. Side-plates 1—3 rounded quadrangular, 4th much deeper, acutely triangular below, 5th and 6th with hind lobe acutely produced downward. Pleon segment 3, postero-lateral corners obtusely quadrate, segments 4—6 flattened. Eyes round, prominent, lateral, dark. Antenna 1, 1st joint longer than 2d or 3d, all stout, flagellum as long as peduncle, tapering, 6 joints with sensory fascicles. Antenna 2 shorter, much more slender, ultimate joint of peduncle longer than penultimate, flagellum rather longer than peduncle, 8-jointed. Gnathopod 1 feeble, almost unarmed, 5th joint shorter than narrowly oblong 6th, finger long, .weak, nearly membranous. Gnathopod 2 stouter, 5th joint subequal to nearly quadrangular 6th, the small palm slightly produced, much overlapped by the rather stout and curved, bidentate finger. Peraeopods 1—5, 4th joint rather wide, not long, finger strong, hooked. Peraeopods 1 and 2, 6th joint stout. Peraeopods 3—5, 2d joint oval quadrangular, in peraeopod 5 angled at lower hind corner. Uropod 1, rami subequal, uropods 2 and 3, outer ramus shorter than inner; uropod 3, rami unarmed, ending obtusely. Telson rounded oval. Colour white, more or less pellucid. L. 7 mm.

North-Atlantic and Mediterranean (from the Trondhjemsfjord to the Gulf of Salerno). On various fishes, Raja batis L., Lophius piscator L., Gadus morhua L., Acipenser sturio L., Galeus galeus (L.).

15. Fam. Laphystiopsidae

1899 Laphystiopsidae, T. Stebbing in: Ann. nat. Hist., ser. 7 v. 4 p. 211.

Body depressed. Head, rostrum broad, not tapering. Side-plates all shallow, 5th the deepest. Eyes rudimentary. Antenna 1 slender, longer than antenna 2, no accessory flagellum. Upper lip apically broad, bilobed. Lower lip with small inner lobes. Mandible, cutting edge denticulate, accessory plate on left and right, no spine-row, molar weak, conical, palp slender, 3d joint longest. Maxilla 1, inner plate with 1 seta, outer with few spines, palp large, 1-jointed. Maxilla 2, inner and outer plates small, subequal, only setose at apex. Maxillipeds, inner plates short, outer small, not overlapping 1st joint of 4-jointed palp. Gnathopods 1 and 2 small, simple. Peraeopods 1—5 not robust, 2d joint in all somewhat expanded, most in peraeopod 5, which is notably the longest. Uropod 3, peduncle very short, rami lanceolate. Telson entire.

Marine.

1 genus, 1 species.

1. Gen. Laphystiopsis O. Sars

1893 Laphystiopsis (Sp. un.: L. planifrons), G. O. Sars, Crust. Norway, v. 1 p. 386.

With the characters of the family.

1 species.

1. L. planifrons O. Sars 1893 L. p., G. O. Sars, Crust. Norway. v. 1 p. 386 t. 135.

Body rather slender, depressed anteriorly, segments imbricated. Pleon segments 3 and 4 with compressed gibbous dorsal projection, preceded in segment 4 by a deep depression. Head depressed, rostrum horizontal, lamellar, broadly truncate, at its base forming lateral ridges on head. Side-plates 1—4 small, not contiguous, simple, 5th and 6th bilobed, 5th larger than any. Pleon segment 3 less deep than 2^d, narrower below, postero-lateral corners rounded. Eyes represented by 2 irregular patches of reddish pigment, coated with chalky white. Antenna 1, 1st joint distally widened, rather longer than 2^d, 3^d short, flagellum more than thrice as long as peduncle, with 24 joints, 1st longest, with 4 sensory fascicles. Antenna 2 much shorter, ultimate and penultimate joints of peduncle subequal, flagellum twice as long as peduncle. Gnathopods 1 and 2, 5th joint about as long as and rather wider than simply cylindric slightly curved 6th, finger small, strong, hooked. Peraeopods 1 and 2 with 2^d joint broad, and 4th a little widened, otherwise very like the much smaller gnathopods 1 and 2. Peraeopods 3—5 successively longer. Uropods 1 and 2, outer ramus much the shorter, uropod 3, rami spinulose, acute, outer little shorter than inner. Telson nearly semicircular. Colour whitish. L. ♀ 8 mm.

Christianiafjord, Trondhjemsfjord, Arctic Ocean (Selsövig [Nordland]). Depth 188–752 m.

---

16. Fam. Acanthonotozomatidae

1871 Subfam. *Iphimedinae*, A. Boeck in: Forh. Selsk. Christian., 1870 p. 178 | 1888 *Iphimedidae*, T. Stebbing in: Rep. Voy. Challenger, v. 29 p. 882 | 1893 *I.*, G. O. Sars. Crust. Norway, v. 1 p. 372 | 1893 *Dexaminidi*, A. Della Valle in: F. Fl. Neapel, v. 20 p. 556.

Integument more or less indurated, processiferous. Head rostrate. Side-plates well developed, 1st—4th usually acuminate. Eyes well developed. Antennae 1 and 2 seldom elongate or very different in length, accessory flagellum absent or rudimentary. Mouth-parts projecting downward, and drawn out as if for piercing rather than biting. Gnathopod 1 very slender and feeble, simple or chelate; gnathopod 2 seldom strong. Peraeopods 3—5, 2^d joint expanded, usually acute· at one or more points of hind margin. Uropod 3, rami lanceolate. Telson unarmed, apically emarginate.

Marine.

4 genera, 11 accepted species and 3 doubtful.

Synopsis of genera:

1 { Maxilla 1, palp 1-jointed . . . . 1. Gen. **Odius** . . . . p. 210
{ Maxilla 1, palp 2-jointed — **2**.

2 { Maxilla 1, palp not reaching apex of outer
{   plate . . . . 2. Gen. **Panoploea** . . . . . p. 211
{ Maxilla 1, palp reaching beyond apex of
{   outer plate — **3**.

3 { Gnathopod 1 minutely chelate . . . . . 3. Gen. **Iphimedia** . . . . . p. 214
{ Gnathopod 1 simple, wholly unchelate . 4. Gen. **Acanthonotozoma** . p. 218

1. Gen. Odius Lillj.

1862 *Otus* (Sp. un.: *O. carinatus*) (non Jac. Hübner 1816. Lepidoptera!), Bate, Cat. Amphip. Brit. Mus., p. 125 | 1865 *Odius* (nom. nov.), W. Lilljeborg in: N. Acta Soc. Upsal., ser. 3 v. 6 nr. 1 p. 18 (tabella), 19 | 1893 *O.*, G. O. Sars, Crust. Norway, v. 1 p. 380 | 1893 *O.*, A. Della Valle in: F. Fl. Neapel, v. 20 p. 581.

Integument indurated, body compressed, carinate. Head, lateral corners well developed. Side-plates 1—4 rather large. Antennae 1 and 2 very short. Upper lip narrow, tapering, minutely bifid. Lower lip without inner lobes, outer acutely produced. Mandible produced to acute cutting edge, spine-row long,· spines short, molar distinct, small, palp slender, almost unarmed. Maxilla 1, inner plate small, tapering, outer elongate, tapering to acute point, palp a minute conical joint. Maxilla 2, both plates long, narrow, obliquely truncate, inner the shorter. Maxillipeds, inner plates long, narrow, outer reaching much beyond 2^d joint of palp, palp not long, finger minute. Gnathopod 1 feeble, slender, minutely chelate. Gnathopod 2 rather robust, subchelate. Peraeopods 1—5 stout, subequal, peraeopods 3—5, 2^d joint expanded. Uropod 3, rami very unequal. Telson oblong, minutely incised.

1 species.

1. O. carinatus (Bate) 1862 *Otus c.*, Bate, Cat. Amphip. Brit. Mus., p. 126 t. 23 f. 2 | 1871 *Odius c.*, A. Boeck in: Forh. Selsk. Christian., 1870 p. 182 | 1893 *O. c.*, G. O. Sars, Crust. Norway, *v.* 1 p. 381 t. 133 f. 2 | 1893 *O. c.*, A. Della Valle in: F. Fl. Neapel, *v.* 20 p. 582 t. 58 f. 86, 87.

Short and stout but highly compressed, carinate from peraeon segment 1 to pleon segment 3, on pleon segment 2 with obtusely rounded projection, on segment 3 with erect acute process. Head. rostrum acute, vertical, well overlapping 1st joint of antenna 1, lateral corners forming a large triangular lobe. Side-plate 1 tapering to obtuse point, 2^d and 3^d slightly narrowed below, 4th broad below, hind margin oblique below emargination, 5th and 6th with hind lobe truncate. Pleon segment 3, postero-lateral angles acutely but slightly produced, a sharp curved tooth higher up below a deep sinus. Eyes reniform, red. Antenna 1, 1st joint subequal to 2^d and 3^d combined, flagellum much shorter than peduncle, with 7 joints, last 3 very small, with long sensory filaments. Antenna 2 shorter, flagellum scarcely longer than ultimate joint of peduncle. Gnathopod 1, 2^d joint widened proximally, rest of limb very slender, 5th joint narrow, much longer than 6th, which is curved and forms a very small chela. Gnathopod 2, 4th joint produced into a narrow lobe, 5th short, cuplike, but narrowly produced, 6th broad, widening distally, palm nearly transverse, finely denticulate. Peraeopods 1—5, 4th joint strongly produced over 5th, finger strong, curved, peraeopods 3—5, 2^d joint quadrate, lower hind angle broadly emarginate or slightly bilobed. Uropod 3, outer ramus much narrower than inner, and little more than $^1/_2$ as long. Telson more than twice as long as broad, tapering to minutely incised apex. Colour whitish or variegated with dark brown shadings. L. ♀ about 5 mm.

Arctic Ocean, North-Atlantic, North-Sea and Skagerrak (Greenland, Spitzbergen. Arctic America, Norway, Shetland Isles). Moderate depths to 113 m; on rocky bottom, among Algae and Hydroids.

2. Gen. **Panoploea** G. M. Thoms.

1880 *Panoploea* (part.), G. M. Thomson in: Ann. nat. Hist., ser. 5 *v.* 6 p. 2 | 1881 *Panoplaea*, G. M. Thomson in: Tr. N. Zealand Inst., *v.* 13 p. 212 | 1893 *Iphimediopsis* (Sp. un.: *I. eblanae*), A. Della Valle in: F. Fl. Neapel, *v.* 20 p. 585.

Back broadly rounded, some segments produced into teeth. Rostrum acute. Side-plates 1—3 more or less acutely tapering, 4th with projecting point of hind margin. Upper lip somewhat narrowed distally. Lower lip without inner lobes, outer incised on inner margin near apex. Mandible narrowly tapering to cutting edge, accessory plate narrow, no spine row, molar exceedingly

14*

feeble, 3^d joint of palp not very long. Maxilla 1, inner plate with several setae, outer with 10(?) spines, palp 2-jointed, not reaching extremity of outer plate. Maxilla 2, outer plate the longer, rather the narrower, obliquely truncate, inner still more so and fringed for half its length. Maxillipeds, inner and outer plates long and narrow, outer fringed on distal part of . outer margin, 1st joint of palp not as long as 2^d and 3^d combined, 2^d much produced along inner margin of 3^d, finger wanting. Gnathopods 1 and 2 very slender, gnathopod 1 with very small chela, gnathopod 2 more or less chelate. Peraeopods 3—5, 2^d joint well expanded. Uropod 3, rami narrowly lanceolate. Telson broadly incised at apex.

3 species accepted, 1 doubtful.

Synopsis of accepted species:

1 { Peraeopods 4 and 5, 2^d joint with lower hind corner rounded 1. **P. spinosa** . p. 212
{ Peraeopods 4 and 5, 2^d joint with lower hind corner acute — **2.**

2 { Peraeopod 5, 2^d joint with an upper tooth on hind margin 2. **P. eblanae** . p. 212
{ Peraeopod 5, 2^d joint without upper tooth on hind margin 3. **P. minuta** . p. 213

1. P. spinosa G. M. Thoms. 1880 *P. s.*, G. M. Thomson in: Ann. nat. Hist., ser. 5 *v.* 6 p. 3 t. 1 f. 2 | 1881 *Panoplaea s.*, G. M. Thomson in: Tr. N. Zealand Inst., *t.* 13 p. 213 | 1888 *Iphimedia s.*, T. Stebbing in: Rep. Voy. Challenger, *v.* 29 p. 524 | 1893 *I. s.*, A. Della Valle in: F. Fl. Neapel, *v.* 20 p. 585.

Peraeon broad, smooth, 7th segment and pleon segments 1 and 2 dorsally produced into 2 teeth. Rostrum nearly reaching apex of 1st joint of antenna 1. Side-plates 5 and 6, hind lobes rounded. Pleon segments 2 and 3, postero-lateral corners acute, scarcely produced. Eyes long, narrow, curved, pale reddish. Antenna 1, 1st joint longer than 2^d and 3^d combined, with an apical tooth, 2^d with 2 apical teeth, flagellum with 22 rather long joints, many with sensory filaments, 1st joint longest. Antenna 2 shorter, ultimate joint of peduncle rather shorter than penultimate, flagellum with 44 joints, all short except 1st. Upper lip slightly narrowing to insinuate apex. Mandible, apical tooth obscurely subdivided, to the rear a 2^d tooth probably representing the accessory plate, but apparently not free, molar a thin expansion slightly fringed with minute spinules, 3^d joint of palp short, scarcely longer than 1st; its blunt apex obliquely truncate, slightly fringed, 2^d joint little longer than 3^d. Maxilla 1, inner plate with 9 setae on oblique distal margin, outer with 10 spines, palp not rudimentary, but not reaching end of outer plate. Triturating lobes of stomach very small, with minute spine-teeth. Gnathopod 1, 5th joint tapering, much longer than 6th, otherwise as in Iphimedia (p. 214), with small finger opposed to very slender thumb. Gnathopod 2 rather stouter, 5th joint shorter than 6th, which widens a little distally and forms a little distinct chela with the small finger. Peraeopods 3—5, 2^d joint strongly expanded, rounded oval, hind margin strongly serrate. Uropod 2, rami very unequal, uropod 3, peduncle with acute points, rami only a little unequal. Telson longer than broad, slightly tapering, apex widely not deeply emarginate. Marsupial plates large and long. Colour varying from light to dark brown, thickly covered with black stellate markings. L. 11 mm.

Dunedin Harbour [New Zealand]. Depth 7—9 m.

2. P. eblanae (Bate) 1857 *Iphimedia e.*, Bate in: Nat. Hist. Rev., *v.* 4 P. Soc. p. 229 t. 16 f. 1—7 | 1893 *Iphimediopsis e.*, A. Della Valle in: F. Fl. Neapel, *v.* 20 p. 586 t. 6 f. 5 (inaccurate); t. 32 f. 1—19; t. 58 f. 93 | 1864 *Iphimedia multispinis*, E. Grube in: Arch. Naturg., *v.* 30ı p. 202 t. 5 f. 1 | 1864 *I. m.*, E. Grube in: Jahresber. Schles. Ges., *v.* 41 p. 58 | 1866 *I. eblanae* + *I. carinata.* Cam. Heller in: Denk. Ak. Wien, *v.* 26ıı p. 28, 29 ! ?1875 *I. corallina*, Catta in: Rev. Sci. nat., *v.* 4 p. 164.

Body stout. Peraeon segment 7, pleon segments 1—3 dorsally produced into a pair of strong teeth, pleon segments 1—3 also armed with a medio-dorsal carina-tooth. Rostrum flat above, narrow, reaching end of 1st joint of antenna 1. Side-plates 1—3 pointed below, 4th with 2 emarginations, 5th—7th with a backward directed point. Pleon segments 2 and 3, postero-lateral corners acute, segments 1—3 with a medio-lateral tooth, in segment 3 all the teeth rather upturned. Eyes subreniform, bright red. Antenna 1, 1st joint longer than 2^d and 3^d combined, produced into 2 strong unequal apical teeth, 2^d produced into 1 apical tooth, flagellum longer than peduncle, 16-jointed. Antenna 2 rather longer, ultimate and penultimate joints of peduncle subequal, flagellum 20-jointed. Upper lip figured by Della Valle with truncate apex. Mandible, accessory plate doubtfully free, 2^d joint of palp much the longest (as figured). Maxilla 1, palp very small, joints equal, inner plate with only 4 setae (Della Valle). Gnathopod 1 very slender, 5th joint longer than 6th, chela very delicate with 7 setae. Gnathopod 2 stronger, 5th joint subequal to 6th, which widens distally and at apex is produced into a rounded lobe against which the small finger impinges more or less chelately, hind margin fringed with setules. Peraeopods 1—5, 6th joint spinose, finger strong. Peraeopods 3—5, 2^d joint quadrate, produced above into a strong tooth, lower hind angle acute. Uropod 1, rami as long as peduncle, uropod 2, rami rather shorter than peduncle, uropod 3, rami much longer than peduncle. Telson short, boat-shaped, apex emarginate. Colour variable, pinkish-white, with transverse bands of orange spots (Grube), sometimes entirely orange-yellow (Della Valle). L. about 7 mm.

North-Atlantic (Ireland), Mediterranean. Depth 50 m, on coralline bottoms; also from Rhizostoma cuvieri Pér. & Les.

3. **P. minuta** (O. Sars) 1874 *Iphimedia eblanae var.*, T. Stebbing in: Ann. nat. Hist., ser. 4 v. 14 p. 11 t. 2 f. 4 | 1882 *I. minuta*, G. O. Sars in: Forb. Selsk. Christian., nr. 18 p. 100 t. 5 f. 2 | 1893 *I. m.*, G. O. Sars, Crust. Norway, v. 1 p. 379 t. 133 f. 1 | 1893 *I. obesa* (part.), A. Della Valle in: F..Fl. Neapel, v. 20 p. 584.

Peraeon segment 7, pleon segments 1—3 dorsally produced into a pair of strong teeth, with no medio-dorsal tooth. Rostrum strongly curved, about reaching apex of 1st joint of antenna 1. Side-plates 1—4 as in P. eblanae, 5th and 6th subtruncate on hind lobe, 7th rounded. Pleon segments 2 and 3 as in P. eblanae. Eyes large, reniform, dark red. Antenna 1, 1st joint produced to apical tooth not very large or acute, flagellum about twice as long as peduncle. Gnathopod 1, 5th joint subequal to 6th, chela very delicate, with a few setae. Gnathopod 2, 6th joint produced so as to make an almost perfect chela. Peraeopods 3—5, hind margin of 2^d joint rounded above, not produced into a tooth, quadrate below, and in peraeopod 5 sharply produced, 4th joint decurrent. Uropod 3, inner ramus considerably longer than outer. Telson oblong quadrangular, scarcely tapering, broadly and angularly emarginate between acute apical points. Colour variable with bright bands, or dark brown shadows, or nearly black. L. ♀ 5—6 mm.

North-Atlantic, North-Sea and Skagerrak (South- and West-Norway, British Isles); Mediterranean.

P. ambigua (HasW.) 1880 *Iphimedia? a.*, Haswell in: P. Linn. Soc. N. S. Wales, v. 4 p 327 t. 24 f. 2 | 1893 *I. obesa* (part.), A. Della Valle in: F. Fl. Neapel, v. 20 p. 584.

Intermediate between P. eblanae and P. minuta.

L. 2·5—3·75 mm.

Port Jackson [East-Australia].

3. Gen. **Iphimedia** H. Rathke

1843 *Iphimedia* (Sp. un.: *I. obesa*), H. Rathke in: N. Acta Ac. Leop., *v.*20ɪ p.85 ¦
1888 *I.*, T. Stebbing in: Rep. Voy. Challenger, *v.*29 p.889 | 1893 *I.*, G. O. Sars, Crust.
Norway, *v.*1 p.376 | 1893 *I.*, A. Della Valle in: F. Fl. Neapel, *v.*20 p.582 | 1846 *Micro-cheles* (Sp. un.: *M. armata*). Krøyer in: Naturh. Tidsskr., ser.2 *v.*2 p.58, 66.

Back broadly rounded, some segments produced into teeth. Rostrum
acute. Side-plates 1--3, at apex simply acute or bidentate, 4ᵗʰ with projecting
tooth between 2 emarginations. Eyes well developed. Antennae 1 and 2 not
greatly differing in length. Upper lip little or not emarginate. Lower lip
having outer lobes incised on inner margin, thus forming an inner process,
which may or may not represent the inner lobe. Mandible rather broadly
tapering to blunt obscurely dentate cutting apex, accessory plate rather long,
spine-row wanting, molar feeble (or sometimes wanting?), palp rather strong.
Maxilla 1, inner plate with several setae, outer with (always?) 11 spines,
palp 2-jointed, reaching beyond apex of outer plate. Maxilla 2 and maxillipeds
as in Panoploea (p. 211). Gnathopods 1 and 2 very slender, delicately chelate,
3ᵈ joint not very short; 2ᵈ joint in gnathopod 1 sinuous. Peraeopods 3—5,
2ᵈ joint well expanded. Uropod 3, rami narrowly lanceolate. Telson broadly
incised at apex.

4 species accepted, 2 doubtful.

Synopsis of accepted species:

1 {	Side-plates 5—7 not acutely produced backward	1. **I. obesa** p.214
	Side-plates 5—7 acutely produced backward — **2.**	
2 {	Side-plates 5—7 forming each 2 acute processes	2. **I. pulchridentata** . p.215
	Side-plates 5—7 forming each 1 acute process — **3.**	
[3 {	Of peraeon segments only 7ᵗʰ with a pair of dorsal teeth.	3. **I. pacifica** p.215
	Of peraeon segments 4ᵗʰ—7ᵗʰ each with a pair of dorsal teeth	4. **I. nodosa** p.216

1. **I. obesa** H. Rathke 1843 *I. o.*, H. Rathke in: N. Acta Ac. Leop., *v.*20ɪ p.85
t.3 f.1A—Q | 1876 *I. o.*, A. Boeck, Skand. Arkt. Amphip., *v.*2 p.245 t.18 f.11d—k
1889 *I. o.*, Hoek in: Tijdschr. Nederl. dierk. Ver., ser.2 *v.*2 p.194 t.7 f.6 | 1893 *I. o.*,
G. O. Sars, Crust. Norway, *v.*1 p.377 t.132 | 1893 *I. o.* (part.), A. Della Valle in: F. Fl.
Neapel, *v.*20 p.584 t.58 f.92 | 1846 *Microcheles armata*, Krøyer in: Naturh. Tidsskr.,
ser.2 *v.*2 p.58 | 1846 *M. a.*, Krøyer in: Voy. Nord, Crust. t.11ʙ f.2a—v.

Peraeon segment 6, pleon segments 1—3 dorsally produced into
2 teeth. Head, rostrum strongly curved, overlapping 1ˢᵗ joint of antenna 1,
lateral corners deflexed, acute. Side-plates 1—3 tapering to acute apex, 5ᵗʰ
and 6ᵗʰ with hind lobe the deeper, subtruncate at apex. Pleon segments 2 and 3,
postero-lateral angles acute, segment 3 having an upper tooth subequal to the corner
one. Eyes large, reniform, dark red, scarlet, or purplish. Antenna 1, 1ˢᵗ joint longer
than 2ᵈ and 3ᵈ combined, forming a large apical tooth. 2ᵈ with small tooth.
flagellum nearly thrice as long as peduncle, with about 24 joints, each with fascicle
of sensory setae, more strongly developed in ♂. Antenna 2 scarcely longer, ultimate joint of peduncle longer than penultimate, flagellum about twice as long as
peduncle. Upper lip distally narrowed, slightly emarginate. Mandible, accessory
plate distinct on both left and right, molar almost obsolete, palp sturdy, 3ᵈ joint
curved, much longer than 1ˢᵗ. Maxillipeds, 1ˢᵗ joint of palp as long as 2ᵈ and 3ᵈ
combined, finger wanting (Sars) or rudimentary (Boeck). Gnathopod 1 very
slender, 5ᵗʰ joint rather shorter than 6ᵗʰ, chela about $1/3$ length of 6ᵗʰ joint,
thumb not stouter than apically denticulate finger. Gnathopod 2 rather stouter,

5th and 6th joints subequal, 6th setose, very slightly expanding to the chela, which is less than $^1/_4$ of the whole length, thumb much stouter than finger. Peraeopods 3—5, 2^d joint oval quadrangular, the lower hind angle a little produced only in peraeopod 5, 4th joint decurrent. Uropod 3, rami long, nearly equal. Telson oblong, little tapering, angularly incised between 2 acute apices. Colour variable, whitish, striped with rosy pink, or lemon-coloured, or purplish grey. L. ♀ reaching 12 mm, ♂ less.

Arctic Ocean, North-Atlantic, North-Sea, Skagerrak and Kattegat (Scandinavia, Great Britain, France, Holland).

2. **I. pulchridentata** Stebb. 1883 *I. p.*, T. Stebbing in: Ann. nat. Hist., ser. 5 *v.* 11 p. 208 | 1888 *I. p.*, T. Stebbing in: Rep. Voy. Challenger, *v.* 29 p. 894 t. 72 | 1893 *I. p.*, A. Della Valle in: F. Fl. Neapel, *v.* 20 p. 583 t. 58 f. 88.

Peraeon broad rounded, pleon rather compressed. Peraeon segments 6 and 7, pleon segments 1 and 2, each with 2 large acute backward directed processes, pleon segment 3 with 1 such process, on either side of medio-dorsal line, pleon segments 1—4 carinate, the carina forming in each a large tooth, with an accessory tooth in segments 2 and 3. Peraeon segments 1—7 with postero-lateral angles sharply produced, in 6th and 7th forming long downward curving processes. Head, rostrum long, acute, downward curving, nearly reaching apex of 1st joint of antenna 1, lateral angles produced into 2 sharp processes curving one towards the other. Side-plates 1—3 ending below in 2 acute processes, side-plate 2 rather narrower than 1st or 3^d; side-plate 4 narrow, very acute below, the lower emargination much longer than the upper; side-plate 5 bilobed, hind lobe produced backward in 2 large processes, side-plate 6 not bilobed, produced into 2 long processes, side-plate 7 into an upper long and lower short process. Pleon segments 1 and 2, postero-lateral angles acute, little produced, a very large process on the margin above, in pleon segment 3, the angle strongly produced in upcurving tooth with similar one above. Eyes round, rather prominent. Antenna 1, 1st joint with tooth near base and 3 unequal distal teeth, 2^d joint with 1 short and 1 very long distal tooth; possibly a 1-jointed accessory flagellum is present. Antenna 2, joints of peduncle dentate, ultimate and penultimate joints equal, flagellum with about 30 joints or more. Upper lip not narrowed, slightly convex. Lower lip, inner edge only a little emarginate. Mandibles, cutting apex undivided, accessory plate dentate on both mandibles, more strap-like on right, molar not dentate, 3^d joint of palp rather longer than 1st. Maxilla 1, 10 setae on inner plate, 11 spines on outer, 2^d joint of palp reaching much beyond outer plate. Maxillipeds, 3 joints of palp nearly equal, 1st rather the longest. Gnathopods 1 and 2 closely resembling those of I. pacifica. Peraeopods 3—5, 2^d joint smooth in front, behind cut into various unequal large acute processes, largest on peraeopod 5, 4th joint strongly and very acutely decurrent. Uropod 3, peduncle short, with 3 unequal apical teeth, rami rather broad lanceolate, unequal. Telson longer than broad, tapering pretty strongly, emarginate more than a quarter the length, between 2 spine-tipped apices. L. 13 mm.

Southern Indian Ocean (Heard Island). Depth 136 m.

3. **I. pacifica** Stebb. 1883 *I. p.*, T. Stebbing in: Ann. nat. Hist., ser. 5 *v.* 11 p. 207 | 1888 *I. p.*, T. Stebbing in: Rep. Voy. Challenger, *v.* 29 p. 890 t. 71 | 1893 *I. p.*, A. Della Valle in: F. Fl. Neapel, *v.* 20 p. 583 t. 58 f. 89.

Peraeon segment 7, pleon segments 1 and 2 dorsally produced into 2 strong teeth. All peraeon segments with lower margin produced acutely backward,

7th strongly. Pleon segments 1—3 with rudimentary dorsal carina. Head, rostrum long, nearly reaching apex of 1st joint of antenna 1, lateral angles acutely bidentate. Side-plates 1—4 as in I. obesa (p. 214), 5th—7th with a strong backward pointing acute process. Pleon segments 1—3, postero-lateral angles strongly

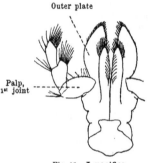

Outer plate

Palp,
1st joint

Fig. 55. I. pacifica.
Maxillipeds.

and acutely produced, each also with an acute upper tooth, that on segment 3 bent upward, serrulate below, segment 6 acutely produced on each side of the telson. Eyes small, oval. Antenna 1, 1st joint longer than 2d and 3d combined, apically produced to a long acute tooth and a shorter bifid one, 2d also produced to a bifid tooth, flagellum longer than peduncle, about 20-jointed. Antenna 2, penultimate joint of peduncle with short apical teeth, rather longer than ultimate, flagellum longer than peduncle, 35-jointed. Upper lip distally narrowed, truncate. Mandible, dentation of cutting apex rather distinct, accessory plate on left mandible strap-shaped, distinct, its presence on right indistinct and doubtful, molar on left feeble, distinct, on right without perceptible denti-culation, palp with 1st joint rather long, yet not as long as 3d. Maxilla 1, inner plate with 7 or 8 setae, outer with 11 spines, 1st joint of palp more than $\frac{1}{2}$ as long as 2d. Maxillipeds (Fig. 55), 1st joint of palp not as long as 2d and 3d combined, finger wanting. Gnathopod 1, 5th joint slightly longer than 6th, both very narrow, chela very small, beset with setae, finger with hooked apex and 2 retroverted teeth on inner margin. Gnathopod 2 rather stouter, 5th and 6th joints subequal, 6th setose, chela minute. Peraeopods 3—5, 2d joint with sinuous strongly serrate hind margin, lower margin also serrate, meeting in sharp tooth. strongest in peraeopod 5, 4th joint very decurrent. Uropod 3, peduncle short, with 3 acute distal processes, one of the rami long, the other at present unknown. Telson concave above, oblong, distal border with small emargination at centre, lateral apices acutely produced. L. about 8 mm.

Southern Indian Ocean (Heard Island, depth 273 m; Kerguelen Island, depth 231 m).

4. I. nodosa Dana 1852 Amphitoë n., I. n., Acanthosoma n., J. D. Dana in: P. Amer. Ac., v.2 p.217 | 1853 & 55 I. n., J. D. Dana in: U. S. expl. Exp.. v.13 ıı p.928; t.63 f.3a, b | 1893 I. n., A. Della Valle in: F. Fl. Neapel, v.20 p.583 t.58 f.89, 90.

Integument crustaceous, peraeon segments 4—7 each with a pair of dorsal teeth to the hind margin, small, tubercular on segment 4, successively larger on following segments, with a subsidiary tooth at each side; there is also a little above the side-plates a lateral carina, forming a backward produced tooth to each of these segments. Pleon segments 1—3 have also the pair of dorsal teeth, very large, upturned at tip on segment 3, accompanied by accessory tooth on segments 1 and 2, in all 3 preceded by a large dorsal central upstanding tooth or process, segment 4 has a little medio-dorsal triangular hump. Head, rostrum produced to end of 1st joint of antenna 1, lateral corners produced not quite so far, with apical notch. Side-plates 1 and 2 narrowly oblong, 3d deeper, similar, 4th axe-like, 5th—7th each with backward pointing tooth. Pleon segments 1—3, each with lateral tooth, rather upward pointing, in addition to the dorsal teeth above mentioned, segment 3 with postero-lateral corners rounded, 2^d and 3^d with small acute tooth. Eyes round, prominent, with reddish tint in spirit. Antenna 1 the shorter, 1st joint much longer and broader than 2^d, .with blunt process and 2 or 3 distal teeth, 2^d with rounded end overlapping the short 3^d, flagellum with over 30 unequal joints. Antenna 2, penultimate joint of peduncle with rounded distal process overlapping the narrower and shorter ultimate, flagellum about 38-jointed. Upper lip rather deeply and almost symmetrically bilobed. Lower lip, principal lobes broad, inner almost obsolete, mandibular processes blunt. Mandible, cutting edge angular with several unequal teeth, accessory plate similar to primary on one mandible, on the other wanting or obsolete, spine-row and molar wanting, 3^d joint of palp rather shorter than 2^d, broad-ended. Maxilla 1, inner plate oval, with 8 or 9 setae along distal margin, outer plate with 9 curved rather slender spines, not much dentate, palp with long setose 2^d joint. Maxilla 2, inner plate with setae on distal part of inner margin. Maxillipeds, inner plates large, prominent, inner and truncate distal margin closely set with setae or slender spines, outer plates closely set with slender seta-like spines round all distal half, the 3 joints of palp subequal in length, 1st and 2^d broad, 3^d long oval. Gnathopod 1, 2^d joint sinuous, narrowest distally, 3^d longer than acute 4th, 5th narrow, subequal to the slightly tapering 6th, of which the little chela-forming thumb is hard to distinguish from a spine; the acute tip of the short finger bends on to its apex; finger thick at base, with a long setule on outer margin. Gnathopod 2 stouter, 2^d joint with sinuous front but straight hind margin, 3^d rather longer than 4th, 5th slightly longer than narrowly oblong 6th, which is distally produced to a blunt end, forming a sort of chela with the closely applied, slender, not very obliquely placed finger. Peraeopods 1 and 2, all the joints rather stout, not expanded, almost bare. Peraeopods 3—5, 2^d joint expanded, hind margin produced into 2 teeth widely apart, the lower one strongly developed only in peraeopod 5, hinder apex of 4th and 5th joints pretty strongly produced. Pleopods, peduncles fringed on outer margin, rami long, 1st joint of inner with several cleft spines. Uropod 1, rami slightly shorter than peduncle, outer ramus slightly shorter than inner. Uropod 2 much shorter than uropod 1, outer ramus much shorter than inner. Uropod 3, rami much longer than peduncle, lanceolate, outer ramus little shorter than inner. Telson oblong oval, apical border very faintly emarginate. L. 8--17 mm.

South-Atlantic (Hermite Island [Tierra del Fuego], Stanley Harbour [Falkland Islands]).

I. normani R. O. Cunningh. 1871 *I. n.*, R. O. Cunningham in: Tr. Linn. Soc. London, *v.* 27 p. 498 t. 59 f. 7 | 1888 *I. n.*, T. Stebbing in: Rep. Voy. Challenger, *v.* 29 p. 405 | 1893 *I. n.*, A. Della Valle in: F. Fl. Neapel, *v.* 20 p. 585.

Head produced into a sharp-pointed rostrum. Pleon segments 1—3 having a sharp-pointed tooth on each lateral margin. Eyes subreniform. Antennae 1 and 2 of nearly equal length. Colour purplish. L. about 8 mm.

„Elizabeth Island".

I. stimpsoni Bate 1855 *I. obesa* (non H. Rathke 1843!). Stimpson in: P. Ac. Philad., *v.* 7 p. 393 | 1862 *I. stimpsoni*, Bate, Cat. Amphip. Brit. Mus., p. 374 | 1888 *I.? s.*, T. Stebbing in: Rep. Voy. Challenger, *v.* 29 p. 288 | 1893 *I.? s.*, A. Della Valle in: F. Fl. Neapel, *v.* 20 p. 585.

Robust, thick. Eyes very large, subreniform, black. Antenna 1 the longer, as long as ²/₃ length of body, basal joints thick. Gnathopods 1 and 2 with equal subcheliform hands of moderate size. Uropods 1—3 slender, smooth, uropod 3 with 2 rami. Telson an elongated scale. Colour crimson with flake-white blotches. L. 6·7 mm.

Port Jackson [East-Australia]. In circumlittoral zone, on weedy and sandy bottoms.

4. Gen. **Acanthonotozoma***) Boeck

1802 *Talitrus* (part.), Latreille, Hist. Crust. Ins., *v.* 3 p. 38 | 1835 *Acanthonotus* (Sp. un.: *A. cristatus*) (non G. Cuvier 1800, Pisces!), (Rich. Owen in MS.) J. C. Ross in: John Ross, App. sec. Voy., nat. Hist. p. 90 | 1876 *Acanthonotozoma* (Sp. typ.: *Acanthonotus cristatus*), *Acanthozoma* (laps.), A. Boeck, Skand. Arkt. Amphip., *v.* 2 p. 237; p. 229, 712 | 1888 *Acanthonotozoma*, T. Stebbing in: Rep. Voy. Challenger, *v.* 29 p. 162 | 1894 *A.*, T. Stebbing in: Bijdr. Dierk., *v.* 17 p. 31 | 1880 *Acanthonotosoma*, D'Urban in: Ann. nat. Hist., ser.·5 *v.* 6 p. 255 | 1893 *A.*, G. O. Sars, Crust. Norway, *v.* 1 p. 372 | 1893 *A.* (part.), A. Della Valle in: F. Fl. Neapel, *v.* 20 p. 674.

Back strongly curved. Rostrum acute, reaching apex of 1ˢᵗ joint of antenna 1. Side-plates 1—3 tapering to acute apex, 4ᵗʰ large with 2 emarginations, 5ᵗʰ and 6ᵗʰ with hind lobe the deeper. Antennae 1 and 2 not elongate. Upper lip rather long, apically narrowed, unsymmetrically incised. Lower lip, lobes nearly contiguous, tapering towards apex, inner lobes almost obsolete. Mandible elongate, tapering to narrow dentate cutting edge, accessory plate narrow on both mandibles, spine-row evanescent, molar obsolete, palp slender. Maxilla 1, inner plate triangular, with many setae, outer obliquely truncate, with 11 spines, palp slender, its 1ˢᵗ joint not very short. Maxilla 2, inner plate the shorter, fringed along most of inner margin, both plates obliquely truncate. Maxillipeds, inner plates long, outer scarcely larger, reaching apex of 2ᵈ joint of palp, palp rather small, finger minute. Gnathopod 1 feeble, very slender, gnathopod 2 shorter and stouter, both pairs simple. Peraeopods 1—5 rather stout. Peraeopods 3—5, 2ᵈ joint quadrangular, lower hind corner acute. Uropod 3, outer ramus the shorter. Telson tapering to a slightly incised apex.

3 species.

Synopsis of species:

1 { Carina absent from peraeon, not dentate on pleon . 3. **A. inflatum** . . p. 219
{ Carina present on peraeon, dentate on pleon — 2.
2 { Carina not forming acute processes on pleon segments 3 and 4 . . . . 1. **A. serratum** . p. 218
{ Carina forming acute processes on pleon segments 3 and 4 . . . . 2. **A. cristatum** . p. 219

1. **A. serratum** (O. Fabr.) 1780 *Oniscus serratus*, O. Fabricius, Fauna Groenl., p. 262 | 1802 *O. s., Talitrus* (part.), Latreille, Hist. Crust. Ins., *v.* 3 p. 39 | 1853 *Acanthonotus s.*, Stimpson in: Smithson. Contr., *v.* 6 nr. 5 p. 52 | 1866 *Vertumnus s.*, Goës in: Öfv. Ak. Förh.. *v.* 22 p. 522 | 1876 *Acanthonotozoma serratum*, A. Boeck, Skand. Arkt. Amphip., *v.* 2 p. 240 | 1893 *Acanthonotosoma s.*, G. O. Sars, Crust. Norway, *v.* 1 p. 374 t. 131 f. 1 | 1893 *A. s.* (part.), A. Della Valle in: F. Fl. Neapel, *v.* 20 p. 675 t. 59 f. 83, 84 | 1838 *Amphithoe serra*, Kröyer in: Danske Selsk. Afh., *v.* 7 p. 266 t. 2 f. 8 | 1840 *Acanthonotus s.*, H. Milne Edwards, Hist. nat. Crust., *v.* 3 p. 25.

Body stout, carina low on peraeon segments 1—4, on segments 5—7 and pleon segments 1 and 2 produced into well marked, not very large, backward pointing teeth, on pleon segment 3 into a laminar obtusely ending process. Head, lateral corners small, narrowly rounded. Side-plate 4 with an acute

---

*) According to the rules of transcription this name should be written *Acanthonotosoma*.   The Editor.

point between the 2 emarginations, 5th and 6th with hind lobe much produced downward, obtuse. Pleon segments 1 and 2, postero-lateral corners forming a small tooth, margin above slightly serrate, in 3^d 2 diverging narrow lobes, the lower coarsely serrate on upper edge and apex. Eyes small, oval, narrow above, bright red. Antenna 1, 1st joint shorter than 2^d and 3^d combined, not dentate, flagellum scarcely longer than peduncle, with about 15 joints carrying sensory filaments. Antenna 2 considerably shorter, ultimate joint of peduncle rather shorter than penultimate, flagellum shorter than ultimate and penultimate combined. Gnathopod 1, 2^d joint setose on both margins, 5th tapering, longer than the very narrow 6th, finger awl-shaped, not dentate, beset with curved setae, the apical one densely ciliated on one edge. Gnathopod 2, 5th joint longer than 6th, both spinose, finger short, broad, sublaminar with acute tip. Feraeopods 3—5, 2^d joint quadrangular, slightly produced at lower hind corner. Uropod 3, outer ramus about $^3/_4$ as long as inner. Telson oblong oval, distally tapering. incision very short. Colour whitish, each segment with 2 or 3 narrow transverse bands of brilliant crimson, which extends also to bases of limbs and antennae. L. ♀ reaching 12 mm.

Arctic Ocean, North-Atlantic, North-Sea and Skagerrak (Greenland, Spitzbergen, Barents Sea, Murman Coast, Kara Sea, North-America, Bohuslän, Norway from Haugesund northward). Depth 19—300 m.

2. A. cristatum (J. C. Ross) 1835 *Acanthonotus cristatus*, J. C. Ross in: John Ross, App. sec. Voy., nat. Hist. p. 90 t. B f. 8—12 | 1838 *Amphithoe cristata*, Kröyer in: Danske Selsk. Afh., *v.* 7 p. 322 | 1866 *Vertumnus cristatus*, Goës in: Öfv. Ak. Förh., *v.* 22 p. 522 | 1876 *Acanthonotozoma cristatum*, A. Boeck, Skand. Arkt. Amphip., *c.* 2 p. 238 | 1894 *A. c.*, T. Stebbing in: Bijdr. Dierk., *v.* 17 p. 32 | 1893 *Acanthonotosoma c.*, G. O. Sars, Crust. Norway, *v.* 1 p. 375 t. 131 f. 2 | 1862 *Acanthonotus serratus* (part), Bate, Cat. Amphip. Brit. Mus., p. 127 | 1893 *Acanthonotosoma serratum* (part.), A. Della Valle in: F. Fl. Neapel, *v.* 20 p. 675.

Body rather compressed, carina conspicuous from peraeon segment 1 to pleon segment 4, forming a large elevated laminar tooth on each segment, except peraeon segments 1—4. Head, lateral corners small, narrowly rounded. Side-plate 4 with an obtuse point between the 2 emarginations, 5th and 6th produced acutely backward. Pleon segments 1—3, postero-lateral corners acute, in segment 3 strongly produced, not bilobed. Eyes very small, round, bright red. Antenna 1, 1st joint subequal to 2^d and 3^d combined, 1st and 2^d each with apical tooth, flagellum subequal to peduncle, setose. Antenna 2 scarcely shorter. Gnathopod 1, 2^d joint not setose, 5th little longer than linear 6th, finger with setae as in A. serratum and distally armed with 7 curved denticles. Gnathopod 2, 5th joint scarcely longer than 6th, both setose, finger small, curved. Feraeopods 1—5 rather robust. Peraeopods 3—5, 2^d joint oblong quadrangular, produced below into two acute processes. Uropod 3, outer ramus about $^2/_3$ as long as inner. Telson scarcely longer than broad, apical incision deeper than in A. serratum. Colour reddish white. ⊸L. ♀ 12—18·5 mm.

Arctic Ocean (Arctic America, Spitzbergen, Barents Sea. Kara Sea, North-Norway). Depth 94—246 m.

3. A. inflatum (Kroyer) 1842 *Acanthonotus inflatus*, Krøyer in: Naturh. Tidsskr., *v.* 4 p. 161 | 1866 *Vertumnus i.*, Goës in: Öfv. Ak. Förh., *v.* 22 p. 523 t. 38 f. 11 | 1876 *Acanthonotozoma inflatum*, A. Boeck, Skand. Arkt. Amphip., *v.* 2 p. 242 | 1894 *A. i.*, T. Stebbing in: Bijdr. Dierk., *v.* 17 p. 32 t. 6 | 1880 *Acanthonotosoma i.*, D'Urban in: Ann. nat. Hist., ser. 5 *v.* 6 p. 255 | 1887 *A. i.*, H. J. Hansen in: Vid. Meddel., ser. 4 *v.* 9 p. 127 | 1893 *A.? i.*, A. Della Valle in: F. Fl. Neapel, *v.* 20 p. 676.

Integument scabrous in parts. Peraeon dorsally broad, rounded, without carina or processes. Pleon segments 1—3 with slight median carina. Head, lateral corners bluntly produced. Side-plates 1—7 and postero-lateral corners of pleon segments 1—3, antennae 1 and 2, mouth-parts, gnathopods 1 and 2, peraeopods 1—5, uropods 1—3 very similar to those of A. serratum (p. 218), telson agreeing rather with that of A. cristatum (p. 219). Eyes rounded oval. Antenna 1, flagellum in ♀ scarcely as long as peduncle, 23-jointed; microscopic rudiment of accessory flagellum. Upper lip less narrow than in Sars' fig. of A. serratum. Mandible, 3^d joint of palp much curved. Gnathopod 1, finger in addition to setae like those in A. serratum having 11 backward directed teeth in correspondence with A. cristatum and a stout curved nail. Gnathopod 2, finger stout, distally fringed with 9 denticles, ending in a stout nail. Peraeopods 3—5, the quadrangular 2^d joint less acute at lower hind corner than in A. serratum. Uropod 3, outer ramus about $^4/_5$ as long as inner. Telson $1^1/_2$ as long as broad, triangular incision equal to $^1/_4$ of total length. L. 6·5—18·5 mm (measurement made perhaps sometimes on specimens coiled up, sometimes on extended ones).

Arctic Ocean (Greenland, Spitzbergen, Franz Joseph Land, Kara Sea, White Sea, Matotschkin Schar). Depth 10—300 m.

---

17. Fam. **Pardaliscidae**

1871 Subfam. *Pardaliscinae*, A. Boeck in: Forb. Selsk. Christian., 1870 p. 150 | 1882 *Pardaliscidae*, G. O. Sars in: Forh. Selsk. Christian., nr. 18 p. 28 | 1888 *P.*, T. Stebbing in: Rep. Voy. Challenger, *v.* 29 p. 990 | 1893 *P.*, G. O. Sars, Crust. Norway, *v.* 1 p. 401.

Body not indurated. Head, rostrum usually small. Side-plates small, 4th like 3^d, 5th with front lobe the deeper. Eyes present or obsolete, never(?) coalescent. Antenna 1 usually with accessory flagellum, which is unlike in ♂ and ♀. Mouth-parts projecting, strongly developed. Mandible without molar (Fig. 56, 58, 60). Maxillipeds, inner plates small (Fig. 59). Gnathopods 1 and 2 subequal, either small and simple, or powerful and imperfectly subchelate (Fig. 57). Peraeopods 1 and 2 unlike peraeopods 3—5, which are rather long, with 2^d joint little expanded. Uropod 3 rather large, rami foliaceous. Telson deeply cleft.

Marine.

7 genera, 11 species accepted and 1 doubtful.

Synopsis of genera:

1. Gen. **Halicoides** A. Walker

1896 *Halicoides* (Sp. un.: *H. anomala*), A. O. Walker in: Ann. nat. Hist., ser. 6 *v.* 17 p. 344.

Rostrum large. Eyes obsolete. Antenna 1 without accessory flagellum, 1st joint of flagellum in ♂ elongate. Antenna 2 longer than antenna 1. Mandible, 3^d joint of palp $^2/_3$ as long as 2^d. Gnathopods 1 and 2 simple, finger rather long. Peraeopods 1 and 2, 4th and 5th joints much widened. Peraeopods 3—5 elongate, slender. Uropod 3, inner ramus rather shorter than outer. Telson cleft to base, apices acute, notched.

1 species.

1. **H. anomalus** A. Walker 1896 *H. anomala*, A. O. Walker in: Ann. nat. Hist., ser. 6 *v.* 17 p. 344 t. 16.

Rostrum fully reaching apex of 1st joint of antenna 1. Side-plates small. Pleon segments 1—3, postero-lateral angles forming a small tooth. Eyes indicated by a clear almost circular space on the top of the head. Antenna 1, 1st joint large, distally expanded, longer than 2^d and 3^d combined (supposed appendage to joint 2 not existing), flagellum with about 28 joints, 1st much the longest, setose. Antenna 2, ultimate joint of peduncle longer than penultimate, both setulose on upper margin, flagellum slender, about 28-jointed. Maxilla 1, outer plate represented with 6 strong spines, 2^d joint of palp with 5 spine-teeth. Maxilla 2, inner plate fringed on inner margin. Gnathopod 1, 5th joint rather more (in fig. much more) than half 6th, 6th narrow, tapering to base of finger, which is half as long. Gnathopod 2 similar, but hind margin of 5th and 6th joints densely setose, the longest setae at the distal ends. Peraeopods 1 and 2, 4th joint obcordate, 5th broadly oval, 6th as long but much narrower, all the 3 fringed with setae, finger $^2/_3$ as long as 6th joint. Peraeopods 3—5 successively longer, 2^d joint shorter than 4th. Uropod 1, peduncle as long as the narrow, spinose, equal rami; uropod 2 shorter than uropod 1; uropod 3, inner ramus coarsely serrate on inner margin and ending in a short strong spine. Telson longer than peduncle of uropod 3. L. 7 mm.

Bay of Biscay (Isle de Yeu). Depth 31—58 m.

2. Gen. **Pardalisca** Krøyer

1842 *Pardalisca* (Sp. un.: *P. cuspidata*), Krøyer in: Naturh. Tidsskr., *v.* 4 p. 153 ⋅ 1888 *P.*, T. Stebbing in: Rep. Voy. Challenger, *v.* 29 p. 991 | 1893 *P.*, G. O. Sars, Crust. Norway, *v.* 1 p. 402 | 1893 *P.* (part.), A. Della Valle in: F. Fl. Neapel, *v.* 20 p. 691.

Body nearly cylindric, back broadly vaulted. Head with small rostrum, lateral margins strongly curved. Side-plates 1—4 not deep, quadrate, 5th with front lobe nearly as deep as side-plate 4, 6th and 7th small. Eyes imperfectly developed, when present, near lateral margins of head. Antenna 1 the shorter, peduncle short, in ♂ 1st joint of flagellum large, bearded, and 1st joint of accessory flagellum large, laminar. Upper lip lamellar, very unsymmetrically bilobed. Lower lip with inner lobes, outer wide apart, mandibular processes elongate. Mandible (Fig. 56, 58) strong, left with broad denticulate cutting edge and accessory plate, spine-row of 2 spines, palp densely setose; right with cutting edge of 4 strong teeth, no accessory plate, 2 spines in spine-row. Maxilla 1, inner plate small, with 1 seta, outer with 7(8?)—10 spines, 1 setiform, the rest smooth or with 1 tooth near the base, 2^d joint of palp distally greatly expanded, with many spine-teeth. Maxilla 2, inner plate rather the broader, both slender. Maxillipeds (Fig. 59), inner plates rudimentary, outer broad, densely fringed, springing from a large base, finger of palp small. Gnathopods 1 and 2 (Fig. 57) not large, simple, 5th joint subfusiform, long, 6th small, spinose, finger beset with spines or spinules. Peraeopods 1 and 2, 4th and 5th joints laminar, spinose, 5th the longer. Peraeopods 3—5 long, finger short. Uropods 1 and 2 spinose, rami subequal, uropod 3 projecting, rami subequal, foliaceous. Telson deeply cleft, armed with marginal spines, apices divergent.

4 species.

Synopsis of species:

$1 \begin{cases}$ Gnathopods 1 and 2, finger unguiform . . . . . . 1. **P. abyssi** . . . p. 222
Gnathopods 1 and 2, finger laminar — **2.** $\end{cases}$

$2 \begin{cases}$ Gnathopods 1 and 2, finger narrowly laminar . . . 2. **P. tenuipes** . . p. 223
Gnathopods 1 and 2, finger broadly laminar — **3.** $\end{cases}$

$3 \begin{cases}$ Maxilla 1, outer plate with 6 (or 7) dentate spines
and a seta . . . . 3. **P. cuspidata** . p. 223
Maxilla 1, outer plate with 9 dentate spines and a seta 4. **P. marionis** . . p. 224 $\end{cases}$

1. **P. abyssi** Boeck 1871 *P. a.*, A. Boeck in: Forh. Selsk. Christian., 1870 p. 152 | 1888 *P. a.,* T. Stebbing in: Rep. Voy. Challenger, *v.* 29 p. 992 t. 93 | 1893 *P. a.*, G. O. Sars, Crust. Norway, *v.* 1 p. 406 t. 143 f. 1 | 1893 *P. a.* (part.), A. Della Valle in: F. Fl. Neapel. *v.* 20 p. 692 t. 59 f. 93 | 1874 *P. cuspidata* (err., non Krøyer 1842!), Buchholz in: Zweite D. Nordpolarf., *v.* 2 p. 306 Crust. t. 1 f. 3; t. 2 f. 1.

Body rather slender, pleon segments 3 and 4 with 2 dorsal teeth, 5th with 1 such tooth. Head, rostrum very small. Pleon segments 1—3, postero-lateral corners minutely acute, or in segment 3 quadrate or slightly rounded. Eyes large, sigmoid, expanded below, near front lateral margin of head, light red. Antenna 1 in ♀, 1st joint as long as 2^d and 3^d combined, flagellum more than twice as long as peduncle, with 40—50 short joints, accessory flagellum 5- or 6-jointed. Antenna 2 considerably longer, ultimate joint of peduncle rather shorter than penultimate, flagellum longer than peduncle, about 40-jointed. Antenna 1 in ♂, 1st joint of accessory flagellum very large, laminar, more than twice as long as 4 small remaining joints combined. Maxilla 1, outer plate with 7 simple spines and a seta, broad apex of palp with 20 small spine-teeth. Maxilla 2, inner plate rather longer and broader than outer. Maxillipeds, outer plates comparatively smaller than in P. cuspidata (p. 223), palp more slender, 2^d joint very narrow. Gnathopod 1, 5th joint twice as broad, not

Fig. 56.
P. abyssi.
Mandible.

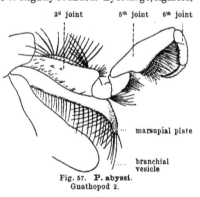

2ᵈ joint   5ᵗʰ joint   6ᵗʰ joint

marsupial plate

branchial
vesicle

Fig. 57. **P. abyssi.**
Gnathopod 2.

twice as long as 6th, fringed with long spines, 6th joint narrow, fringed with pectinate spines, finger a little shorter, curved, slender, hind margin fringed with spinules. Gnathopod 2 (Fig. 57) similar, but 5th joint longer, more fusiform, more than twice as long as 6th. Peraeopods 1 and 2 rather densely spinose, 2^d joint expanded distally, 5th narrowly oval, longer than 4th or linear 6th, finger less than half 6th. Peraeopods 3—5, 2^d joint narrow oblong, 4th, 5th and 6th joints long and narrow, finger short and straight, with short curved nail. Uropod 3, rami broadly foliaceous, subequal, or outer rather longer than inner, scarcely acute. Telson not twice as long as broad, 4 spinules along each margin, cleft nearly $^3/_4$ length, notched apices a little divergent. Colour semipellucid, dorsally tinted with orange, ova rose-coloured. L. reaching 28 mm.

Arctic Ocean, North-Atlantic and North-Sea (Greenland, Spitzbergen, NoVa Scotia, NorWay). Depth 56—282 m.

2. **P. tenuipes** O. Sars 1893 *P. t.*, G. O. Sars, Crust. NorWay, *v.* 1 p. 404 t. 142 f. 2.

Body not very slender, pleon segment 3 with 2 dorsal prominences, rounded, short, segment 4 with 2 small dorsal teeth, 5th with 1 rather larger tooth. Head, rostrum rather more produced than in P. cuspidata. Pleon segment 3, postero-lateral corners acute angled, not forming a tooth. Eyes rather small, sublinear, chalky white. Antenna 1 in ♀, 1st joint as long as 2^d and 3^d combined, flagellum slender, 4 times as long as peduncle, accessory flagellum 6-jointed. Antenna 2 considerably longer, ultimate and penultimate joints of peduncle subequal, flagellum 1$^1/_2$ length of peduncle. Antennae 1 and 2 in ♂ longer, antenna 1 as usual having 1st joint of flagellum very large, densely fringed; accessory flagellum large, with 3 joints, 1st representing 4 coalesced into a large flattened .joint. Gnathopods 1 and 2 as in P. cuspidata, except that the finger is more narrowly laminar, and has the terminal spine much stronger than the others. Feraeopods 1 and 2 nearly as in P. cuspidata, but peraeopods 3—5 more slender, more densely setose, 6th joint very narrow, finger small, narrow, awl-shaped. Uropod 3, rami less broad, rather bluut, outer rather longer than inner. Telson nearly twice as long as broad, with a group of 3 spinules on each side near the base, and one near the apex, cleft more than $^3/_4$ length, dehiscent all along, apices bidentate. Colour semipellucid, yellowish grey; ova orange-coloured. L. ♀ scarcely attaining 11 mm.

North-Atlantic (NorWay). Depth 94—188 m.

3. **P. cuspidata** Krøyer 1842 *P. c.*, Krøyer in: Naturh. Tidsskr., *v.* 4 p. 153 | 1876 *P. c.*, A. Boeck, Skand. Arkt. Amphip., *v.* 2 p. 482 t. 12 f. 5, except 5g | 1893 *P. c.*, G. O. Sars, Crust. NorWay, *v.* 1 p. 403 t. 141, 142 f. 1 | 1893 *P. c.* (part.), A. Della Valle in: F. Fl. Neapel, *v.* 20 p. 692 t. 59 f. 92.

Body rather slender, cylindric, smooth, back broadly rounded, pleon segments 3 and 4 with 2 dorsal teeth, 5th with 1 such tooth. Head, rostrum very small. Pleon segment 3, postero-lateral corners sharply quadrate. Eyes very narrow, slightly sigmoid, along front lateral margins of head, bright red. Antenna 1 in ♀, 1st joint as long as 2^d and 3^d combined, flagellum more than thrice as long as peduncle, with many short joints, accessory flagellum with 5 joints, 1st nearly as long as the other 4 combined. Antenna 2 considerably longer, ultimate and penultimate joints of peduncle subequal, long, flagellum rather longer than peduncle. Antenna 1 in ♂, 1st joint of flagellum very large, clothed with transverse rows of sensory setae, 1st joint of accessory flagellum laminar, more than twice as long as the other 4 combined. Maxilla 1, outer plate with 6(7?) dentate spines and a seta, broad apex of palp with many spinules. Maxilla 2, inner plate rather longer and broader than outer. Maxillipeds, 2ᵈ joint of palp stout, not much longer than 3ᵈ. Gnathopod 1, 5ᵗʰ joint subfusiform, densely spinose, much broader than 6ᵗʰ and fully twice as long, 6ᵗʰ narrow, fringed with pectinate spines, finger laminar, broadly oval, armed on apex and along hind margin with spines, one rather larger than the rest serving as a nail. Gnathopod 2 similar, but longer, especially in 2ᵈ and 5ᵗʰ joints. Peraeopods 1—5 as in P. abyssi (p. 222). Uropod 3, rami equal, broadly foliaceous, spinules on outer, plumose setae on inner margin. Telson 1¹/₂ times as long as broad, margins with 3 separate spinules, and a group of 3, cleft ³/₄ length, dehiscent, apices rather broadly notched. Colour orange, darkest on back; ova rose-coloured. L. 12 mm.

Arctic Ocean, North-Atlantic and Skagerrak (Greenland, Spitzbergen, Matotschkin Schar, Norway southward to Bergen, Bohuslän). Depth 18—72 m.

4. P. marionis Stebb. 1888 *P. m.*, T. Stebbing in: Rep. Voy. Challenger, *v.* 29 p. 996 t. 94 | 1893 *P. cuspidata* (part.), A. Della Valle in: F. Fl. Neapel, *v.* 20 p. 692.

Rostrum very small. Eyes not observed. Antenna 1, 1ˢᵗ joint longer than 2ᵈ and 3ᵈ combined, flagellum longer than peduncle, 18 joints remaining, accessory

Palp

flagellum with 5 joints, 1ˢᵗ longest, as long as 1ˢᵗ of primary. Antenna 2, ultimate joint of peduncle a little shorter than penultimate, flagellum not much longer than peduncle, 29-jointed. Mouth-parts nearly as in P. cuspidata (p. 223), but mandibular palp (Fig. 58) rather shorter;

Fig. 58. Mandible.

maxilla 1, outer plate with 9 dentate spines and a seta, broad apex of palp with 15 little distant spine-teeth;

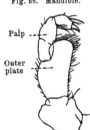

Palp

Outer plate

maxilla 2, inner plate broader but rather shorter than outer; maxillipeds (Fig. 59) broader, 2ᵈ joint of palp little longer than 1ˢᵗ. Gnathopods 1 and 2 nearly as in P. cuspidata, but 2ᵈ joint shorter, 5ᵗʰ much more than twice as long and broad as 6ᵗʰ, finger not much longer than broad, beset with 14 unequal spines and having a spine-like nail as long as itself. Peraeopods 1 and 2, 4ᵗʰ joint not much shorter than the oval 5ᵗʰ, finger fusiform, more than half as long as 6ᵗʰ joint, with a very small curved nail. Peraeopod 4, 2ᵈ joint

Fig. 59. Maxilliped.
Fig. 58 and 59.
P. marionis.

moderately expanded, oblong, 4ᵗʰ, 5ᵗʰ and 6ᵗʰ joints narrow, long. All the pleon and some of the peraeopods missing. L. (from rostrum to end of peraeon) about 4 mm.

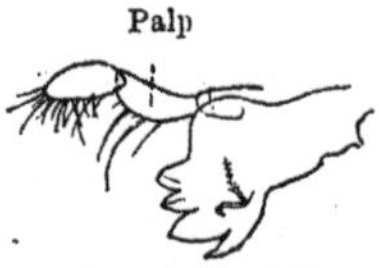

Southern Indian Ocean (Marion Island). Depth 183 m.

3. Gen. **Pardaliscoides** Stebb.

1888 *Pardaliscoides* (Sp. un.: *P. tenellus*), T. Stebbing in: Rep. Voy. Challenger, *v.* 29 p. 1725 | 1897 *P.*, T. Stebbing in: Tr. Linn. Soc. London, ser. 2 *v.* 7 p. 38 | 1893 *Pardalisca* (part.), A. Della Valle in: F. Fl. Neapel, *v.* 20 p. 691.

Antenna 1 the longer; peduncle long, primary and accessory flagella many-jointed. Mandible, left with 4 unequal teeth on cutting edge, crenulate accessory plate and 2 spines; right with 3 teeth on cutting edge, no accessory plate, 3 spines; palp slender, fringed with setiform spines. Maxilla 1, inner plate small, with 1 seta, outer with 7 spines, 1 setiform, 2ᵈ joint of palp wide, carrying many spinules. Maxilla 2, inner plate with 3 setae, outer with 6, both slender. Maxillipeds, inner plates very small, outer on small base, short, narrow, spinulose; 2^d and 3^d joints of palp long, finger long, with setules on inner margin. Gnathopods 1 and 2 similar, simple, 5th joint robust, fusiform, 6th slender, finger fringed with setules. Peraeopods 1—5 slender, long. Uropod 2, rami unequal; uropod 3, rami foliaceous. Telson deeply cleft.

1 species.

1. P. tenellus Stebb. 1888 *P. t.*, T. Stebbing in: Rep. Voy. Challenger. *v.* 29 p. 1725 | 1897 *P. t.*, T. Stebbing in: Tr. Linn. Soc. London, ser. 2 *v.* 7 p. 38 t. 12 | 1893 *Pardalisca abyssi* (part.), A. Della Valle in: F. Fl. Neapel, *v.* 20 p. 692.

Pleon segments 1—3 perhaps with dorsal denticle. Head, rostrum acute. Antenna 1, 1st joint stout, 2^d longer, 3^d half as long as 2^d, flagellum with more than 13 joints, 1st much the longest, accessory flagellum with more than 7 joints, 1st as long as 1st of primary. Antenna 2, ultimate and penultimate joints of peduncle long, ultimate the shorter, flagellum half as long as peduncle, 12-jointed. Gnathopods 1 and 2 spinose; short plumose spines on hind margin of 5th and 6th joints, beginning at narrowed distal end of 5th. Setules on finger's inner margin very small. Peraeopods 1 and 2, 5th joint longer than 4th or 6th; peraeopod 2 rather longer than peraeopod 1, and having about a dozen blunt spines on hind margin of 6th joint. Peraeopod 3, 2^d joint very slightly expanded, 4th the longest, finger slender, acute. Peraeopod 4 like 3^d, but joints rather longer. Peraeopod 5 considerably longer than 4th, 2^d joint expanded above, narrowing downward, 4th joint very long. Branchial vesicle of this limb small, narrowly oval. Telson much longer than broad, cleft $^3/_4$ length, dehiscent, with a spinule at each apex and a setule on each lateral margin. L. 8 mm.

South-Pacific (lat. 37° S., long. 83° W.). Depth 3246 m.

4. Gen. **Nicippe** Bruz.

1859 *Nicippe* (Sp. un.: *N. tumida*), R. M. Bruzelius in: Svenska Ak. Handl., n. ser. *v.* 3 nr. 1 p. 99 | 1893 *N.*, G. O. Sars, Crust. Norway, *v.* 1 p. 409 | 1893 *N.* (part.), A. Della Valle in: F. Fl. Neapel, *v.* 20 p. 657.

Head with very small rostral projection, lateral corners produced. Side-plates very shallow, not overlapping, 4th not emarginate, 5th with front lobe the deeper. Antenna 1 in ♀ much longer than antenna 2, accessory flagellum small. Antenna 1 in ♂ rather shorter than antenna 2, 1st joint of flagellum long, bearded, 1st joint of accessory flagellum long, laminar. Upper lip moderately bilobed. Lower lip, outer lobes small, separated by a single inner lobe. Mandible, cutting edge broad, dentate, accessory plate broad, only present on left mandible, spine-row with small spines, 3^d joint of palp not much shorter than 2^d. Maxilla 1, inner plate with 1 seta, outer with 8 (?) spines, palp with 9 spine-teeth on expanded apex. Maxilla 2, inner plate a little shorter than outer, both slender. Maxillipeds, inner plates conical, small but not rudimentary, outer fringed with spines, narrow, not reaching end of palp's 1st joint, palp very large, 2^d joint long, oval, 3^d not much shorter, finger spinulose on inner margin. Gnathopods 1 and 2, 5th joint rather short, distal expansion densely setose, 6th long oval, finger long, curved, acute. Peraeopods 1 and 2, 4th and 5th joints equal, slightly widened, 6th rather longer. Peraeopods 3—5 successively longer, slender, 2^d joint oblong, little expanded, finger stiliform. Uropods 1 and 2 strong, spinose, outer ramus little shorter than inner; uropod 3 reaching beyond the others, rami subequal, setose on inner margin. Telson narrow, deeply cleft.

1 species.

1. **N. tumida** Bruz. 1859 *N. t.*, R. M. Bruzelius in: Svenska Ak. Handl., n. ser. *v.* 3 nr. 1 p. 99 t. 4 f. 19 | 1868 *N. t.*, A. M. Norman in: Ann. nat. Hist., ser. 4 *v.* 2 p. 414 t. 21 f. 4—6 | 1876 *N. t.*, A. Boeck, Skand. Arkt. Amphip., *v.* 2 p. 492 | 1893 *N. t.*, G. O. Sars, Crust. Norway, *v.* 1 p. 410 t. 144; t. 145 f. 1 | 1893 *N. t.*, A. Della Valle in: F. Fl. Neapel, *v.* 20 p. 658 t. 59 f. 66, 67.

Back broadly vaulted, subdepressed, pleon segment 4 with 2 contiguous very small dorsal teeth. Head, lateral corners angularly produced, their lower margins continued backward as lateral projecting ridge. Side-plate 1 larger than any of the 3 following. Pleon segment 3, postero-lateral corners quadrate. Eyes represented by yellowish pigment near lateral margins of head. Antenna 1 in ♀ long, slender, 1st joint as long as 2^{d} and 3^{d} combined, flagellum nearly 4 times as long as peduncle, 48-jointed, accessory flagellum with 3 or 4 joints, 1st as long as others combined. Antenna 2 in ♀, ultimate and penultimate joints of peduncle equal, flagellum as long as peduncle, 18-jointed. Antenna 2 in ♂, flagellum long, filiform. Gnathopod 1, 6th joint very large and tumid, palm not defined, but subchelate by help of the long clasping finger. Gnathopod 2 like gnathopod 1, except that the distal expansion of 5th joint is somewhat more prominent and very densely spinose. Peraeopod 5 very elongate, 2^{d} joint behind submarginally armed with 2 (Bruzelius; 1 Norman; not noticed by Boeck and Sars) long strong setae, plumose on the distal half. Uropod 3 large, peduncle thick, setose on outer margin, rami oblong lanceolate, inner margin fringed with long setae, outer a little the longer. Telson nearly 3 times as long as broad, cleft nearly to the base, the division dehiscent more at the 2 extremities than in the middle, apices bidentate. Colour semipellucid, tinged yellowish orange. L. ♀ 14 mm, ♂ rather less.

Arctic Ocean, North-Atlantic, North-Sea and Skagerrak (Norway, Shetland Isles, Sound of Skye). Depth 113—565 m.

5. Gen. **Synopioides** Stebb.

1888 *Synopioides*, T. Stebbing in: Rep. Voy. Challenger, *v.* 29 p. 999 | 1893 *S.*, A. Della Valle in: F. Fl. Neapel, *v.* 20 p. 851.

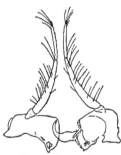

Fig. 60. **S. macronyx.**
Mandibles and upper lip.

Side-plates very shallow. Antenna 1, peduncle short, flagellum long, accessory flagellum rather long. Antenna 2 longer than antenna 1. Mandible (Fig. 60), cutting edge broad, denticulate, accessory plate on both mandibles, spine-row of 2 spines, palp very long, 3^{d} joint linear, much shorter than 2^{d}. Maxilla 1, outer plate with 8 long spines, palp with 7 spine-teeth on distally widened 2^{d} joint. Maxilla 2, inner plate fringed on inner margin, rather shorter than outer. Maxillipeds, inner plates small, conical, setiferous, outer small on a large base, with few fringing spines, palp long. Gnathopods 1 and 2 simple, 5th and 6th joints long, narrow. Peraeopods 1 and 2 as in Pardalisca (p. 221), but rather more slender. Peraeopods 3—5 elongate, finger stiliform. Uropod 1, rami equal; uropod 2, outer ramus much the shorter; uropod 3, rami subequal, foliaceous. Telson deeply cleft.

1 species accepted, 1 doubtful.

1. **S. macronyx** Stebb. 1888 *S. m.*, T. Stebbing in: Rep. Voy. Challenger, *v.* 29 p. 1000, 1223 t. 94 A | 1893 *S. m.*, A. Della Valle in: F. Fl. Neapel, *v.* 20 p. 852.

Head apparently angled in front as in Synopia schéeleana (p. 272). Side-plates 5 and 6 broader than the preceding. Pleon segments 1—3, postero-lateral angles acute, chiefly in segment 3. Eyes not perceived. Antenna 1, 1st joint short, as long as 2^d and 3^d combined, flagellum in ♀ 5 times as long as peduncle, with 33 joints, 1st very long, first 6 or 7 very stout, then tapering, terminal 20 slender, the proximal joints armed with slender spines and a large brush of long broad sensory filaments, accessory flagellum with 3 joints, together as long as the first 8 of primary. Antenna 2 longer, ultimate joint of peduncle elongate, shorter than the spinose penultimate, flagellum longer than peduncle, with about 30 unequal joints. Triturating lobes of stomach of peculiar character. Gnathopod 1, 5th joint considerably shorter than the long, narrow, slightly curved 6th, finger about half 6th, setulose, with tooth at base of the nail. Gnathopod 2, 5th joint rather stouter and longer than the tapering 6th, both densely furnished with setiform spines, some on the 5th of great length, finger more than $^1/_2$ as long as 6th, nearly as in gnathopod 1. Peraeopods 1 and 2 spinose. Peraeopod 3, 2^d joint oblong, rami not very wide, 6th joint straight, slender, finger needle-like. Peraeopod 5, 2^d joint piriform, very broad at the top. Uropods 1 and 2, peduncle and rami very spinose, in uropod 1 peduncle subequal to the rami, in uropod 2 peduncle as long as the shorter ramus, inner rather longer than rami of uropod 1. Uropod 3, peduncle much shorter than the broad lanceolate rami, which are setose on inner margin, outer the longer by the extent of its small 2^d joint. Telson about once and a half as long as broad, cleft nearly to base, apices acute, bidentate, the inner points rather divergent, the outer considerably produced. L. 10 mm.

South-Pacific (lat. 38° S., long. 94° W.). Depth 2743 m.

S. secundus Stebb. 1888 *S. s.*, T. Stebbing in: Rep. Voy. Challenger, *v.* 29 p. 1224.

Perhaps the ♂ of S. macronyx. L. 12 mm.

South-Pacific (lat. 39° S., long. 105° W.). Depth 3703 m.

6. Gen. **Pardaliscella** O. Sars

1893 *Pardaliscella* (Sp. un.: *P. boeckii*), G. O. Sars, Crust. Norway, *v.* 1 p. 407.

Body slender, cylindric, without dorsal processes. Head not rostrate. Side-plates as in Pardalisca (p. 221). Antennae 1 and 2 in ♀ short, subequal, accessory flagellum small. Lower lip, outer lobes small, separated by a single inner lobe. Mandible not very strong; as to cutting edge, accessory plate and spine-row similar to those of Pardalisca, but weaker, palp slender, almost naked except at apex. Maxilla 1, inner plate with 1 seta, outer with 6 spines, 2 very stout, apex of palp moderately expanded. Maxilla 2 with few setae, inner plate a little shorter and wider than outer. Maxillipeds, inner plates conical, rather, not very, small, outer moderately broad, not reaching end of palp's 1st joint, palp of moderate size. Gnathopods 1 and 2 stronger than in Pardalisca, 5th joint wider, scarcely longer than 6th, which narrows distally, finger simple, unguiform. Peraeopods 1—5 nearly as in Pardalisca. Uropods 1 and 2, rami nearly equal, with few spinules; uropod 3, rami equal in length, quite naked, outer the narrower, with small 2^d joint. Telson unarmed, deeply cleft.

2 species.

15*

1. **P. boeckii** (Malm) 1871 *Pardalisca b.*, A. W. Malm in: Öfv. Ak. Förh., *v.* 27 p. 547
t. 5 f. 2 | 1893 *Pardaliscella b.*, G. O. Sars, Crust. Norway, *v.* 1 p. 408 t. 143 f. 2 | 1893
Pardalisca abyssi (part.), A. Della Valle in: F. Fl. Neapel, *v.* 20 p. 692.

Peraeon long. Head, lateral corners slightly projecting, evenly rounded.
Side-plate 1, front corner acute, 5th in front as deep as 4th. Pleon segment 3,
postero-lateral corners evenly rounded, little prominent. Eyes inconspicuous
in specimens in spirit. Antenna 1 in ♀, 1st joint longer than 2^d and 3^d combined,
flagellum rather longer than peduncle, about 12-jointed, accessory flagellum small,
2- or 3-jointed. Antenna 2 scarcely longer, ultimate joint of peduncle subequal to
penultimate, flagellum subequal to the 2 combined. Gnathopods 1 and 2, 5th joint
not longer than 6th, finger with small denticle at middle of concave margin.
Peraeopods 1 and 2 less setose than in Pardalisca (p. 221). Peraeopod 3, 2^d joint
rather narrow, 6th linear, finger as long as 6th. Peraeopod 5, 2^d joint more
widened, finger not so long as 6th. Uropods 1 and 2, rami lanceolate, each
with a single spinule at about middle of each margin. Telson not quite
twice as long as broad, distally tapering rapidly to acute apices, cleft rather
beyond middle. L. (a presumably young specimen) 2·5 mm (Malm), ♀ adult,
scarcely exceeding 4 mm (Sars).

Skagerrak (Bohuslän), Christianiafjord. Depth 188 m.

2. **P. axeli** Stebb.*) 1871 *Pardalisca boeckii* (err., non Malm 1871!), A. Boeck
in: Forh. Selsk. Christian., 1870 p. 152 | 1876 *P. b.*, A. Boeck, Skand. Arkt. Amphip.,
v. 2 p. 485 t. 10 f. 4.

Differing from P. boeckii by following characters. Side-plate 1, front
corners rounded (in fig.) Pleon segment 3, postero-lateral corners broadly
rounded and (in fig.) strongly outdrawn. Antenna 1, flagellum longer than
peduncle, 16-jointed, accessory flagellum 2-jointed. Antenna 2 shorter, ultimate
joint of peduncle shorter than penultimate, flagellum shorter than peduncle,
10—12-jointed. Gnathopods 1 and 2, 5th joint a little longer than 6th. Telson
lanceolate, cleft to the base. L. 8—10 mm.

Christianiafjord, North-Sea (Karmöe).

7. Gen. **Halice** Boeck

1871 *Halice*, A. Boeck in: Forh. Selsk. Christian., 1870 p. 152 | 1893 *H.*, G. O.
Sars, Crust. Norway, *v.* 1 p. 411 | 1893 *H.*, A. Della Valle in: F. Fl. Neapel, *v.* 20 p. 661.

Head, rostrum small but distinct. Side-plates very shallow. Eyes
wanting. Antennae 1 and 2 slender, much stronger in ♂. Antenna 1 the
shorter. Mandible, cutting edge moderately wide, a small accessory plate
on both mandibles, palp not very large, 3^d joint very short. Maxilla 1,
inner plate with 1 seta, outer with 6(?) spines, 2^d joint of palp not distally
expanded. Maxilla 2 comparatively large. Maxillipeds, inner plates almost
obsolete, outer as long as their base, overtopping 1st joint of palp, fringed
with a few spines, palp not very large. Gnathopods 1 and 2 not very powerful,
simple. Peraeopods 1 and 2 as in Pardalisca (p. 221). Peraeopods 3—5 long,
slender, 2^d joint rather narrow, finger stiliform. Uropod 3, rami extended

---

*) Sp. nov. After Axel Boeck.

considerably beyond the others, subfoliaceous, inner margin setose, outer ramus with small 2^d joint. Telson narrow, unarmed, deeply cleft.

1 species.

1. **H. abyssi** Boeck 1871 *H. a.* $+$ *H. grandicornis* ($\male$), A. Boeck in: Forb. Selsk. Christian., 1870 p. 152, 153 | 1876 *H. a.*, A. Boeck, Skand. Arkt. Amphip., *v.* 2 p. 488 t. 10 f. 2 | 1893 *H. a.*, G. O. Sars, Crust. Norway, *v.* 1 p. 412 t. 145 f. 2 | 1893 *H. a.*, A. Della Valle in: F. Fl. Neapel, *v.* 20 p. 661 t. 59 f. 69—71.

Body moderately slender, peraeon in $\female$ much longer than in $\male$, pleon segments 4 and 5 produced to a dorsal acute subdeflexed tooth, larger in segment 5. Head, rostrum nearly reaching middle of 1st joint of antenna 1, lateral corners somewhat produced, narrowly rounded. Side-plate 1, front corners acutely produced, 5th more than twice as broad as deep, front lobe scarcely produced. Pleon segment 3, postero-lateral corners subquadrate. Eyes not represented by any trace of pigment. Antenna 1 in $\female$, 1st joint nearly twice as long as 2^d and 3^d combined, flagellum scarcely thrice as long as peduncle, 1st joint as long as the next 8 combined; accessory flagellum nearly as long as peduncle, with 3 joints, 1st much the longest. Antenna 2 little longer, ultimate joint of peduncle shorter than penultimate, flagellum subequal to the peduncle. In $\male$ antennae 1 and 2 nearly twice as long as in $\female$. Antenna 1, 1st joint of flagellum extremely large, dense with sensory setae, about as long as laminar accessory flagellum. Gnathopods 1 and 2, 5th joint little expanded distally, densely setose, 6th joint much longer, rather narrow, hind margin densely setose, finger slender, smooth, a little curved. Peraeopods 1 and 2 scarcely longer than gnathopod 2, 4th joint rather shorter than 5th. Peraeopods 3—5 long, slender; in peraeopod 3, 2^d joint rather small and narrow, in 4th and 5th successively larger, 4th joint considerably longer than 5th or 6th, finger short. Uropod 3, outer ramus rather longer than inner. Telson more than twice as long as broad, tapering, cleft nearly to base, apices bidentate. Colour pellucid with faint yellowish tinge. L. $\female$ 8, $\male$ 9 mm.

Arctic Ocean, North-Atlantic, North-Sea and Skagerrak (Whole Norway). Depth 188—752 m.

---

18. Fam. **Liljeborgiidae**

1899 *Liljeborgiidae*, T. Stebbing in: Ann. nat. Hist., ser. 7 *v.* 4 p. 211.

Pleon with one or more of the segments dorsally dentate. Head, rostrum seldom large. Side-plate 1 produced forward, 4th emarginate behind. Antenna 1 usually shorter than antenna 2, accessory flagellum well developed. Upper lip slightly or not bilobed. Lower lip without-inner lobes. Mandible, molar feeble. Maxillipeds, inner and outer plates rather small, palp elongate, Gnathopods 1 and 2 strong, subchelate, gnathopod 2 the larger, often with sexual variation. Peraeopods 1 and 2 very slender. Peraeopods 3—5, 2^d joint expanded, 5th pair the longest. Uropod 3, rami subequal in length. Telson cleft.

Marine.

2 genera, 10 accepted species and 2 doubtful.

Synopsis of genera:

Gnathopods 1 and 2, 5th joint produced into a narrow lobe 1. Gen. **Liljeborgia** p. 230
Gnathopods 1 and 2, 5th joint little or not at all produced 2. Gen. **Idunella** . p. 234

1. Gen. **Liljeborgia** Bate

1861 *Iduna* (non Keyserling & H. Blasius 1840, Aves!), A. Boeck in: Forb. Skand. Naturf., Møde 8 p. 656 | 1862 *Liljeborgia* (part.), Bate (& Westwood), Brit. sess. Crust., *v.* 1 p. 202 | 1862 *L.*, Bate, Cat. Amphip. Brit. Mus , p. 118 | 1876 *L.*, A. Boeck, Skand. Arkt. Amphip., *v.* 2 p. 496 | 1888 *L.*, T. Stebbing in: Rep. Voy. Challenger, *v.* 29 p. 980 | 1871 *Lilljeborgia*, A. Boeck in: Forb. Selsk. Christian., 1870 p. 154 | 1894 *L.*, G. O. Sars, Crust. Norway, *v.* 1 p. 529 | 1865 *Microplax* (non Fieber 1861, Hemiptera!). W. Lilljeborg in: N. Acta Soc. Upsal., ser. 3 *v.* 6 nr. 1 p. 19.

Pleon segment 4 dorsally dentate. Antenna 1 the shorter, accessory flagellum strongly developed. Mandible, cutting edge dentate, accessory plate on· both mandibles, spines of spine-row short, palp slender, with less difference in length than usual between the joints. Maxilla 1, inner plate small, with 1 or 2 setae, outer plate with 10 spines, palp large. Maxilla 2, inner plate the wider. Maxillipeds, outer plates narrow, reaching little beyond 1st joint of elongate palp. Gnathopods 1 and 2, 5th joint produced into a considerable lobe, 6th joint large, oval, finger long, more or less serrate on inner margin. Uropod 3, rami 1-jointed.

8 species accepted, 2 doubtful.

Synopsis of accepted species:

1 { Pleon segments 1 and 2 each with a single dorsal tooth — **2.**
{ Pleon segments 1 and 2 each with. more than 1 dorsal tooth — **6.**

2 { Pleon segment 3 without dorsal tooth; telson cleft nearly to base — **3.**
{ Pleon segment 3 with dorsal tooth; telson not cleft nearly to base — **5.**

3 { Pleon segment 3, postero-lateral angles upturned 1. **L. pallida** . . . . p. 230
{ Pleon segment 3, postero-lateral angles not upturned — **4.**

4 { Eyes conspicuous; peraeopods 4 and 5, finger less than half as long as 6th joint . . . . . . 2. **L. brevicornis** . . p. 231
{ Eyes inconspicuous; peraeopods 4 and 5, finger more than half as long as 6th joint . . . . 3. **L. macronyx** . . . p. 231

5 { Peraeopods 3—5, 2d joint narrowly oblong . . . 4. **L. fissicornis** · . p. 231
{ Peraeopods 3—5, 2d joint broadly ovate . . . . 5. **L. consanguinea** . p. 232

6 { Pleon segments 1 and 2 quinquedentate . . . . 6. **L. dubia** . . . . . p. 233
{ Pleon segments 1 and 2 tridentate — **7.**

7 { Pleon segment 3 not dentate, segment 4 unidentate 7. **L. kinahani** . . . p. 233
{ Pleon segment 3 unidentate, segment 4 bidentate 8. **L. dellavallei** . . p. 234

1. L. pallida (Bate) 1856 *Gammarus? pallidus* (nom. nud.), Bate in: Rep. Brit. Ass., Meet. 25 p. 55 | 1857 *G.? p.*, Bate in: Ann. nat. Hist., ser. 2 *v.* 19 p. 145 | 1862 *Liljeborgia pallida* (part.), Bate & Westwood, Brit. sess. Crust., *v.* 1 p. 203 f. | 1862 *L. p.*, Bate, Cat. Amphip. Brit. Mus., p. 118 t. 20 f. 5.

In many respects agreeing with L. brevicornis; but: Pleon segment 3, postero-lateral angles acutely toothed, with a distinct sinus above the upturned point. Antenna 2, flagellum shorter than ultimate joint of peduncle (a character probably variable and less important). Uropod 3, rami much shorter than peduncle. Colour in life very white, back stained with a rich crimson blotch, gnathopods 1 and 2 with a rosy hue on 6th joint, and the same tint on parts of the peraeopods. L. 5 mm.

Plymouth Sound; off Torquay.

2. L. brevicornis (Bruz.) 1859 *Gammarus b.*, R. M. Bruzelius in: Svenska
Ak. Handl., n. ser. *v.* 3 nr. 1 p. 62 t. 3 f. 11a—o | 1861 *G. b.*, *Iduna* (part.), A. Boeck
in: Forb. Skand. Naturf., Møde 8 p. 657 | 1862 *Liljeborgia pallida* (part.) (err., non *Gammarus pallidus* Bate 1857!), Bate & Westwood, Brit. sess. Crust., *v.* 1. p. 205 | 1871
Lilljeborgia p., A. Boeck in: Forh. Selsk. Christian., 1870 p. 155 | 1876 *L. p.*, A. Boeck,
Skand. Arkt. Amphip., *v.* 2 p. 497 t. 18 f. 9 | ? 1889 *L. p.*, A. M. Norman in: Ann. nat.
Hist., ser. 6 *v.* 4 p. 116 t. 10 f. 10 | 1894 *L. p.*, G. O. Sars, Crust. Norway, *v.* 1 p. 530 t. 187.

Pleon segments 1, 2 and 4 (Sars) (Bruzelius and Boeck: segments 1—5)
each produced to a small dorsal tooth. Head, rostrum distinct but short,
lateral corners projecting, obliquely rounded. Side-plate 1 distally expanded,
1st—3d with small tooth at lower hind corner. Pleon segment 3, postero-lateral
angles acute, lateral margin straight. Eyes large, oval quadrangular, very dark.
Antenna 1, 1st joint nearly twice as long as 2d and 3d combined, 3d very small,
flagellum nearly twice as long as peduncle, about 20-jointed, accessory peduncle
more than $\frac{1}{2}$ as long as primary, 13-jointed. Antenna 2 about $\frac{1}{3}$ longer, ultimate
joint of peduncle rather longer than penultimate, flagellum longer than ultimate
joint of peduncle. Gnathopod 1, 4th joint apically acute, 5th produced to a narrow
setose lobe, 6th large, ovate, palm curved, oblique, spinulose and setulose,
defined from the shorter hind margin by an angle and spines, finger curved,
with 6 strong teeth in proximal half. Gnathopod 2 similar, but larger,
and larger in ♂ than in ♀, palm longer and more oblique, finger serrate
almost throughout, with about 12 teeth. Peraeopods 1 and 2, 5th joint
much shorter than 6th, finger about $\frac{1}{3}$ length of 6th. Peraeopods 3—5, 2d joint
oval, serrate behind. Peraeopod 5, 6th joint rather longer than 5th, tapering,
with fascicles of setae on hind margin, finger about $\frac{1}{3}$ as long, stiliform.
Uropod 3, rami longer than peduncle, lanceolate, subequal in length, but
inner much broader than outer, with 3 or 4 spinules on inner margin.
Telson nearly twice as long as broad, cleft nearly to base, apices a little
dehiscent, bidentate, outer point more produced than inner, a long spine in
each notch. Colour pale orange. L. ♀ 8, ♂ nearly 10 mm.

Arctic Ocean, North-Atlantic, North-Sea and Skagerrak (whole Norway, depth
75—564 m; Bohuslän; Scotland; France).

3. L. macronyx (O. Sars) 1894 *Lilljeborgia m.*, G. O. Sars, Crust. Norway,
v. 1 p. 533 t. 188 f. 2.

Body slender, pleon segments 1, 2 and 4 each produced to a small dorsal
tooth. Head, rostral projection well marked, lateral corners slightly produced,
obtusely acuminate. Pleon segment 3, postero-lateral corners almost quadrate,
produced to a very small tooth. Eyes inconspicuous in spirit. Antenna 1, flagellum
nearly twice as long as peduncle, about 12-jointed, accessory flagellum more than
half as long, 8-jointed. Antenna 2, ultimate and penultimate joints of peduncle
subequal, flagellum less than both combined. Gnathopods 1 and 2 nearly as
in L. brevicornis, but finger in gnathopod 1 with–only 3 teeth at base,
and in gnathopod 2 with 7 on proximal half. Peraeopods 1 and 2, finger half as
long as 6th joint. Peraeopods 3—5, 2d joint large and laminar. Peraeopod 5 very
long, 6th joint smooth, finger nearly as long. Uropod 3, both rami rather
narrow, inner the wider, with 3 small spinules on inner margin. Telson
very narrow, nearly thrice as long as broad, cleft nearly to base, apices
minutely bidentate, with spinule in each notch. L. ♀ 6 mm.

Christianiafjord and Trondhjemsfjord. Depth 376—752 m.

4. L. fissicornis (Sars) 1858 *Gammarus f.*, M. Sars in: Forb. Selsk. Christian.,
p. 147 | 1861 *G. f.*, *Iduna f.*, A. Boeck in: Forh. Skand. Naturf., Møde 8 p. 657 |

1871 *Lilljeborgia f.*, A. Boeck in: Forb. Selsk. Christian., 1870 p. 155 | 1876 *L. f.*, A. Boeck, Skand. Arkt. Amphip., *v.* 2 p. 499 t. 18 f. 10 | 1889 *L. f.*, A. M. Norman in: Ann. nat. Hist., ser. 6 *v.* 4 p. 118 t. 10 f. 11 | 1894 *L. f.*, G. O. Sars, Crust. Norway, *v.* 1 p. 534 t. 189 | 1866 *Gammarus pallidus* (err., non Bate 1857!), Goës in: Öfv. Ak. Förh., *v.* 22 p. 529 t. 40 f. 27 | 1893 *Nicippe pallida* (part.), A. Della Valle in: F. Fl. Neapel. *v.* 20 p. 658.

Body rather slender, pleon segments 1—5 each produced into a dorsal tooth, segments 4 and 5 carinate and the tooth on each large, especially that on segment 4. Head, rostrum small, lateral corners subangular. Side-plate 1 much expanded distally. Pleon segment 3, postero-lateral corners produced to a small tooth. Eyes wholly wanting. Antenna 1, flagellum nearly twice as long as peduncle, 18—20-jointed, accessory flagellum more than half as long, 10—12-jointed. Antenna 2 longer by almost the whole flagellum. Gnathopods 1 and 2 in ♀, palm curved, oblique, defined by a projecting angle and strong palmar spines, finger in gnathopod 1 with 4 teeth at base, in gnathopod 2 with about 9 covering the proximal $^2/_3$ of margin. Gnathopod 1 in ♂ as in ♀, but gnathopod 2 with 6th joint very large, dilated at base, then tapering, palm long, straight or slightly concave, finely ciliated, finger very large, greatly curved, without teeth. Peraeopods 3—5, 2^d joint narrowly oblong, 4th pair much longer than 3^d, and 5th than 4th. In peraeopod 5, 6th joint with small setae on both margins, finger rather short. Uropod 3, rami equal in length, inner the broader, with 3 spines on inner margin. Telson longer in ♂ than in ♀, cleft not much beyond the middle, apices contiguous at the inner points, the outer considerably more produced, a long spine in each notch. Colour uniformly yellow. L. ♀ 10, ♂ 11 mm, but arctic specimens reaching 20 mm.

Arctic Ocean and North-Atlantic (Greenland, Spitzbergen, Trondhjemsfjord, Finmark); Skagerrak (Bohuslän). Depth 94—376 m.

5. **L. consanguinea** Stebb. 1888 *L. c.*, T. Stebbing in: Rep. Voy. Challenger, *v.* 29 p. 980 t. 91 | 1893 *Nicippe pallida* (part.), A. Della Valle in: F. Fl. Neapel, *v.* 20 p. 658.

Pleon segments 1—5 carinate, each produced to a small but pronounced dorsal tooth. Head, rostrum small, well marked, lateral corners narrowly obtuse. Side-plate 1 broadly produced forward, 1st—3^d with a denticle at the front and hind corners, 4th with the hind margin below the excavation straight, serrate into 4 teeth. Pleon segments 1—3, postero-lateral corners acute, in segment 3 slightly upturned with small sinus above. Eyes doubtful. Antenna 1, 2^d joint fully half as long as 1st, flagellum rather longer than peduncle, 13-jointed, accessory flagellum more than half as long, 8- or 9-jointed. Antenna 2 not much longer than antenna 1, ultimate joint of peduncle rather longer than penultimate, flagellum shorter than both combined, 13-jointed. Upper lip very slightly bilobed. Mandible, accessory plate strongly dentate on left, feebly denticulate on right, spine-row of 6 spines, molar almost evanescent, represented by a row of spines, 2^d joint of palp a little longer than 1st, 1st than 3^d. Maxilla 1, inner plate with 2 setae instead of 1 as in the European species, outer plate with 10 spines. Gnathopods 1 and 2 in ♀ as in L. fissicornis (p. 231), but 4th joint of gnathopod 1 distally rounded (?). Peraeopods 3—5 distinguished from those of L. fissicornis by the broadly oval 2^d joint, which also in peraeopod 5 has the hind margin very convex. Uropod 3, rami rather longer than peduncle, inner little broader and rather shorter than outer, with 2 spinules on the inner margin. Telson as in L. fissicornis. L. ♀ 11 mm.

Southern Indian Ocean (Kerguelen Island, depth 36 m; Heard Island, depth 137 m).

6. **L. dubia** (Hasw.) 1880 *Eusirus dubius*, HasWell in: P. Linn. Soc. N.S.Wales, *v.*4 p 331 t. 20 f. 3 | 1885 *E. affinis*, HasWell in: P. Linn. Soc. N. S.Wales, *v.*10 p. 101 t. 14 f. 2—4 | 1888 *Liljeborgia haswelli*, T. Stebbing in: Rep. Voy. Challenger, *v.*29 p. 985 t. 92 | 1893 *Nicippe h.*, A. Della Valle in: F. Fl. Neapel, *v.*20 p. 661 t. 59 f. 68.

Pleon segments 1—5 dorsally dentate, 1st and 2d having 5 teeth each, 3d a very minute tooth between two rounded lobes, 4th and 5th carinate each with a strong tooth; peraeon segments 6 and 7 with a dorsal tooth(?). Head, rostrum very distinct. narrow, acute, lateral corners broadly rounded. Side-plates 1—3 with very small denticles at the corners, 4th with 2 teeth on the hind margin. Pleon segments 2 and 3, postero-lateral corners acute, very slightly produced. Eyes oval, not large, remaining dark in spirit. Antenna 1, 1st joint fully twice as long as 2d and 3d combined, flagellum much longer than peduncle, 34-jointed, accessory flagellum more than half as long, 18-jointed. Antenna 2 considerably longer, penultimate joint of peduncle long, but shorter than ultimate, flagellum shorter than both combined, 24-jointed. Maxilla 1, inner plate with 1 seta. Gnathopods 1 and 2 in ♀ nearly as in L. consanguinea, but 6th joint more broadly oval. In the ♂ gnathopod 2 has the palm of 6th joint well defined, but rather concave than convex, the joint distally narrowed, with the palm produced into a blunt tooth, strong but variable in length, with a cavity between it and the lobe adjoining the hinge of the finger; inner margin of the finger somewhat convex at the middle, cut into teeth. Peraeopods 1—5, finger short. Peraeopods 3—5, 2d joint rather broad, hind margin not very convex, strongly serrate. Uropod 3; rami lanceolate, inner rather the broader and longer, with 4 spines on the inner margin. Telson long and narrow, cleft nearly to the base, outer point of bidentate apices produced much beyond the inner, a long spine in each notch. Colour in spirit showing a lustrous green on many parts. L. 15 mm, but often much smaller.

Bass Strait and South-Pacific (Tasmania; Port Jackson, Port Stephens and Jervis Bay [East-Australia]; East Moncoeur Island, depth 69 m).

7. **L. kinahani** (Bate) 1862 *Phaedra k.*, Bate (& WestWood), Brit. sess. Crust., *v.*1 p. 211 f. | 1862 *P. k.*, Bate, Cat. Amphip. Brit. Mus., p. 119 t. 21 f. 1 | 1876 *Liljeborgia k.*, A. Boeck, Skand. Arkt. Amphip., *v.*2 p. 497 | 1888 *Lilljeborgia k.*, Chevreux in: C.-R. Ass. Pranç., Sess. 16 *v.*2 p. 666 | 1894 *L. k.*, G. O. Sars, Crust. NorWay. *v.*1 p. 532 t. 188 f. 1.

Body very short and stout. Pleon segments 1, 2, 4 and 5 each produced into a well marked dorsal tooth, this in segments 1 and 2 (Sars) accompanied by a smaller tooth on each side. Peraeon segments 5—7 and(?) pleon segment 3 dorsally dentate (Bate). Head, rostrum distinct, lateral corners obtusely acuminate. Side-plate 1 broader than 2d or 3d, but scarcely expanded distally, Pleon segment 3, postero-lateral corners acute, slightly upturned, with very small sinus above. Eyes small, round, black. Antenna 1 short, flagellum not nearly twice as long as peduncle, 12-jointed, accessory flagellum more than half as long, 7-jointed. Antenna 2, ultimate joint of peduncle scarcely longer than penultimate. Gnathopod 1 as in L. brevicornis (p. 231), but 2d joint abruptly dilated near base, finger with only 3 teeth at base. Gnathopod 2, finger with 6 teeth on proximal half. Peraeopod 5, 6th joint setose on outer margin, finger about half as long, very slender. Uropod 3, outer ramus quite unarmed, inner with a single spinule on inner margin. Telson cleft nearly to base, lobes divergent, a long spine in the notch of each apex, the points of the apices level. Colour whitish, pellucid (Sars), in front tinted with violet, behind rosy white (Chevreux). L. 3 mm.

North-Atlantic and North-Sea (NorWay, depth 11—19 m; France; Great Britain).

8. **L. dellavallei** Stebb.*) 1893 *Nicippe pallida* (part.) (err., non *Gammarus pallidus* Bate 1857!), A. Della Valle in: F. Fl. Neapel, *v.*20 p. 658 t. 1 f. 1; t. 19 f. 35—52.

Rather robust. Pleon segments 1 and 2 each produced into 3 dorsal teeth, 3 and 5 each into 1 tooth, segment 4 into 2 teeth. Head, rostrum small, acute, slightly deflexed, lateral corners rounded. Side-plate 1 distally expanded, corners rounded. Pleon segments 2 and 3, postero-lateral angles acute, upturned, with a small sinus above. Eyes white, in ♀ rather small, rounded oblong, in ♂ larger, irregular, approximating above. Antenna 1, 1st joint longer than 2^d and 3^d combined, flagellum twice as long as peduncle, with about 20 short joints, accessory flagellum with more than 12 joints. Antenna 2, ultimate joint of peduncle longer than penultimate, rather shorter than 15-jointed flagellum. Gnathopod 1, finger with 7 teeth on proximal half. Gnathopod 2, 11 teeth covering about $2/_3$ length of inner margin of finger. Peraeopods 3—5, 2^d joint oval, broadly so in peraeopod 5, hind margin strongly serrate, finger small and delicate. Uropod 3, rami broadly lanceolate, little longer than peduncle. Telson cleft about $2/_3$ of the length, each apex seemingly evenly bidentate. Colour variable, sometimes uniformly grey, sometimes wine-red to peraeon segment 6, then yellowish grey. L. 5—6 mm.

Gulf of Naples.

L. bispinosa (A. Costa) 1857 *Gammarus bispinosus*, A. Costa in: Mem. Acc. Napoli, *v.*1 p. 223 t. 3 f. 9 | 1876 *Liljeborgia b.*, A. Boeck, Skand. Arkt. Amphip., *v.*2 p. 497 | ?1893 *Gammarus locusta* (part.), A. Della Valle in: F. Fl. Neapel, *v.* 20 p. 660.

L. 5 mm.

Gulf of Naples.

L. pugettensis (Dana) 1853 & 55 *Gammarus p.*, J. D. Dana in: U. S. expl. Exp., *v.* 13 II p. 957; t. 66 f. 1 a—g | 1876 *G. pugetensis*, *Cheirocratus* (part.), A. Boeck, Skand. Arkt. Amphip., *v.* 2 p. 395 | 1894 *G. p.*, *C.* (part.), G. O. Sars in: Crust. NorWay, *v.* 1 p. 523 | 1893 *G. locusta* (part.), A. Della Valle in: F. Fl. Neapel, *v.* 20 p. 760.

Third joint of palp of mandible seems to be longer than in other species of Liljeborgia. L. 18 mm.

Puget Sound [Western North-America].

2. Gen. **Idunella** O. Sars

1894 *Idunella* (Sp. un.: *I. aeqvicornis*), G. O. Sars, Crust. Norway, *v.* 1 p. 536 | 1896 *Idurella*, Pocock in: Zool. Rec., *v.* 32 Crust. p. 41.

Head, rostrum not large. Accessory flagellum of antenna 1 not very large. Mouth-parts probably agreeing nearly with those of Liljeborgia (p. 230). Gnathopods 1 and 2, 5th joint not or very slightly produced.

2 species.

Synopsis of species:

1. **I. aequicornis** (O. Sars) 1876 *Lilljeborgia aeqvicornis*, G. O. Sars in: Arch. Naturv. Kristian., *v.*2 p. 255 | 1894 *Idunella a.*, G. O. Sars, Crust. NorWay, *v.*1 p. 537 t. 190.

Rather slender. Pleon segment 2 produced to a small dorsal tooth. Head, rostrum small, lateral corners narrowly rounded. Side-plate 1 much expanded,

---

*) Sp. nov. After Antonio Della Valle.

broadly rounded in front, 1st—3^d with denticle at lower hind corner, 4th large, quadrate, emarginate above. Pleon segment 3, postero-lateral corners acutely upturned, a small sinus between the point and the bulging hind margin. Eyes represented by small ovoid patches of whitish pigment. Antenna 1, 1st joint scarcely longer than 2^d and 3^d combined, flagellum subequal to the peduncle, about 12-jointed, accessory flagellum not half as long, slender, 5-jointed. Antenna 2 scarcely as long as antenna 1, ultimate joint of peduncle a little shorter than penultimate, flagellum shorter than the 2 combined. Mandibular palp more fully developed than in Liljeborgia, 3^d joint long and falciform. Maxillipeds, palp smaller than in Liljeborgia. Gnathopod 1 in ♀, 2^d joint somewhat fusiform, 5th very slightly produced, 6th very large, gradually widened distally, palm nearly transverse, slightly convex, defined by a distinct angle and 2 unequal spines, finger slender and curved. Gnathopod 2 in ♀ similar, but with 6th joint a little smaller. Gnathopod 1 in ♂ very large, 6th joint longer than joints 2—5 combined, palm long, concave, very oblique, defined by an acute tooth, the joint greatly narrowed towards the hinge near which the palm ends in a small denticle, finger long, slender, greatly curved. Gnathopod 2 in ♂ as in ♀, much smaller than gnathopod 1. Peraeopods 1 and 2 very slender. Peraeopods 3—5, 2^d joint expanded, oval, peraeopod 5 the longest, its 6th joint a little longer than 5th, sparingly setose, finger rather shorter and thin. Uropod 3, rami subequal, inner lanceolate, outer much narrower, 2-jointed, 2^d joint spiniform. Telson about twice as long as broad, little tapering, cleft nearly $^3/_4$ length, inner point of bidentate apices the longer, each with 2 spines in the notch. Colour uniformly pale yellow. L. scarcely 7 mm.

Arctic Ocean (Varangerfjord, depth 94—188 m; Storeggen; Jan Mayen; Barents Sea).

2. **I. picta** (Norm.) 1889 *Lilljeborgia p.*, A. M. Norman in: Ann. nat. Hist., ser. 6 *v.* 4 p. 116 t. 10 f. 5—9.

Pleon segments 1, 2, 4 and 5 produced each into a dorsal tooth. Head, rostrum small, lateral corners scarcely produced. Eyes (in fig.) oval, dark. Antenna 1 shorter than peduncle of antenna 2, flagellum with about 7 joints, accessory flagellum with only 3, the last very minute. Antenna 2, ultimate joint of peduncle $^2/_3$ as long as penultimate, flagellum a little longer than ultimate, 8-jointed. Mouthparts not described. Gnathopod 1, 5th joint not produced, 6th widening gradually from base, palm oblique, well arched, scarcely more than $^1/_3$ of total length of joint. Gnathopod 2, 5th joint not produced, 6th ovate, greatest width at commencement of palm, which is slightly defined, a little longer than hind margin. Peraeopods 1 and 2, finger little more than $^1/_4$ length of 6th joint. Peraeopod 5, 6th joint slightly longer than 5th, finger scarcely more than $^1/_4$ as long. Uropod 3, rami very broad and foliaceous. Colour pale with markings of deep purple, a spot on ultimate joint of peduncle in antennae 1 and 2, the purple suffused over part of head, and most of peraeon, but on peraeopods and the other parts in spots and blotches. L. 6 mm.

English Channel (Guernsey).

19. Fam. **Oedicerotidae**

- 1865 Subfam. *Oedicerina*, W. Lilljeborg in: N. Acta Soc. Upsal., ser. 3 *v.* 6 nr. 1 p. 18 | 1871 Subfam. *Oedicerinae*, A. Boeck in: Forh. Selsk. Christian., 1870 p. 160 ! 1883 *Oediceridae*, J. S. Schneider in: Tromsø Mus. Aarsh., *v.* 6 (p. 1) | 1888 *O.*, T. Stebbing in: Rep. Voy. Challenger, *v.* 29 p. 835 | 1892 *O.*, G. O. Sars, Crust. Norway, *v.* 1 p. 286 | 1893 *Oediceridi*, A. Della Valle in: F. Fl. Neapel, *v.* 20 p. 531.

Side-plates of moderate size, fringed with setae. Pleon segments 1—3, postero-lateral corners usually rounded. Eyes (Fig. 61, 62), when present and distinctly developed, usually contiguous, approximate or wholly confluent, dorsal and more or less frontal. Antenna 1 (Fig. 61, 67) with accessory flagellum absent or rudimentary. Epistome not projecting. Upper lip not bilobed. Lower lip with inner lobes separate or coalesced. Mandible, molar variable, palp usually large. Maxillipeds, plates well developed, but seldom large. Gnathopod 1 (Fig. 65) subchelate. Gnathopod 2 (Fig. 66) subchelate or rarely chelate. Peraeopods 1—4 (Fig. 63), 4th joint moderately large. Peraeopods 3 and 4, 2^d joint elliptic, fringed with plumose setae. Peraeopod 5 (Fig. 64) very long, 2^d joint expanded, 7th stiliform. Branchial vesicles usually simple, large. Uropods 1—3 commonly extending equally far back. Telson small, entire.

Marine; many species are known as sand-burrowers.

18 genera, 56 accepted species and 6 doubtful.

Synopsis of genera:

1. Gen. **Perioculodes** O. Sars

1892 *Perioculodes* (Sp. un.: *P. longimanus*), G. O. Sars, Crust. Norway, *v.* 1 p. 312 |
1895 *P.,* A. M. Norman in: Ann. nat. Hist., ser. 6 *v.* 15 p. 486.

Side-plates moderately deep. Rostral projection short, deflexed. A single eye (Fig. 61) encircling front of head, lenses highly refractive. Antenna 1 in ♀ rather longer than antenna 2, 1ˢᵗ, 2ᵈ and 3ᵈ joints subequal, in ♂ (Fig. 61) 2ᵈ shorter than 1ˢᵗ, 3ᵈ than 2ᵈ, flagellum stronger than in ♀. Antenna 2 in ♂ (Fig. 61), flagellum long, filiform. Upper lip broadly truncate. Lower lip, inner lobes coalesced. Mandible not strong, molar feeble, conical, tipped with 3 spinules, palp in ♀ rather small, 3ᵈ joint short; in ♂ larger. Maxillae 1 and 2 nearly as in Monoculodes (p. 258). Maxillipeds, inner plates small, outer nearly reaching apex of 2ᵈ joint of palp, armed with spaced spines. Gnathopods 1 and 2, process of 5ᵗʰ joint very long, stiliform, adjacent to and overtopping hind margin of the narrow 6ᵗʰ joint, which is long with short palm. Peraeopods 1—5 and uropods 1—3 nearly as in Monoculodes. Telson oblong, apically rounded.

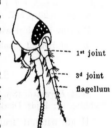

──── 1ˢᵗ joint

──── 3ᵈ joint
── flagellum

Fig. 61.
P. longimanus, ♂.
Head and antennae.
[After G. O. Sars.]

1 species accepted, 2 doubtful.

1. **P. longimanus** (Bate & Westw.) 1868 *Monoculodes l.,* Bate & Westwood, Brit. sess. Crust., *v.* 2 p. 507 | 1892 *Perioculodes l.,* G. O. Sars, Crust. Norway, *v.* 1 p. 313 t. 110 f. 2; t. 111 f. 1 | 1895 *P. l.,* A. M. Norman in: Ann. nat. Hist., ser. 6 *v.* 15 p. 486 | 1893 *Oediceros l.* (part.). A. Della Valle in: F. Fl. Neapel, *v.* 20 p. 547 t. 4 f. 9; t. 33 f. 32—36 | 1871 *Monoculodes grubei,* A. Boeck in: Forh. Selsk. Christian.. 1870 p. 165 | 1888 *M. aequimanus* (err., non *Oedicerus a.* Kossmann 1880!), (A. M. Norman in MS.) D. Robertson in: P. nat. Hist. Soc. Glasgow, n. ser. *v.* 2 p. 30.

Head, rostral projection short, broad, triangular, lateral corners rounded. Side-plate 1 rather expanded and obliquely truncate, 2ᵈ and 3ᵈ oblong, with small tooth at lower hind corner, 4ᵗʰ very large, lower hind corner a little produced, obtuse, 5ᵗʰ broader, less deep. Eyes broadest at top, bright scarlet with about 12 brilliantly iridescent lenses on each side. Antenna 1 in ♀, flagellum little over half length of peduncle, with 6 joints, in ♂ flagellum as long as peduncle, 1ˢᵗ joint very large, densely clothed (the sensory hairs not shown in Fig. 61). Antenna 2 in ♀, ultimate and penultimate joints of peduncle subequal, together as long as flagellum. Antenna 2 in ♂ twice as long as in ♀. Gnathopod 1, process of 5ᵗʰ joint rather flexuous, nearly reaching end of 6ᵗʰ, which is about thrice as long as broad, palm nearly as long as hind margin. Gnathopod 2, process of 5ᵗʰ joint extending much beyond 6ᵗʰ joint, which is 4 times (Sars) or about 5 times (Norman) as long as broad, palm much shorter than hind margin. Peraeopods 1—4 moderately slender, densely setose, in 1ˢᵗ and 2ᵈ finger narrowly acuminate, not half length of broad 6ᵗʰ joint, in 3ᵈ and 4ᵗʰ lanceolate, compressed. Peraeopod 5, 2ᵈ joint quadrangular oval, 5ᵗʰ and 6ᵗʰ subequal, each longer than 4th. Uropod 3, rami as long as peduncle, very narrow, unarmed. Telson nearly twice as long as broad, apex evenly rounded. Colour pale orange; ova in pouch dark bluish. L. 4 mm.

Arctic Ocean, North-Atlantic, North-Sea and Skagerrak (Norway, France, Great Britain); Kattegat; Mediterranean. Shallow water.

P. aequimanus (Kossm.) 1880 *Oedicerus a.*, Kossmann, Reise Roth. Meer., *v.* 21 Malacost. p. 130 t. 13 f. 6—8 | 1893 *Oediceros longimanus* (part.), A. Della Valle in: F. Fl. Neapel, *v.* 20 p. 547.

Probably young ♀ (doubtful marsupial plates on peraeopods 3—5).

Red Sea.

P. megapleon (Giles) 1888 *Monoculodes m.*, G. M. Giles in: J. Asiat. Soc. Bengal, *v.* 57 p. 235 t. 7 f. 12 | 1893 *Oediceros longimanus* (part.), A. Della Valle in: F. Fl. Neapel, *v.* 20 p. 547.

Perhaps identical with P. longimanus (p. 237). L. 3·2 mm.

Bay of Bengal (banks off Chittagong). Surface.

2. Gen. **Gulbarentsia** Stebb.

1894 *Barentsia* (Sp. un.: *B. hoeki*) (non T. Hincks 1880, Bryozoa!), *Gulbarentsia*, T. Stebbing in: Bijdr. Dierk., *v.* 17 p. 25, 2.

Head without frontal process or rostrum (Fig. 62). Side-plates not large, 1st distally expanded, 4th not excavate behind. Eye single, prominent, forming a semicircular ring (Fig. 62). Antenna 1 shorter than antenna 2, flagellum in both many-jointed. Upper lip with rounded sides, distally narrowed. Lower lip, inner lobes small, distinct. Mandible normal, but molar small, palp elongate, 3[d] joint longer than 2[d]. Maxilla 1, inner plate with 2 setae, outer with 7 spines, 2[d] joint of palp apically narrowed. Maxilla 2, plates subequal. Maxillipeds, inner plates small, outer reaching half way along the large and long 2[d] joint of palp, closely fringed with slender spines. Gnathopods 1 and 2 similar, 5th joint much shorter than 6th, with strongly projecting setose process, 6th long oval, the long spinose palm moderately defined. Peraeopods 1 and 2, finger much shorter than 6th joint. Peraeopods 3 and 4, 2[d] joint little expanded, finger longer than 6th joint, with a cap over the nail.

1 species.

1. G. hoeki Stebb. 1894 *Barentsia h.*, *Gulbarentsia*, T. Stebbing in: Bijdr. Dierk., *v.* 17 p. 25 t. 5.

Head, front broadly convex, lateral angles acute, a dorsal carina behind the eye (Fig. 62). Peraeon segments 1—6 transversely ridged on front and

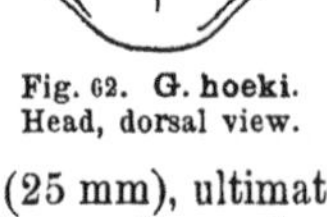

Fig. 62. **G. hoeki.** Head, dorsal view.

hind margins, 5th and 6th with median carinate tooth and postero-lateral angles acute. Side-plates partially serrate and spinulose, 5th and 6th with hind angles of lobes acute. Eye large, collar-like, with many lenses, separated from front of head by a narrow groove. Antennae 1 and 2 little armed. Antenna 1 (18 mm), 1st joint twice as long as 2[d], flagellum nearly as long as peduncle, 36-jointed. Antenna 2 (25 mm), ultimate and penultimate joints of peduncle subequal, flagellum rather longer than peduncle, with 160 joints, many very short and broad. Gnathopod 2, side-plates apically narrowed, 2[d] and 6th joints and lobe of 5th longer than in gnathopod 1, palm rather better defined in gnathopod 1. Peraeopods 1—4, 6th joint curved, spinose on convex margin, the shorter 5th joint armed on the opposite margin. Specimen defective after peraeon segment 6. L.?

Kara Sea (lat. 72° N., long. 65° E.).

3. Gen. **Exoediceros** Stebb.

1899 *Exoediceros* (Sp. typ.: *Oedicerus fossor*), T. Stebbing in: Ann. naf. Hist., ser. 7 *v.* 4 p. 208.

Distinguished from Oediceros (p. 243) by having the rostrum little pronounced, the eyes not contiguous though well developed, antenna 1 with rudiment of accessory flagellum, mandible with well developed molar and the palp's 2d and 3d joints broad, maxilla 1 with numerous setae fringing the inner plate, maxillipeds with inner plates broad, gnathopods 1 and 2 with 5th joint at least as large as 6th, peraeopods 1 and 2 without finger, peraeopods 3 and 4 with minute upturned finger, uropod 2 not reaching so far back as uropod 1 or uropod 3.

1 species.

1. **E. fossor** (Stimps.) 1855 *Oedicerus f.*, Stimpson in: P. Ac. Philad., *v.*7 p. 393 | 1899 *O. f.*, *Exoediceros sp. typ.*, T. Stebbing in: Ann. nat. Hist., ser. 7 *v.* 4 p. 208 | 1880 *O. arenicola*, *O. f.?*, Haswell in: P. Linn. Soc. N.S.Wales, *v.*4 p.325 t.24 f.3 | 1893 *Oediceros a.*, *Halimedon sp.?*, A. Della Valle in: F. Fl. Neapel, *v.*20 p.556.

Eyes small, rounded oval. Antenna 1, joints of peduncle successively a little shorter, flagellum attaining to about 24 distally widened joints carrying small calceoli, accessory flagellum a little blunt joint tipped with long setae. Antenna 2, flagellum as in antenna 1, at least in ♀. Upper lip rounded. Maxilla 2, with outer lobe the broader, inner with row of setae curving from surface along inner margin. Gnathopods 1 and 2 rather larger in ♂ than in ♀ (Haswell), 2d gnathopod larger than 1st, but similar, 5th joint distally wide, setose, lobed, not produced along the oval 6th. Peraeopods 1—4 have the 4th and 5th joints distally widened, 4th much longer than 5th, both strongly setose, 6th narrow, with rows of setiform spines and spinules. Peraeopods 3—5 have the 2d joint oval, with numerous setae projecting from within the expansion. Peraeopod 5 elongate as in Oediceros. Uropods 1 and 2 with few spines on the slender subequal rami, which are much longer in the 1st than in the 2d pair. Uropod 3, rami equal, laminar, as long as peduncle, truncate apically, with numerous setae or spines. Telson at least as broad as long, narrowing slightly to broadly rounded apex, length less than peduncle of uropod 3. Colour white with a few blackish spots. L. 5—7·5 mm.

Shark Island and Port Jackson [East-Australia]. Burrowing in sand above high-water mark.

4. Gen. **Pontocrates** Boeck

1871 *Pontocrates* (part.), A. Boeck in: Forh. Selsk. Christian., 1870 p. 171 | 1888 *P.*, T. Stebbing in: Rep. Voy. Challenger, *v.* 29 p. 307 | 1892 *P.*, G. O. Sars, Crust. Norway, *v.* 1 p. 315.

Head with short deflexed rostrum. Side-plates 4 and 5 rather large. Eyes contiguous, at base of rostrum. Antenna 2 in ♀ much or little longer than antenna 1, in ♂ elongate. Upper lip nearly quadrate. Lower lip, inner lobes separate from one another, but coalesced with outer lobes. Mandible normal. Maxilla 1, inner plate with 2 setae, outer with 9 spines. Maxillipeds, outer plates scarcely reaching beyond middle of large 2^d joint of palp. Gnathopods 1 and 2 very dissimilar. Gnathopod 1, 5th joint with short base and long process, 6th oval, palm well defined. Gnathopod 2, 5th joint with short base and process produced beyond the chela which is formed between produced end of narrow 6th joint and short finger (the process appears to be partially coalesced with the 6th joint, though less so than in Synchelidium (p. 241). Peraeopods 1—4 rather stout, densely setose, finger minute. Peraeopod 3 very short. Peraeopod 5, 2^d joint broad oval, subequal in length to 4th, 5th, 6th or 7th. Uropod 3, rami narrow, longer than peduncle. Telson small.

3 species.

Synopsis of species:

1 { Antenna 2 in ♀ much longer than antenna 1 . . . 1. **P. altamarinus** . p. 240
{ Antenna 2 in ♀ little longer than antenna 1 — **2.**

2 { Gnathopod 2, thumb of chela much broader than finger, process of 5th joint little produced beyond 6th . . . . 2. **P. arcticus** . . . p. 240
{ Gnathopod 2, thumb of chela scarcely broader than finger, process of 5th joint much produced beyond 6th . . . . 3. **P. arenarius** . . p. 241

1. **P. altamarinus** (Bate & Westw.) 1862 *Kroyera altamarina*, Bate & Westwood, Brit. sess. Crust., *v.* 1 p. 177 f. | 1893 *Kröyera a.*, A. Della Valle in: F. Fl. Neapel, *v.* 20 p. 554 | 1895 *Pontocrates altamarinus*, G. O. Sars, Crust. Norway, *v.* 1 p. 695 t. vii f. 2. ·

Moderately robust. Rostrum a little projecting, moderately deflexed. Side-plate 1 expanded and rounded distally, 2^d and 3^d slightly narrowed distally, 4th rather produced at lower hind corner. Eyes rather large, yellowish red. Antenna 1 much shorter than antenna 2 in both sexes, 2^d joint fully as long as 1st, flagellum shorter than peduncle, 8-jointed. Antenna 2 in ♀, ultimate joint of peduncle longer than penultimate, flagellum longer than peduncle. Gnathopod 1, process of 5th joint not reaching much beyond hind margin of 6th, which is rather large, oval fusiform, palm obliquely curved, much longer than hind margin, densely fringed with minute curved spinules. Gnathopod 2, process of 5th joint produced very little beyond chela, 6th joint slightly tapering, thumb of chela little broader than finger, fringed as in palm of gnathopod 1. Peraeopods 1 and 2 moderately robust, 6th joint narrower than 5th, but subequal in length. Uropod 3, peduncle rather long, rami unarmed (Sars, but?). Telson oval, apex a little insinuate. Colour semipellucid, dorsally marbled with opaque light yellowish, antennae banded with orange; ova in pouch reddish brown. L. ♀ nearly 7 mm.

North-Atlantic and North-Sea (Skudesnes, depth 38 m; France; British Isles).

2. **P. arcticus** O. Sars 1883 *P. norvegicus* (err., non *Oediccrus n.* A. Boeck 1861!), J. S. Schneider in: Tromsø Mus. Aarsh., *v.* 6 p. 17 t. 2 f. 15; t. 3 f. 21, 22 | 1889 *P. n.*, Hoek in: Tijdschr. Nederl. dierk. Ver., ser. 2 *v.* 2 p. 193 t. 9 f. 8 | 1892 *P. n.*, G. O. Sars, Crust. Norway, *v.* 1 p. 315 t. 111 f. 2 | 1895 *P. arcticus*, G. O. Sars, Crust. Norway, *v.* 1 p. 693.

Rather stout, tumid in front. Rostrum short, much deflexed. Side-plate 1 with lower margin flattened, slightly insinuate, 2^d rather narrowed distally, 3^d not so, 4th rather produced at lower hind corner. Eyes rounded, light red with whitish coating. Antenna 1 in ♀ nearly as long as antenna 2, 2^d joint shorter than 1st, flagellum nearly as long as peduncle, 8- or 9-jointed. Antenna 2, ultimate and penultimate joints of peduncle subequal, flagellum in ♀ a little, in ♂

very greatly, longer than peduncle. Gnathopod 1, process of 5th joint reaching much beyond hind margin of 6th, which is oval, constricted at base, considerably expanded distally, palm obliquely curved, somewhat longer than hind margin, edge smooth with row of minute curved spinules. Gnathopod 2, process of 5th joint produced very little beyond chela, 6th joint scarcely tapering, nearly 6 times (Sars) or 5 times (Schneider) as long as broad, thumb of chela much broader than finger. Peraeopods 1 and 2 rather stout, 6th joint shorter than 5th, finger extremely small. Uropod 3, peduncle rather long, rami each with 2 spinules. Telson rounded, apex obtusely truncate, beset with 8 spinules (Schneider). Colour yellowish, antero-dorsally tinged with orange. L. ♀ 6 mm.

Arctic Ocean (Finmark). Depth 11—19 m.

3. **P. arenarius** (Bate) 1858 *Kroyera arenaria*, Bate in: Nat. Hist. Northumb., v. 41 p. 15 t. 2 f. 1 | 1862 *Kröyera a.*, *Kroyea* (laps.) *a.*, Bate, Cat. Amphip. Brit. Mus., p. 106 t. 17 f. 4 | 1893 *Kröyera a.*, A. Della Valle in: F. Fl. Neapel, v. 20 p. 554 | 1888 *Pontocrates arenarius*, T. Stebbing in: Rep. Voy. Challenger, v. 29 p. 307 | 1889 *P. a.*, Hoek in: Tijdschr. Nederl. dierk. Ver., ser. 2 v. 2 p. 192 t. 9 f. 7, 8 | 1861 *Oedicerus norvegicus*, A. Boeck in: Forh. Skand. Naturf., Møde 8 p. 650, 652 | 1871 *Pontocrates n.*, A. Boeck in: Forh. Selsk. Christian., 1870 p. 171 | 1876 *P. n.* + *P. n. var.*, A. Boeck, Skand. Arkt. Amphip., v. 2 p. 288 t. 16 f. 4; t. 15 f. 7 | 1895 *P. n.*, G. O. Sars, Crust. Norway, v. 1 p. 693 t. VI f. 2; t. VII f. 1.

Robust, segments dorsally sharply defined. Rostrum short, slightly curved. Side-plate 1 with distal part transversely truncate, 2^d tapering to an obtuse point, 3^d with oblique slightly insinuate distal margin, 4th much deeper than the preceding, little produced at lower hind corner. Eyes rounded, light red with no whitish coating. Antenna 1 in ♀ not much shorter than antenna 2, 2^d joint a little shorter than 1st, flagellum subequal to peduncle. with 9 joints, in ♂ stouter, flagellum with 11 joints, first 6 or 7 densely hirsute. Antenna 2, ultimate and penultimate joints of peduncle subequal, flagellum in ♀ as long as peduncle, in ♂ very much longer. filiform. Gnathopod 1, process of 5th joint reaching beyond 6th, which gradually widens distally, palm nearly transverse, not as long as hind margin, densely fringed with long stiff setae. Gnathopod 2 very slender, process of 5th joint produced considerably beyond chela, 6th joint very narrow, thumb of chela little broader than finger. Peraeopods 1 and 2, 5th joint broad, scarcely longer than 6th, finger almost obsolete. Peraeopod 5, 2^d joint oval quadrangular, with straight hind margin. Uropod 3, peduncle shorter than in the 2 preceding species, rami with a few spinules. Telson oval, apex evenly rounded. Colour semipellucid, whitish, slightly tinged with yellowish. L. ♀ 6, ♂ 6·5 mm.

North-Sea (Skudesnes, depth 19 m; Dutch and British coasts); Christianiafjord.

5. Gen. **Synchelidium** O. Sars

1871 *Pontocrates* (part.), A. Boeck in: Forh. Selsk. Christian., 1870 p. 171 | 1892 *Synchelidium*, G. O. Sars, Crust. Norway, v. 1 p. 317.

Head with more or less strongly deflexed rostrum. Side-plate 1 distally expanded, with straight lower margin, 4th and 5th rather large. Eyes contiguous, at base of rostrum. Antenna 1 in ♀ longer than antenna 2, 3^d joint rather long. Antenna 2 in ♂ very elongate. Mandible, molar small, conical, tipped with 1 spine, 3^d joint of palp in ♀ small. Maxilla 1 as in Pontocrates (p. 239). Maxilla 2, plates short, outer apically truncate. Maxillipeds, outer plates with only a few strong spines. Gnathopod 1 powerful, nearly as in Pontocrates, but process of 5th joint tipped by a strong spine, palm coarsely dentate. Gnathopod 2, process of 5th joint coalesced with 6th, except near the small chela, in which thumb and finger are both narrow. Other parts in agreement with Pontocrates.

4 species.

Synopsis of species:

1 { Blotched with brown — **2.**
  { Pellucid, without pigmentary ornament — **3.**

2 { Gnathopod 1, 6th joint broadly oval . . . . . . 1. **S. haplocheles** . . p. 242
  { Gnathopod 1, 6th joint elongate oval . . . . . 2. **S. maculatum** . . p. 242

3 { Rostrum much deflexed, lateral corners of head
  {   subacute . . . . 3. **S. tenuimanum** . p. 243
  { Rostrum little deflexed, lateral corners of head
  {   rounded . . . . 4. **S. intermedium** . p. 243

1. S. haplocheles (Grube) 1864 *Kroyeria h.*, E. Grube, Lussin, p. 72 | 1893 *Kröyera h.* (part.), A. Della Valle in: F. Fl. Neapel, *v.* 20 p. 553 t. 3 f. 15; t. 34 f. 35—39, 1868 *Kroyera brevicarpa*, Bate & Westwood, Brit. sess. Crust., *v.* 2 p. 508 f. | 1892 *Synchelidium brevicarpum*, G. O. Sars, Crust. Norway, *v.* 1 p. 318 t. 112 f. 1.

Not robust, but back broadly rounded. Rostrum short, evenly curved, scarcely reaching beyond middle of 1st joint of antenna 1. Eyes rather large, rounded, bright red, with whitish coating. Antenna 1 in ♀ slender, 1st and 2^{d} joints subequal, 3^{d} not much shorter, flagellum much shorter than peduncle, 5-jointed, in ♂ 2^{d} and 3^{d} joints shorter and thicker, flagellum as long as peduncle, its 1st joint large, densely clothed. Antenna 2 in ♀ shorter than antenna 1, ultimate and penultimate joints of peduncle subequal, together as long as flagellum. Gnathopod 1, process of 5th joint a little dilated, then tapering, produced considerably beyond hind margin of 6th, which is wide at the middle, fusiform, palm much longer than hind margin, defined by an angular projection and strong spine, its edge denticulate, 6 of the denticles large and blunt (Sars, but scarcely so in Della Valle's Fig.). Gnathopod 2 rather slender, 6th joint thickened at base, thence tapering, chela about $^{1}/_{5}$ of its length, a little overtopped by process of 5th joint. Peraeopods 1 and 2 moderately strong, 5th and 6th joints subequal, finger minute. Peraeopods 3 and 4, finger less insignificant. Peraeopod 5, 5th joint the longest. Uropod 3, rami very slightly longer than peduncle, slender, unarmed. Telson rather large, oblong oval, apex slightly insinuate. Colour whitish, variegated with dark brown, this forming transverse bands on some segments. L. 4—5·5 mm.

North-Atlantic, North-Sea and Skagerrak (South- and West-Norway, Trondhjems-fjord, depth 19—56 m; Great Britain); Mediterranean.

2. S. maculatum Stebb.*) 1893 *Kröyera arenaria* (err., non *Kroyera a.* Bate 1858!), A. Della Valle in: F. Fl. Neapel, *v.* 20 p. 554 t. 4 f. 1; t. 34 f. 18- -34.

Slender. Eyes red. Antenna 1, 1st joint a little longer than 2^{d}, 2^{d} than 3^{d}, flagellum as long as 2^{d} and 3^{d} combined, with 4 rather elongate joints. Antenna 2, ultimate joint of peduncle shorter than penultimate, flagellum 6-jointed. Gnathopod 1, process of 5th joint reaching end of hind margin of 6th, which is ellipsoidal, more than twice as long as broad, palm longer than hind margin,

---

*) Sp. nov.

defined by a stout spine, divided into numerous little teeth, of which 5 interspersed are more conspicuous than the rest. Gnathopod 2, 2^d joint distally constricted, process of 5th showing separation or suture at base, then coalesced till near the chela, which it a little overtops, long 6th joint rather wide at base, the narrow thumb fringed with minute denticle and 4 larger hooked ones at apex, finger apically hooked. Telson subquadrate, not longer than broad. Colour reddish yellow, with large brown scars of various shapes on head and trunk, all the appendages pellucid, violet-grey. L. 4—5 mm.

Mediterranean.

3. **S. tenuimanum** Norm. 1871 *Pontocrates haplocheles* (err., non *Kroyeria h.* E. Grube 1864!), A. Boeck in: Forh. Selsk. Christian., 1870 p.172 | 1876 *P. h.*, A. Boeck, Skand. Arkt. Amphip., *v.*2 p.289 t.16 f.3 | 1892 *Synchelidium h.*, G. O. Sars, Crust. Norway, *v.*1 p.319 t.112 f.2 | 1893 *Kröyera h.* (part.), A. Della Valle in: F.Fl. Neapel, *v.*20 p.553 | 1895 *Synchelidium tenuimanum*, A. M. Norman in: Ann. nat. Hist., ser. 6 *v.* 15 p. 486.

Moderately stout. Head strongly arched, rostrum vertically deflexed, acute, nearly reaching end of 1st joint of antenna 1; lateral corners angularly produced. Eyes rather small, lenses few, bright red with chalky white coating. Antenna 1 in ♀, 1st joint longer than 2^d, flagellum subequal to peduncle, 7-jointed. Antenna 2 in ♀ shorter than antenna·1, flagellum subequal to peduncle. Gnathopod 1, 5th joint as in S. haplocheles, 6th much expanded, oval triangular, palm divided into 7 or 8 coarse denticles. Gnathopod 2 extremely slender, 6th joint narrow at base, but much more so at the chela, which is scarcely more than $1/_9$ of its length, overtopped very slightly by a narrow point of the process of the 5th joint. Peraeopods 1 and 2, 6th joint shorter than 5th, finger minute. Peraeopod 5, 5th and 6th joints subequal. Telson rather small, scarcely longer than broad, insinuate slightly at apex. Colour pellucid, whitish, not at all diversified; ova in pouch orange. L. ♀ 4 mm.

North-Atlantic and North-Sea (Norway). Depth 94—564 m.

4. **S. intermedium** O. Sars 1892 *S. i.*, G. O. Sars, Crust. Norway, *v.*1 p.320 t. 113 f. 1.

Head, rostrum slightly deflexed, scarcely reaching beyond middle of 1st joint of antenna 1; lateral corners obtusely rounded. Side-plate 1 with lower front corner more angular than in the 2 preceding species. Eyes rather small, lenses few, light red with whitish coating. Antenna 1 in ♀, 2^d joint fully as long as 1st, 3^d rather long, flagellum much shorter than peduncle, 6-jointed. Antenna 2 in ♀ much shorter than antenna 1, flagellum scarcely as long as peduncle. Gnathopod 1 nearly as in S. haplocheles, but process of 5th joint less produced. Gnathopod 2 more slender than in S. haplocheles, but less elongate than in S. tenuimanum, chela about $1/_6$ as long as 6th joint. Feraeopods 1 and 2, 5th and 6th joints subequal, otherwise peraeopods 1—5 about as in S. tenuimanum. Telson oval, decidedly longer than broad, apically transversely truncate. Colour pellucid, whitish, not diversified. L. ♀ 4 mm.

Trondhjemsfjord. Depth 282—750 m..

6. Gen. **Oediceros** Krøyer

1842 *Oediceros* (Sp. un.: *O. saginatus*), Krøyer in: Naturh. Tidsskr., *v.*4 p. 155 ; 1892 *O.*, G. O. Sars, Crust. Norway, *v.*1 p.287 | 1893 *O.* (part.), A. Della Valle in: F.Fl.Neapel, *v.*20 p.541 | 1853 *Oedicerus* (non Kollar & L. Redtenbacher 1844, Coleoptera!), J. D. Dana in: U.S. expl. Exp., *v.*13 II p.933 | 1882 *Aedicerus* (laps.), Haswell, Cat. Austral. Crust., p.238, 315.

Back of peraeon broad. Rostrum deflexed, acute. Eyes distinct, contiguous. at base of rostrum. Antenna 1 shorter than antenna 2, peduncle of each elongate. Upper lip rounded. Mandible, molar without triturating surface, palp robust. Maxilla 1, inner plate with a few setae, outer plate narrow. Maxilla 2, plates oval, subequal. Maxillipeds, outer plates strongly fringed with spines. Gnathopods 1 and 2 similar and nearly equal, 5th joint produced to a setose lobe, 6th oval, palm very oblique. Peraeopods 1—4 short, stout, setose, finger lanceolate. Telson rounded quadrangular.

2 species accepted, 1 doubtful.

Synopsis of accepted species:

Rostrum geniculate . . . . 1. **O. saginatus** . p. 244
Rostrum gently curved . . . . 2. **O. borealis** . . p. 244

1. **O. saginatus** Krøyer 1842 *O. s.*, Krøyer in: Naturh. Tidsskr., *v.* 4 p. 156 | 1883 *O. s.*, J. S. Schneider in: Tromsø Mus. Aarsh., *v.* 6 p. 11 t. 2 f. 10 | 1892 *O. s.*, G. O. Sars, Crust. Norway, *v.* 1 p. 288 t. 102 | 1893 *O. s.* (part.), A. Della Valle in: F. Fl. Neapel, *v.* 20 p. 551 t. 58 f. 71, 72.

Head, rostrum abruptly deflexed, acute, nearly reaching apex of 1st joint of antenna 1; lateral corners quadrate. Side-plates not deep, 4th scarcely emarginate behind, 5th broader and nearly as deep. Eyes within a rounded prominence, dark red. Antenna 1, 1st and 2^{d} joints equal in length, flagellum scarcely as long as peduncle, 13-jointed. Antenna 2, ultimate joint of peduncle as long as penultimate and antepenultimate combined, flagellum rather shorter than peduncle. Gnathopods 1 and 2, process of 5th joint slightly curved and a little dilated in the middle, densely tufted, 6th joint oblong oval, more than twice as long as broad, palm denticulate, long, gently curved, defined by a little tooth, finger long and curved. Peraeopods 1—4, finger rather short. Peraeopod 5 more than half length of body, 2^{d} joint with sinuous hind margin. Uropod 3, inner ramus rather the longer, with 4 spinules. Telson little longer than broad, apex transversely truncate. Colour whitish, tinged with orange, head reticulate and back banded with brown violet. L. 20—30 mm.

Arctic Ocean and North-Atlantic (Greenland, Iceland, Spitzbergen, Murman coast, Siberian Polar Sea, Norway).

2. **O. borealis** Boeck 1871 *O. b.*, A. Boeck in: Forh. Selsk. Christian., 1870 p. 162 | 1876 *O. b.*, A. Boeck, Skand. Arkt. Amphip., *v.* 2 p. 261 t. 14 f. 1 | 1892 *O. b.*, G. O. Sars, Crust. Norway, *v.* 1 p. 290 t. 103 f. 1 | 1893 *O. saginatus* (part.), A. Della Valle in: F. Fl. Neapel, *v.* 20 p. 551.

Head, rostrum evenly curved downward, reaching little beyond middle of 1st joint of antenna 1; lateral corners rounded. Side-plates rather large, 1st somewhat pentagonal, 4th distinctly emarginate, deeper than 5th. Eyes rather small, contiguous. Antenna 1, flagellum 10-jointed. Antenna 2, ultimate joint of peduncle scarcely longer than penultimate, flagellum subequal to peduncle. Gnathopods 1 and 2 very strong, process of 5th joint narrow, sparingly setose, 6th tumid, oval, palm not very long, defined by a little tooth, finger to match the palm. Peraeopods 1—4 less stout than in O. saginatus, finger in peraeopods 3 and 4 rather long. Peraeopod 5, 2^{d} joint rounded oval, hind margin scarcely sinuous. Uropod 3 rather small, rami very narrow, inner with 1 spinule. Telson much longer than broad, apex broadly rounded. L. ♀ 9 mm.

Arctic Ocean (Greenland, Finmark).

O. latrans (HasW.) 1880 *Oedicerus l.*, Haswell in: P. Linn. Soc. N.S.Wales, *v.* 4 p. 324 t. 19 f. 1 | 1893 *Oediceros l.*, A. Della Valle in: F. Fl. Neapel, *v.* 20 p. 556. Bondi Bay [New South Wales]. Burrowing in the sand between tide marks.

7. Gen. **Paroediceros** O. Sars

1892 *Paroediceros*, G. O. Sars, Crust. Norway, *v.* 1 p. 291.

Front of head produced to a process bearing the contiguous eyes, when present, at its distal end. Antenna 1 much shorter than antenna 2. Upper lip with flattened margin. Mandibular palp slender. Gnathopod 1, process of 5th joint insignificant, 6th joint elongate, distally widened. Peraeopods comparatively slender. Other characters agreeing with Oediceros (p. 243).

5 species.

Synopsis of species:

1 { Gnathopod 1, 6th joint extremely elongate — **2**.
{ Gnathopod 1, 6th joint not extremely elongate — **3**.

2 { Frontal process small, without eyes . . . . 1. **P. macrocheir** . p. 245
{ Frontal process well developed, with eyes . . . . 2. **P. intermedius** . p. 245

3 { Frontal process deflexed . . . . 3. **P. curvirostris** . p. 246
{ Frontal process horizontal — **4.**

4 { Eyes large, transverse bands of colour on body . 4. **P. lynceus** . . . p. 246
{ Eyes small, no colour bands on body . . . . . . 5. **P. propinquus** . p. 246

1. **P. macrocheir** (O. Sars) 1879 *Oedicerus m.*, G. O. Sars in: Arch. Naturv. Kristian., *v.* 4 p, 449 | 1885 *Oediceros m.*, G. O. Sars in: Norske Nordbavs-Exp., *v.* 6 Crust. I p. 170 t. 14 f. 4 | 1887 *O. m.*, H. J. Hansen in: Dijmphna Udb., p. 221 | 1892 *O. m.*, Paroediceros (part.), G. O. Sars, Crust. Norway, *v.* 1 p. 291 | 1893 *O. lynceus* (part.), A. Della Valle in: F. Fl. Neapel, *v.* 20 p. 546.

Head arched, frontal process very small, not tumid. Side-plates 1—4 large and broad, densely fringed, 4th scarcely so deep as the others, 5th distinctly bilobed, 6th and 7th small. Eyes wanting. Antenna 1 very short, 1st and 2^{d} joints equal in length, flagellum shorter than peduncle, 8-jointed. Antenna 2 more than twice length of antenna 1, ultimate and penultimate joints of peduncle robust, ultimate the shorter, flagellum slender, rather long. Gnathopod 1 very long, 2^{d} joint large and muscular, 3^{d}—5th very short, 6th more than 4 times as long as broad, dilated distally, palm arcuate, oblique, defined by obtuse projection and strong spine. Gnathopod 2 shorter, 5th joint narrowly produced, 6th shorter and broader than in gnathopod 1, palm very oblique, about half length of the joint. Peraeopods 1 and 2 rather slender, very hirsute, finger large, falciform. Peraeopods 3—5, 2^{d} joint narrowly ovate; 4th peraeopod longer than 3^{d}, both very hirsute, with long falciform finger; 5th with short simple setae and long needle-like finger. Uropod 3 shorter than uropod 2, and uropod 2 than uropod 1. Telson short, apically rounded. Skin pellucid, but with brown-violet arborescent markings. L. ♀ 18 mm.

Arctic Ocean (between Jan Mayen and Iceland). Depth 1890 m.

2. **P. intermedius** Stebb.*) 1887 *Oediceros microps* (err., non O. Sars 1882!), H. J. Hansen in: Dijmphna Udb., p. 220 t. 21 f. 12 | 1892 *Paroediceros* (part.), G. O. Sars, Crust. Norway, *v.* 1 p. 291, 294 | 1893 *Oediceros lynceus* (part.), A. Della Valle in: F. Fl. Neapel, *v.* 20 p. 546.

Intermediate between P. propinquus (p. 246) and P. macrocheir. Frontal process produced to end of 1st joint of antenna 1, scarcely deflexed. Side-plates

---

*) Sp. nov.

distally expanded. Eyes placed at apex of frontal process. Antenna 1
reaching about to end of peduncle of antenna 2, both agreeing in general
with those of P. propinquus. Gnathopod 1 agreeing closely with P. macrocheir.
L. 11·5 mm.

    Kara Sea. Depth 113 m.

    3. **P. curvirostris** (H. J. Hansen) 1876 *Oediceros lynceus* (part.), A. Boeck,
Skand. Arkt. Amphip., *v.* 2 p. 259 t. 13 f. 4 | 1893 *O. l.* (part.), A. Della Valle in: F. Fl.
Neapel, *v.* 20 p. 546 | 1887 *O. curvirostris*, H. J. Hansen in: Vid. Meddel., ser. 4 *v.* 9
p. 107 t. 4 f. 4 | 1892 *O. c., Paroediceros* (part.), G. O. Sars, Crust. Norway, p. 291.

    Head, rostrum deflexed, reaching slightly beyond 1st joint of antenna 1,
lateral corners somewhat produced, with rounded apex. Eyes large, somewhat remote from downcurved apex of rostrum. Antennae a little more
slender than in P. lynceus. Peraeopod 1, finger much longer than 6th joint;
in peraeopod 2 rather shorter than it. Peraeopods 3 and 4, finger very little
longer than 6th joint. Other points agreeing with P. lynceus, unless the colour
be fainter. L. ♀ 12·5 mm.

    Davis Strait (lat. 64⁰ N., long. 53⁰ W., depth 81 m; Godthaab, depth 11–19 m).

    4. **P. lynceus** (Sars) 1858 *Oediceros l.*, M. Sars in: Forh. Selsk. Christian.,
p. 143 | 1876 *O. l.* (part.), A. Boeck, Skand. Arkt. Amphip., *v.* 2 p. 259 | 1883 *O. l.*, J. S.
Schneider in: Tromsø Mus. Aarsh., *v.* 6 p. 14 t. 2 f. 12 | 1888 *O. l.*, T. Stebbing in: Rep.
Voy. Challenger, *v.* 29 p. 837 t. 137в | 1893 *O. l.* (part.), A. Della Valle in: F. Fl. Neapel,
v. 20 p. 546 t. 58 f. 65, 66 | 1892 *Paroediceros l.*, G. O. Sars, Crust. Norway, *v.* 1 p. 292
t. 103 f. 2; t. 104 f. 1 | 1867 *Monoculodes nubilatus*, Packard in: Mem. Boston Soc., *v.* 1
p. 298 t. 8 f. 4.

    Head produced horizontally a little beyond 1st joint of antenna 1,
lateral corners produced, acute. Side-plate 1 pentagonal. Pleon segments 1—3
slightly carinate dorsally. Eyes rather large, contiguous, at extremity of
frontal process, leaving a minute rostral tip, yellowish red. Antenna 1 in ♀
shorter than peduncle of antenna 2, 1st joint longest, flagellum much shorter
than peduncle, 10-jointed. Antenna 2, penultimate joint of peduncle scarcely
longer but much thicker than ultimate, very setose, flagellum slender, longer
than peduncle, in young ♂ still longer, with very many short joints. Gnathopod 1 in length subequal to gnathopod 2, 2d joint moderately dilated, 6th
a little widened at the spine which defines the curved oblique palm, longer
than half the joint, finger longer than the palm. Gnathopod 2, 6th joint
oblong oval, palm more than half its length. Peraeopods moderately slender,
finger in peraeopods 1 and 2 shorter than 6th joint, in the others about equal to it
in length; in peraeopod 5, 2d joint rather broad, hind margin setose. Uropod 3,
rami as long as or longer than peduncle, inner one with 1 spinule. Telson
oval quadrangular, apex truncate, carrying 2 setules. Colour whitish, with
brownish-violet bands on back and reticulated on head. L. 22—25 mm.

    Arctic Ocean (Greenland, Iceland, Spitzbergen, Murman coast, Siberian Polar
Sea, Labrador, Finmark). Depth 9—94 m. — No doubt often confused with P. propinquus,
so that notices of localities and depths are exposed to some uncertainty.

    5. **P. propinquus** (Goës) 1866 *Oediceros p.*, Goës in: Öfv. Ak. Förh., *v.* 22 p. 526
t. 39 f. 19 | 1892 *Paroediceros propinqvus*, G. O. Sars, Crust. Norway, *v.* 1 p. 293 t. 104 f. 2 |
1882 *Oediceros microps*, G. O. Sars in: Forh. Selsk. Christian., nr. 18 p. 95 t. 4 f. 8, 8a |
1883 *O. m.*, J. S. Schneider in: Tromsø Mus. Aarsh., *v.* 6 p. 15 t. 2 f. 14 | 1893 *O. lynceus*
(part.), A. Della Valle in: F. Fl. Neapel, *v.* 20 p. 546.

Near to P. lynceus, but frontal process less tumid, less long, lateral corners of head less produced, side-plate 1 foursided, eyes much smaller, on apex of frontal process, bright red. Antenna 1 in ♀, 1st joint scarcely longer than 2^d, flagellum with 8 joints, shorter than peduncle, but in ♂ twice length of peduncle, with 11 joints, many setose; antenna 2, flagellum in ♀ as long as peduncle, in ♂ reaching end of body. Gnathopod 1, 2^d joint fusiform, 6th about 3 times as long as broad, palm about half its length. Gnathopod 2 as in P. lynceus, peraeopods more slender, finger long and (in 1st—4th) falciform; 2^d joint in peraeopod 5 less dilated. Uropod 3, rami as long as peduncle, narrow, quite unarmed. Telson similar, comparatively smaller. Colour light yellowish, tinged with rose, with no brown markings. L. 10—11 mm.

Arctic Ocean (Varangerfjord (Vadsö), depth 36—146 m; Tromsö, depth 146 m; Kvalö [Nordland]; Spitzbergen, depth 7—55 m; Greenland; Iceland).

8. Gen. **Halicreion** Boeck

1871 *Halicreion* (Sp. un.: *H. longicaudatus*). A. Boeck in: Forh. Selsk. Christian., 1870 p. 173 | 1876 *H.*, A. Boeck, Skand. Arkt. Amphip., v. 2 p. 294 | 1892 *H.*, G. O. Sars, Crust. NorWay, v. 1 p. 321 | 1873 *Halicrion*, E. v. Martens in: Zool. Rec., v. 8 p. 190.

Head produced to an acute deflexed rostrum. Side-plates of moderate size. Eyes contiguous at base of rostrum. Antenna 1 rather shorter than antenna 2, slender in ♀, in ♂ robust with tumid short-jointed peduncle and 1st joint of flagellum enormous. Mouth-parts as in Monoculodes (p. 258), except that outer plates of maxillipeds are rather shorter and broader. Gnathopods 1 and 2 nearly alike, 6th joint ovate, process of 5th rather narrower and longer in gnathopod 2 than in gnathopod 1. Peraeopods 1—5 as in Monoculodes; 5th much (not a little, as Boeck says) longer than 3^d and 4th. Uropod 3 greatly elongate, projecting far beyond uropods 1 and 2. Telson very small.

1 species.

1. H. aequicornis (Norm.) 1869 *Oediceros a.*, A. M. Norman in: Rep. Brit. Ass., Meet. 38 p. 278 | 1893 *O. a.* (part.), A. Della Valle in: F. Fl. Neapel, v. 20 p. 545 t. 58 f. 63, 64 | 1889 *Monoculodes a.*, A. M. Norman in: Ann. nat. Hist., ser. 6 v. 3 p. 453 t. 20 f. 1—5 | 1871 *Halicreion longicaudatus*, A. Boeck in: Forh. Selsk. Christian., 1870 p. 173 | 1876 *H. l.*, A. Boeck, Skand. Arkt. Amphip., v. 2 p. 295 t. 21 f. 3 | 1892 *H. l.*, G. O. Sars, Crust. NorWay, v. 1 p. 322 t. 113 f. 2.

Rather slender and compressed. Rostrum somewhat compressed, reaching a little beyond 1st joint of antenna 1. Side-plate 1 scarcely expanded, 4th not very broad. Eyes oval, ill developed, light red, at base of rostrum. Antenna 1 in ♀ reaching end of peduncle of antenna 2, 2^d joint a little shorter than 1st, 3^d than 2^d, each with stiff diverging apical setae, flagellum shorter than peduncle, 5-jointed. Antenna 1 in ♂, 2^d and 3^d joints very short and thick, flagellum twice as long as peduncle, 1st joint longer than other 4 combined, with 2 rows of long sensory setae in dense tufts. Antenna 2 alike in ♂ and ♀, ultimate and penultimate joints of peduncle long, slender, subequal, each subequal to 5-jointed flagellum. Gnathopod 1, process of 5th joint scarcely reaching palm, which is rather oblique, longer than hind margin. Gnathopod 2 subequal to gnathopod 1, process of 5th joint narrower, reaching palm, 6th joint also narrower, otherwise similar. Peraeopods not very setose. Peraeopods 1 and 2, 4th joint little expanded, in length equal to 6th, a little longer than the finger. Peraeopods 3 and 4, 2^d joint oval, with setae projecting from within distal half of expansion, finger nearly equal to 4th and fully equal to 6th joint. Peraeopod 5, very long, 2ᵈ joint piriform, 4ᵗʰ rather shorter than 5ᵗʰ, 5ᵗʰ than either 6ᵗʰ or finger. Uropods 1 and 2 scarcely reaching beyond peduncle of uropod 3. Uropod 3, rami narrow, lanceolate, unarmed, longer than the long peduncle. Telson scarcely longer than broad, apical margin slightly concave. Colour whitish tinged with yellow. L. ♀ 5, ♂ 4 mm.

North-Atlantic, North-Sea and Christianiafjord (Norway, depth 94—188 m; St. Magnus Bay [Shetland Islands], depth 55—110 m).

9. Gen. **Arrhis** Stebb.*)

1861 *Aceros* (Sp. typ.: *Oedicerus obtusus*) (non C. L. Bonaparte 1850, Aves!), A. Boeck in: Forb. Skand. Naturf., Møde 8 p. 651 | 187ŀ *A.*, A. Boeck in: Forb. Selsk. Christian., 1870 p. 172 | 1892 *A.*, G. O. Sars, Crust. Norway, *v.* 1 p. 338.

Body slender. Head without frontal process. Side-plates not deep, 1ˢᵗ produced forward. Eyes inconspicuous, their pigment patches not contiguous. Antenna 1 in ♀ longer than antenna 2, or as long, peduncle very long, much longer than in antenna 2, flagellum short. Lower lip, inner lobes separate. Mandible strong, cutting edge little dentate, molar strong, 2ᵈ joint of palp arcuate. .Maxilla 1, outer plate with 7 spines (Schneider). Maxillipeds, inner plates not reaching apex of 1ˢᵗ joint of palp, outer not nearly reaching apex of the distally widened 2ᵈ joint. Gnathopods 1 and 2, 5ᵗʰ joint long, produced to narrow setose process, 6ᵗʰ oval, palm oblique, pretty well defined. Peraeopods 1 and 2 rather longer than peraeopods 3 and 4, finger in all lanceolate. Peraeopod 5, uropods 1—3 and telson not generically distinctive.

1 species.

1. A. phyllonyx (Sars) 1858 *Leucothoë p.*, M. Sars in: Forh. Selsk. Christian., p. 148 | 1862 *Montagua p.*, Bate, Cat. Amphip. Brit. Mus., p. 369 | 1871 *Aceros p.*, A. Boeck in: Forh. Selsk. Christian., 1870 p. 172 | 1887 *A. p.*, H. J. Hansen in: Vid. Meddel., ser. 4 *v.* 9 p. 117 t. 4 f. 7 | 1892 *A. p.*, G. O. Sars, Crust. Norway, *v.* 1 p. 338 t. 119; t. 120 f. 1 | 1893 *Halimedon p.*, A. Della Valle in: F. Fl. Neapel, *v.* 20 p. 535 t. 58 f. 23—27 | 1859 *Oediceros obtusus*, R. M. Bruzelius in: Svenska Ak. Handl., n. ser. *v.* 3 nr. 1 p. 92 t. 4 f. 17 | 1861 *Oedicerus o.*, *Aceros sp. typ.*, A. Boeck in: Forh. Skand. Naturf., Møde 8 p. 651.

Tumid anteriorly. Head tumid with swollen cheeks, front truncate. Side-plate 1 produced along lower side margins of head, 3ᵈ with insinuate lower margin, 5ᵗʰ, 6ᵗʰ and 7ᵗʰ unusually broad. Eyes separate, on sides of head near the front, represented by 2 small patches of whitish pigment. Antenna 1 in ♀, 2ᵈ joint nearly twice as long as 1ˢᵗ, setose on both edges, flagellum very short, 11-jointed, in (young) ♂ somewhat longer. Antenna 2, ultimate and penultimate joints of peduncle subequal, and together as long as flagellum in ♀, flagellum longer in ♂. Gnathopod 1, 5ᵗʰ joint longer than 6ᵗʰ, its setose process not lying near hind margin of 6ᵗʰ, palm longer than the hind margin. Gnathopod 2 nearly resembling gnathopod 1, but longer, and process of 5ᵗʰ joint narrow, lying near margin of 6ᵗʰ. Peraeopods 1 and 2, 2ᵈ joint bent, 4ᵗʰ large, very setose, 5ᵗʰ and 6ᵗʰ equal, finger shorter, compressed, lanceolate. Peraeopods 3 and 4, 2ᵈ joint long oval, finger as long as 6ᵗʰ joint. Peraeopod 5 elongate, 2ᵈ joint large, elongate piriform, 5ᵗʰ joint shorter than 4ᵗʰ, much shorter than 6ᵗʰ. Uropods 1 and 2, rami densely spinose, uropod 3,

---

*) Nom. nov. ἄῤῥις, without nose. — The name *Aceros* is preoccupied (1850, C. L. Bonaparte, Consp. Av., *v.* 1 p. 90).

rami spinulose, about as long as peduncle. Telson almost a square. Colour whitish, with tinge of flesh colour. L. 15—20 mm.

Arctic Ocean, North-Atlantic, North-Sea and Skagerrak (Greenland, Iceland, Spitzbergen etc.; Norway, depth 90—753 m; Bohuslän); German Ocean?

10. Gen. **Westwoodilla** Bate

1856 *Westwoodea* (nom. nud.), Bate in: Rep. Brit. Ass., Meet. 25 p. 58 | 1857 *Westwoodia* (Sp. un.: *W. caecula*) (non Brullé 1846, Hymenoptera!), Bate in: Ann. nat. Hist., ser. 2 *v.* 19 p. 139 | 1862 *Westwoodilla*, Bate, Cat. Amphip. Brit. Mus., p. 102 | 1871 *Halimedon* (part.), A. Boeck in: Forb. Selsk. Christian., 1870 p. 169 | 1892 *H.*, G. O. Sars, Crust. Norway, *v.* 1 p. 326 | 1893 *H.* (part.), A. Della Valle in: F. Fl. Neapel, *v.* 20 p. 533.

Head with frontal process ending in acute rostrum, usually short. Side-plates of moderate size, 4th with lower hind corner not produced, narrower than 5th. Eyes contiguous on frontal process. Antenna 1 shorter than antenna 2, flagellum of antenna 2 in ♂ very elongate. Upper lip with angular sides. Lower lip, inner lobes well defined. Mandible strong, cutting edge indistinctly dentate, molar well defined, palp slender, 2^d joint longest, much curved. Maxilla 1, outer plate carrying 8 spines (Schneider). Maxillipeds, inner plates rather short, outer reaching apex of 2^d joint of moderate-sized palp. Gnathopods 1 and 2 nearly alike, rather feeble, 5th joint long, ending in short setose expansion, 6th rather small, oblong oval. Peraeopods 1—5, uropods 1—3 and telson as in Monoculodes (p. 258).

5 species.

Synopsis of species:

$$
1 \begin{cases} \text{Frontal process deflexed} - 2. \\ \text{Frontal process horizontal} - 4. \end{cases}
$$

$$
2 \begin{cases} \text{Rostral apex deflexed} \dots \dots \dots \dots 1. \text{ W. brevicalcar . p. 249} \\ \text{Rostral apex not deflexed} - 3. \end{cases}
$$

$$
3 \begin{cases} \text{Eyes oval, of moderate size} \dots \dots \dots 2. \text{ W. caecula . . . p. 250} \\ \text{Eyes rounded, unusually large} \dots \dots \dots 3. \text{ W. megalops . . p. 250} \end{cases}
$$

$$
4 \begin{cases} \text{Eyes not prominent, at centre of tapering frontal} \\ \quad \text{process} \dots \dots \dots \dots \dots \dots 4. \text{ W. acutifrons . . p. 251} \\ \text{Eyes prominent, at distal end of elongate frontal} \\ \quad \text{process} \dots \dots \dots \dots \dots \dots 5. \text{ W. rectirostris . p. 251} \end{cases}
$$

1. **W. brevicalcar** (Goës) 1866 *Oediceros b.*, Goës in: Öfv. Ak. Förh., *v.* 22 p. 527 t. 39 f. 22 | 1871 *Halimedon b.*, A. Boeck in: Forh. Selsk. Christian., 1870 p. 171 | 1876 *H. b.*, A. Boeck, Skand. Arkt. Amphip., *v.* 2 p. 286 t. 15 f. 3 | 1883 *H. b.*, J. S. Schneider in: Tromsø Mus. Aarsh., *v.* 6 p. 37 t. 2 f. 11 | 1892 *H. b.*, G. O. Sars, Crust. Norway, *v.* 1 p. 331 t. 116 f. 3 | 1893 *H. b.* (part.), A. Della Valle in: F. Fl. Neapel, *v.* 20 p. 539.

Rather short and stout. Head frontal process short, evenly curved, the acute deflexed rostrum reaching a little beyond middle of 1st joint of antenna 1. Side-plate 1 little expanded distally. Eyes of moderate size, rounded oval, bright red, at base of frontal process. Antenna 1 in ♀ reaching beyond peduncle of antenna 2, 2^d joint shorter than 1st, flagellum a little shorter than peduncle, 7-jointed. Antenna 2, ultimate joint of peduncle shorter than penultimate, flagellum about as long as the 2 combined. Gnathopod 1, 5th joint not much longer than its width at the somewhat projecting process, 6th joint

as long as 5th, oval, palm oblique, rather longer than hind margin. Gnathopod 2 more elongate, process of 5th joint little projecting, palm of 6th
scarcely longer than hind margin. Peraeopods 1—4, finger nearly as long
as the narrow linear 6th joint; in peraeopods 3 and 4, 2^d joint oval, with long
setae projecting from within most of the expansion. Peraeopod 5, 2^d joint
piriform, 4th, 5th and 6th subequal. Uropod 3, rami narrow, longer than
peduncle. Telson decidedly longer than broad, quadrangular, apically truncate,
armed with spinules. Colour whitish, tinged with yellow. L. ♀ 6 mm.

Arctic Ocean (Iceland, Greenland, Spitzbergen; Norway, depth 9—55 m).

2. **W. caecula** (Bate) 1856 *Westwoodea caeculus* (nom. nud.), Bate in: Rep.
Brit. Ass., Meet. 25 p. 58 | 1857 *Westwoodia caecula*, Bate in: Ann. nat. Hist., ser. 2 *v.*19
p. 140 | 1862 *Westwoodilla c.* + *W. hyalina*, Bate, Cat. Amphip. Brit. Mus., p. 102 t. 16
f. 5; p. 103 t. 17 f. 5 | 1862 *Oediceros parvimanus*, Bate & Westwood, Brit. sess. Crust.;
v. 1 p. 161 f. | 1889 *Halimedon p.*, A. M. Norman in: Ann. nat. Hist., ser. 6 *v.* 3 p. 455
t. 20 f. 10—14 | 1893 *H. p.*, A. Della Valle in: F. Fl. Neapel, *v.* 20 p. 539 t. 58 f. 38—40 |
1871 *H. mølleri*, A. Boeck in: Forh. Selsk. Christian., 1870 p. 169 | 1876 *H. mülleri*,
A. Boeck, Skand. Arkt. Amphip., *v.* 2 p. 281 t. 13 f. 5 | 1883 *H. m.*, J. S. Schneider in:
Tromsø Mus. Aarsh., *v.* 6 p. 33 t. 3 f. 17 | 1892 *H. m.*, G. O. Sars, Crust. Norway, *v.* 1
p. 327 t. 115.

Rather slender, but tumid anteriorly. Frontal process rather large, strongly
arched distally, but the small rostral apex projected horizontally. Side-plate 1
much expanded and produced forward, 5th nearly as deep as 4th. Eyes of moderate
size, oval, light red, at distal end of frontal process. Antenna 1 in ♀
little longer than peduncle of antenna 2, 2^d joint rather longer than 1st, flagellum
as long as 2^d and 3^d combined, 10-jointed. Antenna 1 in ♂, 1st and 2^d joints
equal, flagellum longer than peduncle, with 14 joints, first 10 thickened, hirsute.
Antenna 2 in ♀, ultimate joint of peduncle longer than penultimate, both densely
setose, together longer than flagellum. Antenna 2 in ♂, peduncle nearly bare,
flagellum very long, filiform. Gnathopod 1 rather feeble, process of 5th joint
broadly rounded and setose, 6th joint subequal to 5th, oblong oval, palm ill-
defined, nearly twice as long as hind margin. Gnathopod 2 similar to gnathopod 1, but longer and more slender. Peraeopods 1—4 densely setose, the
finger about as long as 6th joint, the latter in peraeopods 1 and 2 subfusiform,
with dense tufts on distal half of front margin. Peraeopod 5, 2^d joint
rounded oval, 4th and 5th equal, 6th rather longer than either. Uropod 3,
rami spinulose, narrowly lanceolate, longer than peduncle. Telson oval, apex
broadly rounded, carrying several setules. Colour pellucid whitish, lightly
tinged with red; ova orange-coloured. L. 8 mm.

Arctic Ocean, North-Atlantic, North-Sea and Skagerrak (Norway, depth 36—376 m;
France, Great Britain; West Greenland, depth 47—400 m); Kattegat.

3. **W. megalops** (O. Sars) 1882 *Halimedon m.*, G. O. Sars in: Forh. Selsk.
Christian., nr. 18 p. 96 t. 4 f. 9, 9 a | 1883 *H. m.*, J. S. Schneider in: Tromsø Mus. Aarsh.,
v. 6 p. 38 t. 2 f. 9 | 1892 *H. m.*, G. O. Sars, Crust. Norway, *v.* 1 p. 330 t. 116 f. 2 | 1893
H. brevicalcar (part.), A. Della Valle in: F. Fl. Neapel, *v.* 20 p. 539.

Very tumid anteriorly. Frontal process broad, strongly arched, the small
rostral apex jutting out nearly to end of 1st joint of antenna 1. Side-plate 1 a
little expanded distally, 5th decidedly less deep than 4th. Eyes unusually large,
rounded, prominent, dark red, occupying most of frontal process. Antenna 1
in ♀ short, little longer than peduncle of antenna 2, 1st joint as long as 2^d and
3^d combined, flagellum shorter than peduncle, 5-jointed. Antenna 2, ultimate
and penultimate joints of peduncle subequal, flagellum nearly as long as peduncle.

Maxilla 1, outer plate with 8 spines, 2^d joint of palp narrower than 1st (Schneider). Gnathopods 1 and 2, 5th joint distally widened, alike in both pairs, 6th joint oval, longer than 5th (Sars) or equal to it (Schneider), palm in gnathopod 1 much longer than hind margin, in gnathopod 2 little longer. Peraeopods 1—4, 6th joint narrow linear, sparingly setose, finger about as long. Peraeopod 5, 2^d joint broad, obliquely quadrate, 4th and 5th subequal, 6th rather longer. Uropod 3, rami a little longer than peduncle, each carrying 2 spinules. Telson quadrate, nearly as broad as long, slightly narrowed at truncate apex, which has 2 setules. Colour yellowish orange mottled with reddish brown; antennae banded with orange. L. ♀ 6 mm.

Arctic Ocean and North-Atlantic (North-Norway). Depth 36—55 m.

4. **W. acutifrons** (O. Sars) 1892 *Halimedon a.*, G. O. Sars, Crust. Norway, *v.* 1 p. 329 t. 116 f. 1.

Rather slender. Frontal process rather long, little arched, gradually tapering to an acute horizontally projected rostral apex reaching beyond 1st joint of' antenna 1. Side-plate 1 much expanded and produced forward, 5th less deep than 4th. Eyes oblong oval, light red, covering centre of frontal process. Antenna 1 in ♀, 2^d joint longer than 1st. flagellum shorter than 2^d and 3^d combined, 10-jointed. Antenna 2, ultimate and penultimate joints of peduncle subequal, flagellum less than the 2 combined. Gnathopods 1 and 2 feeble. Gnathopod 1, 5th joint distally much expanded, 6th rather longer, more than twice as long as broad, palm very oblique, longer than hind margin. Gnathopod 2, 5th joint less expanded, 6th narrower than in gnathopod 1, equal in length to 5th joint, palm scarcely longer than hind margin. Peraeopods 1 and 2, 6th joint slightly expanded at setose distal end, finger nearly as long. Peraeopods 3 and 4, finger longer than 6th joint. Peraeopod 5 very long, 2^d joint broad above, 5th and 6th joints subequal, longer than 4th. Uropod 3, rami narrow, spinuliferous, longer than peduncle. Telson oval, apex not very broadly rounded. Colour whitish, faintly tinged with yellow; ova rose-coloured. L. ♀ 8 mm.

North-Atlantic (Apelvaer [Namdal], Trondhjemsfjord). Depth 90—270 m.

5. **W. rectirostris** (Della Valle) 1893 *Halimedon r.*, A. Della Valle in: F. Fl. Neapel, *v.* 20 p. 537 t. 4 f. 6; t. 33 f. 1—15.

Rather slender. Frontal process elongate, horizontally projected. Side-plate 1 greatly expanded distally and produced forward, 5th much broader than 4th and nearly as deep. Eyes small, rounded, prominent, bright vermilion, at apex of frontal process, leaving a very minute rostral point (a single eye is represented on t. 33, with no dividing line). Antenna 1 in ♀, 2^d joint as long as 1st, 3^d more than half length of 2^d, flagellum shorter than peduncle, 10-jointed. Antenna 2, ultimate and penultimate joints of peduncle subequal, flagellum in ♀ as long as the 2 combined, 12-jointed. Gnathopods 1 and 2 said to be almost exactly alike, 5th joint triangular, as long as 6th, hinder distal angle dilated, not produced, 6th joint almond-shaped. In the figures the angle of 5th joint is as usual more broadly dilated in gnathopod 1 than in gnathopod 2, and the palm of 6th joint is much longer than hind margin in gnathopod 1, and little longer than it in gnathopod 2, but the 6th joint of gnathopod 1 is $2^1/_2$ times as long as broad, and in gnathopod 2 only twice. Peraeopods 1, 2 and 4, finger subequal to 6th joint, in peraeopod 3 longer than it. Peraeopod 5, 2^d joint broad, 4th longer than 5th or 6th, finger shorter than 6th. Uropod 3, rami longer than peduncle.   Telson longer than broad, sides insinuate, distal margin flatly rounded. Colour orange-yellow blotched with lemon-yellow. L. 5—6 mm.

Gulf of Naples. In muddy sand, depth 12—20 m.

11. Gen. **Carolobatea** Stebb.

1899 *Carolobatea* (Sp. typ.: *Halimedon schneideri*), T. Stebbing in: Ann. nat. Hist., ser. 7 *v.* 4 p. 208.

Frontal process of head apically subacute. Side-plate 1 distally expanded, 4th much deeper than the rest, 5th nearly as broad as 4th. Eyes contiguous, on frontal process. Antenna 1 in ♀ shorter than antenna 2, flagellum in both many-jointed. Upper lip with sides angular. Lower lip with inner lobes separate. Mandible, cutting edge bluntly dentate, molar rather weak, 2d joint of palp very slightly bent. Maxilla 1, outer plate with 9 spines; maxilla 2, inner plate the broader. Maxillipeds, inner plates small, not reaching apex of 1st joint of palp, outer plates not nearly reaching apex of distally expanded 2d joint of palp. Gnathopods 1 and 2, 5th joint subequal to 6th, distally widened and slightly produced, 6th oblong, a little widened distally, palm well defined, very slightly oblique, much shorter than hind margin. Peraeopods 1—4, finger narrowly boat-shaped, and as in gnathopods 1 and 2 having a membranous cap over the tip; in 1st, 2d and 4th pairs finger as long as 6th joint. Uropod 3, rami subequal to peduncle. Telson rather longer than broad.

1 species.

1. **C. schneideri** (Stebb.)   1888 *Halimedon s.*, T. Stebbing in: Rep. Voy. Challenger, *v.* 29 p. 839 t. 59 | 1899 *H. s.*, *Carolobatea sp. typ.*, T. Stebbing in: Ann. nat. Hist., ser. 7 *v.* 4 p. 209 | 1893 *H. brevicalcar* (part.), A. Della Valle in: F. Fl. Neapel, *v.* 20 p. 539 t. 58 f. 41, 42.

. Frontal process deflexed, reaching end of 1st joint of antenna 1. Eyes dark, elongate, widened distally (apparently not always persistent in alcohol). Antenna 1 in ♀, 2d joint rather longer than 1st, flagellum shorter than peduncle, 17-jointed. Antenna 2, ultimate joint of peduncle shorter than penultimate, flagellum shorter than the 2 combined, 24-jointed. Gnathopod 2 longer and more slender than gnathopod 1, the palm a little more oblique, but in both defined by a well marked angle with palmar spine. Peraeopods 1—4, 6th joint with brush of setae on lower half of the convex margin; in peraeopods 3 and 4, hind margin of 2d joint sinuous. Peraeopod 5 imperfect, but with the family character indicated. Telson narrowing distally, apical margin very slightly insinuate. L. ♀ (from front of head to end of outstretched uropod) 13 mm.

Southern Indian Ocean (Kerguelen Island).

12. Gen. **Oediceropsis** Lillj.

1865 *Oediceropsis* (Sp. un.: *O. brevicornis*), W. Lilljeborg in : N. Acta Soc. Upsal., ser. 3 *v.* 6 nr. 1 p. 18, 19 | 1871 *O.*, A. Boeck in: Forh. Selsk. Christian., 1870 p. 174 ! 1892 *O.*, G. O. Sars, Crust. Norway, *v.* 1 p. 324.

Head without frontal process, but narrowed anteriorly. Side-plates 1—4 rather large, 1st expanded distally, 4th excavate behind, 5th small. Eyes lateral. Antenna 1 very small. Antenna 2 large, penultimate joint of peduncle very large, flagellum of many short calceoliferous joints both in ♀ and ♂. Upper lip with margin evenly convex. Mandibular palp elongate. Maxilla 2, inner plate much wider than outer. Plates of maxillipeds rather broad in proportion to length. Gnathopods 1 and 2, 5th joint short with rather small setose lobe, 6th long ovate, palm much longer than hind margin. Peraeopods 1 and 2 feeble. Peraeopods 3 and 4 strong. Uropod 3, rami longer than peduncle. Telson very small.

1 species.

1. **O. brevicornis** Lillj. 1865 *O. b.*, W. Lilljeborg in: N. Acta Soc. Upsal., ser. 3 *v.*6 nr. 1 p. 19 | 1876 *O. b.*, A. Boeck, Skand. Arkt. Amphip., *v.* 2 p. 297 t. 13 f. 2 | 1892 *O. b.*, G. O. Sars, Crust. Norway, *v.*1 p. 325 t. 114 | 1893 *Oediceros b.*, A. Della Valle in: F. Fl. Neapel, *v.* 20 p. 543 t. 58 f. 50—52.

Head with minute rostrum and sinuous margin on each side below it. Side-plate 1 densely setose, 4th very broad distally. Eyes rounded triangular, light red, no distinct lenses. Antenna 1 only reaching middle of penultimate joint of peduncle of antenna 2, 1st joint longer than 2^d, flagellum shorter than peduncle, 10-jointed. Antenna 2, penultimate joint of peduncle with conspicuous spine at apex, ultimate only about half as long, with 5 large spines, flagellum shorter than peduncle, very flexible. Mandibular palp nearly twice as long as trunk of mandible. Gnathopods 1 and 2, palm nearly 4 times as long as hind margin, slightly defined, set with setae and spinules; palm and finger rather longer in gnathopod 2. Peraeopods 1—4 very setose, finger slender, long. Peraeopods 1 and 2, 5th joint longer than 4th or 6th. Peraeopods 3 and 4, 5th joint shorter than 4th; peraeopod 4, 6th joint and finger elongate. Peraeopod 5, 2^d joint expanded, tufts of spinules on others. Uropod 3 not reaching beyond the others. Telson oval quadrangular. Colour pale flesh-tinge, pellucid. L. ♀ 11 mm.

North-Atlantic and Arctic Ocean (West-Norway, to Lofoten Islands). Depth 94—564 m.

13. Gen. **Acanthostepheia** Boeck

1871 *Acanthostepheia* (Sp. un.: *A. malmgreni*), A. Boeck in: Forb. Selsk. Christian., 1870 p. 163 | 1876 *A.*, A. Boeck, Skand. Arkt. Amphip., *v.*2 p. 262 | 1894 *A.*, T. Stebbing in: Bijdr. Dierk., *v.* 17 p. 24 | 1873 *Acanthostephia*, E. v. Martens in: Zool. Rec., *v.*8 p. 190.

Carinate, peraeon broad, pleon compressed. Rostrum long, tapering, carinate, about reaching apex of 1st joint of antenna 1. Side-plate 1 distally expanded, 5th as large as the 4th. Eyes separated by carina of head, large, oval or reniform, prominent. Antenna 1 shorter than antenna 2, 2^d joint shorter than 1st. Mouth-parts nearly as in Monoculodes (p. 258). Gnathopods 1 and 2 similar, 5th joint rather short, with prominent round-ended process, 6th large oval, the oblique palm much longer than hind margin, finger as long as palm. Peraeopods 1 and 2, finger small. Peraeopods 3 and 4, finger rather large, lanceolate. Uropod 3, rami not longer than peduncle. Telson short, quadrate.

2 species accepted, 1 doubtful.

Synopsis of accepted species:

1. A. malmgreni (Goës) 1866 *Amphithonotus m.*, Goës in: Öfv. Ak. Pörh., *v.* 22 p. 526 t. 39 f. 17 | 1871 *Acanthostepheia m.*, A. Boeck in: Forh. Selsk. Christian., 1870 p. 163 | 1876 *A. m.*, A. Boeck, Skand. Arkt. Amphip., *v.* 2 p. 263 | 1894 *A. m.*, T. Stebbing in: Bijdr. Dierk., *v.* 17 p. 24 | 1887 *A. malmgrenii*, H. J. Hansen in: Vid. Meddel., ser. 4 *v.* 9 p. 104 | 1893 *Oediceros m.* (part.), A. Della Valle in: F. Fl. Neapel, *v.* 20 p. 544 t. 58 f. 55.

Head, rostrum not always reaching end of 1st joint of antenna 1, lateral angles acute, carina not quite extending to hind margin. Peraeon segments strongly ridged transversely, carina rudimentary on first 5, on 7th and on pleon segments 1—4 longitudinally bidentate. Postero-lateral angles of peraeon segments 5—7 and pleon segments 1—3 acute, intermediate ridges on the latter. Side-plates 5 and 6 with 2 acute lobes, 7th with one acute lobe. Eyes reniform, diverging backward from base of rostrum. Antenna 1, 1st joint very long, 2^d rather shorter, flagellum subequal to peduncle, 40-jointed. Antenna 2, ultimate and penultimate joints of peduncle long, subequal. Gnathopod 2, process of 5th joint rather longer and narrower, and 6th joint rather longer than in gnathopod 1. Peraeopods 1—5 not very hirsute, finger shorter than 6th joint in peraeopods 1 and 2, longer than it in peraeopods 3 and 4. Peraeopod 5, 2^d joint longitudinally carinate, 5th and 6th joints much longer than 4th, 5th a little longer than 6th. Uropod 3, peduncle carinate, decidedly longer than rami. Telson little longer than broad, distally a little emarginate. Colour dull yellow, dorsally red. L. sometimes 45 mm.

Arctic Ocean (Spitzbergen, Franz Joseph Land, Siberia, Barents Sea, Kara Sea, Greenland). Depth 10—300 m.

2. A. pulchra Miers 1881 *A. p.*, Miers in: Ann. nat. Hist., ser. 5 *v.* 7 p. 47 t. 7 f. 1, 2 | 1894 *A. p.*, T. Stebbing in: Bijdr. Dierk., *v.* 17 p. 25 | 1882 *Acanthostephia malmgreni* (err., non *Amphithonotus m.* Goës 1866!), Stuxberg in: Vega-Exp., *v.* 1 p. 724 f. (no description) | 1893 *Oediceros malmgrenii* (part.), A. Della Valle in: F. Fl. Neapel *v.* 20 p. 544.

Head, rostrum reaching fully to or beyond apex of 1st joint of antenna 1, lateral angles acute, but almost quadrate, carina not quite extending to hind margin. Ridges and carina of peraeon and pleon similar to those in A. malmgreni, but far less strongly developed, carina of pleon segment 3 scarcely bidentate. Postero-lateral angles of peraeon segments 5—7 and pleon segments 1—3 rounded. Side-plates 5 and 6 with rounded lobes, the hinder produced much below the front one. Eyes oval, parallel. Antenna 1 shorter than in A. malmgreni, 2^d joint considerably shorter than 1st, flagellum about 20-jointed. Antenna 2, ultimate joint of peduncle considerably longer than penultimate. Peraeopods 1—4, finger not longer than 6th joint. Peraeopod 5, 2^d joint not distinctly carinate, 5th joint longer than 4th, 6th longer than 5th. Uropod 3, rami nearly as long as peduncle. Telson as in A. malmgreni. L. 37 mm.

Arctic Ocean (Franz Jóseph Land, Siberia). Depth 6—192 m.

A. behringiensis (Lockington) 1877 *Oedicerus b.*, Lockington in: P. Calif. Ac., *v.* 7 p. 47 | 1893 *O. b.*, A. Della Valle in: F. Fl. Neapel. *v.* 20 p. 556.

Perhaps identical with A. pulchra. L. about 32 mm.

North of Behring Strait (West Alaska).

14. Gen. **Aceroides** O. Sars

1892 *Aceroides, Aceropsis* (Sp. un.: *A. latipes*) (non *Aceropsis* Stuxberg 1880, nom. nud.), G. O. Sars, Crust. Norway, *v.* 1 p. 340.

Head with small but distinct rostrum. Side-plates not deep, 1st not expanded. Eyes absent. Antenna 1 in ♀ longer than antenna 2, peduncle rather short, flagellum much longer. Upper lip broad, with margin scarcely insinuate. Lower lip with inner lobes separate. Mandible less strong than in Arrhis (p. 248), palp straight. Maxillae 1 and 2, maxillipeds and gnathopods 1 and 2 as in that genus. Peraeopods 1 and 2 (Fig. 63) large, 4th to 7th joints much expanded. Peraeopods 3 and 4 much smaller, 2d joint little expanded, finger slender. Peraeopod 5 characteristic of the family. Uropods 1—3 and telson normal. 1 species.

1. **A. latipes** (O. Sars) 1866 *Oediceros obtusus var.*, Goës in: Öfv. Ak. Förh., *v.* 22 p. 527, 536 t. 40 f. 24, 24' | 1882 *Halicreion latipes*, G. O. Sars in: Forh. Selsk. Christian., nr. 18 p. 97 t. 4 f. 10 | 1883 *H.? l.*, J. S. Schneider in: Tromsø Mus. Aarsh., *v.* 6 p. 43 | 1892 & 93 *Aceropsis l.*, *Aceroides l*, G. O. Sars, Crust. Norway, *v.* 1 t. 120 f. 2; p. 341 | 1887 *Aceros distinguendus*, H. J. Hansen in: Vid. Meddel., ser. 4 *v.* 9 p. 118 t. 4 f. 8 | 1893 *Halimedon d.*, A. Della Valle in: F. Fl. Neapel, *v.* 20 p. 916.

Neither very slender nor in front very tumid. Rostrum very small, acute. Side-plate 1 obliquely quadrate, 3d slightly emarginate below, subequal to 4th, 5th—7th broad. Antenna 1 in ♀, 1st joint as long as 2d and 3d combined, flagellum nearly twice as long as peduncle, with 12 joints, each with 2 long setae. Antenna 2 in ♀ much shorter, ultimate joint of peduncle shorter than penultimate, flagellum as long as the 2 combined. Gnathopod 1, 5th joint shorter than 6th, process rather large, pointed obliquely forward, 6th joint oblong oval widening to oblique palm, which is longer than hind margin. Gnathopod 2, 5th joint with very narrow process, 6th elongate, palm shorter than hind margin. Peraeopods 1 and 2 (Fig. 63) very large and powerful, 4th joint much expanded distally, a curved ridge on the outside fringed with long setae, 5th joint short, heart shaped, setose, 6th expanded distally, finger large, foliaceous. Peraeopods 3 and 4 much more feebly constructed, subequal to one another, 2d joint not oval, finger slender. Peraeopod 5,

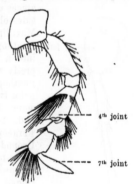

--- 4th joint

--- 7th joint

Fig. 63. **A. latipes.**
Peraeopod 1.
[After G. O. Sars.]

2d joint large, oval, 5th much shorter than 4th or 6th. Uropod 3, rami with a few spinules, narrowly lanceolate, longer than peduncle. Telson nearly square. L. ♀ (young?) 5 mm.

Arctic Ocean (Spitzbergen; Kara Sea; Greenland; Varangerfjord [Norway], depth 144—188 m).

15. Gen. **Bathymedon** O. Sars

1871 *Halimedon* (part.), A. Boeck in: Forb. Selsk. Christian., 1870 p. 169 | 1892 *Bathymedon*, G. O. Sars, Crust. Norway, *v.* 1 p. 332.

Near to Westwoodilla (p. 249). Front of head produced to a short rostrum, lateral angles deep, quadrate. Side-plates not very large, sparingly setose. Eyes poorly developed or wanting. Mandible powerful, 2d joint of palp only a little or not arcuate. Maxillipeds, outer plates rather large, though sometimes not reaching apex of 2d joint of palp. Gnathopods 1 and 2, 5th joint usually long, with or without distal process, 6th joint usually oval, not very large, palm oblique. Gnathopod 2 longer and more slender than gnathopod 1. Other characters nearly as in Westwoodilla (p. 249). 4 species.

Synopsis of species:

1 { Gnathopod 2, 5th joint well expanded distally . . 1. **B. obtusifrons** . p. 256
  { Gnathopod 2, 5th joint scarcely at all expanded — **2**.

2 { Gnathopod 2, 5th joint much shorter than 6th . . 2. **B. saussurei** . . p. 256
  { Gnathopod 2, 5th joint longer than 6th — **3**.

3 { Eyes represented by white pigment . . . . 3. **B. longimanus** . p. 257
  { Eyes entirely wanting . . . . 4. **B. acutifrons** . . p. 257

1. B. obtusifrons (H. J. Hansen) ? 1883 *Halimedon saussurei* (err., non A. Boeck 1871?), J. S. Schneider in: Tromsø Mus. Aarsh., *v.* 6 p. 35 t. 2 f. 13 | 1887 *H. obtusifrons*, H. J. Hansen in: Vid. Meddel., ser. 4 *v.* 9 p. 116 t. 5 f. 1—1e | 1893 *H. o.*, A. Della Valle in: F. Fl. Neapel, *v.* 20 p. 536 t. 58 f. 30—32 | 1892 *Bathymedon o.*, G. O. Sars, Crust. Norway, *v.* 1 p. 336 t. 118 f. 2.

Body tumid anteriorly. Rostrum very short and blunt. Side-plate 1 expanded. Eyes imperfectly developed, rounded, light red, dorsal behind rostrum. Antenna 1 in ♀ a little longer than peduncle of antenna 2, 1st joint longer than 2^d, flagellum longer than peduncle. Antenna 2, ultimate and penultimate joints of peduncle subequal, together as long as flagellum. Divergent setae on

apices of joints of peduncle in antennae 1 and 2. Upper lip, sides obtusely angled. Mandible, 2^d joint of palp straight. Maxillipeds, outer plates reaching end of palp's 2^d joint. Gnathopod 1, 5th joint not longer than 6th, distally expanded into a much projecting narrow lobe, 6th rather broad at pretty well defined palm. Gnathopod 2 much longer than gnathopod 1, 5th joint similarly produced, 6th long oval, palm twice as long as hind margin with very long palmar spine, finger very long, curved, serrate within. Peraeopods 1—4 strongly built, setose, finger lanceolate, short. Peraeopod 5, 5th joint rather shorter than 4th or 6th. Uropod 3, rami narrowly lanceolate, spinuliferous, nearly twice length of peduncle. Telson rather longer than broad, narrowed distally, apical margin truncate. Colour whitish tinged with yellowish; ova orange-coloured. L. ♀ 5 mm.

Arctic Ocean (Greenland; Tromsö, Varangerfjord, depth 188 m).

2 B. saussurei (Boeck) 1871 *Halimedon s.*, A. Boeck in: Forh. Selsk. Christian., 1870 p. 170 | 1876 *H. s.*, A. Boeck, Skand. Arkt. Amphip., *v.* 2 p. 283 t. 15 f. 1 | 1893 *H. s.*, A. Della Valle in: F. Fl. Neapel, *v.* 20 p. 535 t. 58 f. 28, 29 | 1892 *Bathymedon s.*, G. O. Sars, Crust. Norway. *v.* 1 p. 335 t. 118 f. 1.

Body slender. Rostrum acute, reaching a little beyond middle of 1st joint of antenna 1. Side-plate 1 distally expanded, 5th broad. Eyes represented by irregular whitish pigment not extending beyond the head. Antenna 1 in ♀, 2^d joint considerably longer than 1st, flagellum as long as peduncle, 14-jointed. Antenna 2 in ♀ not longer than antenna 1, ultimate and penultimate joints of peduncle nearly equal, together as long as flagellum. Gnathopod 1, 5th joint narrowly produced a little forwards, 6th joint subequal to 5th, oblong oval, palm subequal to hind margin. Gnathopod 2 slender and feeble, 5th joint with very slight process, 6th much longer than 5th, almost linear, palm short and ill-defined. Peraeopods 1 and 2, finger long and slender. Peraeopod 3 much shorter than peraeopod 4, finger as long as 6th joint. Peraeopod 4, 6th joint much longer

2^d joint

7th joint

Fig. 64.
B. saussurei.
Peraeopod 5.
[After
G. O. Sars.]

than the long finger. Peraeopod 5 (Fig. 64) slender, unusually long, very fragile, 2^d joint broad oval, 5th longer than 6th, 6th than 4th. Uropod 3, peduncle longer than rami, inner ramus with dense row of minute spinules on inner margin. Telson small, as broad as long, rounded quadrangular, apical border insinuate. Colour pale yellowish, peraeon transversely banded with faint whitish stripes. L. ♀ 5 mm.

Skagerrak, North-Sea and North-Atlantic (Christianiafjord to Trondhjemsfjord). Depth 94—564 m.

3. **B. longimanus** (Boeck) 1871 *Halimedon l.*, A. Boeck in: Forh. Selsk. Christian., 1870 p. 170 | 1876 *H. l.*, A. Boeck, Skand. Arkt. Amphip., *v.* 2 p. 284 t. 13 f. 6 | 1883 *H. l.*, J. S. Schneider in: Tromsø Mus. Aarsh., *v.* 6 p. 34 t. 3 f. 16 | 1893 *H. l.*, A. Della Valle in: F. Fl. Neapel, *v.* 20 p. 538 t. 58 f. 36, 37 | 1892 *Bathymedon l.*, G. O. Sars, Crust. Norway, *v.* 1 p. 333 t. 117.

Body rather slender and compressed, thin-skinned. Small acute rostrum reaching about middle of 1st joint of antenna 1. 1st side-plate distally expanded and produced, 5th broad. Eyes represented by chalky white pigment irregularly distributed within head and front of peraeon. Antenna 1, 2^d joint longer than 1st, flagellum in ♀ short, with 10 joints, in ♂ longer, with 22 setose joints. Antenna 2 in ♀, ultimate and penultimate joints of peduncle subequal, together as long as flagellum, in ♂ ultimate rather longer than penultimate, flagellum filiform, elongate. Mandible, 2^d joint of palp slightly arcuate. Maxillipeds, outer plates not reaching apex of 2^d joint of palp. Gnathopod 1, 5th joint with short broad distal expansion, 6th rather shorter, oval, palm longer than hind margin, defined by a long spine. Gnathopod 2 much more slender, 5th joint scarcely at all expanded, elongate, 6th joint much shorter, palm twice as long as hind margin, defined by a spine. Peraeopods 1—4 setose, finger longer in peraeopods 1 and 2 than in peraeopods 3 and 4.. Peraeopod 5, 2^d joint large, piriform, 5th shorter than 4th, 4th than 6th. Uropod 3, peduncle stout, rami rather longer, spinuliferous. Telson scarcely as long as breadth at base, narrowing to somewhat emarginate apical border. Colour pale yellowish; ova yellowish. L. 6 mm.

Arctic Ocean, North-Atlantic, North-Sea and Skagerrak (South- and West-Norway northward to Lofoten Isles). Depth 94—376 m.

4. **B. acutifrons** Bonnier 1896 *B. a.*, J. Bonnier in: Ann. Univ. Lyon, *v.* 26 p. 643 t. 38 f. 2.

Nearly akin to B. longimanus. Rostrum very small, acute. Side-plate 1 distally expanded and produced, 5th broad. Eyes and ocular pigment entirely wanting. Antenna 1, 1st and 2^d joints equal, 3^d very short, flagellum multiarticulate. Antenna 2 in ♂ much longer, ultimate and penultimate joints of peduncle subequal, long. Mandible, 2^d joint of palp not arcuate. Maxillipeds, outer plates not reaching apex of 2^d joint of palp. Gnathopods 1 and 2 as in B. longimanus. Peraeopods 1—5 also nearly as in that species, but differing by having in peraeopods 1—4 a rather bluntly lanceolate, instead of an arcuate finger. Telson quadrangular oval, with a couple of long setae near the lower angles, apical margin slightly concave. L. ♂ 5 mm.

Bay of Biscay. Depth 950 m.

16. Gen. **Monoculopsis** O. Sars

1892 *Monoculopsis* (Sp. un.: *M. longicornis*), G. O. Sars, Crust. Norway, *v.* 1 p. 310.

Head produced to a short obtuse rostrum. Side-plates 4 and 5 in ♀ very large. Eyes contiguous within the front of head. Antenna 1 in ♀ longer than antenna 2, 3^d joint long, flagellum short. Upper lip rounded quadrangular Lower lip with inner lobes separate. Mandible, molar well developed, palp slender, 3^d joint shorter than 2^d. Gnathopod 1, process of 5th joint very large, projecting. Gnathopod 2 slender, process of 5th joint long, slender, 6th elongate, tapering distally. Peraeopods 1—4 short. stout, very hirsute. Uropods and telson as in Monoculodes (p. 258).

1 species.

1. **M. longicornis** (Boeck) 1871 Monoculodes l., A. Boeck in: Forh. Selsk. Christian., 1870 p. 165 | 1876 M. l., A. Boeck, Skand. Arkt. Amphip., v. 2 p. 273 t. 16 f. 2 | 1883 M. l., J. S. Schneider in: Tromsø Mus. Aarsh., v. 6 p. 24 t. 1 f. 7; t. 3 f. 18, 23 | 1892 Monoculopsis l., G. O. Sars, Crust. Norway, v. 1 p. 311 t. 110 f. 1 | 1893. Oediceros affinis (part.), A. Della Valle in: F. Fl. Neapel, v. 20 p. 548.

Tumid in front. Head, rostrum very short, triangular, lateral corners subrectangular. Side-plates 1—3 small, 4th as large as 1st—3^d combined, 5th still broader, not so deep. Eyes rounded, light red with whitish coating. Antennae 1 and 2 densely setose. Antenna 1, 2^d joint much longer than 1st; 3^d subequal to 1st, and to 8-jointed flagellum. Antenna 2 in ♀ a little shorter than antenna 1, ultimate joint of peduncle longer than penultimate, flagellum shorter than those 2 combined. Gnathopod 1 not strong, process of 5th joint narrowly linguiform, reaching beyond hind margin of oblong oval 6th, which has palm very oblique, subequal to hind margin. Gnathopod 2 long, slender, process of 5th joint stiliform, reaching beyond palm, 6th joint nearly 5 times as long as broad, distally narrowed, palm small, oblique, well defined. Peraeopods 1 and 2, 4th joint rather expanded, as long as 5th and 6th combined, finger small. Peraeopods 3 and 4 not long, 2^d joint much expanded, 4th broad, finger short and broad. Peraeopod 5 long, 2^d joint broadly oval, 4th and 5th subequal, 6th and 7th subequal. Uropod 3, rami narrowly lanceolate, spinuliferous, a little longer than peduncle. Telson little longer than broad, slightly narrowed to slightly insinuate apical margin. Colour whitish, pale brownish violet on back. L. ♀ 9 mm.

Arctic Ocean, North-Atlantic and North-Sea (Haugesund to Vadsö [Norway], depth 18—36 m; Jan Mayen).

17. Gen. **Monoculodes** Stimps.

1853 Monoculodes (Sp. un.: M. demissus), Stimpson in: Smithson. Contr., v. 6 nr. 5 p. 54 | 1892 M., G. O. Sars, Crust. Norway, v. 1 p. 294 | 1857 Krüyera (Sp. un.: K. carinata), Bate in: Ann. nat. Hist., ser. 2 v. 19 p. 140 | 1858 Kroyera, Bate in: Nat. Hist. Northumb., v. 4 p. 15 | 1862 Kroyea, Bate, Cat. Amphip. Brit. Mus., t. 17 | 1864 Kroyeria, E. Grube, Lussin, p. 72 | 1871 Krøyeria, A. Boeck in: Forh. Selsk. Christian., 1870 p. 171.

Head produced to a rostrum, which is usually deflexed, acute. Sideplates 4 and 5 rather large. Eyes almost always contiguous, at base of rostrum. Antenna 1 generally much shorter than antenna 2, flagellum of latter in ♂ filiform. Mandible, molar with triturating surface. Maxilla 1, with 9 (6?, 7?) spines on outer plate. Gnathopod 1 (Fig. 65) usually shorter and stouter than gnathopod 2 (Fig. 66), 5th joint of latter produced into a long slender process. Peraeopods, uropods and telson nearly as in Oediceros (p. 243).

18 species accepted, 1 obscure.

Synopsis of accepted species:

1 { Eyes separated by carina of head . . . . .   1. **M. gibbosus** . . . . p. 259
    Eyes contiguous -- **2.**

2 { Head narrowly produced behind the eyes -- **3.**
    Head not narrowly produced behind the eyes — **5.**

3 { Antenna 1, 2^d joint of peduncle much longer
        than 1st . . . .   2. **M. hanseni** . . . . p. 260
    Antenna 1, 2^d joint of peduncle not longer
        than 1st — **4.**

4 { Produced part of head long . . . .   3. **M. longirostris** . . p. 260
    Produced part of head short . . . .   4. **M. kröyeri** . . . . p. 261

5 { Rostrum short — **6.**
    Rostrum long — **14.**

6 { Telson transversely truncate. — **7.**
    Telson a little emarginate -- **9.**

7 { Antenna 1 in ♀, 2^d joint longer than 1st . .   5. **M. pallidus** . . . . p. 261
    Antenna 1 in ♀, 1st and 2^d joints subequal — **8.**

8 { Gnathopod 2, process of 5th joint overreaching
        hind margin of 6th; pleon carinate . . .   6. **M. carinatus** . . . . p. 261
    Gnathopod 2, process of 5th joint not over-
        reaching hind margin of 6th; pleon smooth   7. **M. griseus** . . . . . p. 262

9 { Frontal process abruptly deflexed — **10.**
    Frontal process not abruptly deflexed — **11.**

10 { Eyes not very large or prominent; side-plate 1
        much expanded . . . .   8. **M. borealis** . . . . p. 262
     Eyes large, prominent; side-plate 1 little
        expanded . . . .   9. **M. schneideri** . . . p. 263

11 { Peraeopods 1 and 2, 7th joint short — **12.**
     Peraeopods 1 and 2, 7th joint rather long — **13.**

12 { Peraeopod 5, 2^d joint evenly expanded . . .   10. **M. crassirostris** . p. 263
     Peraeopod 5, 2^d joint piriform . . . .   11. **M. latimanus** . . . p. 264

13 { Colouring tessellated; telson apically insinuate   12. **M. tesselatus** . . . p. 264
     Colouring not tessellated; telson with apex
        decidedly emarginate . . . .   13. **M. simplex** . . . . p. 264

14 { Antenna 1, 2^d joint apically tuberculate. . .   14. **M. tuberculatus** . . p. 265
     Antenna 1, 2^d joint not apically tuberculate — **15.**

15 { Eyes large — **16.**
     Eyes not large — **17.**

16 { Side-plate 1 little expanded; gnathopod 1,
        process of 5th joint broad . . . .   15. **M. norvegicus** . . . p. 265
     Side-plate 1 much expanded; gnathopod 1,
        process of 5th joint narrow . . . .   16. **M. subnudus** . . . p. 266

17 { Gnathopods 1 and 2 rather weak; telson broadly
        rounded at apex . . . .   17. **M. packardi** . . . . p. 266
     Gnathopods 1 and 2 rather strong; telson nar-
        rowed at apex . . . .   18. **M. tenuirostratus** . p. 267

1. **M. gibbosus** Chevreux 1888 *M. g.*, Chevreux in: Bull. Soc. zool. France,
v. 13 p. 41 | 1893 *Oediceros affinis* (part.)?, A. Della Valle in: F. Fl. Neapel, *v.* 20 p. 556.

Head with rostrum and pleon carinate. Rostrum large, evenly curved
with high narrow carina, shortly acuminate, reaching apex of 1st joint of

17*

antenna 1. Side-plate 1 expanded into a lobe in front. Pleon segments 1 and 2 ending dorsally in a median tubercle, segment 3 with a very high gibbous carina. Eyes very large and prominent. oval, at the base of the rostrum, separated by the cephalic carina. Antennae 1 and 2 short. Antenna 1 a little shorter than peduncle of antenna 2, flagellum 5-jointed. Antenna 2, ultimate joint of peduncle much longer than penultimate. Gnathopod 1, process of 5th joint broad, 6th joint large, ovate, palm not defined. Gnathopod 2, process of 5th joint narrow, reaching beyond the hind margin of 6th, which is almost 3 times as long as broad, distally widened. Peraeopods 1 and 2 very hirsute, finger short. Peraeopod 5, 2^d joint moderately dilated, hind margin serrate. Uropods 1 and 2 elongate, rami shorter than peduncle. Colour yellowish white, marbled with tawny red. L. ♀ 8 mm.

North-Atlantic (lat. 46° N., long. 7° W.). Depth 180 m.

2. **M. hanseni** Stebb. 1894 *M. h.*, T. Stebbing in: Bijdr. Dierk., *v.* 17 p. 23 t. 4.

Head, frontal process long, horizontal, lateral angles little produced, obtuse. Side-plate 1 distally expanded, 3^d with lower margin insinuate, 4th with concave hind margin. Last peraeon segment with faint trace of carina. Pleon segments 1—3 discontinuously carinate; 1st and 2^d with many adpressed surface spinules. Eyes contiguous, forming an oval prominence at apex of frontal process, leaving a very small deflexed rostral apex. Antenna 1, 1st joint reaching beyond cephalic process, 2^d much longer, flagellum about 11-jointed. Antenna 2, ultimate joint of peduncle shorter than penultimate. Upper lip with slightly concave margin. Gnathopod 1, 5th joint with broad process, scarcely reaching the palm of 6th joint, which is widest at the point where a palmar spine separates the oblique convex palm from the shorter concave hind margin. Gnathopod 2 much longer than gnathopod 1, 5th joint with narrow process, long, but scarcely reaching palm of 6th joint, which widens gradually to the palmar spine, the palm convex, much shorter than the faintly concave hind margin. Peraeopods 1 and 2 very hirsute, 4th joint much widened distally, finger smooth, curved, as long as 6th joint. Peraeopods 3 and 4 very like 1st and 2^d, except that the 2^d joint is piriform, the expansion being almost confined to the upper part. Peraeopod 5, 2^d joint oval, other joints slender, with small spines. Uropods 1—3, in all peduncle longer than rami, in uropod 3 outer ramus rather shorter than inner. Telson small, a little longer than broad, sides and apex slightly insinuate, some setules on rounded corners of apical margin. L. ♀ 21 mm.

Kara Sea (lat. 72° N. long. 65° E.; Varna, depth 104 m).

3. **M. longirostris** (Goës) 1866 *Oediceros l.*, Goës in: Öfv. Ak. Förh., *v.* 22 p. 526 t. 39 f. 20 | 1893 *O. l.*, A. Della Valle in: F. Fl. Neapel, *v.* 20 p. 545 t. 58 f. 61, 62 | 1876 *Monoculodes l.*, A. Boeck, Skand. Arkt. Amphip., *v.* 2 p. 270 | 1883 *M. l.*, J. S. Schneider in: Tromsø Mus. Aarsh., *v.* 6 p. 26 t. 1 f. 1 | 1892 *M. l.*, G. O. Sars, Crust. Norway, *v.* 1 p. 306 t. 108 f. 3 | 1894 *M. l.*, T. Stebbing in: Bijdr. Dierk., *v.* 17 p. 24.

Head, frontal process long, horizontal, and the species in general in close agreement with M. hanseni, but without carina, 1st joint of antenna 1 not reaching end of frontal process, 2^d joint not longer than 1st, palm of gnathopod 1 shorter than the hind margin, peduncle of uropod 3 shorter than rami, telson with sides slightly convex and apical margin straight. Antenna 2 in ♂ with 60—70 joints in flagellum (Boeck). Colour semi-pellucid, whitish, rarely (in adults?) with brownish marbling. L. ♀ 12 mm.

Arctic Ocean (Spitzbergen, Finmark. Tromsö); Kattegat. Depth 9—150 m.

4. **M. kröyeri** Boeck 1871 *M. k.*, A. Boeck in: Forh. Selsk. Christian., 1870
p. 166 | 1876 *M. k.* (part.), A. Boeck, Skand. Arkt. Amphip., *v.*2 t. 15 f. 5 and 6 in part.
(5 l, m, 6 i, k, n, r; the rest referring to other species) | 1892 *M. k.*, G. O. Sars, Crust.
Norway, *v.*.1 p. 305 t. 108 f. 2 | 1894 *M. k.*, T. Stebbing in: Bijdr. Dierk., *v.* 17 p. 24 |
1893 *Oediceros nubilatus* (part.), A. Della Valle in: F. Fl. Neapel, *v.* 20 p. 550.

Body robust, tumid in front. In general like M. hanseni, but without
carina, frontal process much shorter, the rostral tip reaching little beyond
middle of 1st joint of antenna 1, eyes not very large, 2^d joint of antenna 1
not longer than 1st, in antenna 2 ultimate joint of peduncle as long as penultimate. Gnathopod 1, palm of 6th joint about equal to hind margin, in gnathopod 2 palm very oblique, only a little shorter than hind margin. Peraeopods 1—4 unusually expanded, finger rather broad, laminar, shorter than
the 6th joint. Telson with convex sides, a little tapering to the truncate
apex. L. ♀ 17 mm.

Arctic Ocean (Greenland, depth 11—47 m); North-Sea (Haugesund [Norway],
depth 113 m).

5. **M. pallidus** O. Sars 1892 *M. p.*, G. O. Sars, Crust. Norway, *v.*1 p. 299 t. 106
f. 3 | 1893 *Oediceros affinis* (part.), A. Della Valle in: F. Fl. Neapel, *v.* 20 p. 938.

Body rather compressed, but with no distinct carina. Frontal process
short with narrow, acute, evenly curved rostrum not reaching apex of 1st joint
of antenna 1; lateral corners of head obtuse. Side-plate 1 scarcely expanded.
Eyes small, round, imperfectly developed, light red. Antenna 1 in ♀ longer
than peduncle of antenna 2, 2^d joint longer than 1st, flagellum shorter than
peduncle, 9-jointed. Antenna 2, ultimate and penultimate joints of peduncle subequal, flagellum little shorter than peduncle. Gnathopod 1, process of 5th joint
narrow, produced beyond hind margin of oval 6th joint, palm well defined,
moderately oblique. Gnathopod 2, slender process of 5th joint produced
much beyond hind margin of narrow 6th joint, which is more than 3 times
as long as broad. Peraeopods 1 and 2, 5th joint tapering distally, densely
setose on convex hind margin, finger short. Peraeopods 3 and 4, finger short,
compressed. Peraeopod 5, 2^d joint rather broadly oval, 5th little longer than 4th.
Telson quadrangular, apical margin straight. Colour whitish, pellucid.
L. ♀ 8 mm.

North-Atlantic (Norway). Depth 110--360 m.

6. **M. carinatus** (Bate) 1856 *Westwoodea c.* (nom. nud.), Bate in: Rep. Brit.
Ass., Meet. 25 p. 58 | 1857 *Kröyera carinata*, Bate in: Ann. nat. Hist., ser. 2 *v.* 19
p. 140 | 1883 *Monoculodes carinatus*, J. S. Schneider in: Tromsø Mus. Aarsh., *v.* 6
p. 19 t. 2 f. 4 | 1889 *M. c.*, A. M. Norman in: Ann. nat. Hist., ser. 6 *v.* 3 p. 447 t. 19
f. 1—5 | 1892 *M. c.*, G. O. Sars, Crust. Norway, *v.*1 p. 295 t. 105 | 1859 *Oediceros affinis*,
R. M. Bruzelius in: Svenska Ak. Handl., n. ser. *v.* 3 nr. 1 p. 93 t. 4 f. 18 | 1893 *O. a.* (part.),
A. Della Valle in: F. Fl. Neapel, *v.* 20 p. 548 t. 4 f. 3; t. 33 f. 27—31 | 1871 *Monoculodes a.*,
A. Boeck in: Forh. Selsk. Christian., 1870 p. 164 | 1862 *M. carinatus* + *M. stimpsoni*,
Bate, Cat. Amphip. Brit. Mus., p. 104 t. 17 f. 2; p. 105 t. 17 f. 3.

Robust, rather tumid in front. Frontal process strongly vaulted, the
short curved rostrum barely reaching apex of 1st joint of antenna 1. Side-
plate 1 distally expanded. Pleon segments 1—3 dorsally carinate. Eyes
very large, elliptical, purplish with yellowish coating. Antenna 1 in ♀, 1st and
2^d joints subequal, flagellum shorter than peduncle, 8-jointed. Antenna 1 in ♂,
2^d joint much shorter than 1st, flagellum thickened at base, with some
joints densely setose. Antenna 2 in ♀ nearly twice length of antenna 1,
ultimate joint of peduncle much longer than penultimate, flagellum equal to the

2 combined; antenna 2 in ♂ less elongate than usual. Gnathopod 1 (Fig. 65), process of 5th joint densely setose, produced beyond hind margin of oval 6th joint, which has palm very oblique, longer than hind margin, not very sharply defined. Gnathopod 2

(Fig. 66), process of 5th joint very long and slender, extending a little beyond hind margin of 6th joint, which is more than 4 times as long as broad, palm

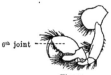

Fig. 65.
M. carinatus.
Gnathopod 1.
[After G. O. Sars.]

short, rather oblique, well defined. Peraeopods 1—4 rather short and stout, densely setose, finger ex-

6th joint   5th joint
Fig. 66. **M. carinatus.**
Gnathopod 2.
[After G. O. Sars.]

tremely small; in peraeopods 1 and 2 6th joint longer than 5th. Peraeopod 5, 2d joint oval, hind margin irregularly curved, 5th, 6th and 7th joints subequal, each longer than 4th. Uropod 3, rami narrowly lanceolate, longer than peduncle, each with about 5 spines. Telson not much longer than broad at the middle, which is somewhat dilated, apical margin straight. Colour yellowish white, mottled with dark brown in transverse bands, the most marked usually on 1st peraeon segment. Ova in pouch rose-coloured. L. ♀ 11, ♂ 9; 6—7 mm (Della Valle).

North-Atlantic, North-Sea and Skagerrak (Great Britain, France, Bohuslän, Norway); Kattegat; Gulf of Naples.

7. **M. griseus** (Della Valle) 1893 *Oediceros g.*, A. Della Valle in: F. Fl. Neapel, *v.* 20 p. 551 t. 33 f. 16—26.

Back smooth. Frontal process short, deflexed. Side-plate 1 little expanded distally. Eyes oval, occupying almost all the frontal process. Antenna 1 longer than peduncle of antenna 2, 1st and 2d joints subequal, flagellum shorter than either, 5-jointed. Antenna 2, ultimate and penultimate joints of peduncle equal. Gnathopod 1, process of 5th joint short, broad, 6th joint oval, palm very oblique, much longer than the obscurely defined hind margin, finger matching palm. Gnathopod 2, process of 5th joint rather long, just reaching the well defined, rather short and oblique palm; the 6th joint a little widened at the palm. Peraeopods 1 and 2, finger described as little, not very weak, for peraeopod 2 figured nearly as long as 6th joint. Peraeopods 3—5, finger straight, in peraeopod 4 figured as long as 6th joint, in peraeopods 3 and 5 longer than 6th. In other respects agreeing with M. carinatus (p. 261). L. 5—6 mm.

Gulf of Naples.

8. **M. borealis** Boeck 1866 *Oediceros affinis* (part.) (err., non Bruzelius 1859!), Goës in: Öfv. Ak. Förh., *v.* 22 p. 527 t. 39 f. 21' | 1871 *Monoculodes borealis*, A. Boeck in: Forh. Selsk. Christian., 1870 p. 168 | 1876 *M. b.* (part.), A. Boeck, Skand. Arkt. Amphip., *v.*2 p. 278 t. 15 f. 4 & ?6 | 1883 *M. b.*, J. S. Schneider in: Tromsø Mus. Aarsh., *v.* 6 p. 22 t. 1 f. 3 | 1892 *M. b.*, G. O. Sars, Crust. NorWay, *v.* 1 p. 298 t. 106 f. 2 | 1893 *Oediceros nubilatus* (part.), A. Della Valle in: F. Fl. Neapel, *v.* 20 p. 550.

Body not very tumid in front, pleon segments 1—3 scarcely carinate. Frontal process moderately large with abruptly deflexed rostrum, apex acute, reaching apex of 1st joint of antenna 1. Side-plate 1 distally expanded. Eyes moderate, rounded, dark red, on geniculation of head. Antenna 1 in ♀ reaching just beyond peduncle of antenna 2, 1st joint rather longer than 2d, flagellum much shorter than peduncle, 8-jointed. Antenna 2, ultimate joint of peduncle shorter than penultimate, flagellum in ♀ as long as those 2 combined. Gnathopod 1, process of 5th joint short, broad, scarcely reaching defining point of oblique palm, which is scarcely longer than hind margin of oblong oval 6th joint. Gnathopod 2, process of 5th joint about or scarcely reaching palm of 6th joint, which is about 3 times as long as broad, a little expanded at palm. Peraeopods 1 and 2, 2[d] joint curved, 2[d], 4th and 5th distally expanded, 5th short, setose, finger not quite so long as 6th. Peraeopods 3 and 4, finger as long as 6th joint. Peraeopod 5 very long, 2[d] joint piriform, 5th joint longer than 4th. Telson rounded, quadrangular, apical margin slightly insinuated. Colour yellowish white, head reticulated and back transversely striped with dark brown. L. ♀ 10 mm.

Arctic Ocean and North-Atlantic (Spitzbergen, Greenland, Siberian Polar Sea, Norway, Scotland).

9. M. schneideri O. Sars 1895 *M. s.*, G. O. Sars, Crust. Norway, *v.* 1 p. 692 t. VI f. 1.

Intermediate between M. tesselatus (p. 264) and M. borealis. Body rather slender, pleon segments 1—3 scarcely carinate. Frontal process moderately large with abruptly deflexed rostrum, apex acute, not reaching apex of 1st joint of antenna 1. Side-plates rather small, 1st pair little expanded, lower hind corner of 4th well produced. Eyes very large and protuberant, red, placed at geniculation of head. Antenna 1 reaching end of peduncle of antenna 2, 1st joint of peduncle a little longer than 2[d], flagellum shorter than peduncle, 8-jointed. Antenna 2 in ♀, ultimate and penultimate joints of peduncle subequal, together as long as flagellum. Gnathopod 1, process of 5th joint broad, reaching ill-defined palm of rather small oval 6th joint. Gnathopod 2, process of 5th joint reaching palm of 6th, which is scarcely 3 times as long as it is broad. Peraeopods 1 and 2 rather stout, 5th joint little expanded, subequal in length to 6th, finger nearly as long. Peraeopods 3 and 4, finger as long as 6th joint. Peraeopod 5 very long, 2[d] joint very broad, irregularly round, 5th joint shorter than 4th, 6th or 7th. Telson quadrangular, apical margin very slightly insinuated. Colour variegated with alternating irregular patches of dark brown. L. ♀ 6 mm.

Arctic Ocean (Tromsö). Depth 13—19 m.

10. M. crassirostris H. J. Hansen 1887 *M. c.*, H. J. Hansen in: Vid. Meddel., ser. 4 *v.* 9 p. 108 t. 4 f. 5—5f | 1889 *M. carinatus?*, A. M. Norman in: Ann. nat. Hist., ser. 6 *v.* 3 p. 450 | 1893 *Oediceros affinis* (part.), A. Della Valle in: F. Fl. Neapel, *v.* 20 p. 548.

Frontal process short, arched as to upper margin, lower nearly straight, apex acute, not reaching apex of 1st joint of antenna 1. Side-plate 1 distally somewhat expanded. Eyes on the bend of the head, large. Antenna 1 short, 2[d] joint a little shorter than 1st, flagellum short, about 8-jointed. Gnathopod 1 robust, process of 5th joint broad and rather long, 6th joint quadrangular oval, palm ill-defined, furnished with simple setae, finger touching process of 5th joint. Gnathopod 2, the long slender process of 5th joint produced beyond the short palm of 6th joint, which is 5 times as long as its middle width, broadest near base. Peraeopods 1 and 2 short, setose, 2[d], 4th and 5th joints rather dilated, 6th scarcely as long as 5th, finger short, very slender. Peraeopod 3, 2[d] and 4th joints broad, 5th much shorter than 4th, finger only half length of 6th. Peraeopod 5, 2[d] joint much expanded, as broad as long. Telson subrectangular, corners rounded, apical margin insinuate. Colour whitish, each segment transversely banded with red-brown. L. 8 mm.

Arctic Ocean (Greenland).

11. **M. latimanus** (Goës) 1866 *Oediceros l.*, Goës in: Öfv. Ak. Förh., *v.* 22 p. 527 t. 39 f. 23 | 1893 *O. l.*, A. Della Valle in: F. Fl. Neapel, *v.* 20 p. 549 t. 58 f. 67, 68 | 1871 *Monoculodes l.*, A. Boeck in: Forh. Selsk. Christian., 1870 p. 168 | 1876 *M. l.*, A. Boeck, Skand. Arkt. Amphip., *v.* 2 p. 279 t. 14 f. 2 | 1883 *M. l.*, J. S. Schneider in: Tromsø Mus. Aarsh., *v.* 6 p. 31 t. 1 f. 2 | 1892 *M. l.*, G. O. Sars, Crust. Norway, *v.* 1 p. 304 t. 108 f. 1.

Rather slender and compressed. Frontal process short, not strongly deflexed, apex of rostrum obtuse, scarcely reaching beyond middle of 1st joint of antenna 1. Side-plate 1 distally expanded. Eyes rather small, rounded, dark red, at base of frontal process. Antenna 1 rather short, but longer than peduncle of antenna 2, 1st joint longer than 2^{d}, flagellum subequal to peduncle, 10-jointed. Antenna 2, ultimate joint of peduncle slightly shorter than penultimate, flagellum subequal to peduncle. Maxilla 1, outer plate with 9 spines. Gnathopod 1, process of 5th joint very short, 6th joint much widened towards the well defined palm, which is much longer than hind margin. Gnathopod 2, process of 5th joint unusually small, scarcely reaching beyond middle of hind margin of large, oblong oval 6th joint, which widens towards well defined large oblique palm. Peraeopods 1 and 2, 2^{d} joint bent, 6th subequal to 5th, finger short. Peraeopods 3 and 4, finger rather more elongate, but not as long as 6th joint. Peraeopod 5, 2^{d} joint piriform, last 4 joints subequal to one another. Telson a little longer than broad, apical margin nearly straight or slightly rounded, carrying 4 spinules. Colour pale yellowish; ova dark violet. L. 7 mm.

Arctic Ocean and North-Atlantic (Norway from Namdal northwards, depth 19—94 m; Spitzbergen; Greenland, depth 4—132 m).

12. **M. tesselatus** J. S. Schn. 1883 *M. norvegicus* (err., non *Oediceros n.* A. Boeck 1861!) J. S. Schneider in: Tromsø Mus. Aarsh., *v.* 6 p. 21 t. 1 f. 5; t. 3 f. 20 | 1884 *M. tesselatus*, J. S. Schneider in: Tromsø Mus. Aarsh, *v.* 7 p. 81 | 1892 *M. tessellatus*, G. O. Sars, Crust. Norway, *v.* 1 p 297 t. 106 f. 1 | 1893 *Oediceros nubilatus* (part.), A. Della Valle in: F. Fl. Neapel, *v.* 20 p. 550.

Frontal process not greatly produced, evenly curved, rostral apex reaching apex of 1st joint of antenna 1. Side-plate 1 subrectangular, little expanded. Pleon segments 1—3 scarcely carinate. Eyes large, rounded oval, bright red, at base of frontal process. Antenna 1 in ♀ reaching a little beyond peduncle of antenna 2, 1st joint rather longer than 2^{d}, flagellum nearly as long as peduncle, 11-jointed; in ♂ 1st joint of peduncle twice length of 2^{d}, flagellum densely setose. Antenna 2 in ♂ as long as body, ultimate joint of peduncle shorter than penultimate (equal to it according to Schneider). Gnathopod 1, process of 5th joint short, broad, not reaching palm, 6th joint oval, palm well defined, about as long as hind margin. Gnathopod 2, process of 5th joint not more than reaching palm, 6th joint of even width, scarcely more than thrice as long as broad. Peraeopods 1—4, finger as long as 6th joint; in peraeopods 1 and 2 6th joint scarcely longer than 5th. Feraeopod 5, 5th joint shorter than 4th or 6th. Uropod 3, peduncle equal to rami in length. Telson shortly oval, apical margin distinctly insinuate. Colour whitish, tessellated with dark brown patches. L. 8 mm.

Arctic Ocean (Norway from Kvalö northwards). Depth 38—94 m.

13. **M. simplex** H. J. Hansen 1887 *M. s.*, H. J. Hansen in: Vid. Meddel., ser. 4 *v.* 9 p. 114 t. 4 f. 6—6 h | 1893 *Oediceros nubilatus* (part.), A. Della Valle in: F. Fl. Neapel, *v.* 20 p. 550.

Near to M. tesselatus and M. borealis (p. 262). Frontal process not greatly produced, pretty evenly curved, little deflexed, rostral apex acute, about reaching apex of 1st joint of antenna 1. Side-plate 1 a little expanded distally. Eyes of moderate size, at base of frontal process. Antenna 1 reaching considerably beyond peduncle of antenna 2, 1st joint rather longer than 2^d, joints of flagellum elongate. Antenna 2. ultimate joint of peduncle rather shorter than penultimate. Gnathopod 1, 2^d joint not dilated, process of 5th joint apically narrowed, scarcely reaching palm of 6th joint, palm minutely pectinate and set with setules and numerous little hooks. Gnathopod 2, process of 5th joint slender and long, yet not quite reaching palm, which is oblique and furnished as in gnathopod 1, 6th joint of uniform width. Peraeopods 1 and 2, joints little dilated, finger a little shorter than 6th joint. Peraeopods 3 and 4, 2^d joint well expanded, the rest not, finger long, nearly straight. Peraeopod 5, 2^d joint broad, narrowed distally, 5th rather longer than 4th, 6th than 5th, finger longer than 4th. Telson a little longer than broad, a little tapering, apical border emarginate more deeply than usual, lateral angles subacute. Colour pale tinged with brown. L. 7·5 mm (not full grown).

Arctic Ocean (Greenland). Depth 19—48 m.

14. **M. tuberculatus** Boeck 1866 *Oediceros affinis* (part.) (err., non Bruzelius 1859!), Goës in: Öfv. Ak. Förh., v.22 p.527 t.39 f.21 | 1871 *Monoculodes tuberculatus*, A. Boeck in: Forh. Selsk. Christian., 1870 p.167 | 1876 *M. t.*, A. Boeck, Skand. Arkt. Amphip., v.2 p.277 t.15 f.2 | 1883 *M. t.*, J. S. Schneider in: Tromsø Mus. Aarsh., v.6 p.29 t.1 f.8 | 1892 *M. t.*, G. O. Sars, Crust. Norway, v.1 p.303 t.107 f.3 | 1893 *Oediceros nubilatus* (part.), A. Della Valle in: F. Fl. Neapel, v.20 p.550.

Rather stout. Frontal process strongly convex, rostral part large, deflexed, acute, reaching beyond 1st joint of antenna 1. Side-plate 1 distally expanded. Eyes of moderate size, round, bright red, at base of frontal process. Antenna 1 without the plumose setae so commonly present, 1st joint longer than 2^d, apex of 2^d joint produced uniquely into a setose quadrate lobe or tubercle, flagellum rather longer than peduncle, 8-jointed. Antenna 2 scarcely longer than antenna 1, ultimate joint of peduncle shorter than penultimate (equal, according to Schneider), flagellum 16-jointed (Boeck), 11-jointed (Sars in fig.). Maxilla 1, outer plate according to Schneider carrying 7 spines. Gnathopod 1, process of 5th joint not reaching palm, 6th joint oblong oval, rather curved, palm as long as hind margin. Gnathopod 2 scarcely longer than gnathopod 1, though more slender, process of 5th joint slender, not reaching palm, 6th joint thrice as long as broad, not widened distally, palm oblique. Peraeopods 1 and 2, 5th joint distally widened, finger about $^2/_3$ length of 6th. Peraeopods 3 and 4, 2^d joint broad, 6th much longer than finger. Peraeopod 5, 2^d joint piriform, 4th shorter than 5th, 5th than 6th. Uropod 3, rami much longer than peduncle. Telson quadrangular, slightly tapering, apical margin insinuate, carrying 2 spinules, angles rounded. Colour yellowish, with shadows of diffuse orange; ova dark violet. L. 6—8 mm.

Arctic Ocean and North-Atlantic (Spitzbergen; Greenland; Norway, depth 94—188 m; Firth of Clyde).

15. **M. norvegicus** (Boeck) 1861 *Oedicerus n.* (part.), A. Boeck in: Forh. Skand. Naturf., Møde 8 p.650 | 1871 *Monoculodes n.*, A. Boeck in: Forh. Selsk. Christian., 1870 p.164 | 1876 *M. n.*, A. Boeck, Skand. Arkt. Amphip., v.2 p.267 t.14 f.5 | 1892 *M. n.*, G. O. Sars, Crust. Norway, v.1 p.301 t.107 f.1 | 1893 *Oediceros nubilatus* (part.), A. Della Valle in: F. Fl. Neapel, v.20 p.550.

Rather short and stout, though somewhat compressed. Frontal process deflexed, rostral apex acute, reaching beyond 1st joint of antenna 1. Side-plate 1 little expanded distally. Eyes rather large, oval, dark red, placed at base of frontal process, some way from apex. Antenna 1 reaching beyond peduncle of antenna 2, 1st joint longer than 2d, flagellum longer than peduncle, about 12-jointed. Antenna 2 in ♀, ultimate and penultimate joints of peduncle subequal, together longer than flagellum. Gnathopod 1, process of 5th joint broad, reaching palm, 6th joint oblong oval, slightly curved, expanded distally, palm well defined, scarcely as long as hind margin. Gnathopod 2, process of 5th joint reaching palm, 6th joint nearly 4 times as long as broad. Peraeopods slender. Peraeopods 1 and 2, 5th joint little expanded, shorter than 6th, finger long, but rather shorter than 6th joint. Peraeopod 5, 2d joint piriform, 5th longer than 4th and slightly longer than 6th. Telson quadrangular, little longer than broad, apical margin scarcely insinuate. Colour persistent, whitish, with transverse bands of reddish brown. L. ♀ scarcely more than 6 mm.

Arctic Ocean, North-Atlantic, North-Sea and Skagerrak (South- and West-Norway northwards to Vadsö); Kattegat. Depth 90—376 m.

16. **M. subnudus** Norm. 1889 *M. s.*, A. M. Norman in: Ann. nat. Hist., ser. 6 v. 3 p. 450 | 1892 *M. falcatus*, G. O. Sars, Crust. Norway, v. 1 p. 302 t. 107 f. 2 | 1893 *Oediceros nubilatus* (part.), A. Della Valle in: F. Fl. Neapel, v. 20 p. 550.

Near to M. borealis, M. simplex (p. 262, 264) and to M. norvegicus (p. 265). Frontal process evenly curved downward, apex of large rostrum reaching beyond apex of 1st joint of antenna 1. Side-plate 1 considerably expanded distally. Eyes large, rounded oval, bright red, at base of frontal process. Antenna 1 in ♀ reaching much beyond peduncle of antenna 2, 1st joint as long as 2d and 3d combined, flagellum longer than peduncle, about 13-jointed (Sars). Antenna 1 in ♀, 2d joint as long as 1st, 3d half length of 2d, flagellum with 11—14 joints (Norman in text, but in figure the 3 joints of peduncle, measure respectively 9, 7, 5 sixteenths of inch). Antenna 2, ultimate joint of peduncle shorter than penultimate (Sars), subequal (Norman), flagellum as long as both combined. Gnathopod 1, process of 5th joint reaching palm (Sars), not reaching it (Norman; their figures scarcely distinguishable), 6th joint narrow oblong, somewhat curved, palm shorter than hind margin. Gnathopod 2, process of 5th joint not nearly reaching palm (Sars), reaching palm (Norman; in figures the difference is trifling), 6th joint narrow sublinear, more than 4 (Sars; Norman: 5) times as long as broad. Peraeopods less setose than usual (Norman). Peraeopods rather less slender than in M. norvegicus, 5th joint in 1st and 2d slightly expanded, finger comparatively shorter. Telson subquadrate, apical border emarginate, and according to Norman each rounded angle carrying 5 spinules. Colour more or less dark orange (Sars). L. 10 mm.

North-Atlantic and Arctic Ocean (Norway to arctic circle, depth 94—188 m; Shetland Islands; Sleat Sound [Isle of Skye]).

17. **M. packardi** Boeck 1871 *M. p.*, A. Boeck in: Forh. Selsk. Christian., 1870 p. 166 | 1876 *M. p.*, A. Boeck, Skand. Arkt. Amphip., v. 2 p. 274 t. 14 f. 3 | 1883 *M. p.*, J. S. Schneider in: Tromsø Mus. Aarsh., v. 6 p. 27 t. 1 f. 6 | 1892 *M. p.*, G. O. Sars, Crust. Norway, v. 1 p. 307 t. 109 f. 1 | 1893 *Oediceros nubilatus* (part.), A. Della Valle in: F. Fl. Neapel, v. 20 p. 550.

Rather slender. Frontal process rather convex, rostral part long, narrow, not reaching end of 1st joint of antenna 1. Side-plate 1 distally expanded. Side-plates in ♂ smaller than in ♀. Eyes small, rounded oval, light red, at base of frontal process. Antennae 1 and 2, peduncle elongate. Antenna 1 longer than peduncle of antenna 2, 2^d joint as long as 1st or longer, flagellum in ♀ rather shorter than peduncle, 9-jointed. Antenna 2, ultimate joint of peduncle nearly as long as penultimate, both densely setose in ♀, not in ♂, flagellum in ♀ nearly as long as peduncle, in ♂ very long, filiform. Lower lip with a single central lobe (Boeck). Maxilla 1, outer plate with 6 spines (Schneider). Gnathopod 1, process of 5th joint short, broad, not reaching palm, 6th joint oblong oval, little expanded distally, palm ill-defined, about as long as hind margin. Gnathopod 2, process of 5th joint about reaching palm, 6th joint more than twice as long as broad, a little expanded distally, palm ill-defined. Feraeopods 1—4, finger, especially in 3^d and 4th, very long, longer than 6th joint. Peraeopod 5, 2^d joint broad above, fringed with delicate setae, 5th joint shorter than 4th. Uropod 3, rami unarmed, about as long as peduncle. Telson oblong oval, little tapering, apically rounded with 2 adjacent spinules. Colour whitish pellucid, orange on back, antennae banded with orange; ova rose-coloured, L. 7—12 mm.

Arctic Ocean, North-Atlantic, North-Sea and Skagerrak (Norway, depth 19—188 m; Firth of Clyde, depth 75 m).

18. M. tenuirostratus Boeck 1871 *M. t.*, A. Boeck in: Forh. Selsk. Christian., 1870 p. 167 | 1876 *M. t.*, A. Boeck, Skand. Arkt. Amphip., *v.* 2 p. 276 t. 14 f. 4 | 1892 *M. t.*, G. O. Sars, Crust. Norway, *v.* 1 p. 309 t. 109 f. 2 | 1893 *Oediceros aequicornis* (part.), A. Della Valle in: F. Fl. Neapel, *v.* 20 p. 545.

Nearly allied to M. packardi. Frontal process evenly curved, rostral part very long and narrow, nearly reaching apex of 1st joint of antenna 1. Side-plate 1 distally expanded. Eyes at base of frontal process, faint in alcoholic specimens. Antennae 1 and 2 elongate. Antenna 1, 2^d joint much longer than 1st, flagellum shorter than peduncle, 12-jointed. Antenna 2 in ♀ scarcely longer than antenna 1, ultimate joint of peduncle shorter than penultimate, flagellum shorter than peduncle. Gnathopods 1 and 2 much stronger than in M. packardi. Gnathopod 1, process of 5th joint large, broad, very setose, reaching palm, 6th joint widened distally, palm well defined. Gnathopod 2, process of 5th joint reaching palm, 6th joint scarcely more than twice as long as broad, widened distally, palm about as long as hind margin. Peraeopods nearly as in M. packardi. Telson oval triangular, strongly tapering to obtuse point with 2 spinules. L. ♀ 8 mm.

Christianiafjord [Norway], North-Atlantic.

M. demissus Stimps. 1853 *M. d.*, Stimpson in: Smithson. Contr., *v.* 6 nr. 5 p. 54 | 1893 *M. d.*, A. Della Valle in: F. Fl. Neapel, *v.* 20 p. 556.

L. 9 mm.

Fundy Bay (Duck Island, Grand Manan). Depth 7 m.

18. Gen. Oediceroides Stebb.

1888 *Oediceroides*, T. Stebbing in: Rep. Voy. Challenger, *v.* 29 p. 843 | 1892 *O.*, G. O. Sars, Crust. Norway, *v.* 1 p. 287.

Head with frontal process conspicuously developed (Fig. 67). Side-plates not very large, 1st expanded distally, produced forward. Eyes, when present, elongate, contiguous, on frontal process. Antenna 1 much shorter than antenna 2 (Fig. 67). Mouth-parts nearly as in Oediceropsis (p. 252), but inner plate of maxilla 1 more strongly developed, carrying from 3—8 plumose setae, outer plate with 9 spines. Gnathopods 1 and 2 nearly alike, powerful, 5th joint distally expanded into a broad setose process, 6th elongate oval or squarish, palm longer than hind margin.

4 species.

Synopsis of species:

$$
1 \begin{cases} \text{Eyes represented only by a pigment mass} \dots \dots \text{ 1. } \textbf{O. rostratus } \text{ . p. 268} \\ \text{Eyes not represented} \dots \dots \dots \dots \dots \text{ 2. } \textbf{O. proximus } \text{ . p. 269} \\ \text{Eyes distinct — 2.} \end{cases}
$$

$$
2 \begin{cases} \text{Back without tubercles} \dots \dots \dots \dots \text{ 3. } \textbf{O. cinderella } \text{ . p. 269} \\ \text{Back studded with tubercles} \dots \dots \dots \text{ 4. } \textbf{O. ornatus } \text{ . . p. 270} \end{cases}
$$

1. O. rostratus (Stebb.) 1883 *Oediceropsis rostrata*, T. Stebbing in: Ann. nat. Hist., ser. 5 *v.* 11 p. 204 | 1888 *Oediceroides r., O. conspicua,* T. Stebbing in: Rep. Voy. Challenger, *v.* 29 p. 844; p. 547 t. 60, 61 | 1893 *Halimedon rostratus,* A. Della Valle in: F. Fl. Neapel, *v.* 20 p. 540 t. 58 f. 46—49.

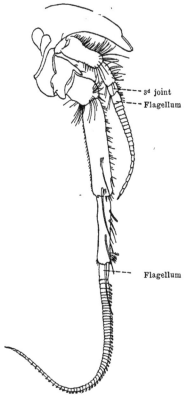

Fig. 67. O. rostratus.
Rostrum and antennae.

Body finely pubescent, peraeon segments 1—6 transversely furrowed, segment 7 and pleon segments 1—4 tuberculately keeled. Side-plate 4 deeply emarginate behind. Frontal process (Fig. 67) arcuate, reaching middle of 2d joint of antenna 1, prismatic in cross section, with a little boat-shaped rostral apex, lateral corners of head acute, postero-antennal rounded. In place of eyes, frontal process filled with a bifid mass of granular bright red pigment. Antenna 1 (Fig. 67) reaching little beyond penultimate joint of peduncle of antenna 2; 1st joint as long as 2d and 3d combined, flagellum 21-jointed, hirsute, tapering. Antenna 2 (Fig. 67), ultimate joint of peduncle shorter than penultimate, both stout, setulose, armed with long spines, flagellum shorter than peduncle, tapering, of 65 joints, most with calceoli, or peduncle without long spines, flagellum longer than peduncle, of 74 joints, without calceoli Upper lip transversely oval, a little projecting at centre. Mandible, cutting plates distinctly denticulate. Maxilla 1, inner plate with 8 plumose setae, 2d joint of palp with many spiniform setae about apex. Maxilla 2, inner margin of inner plate entirely fringed with setae. Gnathopods 1 and 2, 5th joint shorter but broader than oval 6th, considerably so in gnathopod 2, in which the 6th joint is little broader, but considerably longer than in gnathopod 1, palm long, feebly defined. Peraeopods 1—4 setose, finger nearly as long as 6th joint, with cap over the nail. Branchial vesicles in peraeopods 1 and 2 very large, lightly crumpled. Peraeopods 3 and 4, 4th joint as long as 2d or longer, both very setose. Peraeopod 5, 2^d joint piriform, large, but not longer than 4th, which is a little longer than 5th or 6th, all the three spinose; the finger straight, as long as 5th or 6th joint, with slender spines or setae fringing both margins. Uropod 3, rami rather longer than peduncle, subequal, lanceolate, spinulose and partly pectinate. Telson small, subquadrate, sides convex, apical border emarginate. L. 18—31 mm.

Southern Indian Ocean (Cumberland Bay [Kerguelen Island], depth 223 m; Heard Island, depth 274 m).

2. **O. proximus** Bonnier 1896 *O. proxima*, J. Bonnier in: Ann. Univ. Lyon, v. 26 p. 640 t. 38 f. 1.

Pleon segments 1—3 dorsally covered with little conical tubercles, close set, producing a downy appearance. Head, frontal process or rostrum acute, scarcely longer than 1st joint of antenna 1, lateral corners rounded. Side-plate 1 distally expanded and produced, 4th dilated in front, hind margin almost straight. Eyes entirely wanting. Antenna 1 rather longer than peduncle of antenna 2, 1st joint longer than 2^d and 3^d combined, flagellum longer than peduncle, with 17 short thick joints. Antenna 2, ultimate joint shorter than thick penultimate, flagellum long, 64-jointed. Mouth-parts nearly as in O. rostratus; maxilla 1 with 6 setae on inner plate, outer with 8 spines. Gnathopod 1, 5th joint distally expanded, spinose, shorter than oval 6th, in which palm is finely crenulate, longer than hind margin, defined by a palmar spine, finger as long as palm. Gnathopod 2, 5th joint distally produced, though not to end of hind margin of 6th, which is longer not wider than in gnathopod 1. Peraeopod 1, 6th joint narrow at the extremities, dilated at middle of setose front margin. Peraeopods in general like those of O. rostratus. Uropod 3 not known. Telson oval, with 2 spinules and 2 setules at the rounded apex. L. ♂ 7 mm.

Bay of Biscay. Depth 650—950 m.

3. **O. cinderella** Stebb. 1888 *O. c.*, T. Stebbing in: Rep. Voy. Challenger, v. 29 p. 850 t. 62, 63 | 1893 *Halimedon c.*, A. Della Valle in: F. Fl. Neapel, v. 20 p. 540 t. 58 f. 43—45.

Back a little imbricated. Side-plate 1 greatly expanded and produced forward, 4th slightly excavate behind. Frontal process deflexed, nearly reaching apex of 1st joint of antenna 1, prismatic in section; lateral corners of head subacute, postero-antennal corners rounded. Eyes long, narrow, nearly reaching blunt apex of frontal process. Antenna 1, 1st joint longer than 2^d, 2^d nearly twice as long as 3^d, all with plumose setae, flagellum slender, probably not long. Antenna 2, ultimate joint of peduncle much narrower than penultimate, about as long, armed with large spines, flagellum of 54 joints, most with calceoli. Upper lip smoothly rounded. Mandibular palp, 2^d joint decidedly longer than 3^d. Maxilla 1, inner plate with 3 plumose setae. Maxilla 2, inner plate not wholly fringed with setae. Gnathopods 1 and 2, 5th joint as broad as long, shorter than 6th, 6th longer than broad, the long convex palm forming a very decided angle with the much shorter hind margin, this joint being longer, but not broader in gnathopod 2 than in gnathopod 1, finger long and curved. Peraeopods 1—4 moderately setose, finger boat-shaped, in peraeopods 1 and 2 a little shorter, in peraeopods 3 and 4 a little longer than the 6th joint, a cap over the nail. Branchial vesicles with accessory lobe at base. Peraeopods 3 and 4, 2^d joint not broadly oval, in peraeopod 4 shorter than 4th joint. Peraeopod 5, 2^d joint broad oval, longer than 4th, 5th or 6th, which are subequal. Telson small, rounded oval, with spinules and plumose setules on rounded apex. L. 15 mm.

South-Atlantic (Falkland Islands). Depth 1893 m.

4. O. ornatus (Stebb.) 1883 *Acanthostepheia ornata*, T. Stebbing in: Ann. nat. Hist., ser. 5 v. 11 p. 203 | 1888 *Oediceroides o.*, T. Stebbing in: Rep. Voy. Challenger, v. 29 p. 855 t. 64 | 1893 *Halimedon ornatus*, A. Della Valle in: F. Fl. Neapel, v. 20 p. 536 t. 58 f. 33—35.

̄   Back imbricate and discontinuously carinate, with large medio-dorsal tubercles, and on peraeon and pleon segments 1 and 2 smaller lateral ones, hind margin of peraeon segments 1—7 and pleon segment 1 fringed with small tubercles. Side-plate 1 much produced forward, 4th excavate behind. Head dorsally ridged, with conspicuous frontal process, constricted at base, prismatic in cross section; lateral corners rounded. Eyes long, separated only by the narrow carina, occupying frontal process to its acute tip, remaining dark in spirit, ocelli numerous. Antenna 1, peduncle slender, short, 1st joint longer than 2d, 3d feeble, suggesting that the missing flagellum is small. Antenna 2, penultimate joint of peduncle much longer than ultimate, 1st joint of flagellum carrying a calceolus, rest missing. Upper lip, distal margin broad. Lower lip, outer lobes very broad, character of inner not clearly ascertained. Mandible, 2d joint of palp a little longer than 3d. Maxilla 1, inner plate very large, with 5 setae. Maxilla 2, inner margin of inner plate wholly fringed with setae. Gnathopods 1 and 2, 5th joint shorter than 6th, with a rather massive setose or spinose lobe, 6th oval, broadest at base, the very long setose palm moderately defined from the short hind margin. Peraeopods 1—5 all more or less broken, what remained showing the characters usual in the family. Uropod 3 not as long and not reaching so far back as uropods 1 or 2; rami lanceolate, spinose, a little longer than peduncle. Telson small, oblong, slightly narrowing to the scarcely emarginate apical border. L. ♀ 15 mm.

Bass Strait (East Moncoeur Island). Depth 70 m.

Oedicerotidarum species incertae sedis.

Oedicerus novi-zealandiae Dana 1853 & 55 *O. n.*, J. D. Dana in: U. S expl. Exp., v. 13 II p. 934 t. 63 f. 7 a —h | 1861 *O. n.*, A. Boeck in: Forh. Skand. Naturf., Møde 8 p. 650 | 1862 *O. novae-zealandiae, O. novi-zealandiae*, Bate, Cat. Amphip. Brit. Mus., p. 104 t. 17 f. 1 | 1876 *O. novae-zealandiae*, Miers, Cat. Crust. N. Zealand, p. 126 | 1886 *O. neo-zelanicus*, (G. M. Thomson &) Chilton in: Tr. N. Zealand Inst., v. 18 p. 149.

Generically near to Exoediceros (p. 239), but with finger on peraeopods 1 and 2 well developed. Colour greenish. L. 4 mm.

South-Pacific (New Zealand). Rock-pools.

---

20. Fam. **Synopiidae**

1853 Subfam. *Synopinae*, J. D. Dana in: U. S. expl. Exp., v. 13 II p. 981 | 1886 *Synopidae*, Bovallius in: N. Acta Soc. Upsal., ser. 3 v. 13 nr. 9 p. 3 | 1888 *S.*, T. Stebbing in: Rep. Voy. Challenger, v. 29 p. 798.

Integument thin, pellucid. Head triangular, not inflated, produced over bases of antennae 1 and 2. Side-plate 3 the largest. Eyes large, coalesced at top of head, usually with small lateral pair below. Antenna 1 shorter than antenna 2, peduncle short, 1st joint of flagellum large, accessory flagellum rather long. Upper lip bilobed. Mandible normal, but 2^d joint of palp large, 3^d minute. Maxilla 1, inner plate with several setae, outer with 8 spines, 2^d joint of palp with spine-teeth on apex. Maxilla 2, inner plate broader than outer, fringed on inner margin. Maxillipeds strongly setose, inner plates rather short, rounded at apex, outer long, narrow, 4th joint of palp very small. Gnathopod 1 scarcely subchelate (Fig. 68), gnathopod 2 simple (Fig. 69). Peraeopods 1 and 2 with 4th and 5th joints dilated. Feraeopods 3—5, 2^d joint expanded. Uropods 1 and 2, outer ramus shorter than inner, uropod 3 reaching furthest back, outer ramus the longer, 2-jointed. Telson cleft.

Marine.

1 genus, 2 species accepted, 3 doubtful.

1. Gen. **Synopia** Dana

1852 *Synopia*, J. D. Dana in: Amer. J. Sci., ser. 2 *v.* 14 p. 315 | 1853 *S.*, J. D. Dana in: U. S. expl. Exp., *v.* 13 II p. 994 | 1871 *S.*, Claus in: Nachr. Ges. Götting., p. 157 | 1886 *S.*, Bovallius in: N. Acta Soc. Upsal., ser. 3 *v.* 13 nr. 9 p. 4 | 1888 *S.*, T. Stebbing in: Rep. Voy. Challenger, *v.* 29 p. 799 | 1893 *S.*, A. Della Valle in: F. Fl. Neapel, *v.* 20 p. 850.

With the character of family.

2 species accepted, 3 doubtful.

Synopsis of accepted species:

Telson somewhat triangular . . . . 1. **S. ultramarina** . p. 271
Telson oval . . . . 2. **S. schéeleana** . . p. 272

1. **S. ultramarina** Dana 1853 & 55 *S. u.* $+$ *S. gracilis*, J. D. Dana in: U. S. expl. Exp., *v.* 13 II p. 995; t. 68 f. 6a—h, 7a—o | 1862 *S. u.*, Bate, Cat. Amphip. Brit. Mus., p. 341 t. 54 f. 1 | 1886 *S. u.*, Bovallius in: N. Acta Soc. Upsal., ser. 3 *v.* 13 nr. 9 p. 6 t. 1 | 1893 *S. u.* (part.), A. Della Valle in: F. Fl. Neapel, *v.* 20 p. 851.

Body and front of head sharp-edged, peraeon segments 1—7 short, pleon segments 1—3 long and large. Front of head at right angles to dorsal line. Side-plates 1 and 2 small, slightly bent forward, 3^d far larger, quadrangular, emarginate behind, 4th not large, triangular, with the longest side arcuate, 5th—7th small; pleon segments 1—3, postero-lateral corners subquadrate. Eye seemingly variable, round or oval, especially perhaps in ♀, the little lateral eyes having 3 or 4 lenses, colour continuing dark in spirit. Antenna 1, flagellum in ♀ with 5 joints, in ♂ with 10—14, in both 1st long, tapering, more or less fringed, accessory flagellum 2-(Dana: faintly 3-)jointed, a little longer or shorter than 1st of primary, the terminal joint minute. Antenna 2, ultimate joint of peduncle much shorter than penultimate, flagellum in ♀ with 12—14 joints, in ♂ with 18. Triturating lobes of stomach fringed with very slender spines. Gnathopod 1, 2^d joint rather widened distally, 5th joint large and long, narrow at the extremities, fringed on hind margin with long plumose spines or setae. 6th very much smaller, broadly ovate, fringed on part answering to palm with plumose spines or setae, finger nearly as long as 6th joint, slender, slightly curved, scarcely intended to close against any palm margin, since the equivalent of the palm is clothed as above mentioned. Gnathopod 2 much more slender, 5th joint long and narrow 6th also narrow, more than half as long, both fringed on hind margin with long spines, finger small, with a nail nearly as long as the base. Peraeopod 1, 2^d joint widened distally. 4th and 5th joints broad, subequal, the 5th oval, fringed on hind margin, 6th sublinear, rather shorter than 5th, finger more than half length of 6th, the base more than twice as long as the little nail. Peraeopod 2 differing from peraeopod 1, 4th joint broad, rather cup-like, much shorter than the broadly oval densely fringed 5th, 6th narrowly oval, about half as long and broad as 5th, slightly armed, finger as in peraeopod 1. Peraeopods 3 and 4, 2^d joint a very broad oval, the hind wing so transparent as easily to escape observation, 6th joint subequal in length to 2^d, longer than the other joints, finger long and straight. Feraeo-pod 4 rather longer than peraeopods 3 or 5. Peraeopod 5, 2^d joint with straight front, hind margin produced downward, forming a subangular lobe, finger shorter than in peraeopods 3 and 4. Pleopods have 2 cleft spines on 1st joint of inner ramus. Uropod 1 much longer than uropod 2, its peduncle much longer than peduncle of uropod 3, but its rami shorter. Telson reaching distinctly beyond peduncle of uropod 3; triangular, obliquely truncated behind, divided beyond the middle (Bovallius), or, triangular with a tendency to oval, cleft to the middle, the apices notched, with the outer tooth rather more produced than the inner. Colour rich blue to hyaline, tinted in parts with blue. L. 4—6 mm.

Tropical Atlantic (lat. 4—12° S., long. 11—25° W.).

2. **S. schéeleana** Bovall. 1886 *S. s.*, Bovallius in: N. Acta Soc. Upsal., ser. 3 *v.* 13 nr. 9 p. 16 t. 2 f. 22—29 | 1888 *S. s.*, T. Stebbing in: Rep. Voy. Challenger, *v.* 29 p. 799 t. 52 | 1893 *S. ultramarina* (part.), A. Della Valle in: F. Fl. Neapel, *v.* 20 p. 851.

Very near to S. ultramarina (p. 271), but antenna 1 more elongate and telson differently shaped. Eye large, oval, sometimes narrowly prolonged backward, about 30 lenses; lateral eyes of 4 lenses. Antenna 1, flagellum in ♂ with 16—20 joints, 1st long, closely fringed, accessory flagellum equal to it or a little shorter, its 2^d joint minute, tipped with a setule. Antenna 2, flagellum 20-jointed. Uropod 2, outer ramus totally smooth along both margins (Bovallius) or finely pectinate along upper margin (Stebbing), inner ramus with 2 spines (Bovallius) or 1 spine (Stebbing) on inner margin. Telson broadly ovate, reaching a little beyond peduncle of uropod 3, cleft about to centre, sometimes beyond, sometimes falling short of it, apices rounded, slightly notched, inner point equally advanced with outer or falling short of it. Colour hyaline. L. 4—6 mm.

Fig. 68.
S. schéeleana.
Gnathopod 1.

Fig. 69.
S. schéeleana.
Gnathopod 2.

Tropical Atlantic and Pacific.

S. angustifrons Dana 1853 & 55 *S. a.*, J. D. Dana in: U. S. expl. Exp., *v.* 13 II p. 998; t. 68 f. 8a—d | 1886 *S. a.*, Bovallius in: N. Acta Soc. Upsal., ser. 3 *v.* 13 nr. 9 p. 20 t. 2 f. 36—39 | 1893 *S. ultramarina* (part.), A. Della Valle in: F. Fl. Neapel, *v.* 20 p. 851.

Antenna 1 short, flagellum 5-jointed. Antenna 2, flagellum 10-jointed. L. 3 mm.

Tropical Pacific (lat. 18° S., long. 122° W.).

S. caraibica Bovall. 1886 *S. c.*, Bovallius in: N. Acta Soc. Upsal., ser. 3 *v.* 13 nr. 9 p. 14 t. 2 f. 30 | 1893 *S. ultramarina* (part.), A. DellaValle in: F. Fl. Neapel, *v.* 20 p. 851.

L. 5 mm.

Caribbean Sea.

S. orientalis Kossm. 1880 *S. o.*, Kossmann in: Reise Roth. Meer, *v.* 21 Malacost. p. 137 t. 15 f. 11—13 | 1886 *S. o.*, Bovallius in: N. Acta Soc. Upsal., ser. 3 *v.* 13 nr. 9 p. 21 | 1893 *S. ultramarina* (part.), A. DellaValle in: F. Fl. Neapel, *v.* 20 p. 851.

L. 3 mm.

Red Sea.

21. Fam. **Tironidae**

1871 Subfam. *Syrrhoinae*, A. Boeck in: Forh. Selsk. Christian., 1870 p. 146 | 1888 *Syrrhoidae*, T. Stebbing in: Rep. Voy. Challenger. *v.* 29 p. 787 | 1893 *S.*, G. O. Sars, Crust. NorWay, *v.* 1 p. 388.

Pleon well developed. Head usually produced into a deflexed rostrum. Side-plate 4 (except in Argissa, Fig. 70 p. 276) not conspicuously large, often smaller than 3^d. Eyes of various characters. Antenna 1 with accessory flagellum, peduncle longer, flagellum shorter in ♀ than in ♂; antenna 2 little or not longer than antenna 1 in ♀, considerably longer in ♂. Lower lip (Fig. 71 p. 278), with inner lobes. Mandible robust, but palp slight, usually with very short 3^d joint. Maxilla 1, inner plate with several setae, outer with (so far as known) 11 spines, 2^d joint of palp long. Maxilla 2, inner plate rather the broader, fringed on inner margin. Maxillipeds normal. Gnathopods 1 and 2 feeble, slender, not very unequal, 5th joint long, 6th shorter, subchelate or simple. Peraeopods 1 and 2 slight. Peraeopod 5 usually the longest. Uropods 1 and 2, outer ramus the shorter; uropod 3, rami subequal. Telson long, cleft (except in Bruzelia, p. 274).

Marine.

7 genera, 13 accepted species. 1 doubtful.

Synopsis of genera:

1	{ Telson undivided	1. Gen. **Bruzelia** . . . p. 274
	{ Telson deeply cleft — **2**.	
2	{ Four eyes	2. Gen. **Tiron** . . . . p. 275
	{ Eyes two, coalesced or separate, or none — **3**.	
3	{ Side-plate 4 larger than 3^d (Fig. 70) — **4**.	
	{ Side-plate 4 not larger than 3^d — **5**.	
4	{ Mandible, molar large; maxillipeds, outer plates of moderate size	3. Gen. **Argissa** . . . p. 276
	{ Mandible, molar Very small; maxillipeds, outer plates Very large	4. Gen. **Astyra** . . . . p. 278
5	{ Side-plate 3 not greatly expanded distally . .	5. Gen. **Syrrhoites** . . p. 279
	{ Side-plate 3 greatly expanded distally — **6**.	
6	{ Gnathopods 1 and 2 subchelate	6. Gen. **Syrrhoe** . . . p. 281
	{ Gnathopods 1 and 2 simple	7. Gen. **Pseudotiron** . p. 284

1. Gen. **Bruzelia** Boeck

1871 *Bruzelia* (Sp. un.: *B. typica*), A. Boeck in: Forh. Selsk. Christian., 1870
p. 149 | 1876 *B.*, A. Boeck, Skand. Arkt. Amphip., *v.* 2 p. 477 | 1893 *B.*, G. O. Sars,
Crust. Norway, *v.* 1 p. 394 | 1893 *B.*, A. Della Valle in: F. Fl. Neapel, *v.* 20 p. 667.

Body stout, indurated. Head, rostrum deflexed, lateral corners broadly
rounded. Side-plate 3 slightly expanded below, 4th shorter, deeply excavate
behind. Pleon segments 1—3 large, 2^d and 3^d, postero-lateral angles acute.
Eyes rudimentary. Antenna 1 rather short, accessory flagellum 2- or 3-jointed.
Antenna 2 in ♀ rather, in ♂ much longer. Upper lip carinate, broad,
narrowing to truncate apex. Lower lip, inner lobes partially coalesced with
outer. Mandible compact, cutting edge obscurely dentate, little prominent,
molar large, not prominent, palp slender, 3^d joint not or scarcely longer
than 1st. Maxilla 1, inner plate with 10 setae, outer with 11(?) spines,
palp long. Maxilla 2, inner plate broader than outer, fringed on inner margin.
Maxillipeds, inner plates broad, outer reaching beyond middle of palp's
2^d joint, fringed with spines on inner margin, finger of palp small. Gnathopods 1 and 2 similar, but 2^d longer and thinner than 1st, 5th joint long,
6th short, palm oblique, armed with a strong denticulate spine. Peraeopods 1
and 2 very slender. Peraeopods 3—5 stronger, 5th subequal to 4th. Uropod 1, rami
narrow, unequal; uropod 2, peduncle short, outer ramus narrow and short,
inner long, broadly lanceolate, unarmed; uropod 3, rami narrow outer with
spiniform terminal joint. Telson large, broadly lanceolate, entire.

2 species.

Synopsis of species:

1. B. typica Boeck   1871 *B. t.*, A. Boeck in: Forh. Selsk. Christian., 1870
p. 150 | 1876 *B. t.*, A. Boeck, Skand. Arkt. Amphip., *v.* 2 p. 478 t. 10 f. 3 | 1893 *B. t.*,
G. O. Sars, Crust. Norway, *v.* 1 p. 395 t. 138, t. 139 f. 1 | 1893 *B. t.*, A. Della Valle in:
F. Fl. Neapel, *v.* 20 p. 668 t. 59 f. 76.

Body extremely broad, subdepressed, back of peraeon vaulted above, lateral
edges forming abrupt angle with side-plates. Pleon segments 1—3 dorsally
carinate, produced to upturned point on segment 3, segments 4—6 smooth in ♀,
in ♂ segments 5 and 6 armed with a dorsal projection, that on segment 5
narrow linguiform, abruptly deflexed, ending in 2 setules. Head, rostrum
apically abruptly deflexed and obtusely rounded, not reaching apex of 1st joint of
antenna 1. Side-plates 1—3 slightly widened distally, 5th and 6th much broader
than deep, hind lobe angular below. Eyes represented by an irregular patch of
chalky white pigment. Antenna 1 in ♀, 1st joint bent, stout, little longer
than 2^d, 3^d more than half as long as 2^d, flagellum scarcely as long as peduncle,
with 8 joints, accessory flagellum small, 2-jointed; in ♂ 1st joint of flagellum
very large, densely filamented, accessory flagellum large, 3-jointed. Antenna 2
in ♀ rather longer than antenna 1, ultimate and penultimate joints of peduncle
subequal, long, flagellum little more than half as long as peduncle, 7-jointed, in ♂
filiform, very long, with about 12 long joints. Gnathopod 1, 5th joint nearly
twice as long as 6th, 6th oblong oval, palm subequal to hind margin, defined
by obtuse angle, the denticulate spine as in Syrrhoe crenulata (p. 282). Gnathopod 2, 5th and 6th joints narrower and longer. Peraeopods 1 and 2, 5th joint
longer than 4th or 6th, fringed distally with long setae. Peraeopods 3—5,

2^d joint narrowly oblong, hind margin smooth. 4th joint rather large, wide above, acutely produced. Uropod 3, outer ramus 2-jointed, a little shorter than inner, in ♀ unarmed, in ♂ both rami fringed with plumose setae on inner margin. Telson about twice as long as broad at base, distally tapering rapidly, subacute apex minutely bidentate. Colour dark yellowish. L. about 6 mm.

Arctic Ocean, North-Atlantic, North-Sea and Skagerrak (Norway northward to Lofoten Islands). Depth 150—564 m.

2. B. tuberculata O. Sars 1882 *B. t.*, G. O. Sars in: Forb. Selsk. Christian., nr. 18 p. 95 t. 4 f. 7 | 1893 *B. t.*, G. O. Sars, Crust. Norway, *v.* 1 p. 397 t. 139 f. 2 | 1893 *B. t.*, A. Della Valle in: F. Fl. Neapel, *v.* 20 p. 668 t. 59 f. 75.

Body rather stout and compact. Peraeon segments 3—7 and pleon segments 1—3 dorsally raised to obtuse tuberculiform processes. Lateral margins of peraeon segments .forming keels, which are continued on pleon segments 1 and 2, and there produced into small teeth. Pleon segments 4—6 dorsally smooth. Head, rostrum almost vertically deflexed, acuminate, over-lapping 1st joint of antenna 1. •Side-plates smaller than in B. typica. Pleon segment 3, postero-lateral angles more hook-like, with lower margin coarsely serrate. Eyes represented by rounded patch of whitish pigment at base of rostrum. Antenna 1 in ♀, 1st joint slightly curved, flagellum shorter than peduncle, 6-jointed, accessory flagellum as long as the 1st joint of flagellum, 2-jointed. Antenna 2 in ♀ rather longer than antenna 1, ultimate and penultimate joints of peduncle subequal, flagellum about half as long as peduncle, 6-jointed. Gnathopods 1 and 2, peraeopods 1 and 2 as in B. typica. Peraeopods 3—5 rather strong, not very unequal, 2^d joint serrate on hind margin and in peraeopod 5 very broadly oval, 4th joint also much expanded above. Uropods 1 and 2 nearly as in B. typica; uropod 3, rami scarcely longer than peduncle, both unarmed. Telson more than twice as long as broad, slightly constricted near base, distally tapering to an acute simple point. Colour light yellowish with large irregular patches of diffuse brick-red. L. ♀ 6 mm.

Arctic Ocean and North-Atlantic (Lofoten Isles, Bejan in Trondhjemsfjord). Depth 188—564 m.

2. Gen. **Tiron** Lillj.

1865 *Tiron* (Sp. un.: *T. acanthurus*), W. Lilljeborg in: N. Acta Soc. Upsal., ser. 3 *v.* 6 nr. 1 p. 19 | 1876 *T.*, A. Boeck, Skand. Arkt. Amphip., *v.* 2 p. 475 | 1893 *T.*, G. O. Sars, Crust. Norway, *v.* 1 p. 398 | 1893 *T.*, A. Della Valle in: F. Fl. Neapel, *v.* 20 p. 693 | 1866 *Syrrhoë* (part.)?, Goës in: Öfv. Ak. Förh., *v.* 22 p. 527 | 1868 *Tessarops* (Sp. un.: *T. hastata*) (non Rafinesque 1821, Arachnoidea!), A. M. Norman in: Ann. nat. Hist., ser. 4 *v.* 2 p. 412.

Body not robust or indurated. Head, rostrum of moderate length. Side-plate 4 less deep than 3^d. Eyes 4, a subdorsal pair, and a minute lateral pair. Antenna 1 with long accessory flagellum: Antenna 2 longer than antenna 1. Upper lip rounded, bilobed, carinate. Lower lip with inner lobes. Mandible normal, molar large and prominent, palp slender, 3^d joint scarcely longer than 1st. Maxilla 1, inner plate with many setae, outer with 11 (?) spines, palp long. Maxilla 2, inner plate fringed on inner margin, outer rather narrower. Maxillipeds normal. Gnathopods 1 and 2 long, setose, 5th joint narrow, much longer than the linear simple 6th, finger small. Peraeopods short and stout, 2^d joint much expanded in peraeopods 3—5, especially in

18*

peraeopod 5. Uropods 1 and 2, outer ramus the shorter; uropod 3, rami long, lanceolate, subequal, outer with small 2^d joint. Telson narrow, very long, deeply cleft.

1 species.

1. **T. acanthurus** Lillj. 1865 *T. a.*, W. Lilljeborg in: N. Acta Soc. Upsal., ser. 3 *r*. 6 nr. 1 p. 19 | 1876 *T. a.*, A. Boeck, Skand. Arkt. Amphip., *v*. 2 p. 475 t. 9 f. 8; t. 13 f. 1 | 1893 *T. a.*, G. O. Sars, Crust. NorWay, *v*. 1 p. 399 t. 140 | 1893 *T. a.*, A. Della Valle in: F. Fl. Neapel, *v*. 20 p. 693 t. 60 f. 1 | 1866 *Syrrhoë [?] bicuspis*, Goës in: Öfv. Ak. Förh., *v*. 22 p. 528 t. 40 f. 26 | 1868 *Tessarops hastata*, A. M. Norman in: Ann. nat. Hist., ser. 4 *v*. 2 p. 412 t. 22 f. 4—7.

Pleon segments 1—3 dorsally serrate, with about 10 teeth, segments 4—6 dorsally produced to conically pointed process, smallest on segment 6. Head, rostrum reaching about middle of 1st joint of antenna 1, lateral corners nearly quadrate. Side-plates 1—3 rather narrow, distally crenulate, setose, 4th slightly emarginate, obtusely triangular below, 5th and 6th, hind lobe deep, rounded. Pleon segment 3, postero-lateral angles minutely produced. Upper eyes rounded, dark red. Antenna 1 in ♀, 1st joint nearly as long as 2^d and 3^d combined, flagellum longer than peduncle, 10-jointed, accessory flagellum half as long, 5-jointed. Antenna 2 in ♀ much longer, ultimate joint of peduncle shorter than penultimate, flagellum half as long as peduncle, 9-jointed. Antenna 1 in ♂, 1st joint longer than 2^d and 3^d combined, flagellum more than twice as long as peduncle, 1st joint very long, densely fringed. Antenna 2 in ♂, flagellum longer than peduncle, 15-jointed. Gnathopods 1 and 2, 5th joint distally narrowed, more than twice as long as 6th, the small finger 2-jointed. Peraeopods 1 and 2, 4th joint longer than 5th or 6th, finger small. Peraeopods 3—5 much stouter, 4th joint much larger than 6th, especially in the 5th pair. Uropod 3 reaching much beyond uropods 1 and 2, spinules and setae on the rami more developed in ♂ than in ♀. Telson cleft nearly to the base, gradually tapering, apices acute, contiguous. Colour dark reddish brown, antennae banded with brown. L. ♀ 8, ♂ 9 mm.

Arctic Ocean, North-Atlantic, North-Sea and Skagerrak (Greenland, NorWay, Great Britain); Kattegat. Depth 38—113 m.

3. Gen. **Argissa** Boeck

1871 *Argissa* (Sp. un.: *A. typica*), A. Boeck in: Forh. Selsk. Christiau., 1870 p. 125 | 1891 *A.*, G. O. Sars, Crust. Norway, *v*. 1 p. 140 | 1893 *A.*, A. Della Valle in: F. Fl. Neapel, *v*. 20 p. 686 | 1890 *Chimacropsis* (Sp. un.: *C. danica*) (non Zittel 1887, Pisces!), Meinert in: Udb. Hauchs, *v*. 8 p. 167.

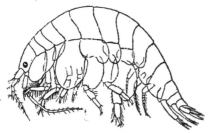

Fig. 70. **A. hamatipes**, ♀.
Lateral view. [After G. O. Sars.]

Body (Fig. 70) compressed. Head not or little rostrate. Sideplates decreasing from 1st to 3^d, 4th large, shield-like, little excavate. Antenna 1 shorter than antenna 2, accessory flagellum small, with 2 joints, 2^d minute. No calceoli. Upper lip slightly notched. Mandible, cutting edge and accessory plate dentate, many spines in spine-row, molar strong, palp slight, 3^d joint longer than 2^d. Maxilla 1, inner plate with 2 or 3 setae, 2^d joint of palp distally widened. Maxilla 2, inner plate with lateral armature. Maxillipeds, plates moderate, outer with apical spines, inner margin simple, 4th joint of palp spiniform. Gnathopods similar, not subchelate, 6th joint slender, much shorter and narrower than 5th, finger slender, feeble. Feraeopods 1 and 2 very small. Peraeopods 3 and 4, 2^d joint moderately expanded, smaller in 4th pair than in 3^d. Peraeopod 5, 2^d joint large, descending much below 3^d, 4th and 5th joints laminar, setose; all peraeopods with 6th and 7th joints small. Uropod 3, rami foliaceous, inner margin setose. Telson large, tapering, deeply cleft.

2 species.

Synopsis of species:

1. **A. hamatipes** (Norm.) 1869 *Syrrhoë h.*, A. M. Norman in: Rep. Brit. Ass., Meet. 38 p. 279 | 1893 *Argissa h.*, T. Scott in: Rep. Fish. Board Scotl., v. 11 p. 213 t. 5 f. 30, 31 | 1871 *A. typica*, A. Boeck in: Forh. Selsk. Christian., 1870 p. 125 | 1872 & 76 *A. t.*, A. Boeck, Skand. Arkt. Amphip., v. 1 t. 7 f. 2; v. 2 p. 206 | 1891 *A. t.*, G. O. Sars, Crust. NorWay, v. 1 p. 141 t. 48 | 1893 *A. t.*, A. Della Valle in: F. Fl. Neapel, v. 20 p. 687 t. 59 f. 91 | 1890 *Chimaeropsis danica*, Meinert in: Udb. Hauchs, v. 3 p. 167 t. 2 f. 42—47.

Head longer than peraeon segments 1 and 2. Side-plate 1 distally widened and rounded, 2^d and 3^d triangular, 4th large, especially in ♀, 5th and 6th with hind lobe, much deeper than front. Pleon segments 4 and 5 dorsally smooth in ♀ (Fig. 70), in ♂ dorsally produced, in segment 5 the expansion large, overarching segment 6. Eyes, each containing 4 small bigeminous lenticular bodies imbedded at the periphery of a common pigmentary mass (Sars). Antenna 1, 1st joint of peduncle as long as 2^d and 3^d combined, flagellum with 7 joints, in ♀ as long as peduncle, in ♂ much longer, in ♂ the first 2 joints of flagellum fused and fringed with cilia. Antenna 2, ultimate joint of peduncle as long as antepenultimate, much shorter than penultimate, flagellum with 7 joints, in ♀ subequal to last 2 joints of peduncle, in ♂ to the whole peduncle. Gnathopods setose. Peraeopods 1—5, finger held at an angle to preceding joint. Peraeopods 3—5 with 2^d joint glandular. Uropod 3, outer ramus with minute 2^d joint. Telson in ♀ subtriangular, in ♂ more regularly oval. Colour whitish, pellucid, orange-tinted on antennae and legs. L. ♀ 5, ♂ 6 mm.

Arctic Ocean, North-Atlantic, North-Sea and Skagerrak (Greenland; Norway, depth 38—188 m; Shetland; Firths of Forth and Clyde; S.W. of Belle Ile en Mer, depth 100 m); Kattegat.

2. **A. stebbingi** Bonnier 1896 *A. s.*, J. Bonnier id: Ann. Univ. Lyon. r. 26 p. 626 t. 36 f. 4.

Pleon segment 5 in ♂ prolonged over 6th in a salient apically rounded lamella. Head, lateral corners a little produced, obtuse; a vertical crest separates bases of antennae 1 and 2 of one side from antennae 1 and 2 of the other, the crest being divided by 2 emarginations into a bluntish rostrum, a truncate prominence, and a subacute projecting epistome reaching halfway along antepenultimate joint of peduncle of antenna 2. Pleon segment 3, posterolateral corners quadrate. Eyes wanting. Antenna 1 in ♂, 3^d joint fully half as long as 2^d, flagellum with 5 joints, together rather longer than peduncle. 1st densely clothed, longer than 2^d and 3^d joints of peduncle combined, accessory flagellum less than half its length, 2-jointed. Antenna 2 in ♂ twice as long. penultimate joint of peduncle more than thrice as long as antepenultimate, nearly twice as long as ultimate, flagellum with 8 slender joints. Antenna 1 in ♀, 3^d joint less than half 2^d, accessory flagellum as long as 1st joint of primary. Antenna 2 in ♀, penultimate joint of peduncle much shorter than in ♂. Mandibular palp proportionally longer than in A. hamatipes (p. 277), 3^d joint twice as long as 2^d. Maxilla 1, inner plate small, with 2 setae, outer with 7 spines, 2^d joint of palp much dilated towards the apex. Maxillipeds, finger ending in a long and strong spine. Gnathopods 1 and 2, 6th joint very narrowly oval, hind margin carrying serrate spines and some long setae, finger ending in a little spine. Feraeopods 1—5 nearly as in A. hamatipes, but without glandular bodies in 2^d joint. Uropods 1—3 and telson also similar, but outer ramus of uropod 3 without spiniform 2^d joint. L. ♂ nearly 5 mm, ♀ a little smaller.

Bay of Biscay. Depth 940 m.

4. Gen. **Astyra** Boeck

1871 *Astyra* (Sp. un.: *A. abyssi*), A. Boeck in: Forh. Selsk. Christian., 1870 p.133 | 1876 *A.*, A. Boeck, Skand. Arkt. Amphip., v.2 p.442 | 1892 *A.*, G. O. Sars, Crust. NorWay, v. 1 p.213 | 1893 *A.*, A. Della Valle in: F. Fl. Neapel, v. 20 p. 693.

Body tumid in front, integument thin. Head little produced in front. Side-plate 1 wider than 2^d or 3^d, 4th obliquely truncate below emargination. Antennae 1 and 2 rather strong, not elongate. 1st with small accessory flagellum, 2^d rather the longer. Mouth-parts projecting below. Upper lip rounded, slightly emarginate. Lower lip (Fig. 71), inner lobes minute, contiguous, between widely separated outer lobes. Mandible, cutting edge rather wide, transversely dentate, accessory plate on the left mandible, spines of spine-row numerous, molar weak, conical, 2^d joint of palp much longer than 3^d. Maxilla 1, inner plate with several setae, 2^d joint of palp long. Maxilla 2, both plates short, broad, densely setose. Maxillipeds, inner plates moderate. outer large, reaching much beyond 2^d joint of palp, inner margin spinose, palp rather short, stout, finger small. Gnathopods 1 and 2 not robust, subequal, scarcely subchelate. 5th joint much wider than 6th. Peraeopods 1 and 2 rather slender. Peraeopods 3—5 rather robust, 2^d joint oblong, not widely expanded. Uropods 1—3, rami narrowly lanceolate, spinulose, peduncle very short and rami long in uropod 3. Telson small, deeply cleft.

1 species.

1. **A. abyssi** Boeck 1871 *A. a.*, A. Boeck in: Forh. Selsk. Christian., 1870 p. 133 | 1876 *A. a*, A. Boeck, Skand. Arkt. Amphip., v. 2 p. 443 t. 9 f. 4 | 1892 *A. a.*, G. O. Sars, Crust. NorWay, v. 1 p.214 t. 73 | 1893 *A. a.*, A. Della Valle in: F. Fl. Neapel, v. 20 p. 694 t. 60 f. 2.

Back quite smooth. Head, rostrum short, blunt, lateral angles almost obsolete. post-antennal linguiform. deflexed. Side-plate 1 distally expanded, 2^d—4th distally narrowed, 5th rounded quadrate, slightly bilobed. Pleon segment 3, postero-lateral corners nearly quadrate, segment 4 with deep dorsal depression. Eyes wanting. Antenna 1. 1st joint

Fig. 71. **A. abyssi.**
Lower lip.
[After G. O. Sars.]

rather flattened, as long as 2^d and 3^d combined, flagellum more than twice as long as peduncle, with 20—30 joints, proximal 5—8 inflated, armed with sensory hairs, accessory flagellum delicate, 1-jointed. Antenna 2, peduncle rather stout, ultimate and penultimate joints subequal, carrying fascicles of setules, flagellum once and a half as long as peduncle. Gnathopod 1 rather stronger than gnathopod 2, 5th joint rather longer and much wider than 6th, ending in a short rounded setose lobe, 6th joint narrow oblong, with no distinct palm, finger short, spinose on inner margin. Peraeopods 3—5, 2^d joint having outside an elevated diagonal ridge, finger lanceolate, spinulose at the edges. Uropod 1, rami nearly equal; uropod 2, rami rather unequal; uropod 3 with the shortest peduncle, the longest rami, inner a little longer than the outer and longitudinally ridged. Telson rather longer than broad, cleft beyond middle, lobes distally diverging. Colour pale yellowish. L. ♀ 8 mm.

Arctic Ocean, North-Atlantic and North-Sea (North- and West-Norway). Depth 188—564 m.

5. Gen. **Syrrhoites** O. Sars

1893 *Syrrhoites* (Sp. un.: *S. serrata*), G. O. Sars, Crust. Norway, *v.* 1 p. 391.

Body rather stout, compressed, carinate. Head, rostrum acute, deflexed, cheeks deep, inflated. Side-plates 1—3 narrowly quadrangular. Eyes wanting. Antennae 1 and 2 subequal in ♀, in ♂ antenna 2 the longer. Upper lip short, broad, distally truncate. Lower lip. inner lobes rather small, outer incised near apex on inner margin. Mandible stout, compact, cutting edge conical. not dentate, or only bidentate, spine-row wanting, molar a very broad, triturating surface, palp extremely slender. 3^d joint not longer than 1st. Maxilla 1, inner plate with several setae. outer with 11 spines, palp long. Maxilla 2, inner plate the broader, fringed on inner margin. Maxillipeds, inner plates broad, outer broad, armed with unusually strong spines, 3^d joint of palp linear, finger very small. Gnathopods 1 and 2 feeble, 5th joint twice as long as 6th, which is small, imperfectly subchelate, palm very oblique, with no strong defining spine. Peraeopods 1 and 2 very slender and feeble. Peraeopods 3—5 much stronger. successively longer, 2^d joint expanded oval. Uropods 1 and 2, outer ramus the shorter; uropod 3, rami lanceolate, subequal. Telson long. distally narrowed. deeply cleft.

3 species accepted, 1 doubtful.

Synopsis of accepted species:

1	{ Pleon segment 3, margin above postero-lateral tooth serrate	1. **S. serratus** . .	p. 279
	{ Pleon segment 3, margin above postero-lateral tooth simple — 2.		
2	{ Pleon segments 1—3 with dorsal teeth inconspicuous	2. **S. fimbriatus** .	p. 280
	{ Pleon segments 1—3 with dorsal teeth conspicuous	3. **S. walkeri** . .	p. 281

1. **S. serratus** (O. Sars) 1879 *Bruzelia serrata*, G. O. Sars in: Arch. Naturv. Kristian., *c.* 4 p. 447 | 1885 *B. s.*, G. O. Sars in: Norske Nordhavs-Exp., *v.* 6 Crust. I p. 182 t. 15 f. 3 1893 *B. s.*, A. Della Valle in: F. Fl. Neapel, *v.* 20 p 668 t. 59 f. 77, 78 1893 *Syrrhoites s.*, G. O. Sars, Crust. Norway, *v.* 1 p. 392 t. 137.

Carinate throughout. peraeon segments 4—7 with carina raised to simple triangular processes, pleon segments 1—3 to a small and a large point, hinder on segment 3 upturned. on segments 4—6 likewise produced, on segment 4 with 2 points in ♀, with only 1 in ♂, on segment 5 upturned in ♀. much stronger and deflexed in ♂. Head. rostrum strongly deflexed, reaching apex of 1st joint of antenna 1. lateral corners rounded. Side-plate 3 scarcely expanded below. 4th rather less deep than 3^d, slightly emarginate, triangular below. Pleon segments 1—3 very large, lateral corners of 1st rounded, of 2^d a little produced, of 3^d very acutely produced below a deep sinus, above which the margin is serrate with 5 upturned teeth. Eyes not even represented by pigment. Antenna 1 in ♀, 1st joint a little longer than 2^d, 2^d than 3^d, flagellum less than peduncle, 9-jointed, accessory flagellum small, 2-jointed. Antenna 2 in ♀ scarcely longer, ultimate joint of peduncle slightly shorter than penultimate, flagellum about half as long as peduncle. Antenna 1 in ♂, 1st joint as long as 2^d and 3^d combined, 1st joint of flagellum very long, with long sensory filaments, accessory flagellum much longer than in ♀. Antenna 2 in ♂ much longer, flagellum longer than peduncle. Maxillipeds, outer plates fringed with 8 unusually strong spines. Gnathopods 1 and 2 very small, 5th joint narrow, not tapering, palm of 6th subequal to and making very obtuse angle with hind margin, finger short, 2-jointed. Peraeopods 3—5, 2^d joint serrate on hind margin, 6th joint very slender and long, especially on peraeopod 5. Uropod 3, rami intermediate in length between outer and inner ramus of uropods 1 and 2, outer ramus with short 2^d joint. Telson oblong triangular, cleft more than $^2/_3$ of length. apices narrow, bidentate, contiguous. Colour yellowish grey. L. 8 mm.

Arctic Ocean, North-Atlantic and North-Sea (West-Norway). Depth 282—658 m.

2. **S. fimbriatus** (Stebb. & D. Roberts.) 1891 *Syrrhoë f.*, T. Stebbing & D. Robertson in: Tr. zool. Soc. London, *v.*13 r p. 34 t. 5 в | 1893 *S. f.*, A. Della Valle in: F. Fl. Neapel, *v.*20 p. 662 | 1895 *S. f.*, A. O. Walker in: P. Liverp. biol. Soc., *v.* 9 p. 304.

Body faintly carinate, integument not indurated, peraeon segment 7 and pleon segments 1 and 2 apparently produced dorsally to a small tooth; in the ♂ only, pleon segment 5 has a tooth projecting over segment 6, and segment 6 is dorsally fringed with close-set spinules. Head curved in front to a strongly deflexed carinate rostrum. Side-plate 3 rather widened below, lower front corner rather acutely produced, 4th less deep, rounded below. Pleon segment 3, postero-lateral corners acutely upturned. Antenna 1 in ♀, 1st joint longer than 2^d, 2^d little longer than 3^d, flagellum not longer than peduncle, with 6 joints, 1st not very long, accessory flagellum with 2 joints, 2^d minute. Antenna 2 in ♀ subequal, ultimate and penultimate joints of peduncle subequal, flagellum much shorter than peduncle, 5-jointed. Antenna 1 in ♂, 1st joint longer than 2^d and 3^d combined, flagellum rather longer than peduncle, with 7 joints, 1st setose, as long as the other 6 combined, accessory flagellum with 3 joints, 1st long, 3^d minute. Antenna 2 in ♂ much longer, ultimate joint of peduncle long, a little longer than penultimate, flagellum longer than peduncle, about 12-jointed. Lower lip slightly incised on inner margin of outer lobes, mandibular processes wide-spread. Mandible, cutting edge seemingly not dentate in ♀, bidentate in ♂, 3^d joint of palp about as long as 1st, tipped with 5 long setae. Maxilla 1, inner plate with 8 setae, outer with 11 spines, spine-like setae on tip of palp's 2^d joint. Maxillipeds, inner plates broad, outer fringed with 8 spines of moderate size. Gnathopod 1. 2^d joint distally widening, 5th not very slender, not twice as long as 6th, spinose on hind margin, 6th broadest where the short hind margin forms very obtuse angle with the very oblique palm, finger short, abruptly narrowed at the centre. Gnathopod 2 more slender, 5th joint twice as long as 6th. 6th in ♂ with margins parallel all along, forming a minute palm much overlapped by the small curved finger; in ♀ palm oblique. corresponding with an oblique row of spines on the hand of the ♂. Peraeopods 1 and 2 slight. Peraeopods 3 and 4, 2^d joint narrowly oblong oval, hind margin serrate. Peraeopod 5, 2^d joint broadly oval, hind margin serrate. Uropod 1, peduncle shorter than inner ramus, more than twice as long as outer, which ends in a long spine. Uropod 2, peduncle a little longer than outer ramus, about half as long as the broadly lanceolate spinulose inner. Uropod 3, peduncle rather shorter than the subequal rami, outer ramus 2-jointed, both in $\male$ armed on inner margin. Telson nearly twice as long as broad, cleft beyond centre, apices acute, minutely notched. L. 1·5 mm.

Firth of Clyde; Irish Sea (Calf of Man).

3. **S. walkeri** Bonnier 1896 *S. w.*, J. Bonnier in: Ann. Univ. Lyon, *v.*26 p.647 t. 38 f. 4.

Carinate (apparently) throughout, hind segments of peraeon projecting, pleon segments 1—6 each prolonged into dorsal tooth, large on 1st—3^d, on 3^d upturned, in $\male$ tooth on 5th produced over 6th, 6th fringed with close-set spinules. Head carinate, rostrum strongly deflexed, acute, reaching apex of 1st joint of antenna 1, lateral corners forming straight-fronted lobe. Side-plates as in S. serratus (p. 279), unless 4th be a little deeper than 3^d. Pleon segment 3, postero-lateral corners strongly produced, acute, margin above not denticulate. Eyes wanting. Antenna 1 in ♀, 1st joint ending in a small hook, 2^d and 3^d equal, flagellum subequal to peduncle, 9-jointed, accessory flagellum small, 2-jointed. Antenna 2 in ♀ equal to antenna 1, flagellum short, 7-jointed. Antenna 1 in $\male$, 3^d joint shorter than 2^d, flagellum with 9 joints, 1st as long as peduncle, longer than the other 8 combined, densely fringed with sensory filaments, accessory flagellum 3-jointed. Antenna 2 in $\male$ much longer, ultimate joint of peduncle rather longer than penultimate, flagellum longer than peduncle. Epistome salient. Upper lip long, rounded. Other mouth-parts nearly as in S. serratus, but maxilla 1, outer plate said to have 12 spines, and maxillipeds have 11 spine-teeth of moderate size on outer plates, which reach apex of palp's rather short 2^d joint. Triturating lobes of stomach with 2 enormous spines in addition to the ordinary armature. Gnathopods 1 and 2 as in S. serratus, but palm defined by a strong spine. Peraeopods 1—5 similar to those of S. serratus but more slender, and 6th joint in peraeopods 3—5 more elongate. Uropods 1—3 as in S. serratus. Telson elongate triangular, cleft to centre, apices notched. L. 7 mm.

Bay of Biscay. Depth 950 m.

S. levis (Boeck) 1871 *Syrrhoë l.*, A. Boeck in: Forh. Selsk. Christian, 1870 p.148 | 1876 *S. l.*, A. Boeck, Skand. Arkt. Amphip., *v.*2 p.473 | 1893 *S. l.*, A. Della Valle in: F. Fl. Neapel, *v.*20 p.662.

L. 5 mm.

Bömmelfjord (Moster [West-Norway]). Depth 282 m.

6. Gen. **Syrrhoe** Goës

1866 *Syrrhoë* (part.), Goës in: Öfv. Ak. Förh., *v.*22 p.527 | 1871 *S.*, A. Boeck in: Forb. Selsk. Christian., 1870 p.147 | 1888 *S.*, T. Stebbing in:, Rep. Voy. Challenger, *v.*29 p.788 | 1893 *S.*, G. O. Sars, Crust. Norway, *v.*1 p.389 | 1893 *Syrrhoe* (part.), A. Della Valle in: F. Fl. Neapel, *v.*20 p.662.

Body rather slender, integument thin. Head produced to sharp strongly deflexed rostrum. Side-plates rather small, 3^d much the largest, distally expanded, 4th very small. Eyes, when present, dorsal, coalesced. Antennae 1 and 2 slender, antenna 2 the longer, in ♂ much the longer. Upper lip distally narrowed, rounded. Lower lip, inner lobes well developed, mandibular processes widely spread. Mandible, cutting edge and accessory plate dentate, spine-row and molar normal, palp slender, 3[d] joint minute, tipped with long setae. Maxilla 1, inner plate with many setae, outer with 11 spines, palp rather long. Maxilla 2, plates subequal, inner fringed on inner margin. Maxillipeds, inner plates rather broad, outer not very, fringed with spines, palp normal. Gnathopods 1 and 2 feeble, slender, subchelate, palm transverse, armed with thumb-like denticulate spine. Peraeopods 1—5 slender; peraeopods 3—5 successively longer, with 2[d] joint expanded, serrate on hind margin. Uropods 1 and 2, outer ramus much the shorter; uropod 3, outer ramus a little the shorter, 2-jointed. Telson long, deeply cleft.

3 species.

Synopsis of species:

1 { Pleon segment 3, postero-lateral margin not serrate 3. **S. papyracea** . . p. 283
  { Pleon segment 3, postero-lateral margin serrate —· **2.**

2 { Pleon segment 3, postero-lateral margin Wholly
    serrate . . . . 1. **S. crenulata** . . p. 282
  { Pleon segment 3, postero-lateral margin partly serrate 2. **S. semiserrata** . p. 283

1. S. crenulata Goës 1866 *S. c.*, Goës in: Öfv. Ak. Förh., *v.* 22 p. 527 t. 40 f. 25 | 1876 *S. c.*, A. Boeck, Skand. Arkt. Amphip., *v.* 2 p. 471 t. 9 f. 5; t. 12 f. 4 | 1893 *S. c.*, G. O. Sars, Crust. Norway, *v.* 1 p. 390 t. 136 | 1893 *S. c.* (part.), A. Della Valle in: F Fl. Neapel, *v.* 20 p. 663 t. 59 f. 74.

Body nearly cylindrical, peraeon segment 7 and pleon segments 1—3 dorsally serrate. Head strongly vaulted in front, rostrum reaching beyond middle of 1st joint of antenna 1. Side-plates 1 and 2 narrow, directed forward, 3[d] nearly securiform, front margin sloping forward, hind lobe narrowly truncate beneath deep emargination which receives side-plate 4. Pleon segments, postero-lateral angles in 2[d] a little produced, acute, in 3[d] nearly quadrate, hind margin above sinuous, serrate with upturned teeth, nearly meeting the dorsal serration. Eyes forming a large dark red dorsal mass. Antenna 1 in ♀, 1st joint with apical tooth, 2[d] a little shorter than 1st, 3[d] than 2[d], flagellum slender, rather longer than peduncle, 18-jointed, accessory flagellum with 3 joints, 3[d] minute. Antenna 2 in ♀ a little longer, ultimate joint of peduncle rather longer than long penultimate, both setulose, flagellum half as long as peduncle. Antenna 1 in ♂, 1st joint fully as long as 2[d] and 3[d] combined, flagellum very long, 1st joint very large, densely fringed, others rather long, accessory flagellum with laminar 1st joint. Antenna 2 in ♂, flagellum very long, filiform, joints short, numerous. Upper lip, partly carinate. Gnathopod 1, 5th joint about twice as long as 6th, which is oblong triangular, widening distally and dense with delicate setae, tip ˚of finger crossing the palmar spine. Gnathopod 2 similar, but longer and more slender, 5th joint about once and a half as long as the narrow 6th. Peraeopods 3—5, 2[d] joint coarsely serrate on hind margin, largest and produced downward in peraeopod 5, which has 5th and 6th joints elongate. Uropod 3, the lanceolate rami setose on inner margin. Telson long, lanceolate, nearly thrice as long as broad, cleft $^2/_3$—$^3/_4$ of length, apices acute (bidentate Goës), contiguous. Colour variable, variegated with yellow, white and pink, side-plates and other parts coral-red, antennae and limbs banded with orange. L. ♀ 10 mm, ♂ less.

Arctic Ocean, North-Atlantic, North-Sea and Skagerrak (Greenland, Spitzbergen, Norway). Depth 38—188 m.

2. **S. semiserrata** Stebb. 1888 *S. s.*, T. Stebbing in: Rep. Voy. Challenger, *v.* 29 p. 793 t. 51 | 1893 *S. s.*, A. Della Valle in: F. Fl. Neapel, *v.* 20 p. 663 t. 59 f. 72, 73.

♀ unknown. — ♂. Peraeon segment 7 and pleon segments 1 and 2 dorsally produced into a small tooth. Head strongly vaulted in front, rostrum carinate, reaching beyond middle of 1st joint of antenna 1. Side-plates 1—4 nearly as in S. crenulata. Pleon segments, postero-lateral angles in 2^d a little produced, acute, in 3^d not produced, hind margin above cut into 8 upturned teeth not nearly reaching the unserrated dorsum. Eyes as in S. crenulata, colour not known. Antenna 1, 1st joint thick, longer than 2^d and 3^d combined, 1st joint of flagellum nearly as long as peduncle, densely fringed, accessory flagellum about as long as 1st joint of primary, slender, serrate, setae along one side and spines along the other. Antenna 2, ultimate joint of peduncle much more than twice as long as penultimate, penultimate nearly twice as long as antepenultimate, flagellum with at least 20 joints, which are slender, not short. Mandible, 3 spines in spine-row. Maxillipeds, outer plates fringed with 11 spines. Gnathopod 1, 5th joint fusiform, nearly twice as long as 6th, which widens a little distally, spinulose along hind margin, palm short, setose, the palmar spine with blunt process near the base and 7 oblique denticles on its concave margin, finger short, hooked, stout at base. Gnathopod 2, 3^d joint longer than 4th, 5th narrow, slightly bent, nearly twice as long as 6th, which scarcely widens distally, hind margin spinulose and finely pectinate, palmar spine without the blunt process. Peraeopods 1 and 2 very slight. Peraeopods 3—5, 2^d joint not very large, the serration not coarse. Uropod 1, peduncle longer than rami, outer ramus much shorter than inner; uropod 2, outer ramus narrow, longer than inner ramus of uropod 1, inner ramus very long, broad, spinose on both margins; uropod 3, rami long, lanceolate, fringed with setae on inner margin, outer a little the shorter, with small 2^d joint. Telson long, deeply cleft, character of apices uncertain. L. 7·5 mm.

Bass-Strait. Depth 60 m.

3. **S. papyracea** Stebb. 1888 *S. p.*, T. Stebbing in: Rep. Voy. Challenger, *v.* 29 p. 789 t. 50 | 1893 *S. p.*, G. O. Sars, Crust. Norway, *v.* 1 p. 390 ! 1893 *S. crenulata* (part.), A. Della Valle in: F. Fl. Neapel, *v.* 20 p. 663.

Peraeon segment 7 and pleon segments 1—4 dorsally serrate, the central tooth most prominent on pleon segment 2. Head produced and sharply curved at base of sharp-edged rostrum. Side-plates 1—4 nearly as in S. crenulata, but 3^d with a more quadrate hind lobe. Pleon segments 2 and 3, postero-lateral angles acutely produced, more strongly in 2^d than 3^d, margin above not denticulate. Eyes not perceived. Antenna 1 in ♀, 1st joint with no large apical tooth, 2^d joint rather shorter than 1st, 3^d than 2^d, flagellum longer than peduncle, with 15 joints or more, 1st longest, accessory flagellum with 2 long joints and 1 very short. Antenna 2 much longer, ultimate joint of peduncle as long as the short antepenultimate and long penultimate combined, flagellum shorter than peduncle, with 18 unequal joints. Mandible, spine-row of 6 spines. Maxillipeds, outer plates fringed with 16 spines, 3^d joint of palp longer than in the other species, finger with a nail as long as the basal part. Gnathopod 1, 5th joint fusiform, more than twice as long as 6th, hind margin densely fringed, 6th short. widening distally, hind margin pectinate, palm slightly oblique, sinuous, setose, the palmar spine with 7 or 8 denticles, finger reaching a little beyond the palm. Gnathopod 2, 3^{d} joint shorter than 4th, 5th long, narrow, spinose, more than twice as long as 6th, which is longer, and narrower than in gnathopod 1, scarcely widened distally, armature similar. Peraeopods 1—5 much as in S. crenulata (p. 282), but 2^{d} joint in peraeopods 3—5 less strongly serrate. Pleopods, peduncles with small interlocking process at apex, coupling spines large, 4 cleft spines on 1st joint of inner ramus. Telson not twice as long as broad, cleft a little beyond middle, apices acute, with small lateral tooth above. L. 12 mm.

Tropical Atlantic (Virgin Island Culebra). Depth 713 mm.

7. Gen. **Pseudotiron** Chevreux

1895 *Pseudotiron* (Sp. un.: *P. bouvieri*), Chevreux in: Bull. Soc. zool. France, *v.* 20 p. 166.

Body compressed, not indurated. Head triangular (as in Synopia, p. 271), more advanced in ♂ than in ♀. Side-plate 3 the largest, greatly excavate (as in Syrrhoe, p. 281). Eyes (as in Argissa, p. 276) small, imperfectly developed. Antenna 1 in ♂ much shorter than antenna 2; accessory flagellum 3-jointed. Mouth-parts and gnathopods nearly as in Tiron (p. 275). Mandible normal, cutting edge and accessory plate dentate, spine-row of several spines, prominent molar, but palp very slender, 3^{d} joint scarcely longer than broad. Maxilla 1, inner plate with several setae, outer with usual spines, 2^{d} joint of palp broad at apex. Maxilla 2, inner plate the broader with inner margin partially fringed. Maxillipeds, inner plates normal, outer armed with 3 long and 5 stout spine-teeth, palp rather long, finger long. Gnathopods 1 and 2 long, slender, simple, setose. Peraeopods 1 and 2 short, stout, finger short, hooked. Peraeopods 3—5, 2^{d} joint expanded, 6th very short, finger small, hooked. Uropods 1 and 2, rami unequal; uropod 3, peduncle short, rami long, broad, subequal, spinulose. Telson long, lanceolate, cleft nearly to base.

1 species.

1. **P. bouvieri** Chevreux 1895 *P. b.*, Chevreux in: Bull. Soc. zool. France, *v* 20 p. 166 f. 1—14.

Peraeon smooth. Pleon segments 1—3, hind margin denticulate, segment 4 with 3 denticles, segment 5 with 1 long tooth extending over segment 6. Side-plate 1 distally expanded, 3^{d} with straight front margin, hind part rectangular below emargination, 4th much smaller, obliquely produced and rounded below a slight emargination. Pleon segment 3, postero-lateral angle a little produced, acute. Eyes in ♀, a group of 4 small lenses near front margin of head (compare Argissa, p. 276), not observed in ♂. Antenna 1 in ♀, 1st joint rather longer than 3^{d}, 2^{d} much shorter than either, flagellum a little longer than peduncle, 9- or 10-jointed, accessory flagellum small. Antenna 1 in ♂, 1st joint much longer than 2^{d} and 3^{d} combined, flagellum considerably longer than peduncle. 1st joint much the longest, fringed, longer than the 3-jointed accessory flagellum. Antenna 2 in ♂, ultimate joint of peduncle shorter than penultimate, flagellum much longer, 23-jointed. Gnathopods 1 and 2, 3^{d} and 4th joints very short, 5th joint long and slender, 6th linear, in gnathopod 1 half as long as 5th, in gnathopod 2 $^2/_3$ as long as 5th, finger rather long, nearly straight. Uropod 2 the smallest. Uropods 1 and 2, the inner ramus the shorter (Chevreux, but?). Uropod 3, outer ramus armed with 3 rows of spinules, inner with a row of setules. Telson, apices subacute, in figure a little divergent. Colour pellucid. L. ♀ 5 mm.

Mediterranean (North of Tunis). Depth 170 m.

˙22. Fam. Calliopiidae

1893 *Calliopiidae*, G. O. Sars, Crust. Norway, *v.* 1 p. 431.

Body compressed or with broadly vaulted peraeon, with or without dorsal teeth. Head, rostrum unimportant. Side-plates small or of moderate size. Antenna 1, peduncle usually short, accessory flagellum 1-jointed or wanting. Upper lip rounded or faintly bilobed. Lower lip with inner lobes small or wanting. Other mouth-parts varying little from the normal. In Laothoes (p. 286), however, the palp of maxilla 1 is abnormally small, and the outer plates of maxillipeds are abnormally large. In Sancho (p. 288), and perhaps also in Paraleptamphopus (p. 294), the gnathopods (Fig. 72, 73) show striking sexual differences. Gnathopods 1 and 2 usually feeble, subchelate. Marsupial plates large, broad. Uropod 3, inner ramus much, or more usually little, longer than outer. Telson entire, sometimes a little notched or emarginate.

Marine, generally distributed; rarely in Wells.

15 genera accepted, 1 imperfectly known, 29 accepted species, 8 obscure.

Synopsis of accepted genera:

1 { Lower lip without inner lobes — **2.**
  { Lower lip with inner lobes.— **8.**

2 { Maxilla 1, palp very small; maxillipeds, outer plates very large . . . . . . 1. Gen. **Laothoes** . . . . . . p. 286
  { Maxilla 1, palp not very small; maxillipeds, outer plates not very large — **3.**

3 { Back broadly vaulted — **4.**
  { Back not broadly vaulted — **6.**

4 { Antenna 1 without accessory flagellum; telson notched . . . . 2. Gen. **Chosroes** . . . . . . p. 287
  { Antenna 1 with accessory flagellum; telson not notched — **5.**

5 { Maxillipeds, outer plates smaller than inner . . . . 3. Gen. **Sancho** . . . . p. 288
  { Maxillipeds, outer plates larger than inner . . . . 4. Gen. **Amphithopsis** . . . p. 289

6 { Body with some segments dorsally dentate 5. Gen. **Halirages** . . . . . . p. 290
  { Body without dorsally dentate segments — 7.

7 { Antenna 1 without accessory flagellum; uropod 3, rami unequal . . . . . . 6. Gen. **Leptamphopus** . . . p. 293
  { Antenna 1 with accessory flagellum; uropod 3, rami equal . . . . . . ˙. . 7. Gen. **Paraleptamphopus** . p. 294

8 { Telson at apex smoothly rounded —˙ **9.**
  { Telson at apex acute or notched — **10.**

9	Peraeopod 5 of moderate length, finger curved	8. Gen. **Calliopius**	p. 295
	Peraeopod 5 very elongate, finger straight	9. Gen. **Paracalliope**	p. 297
10	Telson notched — **11.**		
	Telson not notched — **12.**		
11	Side-plates 1—4 rather deep; mandible with cutting edge outdrawn	10. Gen. **Harpinioides**	p. 298
	Side-plates 1—4 rather shallow; mandible with cutting edge not outdrawn . .	11. Gen. **Atylopsis**	p. 299
12	Antenna 1 longer than antenna 2 — **13.**		
	Antenna 1 not longer than antenna 2 — **14.**		
13	Body with some segments dorsally dentate	12. Gen. **Cleippides**	p. 300
	Body without dorsally dentate segments	13. Gen. **Stenopleura**	p. 302
14	Side-plates shallow, 4th not deeper than 1st	14. Gen. **Haliragoides**	p. 303
	Side-plates not very shallow, successively deeper to the 4th	15. Gen. **Apherusa**	p. 304

1. Gen. **Laothoes** Boeck

1871 *Laothoës* (Sp. un.: *L. meinerti*), A. Boeck in: Forb. Selsk. Christian., 1870 p. 202 | 1876 *L.*, A. Boeck, Skand. Arkt. Amphip., *v.* 2 p. 360 | 1890 *L.*, T. Stebbing in: Ann. nat. Hist., ser. 6 *v.* 5 p. 194 | 1893 *Laothoë* (non J. C. Fabricius 1807, Lepidoptera!), G. O. Sars. Crust. Norway, *v.* 1 p. 453 | 1893 *Thoelaos*, A. Della Valle in: F. Fl. Neapel. *v.* 20 p. 592.

Side-plates not very large. Antennae 1 and 2 slender, antenna 1 the shorter, peduncle short. no accessory flagellum. Mouth-parts much projecting. Upper lip rounded triangular. Lower lip without inner lobes. Mandible normal, strong, accessory plate on both, 3d joint of palp shorter than 2d. Maxilla 1, inner plate with 5 setae. outer with strong spines, palp very small, 2d joint rudimentary, nodiform. Maxillipeds, inner plates normal, outer very large, broad, reaching almost to extremity of palp, with dentate inner margin. Gnathopods 1 and 2 feeble, 5th and 6th joints narrow, rather long, palm very short. Peraeopods 3—5, 2d joint oval, largest in peraeopod 5. Uropod 2, outer ramus much the shorter; uropod 3. rami long, subequal, lanceolate. Telson entire.

1 species accepted, 1 uncertain.

1. **L. meinerti** Boeck 1871 *L. m.*, A. Boeck in: Forh. Selsk. Christian., 1870 p. 202 | 1876 *L. m.*, A. Boeck, Skand. Arkt. Amphip., *v.* 2 p. 361 | 1893 *L. m.*, *Laothoë m.*, G. O. Sars, Crust. Norway, *v.* 1 p. 454 t. 160 | 1893 *Thoelaos meinertii*, A. Della Valle in: F. Fl. Neapel, *v.* 20 p. 592.

Body rather long and cylindric. Head, rostrum very small, lateral corners deflexed, rounded, post-antennal quadrate. Pleon segment 3, postero-lateral corners almost quadrate, the margins slightly convex. Eyes oblong oval, widened below, light red. Antenna 1 in ♀ not half as long as the body, 1st joint thick, much longer than 2d and 3d combined, flagellum setiferous. of many short joints, about 4 times as long as peduncle. Antenna 2 considerably longer, ultimate and penultimate joints of peduncle subequal, flagellum fully thrice as long as peduncle. Gnathopod 1, 5th joint fusiform, as long as narrowly oblong 6th, palm nearly transverse, scarcely half as long as hind margin. Gnathopod 2 rather longer, 5th joint widest distally, as long as the narrow, almost linear 6th. Peraeopods 1—5 moderately slender, 2d joint in peraeopods 3—5 with smooth hind margin. Uropod 3, rami about twice as long as peduncle, fringed with spinules only. Telson oval quadrangular, with apex broad, faintly emarginate. Colour pellucid, yellowish, indistinctly banded with orange; ova in pouch pale violet. L. ♀ 8 mm.

Arctic Ocean, North-Atlantic and North-Sea (Hardangerfjord, depth 565—942 m; Trondhjemsfjord; Nordland).

L. carcinophilus (Chevreux) 1889 *Paramphithoe carcinophila*, Chevreux in: Bull. Soc. zool. France, v.14 p.288 f. | 1893 *P. c., Acanthozone?*, A. Della Valle in: F. Fl. Neapel, v.20 p.619.

L. 7 mm.

North-Atlantic (Azores, between Pico and San Jorge, depth 620 m, and off Florès, depth 1386 m). From the carapace of Geryon sp.

2. Gen. **Chosroes** Stebb.

1888 *Chosroës* (Sp. un.: *C. incisus*), T. Stebbing in: Rep. Voy. Challenger, v.29 p.1208.

Back of peraeon broadly vaulted in front, with side-plates laterally out-spread, pleon segments 4—6 folded under, 5th and 6th not very short. Head, rostrum minute. Side-plate 1 small. Antenna 1 the shorter, without accessory flagellum; both 1st and 2d antennae calceoliferous. Upper lip rounded. Lower lip without inner lobes. Mandible normal, palp very large, 3d joint longer but narrower than 2d. Maxilla 1, inner plate with 2 or 3 setae, outer with 11 spines. Maxilla 2, inner plate only partially fringed on inner margin. Maxillipeds, outer plates scarcely reaching middle of palp's 2d joint, inner margin fringed with submarginal spinules, palp rather elongate, finger short with a short sharp nail. Gnathopods 1 and 2 small, similar, 5th joint rather elongate, palm rather short, slightly oblique. Peraeopods 1 and 2, 2d joint partially dilated; peraeopods 3—5, expanded 2d joint notched, 4th joint in all broad, not long. Uropods 1 and 2, outer ramus the shorter; uropod 3, rami subequal, lanceolate. Telson slightly notched.

1 species.

1. C. incisus Stebb. 1888 *C. i.*, T. Stebbing in: Rep. Voy. Challenger, v.29 p.1209 t.134, 135 | 1893 *Acanthozone incisa*, A. Della Valle in: F. Fl. Neapel, v.20 p.614 t.59 f.26.

Body dorsally very wide at middle of peraeon, pleon segment 6 considerably longer than 5th. Side-plate 4 deeply emarginate, 5th broad, the lobes about equal in depth. Eyes broadly oval, wide apart, lenses numerous. Antenna 1, 1st joint stout, a little longer than 2d, and 2d than 3d, flagellum about twice as long as peduncle, with more than 34 joints, short, distally a little widened, these and last 2 joints of peduncle with many small calceoli. Antenna 2 decidedly longer, ultimate joint of peduncle longer than penultimate, flagellum longer than peduncle, with more than 24 joints, short, broad, calceoliferous. Triturating lobes at entrance of stomach with short curved and long straight spines. Gnathopod 1, 5th joint somewhat expanded in the middle, 6th equally long, broader, oval, palm slightly convex, shorter than hind margin, defined by a group of palmar spines. Gnathopod 2 similar but more elongate. Peraeopods 1 and 2, 2d joint expanded into a broad lobe near but not at the apex, 5th joint longer than 4th, 6th than 5th, finger short. Peraeopods 3—5, 2d joint broad, hind margin deeply notched below the middle. Marsupial plates large. Uropods 1 and 2, outer ramus shorter

than inner; uropod 3, rami equal, broadly lanceolate, inner with spines and
setae on both margins, outer with spines and setae on inner, groups of spines
on outer margin. Telson as long as peduncle of uropod 3, not twice as
long as broad, a short triangular notch separating 2 rounded apices, sides
convex in young, slightly concave in adult. L. 10 mm.

Strait of Magellan (Cape Virgins). Depth 100—128 m.

3. Gen. **Sancho** Stebb.

1897 *Sancho* (Sp. un.: *S. platynotus*), T. Stebbing in: Tr. Linn. Soc. London,
ser. 2 *v.* 7 p. 42.

Back of peraeon broadly vaulted, with side-plates laterally outspread,
pleon narrow, segments 3—6 folded under. Head with angular front. Eyes
dorsal, separate. Antenna 1 with short peduncle, accessory flagellum small,
1-jointed. Antenna 2 not much longer than antenna 1. Upper lip broader
than long. Lower lip without inner lobes. Mandible, cutting edge, accessory
plate and spine-row small, molar and palp powerful. Maxilla 1, inner plate
with 2 setae, outer with 11 spines, small and inconspicuously denticulate.
Maxilla 2, inner plate broader than outer, not fringed on inner margin.
Maxillipeds, inner plates broad, outer smaller than, and scarcely reaching
beyond, the inner, palp well developed, finger small. Gnathopod 1 (Fig. 72)
feeble, subchelate. Gnathopod 2 in ♀ similar to gnathopod 1, in ♂ (Fig. 73)
very elongate, with 6th joint massive. Peraeopods 1—5 normal, 5th the
longest. Uropods 1—3, outer ramus shorter than inner. Telson small, entire.

1 species.

1. **S. platynotus** Stebb.   1897 *S. p.*, T. Stebbing in: Tr. Linn. Soc. London,
ser. 2 *v.* 7 p. 42 t. 9 A.

Peraeon segment 1 very short, segment 7 longer; flexing of pleon
begins at the 2d segment. Head dorsally flattened. Side-plate 1 distally a
little widened, 4th the largest, emarginate behind, serrulate below the excavation.

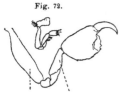

Fig. 72.

2d joint   5th joint

Fig. 73.

Fig. 72 & 73. **S. platynotus**, ♂.
Gnathopods 1 and 2.

Pleon segments 1 and 2, postero-lateral corners
minutely produced. Eyes round, wide apart on top
of head, light pinkish in spirit. Antenna 1, 1st joint
longer than 2d and 3d combined, flagellum in ♂
4 times as long as peduncle, with 41 joints, the 1st
slightly longer than 3d joint of peduncle, and longer
than accessory flagellum. Antenna 2, ultimate
joint of peduncle longer and thinner than penultimate, both spinose, flagellum nearly as in
antenna 1, but setuliferous. Upper lip, distal
margin almost straight. Mandible, 3d joint of
palp as long as 2d, wide, with spines on oblique
apex. Maxillipeds, outer plates fringed with slender spines on the inner
margin, 3d joint of palp narrower but not much shorter than 2d, finger
short, tipped with spinules. Gnathopod 1 in ♂ (Fig. 72), 5th joint considerably
longer and narrower than the 6th, which is rather longer than broad, palm
short, nearly transverse, with small finger to match; in the ♀ 5th and
6th joints narrow, not very long, subequal, 6th rather the longer. Gnathopod 2 in ♂ (Fig. 73), 2d joint narrower at base, a little widened near the
middle, 5th joint subfusiform, shorter and strikingly narrower than the massive

rotundo-quadrate 6th, which has the front margin straight, the hinder curved, slightly crenulate, palm broad, almost transverse, divided into 3 or 4 irregular teeth, finger smooth, curved, closing down into a pocket excavate in thickness of 6th joint; in the $\mathcal{Q}$ 5th and 6th joints almost as in gnathopod 1, feeble, but rather more elongate. Peraeopods 1—5, 4th joint a little widened and produced, finger small, with spinule on inner margin. Peraeopods 1 and 2 rather short; peraeopod 3, 2[d] joint tending to oval; peraeopods 4 and 5, 2[d] joint more oblong, fringed on both margins with long setae, hind margin nearly straight, produced downward, especially in peraeopod 5, which has numerous spines and spinules on the narrow, elongate, subequal 5th and 6th joints. Uropod 2, peduncle as long as in uropod 1 and rami longer than in uropod 1. Uropod 3, peduncle short, apically dentate, outer ramus much shorter and narrower than inner. Telson scarcely as long as peduncle of uropod 3, as broad as long, margins strongly convex, rather abruptly converging to the obtuse apex. L. (without flexed part of pleon) about 3 mm.

Port Jackson [East-Australia].

4. Gen. **Amphithopsis** Boeck

1861 *Amphithopsis* (part.), A. Boeck in: Forh. Skand. Naturf., Møde 8 p. 661 | 1876 *A.*, A. Boeck, Skand. Arkt. Amphip., *v.* 2 p. 349 | 1893 *A.*, G. O. Sars, Crust. Norway, *v.* 1 p. 455 | 1894 *A.*, T. Stebbing in: Bijdr. Dierk., *v.* 17 p. 36.

Back of peraeon broadly vaulted, with side-plates laterally outspread, pleon segments 5 and 6 dorsally not very short. Head, rostrum small, post-antennal corners not produced. Antennae 1 and 2 long, slender, subequal, accessory flagellum very small. Upper lip distally narrowed, rounded. Lower lip without inner lobes. Mandible normal, 3[d] joint of palp large. Maxilla 1, inner plate with several setae. Maxilla 2, inner plate only partially fringed on inner margin. Maxillipeds normal, rather robust. Gnathopods 1 and 2 small, subequal, 5th joint rather elongate, palm short, transverse. Peraeopods 3—5, 2[d] joint oval, peraeopods 4 and 5 much longer than peraeopods 1—3. Uropods 1—3, outer ramus much shorter than inner. Telson small, entire.

1 species.

1. **A. longicaudata** Boeck 1861 *A. l.*, A. Boeck in: Forh. Skand. Naturf., Møde 8 p. 663 | 1876 *A. l.*, A. Boeck, Skand. Arkt. Amphip., *v.* 2 p. 351 t. 22 f. 3 | 1893 *A. l.*, G. O. Sars, Crust. Norway, *v.* 1 p. 456 t. 161 | 1893 *Acanthozone l.* (part.), A. Della Valle in: F. Fl. Neapel, *v.* 20 p. 605 t. 59 f. 17.

Peraeon segment 7 longest. Head, rostrum rather prominent, lateral corners narrowly rounded. Side-plates close set, 1st distally expanded. Pleon segment 3, postero-lateral angles quadrate, produced point scarcely perceptible. Eyes irregularly round, light red. Antenna 1 rather longer than the body, 1st joint a little longer than 2[d], 2[d] about twice as long as 3[d], flagellum 5 times as long as peduncle, brittle, many-jointed, setulose, accessory flagellum linear, 1-jointed, about as long as 1st joint of primary. Antenna 2 subequal to antenna 1, less slender, ultimate joint of peduncle longer than penultimate, both of these together with proximal part of flagellum densely setose, flagellum more than twice as long as peduncle. Gnathopod 1, 5th joint somewhat expanded in the middle, 6th subequal in length, widening distally to transverse palm, which is about half as long as hind margin. Gnathopod 2 scarcely longer, more slender,

6th joint narrower, with shorter palm. Peraeopods 1 and 2 rather slender.
Peraeopods 3—5 rather robust, 2^d joint with hind margin smooth, finger in
all peraeopods large, with 2 denticles near the curved apex. Uropod 2
reaching beyond uropod 1, and about as far as uropod 3, which has outer
ramus scarcely $^1/_3$ as long as inner. Telson oval triangular, apex obtusely
pointed. Colour pellucid, whitish, mottled in hands with light red; ova in
pouch bluish green. L. ♀ 7 mm.

Arctic Ocean. North-Atlantic and North-Sea (Norway). Depth 94—282 m.

5. Gen. **Halirages** Boeck

1871 *Halirages* (part.), A. Boeck in: Forh. Selsk. Christian., 1870 p. 194 | 1876
H., A. Boeck. Skand. Arkt. Amphip., c. 2 p. 337 ¡ 1893 *H.*, G. O. Sars, Crust. Norway,
r. 1 p 435.

Body slender, with some segments dentate. Head, post-antennal
corners acute, not strongly produced. Side-plates 1—4 rather shallow.
Antennae 1 and 2 elongate, with small calceoli in several rows. No accessory
flagellum. Upper lip rounded. Lower lip without inner lobes. Mandible
normal, palp large, 3^d joint curved. Maxilla 1, inner plate with several
(5 or 6) setae, outer with 11 spines. Maxilla 2, inner plate fringed on
inner margin. Maxillipeds normal, palp robust. Gnathopods 1 and 2 feeble,
5th joint elongate, 6th oblong, palm shorter than hind margin. Peraeopods 1—5
rather slender, finger short. Peraeopods 3—5, 2^d joint oval. Uropod 3
reaching beyond the others. Telson entire, with notch or sculpturing.

4 species accepted, 2 obscure.

Synopsis of accepted species:

1 { Peraeon segment 7 and pleon segments 1 and 2
dorsally tridentate . . . . 1. **H. nilssoni** . . . . p. 290
Peraeon segment 7 and pleon segments 1 and 2
dorsally unidentate — 2.

2 { Telson transversely truncate, a little emarginate 2. **H. fulvocinctus** . . p 291
Telson with a little apical notch . . . . 3. **H. huxleyanus** . . p. 291
Telson apically tridentate . . . . 4. **H. quadridentatus** . p 292

1. **H. nilssoni** Ohlin 1895 *H. n.*, Ohlin in: Acta Univ. Lund., r. 31 nr. 6 p. 44
t. f. 1—6.

Peraeon segment 7 and pleon segments 1 and 2 each produced dorsally
into a large central tooth, with a feebler one on either side of it. Pleon
segment 3 with an elevated keel, not dentate but strongly rounded behind.
Side-plates 1—3 (in figure) very much smaller than the 4th. Pleon segment 3,
postero-lateral corners produced to an acute tooth, the lateral margin above
forming a truncate lobe with serrate margin, ending in a tooth above. Eyes
large, oval or slightly reniform, dark with a lighter rim. Antenna 1, 1st joint
rather longer than 2^d, 3^d about $^1/_3$ as long as 2^d, outdrawn in a laminar calceoliferous process to middle of 1st joint of flagellum, which joint is as long as 6 or
7 of the following short joints. Antenna 2 as long as the body, twice as long as
antenna 1, penultimate joint of peduncle as long as ultimate, outdrawn at apex.
Mouth-parts as in H. fulvocinctus, but 3^d joint of mandibular palp even
more strongly developed. Gnathopods 1 and 2, 6th joint as long as 5th,
but considerably wider, widening towards the oblique palm. Gnathopod 2
rather weaker than gnathopod 1. Uropod 3. rami without setae. Telson

not quite twice as long as broad, strongly tapering to an emarginate apex bounded by sharp points. L. 13 to about 18 mm.

Baffin Bay. Depth 9–27 m.

2. **H. fulvocinctus** (Sars) 1858 *Amphithoë fulvocincta*, M. Sars in: Forh. Selsk. Christian., p. 141 | 1866 *Paramphithoë f.*, Goës in: Öfv. Ak. Förh., v. 22 p. 525 t. 39 f. 15 | 1871 *Halirages fulvocinctus*, A. Boeck in: Forb. Selsk. Christian,, 1870 p. 196 | 1876 *H. f.*, A. Boeck, Skand. Arkt. Amphip., v. 2 p. 342 t. 23 f. 11f. | 1888 *H. f.*, T. Stebbing in: Rep. Voy. Challenger, v. 29 p. 90 | 1893 *H. f.*, G. O. Sars, Crust. Norway, v. 1 p. 436 t. 154 | 1893 *Acanthozone fulvocincta* (part.), A. Della Valle in: F. Fl. Neapel, v. 20 p. 614 t. 59 f. 27 | 1864 *Pherusa tricuspis*, Stimpson in: P. Ac. Philad.. 1863 p. 139.

Peraeon segment 7 and pleon segments 1 and 2 each produced to a dorsal tooth. Side-plates 1—3 quadrate with rounded corners, 4th broader, more rounded, emarginate behind. Pleon segment 3, postero-lateral corners and lateral margin as in H. nilssoni. Eyes large, oval, bright red. Antenna 1 in ♀ more than ²/₃ length of body, flagellum more than 4 times as long as peduncle, otherwise as in H. nilssoni: antenna 2 longer, ultimate and penultimate joints of peduncle subequal, with many calceoli, flagellum nearly 4 times as long as peduncle, with calceoli in double row. Antennae 1 and 2 in ♂ still more elongate. Mandible, 3d joint of palp longer than 2d, curved. closely fringed. Maxillipeds, outer plates with smooth inner margin, fringed with submarginal spine-teeth, 3d joint of palp apically a little produced. Gnathopod 1, 5th joint slightly widening distally, 6th subequal, rather narrowly oblong, palm much shorter than hind margin, not very oblique. but defined by an obtuse angle, armed with short spines, finger short. Gnathopod 2 similar, rather longer. Peraeopods 1—5 moderately long. fringed with fascicles of spinules. Peraeopod 5, 2d joint broadly oval, much larger than that of peraeopods 3 and 4, hind margin smooth, roundly produced below. Uropod 3, peduncle rather long, rami nearly twice as long as peduncle, densely fringed with spinules and on the inner margin also with setae. Telson about twice as long as broad, slightly tapering, apex very slightly emarginate and serrulate. Colour pellucid, yellowish, transversely banded with orange. L. ♀ reaching 19 mm.

Arctic Ocean (widely distributed, reaching lat. 82° 27′ N.); North-Atlantic (South of Halifax [Nova Scotia], whole West-Norway). Depth 2–470 (659?) m.

3. **H. huxleyanus** (Bate) 1862 *Atylus h.*, Bate, Cat. Amphip. Brit. Mus., p. 135 t. 25 f. 4 | 1888 *Halirages h.*, T. Stebbing in: Rep. Voy. Challenger, c. 29 p. 902 t. 73 · 1893 *Acanthozone huxleyana*, A. Della Valle in: F. Fl. Neapel, v. 20 p. 612 t. 59 f. 23.

Back rounded to end of peraeon segment 6. imbricated, peraeon segments 5—7 with postero-lateral angles produced acutely backward. Peraeon segment 7 and pleon segments 1—3 each produced into a large tooth, and at least pleon segments 1—3 carinate, peraeon segment 6 . slightly toothed. Head, rostrum acute but very small. Side-plates 1—4 small, of rather irregular shape, 4th not much larger than 3d, 5th strongly bilobed. Pleon segments 1—3, postero-lateral corners produced into short acute points. Eyes round, not very large, dark in spirit, with a light rim. Antenna 1, 1st joint not as long as 2d and 3d combined, 3d more than half 2d, flagellum 3 or 4 times as long as peduncle, with 62 joints, at intervals much widened distally. Antenna 2, peduncle rather longer than in antenna 1,

19*

flagellum about equal; ultimate joint of peduncle rather shorter than penultimate, flagellum evenly tapering, with 56 joints, some of the proximal having
as many as 6 calceoli, and the last 3 joints of peduncle having 2 or 3
rows of them. Mandible, 3^d joint of palp very spinose, scarcely as long
as 2^d. Maxillipeds, outer plates with smooth inner margin, fringed with
submarginal spine-like setae. Gnathopod 1, 5th joint shorter and narrower
than 6th, 6th oval, narrowest at hinge of finger, palm very oblique and
ill-defined. Gnathopod 2 similar, but rather longer, not wider. Peraeopods 1—5 not especially slender, finger short, curved. Peraeopod 5, 2^d joint
somewhat piriform, much larger than in peraeopods 3 and 4. Uropods 1
and 2 with few marginal spinules, each with a stout apical spine, outer ramus
shorter than inner; uropod 3, rami subequal, about twice as long as peduncle,
lanceolate, with several spinules and setae. Telson longer than peduncle
of uropod 3, more than twice as long as its greatest breadth, sides a little sinuous,
not strongly tapering, apex triangularly notched. L. 22·5 mm.

South-Atlantic (Stanley Harbour [Falkland Islands], lat. 52° S., from kelp;
Hermit Island, lat. 56° S.); Strait of Magellan.

4. H. quadridentatus O. Sars 1876 *H. qvadridentatus*, G. O. Sars in: Arch.
Naturv. Kristian., *v.* 2 p. 257 | 1885 *H. q.*, G. O. Sars in: Norske Nordhavs-Exp., *v.* 6
Crust. I p. 172 t. 14 f. 4 II, 4 a —f | 1893 *Acanthozone quadridentata* (part.), A. Della
Valle in: F. Fl. Neapel, *v.* 20 p. 611 t. 59 f. 22 | 1882 *Halirages elegans*, A. M. Norman
in: P. R. Soc. Edinb., *v.* 11 p. 688.

Peraeon segments 6 and 7 and pleon segments 1 and 2 each produced
into a dorsal adpressed tooth. Head, rostrum obsolete, lateral corners narrowly
rounded, post-antennal produced to an acute deflexed lobe. Side-plates 1—4
described as simple rounded, and finely serrate along the lower margin
(but Sars fig. 4 d shows side-plate 1 with an acute front angle, in accordance
with Norman). Pleon segment 1, postero-lateral angles rounded, in segments 2
and 3 quadrate, in segment 3 lateral margin above the angle slightly arcuate,
very finely crenulated. Eyes very large, irregular oval, almost meeting
above, brilliant red. Antenna 1 as long as body, 3^d joint of peduncle very
short, flagellum 5 times as long as peduncle, joints numerous, short, except 1st.
Antenna 2 considerably longer, ultimate and penultimate joints of peduncle
subequal. Mandible, 3^d joint of palp rather shorter than 2^d. Gnathopods 1
and 2, 5th joint slender, considerably longer than 6th, 6th oblong, little
widened distally, palm short, scarcely oblique. Peraeopods 1—5 very slender,
1st and 2^d almost filiform, 3^d—5th with 2^d joint piriform, produced at lower
hind corner to a sharp point. Uropods 1 and 2, outer ramus much shorter
than inner; uropod 3, outer ramus little shorter than inner, both long,
lanceolate, with pectinate edges. Telson reaching beyond peduncle of uropod 3,
twice as long as broad, triangular, apex tridentate, central tooth the largest,
but all 3 small. Colour whitish, translucent, with a few faint red bands.
L. 25 mm.

North-Atlantic (West-Norway, depths 659 and 960 m; lat. 60° N., long. 6° W.,
depth 988 m).

H. batei (R. O. Cunningh.) 1871 *Atylus? b.*, R. O. Cunningham in: Tr. Linn.
Soc. London, *v.* 27 p. 498 t. 59 f. 9 | 1888 *Halirages huxleyanus?*, T. Stebbing in: Rep.
Voy. Challenger, *v.* 29 p. 902.

L. 16 mm.

Strait of Magellan (Possession Bay).

H. megalops (Buchh.) 1874 *Paramphithoë m.*, Buchholz in: Zweite D. Nord-polarf., v. 2 p. 369 Crust. t. 12 | 1893 *Acanthozone fulvocincta* (part.), A. Della Valle in: F. Fl. Neapel, v. 20 p. 614.

Strongly separated from the genus Halirages by the non-dentated segments. the long rostrum, short 3ᵈ joint of mandibular palp and small size; but in other respects close to H. fulvocinctus (p. 291). Also like. but not reconcilable with Apherusa megalops (p. 306). L. 5—7 mm.

Arctic Ocean (Sabine Island, Germania Harbour. Shannon). Depth 19 m.

6. Gen. **Leptamphopus** O. Sars

1880 *Panoploea* (part.), G. M. Thomson in: Ann. nat. Hist.. ser. 5 v. 6 p. 2 1893 *Leptamphopus* (Sp. un.: *L. longimanus*). G. O. Sars, Crust. Norway, c. 1 p 458.

Body not acutely dentate. Head, rostrum not very pronounced. Side-plates of moderate size, 4ᵗʰ the largest, distinctly emarginate. Antennae 1 and 2 elongate, antenna 1 the longer, no accessory flagellum. Upper lip rounded. Lower lip without inner lobes. Mandible. cutting edge much produced, molar large, palp moderate. Maxilla 1, inner plate fringed with many setae. outer with spines very closely set, palp rather elongate. Maxilla 2, inner plate fringed on inner margin, both plates with the apex densely fringed. Maxillipeds not very large, outer plates not reaching end of palp's 2ᵈ joint. fringed with spine-teeth on inner margin, palp comparatively small. Gnatho-pods 1 and 2 slender, unequal, the 2ᵈ much the longer, 6ᵗʰ joint nearly linear, palm very short and nearly transverse. Peraeopods 3--5. 2ᵈ joint oval. Uropod 3, rami narrow, unequal. Telson entire.

2 species.

Synopsis of species:

Noue of the segments dorsally produced . . . . . 1. **L. longimanus** . . . . p. 293
Peraeon segment 7 and pleon segments 1 and 2
  dorsally produced . . . . 2. **L. novaezealandiae** . . p. 294

1. L. longimanus (Boeck) 1871 *Amphithopsis longimana*, A. Boeck in: Forh. Selsk. Christian., 1870 p. 200 | 1876 *A. l.*, A. Boeck, Skand. Arkt. Amphip., v. 2 p. 353 t. 22 f. 2 | 1893 *Leptamphopus longimanus*, G. O. Sars, Crust. Norway, c. 1 p. 459 t. 162 | 1893 *Acanthozone longimana* (part.), A. DellaValle in: F. Fl. Neapel, v. 20 p 604 t. 59 f. 16.

Body slender, back evenly rounded. Head, rostral projection very small. lateral corners defined below by a small notch. post-antennal corners forming a triangular lobe. Pleon segments 2 and 3. postero-lateral corners sub-quadrate. Eyes faintly traceable in spirit. Antenna 1 in ♀ nearly as long as body, 1ˢᵗ joint subequal to 2ᵈ and 3ᵈ combined, flagellum slender, about 5 times as long as peduncle. Antenna 2 much shorter, ultimate joint of peduncle longer than penultimate, flagellum not quite twice as long as peduncle. Mandible, said to have no accessory plate on right mandible. Gnathopod 1. 5ᵗʰ joint somewhat widened distally, forming a short setose lobe, 6ᵗʰ as long as 5ᵗʰ, but much narrower. sublinear. fringed behind with fascicles of setules, finger exceedingly small. Gnathopod 2 much longer, 3ᵈ joint not twice as long as broad, 5ᵗʰ and 6ᵗʰ joints subequal, very long and narrow, fringed behind with setules. finger very small. Peraeopods 1—5 slender and brittle. Peraeopods 3—5, 2ᵈ joint oblong oval, hind margin perfectly smooth, that of peraeopod 5 much the largest. Uropod 1 much larger than uropod 2: uropod 3 reaching about to apex of uropod 2. outer ramus about half as long as inner. Telson oval quadrangular, with small notch in broadly truncate apex. L. ♀ 10 mm.

Arctic Ocean, North-Atlantic and North-Sea (Greenland; Norway, depth 282—752 m).

2. **L. novaezealandiae** (G. M. Thoms.) 1879 *Pherusa n.-z.*, G. M. Thomson in: Tr. N. Zealand Inst., v. 11 p. 239 t. 10c f. 2, 2a—c | 1886 *P. neo-zelanica*, (G. M. Thomson &) Chilton in: Tr. N. Zealand Inst., v. 18 p. 148 | 1880 *Panoploea debilis*, G. M. Thomson in: Ann. nat. Hist., ser. 5 v. 6 p. 3 t. 1 f. 3 | 1893 *Acanthozone longimana* (part.), A. Della Valle in: F. Fl. Neapel, v. 20 p. 604, 620.

Back rounded, peraeon segment 7 and pleon segments 1 and 2 produced each dorsally into a flat obtuse tooth. Head with triangular front, no proper rostrum. Pleon segments 2 and 3, postero-lateral angles produced to a small tooth, smaller on 3^d than 2^d, margin above smooth. Antenna 2 nearly as long as antenna 1. Right mandible perhaps with small accessory plate. Gnathopod 2. 3^d joint at least 3 times as long as broad. Peraeopods 3—5, 2^d joint with hind margin slightly serrate. First 3 marsupial plates very long and broad, 4th narrow. Apparently in uropods 1 and 2 the outer ramus is much the shorter, but in uropod 3 not very much shorter than the inner. Telson slightly tapering to a broadly rounded unnotched extremity. In other respects, the species shows close agreement with L. longimanus. Colour light brown, made up of stellate markings. L. 9 mm.

Dunedin Harbour [New Zealand].

7. Gen. **Paraleptamphopus** Stebb.

1899 *Paraleptamphopus*, T. Stebbing in: Ann. nat. Hist., ser. 7 v. 4 p. 209.

Body not dentate. Antenna 1 the longer, with small accessory flagellum. Upper lip flatly rounded. Lower lip without inner lobes. Mandible normal, 3^d joint of palp shorter than 2^d. Maxilla 1, inner plate with many setae, outer with 10 or 11 spines, 2^d joint of palp with longer spines on one maxilla than on the other. Maxilla 2, inner plate fringed on inner margin. Maxillipeds, outer plates not reaching apex of palp's 2^d joint, fringed with spine-teeth on inner margin. Gnathopods 1 and 2 subchelate, weak, 5^d joint in gnathopod 2 rather long. Peraeopods 1—5, 6th joint longer than 5th, finger short. Peraeopods 3—5, 2^d joint oval. Uropods 1 and 2, outer ramus the shorter; uropod 3, rami equal. Telson entire.

2 species.

Synopsis of species:

Eyes wanting, colour of body white, semipellucid . . . 1. **P. subterraneus** . p. 294
Eyes small, colour of body dark indigo blue . . . . . 2. **P. caeruleus** . . . p. 295

1. **P. subterraneus** (Chilton) 1882 *Calliope subterranea*, Chilton in: N. Zealand J. Sci., v. 1 p. 44 | 1882 *C. s.*, Chilton in: Tr. N. Zealand Inst., v. 14 p. 177 t. 9 f. 1—10 | 1889 *C. s.*, Moniez in: Rev. biol. Nord France, v. 1 p. 253 | 1890 *C. s.*, Wrześniowski in: Z. wiss. Zool., v. 50 p. 611 | 1886 *Calliopius subterraneus*, (G. M. Thomson &) Chilton in: Tr. N. Zealand Inst., v. 18 p. 148 | 1894 *C. s.* ♀, Chilton in: Tr. Linn. Soc. London, ser. 2 v. 6 p. 234 t. 23 f. 10—18; (?♂ t. 22 f. 1—15, t. 23 f. 1—9) | 1893 *Acanthonotosoma subterraneum* (part.), A. Della Valle in: F. Fl. Neapel, v. 20 p. 678 | 1899 *Calliope subterranea, Paraleptamphopus* (part.), T. Stebbing in: Ann. nat. Hist., ser. 7 v. 4 p. 210.

♀. Body slender. Head, rostral projection very small. Side-plates not very deep. Pleon segment 3, postero-lateral corners rounded. Eyes wanting. Antenna 1, 1ˢᵗ joint rather longer than 2ᵈ, 3ᵈ short, flagellum many-jointed, accessory flagellum 1-jointed. Antenna 2, ultimate joint of peduncle slightly shorter than penultimate, flagellum much shorter than that of antenna 1. Maxilla 1, inner plate with 10 setae. Maxilla 2, inner plate strongly fringed on the inner margin. Maxillipeds, outer plates nearly reaching end of palp's 2ᵈ joint, 3ᵈ joint of palp a little produced over the short finger. Gnathopod 1, 5ᵗʰ joint distally widened, rather shorter than 6ᵗʰ, which is oblong oval, palm slightly oblique and serrulate, fairly well defined, finger with setules on the concave margin. Gnathopod 2 much longer and thinner. 5ᵗʰ joint considerably longer than 6ᵗʰ, both nearly parallel-sided, palm transverse, very short, well defined, finger very short. Uropod 1, peduncle longer than rami, outer ramus not much shorter than inner; uropod 2, outer ramus considerably shorter than inner; uropod 3, rami not very much longer than peduncle. Telson short, not tapering, broader than long, sides convex, apical border faintly concave. Colour white, semipellucid. L. 5 mm. — The supposed ♂ is uncertain in respect to sex and to identity with the species. It has the flagellum calceoliferous in both 1ˢᵗ and 2ᵈ antennae, and on some of the limbs a plate additional to the normal branchia.

New Zealand (Eyreton, Lincoln, Ashburton and Winchester). In wells.

2. P. caeruleus (G. M. Thoms.) 1885 *Pherusa caerulea*, G. M. Thomson in: N. Zealand J. Sci., v.2 p 576 | 1887 *P. c.*, T. Stebbing in: Tr. zool. Soc. London, v.12 vi p 206 t.39 f. ʙ | 1886 *Amphithopsis c.*, T. Stebbing in: P. zool. Soc. London, p.5 | 1899 *Pherusa c.*, *Paraleptamphopus* (part.), T. Stebbing in: Ann. nat. Hist., ser.7 v.4 p.210 | 1893 *Acanthonotosoma subterraneum?*, A. Della Valle in: F. Fl. Neapel, v.20 p.678.

Body rather compact, shining. Head, rostrum short, lateral corners narrowly obtuse. Side-plates 1—4 rather deep, 1ˢᵗ—3ᵈ oblong, 4ᵗʰ much the largest, rounded below, emarginate behind. Pleon segment 3, postero-lateral corners rounded. Eyes very small, dark. Antenna 1, 1ˢᵗ joint much thicker than 2ᵈ and rather longer, 2ᵈ much longer than 3ᵈ, flagellum about thrice as long as peduncle, 33-jointed, accessory flagellum 1-jointed. Antenna 2 much shorter, ultimate joint of peduncle rather shorter than penultimate, flagellum 19-jointed. Maxilla 1, inner plate with 15 setae, outer with 10 spines. Maxilla 2, fully fringed on inner margin of inner plate. Maxillipeds, outer plates nearly reaching end of palp's 2ᵈ joint. Gnathopods 1 and 2 feeble and slender, 5ᵗʰ joint distally a little widened, longer than 6ᵗʰ, which is also a little widened distally, palm oblique, short, in gnathopod 2 defined by a prominence, against which the setulose end of the blunt finger impinges. Uropod 1, peduncle longer than rami, outer ramus not much shorter than inner. Uropod 3, rami not very long, but longer than peduncle. Telson rounded, shorter than the short peduncle of uropod 3. Colour deep indigo-blue, appearing black when alive, persistent in spirit, but flagellum of antennae 1 and 2, and appendages of peraeon and pleon light. L. 5 mm.

New Zealand (Otago). In streamlet, at height of about 915 m above sea-level.

8. Gen. **Calliopius** Lillj.

1856 *Calliope* (Sp. un.: *C. leachii*) (non J. Gould 1836. Aves!), Bate in: Rep. Brit. Ass., Meet.25 p.58 | 1857 *C.*, Bate in: Ann. nat. Hist., ser.2 v.19 p.142 | 1859 *Paramphithoe* (part.), R. M. Bruzelius in: Svenska Ak. Handl., n. ser. v.3 nr.1 p.68 | 1861 *Amphithopsis* (part.), A. Boeck in: Forb. Skand. Naturf., Møde 8 p 661 | 1865

Calliopius, W. Lilljeborg in: N. Acta Soc. Upsal., ser. 3 *v.* 6 nr. 1 p. 18, 19 | 1876 *C..*
A. Boeck, Skand. Arkt. Amphip., *v.* 2 p. 344 | 1893 *C.*, G. O. Sars, Crust. Norway. *f* 1
p. 446 | 1866 *Paramphithoë*, Goës in: Öfv. Ak. Förh., *v.* 22 p. 523.

Body rather strongly built, without dorsal teeth. Side-plates of moderate
size. Antennae 1 and 2 not very slender, subequal, 3^d joint of peduncle
of antenna 1 apically produced, no accessory flagellum, joints in flagella
sharply defined, calceoli in both sexes. Mouth-parts nearly as in Apherusa (p. 304),
but palp of mandible larger, with 3^d joint as long as 2^d. Gnathopods 1
and 2, 5th joint short, cup-shaped, 6th large, oval, palm longer than hind
margin, weakly defined. Peraeopods 1—5 rather strongly built, finger curved.
not elongate, 3^d—5th pairs with 2^d joint oval. Uropod 3 scarcely reaching
beyond the others, peduncle short. rami lanceolate, subequal. Telson entire.
linguiform.

     2 species accepted, 1 obscure.

     Synopsis of accepted species:

Antenna 1, 3^d joint produced into a large lobe . . . . 1. **C. laeviusculus** . . p. 296
Antenna 1, 3^d joint produced into a small lobe . . . . 2. **C. rathkii** . . . . p. 296

     1. **C. laeviusculus** (Kröyer) 1838 *Amphithoe laeviuscula*, Kröyer in: Dansko
Selsk. Afh., *v.* 7 p. 281 t. 3 f. 13 a—h | 1859 *Paramphithoe l.*, R. M. Bruzelius in: Svenska
Ak. Handl., n. ser. *v.* 3 nr. 1 p. 76 | 1861 *Amphithopsis l.*, A. Boeck in: Forh. Skand.
Naturf.. Møde 8 p. 662 | 1862 *Calliope l.*, Bate, Cat. Amphip. Brit. Mus., p. 148 t. 28 f. 2
1871 *Calliopius laeviusculus*, A. Boeck in: Forh. Selsk. Christian., 1870 p. 197 | 1876 *C. l.*
(part.), A. Boeck, Skand. Arkt. Amphip., *v.* 2 p. 345 | 1893 *Acanthozone laeviuscula* (part.),
A. Della Valle in: F. Fl. Neapel, *v.* 20 p. 602 t. 59 f. 12 | 1856 *Calliope leachii*, Bate in:
Rep. Brit. Ass., Meet. 25 p. 58 t. 17 f. 3 | 1857 *C. l.*, Bate in: Ann. nat. Hist., ser. 2 *r.* 19
p 142 | 1858 *Amphithoe serraticornis*, M. Sars in: Forh. Selsk. Christian., p 140.

Pleon segments 1—3 dorsally raised at end, giving an imbricated
appearance. Head, rostrum obsolete, lateral corners obtusely rounded. Pleon
segment 3, postero-lateral corners subquadrate. Eyes oval, reniform, very
dark, larger in ♂ than in ♀. Antenna 1 of $^1/_3$ length of body, 1st joint as
long as 2^d and 3^d combined, 3^d produced beyond 1st joint of flagellum.
with row of 8 calceoli on the triangular lobe, flagellum rather longer than
peduncle, with 20—30 joints sharply defined, as though the flagellum were
serrate. Antenna 2 scarcely longer, flagellum similar in appearance. subequal to peduncle. Gnathopods 1 and 2 strongly built. 5th joint with setose
lobe, 6th wide at base, narrowing from commencement of oblique palm to
hinge of finger. Peraeopods 1—5 strong, 2^d joint of 5th pair rather large,
especially in the ♂. Uropod 3, both rami plumose and spinose. Telson
nearly twice as long as broad. a little dilated at base, smoothly rounded
at apex. Colour uniformly light green, with whitish area on front of back.
L. ♀ 12, ♂ 13—14 mm.

     Arctic Ocean, North-Atlantic, North-Sea and Skagerrak (Greenland; Spitzbergen:
Labrador; Norway, depth 6—56 m; British Isles); North-Pacific.

     2. **C. rathkii** (Zadd.) 1844 *Amphithoe r.*, Zaddach, Syn. Crust. Pruss.. p 6 |
1893 *Calliopius rathkei*, G. O. Sars, Crust. Norway, *v.* 1 p. 447 t. 157 f. 2 | ?1847 *Amphi-*
thoe gibba, (H. Frey &) R. Leuckart, Wirbell. Th., p. 162 | 1862 *Calliope grandoculis*,
Bate. Cat. Amphip. Brit. Mus., p. 149 t. 28 f. 4 | 1879 *Calliopius laeviusculus* (err.. non
Amphithoe laeviuscula Kröyer 1838!), Hoek in: Tijdschr. Nederl. dierk. Ver., *v.* 4 p. 138
t. 6 f. 4, 6, 7, 10—12; t. 10 f. 7.

Closely approaching C. laeviusculus. Segments rather sharply defined. Head, rostrum short but distinct, lateral corners subtruncate. Pleon segment 3, postero-lateral corners produced to a small but distinct upturned point, margin above convex. Eyes obliquely oval, dark brown, larger in ♂ than in ♀. Antenna 1, 1st joint not as long as 2^d and 3^d combined, 3^d about $^1/_2$ as long as 2^d, produced lobe quite short, with only 2 or 3 calceoli. Gnathopods 1 and 2, lobe of 5th joint narrow, sparingly setose, 6th less stout than in preceding species. Peraeopods 1—5 not so stoutly built as in C. laeviusculus. Uropod 3, both rami spinose, only the inner plumose. Telson fully twice as long as broad, of uniform width, smoothly rounded at apex. Colour yellowish violet, with irregular orange specks, segments banded behind with reddish brown, on dorsum of peraeon segments 3 and 4 a shield of silvery lustre. L. ♀ scarcely over 6 m.

Arctic Ocean, North-Atlantic, North-Sea and Skagerrak (Norway, Bohuslän, France, Holland, Great Britain); Kattegat.

C. pictus (Giles) 1890 *Parapleustes p.*, G. M. Giles in: J. Asiat. Soc. Bengal, v. 59 p. 70 t. 2 f. 6 | 1893 *Acanthozone laeviuscula?*, A. Della Valle in: F. Fl. Neapel, r. 20 p. 895.

Very like C. laeviusculus. Pleon segment 3, postero-lateral angles rounded. Eyes very large. Colour resembling that of the host. L. reaching 7 mm.

Bay of Bengal (Andaman Isles). Depth 55 m. On Pennatula sp.

9. Gen. **Paracalliope** Stebb. ✦

1899 *Paracalliope* (Sp. typ.: *Calliope fluviatilis*), T. Stebbing in: Ann. nat. Hist., ser. 7 v. 4 p. 210.

Body without dorsal teeth. Side-plates not very large. Antenna 1 the shorter, without accessory flagellum or calceoli. Upper lip rounded. Lower lip with inner lobes. Mandible slightly constructed, normal, 3^d joint of palp at least as long as 2^d. Maxilla 1, inner plate with many (11) setae, outer with (apparently) 11 spines, 2^d joint of palp carrying spine-teeth and slender spines. Maxilla 2, inner plate the narrower, fringed along inner margin. Maxillipeds, inner plates rather broad, outer reaching apex of palp's 2^d joint, fringed with spines on inner margin, the series not continued on the apical border, finger of palp acute. Gnathopod 1, 5th joint rather shorter than oval 6th. Gnathopod 2, 5th joint cup-shaped, 6th quadrangular oval, with hollow well-defined palm. Peraeopods 1—4, finger curved. Peraeopod 5 much longer than the others, huger long, straight, spinose. Uropods 1—3, rami slender, with few spines. Uropod 3, rami equal, not longer than peduncle. Telson short, entire.

1 species.

1. P. fluviatilis (G. M. Thoms.) 1879 *Calliope f.*, G. M. Thomson in: Tr. N. Zealand Inst., v. 11 p. 240 t. 10c f. 4, 4a—c | 1886 *Calliopius f.*, (G. M. Thomson &) Chilton in: Tr. N. Zealand Inst., v. 18 p. 148 1899 *Calliope f.*, ? *Oediceros novae-zealandiae*, *Paracalliope* sp. typ., T. Stebbing in: Ann. nat. Hist., ser. 7 v. 4 p 210 | 1880 *Pherusa australis*, Haswell in: P. Linn. Soc. N. S. Wales, v. 5 p. 103 t. 7 f. 1.

Peraeon segments 1—6 rather short. Head, front obtuse-angled. Side-plate 1 a little widened distally. Pleon segment 3, postero-lateral corners subquadrate. Eyes oblong oval, large, black. Antenna 1, 1st joint as long as 2^d and 3^d combined, but much stouter, 3^d nearly $^1/_2$ as long as 2^d, and (contrary to Haswell and Thomson) quite distinguishable from the flagellum, which is slender, about twice as long as peduncle, 21-jointed. Antenna 2 longer, ultimate and penultimate joints of peduncle subequal, flagellum about twice as long as peduncle, 24-jointed. Upper lip transversely oblong, with distal margin not strongly convex. Mandible, cutting edge and accessory plate slight in structure, spine-row with few spines, molar very prominent, 3^d joint of palp curved, spinose, narrow at apex. Gnathopod 1. 5th joint elongate triangular but not very long, 6th with oblique palm. Gnathopod 2 considerably stronger, lobe of 5th joint fringed with slender spines, 6th broad at base, widening a little to the moderately oblique palm, which on one margin is evenly concave, on the other sinuous and spinose, the rather strong finger closing down in the hollow between them; the reversion of gnathopod 2, beginning at 5th joint, recalls the curious torsion in gnathopod 1 of Trischizostoma nicaeense (p. 13). Peraeopods 1—5, 6th joint longer than either 4th or 5th, the 4th—6th joints spinose, 2^d joint in peraeopods 3—5 expanded, hind margin little serrate. Peraeopod 5. 6th joint elongate, but shorter than the long, stiliform, spinose finger. Uropod 1, rami shorter than peduncle, outer ramus little shorter than inner. Uropod 2, outer ramus much shorter than inner. Uropod 3, peduncle rather elongate. Telson little longer than broad, of uniform width to the smoothly rounded apex. Colour greyish, more or less marked with dark spots. L. 5 mm.

This species is not improbably identical with *Oedicerus novi-zealandiae* Dana (p. 270), which Boeck, in 1861, includes into his genus *Aceros* (p. 248).

New Zealand. In fresh water, in brackish or nearly salt water. — Botany Bay [East-Australia].

10. Gen. **Harpinioides** Stebh.

1888 *Harpinioides* (Sp. un.: *H. drepanocheir*), T. Stebbing in: Rep. Voy. Challenger, r. 29 p. 936 | 1890 *Harpinioides*, Warburton in: Zool. Rec., v. 25 Crust. p. 19.

Back rounded. Head, rostrum obsolete. Side-plates 1—4 rather deep, 1st somewhat expanded distally. Antenna 1 the longer, peduncle short, accessory flagellum minute. Upper lip, distal margin broad, flat. Lower lip with inner lobes. Mandible, cutting edge very oblique, long, cut into numerous denticles, accessory plate on left similar but shorter, on right represented by a short prickly spine, 10 spines in spine-row, molar small and slender, not denticulate. 2^d joint of palp dilated, 3^d equally long, but narrower. Maxilla 1, inner plate with 1 seta, outer with 9 spines, mostly smooth, 2^d joint of palp somewhat dilated. Maxilla 2, inner plate shorter and narrower than outer, partially fringed on inner margin. Maxillipeds, inner and outer plates narrow, outer prolonged much beyond inner, but not reaching end of palp's 2^d joint, fringed on inner margin with slender spine-teeth, palp rather long, finger as long as 3^d joint. Gnathopods 1 and 2 similar, 5th joint short, 6th long, tapering. Peraeopods 3—5. 2^d joint greatly expanded. Uropods 1 and 2, outer ramus the shorter; uropod 3, rami subequal, lanceolate. Telson almost entire, feebly notched.

1 species.

1. **H. drepanocheir** Stebh. 1888 *H. d.*, T. Stebbing in: Rep. Voy. Challenger, v. 29 p. 937 t. 82 | 1893 *Acanthonotosoma d.*, A. Della Valle in: F. Fl. Neapel, v. 20 p. 677 t. 59 f. 87.

Pleon segment 3, postero-lateral corners strongly rounded. Eyes not observed. Antenna 1. 1st joint stout, longer than 2^d and 3^d combined. flagellum twice as long as peduncle. with 24 joints, 1st the longest; accessory flagellum shorter than 1st joint of primary, consisting of 1 narrow, truncate joint. Antenna 2. ultimate joint of peduncle rather shorter than penultimate, flagellum shorter than peduncle, 14-jointed. Gnathopods 1 and 2, 5th joint cup-like. about $^1/_3$ length of 6th, 6th broadest at base, tapering with a curve to a narrow apex, without definite palm, finger long. slender, slightly curved. about half length of 6th joint and closing over the concave part of its margin. Gnathopod 2 rather the longer. Peraeopods 1—5, 5th joint shorter than 4th or 6th, finger curved. Peraeopod 3, 2^d joint almost as broad as long, and of nearly equal breadth throughout. Peraeopod 5 longer than peraeopod 4, and with 2^d joint broader. Telson longer than peduncle of uropod 3, once and a half as long as broad, slightly tapering to a slight triangular emargination between minutely notched apices. L. 6 mm.

Cumberland Bay [Kerguelen Island]. Depth 232 m.

11. Gen. Atylopsis Stebb.

1888 *Atylopsis* (part.), T. Stebbing in: Rep. Voy. Challenger, *c.* 29 p. 924.

Side-plates rather shallow. Antenna 1, peduncle short. Upper lip slightly bilobed. Lower lip with inner lobes. Mandible normal, palp strong, 3^d joint as long as 2^d. Maxilla 1, inner plate with few setae, outer with 11 spines. Maxilla 2, inner plate with few setae on inner margin. Maxillipeds, outer plates without spine-teeth on inner margin, 3^d joint of palp produced over base of finger. Gnathopods 1 and 2 not very powerful. subchelate, 2^d larger than 1st. Uropods 1—3, outer ramus shorter than inner. Telson short, with emarginate apex.

2 species.

Synopsis of species:

1. **A. dentata** Stebb. 1888 *A. dentatus*, T. Stebbing in: Rep. Voy. Challenger, *c.* 29 p. 929 t. 80 | 1893 *Acanthonotosoma?, Acanthozone?*, A. Della Valle in: F. Fl. Neapel, *v.* 20 p. 910.

Like, but quite distinct from Apherusa tridentata (p. 305). Peraeon segment 7 and pleon segments 1 and 2 each produced to a dorsal tooth, surface of body rather hairy. Head, rostrum small, obtuse, lateral margin forming a rounded lobe, post-antennal corners not produced. Side-plate 1, lower front corner rounded, slightly outdrawn. Pleon segment 3, postero-lateral corners produced into a small tooth. Eyes broadly oval. Antenna 1, 1st joint not twice as long as 2^d. 2^d scarcely longer than broad, remainder missing. Antenna 2, penultimate joint of peduncle not elongate, after-part missing. Maxilla 1, inner plate with 2 setae; on outer plate only 10 spines were actually observed. Maxilla 2, inner plate rather shorter and narrower than outer, with

2 setae on the inner margin. Maxillipeds, outer plates small, feebly armed.
finger of palp short, with spine-like nail. Gnathopod 1, 5th and 6th joints
equal, 6th very little widened distally, palm short, oblique, much shorter
than hind margin, with defining spine at the curve in which they meet.
Gnathopod 2 longer, 5th joint rather shorter than 6th, which is a little
more widened towards the more oblique palm than in gnathopod 1.
Peraeopod 1, 6th joint longer than 4th or 5th, finger rather strong. with
2 or 3 setules on convex margin. Peraeopods 3—5, 2^d joint broadly
oval, serrate on hind margin, largest in peraeopod 5. Uropods 1—3.
outer ramus much shorter than inner; uropod 3, peduncle very short.
outer ramus narrower than inner and about half its length. both acute
and carrying spines. Telson as long as peduncle of uropod 3. a little
longer than broad, ending in a small triangular emargination between obtuse
apices, sides convex. L. 6 mm.

   Strait of Magellan (Cape Virgins). Depth 100 m.

   2. A. emarginata Stebb. 1888 *A. emarginatus*, T. Stebbing in: Rep. Voy.
Challenger, *v.* 29 p. 932 t. 81 | 1893 *Acanthonotosoma emarginatum*, A. Della Valle in:
F. Fl. Neapel, *v.* 20 p. 678 t. 59 f. 86.

   Body not dentate. Head, rostrum small, obtuse, post-antennal corners
slightly produced, rounded. Side-plate 1, lower front corner slightly outdrawn.
Pleon segment 3, postero-lateral corners rounded. Eyes seemingly large,
reniform. Antenna 1, 1st joint little longer than 2^d, 2^d than 3^d, flagellum not
long, but nearly twice as long as peduncle. 30-jointed, accessory flagellum
minute, 1-jointed. Antenna 2 longer. ultimate joint of peduncle rather
longer than penultimate, flagellum with 33 joints, 1st longest. Maxilla 1, inner
plate with 4 or 5 setae, outer with 11 spines. Maxilla 2, inner plate as
broad as outer, with 4 setae on inner margin. Maxillipeds, outer plates
moderately large, with long spines on distal margin and slender submarginal
spines near the inner border, finger rather large, acute. Gnathopods 1 and 2.
5th joint much shorter than 6th, distally rather wide and cup-like. 6th joint
oblong oval, palm oblique, with a tooth near the hinge, then sinuous and
closely pectinate, in gnathopod 1 as long as the hind margin, but shorter than
it in gnathopod 2; finger with inner margin cut into many adpressed teeth.
Peraeopod 1, 6th joint longer than 4th or 5th, finger curved, strong. unarmed.
Peraeopods 3—5, 2^d joint rather wider above than below. not strongly
expanded, hind margin straight. Uropod 1, rami long; uropod 2. outer
ramus much shorter than inner; uropod 3, peduncle short, rami long. outer
ramus in ♂ not greatly shorter than inner, but more decidedly so in ♀.
Telson longer than peduncle of uropod 3, a little longer than broad, distal
emargination not as deep as wide, the triangular apices a little serrate on
outer margin. L. 7 mm.

   Southern Indian Ocean (Marion Island). Depth 567 m.

12. Gen. **Cleippides** Boeck

   1871 *Cleïppides* (Sp. un.: *C. tricuspis*), A. Boeck in: Forh. Selsk. Christian,. 1870
p. 201 | 1876 *C.*, A. Boeck, Skand. Arkt. Amphip., *v.* 2 p. 357 | 1875 *Cleippides*, Cam.
Heller in: Denk. Ak. Wien, *v.* 35 p 32.

   Back rounded in front, some of the hinder segments produced to a
dorsal tooth. Side-plates 1—4 not large. Antenna 1 the longer,. without

accessory flagellum. Upper lip little or not at all emarginate. Lower lip
with inner lobes obsolescent. Mandible, cutting edge not dentate, accessory
plate denticulate, spine-row small, or accessory plate and spine-row wanting
(Heller), molar with a circular ridge, 2^d and 3^d joints of palp distally
widened. Maxilla 1, inner plate with several setae, outer with 11 spines,
1st joint of palp about half as long as 2^d. Maxilla 2, inner plate rather
the shorter, fringed on inner margin. Maxillipeds, inner plates of moderate
size, outer not nearly reaching end of palp's 2^d joint (if Krøyer's figure
can be trusted against his text), fringed with spine-teeth on inner margin,
palp stout, $1^{st}-3^d$ joints subequal in length, 4th small. Gnathopod 1
weakly subchelate, 5th joint much longer than 6th. Peraeopods 1—5 slender,
3^d-5th with 2^d joint little expanded. Uropod 3, rami lanceolate, equal.
Telson small, entire.

2 species.

Synopsis of species:

1. **C. tricuspis** (Krøyer) 1846 *Acanthonotus t.*, Krøyer in: Naturh. Tidsskr.,
ser. 2 *v.* 2 p. 115 | 1846 *A. t.*, Krøyer in: Voy. Nord, Crust. t. 18 f. 1 a—v | 1862 *Dexamine t.*,
Bate, Cat. Amphip. Brit. Mus., p. 133 t. 24 f. 5 | 1866 *Paramphithoë t.*, Goës in: Öfv.
Ak. Förh., *v.* 22 p. 525 | 1871 *Cleïppides t.*, A. Boeck in: Forh. Selsk. Christian., 1870
p. 201 | 1876 *C. t.*, A. Boeck, Skand. Arkt. Amphip., *v.* 2 p. 358 | 1893 *Acanthozone t.*
(part.), A. Della Valle in: F. Fl. Neapel, *v.* 20 p. 603 t. 59 f. 13, 14.

Without carina. Peraeon segment 7 and pleon segments 1 and 2
each produced into a large flat dorsal tooth. Head, rostral projection
small. Side-plates 1—3 deeper than broad, lower corners rounded. Side-
plate 4 rather broader. slightly emarginate behind. Pleon segment 3,
postero-lateral angles outdrawn into a tooth, above which is another,
separated by a rather large sinus. Eyes narrowly reniform, dark, the
colour disappearing in spirit. Antenna 1 long, thin, 1st joint as long as
2^d and 3^d combined. flagellum thrice as long as peduncle, 50—60-jointed.
Antenna 2 much shorter, ultimate joint of peduncle rather shorter than
penultimate, flagellum rather longer than peduncle, 40-jointed. Gnathopod 1, 5th joint much longer than the narrowly oval 6th, palm undefined,
finger small, but strong, curved, bidentate. Gnathopod 2 rather longer,
5th joint narrow, scarcely longer than the narrowly quadrangular 6th.
Peraeopods 3—5, 2^d joint larger in peraeopod 4 than in 3^d, in 5th than
in 4th, narrowing downward, the hind margin serrate, tending to concave.
Uropod 1, outer ramus little shorter than inner; uropod 2, outer ramus much
shorter than inner. Telson oval, feebly acuminate. Colour chestnut-brown.
L. reaching 16 mm or more.

Arctic Ocean and North-Atlantic (South-Greenland, Spitzbergen).

2. **C. quadricuspis** Heller 1875 *C. q.*, Cam. Heller in: Denk. Ak. Wien,
v. 35 p. 32 t. 3 f. 1—16 | 1885 *C. q.*, G. O. Sars in: Norske Nordhavs-Exp., *v.* 6 Crust I
p. 174 t. 14 f. 5 | 1893 *Acanthozone tricuspis* (part.). A. Della Valle in: F. Fl. Neapel,
v. 20 p. 603.

Peraeon segment 7 and pleon segments 1—3 bluntly carinate and
each produced into a long dorsal tooth. Head, rostrum small, acute, lateral

corners angular, scarcely produced, post-antennal corners acutely produced
downward. Side-plates 1—4 subquadrate, produced ·below into a minute⁓
pointed tooth, in front of 1ˢᵗ, behind in 2ᵈ—4ᵗʰ; 5ᵗʰ and 6ᵗʰ each produced
into 2 teeth, 5ᵗʰ pair much wider than the rest. Pleon segment 3, postero-
lateral corners with 2 teeth as in C. tricuspis. Eyes small, reniform, almost
colourless. Antenna 1, 1ˢᵗ joint as long as 2ᵈ and 3ᵈ combined, flagellum more
than thrice as long as peduncle, many-jointed. Antenna 2 scarcely half as long.
Gnathopods 1 and 2 alike, but the 2ᵈ rather longer than the 1ˢᵗ; 5ᵗʰ joint
long, distally wide, 6ᵗʰ abruptly narrower, oval, about half as long, palm
undefined, finger curved. Peraeopods 3—5, 2ᵈ joint scarcely expanded,
except a little at the base; peraeopod 5 with a tooth, the lower hind
angle produced to a point. Uropods 1—3 and telson as in C. tricuspis.
Colour yellowish white shading into violet and rose-red; mouth-parts and
most of the limbs of a rich carmine. L. 42—52 mm.

Arctic Ocean (lat. 67⁰—80⁰ N., long. 16⁰ E.—12⁰ W., depth 663—1890 m; Spitzbergen, depth 160—265 m).

13. Gen. **Stenopleura** Stebb.

1888 *Stenopleura* (Sp. un.: *S. atlantica*), T. Stebbing in: Rep. Voy. Challenger,
c. 29 p. 949.

Body slender, not dentate. Head, post-antennal corners not produced.
Side-plates 1—4 very shallow. Antenna 1 the longer, without accessory
flagellum. Lower lip with inner lobes. Mandible with accessory plate on
left, spine-row with very few spines, molar strong, palp robust, ·3ᵈ joint as
long as 2ᵈ. Maxilla 1, inner plate with 1 seta, outer with 10 spines,
2ᵈ joint of palp tipped with spine-teeth. Maxilla 2, inner plate shorter than
outer, each with a few apical spines, inner margin unarmed. Maxillipeds,
inner plates with distal margin sloping outward, outer plates not reaching
beyond apex of palp's 1ˢᵗ joint, with 3 spines on apical margin and a few
submarginal setae on inner, finger as long as 3ᵈ joint of palp. Gnathopods 1
and 2 alike, subequal, 5ᵗʰ joint short, cup-like, 6ᵗʰ oval, rather large, finger
long. Peraeopods 1 and 2, 2ᵈ and 4ᵗʰ joints widened. Peraeopods 3—5,
2ᵈ joint expanded, finger in all peraeopods smooth, curved. Uropods 1
and 2, outer ramus much shorter than inner; uropod 3, rami long, lanceolate, outer nearly as long as inner. Telson entire, apically sculptured.

1 species.

1. S. atlantica Stebb. 1888 *S. a.*, T. Stebbing in: Rep. Voy. Challenger, v. 29
p 950 t. 84 | 1893 *Acanthozone a.*, A. Della Valle in: F. Fl. Neapel, v. 20 p. 601 t. 59 f. 11.

Head, rostrum and lateral corners little prominent. Side-plate 1
distally a little produced forward. · Pleon segment 3, postero-lateral corners
subquadrate. Eyes large, oblong oval. Antenna 1, peduncle short, 1ˢᵗ joint stout,
longer than 2ᵈ and 3ᵈ combined, flagellum about 5 times as long as peduncle,
soon becoming filiform, with 33 joints, 1ˢᵗ nearly as long as 2ᵈ of peduncle.
Antenna 2 rather shorter, ultimate joint of peduncle shorter than penultimate,
flagellum more than 4 times as long as peduncle, filiform, 35-jointed. Gnathopods 1 and 2, 5ᵗʰ joint triangular, with serrate spines on the slightly produced
lobe, 6ᵗʰ joint triangular oval, tapering to the hinge of the curved finger;
to judge by the length of the latter, the palm may be considered to form
the hind margin. Gnathopod 2 is slightly the larger, with the lobe of the

5th joint rather narrower and more decidedly produced. Peraeopods 1 and 2, 2d joint not twice as long as broad, narrowest at the base, 4th shorter than either 5th or 6th, broad, with convex outer margin. Peraeopods 3—5, 2d joint widely oblong oval, but in peraeopod 5 with straightened hind margin and produced below the 3d joint. Uropods 1—3, all the rami spinulose. in uropod 2 outer ramus only half as long as inner, in uropod 3 peduncle very short, outer ramus a little shorter and more slender than inner. Telson rather longer than peduncle of uropod 3, a little longer than broad, the central piece of the tridentate apex much larger than the lateral teeth. L. 7·5 mm.

Tropical Atlantic (lat. 2—3° N., long. 8—24° W., depths 91 and 3383 m); South-Atlantic (near Tristan da Cunha).

14. Gen. **Haliragoides** O. Sars

1893 Haliragoides (Sp. un.: H. inermis), G. O. Sars, Crust. Norway. v. 1 p. 432.

Body slender, not dentate. Head, post-antennal corners strongly produced. Side-plates 1—4 shallow, 4th not deeper. than 1st—3d. Antenna 1 the shorter, without accessory flagellum. Upper lip with rounded, very slightly insinuate margin. Lower lip, inner lobes distinct, small. Mandible normal, well developed, 3d joint of palp much shorter than 2d. Maxilla 1, inner plate with several setae. Maxilla 2, inner plate fringed on inner margin. Maxillipeds normal, outer plates fringed with spine-teeth, palp rather robust than elongate. Gnathopods 1 and 2 subequal, rather feeble. 5th joint elongate, 6th oval, palm very oblique. Peraeopods 1—5 very slender, long, finger long and thin. Peraeopod 5, 2d joint much larger than in peraeopods 3 and 4. Uropod 3 reaching much beyond the others. Telson entire, apically sculptured.

1 species.

1. **H. inermis** (O. Sars) 1882 Halirages i., G. O. Sars in: Forb. Selsk. Christian., nr. 18 p. 103 t. 5 f. 5 | 1893 Haliragoides i., G. O. Sars, Crust. Norway. v. 1 p. 433 t. 153 | 1893 Acanthozone quadridentata (part.), A. Della Valle in: F. Fl. Neapel, v. 20 p. 611.

Back rounded, not at all carinate. Head, rostrum and lateral corners very little produced, post-antennal lobes linguiform, deflexed, serrate in front. Side-plate 1 obtusely produced, 4th a little emarginate. Pleon segment 3, postero-lateral corners produced to a small acute tooth, above which the margin is convex, smooth. Eyes large, oval, feebly developed, light red. Antenna 1, 1st joint rather stout, apically dentate, nearly as long as 2d and 3d combined, flagellum about thrice as long as peduncle, with 28 joints, 1st elongate. Antenna 2 much longer, longer than the body, ultimate and penultimate joints of peduncle subequal, flagellum 5 or 6 times as long as peduncle. 1st joint long, slender, the others short. Gnathopods 1 and 2, 5th joint elongate triangular, being a little widened distally, 6th shorter, widest at beginning of palm, which is much longer than the hind margin, defined from it by an obtuse angle, finger about as long as palm. Peraeopods 1—5 very brittle, fringed with fascicles of setules, 4th joint shorter than 5th, 6th much longer than either. Peraeopods 3 and 4, 2d joint rather small, narrowly oval. Peraeopod 5, 2d joint much wider above than below. Uropod 2 small. Uropod 3 long, peduncle well developed, rami narrowly lanceolate, outer a little the shorter, both densely spinulose and with setae

on inner margin. Telson shorter than peduncle of uropod 3, oval triangular.
tridentate, the central tooth much the largest. Colour brilliant with red,
orange and pure white. L. 10—14 mm.

Arctic Ocean, North-Atlantic and North-Sea (Norway). Depth 188—564 m.

15. Gen. **Apherusa** A. Walker .

1859 *Paramphithoe* (part.), R. M. Bruzelius in: Svenska Ak. Handl., n. ser. *v.* 3
nr. 1 p. 68 | 1861 *Amphithopsis* (part.), A. Boeck in: Forh. Skand. Naturf., Møde 8
p. 661 | 1862 *Gossea* (Sp. un.: *G. microdeutopa*) (non L. Agassiz 1862, Coelenterata!), Bate
(& Westwood), Brit. sess. Crust., *v.* 1 p. 276 | 1871 *Halirages* (part.), A. Boeck in: Forh. Selsk.
Christian., 1870 p. 194 | 1891 *Apherusa*, A. O. Walker in: Ann. nat. Hist., ser. 6 *v.* 8
p. 83 | 1893 *A.*, G. O. Sars, Crust. Norway, *v.* 1 p. 438.

Head, post-antennal corners more or less produced. Side-plates of
moderate size, encreasing in depth to the 4th. Antenna 1 shorter than
antenna 2, without calceoli, no accessory flagellum. Upper lip rounded.
Lower lip with small inner lobes. Mandible, molar strong, 3^d joint of palp
shorter than 2^d. Maxilla 1, inner plate with variable number of setae,
outer with variable number of spines. Maxilla 2, inner plate the narrower,
fringed on inner margin. Maxillipeds, inner plates large as compared with
outer, outer fringed on inner margin with spine-teeth or slender spines, palp
of moderate size. Gnathopods 1 and 2 not strong or differing much in
size. Peraeopods 1—5 usually rather strongly built. Uropod 3, rami usually
projecting beyond the others. Telson not large, entire.

7 species accepted, 3 doubtful.

Synopsis of accepted species:

1 { Back with some segments dentate — **2.**
  { Back with no segments dentate — **5.**

2 { Postero-lateral margin of pleon segment 3 smooth . 1. **A. cirrus** . . . p. 304
  { Postero-lateral margin of pleon segment 3 denti-
  {   culate — **3.**

3 { Peraeon segment 7 dentate. . . . . 2. **A. tridentata** . p. 305
  { Peraeon segment 7 not dentate — **4.**

4 { Head, post-antennal corners slightly produced; telson
  {   tridentate . . . . 3. **A. bispinosa** . p. 305
  { Head, post-antennal corners strongly produced; telson
  {   bidentate . . . . 4. **A. megalops** . p. 306

5 { Gnathopods 1 and 2, 5th joint much longer than 6th 5. **A. glacialis** . . p. 307
  { Gnathopods 1 and 2, 5th joint not longer than 6th — **6.**

6 { Antenna 1, 3^d joint much shorter than 2^d . . . . . 6. **A. jurinei** . . . p. 307
  { Antenna 1, 3d joint not much shorter than 2^d . . . 7. **A. georgiana** . p. 308

1. A. cirrus (Bate) 1862 *Pherusa bicuspis* (*P. cirrus* Bate) (err., non *Amphithoe b.*
Kröyer 1838!), Bate & Westwood, Brit. sess. Crust., *v.* 1 p. 253 f. | 1862 *P. cirrus* +
P. b., Bate, Cat. Amphip. Brit. Mus., p. 143 t. 27 f. 6; p. 144 t. 27 f. 7 | 1871 *Halirages
borealis*, A. Boeck in: Forh. Selsk. Christian., 1870 p. 196 | 1876 *H. b.*, A. Boeck,
Skand. Arkt. Amphip., *v.* 2 p. 340 t. 23 f. 6 | 1893 *Apherusa b.*, G. O. Sars, Crust. Norway,
v. 1 p. 441 t. 155 f. 2 | 1895 *A. b.?*, A. O. Walker in: Ann. nat. Hist., ser. 6 *v.* 15 p. 468 |
1893 *Acanthozone laeviuscula?*, A. Della Valle in: F. Fl. Neapel, *v.* 20 p. 619.

Pleon segments 1 and 2 each produced into a dorsal tooth. Head
somewhat produced in front, but not rostrate, lateral and post-antennal

corners of little prominence. Side-plates 1—3 rounded quadrangular, 4th obliquely rounded and little widening distally. Pleon segment 3, postero-lateral corners ending in a short blunt point, margin above smooth. Eyes small, round, black. Antenna 1, 1st joint longer than 2^d and 3^d combined, flagellum 2—3 times as long as peduncle, 22-jointed. Antenna 2 longer, ultimate and penultimate joints of peduncle subequal, flagellum 2—3 times as long as peduncle. In ♂ the underside of peduncle of antenna 1 and upperside of peduncle of antenna 2 have fascicles of setae (Bate's figure; but flagellum of antenna 2 much longer than he represents). Gnathopods 1 and 2, 5th joint shorter than the narrow oblong 6th, which has a short rather oblique palm. Gnathopod 2 larger than gnathopod 1, having the 5th joint stouter and the 6th longer. Peraeopods 1—5, 4th joint a little widened, shorter than 6th, finger strong. Peraeopods 3—5, 2^d joint oval, in peraeopod 5 oblong oval, much larger than in the other peraeopods. Uropod 3, rami not greatly longer than peduncle, extending much beyond uropod 2. Telson triangular, scarcely twice as long as broad, tip pointed, simple. Colour uniformly claret-red. L. 7·5 mm.

Arctic Ocean, North-Atlantic, North-Sea and English Channel (North-Norway, depth 11—19 m; Great Britain, among shore Weeds).

2. **A. tridentata** (Bruz.) 1859 *Paramphithoe t.*, R. M. Bruzelius in: Svenska Ak. Handl., n. ser. *v.* 3 nr. 1 p. 74 t. 3 f. 13 | 1861 *Amphithopsis t.*, A. Boeck in: Forh. Skand. Naturf.. Møde 8 p. 662 ¦ 1862 *Dexamine t.*, Bate. Cat. Amphip. Brit. Mus.. p. 376 | 1871 *Halirages tridentatus*, A. Boeck in: Forh. Selsk. Christian , 1870 p. 196 | 1876 *H. t.*, A. Boeck, Skand. Arkt. Amphip., *v.* 2 p. 341 | 1893 *Apherusa tridentata*, G. O. Sars, Crust. Norway, *v.* 1 p. 442 t. 156 f. 1 | 1893 *Acanthozone fulvocincta* (part.), A. Della Valle in: F. Fl. Neapel, *v.* 20 p. 614.

Peraeon segment 7 and pleon segments 1 and 2 each produced into a dorsal tooth. Head, rostrum small, lateral corners rounded, sharply defined below, post-antennal corners prominent, acute. Side-plate 1, front corner rather angular, 1st—3^d with lower margin minutely serrate, 4th much widened distally. Pleon segment 2, lower part of hind margin serrate, segment 3, postero-lateral corners sharply produced, margin above broadly lobed, cut into 12 upturned teeth. Eyes oval reniform, dark, larger in ♂ than in ♀. Antenna 1 in ♀, about ⅓ length of body, 1st joint as long as 2^d and 3^d combined, flagellum rather over twice as long as peduncle, with 40 joints, 1st the largest. Antenna 2 nearly twice as long, ultimate and penultimate joints of peduncle subequal, flagellum slender, about thrice as long as peduncle. Mandible, molar strong. Gnathopods 1 and 2, 5th and 6th joints slender, subequal, 6th slightly widening distally, palm short, defined by an obtuse angle. Peraeopods 1—5 moderately robust, 2^d joint in peraeopods 3—5 oval, with serrate hind margin, largest in 5th pair. Uropod 3 reaching not greatly beyond the others; inner ramus rather the longer, outer edge bulging at the base. Telson oblong oval, not quite twice as long as broad, apical border irregularly serrate, with some of the teeth double. Colour mottled with a magnificent carmine red. L. ♀ attaining 14 m.

Arctic Ocean (Norway southWard to Lofoten Isles).

3. **A. bispinosa** (Bate) 1857 *Dexamine b.*, Bate in: Ann. nat. Hist., ser. 2 *v.* 19 p. 142 | 1871 *Halirages bispinosus*, A. Boeck in: Forh. Selsk. Christian., 1870 p. 195 | 1876 *H. b.*, A. Boeck, Skand. Arkt. Amphip., *v.* 2 p. 338 t. 23 f. 9 | 1881 *Pherusa bispinosa*, Nebeski in: Arb. Inst. Wien. *v.* 3 p. 146 | 1893 *Acanthozone b.*, A. Della Valle in: F. Fl. Neapel, *v.* 20 p. 609 t. 3 f. 5; t. 17 f. 22—36 | 1893 *Apherusa b.*, G. O. Sars,

Crust. Norway, *v.* 1 p. 439 t. 155 f. 1 | 1858 *Amphithoë macrocephala*, M. Sars in: Forb.
Selsk. Christian., p. 142 | 1859 *Paramphithoe elegans*, R. M. Bruzelius in: Svenska
Ak. Handl., n. ser. *v.* 3 nr. 1 p. 75 t. 3 f. 14 | 1861 *Amphithopsis e.*, A. Boeck in: Forh.
Skand. Naturf., Møde 8 p. 662 | 1862 *Atylus bispinosus* + *Pherusa e.*, Bate, Cat. Amphip.
Brit. Mus., p. 140 t. 27 f. 1; p. 377 | 1868 *P. pontica*, Czerniavski in: Syezda Russ. Est.,
Syezda 1 Zool. p. 110 t. 8 f. 15.

Pleon segments 1 and 2 each produced into a dorsal tooth. Head.
rostrum distinct, acute, lateral corners small, angular, post-antennal forming
a short deflexed point. Side-plate 1 subangular in front, a little expanded
distally. Pleon segment 2, postero-lateral corners acutely produced, segment 3,
postero-lateral corners acute, margin above serrate, upper corner forming
a sharp upturned point, followed by a deep sinus with a bidentate projection
in it or margin above with only 4 teeth (Czerniavski). Eyes large, roundish,
dark brown. Antenna 1 in ♀ scarcely more than $^1/_3$ as long as body, 1st joint
longer than 2^d and 3^d combined, flagellum nearly 4 times as long as peduncle,
30-jointed. Antenna 2 in ♀ nearly twice as long, ultimate joint of peduncle
rather longer than penultimate, flagellum long and slender. Antennae 1
and 2 in ♂ longer than in ♀, and setose on confronting edges of peduncle;
antenna 2, flagellum with 60—70 joints (Boeck). Gnathopods 1 and 2 in ♀
slender, feeble, 5th joint in gnathopod 1 longer than 6th, in gnathopod 2
equal to it, 6th joint narrow oblong, palm oblique, much shorter than hind
margin. In ♂ the hand is very often elongate piriform, narrowed at the
hinge of the long weak finger, and the hand is longer than the wrist or
5th joint (Czerniavski). Peraeopods rather slender, spinulose, 2^d joint in
peraeopods 3—5 oval, hind margin slightly serrate, largest in peraeopod 5.
Uropod 3 reaching much beyond the others, rami spinulose, inner rather the
longer, its outer edge bulging at the base. Telson triangular, fully twice
as long as broad, apex very minutely tridentate. Colour variable, brightly
mottled. L. ♀ 6 mm, longer in greater depths.

Arctic Ocean, North-Atlantic, North-Sea and Skagerrak (Norway, France, Great
Britain); Kattegat; Mediterranean.

4. **A. megalops** (O. Sars) 1882 *Halirages m.*, G. O. Sars in: Forh. Selsk. Christian.,
nr. 18 p. 102 t. 5 f. 4 | 1893 *Apherusa m.*, G. O. Sars, Crust. Norway, *v.* 1 p. 443 t. 156
f. 2 | 1893 *Acanthozone fulvocincta* (part.), A. Della Valle in: F. Fl. Neapel, *v.* 20 p. 614.

Pleon segments 1 and 2 each produced into a dorsal tooth. Head,
rostrum obsolete, lateral corners little prominent, transversely truncate, post-
antennal strongly produced, acute, upper edge serrulate. Side-plates 1—3
rather small, rounded quadrangular, slightly serrate below. Pleon segment 2,
postero-lateral corners a little produced, segment 3, postero-lateral corners
quadrate, margin above obliquely truncate, regularly serrate, the deep sinus
above containing a simple projection within. Eyes very large, obliquely oval,
nearly contiguous above, dark brown. Antenna 1 scarcely $^1/_3$ as long as
body, 1st joint as long as 2^d and 3^d combined, apically toothed, flagellum
twice as long as peduncle. Antenna 2 twice as long, ultimate and penultimate joints of peduncle equal, 1st joint of flagellum much the longest.
Gnathopods 1 and 2 nearly as in A. tridentata (p. 305), peraeopods 1—5
rather more slender, uropod 3 also similar but rather more elongate, telson
differing in the apical border, which has a rounded notch between two acute
points. Colour semipellucid, tinged with orange, transversely banded with
light red. L. ♀ 11 mm.

Varangerfjord [North-Norway]. Depth 94—123 m.

5. **A. glacialis** (H. J. Hansen) 1887 *Amphithopsis g.*, H. J. Hansen in: Vid.
Meddel., ser. 4 *v.* 9 p. 137 t. 5 f. 6—6c | 1894 *A. g.*, T. Stebbing in: Bijdr. Dierk., *v.* 17
p. 35 | 1895 *Apherusa g.*, Ohlin in: Acta Univ. Lund., *v.* 31 nr. 6 p. 46 | 1893 *Acanthozone
longimana* (part.), A. Della Valle in: F. Fl. Neapel, *v.* 20 p. 604.

Body slender, back without dorsal teeth, except a pair of minute points
on pleon segment 6. Head, rostrum obsolete, lateral corners rounded. Pleon
segment 3, postero-lateral corners showing only an incipient tooth, margin
above slightly convex. Eyes rather large, rounded, dark with light rim.
Antenna 1 scarcely $\frac{1}{3}$ as long as body, 1st joint as long as 2^d and 3^d combined,
flagellum nearly thrice as long as peduncle. Antenna 2 rather longer, ultimate and
penultimate joints of peduncle subequal. Maxilla 1, inner plate with 9 setae,
outer narrow with 7 or 8 spines, slender, crowded. Maxillipeds, inner plates
as large as outer and reaching as far, outer plates fringed on inner margin
with spine-like setae, 1st joint of palp short, 2^d broad, 4th short, with minute
nail on the blunt apex. Gnathopods 1 and 2, 5th joint long and narrow,
but wider as well as much longer than the sublinear 6th, which has a very
short transverse palm and finger to match. Gnathopod 1 is longer than
gnathopod 2, the extra length being in the 5th and 6th joints. Peraeopods 1—5
not very elongate, 5th considerably the longest, 2^d joint in 3^d—5th pairs
oblong ovate. Uropods 1 and 2, outer ramus the shorter; uropod 3, rami
not greatly longer than peduncle, outer ramus very little shorter than inner.
Telson little longer than broad, slightly narrowed distally, apex entire,
furnished with a pair of setules. Colour yellowish red. L. 7—10 mm.

Arctic Ocean, reaching lat. 76° 30′ N.

6. **A. jurinei** (M.-E.) 1830 *Amphithoe j.*, H. Milne Edwards in: Ann. Sci. nat.,
v. 20 p. 376 | 1840 *A. jurinii*, H. Milne Edwards, Hist. nat. Crust., *v.* 3 p. 30 t. 1 f. 2 |
1891 *Pherusa j.*, A. O. Walker in: Ann. nat. Hist., ser. 6 *v.* 7 p. 421 | 1893 *Apherusa j.*,
G. O. Sars, Crust. Norway, *v.* 1 p. 445 t. 157 f. 1 | 1895 *A. j.*, A. O. Walker in: P.
Liverp. biol. Soc., *v.* 9 p. 305 | 1843 *Amphithoë norvegica*, H. Rathke in: N. Acta Ac.
Leop., *v.* 20₁ p. 83 t. 4 f. 6 | 1859 *Paramphithoe n.*, R. M. Bruzelius in: Svenska Ak.
Handl., n. ser. *v.* 3 nr. 1 p. 77 | 1871 *Calliopius norvegicus*, A. Boeck in: Forh. Selsk.
Christian., 1870 p. 198 | 1876 *C. n.*, A. Boeck, Skand. Arkt. Amphip., *v.* 2 p. 348 t. 22
f. 6 | 1862 *Pherusa fucicola* (part.) (err., non Leach 1813/14!) + *?Gossea microdeutopa*
(Bate), Bate & Westwood. Brit. sess. Crust., *v.* 1 p. 255; p. 277 f. | 1862 *P. f.* (part.) +
Calliope norvegica + *?G. microdentopa*, *G. microdeutopa*, Bate, Cat. Amphip. Brit. Mus.,
p. 145; p. 150; p. 160, 387. 396 t. 29 f. 6 | 1891 *Apherusa*, A. O. Walker in: Ann. nat.
Hist., ser. 6 *v.* 8 p. 83 | 1893 *Acanthozone laeviuscula* (part.), A. Della Valle in: F. Fl.
Neapel, *v.* 20 p. 602.

Back without any dorsal teeth. Head, rostrum very small, lateral corners
rounded, post-antennal acute, neither strongly produced. Side-plates 1—3
rounded quadrangular. Pleon segment 3, postero-lateral angles forming a
short acute projection, the margin above a triangular smooth lobe (Milne
Edwards). Eyes oval reniform, dark brown. Antenna-1 about $\frac{1}{3}$ as long as
body, 1st joint not quite as long as 2^d and 3^d combined, flagellum twice as
long as peduncle, 26—28-jointed. Antenna 2 about once and a half as long as
antenna 1 (Milne Edwards: longer than antenna 1), ultimate and penultimate
joints of peduncle equal, flagellum twice as long as peduncle, 1st joint longest.
Gnathopods 1 and 2 rather feeble, 5th joint shorter than 6th, which is subfusiform, widest at beginning of palm, palm oblique, nearly equal to hind
margin. Peraeopods 1—5 rather stout, finger curved, 2^d joint in 3^d—5th
peraeopods oval, hind margin quite smooth. Uropod 3 extending but little (Milne
Edwards: much) beyond the others, rami narrowly lanceolate. Telson
triangular, not quite twice as long as broad, apex obtusely pointed, having

20*

on each side 2 minute spinules. Colour light straw-yellow, variegated with reddish orange. L. ♀ 8 mm.

North-Atlantic and North-Sea (Norway, France, Great Britain); Kattegat; Mediterranean.

7. A. georgiana (Pfeff.) 1888 *Calliopius georgianus*, Pfeffer in: Jahrb. Hamburg. Anst., v.5 p.116 t.2 f.6 | 1893 *Atylus?*, A. Della Valle in: F. Fl. Neapel. v.20 p.704.

Except in size closely resembling A. jurinei (p. 307). Back of peraeon broad, rounded, hinder segments of pleon compressed, no dorsal teeth. Head, front obtuse angled, lateral corners transversely truncate. Side-plates 1—4 not very large, but size encreasing rapidly from 1st to 4th. Pleon segment 3, postero-lateral corners subquadrate. Eyes tending to rounded oval. Antenna 1, peduncle stout, 1st—3d joints nearly equal in length, 2d and 3d distally somewhat outdrawn, flagellum (in figure) shorter than peduncle, with about 25 joints, alternately weakly and strongly distally outdrawn. Antenna 2 much longer, ultimate joint of peduncle rather longer than penultimate, flagellum twice as long as that of antenna 1, with about 38 joints. No calceoli mentioned. Mandible with prominent molar, palp rather short, 3d joint shorter than 2d, curved, acute at apex. Maxilla 1, inner plate with 3 setae. Maxilla 2, inner plate narrower but rather longer than outer. Gnathopods 1 and 2, 5th joint nearly as long as 6th, subdistally lobed, more strongly in gnathopod 2, 6th joint oval, narrow, palm seemingly weakly defined. Peraeopods 1 and 2 more slender than peraeopods 3—5, otherwise like them, 4th joint in all outdrawn. Uropods 1 and 2, rami small, outer ramus the shorter; uropod 3, peduncle strong, smooth, rami serrate, spinose. reaching back much beyond the other uropods. Colour greenish grey. L. 17 mm.

South-Atlantic (South Georgia). Under stones, in Florideae, etc.

A. barretti (Bate) 1862 *Pherusa b.*, Bate, Cat. Amphip. Brit. Mus., p.146 t.27 f.9 (10 in text) | 1893 *Acanthozone fulvocincta* (part.), A. Della Valle in: F. Fl. Neapel, v. 20 p 943.

L. 12·5 mm.

North-Atlantic.

A. laevis (HasW.) 1880 *Pherusa l.*, Haswell in: P. Linn. Soc. N. S. Wales, v. 4 p. 260 t. 9 f. 4.

L. 9 mm.

South-Pacific (Kiama [New South Wales]).

A. translucens (Chilton) 1884 *Panoplaea t.*, Chilton in: Tr. N. Zealand Inst., v. 16 p. 263 t. 21 f. 3a—c | 1893 *Acanthozone longimana* (part.)?, A. Della Valle in: F. Fl. Neapel, v. 20 p. 619.

Closely related to *Panoploea debilis* (Leptamphopus novaezealandiae, p. 294): Chilton. L. 14 mm.

Lyttelton Harbour [New Zealand].

Gen. **Schraderia** Pfeff.

1888 *Schraderia* (Sp. un.: *S. gracilis*), Pfeffer in: Jahrb. Hamburg. Anst., v.5 p.141. 1 species.

S. gracilis Pfeff. 1888 *S. g.*, Pfeffer in: Jahrb. Hamburg. Anst., v.5 p.141 t.2 f.5 | 1893 *Acanthozone?, Pontogeneia?*, A. Della Valle in: F. Fl. Neapel, v.20 p.904.

No description; only one figure.

South-Atlantic (South Georgia).

23. Fam. Pleustidae

1874 Subfam. *Pleustinae*, Buchholz in: Zweite D. Nordpolarf, *v.*2 p. 333 | 1888 *Pleustidae*, T. Stebbing in: Rep. Voy. Challenger, *v.*29 p. 870 | 1893 *Paramphithoidae*, G. O. Sars, Crust. Norway, *v.*1 p. 343.

Rostrum more or less prominent. Side-plates 5—7 small. Antenna 1 without accessory flagellum, longer than antenna 2, flagellum in both many-jointed. Upper lip unsymmetrically bilobed. Lower lip, inner lobes continuous with outer, little prominent. Mandible, accessory plate sometimes wanting on right, palp well developed, 3^d joint falciform. Maxilla 1, inner plate small, with 1—4 setae. Maxillipeds, inner and outer plates rather small, not strongly armed, palp long. Gnathopods 1 and 2 often alike, subchelate. Peraeopods 3—5, 2^d joint expanded. Branchial vesicles rather small, simple. Uropod 3. rami longer than peduncle, slender, lanceolate, spinulose, outer shorter than inner. Telson small, entire or (very rarely) notched, boat-shaped. Sexual difference very slight.

Marine.

5 accepted genera and 1 doubtful, 18 accepted species and 4 doubtful.

Synopsis of accepted genera:

1 { Mandible, molar feebly deVeloped — **2.**
{ Mandible, molar strongly deVeloped — **3.**

2 { Mandibular palp of moderate size; gnathopods 1
{ and 2 poWerful . . . . 1. Gen. **Pleustes** . . . . p. 309
{ Mandibular palp large; gnathopods 1 and 2. of
{ moderate size . . . . 2. Gen. **Neopleustes** . . p. 311

3 { Maxillipeds, finger of palp strong . . . . . . 3. Gen. **Mesopleustes** . p. 315
{ Maxillipeds, finger of palp slight — **4.**

4 { Mandible, molar compressed; maxilla 1, inner
{ plate With 1 seta . . . . 4. Gen. **Stenopleustes** . p. 316
{ Mandible, molar cylindric; maxilla 1, inner
{ plate With 2 setae . . . . 5. Gen. **Sympleustes** . . p. 317

1. Gen. Pleustes Bate

1858 *Pleustes* (Sp. un.: *P. tuberculata*), Bate in: Ann. nat. Hist., ser. 3 *v.*1 p. 362 1876 *P.* (part.). A. Boeck, Skand. Arkt. Amphip., *v.*2 p. 299 | 1893 *P.*, G. O. Sars, Crust. Norway, *v.*1 p. 343 | 1859 *Paramphithoe* (part.), R. M. Bruzelius in: Svenska Ak. Handl., n. ser. *v.*3 nr. 1 p. 68 | 1871 *P.*, A. Boeck in: Forh. Selsk. Christian., 1870 p. 174.

Coating indurated, with keels or tubercles. Rostrum large, flat. Side-plates 1—4 large, 4th emarginate behind. Antennae 1 and 2 not elongate. Upper lip, lobes subequal, finely hirsute. Mandible, molar feeble, palp of moderate size. Maxilla 1, inner plate with 1 seta, outer with 9 spines. Maxilla 2, inner plate the wider. Maxillipeds, outer plates scarcely reaching beyond 1st joint of elongate palp. Gnathopods 1 and 2 powerful, 5th joint short, produced to a narrow lobe, 6th large, oval, palm oblique, well defined, tufted with spines. Peraeopod 4 little longer than 3^d and 5th little longer than 4th. Telson broadly rounded at apex.

2 species accepted, 3 obscure.

Synopsis of accepted species:

Peraeon tricarinate, carinae medio-dorsal and latero-
marginal . . . . 1. **P. panoplus** . . . p. 310
Peraeon quinquecarinate, With 5 longitudinal roWs of
teeth or tubercles . . . . 2. **P. cataphractus** . p. 310

1. **P. panoplus** (Kröyer) 1838 *Amphithoe panopla*, Kröyer in: Danske Selsk. Afh.,
v. 7 p. 270 t. 2 f. 9 a—i | 1846 *Amphitoe p.*, Krøyer in: Voy. Nord, Crust. t. 11 f. 2 a—x |
1859 *Paramphithoe p.*, R. M. Bruzelius in: Svenska Ak. Handl., n. ser. *v.* 3 nr. 1 p. 69 |
1862 *Pleustes panoplus*, Bate, Cat. Amphip. Brit. Mus., p. 63 t. 9 f. 9 | 1874 *P. p.*,
Buchholz in: Zweite D. Nordpolarf., *v.* 2 p. 334 Crust. t. 6 | 1893 *P. p.*, G. O. Sars. Crust.
NorWay, *v.* 1 p. 344 t. 121 | 1894 *P. panopla*, T. Stebbing in: Bijdr. Dierk., *v.* 17 p. 28
(synonymy) | 1893 *Acanthozone p.* (part.), A. Della Valle in: F. Fl. Neapel, *v.* 20 p. 607
t. 59 f. 19 | 1867 *Amphithonotus cataphractus* (err., non Stimpson 1853!), Packard in:
Mem. Boston Soc., *v.* 1 p. 298.

Back dorsally carinate, strongly on peraeon, which also has lateral
margins carinate; pleon segments 1—4 and 6 with a pair of dorsal projections,
1st—3d with a tubercle on each side below these. Rostrum large, obtuse or
subacute, slightly hollowed dorsally, carinate below; lateral corners of head
subquadrate. Side-plate 4 acute angled below emargination, 5th and 6th
with quadrate hind lobe. Pleon segment 3 with postero-lateral angles acute,
slightly produced. Eyes rather small, rounded oval, prominent, dark red.
Antenna 1 short, 1st joint about as long as 2d and 3d combined, flagellum
scarcely thrice as long as peduncle, with 26—45 joints. Antenna 2 still
shorter, ultimate and penultimate joints of peduncle subequal, flagellum
subequal to peduncle, with 20—26 joints. Gnathopod 1, palm much longer
than hind margin of 6th joint, finger only reaching 1st of 3 transverse rows
of palmar spines. Gnathopod 2 more powerful, narrow lobe of 5th joint
more projecting, 6th much more dilated at junction of palm with hind
margin (Sars appears to reckon the palms not from this angle, but from the
point reached by the finger). Peraeopods 3—5, 2d joint oblong quadrangular.
Uropod 3 not very long, inner ramus not twice length of outer. Telson
at base constricted and having on the upper face a triangular prominence,
apical margin broadly rounded. Colour dark brown. or whitish, or variegated
with shades of brown. L. ♀ reaching 18—27 mm.

Arctic Ocean (Widely distributed, depth 5—113 m); North-Atlantic (West-NorWay
from Bergen northWards, depth 19—56 m; South of Halifax [NoVa Scotia], depth 155 m).

2. **P. cataphractus** (Stimps.) 1853 *Amphithonotus c.*, Stimpson in: Smithson.
Contr., *v.* 6 nr. 5 p. 52 | 1876 *Paramphithoe cataphracta*, S. I. Smith & Harger in: Tr.
Connect. Ac., *v* 3 p. 31 | 1876 *Tritropis cataphractus*, A. Boeck in: Skand. Arkt. Amphip.,
v. 2 p. 510 | 1888 *Rhachotropis c.*, T. Stebbing in: Rep. Voy. Challenger, *v.* 29 p. 278, 1720.

Body robust. Rostrum very large, elongate triangular. pointed,
curving downwards, concave above and with a sharp median ridge below.
One strong median dorsal carina commencing on peraeon segment 1,
becoming strongly dentate on the last peraeon segment, and ceasing on
pleon segment 2; the next 2 carinae (proceeding outwards) are developed
in the form of strong teeth on peraeon segments 6 and 7 and all pleon
segments, being spine-like on pleon segment 2, and almost lamelliform on
the last 4; the next carinae are sharp ridges, extending along the bases
of the side-plates, and slightly continued on pleon segments 1 and 2; and
the last or outer carinae are very short, extending only along the bases of

uropods 1—3. Side-plates large, angular. Eyes very large, rounded, prominent, yellowish or vermilion, with a black dot in the middle. Antennae 1 and 2 short, slender, subequal. Gnathopods 1 and 2, 5th joint slenderly produced, 6th large, ovate, dentate below, finger about $^2/_3$ length of 6th joint. Feraeopods 1—5 slender, peraeopods 3—5, 2^d joint but slightly expanded. Uropods 2 and 3, outer ramus shorter than inner. Telson subquadrate. Colour very variable, generally dark reddish or brown, variegated and mottled with white, sometimes a uniform deep purple or pure white. When disturbed, rolls itself up, as if feigning death. L. 12·5 mm.

Fundy Bay (Grand Manan). Depth 18 m.

P. medius (Goës) 1866 *Paramphithoë media*, Goës in: Öfv. Ak. Förh., *v.* 22 p. 523 t. 38 f. 13 | 1871 *P. m.*, A. Boeck in: Forb. Selsk. Christian., 1870 p. 176 | 1876 *Pleustes medius*, A. Boeck, Skand. Arkt. Amphip., *v.* 2 p. 302 | 1893 *Acanthozone pulchella* (part.)?, A. Della Valle in: F. Fl. Neapel, *v.* 20 p. 605.

L. about 8 mm.

Arctic Ocean (Spitzbergen). Depth 38 m.

P. occidentalis (Stimps.) 1864 *Amphithonotus o.*, Stimpson in: P. Ac. Philad., p. 158.

Closely allied to P. panoplus and P. cataphractus. L. 19 mm.

North-Pacific (North-America).

P. tuberculatus Bate 1858 *P. tuberculata*, Bate in: Ann. nat. Hist., ser. 3 *v.* 1 p. 362 | 1893 *P. tuberculatus*, G. O. Sars, Crust. Norway, *v.* 1 p. 344 | 1871 *Paramphithoe panopla* (err., non *Amphithoe p.* Kröyer 1838!), A. Boeck in: Forb. Selsk. Christian., 1870 p. 176 | 1876 *Pleustes panoplus*, A. Boeck, Skand. Arkt. Amphip., *v.* 2 p. 303 | ? 1894 *P. panopla*, T. Stebbing in: Bijdr. Dierk., *v.* 17 p. 28 | 1893 *Acanthozone p.* (part.), A. Della Valle in: F. Fl. Neapel, *v.* 20 p. 607 | ? 1858 *Amphithoë panoploides*, M. Sars in: Forb. Selsk. Christian., p. 138.

Perhaps only an aged form of P. cataphractus. L. 21 mm.

Arctic Ocean.

2. Gen. **Neopleustes** Stebb.*)

1859 *Paramphithoe* (part.), R. M. Bruzelius in: Svenska Ak. Handl., n. ser. *v.* 3 nr. 1 p. 68 | 1866 *Paramphithoë* (part.), Goës in: Öfv. Ak. Förh., *v.* 22 p. 523 | 1893 *P.*, G. O. Sars, Crust. Norway, *v.* 1 p. 346 | 1861 *Amphithopsis* (part.) + *P.* (part.), A. Boeck in: Forb. Skand. Naturf., Møde 8 p. 661, 662.

Body slender, not indurated. Head more or less produced over antenna 1, post-antennal corners usually projecting acutely. Side-plates not powerfully developed. Antenna 1 usually long and much longer than antenna 2. Upper lip unsymmetrically bilobed, incision oblique. Mandible, molar weak, palp very large. Maxillae 1 and 2 nearly as in Pleustes. Maxillipeds differing by joint 3 of palp distally attenuated. 4th spiniform. Gnathopods 1 and 2 feeble to moderately strong. Peraeopods more or less elongated. Telson hollowed above, carinate below.

7 species.

---

*) Gen. nov. νέος, new, and *Pleustes*. Type is *Amphitoe pulchella* Kröyer (1846).

Synopsis of species:

1 { Dorsum with teeth — **2.**
  { Dorsum without teeth — **3.**

2 { Dorsum with 5 or more teeth . . . . 1. **N. pulchellus**. . p. 312
  { Dorsum with 3 teeth . . . . 2. **N. boeckii** . . . p. 312
  { Dorsum with 2 teeth . . . . 3. **N. bicuspis**. . . p. 313
  { Dorsum with 1 tooth . . . . 4. **N. monocuspis** . p. 313

3 { Antenna 1 short . . . . 5. **N. brevicornis** . p. 313
  { Antenna 1 elongate — **4.**

4 { Gnathopods 1 and 2 feeble . . . . 6. **N. assimilis** . . p. 314
  { Gnathopods 1 and 2 moderately strong . . . . . 7. **N. bairdi** . . . . p. 314

1. **N. pulchellus** (Krøyer) 1846 *Amphitoe pulchella.*, Krøyer in: Voy. Nord, Crust. t. 10 f. 2 a—r | 1859 *Paramphithoe p.*, R. M. Bruzelius in: Svenska Ak. Handl., n. ser. *v.* 3 nr. 1 p. 70 | 1887 *P. p.*, H. J. Hansen in: Vid. Meddel., ser. 4 *v.* 9 p. 119 t. 5 f. 2—2 b | 1893 *P. p.*, G. O. Sars, Crust. Norway, *v.* 1 p. 346 t. 122 f. 1 | 1862 *Pherusa p.*, Bate, Cat. Amphip. Brit. Mus., p. 143 t. 27 f. 5 | 1876 *Pleustes pulchellus* (part.), A. Boeck, Skand. Arkt. Amphip., *v.* 2 p. 306 (not fig.) | 1893 *Acanthozone pulchella* (part.), A. Della Valle in: F. Fl. Neapel, *v.* 20 p. 605 t. 59 f. 18 | 1876 *Pleustes euacanthus*, G. O. Sars in: Arch. Naturv. Kristian., *v.* 2 p. 256 | 1885 *Paramphithoe euacantha*, G. O. Sars in: Norske Nordhavs-Exp., *v.* 6 Crust. I p. 168 t. 14 f. 3.

Sharp compressed teeth on peraeon segments 5—7, sometimes also on 4th and others; similar teeth on pleon segments 1 and 2, an upturned lamellar expansion on segment 3. Head produced to rather broad blunt rostrum, post-antennal corners spiniform. Side-plate 1 tapering to blunt point, a little serrate behind, 2d more rounded apically, also serrate, hind margin of 4th very oblique, acute below emargination. Pleon segment 3, postero-lateral angles acutely produced, slightly upturned, hind margin straight, finely serrulate. Eyes moderate, irregular oval, dark red. Antenna 1 very long, 1st joint as long as 2d and 3d combined, flagellum 4 to 5 times as long as peduncle, with 90 joints, 1st longer than 3d joint of peduncle. Antenna 2 scarcely $\frac{1}{2}$ as long as antenna 1, ultimate and penultimate joints of peduncle equal, flagellum about twice as long as peduncle. Gnathopods 1 and 2 rather feeble, 5th joint rather long, little expanded, 6th joint longer, oblong oval, widening distally a little. palm shorter than hind margin, forming with it a very obtuse angle. Gnathopod 2 the stronger. Feraeo-pods 1—5, finger strong, curved; peraeopods 3—5, 2d joint oblong oval, serrulate behind. Uropod 3, inner ramus long, outer more than $\frac{1}{2}$ as long, much longer than peduncle. Telson very small, oval, with small dentiform projection on each side near apex. Colour whitish, pellucid, mottled with dark brown. L. ♀ reaching 17 mm.

Arctic Ocean, North-Atlantic, North-Sea and Skagerrak (Greenland, Iceland, Spitzbergen, Norway, Bohuslän). Depth 75—282 m.

2. **N. boeckii** (H. J. Hansen) 1876 *Pleustes pulchellus* (part.) (err., non *Amphitoe pulchella* Krøyer 1846!), A. Boeck, Skand. Arkt. Amphip., *v.* 2 t. 23 f. 1 (not text) | 1893 *Acanthozone pulchella* (part.), A. Della Valle in: F. Fl. Neapel, *v.* 20 p. 605 | 1887 *Paramphithoë boeckii*, H. J. Hansen in: Vid. Meddel., n. ser. 4 *v.* 9 p. 121 t. 5 f. 3—3 b | 1893 *P. b.*, G. O. Sars, Crust. Norway, *v.* 1 p. 348 t. 122 f. 2.

Sharp compressed teeth only on peraeon segment 7 and pleon segments 1 and 2. Head produced to strongly carinate acute rostrum, post-antennal corners little produced. Side-plates not serrate, 1st apically rounded. Pleon segment 3, postero-lateral angles acutely produced, hind margin convex, smooth. Eyes very large, rounded triangular. Antenna 1, 1st joint of flagellum shorter than 3d of peduncle. Gnathopods 1 and 2 less slender than in N. pulchellus, 6th joint more regularly oblong oval. Peraeopods 1—5 less robust; peraeopods 3—5, 2d joint with hind margin smooth. Uropod 3, outer ramus little longer than $\frac{1}{2}$ inner. Telson about twice as long as broad, apex without dentiform projections. Colour pale red (Holböll). In other points agreeing with N. pulchellus. L. ♀ scarcely more than 8 mm.

Arctic Ocean (Greenland). Depth 9—113 m.

3. **N. bicuspis** (Kröyer) 1838 *Amphithoe b.*, (Reinhardt in MS.) Kröyer in: Danske Selsk. Afh., *v.*7 p.273 t.2 f.10a—e | 1859 *Paramphithoe b.*, R.M. Bruzelius in: Svenska Ak. Handl., n. ser. *v.*3 nr.1 p.73 | 1887 *P. b.*, H.J. Hansen in: Vid. Meddel., ser.4 *v.*9 p.122 | 1893 *P. b.*, G.O. Sars, Crust. Norway, *v.*1 p.349 t.123 f.1 | 1895 *P. b.* (part), A.O. Walker in: P. Liverp. biol. Soc., *v.*9 p.303 | 1861 *Amphithopsis b.*, A. Boeck in: Forh. Skand. Naturf., Møde 8 p.662 | 1876 *Pleustes b.*, A. Boeck, Skand. Arkt. Amphip, *v.*2 p.308 | 1893 *Acanthozone pulchella* (part.)?, A. Della Valle in: F. Fk. Neapel. *v.*20 p.605.

Not carinate, but dorsally toothed on pleon segments 1 and 2. Rostrum small, obtuse, post-antennal corners little produced. Side-plate 1 scarcely expanded distally. Pleon segment 3, postero-lateral angles forming a small tooth with sinus above. Eyes rather small, oval triangular, dark red. Antenna 1, 1st joint longer than 2d and 3d combined, flagellum 4 or 5 times as long as peduncle, 90-jointed. Antenna 2 much shorter, ultimate joint of peduncle scarcely as long as penultimate, flagellum about twice as long as peduncle, 60-jointed. Maxilla 1 with 1 seta (Bruzelius; Boeck: 3 setae) on inner plate. Gnathopods 1 and 2, 5th joint short, lobe setose, 6th rather large, as long as 2d joint, tapering a little, palm spinose, leaving scarcely any hind margin. Peraeopods 1—5 rather slender, spinulose; peraeopods 3—5 subequal. 2d joint oval, serrulate behind. Uropod 3, outer ramus scarcely more than half as long as inner, longer than peduncle. Telson nearly twice as long as broad, apex evenly rounded. Colour whitish, tinged with yellowish, mottled with reddish brown spots, but sometimes almost pure white. L. ♀ 12—14 mm.

Arctic Ocean, North-Atlantic, North-Sea and Skagerrak (Greenland, Spitzbergen, Iceland, Labrador, Norway, Bohuslän, France, Great Britain); Kattegat. Depth 6—113 m.

4. **N. monocuspis** (O Sars) 1893 *Paramphithoë m.*, G.O. Sars, Crust. Norway, *v.*1 p.351 t.123 f.2 | 1895 *P. bicuspis* (part.). A.O. Walker in: P. Liverp. biol. Soc., *v.*9 p.303.

Dorsal tooth only on pleon segment 2. Rostrum small, obtuse, post-antennal corners sharply produced. Side-plate 1 broad, distinctly expanded distally, with denticle at hind corner. Pleon segment 3, produced angles without sinus above. Eyes large, almost regularly oval, dark red. Antenna 1, 1st joint subequal to 2d and 3d combined. Gnathopods 1 and 2 stouter than in N. bicuspis. Peraeopods 1—5 rather strong, 4th longer than 3d, 5th than 4th. Uropod 3, outer ramus much longer than $\frac{1}{2}$ inner. Telson oval, not nearly twice as long as broad. Colour whitish, sparingly mottled, except in front of head and dark peduncles of antennae. L. ♀ 11 mm.

Arctic Ocean (Greenland. Hammerfest [Finmark]). Depth 56—94 m.

5. **N. brevicornis** (O. Sars) 1882 *Paramphithoë b.*, G.O. Sars in: Forh. Selsk. Christian., nr.18 p.98 t.4 f.11, 11a | 1893 *P. b.*, G.O. Sars, Crust. Norway, *v.*1 p.353 t.124 f.2 | ?1887 *P. gracilis*, H.J. Hansen in: Vid. Meddel., ser.4 *v.*9 p.124.

Stout, compact, no dorsal projections. Rostrum very slight, post-antennal corners little produced. Side-plates 1—4 deep, 1st rather expanded distally, 1st—3^d with denticle at hind corner, 4th subquadrate below emargination. Pleon segment 3, postero-lateral corners subquadrate, point obtuse. Eyes small, rounded, dark red. .Antennae 1 and 2 unusually short. Antenna 1, 1st joint rather thick, as long as 2^d and 3^d combined, flagellum not twice as long as peduncle, 12-jointed. Antenna 2 a little shorter, ultimate and penultimate joints of peduncle subequal, together as long as flagellum. Gnathopods 1 and 2 moderately strong, closely alike, 5th joint very short, lobe linguiform, 6th very large, as long as 2^d joint, widest at palm, which is as long as hind margin, convex, defined by projecting angle and palmar spines. Peraeopods 1—5 moderately long and spinulose, very slender. Peraeopods 3—5, 2^d joint oval, largest in peraeopod 5 with hind margin nearly straight. Uropod 3 comparatively short, outer ramus $^2/_3$ as long as inner. Telson oblong triangular, not nearly twice as long as broad at base, apex obtuse. Colour whitish, more or less darkened with dark brownish violet. L. ♀ scarcely 4 mm.

Arctic Ocean (North-Norway; Greenland?).

6. **N. assimilis** (O. Sars) 1882 *Paramphithoë a.*, G. O. Sars in: Forb. Selsk. Christian., nr.18 p. 99 t. 5 f. 1, 1a | 1887 *P. a.*, H. J. Hansen in: Vid. Meddel., ser. 4 v. 9 p. 124 | 1893 *P. a.*, G. O. Sars, Crust. Norway, v. 1 p. 352 t. 124 f. 1 | 1893 *Acanthozone pulchella* (part.)?, A. Della Valle in: F. Fl. Neapel, v. 20 p. 605.

With no dorsal projections. Rostrum short, obtuse, post-antennal corners little produced. Side-plate 1 slightly expanded distally, 2^d and 3^d large, 1st—3^d with denticle at hind corner, 4th acute below emargination, narrow distally. Pleon segment 3, postero-lateral corners acutely produced, but not into a tooth. Eyes rather large, rounded triangular. Antenna 1 very long, 1st joint longer than 2^d and 3^d combined, flagellum more than 4 times as long as peduncle. Antenna 2 much shorter, ultimate joint of peduncle longer than penultimate, flagellum about twice as long as peduncle. Gnathopods 1 and 2 rather feeble, 5th joint rather long and little widened, 6th longer than 5th, not nearly so long as 2^d, widest at palm, which is not nearly so long as spinose hind margin. Peraeopods 1—5 long and slender. Peraeopods 3—5. 2^d joint oval, serrulate behind, in peraeopod 5 its hind margin very convex. Uropod 3, rami spinose, outer longer than peduncle, little over half as long as inner. Telson fully twice as long as broad, apex evenly rounded. L. ♀ 8 mm.

Arctic Ocean and North-Atlantic (West-Norway; Greenland, depth 19—113 m).

7. **N. bairdi** (Boeck) 1872 *Paramphithoë b.*, A. Boeck in: Forh. Selsk. Christian., 1871 p. 45, 50 t. 1 f. 3.

Dorsum without keel or teeth. Head a little outdrawn, forming no proper rostrum; post-antennal corners blunt. Side-plates 1—4 large, distally rounded, feebly serrulate on lower hind margin. Pleon segment 3, postero-lateral corners subquadrate. Eyes moderately large, oval, black. Antenna 1, 1st joint stout, longer than 2^d and 3^d combined, flagellum about thrice as long as peduncle, 40—45-jointed. Antenna 2 much shorter, ultimate joint of peduncle longer than penultimate, flagellum much longer than peduncle. about 24-jointed. Mandibular palp, 3^d joint much longer than 2^d. Maxilla 2, inner and outer plates small. Maxillipeds, outer plates not reaching middle of palp's 2^d joint, palp's 3^d joint tapering distally, finger spine-like. Gnathopods 1 and 2, 5th joint short, triangular, lobe rounded, 6th joint very large, oval, widest at junction of subequal palm and hind margin, larger in gnathopod 2 than in gnathopod 1. Peraeopods 3—5, 2^d joint large, feebly serrate, 4th joint produced downward. Uropod 3, peduncle very short, rami spinose, outer about $^3/_4$ as long as inner. Telson oval. L. ?.

North-Pacific (California).

3. Gen. **Mesopleustes** Stebb.

1899 *Mesopleustes* (Sp. typ.: *Pleustes abyssorum*), T. Stebbing in: Ann. nat. Hist., ser. 7 v. 4 p. 209.

Coating indurated, carinate. Rostrum large. Side-plates 1—4 distally narrowed, 5th and 6th with hind lobe deeper than front. Upper lip with small oblique incision. Mandible, accessory plate on both left and right, molar prominent, strong, oval, 3^d joint of palp longer than 2^d. Maxilla 1, inner plate with 2 short and 2 long plumose setae, outer plate with 10 spines. Maxillipeds, inner plates with 3 spine-teeth in a group at apical angle, outer plates scarcely reaching beyond 1st joint of palp, inner margin unarmed, apical with 6 close-set spines, 1st and 2^d joints of palp equal, 3^d rather shorter, as long as strong finger. Gnathopods 1 and 2 strong, 5th joint cup-like, 6th broad oblong oval, palm well defined. Gnathopod 2 larger than gnathopod 1. Peraeopods 1—5 robust, subequal. Telson subrotund.

1 species.

1. M. abyssorum (Stebb.) 1888 *Pleustes a.*, T. Stebbing in: Rep. Voy. Challenger, v. 29 p. 872 t. 67 | 1893 *Acanthozone a.*, A. Della Valle in: F. Fl. Neapel, v. 20 p. 609 t. 59 f. 21 | 1899 *Pleustes a.*, *Mesopleustes sp. typ.*, T. Stebbing in: Ann. nat. Hist., ser. 7 v. 4 p. 209.

In general appearence resembling Pleustes panoplus (p. 310). Peraeon and pleon segments all carinate except 4th of pleon; segments imbricate, pleon segment 3 with erect dorsal tooth. Rostrum long, narrow, carinate below, lateral and post-antennal corners little projecting. Side-plate 1 narrowly produced forward, 2 denticles on hind margin, 2^d—4th successively encreasing in depth, vertically ridged, 4th with very oblique hind margin below emargination, 5th with distally narrowed hind lobe. Pleon segment 3, postero-lateral corners produced into a small tooth. Eyes small, oval. Antenna 1, 1st joint a little longer than 2^d, 3^d more than $^1/_2$ as long as 2^d, flagellum with 44 joints, 1st longest. Antenna 2 much shorter, ultimate joint of peduncle rather longer than penultimate, flagellum with 24 joints, 1st longest. Gnathopods 1 and 2 spinose, 5th joint as broad as long; gnathopod 1, 6th joint with palm divided between 2 broad shallow emarginations, finger reaching beyond them. Gnathopod 2 much larger, palm with deep cavity between its blunt end and a large tooth or process near the finger hinge. Feraeopods 1—5 nearly as in Sympleustes latipes (p. 317), but 2^d joint oblong, hind margin sinuous, with much produced lobe. Uropod 3, outer ramus little longer than peduncle, only half length of inner. Telson short, as broad as long. L. about 17 mm (if fully extended).

Southern Indian Ocean (near Marion Island). Depth 3013 m.

4. Gen. **Stenopleustes** O. Sars

1893 *Stenopleustes*, G. O. Sars, Crust. Norway, *v.* 1 p. 354.

Body slender, smooth, thin-skinned. Rostrum and post-antennal corners little produced. Side-plates not very large. Antenna 1 slender, much longer than antenna 2. Lips nearly as in Neopleustes (p. 311). Mandible, molar powerful, compressed, palp moderate. Maxilla 1, inner plate with 1 seta. Maxillipeds, inner plates rather broad, outer plates small, with a few hairs, 3^d joint of palp apically much produced. Gnathopods 1 and 2 imperfectly subchelate, not strong. Peraeopods 1—5 rather slender. Telson nearly as in Neopleustes.

2 species.

Synopsis of species:

Pleon smooth . . . . 1. **S. malmgreni** . p. 316
Pleon nodiferous . . . . 2. **S. nodifer** . . p. 316

1. **S. malmgreni** (Boeck) 1871 *Amphithopsis m.*, A. Boeck in: Forb. Selsk. Christian., 1870 p. 199 | 1876 *A. m.*, A. Boeck, Skand Arkt. Amphip., *v.* 2 p. 350 t. 23 f. 7 | 1893 *Stenopleustes m.*, G. O. Sars, Crust. Norway, *v.* 1 p. 355 t. 125 f. 1 | 1893 *Acanthozone longicaudata* (part.), A. Della Valle in: F. Fl. Neapel. *v.* 20 p. 605.

Pleon without nodules. Head, rostrum short, blunt, lateral corners angular. Pleon segment 3, postero-lateral corners quadrate. Eyes very large, reniform, light red. Antenna 1, 1st joint subequal to 2^d and 3^d combined, flagellum nearly 6 times as long as peduncle, 1st joint elongate. Antenna 2, ultimate joint of peduncle longer than penultimate, flagellum longer than peduncle. Gnathopod 1, 5th joint rather longer than oblong oval 6th, palm very oblique, spinose. Gnathopod 2 rather larger, 5th joint rather shorter than 6th. Peraeopods 3—5, 2^d joint large oval. Uropod 3, inner ramus 3 times as long as peduncle, outer much the shorter, both spinose. Telson triangular oval. Colour whitish, sometimes faintly carneous. L. ♀ 7 mm.

Arctic Ocean, North-Atlantic, North-Sea and Skagerrak (Norway). Depth 150—282 m.

2. **S. nodifer** (O. Sars) 1882 *Amphithopsis nodifera*, G. O. Sars in: Forb. Selsk. Christian., nr. 18 p. 103 t. 5 f. 6a—b | 1893 *Stenopleustes nodifer*, G. O. Sars, Crust. Norway, *v.* 1 p. 356 t. 125 f. 2 | 1895 *S. n.*, A. O. Walker in: P. Liverp. biol. Soc , *v.* 9 p. 303 | 1893 *Acanthozone nodifera*, A. Della Valle in: F. Fl. Neapel, *v.* 20 p. 604 t. 59 f. 15.

Pleon segments 1 and 2 ending dorsally in pair of juxtaposed nodiform projections (often difficult to make out). Head, rostrum a little produced, lateral corners rounded. Pleon segment 3, postero-lateral corners somewhat produced. Eyes large, reniform, dark red. Antenna 1, 1st joint rather longer than 2^d and 3^d combined. Gnathopods 1 and 2, palm rather shorter than in S. malmgreni. Telson longer, triangular with narrowly rounded apex. Other points closely agreeing with S. malmgreni. Colour whitish, pellucid, speckled yellowish or brownish; peduncle of antenna 1 and uropods 1—3 generally dark brown (Sars); limbs and specially peraeon segments 4—6 speckled with dark red (Walker). L. ♀ 3—5 mm.

North-Atlantic, North-Sea and Skagerrak (Norway, depth 56–188 m; Great Britain, depth 57 m).

5. Gen. **Sympleustes** Stebb.

1861 *Amphithopsis* (part.), A. Boeck in: Forh. Skand. Naturf., Møde 8 p. 661 |
1899 *Sympleustes*, T. Stebbing in: Ann. nat. Hist., ser. 7 v. 4 p. 209.

Not very slender. Rostrum small, post-antennal corners more or less projecting. Side-plates moderate. Upper lip as in Neopleustes (p. 311). Molar of mandible strong, not compressed but cylindrical, palp rather large. Maxilla 1, inner plate with 2 setae. Other mouth-parts nearly as in Stenopleustes, but palp of maxillipeds variable, finger slight. Gnathopod 2 usually much stronger than gnathopod 1, and more distinctly subchelate. Peraeopods 1—5, stoutness varying. Telson entire or (in S. megacheir) notched.

6 species.

Synopsis of species:

1 { Telson With notched apex . . . . 1. **S. megacheir** . . . p. 317
  { Telson With apex entire — **2.**

2 { Back not entirely smooth — **3.**
  { Back entirely smooth — **4.**

3 { With obtuse dorsal projection . . . . 2. **S. latipes** . . . . p. 317
  { With partial carina . . . . 3. **S. grandimanus** . p. 318

4 { Antenna 1, 1st joint with spiniform process . . . 4. **S. glaber** . . . . p. 318
  { Antenna 1, 1st joint Without spiniform process — **5.**

5 { Gnathopods 1 and 2, palm not incised . . . . . 5. **S. pulchellus** . . . p. 319
  { Gnathopods 1 and 2, palm deeply incised . . . 6. **S. olrikii** . . . . . p. 319

1. **S. megacheir** (A. Walker) 1897 *Parapleustes m.*, A. O. Walker in: J. Linn. Soc., v. 26 p. 230 t. 18 f. 4—4 c.

Pleon segment 2 having a small dorsal tooth. Side-plates 1 and 2 small, front angle of 1st acute. Pleon segment 3, postero-lateral corners obtusely quadrate. Eyes wanting. Antenna 1 about ⅔ as long as body, 1st joint as long as 2d and 3d combined. Antenna 2 shorter, ultimate joint of peduncle rather shorter than penultimate. Maxillipeds strong, finger longer than 3d joint of palp; other mouth-parts not described. Gnathopod 1, 2d joint not longer than 6th, 5th shorter than ovate 6th, both setose on hind margin, finger with distal half serrate. Gnathopod 2, 5th joint short, cup-shaped, 6th very large, the long oblique palm having three crenate lobes and two sinuses, and being defined by a small tooth. Peraeopods 1—5 and uropod 3 as in S. pulchellus (p. 319). Telson spoon-shaped, apically notched for one quarter of its length. L. 8 mm.

North-Atlantic (South-West of Ireland). Depth 1371 m.

2. **S. latipes** (Sars) 1858 *Amphithoë l.*, M. Sars_ in: Forh. Selsk. Christian., p. 139 | 1871 *Amphithopsis l.*, A. Boeck in: Forh. Selsk. Christian., 1870 p. 200 | 1876 *A. l.*, A. Boeck, Skand. Arkt. Amphip., v. 2 p. 355 t. 22 f. 4 | 1887 *A. l.*, H. J. Hansen in: Vid. Meddel., ser. 4 v. 9 p. 135 t. 5 f. 4 | 1893 *Parapleustes l.*, G. O. Sars, Crust. NorWay. v. 1 p. 360 t. 127 | 1897 *P. l.*, J. Bonnier in: Ann. Univ. Lyon, v. 26 p. 645 t. 38 f. 3 | 1893 *Acanthozone l.* (part.), A. Della Valle in: F. Fl. Neapel, v. 20 p. 608 t. 59 f. 20 | 1899 *Amphithoe l.*, *Sympleustes* (part.), T. Stebbing in: Ann. nat. Hist., ser. 7 v. 4 p. 209 | 1862 *Calliope ossiani* (Bate) + *C. fingalli*, Bate & WestWood, Brit. sess. Crust., v. 1 p. 261 f.; p. 263 f. | 1862 *C. o.* + *C. f.*, Bate, Cat. Amphip. Brit. Mus., p. 149 t 28 f. 3; p. 377 | 1869 *Calliopius o.* + *C. f.*, A. M. Norman in: Rep. Brit. Ass., Meet. 38 p. 280, 281.

Robust; peraeon segment 7 and pleon segments 1—3 each with obtuse dorsal projection, that in pleon segment 3 compressed gibbous. Head, rostrum small, lateral corners emarginate, post-antennal corners little produced. Side-plates 1—3 without denticle, 1st acute at lower front corner, 4th deep, narrow below, 5th and 6th with hind lobe much deeper than front. Pleon segment 3, postero-lateral corners subquadrate, obscurely produced. Eyes rather large, oblong reniform, dark red. Antenna 1, 1st joint longer than 2[d] and 3[d] combined, apically somewhat produced, flagellum about thrice as long as peduncle, 1st joint longer than 3[d] of peduncle, accessory flagellum rudimentary (Bonnier). Antenna 2, ultimate joint of peduncle longer than penultimate, flagellum little longer than peduncle. Upper lip very unsymmetrical. Mandible, 3[d] joint of palp much longer than 2[d]. Maxillipeds, 3[d] joint of palp fully as long as 2[d], apically acutely produced. Gnathopod 1 rather feeble, 5th joint distally broad, as long as 6th, palm well defined, equal to hind margin and almost at right angles to it, so that the joint is sub-triangular. Gnathopod 2 powerful, 5th joint short, cup-like, 6th very large, expanded distally, palm excavate between a small projection and a broadly truncate lobe armed with 5 stout spines, the point of the finger closing into a groove between these and a similarly armed surface ridge. Peraeopods 3—5 very strong, 2[d] joint oval, hind margin quite smooth, nearly straight, 4th and 5th expanded, produced downward; all peraeopods spinose, with finger strong. Uropod 3, outer ramus $^2/_3$ as long as inner. Telson distinctly boat-shaped, very small, apex evenly but not broadly rounded. Colour whitish, transversely banded with patches of dark brown, found also on peduncle of antennae and on legs. L. ♀ reaching 12 mm.

Arctic Ocean, North-Atlantic, North-Sea and Skagerrak (Greenland, Norway, Great Britain), depth 56—330 m; Bay of Biscay, depth 1410 m.

3. **S. grandimanus** (Chevreux) 1887 *Amphithopsis grandimana*, Chevreux in: Bull. Soc. zool. France, *v.* 12 p. 570 | 1899 *A. g., Sympleustes* (part.), T. Stebbing in: Ann. nat. Hist., ser. 7 *v.* 4 p. 209 | 1893 *Acanthozone latipes* (part.), A. Della Valle in: F. Fl. Neapel, *v.* 20 p. 608.

Said to be near to S. pulchellus. Compressed, pleon segments 1—3 slightly carinate. Head, rostrum very short, lateral corners rounded. Side-plates not deep. Pleon segment 3, postero-lateral corners almost quadrate. Eyes large, oval. Antenna 1 long, 2[d] joint almost as long as 1st, 3[d] half as long as 2[d], flagellum with 53 joints, 1st long. Antenna 2, $^2/_3$ as long as antenna 1, ultimate joint of peduncle a little shorter than penultimate, flagellum 29-jointed. Mandible, 3[d] joint of palp very long. Gnathopod 1, 5th joint rather shorter than the broadly oval 6th. Gnathopod 2, 6th joint very large, elongate oval, front margin with strong sharp apical tooth, palm slightly crenulate, spinose, well defined by a strong tooth. Peraeopods 1—5 large, robust; peraeopods 3—5, 2[d] joint broadly oval, slightly produced downward, smooth behind. Uropods 1—3 elongate, outer ramus only a little shorter than inner. Telson very short, rounded. L. 8 mm.

North-Atlantic (Cape Finisterre). Depth 510. m.

4. **S. glaber** (Boeck) 1861 *Amphithopsis g.*, A. Boeck in: Forh. Skand. Naturf., Møde 8 p 662 | 1871 *Paramphithoë glabra*, A. Boeck in: Forh. Selsk. Christian., 1870 p. 175 | 1876 *Pleustes glaber*, A. Boeck, Skand. Arkt. Amphip., *v.* 2 p. 300 t. 21 f. 1 | 1893 *Parapleustes g.*, G. O. Sars, Crust. Norway, *v.* 1 p. 358 t. 126 f. 1 | 1899 *Amphithopsis g., Sympleustes* (part.), T. Stebbing in: Ann. nat. Hist., ser. 7 *v.* 4 p. 209 | 1866 *Paramphithoë exigua*, Goës in: Öfv. Ak. Förh., *v.* 22 p. 523 t. 38 f. 12 | 1893 *Acanthozone pulchella* (part.)?, A. Della Valle in: F. Fl. Neapel, *v.* 20 p. 605.

Rather slender, smooth. Head, rostrum not large but distinct, lateral corners acute, post-antennal corners projecting spiniform. Side-plates 1—3 with tooth at lower hind corner, 5th with hind lobe little deeper than front. Pleon segment 3, postero-lateral angles forming a little recurved tooth. Eyes moderate, rounded oval, dark red. Antenna 1, 1st joint much longer than 2[d] and 3[d] combined, with spiniform apical process, flagellum more than thrice as long as peduncle, 1st joint long. Antenna 2 much shorter, ultimate joint of peduncle longer than penultimate, flagellum not twice as long as peduncle. Upper lip not very unsymmetrical. Mandible, 3[d] joint of palp not greatly longer than 2[d]. Maxillipeds, 3[d] joint of palp shorter than 2[d], not apically produced. Gnathopods 1 and 2 not large, nor very unequal, 5th joint much shorter than 6th, lobe narrowly rounded, densely setose, 6th oblong oval, palm very oblique, much longer than hind margin, defined by fascicles of strong spines. Peraeopods 3—5, 2[d] joint regularly oval, serrulate behind. Uropod 3, outer ramus more slender than inner, less than $^2/_3$ of its length. Telson oval, nearly twice as long as broad. Colour generally whitish, marbled with reddish brown patches. L. $\female$ about 6 mm.

Arctic Ocean, North-Atlantic, North-Sea and Skagerrak (Greenland, Spitzbergen, Iceland, Murmen Coast, NorWay); Kattegat.

5. **S. pulchellus** (O. Sars) 1876 *Amphithopsis pulchella*, G. O. Sars in: Arch. Naturv. Kristian., *v.* 2 p. 258 | 1885 *A. p.*, G. O. Sars in: Norske Nordhavs-Exp., *v.* 6. Crust. I p. 175 t. 14 f. 6 | 1893 *Parapleustes pulchellus*, G. O. Sars, Crust. NorWay, *v.* 1 p. 359 t. 126 f. 2 | 1899 *Amphithopsis pulchella, Sympleustes* (part.), T. Stebbing in: Ann. nat. Hist., ser. 7 *v.* 4 p. 209 | 1893 *Acanthozone latipes* (part.), A. Della Valle in: F. Fl. Neapel, *v.* 20 p. 608.

Not very slender, smooth. Head, rostrum very small, lateral corners angular, post-antennal corners little produced. Side-plates 1—3, denticle nearly obsolete. 4th narrow below, 5th and 6th, hind lobe much deeper than front. Pleon segment 3, postero-lateral corners obtusely quadrate. Eyes moderate, reniform, bright red. Antenna 1, 1st joint scarcely longer than 2[d] and 3[d] combined, not produced, flagellum about thrice as long as peduncle, 1st joint long. Antenna 2, ultimate and penultimate joints of peduncle subequal, flagellum not nearly twice as long as peduncle. Gnathopod 1 rather slender, 5th joint nearly as long as 6th, lobe broad, setose, 6th slightly widening distally, palm much shorter than hind margin, imperfectly defined between 2 sets of 3 small spines. Gnathopod 2 much stronger, not much longer, 5th joint short, lobe narrow, setose, 6th slightly widening distally, palm rather sinuous, slightly oblique, much shorter than hind margin, well defined between 2 sets of 3 strong spines. Peraeopods 3—5, 2[d] joint large oval, smooth behind. Uropod 3, outer ramus about $^2/_3$ as long as inner, both with few marginal spines. Telson rather small, distinctly boat-shaped, apex narrowly obtuse. Colour whitish with small pinkish spots all over. L. $\female$ 7 mm.

Arctic Ocean (Greenland, Spitzbergen, Iceland, Varangerfjord). Depth 94—377 m.

6. **S. olrikii** (H. J. Hansen) 1887 *Amphithopsis o.*, H. J. Hansen in: Vid. Meddel., ser. 4 *v.* 9 p. 136 t. 5 f. 5—5 b | 1892 *A. olriki*, *Parapleustes* (part.), G. O. Sars, Crust. NorWay, *v.* 1 p. 357 | 1899 *A. olrikii, Sympleustes* (part.), T. Stebbing in: Ann. nat. Hist., ser. 7 *v.* 4 p. 209 | 1893 *Acanthozone latipes* (part.), A. Della Valle in: F. Fl. Neapel, *v.* 20 p. 608.

Rather robust, smooth. Head, lateral corners obtuse. Side-plate 1 shallow, 2^d—4th successively deeper, all 4 narrow distally, 5th with hind lobe much deeper than front. Pleon segment 3, postero-lateral corners obtusely quadrate. Eyes large, subreniform. Antenna 1, 1st joint longer than 2^d and 3^d combined. Antenna 2 little more than $^1/_2$ length of antenna 1. Gnathopod 1 rather stout, 5th joint a little shorter and distally wider than 6th, 6th oblong, palm very oblique but well defined, near middle deeply and irregularly incised, incision armed with 2 spines. Gnathopod 2 rather longer and stronger, 5th joint like that of gnathopod 1, but 6th as wide as 5th and much longer, palm rather oblique, defined by a tooth and palmar spines, and having a wide and deep almost semicircular incision. Peraeopods 3—5 subequal in length, 2^d joint oblong with rounded corners, 4th long, not broad. Uropod 3, outer ramus $^2/_3$ as long as inner. L. about 8 mm (mouth-organs and telson not described).

Arctic Ocean (Greenland). Depth 94 m.

Gen. **Parapleustes** Buchh.

1874 *Parapleustes* (Sp. un.: *P. glacilis*), Buchholz in: Zweite D. Nordpolarf., v. 2 p. 337.

1 species.

P. gracilis Buchh. 1874 *P. glacilis, P. gracilis*, Buchholz in: Zweite D. Nordpolarf., v. 2 p. 269. 337 Crust. t. 7 f. 1 | ? 1887 *Paramphithoë gracilis*, H. J. Hansen in: Vid. Meddel., ser. 4 v. 9 p. 124 | 1893 *Acanthozone pulchella* (part.)?, *A. bicuspis*, A. Della Valle in: F. Fl. Neapel, v. 20 p. 605, 942.

In many respects agreeing closely with Neopleustes brevicornis (p. 313). Mandible with no molar tubercle, excluding the species from Gen. Sympleustes (p. 317). L. 5 mm.

Arctic Ocean (Sabine Island). Depth 19 m.

---

24. Fam. **Paramphithoidae**

1871 Subfam. *Epimerinae*, A. Boeck in: Forh. Selsk. Christian., 1870 p. 183 | 1888 *Epimeridae*, T. Stebbing in: Rep. Voy. Challenger, v. 29 p. 876 | 1893 *E.*, G. O. Sars, Crust. Norway. v. 1 p. 362.

Integument indurated, processiferous (Fig. 74, p. 327). Side-plates rigid, some acute. Eyes, when present, prominent. Antenna 1 shorter than antenna 2, flagellum in both many-jointed, accessory flagellum rudimentary or absent. Upper lip not deeply or not incised. Lower lip, inner lobes coalescing with outer or absent. Mandible, accessory plate on right as well as on left. Maxillipeds, outer plates broad, not long, scantily armed on inner margin, finger small. Mouth-parts in general strongly developed. Gnathopods 1 and 2 (Fig. 75, 76, p. 327) not stout in structure, 5th and 6th joints narrow, finger small. Peraeopods 3—5, 2^d joint not widely expanded. Uropod 3, rami lanceolate, longer than peduncle. Telson not large, entire or distally insinuate. Sexual difference very slight.

Marine. Free swimming, rarely semiparasitic.

3 genera, 8 accepted species and 1 obscure.

Synopsis of genera:

1 {
Side-plates 4 and 5 together forming a crescentic
curve below . . . . 1. Gen. **Epimeria** . . p. 321
Side-plates 4 and 5 not combining to form a
curve — **2.**

2 {
Gnathopods 1 and 2, 5th joint rather shorter
than 6th . . . . 2. Gen. **Paramphithoe** p. 324
Gnathopods 1 and 2 (Fig. 75, 76), 5th joint greatly
longer than 6th . . . . 3. Gen. **Actinacanthus** p. 326

1. Gen. **Epimeria** A. Costa

1793 Subgen. *Gammarellus* (part.), J. F. W. Herbst, Naturg. Krabben Krebse, *v.* 2
p. 106 | ?1847 *Vertumnus* (Sp. un.: *V. cranchii*) (nom. nud.), (Leach in MS.) A. White,
Crust. Brit. Mus., p. 89 | 1851 *Epimeria* (Sp. un.: *E. tricristata*), (A. Costa in:) F. W.
Hope, Cat. Crost. Ital., p. 46 | 1888 *E.*, T. Stebbing in: Rep. Voy. Challenger, *v.* 29
p. 877 | 1893 *E.*, G. O. Sars, Crust. Norway, *v.* 1 p. 363.

Body rather stout, with dorsal projections. Head with acute curved
rostrum, lateral corners weak, post-antennal deflexed, obtuse. Side-plates 1—3
strongly tapering, grooved, 4th deep, with large upper emargination and
larger one below defined by 2 acute points, 5th produced acutely or subacutely backward, its lower margin prolonging the arch of side-plate 4,
6th rather acute at hind corner, 7th rounded. Eyes prominent, not strongly
developed. Antenna 2 not elongate. Mandible, molar small, normal, palp
not very large. Maxilla 1, inner plate with many setae. Maxilla 2, inner
and outer plates broad, short, obliquely truncate, densely fringed, inner shorter,
not narrower than outer. Maxillipeds, inner and outer plates pretty well
developed, outer serrulate on inner margin, fringed with spines and setae,
palp not very large. Gnathopods 1 and 2 similar, subequal, rather small,
palm of 6th joint short. Peraeopods 1—5 not robust; in peraeopod 3
2d joint narrow, in peraeopods 4 and 5 not greatly expanded; peraeopod 5
a little shorter than peraeopod 4. Uropods 1—3, rami lanceolate, spinulose, rami of uropod 3 large, equal. Telson rounded quadrangular, apex
emarginate.

4 species.

Synopsis of species:

1 {
Pleon segment 3, postero-lateral angles strongly
produced, no accessory tooth . . . . 1. E. parasitica . . p. 321
Pleon segment 3, postero-lateral angles weakly
produced, an accessory tooth above — **2.**

2 {
The whole peraeon carinate . . . . 2. E. loricata . . . p. 322
Only hinder part of peraeon carinate — **3.**

3 {
Side-plates 1—5 with acute apices . . . . . . 3. E. cornigera . . p. 323
Side-plates 1—5 with obtuse apices . . . . . . 4. E. tuberculata . p. 323

    1. E. parasitica (Sars) 1858 *Amphithoë p.*, M. Sars in: Forb. Selsk. Christian.,
p. 131 | 1861 *Acanthosoma p.*, ?*A. tricristata*, A. Boeck in: Forh. Skand. Naturf.,
Møde 8 p. 665, 666 | 1862 *Acanthonotus parasiticus*, Bate, Cat. Amphip. Brit. Mus.,
p. 375 | 1893 *Epimeria parasitica*, G. O. Sars, Crust. Norway, *v.* 1 p. 366 t. 129 f. 1 | 1871
E. cornigera (err., non *Gammarus corniger* J. C. Fabricius 1779!), A. Boeck in: Forh.

Selsk. Christian., 1870 p. 185 | 1888 *E. c.*, T. Stebbing in: Rep. Voy. Challenger, *v.* 29 p. 318 | 1893 *Acanthonotosoma cornigerum* (part.), A. Della Valle in: F. Fl. Neapel, *v.* 20 p. 676.

Dorsal carina from peraeon segment 5 to pleon segment 4, the dentiform projection encreasing in strength on successive segments, subdorsal carinae ending in a small tooth on each side of the segments. Head, rostrum reaching slightly beyond 1st joint of antenna 1, post-antennal corners subacute. Pleon segments 1 and 2, postero-lateral corners with 2 projecting angular points, segment 3, postero-lateral corners acute, strongly produced, without upper projecting point. Eyes rather large, piriform, brilliant carmine. Antenna 1 not very long, 1st joint longer than 2^d and 3^d combined, flagellum about twice as long as peduncle, 20-jointed. Antenna 2 not much longer, ultimate and penultimate joints of peduncle subequal, flagellum twice as long as peduncle. Gnathopods 1 and 2, 5th joint as long as 6th, 6th slightly widening to not very oblique palm. Peraeopod 5, 2^d joint as long as 3^d—6th combined, widened except at distal end, which is produced to small acute lobe. Uropod 3, rami long, not very broad, spinules few. Telson oblong oval, apical margin insinuate, not deeply. Colour dark red, especially at hind margin of segments. L. ♀ about 9 mm.

North-Atlantic, North-Sea and Skagerrak (South- and West-Norway). Semiparasitic on Holothuria tremula Gunn.

2. **E. loricata** O. Sars   1872 *E. coniger* (nom. nud.), Whiteaves in: Ann. nat. Hist., ser. 4 *v.* 10 p. 347 | 1874 *E. cornigera* (err., non *Gammarus corniger* J. C. Fabricius 1779!), Whiteaves in: Amer. J. Sci., ser. 3 *v.* 7 p. 213 | 1893 *Acanthonotosoma cornigerum* (part.), A. Della Valle in: F. Fl. Neapel, *v.* 20 p. 676 | 1879 *Epimeria loricata*, G. O. Sars in: Arch. Naturv. Kristian., *v.* 4 p. 450 | 1885 *E. l.*, G. O. Sars in: Norske Nordhavs-Exp., *v.* 6 Crust. I p. 166 t. 14 f. 2 | 1888 *E. l.*, T. Stebbing in: Rep. Voy. Challenger, *v.* 29 p. 878 t. 68 | 1893 *E. l.*, G. O. Sars, Crust. Norway, *v.* 1 p. 368 t. 129 f. 3 | 1883 *E. conspicua*, T. Stebbing in: Ann. nat. Hist., ser. 5 *v.* 11 p. 204.

Much indurated, carinate from peraeon segment 1 to pleon segment 4, laminar, obtuse, backward produced processes, encreasing in size successively to pleon segment 1, on 2^d—4th more acute, on 4th diminished in size; peraeon segments 1—7 with subdorsal lateral tubercle, pleon segments 1—3 with 4 lateral tubercles, 4th with one and a ridge, 5th and 6th centrally faintly carinate. Head not carinate, rostrum very long, acute, reaching end of peduncle of antenna 1. Pleon segments 1—3, postero-lateral angles acute, with another tooth above the produced point, pleon segment 2 with tooth at antero-lateral corner. Eyes very prominent, rounded or oval, rich vermilion. Antenna 1, 1st joint longer than 2^d and 3^d combined, flagellum nearly thrice as long as peduncle, with about 30 joints, 1st longest, accessory flagellum 1-jointed, almost rudimentary. Antenna 2 not much longer, ultimate joint of peduncle shorter than penultimate, flagellum with 50 joints or more, 1st longest. Upper lip faintly insinuate at narrowed apex. Maxilla 1, inner plate with 9 setae, outer with 11 spines. Gnathopods 1 and 2 stronger than in other species of the genus. Gnathopod 1, 5th joint nearly as long as 6th, twice as long as broad, 6th oblong, a little widened at the short, serrate, almost transverse palm, finger armed on inner margin with 12 spines. Gnathopod 2 like gnathopod 1, but 6th joint longer. Peraeopods 3 and 4, 2^d joint deeply grooved behind; peraeopod 5, 2^d joint not equal in length to 3^d—6th combined, expanded above, produced into a narrow lobe below. Uropod 3, rami long, broad, ending very acutely. Telson little longer than broad, triangularly incised but not deeply between 2 triangular apices. Colour magnificent coral-red, more vivid on hind margin of segments. L. reaching 40 mm.

Arctic Ocean and North-Atlantic (Greenland; Spitzbergen; North-America; Hasvig [West-Finmark], depth 150—282 m).

3. **E. cornigera** (F.) 1779 *Gammarus corniger*, J. C. Fabricius, Reise Norweg., p. 383 | 1788 *Cancer (G.) c.*, J. F. Gmelin, Syst. Nat., v. 5 p. 2992 | 1793 *C. (Gammarellus) c.*, J. F. W. Herbst, Naturg. Krabben Krebse, v. 2 p. 141 | 1871 *Epimeria cornigera*, A. Boeck in: Forb. Selsk. Christian., 1870 p. 185 | 1893 *E. c.*, G. O. Sars, Crust. Norway, v. 1 p. 364 t. 128 | 1893 *Acanthonotosoma cornigerum* (part.), A. Della Valle in: F. Fl. Neapel, v. 20 p. 676 t. 59 f. 85 | ? 1847 *Vertumnus cranchii* (nom. nud.), (Leach in MS.) A. White, Crust. Brit. Mus., p. 89 | ? 1850 *Acanthonotus testudo*, A. White in: P. zool. Soc. London, v. 18 p. 97 t. 16 | 1851 *Epimeria tricristata*, (A. Costa in:) F. W. Hope, Cat. Crost. Ital., p. 46 | 1857 *Acanthonotus owenii*, Bate in: Ann. nat. Hist., ser. 2 v. 19 p. 141.

Front part broadly arched, back carinate from peraeon segment 5 to pleon segment 4, the keel forming in each segment an acute backward directed tooth, in pleon segment 4 the back indented in front of the tooth. There is also a similar but much slighter subdorsal keel on either side from peraeon segment 6 to pleon segment 3. Rostrum reaching beyond 1st joint of antenna 1. Side-plates 1—5 with acute apices. Pleon segments 1—3 each having two rather distant teeth at postero-lateral corners. Eyes very prominent, rounded, bright carmine. Antenna 1, 1st joint longer than 2d and 3d combined, flagellum about thrice as long as peduncle, 40-jointed, accessory flagellum 1-jointed, almost rudimentary. Antenna 2 not much longer, ultimate and penultimate joints of peduncle subequal, flagellum about twice as long as peduncle. Gnathopods 1 and 2 rather feeble, 5th joint longer than 6th, palm small, joining hind margin by a rounded angle. Feraeopods 1—5, uropods 1—3 and telson nearly as in E. loricata, but telson scarcely at all longer than broad. Colour whitish with reddish tinge, hind rim of each segment and tips of 4th and 5th side-plates pink. L. reaching 16 mm.

North-Atlantic and North-Sea (West-Norway at least up to Trondhjemsfjord, depth 94—282 m; France; Great Britain); Mediterranean.

4. **E. tuberculata** O. Sars 1893 *E. t.*, G. O. Sars, Crust. Norway, v. 1 p. 367 t. 129 f. 2.

Front part broadly arched, carina from peraeon segment 6 to pleon segment 4 with rather blunt projections, subdorsal keels very slight without dentiform projections, pleon segment 4 with a lateral ridge. Rostrum little overlapping 1st joint of antenna 1. Side-plates 1—3 and 5 with blunt apices, 4th less produced and less distinctly falciform than in other species. Pleon segments 1—3 with 2 distant, not very sharp, teeth at postero-lateral corners. Eyes rounded, light carmine. Antenna 1 short, 1st joint much longer than 2d and 3d combined, flagellum about twice as long as peduncle, 24-jointed. Antenna 2 considerably longer, ultimate joint of peduncle shorter than penultimate, flagellum 2½ times as long as peduncle. Gnathopods 1 and 2, 5th joint considerably longer than 6th, which is narrow, scarcely widening distally. Peraeopod 5, 2d joint not nearly as long as remainder of leg, ending in a blunt lobe. Uropod 3 as in E. cornigera. Telson rather broad, scarcely longer than breadth at base, slightly emarginate between 2 blunt apices. Colour whitish with faint rosy tinge (Sars), clear

21*

white variegated by a peculiar, very pretty shade of chocolate brown (T. Scott, MS.). L. reaching 16 mm.

North-Atlantic and North-Sea (West-Norway, depth 282—376 m; Ailsa Craig in Firth of Clyde).

2. Gen. **Paramphithoe** Bruz.

1835 *Acanthosoma* (Sp. un.: *A. hystrix*) (non J. Curtis 1824, Hemiptera!), (Rich. Owen in MS.) J. C. Ross in: John Ross, App. sec. Voy., nat. Hist. p. 91 | 1859 *Paramphithoe* (part.), R. M. Bruzelius in: Svenska Ak. Handl., n. ser. *v.* 3 nr. 1 p. 68 | 1871 *Acanthozone* (Sp. un.: *A. cuspidata*), A. Boeck in: Forh. Selsk. Christian., 1870 p. 184 | 1893 *A.*, G. O. Sars, Crust. Norway, *v.* 1 p. 369 | 1893 *A.* (part.), A. Della Valle in: F. Fl. Neapel, *v.* 20 p. 599 | 1894 *A.*, T. Stebbing in: Bijdr. Dierk., *v.* 17 p. 29.

Body indurated, beset with acute processes. Head, rostrum very small, post-antennal corners spiniform. Side-plates 1—3, 6 and 7 forming one acute lobe, 4th and 5th with 2 acute lobes. Eyes prominent. Antenna 2 elongate. Upper lip apically narrow, slightly insinuate, symmetrical. Lower lip with well formed inner plates coalescent on their outer margins with the outer plates. Mandible powerful, all parts normal, cutting plates bluntly dentate, spine-row of about 15 slender spines, 3d joint of palp shorter than 2d. Maxilla 1 with about 10 setae on inner plate, 11 spines on outer. Maxilla 2, inner plate fringed along much of inner margin, rather shorter and narrower than outer. Maxillipeds, inner and outer plates rather broad, outer closely fringed with spines on apical margin, 2d joint of palp much the longest, finger small. Gnathopods 1 and 2 rather feeble, narrow, 5th joint shorter than 6th, finger small. Peraeopods 3—5, 2d joint with acute processes on hind margin. Uropod 2, rami unequal, shorter than in uropods 1 and 3; uropod 3, rami long, lanceolate. Telson rather elongate and boat-shaped.

3 accepted species, 1 doubtful.

Synopsis of accepted species:

1 { Rostrum elongate . . . . 1. **P. buchholzi** . . . p. 324
  { Rostrum short — **2.**

. 2 { Peraeon segments 1—4 with elevated tooth on
    {   hind margin . . . . 2. **P. hystrix** . . . . p. 325
    { Peraeon segments 1—4 without elevated tooth on
    {   hind margin . . . . 3. **P. polyacantha** . p. 325

1. **P. buchholzi** (Stebb.) 1874 *Acanthozone hystrix* (err., non *Acanthosoma h.* J. C. Ross 1835!), Buchholz in: Zweite D. Nordpolarf., *v.* 2 p. 362 Crust. t. 11 | 1888 *A. buchholzi*, T. Stebbing in: Rep. Voy. Challenger, *v.* 29 p. 162, 467 | 1894 *A. b.*, T. Stebbing in: Bijdr. Dierk., *v.* 17 p. 31 | 1893 *A. cuspidata* (part.), A. Della Valle in: F. Fl. Neapel, *v.* 20 p. 613.

Peraeon segment 1 with 7 acute processes in front, the medio-dorsal largest, pointing upward and forward, and 5 on hind margin; segments 2—7 with 5 processes of hind margin, those on segments 5—7 pointing backward, the middle one with a subsidiary tooth near the top. Pleon segments 1—3 with a similar central process, and on either side of it in segment 1 with 4 sharp processes on the hind margin, in segment 2 with 6, in segment 3 with 3; segment 4 has a central lobe not acutely produced and at the postero-lateral corners a sharp process, which is repeated on segment 5. Head, rostrum straight, pointing slightly upward, overlapping 1st joint of antenna 1,

lateral and post-antennal angles alike sharply outdrawn. Side-plates 1
and 3 apically serrate in 4 teeth, plates 2, 5 and 7 each with 2 denticles,
4th and 5th irregularly sculptured below. Eyes round. Maxilla 2, inner plate
rather broader than outer(?). Gnathopods 1 and 2 feeble, subequal. Feraeopods 3—5, 2[d] joint with 3 acute processes of the hind margin in peraeopod 3, with 4 in peraeopods 4 and 5. Telson longer than breadth at base,
tapering, with a triangular emargination between 2 triangular apices. Colour
uniformly pale reddish yellow. L. 22·5 mm (2[d] antenna 23 mm long).

Arctic Ocean (Greenland). Depth 56 m.

2. **P. hystrix** (J. C. Ross) 1835 *Acanthosoma h.*, (Rich. Owen in MS.) J. C.
Ross in: John Ross. App. sec. Voy., nat. Hist. p. 91 t. B f. 4—7 | 1838 *Amphithoe h.*,
Kröyer in: Danske Selsk. Afh.. *v.* 7 p. 259 t. 2 f. 7 | 1859 *Paramphithoe h.*, R. M.
Bruzelius in: Svenska Ak. Handl., n. ser. *v.* 3 nr. 1 p. 71 | 1877 *Acanthozone h.*, Miers
in: Ann. nat. Hist., ser. 4 *v.* 19 p. 136; *r.* 20 p. 100 | 1888 *A. h.*, T. Stebbing in: Rep.
Voy. Challenger. *c.* 29 p. 50, 318, 466 | 1894 *A. h.*, T. Stebbing in: Bijdr. Dierk., *v.* 17
p. 29 | 1871 *A. cuspidata* (err.?, non *Oniscus cuspidatus* Lepechin 1780!), A. Boeck in:
Forh. Selsk. Christian., 1870 p. 184 | 1876 *A. c.*, A. Boeck, Skand. Arkt. Amphip., *v.* 2
p. 229 t. 20 f. 3 | 1893 *A. c.*, G. O. Sars, Crust. Norway. *v.* 1 p. 370 t. 130 | 1893 *A. c.*
(part.), A. Della Valle in: F. Fl. Neapel, *v.* 20 p. 613 t. 59 f. 25.

Body nearly cylindric. Peraeon segment 1 with large laminar carinate
process completely overarching the head, a smaller subdorsal process on either
side, segments 1—7 with 5 processes from transverse ridge of hind margin
as in A. buchholzi. Pleon segments 1—3 with similar central process, and
on either side of it in segments 1 and 2, 4 sharp processes, the 2 upper
much larger than the 2 lower, in segment 3 two on each side which are
subequal, segment 4 with sharp median and postero-lateral processes, segments 5
and 6 with acute process at postero-lateral angles but no dorsal process. Head,
rostrum very small, horizontal, lateral corners forming a subtruncate lobe,
post-antennal corners sharply outdrawn. Side-plates 1—3 acutely produced
at apex, not serrate, 4th with 2 acute processes pointing downward, 5th
with 2 pointing backward, 6th and 7th with 1 acute process. Eyes prominent,
rounded, dark brown (Sars) or white (Bate). Antenna 1, 1st joint as long as
2[d] and 3[d] combined, with large apical process, flagellum about 3 times as
long as peduncle. Antenna 2 much longer, joints of peduncle produced to
dentiform projections, ultimate and penultimate joints of peduncle subequal,
flagellum nearly 4 times as long as peduncle. Gnathopods 1 and 2 subequal,
5th joint slightly shorter than 6th, 6th widening a little distally, palm nearly
transverse, finger small. Peraeopods 3—5, 2[d] joint with only 2 acute processes
of hind margin. Telson as long as once and a half breadth at base, tapering,
slightly (more than in figure by Sars 1893) emarginate. Colour light straw-
yellow, mottled with brown. L. reaching over 30 mm.

Arctic Ocean, widely distributed; North-Atlantic (North-America; Norway,
Trondhjemsfjord and northward).

3. **P. polyacantha** (J. Murdoch) 1885 *Acanthozone p.*, J. Murdoch in: P. U. S.
Mus., *v.* 7 p. 520 | 1893 *A. cuspidata* (part.)?, A. Della Valle in: F. Fl. Neapel, *v.* 20 p. 613.

Peraeon segment 1, anterior margin raised into a ridge curving forward over the head, segments 1—5, hind margin raised into a rounded
ridge, developing into a median tooth on segment 5. Peraeon segments 6
and 7 and pleon segments 1—4 having on hind margin broad median tooth
pointing backward, largest on pleon segment 3, nearly obsolete on segment 4,

and postero-lateral angles produced acutely backward; peraeon segments 6 and 7 and pleon segments 1 and 2 with small intermediate tooth on each side, peraeon segments 1—5 with lateral margin deeply carinate. Head, rostrum very short, sharp, post-antennal corners spiniform. Side-plates with spiniform process. Pleon segment 2, postero-lateral corners bidentate. Eyes round, prominent. Antenna 1 about $^2/_3$ as long as antenna 2. Gnathopods 1 and 2 slender, subchelate. Telson rather long, entire.

Arctic Ocean (Point Franklin [arctic Alaska]). Depth 24 m.

P. cuspidata (Lepech.) 1780 *Oniscus cuspidatus*, Lepechin in: Acta Ac. Petrop., 1778 *v.* I t. 8 f. 3 | 1877 *Acanthozone cuspidata*, Miers in: Ann. nat. Hist., ser. 4 *v.* 19 p. 136; *v.* 20 p. 100 | 1888 *A. c.*, T. Stebbing in: Rep. Voy. Challenger, *v.* 29 p. 49 (with copy of Lepechin's fig.).

Perhaps identical with P. hystrix (p. 325). L. 20 mm.

White Sea.

3. Gen. **Actinacanthus** Stebb.*)

1888 *Acanthechinus* (Sp. un.: *A. tricarinatus*) (non P. M. Duncan & P. Sladen 1882, Echinodermata!), T. Stebbing in: Rep. Voy. Challenger, *v.* 29 p. 883.

Body indurated, beset with acute processes (Fig. 74). Head, rostrum very small. Side-plates 1—4 acute, 3^d—7th furnished with acute backward pointing lobes. Antennae 1 and 2, some joints of peduncle with acute projections. Upper lip apically broad, rather unsymmetrically bilobed. Mandible, spine-row abnormal, molar very prominent, palp elongate. Maxilla 1, inner plate with 3 setae, outer with 7 spines, palp long. Maxilla 2, plates broad, outer the broader. Maxillipeds, inner plates well developed, outer rather broad, reaching a little beyond 1st joint of palp, thinly furnished with setae, 1st joint of palp rather longer than 2^d, finger slight, short. Gnathopods 1 and 2 (Fig. 75, 76) very long and thin. 5th joint much longer than subchelate 6th, finger very small. Peraeopods 3—5, 2^d joint channelled, others decurrent. Uropod 3, rami long, lanceolate. Telson short, entire. Branchial vesicles simple. Marsupial plates long, narrow.

1 species.

1. **A. tricarinatus** (Stebb.) 1883 *Acanthozone tricarinata*, T. Stebbing in: Ann. nat. Hist., ser. 5 *v.* 11 p. 205 | 1893 *A. t.*, A. Della Valle in: F. Fl. Neapel, *v.* 20 p. 601 t. 59 f. 11 | 1888 *Acanthechinus tricarinatus*, T. Stebbing in: Rep. Voy. Challenger, *v.* 29 p. 884 t. 69, 70.

Body cylindric, except hinder part of pleon which is a little depressed and strongly flexed. Peraeon segments 1—7 (Fig. 74) each with central and 2 lateral acute, 3-sided, serrate, processes connected by transverse ridge on hinder part of segment; central process on segment 1 bifurcate, one arm pointing forward, the other backward. Pleon segments 1 and 2 with large central process flanked by 2 tubercles on each side, 3^d with smaller central process and 1 tubercle, and having a tridentate process rising from hind margin; 4th with hump, depression, and 2 serrate processes in succession; 5th unarmed; 6th with central and 2 lateral processes. Side-plates 1 and 2 pointing forward, tapering, carinate, serrate; 3^d and 4th with both lobes acute, pointing backward; 7th with the largest lobe. Eyes not perceived. Antenna 1, 1st joint

---

*) Nom. nov. ἀκτίς, ray, ἄκανθα, spine. — The name *Acanthechinus* is preoccupied (1882, P. M. Duncan & W. P. Sladen in: Pal. Ind., ser. 14 p. 34).

distally dilated, produced into 3 long processes, 2^d joint subequal to 1st, produced into 2 long processes, 3^d joint short, flagellum with 1st joint long, longer than shaft of 1st or 2^d joint of peduncle. Antenna 2 with 2 large processes on basal joint, with 1 long and 2 short on antepenultimate, 2 long and 2 short on penultimate, 2 very small on ultimate joint, which is longer than shaft of penultimate, flagellum rather longer than peduncle, with about 12 joints, 1st very long. Mandible, spines of spine-row extremely unequal, 1st joint of palp with tooth-like process, which seems to be movable, 3^d joint rather longer than 2^d, spinose. Gnathopod 1 (Fig. 75), 2^d joint the thickest part of the limb, 5th joint stouter and much longer than 6th, which is about 5 times as long as its greatest breadth, tapering to minute oblique spinose palm, finger very small, with tooth and

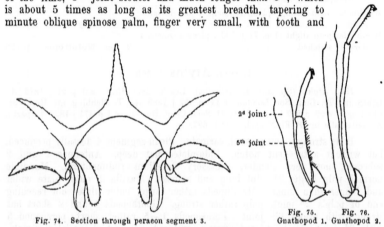

2^d joint

5th joint

Fig. 74. Section through peraeon segment 3.

Fig. 75. Fig. 76.
Gnathopod 1. Gnathopod 2.

Fig. 74—76. A. tricarinatus.

slender nail overlapping palm. Gnathopod 2 (Fig. 76) rather similar to gnathopod 1, but much longer, 5th joint not thicker than 6th but 3 times as long, 6th of uniform width, otherwise nearly as in gnathopod 1, palm less spinose. Peraeopods 1—5 with finger short, broad, spinulose. Peraeopods 3—5, 2^d joint with 4 longitudinal ridges, below overlapping 3^d joint with rounded lobe. 3^d—6th joints carinate in front, 4th and 5th sharply decurrent. Peraeopod 5 rather longer than peraeopod 4. Uropods 1—3, peduncle channelled above, rami equal, lanceolate, in uropod 1 rather shorter than peduncle, in uropod 2 a little longer, in uropod 3 much longer. Telson rather longer than broad, somewhat boat-shaped, broad distal margin convex. L. about 16 mm.

Southern Indian Ocean (Heard Island). Depth 282 m.

25. Fam. Atylidae

1865 Subfam. *Atylina* (part.), W. Lilljeborg in: N. Acta Soc. Upsal., ser. 3 v. 6 nr. 1 p. 18 | 1876 Subfam. *Atylinae* (part.), A. Boeck, Skand. Arkt. Amphip., v. 2 p. 320 | 1882 *Atylidae*, G. O. Sars in: Forh. Selsk. Christian., nr. 18 p. 26 | 1888 *A.*, T. Stebbing in: Rep. Voy. Challenger, v. 29 p. 899.

Body strongly compressed, carinate. Pleon segments 5 and 6 coalesced. Antenna 1, accessory flagellum rudimentary or absent. Upper lip rounded.

Lower lip with inner lobes obsolescent. Mandible (Fig. 77 p. 333) with palp.
Maxilla 1 (Fig. 78 p. 333), inner plate with several setae. Maxilla 2, inner plate
partially fringed on inner margin. Maxillipeds, palp well developed. Gnathopods 1 and 2 subchelate, gnathopod 2 generally the more slender. Feraeopods 1—5, finger usually pointing backward. Uropod 3, rami subequal,
extending beyond those of uropod 2. Telson short, deeply cleft.

Marine.

2 genera, 8 accepted species and 1 obscure.

Synopsis of genera:

Mandibular palp strong; pleon segment 4 not dorsally
notched . . . . 1. Gen. **Atylus** . . . p. 328
Mandibular palp slight (Fig. 77 p. 333); pleon segment 4
dorsally notched . . . . 2. Gen. **Nototropis** . p. 329

1. Gen. **Atylus** Leach

1815 *Atylus* (Sp. un.: *A. carinatus*), Leach, Zool. Misc., *v.* 2 p. 21 | 1815 *A.*,
Leach in: Tr. Linn. Soc. London, *v.* 11 p. 357 | 1888 *A.*, T. Stebbing in: Rep. Voy.
Challenger. *v.* 29 p. 907 | 1893 *A.*, G. O. Sars, Crust. Norway, *v.* 1 p. 471 | 1893 *A.* (part.),
A. Della Valle in: F. Fl. Neapel, *v.* 20 p. 697.

Body strongly compressed, carinate. Pleon segment 4 dorsally produced,
but without antecedent notch. Side-plates not deep. Antennae 1 and 2
subequal, not very slender, accessory flagellum rudimentary. Mandible
normal, palp large, 3^d joint long and spinose. Maxilla 2, outer plate wider
and longer than inner. Maxillipeds rather large, outer plates not reaching
end of palp's 2^d joint, palp rather strong. Gnathopods 1 and 2 short but
strong, with broad 6th joint. Peraeopods 1—5 rather short, peraeopod 5
with 2^d joint much dilated, finger reversed. Branchial vesicles simple.
Uropod 3, rami long, nearly equal, not reaching beyond uropod 1. Telson
deeply cleft.

1 species.

1. **A. carinatus** (F.) 1793 *Gammarus c.*, J. C. Fabricius, Ent. syst., *v.* 2 p. 515 |
1815 *Atylus c.*, Leach, Zool. Misc., *v.* 2 p. 22 t. 69 | 1874 *A. c.*, Buchholz in: Zweite D.
Nordpolarf., *v.* 2 p. 357 Crust. t. 10 | 1876 *A. c.*, A. Boeck, Skand. Arkt. Amphip., *v.* 2
p. 324 | 1893 *A. c.*, G. O. Sars, Crust. Norway, *v.* 1 p. 471 t. 166 f. 1 | 1893 *A. c.*, A. Della
Valle in: F. Fl. Neapel, *v.* 20 p. 701 t. 60 f. 3 | 1838 *Amphithoe carinata*, Kröyer in:
Danske Selsk. Afh., *v.* 7 p. 256 t. 2 f. 6 a—k | 1838 *A. c.*, Kröyer in: Naturh. Tidsskr.,
v. 2 p. 259 | 1866 *Paramphithoë c.*, Goës in: Öfv. Ak. Förh., *v.* 22 p. 523.

Body elongate, carinate throughout, the keel produced into a tooth in
peraeon segment 7 and pleon segments 1—4. Head, rostrum rather large,
compressed, distally somewhat expanded, ending obtusely; lateral corners
subtruncate. Side-plates 1—4 almost quadrangular, 4th broader, scarcely
deeper than the preceding, little emarginate, 5th with front lobe the deeper.
Pleon segment 3, postero-lateral corners subquadrate. Eyes very small,
rounded oval, prominent, very dark. Antenna 1 not nearly $^1/_3$ as long as
body, 1st joint as long as 2^d, 3^d about half 2^d, flagellum scarcely longer
than peduncle, 22-jointed, accessory flagellum 1-jointed, very small. Antenna 2
not (Boeck: a little) longer, ultimate and penultimate joints of peduncle subequal, together longer than the 20-jointed flagellum. Gnathopod 1, 5th joint
not very long, but nearly as long as 6th, 6th rounded oval, front margin
densely setose, palm shorter than hind margin, with which it forms an

obtuse angle, finger rather short and stout. Gnathopod 2 similar, rather longer. Peraeopod 1 longer than peraeopod 2, 4th joint in both rather stout. Peraeopods 3 and 4, 2^d joint not very wide, narrowing distally. Peraeopod 5, 2^d joint broader than long, hind lobe produced, defined by a notch in front, finger short and stout as in peraeopods 3 and 4, in all 3 easily reversed. Uropod 2 much shorter than uropods 1 and 3, the outer ramus a little shorter (Boeck: a little longer) than the inner. Uropod 3, peduncle very short, rami long, fringed with slender setae. Telson oval triangular, small, longer than broad, cleft nearly to base, tapering to obtuse apices, each carrying a spinule. Colour whitish with red spots? L. 32—43 mm.

Arctic Ocean (widely distributed; north of Norway, in stomach of Liparis montagui Donov.).

2. Gen. Nototropis A. Costa

1853 *Nototropis, Notrotopis* (Sp. un.: *N. spinulicauda*), A. Costa in: Rend. Soc. Borbon., n. ser. *v*.2 p.170, 173 | 1859 *Paramphithoe* (part.), R. M. Bruzelius in: Svenska Ak. Handl., n. ser. *v*.3 nr.1 p.68 | 1861 *Epidesura* (Sp. typ.: *Amphithoë compressa*), A. Boeck in: Forh. Skand. Naturf., Møde 8 p.659 | 1893 *Paratylus*, G. O. Sars, Crust. Norway, *v*.1 p.462.

Body strongly compressed, carinate. Pleon segment 4 dorsally produced with antecedent notch. Head rostrate, lateral margins sinuous. Side-plates 1—4 of moderate size, generally smaller in ♂ than in ♀, 5th with front lobe the deeper. Antenna 1 the shorter, without accessory flagellum, both pairs elongate in ♂. Mandible (Fig. 77 p. 333) normal, palp slender, feebly armed. Maxilla 1 see fig. 78 (p. 333), maxilla 2 as in Atylus. Maxillipeds, outer plates fully reaching or passing end of palp's 2^d joint, palp slender. Gnathopod 1 rather stouter and shorter than gnathopod 2, 6th joint oval, densely spinose in front; 5th joint in both pairs elongate triangular, palm oblique. Peraeopod 1 longer than peraeopod 2, peraeopod 3 short, with 2^d joint piriform, peraeopods 4 and 5 longer, subequal, 2^d joint in peraeopod 5 widely expanded, finger in peraeopods 3—5 sometimes reversed. Branchial vesicles often pleated. Uropods 1 and 2, outer ramus the shorter. Uropod 3, rami subequal, lanceolate, spinulose. Telson short, deeply cleft, apices subtruncate.

7 species accepted, 1 obscure.

Synopsis of accepted species:

<table>
<tr><td>1</td><td>{ Peraeopod 1, finger very large . . . .
{ Peraeopod 1, finger not very large — 2.</td><td>1. N. falcatus . . . . . p. 330</td></tr>
<tr><td>2</td><td>{ Pleon segments 1—3 without dorsal teeth . .
{ Pleon segments 1—3 with dorsal teeth — 3.</td><td>2. N. swammerdamei . p. 330</td></tr>
<tr><td>3</td><td>{ Peraeopod 5, 2^d joint with lower hind corner
   triangularly produced — 4.
{ Paraeopod 5, 2^d joint with lower hind corner
   not triangularly produced — 6.</td><td></td></tr>
<tr><td>4</td><td>{ Branchial vesicles simple . . . .
{ Branchial vesicles not simple — 5.</td><td>3. N. vedlomensis . . . p. 331</td></tr>
<tr><td>5</td><td>{ Eyes large, reniform; a spinulose hump on
   composite segment of pleon . . . .
{ Eyes small, round; no spinulose bump on
   composite segment of pleon . . . . . .</td><td>4. N. guttatus . . . . . p. 331

5. N. nordlandicus . . . p. 332</td></tr>
<tr><td>6</td><td>{ Post-antennal corners not produced; uro-
   pod 3, rami very long . . . .
{ Post-antennal corners produced; uropod 3,
   rami not very long . . . .</td><td>6. N. smitti . . . . . . p. 332

7. N. homochir . . . . p. 333</td></tr>
</table>

1. **N. falcatus** (Metzg.) 1871 *Atylus f.*, Aug. Metzger in: Jahresber. Ges.
Hannover, *v.* 21 p. 28 | 1889 *A. f.*, Hoek in: Tijdschr. Nederl. dierk. Ver., ser. 2 *v.* 2
p. 195 t. 8 f. 2, 21 | 1893 *A. f.*, A. Della Valle in: F. Fl. Neapel, *v.* 20 p. 703 | 1888 *A. f.*,
Tritaeta?, T. Stebbing in: Rep. Voy. Challenger, *v.* 29 p. 408 | 1882 *A. uncinatus*, G. O.
Sars in: Forh. Selsk. Christian., nr. 18 p. 101 t. 5 f. 3, 3 a | 1893 & 95 *Paratylus falcatus*,
P. u., G. O. Sars, Crust. Norway, *v.* 1 p. 465 t. 164 f. 1; p. 697 | 1895 *P. f.* + *P. u.*,
A. O. Walker in: P. Liverp. biol. Soc., *v.* 9 p. 306.

Resembles N. swammerdamei except in the following points. Pleon
segments 1—3 produced (not so in *Atylus uncinatus* O. Sars) to a small dorsal
tooth. Side-plate 1 tapering distally, 4th little broader than 3^d. Antenna 1 in ♀
scarcely $^1/_8$ as long as body, flagellum little longer than peduncle. Antenna 2
much longer, ultimate joint of peduncle not nearly twice as long as penultimate, flagellum not as long as both combined. Gnathopod 1 less strongly
built. Gnathopod 2, 6th joint narrowly oblong, palm very oblique, subequal to
hind margin. Peraeopod 1 strong, 5th joint very short, cup-shaped, 6th curved,
with strong spines at base, finger huge, falciform, fitted for clasping. Peraeopod 2 very small, 5th—7th joints almost rudimentary. Peraeopods 3—5, 5th
joint much longer than 6th, finger reversed, 2^d joint in peraeopod 5 broader
and with more convex hind margin than in the other species. Colour rather
pellucid, with scattered brownish patches (Sars) or white with a red spot on the
back of each segment except pleon segment 4 (Walker). L. ♀ 5—7 mm.

Arctic Ocean, North-Atlantic and North-Sea (Norway, East Frisian coast, Holland,
France, North-West England, Wales).

2. **N. swammerdamei** (M.-E.) 1830 *Amphithoe s.*, H. Milne Edwards in: Ann.
Sci. nat., *v.* 20 p. 378 | 1840 *Amphitoe swammerdamii*, H. Milne Edwards, Hist. nat. Crust.,
v. 3 p. 35 | 1876 *Atylus s.*, A. *schwammerdamii*, A. Boeck, Skand. Arkt. Amphip., *v.* 2
p. 328 t. 21 f. 5; t. 22 f. 1 | 1879 *A. swammerdammii*, Hoek in: Tijdschr. Nederl. dierk.
Ver., *v.* 4 p. 134, 152 t. 10 f. 1—6 | 1893 *A. swammerdamii* (part.), A. Della Valle in:
F. Fl. Neapel, *v.* 20 p. 698 t. 3 f. 12; t. 17 f. 1—21 | 1893 *Paratylus swammerdami*,
G. O. Sars, Crust. Norway, *v.* 1 p. 463 t. 163 | 1895 *P. swammerdamii*, A. O. Walker in:
P. Liverp. biol. Soc., *v.* 9 p. 305 | 1852 *Amphithoë compressa*, W. Liljeborg in: Öfv. Ak.
Förh., *v.* 9 p. 8 | 1859 *Paramphithoe c.*, R. M. Bruzelius in: Svenska Ak. Handl., n. ser.
v. 3 nr. 1 p. 72 | 1861 *Amphithoë c.*, *Epidesura sp. typ.*, A. Boeck in: Forh. Skand.
Naturf., Møde 8 p. 659 | 1857 *Dexamine gordoniana*, Bate in: Ann. nat. Hist., ser. 2
v. 19 p. 142 | 1862 *D. loughrini* + *Atylus swammerdamii* + *A. compressus*, Bate, Cat.
Amphip. Brit. Mus., p. 132 t. 24 f. 3; p. 136 t. 26 f. 2; p. 142.

None of the segments produced to a dorsal tooth except 4th of pleon,
the following composite segment ending in a little spinulose hump. Head,
rostrum rather short, lateral margin bilobate, lobes rounded, post-antennal
corners not produced. Side-plates 1—4 much larger and deeper in ♀ than
in ♂, 1st not tapering, 4th slightly emarginate, front lobe of 5th rounded.
Pleon segment 3, postero-lateral corners quadrate, with small acutely produced
point. Eyes oblong reniform, larger in ♂ than in ♀, dark brown with
whitish coating. Antenna 1 in ♀ more than $^1/_3$ length of body, 1st joint
as long as 2^d and 3^d combined, flagellum not quite twice as long as
peduncle, about 24-jointed. Antenna 2 in ♀ little longer, ultimate joint of
peduncle nearly twice as long as penultimate, flagellum about equal to both
combined. Antennae 1 and 2 in ♂ much longer, with fascicles of setules
on confronting edges of peduncles. Gnathopod 1, 5th joint shorter than
the 6th, which is rather stout, ovate, front margin with many series of
spinules; palm very oblique, not strongly defined. Gnathopod 2 longer, more
slender, 5th joint nearly as long as 6th, which is narrow, with short, moderately

·definite palm. Peraeopods 1—5, 5th joint shorter than 6th, finger small. Peraeopods 3 —5, 2^d joint with subacutely produced lower hind corner (production very slight in peraeopod 4), finger reversed; in peraeopods 3 and 4 front margin of 2^d joint setose, in peraeopod 5 2^d joint broadly oval. Uropod 3, rami more than twice as long as peduncle. Telson rather longer than broad, cleft nearly to base, apices truncate, each with a spinule. Colour pellucid, whitish, with small patches of chestnut-brown. L. ♀ 8 mm, ♂ rather less (Sars), adult ♂ and ♀ 4—9·5 mm (Walker).

Arctic Ocean, North-Atlantic and Mediterranean (Europe from Vadsö [Finmark] round to Naples, also at the Azores). From between tide-marks to 37 m.

3. N. vedlomensis (Bate & Westw.) 1862 *Dexamine v.*, Bate & Westwood, Brit. sess. Crust., *v.*1 p.242 f. | 1871 *Atylus v.*, A. Boeck in: Forh. Selsk. Christian., 1870 p.192 | 1876 *A. v.*, A. Boeck, Skand. Arkt. Amphip., *v.*2 p.330 t.9 f.9; t.11 f.6 | 1893 *Paratylus v.*, G. O. Sars, Crust. Norway, *v.*1 p.466 t.164 f.2 | 1866 *Atylus costae*, Cam. Heller in: Denk. Ak. Wien., *v.*26 n p.31 | 1893 *Dexamine spinosa* (part.), A. Della Valle in: F. Fl. Neapel, *v.*20 p.579.

Body slender, very distinctly carinate, peracon segment 7 and pleon segments 1—3, as well as the 4th, produced to compressed acute dorsal teeth, the following composite segment being apparently smooth, though with slight dorsal depression. Head, rostrum rather prominent, lateral margin bilobate, upper lobe acute, lower rounded, post-antennal corners not produced. Side-plates 1 and 2 slightly tapering, 3^d and 4th broader, not very unequal. Pleon segment 3, postero-lateral corners rotundo-quadrate with obsolete point. Eyes small, oval, dark brown with whitish coating (Sars; perhaps only applying to ♀). Antenna 1 in ♀ much more than ¹/₃ as long as body, 1st joint thicker than 2^d, scarcely as long, produced to a denticle varying from obtuse to acute, 3^d joint rather small, flagellum rather longer than peduncle, about 28-jointed. Antenna 2 in ♀ longer, ultimate joint of peduncle longer than penultimate, flagellum about as long as peduncle. In ♂ antennae 1 and 2 attain greater length, 2^d joint of peduncle of antenna 1 and ultimate joint of peduncle of antenna 2 being more elongate than in ♀. Gnathopods 1 and 2, peraeopods 1 and 2, uropods 1—3 and telson nearly as in N. swammerdamei, but gnathopods 1 and 2 rather more slender. Feraeopods 3—5, 5th joint much longer than 6th, finger recurved. Peraeopod 3, 2^d joint in ♀ uncinately produced at lower hind apex. Peraeopod 5, 2^d joint very broad, lower hind corner rather sharply produced. Branchial vesicles not pleated. Colour semipellucid, yellowish, with a row of rounded orange spots along the back; sides, antennae, legs streaked and mottled with orange, mouth-parts and uropods tinged with dark brown; ova in pouch bluish-green. L. ♀ 8 mm.

North-Atlantic, North-Sea and Skagerrak (South—and West-Norway, depth 19—94 m; Shetland Islands and southward to France); Kattegat.

4. N. guttatus (A. Costa) 1851 *Acanthonotus g.*, *Amphithonotus? g.*, (A. Costa in:) F. W. Hope, Cat. Crost. Ital., p.46 | 1857 *Nototropis g.*, A. Costa in: Mem. Acc. Napoli, *c.*1 p.194 t.1 f.7 | 1885 *Amphitonotus g.*, J. V. Carus, Prodr. F. Medit., *v.*1 p.408 | ?1853 *Notrotopis spinulicauda*, A. Costa in: Rend. Soc. Borbon., n. ser. *v.*2 p.173 | ?1857 *Nototropis s.*, A. Costa in: Mem. Acc. Napoli, *v.*1 p.194 t.1 f.8 | 1893 *Dexamine spinosa* (part.), A. Della Valle in: F. Fl. Neapel, *v.*20 p.573 | ?1895 *Atylus andrusowi*, Sowinski in: Mém. Soc. Kiew. *v.*14 t.4 f.7--14.

In general appearance strikingly in agreement with N. vedlomensis, but clearly distinct. The composite pleon segment ends in a well marked

little spinulose hump. Rostrum rather long, with rounded end. Eyes reniform, of moderate size in ♀, very large in ♂, almost meeting at the top
of the head. In peraeopod 3 the lower hind corner much less distinctly
uncinate than in N. vedlomensis (p. 331); in peraeopod 5 the 2^d joint not
very broad, with the produced point not acute. Branchial vesicles of peraeopod 5 and to some extent those of peraeopod 4 simple, but those of gnathopod 2
and peraeopods 1—3 remarkably developed; their numerous distinct lobes
approaching to the phyllobranchiate structure. Colour reddish white or
yellowish, with 3 longitudinal series of milky white spots on each side,
and with various little broken snuff-coloured lines. L. about 10 mm.

Mediterranean; Bay of Biscay (Vicero [North of Spain]); Black Sea?

5. N. nordlandicus (Boeck) 1871 *Atylus n.*, A. Boeck in: Forh. Selsk. Christian.,
1870 p. 193 | 1876 *A. n.*, A. Boeck, Skand. Arkt. Amphip., *v.* 2 p. 332 t. 23 f. 2 | 1893
Paratylus n., G. O. Sars, Crust. NorWay, *v.* 1 p. 469 t. 165 f. 2 | 1893 *Atylus swammer-
damii* (part.), A. Della Valle in: F. Fl. Neapel, *v.* 20 p. 698.

Body as in N. vedlomensis (p. 331), but with peraeon segment 6 also toothed;
head also similar, but rostrum more prolonged, reaching beyond middle of
1st joint of antenna 1, nearly horizontal. Side-plates 1—4 rather deep, 1st and
2^d rather narrow, distal margin serrate and setose, 3^d and 4th distally rounded,
serrate, 4th not emarginate, 5th with front lobe narrowly rounded. Pleon
segment 3, postero-lateral corners subquadrate, with small produced point
and margin above slightly crenulate. Eyes very small, rounded, red.
Antenna 1 in ♀ about $^1/_3$ as long as body, 1st joint a little shorter than 2^d,
flagellum little longer than peduncle, 12-jointed. Antenna 2 rather longer,
ultimate joint of peduncle little longer than penultimate, flagellum as long
as both combined. Gnathopod 1, 5th and 6th joints equal, 6th oblong oval,
palm ill-defined. Gnathopod 2 much longer, 6th joint shorter than elongate 5th,
slightly widened distally. Peraeopods 1 and 2, 4th joint as long as 5th and
6th combined, 5th much shorter than 6th. Peraeopod 3, 2^d joint with lower
hind angle acutely produced, 5th rather longer than 4th or 6th. Peraeopod 5,
2^d joint very wide, lower hind angle acutely produced. Branchial vesicles
pleated. Uropod 3, rami nearly 3 times as long as peduncle. Telson
longer than broad, cleft $^2/_3$ length, slightly dehiscent, apices obliquely truncate,
armed with spinules and 2 setules. Colour dark yellowish grey, with row
of obscure orange spots. L. ♀ scarcely 8 mm.

Arctic Ocean, North-Atlantic, North-Sea and Skagerrak (NorWay northWard to
Hasvig [West Finmark]). Depth 94—188 m.

6. N. smitti (Goës) 1866 *Paramphithoë s.*, Goës in: Öfv. Ak. Förh., *v.* 22 p. 524
t. 38 f. 14 | 1871 *Atylus s.*, A. Boeck in: Forb. Selsk. Christian., 1870 p. 190 | 1876 *A. s.*,
A. Boeck, Skand. Arkt. Amphip., *v.* 2 p. 326 | 1887 *A. smittii*, H. J. Hansen in: Dijmphna
Udb., p. 223 | 1894 *A. smitti*, T. Stebbing in: Bijdr. Dierk., *v.* 17 p. 35 | 1893 *Para-
tylus s.*, *P. smithi*, G. O. Sars, Crust. NorWay, *v.* 1 p. 468; t. 165 f. 1 | 1893 *Atylus
swammerdamii* (part.), A. Della Valle in: F. Fl. Neapel, *v.* 20 p. 698.

Body highly compressed, very distinctly carinate, peraeon segments 5—7
and pleon segments 1—4 produced to compressed teeth, the size at first
small but encreasing successively to that on segment 4, which is preceded
by the usual notch. Head, rostrum sometimes reaching end of 1st joint of
antenna 1, lateral margin sinuous, forming below a deflexed lobe defined
by a notch, post-antennal corners not produced. Side-plate 1 narrow,
bent forward, distally a little widened and serrate, 2^d narrow, 3^d and 4th

broader, less deep, distal border slightly emarginate, 5th with front lobe subacute. Pleon segment 3, postero-lateral corners rotundo-quadrate with minutely produced point. Eyes small, rounded, carmine. Antenna 1 very slender, nearly $\frac{1}{2}$ as long as body, 1st joint much shorter than 2^d, 3^d very small, flagellum about twice as long as peduncle, with about 45 joints. Antenna 2 little longer, ultimate and penultimate joints of peduncle subequal, flagellum about as long as peduncle, with 50 short joints. Gnathopods 1 and 2 very slender, 5th joint longer, little narrower than 6th, both densely setose, palm rather oblique, shorter than hind margin. Gnathopod 2 longer and more slender than gnathopod 1. Peraeopods 1 and 2, 4th joint scarcely as long as 5th and 6th combined, 5th much shorter than 6th. Feraeopods 3—5 rather long, 5th joint much longer than 6th, finger scarcely reversed, 2^d joint in peraeopod 3 with lower hind corner acute but scarcely produced, 2^d joint in peraeopod 5 very wide, and distally abruptly narrowed, scarcely forming a lobe and not ending acutely. Branchial vesicles not pleated. Uropod 3, rami extending much or little beyond uropods 1 and 2, 4 or 5 times as long as peduncle, densely spinose. Telson scarcely longer than broad, cleft nearly to the base (Boeck: much longer than broad; cleft not so far extending), apices not dehiscent, transversely truncate, each with a spinule. L. ♀ 23—35 mm.

Arctic Ocean (Finmark, Greenland, Spitzbergen, Siberia).

7. **N. homochir** (HasW.) 1885 *Atylus h.*: HasWell in: P. Linn. Soc. N. S. Wales. *v.* 10 p. 101 t. 13 f. 5—7 (juv.) | 1888 *A. h.*, T. Stebbing in: Rep. Voy. Challenger, *r.* 29 p. 908 t. 74 | 1893 *A. swammerdamii* (part.). A. Della Valle in: F. Fl. Neapel, *v.* 20 p. 698.

Body highly compressed, very distinctly carinate, peraeon segment 7 and pleon segments 1—3 produced to minute teeth, the hinder tooth on segments 4 and 6 more pronounced, the 4th with the usual notch, the coalesced 5th and 6th with a small anterior tooth. Head, rostrum slender, acute, reaching middle of 1st joint of antenna 1, lateral margins forming

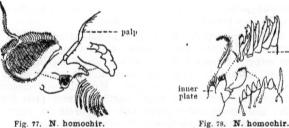

Fig. 77. **N. homochir.** Mandible.

Fig. 78. **N. homochir.** Maxilla 1.

below a deflexed lobe, post-antennal corners acutely produced. Side-plate 1 narrow, bent forward, spiniferous, 2^d—4th successively wider, also crenulate and spiniferous, 4th with lower border a little emarginate, 5th with narrowly produced front lobe. Pleon segments 1—3, postero-lateral corners produced to a small acute point, margin above convex. Eyes oval reniform. Antennae 1 and 2 elongate, peduncle carinate. Antenna 1, 1st joint distally produced into a short tooth, 2^d joint rather longer, 3^d about $\frac{1}{4}$ length of 2^d, with a little tubercle looking like rudiment of accessory flagellum, flagellum considerably longer than peduncle, 40-jointed. Antenna 2 rather longer, ultimate joint of peduncle longer than penultimate, both long, spinose, flagellum slender, rather longer than ultimate joint of peduncle, 35-jointed. Mandible (Fig. 77), 3^d joint of slender palp rather longer than 2^d. Maxilla 1 (Fig. 78), inner plate with 6 setae, outer with 11 spines, 2^d joint of palp with slender spines on apex of one maxilla and stouter spines on the other, characters which probably apply to all the species, but have not been definitely recorded for all. Maxillipeds, outer plates reaching end of 2^d joint of slender palp. Gnathopod 1, 5th joint slightly shorter and narrower than 6th, 6th narrowly oval, widening a little distally, palm oblique, moderately defined. Gnathopod 2, 5th and 6th joints subequal, longer and less spinose than in gnathopod 1. Peraeopods 1 and 2 attached as in N. smitti (p. 332) very low down on the sideplates, 4th joint as long as 5th and 6th combined, 5th very short, shorter than the finger. Peraeopod 3, 2^d joint piriform, lower hind corner not produced, 4th longer than 5th, 5th than 6th. Peraeopod 5, 2^d joint very broad, scarcely forming a lobe below, not produced, 5th joint as in peraeopod 4 longer than 4th or 6th, finger reversed. Branchial vesicles simple. Uropod 3, rami about thrice as long as peduncle, not reaching beyond uropod 1. Telson short, not longer than peduncle of uropod 3, rather longer than broad, cleft $^3/_4$ length, apices subtruncate, each with a spinule. L. 14 mm.

Port Stephens [East-Australia], Port Phillip [southern Australia]. Depth 62 m.

N. villosus (Bate) 1862 *Atylus v.*, Bate, Cat. Amphip. Brit. Mus., p. 135 t. 26 f. 1 | 1893 *A. smammerdamii* (part.), A. Della Valle in: F. Fl. Neapel, *v.* 20 p. 698.

L. 23 mm.

South-Atlantic (Hermit Island, lat. 56° S.).

---

26. Fam. **Melphidippidae**

1899 *Melphidippidae*, T. Stebbing in: Ann. nat. Hist., ser. 7 *v.* 4 p. 210 | 1899 *M.*, T. Stebbing in: Tr. Linn. Soc. London, ser. 2 *v.* 7 p. 422.

Some of the segments dentate, body extremely slender. Head without distinct rostrum. Side-plates very shallow, 4th not emarginate, 5th with front lobe the deeper. Antenna 1 slender, with accessory flagellum. Antenna 2 not longer than antenna 1. Upper lip rounded, with slight emargination. Lower lip with inner lobes well developed. Mandible with slender palp; maxilla 1, inner plate with many setae; maxilla 2, inner plate fringed on inner margin; maxillipeds, outer plates with spine-teeth on inner margin. Gnathopods 1 and 2 weakly subchelate, 5th joint rather elongate. Feraeopods 1—5 long and slender. Peraeopods 3—5, 2^d joint little expanded. Branchial vesicles simple. Marsupial plates narrow. Uropods 1 and 2, outer ramus the shorter. Uropod 3 very elongate. Telson cleft.

Marine.

2 genera, 5 species.

Synopsis of the genera:

1. Gen. **Melphidippa** Boeck

1871 *Melphidippa* (part.), A. Boeck in: Forb. Selsk. Christian., 1870 p.218 | 1876
M., A. Boeck, Skand. Arkt. Amphip., *v.*2 p.413 | 1894 *M.*, G. O. Sars, Crust. Norway,
*v.*1 p.482.

Side-plate 1 triangularly produced, lower margin fringed with setules.
Eyes imperfectly developed. Antenna 1, peduncle rather long, accessory
flagellum well developed. Antenna 2 with peduncle long and densely setose.
Mandible, 3^d joint of palp little·shorter than 2^d. Maxillipeds, palp elongate.
Gnathopod 1, 5th joint expanded proximally. Peraeopods 1 and 2 setose,.
finger long, with setae on both margins. Peraeopods 3—5, finger small,
in peraeopods 4 and 5 reversed. Uropod 3, peduncle elongate, narrow,
rami subequal, linear, spinose. Telson triangular, longer than broad, apices.
bidentate.

4 species.

Synopsis of species:

1 { Palm of gnathopod 2 excavate and rather long . . 4. **M. serrata** . . p.337
   { Palm of gnathopod 2 not excavate, short — 2.

2 { Antenna 1, 3^d joint short . . . . 1. **M. goësi** . . . p.335.
   { Antenna 1, 3^d joint long ·— 3.

3 { Antenna 1, 2^d joint much longer than 1st; gnatho-
       pod 2, 5th joint much longer than 6th . . . . . 2. **M. macrura** . . p.336
   { Antenna 1, 2^d joint not much longer than 1st; gnatho-
       pod 2, 5th joint not longer than 6th . . . . . . 3. **M. borealis** . . p.336

1. **M. goësi** Stebb. 1866 *Gammarus spinosus* (non *Cancer (G.) s.*, Montagu
1813!), Goës in: Öfv. Ak. Förh., *v.*22 p.530 t.40 f.30 | 1871 *Melphidippa spinosa*,
A. Boeck in: Forb. Selsk. Christian., 1870 p.219 | 1876 *M. s.*, A. Boeck, Skand. Arkt.
Amphip., *v.*2 p.417 t.23 f.4 | 1894 *M. s.*, G. O. Sars, Crust. Norway, *v.*1 p.483 t.169 |
1893 *Ceradocus spinosus*, A. Della Valle in: F. Fl. Neapel, *v.*20 p.719 | 1899 *Melphidippa.
goësi*, T. Stebbing in: Tr. Linn. Soc. London, ser.2 *v.*7 p.422.

Pleon segments 1—3 each produced into 3 sharp dorsal teeth, with
small denticles between them, segments 4 and 5 each produced into a large
simple tooth. Head, lateral corners obtusely truncate. Pleon segments 1—3,
postero-lateral margins irregularly dentate, in segment 3 postero-lateral
corners acutely produced. Eyes oval, narrowed above, red with whitish·
coating. Antenna 1 long, 2^d joint much longer than 1st, 3^d scarcely $^1/_4$ as long
as 2^d, flagellum nearly thrice as long as peduncle, 24-jointed; accessory
flagellum not very large, 2-jointed. Antenna 2 rather shorter, ultimate joint
of peduncle shorter than penultimate, flagellum about half length of peduncle.
Gnathopod 1, 5th joint broad, laminar, densely setose at base, narrowed
distally, 6th much shorter, subfusiform, setose, palm ill-defined, finger slender,
curved. Gnathopod 2 much more slender, 5th and 6th joints narrow, subequal,
palm very short, finger extremely small. Peraeopods 1. and 2, 5th joint
longer than 4th or 6th. Peraeopod 3, 2^d joint acutely produced at hind angle. Peraeopods 3—5, 4th and 5th joints subequal, 6th much shorter.
Uropod 2, outer ramus about half length of inner, longer than peduncle.
Uropod 3, inner ramus rather the longer, nearly as long as the elongate
peduncle. Telson nearly twice as long as broad, with a row of spinules
along each margin, cleft nearly to middle, with a pair of plumose (auditory)
setae near top of cleft, a long spine in each apical notch, of which the

outer tooth is the longer. Colour whitish, varied with reddish brown.
L. ♀ 9 mm.

Arctic Ocean, North-Atlantic and North Sea (Spitzbergen, Kara Sea, Norway,
Scotland). Depth 9—94 m.

2. **M. macrura** O. Sars 1894 *M. m.*, G. O. Sars, Crust. Norway, *v.* 1 p. 484
t. 170 f. 1.

Pleon segments 1—3 each produced to a dorsal tooth, with the hind
margin on either side of it finely denticulate, segments 4 and 5 each
produced to a long somewhat curved tooth. Head, lateral corners narrowly
rounded. Pleon segments 1—3, postero-lateral margins smooth, except in
segment 3, which has the postero-lateral corners produced to a small tooth.
Eyes about as in M. goësi (p. 335). Antenna 1, 2^d joint nearly twice as long as
1st, 3^d nearly $^2/_3$ length of 2^d, flagellum little longer than peduncle, about
15-jointed, accessory flagellum rather long, 4-jointed. Antenna 2 scarcely
shorter, ultimate joint of peduncle rather longer than penultimate, and longer
than flagellum. Gnathopod 1, 5th joint less expanded than in M. goësi.
Gnathopod 2, 5th joint expanded in an oblong form between the narrow
extremities, 6th much smaller, subfusiform as in gnathopod 1, finger rather
elongate. Peraeopods 1 and 2, 5th joint subequal to 6th, and scarcely
longer (shorter in figure) than the 4th. Peraeopods 3—5 as in M. goësi,
except that the 2^d joint of peraeopod 3 is not produced. Uropods 1—3
nearly as in M. goësi, except that uropod 3 is still more elongate, with
the peduncle considerably longer than the rami. Telson with a single pair
of lateral spinules close to the base, 2 pairs of setae near top of cleft,
which extends beyond the middle, inner tooth of each bidentate apex very
short, the spine in the notch not elongate. L. ♀ 8 mm.

North-Atlantic (West-Norway).

3. **M. borealis** Boeck 1871 *M. b.*, A. Boeck in: Forh. Selsk. Christian., 1870
p. 219 | 1876 *M. b.*, A. Boeck, Skand. Arkt. Amphip., *v.* 2 p. 415 t. 23 f. 3 | 1894 *M. b.*,
G. O. Sars, Crust. Norway, *v.* 1 p. 486 t. 170 f. 2 | 1893 *Ceradocus b.*, A. Della Valle in:
F. Fl. Neapel, *v.* 20 p. 720.

Pleon segments 1—5 each produced into a small dorsal tooth, with
denticulation of the hind margin on either side of it in segments 2—5.
Head, lateral corners broadly rounded. Pleon segments 1—3, postero-
lateral margins smooth. postero-lateral corners of segment 3 quadrate. Eyes
rather small, rounded, red with whitish coating. Antenna 1, 2^d joint longer
than 1st, 3^d about half as long as 2^d, flagellum twice as long as peduncle,
19-jointed, accessory flagellum longer than 3^d joint of peduncle, 5-jointed.
Antenna 2 scarcely shorter, ultimate and penultimate joints of peduncle
subequal, flagellum half as long as peduncle, 4-jointed. Gnathopod 1 about
as in the preceding species. Gnathopod 2, 5th joint a little expanded, 6th fully
as long as 5th, long oval, palm very short, finger small. Peraeopods 1
and 2, 5th joint nearly as long as 4th, longer than 6th. Peraeopods 3—5
as in M. macrura. Uropods 1—3 as in M. goësi (p. 335). Telson very
narrow, nearly thrice as long as broad, cleft beyond middle, lateral margins
smooth, a small spine in each apical notch. Colour whitish, broadly
banded across with chestnut brown patches; ova in pouch dark bluish.
L. ♀ 7 mm.

Arctic Ocean, North-Atlantic, North-Sea and Skagerrak (Norway from Christianiafjord to Vadsö). Depth 54—188 m.

4. **M. serrata** (Stebb.) 1888 *Neohela s.*, T. Stebbing in: Rep. Voy. Challenger, v. 29 p. 1215 t. 136 | 1893 *N. s.*, A. Della Valle in: F. Fl. Neapel, v. 20 p. 343 | 1894 *Melphidippa s.*, (T. Stebbing in:) G. O. Sars, Crust. Norway, v. 1 p. 624 | 1899 *M. s.*, T. Stebbing in: Tr. Linn. Soc. London, ser. 2 v. 7 p. 422.

Pleon segments 1—5 agreeing with those of M. borealis, but the 1st also dorsally denticulate and the 3d with postero-lateral corners produced into an acute tooth. Head with rostrum produced beyond the narrow lateral corners. In general agreeing with M. borealis, but gnathopod 2 with 6th joint not quite so long as 5th, and having a rather long, oblique, well excavate palm, and the finger not very small, the telson deeply cleft, with somewhat divergent lobes. Peraeopods 3—5, uropods 1—3, much of antennae 1 and 2 and apices of telson missing. Triturating organs of stomach seemingly devoid of strong spines. L. 8 mm.

Cumberland Bay [Kerguelen Island]. Depth 239 m.

2. Gen. **Melphidippella** O. Sars

1871 *Melphidippa* (part.), A. Boeck in: Forh. Selsk. Christian., 1870 p. 218 | 1894 *Melphidippella* (Sp. un.: *M. macera*), G. O. Sars, Crust. Norway, v. 1 p. 487.

Side-plate 1 subquadrate. Eyes well developed. Antenna 1, peduncle not very long, accessory flagellum rudimentary. Antenna 2 shorter than antenna 1 in ♀, but not in ♂. Mandible, 3d joint of palp much shorter than 2d. Maxillipeds, palp not elongate. Gnathopod 1, 5th joint scarcely expanded proximally. Peraeopods 1 and 2 extremely slender, finger very short. Peraeopods 3—5, finger reverted. Uropod 3, peduncle elongate, sublaminar, rami subequal, narrowly lanceolate. Telson triangular, longer than broad, apices bidentate.

1 species.

1. **M. macra** (Norm.) 1869 *Atylus macer*, A. M. Norman in: Rep. Brit. Ass., Meet. 38 p. 280 | 1888 *A. ? m.*, T. Stebbing in: Rep. Voy. Challenger, v. 29 p. 1628 | 1889 *Melphidippa macra*, A. M. Norman in: Ann. nat. Hist., ser. 6 v. 4 p. 121 t. 10 f. 14; t. 12 f. 4—7 | 1893 *Ceradocus macer*, A. Della Valle in: F. Fl. Neapel. v. 20 p. 720 | 1894 *Melphidippella macera*, G. O. Sars, Crust. Norway, v. 1 p. 488 t. 171 | 1871 *Melphidippa longipes*, A. Boeck in: Forh. Selsk. Christian., 1870 p. 219 | 1876 *M. l.*, A. Boeck, Skand. Arkt. Amphip., v. 2 p. 414 t. 24 f. 5.

Peraeon in ♀ rather tumid. Pleon segments 1—5 each produced dorsally to a tooth, with the hind margin on either side of it finely denticulate. Head short, but broad and deep, rostrum small, lateral lobes swollen and produced to an acute deflexed point. Side-plate 1, lower front corner minutely produced, 2d obliquely truncate below, 5th in front as deep as 4th. Pleon segment 3, postero-lateral corners acutely produced, postero-lateral margin minutely serrate. Eyes large, semi-globose, red, on lateral lobes of head. Antenna 1 long, 1st joint thick, subequal to 2d and 3d combined, 3d very small, flagellum more than 4 times as long as peduncle, slender, with about 22 joints, accessory flagellum a minute setulose nodule. Antenna 2, penultimate joint of peduncle longer than ultimate, in ♂ somewhat laminar, flagellum in ♀ shorter than peduncle, with about 7 joints, in ♂ fully as long as peduncle, with about 12 joints. Gnathopod 1, 5th joint rather compressed, densely setose on hind margin, narrowing distally, 6th much shorter, narrowly fusiform, finger slender. Gnathopod 2 rather longer, 5th joint rather narrow, in length subequal to the sublinear 6th, which has a very short palm and finger to match. Peraeopods 1 and 2, 4th, 5th and 6th joints subequal. Peraeopods 3—5 longer and somewhat stronger. Uropod 3, rami a little longer than peduncle. Telson cleft much beyond the middle, slender spinules along lateral margins, a long spine in each apical notch. Colour brick-red, mottled with opaque white. L. 6 mm.

North-Atlantic, North-Sea and Skagerrak (Norway, Shetland Isles); Kattegat. Depth 11—54 m.

---

27. Fam. Eusiridae

1888 *Eusiridae*, T. Stebbing in: Rep. Voy. Challenger, *v.* 29 p. 953 | 1893 *E.*, G. O. Sars, Crust. Norway, *v.* 1 p. 414.

Pleon strongly developed. Head, margins incised for base of antenna 2. Side-plate 1 usually distally expanded (Fig. 79 p. 341). Antennae 1 and 2 (Fig. 82 p. 347), peduncle long; antenna 1 usually with small accessory flagellum. Upper lip not bilobed. Lower lip with small inner lobes. Mandible, molar generally well developed, 3^d joint of palp elongate. Maxillae 1 and 2 and maxillipeds normal; maxillipeds with strong palp. Gnathopods 1 and 2 (Fig. 80 p. 344) subequal, with large subchelate hands. Peraeopods 1 and 2 slender, shorter than the hinder peraeopods. Uropod 3 usually with subequal rami (Fig. 81 p. 344). Telson large, partly cleft.

Marine.

6 genera, 26 accepted species and 1 obscure.

Synopsis of genera:

1 { Gnathopods 1 and 2, 6th joint attached to produced apex of 5th — 2.
Gnathopods 1 and 2, 6th joint with normal attachment to 5th — 3.

2 { Gnathopods 1 and 2 (Fig. 79), 5th joint with hind lobe prominent . . . . 1. Gen. **Eusirus** . . . . . p. 338
Gnathopods 1 and 2, 5th joint with hind lobe not prominent . . . . 2. Gen. **Eusiropsis** . . . p. 843

3 { Peraeopods stout . . . . 3. Gen. **Eusiroides** . . . p. 345
Peraeopods (Fig. 83 p. 347) slender — 4.

4 { Body smooth . . . . 4. Gen. **Cleonardo** . . . p. 346
Body spiny and carinate — 5.

5 { Telson elongate . . . . 5. Gen. **Rhachotropis** . p. 347
Telson short . . . . 6. Gen. **Rozinante** . . . p. 854

1. Gen. Eusirus Krøyer

1845 *Eusirus* (Sp. un.: *E. cuspidatus*), Krøyer in: Naturh. Tidsskr., ser. 2 *v.* 1 p. 501, 511 | 1861 *E.*, A. Boeck in: Forh. Skand. Naturf., Møde 8 p. 655 | 1888 *E.*, T. Stebbing in: Rep. Voy. Challenger, *v.* 29 p. 964 | 1893 *E.*, A. Della Valle in: F. Fl. Neapel, *v.* 20 p. 669 | 1893 *E.*, G. O. Sars, Crust. Norway, *v.* 1 p. 415 | 1873 *Eusinus*, A. Marschall, Nomencl. zool., p. 409.

Body compressed, more or less carinate, with dorsal projections. Head, rostrum small, lateral corners short, broad. Side-plate 4 the largest, emarginate behind. Pleon segment 3, postero-lateral angles not produced, lateral margin serrate. Eyes distinct. Antenna 1, 3^d joint short, more or less covered by 2 dentate lobes of 2^d joint, accessory flagellum a small linear joint with minute 2^d. Upper lip rounded. Mandible, cutting edge almost undivided, accessory plate dentate, spines of spine-row small, 3^d joint of palp as long as 1st and 2^d combined. Maxilla 1, inner plate with 1 or 2 setae, outer at least sometimes with 11 spines, palp not widened distally. Maxilla 2, both plates short, broad, apically rounded, inner the broader. Maxillipeds, inner plates sometimes partly coalesced, outer of moderate size, fringed with setiform spines, palp robust, finger unguiform. Gnathopods 1 and 2 (Fig. 79), 5th joint prolonged in front, behind produced to a narrow setose lobe, 6th very large, subquadrate or transversely elliptical, attached to the 5th joint only by its proximal front corner, palm nearly transverse, defined by a tooth and palmar spines, its edge sharp, minutely setose, finger long, slender, curved. Peraeopods 1 and 2 generally very slender. Peraeopods 3—5 long, 2^d joint oblong oval or piriform. Uropod 3, rami lanceolate, spinulose. Telson elongate, tapering, cleft or notched.

8 species.

Synopsis of species:

1 { Telson cleft beyond one third of length — 2.
  { Telson not cleft beyond one third of length — 4.

2 { Peraeopods 1 and 2, 4th joint not nearly twice
     as long as 5th . . . . 1. E. cuspidatus . . p. 339
  { Peraeopods 1 and 2, 4th joint twice as long
     as 5th — 3.

3 { Antenna 1, 1st joint much longer than 2^d . . . 2. E. propinquus . . p. 340
  { Antenna 1, 1st joint not much longer than 2^d . 3. E. antarcticus . . p. 340

4 { Only 2 segments are dorsally produced into a tooth 4. E. longipes . . . p. 341
  { 3 segments are dorsally produced into a tooth — 5.

5 { Pleon segment 3 dorsally dentate — 6.
  { Pleon segment 3 not dorsally dentate — 7.

6 { Telson cleft nearly one third of the length . . . 5. E. leptocarpus . . p. 341
  { Telson with minute apical notch . . . . 6. E. biscayensis . . p. 342

7 { Peraeopods 3—5, 6th joint not twice as long as 2^d 7. E. minutus . . . p. 342
  { Peraeopods 3—5, 6th joint nearly thrice as long as 2^d 8. E. holmii . . . . p. 342

1. **E. cuspidatus** Krøyer 1845 *E. c.*, Krøyer in: Naturh. Tidsskr., ser. 2 v. 1 p. 501 t. 7 f. 1 | 1874 *E. c.*, Buchholz in: Zweite D. Nordpolarf.. v. 2 p. 313 Crust. t. 3 f. 2, 2b | 1893 *E. c.*, G. O. Sars, Crust. Norway, v. 1 p. 416 t. 146 | 1893 *E. c.* (part.), A. Della Valle in: F. Fl. Neapel, v. 20 p. 669 t. 18 f. 41—50; t. 59 f. 79—82.

Body rather robust, carina produced to acute dorsal tooth in peraeon segments 6 and 7 and pleon segments 1 and 2; pleon segment 4 partly carinate. Head, rostrum short, obtuse, lateral corners slightly bilobed. Side-plate 1 moderately expanded distally, front corner rounded, hinder serrate with 3 denticles, as in the rather tapering 2^d and 3^d. Pleon segment 1, postero-lateral corners evenly rounded, convex lateral margin minutely serrate. Eyes rather large, oval reniform, red. Antenna 1, 1st and 2^d joints subequal, with dentate apical lobes, flagellum about twice as long as peduncle, with about 50 short joints, accessory flagellum very small, scarcely as long as 3^d joint of peduncle. Antenna 2 considerably shorter, ultimate and penultimate joints of peduncle subequal, together fully as long as flagellum. Gnathopods 1 and 2 rather robust, lobe of 5th joint prominent, 6th not much broader than long,

22*

rounded quadrangular, palm little arched, finger powerful. Peraeopods 1—5 more robust than in other species, very spinulose. Peraeopods 1 and 2, 4ᵗʰ joint not much longer than 5ᵗʰ. Peraeopod 5, 2ᵈ joint not nearly twice as long as broad, in length subequal to 6ᵗʰ joint. Uropod 3, inner ramus rather longer than outer. Telson large, slightly tapering, cleft nearly to centre, fissure not dehiscent till near to the acute divergent apices. Colour uniformly yellowish, or whitish yellow passing over to the colour of the face. L. reaching 39 mm.

North-Atlantic and Arctic Ocean (Greenland, Spitzbergen, Finmark). Depth 38—113 m.

2. **E. propinquus** O. Sars 1893 *E. propinqvus*, G. O. Sars, Crust. Norway, *v.* 1 p. 417 t. 147 f. 1.

Body moderately slender, not strongly carinate, peraeon scarcely carinate till segment 7, which has the dorsal tooth almost obsolete, only pleon segments 1 and 2 produced into a prominent dorsal tooth. Head, rostrum small, lateral corners nearly transversely truncate. Side-plate 1 well expanded distally, front corner rounded, hinder with 1 denticle. Pleon segment 3, postero-lateral corners narrowly rounded, lateral margin finely serrate. Eyes large, oval reniform, light red. Antenna 1 in ♀, 1ˢᵗ joint fully as long as 2ᵈ and 3ᵈ combined, with dentiform apical lobe, flagellum scarcely as long as peduncle, 3ᵈ joint of peduncle and the flagellum calceoliferous on lower margin, accessory flagellum linear, longer than 3ᵈ joint of peduncle. Antenna 2 little shorter, ultimate joint of peduncle rather shorter than penultimate, both with flagellum calceoliferous on upper margin, flagellum about half as long as peduncle. Gnathopods 1 and 2 less robust than in E. cuspidatus (p. 339), lobe of 5ᵗʰ joint very narrow, 6ᵗʰ nearly elliptical, much broader than long, palm rather arched, finger very slender. Peraeopods 1 and 2 very slender, 4ᵗʰ joint twice as long as 5ᵗʰ. Peraeopods 3—5 rather slender, 6ᵗʰ joint narrow, long, finger slight; 2ᵈ joint in peraeopod 5 nearly twice as long as broad, hind margin insinuate. Uropod 3, rami narrowly lanceolate, inner little longer than outer, both with setae as well as spinules on inner margin. Telson narrow, tapering, margins finely spinulose, cleft not quite to middle, apices acute, scarcely divergent. Colour semipellucid, orange-tinted, reddish brown on hinder margins of segments, antennae 1 and 2 banded with red. L. ♀ 12 mm.

North-Atlantic and Arctic Ocean (Trondhjemsfjord, depth 188—282 m; Selsövig [Nordland]; Finmark).

3. **E. antarcticus** G. M. Thoms. 1880 *E. cuspidatus var. a.*, G. M. Thomson in: Ann. nat. Hist., ser. 5 *v.* 6 p. 4 | 1888 *E. longipes* (err., non A. Boeck 1861!), T. Stebbing in: Rep. Voy. Challenger, *v.* 29 p. 965 t. 87 | 1893 *E. l.?*, G. O. Sars, Crust. Norway, *v.* 1 p. 421 | 1893 *E. cuspidatus* (part.), A. Della Valle in: F. Fl. Neapel, *v.* 20 p. 669, 671.

Exceedingly like E. propinquus; differs in following points. Side-plate 1 more produced forward. Peraeon segment 7 with dorsal tooth more pronounced. Eyes rather dark brown (in spirit), visual elements well developed. Antennae 1 and 2 apparently not calceoliferous. Antenna 1, 1ˢᵗ joint much thicker, scarcely longer than 2ᵈ, in ♂ transversely banded below with setules, accessory flagellum linear, length variable, but scarcely longer than 3ᵈ joint of peduncle, flagellum of many short joints, the 1ˢᵗ sometimes long. Antenna 2 in ♂, upper margin of ultimate and penultimate joints of peduncle transversely banded with setules. Mandible, molar prominent, but smaller than that figured by Sars for E. propinquus.

Maxillipeds, finger well formed, ending in a spine or nail, its concave margin armed with 4 spinules. Gnathopods 1 and 2 (Fig. 79) agreeing with E. longipes more nearly than E. propinquus. Uropod 3 in young ♂ without, in adult ♂ probably with a fringe of setae as in the two species compared. Telson cleft for barely ²/₅ of length, apices a little divergent. L. 9 mm.

South Pacific (New Zealand); Southern Indian Ocean (Kerguelen Island, Heard Island). Depth 282 m.

4. **E. longipes** Boeck 1861 *E. l.*, A. Boeck in: Forb. Skand. Naturf., Møde 8 p. 656 | 1876 *E. l.*, A. Boeck, Skand. Arkt. Amphip., *v.* 2 p. 504 t. 19 f. 4 | 1893 *E. l.*, G. O. Sars, Crust. Norway, *v.* 1 p. 420 t. 148 f. 1 | ?1862 *E. helvetiae*, Bate (& Westwood), Brit. sess Crust, *v.* 1 p. 267 f. | ?1862 *E. h.*, *E. helvetioe*, Bate. Cat. Amphip. Brit. Mus., p. 155 t. 29 f. 1 | 1866 *E. bidens*, Cam. Heller in: Denk. Ak. Wien, *v.* 26 ɪɪ p. 82 t. 3 f. 19 | 1893 *E. cuspidatus* (part.), A. Della Valle in: F. Fl. Neapel, *v.* 20 p. 669.

Body moderately slender, dorsal carina only slightly indicated, only pleon segments 1 and 2 produced into a dorsal tooth. Head, lateral corners obtusely truncate. Side-plate 1 well produced and rounded in front, side-plates 1—3 with 2 denticles at lower hind corner. Pleon segment 3, postero-lateral corners narrowly rounded, convex lateral margin coarsely serrate. Eyes very large, oblong reniform, bright red. Antenna 1, 1ˢᵗ joint subequal to 2ᵈ, with several dentiform apical projections, 2ᵈ with the usual 2, of which the inner is 5-dentate, flagellum nearly twice as long as peduncle, with 40—50 joints, calceoli-ferous, accessory flagellum very narrow, subequal to 3ᵈ joint of peduncle. Antenna 2 considerably shorter, ultimate joint of peduncle longer than penultimate, together as long as flagellum, ultimate joint of peduncle and flagellum calceoliferous.

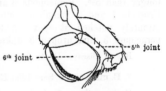

Fig. 79. **E. antarcticus.** Gnathopod 1.

Gnathopods 1 and 2 moderately strong, process of 5ᵗʰ joint slender, 6ᵗʰ sub-quadrate, little broader than long, palm slightly arcuate, finger slender. Peraeopods 1—5 very slender and long. Peraeopods 1 and 2, 4ᵗʰ joint subequal to 6ᵗʰ, not nearly twice as long as 5ᵗʰ. Peraeopods 3—5, 2ᵈ joint not broadly oval, strongly serrate behind. Uropod 3, rami subequal, narrowly lanceolate. Telson comparatively small, tapering, cleft scarcely ¹/₃ of length, apices acute, very little divergent. Colour straw-yellow, mottled all over with brick-red specks; ova dark bluish green. L. ♀ reaching 13 mm.

Arctic-Ocean, North-Atlantic, North-Sea and Skagerrak (Norway, depth 56—188 m; Shetland Isles; Firth of Clyde; France); Adriatic.

5. **E. leptocarpus** O. Sars 1893 *E. l.*, G. O. Sars, Crust. Norway, *v.* 1 p. 422 t. 148 f. 2.

Hind part of body distinctly carinate, pleon segments 1—3 produced to dorsal tooth. Head, lateral corners truncate. Side-plate 1 well expanded, narrowly rounded in front, hind corner smooth. Pleon segment 3, postero-lateral corners narrowly rounded, fine serration of lateral margin turning the corner as in E. propinquus. Eyes faintly traced in specimens in spirit. Antenna 1, 1ˢᵗ joint subequal to 2ᵈ, both long, flagellum shorter than peduncle, without calceoli, accessory flagellum linear, fully as long as 3ᵈ joint of peduncle. Antenna 2 not much shorter, ultimate joint of peduncle shorter than penultimate, flagellum about ¹/₂ as long as peduncle, without calceoli.

Gnathopods 1 and 2 rather large, 5th joint long, distally attenuated, 6th fully twice as broad as long, palm evenly curved, finger narrow, long. Peraeopods 1—5 rather slender. Peraeopods 1 and 2, 4th joint rather longer than 6th, nearly twice as long as 5th. Peraeopods 3—5, 2^d joint narrowly oval, finely serrate behind. Uropod 3, rami subequal, narrowly lanceolate. Telson as in E. longipes (p. 341), but comparatively larger. L. ♀ 8 mm.

Hardangerfjord, depth 564—752 m; Trondhjemsfjord.

6. **E. biscayensis** Bonnier 1896 *E. b.*, J. Bonnier in: Ann. Univ. Lyon, *v.* 26 p. 651 t. 39 f. 1.

Peraeon segment 7 with slight carina ending in denticle, pleon segment 3 with carina ending in tooth. Head, rostrum narrow, as long as 1st joint of antenna 1. Side-plate 1 quadrate. Pleon segment 3, postero-lateral margin serrate. Eyes reniform. Antenna 2, ultimate and penultimate joints of peduncle equal, flagellum calceoliferous. Mouth-parts in most respects normal, but inner plate of maxilla 1 with 2 setae, outer with 8 spines. Gnathopods 1 and 2, 5th joint elongate, hind lobe small, acute, with only 2 setae, 6th more than twice as broad as long; the long, curved palm defined by a little hollow with palmar spines. Peraeopods 1 and 2 very slender, 4th joint not longer than 5th. Peraeopods 3 and 4, finger nearly half as long as 6th joint. Uropods 1 and 2 spinulose, outer ramus little shorter than inner. Telson minutely notched at apex. Antenna 1 except 1st joint, terminal joints of peraeopod 5, and uropod 3 unknown. L. ♀ 12 mm.

Bay of Biscay. Depth 940 m.

7. **E. minutus** O. Sars 1893 *E. m.*, G. O. Sars, Crust. Norway, *v.* 1 p. 419 t. 147 f. 2.

Body rather stout, hind part distinctly carinate, peraeon segment 7 and pleon segments 1 and 2 produced to a dorsal tooth. Head, rostrum pretty well developed, lateral corners transversely truncate. Side-plate 1 well expanded, both corners smoothly rounded. Pleon segment 3, postero-lateral corners narrowly rounded, margin finely serrate only a little way up, serration turning the corner. Eyes faintly traced in specimens in spirit. Antenna 1 in ♀ short, 1st joint thick, longer than 2^d and 3^d combined, with acute apical lobe, inner lobe of 2^d joint bidentate, flagellum rather longer than peduncle, 10-jointed, without calceoli, accessory flagellum very minute. Antenna 1 in ♂ much longer, flagellum about thrice as long as peduncle, with many joints, several of the proximal densely clothed with sensory setae. Antenna 2 alike in ♂ and ♀, ultimate joint of peduncle shorter than penultimate, flagellum about $^1/_2$ as long as peduncle. Gnathopods 1 and 2 not very strong, process of 5th joint long, narrow, 6th scarcely shorter than breadth at base, distally somewhat widened, palm little curved. Peraeopods 1—5 slender, less elongate than in E. propinquus (p. 340). Peraeopods 1 and 2, 4th joint as long as 6th, not nearly twice as long as 5th. Peraeopods 3—5, 2^d joint strongly serrate behind, in peraeopod 5 broadly oval, scarcely at all tapering distally as in other species. Uropod 3, outer ramus much shorter than inner. Telson rather long, tapering throughout, cleft scarcely $^1/_6$ of length, apices acute, standing well apart. L. ♀ about 6 mm.

Trondhjemsfjord. Depth 752 m.

8. **E. holmii** H. J. Hansen 1887 *E. h.*, H. J. Hansen in: Dijmphna Udb., p. 224 t. 22 f. 1—1b | 1893 *E. h.*, G. O. Sars, Crust. Norway, *v.* 1 p. 416 | 1893 *E. cuspidatus* (part.), A. Della Valle in: F. Fl. Neapel, *v.* 20 p. 669.

Body rather slender, peraeon segment 6 a little thickened and rounded, segment 7 and pleon segments 1 and 2 produced into a prominent dorsal tooth, these 3 and 2 following segments carinate. Side-plate 1 produced subacutely in front. Pleon segment 3, postero-lateral corners quadrate, above the acute point finely serrate on lateral margin. Antenna 1 long, 1st joint as long as 2d, flagellum more than twice as long as peduncle, in appearance serrate below, by distal widening of the proximal joints, accessory flagellum spine-like. Antenna 2 scarcely half as long, much more slender, ultimate joint of peduncle in 1 specimen considerably longer than penultimate, only a little in a smaller specimen. Gnathopods 1 and 2 long, slender, 6th joint transversely oblong ovate. Peraeopods 1—5 extremely slender and elongate. Peraeopods 1 and 2, 4th joint not nearly twice as long as 5th. Peraeopods 3—5, 2d joint narrowly oval, distally narrowed, 4th shorter than 5th, 5th much shorter than 6th, which is $2^1/_2$—3 times as long as 2d, finger small, straight. Uropods 1—3 long, but strong. Telson narrow, tapering, cleft very short, widely separating the acute apices. Colour light rose-red. L. ♀ 53 mm.

Kara Sea. Depth 172—176 m.

2. Gen. **Eusiropsis** Stebb.

1897 *Eusiropsis* (Sp. un.: *E. riisei*), T. Stebbing in: Tr. Linn. Soc. London, ser. 2 *v.* 7 p. 39.

Body (Fig. 80 p. 344) without dorsal projections. Head rostrate. Side-plates shallow. Pleon segment 3, postero-lateral corners not produced, lateral margin not serrate. Antenna 1 the shorter, accessory flagellum minute, 1-jointed. Antenna 2, ultimate joint of peduncle in ♂ elongate. Mouth-parts nearly as in Eusirus (p. 338), but mandible with molar feeble, maxilla 1, inner plate unarmed, outer with 10 spines, 2d joint of palp narrower and scarcely longer than 1st. Gnathopods 1 and 2 as in Eusirus, but hind lobe of 5th joint almost obsolete. Peraeopods 1 and 2 slender, finger ending obtusely and tipped with long setae. Peraeopods 3—5 slender, elongate, finger straight. acute, armed like the preceding joints with long setae. Branchial vesicles with accessory lobes. Uropod 3, rami lanceolate, spinulose, in ♂ very plumose (Fig. 81 p. 344). Telson narrow, apically cleft.

1 species.

1. **E. riisei** Stebb. 1897 *E. r.*, (nom. sp. Lütken in MS.) T. Stebbing in: Tr. Linn. Soc. London, ser. 2 *v.* 7 p. 39 t. 13, 14.

Peraeon segments 2, 3 and 4 shorter than the others, pleon segments 1—3 large. Head, rostrum triangular, longer than broad. Side-plates 1 and 2 very shallow. 2d distally narrowed, 3d securiform, front corner rounded, hinder acutely produced. 4th nearly transversely oblong, slightly emarginate, 5th and 6th with hind lobe produced. Pleon segments 1—3, postero-lateral corners rounded (Fig. 80). Eyes wanting. Antenna 1 in ♂, 1st joint stout, with tufts of setules below, 2d a little shorter, calceoliferous, 3d short, flagellum rather stout, with at least 30 joints, calceoliferous. Antenna 2, penultimate joint of peduncle with tufts of setules above, ultimate joint much longer, longer than peduncle of antenna 1, calceoliferous, flagellum longer than peduncle, with 42 joints, 27 rather stout, with calceoli, rest more slender, with sensory filaments. Upper lip apically rounded. Lower lip, inner lobes small. Mandible with 5 or 6 small spines in spine-row. Maxilla 2, inner plate rather broader than outer, scantily armed. Maxillipeds, outer plates not very large, finger of palp as long as 3ᵈ joint. Gnathopods 1 and 2, 5ᵗʰ joint not very elongate, but having the attachment of the 6ᵗʰ as in Eusirus, the hind lobe very

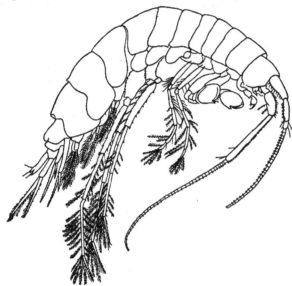

Fig. 80. E. riisei. Lateral View.

slight, 6ᵗʰ joint massive, distally widened, palm long, convex, carrying setules and ending in a spinigerous pocket, into which the long curved finger inserts its

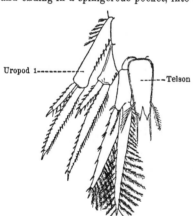

Uropod 1- - - - - - - - -

- - - - -Telson

Fig. 81. E. riisei.
Uropods 1—3 and telson.

apex. Peraeopods 1 and 2, finger not unguiform, less than half as long as 6ᵗʰ, apical margin not acute, fringed with 6 plumose setae, mostly of great length. Peraeopod 3, 2ᵈ joint expanded, hind margin serrate, 4ᵗʰ, 5ᵗʰ and 6ᵗʰ joints all longer than 2ᵈ, finger as long as 2ᵈ joint, the limb copiously supplied with spines and plumose setae. Peraeopod 4, 2ᵈ joint larger than in peraeopod 3, 5ᵗʰ joint shorter than 6ᵗʰ. Peraeopod 5. 2ᵈ joint larger than in peraeopod 4, 5ᵗʰ and 6ᵗʰ not quite so long, plumose setae of striking length. In uropod 1 outer ramus a little more than half as long as inner, in uropod 2 a little less than half; in uropod 3 peduncle a little longer than in uropod 2, rami long, outer a little the shorter (Fig. 81). Telson longer than peduncle of uropod 3, cleft about a quarter of length, apices acute, lateral margins with 2 notches. L. about 10 mm.

North-Atlantic (lat. 22° N., long. 36° W.).

3. Gen. **Eusiroides** Stebb.

1888 *Eusiroides*, T. Stebbing in: Rep. Voy. Challenger, *v.*29 p. 969 | 1893 *E.*, G. O. Sars, Crust. Norway, *v.*1 p. 414 | 1893 *E.*, A. Della Valle in: F. Fl. Neapel, *v.*20 p. 671.

Body with rounded back, not carinate. Side-plate 1 distally widened. Calceoli present on antennae 1 and 2 in ♀; accessory flagellum of antenna 1 consisting of 1 narrow short joint. Maxilla 1 with 1—3 setae on inner plate, 10 spines on outer. Maxilla 2 with varying proportions of the 2 plates. Gnathopods 1 and 2, 5ᵗʰ joint short, cup-shaped, attachment of 6ᵗʰ normal, 6ᵗʰ large, oval, palm ill-defined except by spines, finger long, curved. Peraeopods 1—5 rather robust, 3ᵈ—5ᵗʰ with 2ᵈ joint large, broadly oval. In other respects scarcely differing from Eusirus (p. 338).

3 species accepted, 1 obscure.

Synopsis of accepted species:

1 { Pleon segment 3, postero-lateral margin not serrate 3. **E. crassi** . . . . . p. 346
  { Pleon segment 3, postero-lateral margin serrate — **2.**

2 { Eyes dark-coloured . . . . 1. **E. monoculoides** . p. 345
  { Eyes light-coloured . . . . 2. **E. dellavallei** . . p. 346

1. **E. monoculoides** (HasW.) 1880 *Atylus m.*, HasWell in: P. Linn. Soc. N.S.Wales, *v.*4 p. 327 t. 18 f. 4 | 1888 *A. m.* + *Eusiroides caesaris* + *E. pompeii*, T. Stebbing in: Rep. Voy. Challenger, *v.*29 p. 969; p. 970 t. 88; p. 974 t. 89 | 1893 *E. m.* + *E. c.* (part.), A. Della Valle in: F. Fl. Neapel, *v.* 20 p. 674, 672.

Pleon segments 1 and 2 dorsally produced to a small acute tooth, which is sometimes obsolete. Head, rostrum small, lateral corners rounded. Pleon segments 1 and 2 sharply quadrate or a little produced, segment 3 with postero-lateral margin very convex, serrate with 10—18 upturned teeth. Eyes very large, reniform, nearly meeting above, blue-black. Antenna 1, 1ˢᵗ joint as long as or longer than 2ᵈ and 3ᵈ combined, 3ᵈ less than half as long as 2ᵈ, both these and flagellum calceoliferous, flagellum nearly twice as long as peduncle, attaining 74 joints, the proximal broader than long, closely set with calceoli, many at intervals distally widened, terminal joints slender, without calceoli. Antenna 2 rather shorter, ultimate joint of peduncle calceoliferous, a little or scarcely shorter than penultimate, flagellum sometimes as long as peduncle, attaining 54 joints, arranged and armed as in antenna 1. Maxilla 1, inner plate carrying 2 setae, 1ˢᵗ joint of palp more than half as long as 2ᵈ, which carries many setae on the outer margin. Maxilla 2, inner plate broader than outer, and not shorter. Gnathopods 1 and 2, 5ᵗʰ joint spinose, 6ᵗʰ broadly oval, narrowest at hinge of finger, with transverse groove on outer surface near the base, palm long, convex, defined by transverse line of strong palmar spines on inner surface, palm border striated, set on both sides with spines and spinules, finger large, curved, with channelled inner margin. Gnathopod 2 rather the larger. Peraeopods 1—5 rather robust, carrying many short spines, finger short and stout. Peraeopods 3—5, 2ᵈ joint well expanded, peraeopods 4 and 5 rather longer than 3ᵈ. Uropods 1—3 spinose, in uropod 1 outer ramus a little, in uropod 2 much shorter than the inner, which exceeds all the rest in length. Uropod 3, rami subequal. Telson rather long, tapering, cleft about to the middle, lateral margins slightly sinuous near base, apices narrow, bidentate. Colour light olive, with a few red spots on the antennae. L. 6—18 mm.

South-Pacific and southern Indian Ocean (Port Jackson, JerVis Bay and Melbourne, [Australia]; Heard Island).

2. **E. dellavallei** Chevreux 1893 *E. caesaris* (part.) (err., non T. Stebbing 1888!),
A. Della Valle in: F. Fl. Neapel, *v.* 20 p. 672 t. 3 f. 8; t. 17 f. 37—48 | 1899 *E. dellavallei*,
Chevreux in: C.-R. Ass. Franç., Sess. 27 *v.* 2 p. 479.

In close general agreement with E. monoculoides (p. 345), but the large,
reniform eyes are white with a faint rose tint; in antenna 1 the 3^d joint is
more than half as long as the 2^d; maxilla 1 has only one seta on the·
inner plate and only one on outer margin of the palp's 2^d joint; maxilla 2
has the inner plate shorter and narrower than the outer. Colour grey, tinged
with yellowish red. L. about 13 mm.

Mediterranean (Naples, France); Bay of Cadiz.

3. **E. crassi** Stebb. ·1888 *E. c.*, T. Stebbing in: Rep. Voy. Challenger, *v.* 29
p. 977 t. 90 | 1893 *E. caesaris* (part.), A. Della Valle in: F. Fl. Neapel, *v.* 20 p. 672.

Differing little from E. monoculoides (p. 345). Pleon segments 1 and 2
not produced into teeth dorsally, but forming a small acute tooth at postero-
lateral corners, segment 3 with those corners almost quadrate, with no
serration. Eyes very large, reniform, almost contiguous at top of head.
Maxilla 1, inner plate long and narrow, with 3 setae, 1st joint of palp half as long
as 2^d, 2^d with 1 seta on outer margin. Gnathopods 1 and 2, 6th joint dilated
about the middle, so that the palm instead of being continuous with the
hind margin forms a well-marked obtuse angle with it. Telson shorter
than in preceding species, cleft for less than half its length, evenly tapering
to subacute apices, which in the single specimen were simple, without any
notch. L. about 15 mm.

South-Atlantic (lat. 37° S., long. 54° W.). Depth 1092 m.

E. lippus (Hasw.) 1880 *Atylus l.*, Haswell in: P. Linn. Soc. N.S. Wales, *v.* 4
p. 328 t. 20 f. 1 | ?1885 *A. l.*, Chilton in: P. Linn. Soc. N.S. Wales, *v.* 9 p. 1037 | 1888
A. l., T. Stebbing in: Rep. Voy. Challenger, *v.* 29 p. 969.

L. 6 mm.

Port Jackson (Clark Island [East-Australia]); Sydney Harbour?

4. Gen. **Cleonardo** Stebb.

1888 *Cleonardo*, T. Stebbing in: Rep. Voy. Challenger, *v.* 29 p. 959 | 1893 *C.*,
G. O. Sars, Crust. Norway, *v.* 1 p. 414.

Body without dentate or serrate segments. Head rostrate. Side-plates 1—4
not large but notably deeper than the following, side-plate 1 not produced
forward, 4th moderately emarginate. Antenna 1 longer than antenna 2 (Fig. 82),
at least in ♂, both calceoliferous. Upper lip apically convex. Mandible normal,
3^d joint of palp longest, slender. Maxilla 1, inner plate with 2 setae, outer
with 11 spines, 2^d joint of palp not widened distally. Maxilla 2, inner plate
broader than outer. Maxillipeds, inner and outer plates not very large,
armature normal, palp long. Gnathopods 1 and 2, 5th joint short, lobed,
6th elongate oval, finger long, slender, curved. Peraeopods 1—5 (Fig. 83)
very slender, finger long. Peraeopods 3—5 long, not very unequal. Uropod 3,
rami lanceolate. Telson long, tapering, deeply cleft.

2 species.

Synopsis of species:

Uropods with the rami unequal, spinulose . . . . . . 1. **C. longipes** . . . p. 347
Uropods with the rami equal, almost unarmed . . . . 2. **C. appendiculatus** p. 347

1. **C. longipes** Stebb. 1888 *C. l.*, T. Stebbing in: Rep. Voy. Challenger, *v.* 29 p. 959 t. 86 | 1893 *Pontogeneia l.*, A. Della Valle in: F. Fl. Neapel, *v.* 20 p. 618 t. 59 f. 31.

♀ unknown. — ♂. Head, rostrum short and broad. Side-plates 1—4 rounded below. Pleon segments 1—3, postero-lateral corners more or less acutely produced. Eyes not perceived. Antenna 1, 1st joint stout, with obtuse apical tooth, 2d rather longer, 3d very short, flagellum longer than peduncle, with 46 joints, only 1st joint long, most with 2d and 3d joints of peduncle calceoliferous. Antenna 2 (Fig. 82) rather shorter, ultimate and penultimate joints of peduncle long, ultimate the shorter and much narrower, flagellum shorter than peduncle, with 35 joints, many with ultimate joint of peduncle calceoliferous. Gnathopods 1 and 2 rather strong, 5th joint cup-shaped, with spinose hind lobe, 6th oval, apically narrowest, palm long, oblique, curved, ciliated, defined by a tooth with palmar spines. Peraeopods 1 and 2, finger as long as 6th joint, with setules on the convex margin. Peraeopods 3—5, 2d joint oval, in peraeopod 5 (Fig. 83) produced somewhat downward, 5th—7th joints very long and slender. Pleopods 1—3, each pair closely united by very short coupling spines and numerous cleft spines. Uropods 1—3 very spinulose, outer ramus shorter and narrower than inner. Telson nearly reaching end of uropod 3, cleft for more than ³/₄ of length, margins finely pectinate, apices acute, not dehiscent. L. 10 mm.

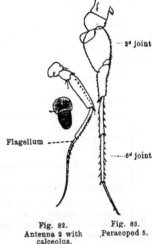

... 2d joint

Flagellum ----

......6d joint

Fig. 82.
Antenna 2 with calceolus.

Fig. 83.
Peraeopod 5.

Fig. 82 & 83. C. longipes, ♂.

South-Pacific (lat. 37° S., long. 83° W.). Depth 3244 m.

2. **C. appendiculatus** (O. Sars) 1879 *Tritropis? appendiculata*, G. O. Sars in: Arch. Naturv. Kristian., *v.* 4 p. 451 | 1885 *T. a.*, G. O. Sars in: Norske Nordhavs-Exp., *v.* 6 Crust. I p. 194 t. 16 f. 3, 3 a | 1888 *T. a.*, *Cleonardo* (part.), T. Stebbing in: Rep. Voy. Challenger, *v.* 29 p. 498, 959 | 1893 *Pontogeneia?*, A. Della Valle in: F. Fl. Neapel, *v.* 20 p. 620.

Peraeon inflated, back rounded, pleon segments 1—3 with distinct but low central carina. Head, rostrum exceedingly small, slightly curved, lateral corners slightly produced, narrowly rounded. Pleon segments 1—3, postero-lateral corners acute, postero-lateral margins quite smooth. Eyes very small, oval, placed low down, pigment light, whitish. Antennae 1 and 2 nearly as in C. longipes, but 2d joint of antenna 1 rather shorter than 1st. Mouth-parts undescribed. Gnathopods 1 and 2, peraeopods 1—5 closely resembling those in C. longipes. Uropods 1—3 each with lanceolate, equal rami, without distinct spines or setae. Telson unknown. Colour whitish, translucent. L. 13·5 mm.

Arctic Ocean (lat. 71° N., long. 13° E.). Depth 2423 m.

5. Gen. **Rhachotropis** S. I. Sm.

1871 *Tritropis* (non Fitzinger 1843, Reptilia!) (part.), A. Boeck in: Forh. Selsk. Christian., 1870 p. 158 | 1883 *Rhachotropis*, S. I. Smith in: P. U. S. Mus., *v.* 6 p. 222 | 1888 *R.*, T. Stebbing in: Rep. Voy. Challenger, *v.* 29 p. 954 | 1893 *R.*, G. O. Sars, Crust. Norway, *v.* 1 p. 423 | 1896 *Rachotropis*, J. Bonnier in: Ann. Univ. Lyon, *v.* 26 p. 653.

More or less of the body dorsally and subdorsally carinate. Head rostrate. Side-plates rather small, 1st narrowly produced forward, 4th very slightly emarginate. Pleon segment 3, postero-lateral corners rounded, serrate above. Antenna 1 generally shorter than antenna 2, and sometimes with small 2-jointed accessory flagellum. Antennae 1 and 2 larger in ♂ than in ♀. Mouth-parts nearly as in Eusirus (p. 338). Maxilla 1, outer plate with 9 spines (sometimes 8?). Gnathopods 1 and 2 strong, subequal, 5th joint short, cup-shaped, 6th large, oblong oval, palm very oblique. Peraeopods 1—5 slender, long, 5th very long, its 2d joint larger than that in 3d and 4th pairs. Uropod 3, rami foliaceous. Telson long, more or less cleft.

11 species.

Synopsis of species:

1 { Peraeopod 5, 2d joint acute at lower hind angle — 2.
    Peraeopod 5, 2d joint not acute at lower hind angle — 3.

3 { Pleon segments 1—3 having some of the dorsal teeth upturned — 4.
    Pleon segments 1—3 not having some of the dorsal teeth upturned — 5.

5 { Eyes well developed — 6.
    Eyes not well developed — 9.

1. R. aculeata (Lepech.) 1780 *Oniscus aculeatus*, Lepechin in: Acta Ac. Petrop., 1778ı p. 247 t. 8 f. 1 | 1866 *Amphithonotus a.*, Goës in: Öfv. Ak. Förh., *v.* 22 p. 526 | 1874 *A. a.* (part.), Buchholz in: Zweite D. Nordpolarf., *v.* 2 p. 316 Crust. t. 4 | 1871 *Tritropis aculeata*, A. Boeck in: Forh. Selsk. Christian., 1870 p. 158 | 1872 & 76 *Acanthonotus aculeatus + T. aculeata*, A. Boeck, Skand. Arkt. Amphip., *v.* 1 p. 37; *v.* 2 p. 511 | 1883 *Rhachotropis aculeata, R. aculata*, S. I. Smith in: P. U. S. Mus., *v.* 6 p. 229, 222 | 1893 *R. aculeata*, G. O. Sars, Crust. Norway, *v.* 1 p. 424 t. 149 | 1894 *R. aculeatus*, T. Stebbing in: Bijdr. Dierk., *v.* 17 p. 37 | 1893 *Pontogeneia aculeata* (part.), A. Della Valle in: F. Fl. Neapel, *v.* 20 p. 616 t. 59 f. 28 | 1821 *Talitrus edvardsii*, E. Sabine, An. north. Exp., p. 55 t. 2 f. 1—4 | 1835 *Amphithoe edvardsi*, J. C. Ross in: John Ross, App. sec. Voy., nat. Hist. p. 90 | 1846 *A. edwardsii*, Krøyer in: Naturh. Tidsskr., ser. 2 *v.* 2 p. 76 | 1862 *Amphithonotus e.*,

Bate, Cat. Amphip. Brit. Mus., p.151 t.28 f.5 | 1867 *Amphitonotus e.*, Packard in: Mem. Boston Soc., *v.*1 p.297 | 1882 *Tritropis avirostris*, G. O. Sars in: Forb. Selsk. Christian., nr. 18 p.105 t. 5 f. 8 | 1888 *Rhachotropis aculeatus* (part.) + *R. avirostris* (part.), T. Stebbing in: Rep. Voy. Challenger, *v.*29 p. 49, 954; p. 540.

Peraeon segments 1—5, dorsal carina faintly indicated in large specimens, segment 6 with 3 teeth, segment 7 and pleon segments 1—4 triply carinate, carinae produced into sharp teeth except the lateral of segment 4, which, however, like the 5th and 6th, has a small lateral tooth. Peraeon segment 7 has lateral corners acute. Central carina forms 2 successive teeth on each of pleon segments 1—4. Head with dorsal bump, rostrum nearly reaching end of 1st joint of antenna 1, acute, channelled above, carinate below, lateral corners with a triangular, somewhat deflexed lobe. Side-plate 1 concave below, front corner acute, 1st—3d with hind corner acute, 5th and 6th slightly ridged behind. Pleon segment 3, lateral margin very finely serrate. Eyes large, convex, rounded triangular, dark brown. Antenna 1, 1st joint large, as long as 2d and 3d combined, with apical tooth, 3d very short, 2d and 3d calceoliferous, flagellum longer than peduncle, with about 65 short joints, each with calceolus; accessory flagellum 2-jointed. Antenna 2 rather longer, ultimate and penultimate joints of peduncle subequal, calceoliferous, flagellum longer than peduncle, with about 85 calceoliferous joints. Gnathopods 1 and 2 nearly equal, 6th joint powerful, palm defined by obtuse spiniferous tooth, finger long, evenly curved. Peraeopods 1—5 moderately slender. Peraeopods 3 and 4, 2d joint rather small, hind margin proximally produced to acute triangular process. Peraeopod 5, 2d joint broad above, hind margin angular above the middle, acutely produced at apex. Uropods 1 and 2, outer ramus much the shorter; uropod 3, outer ramus a little the shorter, both narrowly lanceolate; all the rami closely spinulose. Telson very long, tapering, closely margined with spinules, cleft ¹/₃ of length, apices approximate, acute. Colour variegated with brown, orange and pure white, gnathopods 1 and 2 carmine. L. 30—44 mm.

Arctic Ocean (widely distributed, circumpolar, reaching lat. 82° 27' N.). Depth 19—263 m.

2. **R. kergueleni** Stebb. 1888 *R. k.*, T. Stebbing in: Rep. Voy. Challenger, *v.*29 p.955 t.85 | 1893 *R. k.*, G. O. Sars, Crust. Norway, *v.*1 p.424 | 1893 *Acanthozone k.*, A. Della Valle in: F. Fl. Neapel, *v.*20 p.612 t. 59 f.24.

Peraeon very short, pleon segments 1—4 long, tricarinate, each produced into a dorsal tooth, largest on segment 2, segments 1 and 2 with smaller tooth on each side. Head, rostrum very long, narrow, subdepressed, lateral corners narrow, prominent. Side-plate 1 produced linguiform, 4th moderately excavate, 5th and 6th slightly serrate behind. Pleon segment 3, lateral margin firmly serrate. Eyes not perceived. Antenna 1, 1st and 2d joints subequal, 3d not ¹/₃ as long as 2d, flagellum much longer than peduncle, with 34 joints, those of distal half very slender, accessory flagellum not perceived. Antenna 2 longer, penultimate joint of peduncle long, fringed with plumose cilia, ultimate much longer, flagellum longer than peduncle, abruptly narrower, with 37 slender joints. Mandible, molar rather small, palp very large. Gnathopods 1 and 2 similar, gnathopod 2 the larger; 5th joint with distal lobe well produced, very spinose, 6th large, broadly oval, distally narrowed, palm well defined by spinose tooth, finger long, much curved. Peraeopods 1—5 very slender. Peraeopods 3 and 4, 2d joint very small, in peraeopod 5

much larger, but still small, in all 3 oval, with lower hind corner acute.
Uropod 1, rami nearly equal; uropod 2, outer ramus somewhat shorter than
inner; uropod 3, rami nearly equal, broad and long, narrowing rather
abruptly to the acute apex; all the rami spinuliferous. Telson long, narrow,
spinuliferous, more than 3 times as long as broad, tapering little till near
the acute apices, which are separated by an exceedingly short slightly debis-
cent slit. L. 11 mm.

Southern Indian Ocean (Kerguelen Island).

3. **R. grimaldii** (Chevreux) 1887 *Tritropis g.*, Chevreux in: Bull. Soc. zool.
France, *v.* 12 p. 571 | 1888 *Rhachotropis g.*, T. Stebbing in: Rep. Voy. Challenger, *v.* 29
p. 1641 | 1893 *R. g.*, G. O. Sars, Crust. Norway, *v.* 1 p. 424.

Peraeon inflated, slightly carinate but without teeth, pleon segments 1—3
tricarinate, the carinae produced into long sharp teeth, upturned, especially
those on segment 3. Head. rostrum short, straight, forming an angle with
upper surface of head, lateral corners elongated, slightly rounded at apex.
Side-plate 1 produced into a narrow angular lobe, lower margin behind
ending in an obtuse tooth. Pleon segment 3, postero-lateral lobe prolonged
backward, broadly rounded, firmly serrate. Antenna 1, 1st joint ending in
2 denticles, 2d rather longer, 3d not quite half as long as 2d, flagellum 11-jointed;
no trace of accessory flagellum. Antenna 2 longer, ultimate and penultimate
joints of peduncle long, equal, flagellum 14-jointed. Gnathopod 1, 5th joint
with long narrow lobe. 6th large, elongate oval, finger shorter than palm,
thin, curved. Gnathopod 2 unknown. Peraeopods 3 and 4, 2d joint narrow,
smooth. Peraeopod 5, 2d joint with hind margin broadly rounded, strongly serrate.
Uropods 1—3 long and equal(?). Telson very long, cleft $^1/_3$ of length.
L. 11 mm.

North-Atlantic (Cape Finisterre). Depth 510 m.

4. **R. elegans** (Bonnier) 1896 *Rachotropis e.*, J. Bonnier in: Ann. Univ. Lyon,
v. 26 p. 658 t. 39 f. 4.

Perhaps identical with R. grimaldii. Slightly compressed, peraeon
segments 1—7 dorsally raised behind, producing a regularly undulating
profile, pleon segments 1—3 tricarinate, carinae produced into teeth, some
slightly upward turned, the central preceded by a small hump, segments 4—6
without any crest. Head with dorsal hump, rostrum very small. Pleon
segment 3, postero-lateral margins rather produced, firmly serrate. Eyes
wanting. Antenna 1 in ♂, 1st joint long, little longer than 2d, 3d very
short, flagellum longer than peduncle, accessory flagellum a single very
short joint. Antenna 2 much longer, ultimate joint of peduncle rather
longer than penultimate, flagellum much longer than peduncle. No calceoli,
but many long sensory setae. Peraeopod 5, 2d joint figured as oblong
oval, not narrowed below. Telson reaching extremity of uropod 3 (in figure
surpassing it), cleft $^2/_5$ of length. Many parts figured, but not described.
L. a little over 9 mm.

Bay of Biscay. Depth 950 m.

5. **R. oculata** (H. J. Hansen) 1887 *Tritropis o.*, H. J. Hansen in: Vid. Meddel.,
ser. 4 *v.* 9 p. 140 t. 5 f. 7—7 e | 1888 *Rhachotropis o.*, T. Stebbing in: Rep. Voy. Challenger,
v. 29 p. 1644, 1721 | 1893 *R. o.*, G. O. Sars, Crust. Norway, *v.* 1 p. 424 | 1893 *Pontogeneia
aculeata* (part.), A. Della Valle in: F. Fl. Neapel, *v.* 20 p. 616.

Body thick, broad, peraeon unarmed, except for dorsal denticle on segment 7, which has the hind corners acute, pleon segments 1—3 tricarinate, teeth acute, except dorsal tooth on segment 3, which is subacute. Head, rostrum rather small, lateral corners little produced, slightly rounded. Side-plate 1 nearly twice as broad as deep. Pleon segment 3, postero-lateral margin firmly serrate. Eyes very large, triangular, occupying major part of head's surface, nearly meeting at top, black. Antenna 1, 1st joint very stout, not very long, but as long as 2^d and 3^d combined, 3^d very short, flagellum rather longer than peduncle, 23-jointed; accessory flagellum apparently wanting. Antenna 2 subequal to antenna 1, flagellum rather shorter than peduncle. Gnathopods 1 and 2, 6th joint large, broadly oval, defining tooth of palm not very prominent. Peraeopods 1—5 long, slender, successively longer. Peraeopods 3—5, 2^d joint piriform, very small in peraeopods 3 and 4, much larger but still small in peraeopod 5. Telson more than twice as long as broad, tapering to the acute, little separated apices, cleft $^2/_5$ of length. L. ♀ 9·5, ♂ 11·5 mm.

Arctic Ocean (Greenland). Depth 19—81 m.

6. **R. inflata** (O. Sars) 1882 *Tritropis i.*, G. O. Sars in: Forb. Selsk. Christian., nr. 18 p. 104 t. 5 f. 7 | 1888 *Rhachotropis i.*, T. Stebbing in: Rep. Voy. Challenger, *v.* 29 p. 540 | 1893 & 95 *R. tumida*, *R. i.*, G. O. Sars, Crust. NorWay, *v.* 1 p. 430 t. 152; p. 697 | 1893 *Pontogeneia aculeata* (part.)?, A. Della Valle in: F. Fl. Neapel, *v.* 20 p. 617.

Body rather short, peraeon much inflated, dorsally smooth, pleon segments 1—3 tricarinate, carinae produced into small teeth, except central of segment 3, segments 4—6 quite smooth. Head very broad, rostrum rather prominent, lateral corners linguiform, slightly deflexed. Side-plates nearly as in R. helleri. Pleon segment 3, postero-lateral margins strongly serrate, serration turning the corner. Eyes very large, convex, rounded oval, or reniform, dark brown. Antenna 1 in ♀ short, without distinct calceoli, 1st joint longer than 2^d, 3^d scarcely half as long as 2^d, flagellum longer than peduncle, 9-jointed; accessory flagellum 2-jointed. Antenna 2 in ♀ little longer, ultimate and penultimate joints of peduncle subequal, flagellum half as long as peduncle, 8- or 9-jointed. Antennae 1 and 2 in ♂ much longer; antenna 1, flagellum with 12 joints, 1st very long, densely clothed with sensory setae. Gnathopod 1, 6th joint oval, breadth much more than half length; gnathopod 2, 6th joint twice as long as broad; in both the palm evenly curved, defined by slight angle carrying 1 spinule, finger long, curved. Peraeopods 1—5 slender, of moderate length, 2^d joint in peraeopod 3 very slightly expanded, in peraeopod 5 much broader, narrowing distally. Uropods 1—3, outer ramus notably shorter than inner, rami of uropod 3 narrowly lanceolate. Telson less elongate than usual. carrying a large sensory setule above the middle on each side, thence rapidly tapering to the acute contiguous apices, cleft almost to the centre. Colour generally variegated with patches of carmine and specks of golden yellow, antennae 1 and 2 and limbs banded. L. ♀ scarcely more than 6 mm, ♂ somewhat less.

Arctic Ocean, North-Atlantic, North-Sea and Skagerrak (Greenland; Kara Sea; NorWay, depth 19—94 m).

7. **R. helleri** (Boeck) 1871 *Tritropis h.*, A. Boeck in: Forh. Selsk. Christian., 1870 p. 159 | 1876 *T. h.*, A. Boeck, Skand. Arkt. Amphip., *v.* 2 p. 513 t. 20 f. 6 | 1893 *Rhachotropis h.*, G. O. Sars, Crust. NorWay, *v.* 1 p. 426 t. 150 | 1888 *R. aculeatus* (part.), T. Stebbing in: Rep. Voy. Challenger, *v.* 29 p. 954 | 1893 *Pontogeneia aculeata* (part.), A. Della Valle in: F. Fl. Neapel, *v.* 20 p. 616.

Body rather slender, peraeon segments 1—3 dorsally a little raised behind, segment 7 produced to a very small dorsal tooth, pleon segments 1—3 tricarinate, carinae ending in teeth, the central on segment 3 much smaller and more obtuse than the preceding, segment 4 with dorsal tooth, lateral ridges not produced. Head, rostrum short, lateral corners narrow triangular. Side-plate 1 linguiform, lower margin straight, corners not acute, 4th slightly emarginate, 5th and 6th without lateral ridge. Pleon segment 3, postero-lateral margins firmly serrate. Eyes rather large, rounded oval, light brown. Antenna 1 in ♀ not very long, 1st joint not as long as 2d and 3d combined, 3d about half as long as 2d, both calceoliferous, flagellum subequal to peduncle, 12—15-jointed, calceoliferous; accessory flagellum 2-jointed. Antenna 2 in ♀ a little longer, ultimate and penultimate joints of peduncle subequal, fringed above with calceoli, together as long as flagellum of about 14 joints. Antenna 1 in ♂ more than twice as long as in ♀, flagellum at base carrying fascicles of sensory filaments. Antenna 2 in ♂ rather longer, flagellum nearly twice as long as peduncle. Gnathopod 1 rather strong, somewhat smaller than otherwise similar gnathopod 2, 6th joint oblong oval, about twice as long as broad, elongate palm defined by very slight spiniferous angle. Peraeopods 1—5 more slender and with more elongate fingers than in R. aculeata (p. 348); peraeopods 3—5 rapidly encreasing in length. 2d joint simple, oblong oval, a little narrowed below, 6th joint very long in peraeopod 5, which is about 3/4 as long as the body. Uropods 1—3, outer ramus not much shorter than inner, in uropod 3 both rami subfoliaceous, abruptly tapering to the apex, marginal spinules few. Telson unarmed, not strongly tapering till near the acute contiguous apices, cleft nearly to the middle. Colour semipellucid, whitish, blotched with light red. L. ♀ 12, ♂ 10 mm.

Arctic Ocean, North-Atlantic, North-Sea and Skagerrak (South- and West-Norway). Depth 94—188 m. Other localities doubtful.

8. **R. macropus** O. Sars ?1874 *Amphithonotus aculeatus* (part.), Buchholz in: Zweite D. Nordpolarf., *v.* 2 p. 316 | 1882 *Tritropis helleri* (err., non A. Boeck 1871!), Hoek in: Niederl. Arch. Zool., suppl. 1 nr. 7 p. 58 | 1894 *Rhachotropis h.*, T. Stebbing in: Bijdr. Dierk., *v.* 17 p. 37 | 1893 *R. macropus*, G. O. Sars, Crust. Norway, *v.* 1 p. 428 t. 151 f. 1.

Very like R. helleri (p. 351), but peraeon segments not dorsally raised; segment 7 produced to very small tooth, pleon segments 1—3 tricarinate, carinae ending in teeth, central of segment 3 like the preceding, segment 4 with dorsal tooth, lateral ridges not produced. Pleon segment 3, postero-lateral margins strongly serrate, serration turning the corner. Eyes large, rounded, dark brown with whitish coating. Antenna 1 in ♀ moderately long, 1st joint as long as 2d, 3d more than half as long as 2d, both densely calceoliferous, flagellum scarcely as long as peduncle; accessory flagellum 2-jointed. Antenna 2 subequal, ultimate joint of peduncle longer than penultimate, flagellum not nearly equal the 2 joints combined. Peraeopods 1—5 much more elongated and slender, peraeopod 5 longer than the whole body. Uropods 1—3 and telson not notably differing from those of R. helleri, but telson large, extending beyond uropod 3, cleft to the centre, apices only subacute. Colour whitish, blotched with reddish orange, patches generally less confluent than in R. helleri. L. ♀ 16 mm.

Arctic Ocean, North-Atlantic, North-Sea and Skagerrak (Norway, depth 188—752 m; Barents Sea; Kara Sea; Spitzbergen.)

9. **R. leucophthalma** O. Sars 1893 *R. l.*, G. O. Sars, Crust. Norway, *v.* 1 p. 429 t. 151 f. 2.

Rather less slender than R. macropus, otherwise like it in general. Peraeon segments 1—7 without teeth, pleon as in R. macropus, but tooth of 4th segment unusually large. Head, lateral corners very narrow. Side-plates as in R. macropus. Eyes quite rudimentary, represented by 2 irregular yellowish white patches of pigment. Antenna 1 not very long, 1st joint subequal to 2[d], 3[d] more than half as long as 2[d], flagellum subequal to peduncle; accessory flagellum 2-jointed. Antenna 2 a little longer, ultimate joint of peduncle longer than penultimate, flagellum not nearly as long as both combined. Gnathopods 1 and 2 rather powerful, similar to those of R. macropus, as also peraeopods 1—5, but peraeopod 5 shorter than the body. Uropods 1—3 as in R. macropus, telson shorter, not extending beyond uropod 3, more tapering distally, cleft to the centre. Colour light yellowish, tinged with reddish orange, and pinkish on mouth-parts and gnathopods 1 and 2. L. ♀ 14 mm.

North-Atlantic, North-Sea and Skagerrak (Norway). Depth 188—752 m.

10. **R. rostrata** (Bonnier) 1896 *Rachotropis r.*, J. Bonnier in: Ann. Univ. Lyon, *v.* 26 p. 653 t. 39 f. 2.

Peraeon segments 1—7 smooth, not produced, pleon segment 1 slightly carinate behind, without lateral carinae, segments 2, 3 and 4 tricarinate, carinae produced into teeth on segment 2, ending obtusely on segment 3, the central produced into a tooth on segment 4. Head, rostrum reaching along ²/₃ length of 1st joint of antenna 1, slightly deflexed, lateral corners produced to a truncate lobe. Side-plates of the usual form; pleon segment 3, postero-lateral margins cut into 10 sharp teeth. Eyes entirely wanting. Antenna 1 in ♂, 1st joint as long as 2[d] and 3[d] combined, 2[d] calceoliferous, 3[d] less than half as long as 2[d], flagellum much longer than peduncle, with 20 joints, 1st very long, densely clothed, accessory flagellum wanting. Antenna 2 much longer, ultimate and penultimate joints of peduncle subequal, calceoliferous, flagellum longer than peduncle, with 40 elongate joints. Maxilla 1, inner plate with only 1 seta, outer with 8 spines. Gnathopods 1 and 2 nearly as in R. helleri (p. 351), but 5th joint with lobe apparently very little produced, palmar angle set with 3 strong spines. Peraeopods 1—5 slender, elongate, 2[d] joint of peraeopod 5 broad proximally. Uropods 1 and 2, outer ramus rather the shorter; uropod 3, rami lanceolate, equal; all rami minutely spinulose. Telson tapering, not reaching extremity of uropod 3, carrying a pair of plumose cilia above the centre of lateral margins, cleft less than ¹/₄ of length. L. 13 mm.

Bay of Biscay. Depth 950 m.

11. **R. gracilis** (Bonnier) 1896 *Rachotropis g.*, J. Bonnier in: Ann. Univ. Lyon, *v.* 26 p. 657 t. 39 f. 3.

Body compressed, dorsally carinate on peraeon segments 1—7 and pleon segments 1—4, carina produced into a small tooth only on pleon segments 1—4 (in text: on segments 1—3), no lateral carinae. Head, rostrum very short, lateral angles apparently acute, little produced. Side-plate 1 (in figure) not produced forward. Pleon segment 3, postero-lateral margins with only 3 scarcely visible denticles. Eyes wanting. Antenna 1 in ♂ short, 1st and 2[d] joints long, equal, 3[d] much less than half as long as 2[d], flagellum shorter than peduncle, accessory flagellum 2-jointed, nearly as long as 1st joint of flagellum. Antenna 2 in adult ♂ longer than the whole body, ultimate joint of peduncle longer than penultimate, both elongate. Upper lip, apex produced, narrowly rounded. Maxilla 1, inner plate with 1 seta, outer with 9 spines. Maxillipeds with elongate palp. Gnathopods 1 and 2 as in R. inflata (p. 351), with which also peraeopods 1—5 appear nearly to agree, but in peraeopod 5 the 5th joint is rather longer than the 6th instead of the reverse. Uropod 3, rami foliaceous, rather blunt. Telson nearly reaching extremity of uropod 3, tapering but slightly, ending in a very small triangular emargination. L. about 10 mm.

Bay of Biscay. Depth 950 m.

6. Gen. **Rozinante** Stebb.

1871 *Tritropis* (part.), A. Boeck in: Forh. Selsk. Christian., 1870 p. 158 | 1894 *Rozinante* (Sp. typ.: *R. fragilis*), T. Stebbing in: Bijdr. Dierk., *v.* 17 p. 88 | 1899 *R.,* T. Scott in: J. Linn. Soc., *v.* 27 p. 77.

Front part of pleon carinate. Head minutely rostrate. Side-plates small, 1st narrowly produced forward. Pleon segment 3, postero-lateral margins denticulate. Antennae 1 and 2 sexually varying in length. Antenna 1 shorter than antenna 2, without accessory-flagellum. Upper lip slightly emarginate. Mandible, cutting edge dentate, accessory plate on both mandibles, many spines in spine-row, molar prominent, 3d joint of palp not longer than 2d. Maxilla 1, inner plate with 4 setae, outer with 11 spines, 2d joint of palp distally widened, fringed with numerous spines. Maxilla 2, plates moderately long. Maxillipeds, inner and outer plates moderately long, finger of palp attached below the abruptly narrowed apex of 3d joint. Gnathopods 1 and 2 subchelate, 5th joint elongate, unlobed, 6th narrow, finger short. Peraeopods 1—5 slender, 5th the longest. Uropod 3 reaching much beyond the others, rami long, lanceolate, subequal, spinulose. Telson short, partially cleft.

1 species.

1. **R. fragilis** (Goës) 1866 *Paramphithoë f.,* Goës in: Öfv. Ak. Förh., *v.* 22 p. 524 t. 39 f. 16 | 1871 *Tritropis f.,* A. Boeck in: Forh. Selsk. Christian., 1870 p. 160 | 1874 *T. f.,* Buchholz in: Zweite D. Nordpolarf., *v.* 2 p. 320 Crust. t. 3 f. 1 | 1876 *T. f.,* A. Boeck, Skand. Arkt. Amphip., *v.* 2 p. 515 | 1888 *Rhachotropis f.,* T. Stebbing in: Rep. Voy. Challenger, *v.* 29 p. 856 | 1894 *Rozinante f.,* T. Stebbing in: Bijdr. Dierk., *v.* 17 p. 39 | 1899 *R. f.,* T. Scott in: J. Linn. Soc., *v.* 27 p. 77 | 1893 *Pontogeneia aculeata* (part.)?, A. Della Valle in: F. Fl. Neapel, *v.* 20 p. 617.

Pleon segments 1—3 obtusely tricarinate, carinae not produced into dorsal teeth, but notched at end of lateral keels. Head, lateral corners slightly produced, obtuse, post-antennal corners very acutely produced, with serrate upper margin. Side-plates 5—7 serrate on hind margin. Pleon segment 2, postero-lateral corners a little acutely produced, segment 3 rotundo-quadrate, postero-lateral margins of segments 1—3 finely serrate (or strongly, Buchholz). Eyes large, subglobose or subreniform, black. Antenna 1, 1st joint longer than 2d and 3d combined, all with apical teeth, 2d joint only sometimes longer than 3d, flagellum thrice as long as peduncle, of many short joints. Antenna 2, ultimate joint of peduncle as long as apically dentate penultimate, flagellum much longer than peduncle. Gnathopods 1 and 2, 5th joint as long as or rather longer than 6th, widened a little distally,

6th joint two and a half times as long as broad, widening slightly to the not very oblique palm, which is overlapped by the denticulate finger. Feraeopods 3—5; 2^d joint narrowed distally, serrate on hind margin, finger long, slender. Uropods 1 and 2, outer ramus much shorter than inner, neither very long. Uropod 3; outer ramus very little shorter than inner. Telson about once and a half as long as broad, cleft from $^1/_6$—$^1/_3$ of length, lateral margins serrate and also obtuse apices. Colour pale reddish yellow, with faint red markings. L. reaching 22·5 mm.

Arctic Ocean (Greenland; Spitzbergen; Cape Wynn, depth 6 m; Barents Sea, depth 372 m).

28. Fam. Bateidae

Head strongly rostrate. Side-plate 1 rudimentary. Antenna 1 without accessory flagellum. Mandible with palp. Gnathopod 1 degraded, without hand. Telson cleft.

Marine.

1 genus, 1 species.

1. Gen. Batea Fr. Müll.

1865 *Batea* (Sp. un.: *B. catharinensis*), Fritz Müller in: Ann. nat. Hist., ser. 3 *v.* 15 p. 276.

Body not dentate. Side-plate 4 rather large and deeply excavate behind. Antenna 1 little shorter than antenna 2. Maxillipeds, outer plates fully reaching apex of palp's 2^d joint, fringed with spine-teeth on inner margin. Gnathopod 1 degraded, ending with a feeble linear 2^d joint, which is longer in the ♀ than in the ♂. Gnathopod 2 subchelate. Peraeopods 3—5, 2^d joint expanded. Uropod 2 shorter than uropod 1 or 3; uropod 3, peduncle short, rami subfoliaceous. Telson rather short, deeply cleft.

1 species.

1. **B. catharinensis** Fr. Müll. 1865 *B. c.*, Fritz Müller in: Ann. nat. Hist., ser. 3 *v.* 15 p. 276 t. 10.

Segment 7 of peraeon much the longest. Head, rostrum (in figure) nearly as long as 1st joint of antenna 1, very blunt. Pleon segment 3, posterolateral corners subquadrate, seemingly with spinules on the straight part of hind margin. Eyes large, dark, reniform, larger in ♂ than in ♀. Antennae 1 and 2 longer in ♂ than in ♀. Antenna 1, 2^d joint narrower and a little shorter than 1st, flagellum many-jointed, setae of alternate joints directed downward in ♀, backward in ♂. Antenna 2, ultimate and penultimate joints of peduncle subequal, with fascicles of hairs on upper side, but only in ♂, flagellum many-jointed, with long upward directed setae on alternate joints, but only in ♀. Gnathopod 1 with more hairs in ♀ than in ♂. Gnathopod 2, 5th joint shorter than 6th, distally widened, 6th oval, palm subequal to hind margin, defined, but not very sharply. Peraeopods 1 and 2 with long hairs on hind margin of 5th and 6th joints, but only in ♂. Uropod 1, peduncle much longer than rami; uropod 2, peduncle little longer than rami; uropod 3,

23*

rami subequal in length, one stouter, both fringed with plumose setae. Telson with rounded apices. L.?

South-Atlantic (Desterro [Brazil]).

---

29. Fam. **Pontogeneiidae**

Body compressed, with or without dorsal teeth. Head, rostrum unimportant. Side-plates 1—4 rounded. Antenna 1, peduncle not elongate, accessory flagellum usually wanting, 1-jointed when present. Upper lip rounded. Lower lip with inner lobes weakly developed or wanting. Mouth-parts in general normal. Gnathopods 1 and 2, hands not powerful, subchelate. Uropod 3, rami subequal, of moderate size. Telson deeply cleft.

Marine.

7 genera, 9 accepted species, 4 obscure.

Synopsis of genera:

1 { Antenna 1 without accessory flagellum — **2.**
  { Antenna 1 with accessory flagellum — **6.**

2 { Body very robust, middle side-plates very deep — **3.**
  { Body not very robust, middle side-plates not
  {   very deep — **4.**

3 { Uropod 3 very short . . . . 1. Gen. **Eurymera** . . . p. 356
  { Uropod 3 not short . . . . 2. Gen. **Bovallia** . . . . p. 357

4 { Antenna 1 longer than antenna 2 . . . . . . 3. Gen. **Stebbingia** . . . p. 358
  { Antenna 1 shorter than antenna 2 — **5.**

5 { Peraeopods 3—5, 4th joint and finger normal 4. Gen. **Pontogeneia** . . p. 359
  { Peraeopods 3—5, 4th joint much dilated, finger
  {   with membranous cap . . . . 5. Gen. **Zaramilla** . . . p. 361

6 { Antenna 1 longer than antenna 2 . . . . . 6 Gen. **Atyloides** . . . p. 362
  { Antenna 1 not longer than antenna 2. . . . 7. Gen. **Paramoera** . . p. 363

1. Gen. **Eurymera** Pfeff.

1888 *Eurymera* (Sp. un.: *E. monticulosa*), Pfeffer in: Jahrb. Hamburg. Anst., *v.* 5 p. 102.

Peraeon dorsally broad. Head not rostrate. Side-plates 1—4 very large, 4th emarginate. Antennae 1 and 2 short; no accessory flagellum. Upper lip rounded. Lower lip not known. Mandible normal, well developed. Maxilla 1, inner plate with many setae, outer with 9 spines (?), 2d joint of palp broad, with many spines. Maxilla 2, inner plate rather the smaller, strongly fringed on inner margin. Maxillipeds, outer plates not reaching apex of palp's 2d joint, fringed with spine-teeth on inner and setae on apical margin, palp not very long, finger acute, with setules on inner margin. Gnathopods 1 and 2 feeble, slender, subchelate, 5th and 6th joints nearly equal in length. Peraeopods 1—5, 4th joint not especially dilated, but, especially in 3d—5th pairs, 4th and 5th joints distally produced; finger seemingly in all short, curved. Uropods 1—3, outer ramus shorter than inner; uropod 3 seemingly very short. Telson deeply cleft.

1 species.

1. E. monticulosa Pfeff. 1888 *E. m.*, Pfeffer in: Jahrb. Hamburg. Anst., *v.* 5 p. 103 t. 1 f. 3.

Back extremely broad at peraeon segments 4 and 5, hinder part of pleon compressed, peraeon and pleon segments 1—3 with transverse dorsal ridges and longitudinal lateral tubercular elevations. Head, lateral corners weakly convex, defined below by a small sharp incision, post-antennal corners strongly produced, obtusely acuminate. Pleon segment 3, postero-lateral corners rounded quadrangular. Eyes small, reniform, very bright. Antenna 1, 1ˢᵗ joint as long as 2ᵈ and 3ᵈ combined, flagellum rather longer than peduncle, with about 34 short joints, the alternate distally outdrawn. Antenna 2 subequal to antenna 1, ultimate joint of peduncle shorter than penultimate, both distally outdrawn into several lobes, flagellum a little longer than peduncle, 1ˢᵗ joint as long as the next 3 combined. Mandible. spine-row of about 12 simple spines, 2ᵈ joint of palp rather longer than 3ᵈ. Maxilla 1, inner plate with more than 20 plumose setae, 2ᵈ joint of palp twice the length of 1ˢᵗ, fringed with many setae and spine-teeth. Gnathopod 1, 5ᵗʰ joint twice as long as broad. subequal to 6ᵗʰ, which widens a little distally, the palm small, defined by an obtuse angle. Gnathopod 2 similar, but 5ᵗʰ joint more than twice as long as broad and 6ᵗʰ not widening distally; finger small in both gnathopods. Peracopods 3—5, 2ᵈ joint well expanded, 5ᵗʰ joint (in figure) shorter than 6ᵗʰ. Telson not quite twice as long as broad, triangular oval, cleft for ²/₃ of length, the lobes not dehiscent, their apices rounded. L. 27 mm.

South-Atlantic (South Georgia).

2. Gen. **Bovallia** Pfeff.

1888 *Bovallia* (Sp. un.: *B. gigantea*), Pfeffer in: Jahrb. Hamburg. Anst., *v.* 5 p. 95.

Some of the segments carinate and elevated. Head, rostrum small. Side-plates 1—4 very large, not setiferous. Antennae 1 and 2 not elongate, antenna 1 with strong peduncle, no accessory flagellum. Upper lip semi-oval, outdrawn to a little point. Mandible normal, well developed. Maxilla 1, inner plate with many setae, outer with about 10 spines, 2ᵈ joint of palp with many slender spines. Maxilla 2, inner plate rather the broader, fringed on inner margin. Maxillipeds strongly developed. outer plates closely fringed on inner margin with short spine-teeth, finger of palp acute, almost concealed among setae of 3ᵈ joint. Gnathopods 1 and 2 similar, moderately large, subchelate, 5ᵗʰ joint cup-shaped, 6ᵗʰ oval, finger strong. Peraeopods 1 and 2 slender. Peraeopods 3—5, 2ᵈ joint long oval, laminar, the rest slender. Uropods 1 and 2, outer ramus the shorter; uropod 3, rami equal, elongate, lanceolate. Telson slender. cleft to the middle. ‾

1 species.

1. B. gigantea Pfeff. 1888 *B. g.*, Pfeffer in: Jahrb. Hamburg. Anst., *v.* 5 p. 96 t. 1 f. 5.

Body thick and rather short, peraeon segments 1—5 dorsally rounded, 6ᵗʰ and 7ᵗʰ and pleon segments 1—3 carinate, each produced to a dorsal tooth, large in the last four, pleon segment 4 with deep dorsal depression. Head, rostrum a small triangular deflexed tooth, lateral corners quadrate, little produced. Side-plates 1—4 rectangular with corners slightly rounded,

4th much the largest, strongly emarginate. Pleon segment 3, postero-lateral corners subquadrate. Eyes very large, narrowly reniform, with narrow interval at top of head. Antenna 1, 1st joint as long as 2^d and 3^d combined, strong, 1st and 2^d serrately carinate below, 3^d serrate, flagellum rather longer than peduncle, joints except 1st small, the alternate distally outdrawn and in ♂ calceoliferous. Antenna 2 scarcely as long as antenna 1, ultimate joint of peduncle about $^2/_5$ as long as penultimate, flagellum as long as peduncle, similar to flagellum of antenna 1. Gnathopod 1, lobe of 5th joint rounded, 6th rather longer than 5th, broad, palm ill-defined, oblique, convex, finger slender, strongly bent, with dark chitinized point. Gnathopod 2, 6th joint apparently rather slighter than in gnathopod 1, palm straighter and rather better defined. Peraeopod 5 a little longer than peraeopod 4, considerably longer than peraeopod 3, 2^d joint with hind margin straighter than in peraeopods 3 and 4. Colour orange- to purple-red. L. 45 mm.

South-Atlantic (South Georgia).

3. Gen. Stebbingia Pfeff.

1888 *Stebbingia* (Sp. un.: *S. gregaria*), Pfeffer in: Jahrb. Hamburg. Anst., v. 5 p. 110.

Body slender, not carinate or dentate. Head scarcely rostrate. Side-plates 1—4 moderately large. Antenna 1 rather the longer, without accessory flagellum. Lower lip without inner lobes. Mandible normal. Maxilla 1, inner plate with several (about 9) setae, outer with spines close-set, outer corner of 1st joint of palp somewhat outdrawn. Maxilla 2, inner plate rather shorter and broader than outer. Maxillipeds well developed, outer plates fringed on inner margin with short spine-teeth, finger of palp narrow and rather feeble. Gnathopods 1 and 2 subchelate, similar, but 1st much weaker than 2^d. Peraeopods 1 and 2 slender; peraeopods 3—5 rather stouter, 2^d joint not greatly widened, the lower hind corner narrowly rounded. In uropod 1 outer ramus scarcely, in uropod 2 much, in uropod 3 not, shorter than inner. Telson cleft to the centre.

1 species.

1. **S. gregaria** Pfeff. 1888 *S. g.*, Pfeffer in: Jahrb. Hamburg. Anst., v. 5 p. 110 t. 2 f. 7 | 1893 *Pontogeneia g.*, A. Della Valle in: F. Fl. Neapel, v. 20 p. 904.

Head, rostrum obsolete, lateral corners feebly convex. Side-plates 1—3 directed slightly forward, rounded below, 4th much broader, also rounded below, emarginate behind. Pleon segment 3, postero-lateral corners obtusely quadrate. Eyes large, broadly reniform, narrowly separate above. Antenna 1, 1st joint rather strong, but not as long as the much narrower 2^d and 3^d combined, flagellum considerably longer than peduncle, with 56 joints, the proximal ones, except the 1st, shorter than broad, alternate joints distally outdrawn. Antenna 2 shorter, ultimate and penultimate joints of peduncle subequal, flagellum about 46-jointed, alternately calceoliferous. Mandible, spines of the spine-row are described as short, curved, hyaline, and unfeathered, the palp and also maxilla 1 are said to resemble those of Bovallia gigantea (p. 357). Gnatho-pod 1, 5th joint seemingly quadrate, 6th oval, palm moderately oblique, defined by an obtuse angle, finger curved, rather shorter than the palm. Gnatho-pod 2, 5th joint cup-shaped but not very short, 6th longer, said to be similar in structure to that of gnathopod 1, though very much larger and differently figured, with a much more oblique and less defined palm. Uropods 1—3, rami spinulose. Uropod 3, rami equal, lanceolate, acute. Telson about $^2/_5$ as broad as long, very little tapering, cleft to the centre, the lobes gradually dehiscent, with truncate notched apices. Colour dull green. L. 17 mm.

South-Atlantic (South Georgia). Under stones, at low tide.

4. Gen. **Pontogeneia** Boeck

1871 *Pontogeneia* (Sp. un.: *P. inermis*), A. Boeck in: Forh. Selsk. Christian., 1870 p. 193 | 1876 *P.*, A. Boeck, Skand. Arkt. Amphip., *v.* 2 p. 334 | 1893 *P.*, G. O. Sars, Crust. Norway. *v.* 1 p. 451 | 1893 *P.* (part.), A. Della Valle in: F. Fl. Neapel, *v.* 20 p. 615 | 1873 *Pontogenia* (non Claparède 1868, Polychaeta!), E. v. Martens in: Zool. Rec., *v.* 8 p. 190 | 1888 *Atylopsis* (part.), T. Stebbing in: Rep. Voy. Challenger, *v.* 29 p. 924.

Body without dorsal teeth. Head, rostrum small or obsolete, post-antennal corners produced. Side-plates not very large, 4th emarginate. Antennae 1 and 2 in ♂ calceoliferous and longer than in ♀, antenna 1 the shorter, without accessory flagellum. Upper lip rounded or slightly emarginate. Lower lip with small inner lobes. Mandible normal, palp strong. Maxilla 1 with about 6 setae on inner plate, 11 spines on outer, 2[d] joint of palp large. Maxilla 2, inner plate fringed on inner margin. Maxillipeds, outer plates not reaching end of palp's 2[d] joint, apically fringed with slender spines, 3[d] joint of palp produced over base of finger. Gnathopods 1 and 2 similar, feeble, subchelate, palm shorter than hind margin of 6th joint. Peraeo-pods 3—5, 2[d] joint well expanded, the remaining joints normally slender. Uropod 3, peduncle short. Telson deeply cleft.

3 species accepted, 3 doubtful.

Synopsis of accepted species:

1 { Gnathopod 2, 5th joint longer than 6th . . . . . 1. **P. inermis** . . . p. 359
  { Gnathopod 2, 6th joint longer than 5th — **2**.

2 { Apices of telson serrate . . . . 2. **P. magellanica** . p. 360
  { Apices of telson smooth . . . . 3. **P. danai** . . . . p. 360

1. **P. inermis** (Kröyer) ? 1780 *Oniscus abyssinus*, O. Fabricius, Fauna Groenl., p. 261 | 1838 *Amphithoe inermis* (♀) + *A. crenulata* (♂), (Reinhardt in MS.) Kröyer in: Danske Selsk. Afh., *v.* 7 p. 275 t. 3 f. 11 a—g; p. 278 t. 3 f. 12 a—g | 1862 *Atylus i.* + *A. crenulatus*, Bate, Cat. Amphip. Brit. Mus., p. 138 t. 26 f. 5; p. 139 t. 26 f. 6 | 1866 *Paramphithoë i.*, Goës in: Öfv. Ak. Förh., *v.* 22 p. 524 | 1874 *P. i.*, Buchholz in: Zweite D. Nordpolarf., *v.* 2 p. 366 | 1871 *Pontogeneia i.*, A. Boeck in: Forh. Selsk. Christian., 1870 p. 194 | 1876 *P. i.*, A. Boeck, Skand. Arkt. Amphip., *v.* 2 p. 335 t. 21 f. 4 | 1893 *P. i.*, G. O. Sars, Crust. Norway, *v.* 1 p. 451 t. 159 | 1893 *P. i.*, A. Della Valle in: F. Fl. Neapel, *v.* 20 p. 617 t. 59 f. 29.

Back evenly rounded. Head, rostrum very small, lateral corners obtusely rounded, separated by an angular incision from the triangular post-antennal lobes. Side-plates 1—3 rounded quadrangular, lower margin minutely crenulate, 4th considerably larger. Pleon segment 3, postero-lateral corners scarcely produced, lateral margin bulging above abruptly. Eyes oblong reniform, light red. Antenna 1, 1st joint not much longer than 2[d], 2[d] twice as long as 3[d], 3[d] a little produced at apex, flagellum about twice as long as peduncle, many-jointed. Antenna 2, ultimate and penultimate joints of peduncle sub-equal, flagellum about twice as long as peduncle. Antennae 1 and 2 in ♂ with large calceoli on confronting edges of ultimate and penultimate joints of peduncle, but none on the long slender flagella. Maxillipeds, outer plates fringed with spines on inner margin. Gnathopods 1 and 2, 5th joint slender, longer than 6th, 6th slightly widening distally, palm somewhat oblique, slightly defined, finger small; 6th joint rather more widened distally in gnathopod 2 than in gnathopod 1, with the palm rather less oblique and therefore better defined. Peraeopods 3—5, 2[d] joint rounded oval with hind margin smooth, that of peraeopod 5 much the largest. Uropod 2, outer ramus much shorter than inner; uropod 3, rami subequal, lanceolate, about twice as long as peduncle, fringed with spinules and setae. Telson oblong oval, cleft beyond the middle, lobes rather dehiscent, obtusely pointed. Transparent, colourless. L. ♀ 12 mm, ♂ somewhat less.

Arctic Ocean and North-Sea (Greenland, Siberia, West-Norway).

2. **P. magellanica** (Stebb.) 1888 Atylopsis magellanicus, T. Stebbing in: Rep. Voy. Challenger, v. 29 p. 925 t. 79 | 1893 Atylus m., A. Della Valle in: F. Fl. Neapel, v. 20 p. 701.

Head angular in front, not rostrate. Side-plate 1 distally widened, 4th much the widest. Pleon segment 3, postero-lateral angles acutely upturned, and lateral margin bulging above a small sinus. Eyes rather large, reniform. Antenna 1, 1st joint rather longer than 2[d], remainder, and all but the basal joints of antenna 2, unknown. Upper lip slightly emarginate, so as to be unequally bilobed. Maxillipeds, outer plates fringed with submarginal spinules. Mouth-parts otherwise agreeing well with P. inermis (p. 359). Gnathopod 1, 5th joint as long as 6th, 6th more than twice as long as broad, palm only slightly oblique, defined by a group of spines at the obtuse angle, finger small, curved, denticulate on inner margin. Gnathopod 2 rather similar but longer, 5th joint not so long as 6th, which is rather widened distally. Peraeopods 3—5 nearly as in P. inermis, but hind margin of 2[d] joint slightly serrate. Uropod 3, peduncle shorter than one of the rami, the other ramus unknown. Telson not longer than peduncle of uropod 3, not much longer than broad, cleft for about $^2/_3$ of length, lobes a little dehiscent near the broad apices, which form 3 denticles on their slope from the lateral margins. L. about 7 mm.

Strait of Magellan (Cape Virgins). Depth 100 m.

3. **P. danai** (G. M. Thoms.) 1879 Atylus dania, A. danai, G. M. Thomson in: Tr. N. Zealand Inst., v. 11 p. 238, 248 t. 10 f. C 1 | 1879 A. danai, G. M. Thomson in: Ann. nat. Hist., ser. 5 v. 4 p 330 | 1893 A. d., A. Della Valle in: F. Fl. Neapel, v. 20 p. 703.

Head with short rostrum, post-antennal corners blunt, little produced. Eyes large, round, black. Antenna 1 about $^1/_3$ shorter than antenna 2, joints of peduncle short, dentate on lower margin, flagellum with more than 25 joints, every 3[d] or 4th produced to a setose tubercle. Antenna 2, ultimate and penultimate joints of peduncle rather short, equal, flagellum 40—50-jointed. Side-plates, mouth-parts and peraeopods apparently in very close agreement with those of P. magellanica, and gnathopods 1 and 2 not very different, but with palm imperfectly defined and finger smooth. Also gnathopod 1, at least in ♀, has the 6th joint larger and more ovate than that of gnathopod 2. Uropods 1 and 2, peduncle longer than rami, outer ramus shorter than inner. Uropod 3, rami about twice as long as peduncle, thickly studded with short spines and fringed with long cilia, probably in both sexes, certainly in ♀. Telson cleft to nearly half its length, apices rounded, smooth, with a spinule somewhat remote from them on each side. Colour semi-transparent, with dark bluish spots. L. 7·5 mm.

South-Pacific (Dunedin [New Zealand]). Rock pools.

P. capensis (Dana) 1853 &.55 *Iphimedia c.*, J. D. Dana in: U. S. expl. Exp., *v.* 13 п p. 931 t. 63 f. 5 a—g | 1862 *Atylus c.*, Bate, Cat. Amphip. Brit. Mus., p. 141 t. 27 f. 4.

Possibly identical with P. magellanica. L. 8 mm.

Near to Cape of Good Hope.

P. fissicauda (Dana) 1852 *Amphithoe f.*, J. D. Dana in: P. Amer. Ac., *v.* 2 p. 214 | 1853 & 55 *Iphimedia f.*, J. D. Dana in: U. S. expl. Exp., *v.* 13 п p. 929 t. 63 f. 4 | 1862 *Atylus f.*, Bate, Cat. Amphip. Brit. Mus., p. 141 t. 27 f. 3.

South-Pacific (north of Valparaiso).

P. tasmaniae (G. M. Thoms.) 1893 *Atyloides t.*, G. M. Thomson in: P. R. Soc. Tasmania, 1892 p. 21 t. 2 f. 9—15; t. 3 f. 1, 2.

L. 7 mm.

Pirates Bay [Tasmania]. Rock pools.

5. Gen. Zaramilla Stebb.

1888 *Zaramilla* (Sp. un.: *Z. kergueleni*), T. Stebbing in: Rep. Voy. Challenger, *v.* 29 p. 866 | 1890 *Zaramella*, Beddard in: Zool. Rec., *v.* 25 Index p. 17.

Back round, not dentate. Head not rostrate. Side-plates 1—4 not very large, setiferous. Antennae 1 and 2 short; no accessory flagellum. Upper lip rounded. Lower lip, inner lobes not sharply separated from outer. Mandible normal, well developed. Maxilla 1, inner plate with many setae, outer with 9 spines, 2^d joint of palp large. Maxilla 2, inner plate rather the smaller, strongly fringed on inner margin. Maxillipeds, outer plates not nearly reaching apex of palp's 2^d joint, fringed with spine-teeth on inner margin and setae on apical, finger slender, with a sharp nail. Gnathopods 1 and 2 feeble, slender, subchelate, 5th and 6th joints nearly equal in length. Peraeopods 1—5 setose, 4th joint long, dilated, especially in 3^d—5th pairs. finger very short in 1st and 2^d, long in 3^d—5th, in all with a membranous cap over the nail. Uropod 3, rami subequal, lanceolate. Telson not long, deeply cleft.

1 species.

1. **Z. kergueleni** Stebb. 1888 *Z. k.*, T. Stebbing in: Rep. Voy. Challenger, *v.* 29 p. 867 t. 66.

Head, lateral corners rounded, little produced. Side-plates 1—4 rounded below; 4th much wider than 1st, 2^d or 3^d, emarginate above. Pleon segment 3, postero-lateral corners quadrate. Eyes large, nearly meeting at top of head, oval, dark. Antenna 1, 2^d joint a little shorter than 1st, and 3^d than 2^d, flagellum as long as peduncle, with 12 joints, some in ♂ with calceoli. Antenna 2 a little longer, ultimate joint of peduncle slightly longer than penultimate, flagellum subequal to peduncle, with 14 unequal joints, the longer more expanded distally and in ♂ with calceoli. Mandibles, on the left both plates strongly dentate, on the right accessory plate denticulate; spine-row, molar and palp strongly developed. Gnathopod 1, 5th joint in the ♂ a little shorter, in the ♀ a little longer than the 6th; 6th oblong oval, palm oblique, crenulate, defined by a group of spines, finger not long. Gnathopod 2, 5th joint in ♂ considerably, in ♀ a little, shorter than 6th, otherwise nearly as in gnathopod 1. Peraeopods 1 and 2, 4th joint broader and much longer than 5th, 5th broader and a little longer than 6th. Peraeopods 3—5, 2^d joint broadly oval, hind margin serrulate, 4th subequal in length to 2^d, much broader than 5th, somewhat decurrent behind, 5th longer and much broader than 6th, which is subequal to the tapering, minutely pectinate finger. Uropod 1, rami equal, rather shorter than peduncle; uropod 2 shorter and stouter, outer ramus shorter than inner; uropod 3, peduncle much shorter than rami, outer ramus more slender, scarcely shorter than inner, both spinulose. Telson once and a half as long as broad, cleft $^3/_4$ of length, with 3 pairs of dorsal spinules, apices acute, slightly divergent. L. ♀ 9 mm.

Southern Indian Ocean (Kerguelen Island). At the surface.

6. Gen. **Atyloides** Stebb.

1888 *Atyloides* (part.), T. Stebbing in: Rep. Voy. Challenger, *v.* 29 p. 913.

Body not carinate or dentate. Side-plates 1—4 rather deep, 4th emarginate behind. Antenna 1 longer than antenna 2, with 1-jointed accessory flagellum. Upper lip rounded, lower lip without inner lobes. Mandible normal, 3^d joint of palp distally widened, not as long as 2^d. Maxilla 1, inner plate with many setae, outer with 11 spines, 2^d joint of palp with many spine-teeth. Maxilla 2, inner plate fringed on inner margin. Maxillipeds, inner plates rather long, outer not reaching end of palp's 2^d joint, fringed with spine-teeth on inner margin. Gnathopods 1 and 2 slender, unequal, 5th joint elongate, 6th joint with parallel margins and very short palm. Peraeopods 3—5, 2^d joint oval. Branchial vesicles simple. Uropods 1 and 2, outer ramus rather shorter than inner; uropod 3, rami lanceolate, subequal. Telson deeply cleft.

1 species.

1. **A. serraticauda** Stebb. 1888 *A. s.*, T. Stebbing in: Rep. Voy. Challenger, *v.* 29 p. 920 t. 78 | 1893 *Atylus s.*, A. Della Valle in: F. Fl. Neapel, *v.* 20 p. 702.

Body with downy surface. Head, rostral projection minute, lateral margins subtruncate, post-antennal corners not produced. Side-plate 5, hind lobe scarcely deeper than front. Pleon segment 3, postero-lateral corners bidentate, the upper tooth the larger. Eyes large, reniform, nearly meeting at top of head, retaining colour in spirit. Antenna 1 of $^2/_3$ length of body, 1st joint longer than 2^d and 3^d combined, 2^d twice as long as 3^d, flagellum 5 times as long as peduncle, with more than 62 joints. Antenna 2 much shorter, ultimate joint of peduncle a little longer than penultimate, flagellum more than twice as long as peduncle, with 34 joints, 1st longest. Gnathopod 1, 5th joint much longer than 6th, both fringed with varied spines, 6th joint about thrice as long as broad, palm convex, slightly oblique, finger not reaching beyond it. Gnathopod 2 similar, but much longer, 5th and 6th joints equal, 6th 5 times as long as broad. Peraeopods 1 and 2, 4th and 6th joints equal, rather longer than 5th, all spinose, finger short, curved. Marsupial plates very large in gnathopod 2 and peraeopod 1, smaller in peraeopod 2, and small in peraeopod 3. Peraeopods 3—5, 4th joint rather produced behind. Uropod 3, rami about once and a half as long as peduncle, serrate on inner margin, with spines on both margins, apices acute, inner ramus with pectinate inner margin. Telson longer than peduncle of uropod 3, longer than broad, only slightly tapering, cleft $^3/_4$ of length, the broad inward-sloping apices each cut into 5 teeth. L. about 8 mm.

Port Phillip (Melbourne [southern Australia]). Depth 60 m.

7. Gen. **Paramoera** Miers

1875 *Paramoera* (part.), Miers in: Ann. nat. Hist., ser. 4 *v.* 16 p. 75 | 1882 *Paramera*, (Miers in:) Scudder, Nomencl. zool., suppl. L. p. 247 | 1888 *Atyloides* (part.), T. Stebbing in: Rep. Voy. Challenger, *v.* 29 p. 913.

Body not carinate or dentate. Pleon segments 5 and 6 not coalesced. Side-plates 1—4 of moderate size, 4th emarginate behind, 5th with hind lobe the deeper. Antenna 1 not longer than antenna 2, with 1-jointed accessory flagellum. Upper lip rounded. Mandible normal, but 3d joint of palp as long as 2d. Maxilla 1, inner plate with many setae, outer with 11 strong denticulate spines, 2d joint of palp with many spine-teeth. Maxilla 2, inner plate fringed on inner margin. Maxillipeds, inner plates rather large, outer not reaching apex of palp's 2d joint, fringed on inner margin with spine-teeth. Gnathopods 1 and 2, 5th joint shorter than 6th, especially in ♂, 6th joint oblong, much wider in ♂ than in ♀, palm not very oblique, well defined. Peraeopods 1—5, 5th joint shorter than 6th, finger small. Branchial vesicles simple. Uropods 1 and 2, outer ramus the shorter; uropod 3, rami long, lanceolate, subequal, reaching back as far as the 1st. Telson deeply cleft.

1 species accepted. 1 obscure.

1. **P. austrina** (Bate) 1862 *Atylus austrinus*, Bate, Cat. Amphip. Brit. Mus., p. 137 t. 26 f. 4 | 1893 *A. a.*, A. Della Valle in: F. Fl. Neapel, *v.* 20 p. 702 | 1875 *Paramoera australis*, Miers in: Ann. nat. Hist., ser. 4 *v.* 16 p. 75 | 1875 *Atylus a.*, Miers in: Ann. nat. Hist., ser. 4 *v.* 16 p. 117 | 1879 *A. a.*, Miers in: Phil. Tr., *v.* 168 p. 208 t. 11 f. 5, 5 a—g | 1880 *A. megalophthalmus*, Haswell in: P. Linn. Soc. N.S. Wales, *v.* 5 p. 102 t. 6 f. 4 | ?1885 *A. m.*, Chilton in: P. Linn. Soc. N.S. Wales, *v.* 9 p. 1037 | 1888 *Atyloides australis* + *A. assimilis*, T. Stebbing in: Rep. Voy. Challenger, *v.* 29 p. 914 t. 75, 76; p. 918 t. 77.

Body rather compressed. Head, rostrum obsolete (Chilton: about $^4/_5$ as long as 1st joint of antenna 1), lateral margins inconspicuously bilobed, post-antennal corners not produced. Pleon segment 3, postero-lateral corners rounded, segments 1—3 dorsally microscopically scabrous. Eyes rather large, reniform, black. Antenna 1, 1st joint as long as 2d and 3d combined, flagellum tapering, with 16—55 joints, alternate joints widening at apex, accessory flagellum (Miers: not existing) consisting of 1 joint shorter than 1st of primary, tipped with 2 setae. Antenna 2 scarcely longer (length variable, Haswell), ultimate and penultimate joints of peduncle subequal, flagellum 20—56-jointed. Gnathopods 1 and 2, 5th joint in young and ♀ not greatly shorter than the narrowly oblong 6th, in the ♂ much shorter, while the 6th joint is much wider; both these joints in gnathopod 2 are longer than in gnathopod 1, without being wider; the palm is short, slightly oblique, well defined by palmar spines. Peraeopods 3—5, 2d joint oval, widest on the last pair. Uropod 3, rami more than twice as long as peduncle, acute, with spines on each margin. Telson longer than peduncle of uropod 3, about $^2/_3$ as broad as long, cleft $^2/_3$ of length, tapering, with a spinule in each bidentate apex (Chilton: oblong, cleft about to middle, apices rounded on outer margin). L. reaching 17 mm.

Southern Indian Ocean (Kerguelen Island, rocky beaches and surface to 45 m; Cape Agulhas, from screw of Challenger); South-Pacific (Sydney, Port Jackson).

P. simplex (Dana) 1852 *Amphitoe (Iphimedia) s.*, J. D. Dana in: P. Amer. Ac., *v.* 2 p. 217 | 1893 *A. (I.) s.*, A. Della Valle in: F. Fl. Neapel, *v.* 20 p. 585 | 1853 & 55

I. s., J. D. Dana in: U. S. expl. Exp., *v.* 13 π p. 927 t. 63 f. 2 a—i | 1862 *Atylus s.*, Bate, Cat. Amphip. Brit. Mus., p. 140 t. 27 f. 2 | 1880 *A. microdeuteropus*, Haswell in: P. Linn. Soc. N.S. Wales, *v.* 5 p. 102 t. 6 f. 2.

Substantially agreeing and probably identical with P. austrina. L. 6—10 mm.

South-Atlantic (Hermit Island); South-Pacific (Port Jackson, Botany Bay).

---

30. Fam. **Gammaridae**

1813/14 *Gammaridae* (part.), Leach in: Edinb. Enc., *v.* 7 p. 432 | 1882 *G.*, G. O. Sars in: Forh. Selsk. Christian., nr. 18 p. 28 | 1888 *G.*, T. Stebbing in: Rep. Voy. Challenger, *v.* 29 p. 1004 | 1894 *G.*, G. O. Sars, Crust. Norway, *v.* 1 p. 481 | 1871 & 76 Subfam. *Gammarinae*, A. Boeck, Skand. Arkt. Amphip., *v.* 1 p. 74; *v.* 2 p. 362 | 1874 „*Gammariden*", B. Dybowsky in: Horae Soc. ent. Ross., *v.* 10 suppl. p. 6 | 1893 *Gammaridi* (part.), A. Della Valle in: F. Fl. Neapel, *v.* 20 p. 620.

Body (Fig. 84, 85, 89) more or less slender, pleon segments 4—6 usually well defined. Antennae (Fig. 86) generally rather slender, and, as a rule, but little different in the two sexes; accessory flagellum of antenna 1, often greatly developed, may dwindle sometimes to a single joint or tubercle, or disappear entirely. Mouth-parts (Fig. 87, 88) normal. Upper lip with a rounded, entire or only slightly emarginate distal border; lower lip with inner lobes well developed, slightly indicated, or absent. Mandible with dentate cutting edge and accessory plate, spine-row, molar, and 3-jointed palp (only in one genus 2^d joint smaller than 1st). Maxilla 1 with inner and outer plate and 2-jointed palp (only exceptionally 1st joint as large as 2^d). Maxillipeds with inner and outer plates and palp well developed. Gnathopods 1 and 2 generally rather powerful, rarely less than subchelate, seldom both weak; sometimes stronger and larger in ♂ than in ♀. Peraeopods more or less slender, sometimes stout, sometimes 2^d joint in peraeopods 3—5 little expanded. Pleopods lose only by exception one ramus and uropod 3 one or both of the rami. Uropod 3, rami more or less foliaceous, projecting beyond uropods 1 and 2; uropod 3 and telson never hooked. Telson (Fig. 89 p. 481) either with entire margin or cleft to the base, sometimes strikingly different in the two sexes.

The greater part of the family lives in fresh and brackish water of seas, lakes, rivers; also in wells and other subterranean waters; rarely in the ocean. Cosmopolitan.

52 genera, 243 accepted and 48 doubtful species.

Synopsis of genera:

1 ⎰ Pleon (Fig. 84), segments 4—6 coalesced — **2.**
  ⎱ Pleon (Fig. 85), segments 4—6 not coalesced — **3.**

2 ⎰ Eyes wanting, body pellucid . . . . 1. Gen. **Boruta** . . . . p. 367
  ⎱ Eyes present, body not pellucid . . . 2. Gen. **Synurella** . . . . p. 368

3 ⎰ Pleopods with a single ramus . . . . 3. Gen. **Paracrangonyx** . . . . p. 369
  ⎱ Pleopods with two rami — **4.**

4 ⎰ Uropod 3 without rami . . . . . . 4. Gen. **Apocrangonyx** . . . . p. 370
  ⎱ Uropod 3 not without rami — **5.**

5 ⎰ Uropod 3 with a single ramus — **6.**
  ⎱ Uropod 3 with two rami — **7.**

6 {	Peraeopod 5 long, 2^d joint moderately expanded	5. Gen. **Crangonyx**. . . . . . p. 370	
	Peraeopod 5 short, 2^d joint greatly expanded	6. Gen. **Hyalellopsis** . . . . . p. 374	

7 {	Telson sometimes entire, sometimes emarginate or cleft	7. Gen. **Pallasea** . . . . p. 374	
	Telson entire, with or without emargination — **8**.		
	Telson cleft — **15**.		

8 { Telson not emarginate — **9**.
Telson emarginate — **11**.

9 { Uropod 3, rami long . . . . 8. Gen. **Weyprechtia** . . . . . p. 380
Uropod 3, rami short — **10**.

10 { Antenna 1, accessory flagellum short 9. Gen. **Paramicruropus** . . . p. 382
Antenna 1, accessory flagellum long 10. Gen. **Parapherusa** . . . . . p. 383

11 { Uropod 3, inner ramus longer than outer 11. Gen. **Amathillopsis** . . . . p. 384
Uropod 3, inner ramus shorter than outer — **12**.

12 { Uropod 3, inner ramus well developed 12. Gen. **Gammarellus** . . . . p. 386
Uropod 3, inner ramus very short — **13**.

13 { Body smooth . . . . 13. Gen. **Eucrangonyx** . . . . p. 388
Body with dorsal carina — **14**.

14 { Peraeopods 3—5 rather short . . . . 14. Gen. **Axelboeckia** . . . . . p. 391
Peraeopods 3—5 very long . . . . . 15. Gen. **Brachyuropus** . . . . p. 392

15 { Antenna 1 without accessory flagellum 16. Gen. **Macrohectopus** . . . . p. 394
Antenna 1 with accessory flagellum — **16**.

16 { Antenna 1, accessory flagellum 1- or 2-jointed — **17**.
Antenna 1, accessory flagellum of more than 2 joints — **29**.

17 { Maxilla 1, 2^d joint of palp not longer than 1st . . . . 17. Gen. **Cardiophilus** . . . . . p. 395
Maxilla 1, 2^d joint of palp longer than 1st — **18**.

18 { Uropod 3 short — **19**.
Uropod 3 well developed — **20**.

19 { Body more or less carinate . . . . . 18. Gen. **Brandtia** . . . . p. 395
Body not carinate . . . . 19. Gen. **Micruropus** . . . . . p. 398

20 { Uropod 3, outer ramus much longer than inner — **21**.
Uropod 3, outer ramus not much longer than inner — **25**.

21 { Uropod 3, outer ramus 1-jointed . . 20. Gen. **Neoniphargus** . . . . p. 404
Uropod 3, outer ramus 2-jointed — **22**.

22 { Maxilla 1, inner plate with few setae . 21. Gen. **Niphargus** . . . . . . p. 405
Maxilla 1, inner plate with many setae — **23**.

23 { Uropod 3, 2^d joint of outer ramus long . . . . 22. Gen. **Eriopisa** . . . . p. 411
Uropod 3, 2^d joint of outer ramus very short — **24**.

24 ⎰ Gnathopods 1 and 2 subequal . . . 23. Gen. **Gmelina** . . . . p. 412
   ⎱ Gnathopod 1 much stronger than gna-
        thopod 2 . . . . 24. Gen. **Gmelinopsis** . . . . . p. 414

25 ⎰ Gnathopod 2 not larger than gna-
        thopod 1 — **26.**
   ⎱ Gnathopod 2 larger than gnatho-
        pod 1 — **27.**

26 ⎰ Uropod 3, peduncle shorter than outer
        (1-jointed?) ramus . . . . 25. Gen. **Hakonboeckia** . . . . p. 415
   ⎱ Uropod 3, peduncle as long as outer
        (2-jointed) ramus . . . . 26. Gen. **Baikalogammarus** . . p. 416

27 ⎰ Mandible, 2d joint of palp shorter
        than 1st . . . . 27. Gen. **Parelasmopus** . . . . p. 417
   ⎱ Mandible, 2d joint of palp longer
        than 1st — **28.**

28 ⎰ Uropod 3, rami narrowly lanceolate . 28. Gen. **Cheirocratus** . . . . . p. 417
   ⎱ Uropod 3, rami laminar, apically
        rounded . . . . 29. Gen. **Megaluropus** . . . . . p. 420

29 ⎰ Body not at all or scarcely carinate — **30.**
   ⎱ Body distinctly carinate — **46.**

30 ⎰ Body without groups of dorsal spi-
        nules — **31.**
   ⎱ Body with groups of dorsal spinules — **40.**

31 ⎰ Uropod 3 large — **32.**
   ⎱ Uropod 3 not large — **36.**

32 ⎰ Uropod 3, rami very unequal . . . . 30. Gen. **Melita** . . . . p. 421
   ⎱ Uropod 3, rami not very unequal — **33.**

33 ⎰ Antenna 2 stout . . . . 31. Gen. **Paraceradocus** . . . . p. 429
   ⎱ Antenna 2 slender — **34.**

34 ⎰ Maxillae 1 and 2 with inner plate very
        setose . . . . 32. Gen. **Ceradocus** . . . . . . p. 430
   ⎱ Maxillae 1 and 2 with inner plate not
        very setose — **35.**

35 ⎰ Peraeopods 3—5 slender . . . . . . 33. Gen. **Maera** . . . . p. 433
   ⎱ Peraeopods 3—5 robust . . . . . . 34. Gen. **Elasmopus** . . . . . . p. 441

36 ⎰ Uropod 3, rami equal . . . . 35. Gen. **Plesiogammarus** . . . p. 446
   ⎱ Uropod 3, rami unequal — **37.**

37 ⎰ Peraeopods 1—5, 6th joint apically
        widened . . . . 36. Gen. **Iphigenella** . . . . . p. 447
   ⎱ Peraeopods 1—5, 6th joint not apically
        widened — **38.**

38 ⎰ Lower lip without inner lobes . . . 37. Gen. **Pandorites** . . . . . . p. 448
   ⎱ Lower lip with inner lobes — **39.**

39 ⎰ Antenna 1 long, with long peduncle 38. Gen. **Pherusa** . . . . p. 449
   ⎱ Antenna 1 short, with short peduncle 39. Gen. **Niphargoides** . . . . p. 450

40 ⎰ Without eyes . . . . 40. Gen. **Phreatogammarus** . . p. 453
   ⎱ With eyes — **41.**

41 ⎰ Eyes (Fig. 86) of irregular form . . . 41. Gen. **Ommatogammarus** . . p. 454
   ⎱ Eyes of regular form — **42.**

42 ⎰ Side-plate 5 dentately produced . . 42. Gen. **Odontogammarus** . . p. 456
   ⎱ Side-plate 5 not dentately produced — **43.**

43 ⎰ Pleon segments 4 and 5 each raised
        dorsally to a spiniferous tubercle 43. Gen. **Dikerogammarus** . . . p. 458
   ⎱ Pleon segments 4 and 5 not raised
        dorsally to a spiniferous tubercle — **44.**

44 { Groups of dorsal spinules limited to
pleon segments 4—6 . . . . . . 44. Gen. **Gammarus** p.460
& 47. Gen. **Heterogammarus***) . . p.494
Groups of dorsal spinules (Fig. 89)
not limited to pleon segments
4—6 — **45.**

45 { Antenna 1, peduncle longer than pe-
duncle of antenna 2 . . . . . . 45. Gen. **Poekilogammarus** . . p.477
Antenna 1, peduncle shorter than pe-
duncle of antenna 2 . . . . . . 46. Gen. **Echinogammarus** . . . p.479

46 { Pleon segments 1—3 without median
carina . . . . 48. Gen. **Parapallasea** p.497
Pleon segments 1—3 with median
carina — **47.**

47 { Body with median carina only — **48.**
Body with median and lateral carinae — **49.**

48 { Peraeopod 5, 2ᵈ joint with broad
lower hind lobe . . . . 49. Gen. **Amathillina** p.499
Peraeopod 5, 2ᵈ joint without broad
lower hind lobe . . . . 50. Gen. **Carinogammarus** . . . p.501

49 { Rostrum long . . . . 51. Gen. **Gammaracanthus** . . p.507
Rostrum short . . . . 52. Gen. **Acanthogammarus** . . p.508

1. Gen. **Boruta** Wrześn.

1888 *Boruta* (Sp. un.: *B. tenebrarum*) Wrześniowski in: Pam. Fizyjogr., *v.* 8 p.264 | 1888 *B.*, T. Stebbing in: Rep. Voy. Challenger, *v.*29 p.1656 | 1890 *B.*, Wrześniowski in: Z. Wiss. Zool., *v.*50 p. 639 | 1893 *B.*, A. Della Valle in: F. Fl. Neapel, *v.* 20 p. 647 | 1896 *Crangonyx*, Vejdovský in: SB. Böhm. Ges., nr. 10 p. 5.

Close to Synurella (p. 368). Pleon segments 4—6 coalesced (Fig. 84). Eyes wanting. Lower lip without inner lobes. Mandible, molar with a very

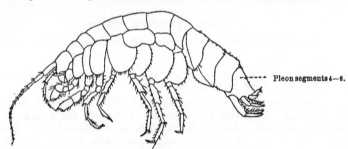

---- Pleon segments 4—6.

Fig. 84. **B. tenebrarum**, ♀.
Lateral View.
[After Wrześniowski.]

short, apically truncate seta. Maxilla 1, inner plate with 7 setae, outer with 7 spines, which are apically broad and pectinate, 2ᵈ joint of palp carrying curved setae on both the right and left maxilla. Maxillipeds with more setae on outer plates than in Synurella. Telson emarginate, not cleft.

1 species.

---

*) For the differences of these two genera see p. 494.

1. **B. tenebrarum** Wrześn. 1888 *B. t.*, Wrześniowski in: Pam. Fizyjogr., *v.* 8
{p. 72) t. 9—16 ; 1888 *B. t.*, T. Stebbing in: Rep. Voy. Challenger, *v.* 29 p. 1656 | 1890 *B. t.*,
Wrześniowski in: Z. Wiss. Zool.. *v.* 50 p. 677 t. 28 f. 6. 7, 9, 13, 15, 16; t. 29 f. 4—6; t. 30
f. 1, 2, 4, 6, 15; t. 31 f. 1, 2, 4, 6—13, 15—17; t. 32 f. 2—7, 9 | 1893 *B. t.*, A. Della Valle
in: F. Fl. Neapel, *v.* 20 p. 647.

Body moderately elongate, pleon segments carrying a few dorsal
setules. Head, front corners rather prominent, truncate. Side-plate 4 the
largest, emarginate. Pleon segments 1—3 (Fig. 84), postero-lateral corners
quadrate (figure), sharply pointed (text). Antenna 1 a third of length of body or
a little more, 1st joint of peduncle subequal to 2^d and 3^d combined, flagellum
once and a half as long as peduncle, 12- or 13-jointed, accessory flagellum 2-
jointed, shorter than first 2 joints of primary. Antenna 2 much shorter, ultimate
joint of peduncle shorter than penultimate, flagellum not equal these 2 combined,
5-or 6-jointed. Mandible, 2^d and 3^d joints of palp subequal. Gnathopod 1,
5th joint little longer than broad, 6th once and a half as long as broad, palm
almost straight, scarcely oblique. Gnathopod 2, 5th and 6th joints longer than in
gnathopod 1, especially in ♀, with palm more oblique; finger in both gnathopods
acute, curved, smooth on concave margin. Peraeopod 4 the longest. Peraeopods 3—5, 2^d joint broadly oval, with serrate margins. Uropod 1, peduncle
rather longer than rami. Uropod 2, peduncle subequal to rami. Uropod 3
subequal to peduncle of uropod 2, ramus between half and a third as long as peduncle,
tipped with a spine. Telson rather longer than broad, emarginate less than
$1/_3$ of length, the broad apices each armed with 4 spines. Branchial vesicles
attached to gnathopod 2 and peraeopods 1—5 by a narrow stalk-like beginning.
On peraeon segments 2, 3, 6 and 7 and between 1st pair of pleopod 1 there
are pairs of tube-like sacs (accessory branchiae), close together on peraeon
segments 7 and 8 and pleon segment 1, but wide apart and close to the
side-plates on peraeon segments 6 and 7, these latter being the largest.
Marsupial plates very large. Colour pellucid. L. ♀ over 7, ♂ under 4 mm.

Northern slopes of Tatry mountains. In Wells.

2. Gen. Synurella Wrzesn.

1877 *Synurella* (Sp. un.: *S. polonica*), (Wrzesniowski in:) Hoyer in: Z. Wiss.
Zool., *v.* 28 p. 403 | 1879 *Goplana*, Wrześniowski in: Zool. Anz., *v.* 2 p. 299 | 1888 *G.*,
T. Stebbing in: Rep. Voy. Challenger, *v.* 29 p. 472, 501 | 1890 *G.*, Wrześniowski in: Z.
Wiss. Zool . *v.* 50 p. 635 | 1890 *G.*, Wrześniowski in: Biol. Centralbl., *v.* 10 p. 151 | 1893
G., A. Della Valle in: F. Fl. Neapel, *v.* 20 p. 645.

Pleon segments 4—6 coalesced. Side-plates 1—4 much deeper than
the following. Eyes with few elements. Antenna 1 the longer, accessory
flagellum 2-jointed. Antenna 2, flagellum in ♂ calceoliferous. Lower lip
with inner lobes. Mandible normal, molar with a long plumose seta.
Maxilla 1, inner plate with several (7) setae, outer with 7 spines, some
furcate, some serrulate, 2^d joint of palp carrying spines and setae. Maxilla 2
with few setae on inner margin of inner plate. Maxillipeds, outer plates
with few setae, palp long. Gnathopods 1 and 2 similar, subchelate, 6th joint
little dilated. Uropod 3 with a single conical ramus. Telson emarginate
or cleft.

2 species.

Synopsis of species:

1. **S. ambulans** (Fr. Müll.) 1846 *Gammarus a.*, Friedr. Müller in: Arch. Naturg., *v.* 12r p. 296 t. 10 f. A—C | 1872 *Crangonyx a.*, A. Boeck, Skand. Arkt. Amphip., *v.* 1 p. 52 | 1879 *Goplana a.*, Wrześniowski in: Zool. Anz., *v.* 2 p. 302 | 1888 *G. a.*, T. Stebbing in: Rep. Voy. Challenger, *v.* 29 p. 217, 502 | 1893 *G. a.*, A. Della Valle in: F. Fl. Neapel, *v.* 20 p. 644.

Close to S. polonica, perhaps identical with it. Back smooth. Head without rostrum. Eyes tending to round. Antenna 1, the 2-jointed accessory flagellum very small. Uropod 3 very small. Telson cleft (?), appendages short, cylindrical, spinulose at apex. L. about 4·5 mm.

Near Greifswald, and near Berlin.

2. **S. polonica** Wrzesn. 1877 *S. p.*, (Wrzesniowski in:) H. Hoyer in: Z. wiss. Zool., *v.* 28 p. 403 | 1879 *Goplana p.*, Wrześniowski in: Zool. Anz., *v.* 2 p. 300 | 1888 *G. p.*, T. Stebbing in: Rep. Voy. Challenger, *v.* 29 p. 472, 501, 503 | 1890 *G. p.*, Wrześniowski in: Z. Wiss. Zool., *v.* 50 p. 639; t. 28 f. 13; t. 30 f. 5, 14; t. 31 f. 3 | 1893 *G. p.*, A. Della Valle in: F. Fl. Neapel, *v.* 20 p. 646.

Body rather stout, back broad, rounded. Head, front corners very prominent. Eyes of moderate size, deep brown black, pigment somewhat irregularly distributed, crystalline cones few. Uropod 3, peduncle broad, the single ramus very short, with 1 spine at the apex. Telson apically emarginate. In both sexes gnathopod 2 and peraeopod 1 have a pair of cylindrical accessory branchiae on the front rim of fleshy part of side-plates, and to peraeopods 4 and 5 and to the front rim of pleon segment 1 similar but simple accessory branchiae are attached. In the ♂ from gnathopod 2 to peraeopod 5 there are lamellar appendages homologous in situation and structure to the marsupial plates of the ♀. Colour, ♂ green, ♀ brownish-yellow, both with an irregular patch of lemon-yellow on top of the head. L. ♀ 6·5 mm, ♂ smaller.

Warsaw. In Weedy ditch.

3. Gen. **Paracrangonyx** Stebb.

1899 *Paracrangonyx* (Sp. typ.: *P. compactus*), T. Stebbing in: Tr. Linn. Soc. London, ser. 2 *v.* 7 p. 422.

Side-plates 1—4 shallow, not deeper than the rest. Eyes rudimentary. Antenna 1 longer than antenna 2, accessory flagellum small. Upper lip faintly emarginate. Lower lip with small inner lobes. Mandible, molar small, 3^d joint of palp subequal to 2^d. Maxilla 1, inner plate with 2 setae, outer with 7 spines, palp 2-jointed. Maxilla 2, outer plate rather longer and broader than inner. Maxillipeds, outer plates short, palp large. Gnathopods 1 and 2 subchelate, 6th joint oval, palm oblique. Peraeopods 1 and 2, 2^d joint narrowly ovate. Peraeopods 3—5, 2^d joint narrowly oblong. Pleopods 1—3 slight, 1-branched. Uropods 1—3 short, biramous. Uropod 3 with inner ramus very small. Telson entire.

1 species.

1. **P. compactus** (Chilton) 1882 *Crangonyx c.*, Chilton in: N. Zealand J. Sci., *v.* 1 p. 44 | 1893 *C. c.*, A. Della Valle in: F. Fl. Neapel, *v.* 20 p. 682 t. 60 f. 14 | 1894 *C. c.*, Chilton in: Tr. Linn. Soc. London, ser. 2 *v.* 6 p. 220 t. 20 | 1899 *Paracrangonyx c.*, T. Stebbing in: Tr. Linn. Soc. London, ser. 2 *v.* 7 p. 422.

Pleon rather deep, segments 1—3 not much longer than those of peraeon. Head not rostrate, lateral corners rounded. Pleon segment 3, postero-lateral corners subquadrate. Eyes represented by 2 or 3 imperfect lenses without pigment. Antenna 1, 1st joint longer than 2d, 2d once and a half as long as 3d, flagellum rather longer than peduncle, 13-jointed, accessory flagellum with 2 joints, 2d minute. Antenna 2, ultimate and penultimate joints of peduncle equal, flagellum shorter than either, 5-jointed. Gnathopod 1, 5th joint short, cup-shaped, 6th twice as long, palm about as long as hind margin, defined by spines, finger closely fitting it. Gnathopod 2 similar, but 5th joint much longer and 6th slightly shorter, with shorter and better defined palm. Peraeopods 1—5, finger short. Pleopods well armed with coupling-spines, the slender single ramus without cleft spines, such as are commonly found on 1st joint of inner ramus. Pleopod 1, ramus with 11 joints, pleopod 2 with 6, pleopod 3 with 3. Uropod 1, peduncle longer than the falciform rami, outer ramus rather the shorter. Uropod 2, peduncle little longer than the subequal rami. Uropod 3, peduncle short, outer ramus thrice as long, with small 2d joint, inner ramus rudimentary. Telson oblong ovate, with spinule at each corner of the broad apex. Colour white, semi-transparent. L. about 8 mm.

New Zealand (from pump at Eyreton, from Leeston, and from Wells at Canterbury).

4. Gen. **Apocrangonyx** Stebb.

1899 *Apocrangonyx* (Sp. typ.: *A. lucifugus*), T. Stebbing in: Tr. Linn. Soc. London, ser. 2 *v.* 7 p. 422.

Eyes wanting. Antenna 1 longer than antenna 2, accessory flagellum small. Gnathopod 2 stouter than gnathopod 1. Peraeopods stout. Uropod 3 rudimentary, without rami. Telson entire.

1 species.

1. **A. lucifugus** (O. P. Hay) 1882 *Crangonyx l.*, O. P. Hay in: Amer. Natural., *v.* 16 p. 144 | 1893 *C. l.*, A. Della Valle in: F. Fl. Neapel, *v.* 20 p. 682 | 1894 *C. l.*, Chilton in: Tr. Linn. Soc. London, ser. 2 *v.* 6 p. 219 | 1899 *Apocrangonyx l.*, T. Stebbing in: Tr. Linn. Soc. London, ser. 2 *v.* 7 p. 422.

♂. Pleon segment 3, postero-lateral corners subquadrate. Antenna 1 scarcely half as long as body, 2d joint $^2/_3$ as long as 1st, 3d joint $^2/_3$ as long as 2d, flagellum 14-jointed, accessory flagellum 2-jointed. Antenna 2, half as long as antenna 1, ultimate and penultimate joints of peduncle long, together longer than flagellum, which is about 5-jointed. Gnathopod 1, 6th joint quadrate, palm smooth-edged, fringed with spines. Gnathopod 2, 6th joint ovate, edge of palm uneven. Peraeopods 4 and 5 the longest, subequal. Uropod 1, rami equal, shorter than peduncle. Uropod 2, outer ramus nearly as long as peduncle, inner nearly twice as long as outer. Uropod 3 consisting of a single joint which is rather longer than the telson, broadly ovate, $^2/_3$ as broad as long, with 2 short spines at apex. Telson a little longer than wide, narrowing a little to the truncate apex, which has 2 stout spines at each corner. Colour pale. L. reaching 6 mm. — In ♀ gnathopod 2 has the palmar margin less oblique.

Illinois (Well in Abingdon).

5. Gen. **Crangonyx** Bate

1859 *Crangonyx* (Sp. un.: *C. subterraneus*), Bate in: P. Dublin Univ. zool. bot. Ass., *v.* 1 p. 237 | 1890 *C.*, Wrześniowski in: Z. Wiss. Zool., *v.* 50 p. 634, 697 | 1893 *C.*, A. Della Valle in: F. Fl. Neapel, *v.* 20 p. 681 | 1894 *C.*, Chilton in: Tr. Linn. Soc. London, ser. 2 *v.* 6 p. 218 | 1896 *C.*, Vejdovský in: SB. Böhm. Ges., nr. 10 p. 5 | 1872 *Stygobromus* (Sp. un.: *S. vitreus*), E. D. Cope in: Amer. Natural., *v.* 6 p. 422 | 1875 *Stygodromus*, E. v. Martens in: Zool. Rec., *v.* 10 p. 189.

Body compressed, not carinate. Side-plates 1—4 deeper than the following. Eyes present or absent. Antenna 1 the longer, accessory flagellum small, 2-jointed. Mouth-parts imperfectly known, probably near to those of Niphargus (p. 405), but (in C. flagellatus) lower lip with very small inner lobes, maxilla 1, inner plate with 6 setae, maxilla 2, inner plate partially fringed on inner margin, maxillipeds, outer plates narrow. Gnathopods 1 and 2 subchelate, 6th joint not strikingly broader than 5th. Peraeopods 3—5, 2d joint moderately expanded. Uropod 3 not elongate, with single 1-jointed ramus. Telson entire or partly cleft.

6 accepted species, 1 doubtful.

Synopsis of accepted species:

1 { Uropod 3, peduncle shorter than ramus . . . . . 1. **C. subterraneus** . p. 371
  { Uropod 3, peduncle longer than ramus — **2.**

2 { Telson not cleft — **3.**
  { Telson more or less deeply cleft — **5.**

3 { Telson more than twice as long as broad . . . . 2. **C. flagellatus** . . p. 371
  { Telson less than twice as long as broad — **4.**

4 { Apex of telson with shallow sinus . . . . 3. **C. vitreus** . . . . p. 372
  { Apex of telson slightly arcuate . . . . 4. **C. tenuis** . . . . p. 372

5 { Uropod 3, ramus more than half length of peduncle 5. **C. bifurcus** . . . p. 373
  { Uropod 3, ramus less than half length of peduncle 6. **C. recurvus** . . . p. 373

1. **C. subterraneus** Bate 1859 *C. s.*, Bate in: P. Dublin Univ. zool. bot. Ass., *v.* 1 p. 237 | 1862 *C. s.*, Bate, Cat. Amphip. Brit. Mus., p. 178 t. 32 f. 6 | 1893 *C. s.*, A. Della Valle in: F. Fl. Neapel, *v.* 20 p. 681 | 1896 *C. s.*, Vejdovský in: SB. Böhm. Ges., nr. 10 p. 5.

Side-plates 1—4 moderately deep, 4th the largest, slightly excavate behind. Pleon segments 1—3, postero-lateral corners seemingly obtusely quadrate. Eyes imperfectly formed, showing small irregular patches of lemon-coloured pigment. Antenna 1 about $\frac{1}{4}$ length of the rest of the animal, 1st joint as long as 2d and 3d combined, flagellum as long as peduncle, about 11-jointed. Antenna 2 more slender. Gnathopod 1 stouter but shorter than gnathopod 2, 5th joint about as broad as 6th, but rather shorter, 6th quadrate, palm slightly convex, not very oblique. Gnathopod 2, 5th joint much longer than broad, 6th longer but scarcely broader, fusiform, palm very oblique, ill-defined. Peraeopods 1 and 2 much more slender than the following, which are successively longer, with well expanded 2d joint. Uropods 1—3 reaching nearly equally far back. Uropod 1, rami equal, rather shorter than peduncle. Uropod 2, rami equal, subequal to peduncle. Uropod 3, ramus about twice as long as the short peduncle. Telson narrow, entire. L. 5 mm.

England (Well in Hampshire).

2. **C. flagellatus** Benedict 1896 *C. f.*, J. E. Benedict in: P. U. S. Mus., *v.* 18 p. 616.

Side-plates 1—4 moderately deep. Pleon segment 3, postero-lateral corners obtusely quadrate. Eyes wanting. Antenna 1, 1st joint rather longer than 2d, 2d nearly thrice as long as 3d, 3d thrice as long as broad, flagellum sometimes as long as body, reaching 61 joints. Antenna 2 shorter, ultimate and penultimate joints of peduncle long, subequal, flagellum reaching 19 joints. Upper lip broadly rounded, longitudinally carinate. Lower lip, inner lobes small. Mandible, cutting edge and accessory plate dentate, accessory plate slighter on right mandible, few spines in spine-row, molar prominent, strong, palp, 2d and 3d joints large, subequal, well armed. Maxilla 1, inner plate

24*

with 6 setae, outer with 7 spines, 2^d joint of palp with a few spinules on apex. Maxilla 2, inner plate with 4 setae on inner margin, both plates armed apically. Maxillipeds, inner plates with 4 spine-teeth on truncate apex, outer plates scarcely so large, not reaching apex of palp's 1st joint, fringed on apex and inner margin with slender spines, palp elongate. Gnathopods 1· and 2 similar, 5th joint triangular, cup-shaped, 6th much longer, large, broadest at base, the long, oblique, rather convex palm fringed with numerous spines, the series continued beyond a large palm-defining spine, which the large curved finger reaches. Peraeopods 1—3 slight. Peraeopods 4 and 5 rather stronger and much longer. Peraeopods 3—5, 2^d joint long oval. Uropods 1 and 2 extending much beyond uropod 3, strongly spinose. Uropod 1, peduncle longer than subequal rami. Uropod 2, peduncle subequal to slightly unequal rami. Uropod 3 short, peduncle rather broader than long, smooth, ramus much narrower and shorter, apically spinose. Telson about thrice as long as broad, slightly tapering to the truncate, very spinose apex. L. 12—14 mm.

Texas (artesian Well at San Marcos).

3. **C. vitreus** (Cope) 1872 *Stygobromus v.*, E. D. Cope in: Amer. Natural., *v.* 6 p.422 | 1888 *Crangonyx v.*, (S. I. Smith in:) Packard in: Mem. Ac. Washington, *v.* 41 p. 34 t. 5 f. 1—4.

Side-plate 4 the deepest, excavate behind, broader than any but the 5th. Pleon segments 1—3, postero-lateral corners obtusely quadrate. Eyes wanting. Antenna 1 about $^1/_3$ length of body, 1st joint as long as 2^d and 3^d combined, flagellum rather longer than peduncle, 10-jointed. Antenna 2 shorter, flagellum about 6-jointed. Gnathopod 1, 5th joint rather short and broad, 6th wider and much longer, palm making a very obtuse angle with the shorter hind margin. Gnathopod 2 much larger, 5th joint short, triangular, cup-shaped, 6th large, widest at commencement of the long palm, thence narrowing to hinge of finger, palm fringed with spines and defined from the much shorter hind margin by 2 spines much longer than the rest, finger strong, curved. Peraeopods 1—3 subequal. Peraeopods 4 and 5 subequal, much longer than peraeopods 1—3; 2^d joint in peraeopods 3—5 oblong oval. Uropods 1 and 2 extending beyond uropod 3; in uropod 1 peduncle considerably longer than the subequal rami, in uropod 2 only a little longer than the rami, of which the outer is a little shorter than the inner. Uropod 3, peduncle short, stout, little longer than broad, single ramus minute, scarcely longer than broad, with 3 spinules at apex. Telson rectangular, nearly as broad as long, apical margin with a very shallow sinus in middle, each side of which is armed with several slender spines. Colour translucent. L. ♀ over 5 mm.

Kentucky (Mammoth Cave). Furrowing in the mud of pools.

4. **C. tenuis** S. I. Sm. 1874 *C. t.*, S. I. Smith in: Rep. U. S. Fish Comm., *v.* 2 p. 656 | 1893 *C. t.*, A. Della Valle in: F. Fl. Neapel, *v.* 20 p. 682.

Slender, elongated. Side-plates very shallow, more as in Niphargus than in Crangonyx. Eyes not perceived. Antenna 1 usually longer than antenna 2, flagellum with 16—19 joints in antenna 1, with 8—10 in antenna 2, but exceptionally large ♂ had 1st antenna 5 mm long; flagellum twice as long as peduncle, with about 22 joints, antenna 2 stout, 6 mm long, flagellum as long as peduncle, 15-jointed. Gnathopods 1 and 2 differing little in the 2 sexes, gnathopod 1 the stouter, with palm more oblique, palm in both pairs armed on each side with a series of stout, obtuse spines, with a notch and a cilium near the tip (this feature appears to be common in the genus). Uropods 1—3 extending to about the same point. Uropod 1, rami subequal, scarcely half as long as peduncle. Uropod 2, rami very unequal, outer half as long as inner. Uropod 3 scarcely as long as telson, the ramus very small, tipped with 4 or 5 setiform spinules. Telson $^2/_3$ as broad as long, tapering slightly to the entire, slightly arcuate apex, which is armed with about 10 slender spinules. L. usual 6—8, largest ♂ 13·5 mm.

Connecticut (Wells at Middletown).

5. **C. bifurcus** O. P. Hay 1882 *C. b.*, O. P. Hay in: Amer. Natural., *v.*16 p.145 | 1894 *C. b.*, Chilton in: Tr. Linn. Soc. London, ser. 2 *v.* 6 p.218.

Pleon segments 1—3, postero-lateral angles drawn backward into a decided tooth. Eyes oval, black. Antenna 1, 1st and 2^d joints subequal, 3^d $^2/_3$ length of 2^d, flagellum 24-jointed. Antenna 2 half as long as antenna 1, ultimate and penultimate joints of peduncle equal, these and the flagellum carrying sensory organs, which in spirit resemble a lanceolate or oblanceolate leaf having a midrib and parallel veinlets running from this to the margin. Gnathopod 2 with 6th joint more elongate and with a more oblique palm than in gnathopod 1. Uropod 3, the ramus about $^2/_3$ as long as peduncle; a process of the peduncle is thought perhaps to represent the missing inner ramus. Telson twice as long as broad, sides nearly parallel, cleft nearly $^3/_4$ of length, each prong armed at apex with 3—5 spines. L. about 9 mm.

Mississippi (at Macon, from rivulet flowing down the limestone hills into the Noxubee-River).

6. **C. recurvus** (Grube) 1861 *Gammarus r.*, E. Grube, Ausfl. Triest, p.137 | 1864 & 66 *Crangonyx r.*, E. Grube in: Arch. Naturg., *v.*301 p.200; *v.*321 p.410 t.10 f.1 | 1893 ʻ*C. r.*, A. Della Valle in: F. Fl. Neapel, *v.*20 p.682 | 1871 *C. recurvatus*, S. I. Smith (& A. E. Verrill) in: Amer. J. Sci., ser.3 *v.*2 p.448.

Back smooth, pleon segments 1—3 not longer than the preceding. Sideplates 1—4 comparatively deep. Pleon segments 1—3, postero-lateral angles (in figure) acute. Eyes small, oval, composed of about 15 elements, pigment dispersed, black. Antenna 1 a little longer than half the body, flagellum considerably longer than peduncle, 14- or 15-jointed. Antenna 2 much shorter, ultimate and penultimate joints of peduncle equal, flagellum 5- or 6-jointed. Gnathopods 1 and 2 similar, 5th and 6th joints about equal in length and breadth, 6th longer than broad, subrectangular, palm scarcely oblique in gnathopod 1, a little more so in gnathopod 2, in neither as long as hind margin. Gnathopod 2, 5th and 6th joints longer than in gnathopod 1. Peracopods 3—5, 2^d joint oblong oval, with crenulate hind margin. Marsupial plates, except the last, very large, with long setae. Uropod 3 very short, not extending so far as 1st and 2^d, the ramus shorter than peduncle, with 1 or 2 spines at apex. Telson rather long, cleft to the middle, apices acute. L. 3—6 mm.

Island of Cherso.

C. ermannii (M.-E.) 1840 *Gammarus e.*, H. Milne Edwards, Hist. nat. Crust., *v.*3 p.49 | 1862 *Crangonyx e.*, Bate, Cat. Amphip. Brit. Mus., p.179 t.32 f.7 | 1893 *C. e.*, A. Della Valle in: F. Fl. Neapel, *v.*20 p.682.

Eyes round, small. Antenna 1 the longer, flagellum longer than peduncle. Gnathopods 1 and 2 subequal, alike, 6th joint ovate, palm oblique, not defined, scarcely leaving any hind margin. Uropod 3 short, scarcely extending beyond preceding. ramus unarmed (Bate). According to Milne Edwards uropod 3 and telson end in two conical unarmed stylets. L. 15 mm.

Warm springs of Kamtchatka.

6. Gen. **Hyalellopsis** Stebb.

1899 *Hyalellopsis* (Sp. typ.: *H. czyrniañskii*), T. Stebbing in: Tr. Linn. Soc. London, ser. 2 *v.* 7 p. 422.

Body (Fig. 85) smooth, pleon segments 4—6 very short. Head, rostrum insignificant. Side-plates 1—4 moderately deep, rounded below, setuliferous, 4th excavate behind, much deeper than 5th. Antennae 1 and 2 short, antenna 1 rather the longer, with 1-jointed accessory flagellum. Mouth-parts perhaps as in Gammarus (p. 460). Gnathopods 1 and 2 subchelate. Peraeopods 1—4, 2^d joint heart-shaped. Peraeopod 5 short, 2^d joint as long as rest of limb, much expanded and distally produced. Uropods 1—3 short, especially uropod 3, which is uniramous. Telson small, rounded.

1 species.

1. **H. czyrniañskii** (Dyb.) 1874 *Gammarus c.*, B. Dybowsky in: Horae Soc. ent. Ross., *v.* 10 suppl. p. 153 t. 9 f. 5 | 1899 *Hyalellopsis c.*, T. Stebbing in: Tr. Linn. Soc. London, ser. 2 *v.* 7 p. 422.

Surface of body (Fig. 85) free from prominences, setae, spines, or hairs. Head, lateral lobes truncate, little produced. Pleon segment 3, postero-lateral corners (in figure) rounded. Eyes small, roundish, very prominent, black. Antenna 1 about $^1/_4$ as long as body, little longer than antenna 2, peduncle

Fig. 85.
H. czyrniañskii.
Lateral view.
[After B. Dybowsky.]

stouter and somewhat longer, its 1st joint stout, flagellum 9—11-jointed, a little longer than peduncle; antenna 2, flagellum 5—8-jointed. Gnathopod 1, 6th joint piriform. more slender in ♂ than in ♀. Gnathopod 2, 6th joint oblong. Peraeopods 3 and 4, 2^d joint having the slightly convex hind margin beset with short setae. Peraeopod 5, 2^d joint having hind margin evenly rounded, produced below into a rounded lobe, and carrying short setae. while the front margin has long stout setae in fascicles. Uropods 1 and 2 are said to reach beyond uropod 3 (scarcely in figure). Uropod 3 very short, rudimentary, ramus set like a stump on the short, broad peduncle, and carrying on its flat broad apex 5 or 6 stout spines. Telson small, leaf-like, semicircular. Colour dark horn-grey. L. ♂ 11, ♀ 7 mm.

Lake Baikal. Depth 10 m.

7. Gen. **Pallasea** Bate

1793 [Subgen.] *Gammarellus* (part.), J. F. W. Herbst, Naturg. Krabben Krebse, *v.* 2 p. 106 | 1862 *Pallasea*, Bate, Cat. Amphip. Brit. Mus., p. 200 | 1867 *P.*, G. O. Sars, Crust. d'Eau douce Norvège, p. 67 | 1876 *P.*, A. Boeck, Skand. Arkt. Amphip., *v.* 2 p. 374 | 1893 *P.*, A. Della Valle in: F. Fl. Neapel, *v.* 20 p. 755 | 1899 *P.*, T. Stebbing in: Tr. Linn. Soc. London, ser. 2 *v.* 7 p. 422 | 1871 *Pallasia* (non Robineau-Desvoidy 1830, Diptera!), A. Boeck in: Forh. Selsk. Christian., 1870 p. 206 | 1894 *Pallasiella*, G. O. Sars, Crust. Norway, *v.* 1 p. 505.

Rows of processes arming the back and sides or sides alone of some of the segments. Head, rostrum obsolete. Eyes prominent. Antenna 1 the longer, peduncle long, accessory flagellum rather short. Mouth-parts normal. Gnathopods 1 and 2 subchelate. Peraeopods 3—5 rather slender, 2^d joint not greatly expanded, most so in peraeopod 5. Uropod 3 reaching to some extent beyond the others, inner ramus the shorter, but often nearly equal to the outer. Telson seldom deeply cleft.

10 species.

Synopsis of species:

1 $\Big\{$ Antenna 1, accessory flagellum 1-jointed . . . . 1. **P. dybowskii** . . . p. 375
   Antenna 1, accessory flagellum with more than one joint — **2.**

2 $\Big\{$ Head with subdorsal dentate ridges . . . . . . 2. **P. reissnerii** . . . p. 376
   Head without dentate ridges — **3.**

3 $\Big\{$ Peraeon segment 5 distinguished by its large pair of subdorsal teeth — **4.**
   Peraeon segment 5 not distinguished by subdorsal teeth — **5.**

4 $\Big\{$ Teeth of peraeon segment 5 simple, bent backward and downward . . . . 3. **P. cancellus** . . . p. 376
   Teeth of peraeon segment 5 uhcinate, raised upward 4. **P. gerstfeldtii** . . . p. 377

5 $\Big\{$ Subdorsal processes wanting on peraeon — **6.**
   Subdorsal processes present on peraeon — **8.**

6 $\Big\{$ Uropod 3, inner ramus less than half outer . . 5. **P. quadrispinosa** . p. 377
   Uropod 3, inner ramus more than half outer — **7.**

7 $\Big\{$ Peraeon segments with median tubercles . . . . 6. **P. kesslerii** . . . . p. 378
   Peraeon segments without median tubercles . . 7. **P. baikali** . . . . . p. 378

8 $\Big\{$ Pleon segments with subdorsal and marginal processes . . . . 8. **P. brandtii** . . . . p. 379
   Pleon segments without marginal processes — **9.**

9 $\Big\{$ Peraeon segments 1—4 with paired tubercles in median line . . . . 9. **P. grubii** . . . . . p. 379
   Peraeon segments 1—4 with single (almost obsolete) tubercles in median line . . . . . . 10. **P. cancelloides** . . p. 380

1. P. dybowskii Stebb. 1874 *Gammarus asper* (non J. D. Dana 1852!), B. Dybowsky in: Horae Soc. ent. Ross., *v.* 10 suppl. p. 155 t. 13 f. 1 | 1893 *Acanthonotosoma?* a., A. Della Valle in: F. Fl. Neapel, *v.* 20 p. 927 | 1899 *Pallasea dybowskii*, T. Stebbing in: Tr. Linn. Soc. London, ser. 2 *v.* 7 p. 422.

Peraeon segments 1—7 with lateral and marginal rows of tooth-like carinae or tubercles, marginal much stronger than the lateral, flattened, the first 5 almost twice as long as last 2, 5th longer and more acute than the preceding; lateral rows consisting of small tubercles, the first 5 the stronger, the lateral rows slightly oblique. Pleon segments 1—3 with lateral rows of small tubercles, obscurely continued on segments 4—6, on segments 1 and 2 a small median tubercle. Head with arched upper profile, rostral margin blunt, little projecting; on the cheek-lobe a thick rounded hump prolonged outward and downward. Side-plates 1—4 with no setae on lower margin, side-plate 4 with a central boss. Eyes not very large but very prominent, roundish, brown. Antenna 1 about $^1/_3$ as long as body, nearly twice as long as antenna 2, peduncle somewhat longer than peduncle of antenna 2, flagellum 15—17-jointed, accessory flagellum 1-jointed. Gnathopods 1 and 2, 6th joint larger in ♂ than in ♀, in gnathopod 1 piriform, in gnathopod 2 oblong. Peraeopods 3—5, 2[d] joint moderately broad, elongate oval, with short, distant setae on hind margin; in peraeopod 5 wing produced downward in a rounded lobe. Uropod 1 reaching end of peduncle of uropod 3. Uropod 2 reaching end of pleon segment 6, which is twice as long as segment 5. Uropod 3 in $\male$ $^{1}/_{7}$, in $\female$ $^{1}/_{9}$ as long as body, rami subequal, fringed with plumose setae. Telson much shorter than peduncle of uropod 3, rounded pentagonal, apical border almost truncate or faintly concave. Colour dark grey brown, with brighter markings and stripes, a dark stripe along the back, carinae darker than the rest of the body, often those to the rear red. L. 17 mm.

Lake Baikal. Depth 10 m.

2. P. reissnerii (Dyb.) 1874 *Gammarus r.*, B. Dybowsky in: Horae Soc. ent. Ross., *v.* 10 suppl. p. 126 t. 3 f. 1; t. 4 f. 7 | 1893 *Pallasea? reissneri*, A. Della Valle in: F. Fl. Neapel, *v.* 20 p. 930 | 1899 *P. reissnerii*, T. Stebbing in: Tr. Linn. Soc. London, ser. 2 *v.* 7 p. 422.

Peraeon segments 1—3 with 2 rather remote median tubercular carinae, which dwindle away on segments 4 and 5, and are replaced by a single carina on segment 6, which gradually encreases in height to pleon segment 3, successively more clearly divided by a saddle-shaped depression of the upper margin into 2 halves, of which the hinder is tooth-like, and on pleon segment 3 bent forward as a hook; pleon segment 4 with a tubercular keel, segments 5 and 6 quite flat. Peraeon segments 1—7 have marginal keels, perhaps representing a coalescence of marginal with subdorsal processes, the processes on segments 1—5 tolerably long and acute, on segments 6 and 7 blunt, tubercular. The swellings on the pleon segments are feebly indicated. Head, rostral point insignificant, upper surface rough, separated from the sides by ridges which have tolerably high, flat teeth at the ends; there is a lateral swelling above the eyes, and below and somewhat behind them a flat, double-pointed process. Eyes small, irregularly reniform, white. Antenna 1 half as long as body, about $^{1}/_{3}$ longer than antenna 2, 1st joint thick, longer than 2d and 3d combined, flagellum 32-jointed, accessory flagellum 3-jointed. Antenna 2, ultimate and penultimate joints of peduncle equal, flagellum 12-jointed. Gnathopod 1, 6th joint slender, piriform, finger short. Gnathopod 2, 5th joint longer than in gnathopod 1, 6th joint slender, oblong, palm rather oblique, finger short. Peraeopods 3 and 4, 2d joint more or less heart-shaped. Peraeopod 5, 2d joint oblong, with rather long setae on hind margin, as in peraeopods 3 and 4. Uropod 3, outer ramus about $^{1}/_{6}$ longer (not shorter, as misprinted in Dybowsky, p. 126) than the inner, 2d joint scarcely perceptible. Telson divided beyond the middle, with blunt apices (in figure). Colour white with shimmer of brown. L. $\male$ 25 mm.

Lake Baikal. Depth 1300 m.

3. P. cancellus (Pall.) 1772 *Oniscus c.* (*Cancer baikalensis* Laxman in MS.), Pallas, Spic. zool., *v.* 9 p. 52 t. 3 f. 18 | 1781 *Gammarus c.*, J. C. Fabricius, Spec. Ins., *v.* 1 p. 515 | 1874 *G. c.*, B. Dybowsky in: Horae Soc. ent. Ross., *v.* 10 suppl. p. 127 | 1793 *Cancer* (*Gammarellus*) *c.*, J. F. W. Herbst. Naturg. Krabben Krebse, *v.* 2 p. 125 t. 35 f. 12 | 1806 *C.* (*Gammarus*) *c.*, Turton in: J. F. Gmelin, Gen. Syst. Nat., *v.* 3 p. 760 | 1825 *Amphithoe c.*. A. G. Desmarest, Consid. gén. Crust., p. 268 | 1830 *A. cancella*, H. Milne Edwards in: Ann. Sci. nat., *v.* 20 p. 377 | 1862 *Pallasea cancellus* (*Gammarus latreillii* Guérin-Méneville in MS.), Bate, Cat. Amphip. Brit. Mus., p. 200 t. 36 f. 1 | 1893 *P. c.* (part.), A. Della Valle in: F. Fl. Neapel, *v.* 20 p. 755 t. 60 f. 26, 27 | 1776 *Oniscus muricatus*, Pallas, Reise Ruß., *v.* 3 p. 709.

Peraeon segments 1—7 and pleon segments 1—4 with a weak median carina formed by a small tubercle or blunt hump, the humps being strongest on the pleon segments. Immediately above the side-plates the margins of the peraeon segments form a weak carina, close above which are backward pointing blunt teeth, successively increasing towards and decreasing from those of segment 5, which are bent slightly backward and downward, and, besides being stronger than the rest, are set nearer to the middle of the back. In pleon segments 1—3 the lateral teeth are acute and rather long, on segments 4 and 5 they form weak carinae occupying about $^2/_3$ length of segment. Head, upper surface smooth, rostrum minute, a flat subacute tooth projecting outward on the under margin of the lateral lobes. Eyes reniform or half-moon-shaped, black. Antenna 1 longer than half the body, twice as long as antenna 2, 1st joint rather long, 3^d shorter than 2^d, flagellum 36—67-jointed, accessory flagellum 4—6-jointed. Antenna 2, flagellum 8—14-jointed. Maxillipeds normal (Bate: having a small squamiform plate to the ischium only, but his figure shows 2^d joint produced as well as the 3^d). Gnathopod 1, 6th joint slender, piriform. Gnathopod 2, 6th joint more rhomboidal, broader, palm oblique. Peraeopods 3—5, 2^d joint heart-shaped, with short setae on the hind margin. Uropod 3, rami subequal. Telson cleft $^1/_3$ of length, each apex armed with a spinule. Colour greenish-brown, or brown with scattered dots. L. $\circ$ attaining 70 mm to end of telson (with antenna 1:108 mm).

Lake Baikal, near the shore; River Angara.

4. **P. gerstfeldtii** (Dyb.) 1874 *Gammarus cancellus var. g.*, B. Dybowsky in: Horae Soc. ent. Ross., v. 10 suppl. p. 129 t. 2 f. 1 | 1893 *Pallasea c.* (part.), A. Della Valle in: F. Fl. Neapel, v. 20 p. 928 | 1899 *P. gerstfeldtii*, T. Stebbing in: Tr. Linn. Soc. London, ser. 2 v. 7 p. 422.

Resembling P. cancellus, but very strongly developed. On peraeon segments 1—4 a median carina of strong blunt teeth, the edge of which is often divided in segments 1 and 4; on segment 5 the carina is reduced to a little point, on segments 6 and 7 it is broad-based, spine-like, but variable in its strength and acuteness. The lateral teeth on peraeon segments 1—4 acute, successively longer, on segment 5 they are nearer to the median line, and appear as reaching high over the body, greatly curved or actually uncinate and backward bent, the distance between their tips being twice the breadth of the body at the base of the preceding pair of teeth; on segments 6 and 7 the lateral teeth are sharp; on pleon segments 1—3 they are somewhat more bent, successively shorter, and on segments 4 and 5 they are simple carinae. At the base of the lateral teeth, or between the latter and the median teeth, there is a small tubercle on peraeon segments 6 and 7 and pleon segments 1—3. As in P. cancellus, there are tubercles and spinules on the pleon segments 4 and 5, or 4—6, the position of which is not very clear. Head, the tooth on the under margin of lateral lobes is more pointed and somewhat longer than in P. cancellus. Antenna 1, 3^d joint as long as 2^d, or somewhat longer. L. attaining 75 mm to end of telson.

Lake Baikal. Depth 20—50 m.

5. **P. quadrispinosa** O. Sars 1861 *Gammarus cancelloides* (err., non Gerstfeldt 1858!), S. Lovén in: Öfv. Ak. Förh., v. 18 p. 287 | 1867 *Pallasea c. var. qvadrispinosa*, G. O. Sars, Crust. d'Eau douce Norvège, p. 68 t. 6 f. 21—34 | 1876 *P. q.*, A. Boeck, Skand. Arkt. Amphip., v. 2 p. 375 | 1871 *Pallasia q.*, A. Boeck in: Forh. Selsk. Christian, 1870 p. 207 | 1894 *Pallasiella q.*, G. O. Sars, Crust. Norway, v. 1 p. 506 t. 178 | 1874 *Gammarus kesslerii var. europaeus*, B. Dybowsky in: Horae Soc. ent. Ross., v. 10 suppl. p. 135 | 1893 *Pallasea cancellus* (part.), A. Della Valle in: F. Fl. Neapel, v. 20 p. 755.

Median carina wanting, body slender, peraeon segments 1—7 each with supramarginal obtuse nodiform prominence; pleon segments 1 and 2 each with a pair of strong, subdorsal, backward directed, parallel teeth, wide apart, segments 4—6 without fascicles of spinules. Head, rostrum minute, lateral lobes smoothly rounded, a boss-like projection overhanging post-antennal corners. Side-plates 1—4 rather narrow, not contiguous below, 4th rather less deep than 3^d. Pleon segments 1—3, postero-lateral corners rounded (as perhaps throughout the genus). Eyes rounded oval, very dark. Antenna 1 about $^1/_8$ as long as body, 1st joint nearly as long as 2^d and 3^d combined, flagellum shorter than peduncle, with about 16 (Sars), with 24—15 (Dybowsky) joints, accessory flagellum with 2 joints, 2^d the longer. Antenna 2 rather shorter, ultimate and penultimate joints of peduncle nearly equal, flagellum 6- or 7-jointed. Gnathopod 1, 6th joint piriform. Gnathopod 2 rather more slender, 6th joint oblong oval, palm nearly transverse; both pairs of gnathopods in ♂ stronger than in ♀. Peraeopods 3 and 4, 2^d joint little expanded, slightly narrowed below. Peraeopod 5, 2^d joint larger than in preceding peraeopods, oval, fringed behind with setae. Uropod 1, rami much shorter than peduncle, smooth. Uropod 2 much shorter than uropod 1, rami with smooth sides. Uropod 3, outer ramus fringed with fascicles of slender setae, 2^d joint minute, inner ramus scarcely more than $^1/_8$ as long as outer, setiferous on inner margin. Telson a little broader than long, 3 spinules on each side, emargination $^1/_4$ of the length, apices rounded, tipped with a spinule. Colour yellowish grey, each segment banded transversely with brownish green. L. ♂ 15—19 mm.

Lakes of Norway (Mjös-Sjö, depth 6—94 m, and others, also River Vorm), Sweden, and Russia (Finland, Onega).

6. **P. kesslerii** (Dyb.) 1874 *Gammarus k.*, B. Dybowsky in: Horae Soc. ent. Ross., *v.* 10 suppl. p. 133 t. 1 f. 7 | 1893 *Amathilla? k.*, A. Della Valle in: F. Fl. Neapel, *v.* 20 p. 929 | 1899 *Pallasea k.*, T. Stebbing in: Tr. Linn. Soc. London, ser. 2 *v.* 7 p. 422.

Near to P. quadrispinosa (p. 377). Median carina represented by small tubercles on peraeon segments, peraeon segments 1—5 each with supramarginal strong acute tooth, segments 6 and 7 each with only a weak hump over the margin; pleon with addition of a little median tubercle on each of segments 1—3, and a flat lateral swelling on segment 3. Head, upper surface rough, with a median groove, narrowing forward; a strong outward and somewhat downward projecting tooth overhanging post-antennal corners. Side-plates carinate. Eyes oval, very prominent, dark brown or black. Antenna 1 half as long as body, nearly twice as long as antenna 2, flagellum 20—28-jointed, accessory flagellum 2—4-jointed. Antenna 2, 7—9-jointed. Peraeopod 3, 2^d joint sometimes widened below. Uropod 3, inner ramus in adults only about $^1/_7$ shorter than outer, both carrying on their margins numerous setae, the majority plumose. Telson very feebly emarginate, and very feebly grooved above distally. Colour bright greenish, with black or brownish spots, the appendages banded. L. reaching over 37 mm.

Lake Baikal. Depth 10—20 m.

7. **P. baikali** Stebb. 1874 *Gammarus lovenii* (non *G. loveni* R. M. Bruzelius 1859!), B. Dybowsky in: Horae Soc. ent. Ross., *v.* 10 suppl. p. 137 t. 13 f. 7 | 1893 *Pallasea cancellus* (part.)?, A. Della Valle in: F. Fl. Neapel, *v.* 20 p. 929 | 1899 *P. baikali*, T. Stebbing in: Tr. Linn. Soc. London, ser. 2 *v.* 7 p. 422.

Median carina wanting, peraeon segments 1—5 with marginal teeth, thinner, somewhat longer and more backward curved than in P. cancelloides (p. 380); weak lateral swellings probably represent the lateral carinae; pleon segments without teeth or humps, segments 4—6 each with a couple of small

.spinules on hind margin. Head, upper surface a little rough, under margin of lateral lobes produced into an acute tooth. Eyes strongly protuberant, brown. Antenna 1 nearly $\frac{1}{2}$ as long as body, about $\frac{1}{3}$ longer than antenna 2, peduncle longer than flagellum, flagellum 26—32-jointed, accessory flagellum 4-jointed. Antenna 2, flagellum 8-jointed. Gnathopods 1 and 2, 6th joint piriform. Peraeopod 5, 2^d joint twice as long as broad, narrower than usual in the genus. Uropod 3, outer ramus about $\frac{1}{10}$ longer than inner, both rami setose. Telson rather long, with many lateral setules and shallowly concave apical margin. Colour brown, with white spots and bands. L. 33 mm.

Lake Baikal. Depth 10—50 m.

8. **P. brandtii** (Dyb.) 1874 *Gammarus b.*, B. Dybowsky in: Horae Soc. ent. Ross., *v.* 10 suppl. p. 136 t. 14 f. 1 | 1893 *Ceradocus? b.*, A. Della Valle in: F. Fl. Neapel, *v.* 20 p. 927 | 1899 *Pallasea b.*, T. Stebbing in: Tr. Linn. Soc. London, ser. 2 *v.* 7 p. 422.

Peraeon and pleon with 2 parallel median rows of tubercles, peraeon segments 1—7 each with marginal carina formed by a low, flattened, rounded tubercle, above which the lateral humps, successively higher, pass over into short. backward directed teeth; corresponding with and continuing the marginal and lateral carinae of the peraeon are ridges and teeth on the sides of pleon segments 1—3, but on segment 3 the median tubercles and the lateral tooth are more feebly, while the ridge is more strongly developed than in segments 1 and 2. Head uncommonly large and tumid, rostrum very short, margin of lateral lobe thickened and carinate, a downward directed boss behind the eyes. Side-plates 3—4 have little tubercles. Eyes round or oval, strongly protuberant, as if stalked. Antenna 1 nearly half as long as body, and twice as long as antenna 2, 1st joint thick, 3^d variable, never less than half as long as 1st, longer or shorter than 2^d, or equal to it, flagellum 30—45-jointed, accessory flagellum 4—6-jointed. Antenna 2, flagellum 9—12-jointed. Gnathopod 1, 6th joint piriform. Gnathopod 2, 6th joint oblong, widening somewhat to the oblique palm. Peraeopod 5, 2^d joint oval, rather wider above than below, marginal setae very short. Uropods 1 and 2 with setae on the margins and 1—3 spinules at the apices of the rami. Uropod 3, inner ramus scarcely shorter than outer, both carrying many setae. Telson cleft to the middle (beyond it in figure), forming 2 tube-like lobes with the apices bent one to meet the other. Colour: head, peraeon segments 6 and 7, pleon segments 4—6 bright green, other segments brownish. L. reaching 35 mm.

Lake Baikal. Depth 10—50 m.

9. **P. grubii** (Dyb.) 1874 *Gammarus g.*, B. Dybowsky in: Horae Soc. ent. Ross., *v.* 10 suppl. p. 132 t. 1 f. 5 | 1893 *Crangonyx? g.*, A. Della Valle in: F. Fl. Neapel, *v.* 20 p. 928 | 1899 *Pallasea g.*, T. Stebbing in: Tr. Linn. Soc. London, ser. 2 *v.* 7 p. 422.

Peraeon segments 1—4 or 1—5 with 2 parallel median rows of tubercles separated by a groove; on the other peraeon segments the tubercles form a single median row, continued by a weak median carina on pleon segments 1—3; on peraeon segments 1—5 the marginal carina is represented by thick, broad, feebly flattened humps, on segments 6 and 7 by quite feeble, small humps; the lateral carinae begin on peraeon segment 1 with very weak humps, encreasing on the following segments, and on 5—7 or 6 and 7 running out into short teeth; pleon segments 1 and 2 have each a pair of long, backward bent teeth, as in P. quadrispinosa (p. 377), and segment 3 has a lateral swelling. Head, upper surface rough, with median groove, bounded behind by 2 little tubercles; the sides below have a boss-like prominence. Eyes roundish oval, very protuberant, dark brown. Antenna 1 less than $^1/_2$ as long as body, nearly twice as long as antenna 2, 1st joint elongate, flagellum shorter than peduncle, 25—29-jointed, accessory flagellum 2- or 3-jointed. Antenna 2, flagellum 8-jointed. Gnathopod 1, 6th joint piriform. Gnathopod 2, 6th joint oblong. Peraeopods 3 and 4, 2d joint narrow, heart-shaped; peraeopod 5, 2d joint oval; long and thick setae on the margin in peraeopods 3—5. Uropod 3, outer ramus $4^1/_2$ times as long as the short, rudimentary inner ramus. Telson with a curved emargination, the upper surface depressed. Colour bright brownish green, with a whitish band over the middle of the back, often segments 2—4 darker than others; brownish spots occur; appendages banded. L. reaching 37 mm.

Lake Baikal.

10. **P. cancelloides** (Gerstf.) 1858 *Gammarus c.*, Gerstfeldt in: Mém. prés. Ac. St.-Pétersb., v. 8 p. 287 t. 9 f. 8 | 1874 *G. c.*, B. Dybowsky in: Horae Soc. ent. Ross., v. 10 suppl. p. 130 t. 13 f. 6 | 1862 *Pallasea c.*, Bate, Cat. Amphip. Brit. Mus., p. 380 | 1893 *P. cancellus* (part.), A. Della Valle in: F. Fl. Neapel, v. 20 p. 755.

Peraeon segments 1—7, median carina weak, tubercular, and on segments 1—3 inconspicuous, marginal carina more strongly developed than the lateral, the teeth directed outward and backward, acute, subequal, about twice as long as those of the lateral carina, which are directed backward, those of the 6th segment longer than the preceding, those of segment 7 nearly as long as the marginal teeth; pleon segments 1 and 2 with rather weak median carina, and on each side a strong, backward and outward bent tooth, longer than the lateral teeth on peraeon, pleon segment 3 with median carina flanked by weak humps, segments 4—6 almost smooth. Head tumid, with a prominence at the back, and a strong acute tooth projecting laterally from each cheek. Side-plate 4 with a weak tubercular projection. Eyes very protuberant, oval or broadly reniform, brownish black. Antenna 1 about $^1/_4$ as long as body, about $^1/_3$ longer than antenna 2, 1st joint twice as long as 2d, 2d twice as long as 3d, flagellum 22—26-jointed, accessory flagellum 3-jointed. Antenna 2, flagellum 8- or 9-jointed. Gnathopods 1 and 2, 6th joint piriform, narrower and smaller in ♀ than in ♂. Peraeopods 3 and 4, 2d joint elongate quadrangular, rather broader above than below, hind margin somewhat sinuous. Peraeopod 5, 2d joint oval, hind margin as in peraeopods 3 and 4 carrying long setae. Uropod 3, inner ramus only about $^1/_8$ shorter than outer, both with numerous setae, the majority plumose. Telson shallowly emarginate. Colour green or whitish green, with brownish black or brown spots. L. reaching 28 mm.

Lake Baikal, under stones close to the shore; River Angara at Irkutsk.

8. Gen. **Weyprechtia** Stuxb.

1880 *Weyprechtia* (Sp. un.: *W. mirabilis*), Stuxberg in: Bih. Svenska Ak., v. 5 nr. 22 p. 27 | 1888 *W.*, T. Stebbing in: Rep. Voy. Challenger, v. 29 p. 523 | 1894 *W.*, T. Stebbing in: Bijdr. Dierk., v. 17 p. 41.

Body dorsally smooth. Pleon segment 3, postero-lateral margins bidentate. Antenna 1 the shorter, accessory flagellum 4—7-jointed. Upper lip not emarginate. Lower lip, inner lobes not distinct. Mandible, cutting edge little or not dentate, accessory plates dentate, spines of spine-row numerous, molar powerful, 3d joint of palp not curved, fully as long as 2d and 3d combined. Maxilla 1, inner plate with setae very numerous, outer with

11 spines, 2^d joint of palp distally expanded and fringed in part with setae. Maxilla 2, inner plate with close-set row of setae near the inner margin. Maxillipeds, inner plates transversely truncate, outer scarcely reaching middle of palp's 2^d joint, which is little longer than 1st or 3^d, the finger short, thick, acute. Gnathopods 1 and 2 similar, slender, 5th joint at least as long as 6th, palm ill-defined, finger small. Peraeopods 3—5, 2^d joint oval. Uropod 2 with unequal rami. Uropod 3, rami single-jointed, lanceolate, subequal. Telson entire, longer than broad.

2 species.

Synopsis of species:

Side-plate 4 deeply and doubly emarginate . . . . 1. **W. heuglini** . p. 381
Side-plate 4 simply emarginate . . . . 2. **W. pinguis** . p. 382

1. **W. heuglini** (Buchh.) 1874 *Amathilla h.*, Buchholz in: Zweite D. Nordpolarf., *v.*2 p.345 | 1893 *A. heuglinii*. A. Della Valle in: F. Fl. Neapel, *v.*20 p.685 | 1894 *Weyprechtia heuglini*, T. Stebbing in: Bijdr. Dierk., *v.*17 p.41 t.7 | 1880 *W. mirabilis*, Stuxberg in: Bih. Svenska Ak., *v.*5 nr.22 p.28.

Integument crustaceous, back rounded, broad; pleon segment 4 with transverse depression, of which also segments 1—3 show traces. Head dorsally broad, rostrum minute, front angles subtruncate, not very prominent. Side-plates 1—3 narrow, truncate below, 1st with subacute front angle, 4th deeper, lower hind angle subacute, a strong downward curved horn dividing the hind margin into 2 great curves. Side-plates 1—5 somewhat projecting laterally, especially the 2 horns or angles of the 4th. Pleon segments 2 and 3, postero-lateral angles acute, segment 3 having a 2^d and somewhat more produced tooth a little way above that of the postero-lateral angle. Eyes reniform, shining, black, curving round from side towards top of head. Antenna 1, 1st joint scarcely as long as 2^d and 3^d combined, flagellum twice as long as peduncle, attaining 32 joints, accessory flagellum 6- or 7-jointed. Antenna 2 about once and a half as long as antenna 1, ultimate and penultimate joints of peduncle equal, flagellum nearly twice as long as peduncle, about 40—59-jointed. Upper lip broadly rounded, almost truncate. Mandible, cutting edge apparently undivided (in younger specimens perhaps dentate), accessory plate of 4 teeth, as usual dissimilar in right and left mandible, 10 spines in spine-row, 3^d joint of palp with one margin straight, longer than 2^d and 3^d combined. Maxilla 1, inner plate with about 30 plumose setae, 2^d joint of palp distally widened, with several slender spines and setae on inner distal margin. Maxilla 2, inner plate with oblique row of about 30 long plumose setae near inner margin. Maxillipeds, outer plates, apical margin crowded with plumose setae passing over into serrate spines on inner margin. Gnathopod 1, 2^d joint the widest, 5th joint narrow, longer than 6th, which is rather long and narrow with small, indistinct palm, finger longer than palm, faintly serrate within, nail curved, acute. Gnathopod 2 similar, but 5th joint more widened, 6th nearly as long, palm even more indistinct. Peraeopods 1 and 2 slender. Peraeopods 3—5 considerably longer, 4th and 5th subequal, longer than 3^d, 2^d joint not very broadly oval, joints 4—7 rather elongate. Uropod 1, rami a little unequal, nearly as long as peduncle. Uropod 2, rami longer than peduncle, inner longer and broader than outer. Uropod 3, rami broad, much longer than peduncle. Telson a little variable, rather longer than broad, tapering to faintly emarginate apex, which carries 2 setules, sometimes producing a slightly trilobed appearance. L. reaching 51 mm (antenna 2: 24 mm).

Arctic Ocean (Spitzbergen; Murman Sea, lat. 69° N., long. 59° E., depth 28 m; glacial sea of Siberia between Cape Vankarema and Behring Strait, depth 7—11 m).

2. **W. pinguis** (Kröyer) 1838 *Gammarus pingvis*, Kröyer in: Danske Selsk. Afh., *v.* 7 p. 252 t. 1 f. 5 | 1862 *Amathia pinguis*, Bate, Cat. Amphip. Brit. Mus., p. 200 | 1871 *Amathilla pingvis*, A. Boeck in: Forb. Selsk. Christian., 1870 p. 218 | 1874 *A. pinguis*, Buchholz in: Zweite D. Nordpolarf., *v.* 2 p. 353 t. 9 f. 2 | 1878 *A. p.*, Miers in: Nares, Voy. Polar Sea, *v.* 2 p. 246 | 1893 *A. p.*, A. Della Valle in: F. Fl. Neapel, *v.* 20 p. 684 t. 59 f. 89 | 1895 *A. p.*, Ohlin in: Acta Univ. Lund., *v.* 31 nr. 6 p. 50 | 1894 *Weyprechtia p.*, T. Stebbing in: Bijdr. Dierk., *v.* 17 p. 41.

Closely resembling W. heuglini. Side-plate 4 with single normal emargination. Upper lip very large, not divided at the end (broadly conical, Buchholz; but ?). Mandible, cutting plate with few teeth on the point (Boeck), or completely without teeth (Buchholz). Gnathopods 1 and 2 with very oblique, straight, fairly well defined palm, longer than in W. heuglini. Peraeopod 4 intermediate in length between peraeopods 3 and 5 (Kröyer: a little smaller than 3^d or 5th). L. 24—29 mm (antenna 2 half this length).

Arctic Ocean (Greenland, depth 6—56 m; Spitzbergen, depth 4—22 m; circumpolar, reaching lat. 82° 24′ N.).

9. Gen. **Paramicruropus** Stebb.

1899 *Paramicruropus*, T. Stebbing in: Tr. Linn. Soc. London, ser. 2 *v.* 7 p. 423.

One of the pleon segments abruptly elevated above the next; pleon segment 6 very small. Antennae 1 and 2 short; antenna 1 the longer, accessory flagellum small. Gnathopod 1, 6th joint not smaller than that of gnathopod 2. Peraeopods 1—5 not elongate. Uropods 1 and 2 not very long. Uropod 3 rudimentary, rami not very unequal. Telson small, entire.

2 species.

Synopsis of species:

Pleon segment 3 overarching the small 4th segment . . 1. **P. solskii** . . . . p. 382
Pleon segment 4 large, overarching the small 5th segment 2. **P. taczanowskii** . p. 382

1. **P. solskii** (Dyb.) 1874 *Gammarus s.*, B. Dybowsky in: Horae Soc. ent. Ross., *v.* 10 suppl. p. 42, 153 t. 3 f. 2 ⌊ 1893 *Amathilla? s.*, A. Della Valle in: F. Fl. Neapel, *v.* 20 p. 930 | 1899 *Paramicruropus s.*, T. Stebbing in: Tr. Linn. Soc. London, ser. 2 *v.* 7 p. 423.

Description incomplete. All segments (in figure) strongly ridged transversely, perhaps with exception of the very small pleon segments 4—6. Head not rostrate. Pleon segment 3 unusually large, overarching the small segment 4, grooved by a deep, horizontal, arched furrow. Eyes small, reniform, black. Antenna 1 a little the longer, $^1/_4$ as long as the body, peduncle shorter than in antenna 2, flagellum 17-jointed, accessory flagellum 3-jointed. Antenna 2, flagellum 12-jointed. Gnathopods 1 and 2 have short 5th joint, 6th almond-shaped. Peraeopod 1 longer than peraeopod 2; peraeopods 3—5 with 2^d joint narrowly oval, longitudinally carinate. Uropods 1 and 2 feebly developed, reaching back almost as far as uropod 3. Uropod 3, rami small, equal. Telson apically rounded (cf. Dybowsky's analytical key). Colour clouded yellow. L. 23 mm.

Lake Baikal.

2. **P. taczanowskii** (Dyb.) 1874 *Gammarus t.*, B. Dybowsky in: Horae Soc. ent. Ross., *v.* 10 suppl. p. 156 t. 14 f. 9 | 1893 *Acanthonotosoma? t.*, A. Della Valle in: F. Fl. Neapel, *v.* 20 p. 930 | 1899 *Paramicruropus t.*, T. Stebbing in: Tr. Linn. Soc. London, ser. 2 *v.* 7 p. 423.

Head and all the segments dorsally granular, on each segment immediately
above the side-plates a strongly developed carina, and encircling the back of
each segment a broad, rib-like, granular, transverse prominence; on pleon
segment 4 the prominence is divided by a deep longitudinal furrow; segments 5
and 6 very small, in the ♀ so overarched by segment 4 as to be invisible in a
side view, in the ♂ more conspicuous. Head with rather long pointed rostrum,
lateral lobes produced into an acute process as long as rostrum. Side-
plates 1—4 ,setose. Eyes point-like, black. Antenna 1 scarcely $\frac{1}{4}$ as long
as body, but much longer than antenna 2, 1st joint stout, flagellum shorter
than peduncle, 12-jointed, accessory flagellum 1-jointed. Antenna 2, flagellum
6-jointed. Gnathopod 1, 6th joint oval, narrowing to finger hinge. Gnathopod 2
not larger, oblong, in ♂ widening a little to the not very oblique palm. Peraeopod 5, 2d joint broader below than above, with long setae on both margins.
Uropods 1 and 2 moderately long (not so long as in figure), reaching beyond
the rudimentary uropod 3. Uropod 3, both rami very short, conical, with 1 long
apical seta, inner ramus in ♂ $\frac{1}{10}$, in ♀ $\frac{1}{8}$ shorter than outer (figure e difficult
to reconcile with description). Telson broader than long, apical margin concave
or shallowly emarginate. Colour wax-yellow. L. 10 mm.

Lake Baikal. Depth 10—50 m.

10. Gen. **Parapherusa** Stebb.[*])

1879 *Harmonia* (Sp. un.: *H. crassipes*), *Harmomia* (laps.) (non *Harmonia* E.
Mulsant 1846, Coleoptera!), Haswell in: P. Linn. Soc. N. S. Wales, v. 4 p. 330 | 1883
Harmonia, Chilton in: Tr. N. Zealand Inst., v. 15 p. 82 | 1880 *Chloris* (non Moehring
1758, Aves!), Haswell in: Ann. nat. Hist., ser. 5 v. 5 p. 33.

Side-plates shallow. Antenna 1 the shorter, accessory flagellum well
developed. Mouth-parts normal. Mandible, 2d joint of palp as long as 3d,
but stouter. Maxilla 1, inner plate with about 10 long setae, outer with
11 spines, 2d joint of palp with 7 or 8 spine-teeth. .Maxilla 2, inner plate
fringed on inner margin. Maxillipeds, inner and outer plates well armed.
Gnathopods 1 and 2 subchelate, 2d much the larger in ♂, but not in ♀.
Peraeopods 3—5 very stout. Uropod 3 very short, rami equal, shorter than
peduncle. Telson simple.

1 species.

1. **P. crassipes** (Hasw.) 1879 *Harmonia c.*, *Harmomia c.*, Haswell in: P. Linn.
Soc. N. S. Wales, v. 4 p. 330, 349; t. 19 f. 3 | 1883 *Harmonia c.*, Chilton in: Tr. N. Zealand
Inst., v. 15 p. 82 t. 2 f. 5 a, h | 1893 *Protomedeia? c.*, A. Della Valle in: F. Fl. Neapel,
v. 20 p. 442 | 1880 *Chloris*, Haswell in: Ann. nat. Hist., ser. 5 v. 5 p. 33.

Pleon segments 5 and 6 very short. Head, lateral lobes rounded, little
produced. Eyes well developed. Antenna 1 scarcely shorter than antenna 2, half
as long as body, 2d joint scarcely longer than 1st, 3d short, flagellum rather longer
than peduncle, about 18-jointed, accessory flagellum 8-jointed. Antenna 2, ultimate
and penultimate joints of peduncle equal, flagellum much shorter than peduncle,
about 10-jointed. Gnathopod 1, 6th joint rather longer than 5th, oval, palm
ill-defined. Gnathopod 2 in ♂, 5th joint short, cup-shaped, 6th very large,
widening to the well-defined, suboblique, undulating palm, which has a square
spinuliferous prominence adjacent to finger hinge and a similar but more
convex one near the middle, finger strong, matching palm. Gnathopod 2

---

*) Nom. nov. Παρά, beside, *Pherusa* (see p. 449). — The name *Harmonia* is
preoccupied (1846, E. Mulsant, Hist. nat. Col. France, Sécuripalpes p. 108), also the
name *Chloris* (1758, P. H. G. Moehring, Gesl. Vog., p. 3, 24).

in ♀, resembling gnathopod 1, but with sborter 5th and longer 6th joint. Peraeopods 1 and 2 not very stout. Peraeopods 3—5, 2d joint oblong, other joints, including finger, very stout, peraeopods 4 and 5 longer than 3d. Uropods 1—3 not elongate, in uropod 1 rami as long as peduncle, in uropods 2 and 3 shorter than peduncle. Telson rather longer than broad, sides converging to a truncate apex, which has a setule at each of the blunt angles. Colour brown. L. 4 mm.

South-Pacific (Port Jackson [New South Wales], Griffith's Point [Victoria]; Lyttelton Harbour, and Timaru [New Zealand]).

11. Gen. **Amathillopsis** Heller

1875 *Amathillopsis* (Sp. un.: *A. spinigera*), Cam. Heller in: Denk. Ak. Wien, v. 35 p. 35 | 1885 *A.*, G. O. Sars in: Norske Nordhavs-Exp., v. 6 Crust. I p. 181 | 1888 *A.*, T. Stebbing in: Rep. Voy. Challenger, v. 29 p. 442, 859 | 1894 *A.*, T. Stebbing in: Bijdr. Dierk., v. 17 p. 27.

Integument crustaceous, body dorsally carinate, with several strongly produced teeth. Head, rostrum short. Side-plates not very deep. Antenna 1 the longer, peduncle long, 2d joint as long as 1st, accessory flagellum very small. Upper lip with margin entire. Mandible normal, 2d and 3d joints of palp long. Maxilla 1, inner plate with 6 or 7 setae, outer with 11 only slightly denticulate spines, 2d joint of palp elongate, armed both on right and left maxilla with spines and setae. Maxilla 2, inner plate setose or only ciliated on inner margin. Maxillipeds, outer plates small, extending not far along the 2d joint of the long palp. Gnathopods 1 and 2 alike, sub-chelate, 5th joint as long as 6th, palm ill-defined. Peraeopod 5, so far as known, the shortest. Uropod 1, rami shorter than peduncle, outer ramus shorter than inner. Uropod 3, rami lanceolate, outer a little shorter than inner. Telson short, slightly emarginate.

3 species.

Synopsis of species:

1 { Gnathopods 1 and 2, 5th joint distally little expanded . 2. **A. affinis** . . p. 385
  { Gnathopods 1 and 2, 5th joint distally much expanded — 2.

2 { Telson much shorter than peduncle of uropod 3 . . . 1. **A. spinigera** . p. 384
  { Telson subequal to peduncle of uropod 3 . . . . . . 3. **A. australis** . p. 385

1. A. spinigera Heller 1875 *A. s.*, Cam. Heller in: Denk. Ak. Wien, v. 35 p. 35 t. 3 f. 17—22; t. 4 f. 1—8 | 1885 & 86 *A. s.*, G. O. Sars in: Norske Nordhavs-Exp., v. 6 Crust. I p. 181 t. 15 f. 2; Crust. II p. 61 | 1888 *A. s.*, T. Stebbing in: Rep. Voy. Challenger, v. 29 p. 442, 865 | 1894 *A. s.*, T. Stebbing in: Bijdr. Dierk., v. 17 p. 28 | 1893 *Amathilla s.* (part.), A. Della Valle in: F. Fl. Neapel, v. 20 p. 684 t. 59 f. 88.

Peraeon broad, pleon compressed; body carinate faintly on head, and thence strongly to pleon segment 4, peraeon segment 1 with 2 upstanding acute processes, the other carinate segments each with one, successively larger to that on pleon segment 2, the next a little, and the next again much smaller than that preceding it. Peraeon segments 1—4 with lateral margins produced outward in acute processes. Head, rostrum very short, slightly deflexed, not projecting beyond the lateral corners which are quadrate, bases of antenna 2 conspicuous in the incision below. Side-plate 1 small, faintly emarginate below, 2d—4th strongly emarginate below between 2 acute points, 5th subequal to 4th in depth and much broader, both lobes acute, 6th with only hind lobe acute. Pleon segments 1—3, postero-lateral corners acutely produced, less in 3^d than in 1st and 2^d. Eyes yellowish white, very small, oval, prominent, adjacent to a cavity in the lateral corners of head. Antenna 1, joints 1 and 2 long, subequal, 3^d about $^1/_3$ as long as 2^d, flagellum rather longer than peduncle, with a great number of very short joints, accessory flagellum spine-like, 2^d joint minute. Antenna 2 rather shorter and more slender, basal joints short and stout, ultimate joint of peduncle shorter than penultimate, flagellum rather longer than peduncle and, as in antenna 1, calceoliferous. Mandible, 3^d joint of palp very little shorter than 2^d. Maxilla 1, outer plate has 12 spines (Heller; but 11?). Maxilla 2, inner plate with setae on inner margin. Maxillipeds, outer plates with short spine-teeth on inner margin, continued in longer forms round the apical margin. Gnathopod 1, 5th joint narrow proximally, then much widened, wider than the oval sub-fusiform 6th, both setose on hind margin, finger long, curved. Gnathopod 2 with 5th joint wider and rather longer, its distal lobe more rounded, 6th joint slightly longer than in gnathopod 1, but scarcely differing. Peraeopods 3—5, 2^d joint scarcely (in peraeopod 5 a little more) expanded, sharply carinate behind and produced into a small distal tooth, peraeopod 4 shorter than 3^d, and 5th shorter than 4th. Uropods 1—3, peduncle and rami with sharp margins, outer ramus shorter than inner, especially in uropod 2. Uropod 3, rami nearly equal to one another and to peduncle. Telson much shorter than peduncle of uropod 3, a little hollowed on surface, rather longer than broad, apically a little widened and emarginate between the rounded corners. Colour straw-yellow, mouth-parts and front legs vivid red. L. reaching 57 mm from rostrum to end of uropod 3; (antenna 1: 43 mm).

Arctic Ocean (Spitzbergen, Franz Josef Land); North-Atlantic (lat. 63°—76° N.). Depth 240—1408 m.

2. **A. affinis** Miers 1881 *A. a.*, Miers in: Ann. nat. Hist., ser. 5 *v.* 7 p. 48 t. 7 f. 3—5 | 1888 *A. a.*, T. Stebbing in: Rep. Voy. Challenger, *v.* 29 p. 865 | 1893 *Amathilla spinigera* (part.)?, A. Della Valle in: F. Fl. Neapel, *v.* 20 p. 684.

Agreeing in many respects with A. spinigera. Side-plates not so distinctly emarginate, 4th much more developed. Antenna 1, 3^d joint relatively shorter than in A. spinigera, flagellum very long and slender, accessory flagellum very minute. Antenna 2 little more than half as long as antenna 1. Gnathopods 1 and 2 slender and feeble, 5th joint gradually widening to the distal end, which forms no lobe, 6th joint as long and wide, oblong, slightly widened at the palm, which is much shorter than hind margin, finger short in proportion. Gnathopod 2 rather the longer. In all the legs 2^d joint oblong oval, much more expanded than in A. spinigera. Telson not distally widened.

Arctic Ocean (Franz Josef Land).

3. **A. australis** Stebb. 1883 *A. a.*, T. Stebbing in: Ann. nat. Hist., ser. 5 *v.* 11 p. 205 | 1888 *A. a.*, T. Stebbing in: Rep. Voy. Challenger, *v.* 29 p. 860 t. 65 | 1893 *Amathilla spinigera* (part.), A. Della Valle in: F. Fl. Neapel, *v.* 20 p. 684.

Peraeon broad, pleon compressed. Carina indicated on head, peraeon segments 1—4, pleon segments 4—6; on peraeon segments 5—7 and pleon segments 1—3, prolonged into acute processes, successively larger. Head, rostrum small, deflexed, not projecting so far as the lateral corners. Side-plates 3 and 4 deeper than the rest, acute at lower front corner. Pleon segments 2 and 3, postero-lateral corners acutely produced. Eyes not perceived.

Antenna 1, 1st and 2^d joints equal in length, 3^d rather longer than $^1/_3$ of 2^d, flagellum broken, the remnant as long as 1st joint of peduncle, containing 50 short joints, accessory flagellum seemingly represented by a strong, flat, incurved spine. Antenna 2 less robust, similar to that of A. spinigera, flagellum incomplete, with 70 joints, as in antenna 1 calceoliferous. Epistome carinate. Upper lip, with the rounded margin flattened at centre. Lower lip probably without inner lobes. Mandible, accessory plate much stronger on left than on right mandible, 3^d joint of palp considerably longer than 1st and 2^d combined. Maxilla 1, inner plate with 7 setae, outer with 11 spines. Maxilla 2, inner plate merely ciliated on inner margin. Maxillipeds, inner plates short, outer reaching little beyond 1st joint of palp, inner margin denticulate, with submarginal spinules, apical margin with long spines, palp elongate. Gnathopods 1 and 2, 2^d joint with distal half much expanded behind, rest of limb as in A. spinigera. Peraeopods 1 and 2, 2^d joint with hind margin a little expanded. Peraeopods 3—5 imperfect, 2^d joint carinate on front and hind margins, a little expanded proximally and in front distally, shorter in 4th than 3^d, and in 5th than 4th, with proximal expansion encreasing. Pleopods with numerous cleft spines on 1st joint of inner ramus. Uropod 1, peduncle longer than rami, outer ramus a little the shorter. Uropod 2, peduncle shorter than inner ramus. Uropod 3, peduncle much shorter than the lanceolate rami, of which outer is slightly the shorter. Telson subequal to peduncle of uropod 3, longer than broad, slightly narrowing to the weakly emarginate apex. L. about 38 mm.

.Coral Sea (between Australia and New Guinea, lat. 12^0 S., long. 145^0 E.). Depth 2560 m.

12. Gen. **Gammarellus** Herbst

1793 [Subgen.] *Gammarellus* (part.), J. F. W. Herbst, Naturg. Krabben Krebse, v. 2 p. 106 | 1899 *G.*, T. Stebbing in: Tr. Linn. Soc. London, ser. 2 v. 7 p. 423 | 1837 *Amathia* (Sp. un.: *A. carinata*) (non Lamouroux 1812, Hydrozoa!), H. Rathke in: Mém. prés. Ac. St.-Pétersb., v. 3 p. 375 | 1862 *Grayia* (Sp. un.: *G. imbricata*) (non Alb. Günther 1858, Reptilia!), Bate (& Westwood), Brit. sess. Crust., v. 1 p. 151 | 1862 *Graya* (non C. L. Bonaparte 1856, Aves!) (part.), Bate, Cat. Amphip. Brit. Mus., p. 101; t. 14 a, 16 | 1895 *Grayia*, A. O. Walker in: Ann. nat. Hist., ser. 6 v. 15 p. 466 | 1862 *Amathilla*, Bate & Westwood, Brit. sess. Crust., v. 1 p. 359 | 1893 *A.* (part.), A. Della Valle in: F. Fl. Neapel, v. 20 p. 683 | 1894 *A.*, T. Stebbing in: Bijdr. Dierk., v. 17 p. 40 | 1894 *A.*, G. O. Sars, Crust. Norway, v. 1 p. 489.

Body dorsally carinate. Antennae 1 and 2 subequal, in ♂ calceoliferous, accessory flagellum 4—7-jointed. Upper lip not emarginate. Lower lip, inner lobes not distinct. Mandible normal, both plates dentate, spines of spine-row numerous, 3^d joint of palp falciform, armed on both margins. Maxilla 1, inner plate with setae numerous, outer with 11 spines, 2^d joint of palp scarcely expanded distally. Maxilla 2, inner plate fringed near inner margin. Maxillipeds, outer plates well armed, not reaching middle of palp's 2^d joint, 3^d joint of palp notably smaller than 2^d, very setose. Gnathopods 1 and 2, 5th joint cup-shaped, 6th much larger, oval, palm feebly defined, finger strong. Peraeopods 1—5 rather stout, 2^d joint oval in last three. Branchial vesicles pleated obliquely. Uropod 3, rami single-jointed, lanceolate, outer a little the longer. Telson feebly emarginate.

2 accepted and 1 doubtful species.

Synopsis of the accepted species:

1. **G. homari** (F.) 1779 *Astacus h.*, J. C. Fabricius, Reise Norweg., p.247 | 1788 *Cancer (A.) h.*, J. F. Gmelin, Syst. Nat., *v.*1 p.2986 | 1798 *Gammarus h.*, J. C. Fabricius, Ent. syst., Suppl. p.418 | 1888 *Amathilla h.*, T. Stebbing in: Rep. Voy. Challenger, *v.*29 p.45 etc. | 1893 *A. h.* (part.), A. Della Valle in: F. Fl. Neapel, *v.*20 p.685 t.59 f.90 | 1894 *A. h.*, T. Stebbing in: Bijdr. Dierk., *v.*17 p.40 | 1894 *A. h.*, G. O. Sars, Crust. Norway, *v.*1 p.490 t.172, t.173 f.1 | 1895 *A. h.*, A. O. Walker in: Ann. nat. Hist., ser.6 *v.*15 p.466, 471 | 1899 *Gammarellus h.*, T. Stebbing in: Tr. Linn. Soc. London, ser.2 *v.*7 p.423 | 1780 *Oniscus arenarius* (non Slabber 1769!), O. Fabricius, Fauna Groenl., p.259 | 1887 *Amathilla arenaria*, H. J. Hansen in: Vid. Meddel.. ser.4 *v.*9 p.149 |,1793 *Cancer (Gammarellus) homari* + *C. (G.) arenarius*, J. F. W. Herbst, Naturg. Krabben Krebse, *v.*2 p.113; p.133 | 1819 *Gammarus sabini*, Leach in: John Ross, Voy. Baffin's Bay, app. p.63 | 1862 *Grayia imbricata* + *Amathilla s.* (part.), Bate & Westwood, Brit. sess. Crust., *v.*1 p.152; p.361 f. | 1874 *A. s.*, Buchholz in: Zweite D. Nordpolarf., *v.*2 p.346 Crust. t.8 f.1, 2; t.9 f.1 | 1827 *Gammarus marinus* (err., non Leach 1815!), G. Johnston in: Zool. J., *v.*3 p.176 | 1828 *G. carinatus* (non J. C. Fabricius 1793!), G. Johnston in: Zool. J., *v.*4 p.52 | 1857 *Amathia carinata*, A. White, Hist. Brit. Crust., p.182 | 1851 *Amphithoë moggridgei*, Bate in: Ann. nat. Hist., ser.2 *v.*7 p.318 t.10 f.10 | 1862 *Grayia imbricata*, Graya i. + *Amathia sabinii* + *A. carino-spinosa* (part.), Bate, Cat. Amphip. Brit. Mus., p.101 t.16 f.4; p.197 t.35 f.9; p.199 t.35 f.11.

Body robust in ♀, more compressed in ♂. Peraeon segments 1—7 and pleon segments 1—3 with compressed dorsal carina produced more or less backward, that of peraeon segment 1 rather broad and also a little acutely produced forward; pleon segment 4 with a transverse dorsal depression followed by a carina. In the young the dorsal prominences are much less marked. Head depressed, rostrum small, lateral corners obtusely truncate. Side-plates 1—4 successively larger, but in young 1st little smaller than 2^d or 3^d. Side-plate 4 emarginate behind. Pleon segment 3, postero-lateral corners quadrate. Eyes large, oblong reniform, dark brown. Antenna 1 in ♀ about ¹/₈ as long as body, 1st joint not as long as 2^d and 3^d combined, 2^d not much longer than 3^d, flagellum once and a half as long as peduncle, with 40—55 setulose joints, accessory flagellum 6- or 7-jointed. Antenna 2 a little shorter, ultimate joint of peduncle shorter than penultimate, flagellum subequal to peduncle, 40-jointed, setulose and calceoliferous. Antenna 2 in ♂ more slender than in ♀, rather longer than antenna 1, peduncle longer than flagellum. Gnathopods 1 and 2, 6th joint oval, narrowing to hinge of finger, palm setose and spinose. Gnathopod 2 the larger. Peraeopod 5, 2^d joint larger than in peraeopods 3 and 4, its hind margin sinuous below. Uropod 3, outer ramus rather the larger, both rami setose on inner margin. Telson rather longer than peduncle of uropod 3, much longer than broad, tapering to a rounded apex, in the centre of which is a little emargination flanked by a pair of spinules. Colour variable, olive-green or white marbled or banded with brown or red. L. variable, reaching 44 mm (antenna 2 ♂: 20 mm).

Arctic Ocean, North-Atlantic, North-Sea and Skagerrak (Greenland, Spitzbergen, Murman Coast, Arctic America, Norway, British Isles, France); Kattegat. Depth 0—206 m.

2. **G. angulosus** (H. Rathke) 1843 *Gammarus a.*, H. Rathke in: N. Acta Ac. Leop., *v.*201 p.72 t.3 f.3 | 1855 *G. a.*, W. Liljeborg in: Vetensk. Ak. Handl., 1853 p.447 | 1871 *Amathilla angulosa*, A. Boeck in: Forb. Selsk. Christian., 1870 p.217 | 1894 *A. a.*, G. O. Sars, Crust. Norway, *v.*1 p.492 t.173 f.2 | 1899 *Gammarellus angulosus*, T. Stebbing in: Tr. Linn. Soc. London, ser.2 *v.*7 p.423 | 1862 *Amathia carino-spinosa* (part.), Bate, Cat. Amphip. Brit. Mus., p.199 | 1862 *Amathilla sabini* (part.), Bate & Westwood, Brit. sess. Crust., *v.*1 p.362 | 1893 *A. homari* (part.), A. Della Valle in: F. Fl. Neapel, *v.*20 p.685 | 1895 *A. h.*, A. O. Walker in: P. Liverp. biol. Soc., *v.*9 p.307.

Often confounded with but distinct from G. homari. Body short and stout, in front angular, behind dorsally carinate, but carina in each segment

25*

truncate, not produced into a tooth, integument very minutely squamose. Head, rostrum short, obtuse, lateral corners evenly rounded. Side-plate 5, hind lobe deeper than front. Pleon segment 3, postero-lateral corners obtusely angular. Eyes oblong, dark brown. Antennae 1 and 2 subequal. Antenna 1, flagellum 21-jointed, accessory flagellum 4- or 5-jointed. Antenna 2, flagellum 22- or 23-jointed. Gnathopods 1 and 2, exactly alike both in size and structure (Sars). Peraeopods 1—5 short and stout, subequal. Peraeopod 5 the longest, 2^d joint with hind margin not sinuous. Uropod 3 less setose than in G. homari (p. 387). Telson oval quadrangular, scarcely longer than broad, little tapering, with a small apical emargination flanked by a pair of spinules. Colour olive-green (Liljeborg), or yellowish, mottled with reddish brown (Sars). L. reaching 12 mm.

Arctic Ocean, North-Atlantic, North-Sea and Skagerrak (Vardö [North-Norway], South- and West-Norway. Bohuslän); Kattegat. French and British localities doubtful.

G. carinatus (H. Rathke) 1837 Amathia carinata, H. Rathke in: Mém. prés. Ac. St.-Pétersb., v. 3 p. 375 t. 5 f. 29—35 | 1862 A. c., Bate, Cat. Amphip. Brit. Mus., p. 198 t. 35 f. 10 | 1868 Amathilla c., Czerniavski in: Syezda Russ. Est., Syezda 1 Zool. p. 131 | 1899 Gammarellus carinatus, G. homari (part.)?, T. Stebbing in: Tr. Linn. Soc. London, ser. 2 v. 7 p. 423.

Probably synonym to G. homari (p. 387). Carina only on peraeon segments 5—7 and pleon segments 1—3. L. about 15 mm.

Black Sea.

13. Gen. **Eucrangonyx** Stebb.

1899 Eucrangonyx, T. Stebbing in: Tr. Linn. Soc. London, ser. 2 v. 7 p. 423.

Like Crangonyx (p. 370) in general, but with a small inner ramus to the 3^d uropod, telson emarginate.

5 species.

Synopsis of species:

1 { Telson elongated in ♂, not in ♀ . . . . 1. **E. mucronatus** . p. 388
  { Telson alike in ♂ and ♀ — **2**.

2 { Eyes wanting . . . . 2. **E. vejdovskyi** . p. 389
  { Eyes devoid of black pigment . . . . 3. **E. packardii** . . p. 389
  { Eyes with black pigment — **3**.

3 { Side-plate 4 not unusually large . . . . 4. **E. gracilis** . . . p. 390
  { Side-plate 4 unusually large . . . . 5. **E. antennatus** . p. 390

1. **E. mucronatus** (S. A. Forb.) 1876 Crangonyx m., S. A. Forbes in: Bull. Illinois Mus., v. 1 p. 6. f. 1—7 | 1888 C. m., Packard in: Mem. Ac. Washington, v. 41 p. 37 t. 5 f. 15 | 1893 C. m., A. Della Valle in: F. Fl. Neapel, v. 20 p. 682 | 1894 C. m., Chilton in: Tr. Linn. Soc. London, ser. 2 v. 6 p. 219 | 1899 Eucrangonyx m., T. Stebbing in: Tr. Linn. Soc. London. ser. 2 v. 7 p. 423.

Pleon segments 1—3, postero-lateral corners broadly rounded. Eyes wanting. Antenna 1 in ♂ sometimes $^4/_5$ as long as body, 1st and 2^d joints subequal, 3^d $^1/_3$ as long as 2^d, flagellum about 5 times as long as peduncle, 30—35-jointed, 2-jointed accessory flagellum a little longer than 1st joint of primary. Antenna 2, ultimate and penultimate joints of peduncle subequal, flagellum 9- or 10-jointed. Antenna 1 in ♀ about half length of body, flagellum thrice as long as peduncle. Upper lip well rounded. Mandible with several spines in spine-row, 1ˢᵗ joint of palp longer than broad, · 3ᵈ fully as long as 2ᵈ, curved. Other mouth-parts agreeing with those of Crangonyx flagellatus (p. 371). Gnathopod 1 in ♂, 5ᵗʰ joint subtriangular, ³/₄ as wide as 6ᵗʰ, 6ᵗʰ broadly ovate, ²/₃ as wide as long, hind margin very short, palm oblique, not sharply defined, fringed with about 15 notched spines. Gnathopod 2, 5ᵗʰ joint as wide as 6ᵗʰ, rather longer than in gnathopod 1, 6ᵗʰ joint also a little longer and narrower. In ♀ gnathopod 1 has hind margin longer, palm shorter, and gnathopod 2 is decidedly smaller than gnathopod 1. Peraeopod 3 about ²/₃ as long as peraeopod 5, peraeopod 4 intermediate in length, 2ᵈ joint in all narrowly oval. Uropod 1 extending beyond uropod 2, and 2ᵈ beyond 3ᵈ. Uropods 1 and 2 stout, each with subequal rami, shorter than peduncle. Uropod 3 short, outer ramus half as long as peduncle, inner ramus rudimentary, unarmed, ¹/₄ or ¹/₅ as long as outer. Telson in ♂ smooth, cylindrical, apically rounded and tipped with spinules, normally about as long as pleon segments 1—3, sometimes half as long as the body. In ♀ it is only little longer than broad, extending to apex of uropod 3, flattened, slightly emarginate, with 2 terminal clusters of 4 or 5 spines. Colourless. L. 9—10 mm.

Illinois. Well at Normal, and in springs.

2. **E. vejdovskyi** Stebb. 1896 *Crangonyx subterraneus* (err., non Bate 1859!), Vejdovský in: SB. Böhm. Ges., nr. 10 p. 12 t. 1—3 | 1899 *Eucrangonyx vejdovskyi*, T. Stebbing in: Tr. Linn. Soc. London, ser. 2 v. 7 p. 423.

♀. Side-plate 4 much deeper than those which follow, broad, emarginate, lower margin crenulate. Pleon segments 1—3, postero-lateral corners rounded. No trace of eyes, but variable orange-or lemon-coloured pigment-masses in their place. Antenna 1, 1ˢᵗ joint not as long as 2ᵈ and 3ᵈ combined, 3ᵈ in figure even longer than 2ᵈ, flagellum somewhat longer than peduncle with 9—13 joints; accessory flagellum, 1ˢᵗ joint twice as long as 2ᵈ. Antenna 2 shorter, ultimate joint of peduncle (in figure) much shorter than penultimate, flagellum 3—5-jointed. Lower lip, inner lobes (in figure) not very distinct, mandibular processes unusually prolonged. Mandible, 2ᵈ joint of palp broad, considerably longer than 3ᵈ. Maxilla 1, inner plate with 4 setae, outer with 6 (7?) spines. Maxilla 2, inner plate with setae near the inner margin. Maxillipeds, outer plates not reaching the middle of palp's 2ᵈ joint, palp's 4ᵗʰ joint with a nail. Gnathopods 1 and 2, 5ᵗʰ joint shorter than · 6ᵗʰ, distally nearly as· wide, 6ᵗʰ rectangular, nearly twice as long as broad in gnathopod 1, and more than twice as long as broad in gnathopod 2, palm convex, not very oblique, much shorter than hind margin, finger with nail. Peraeopods 1 and 2 very slender, said to be slightly longer than peraeopods 3—5, which have the 2ᵈ joint oval, with serrate margins, the 4ᵗʰ rather long and distally somewhat widened. Uropod 1, peduncle as long as the equal rami. Uropod 2, peduncle shorter than the rami. Uropod 3, peduncle half as long as outer ramus, which carries 6 or 7 spines, inner ramus rudimentary, flattened, shorter than peduncle. Telson almost square, apex with rounded emargination, and 5 or 6 spines on either side. Accessory branchiae on peraeon segments 5—7 latero-ventrally inserted somewhat behind the basal joint of the peraeopod, directly behind the primary branchiae. L. 4 mm.

Wells near Prague.

3. **E. packardii** (S. I. Sm.) 1873 *Crangonyx vitreus* (err., non *Stygobromus v.* E. D. Cope 1872!), Packard in: Rep. Peabody Ac., v. 5 p. 95 | 1888 *C. packardii*, (S. I. Smith in:) Packard in: Mem. Ac. Washington, v. 41 p. 35 t. 5 f. 5—11 | 1899 *Eucrangonyx p.*, T. Stebbing in: Tr. Linn. Soc. London, ser. 2 v. 7 p. 423.

Eyes small, round, without dark pigment. Antenna 1, 1st joint little longer than 2d, 3d 3/4 as long as 2d, flagellum not twice as long as peduncle, with 14 joints, accessory flagellum not longer than 1st joint of primary. Antenna 2, ultimate and penultimate joints of peduncle equal, flagellum of few joints. Gnathopod 1 (in a ♀ 5 mm long), 5th joint trapezoidal, 6th oblong, palm little oblique, much shorter than hind margin; gnathopod 2 similar, but 6th joint longer, with more oblique palm. In largest ♀ gnathopod 1 has a more triangular 5th joint and the 6th more expanded, wider than the 5th, and gnathopod 2 also has the palm much more oblique and longer than hind margin. Peraeopods 1—5 differ from those in E. gracilis in having the spines more numerous, longer and more slender. Uropods 1 and 2 have the spines shorter and more obtuse than they usually are in E. gracilis, but otherwise these 2 closely allied species agree in uropods and telson. L. reaching 7·5 mm.

Indiana (wells in Orleans and at New Albany).

. 4. **E. gracilis** (S. I. Sm.) 1871 *Crangonyx g.*, S. I. Smith (& A. E. Verrill) in: Amer. J. Sci., ser. 3 v. 2 p. 453 | 1893 *C. g.*, A. Della Valle in: F. Fl. Neapel, v. 20 p. 682 | 1894 *C. g.*, Chilton in: Tr. Linn. Soc. London, ser. 2 v. 6 p. 218 | 1899 *Eucrangonyx g.*, T. Stebbing in: Tr. Linn. Soc. London, ser. 2 v. 7 p. 423.

Pleon segments 1—3, postero-lateral corners produced and ending in a small tooth. Eyes slightly elongated, composed of a few black facets. Antenna 1 rather more than half as long as body, 1st and 2d joints subequal, 3d 2/3 as long as 2d, flagellum not twice as long as peduncle, with about 20 joints, accessory flagellum as long as 1st joint of primary. Antenna 2 about half as long as antenna 1, ultimate and penultimate joints of peduncle long, subequal, flagellum rather shorter than peduncle, 7- or 8-jointed. Gnathopods 1 and 2 of ♀ as in E. packardii (p. 389). In ♂ gnathopod 1 has palm slightly oblique, strongly spined, gnathopod 2 has 6th joint proportionally shorter than in ♀, increasing considerably in breadth distally. palm much more oblique, slightly arcuate and spinose. Peraeopod 5 slightly longer than peraeopods 3 and 4, 2d joint in all having hind margin serrate. Uropod 3 reaching apex of uropod 2, outer ramus nearly twice as long as peduncle, armed with a few slender spines, inner minute, unarmed, shorter than diameter of outer. Telson scarcely as long as peduncle of uropod 3, a little broader than long, apical margin with a triangular emargination, on either side of which the extremity is truncate, and armed with three spines. L. ♀ 6—7, ♂ 5—14 mm.

North-America (Lake Superior, Lake Huron).

5. .**E. antennatus** (Pack.) 1881 *Crangonyx a.*, Packard in: Amer. Natural., v. 15 p. 880 | 1888 *C. a.*, Packard in: Mem. Ac. Washington, v. 41 p. 36 t. 5 f. 12, 14; textf. | 1893 *C. a.*, A. Della Valle in: F. Fl. Neapel, v. 20 p. 682 | 1899 *Eucrangonyx a.*, T. Stebbing in: Tr. Linn. Soc. London, ser. 2 v. 7 p. 423.

Side-plates 1—4 very unequal, 1st very small, 4th unusually large and nearly square, but seemingly much excavate behind. Pleon segments 1—3, postero-lateral corners subacute. Eyes distinct, well developed, black, but not so distinct and only 1/4 as large as in E. gracilis. Antenna 1 nearly 2/3 as long as the body, 2d joint of peduncle is said to be much longer than in E. gracilis, 3d rather more than half 2d, flagellum 20—24-jointed. Antenna 2 (in figure) not 1/3 as long as antenna 1. Gnathopods 1 and 2 and peraeopods 1—5 apparently as in E. gracilis, as also uropods 1—3 and telson, but rami of uropods are said to be slightly stouter and more polished, and the spinules a little stouter. Colour purplish. L. 6—7 mm.

Tennessee (subterranean stream of Nickajack Cave).

14. Gen. **Axelboeckia** Stebb.

1894 *Boeckia* (Sp. un.: B. spinosa) (non A. W. Malm 1871!), (O. Grimm in MS.)
G. O. Sars in: Bull. Ac. St.-Pétersb., ser. 5 *v.* 1 p. 182 | 1899 *Axelboeckia*, T. Stebbing
in: Tr. Linn. Soc. London, ser. 2 *v.* 7 p. 423.

Integument crustaceous, body robust, with median and marginal carinae,
pleon segments 5 and 6 small. Head rostrate. Side-plates 1 and 4 shorter
than 2^d or 3^d, 4th not very broad, nor much emarginate behind. Antenna 1
the longer, with accessory flagellum. Mouth-parts nearly as in Gammarus
(p. 460), but apex of upper lip scarcely emarginate, lower lip with broad
principal lobes separated by rudiments of inner lobes. Maxilla 1 on right
has apex of palp expanded, denticulate, on left narrower, with a few
slender spines. Maxillipeds, 3^d joint of palp distally expanded. Gnathopods
1 and 2 nearly alike, small, subchelate, stronger in ♂ than in ♀. Peraeopods
1—5 differing little in length. Peraeopods 3 and 4, 2^d joint narrow.
Peraeopod 5, 2^d joint broadly expanded. Branchial vesicles broad, with
narrow attachment. Uropods 1—3 successively smaller. Uropod 3 short, peduncle
short, outer ramus 1-jointed, inner shorter. Telson very small, unarmed, not
deeply cleft.

2 species.

Synopsis of species:

1. **A. spinosa** (O. Sars) 1894 *Boeckia s.* (*B. nasuta, B. hystrix*), (O. Grimm
in MS.) G. O. Sars in: Bull. Ac. St.-Pétersb., ser. 5 *v.*1 p. 183 t. 1, 2 | 1899 *Axelboeckia's.*,
T. Stebbing in: Tr. Linn. Soc. London, ser. 2 *v.* 7 p. 424.

Body extremely stout in ♀, less so in ♂. Back obtusely carinate
throughout, peraeon segments each with spiniform process on each side near
side-plates, in segment 5 very large, acute, the others smaller and rather
obtuse; pleon segments 1—4 with a pair of subdorsal, upturned processes,
segment 4 overlapping 5th and 6th, and ending in a rather large, hooked,
median projection. Young specimens have all processes of head and body
digitiform, ending obtusely, and the dorsal prominences more strongly
elevated, giving the back a serrated appearance. Head slightly keeled,
rostrum horizontal, forming an equilateral triangle, lateral lobes acute, longer
than rostrum, diverging nearly at a right angle. Side-plate 1 setose, curved
forward, 4th obliquely truncate below emargination. Pleon segment 3,
postero-lateral angles obtusely quadrate. Eyes small, rounded, dark, wide
apart. Antenna 1 not long. 1st joint as long as 2^d and 3^d combined, 3^d nearly
as long as 2^d, flagellum twice as long as peduncle, with about 25 short, setose
joints, accessory flagellum an extremely small nodule, carrying 2 setules.
Antenna 2 much shorter, bent back, ultimate joint of peduncle shorter than
penultimate, flagellum 9-jointed. Gnathopods 1 and 2 in ♂, 6th joint much
wider than 5th, hind margin armed with forward curving spines and forming
a broadly rounded lobe where it meets the deeply concave palm, finger
strongly curved, impinging within the lobe. Gnathopods 1 and 2 in ♀
comparatively small, 6th joint about as long as 5th, and scarcely broader,
palm in gnathopod 1 somewhat oblique, in gnathopod 2 nearly transverse,
finger not strong, matching palm. Peraeopods 3 and 4 rather longer than
the rest, 2^d joint little expanded, slightly tapering distally. Peraeopod 5,
2^d joint nearly straight in front, strongly expanded behind with curved hind

margin. Uropod 1 much the longest, rami subequal, slightly longer than peduncle. Uropod 2, rami subequal, longer than peduncle. Uropods 1—3 very slightly armed. Telson extremely small, broadly triangular, length not half the breadth, incision about to the middle, apices rounded. L. ♂ 25, ♀ 20 mm.

Caspian Sea. Depth 13—282 m.

2. **A. carpenterii** (Dyb.) 1874 *Gammarus c.*, B. Dybowsky in: Horae Soc. ent. Ross., *v.* 10 suppl. p. 113 t. 13 f. 2 | 1899 *Axelboeckia c.*, T. Stebbing in: Tr. Linn. Soc. London, ser. 2 *v.* 7, p. 424 | 1893 *Amathilla homari* (part.)?, A. Della Valle in: F. Fl. Neapel, *v.* 20 p. 928.

All segments with a median carina, low and divided by a transverse depression; all segments with hind margin more or less thickened, the thickening on peraeon segments 1—3 and pleon segments 1—3 carrying very small tubercles, representatives of lateral carinae; all segments with marginal swellings, stronger and more carina-like on peraeon segments 1—3 than on any others; the various elevations feebly developed on pleon segments 4—6. Head, rostrum long, acute, slightly deflexed, with carina-like compression below; upper surface of head separated from the cheeks by weak ridge-like elevations, and grooved with a furrow continued along the rostrum. Side-plates 1—4 very deep and obliquely outward and downward directed, owing to the breadth and flatness of the body. Eyes moderately large, prominent, biscuit-shaped, black. Antenna 1 about $^1/_4$ as long as body, and about $^1/_3$ longer than antenna 2, peduncle stouter and usually longer than peduncle of antenna 2, 3^d joint longer than 2^d, in ♂ flagellum with 12 joints, accessory flagellum with 2; antenna 2, flagellum with 4 joints; but in (much longer) ♀ antenna 1, flagellum with 20—29, accessory flagellum with 4 or 5, antenna 2, flagellum with 7 or 8 joints. Gnathopods 1 and 2, 5th joint short, 6th joint broad, oblong, that of gnathopod 1 somewhat larger than that of gnathopod 2 (in figure equal in size), the length little greater than the breadth, which encreases to the palm, this being oblique, well defined, subequal to the hind margin. Peraeopods 3—5, 2^d joint not broad, hind margin feebly convex, with few but rather long setae. Uropods 1 and 2 reaching end of uropod 3 or rather beyond. Uropod 3 $^1/_{15}$ as long as body, inner ramus about $^2/_3$ as long as outer, with simple setae, outer with a few plumose setae on inner margin. Telson cleft almost to centre (in figure less than $^1/_3$ of length). Colour brown or yellow. L. ♂ 11, ♀ 24—31 mm.

Lake Baikal. Depth 50—300 m.

15. Gen. **Brachyuropus** Stebb.

1899 *Brachyuropus*, T. Stebbing in: Tr. Linn. Soc. London, ser. 2 *v.* 7 p. 424.

With median dentate carina. Side-plate 4 with projecting tooth. Antenna 1 much the longer, accessory flagellum long. Gnathopods 1 and 2 similar, subchelate. Peraeopods 3—5 elongate. Uropods 1 and 2 elongate. Uropod 3 rudimentary, rami very unequal. Telson apically emarginate.

2 species.

Synopsis of species:

1. **B. grewingkii** (Dyb.) 1874 *Gammarus g.*, B. Dybowsky in: Horae Soc. ent. Ross., *v.* 10 suppl. p. 150 t. 2 f. 4 | 1893 *Crangonyx? g.*, A. Della Valle in: F. Fl. Neapel, *v.* 20 p. 928 | 1899 *Brachyuropus g.*, T. Stebbing in: Tr. Linn. Soc. London, ser. 2 *v.* 7 p. 424.

Median dorsal carina represented on all segments, except pleon segments 4—6 or 5 and 6, by strong teeth, usually strongest on peraeon segments 5—7 and pleon segments 1 and 2, all, and especially those on pleon segments 1—3, laterally strongly compressed, hinder pleon segments dorsally rounded and completely smooth. Lateral marginal carinae on peraeon segments weak, tubercular (a variety with double series of small tubercles on peraeon segments 6 and 7 and pleon segments 1—3). Head, surface rough, rostrum short, obtusely pointed; behind the eye from the cheek projects a long, pointed, obliquely outward and forward directed tooth. Side-plates 1—4 more or less deeply emarginate below between 2 acute points, the hinder longest. Side-plate 4 on lower part of its surface provided with a downward and backward directed, strong, pointed, flattened tooth, reaching far over its hind angle. Pleon segments 1—3, postero-lateral angles acute (in figure rounded). Eyes weakly developed, white, dimly showing through the pellucid integument, sometimes not discernible at all. Antenna 1 more than half as long as body, about thrice as long as antenna 2, flagellum with more than 60 joints, accessory flagellum 22-jointed. Antenna 2, ultimate joint of peduncle shorter than penultimate, flagellum 14-jointed. Gnathopods 1 and 2, 5th joint short, 6th piriform, narrowing to the finger hinge, palm very oblique, much longer than hind margin, defined by a spine, and having a prominent spine near the middle (in figure); this joint (in figure) is longer in gnathopod 1 than in gnathopod 2. Peraeopods 3—5, 2d joint narrowly heart-shaped (in figure; narrowly oblong, a little widened proximally), without setae on hind margin. Peraeopod 4 as long as body. Peraeopod 5 subequal to peraeopod 4 (in figure). Uropods 1 and 2 strongly developed, reaching about equally far, rami thickly fringed with simple setae. Uropod 3 scarcely $^1/_{25}$—$^1/_{20}$ as long as body, outer ramus fringed with simple setae, 3—5 times as long as inner, which has only an apical seta. Telson broader than long, with convex margins narrowing to a shallowly emarginate apex, with angles only subacute. Colour dark horn-yellow to yellowish white. L. reaching 66 mm.

Lake Baikal. Depth 100—1000 m.

2. **B. reichertii** (Dyb.) 1874 *Gammarus r.*, B. Dybowsky in: Horae Soc. ent. Ross., *v.* 10 suppl. p. 42, 152 t. 13 f. 4 | 1893 *Crangonyx reicherti*, A. Della Valle in: F. Fl. Neapel, *v.* 20 p. 930 | 1899 *Brachyuropus reichertii*, T. Stebbing in: Tr. Linn. Soc. London, ser. 2 *v.* 7 p. 424.

Median dorsal carina as in B. grewingkii, but the teeth higher, more acute, less compressed, longest on pleon segments 1 and 2; peraeon segments 1—4 with single weak lateral carina, segments 5—7 with lateral row of little roundish tubercles, and marginal prominences carina-like, tuberculate in the middle; pleon segments 1—3 with lateral tubercles on hind margin. Head, surface rough, rostrum a short, blunt hump; from the cheek projects forward, outward, and upward a slightly bent tooth. Side-plates 1—4 below shallowly emarginate, the angles blunt and short. From the surface of side-plate 4, and nearly at right angles to it, a very long, rounded tooth is directed horizontally outward. Distance between apices of this pair of teeth equal to $2^1/_2$—3 times greatest breadth of body without the teeth. Pleon segments 1—3, postero-lateral angles rounded. Eyes as in B. grewingkii. Antenna 1 $^2/_3$ as long as body, 3—4 times as long as antenna 2, flagellum 46-jointed, accessory flagellum 12-jointed. Antenna 2, ultimate joint of peduncle shorter than penultimate, flagellum 8-jointed. Gnathopod 1 as in B. grewingkii; gnathopod 2 smaller. narrowing less, palm (in figure) shorter than hind margin. Peraeopod 3 not much shorter than the body. Uropods 1 and 2, rami scantily fringed with setae. Uropod 3 less than $^1/_{30}$ as long as body, outer ramus 2—3 times as long as inner, both having only apical setae. Telson two-pointed, i. e. with acute angles to the emarginate apex. Colour whitish or yellow. L. 32 mm.

Lake Baikal. Depth 200—500 m.

16. Gen. **Macrohectopus** Stebb.*)

1874 *Constantia* (non A. Adams 1860, Mollusca!), *Costantia* (laps.), B. Dybowsky in: Horae Soc. ent. Ross., *v.* 10 suppl. p. 50, 51, 186 | 1888 *Costantia*, T. Stebbing in: Rep. Voy. Challenger, *v.* 29 p 427, 428.

Peraeon smooth, pleon carinate. Head not rostrate. Side-plates all very small and shallow, apparently the 2^d deepest, the 6th widest. Antenna 1 longer and stronger than antenna 2, peduncle very long, 3^d joint longest, accessory flagellum wanting. Mouth-parts (so far as known) as in Gammarus (p. 460). Gnathopods 1 and 2 slender, feebly subchelate, 5th joint longer than 6th. Peraeopods 1—5 very slender, peraeopod 4 much the longest, 2^d joint little expanded. Uropod 1 much the longest. Uropods 1 and 2, outer ramus much shorter than inner. Uropod 3, rami long, lanceolate, equal, 1-jointed, fringed with plumose setae. Telson long, divided.

1 species.

1. **M. branickii** (Dyb.) 1874 *Constantia b.* + *C. b. var. alexandri*, B. Dybowsky in: Horae Soc. ent. Ross., *v.* 10 suppl. p. 186 t. 3 f. 7; p. 187 t. 3 f. 6 | 1888 *Costantia b.*, T. Stebbing in: Rep. Voy. Challenger, *v.* 29 p. 1696 | 1893 *Pontogeneia* (part.)?, A. Della Valle in: F. Fl. Neapel, *v.* 20 p. 891.

Body slender. Pleon segment 1 with weak median carina running out to a long, horn-like, upward projecting and slightly forward bent tooth (in *var. alexandri* directed straight backward). Pleon segments 2 and 3 with median carina much stronger, but tooth directed straight backward and insignificant. Head, front somewhat depressed. Side-plates so small and shallow that the branchial vesicles and marsupial plates in a lateral view are exposed. Pleon segments 1—3, postero-lateral corners (in figure) rounded. Eyes rather large, narrowly reniform, black. Antenna 1 about $^2/_3$ as long as body, flagellum shorter than peduncle, with 23—56 joints. Antenna 2 very slender, ultimate and penultimate joints of peduncle long, subequal, flagellum with 12—22 joints, setiferous as in antenna 1. Gnathopods 1 and 2 subequal. 5th joint almost twice as long as 6th (in figure subequal to it), 6th joint narrowly oval, palm feebly defined. Peraeopod 1 rather longer than peraeopod 2. Peraeopods 3—5, 2^d joint short, narrow, not setiferous on hind margin. Peraeopod 4, 4 times as long as peraeopod 3, and twice as long as peraeopod 5. Uropod 1 reaching beyond uropod 3, peduncle long, but inner ramus longer. 5 times as long as the outer (in figure of *var. alexandri* shorter than peduncle and not much over twice as long as the outer ramus). Uropod 2, inner ramus about twice as long as outer, with plumose setae on inner margin. Colour pellucid. L. ♀ 36 mm.

Lake Baikal (only met with swimming).

---

*) Nom. nov. Μακρός, long, ἔκτος, sixth, πούς, foot. — The name *Constantia* is preoccupied (1860, A. Adams in: Ann. nat. Hist., ser. 3 *v.* 5 p. 300). The name *Costantia* is an accidental misspelling.

17. Gen. **Cardiophilus** O. Sars

1896 *Cardiophilus* (Sp. un.: *C. baeri*), G. O. Sars in: Bull. Ac. St.-Pétersb.. ser. 5
v. 4 p. 474 | 1899 *Caridophilus*, J. V. Carus in: Zool. Anz., Regist. 16—20 p. 89.

Body elongate, smooth. Side-plates 1—4 rather broad. Antenna 1,
accessory flagellum very small. Antenna 2 exceedingly small. Upper lip
small, rounded. Lower lip without inner lobes. Mandible, cutting plates
narrow, spine-row of 3 spines, molar not very powerful, palp large. Maxilla 1,
inner plate with 3 plumose setae, outer with denticulate spines, number not
stated, palp weak, of 2 subequal slender joints, the 2^d tipped with 2 setules.
Maxilla 2 normal, inner plate with setae only on distal half of inner margin.
Maxillipeds normal, except that the finger is rudimentary. tipped with 2
minute setules. Gnathopods 1 and 2 unequal, subchelate, 5th and 6th joints
short in gnathopod 1, narrowly elongate in gnathopod 2. Peraeopods 1
and 2, 4th joint expanded, but not large. Peraeopods 3—5, 2^d joint moderately
expanded. Peraeopods 1—5, finger strong, hooked. Uropod 3 very small,
outer ramus with minute 2^d joint, inner scale-like. Telson very short, cleft
to. the base.

1 species.

1. **C. baeri** O. Sars 1896 *C. b.*, G. O. Sars in: Bull. Ac. St.-Pétersb.. ser. 5 *v.* 4
p. 474 t. 11 f. 13—27.

Body extremely slender and elongate. nearly cylindric, pleon segments
4—6 not long, perfectly smooth. Head not rostrate, lateral lobes rounded,
post-antennal corners subacute. Side plate 1 subquadrate, 4th little emarginate.
Pleon segment 3, postero-lateral corners quadrate, but not very sharply. Eyes
subrotund, dark. Antenna 1 about $^1/_4$ as long as body, peduncle short, 1st joint
longer than 2^d and 3^d combined, flagellum 10-jointed, accessory flagellum
extremely small, with 2 joints, 2^d minute. Antenna 2 scarcely half as long,
ultimate joint of peduncle rather shorter than penultimate, flagellum 4-jointed.
Gnathopod 1, 5th joint subequal to 6th, 6th oval quadrangular, slightly narrowed
distally. palm nearly transverse. Gnathopod 2 much longer, 5th joint fully
as long as 6th and distally wider. 6th narrow, palm very short, transverse.
Peraeopods 3—5, 2^d joint oblong oval, slightly narrowed below, little differing
in peraeopod 5 from that in peraeopod 4. Uropods 1 and 2 rather stout,
rami subequal, having at the apex several spines, some of them hooked.
Uropod 3 scarcely at all projecting beyond uropod 1, outer ramus scarcely
longer than peduncle, furnished with a few setules, inner ramus with 1 apical
seta. Telson twice as broad as long, almost semicircular. each lobe with
a small apical and 2 lateral hairs. L. ♀ 5·5 mm.

Caspian Sea.

18. Gen. **Brandtia** Bate

1862 *Brandtia* (Sp. un.: *B. latissima*), Bate, Cat. Amphip. Brit. Mus., p. 129.

A median tubercular carina on at least some of the segments of
peraeon and pleon; marginal carinae more or less developed. Antenna 1
longer, but seldom much longer and in ♂ sometimes shorter than antenna 2,
accessory flagellum 1-jointed. Mouth-parts nearly as in Gammarus (p. 460), at
least in B. lata. Upper lip rounded. Lower lip without inner lobes. Maxilla 1,
inner plate with numerous setae, outer with 11 spines. Peraeopod 5, 2^d joint
moderately expanded. Uropod 3 short. Telson deeply cleft.

6 species.

Synopsis of species:

1. **B. lata** (Dyb.) 1874 *Gammarus latus*, B. Dybowsky in: Horae Soc. ent. Ross.,
*v.*10 suppl. p.159 t.4 f.5 | 1893 *Atylus? l.*, A. Della Valle in: F. Fl. Neapel, *v.* 20 p. 929 :
1899 *Brandtia lata*, T. Stebbing in: Tr. Linn. Soc. London, ser. 2 *v.*7 p. 424.

Back of peraeon flatly roof-shaped, median carina throughout, tubercular, on segment 1 carrying 1 or 2 backward bent spines; marginal carinae
strongly developed, obtusely projecting immediately over the side-plates;
on pleon segments 1—3 the median tubercular carina carries on segment 1
in ♂ 2, in ♀ 4 spines, on segments 2 and 3 of both sexes 4 forward
directed spines; corresponding to the marginal carinae of peraeon are small,
slightly curved elevations; pleon segments 4—6 arched, without carinae.
Head, front with tip-tilted rostrum, carrying 2 backward pointing spines,
top of head surmounted by 4 humps each carrying a backward directed
spine; an acute spine projecting from lower margin of each cheek. Side-
plates 1—4 setiferous, 1—3 carinate. Lateral compression of body greatest
at lower part of side-plates. Eyes prominent, oval or reniform, black.
Antenna 1 not long, in ♂ little, in ♀ considerably, longer than antenna 2,
1st joint with hump near distal end above, flagellum 17—19 - jointed.
Antenna 2, flagellum 7—9 - jointed. Gnathopods 1 and 2 subequal,
6th joint oblong, but in gnathopod 1 with more oblique palm, which is
armed with 1 large and 2 small spines. Peraeopods 3—5. 2d joint heart-
shaped, with 11—15 rather long setae on hind margin. Uropod 3 smaller
in ♀ than in ♂, rami nearly equal, ending in simple setae longer than the
rami, with delicate plumose setae on the inner margin. Telson broader than
long, cleft nearly to the base, with small setules on the rounded apices.
Colour brownish yellow to brownish green, sometimes with a bright mark
across end of peraeon. L. 17—20 mm.

Lake Baikal. Depth 2—20 m.

2. **B. latissima** (Gerstf.) 1858 *Gammarus latissimus*, Gerstfeldt in: Mém. prés.
Ac. St.-Pétersb., *v.*8 p. 288 | 1862 *Brandtia latissima*, Bate, Cat. Amphip. Brit. Mus.,
p. 129 t. 28 f. 5, 6 | 1893 *Pontogeneia? l.*, A. Della Valle in: F. Fl. Neapel, *v.*20 p. 616. 929
1874 *Gammarus latissimus* + *G. latior*, B. Dybowsky in: Horae Soc. ent. Ross., *v.* 10
suppl. p. 161; p. 158 t. 4 f. 6.

General shape as in B. lata, median carina strongly developed, on pleon
segments 2 and 3 carrying 3—4 pairs of spines, but none on segment 1;
on pleon segment 3 the spines of the last pair almost on the hind margin and almost
twice as far asunder as those of the other pairs, the same being the case
on segment 2 only when the 4th pair of spines is developed. Marginal
carina on each of peraeon segments carrying a short spine, this carina

represented by lateral swellings on pleon segments 1—3; pleon segments 4—6 much longer than in B. lata. Head with spines as in B. lata and 2 additional lateral pairs. Side-plates 1—4 all carinate, but not setiferous. Eyes prominent, oval or reniform, black. Antennae 1 and 2 in general as in B. lata, but 1st joint of antenna 1 with 2 humps in line, the front one armed with 2 backward directed spines, flagellum with 15—27 joints, of which in the ♂ the 3d, in the ♀ the 4th is longest. Gnathopods 1 and 2 subequal, 6th joint slender, with slightly convex margins (Dybowsky), in gnathopod 1 6th joint more elongate, in gnathopod 2 more rhomboidal, with palm more obliquely truncate (Gerstfeldt). Peraeopods 3—5, 2d joint heart-shaped, narrower than in B. lata, with extremely short, distant, setae on hind margin. Uropod 3 shorter in ♀ than in ♂, but never rudimentary. Colour in general greenish brown, parts of head and peraeon brighter. L. 19—27 mm.

Lake Baikal, River Angara.

3. **B. tuberculata** (Dyb.) 1874 *Gammarus tuberculatus*, B. Dybowsky in: Horae Soc. ent. Ross., *v.* 10 suppl. p. 161 | 1893 *Atylus? t.*, A. Della Valle in: F. Fl. Neapel, *v.* 20 p. 931 | 1899 *Brandtia tuberculata*, T. Stebbing in: Tr. Linn. Soc. London, ser. 2 *v.* 7 p. 424.

The median carina weakly developed as longish tubercles on peraeon segments 6 and 7 and pleon segments 1—3; marginal carina represented by a rounded, smooth, not very prominent swelling immediately over the side-plates of all peraeon segments and pleon segments 1—3." Head, rostrum a short obtuse point. Side-plates 1—4 setiferous. Eyes round oval, broader below than above, black. Antenna 1 less than $\frac{1}{2}$ as long as body, twice as long as antenna 2, 1st joint broad, flagellum 24-jointed. Antenna 2, flagellum 8-jointed. Gnathopods 1 and 2 subequal, 6th joint piriform, palm defined by 3 spines, finger long, slightly bent. Peraeopods 3 and 4, 2d joint expanded above, all the hind margin beset with rather long setae. Peraeopod 3, 2d joint greatly widened below. Uropods 1 and 2 reach beyond uropod 3, which is very short, outer ramus twice as long as inner, with 3—7 spines on apex and 2 on outer margin, inner ramus with only 2 spines, both apical. Telson divided. Colour greyish white or clear brown, with narrow dark stripes across the segments. L. 16 mm.

Lake Baikal. Litoral.

4. **B. morawitzii** (Dyb.) 1874 *Gammarus .m.*, B. Dybowsky in: Horae Soc. ent. Ross., *v.* 10 suppl. p. 163 | 1893 *Atylus? m.*, A. Della Valle in: F. Fl. Neapel, *v.* 20 p. 929 | 1899 *Brandtia m.*, T. Stebbing in: Tr. Linn. Soc. London, ser. 2 *v.* 7 p. 424.

Median carina distinct on the convex peraeon segments 6 and 7 and pleon segments 1—3; spines on pleon segments 5 and 6 extremely delicate; whether marginal carinae are present or absent is not stated. Head, rostrum broad, rounded, little advanced. Side-plates 1—4 setiferous. Eyes moderately large, roundish, convex, black or dark red. Antenna 1 rather more than $\frac{1}{3}$ as long as body, and about $\frac{1}{3}$ longer than antenna 2, 1st joint stout, flagellum 14—18-jointed. Antenna 2, flagellum with 7 or 8 joints. Gnathopods 1 and 2 small, slender, 6th joint oval in gnathopod 1, oblong in gnathopod 2. Peraeopod 3, 2d joint on upper margin strongly rounded. Peraeopod 4, 2d joint heart-shaped. Peraeopod 5, 2d joint broader than in peraeopods 3 and 4, hind margin evenly rounded. Uropod 3 extremely short, outer ramus scarcely longer than peduncle, 1-jointed, with 4 spines on apex, 2 on outer margin; inner ramus $\frac{3}{4}$ as long as outer, with 1—3 apical spines. Each half of telson like a cone with broad base. Colour greyish white, sometimes dark longitudinal stripe on the back. L. 10 mm.

Lake Baikal. Depth 20 m.

5. B. smaragdina (Dyb.) 1874 *Gammarus smaragdinus* + *G. s. var. intermedius.* B. Dybowsky in: Horae Soc. ent. Ross., *v.* 10 suppl. p. 164 t. 11 f. 6 | 1893 *Atylus? s.*, A. Della Valle in: F. Fl. Neapel. *v.* 20 p. 930 | 1899 *Brandtia smaragdina*, T. Stebbing in: Tr. Linn. Soc. London, ser. 2 *v.* 7 p. 424.

Shows some affinity to Cardiophilus (p. 395). Sexual variation very marked. Body stouter in ♀ than in ♂. Median carina on peraeon segments 6 and 7 and pleon segments 1—3 consisting of long tubercles of little prominence; pleon segments 4—6 carrying 2 or 3 little groups of very delicate spines; marginal carinae not mentioned. Head, rostrum rounded, depressed. Side-plates 1—4 with marginal setae. Eyes convex, reniform, black. Antenna 1 in ♂ as long as body, 4 times as long as antenna 2, in ♀ half as long as body, twice as long as antenna 2, 1st joint stout, flagellum in ♂ with 44, in ♀ with 34 joints. Antenna 2, flagellum 8-jointed. Gnathopods 1 and 2, 6th joint oval, in ♂ that of gnathopod 1 somewhat the larger. Peraeopods 3—5, 2^d joint moderately broad, most expanded in peraeopod 5, with hind margin and surface setiferous and margin also spinose. Uropod 3 short, longer in ♂ than in ♀, outer ramus $^1/_4$—$^1/_3$ longer than inner, in ♂ with plumose setae on inner margin, on outer and apex with spines and simple setae; in ♀ with 2 spines on outer margin, on inner with 2 or 3 setae, at apex with spines and setae, inner ramus carrying only setae. Telson divided. Colour emerald-green (*var. intermedius* whitish; it has in ♀ smaller eyes, shorter antennae, and shows some minute differences in the uropod 3). L. 11—12 mm.

Lake Baikal. Depth 50—100 m (*var. intermedius* at 15 m).

6. B. fasciata Stebb. 1874 *Gammarus zebra* (non Rathke 1843!), B. Dybowsky in: Horae Soc. ent. Ross., *v.* 10 suppl. p. 166 t. 14 f. 7 | 1893 *Atylus? z.*, A. Della Valle in: F. Fl. Neapel, *v.* 20 p. 931 | 1899 *Brandtia fasciata*, T. Stebbing in: Tr. Linn. Soc. London, ser. 2 *v.* 7 p. 424.

Peraeon segments 5—7 or 6 and 7 and pleon segments 1—3 carinate or provided with little tubercles in the median dorsal line; pleon segments 4—6 very convex, with very delicate spine-points or setae. Head, rostrum short, slightly depressed. Side-plates 1—4 scantily setiferous. Eyes reniform, black. Antenna 1 very little longer than antenna 2, $^1/_4$—$^1/_3$ as long as body, 1st joint rather broad, slightly flattened, flagellum 10—13-jointed. Antenna 2, flagellum 5—7-jointed. Gnathopods 1 and 2 subequal, 6th joint oblong, palm shorter and less oblique in gnathopod 2. Peraeopod 5, 2^d joint trapezoidal, with the corners rounded, broadest below, much broader and longer than the corresponding joint of peraeopods 3 and 4. Uropod 3 short, outer ramus with small 2^d joint, in length about 4 times as long as the inner ramus, with 4 spines and several setae on outer margin, simple setae on the inner, spines and setae at the apex, the inner ramus with 1 long apical seta. Telson divided. Colour greenish brown, with dark brown or black stripe across each segment. L. 17 mm.

Lake Baikal. At river mouths, ascending the streams in swarms.

19. Gen. **Micruropus** Stebb.

1899 *Micruropus*, T. Stebbing in: Tr. Linn. Soc. London, ser. 2 *v.* 7 p. 424.

Without carinae or overarching segments. Antennae 1 and 2 short; antenna 1 usually longer than antenna 2, the peduncles not greatly differing in length, accessory flagellum 1-jointed. Gnathopods 1 and 2, 6th joint not greatly differing in size. Uropod 3 small or very small, rami unequal, outer ramus usually 1-jointed. Telson cleft.

12 species.

Synopsis of species:

1 { All segments dorsally unarmed — **2.**
{ Some segments with dorsal setules or spinules — **3.**

2 { Eyes rather large; uropod 1 not reaching beyond
{ uropod 3 . . . . 1. **M. puella** . . . p. 399
{ Eyes very small; uropod 1 reaching beyond uropod 3 2. **M. inflatus** . . p. 399

3 { All segments anterior to pleon segment 4 dorsally
{ smooth — **4.**
{ Some segments anterior to pleon segment 4 dorsally
{ armed — **5.**

4 { Uropod 1 not reaching beyond middle of uropod 3 3. **M. vortex** . . . p. 400
{ Uropod 1 reaching end of uropod 3 . . . . 4. **M. talitroides** . p. 400

5 { Antenna 1, peduncle shorter than in antenna 2 — **6.**
{ Antenna 1, peduncle not shorter than in antenna 2 — **7.**

6 { Antenna 1 shorter than antenna 2; uropod 3, inner
{ ramus half as long as outer . . . . . . 5. **M. littoralis** . . p. 401
{ Antenna 1 longer than antenna 2; uropod 3, inner
{ ramus scarcely ⅓ as long as outer . . . . . . 6. **M. glaber** . . . p. 401

7 { Peraeon dorsally minutely granular . . . . 7. **M. rugosus** . . p. 402
{ Peraeon dorsally smooth — **8.**

8 { Antennae 1 and 2, in ♂ and ♀, flagellum with calceoli 8. **M. wahlii** . . . p. 402
{ Antennae 1 and 2, flagellum without calceoli — **9.**

9 { Antenna 1 longer than antenna 2 — **10.**
{ Antenna 1 not longer than antenna 2 — **11.**

10 { Uropod 3, inner ramus nearly as long as outer . . 9. **M. fixsenii** . . p. 402
{ Uropod 3, inner ramus only half as long as outer 10. **M. perla** . . . p. 403

11 { Gnathopod 2, 6th joint piriform . . . . 11. **M. klukii** . . . p. 403
{ Gnathopod 2, 6th joint oblong . . . . 12. **M. pachytus** . . p. 404

1. **M. puella** (Dyb.) 1874 *Gammarus p.*, B. Dybowsky in: Horae Soc. ent. Ross.. *v.*10 suppl. p. 48, 175 | 1893 *Atylus? p.*, A. Della Valle in: F. Fl. Neapel, *v.* 20 p. 930 | 1899 *Micruropus p.*, T. Stebbing in: Tr. Linn. Soc. London, ser. 2 v.7 p. 424.

All segments dorsally completely smooth and level. Side-plates 1—4 each with 3—5 setules. Eyes rather large, reniform, ruby-red. Antenna 1 ⅖ as long as body, more than twice as long as antenna 2, peduncle a little longer than peduncle of antenna 2, flagellum 12—14-jointed. Antenna 2, flagellum with 3 or 4 joints. Gnathopod 1, 6th joint piriform, in ♂ larger than the oblong 6th joint of gnathopod 2. Peraeopods 3—5, 2^d joint broad, rather convex in front, strongly widened behind, the almost semicircular wing ending below in a short rounded lobe, setules on both margins. Uropod 1 reaching end of uropod 3. Uropod 3 1/11 or 1/12 as long as body, outer ramus 2-jointed, 2^d joint ⅓ as long as 1st, with 2 apical setae; 1st joint with 1 marginal spine and 2 apical spines; inner ramus ½ as long as outer, with 2 apical setae. Telson cleft. Colour white. L. 5—6 mm.

Lake Baikal. Depth 100 m.

2. **M. inflatus** (Dyb.) 1874 *Gammarus i.*, B. Dybowsky in: Horae Soc. ent. Ross., *v.*10 suppl. p.169 t.12 f.4 | 1893 *Atylus? i.*, A. Della Valle in: F. Fl. Neapel, *v.*20 p.929 | 1899 *Micruropus i.*, T. Stebbing in: Tr. Linn. Soc. London, ser.2 *v.*7 p.424.

All segments dorsally smooth. Side-plates 1—4 with setae. Eyes very small, punctiform, black. Antenna 1 half as long as body, twice as long as antenna 2, longer than in other species of this genus, flagellum more than twice as long as peduncle, 23-jointed. Antenna 2, ultimate joint of peduncle shorter than penultimate, flagellum 9-jointed. Gnathopods 1 and 2, 6th joint piriform, little larger in gnathopod 1 than in gnathopod 2, palm very long and oblique, finger long, curved. Peraeopods 3—5, 2d joint with numerous fascicles of setae at the strongly convex lower front corner, upper hinder part in peraeopods 3 and 4 with broad winglike expansion, hind margin with long setae; peraeopod 5 very broad, evenly expanded behind. setose on both margins. Uropods 1 and 2 reaching much beyond uropod 3. Uropod 3 very short, scarcely $^1/_{20}$ as long as body, outer ramus about twice as long as inner, with 3 apical and 2 pairs of marginal spines, inner with 2 marginal spines and 1 or 2 apical setae. Telson broader than long, cleft to base, lobes conical, divergent. L. about 20 mm.

Lake Baikal. Depth 2—10 m.

3. **M. vortex** (Dyb.) 1874 *Gammarus v.*, B. Dybowsky in: Horae Soc. ent. Ross., v. 10 suppl. p. 178 t. 9 f. 4 | 1899 *Micruropus v.*, T. Stebbing in: Tr. Linn. Soc. London, ser. 2 v. 7 p. 424 | 1893 *Gammarus pungens* (part.)?, A. Della Valle in: F. Fl. Neapel, v. 20 p. 931.

Pleon segments 4—6 with 3 groups of delicate dorsal spines and setae, the number of setae small and variable; other segments smooth. Side-plates 1—4 with isolated setae. Eyes of moderate size, reniform, black. Antenna 1 about $^1/_3$ as long as body, a little longer than antenna 2, peduncle a little shorter than peduncle of antenna 2, but in ♀ sometimes longer, flagellum 9—16-jointed. Antenna 2, flagellum with 5—9 joints, calceoliferous only in the ♂. Gnathopod 1, 6th joint piriform, as long as the oblong 6th joint of gnathopod 2, but (in figure) less bulky. Peraeopods 1 and 2, finger with a strong spine on inner margin adjacent to the nail. Peraeopod 3, 2d joint slightly convex in front, the wing-like expansion produced somewhat downward in a lobe-like rounded corner. Peraeopods 4 and 5, 2d joint heart-shaped, wide above, narrowed below, differing from the form usual in this group, hind margin fringed with long setae. Uropods 1 and 2 scarcely reaching middle of uropod 3. Uropod 3 about $^1/_6$ as long as body, therefore rather distinct from rest of genus, outer ramus 2-jointed, inner only $^1/_3$ or $^1/_4$ as long, both with plumose setae on inner margin. Telson cleft. Colour greenish. L. ♂ 10, ♀ 6—7 mm.

Lake Baikal. Depth 5—30 m; on stony coasts.

4. **M. talitroides** (Dyb.) 1874 *Gammarus t.*, B. Dybowsky in: Horae Soc. ent. Ross., v. 10 suppl. p. 47, 171 t. 14 f. 3 | 1893 *Atylus? t.*, A. Della Valle in: F. Fl. Neapel, v. 20 p. 930 | 1899 *Micruropus t.*, T. Stebbing in: Tr. Linn. Soc. London, ser. 2 v. 7 p. 424.

Pleon segments 5 and 6 with short, delicate, dorsal spinules, other segments smooth. Head, rostrum somewhat prominent, curved, bluntly rounded. Side-plates 1—4 with setae. Eyes not large, reniform, brownish black. Antenna 1 about $^1/_3$ as long as body, about $^1/_3$ longer than antenna 2, but with rather shorter peduncle in ♂, flagellum 14—18-jointed. Antenna 2, flagellum with 7 joints, calceoliferous only in ♂. Gnathopod 1, 6th joint rather shorter and decidedly narrower than the oblong 6th joint of gnathopod 2, described as piriform, but in figure the palm is shorter than hind margin. and, though very oblique, well defined. Peraeopods 3 and 4, 2d joint with convex, strongly setose, lower front corner, hinder expansion setose, rounded

above, narrowed below. Peraeopod 5, 2^d joint setose as in preceding peraeopods, but longer and somewhat broader, hinder expansion narrower than
muscle-bearing part. Uropod 1 reaching end of uropod 3. Uropod 3 $^1/_{13}$—$^1/_{12}$
as long as body, outer ramus 1-jointed, $^2/_7$—$^2/_5$ longer than inner, with
2 pairs of strong spines on outer margin, both rami having long apical setae.
Telson cleft to base, each half (in figure) with 3 apical spinules. Colour
yellowish or greenish grey; front margin of head and side-plates 1—3, side-
plate 4, 2^d joint of peraeopods 3—5, uropod 3 and telson, brownish black.
L. 11 mm.

Lake Baikal.

5. **M. littoralis** (Dyb.) 1874 *Gammarus l.*, B. Dybowsky in: Horae Soc. ent.
Ross., *v.* 10 suppl. p. 168 t. 14 f. 2 | 1893 *Atylus? l.*, A. Della Valle in: F. Fl. Neapel,
v. 20 p. 929 | 1899 *Micruropus l.*, T. Stebbing in: Tr. Linn. Soc. London, ser. 2 *v.* 7 p. 424.

Peraeon segments 6 and 7 and pleon segments 1—3 with dorsal setae,
pleon segments 4—6 with groups of delicate spines arranged on upper curvature
of the segments. Head, rostrum obtusely pointed. Side-plates 1—4 with few
marginal setae. Eyes small, rounded reniform, wide apart, black. Antenna 1
(usually) shorter and with shorter peduncle than antenna 2, scarcely $^1/_4$ as
long as body. 1st joint stout, longer than 2^d and 3^d combined, flagellum
8-jointed, accessory flagellum 1-jointed. Antenna 2, ultimate and penultimate
joints of peduncle subequal, flagellum 5—7-jointed. Gnathopod 1, 6th joint
piriform, longer but narrower than the oblong 6th joint of gnathopod 2,
palm concave, very oblique. Peraeopods 3 and 4 with lower front corner of
2^d joint strongly convex and setose; peraeopod 5 as in M. inflatus (p. 399).
Uropod 1 (not uropod 2) reaching end of uropod 3. Uropod 3 scarcely
$^1/_{15}$ as long as body, outer ramus about twice as long as inner, with 4 or
5 apical and 2 marginal spines and as many setae, inner with only a long
apical seta. Telson broader than long, not cleft to base, lobes conically
divergent, tipped with spinules. Colour white. L. about 10 mm.

Lake Baikal.

6. **M. glaber** (Dyb.) 1874 *Gammarus g.*, B. Dybowsky in: Horae Soc. ent. Ross.,
v. 10 suppl. p. 176 t. 14 f. 6 | 1893 *Atylus? g.*, A. Della Valle in: F. Fl. Neapel, *v.* 20
p 928 | 1899 *Micruropus g.*, T. Stebbing in: Tr. Linn. Soc. London, ser. 2 *v.* 7 p. 424.

Peraeon segments 6 and 7 each with 2 long setae on hind margin,
pleon segments 1—4 with long, isolated setae on the whole upper surface,
segments 5 and 6 with groups of delicate spinules. Head, rostrum a slightly
depressed, rounded point. Side-plates 1—4 with isolated setae. Eyes of
moderate size, rounded reniform, black. Antenna 1 about $^1/_4$ as long as
body, about $^1/_3$ longer than antenna 2, but with peduncle rather shorter,
flagellum 11-jointed. Antenna 2, ultimate and penultimate joints of peduncle
equal, flagellum 4-jointed. Gnathopod 1, 6th joint in $\male$ larger than oblong
6th joint of gnathopod 2, which it much resembles in shape, but that the
much more oblique palm adds to its length, the part bounded by the hind
margin being in both gnathopods oblong. Peraeopods 3—5, 2^d joint with
lower front corner little or not at all inflated, but setose; in peraeopod
5 this joint is trapezoidal, with numerous and long setae on hind margin.
Uropods 1 and 2 reaching last fourth of outer ramus of uropod 3. Uropod 3
about $^1/_8$ as long as body, outer ramus with apical spines and setae, 1 spine
on inner margin, 2 pairs of spines on outer, inner ramus scarcely $^1/_3$
as long as outer, with 1 apical seta. Telson cleft to base, lobes conical,
with some armature. Colour white. L. 9 mm.

Lake Baikal.

7. **M. rugosus** (Dyb.) 1874 *Gammarus r.*, B. Dybowsky in: Horae Soc. ent. Ross., v. 10 suppl. p. 48, 174 t. 14 f. 8 | 1893 *Atylus? r.*, A. Della-Valle in: F. Fl. Neapel, v. 20 p. 930 | 1899 *Micruropus r.*, T. Stebbing in: Tr. Linn. Soc. London. ser. 2 v. 7 p 424.

All peraeon segments and a narrow girdle on head dorsally minutely granular and here and there beset with short setules. Head with short, rounded, rostral point. Side-plates 1—4 setose on lower margin. Eyes small, wide apart, rounded reniform, black. Antenna 1 scarcely $^1/_4$ as long as body, about $^1/_6$ longer than antenna 2, peduncle stouter and a little longer than that of antenna 2, 1st joint exceedingly stout, longer than 2d and 3d combined, flagellum rather shorter than peduncle, 10-jointed; accessory flagellum 1-jointed. Antenna 2, ultimate joint of peduncle shorter than penultimate, flagellum 8-jointed. Gnathopod 1, 6th joint rather longer and broader than that of gnathopod 2 (oval in gnathopod 1, oblong in gnathopod 2, but in figure very similar), the palm little oblique, well defined, much shorter than hind margin. Peraeopods 3—5, 2d joint at lower front corner fringed with fascicles of stiff setae; in peraeopods 3 and 4 the wing-like widening behind chiefly on the upper half, in peraeopod 5 the greatest breadth at the lowest quarter. Uropods 1 and 2 reaching beyond uropod 3. Uropod 3 $^1/_{20}$ as long as body, outer ramus 1-jointed, as long as peduncle, with 3 apical spines and 2 in middle of outer margin, inner ramus about half as long, with 1 long apical seta. Telson (in figure) nearly twice as broad as long, cleft not nearly to the base, a seta on each divergent apex. Colour brownish white. L. 9 mm.

Lake Baikal.

8. **M. wahlii** (Dyb.) 1874 *Gammarus w.* + *G. w. var. platycercus*, B. Dybowsky in: Horae Soc. ent. Ross., v. 10 suppl. p. 179; p. 180 | 1893 *Atylus? w.*, A. Della Valle in: F. Fl. Neapel, v. 20 p. 931 | 1899 *Micruropus w.*, T. Stebbing in: Tr. Linn. Soc. London, ser. 2 v. 7 p. 424.

Pleon segments 4—6 each with 3 groups of spinules, 2 or 3 in each group on segments 4 and 5, only 1 or 2 in each on segment 6. Side-plates 1—4 with setae. Eyes moderately large, reniform, black. Antenna 1 less than $^1/_3$ as long as body, usually rather shorter than antenna 2, in *var. platycercus* rather longer, peduncle stouter but shorter, 1st joint stout, flagellum with 21—27 joints (in *var.* with 43). Antenna 2, flagellum with 21—27 joints (in *var.* with 31—43), as in antenna 1 furnished with calceoli (in antenna 2 smaller in ♀ than in ♂). Gnathopods 1 and 2 with 6th joint of same size, piriform in gnathopod 1, oblong in gnathopod 2. Peraeopods 3 and 4, 2d joint little inflated at the setose lower front corner, hind margin closely set with long setae. Peraeopod 5, 2d joint broad. broadest in the upper half, with long setae on the hind margin. Uropod 3 $^1/_{11}$ as long as body, reaching beyond the others, outer ramus 4 times as long as the inner, with plumose setae on inner margin, 2 pairs of spines and simple setae on the outer, 2d joint short (not found in *var.*); inner ramus with 2 plumose setae on inner margin and a long apical seta. Telson divided. Colour greenish white. L. 11 mm.

Lake Baikal (mouth of the river Sljudianka).

9. **M. fixsenii** (Dyb.) 1874 *Gammarus f.*, B. Dybowsky in: Horae Soc. ent. Ross., v. 10 suppl. p. 172 | 1893 *Atylus? f.*, A. Della Valle in: F. Fl. Neapel, v. 20 p. 928 | 1899 *Micruropus f.*, T. Stebbing in: Tr. Linn. Soc. London, ser. 2 v. 7 p. 424.

Pleon segments 1—6 each with 2 delicate setae on hind margin, otherwise upper surface of body quite smooth and even. Head, upper profile strongly convex, rostrum depressed, beak-like. Side-plates 1--4 with setae.

Eyes moderately large, broadly reniform, not prominent, black. Antenna 1 about $\frac{1}{3}$ as long as body, nearly twice as long as antenna 2, but peduncle, though stouter, scarcely longer, flagellum 16-jointed. Antenna 2, flagellum 4- or 5-jointed. Gnathopods 1 and 2, 6th joint small, piriform in gnathopod 1, oblong in gnathopod 2. Peraeopods 3 and 4, 2d joint narrow, front margin little convex, with fascicles of long setae, hind margin setose, wing-like expansion very weakly developed. Peraeopod 5, 2d joint very broad, in ♂ only $\frac{1}{7}$ longer than broad, lower part of front margin almost monstrously bulging, upper surface of the swelling closely set with long setae. Uropods 1 and 2 moderately long, in ♂ both reaching beyond uropod 3, in ♀ uropod 2 somewhat shorter. Uropod 3, outer ramus as long as peduncle, with 2 or 3 apical spines and 4 pairs on outer margin, inner ramus only about $\frac{1}{8}$ shorter, with 2 or 3 apical spines. Telson cleft. Colour white, with a longitudinal dorsal dark stripe. L. about 9 mm.

Lake Baikal.

10. **M. perla** (Dyb.) 1874 *Gammarus p.*, B. Dybowsky in: Horae Soc. ent. Ross., *v.* 10 suppl. p. 184 | 1893 *Atylus? p.*, A. Della Valle in: F. Fl. Neapel, *v.* 20 p. 929 | 1899 *Micruropus p.*, T. Stebbing in: Tr. Linn. Soc. London, ser. 2 *v.* 7 p. 424.

Peraeon segments smooth, pleon segments with scattered dorsal setules. Head, rostral point short, rounded, slightly depressed. Side-plates 1—4 with few setae. Eyes punctiform or linear, white, becoming invisible in spirit. Antenna 1 about $\frac{1}{4}$ as long as body and $\frac{1}{4}$ longer than antenna 2, peduncle rather longer than peduncle of antenna 2, 1st joint broad, flagellum rather shorter than peduncle, 7- or 8-jointed, accessory flagellum 1-jointed. Antenna 2, flagellum 4- or 5-jointed. Gnathopod 1, 6th joint oval, rather larger than the oblong 6th joint of gnathopod 2. Peraeopods 3—5, lower front corner with fascicles of stiff setae; this joint in peraeopods 3 and 4 very narrow, above broader than below, hind margin sinuous, in peraeopod 5 very broad, wider below than above. Uropod 1 reaching rather beyond uropod 2, both beyond uropod 3. Uropod 3 about $\frac{1}{22}$ as long as body, outer ramus shorter than peduncle or equal to it, with 2 or 3 apical spines, inner ramus half as long, sometimes with an apical seta. Telson cleft. Colour white. L. ♂ 9, ♀ 7 mm.

Lake Baikal. Depth 10 m.

11. **M. klukii** (Dyb.) 1874 *Gammarus k.*, B. Dybowsky in: Horae Soc. ent. Ross., *v.* 10 suppl. p. 181 | 1893 *Atylus? k.*, A. Della Valle in: F. Fl. Neapel, *v.* 20 p. 929 | 1899 *Micruropus k.*, T. Stebbing in: Tr. Linn. Soc. London, ser. 2 *v.* 7 p. 424.

Pleon segments 4—6 with 2 or 3 groups of delicate spinules, back otherwise quite smooth. Side-plates 1—4 setiferous. Eyes narrowly reniform, black. Antenna 1 only $\frac{1}{4}$ as long as body, rather shorter than antenna 2. 1st joint broad, flat, 2d and 3d thin, flagellum with 10 joints, 1st rather long, and as long as 1-jointed accessory flagellum. Antenna 2, flagellum 4- or 5-jointed. Gnathopods 1 and 2 equal, 5th joint piriform. Peraeopod 3, 2d joint broader below than above, hind margin with upper and lower corners rounded. Peraeopod 4, 2d joint wider above than below. Peraeopod 5, 2d joint wider below, hind margin evenly rounded, and as in peraeopods 3 and 4 setose. Uropod 3 $\frac{1}{11}$ as long as body, but reaching beyond uropods 1 and 2, outer ramus 1-jointed, 4 or 5 times as long as inner, with 5 spines and 3 setae on the apex, a spine and a seta on outer margin, inner ramus

26*

with only 1 very long apical seta. Telson divided. Colour greyish white.
L. about 10 mm.

Lake Baikal. Close to the shore.

12. **M. pachytus** (Dyb.) 1874 *Gammarus p.* + *G. p. var. dilatatus*, B. Dybowsky
in: Horae Soc. ent. Ross., *v.* 10 suppl. p. 182; p. 183 | 1893 *Atylus? p.*, A. Della Valle
iu: F. Fl. Neapel, *v.* 20 p. 929 | 1899 *Micruropus p.*, T. Stebbing in: Tr. Linn. Soc.
London, ser. 2 *v.* 7 p. 424.

Segments smooth, except that the hind margin is thickened and overlaps
front margin of following segment. In *var.* dilatatus the surface of the
body covered with short hairs, and the setae of side-plates 1—4 long,
stout, almost spine-like. Eyes small, rounded reniform, black. Antenna 1
$^1/_5$ as long as body, subequal to antenna 2, 1st joint broad, flagellum
11-jointed, with oblique attachment, so that it is always directed outwards.
accessory flagellum 1-jointed. Antenna 2, flagellum 8- or 9-jointed. Gnathopods 1 and 2 almost equally long, 6th joint narrowly piriform in gnathopod 1, obliquely oblong in gnathopod 2, in the *var.* gnathopod 1 is said to
be the larger. Peraeopods 3—5, 2^d joint very setose on hind margin. Peraeopods 3 and 4, 2^d joint broader below than above (in *var.* considerably broader
above than below). Peraeopod 5, 2^d joint broad, strongly and evenly curved
behind, broadest in the upper half (in *var.* trapezoidal, broader below than
above). Uropod 1 reaching at least as far back as uropod 3. Uropod 3
very short, scarcely $^1/_{16}$ as long as body, (in *var.* rudimentary, $^1/_{25}$ as long
as body), outer ramus 2-jointed, less than twice as long as inner, with
3 spines on apex, 2 or 3 on outer margin (in *var.* with 4 on apex, 1 on
outer margin); inner ramus with 1 apical spine. Telson divided (in *var.*
short, divided almost to the middle). Colour greyish white (in *var.* greenish
white). L. 15 mm.

Lake Baikal. Depth 2—10 m.

20. Gen. **Neoniphargus** Stebb.

1899 *Neoniphargus* (Sp. typ.: *N. thomsoni*), T. Stebbing in: Tr. Linn. Soc. London,
ser. 2 *v.* 7 p. 424.

Side-plates 1—4 much deeper than those which follow. Eyes well
developed. Antenna 1 the longer, accessory flagellum very small, 2-jointed.
Mouth-parts nearly as in Niphargus (p. 405), but maxilla 1 said to have
6 spines on outer plate. Gnathopods 1 and 2 similar, subchelate, 5th joint
distally wide, 6th quadrate. Peraeopod 5 shorter than peraeopod 4. Uropod 3
not elongate, outer ramus 1-jointed, inner minute. Telson partly cleft.

1 accepted and 1 doubtful species.

1. **N. thomsoni** Stebb. 1893 *Niphargus montanus* (non *Gammarus m.* A. Costa
1851!), G. M. Thomson in: P. R. Soc. Tasmania, 1892 p. 70 t. 6 f. 1—13 | 1899 *Neoni-
phargus thomsoni*, T. Stebbing in: Tr. Linn. Soc. London, ser. 2 *v.* 7 p. 425.

Head, lateral corners very little prominent. Side-plate 1 rounded below,
4th broad, emarginate. Pleon segments 1—3, postero-lateral corners rounded.
Eyes relatively large, subreniform, close to lateral margin of head. Antenna 1
short, 2^d joint rather shorter than 1st, and 3^d than 2^d, flagellum rather longer
than peduncle, with about 14 joints, accessory flagellum rather longer than
1st joint of primary. Antenna 2 said to be about half as long as antenna 1,
flagellum shorter than peduncle, 8-jointed. Gnathopods 1 and 2, 5th joint triangular,

cup-shaped, distally very wide, 6th nearly square, hind margin rugose, palm transverse, minutely denticulate. Uropods 1 and 2 reach nearly as far back as uropod 3. Uropod 3, outer ramus rather longer than peduncle. Telson as broad as long, cleft to centre, apices broad, each carrying 4 spinules. Colour whitish yellowish to brown. L. 6 mm.

Tasmania (swampy ground near top of Mount Wellington about 1200 m).

N. moniezi (Wrześu.) 1889 *Gammarus puteanus* (err., non C. L. Koch 1835!), Moniez in: Rev. biol. Nord France, v.1 p.245 | 1890 *Niphargus moniezi*, Wrześniowski in: Z. wiss. Zool., v.50 p.672, 675, 676 | 1899 *N. m., Neoniphargus* (part.)?, T. Stebbing in: Tr. Linn. Soc. London, ser.2 v.7 p.425.

Eyes wanting. Gnathopods, 1 and 2, 6th joint oval. Uropod 3 short, inner ramus represented by a scale, oval, broad, „repliée", without apical spines or setae, outer ramus 1-jointed. Telson double. L. 6—7 mm.

North of France. Subterranean Waters.

21. Gen. **Niphargus** Schiödte

1849 *Niphargus* (Sp. un.: *N. stygius*), Schiödte in: Danske Selsk. Skr., ser.5 v.2 p.26 | 1890 *N.*, Wrześniowski in: Z. wiss. Zool., v.50 p.620 | 1893 *N.*, A. Della Valle in: F. Fl. Neapel, v.20 p.704 | 1894 *N.*, Chilton in: Tr. Linn. Soc. London, ser.2 v.6 p.165.

Body compressed, not carinate. Side-plates 1—4 not large. Eyes rudimentary or wanting. Antenna 1 the longer, accessory flagellum not more than 2-jointed. Upper lip rounded. Lower lip, inner lobes well developed. Mandible normal. 3[d] joint of palp longer than 2[d]. Maxilla 1, inner plate with few (2 or 3) setae, outer with 7 spines, palp of right and left maxilla alike in form. Maxilla 2, inner plate not or very slightly fringed along inner margin. Maxillipeds, outer plates not very broad, fringed with spine-teeth on inner margin, palp elongate. Gnathopods 1 and 2 similar, subchelate, 5th joint less wide than 6th, 6th joint dilated, finger with nail. Uropod 3, outer ramus long, 2-jointed, inner very small. Telson more or less deeply cleft.

10 species accepted, 10 doubtful.

Synopsis of accepted species:

1 { Accessory flagellum of antenna 1 with 2 joints — **2.**
  { Accessory flagellum of antenna 1 with only 1 joint — **8.**

2 { Antenna 1 longer than the body; telson differing in ♂ and ♀ . . . . 1. **N. croaticus** . p 406
  { Antenna 1 not longer than the body; telson not differing in ♂ and ♀ — **3.**

3 { Pleon segment 2, postero-lateral corners rounded, in segment 3 acute . . . . 2. **N. forelii** . . p. 406
  { Pleon segments 2 and 3, postero-lateral corners more or less rounded — **4.**
  { Pleon segments 2 and 3, postero-lateral corners acute or quadrate — **5.**

4 { Side-plates 1—4 not deeper than side-plate 5 . . . 3. **N. aquilex** . . p. 406
  { Side-plates 1—4 deeper than side-plate 5 . . . . . 4. **N. stygius** . . p. 407

5 { Gnathopods 1 and 2, 6th joint broader than long — **6.**
  { Gnathopods 1 and 2, 6th joint longer than broad — **7.**

6 { With eyes . . . . 5. **N. puteanus** . p. 407
  { Without eyes . . . . 6. **N. viréi** . . . p 408

1. **N. croaticus** (Jurinac) 1887 *Eriopis croatica*, Jurinac in: Rad JugoslaV. Ak., *v.* 83 p. 96 t. f. 1—12a | 1888 *Niphargus croaticus*, Jurinac, Fauna Kroat. Karst., p. 11—16 f. | 1890 *N. c.*, Wrześniowski in: Z. Wiss. Zool., *v.* 50 p. 610, 670, 676, 716 | 1889 *Gammarus c.*, Moniez in: Rev. biol. Nord France, *v.* 1 p. 252.

Pleon segments 1--6, hind margin armed with a close-set row of backward pointing spinules. Side-plates successively deeper from 1st to 3d, 4th less deep than 3d. Pleon segments 1—3, postero-lateral corners subacute. Eyes wanting. Antenna 1 in ♂ rather longer than the body, flagellum more than 4 times as long as peduncle, 73-jointed, accessory flagellum 2-jointed. Antenna 2 less than $\frac{1}{3}$ as long as antenna 1, ultimate joint of peduncle longer than penultimate. more than twice as long as the 8-jointed flagellum. Gnathopod 1, 6th joint trapezoidal, a little tending to oval, being outdrawn towards the hinge of the finger, palm slightly convex, well defined. In the ♀ gnathopod 1 appears to have the 6th joint almost rectangular. Gnathopod 2, 6th joint rectangular, palm slightly convex. Peraeopod 3 considerably longer ˅than peraeopods 1 and 2, but shorter than the long equal peraeopods 4 and 5. Peraeopods 3—5, 2d joint comparatively narrow, coarsely serrate. Uropod 1, rami equal to peduncle. Uropod 2 shorter, rami longer than peduncle. Uropod 3 reaching much beyond the others, 1st joint of outer ramus 15 times as long as 2d, carrying 8 plumose setae and many spinules, inner ramus very short, with apical spine and seta. Telson in ♂ with a wide arched emargination rather more than $\frac{1}{3}$ of its length, and 2 or 3 spines on the blunt apices, in ♀ cleft more than $\frac{2}{3}$ of length, the incision not rounded but acute, 5—8 spines on the apices. Colour quite white. L. ♀ 18·5, ♂ 20 mm.

Croatia (caVern near Zagorije).

2. **N. forelii** Humb. 1877 *N. puteanus var. f.* + *N. p. var. onesiensis*, A. Humbert n: Bull. Soc. Vaudoise, *v.* 14 p. 278 t. 6, 7 | 1888 *N. p. var. f.*, T. Stebbing in: Rep. Voy. Challenger, *v.* 29 p. 456 | 1890 *N. f.*, Wrześniowski in: Z. wiss Zool., *v.* 50 p. 621, 631, 668, 674 | 1889 *Gammarus puteanus var. f.*, Moniez in: Rev. biol. Nord France, *v.* 1 p. 247 | 1882 *Niphargus p. var. forellii*, G. Joseph in: Berlin. ent. Z., *v.* 26 p. 7.

Side-plates 1—4 deeper than the following. Pleon segment 2, posterolateral corners rotundo-quadrate, segment 3, postero-lateral corners acute. Eyes wanting. Antenna 1 rather less than half as long as body, 1st joint of peduncle nearly as long as 2d and 3d combined, flagellum 16-jointed. Antenna 2, ultimate joint of peduncle rather shorter than penultimate, flagellum 7—9-jointed. Inner lobes of lower lip not mentioned. Gnathopods 1 and 2 similar, 6th joint moderately broad at base, widening towards the straight palm at right angles with the convex hind margin; palm rather longer in gnathopod 2 than in gnathopod 1. Peracopods 1—3 subequal, shorter than peraeopods 4 and 5. Peraeopods 3—5, 2d joint rather broad, according to figure. Uropod 3, outer ramus with 2d joint well developed. Telson cleft about $\frac{3}{4}$ of length. L. 9 mm.

SWitzerland (Lake of GeneVa, well at Onex).

3. **N. aquilex** Schiödte 1841 *Gammarus puteanus* (part.: forma II), C. L. Koch, C. M. A., *v.* 36 nr. 22 | 1889 *G. p.* (part.: forma II), Moniez in: Rev. biol. Nord France, *v.* 1 p. 245 | 1853 *Niphargus stygius* (err., non *Gammarus s.*, Schiödte 1847!), Westwood

in: Ann. nat. Hist., ser. 2 *v.*12 p. 44 | 1855 *N. aquilex,* Schiödte in: Ov. Danske Selsk., p. 350 f. | 1888 *N. a.* (part.), T. Stebbing in: Rep. Voy. Challenger, *v.*29 p.158, 287, 304, 316, 448, 456 | 1890 *N. puteanus var. vejdovskýi,* Wrześniowski in: Z. wiss. Zool., *v.*50 p. 655 t. 27 f. 6, 9—11, 15; t. 28 f. 1—3; t. 30 f. 3 | 1893 *N. subterraneus* (part.), A. Della Valle in: F. Fl. Neapel, *v.*20 p. 704 t. 38 f. 31—34.

Body long and narrow. Side-plates 1—4 not deeper than 5th. Pleon segments 1—3, postero-lateral corners rounded. Eyes wanting. Antenna 1, 1st joint nearly as long as 2[d] and 3[d] combined, 3[d] about half as long as 2[d], flagellum with about 20 joints, accessory flagellum very small, 2-jointed. Antenna 2 much shorter but with longer peduncle, ultimate joint of peduncle a little shorter than penultimate, flagellum 7- or 8-jointed. Gnathopods 1 and 2, 5th joint much narrower than 6th, 6th rounded triangular, widening to the palm, which is long, straight, defined by a spine at the obtuse angle. Peraeopods 1—3 subequal, peraeopods 4 and 5 rather longer. Peraeopods 3—5, 2[d] joint oblong oval. Uropods 1 and 2 very short, uropod 3 very long, especially in ♂, peduncle short, outer ramus very long, 2[d] joint in ♂ nearly as long as 1st, in ♀ much shorter, inner ramus short, rudimentary. Telson short, deeply cleft, apices obtuse. Colour whitish, pellucid. L. ♂ reaching 20, ♀ 12·5 mm.

Europe (south of England), St. Briac, Leyden, Cologne, Wells in Bohemia *(var. vejdovskýi),* and other places).

4. **N. stygius** (Schiödte) 1847 *Gammarus s.,* Schiödte in: Ov. Danske Selsk., p. 81 | 1849 *Niphargus s.,* Schiödte in: Danske Selsk. Skr., ser. 5 *v.*2 p. 26 t. 3 | 1890 *N. s.,* Wrześniowski in: Z. wiss. Zool., *v.*50 p. 603, 620, 665.

Body long and narrow. Side-plate 3 slightly deeper than 4th, 4th than 5th. Pleon segments 1—3, postero-lateral corners in figure slightly rounded. Eyes wanting. Antenna 1, 1st joint rather longer than 2[d], 3[d] half as long as 2[d], flagellum 16—25-jointed, accessory flagellum very small, 2-jointed. Antenna 2 $^2/_3$ as long as antenna 1, ultimate and penultimate joints of peduncle subequal, flagellum 7—9-jointed. Maxilla 1 with 7 spines on the outer plate, but figured with several setae on the inner plate and a great number on the outer margin of the palp's 2[d] joint, peculiarities not to be accepted without corroboration. Gnathopods 1 and 2 very setose according to figure, 5th joint triangular, much shorter than 6th, 6th longer than broad, widest at the somewhat oblique palm, which is well defined by a strong spine at the angle. Gnathopod 2 rather larger than gnathopod 1. Peraeopods 1 and 2 slender, equal, shorter than the 3 following, which successively encrease in length, with oval 2[d] joint. Uropods 1 and 2 short, each with equal rami. Uropod 3, inner ramus in ♂ $^1/_{16}$, in ♀ $^1/_7$ as long as outer, tipped with 2 spinules and a seta, outer ramus with 1st joint in ♂ thrice as long as peduncle and subequal to 2[d] joint, in ♀ twice as long as peduncle and twice as long as 2[d] joint. Telson cleft beyond the middle, apices subacute. Colour snow-white. L. 10—14 mm.

Caves of Adelsberg and Lueg in Krain.

5. **N. puteanus** (C. L. Koch) 1835 *Gammarus p.* (part.: forma I), C. L. Koch, C. M. A.. *v.*5 nr. 2 | 1862 *Niphargus p.,* Bate, Cat. Amphip. Brit. Mus., p. 177 t. 32 f. 4 | 1890 *N. ratisbonensis,* Wrześniowski in: Z. wiss. Zool., *v.*50 p. 662, 673, 696.

Scarcely distinguishable from N. stygius, except in having yellow eyes, gnathopods 1 and 2, 6th joint very large, almost rectangular, broader than long, and the otherwise pellucid body streaked along the edges of the peraeon segments and to the and of pleon with a yellow ochre stripe. L. 12·5 mm.

Regensburg, in Wells; also at Poitiers.

6. **N. viréi** Chevreux 1896 *N. v.*, Chevreux in: Bull. Mus. Paris, *v.* 2 p. 136.

♀.. Body less slender than usual in Niphargus. Pleon segment 5 with 6 very minute dorsal spines, segment 6 with 8. Head, lateral corners very prominent. rounded. Anterior side-plates much deeper than their segments, 4th much deeper than 5th. Pleon segments 1—3, postero-lateral corners quadrate with slightly acute point. Eyes wanting or rudimentary, perhaps represented by irregular blotches of yellowish pigment, which disappear in spirit. Antenna 1 $^2/_3$ as long as body, flagellum 44—56-jointed, accessory flagellum 2-jointed. Antenna 2 very short, flagellum a little longer than ultimate joint of peduncle, 15-jointed. Gnathopod 1, 6th joint almost quadrangular, a little broader than long, finger matching palm. Gnathopod 2, 6th joint considerably larger than in gnathopod 1, subtriangular, tending to quadrate, much broader than long, twice as broad as the 5th joint, finger strongly curved, slightly shorter than palm. Peraeopods 1—5 slender and very elongate. Uropod 3, inner ramus rudimentary, outer very long, 2^d joint less than half as long as 1st. Telson as broad as long, squarely truncate, with a wide emargination occupying $^2/_3$ of the length, each lobe ending in 6 strong spines and carrying a group of 3 spines and a setule towards the middle of the outer margin. L. 25—31 mm.

Jura. Caves.

7. **N. fontanus** Bate 1859 *N. f.*, Bate in: P. Dublin Univ. zool. bot. Ass., *v.* 1 p. 237 f. | 1862 *N. f.*, Bate, Cat. Amphip. Brit. Mus., p. 175 t. 32 f. 2 | 1890 *N. f.*, Wrześniowski in: Z. Wiss. Zool., *v.* 50 p. 665 | 1893 *N. subterraneus* (part.), A. Della Valle in: F. Fl. Neapel, *v.* 20 p. 704.

Comparatively robust. Side-plate 4 the deepest. Pleon segments 1—3, postero-lateral corners acute. Eyes very small. of irregular shape, bright lemon. Antennae 1 and 2 as in N. aquilex (p. 406). Gnathopods 1 and 2 alike, 5th joint narrow, 6th large, the base much wider than the 5th, pear-shaped, narrowing from the palm to the hinge of finger, palm very oblique, much longer than hind margin, defined by a strong palmar spine, finger long, much curved. Peraeopods 1—5 as in N. stygius (p. 407). Uropod 3, 2^d joint of outer ramus shorter than 1st. In the ♀ gnathopods 1 and 2 have the 6th joint shorter and less distinctly pear-shaped, and uropod 3 has the 2^d joint of outer ramus much shorter than 1st. Telson apparently as in N. aquilex. L. 12·5 mm.

South of England. Wells.

8. **N. kochianus** Bate 1859 *N. k.*, Bate in: P. Dublin Univ. zool. bot. Ass., *v.* 1 p. 237 f. | 1862 *N. k.*, Bate, Cat. Amphip. Brit. Mus., p. 176 t. 32 f. 3 | 1890 *N. k.*, Wrześniowski in: Z. Wiss. Zool., *v.* 50 p. 665 | 1893 *N. subterraneus* (part.), A. Della Valle in: F. Fl. Neapel, *v.* 20 p. 704.

In close agreement with N. fontanus except in regard to gnathopods 1 and 2, 6th joint, which approaches in form an imperfect oblong square, abruptly wider than the 5th joint, and widening slightly to the sinuous palm, this being nearly at right angles with the serrate hind margin. Nothing is said about eyes. Uropod 3 figured with a very short 2^d joint to outer ramus. L. about 8 mm.

England (Hampshire and Wiltshire), Ireland (Dublin). Wells.

9. **N. orcinus** Joseph 1869 *N. o.*, G. Joseph in: Jahresber. Schles. Ges., *v.* 46 p. 52 | 1890 *N. o.*, Wrześniowski in: Z. Wiss. Zool., *v.* 50 p. 669. 676 | 1889 *Gammarus o.*, Moniez in: Rev. biol. Nord France, *v.* 1 p. 252 | 1893 *Niphargus subterraneus* (part.)?, A. Della Valle in: F. Fl. Neapel, *v.* 20 p. 705.

Somewhat robust. Head somewhat acutely rostrate. Eyes wanting. Antenna 1 short, in the $\male$ $^1/_3$, in the $\female$ $^1/_4$ of length of body, 2^d joint longer than 1st, 3^d $^2/_3$ as long as 2^d, flagellum twice as long as peduncle, 34-jointed, accessory flagellum represented by a small·tubercle carrying 1 seta. Antenna 2 half as long as antenna 1, ultimate joint of peduncle rather longer than penultimate, flagellum 13-jointed. Mouth-parts said to differ from those of other species. Gnathopods 1 and 2, 5th joint short, triangular, 6th large, rotundo-quadrate, palm defined by a claw-like inward-directed strong tooth (palmar spine?), finger large, curved. Peraeopods 1 and 2 strong, slender. Peraeopod 5 the longest. Uropods 1 and 2 short. Uropod 3, outer ramus in $\male$ 6 times as long as peduncle, its 2^d joint half as long as 1st, inner ramus short, laminar, almost as long as peduncle, more setose in $\male$ than in $\female$. Telson deeply cleft, the conical halves as long as peduncle of uropod 3, armed with spines. Colour clear yellowish grey, tip of finger and tooth of palm in gnathopods 1 and 2 flesh-coloured. L. $\male$ 51, $\female$ 47·5 mm.

Krain. Stalactite-caves.

10. **N. godeti** Wrześn. 1867 *Gammarus puteanus* (err.. non C. L. Koch 1835!), Godet in: Bull. Soc. Neuchatel, *v.*7 p. 424 | 1889 *G. p.*, Moniez in: Rev. biol. Nord France, *v.*1 p. 252 | 1890 *Niphargus godeti*, Wrześniowski in: Z. Wiss. Zool., *v.*50 p. 668, 674.

Side-plate 4 much the largest. Pleon segments 1—3, postero-lateral corners rounded. Eyes wanting. Antenna 1 almost as long as body, flagellum 6 times as long as peduncle, 51-jointed, accessory flagellum spine-like. Antenna 2 very short, little more than $^1/_5$ as long as antenna 1. Gnathopods 1 and 2, 6th joint spoken of as triangular, but in fact figured as rather broad at base though widening to the slightly convex, not oblique palm. Peraeopods 3—5 seemingly subequal in length, 2^d joint rather broad, hind margin coarsely serrate. Uropod 3 more than half as long as body, outer ramus with very long 2^d joint. L. 33 mm.

Neuchâtel. Well.

N. casparianus Wrześn. 1849 *Gammarus puteanus* (err., non C. L. Koch 1835!), Caspary in: Verh. Ver. Rheinlande, *v.*6 p. 39 t. 2 | 1850 *G. p.*, Hosius in: Arch. Naturg., *v.*161 p. 233 | 1890 *Niphargus casparianus*, Wrześniowski in: Z. Wiss. Zool., *v.*50 p.664, 673.

Scarcely distinguishable from N. kochianus.

Germany (Elberfeld, Bonn).

N. caspary (Pratz) 1866 *Gammarus c.*, Pratz, Grundw. Thiere | 1889 *G. c.*, Moniez in: Rev. biol. Nord France, *v.*1 p. 230 | 1890 *Niphargus c.*, Wrześniowski in: Z. Wiss. Zool., *v.*50 p. 666.

Gnathopods 1 and 2 with 6th joint triangular (probably widening from the base). Uropod 3 in $\male$, 1st joint of outer ramus 4—5 times as long as peduncle. 2^d joint short; in $\female$ outer (Moniez: „interne") ramus 1-jointed, the joint scarcely longer than peduncle.

Munich. Well.

N. elegans Garbini 1894 *N. e.* + *N. e. var. imperfectus* + *Gammarus fluviatilis var. manophtalmus*, Garbini in: Mem. Acc. Verona, ser. 3 *v.*70 p 108, 109 | 1895 *N. e.* + *N. e. var. i.* + *G. f. var. m.*, Garbini in: Mem. Acc. Verona, ser. 3 *v.*711 p. 33, 38, 31.

Near to N. tatrensis (p 410), differing by want of spine-rows on pleon segments 1—3, by greater number of joints in antennae 1 and 2, by relative length of joints and armature of uropod 3 etc. Eyes developed to wanting.

Verona.

N. longicaudatus (A. Costa) 1851 *Gammarus longicaudatus*, *G. longicauda* (non F. Brandt 1851!), (A. Costa in:) F. W. Hope, Cat. Crost. Ital., p. 45, 23 | 1857 *G. longicaudatus*, A. Costa in: Mem. Acc. Napoli, *v.* 1 p. 217 t. 4 f. 6 | 1890 *Niphargus l.* (part.), Wrześniowski in: Z. wiss. Zool., *v.* 50 p. 603, 625, 674 | 1893 *N. subterraneus* (part.), A. Della Valle in: F. Fl. Neapel, *v.* 20 p. 704.

Probably the same as N. aquilex (p. 406). L. 12·5 mm.

Naples. In the drinking water.

N. minutus (Gerv.) 1835 *Gammarus pulex m.*, P. Gervais in: Ann. Sci. nat., ser. 2 *v.* 4 p. 128 | 1859 *G. lacteus*, P. Gervais & P. J. Beneden, Zool. méd., *v.* 1 p. 488 | 1890 *G. l.*, Wrześniowski in: Z. wiss. Zool., *v.* 50 p. 602 | 1893 *G. fluviatilis* (part.), A. Della Valle in: F. Fl. Neapel, *v.* 20 p. 763.

L. minute.

Neighbourhood of Paris. Only in wells.

N. montanus (A. Costa) 1851 *Gammarus m.*, (A. Costa in:) F. W. Hope, Cat. Crost. Ital., p. 44 | 1857 *G. m.*, A. Costa in: Mem. Acc. Napoli, *v.* 1 p. 218 t. 4 f. 7, 8 | 1867 *G. longicaudatus var.*, A. Costa in: Annuario Mus. Napoli, *v.* 4 p. 38 | 1890 *Niphargus longicaudatus* (part.), Wrześniowski in: Z. wiss. Zool., *v.* 50 p. 674.

Agreeing nearly with N. longicaudatus but differing in uropods and telson. L. 12·5 mm.

Italy (Lake Matese).

N. ponticus Czern. 1868 *N. p.*, Czerniavski in: Syezda Russ. Est., Syezda 1 Zool. p. 108 t. 8 f. 12—14 | 1888 *N. p.*, T. Stebbing in: Rep. Voy. Challenger, *v.* 29 p. 378 | 1890 *N.? p.*, Wrześniowski in: Z. wiss. Zool., *v.* 50 p. 673 | 1893 *N. subterraneus* (part.)?, A. Della Valle in: F. Fl. Neapel, *v.* 20 p. 705.

L. about 2 mm.

Black Sea.

N. rhipidiophorus (Catta) 1878 *Gammarus r.*, Catta in: Act. Soc. Helvét., Sess. 60 p. 257 | 1889 *G. r.*, Mouiez in: Rev. biol. Nord France. *v.* 1 p. 252 | 1892 *G. r.*, Chevreux and Guerne in: Bull. Soc. zool. France, *v.* 17 p 140 | 1888 *Niphargus r.*, T. Stebbing in: Rep. Voy. Challenger, *v.* 29 p. 475 | 1890 *N.? r.*, Wrześniowski in: Z. wiss. Zool., *v.* 50 p. 607, 673 | 1893 *Gammarus pungens* (part.), A. Della Valle in: F. Fl. Neapel, *v.* 20 p. 764.

Eyes small. Peraeopod 1 with brush of long setae on 5th and 6th joints. Uropod 3 enormous.

France (a well, brackish in summer, at La Ciotat, near the Mediterranean).

N. subterraneus (Leach) 1813/14 *Gammarus s.*, Leach in: Edinb. Enc., *v.* 7 p. 402 | 1893 *Niphargus s.* (part.), A. Della Valle in: F. Fl. Neapel, *v.* 20 p. 704 | 1888 *N. aquilex* (part.), T. Stebbing in: Rep. Voy. Challenger, *v.* 29 p. 84.

London. Well.

N. tatrensis Wrześn. 1888 *N. t.*, Wrześniowski in: Pam. Fizyjogr., *v.* 8 p. 47 t. 6—16 | 1890 *N. t.*, Wrześniowski in: Z. wiss. Zool., *v.* 50 p. 643 t. 27 f. 1—5, 7, 8, 12—14; t. 28 f. 4, 5, 8; t. 29 f. 1—3, 12; t. 30 f. 11—13; t. 31 f. 5, 14, 14a, 18—21; t. 32 f. 1, 8, 8a.

Perhaps identical with N. stygius (p. 407); but said to be distinguished by the postero-lateral corners of the pleon segments 1—3, the 6th joint of the gnathopods 1 and 2, the relative length of peraeopods 3 —5, the relative length of the apical joint in uropod 3, and the setose armature of the mouth-parts.

Tatra Mountain.

22. Gen. **Eriopisa** Stebb.

1859 *Eriopis* (non E. Mulsant 1851, Coleoptera!), R. M. Bruzelius in: Svenska
Ak. Handl., n. ser. *v.* 3 nr. 1 p. 64 | 1890 *Eriopisa*, T. Stebbing in: Ann. nat. Hist., ser. 6
v. 5 p. 193 | 1893 *E.*, A. Della Valle in: F. Fl. Neapel, *v.* 20 p. 705 | 1894 *E.*, G. O. Sars,
Crust. NorWay, *v.* 1 p. 514 | 1890 *Eriopsis*, Wrześniowski in: Z. wiss. Zool., *v.* 50 p. 632.

Body slender and smooth. Head not rostrate. Side-plates all shallow.
Eyes wanting. Antenna 1 much the longer, accessory flagellum small,
2-jointed. Upper lip rounded. Lower lip with inner lobes. Mandible normal,
spine-row with many spines, palp slender, 3^d joint longer than 2^d. Maxilla 1,
inner plate with numerous setae, outer with 7 spines, 2^d joint of palp with
3 or 4 spine-teeth. Maxilla 2, inner plate fringed on $^2/_3$ of inner margin.
Maxillipeds, inner plates squarely truncate, outer fully reaching middle of
palp's 2^d joint, with setiform spines on apical margin, spinules of inner
margin not close to the edge. 3^d joint of palp wide, finger slender, tipped with
a spine. Gnathopods 1 and 2 subchelate, unlike and unequal, finger without
nail or denticle, gnathopod 2 the larger. Peraeopod 5 the longest, with 2^d
joint larger than in the other peraeopods. Uropod 3, inner ramus very
small, outer very long, 2-jointed, the joints laminar. Telson deeply cleft.
1 species.

1. **E. elongata** (Bruz.) 1859 *Eriopis e.*. R. M. Bruzelius in: Svenska Ak. Handl.,
n. ser. *v.* 3 nr. 1 p. 65 t. 3 f. 12 | 1871 *Niphargus elongatus*, A. Boeck in: Forh. Selsk.
Christian., 1870 p. 136 | 1876 *N. e.*, A. Boeck. Skand. Arkt. Amphip., *v.* 2 p. 403 t. 22
f. 5 1890 *Eriopsis elongata*, Wrześniowski in: Z. wiss. Zool., *v.* 50 p. 632 | 1893 *Eriopisa e.*,
A. Della Valle in: F. Fl. Neapel, *v.* 20 p. 706 t. 38 f. 17–30; t. 60 f. 5 | 1894 *E. e.*,
G. O. Sars, Crust. NorWay, *v.* 1 p. 515 t. 181 f. 2 | 1895 *E. e.*, A. M. Norman in: Ann.
nat. Hist., ser. 6 *v.* 15 p. 490.

Body elongate. Head, lateral corners slightly projecting, rounded, post-
antennal corners dentiform. Side-plates scarcely contiguous, 1st angularly
produced forward, 4th little larger than 5th and like it in shape. Pleon
segment 3, postero-lateral corners produced into a small tooth. Antenna 1 as
long as body, 1st joint not longer than 2^d, with slender spine at apex, 3^d little
over $^1/_4$ as long as 2^d, flagellum nearly thrice as long as peduncle, 32-jointed,
accessory flagellum shorter than 1st joint of primary. Antenna 2 little longer
than peduncle of antenna 1, ultimate joint of peduncle shorter than penultimate
and subequal to the 6-jointed flagellum. Gnathopod 1, 2^d joint as in 3 following
limbs very narrow at base, 5th joint as long as 6th but narrower, 6th triangular,
palm and hind margin almost at right angles one to other. Gnathopod 2
much larger, 5th joint triangular, cup-shaped, much shorter than 6th, 6th oblong
oval, narrowing distally, palm very oblique, ill-defined, long, a little notched,
armed with some strong spines, finger ciliated on inner margin. Peraeopods 1
and 2, 5th joint rather expanded. Peraeopod 3, 2^d joint narrowly oval.
Peraeopod 4 longer, 2^d joint considerably wider. Peraeopod 5 with 2^d joint oval,
much larger than in peraeopod 4, hind margin strongly serrate as also in
peraeopods 3 and 4. Uropods 1 and 2 of very moderate size. Uropod 3,
peduncle short, inner ramus not quite equal to it, outer ramus as long as
pleon segments 1—5 combined, both joints laminar, fringed with spinules,
2^d rather the shorter, its truncate apex armed with dense brush of setae.
Telson scarcely as long as peduncle of uropod 1, subtriangular, cleft close to
base, apices acute, with spine and setule in notch formed by external tooth
a little way up lateral margin. Colour yellowish. L. 11 mm.

Arctic Ocean, North-Atlantic, North-Sea and Skagerrak (South- and West-NorWay
to Lofoten Isles, depth 94—752 m; Bohuslän; Great Britain); Kattegat; Mediterranean.

23. Gen. **Gmelina** O. Sars

1894 *Gmelina*, (O. Grimm in MS.) G. O. Sars in: Bull. Ac. St.-Pétersb., ser. 5 *v.* 1 p. 191.

Head with small rostrum, lateral lobes rather small. Side-plates 1—4 larger in ♀ than in ♂, 4th not very wide, nor strongly emarginate. Antennae 1 and 2 subequal, not elongate, slender, accessory flagellum 1-jointed. Mouth-parts nearly as in Gammarus (p. 460), but upper lip with no emargination at apex and with an obtuse prominence in front. Maxilla 2 with outer plate considerably broader than inner, palp of maxillipeds rather short, with 3^d joint distally expanded. Gnathopods 1 and 2 alike, subchelate, feeble in ♀, stronger in ♂. Peraeopods 1—5 not elongate, successively shorter to peraeopod 3 and thence successively longer, 2^d joint rather larger and more laminar in peraeopod 5 than in 3 or 4. Uropod 3 more or less projecting, outer ramus well developed, with very small 2^d joint, inner ramus small, scale-like. Telson deeply cleft.

4 species.

Synopsis of species:

<table>
<tr><td rowspan="2">1</td><td>Many segments With conspicuous projections — 2.</td><td></td><td></td></tr>
<tr><td>No segments With conspicuous projections — 3.</td><td></td><td></td></tr>
<tr><td rowspan="2">2</td><td>With median dorsal carina . . . .</td><td>1. **G. costata** . . .</td><td>p. 412</td></tr>
<tr><td>Without median dorsal carina . . . .</td><td>2. **G. kusnezowi** .</td><td>p. 413</td></tr>
<tr><td rowspan="2">3</td><td>Uropod 3, inner ramus longer than peduncle . . .</td><td>3. **G. laeviuscula** .</td><td>p. 413</td></tr>
<tr><td>Uropod 3, inner ramus not longer than peduncle .</td><td>4. **G. pusilla** . . .</td><td>p. 414</td></tr>
</table>

1. **G. costata** O. Sars 1894 *G. c.* (O. Grimm in MS.), (*Palasiella macera* in MS.) G. O. Sars in: Bull. Ac. St.-Pétersb., ser. 5 *v.* 1 p. 192 t. 3.

Very slender and compressed, especially in ♂, integument crustaceous, median dorsal carina beginning faintly on peraeon segment 1, reaching highest elevation on pleon segment 3, the obtusely triangular laminar process on this and on the next preceding and 2 following segments being very conspicuous, pleon segments 4 and 5 ending in upturned points, segment 6 smooth. On peraeon segments 1—7 prominences near the side-plates form a low obtuse lateral keel. Head, rostrum triangular, not elongate, lateral lobes broadly truncate, post-antennal corners acute. Pleon segment 2, postero-lateral corners acutely produced, in segment 3 acute, subquadrate. Eyes oval reniform, not large, dark. Antenna 1, 1st joint nearly as long as 2^d and 3^d combined, 2^d not twice as long as 3^d, flagellum little longer than peduncle, with about 16 short joints, accessory flagellum with 3 setules at tip. Antenna 2 in ♂ equal to antenna 1, ultimate joint of peduncle as long as penultimate, flagellum 8-jointed, in ♀ rather shorter than antenna 1, ultimate joint of peduncle shorter than penultimate, flagellum 6-jointed. Gnathopods 1 and 2 in ♂, 6th joint much longer than 5th, large, oblong oval, widening a little to the concave palm, which is defined by a nearly rectangular lobe, armed with 2 strong spines, between which the strongly curved finger impinges. Gnathopod 1 in ♀ feeble, 6th joint longer than 5th, oblong quadrangular, palm rather oblique; gnathopod 2 rather more slender, 6th joint as long as 5th, palm nearly transverse. Peraeopods 3 and 4, 2^d joint little expanded, narrowing distally. Peraeopod 5, 2^d joint oblong quadrangular. Uropod 2 the smallest, the rami linear, with spines only at the apex. Uropod 3 large, projecting far beyond the others, peduncle short and thick, outer ramus large, foliaceous, margins setose and spinulose, apex broad, with a diminutive 2^d joint, inner ramus rather longer than broad. Telson longer than broad, projecting beyond peduncle of uropod 3, cleft more than $^3/_4$ length, outer margin finely ciliated, the lobes narrowing to obtuse divergent apices, each armed with a spinule and setules. L. ♀ 12, ♂ 16 mm.

Caspian Sea. Depth 0—38 m.

2. **G. kusnezowi** (Sowinski) 1894 *Gammarus k.*, Sowinski in: Mém. Soc. Kiew, v. 13 p. 383 t. 8 | 1894 *Gmelina k.* (*Pallasiella mammillifera* in MS.), G. O. Sars in: Bull. Ac. St.-Pétersb., ser. 5 v. 1 p. 197 t. 4.

Rather slender and compressed, and more so in ♂ than in ♀, integument crustaceous, with no median dorsal carina, but with a pair of subdorsal prominences on each segment from peraeon segment 1 to pleon segment 3, gradually encreasing from low tubercles to conspicuous knobs; peraeon segments 1—5 (or 1—6) with conspicuous lateral projecting tuberculiform prominences; pleon segments 4—6 dorsally smooth, except that on 5 and 6 there are 2 fascicles of spinules. Head tapering, rostrum short and narrow. lateral lobes narrowly rounded, with a boss-like process above the post-antennal corners. Pleon segments 2 and 3, postero-lateral corners acute, subquadrate. Eyes oval reniform, highly protuberant, very dark. Antenna 1 short, 1st joint little longer than 2^d, 2^d twice as long as 3^d, flagellum shorter than peduncle, with 15 short joints; accessory flagellum with 3 setules. Antenna 2 subequal to antenna 1, ultimate joint of peduncle shorter than penultimate, flagellum 6-jointed. Gnathopods 1 and 2 nearly as in G. costata, but gnathopod 2 in ♀ with palm more oblique. Peraeopods 1—5 and uropods 1 and 2 nearly as in G. costata. Uropod 3 a little projecting beyond uropod 1. outer ramus tapering to a narrowly truncate apex, the small but distinct 2 joint carrying several setules, inner ramus a little longer than broad, tapering. Telson broader than long, scarcely as long as peduncle of uropod 3. cleft deep, scarcely dehiscent, the lobes obtusely rounded, 2 spinules on outer and 4 on apical margin of each. Colour in dark transverse bands. L. ♀ 14, ♂ 18 mm.

Sea of Azov, Caspian Sea.

3. **G. laeviuscula** O. Sars 1896 *G. l.*, G. O. Sars in: Bull. Ac. St.-Pétersb., ser. 5 v. 4 p. 430 t. 2 f. 8—12.

♀ unknown. — ♂. Rather slender, much compressed, without any distinct tubercles or processes, pleon segments 4—6 dorsally tipped with small bristles. Head slightly tapering, rostrum well marked, lateral lobes slightly produced, narrowly rounded. Pleon segment 3, postero-lateral corners slightly produced, 2^d segment simply quadrate. Eyes very small, oval, dark. Antenna 1 rather the longer, flagellum shorter than peduncle, 10-jointed. Antenna 2, ultimate joint of peduncle shorter than ⁻penultimate, flagellum about half as long as peduncle. Gnathopods 1 and 2 rather powerful, these and peraeopods 1—5 seeming to agree with those of the 2 preceding species. Uropod 1 twice as long as uropod 2. Uropod 3 rather robust, projecting somewhat beyond the others. outer ramus very broad, foliaceous, fringed with fascicles of spinules, 2^d joint small, inner ramus nearly half length of outer, not very narrow, nearly thrice as long as broad, tipped with spinules. Telson rather large, oval, cleft nearly to the base, lobes scarcely at all dehiscent, each with 4 or 5 spinules on outer margin, and about 7 on broadly truncate apex. L. 7 mm.

North Caspian Sea.

4. **G. pusilla** O. Sars 1896 *G. p.*, G. O. Sars in: Bull. Ac. St.-Pétersb., ser. 5 *v.* 4 p. 432 t. 2 f. 13—21.

♀. Rather short and stout, somewhat compressed, smooth. Head slightly tapering, rostrum small, distinct. lateral lobes angularly produced. Side-plate 1 expanded below and produced forward. Pleon segments 2 and 3, postero-lateral corners quadrate. Eyes small, rounded oval, dark. Antenna 1 feeble, 2^d joint subequal to 1st, 3^d much smaller, flagellum as long as peduncle, 7-jointed. Antenna 2 stouter, scarcely longer, flagellum small, 4-jointed. Gnathopods 1 and 2 feeble, setose, in gnathopod 1 the 6th joint oblong oval, palm rather oblique, in gnathopod 2 more slender, not longer than 5th, oblong quadrate, palm transverse. Peraeopods 1—5 nearly as in G. laeviuscula (p. 413), but peraeopods 3—5 relatively shorter, more setose, and 2^d joint of peraeopod 5 broader, irregularly oval. Uropod 1 not twice as long as uropod 2. Uropod 3 much less robust than in G. laeviuscula, outer ramus tapering, with few fascicles of spinules, 2^d joint small, distinct, inner ramus narrow, about $^1/_3$ as long as outer, with one apical spinule. Telson longer than broad, cleft nearly to base, lobes each with 2 spinules on outer margin and 2 unequal ones on narrow apices. L. 5 mm. — ♂ unknown.

North Caspian Sea.

24. Gen. **Gmelinopsis** O. Sars

1896 *Gmelinopsis*, G. O. Sars in: Bull. Ac. St.-Pétersb., ser. 5 *v.* 4 p. 434.

Peraeon segment 7 and pleon segments 1—3 dorsally carinate; marginal tubercular carinae on peraeon segments 1—7. Side-plates 1—4 rather deep. Antennae 1 and 2 short, slender. Antenna 1 the longer, accessory flagellum small, 2-jointed. Mouth-parts nearly as in Gmelina (p. 412). Gnathopod 1 stronger than gnathopod 2. Peraeopod 5, 2^d joint strongly expanded. Uropod 3 rather short, outer ramus with small narrow 2^d joint, inner ramus small, scale-like. Telson more or less deeply cleft, lobes apically narrow.

2 species.

Synopsis of species:

1. **G. tuberculata** O. Sars 1896 *G. t.*, G. O. Sars in: Bull. Ac. St.-Pétersb., ser. 5 *v.* 4 p. 435 t. 3 f. 1—19.

Peraeon segment 7 and pleon segments 1—3 with median dorsal carina formed by low obtusely rounded laminar expansions; peraeon segments 1—7 with distinct lateral tubercles. pleon segments 4—6 smooth. Head with small distinct rostrum, lateral lobes small, angular, below each of them a conspicuous boss-like prominence extends obliquely downward. Side-plates 1—4 deep, not very broad. slightly crenulate distally; side-plate 4 slightly emarginate behind. Pleon segments 1—3, postero-lateral corners rounded off. Eyes rather small, oblong oval, slightly protuberant. Antenna 1, 1st joint as long as 2^d and 3^d combined, flagellum longer than peduncle, 12-jointed, 2^d joint of accessory flagellum minute. Antenna 2 little more than half as long as antenna 1, ultimate joint of peduncle much shorter than penultimate, flagellum scarcely longer than ultimate, 4-jointed. Gnathopod 1, 5th joint short, cup-shaped, 6th large, oblong oval, palm rather oblique, defined by an obtuse corner, carrying a strong spine. Gnathopod 2 much feebler, 6th joint shorter and much narrower, oblong quadrangular, palm nearly transverse. Peraeopods 1— 5 rather slender and spinulose, 4th longer than 3^{d} or 5th, in 3^{d} and 4th 2^{d} joint oblong oval, rather narrowed distally, larger in peraeopod 4 than in 3. Peraeopod 5, 2^{d} joint greatly expanded, almost heart-shaped, its greatest breadth being below, hind margin minutely crenulate and setuliferous. Uropods 1 and 2 with subequal rami, apically tufted with strong spines. Uropod 3 rather short and thick, reaching little beyond the others, outer ramus rather thick, slightly curved, apex of 1st joint much broader than the small 2^{d} joint, inner ramus less than twice as long as broad. Telson cleft nearly to base, the lobes tapering, their narrow apices standing wide apart, each armed with a spinule and a hair. L. ♀ 8 mm.

Caspian Sea. Depth 53 m.

2. **G. aurita** O. Sars 1896 *G. a.*, G. O. Sars in: Bull. Ac. St.-Pétersb., ser. 5 *v.* 4 p. 437 t. 3 f. 20—28.

♀. Peraeon segments 6 and 7 and pleon segments 1—3 with median dorsal carina formed by laminar expansions, rounded on pleon segment 3, triangular on the others. Peraeon segments 1—7 with lateral tubercles less prominent than in G. tuberculata. Pleon segments 4—6 short and stout, with small hairs dorsally, segment 6 with a very minute denticle on each side. Head, lateral lobes small, obtuse-angled, lateral protuberances greatly developed, these spiniform projections extending obliquely downward and looking as seen from above like a pair of pointed ears. Side-plates 1—4 rather less deep than in G. tuberculata. Pleon segments 2 and 3, postero-lateral corners subquadrate. Antenna 1, flagellum shorter than peduncle, 10-jointed. Antenna 2 more than half as long as antenna 1. Telson triangular, about as long as broad at base, distally conically tapered, cleft narrow, reaching only to middle, apices acute, each armed with a spinule and 2 hairs. Other parts nearly as in G. tuberculata, but gnathopods 1 and 2 rather less unequal. L. 8 mm. — ♂ unknown.

Caspian Sea. Depth 203 m.

25. Gen. **Hakonboeckia** Stebb.

1899 *Hakonboeckia* (Sp. un.: *H. strauchii*), T. Stebbing in: Tr. Linn. Soc. London, ser. 2 *v.* 7 p. 425.

Near to Axelboeckia (p. 391) and Gmelinopsis. Peraeon segments with margins acutely produced. Head with rostral and lateral projections. Antenna 1 longer than antenna 2, peduncle of antenna 1 equal to peduncle of antenna 2; accessory flagellum very small. Gnathopod 1, 6th joint like that of gnathopod 2, but longer. Peraeopods 3—5, 2^{d} joint broad, not produced downward. Uropod 3, rami subequal, outer (seemingly) 1-jointed. Telson cleft nearly to base.

1 species.

1. **H. strauchii** (Dyb.) 1874 *Gammarus s.*, B. Dybowsky in: Horae Soc. ent. Ross., *v.* 10 suppl. p. 112 t. 12 f. 7 | 1899 *Hakonboeckia s.*, T. Stebbing in: Tr. Linn. Soc. London, ser. 2 *v.* 7 p. 425 | 1893 *Gammarus strauchi*, *G. locusta* (part.)?, A. Della Valle in: F. Fl. Neapel, *v.* 20 p. 930.

Peraeon segments 1—7, like serrate edge of a roof broadly expanded in outward and downward directed pointed angles over the side-plates, most strongly in segments 1—5. Peraeon segments 6 and 7, and pleon segments 1—3 each with mid hind margin a little thickened and raised over following segment. Head, upper profile arched with rostrum projecting helm-like and deflexed; under part of cheek-lobes strongly projecting outward and running down into a free acute point. Eyes very prominent, large, oval, dark brown. Antenna 1 little more than $^1/_3$ as long as body, about twice as long as antenna 2, peduncle stouter than that of antenna 2 but equal in length, flagellum 9- or 10-jointed, accessory flagellum 2-jointed. Gnathopods 1 and 2, 6th joint oblong, widening a little to the not very oblique palm, larger in gnathopod 1. Peraeopods 3—5, 2^d joint broad, heart-shaped, hind margin with 3 or 4 short spine-like setae. Uropod 1 not reaching the end of uropod 3. Uropod 3 $^1/_6$ as long as body, rami subequal, outer with plumose setae only on inner margin, inner ramus having them on both margins. Colour dark grey or dark green brown, antennae bright, under part of body and peraeopods dark blue or dark violet; a variety brightly coloured, side-plates and under body dark blue. L. 10—11 mm.

Lake Baikal. Depth 20—100 m.

26. Gen. **Baikalogammarus** Stebb.

1899 *Baikalogammarus* (Sp. typ.: *B. pullus*), T. Stebbing in: Tr. Linn. Soc. London, ser. 2 v. 7 p 425.

Near to Gammarus (p. 460). Pleon segments 4—6 with a few dorsal setules or spinules. Antenna 1 the longer, peduncle shorter than peduncle of antenna 2, accessory flagellum very short. Gnathopod 1, 6th joint not smaller than 6th joint of gnathopod 2. Peraeopods 3—5, 2^d joint broad, the wing produced downward in a long rounded lobe. Uropod 3 rather elongate, with peduncle as long as the 2-jointed outer ramus. Telson cleft.

1 species.

1. **B. pullus** (Dyb.) 1874 *Gammarus p.*, B. Dybowsky in: Horae Soc. ent. Ross., v. 10 suppl. p. 170 t. 11 f. 4 | 1893 *Atylus? p.*, A. Della Valle in: F. Fl. Neapel, v. 20 p. 930 | 1899 *Baikalogammarus p.*, T. Stebbing in: Tr. Linn. Soc. London, ser. 2 v. 7 p. 425.

Pleon segments from 2^d, 3^d or 4th to the 6th with 2—4 dorsal setules or very fine spines. Head, rostrum slightly deflexed and forming a helm-like prominence, post-antennal corners (in figure) rounded. Side-plates 1—4 with isolated short setae. Eyes large, widened above, black. Antenna 1 about $^1/_2$ as long as body, a little longer than antenna 2, 1st joint stout, not as long as 2^d and 3^d combined, flagellum 15—18-jointed, accessory flagellum 1-jointed. Antenna 2, ultimate joint of peduncle longer than penultimate, curved, flagellum 7—10-jointed. Gnathopod 1, 6th joint described as piriform, but in figure oblong oval, widening a little to the oblique palm which is rather shorter than the hind margin. Gnathopod 2. 6th joint oblong, rather smaller than in gnathopod 1. Uropod 1 reaching end of uropod 3. Uropod 2 only reaching end of peduncle of uropod 3. Uropod 3 $^1/_5$ (less?) of length of body; peduncle as long as outer ramus, which has 2^d joint nearly $^1/_3$ as long as 1st; inner ramus $^2/_3$ as long as outer; both with plumose setae on inner margin. Telson in figure very small, much shorter than peduncle of uropod 3. Colour green, spotted with brown. L. ♀ 6, ♂ 8—9 mm (including uropod 3).

Lake Baikal. Depth 3—10 m.

27. Gen. **Parelasmopus** Stebb.

1888 *Parelasmopus* (Sp. un.: *P. suluensis*), T. Stebbing in: Rep. Voy. Challenger, *c.* 29 p. 1029.

Head without rostrum. Antenna 1 the longer, accessory flagellum small. Upper lip rounded. Lower lip with inner lobes. Mandible normal except as to palp, which is slender and has the 2^d joint smaller than either 1st or 3^d. Maxilla 1, inner plate with few setae, outer with 7 spines. Maxilla 2, inner plate fringed on distal part of inner margin. Maxillipeds normal, but spine-teeth on apex of inner plate uncertain. Gnathopods 1 and 2 subchelate, 2^d with 6th joint in $\male$ very large. Peraeopods 3—5, 2^d joint expanded. Uropod 3 not reaching beyond the others (Dana). Telson small, deeply cleft.

1 species.

·1. **P. suluensis** (Dana) 1852 *Gammarus s.*, J. D. Dana in: P. Amer. Ac., *v.* 2 p. 210 | 1853 & 55 *G. s.*, J. D. Dana in: U. S. expl. Exp., *v.* 13 II p. 947; t. 65 f. 3 | 1862 *Megamoera s.*, Bate, Cat. Amphip. Brit. Mus., p. 230 t. 40 f. 6 | 1888 *Parelasmopus s.*, T. Stebbing in: Rep. Voy. Challenger, *v.* 29 p. 1029 t. 100.

Peraeon segment 7 and pleon segments 1 and 2 each with hind margin dorsally produced into a pair of teeth, segment 3 dorsally emarginate, segment 4 with 2 pairs of upward curved teeth, segment 6 very short. Head with no rostral point, lateral lobes rounded, post-antennal corners acute. Side-plates 1—4 moderate in depth, 1st a little distally outdrawn. Pleon segments 1—3, postero-lateral corners acute, in segment 3 produced upward, with adjacent lower margin serrate in 4 or 5 teeth. Eyes large, oval, close to lateral lobes of head. Antenna 1 much the longer, 2^d joint rather longer than 1st, 3^d less than $^1/_3$ as long as 2^d, flagellum with more than 17 joints, accessory flagellum with 2 slender joints. Antenna 2 slender, gland-cone reaching end of antepenultimate joint, ultimate long. but shorter than penultimate, flagellum 10-jointed. Mandibular palp not so long as the trunk, 1st and 3^d joints equal, each twice or more than twice as long as 2^d, 3^d slightly tapering. Gnathopod 1, 5th joint nearly as long as 6th, 6th oblong, with oblique row of 20 spines on inner surface, palm slightly oblique, finely denticulate. Gnathopod 2 in $\male$, 5th joint short, cup-shaped, broader than long, 6th much broader, twice as long as broad, with oblique, irregular palm, having a spinose prominence near hinge of finger, then 4 emarginations and on either side of the last of them 2 processes of the inner surface adapted to receive the finger-tip. Peraeopods 3 and 4, 2^d joint broader above than below, 4th moderately broad. Peraeopod 5, 2^d joint oval. Uropods 1—3 reaching equally far (Dana). Telson scarcely longer than broad, cleft nearly to base. lobes divergent, each with bidentate apex, outer tooth the longer, 2 unequal spines inserted between the teeth. L. about 8 mm.

Sooloo Sea (between Cape York and the Arrou Islands). Depth 16 m; among sea-weed floating off the shore.

28. Gen. **Cheirocratus** Norm.

1862 *Liljeborgia* (part.). Bate (& Westwood), Brit. sess. Crust., *v.* 1 p. 202 | 1867 *Cheirocratus* (Sp. un.: *C. mantis*), A. M. Norman in: Nat. Hist. Tr. Northumb., *v.* 1 p. 12 | 1876 *C.*, A. Boeck, Skand. Arkt. Amphip., *v.* 2 p 395 | 1893 *C.*, A. Della Valle in: F. Fl. Neapel, *v.* 20 p. 687 | 1894 *C.*, G. O. Sars, Crust. Norway, *v.* 1 p. 523.

Pleon segments 4 and 5 dentate. 4th—6th armed with bristles. Head not rostrate, lateral corners obtusely lobed, post-antennal angles projecting

just below them. Side-plate 2 the deepest, 4th scarcely emarginate, little
or not at all deeper than 5th. Eyes small, round. Antenna 1 much the
shorter, accessory flagellum very small. 2- or 3-jointed. Upper lip rounded.
Lower lip, inner lobes moderately distinct. Mandible normal, 1st joint of
palp not very short, 3[d] shorter than 2[d]. Maxilla 1, inner plate with numerous
setae. Maxilla 2, inner plate fringed along inner margin. Maxillipeds, palp
rather short and slender. Mouth-parts in other respects normal. Gnathopods 1
and 2 in ♀ rather feeble, subequal, 6th joint narrow, without palm. Gnathopod 1
in ♂ as in ♀, gnathopod 2 in ♂ powerful, strongly subchelate. Peraeopods 1
and 2 slight. Peraeopods 3—5 successively longer, 2[d] joint narrowly oblong.
Uropod 3 long, rami narrowly lanceolate, subequal. Telson deeply cleft,
lobes divergent, apically spinose.

4 species.

Synopsis of species:

1 { Peraeopod 5, 4th.—6th joints narrow — **2.**
  { Peraeopod 5, 4th—6th joints broad — **3.**

2 { Gnathopod 2 in ♂, palm not marginal, but formed
    by teeth on inner surface of 6th joint . . . . 1. **C. sundevallii** . . p. 418
  { Gnathopod 2 in ♂, palm marginal, concave . . . 2. **C. intermedius** . p. 419

3 { Gnathopod 2 in ♂, palm not marginal, but formed
    on inner surface of 6th joint . . . . 3. **C. robustus** . . . p. 419
  { Gnathopod 2 in ♂, palm marginal, distal, dentate 4. **C. assimilis** . . . p. 419

1. **C. sundevallii** (H. Rathke) 1843 *Gammarus s.*, H. Rathke in: N. Acta
Ac. Leop., *v.* 20 I p. 65 t. 3 f. 2 | 1888 *Cheirocratus s.*, T. Stebbing in: Rep. Voy. Challenger,
v. 29 p. 204 | 1893 *C. s.*, A. Della Valle in: F. Fl. Neapel, *v.* 20 p. 690 t. 20 f. 3, 4, 24,
25, 27, 30 | 1876 *C. sundevalli*, A. Boeck, Skand. Arkt. Amphip., *v.* 2 p. 396 t. 24 f. 2
(but not 2k) | 1889 *C. s.*, A. M. Norman in: Ann. nat. Hist., ser. 6 *v.* 4 p. 130 t. 11
f. 9, 10; t. 12 f. 1—3 | 1894 *C. sundevalli*, G. O. Sars, Crust. Norway, *v.* 1 p. 524 t. 184; t. 185
f. 1 | 1862 *Liljeborgia shetlandica*, Bate & Westwood, Brit. sess. Crust., *v.* 1 p. 206 f. |
1862 *Protomedeia whitei*, Bate, Cat. Amphip. Brit. Mus., p. 169 t. 31 f. 2 | 1874 *Lilje-
borgia normanni*, T. Stebbing in: Ann. nat. Hist., ser. 4 *v.* 14 p. 10 t. 1 f. 1 | 1879
Cheirocratus brevicornis, Hoek in: Tijdschr. Nederl. dierk. Ver., *v.* 4 p 142 t 10 f. 10—13 |
1884 *C. b.*, H. Blanc in: N. Acta Ac. Leop., *v.* 47 p. 72 t. 8 f. 76. 77.

Body rather slender and compressed, pleon segment 4 produced dorsally
to 3, segment 5 to 4 teeth, with stiff bristles between them, segment 6
with 2 bristles. Head, the lateral lobes separated from the triangular post-
antennal corners by a narrow incision. Side-plate 1 somewhat expanded
below and angularly produced, 2[d] deeper in ♂ than in ♀, 6th as deep as 4th.
Pleon segment 3, postero-lateral corners acutely produced. Eyes reddish
brown. Antenna 1 shorter than peduncle of antenna 2, 1st and 2[d] joints subequal in length, flagellum shorter than peduncle, 12—18-jointed, accessory
flagellum with 3 (Boeck) or 2 (Sars) joints. Antenna 2 nearly twice as long as
antenna 1, ultimate joint of peduncle longer than penultimate, flagellum shorter
than both joints combined, with about 24 joints. Gnathopod 1, 6th joint rather
shorter and much narrower than the elongate 5th, both densely setose, finger very
small, denticulate. Gnathopod 2 in ♀ rather more slender, 6th joint about
as long as 5th. Gnathopod 2 in ♂, 5th joint cup-shaped, much shorter than
the large, tumid, ovate 6th, which is densely setose about hind margin with
long setae and has a sort of sinuous palm on the inner surface armed by
3 or 4 irregularly placed denticles, against which the curved finger impinges.
Peraeopods 3—5 in both sexes very slender and long, with slender spines,

the 2^d joint distally a little tapering. Uropod 3, peduncle long, more than half as long as rami. Telson about $^2/_3$ as long as peduncle of uropod 3, cleft rather wide, $^2/_3$ of length, apices obliquely truncate from within outward, armed with 3 unequal spines. Colour orange, mottled with rose-red; golden yellow, mottled all over with small red specks, an opaque whitish patch on front part of peraeon; segments and side-plates spotted with scarlet, margins of antennae 1 and 2 and basal joint of legs scarlet. L. 8—10 mm.

Arctic Ocean and North-Atlantic with the adjoining seas (Europe from Lofoten Isles round to Naples and Constantinople). Depth 0—112 m.

2. **C. intermedius** O. Sars 1894 *C. i.*, G. O. Sars, Crust. Norway, v. 1 p. 527 t. 186 f. 1 | 1896 *C. i.*, T. Scott in: Rep. Fish. Board Scotl., v. 14 p. 160 t. 4 f. 7.

Closely resembling C. sundevallii. Head, lateral lobes slightly constricted at base, post-antennal corners projecting beyond them. Side-plate 1 little expanded distally, 2^d in $\male$ much deeper. Pleon segment 3, postero-lateral corners acutely and considerably produced. Antenna 1, flagellum subequal to peduncle. Gnathopod 2 in $\male$, 6th joint very large, tumid at base, palm oblique, longer than hind margin, concave, densely setose, defined by an obtuse prominence with a bifid denticle and 2 others, having also 2 distant submarginal denticles, and near the finger-hinge 2 or 3 prominent tubercles, finger curved, shorter than palm and impinging upon its edge, not closing against the surface of the joint. Uropod 3 long, rami little longer than peduncle. Telson scarcely longer than broad, cleft wide, apices armed with 5 spines, 1 longer than the rest. Colour pale yellow with irregular patches of vivid carmine. L. $\male$ reaching 11 mm.

North-Atlantic and North-Sea (South- and West-Norway up to the Trondhjemsfjord, in moderate depths; Firth of Forth).

3. **C. robustus** O. Sars 1894 *C. r.*, G. O. Sars, Crust. Norway, v. 1 p. 526 t. 185 f. 2.

Very like C. sundevallii. Body comparatively stouter, with side-plates smaller, 2^d in $\male$ only a little deeper than the others, pleon segment 3 with postero-lateral corners rather more produced. Antenna 1 as long as peduncle of antenna 2, flagellum longer than peduncle, 22-jointed. Antenna 2, ultimate and penultimate joints of peduncle subequal. Gnathopod 1, 6th joint as long as 5th. Gnathopod 2 in $\male$, joints short except 6th, which is very large, tumid, ovate, inner face densely clothed with short setae, palm on inner surface, slightly concave, armed with 1 tubercle to meet point of finger. Peraeopods 3—5 strongly built, especially in $\male$, peraeopod 5 much the strongest, 4th—6th joints broad, compressed, finger small. Telson broader than long, 5 not very unequal spines at each apex. Colour pale yellow with small red spots. L. $\male$ nearly 10 mm.

Christianiafjord and Trondhjemsfjord. Moderate depths.

4. **C. assimilis** (Lilj.) 1851 *Gammarus a.*, W. Liljeborg in: Öfv. Ak. Förh., v. 8 p. 23 | 1871 *Cheirocratus a.*, A. Boeck in: Forh. Selsk. Christian., 1870 p. 214 | 1876 *C. a.*, A Boeck. Skand. Arkt. Amphip , v. 2 p. 398 t. 24 f. 3 | 1889 *C. a.*, A. M. Norman in: Ann. nat. Hist., ser. 6 v. 4 p. 129 t. 10 f. 13; t. 11 f. 11 | 1893 *C. a.*, A. Della Valle in: F. Fl. Neapel, v. 20 p. 688 t. 20 f. 1, 2, 5—23, 26, 28, 29 | 1894 *C. a.*, G. O. Sars, Crust. Norway, v. 1 p. 528 t. 186 f. 2 | 1867 *C. mantis*, A. M. Norman in: Nat. Hist. Tr. Northumb., v. 1 p. 13 t. 7 f. 14. 15.

Body as in C. intermedius, post-antennal corners of head even more produced, postero-lateral angles of pleon segment 3 rather less. Gnathopod 2

97*

in ♂ much elongated, 6th joint widening distally, with fascicles of spinules
on both edges, palm much shorter than hind margin, scarcely setose,
defined by a slender tooth and spine, divided into 3 unequal teeth, and
minutely trilobed near finger-hinge, finger strong, falciform, dilated in
the middle, impinging against palmar tooth. Peraeopods 3—5 nearly as in
C. robustus (p. 419), but here peraeopod 5 still longer. Uropod 3, rami rather
broad, twice as long as peduncle. Telson much broader than long, each of the
concave apices carrying 3 spines, the central longest. Colour pale yellow,
mottled with small red spots. Gnathopod 2 in ♀, transverse rows of setae of 5th
joint appear to want the hamate character conspicuous in C. sundevallii
(p. 418) (Norman). L. ♂ 13 mm.

Distribution appears to be nearly the same as that of C. sundevallii (p. 418).

29. Gen. **Megaluropus** Hoek

1889 *Megalonoura* (Sp. un.: *M. agilis*) (nom. nud.), (A. M. Norman in MS.) (A. O.
Walker in:) Herdman in: P. Liverp. biol. Soc., *v.* 3 p. 39 | 1889 *Megaluropus* (Sp. un.:
M. agilis) (*Megaloura* A. M. Norman in MS.), (A. M. Norman in MS.) Hoek in: Tijdschr.
Nederl. dierk. Ver., ser. 2 *v.* 2 p. 197, 198 | 1889 *Megaluropus*, A. M. Norman in: Ann.
nat. Hist., ser. 6 *v.* 3 p. 446 | 1889 *M.*, Hoek in: Tijdschr. Nederl. dierk. Ver., ser. 2 *v.* 2
p. 260 | 1893 *M.*, A. Della Valle in: F. Fl. Neapel, *v.* 20 p. 694.

Head with small, acute rostrum and very prominent lateral lobes, on
which the eyes are placed. Antenna 1 much shorter than antenna 2, accessory
flagellum very small. Upper lip broad, apically bilobed. Lower lip with
inner lobes. Mandible, maxillae 1 and 2 and maxillipeds normal. Maxilla 1
with several setae on inner plate, 11 (?) spines on outer. Maxilla 2, inner
plate fringed on inner margin. Gnathopod 1 simple; gnathopod 2 larger,
simple or feebly subchelate. Peraeopod 5 elongate, finger stiliform. Uropod
3 much produced, rami equal. membranaceous, apically rounded. Telson cleft
to the base.

1 species.

1. M. agilis Hoek 1889 *M. a.*, (A. M. Norman in MS.) Hoek in: Tijdschr. Nederl.
dierk. Ver., ser. 2 *v.* 2 p. 197, 260 t. 7 f. 7; t. 8 f. 3, 3d, f, l; t. 9 f. 3g, h, i, k | 1889 *M. a.*,
A. M. Norman in: Ann. nat. Hist., ser. 6 *v.* 3 p. 446 t. 18 f. 1—10; *v.* 4 p. 123 t. 10 f. 15--17 |
1893 *M. a.*, A. Della Valle in: F. Fl. Neapel, *v.* 20 p. 695. t. 3 f. 9; t. 34 f. 1—17 | 1895
M. a., A. O. Walker in: P. Liverp. biol. Soc., *v.* 9 p 309 | 1890 *Cheirocratus drechselii*,
Meinert in: Udb. Hauchs, *v.* 3 p. 170 t. 2 f. 48—52.

Pleon segment 3 (and perhaps also 4 and 5) with hind margin dorsally
serrate. Head, rostrum reaching nearly middle of 1st joint of antenna 1,
lateral lobes rather broad, abruptly sharpened at apex, reaching much
beyond rostrum. Side-plate 1 a little widened distally, 2^d larger, 3^d irregularly
elliptic, with concave hind margin, 4th largest, scarcely emarginate. Pleon
segment 3, postero-lateral corners quadrate, the postero-lateral margin
serrate, with a cilium between every 2 or 3 teeth. Eyes red, occupying all
but the apex of lateral lobes of head, and in the ♂ passing upward behind
base of antenna 1. Antenna 1 rather short, 1st joint not longer, and in ♂
rather shorter than 2^d, 3^d joint very short, flagellum 6—8-jointed, with
sensory filaments but without calceoli, accessory flagellum 2-jointed. Antenna
2, ultimate and penultimate joints of peduncle subequal, very long, flagellum
in ♀ with 8 joints, in ♂ with 16 joints, which are long and slender, without
calceoli or filaments. Gnathopod 1, 5th joint oblong ovate, broader than 6th,
which is of a tapering form, the long palm scarcely distinguished from the
short hind margin, finger strong, curved. Gnathopod 2, 5th joint in ♀ longer

taan 6th, widening distally, 6th joint not more than half as broad, with palm nearly as in gnathopod 1, but more defined and rather shorter. Gnathopod 2 in ♂, 5th joint shorter and not much broader than 6th, which is long oval, palm ill-defined. Peraeopods 1 and 2, 2[d] joint bent, distally widened. Peraeopods 3 and 4, 2[d] joint wider below than above, 4th joint wide, longer than any of the following joints. Peraeopod 5, 2[d] joint well expanded, produced behind in rounded lobe below the 3[d] joint, 4th—7th joints elongate, and except the 4th slender. Uropod 1, rami equal, shorter than peduncle. Uropod 2, rami unequal. Uropod 3, the broad rami setose in ♂. Telson much longer than broad, the convex outer margin of each lobe fringed with sëtules. Colour variegated with crimson, orange, and white. L. 4—5 mm.

Kattegat; North-Sea (Holland, Firth of Forth); Firth of Clyde; Liverpool Bay; English Channel, Bristol Channel (Devon, Jersey); Gulf of Naples. Surface and Various depths to 36 m.

30. Gen. **Melita** Leach

1813/14 *Melita* (Sp. un.: *M. palmata*). Leach in: Edinb. Enc., v. 7 p. 403 | 1876 *M.*, A. Boeck, Skand. Arkt. Amphip., r. 2 p. 384 | 1893 *M.*, A. Della Valle in: F. Fl. Neapel, v. 20 p 707 | 1894 *M.*, G. O. Sars, Crust. Norway, v. 1 p. 507 | 1875 *Paramoera* (part.), Miers in: Ann. nat. Hist., ser. 4 c. 16 p 75.

Body slender, peraeon smooth, pleon usually with 1 or more of the segments dorsally dentate and armed with bristles. Head not rostrate, lateral corners rounded. Side-plate 4 the largest, emarginate behind. Eyes usually distinct, rather small. Antenna 1 slender, longer than antenna 2, 1st and 2[d] joints rather long, 3[d] not short, with accessory flagellum. Mouth-parts, so far as known, normal. Upper lip with small central emargination. Lower lip, inner lobes tolerably distinct. Mandibular palp rather slender. Maxilla 1, inner plate with several setae, outer with 11 spines. Maxilla 2, inner plate sometimes with setae on inner margin. Maxillipeds, outer plates with teeth on inner margin, passing into slender spines on apex. Gnathopod 1 small, subchelate. Gnathopod 2 larger, often unequal, and one in the ♂ sometimes much larger than the other, sometimes approximately chelate. Peraeopods 3—5. 2[d] joint well expanded. Peraeopods 4 and 5 subequal, longer than others. Branchial vesicles simple. Marsupial plates narrow. Uropod 2 the shortest. Uropod 3 projecting much beyond the others, outer ramus long, 2[d] joint wanting or rudimentary, inner ramus very short. Telson small, deeply cleft. Some characters subject to much variation within the species.

12 accepted and 7 doubtful species.

Synopsis of accepted species:

1 { Without eyes . . . . 1. **M. pallida** . . . . . p. 422
{ With eyes — **2.**

2 { Eyes scarcely pigmented . . . . 2. **M. quadrispinosa** . p. 422
{ Eyes darkly pigmented — **3.**

3 { Gnathopod 2 in ♂, one of the pair chelate . 3. **M. fresnolii** . . . . p. 423
{ Gnathopod 2 in ♂ subchelate — **4.**

4 { No pleon segment dorsally dentate — **5.**
{ At least 1 pleon segment dorsally dentate — **6.**

5 { Peraeopods 3—5 slender . . . . 4. **M. nitida** . . . . . p. 424
{ Peraeopods 3—5 rather strongly built . . . . 5. **M. pellucida** . . · . p. 424

6 { Pleon segment 4 not dentate . . . . 6. **M. coroninii** . . . . p. 424
{ Pleon segment 4 dentate — **7.**

1. **M. pallida** O. Sars 1879 *M. p.*, G. O. Sars in: Arch. Naturv. Kristian., *v.* 4 p. 457 | 1885 *M. p.*, G. O. Sars in: Norske Nordhavs-Exp.. *v.* 6 Crust. I p. 179 t. 15 f. 1, 1 a—1 | 1893 *M. p.*, A. Della Valle in: F. Fl. Neapel, *v.* 20 p. 716.

Back rounded, but body greatly compressed, pleon segments 1—5 each produced into a pair of dorsal adpressed teeth, between which there are 2—4 smaller denticles. Side-plates 1—4, lower margin quite smooth. Pleon segments 2 and 3, postero-lateral corners acutely produced, in segment 3 upward curving. Eyes entirely wanting. Antenna 1 nearly as long as body, 2^d joint not longer than 1st, flagellum longer than peduncle, filiform, 24-jointed, accessory flagellum 3-jointed. Antenna 2 about half as long as antenna 1, ultimate and penultimate joints of peduncle equal, flagellum 11-jointed. Mandibular palp short and slender, 3^d joint acute, ending in 2 simple setae. Maxilla 1, inner plate with numerous setae, outer with strong spines, 2^d joint of palp slender, curved, with a few delicate spines at apex. Maxillipeds, 3^d joint of palp distally dilated, setose, finger short, conical. Gnathopod 1, 6th joint as long as 5th, abruptly truncate at palm. Gnathopod 2, 6th joint large, oval, palm rather oblique, defined by a projecting corner, armed with a short spine. Peraeopods 3—5, 2^d joint indistinctly serrate behind. Uropod 3 the longest, outer ramus more than twice as long as peduncle, armed with few spines, rudimentary inner ramus bare. Telson much shorter than peduncle of uropod 3, cleft to base, lobes scarcely dehiscent, narrowly truncate apices each with a setule. Colour uniform white. L. reaching 26 mm.

Arctic Ocean (West of Spitzbergen). In cavities (Work of Teredo?) in Wood dredged from 2510 m.

2. **M. quadrispinosa** Vosseler 1889 *M. q.*, Vosseler in: Arch. Naturg., *v.* 551 p. 157 t. 8 f. 15—24 | 1893 *M. palmata* ♀ ?, A. Della Valle in: F. Fl. Neapel, *v.* 20 p. 716.

Very like M. dentata (p. 427). Pleon segment 4 produced to 1 strong dorsal tooth, segment 5 to 3 weaker teeth. Pleon segments 2 and 3, postero-lateral corners acutely produced. and in segment 3 upturned. Body slender, strongly compressed. Eyes small, scarcely pigmented. Antenna 1, accessory flagellum 3-jointed. Antenna 2, flagellum scarcely longer than penultimate joint of peduncle. Mandibular palp thin, with few setae. Maxilla 1 with few - setae (very numerous in M. dentata) on inner plate. Gnathopod 1, 5th joint much longer (in figure) than the oval 6th, which has a palm, though weakly defined, 5th and 6th joints both setose on hind margin. Gnathopod 2, 5th joint not short, 6th longer and broader, oval, palm weakly defined, finger curved, acute, not especially large, and seemingly not closing against surface of 6th joint. Peraeopods 3—5, 2^d joint broadly oval. Uropod 3, outer ramus long, inner reduced to a small scale, peduncle longer than the telson. L. 11·5 mm.

Arctic Ocean (Spitzbergen).

3. **M. fresnelii** (Aud.) 1826 *Gammarus f.*, Audouin in: Descr. Égypte, *v.* 1 IV p. 93 Crust. t. 11 f. 3 | 1830 *Amphithoe f.*, H. Milne Edwards in: Ann. Sci. nat., *v.* 20 p. 377 | 1875 *Melita f.*, *Paramoera* (part.), Miers in: Ann. nat. Hist., ser. 4 *v.* 16 p. 75 | 1893 *M. f.*, A. Della Valle in: F. Fl. Neapel, *v.* 20 p. 708 t. 60 f. 6 | 1845 *Gammarus anisochir*, Krøyer in: Naturh. Tidsskr., ser. 2 *v.* 1 p. 317 t. 2 f. 1 a—p | 1852 *G. (Maera) validus + G. (M.) pilosus + G. (M.) setipes*, J. D. Dana in: P. Amer. Ac., *v.* 2 p. 212, 213 | 1853 & 55 *M. valida + M. s. + M. anisochir*, J. D. Dana in: U. S. expl. Exp., *v.* 13 II p. 966 t. 66 f. 6; p. 967 t. 66 f. 7; p. 968 t. 66 f. 8 | 1862 *Melita v. + M. s. + M. a. + M. fresnelli*, Bate, Cat. Amphip. Brit. Mus., p. 185 t. 33 f. 7; p. 186 t. 33 f. 8; p. 186 t. 34 f. 1; p. 186 t. 34 f. 2 | 1864 *M. exilii*, Fritz Müller, Für Darwin, p. 6 f. | 1879 *M. australis*, Haswell in: P. Linn. Soc. N. S. Wales, *v.* 4 p. 264 t. 9 f. 6, 7 | 1890 *M. cotesi*, G. M. Giles in: J. Asiat. Soc. Bengal, *v.* 59 p. 64 t. 2 f. 1.

Peraeon quite smooth. Pleon segment 1, a small median tooth flanked on each side by 3 larger teeth and an outer rudimentary tooth; segment 2 similar, but median tooth comparatively smaller, and outer teeth larger; segment 3 with 7 teeth, the outermost on each side furcate; segment 4 with 5 teeth. 2 large, the middle and outer small; segment 5 with 2 rather long teeth. Side-plates 1—4 tolerably large, setose on the lower margin, 1st with denticle at lower hind corner. Pleon segment 3, postero-lateral corners produced to a long tooth, serrate on its upper margin. Eyes dark, nearly circular. Antenna 1 nearly $^4/_5$ length of body, 1st joint shorter than 2^d, 3^d scarcely $^1/_4$ as long as 2^d, flagellum much longer than peduncle, 40-jointed, accessory flagellum more than $^1/_4$ of primary, 6-jointed. Antenna 2 a little shorter, ultimate joint of peduncle slightly shorter than penultimate, flagellum $^2/_3$ as long as peduncle, about 20-jointed. Mandible, 10 spines in spine-row, palp slight, but a little longer than the trunk. Maxilla 1, inner plate with 10 setae. Gnathopod 1 small, 6th joint oval, broader than 5th, both setose, finger short, pointed, strongly curved, seemingly not very freely movable. Gnathopod 2 in ♀ similar to gnathopod 1, but longer and larger, 6th joint longer but not proportionately broader, palm and inner margin of finger finely denticulate (as seen in *M. australis* Haswell). Gnathopod 2 in ♂, the right limb nearly as in ♀, the left quite different, 2^d joint rather dilated, 4th produced to a backward-directed point, 5th very small, 6th and finger forming a sort of chela; the 6th joint, instead of being shorter than 2^d, is more than twice as long, and of great breadth, its hind margin longer than the front, produced to a great spoon-shaped tooth, between which and the finger-hinge are 3 teeth or tubercles, the finger long, massive, slightly sinuous and apically blunt. Peraeopod 1 rather longer than peraeopod 2. Peraeopods 3—5, 2^d joint minutely serrate behind. Peraeopod 4 slightly longer than peraeopod 5. Uropod 1, rami about equal to peduncle and one to other. Uropod 2, rami rather longer than peduncle. Uropod 3, outer ramus nearly twice as long as peduncle, inner ramus small, almost linear. Telson cleft nearly to base, lobes conical, widely dehiscent, apices curved a little inward. Colour clouded yellowish grey, thumb of chela porcelain-white. L. about 13 mm. (Diagnosis taken especially from Krøyer's account).

Red Sea? (Egypt); Atlantic (Rio Janeiro, depth 11—13 m; Desterro); Indian Ocean (Andaman Islands, shallow water; Singapore, depth 3 m); Port Jackson [East-Australia].

4. **M. nitida** S. I. Sm. 1873 *M. n.*, (S. I. Smith in:) A. E. Verrill in: Rep. U. S. Fish Comm., *c.* 1 p. 560 | 1893 *M. n.*, A. Della Valle in: F. Fl. Neapel, *v.* 20 p. 716.

Pleon segments 1—6, none serrate or emarginate, segment 5 with several slender spines on each side near median dorsal line. Eyes small, round, black. Antenna 1 about $^2/_3$ as long as body, 2^d joint slightly longer than 1st, nearly twice as long as 3^d, flagellum longer than peduncle. Antenna 2 shorter, but with longer peduncle, ultimate joint of peduncle rather longer than penultimate. Gnathopod 1, 5th joint longer and broader than 6th, which is oblong, slightly curved, finger very small but stout, curved, attached in notch of apex of 6th joint, not closing on palm but projecting inwards. Gnathopod 2 stout, 5th joint short, triangular, 6th somewhat oval, palm oblique, arcuate, continuous with hind margin, with spinules and stiff hairs, tip of finger resting within the palmary margin. Peraeopod 1 rather longer than peraeopod 2. Peraeopods 3—5, 2^d joint minutely serrate behind. Uropod 2 not reaching quite so far as uropod 1. Uropod 3 very long, with fascicles of spines along the margins. Lobes of telson slender, with spines at the tips. Colour dark greenish-slate. L. 7--9 mm.

North-Atlantic (New Jersey to Cape Cod). Near low-water mark.

5. **M. pellucida** O. Sars 1882 *M. p.*,, G. O. Sars in: Forh. Selsk. Christian., nr. 18 p. 106 t. 5 f. 9, 9 a, b | 1894 *M. p.*, G. O. Sars, Crust. Norway, *v.* 1 p. 511 t. 180 f. 2.

Body rather slender, with no dorsal teeth, but some simple bristles on pleon segments 4 and 5. Head, lateral corners broadly rounded, postantennal corners to the rear but distinct. Side-plate 1 expanded distally, 4th quadrate below the emargination. Pleon segment 3, postero-lateral corners acute, but scarcely produced. Eyes small, rounded oval, black. Antenna 1 about $^2/_3$ as long as body, 1st joint fully as long as 2^d, 3^d more than half as long as 2^d, flagellum nearly twice as long as peduncle, 18-jointed, accessory flagellum very small, 1-jointed. Antenna 2 scarcely more than $^1/_2$ as long as antenna 1, ultimate and penultimate joints of peduncle subequal, together longer than flagellum. Gnathopod 1, 6th joint little more than $^1/_2$ as long as 5th, subquadrate, palm distinctly defined, transverse. Gnathopod 2 in ♀ larger, 6th joint a little longer than 5th and broader, oval, but with palm nearly transverse, defined by a distinct angle and 2 palmar spines. in ♂ similar, but with 5th joint shorter, and 6th considerably larger. Peraeopods 3—5 rather strongly built, 2^d joint rather broadly oval. Uropod 3, peduncle about $^1/_3$ as long as outer ramus, which has apex obliquely truncate, armed with many unequal spines, inner ramus not twice as long as broad. Telson broader than long, not cleft to the base, the dehiscent lobes obtusely rounded, with several apical spines. Colour whitish, pellucid. L. ♂ 6 mm, ♀ considerably less.

South-Norway (brackish basin inside the Listerland).

6. **M. coroninii** Heller 1866 *M. c.*, Cam. Heller in: Denk. Ak. Wien, *v.* 26 n p. 37 t. 3 f. 20, 21 | 1893 *M. palmata* (part.) A. Della Valle in: F. Fl. Neapel, *v.* 20 p. 714.

Pleon segments 1—4 entirely unarmed, segment 5 produced into 3 dorsal denticles with a bristle in each of the intervals (the figured small 7th segment is an error). Eyes small, round, black. Antenna 1 almost as long as body, 2^d joint longer than 1st, flagellum of more than 40 joints, accessory flagellum of 4 rather elongate joints. Antenna 2 shorter, flagellum much shorter than peduncle, about 20-jointed. Gnathopod 1, 6th joint shorter than 5th, narrow proximally, distally a little widened, palm short, defined by a projecting rounded lobe, behind which the point of the finger closes down. Gnathopod 2, 6th joint in ♂ rather tumid, oval. palm weakly defined, finger strongly curved, closing against inner surface of 6th joint. Peraeopods 1 and 2 slender. Peraeopods 3—5 stouter, 2^d joint oval, pretty well expanded. Uropods 1—3 as usual in the genus. Telson with lobes about reaching end of peduncle of uropod 3. L. 10—11 mm.

Adriatic (Lesina).

7. **M. palmata** (Mont.) 1804 *Cancer palmatus*, Montagu in: Tr. Linn. Soc. London, *v.* 7 p. 69 t. 6 f. 4 | 1812 *Astacus p.*, Pennant, Brit. Zool., ed. 5 *v.* 4 p. 35 | 1813/14 *Melita palmata, M. palmeta*, Leach in: Edinb. Enc., *v.* 7 p 403, 432 | 1876 *M. palmata*, A. Boeck, Skand. Arkt. Amphip., *v.* 2 p. 387 t. 24 f. 4 | 1878 *M. p.*, Zaddach in: Schr. Ges. Königsb., *v.* 19 Abh. p. 32 f. 4 | 1889 *M. p.*, A. M. Norman in: Ann. nat. Hist., ser. 6 *v.* 4 p. 132 | 1893 *M. p.* (part.), A. Della Valle in: F. Fl. Neapel, *v.* 20 p. 713 t. 1 f. 6; t. 23 f. 24—40 | 1894 *M. p.*, G. O. Sars, Crust. Norway, *v.* 1 p. 508 t. 179 | 1818 *Gammarus palmatus*, Lamarck, Hist. An. s. Vert., *v.* 5 p. 181 | 1830 *G. dugesii*, H. Milne Edwards in: Ann. Sci. nat., *v.* 20 p. 368 | 1857 *G. inaequimanus*, Bate in: Ann. nat. Hist., ser. 2 *v.* 19 p. 145.

Pleon segment 4 produced dorsally to a compressed tooth, segment 5 to 2 small contiguous denticles, each with a bristle at the base. Head, a small rounded projection in the rostral position, post-antennal corners separated by a small incision from the lateral corners. Side-plate 6 in ♀ (not in ♂, as Boeck asserted) having the front lobe produced below to a spiral process. Pleon segment 3, postero-lateral corners slightly produced, acute. Eyes very small, roundish, dark brownish. Antenna 1 more than ¹/₂ as long as body, 2^d joint rather longer than 1st, 3^d about half 2^d or less, flagellum longer than peduncle, 20-jointed, accessory flagellum 2-jointed. Antenna 2 much shorter, ultimate and penultimate joints of peduncle subequal, flagellum short, sometimes with 10 joints. Maxillipeds, 3^d joint of palp distally widened, finger with setules on inner margin. Gnathopod 1 very small, 6th joint much shorter than 5th, widening distally, palm transverse; (Sars: finger in ♀ well defined, in ♂ obtuse and immovable; but Zaddach and others: 6th joint in ♂ on its front border deeply emarginate and so obliquely that the emargination is larger on the inner than on the outer side, whereby arises an upper and an under process, the finger being placed in the emargination on the outer surface, and moving so obliquely that in repose it rests on the inner side of the under process, its place of insertion being protected by a chitinous process on each side). The finger ends in an acute hook. Gnathopod 2 large in ♀, 5th joint expanded distally to a broad setiferous lobe, 6th broad, ovately oblong, palm rather oblique, with a few spines at the obtuse angle; in ♂ very large, 5th joint short, cup-shaped, 6th much expanded, widest at palm which is somewhat variable, sometimes where it meets hind margin much projecting and rounded off (Sars), at others, slightly oblique and angular, with a strong group of setae on the adjoining surface. finger curved, acute, folding down on to the inner concave surface of 6th joint. Peraeopods 3—5 much stouter than peraeopods 1 and 2, 2^d joint broadly oval, very slightly serrate, peraeopods 4 and 5 much longer than 3^d, terminal joints overturned backward. Finger in maxillipeds and all the limbs ending in a nail, finger short in peraeopods 1—5. Uropods 1—3 of the usual character. Telson about reaching end of peduncle of uropod 3, with lobes rather narrow, widely divergent, each carrying 4 spinules on truncate apex, and 3 stronger spines on projection of inner margin (Sars); divided to the base, each lobe very acute, with slight indent on the outer margin (Della Valle). Colour pale brownish yellow. L. 8—16 mm.

North-Atlantic with the adjoining seas (Europa: Great Britain, West-Norway, Baltic etc.; Azores).

8. **M. obtusata** (Mont.) 1813 *Cancer (Gammarus) obtusatus,* Montagu in: Tr. Linn. Soc. London, *v.* 11 p. 5 t. 2 f. 7 | 1818 [?] *G. o.,* Latreille in: Tabl. enc. méth., Crust. Arach. Ins. t. 336 f. 29 | 1830 *Amphithoe obtusata,* H. Milne Edwards in: Ann. Sci. nat., *v.* 20 p. 377 | 1879 *Melita o.,* Hoek in: Tijdschr. Nederl. dierk. Ver., *v.* 4 p. 140 t. 10 f. 8, 9 | 1889 *M. o.,* A. M. Norman in: Ann. nat Hist., ser. 6 *v.* 4 p. 132 | 1893 *M. o.* (part.). A. Della Valle in: F. Fl. Neapel, *v.* 20 p 711 t. 1 f. 7; t. 23 f. 1—19; & ? juv. p. 713 t. 3 f. 14; t. 23 f. 20—23 | 1894 *M. o.,* G. O. Sars, Crust. Norway, *v.* 1 p. 510 t. 180 f. 1 | 1852 *Gammarus maculatus* (non G. Johnston 1827!), W. Liljeborg in: Öfv. Ak. Förh., *v.* 9 p. 10 | 1853 *G. obtusunguis,* A. Costa in: Rend. Soc. Borbon., n. ser. *v.* 2 p. 176 | 1862 *Melita obtusata* + *M. proxima* + *Megamoera alderi,* Bate, Cat. Amphip. Brit. Mus., p. 183 t. 33 f. 3; p. 184 t. 33 f. 4; p. 228 t. 40 f. 1.

Pleon segments produced to variable number of teeth. Head, lateral corners broadly rounded, post-antennal corners forming a minute tooth. Side-plates 1—4 rather deep, 1st a little widened distally, 1st—3d with denticle at lower hind corner. Pleon segment 3, postero-lateral angles much produced, acute, upturned, margins smooth or slightly serrate. Eyes small, round, dark. Antenna 1 more than ½ as long as body, 2d joint rather longer than 1st, flagellum longer than peduncle, flagellum about 16-jointed (Sars), accessory flagellum 4-jointed, last joint minute. Antenna 2, ultimate joint of peduncle rather shorter than penultimate, flagellum subequal to both combined. Mandibular palp very narrow, joints successively longer. Gnathopod 1, 6th joint as long as 5th, rather broad, somewhat triangular, palm oblique, ill-defined. Gnathopod 2 larger in ♀, 6th joint longer than 5th, oblong oval, palm rather oblique, defined by an obtuse angle; in ♂ 5th joint short, cup-like, 6th very large, tumid, widest at palm, which is irregularly serrated, and hollowed near the triangular defining lobe; finger, which closes into the cavity, is scimitar-shaped, its obtuse apex rounded above and squared below; (Bate: palm sometimes less sinuous and more irregularly denticulate). Peraeopods 3—5, 2d joint oval, hind margin distinctly serrate. Uropod 3, outer ramus ending in a strong spine, which perhaps represents a 2d joint, about once and a half as long as peduncle, inner ramus very small. Telson not nearly reaching end of peduncle of uropod 3, cleft nearly to base, lobes dehiscent, acute, with some spinules on inner margin. Colour pale brown, mottled with rufous brown, especially about the legs (Montagu). L. 6—9 mm.

North-Atlantic with the adjoining seas (from West-Norway round Europe to Italy).

9. **M. amoena** H. J. Hansen 1887 *M. a.,* H. J. Hansen in: Vid. Meddel., ser. 4 *v.* 9 p. 147 t. 6 f. 1, 1 a | 1893 *M. obtusata* (part.), A. Della Valle in: F. Fl. Neapel, *v.* 20 p. 711.

♂. Body somewhat compressed, pleon segment 1 with 1 very little dorsal tooth, segment 2 with 5, middle small, other 4 very minute, segment 3 unarmed, segment 4 with 2 rather small teeth, segment 5 with 2 minute teeth. Head, post-antennal corners not dentiform. Side-plates 1—4, lower margin broadly rounded, 1st the smallest. Pleon segment 2, postero-lateral corners produced to a process rather long, narrow, obtuse. corners of segment 3 acute, upturned. Eyes small, round. Antenna 1 about ½ as long as body, 1st joint rather shorter than 2d. flagellum shorter than peduncle, accessory flagellum 4-jointed, last joint minute. Antenna 2, ultimate and penultimate joints of peduncle subequal. Gnathopod 1, 6th joint shorter than 5th, much longer than broad, palm very oblique. Gnathopod 2, 5th joint very short, cup-shaped. 6th very large, about twice as long as broad, palm very oblique, longer than hind margin, irregularly dentate, finger long, the acute apex closing behind the little triangular lobe which defines the palm. Peraeopods 3—5, 2d joint much expanded, in peraeopods 4 and 5 not ⅓ longer than broad. Uropod 3, outer ramus not quite twice as long as peduncle, inner very minute. Colour whitish. L. 8·5 mm.

Ikertokfjord [West-Greenland]. Depth 56 m.

10. **M. formosa** J. Murdoch 1866 *Gammarus dentatus* (part.), Goës in: Öfv. Ak. Förh., *v.* 22 p. 530 t. 40 f. 29′ | 1885 *Melita formosa*, J. Murdoch in: P. U. S. Mus., *v.* 7 p. 520 | 1887 *M. goësii*, H. J. Hansen in: Dijmphna Udb., p. 228 t. 21 f. 13 | 1887 *M. g.*, H. J Hansen in: Vid. Meddel., ser. 4 *v.* 9 p. 146 t. 5 f. 8 | 1893 *M. obtusata* (part.), A. Della Valle in: F. Fl. Neapel, *v.* 20 p. 711.

Close to M. dentata. Pleon segments 2 and 3 produced each to ·1 dorsal tooth, segment 4 to 3 teeth, segment 5 to 4, all very small. Pleon segments 1—3, postero-lateral corners acute, in segment 3 produced upward. Antenna 1, 1st joint not quite so long as 2d. Gnathopod 1, 6th joint oval, fringed with long hairs on hind margin. Gnathopod 2, 6th joint in ♂ broadly oval, armed on the hind margin with 3 or 4 blunt teeth and running out into a broad blunt tooth, palm minutely denticulate, finger large, curved, acute, shutting on the inside of the palm. Uropod 3, inner ramus ovate. Colour purple, with lighter streak down middle of back (Murdoch). Hansen adds: body depressed, lateral corners of head produced into a long, acute process; side-plates 1—4 shallow; peraeopods 3—5, 2d joint long and narrow, with front and hind margins subparallel; uropod 3, outer (not inner!) ramus very long; colour red. L. 17·5—21 mm.

Arctic Ocean (arctic Alaska; Jugor Schar, depth 19—23 m; West-Greenland, depth 66 m; Spitzbergen, depth 19—75 m).

11. **M. dentata** (Krøyer) 1842 *Gammarus dentatus*, Krøyer in: Naturh. Tidsskr., *v.* 4 p. 159 | 1866 *G. d.* (part.), Goës·in: Öfv. Ak. Förh., *v.* 22 p. 530 t. 40 f. 29 | 1862 *Megamoera dentata*, Bate, Cat. Amphip. Brit. Mus., p. 225 t. 39 f. 4 | 1871 *Melita d.*, A. Boeck in: Forh. Selsk. Christian., 1870 p. 211 | 1876 *M. d.* (part.), A. Boeck, Skand. Arkt. Amphip., *v.* 2 p. 389 t. 23 f. 10 | 1884 *M. d.*, J. S. Schneider in: Tromsø Mus. Aarsh., *v.* 7 p. 113 | 1889 *M. d.*, A. M. Norman in: Ann. nat. Hist.. ser. 6 *v.* 4 p. 135 | 1894 *M. d.* (part.), G. O. Sars, Crust. Norway, *v.* 1 p 513 t. 181 f. 1 | 1853 *Gammarus purpuratus*, Stimpson in: Smithson. Contr.; *v.* 6 nr. 5 p. 55 | 1893 *Melita palmata* (part.). A. Della Valle in: F. Fl. Neapel, *v.* 20 p. 713.

Elongate, compressed, but with rounded back. Pleon segments 1—6 produced dorsally to several teeth, of which median usually the largest, numbers very variable. Head, post-antennal corners short, dentiform. Side-plates 1—4 shallow, 1st scarcely expanded distally, 1st—3d with lower hind corner dentiform. Pleon segment 3, postero-lateral corners acutely produced, slightly upturned. Eyes small, rounded oval, black. Antenna 1 nearly as long as body, 1st joint with spines on lower margin, a long one at apex, 2d joint longer, 3d about $^1/_4$—$^1/_3$ of 2d; flagellum longer than peduncle, 36—46-jointed, accessory flagellum 4—9-jointed. Antenna 2, peduncle long, ultimate and penultimate joints subequal, flagellum short, about 13-jointed. Mandibular palp rather slender, 1st joint produced to a dentiform projection, 3d and 2d subequal, or 3d the longer. Maxilla 1, outer plate with 7—9(?) spines. Gnathopod 1 densely setose, 6th joint (Sars) about as long as 5th, oblong oval, palm rather oblique, defined by an obtuse angle; (Norman: 6th joint ovate, much shorter than 5th, palm defined by a small tooth-like process. very minutely crenulated and spinulose, finger falcate, its inner margin divided up into minute teeth of peculiar form, widening in the middle and apiculate). Gnathopod 2 much stronger, 5th joint rather short, 6th rather large, especially in ♂, oblong, a little widening to the oblique slightly denticulated palm, which is defined by a tooth-like process, and in ♂ is angular near the finger-hinge, finger

large, curved, acute. Peraeopods 1 and 2 slender, short. Peraeopods 3—5
successively longer, 2^d joint large, oblong oval, hind margin serrate, with
the lower hind corners rounded or subangular. Uropod 3, outer ramus very long,
sublinear, fully twice as long as peduncle, ending in 4 spines, the largest perhaps
representing a 2^d joint, inner ramus very small. Telson not nearly reaching
end of peduncle of uropod 3, lobes acute, each with 2 spines inside the point,
1 setule outside. Colour varying with the surroundings (Schneider); uniform
dark purple, never varying (Stimpson). L. 11—22(—27·5?) mm.

Arctic and Scandinavian Waters; North-Atlantic (Halifax [Nova Scotia], Labrador,
Grand Manan, Northumberland). Down to 113 m.

12. **M. gladiosa** Bate 1862 *M. g.*, Bate (& Westwood), Brit. sess. Crust., *v.* 1
p. 346 f. | 1862 *M. g.*, Bate, Cat. Amphip. Brit. Mus., p. 185 t. 33 f. 6 | 1876 *M. g.*,
T. Stebbing in: Ann. nat. Hist., ser. 4 *v.* 17 p. 77 t. 4 f. 2, 2a—d | 1889 *M. g.*, A. M.
Norman in: Ann nat. Hist., ser. 6 *v.* 4 p. 134 | 1893 *M. obtusata* (part), A. Della Valle
in: F. Fl. Neapel, *v.* 20 p. 711.

Near to M. obtusata (p. 426). Pleon segments 1—6 each with 3 dorsal teeth,
but sometimes segments 1, 5 and 6 with more or fewer, or 1st with none; teeth
large on segments 1—4, except central of segment 4, teeth on segment 6 very
small. Pleon segment 3, postero-lateral corners acutely produced, upturned,
their upper and under margins serrate. Gnathopod 1, 5th and 6th joints
equal and similar, broadly oval, palm scarcely defined, hind margin of
4th—6th joints with dense fringe of fine short hairs, as in M. obtusata ♀.
Gnathopod 2 in ♀ without the fur seen in gnathopod 1, not much larger, 5th joint
less broad, 6th equal in breadth but longer, palm defined by a small tooth.
Gnathopod 2 in ♂ much larger, 5th joint short, cup-shaped, 6th large, broad,
palm serrate, strongly sinuous, with an abrupt curve at the middle, finger
scimitar-shaped, broad almost to the end, but with an acute apex.

North-Atlantic (South-West-England; North-East-Scotland; Boulogne and West-
France; Azores).

M. appendiculata (Say) 1818 *Gammarus appendiculatus,* Say in: J. Ac. Philad.,
v. 1 II p. 374 | 1845 *G. a.*, Krøyer in: Naturh. Tidsskr., ser. 2 *v.* 1 p. 326 1862 *G. a.*, *Podocerus*
(part.)?, Bate, Cat. Amphip. Brit. Mus., p. 223, 224 | 1888 *Maera? a.*, T. Stebbing in:
Rep. Voy. Challenger, *v.* 29 p. 103 | 1893 *Ceradocus? a.*, A. Della Valle in: F. Fl. Neapel,
v. 20 p. 765.

Peraeon segments 5—7 and pleon segments dentated on their posterior edges.
L. 7—8 mm.

· United States of America (Georgia).

M. confervicola (Stimps.) 1857 *Maera c.*, Stimpson in: P. Calif. Ac., *v.* 1 p. 99 |
1857 *Gammarus confervicolus*, Stimpson in: Boston J. nat. Hist., *v.* 6 p. 520 | 1862 *G. c.*,
Bate, Cat. Amphip. Brit. Mus, p. 218 t. 38 f. 9 | 1899 *Melita c.*, T. Stebbing in: Tr. Linn.
Soc. London, ser. 2 *v.* 7 p. 425 | 1893 *Gammarus marinus* (part.), A. Della Valle in:
F. Fl. Neapel, *v.* 20 p. 762.

L. 12·5 mm.

North-Pacific (San Francisco, among confervae in salt marshes; Pugets Sound).

M. gayi (Nic.) 1849 *Amphitoe g.*, H. Nicolet in: Gay. Hist. Chile, *v.* 3 p. 236,
Crust. t. 2 f. 6a, b | 1893 *Amphithoe g.*, A. Della Valle in: F. Fl. Neapel, *v.* 20 p. 463 ·
1899 *Melita g.*, T. Stebbing in: Tr. Linn. Soc. London, ser. 2 *v.* 7 p. 425.

Side-plate 5 much shorter than 4th, 6th deeply emarginate so that the anterior
lobe takes the form of a horn much curved backward (compare M. palmata ♀, p. 425).
L. about 9 mm.

South Pacific (off Chili).

M. inaequistylis (Dana) 1852 *Amphitoë (Melita) i.* (♀) + *A. (M.) tenuicornis* (part.: ♂), J. D. Dana in: P. Amer. Ac., *v.* 2 p. 214; p. 215 | 1853 & 55 *M. t.*, J. D. Dana in: U. S. expl. Exp., *v.* 13ıı p. 963 t. 66 f. 5 a—m | 1881 *M. t.*, G. M. Thomson in: Tr. N. Zealand Inst., *v.* 13 p. 218 | 1862 *Moera t.*, Bate, Cat. Amphip. Brit. Mus., p. 195 t. 35 f. 6 | 1875 *Paramoera t.*, Miers in: Ann. nat. Hist., ser. 4 *v.* 16 p. 75 | 1876 *P. t.*, Miers, Cat. Crust. N. Zealand, p. 127 t. 8 f. 8 (♀) | 1893 *Melita palmata* (part.), A. Della Valle in: F. Fl. Neapel, *v.* 20 p. 714.

Dana's and Thomson's statements are in many respects contradictory: whether ♀ and ♂ belong to the same species, is doubtful. •

South-Pacific (New Zealand). Between tidemarks and in rock pools along coast.

M. oxyura Catta 1875 *M. o.*, Catta in: Rev. Sci. nat., *v.* 4 p. 164 | 1893 *M. o.*, A. Della Valle in: F. Fl. Neapel, *v.* 20 p. 716.

Rather near to M. gladiosa; specially distinguished by the strong dentation of the lower hind margin of pleon segments.

Gulf of Lyon (Marseilles). Among Zostera, depth 19—25 m.

M. sp., F. Brandt 1851 *Gammarus longicauda* (non A. Costa 1851!), F. Brandt in: Middendorff, Reise Sibirien, *v.* 2ı p. 141 t. 6 f. 32 | 1862 *Megamoera l.*, Bate, Cat. Amphip. Brit. Mus., p. 229 t. 40 f. 3 | 1893 *Maera? l.*, A. Della Valle in: F. Fl. Neapel, *v.* 20 p. 730 | 1876 *Melita dentata* (part.), A. Boeck, Skand. Arkt. Amphip., *v.* 2 p. 392 | 1894 *M. d.* (part.), G. O. Sars, Crust. Norway, *v.* 1 p. 514.

Perhaps identical with M. dentata (p 427).

Sea of Ochotsk.

M. sp., Stimps. ?1852 *Amphithoe tenuicornis* (part.) (err., non *Amphithoe t.* Rathke 1843!), J. D. Dana in: P. Amer. Ac., *v.* 2 p. 214 | 1855 *Gammarus t.* (sp. nov.!), Stimpson in: P. Ac. Philad., *v.* 7 p. 382 | 1899 *Melita t.*, T. Stebbing in: Tr. Linn. Soc. London, ser. 2 *v.* 7 p. 425.

L. 8 mm.

North-Pacific (Loo Choo Islands).

31. Gen. Paraceradocus Stebb.

1899 *Paraceradocus* (Sp. typ.: *P. miersii*), T. Stebbing in: Tr. Linn. Soc. London, ser. 2 *v.* 7 p. 426.

Side-plates not deep, 1st larger than 4th. Antenna 1 longer than antenna 2, but not stouter and with shorter peduncle. Upper lip transversely elliptic. Lower lip with principal lobes dehiscent. Mandibular palp long, 3^d joint not short. Maxilla 1, inner plate large, with setae only on the apex, palp broad. Maxilla 2, inner plate fringed on inner margin. Peraeopod 1, uropods and telson as in Ceradocus (p. 430).

1 species.

1. **P. miersii** (Pfeff.) 1888 *Megamoera m.*, Pfeffer in: Jahrb. Hamburg. Anst., *v.* 5 p. 121 t. 3 f. 3 | 1893 *Maera? m.*, A. Della Valle in: F. Fl. Neapel, *v.* 20 p. 732 | 1899 *Paraceradocus m.*, T. Stebbing in: Tr. Linn. Soc. London, ser. 2 *v.* 7 p 426.

Body elongate, back rounded as far as peraeon segment 6, segment 7 and pleon segments 1—5 with dorsal angle, on segments 4—5 raised into a carina produced acutely backward, pleon segment 6 dorsally flat. Head, lateral lobes a little produced, narrowly rounded, post-antennal corners not acute. Side-plates 1 and 2 rather shallow, larger than 3^d and 4th, 4th not deeper than 5th. Pleon segment 3, postero-lateral corners strongly

produced backward, acute, with slight upward curve. Eyes narrowly oval.
Antenna 1 about half as long as body, 1st joint long and stout, 2^d thinner
and rather shorter, 3^d small, flagellum rather longer than peduncle, 50-jointed,
accessory flagellum about 7-jointed. Antenna 2 very strong, gland-cone
short, antepenultimate joint of peduncle very stout, penultimate and ultimate
joints of peduncle corresponding in thickness and length respectively to
1st and 2^d joints of antenna 1, flagellum stout, as long as ultimate joint of
peduncle, about 17-jointed. Gnathopod 1, 5th joint rather longer than 6th,
6th oblong, a little widened distally, palm nearly transverse. Gnathopod 2
much larger, 5th joint broad, cup-shaped, 6th large, subquadrate, a little
widened distally, palm transverse, sinuous, defined by a short strong tooth,
not quite reached by the finger-point. Peraeopods 1 and 2 not very strong.
Peraeopods 3—5 successively longer, robust, spinose, 2^d joint well expanded,
but narrowing somewhat downward. Uropods 1—3 stout, in 1st and 2^d outer
ramus rather shorter than inner, 1st not reaching so far back as 2^d, 2^d little
beyond peduncle of uropod 3. Uropod 3, peduncle short. rami forming
long narrow laminae, in ♂ of quite exceptional size, slightly widening to
rounded apices, in ♀ narrowly elliptic, inner ramus in ♂ scarcely, in ♀ considerably longer and somewhat broader than the outer. Telson cleft nearly
to base, breadth at base equal to $^2/_{:;}$ of length, lobes conical, divergent.
Colour orange-red. L. ♀ 35—38, ♂ 46 mm.

South-Atlantic (South Georgia). Low-tide.

32. Gen. **Ceradocus** A. Costa

1853 Ceradocus (Sp. un.: C. orchestiipes), A. Costa in: Rend. Soc. Borbon., n. ser.
v. 2 p. 170 | 1893 C. (part.), A. Della Valle in: F. Fl. Neapel, v. 20 p. 718 | 1862 Mega-
moera (part.), Bate (& Westwood), Brit. sess. Crust., v. 1 p. 400 | 1862 Megamaera, Bate,
Cat. Amphip. Brit. Mus., t. 39.

Head without conspicuous rostrum. Side-plates shallow, 4th not deeper
than 5th, nor excavate behind. Antenna 1 the longer and stouter, accessory
flagellum well developed. Antenna 2, flagellum short. Upper lip with rounded
margin. Lower lip with inner lobes. Mandible, 3^d joint of palp straight,
much shorter than 2^d. Maxilla 1, inner plate large, subacute above, margin
fringed with numerous setae, outer plate with 9 or more spines. Maxilla 2,
inner plate fringed along inner margin. Maxillipeds, gnathopods 1 and 2,
peraeopods 1—5, uropods 1 and 2 and telson as in Maera (p. 433); uropod 3
with rami greatly developed.

4 species.

Synopsis of species:

1 { Some segments dorsally dentate . . . . 1. C. rubromaculatus . p. 430
  { No segments dorsally dentate — **2.**

2 { Side-plates 1—4 with tooth at lower hind corner 2. C. semiserratus . . p. 431
  { Side-plates 1—4 without tooth at lower hind
  { corner — **3.**

3 { Pleon segment 3, postero-lateral corners denti-
  { culate . . . . 3. C. orchestiipes . . . p. 432
  { Pleon segment 3, postero-lateral corners forming
  { a simple hook . . . . 4. C. torelli . . . . . p. 432

1. **C. rubromaculatus** (Stimps.) 1855 Gammarus rubro-maculatus, Stimpson
in: P. Ac. Philad., v. 7 p. 394 | 1885 Moera rubromaculata, Haswell in: P. Linn. Soc.
N. S. Wales, v. 10 p. 105 t. 15 f. 5—12 | 1888 Maera r., T. Stebbing in: Rep. Voy.

Challenger, v. 29 p. 1008 t. 95, 96 | 1893 *Ceradocus rubromaculatus*, A. Della·Valle in: F. Fl. Neapel, v. 20 p. 720 | 1899 *C. r.*, T. Stebbing in: Tr. Linn. Soc. London, ser. 2 v. 7 p. 426 | 1862 *Megamoera serrata*, Bate, Cat. Amphip. Brit. Mus., p. 226 t. 39 f. 5 | 1879 *Melita? ramsayi + Moera rubro-maculata + M. spinosa + M. ramsayi*, HasWell in: P. Linn. Soc. N. S. Wales, v. 4 p. 246 t. 10 f. 1; p. 267 t. 10 f. 4; p. 268 t. 10 f. 5; p. 334 | 1885 *Moera festiva*, Chilton in: P. Linn. Soc. N. S. Wales, v. 9 p. 1037 t. 46 f. 2.

Pleon segments 1—6 dorsally serrate, with setules in the serrations, segment 1 with 15 teeth, 2^d with 17, 3^d with 15, 4th with 9, 5th with 7, 6th with 3, central most prominent, especially in segments 3 and 4, but this ornamentation is very variable. Head, lateral lobes nasiform, post-antennal corners acute. Side-plates 1—4 shallow, not deeper than 5th, 1st a little outdrawn, acute, 2^d—4th subquadrate· with rounded angles. Pleon segments 1—3, postero-lateral corners acute, in segments 1 and 2 and sometimes in 3 lower margin a little serrate, and in segment 3 hind margin of the corner serrate. Eyes dark, oval, close to margin of lateral lobes of head. Antenna 1 half as long as body, 2^d joint more slender than 1st, but as long or a little longer, 3^d about $^1/_5$ as long as 2^d, flagellum shorter than peduncle, 10—33-jointed, accessory flagellum also variable, 4—12-jointed. Antenna 2, gland-cone nearly reaching end of long antepenultimate joint of peduncle, this and ultimate joint much shorter than penultimate, flagellum short, with 18 joints (to match 33 in flagellum of antenna 1). Upper lip with almost semicircular margin. Lower lip, principal lobes dehiscent, with little quadrate corner at top of inner margin. Mandible, 3^d joint of palp much shorter than 2^d, with long apical spines. Gnathopod 1, 5th joint longer than 6th, both spinose, 6th rather broadly ovate, palm oblique. slightly convex, finely denticulate, finger fitting palm. Gnathopod 2 in $\male$, 5th joint short, cup-shaped, much narrower than the massive 6th, which is distally widened, hind margin nearly straight, front very convex, palm defined by a strong tooth between palmar spines, very oblique, with 1 or 2 gaps, or with variable denticulation; finger strong, much or little curved. Peraeopods 1 and 2 slender, 1st rather longer than 2^d. Peraeopods 3—5 rather stout, 3^d much shorter than the following, 4th rather longer than 5th, 2^d joint narrowly oblong in peraeopod 3, broader in the others. Uropods 1 and 2 not or scarcely reaching beyond peduncle of uropod 3. Uropod 3, rami subequal, broad and long, strongly serrate and spined on both margins, apices narrow, not acute. Telson short, broader than long, cleft nearly to base, almost the full width between the acute apices, inner margins of the lobes sinuous. Colour, spotted with crimson above, white below. L. reaching 23 mm.

Australian and Tasmanian Waters, and off Cape Agulhas. Depth 9—282 m, and in sea-weed on sandy beach.

2. **C. semiserratus** (Bate) 1862 *Megamoera semiserrata*, Bate (& Westwood), Brit. sess. Crust., v. 1 p. 401 f. | 1862 *M. s.*, *Megamaera semmiserrata*, Bate, Cat. Amphip. Brit. Mus., p. 226; t. 39 f. 6 | 1869 *Maera semiserrata*, (A. M. Norman in:) G. S. Brady & D. Robertson in: Ann. nat. Hist., ser. 4 v. 3 p. 359 | 1889 *M. s.*, A. M. Norman in: Ann. nat. Hist., ser. 6 v. 4 p. 127 | 1899 *Ceradocus semiserratus*, T. Stebbing in: Tr. Linn. Soc. London, ser. 2 v. 7 p. 426 | 1893 *C. fasciatus* (part.), A. Della Valle in: F. Fl. Neapel. v. 20 p. 721.

Body slender, smooth. Side-plates 1—4 each with a tooth at the lower hind corner. Pleon segment 3, postero-lateral corners rather produced, smooth below, but with the hind margin regularly serrate. Eyes narrowly reniform. Antenna 1 about half as long as body, 2^d joint rather longer than 1st, 3^d less than half as long as 1st, flagellum rather longer than peduncle

Antenna 2 reaching little beyond peduncle of antenna 1. Gnathopods 1 and 2 somewhat similar in shape, but 1st (in figure) has 5th joint longer than oblong oval 6th, both very hirsute; in the considerably larger gnathopod 2 the 5th joint is rather shorter than the 6th, which has the palm oblique, slightly denticulated, defined by short palmar spines. Peraeopods 1 and 2 very slender. Peraeopods 3—5 successively longer, 2^d joint not greatly expanded, serrate behind. Uropod 3 reaching far beyond the others, rami equal. Telson cleft, each lobe obliquely truncate. L. 6—7 mm.

North-Atlantic (British Isles, West of France).

3. C. orchestiipes A. Costa 1844 *Gammarus fasciatus* (non Say 1818!) + Var. *G. f. corallinus* (nom. nud.) + Var. *G. f. violaceus* (nom. nud.), O. G. Costa in: Atti Acc. Borbon., *v.* 5 п p. 73 t. 1 f. 3 | 1893 *Ceradocus f.* (part.), A. Della Valle in: F. Fl. Neapel, *v.* 20 p. 721 t. 6 f. 1; t. 21 f. 17—33 | 1853 *C. orchestiipes*, A. Costa in: Rend. Soc. Borbon., n. ser. *v.* 2 p. 177 | 1899 *C. o.*, T. Stebbing in: Tr. Linn. Soc. London, ser. 2 *v.* 7 p. 426 | 1866 *Maera o.*, Cam. Heller in: Denk. Ak. Wien, *v.* 26 п p. 38 t. 3 f. 22, 23 | 1862 *Melita orchestipes*, Bate. Cat. Amphip. Brit. Mus., p. 187 | 1864 *Megamoera o.*, E. Grube, Lussin, p. 73.

Body dorsally smooth, rounded, but pleon segments 1—6 (variably: all to none) medio-dorsally produced into a tooth. Head without rostrum, lateral lobes narrowly produced. Side-plates 1—4 shallow, subquadrate, 1st produced forward in an acute recurved point. Pleon segment 3, postero-lateral corners acutely outdrawn, with irregular and variable denticulation. Eyes small, oval, black (Heller), circular, yellow-brown (Della Valle). Antenna 1 about half as long as body, 2^d joint thinner than 1st, as long or longer, 3^d short, flagellum sometimes longer than peduncle, 40-jointed, accessory flagellum 7—10-jointed. Antenna 2 much shorter, gland-cone as long as the unusually long antepenultimate joint of peduncle, ultimate shorter than penultimate, flagellum longer than ultimate joint of peduncle and sometimes more than 20-jointed. Gnathopod 1, 2^d joint curved, 5th as broad as 6th and rather longer, 6th oval, hind margin more convex than front, palm ill-defined, finger stout. Gnathopod 2 much larger, especially in ♂. 5th joint small, 6th widening to the very oblique palm, which is defined by a strong tooth, and is regularly denticulate in ♀, irregularly in ♂. In ♂ the hands of gnathopod 2 often unlike in size and shape. Peraeopods 1 and 2 slender. Peraeopods 3—5 successively longer, moderately robust. 2^d joint oblong, not greatly expanded, hind margin smooth, distally produced into a small acuminate process. Uropods 1 and 2 reaching equally far back. Uropod 3 reaching much beyond the others, rami very long, equal, laminar, fringed with spinules, apices narrowly truncate. Telson cleft nearly to base, lobes dehiscent, with 2 spines between acute outer apex and inner angle. Colour, banded pale yellow and crimson, appendages variegated with the same hues. L. 15 mm.

Mediterranean. Depth 17—70 m.

4. C. torelli (Goës) 1866 *Gammarus t.*, Goës in: Öfv. Ak. Förb., *v.* 22 p. 530 t. 40 f. 28 | 1867 *Megamoera t.*, Bate in: Zool. Rec., *v.* 3 p. 232 | 1871 *Maera t.*, A. Boeck in: Forh. Selsk. Christian., 1870 p. 208 | 1876 *M. t.*, A. Boeck, Skand. Arkt. Amphip., *v.* 2 p. 380 | 1893 *Ceradocus t.*, A. Della Valle in: F. Fl. Neapel, *v.* 20 p. 723 | 1899 *C. t.*, T. Stebbing in: Tr. Linn. Soc. London, ser. 2 *v.* 7 p. 426.

Body elongate, smooth, not very compressed. Head, lateral lobes rounded, little prominent. Side-plates 1, 2 and 5 larger than 3^d or 4th, none very deep. Pleon segments 1—3, postero-lateral corners forming a pointed hook. Eyes unknown. Antenna 1 about half as long as body, peduncle very long, 2^d joint longer than 1st, 3^d scarcely $1/_3$ as long as 2^d, flagellum shorter than peduncle, with 35 short joints, accessory flagellum 5-jointed. Antenna 2 much shorter, slender, gland-cone long, ultimate joint of peduncle shorter than penultimate, flagellum subequal to ultimate joint of peduncle, 16-jointed. Gnathopod 1, 5th joint rather smaller than 6th, both hirsute. Gnathopod 2 much larger, 5th joint short, cup-shaped, 6th very large, oblong, widening a little to the oblique, well defined palm, which is much shorter than the hind margin, finger serrate within. Peraeopods 1 and 2 small. Peraeopods 3—5 rather robust, peraeopod 3 very short, 5th long; 2^d joint not greatly expanded, hind margin serrate, forming below an acute corner. Uropods 1—3 as in C. orchestiipes. Telson deeply cleft, lobes divergent, with spinules on outer margin and in the emarginate apices. L. over 50 mm.

Arctic Ocean (Iceland).

33. Gen. **Maera** Leach

1813/14 *Maera* (Sp. un.: *M. grossimana*), Leach in: Edinb. Enc., *v.* 7 p. 403, 432 | 1893 *M.*, A. Della Valle in: F. Fl. Neapel, *v.* 20 p. 724 | 1894 *M.*, G. O. Sars, Crust. Norway, *v.* 1 p. 517 | 1843 *Moera*, L. Agassiz, Nomencl. zool., Crust. p 18 | 1853 *Leptothoe* (Sp. un.: *L. danae*), Stimpson in: Smithson. Contr., *v.* 6 nr. 5 p. 46 | 1862 *Megamoera* (part.), Bate (& Westwood), Brit. sess. Crust., *v.* 1 p. 400.

Body more or less slender. Head without conspicuous rostrum. Side-plates shallow, 4th scarcely or not emarginate behind. Antenna 1 the longer, 2^d joint usually longer than 1st, accessory flagellum well developed. Antenna 2, flagellum short. Upper lip symmetric. Lower lip with inner lobes. Mandible with slender palp, 3^d joint straight. Maxilla 1, inner plate with few setae, outer with (probably) 10 spines. Maxilla 2, inner plate not fringed with setae on inner margin. Maxillipeds normal, but the 3 spine-teeth on apex of inner plate are not certainly present. Gnathopods 1 and 2 subchelate, gnatho-pod 2 usually much the larger in the ♂. Peraeopods slender, peraeopods 4 and 5 longer than the rest, 2^d joint of peraeopods 3—5 variable, sometimes very slender. Uropod 3 often reaching much beyond the others, rami equal or not very unequal, 1-jointed. Telson rather small, deeply cleft.

12 accepted and 8 doubtful species.

Synopsis of accepted species:

$1 \begin{cases} \text{Uropod 3 scarcely or not at all extending} \\ \quad \text{beyond uropod 1 — 2.} \\ \text{Uropod 3 extending much beyond uropod 1 — 5.} \end{cases}$

$2 \begin{cases} \text{Without eyes . . . .} & \text{1. M. tenera . . . . . p. 434} \\ \text{With eyes — 3.} \end{cases}$

$3 \begin{cases} \text{Uropod 3, rami not apically obtuse . . . . .} & \text{2. M. quadrimana . . p. 434} \\ \text{Uropod 3, rami apically obtuse — 4.} \end{cases}$

$4 \begin{cases} \text{Peraeopods 4 and 5, 2^d joint not produced} \\ \quad \text{downward . . . .} & \text{3. M. grossimana . . p. 435} \\ \text{Peraeopods 4 and 5, 2^d joint produced downward} & \text{4. M. inaequipes . . . p. 435} \end{cases}$

$5 \begin{cases} \text{Some segments dorsally dentate — 6.} \\ \text{No segments dorsally dentate — 7.} \end{cases}$

$6 \begin{cases} \text{Pleon segment 1 without dorsal teeth . . . .} & \text{5. M. tenuimana . . . p. 436} \\ \text{Pleon segment 1 with dorsal teeth . . . . .} & \text{6. M. westwoodi . . . p. 436} \end{cases}$

$7 \begin{cases} \text{Body pubescent . . . .} & \text{7. M. furcicornis . . . p. 437} \\ \text{Body not pubescent — 8.} \end{cases}$

8 { Body extremely slender — **9.**
{ Body not extremely slender — **11.**

9 { Gnathopod 2, palm excaVate near defining angle   8. **M. hamigera** . . . . p. 437
{ Gnathopod 2, palm not excaVate — **10.**

10 { Peraeopod 5 shorter than peraeopod 4 . . .   9. **M. tenella** . . . . . p. 438
{ Peraeopod 5 not shcrter than peraeopod 4 .   10. **M. lovéni** . . . . . p. 438

11 { Pleon segment 3, postero-lateral corners serrate
{   aboVe and below, acutely produced . . .   11. **M. othonis** . . . . . p. 438
{ Pleon segment 3, postero-lateral corners serrate
{   only aboVe, quadrate . . . .   12. **M. mastersii** . . . . p. 439

1. **M. tenera** O. Sars 1876 *Moera tenella, Maera t.* (non *Gammarus tenellus* J. D. Dana 1852!), G. O. Sars in: Arch. Naturv. Kristian., *v.* 2 p. 259, 271 | 1885 *Maera tenera,*[1] G. O. Sars in: Norske Nordhavs-Exp., *v.* 6 Crust. I p. 177 t. 14 f. 7 | 1893 *M. t.,* A. Della Valle in: F. Fl. Neapel, *v.* 20 p. 724.

Body very slender, almost cylindric, smooth. Head, lateral lobes rounded, little prominent. Side-plates very shallow, 1st the deepest, rounded in front, produced a little forward. Pleon segment 3, postero-lateral corners quadrate. Eyes entirely wanting. Antenna 1 elongate, 1st and 2d joints subequal, 3d about $\frac{1}{3}$ as long as 2d, flagellum shorter than peduncle, 16-jointed, accessory flagellum 4-jointed. Antenna 2 of $\frac{2}{3}$ length of antenna 1, ultimate joint of peduncle shorter than penultimate, flagellum shorter than peduncle, 8-jointed. Gnathopod 1 small, 6th joint not larger than 5th. Gnathopod 2 powerful, 5th joint (in figure) broad and rather longer than broad, 6th very large, almost quadrate, twice as large as 5th, setose, finger not very large. Peraeopods 1—5 very slender, with scanty armature, 3d—5th with linear 2d joint, peraeopods 4 and 5 elongate. Uropods 1—3, rami slender, lanceolate. Uropod 3 reaching little beyond the others. Telson very small, deeply incised. L. 10 mm.

North-Atlantic (Storeggen [Norway]). Cold area, depth 785 m.

2. **M. quadrimana** (Dana) 1853 & 55 *Gammarus quadrimanus,* J. D. Dana in: U. S. expl. Exp., *v.* 13 ɪɪ p. 955 t. 65 f. 9 | 1862 *Moera q.,* Bate, Cat. Amphip. Brit. Mus., p. 194 t. 35 f. 5 | ? 1882 *M. q.,* G. M. Thomson in: Tr. N. Zealand Inst., *v.* 14 p. 235 t. 17 f. 4 a, 4 b | 1893 *Maera truncatipes* (part.), A. Della Valle in: F. Fl. Neapel, *v.* 20 p. 725.

Body slender, smooth. Side-plates 1—4 shallow, subquadrate. Pleon segment 3, postero-lateral corners said to be dentate (Della Valle). Eyes (in figure) small, rounded. Antenna 1, 1st and 2d joints long, subequal, 3d very short, flagellum rather shorter than peduncle, accessory flagellum rather longer than half the primary. Antenna 2, peduncle shorter than in antenna 1, flagellum very short. Gnathopod 1 small, 6th joint oblong, both 5th and 6th widened distally. Gnathopod 2, 5th joint small, cup-shaped, 6th very large, widened to the palm, which forms a somewhat obtuse angle with the hind margin, is divided into 3 blunt teeth, and defined by a long, acute tooth, separated from the others by a cavity (Thomson: front margin longer than the hind margin; 2 small teeth in place of the long defining tooth). Feraeopods 1 and 2 very slender. Peraeopods 3—5, 2d joint moderately expanded and produced downward, 4th joint rather broad, produced, finger sharply produced behind the short nail. Peraeopod 5 a little shorter or longer than peraeopod 4. Uropods 1—3 (in figure) reaching about equally far back. Uropod 3, rami straight, equal, in figure subacute, tipped with slender setae. Uropod 2 with very long rami (Della Valle). Telson undescribed. Colour uniform yellowish white or dirty green. L. 12—16 mm.

Tropical and South-Pacific (Fiji Islands, coral reefs; Paterson Inlet [New Zealand]).

3. **M. grossimana** (Mont.) 1808 *Cancer (Gammarus) grossimanus*, Montagu in: Tr. Linn. Soc. London, *v.*9 p. 97 t. 4 f. 5 | 1812 *Astacus g.*, Pennant, Brit. Zool.. ed. 5 *v.*4 p. 33 | 1813/14 *Maera grossimana*, Leach in: Edinb. Enc., *v.*7 p.403. 432 | 1889 *M. g.*, A. M. Norman in: Ann. nat. Hist., ser. 6 *v.*4 p. 126 | 1893 *M. g.* (part.), A. Della Valle in: F. Fl. Neapel, *v.*20 p.727 t.2 f.10; t.21 f.1—16; t.41 f.37 | 1818 *Gammarus grossimanus*, Lamarck, Hist. An. s. Vert., *v.*5 p.182 | 1862 *Moera g.*, Bate, Cat. Amphip. Brit. Mus., p. 188 t. 34 f. 3 | 1888 *M. grossimana*, T. Barrois, Cat. Crust. Açores, p. 38 f. | 1830 *Gammarus impostii*, H. Milne Edwards in: Ann. Sci. nat., *v.*20 p. 368 | 1866 *Maera donatoi*, Cam. Heller in: Denk. Ak. Wien, *v.*26 II p. 41 t. 3 f. 26.

Body slender, smooth. Head, rostrum minute. Side-plate 1 acutely produced forward, 2^d—4th quadrate. Pleon segment 3, postero-lateral corners acutely produced, not serrate. Eyes narrow, sometimes reniform, reddish brown. Antenna 1, peduncle elongate, 2^d joint longer, sometimes much longer than 1st, flagellum shorter than peduncle, 22-jointed, accessory flagellum 8-jointed. Antenna 2 much shorter, gland-cone long, but not reaching end of antepenultimate joint of peduncle, penultimate slightly curved, longer than ultimate, flagellum short, 8- or 9-jointed. Mandibular palp slender, 1st joint slightly produced, 3^d joint straight, shorter than 2^d. Maxilla 1, inner plate with 3 setae at apex. Maxillipeds, inner plates with a spine tooth subapical on inner margin, others perhaps on apex, 4th joint of palp with spine-like nail. Gnathopod 1 slender, 5th joint rather longer than 6th, front margin distally acute, 6th joint widened a little to the oblique convex palm. Gnathopod 2, 4th joint with acute hind margin, 5th short, cup-shaped, 6th large, longer than broad, palm defined by a tooth, with its margin irregularly notched in ♂, regularly serrate or crenulate and convex in ♀, finger matching palm, large, curved, outer margin fringed with setules. Peraeopods 1 and 2 slender. Peraeopods 3—5, 2^d joint little expanded, narrowing downward, slightly produced in peraeopod 3, but forming no free hind corner in peraeopods 4 and 5, finger with a short sharp nail on its blunt apex. Uropod 3 not or not conspicuously reaching beyond uropod 1, the rami equal, rather longer than peduncle, apices squarely truncate, tipped with setae. Telson shorter than peduncle of uropod 3, cleft about $^4/_5$ of length, lobes very divergent, each having a tridentate apex, middle tooth longest, outer minute, with intervening spinules. Colour transparent yellowish, with rosy tinting. L. 6—9 mm.

North-Atlantic with English Channel (Devon, Cornwall, Channel Islands; Azores; West France); Mediterranean.

4. **M. inaequipes** (A. Costa) 1847 *Amphithoe truncatipes* (nom. nud.), (Spinola in MS.) A. White, Crust. Brit. Mus., p. 87 | 1893 *Maera t.* (part.), A. Della Valle in: F. Fl. Neapel, *v.* 20 p. 725 t. 1 f. 2; t. 22 f. 26—40 | 1851 *Amphithoe inaequipes*, (A. Costa in:) F. W. Hope, Cat. Crost. Ital., p. 45 | 1857 *Gammarus scissimanus*, A. Costa in: Mem. Acc. Napoli, *v.*1 p. 221 t. 3 f. 7 | 1866 *Maera scissimana + M. integrimana*, Cam. Heller in: Denk. Ak. Wien, *v.*26 II p. 40 t. 3 f. 24; p. 40 t. 3 f. 25 | 1888 *Moera s.*, F. Barrois in: Bull. Soc. zool. France, *v.*13 p. 58 | 1888 *M. s.*, F. Barrois, Cat. Crust. Açores, p. 35 textf. | 1862 *M. truncatipes + M. blanchardi*, Bate, Cat. Amphip. Brit. Mus., p. 189 t. 34 f. 4; p. 190 t. 34 f. 5.

Body rather robust, smooth. Head, lateral lobes rounded. Side-plate 1, front corner a little produced, acute, 2^d squarely rounded, small. Pleon

28*

segment 3, postero-lateral corners not serrate. Eyes small, round, dark. Antenna 1 slender, 2^d joint longer than 1st, 3^d rather short, flagellum rather shorter than peduncle, 15—26-jointed, accessory flagellum 7—11-jointed. Antenna 2 rather longer than peduncle of antenna 1, gland-cone reaching beyond antepenultimate joint of peduncle, ultimate scarcely shorter than penultimate, flagellum subequal to ultimate, 5—10-jointed. Gnathopod 1, 5th joint as long as 6th, 6th subovate, wider than 5th at junction of hind margin with convex, ill-defined palm. Gnathopod 2 much larger, 5th joint small, cup-shaped, 6th broad, a little longer than broad, palm almost transverse, defined by an acute tooth, margin in ♀ convex, serrulate, in ♂ variable but usually with a deep central notch or cavity; finger powerful. Peraeopods 1—5 with tricuspidate finger, that is, having a nail projecting between 2 points. Feraeopods 3—5, 2^d joint moderately expanded, hind margin somewhat produced downward in a rounded lobe, peraeopods 4 and 5 equal. Uropods 1—3 reaching about equally far back. Uropod 3, rami not very long, apically obtuse, tipped with setae. Telson deeply cleft, each lobe bidentate, a spine between the short outer and long inner tooth. Colour, dorsally green bronzed with a little red, gnathopods 1 and 2 tinged with green, other appendages pellucid, pinkish. L. 7 mm.

Mediterranean; North-Atlantic (Azores).

5. **M. tenuimana** (Bate) 1862 *Gammarus tenuimanus* (♀), Bate (& Westwood), Brit. sess. Crust., v. 1 p. 384 f. | 1862 *G. t.*, Bate, Cat. Amphip. Brit. Mus., p. 214 t. 38 f. 2 | 1895 *G. t.*, A. O. Walker in: Ann. nat. Hist., ser. 6 v. 15 p. 471 | 1868 *Maera batei* (♂), A. M. Norman in: Ann. nat. Hist., ser. 4 v. 2 p. 416 t. 22 f. 1—3 | 1889 *M. b.*, A. M. Norman in: Ann. nat. Hist., ser. 6 v. 4 p. 127 | 1893 *M. b.*, A. Della Valle in: F. Fl. Neapel, v. 20 p. 726 | 1895 *Moera b.*, A. O. Walker in: P. Liverp. biol. Soc., v. 9 p. 308 ! 1868 *Megamoera multidentata*, (A. M. Norman in MS.) Bate & Westwood, Brit. sess. Crust., v. 2 p. 515 f.

Pleon segments 2—6 dorsally dentate, 2^d with 3, 3^d with 5, 4th—6th each with 2 teeth, a spinule at inner side of base of each tooth. Head, lateral lobes somewhat produced, rounded. Side-plates 1—4 shallow, 2^d seemingly somewhat large in ♂. Pleon segments 2 and 3, postero-lateral corners acute, not serrate. Eyes ovate, dark. Antenna 1, 2^d joint considerably longer than 1st, 3^d short, flagellum about as long as peduncle, 22—24-jointed, accessory flagellum 4- or 5-jointed. Antenna 2 subequal to peduncle of antenna 1, ultimate and penultimate joints of peduncle subequal, flagellum not longer than ultimate, 8-jointed. Gnathopod 1 slender, 5th and 6th joints parallelsided, 6th the shorter, palm slightly oblique, short. Gnathopod 2 in ♀ very like gnathopod 1, but 6th joint ovate, palm undefined, finger small; in ♂, 5th joint small, cup-shaped, 6th large, subrectangular, palm $^1/_8$ its length, with 3 tubercles, the one which defines the palm being flat-topped and setose, finger strongly curved, leaving a gap when closed. Peraeopods 1 and 2 slender. Peraeopods 3—5, 2^d joint not much expanded, in peraeopods 4 and 5 not forming a free corner below, finger strong. Uropod 1 much longer than uropod 2, but scarcely reaching beyond peduncle of uropod 3. Uropod 3, peduncle long, very stout, rami subequal, very long, laminar, fringed with spines. Telson with 1 spine at apex of each lobe. L. 6—9 (15?) mm.

English Channel (Guernsey); Menai Strait and Liverpool Bay, depth 19—38 m; Moray Firth, at mouth of rivers.

6. **M. westwoodi** Stebb. 1855 *Gammarus kroyeri* (non H. Rathke 1843!), T. Bell in: Belcher, Last arct. Voy., v. 2 p. 405 t. 34 f. 4 | 1862 *Megamoera kröyeri*, Bate, Cat. Amphip. Brit. Mus., p. 229 t. 40 f. 4 | 1893 *Maera? k.*, A. Della Valle in: F. Fl.

Neapel, v. 20 p. 730 | 1871 *Melita dentata* (err., non *Gammarus dentatus* Krøyer 1842!), A. Boeck in: Forb. Selsk. Christian., 1870 p. 211 | 1888 *M. d.*, T. Stebbing in: Rep. Voy. Challenger, v. 29 p. 281, 1710 | 1899 *Maera westwoodi*, T. Stebbing in: Tr. Linn. Soc. London, ser. 2 v. 7 p. 426.

Little removed from M. tenuimana, but: pleon segment 1 dorsally dentate, accessory flagellum of antenna 1 extremely small, antenna 2 with peduncle as long as peduncle of antenna 1, and gnathopod 2 (Bate) with 6th joint ovate, palm oblique, straight (sinuous in figure.) L. 22 mm.

Wellington Channel [arctic Canada]. Depth 66 m.

7. **M. furcicornis** (Dana) 1852 *Gammarus f.*, J. D. Dana in: P. Amer. Ac., v. 2 p. 211 | 1853 & 55 *G. f.*, J. D. Dana in: U. S. expl. Exp., v. 13 II p. 951 t. 65 f. 6 | 1862 *Moera f.*, Bate. Cat. Amphip. Brit. Mus., p. 193 t. 35 f. 2 | 1888 *Maera f.*, T. Stebbing in: Rep. Voy. Challenger, v. 29 p. 1709 | 1893 *M.? f.*, A. Della Valle in: F. Fl. Neapel, v. 20 p. 730.

Body rather slender, sparsely pubescent. Head, lateral lobes not prominent. Pleon segments 2 and 3, postero-lateral corners (in figure) not produced nor acute. Eyes (in figure) small, round. Antenna 1, 1st joint stout, 2^{d} slender, a little longer, flagellum rather longer than peduncle. about 14-jointed, accessory flagellum about half primary, 5-jointed. Antenna 2 much shorter, but with peduncle nearly equal to that of antenna 1, its ultimate joint rather shorter than penultimate, flagellum subequal to penultimate, 7—10-jointed. Gnathopod 1 small, 5th and 6th joints subequal, 6th joint subovate. Gnathopod 2, 5th joint short, cup-shaped. 6th broad, oblong, a little widening to the truncate, or very slightly oblique, and not excavate palm; finger half as long as 6th joint. Peraeopods 3—5 rather long, 5th the longest, 2^{d} joint in all narrow oblong, hind margin serrulate. Uropods 1 and 2 reaching about equally far, uropod 3 much farther, elongate. L. 6 mm.

Sooloo Sea (shores of island off harbour of Soung).

8. **M. hamigera** (Hasw.) 1879 *Moera h.*, Haswell in: P. Linn. Soc. N. S. Wales, v. 4 p. 333 t. 21 f. 1 | 1888 *Maera h.*, T. Stebbing in: Rep. Voy. Challenger, v. 29 p. 1709 | 1893 *M. h.*, A. Della Valle in: F. Fl. Neapel, v. 20 p. 723 | 1885 *Megamoera suensis var.*, Haswell in: P. Linn. Soc. N. S. Wales, v. 10 p. 103 t. 15 f. 1—4.

Body slender, smooth. Head, lateral lobes not produced. Side-plates 1—4 not deep, 2^{d} the largest. Pleon segment 3, postero-lateral corners apparently quadrate with a few teeth(?) on their posterior margin. Eyes long, narrow. Antenna 1 elongate, 2^{d} joint longer than 1st; flagellum about as long as peduncle, 33-jointed; accessory flagellum 4—6-jointed. Antenna 2 about as long as peduncle of antenna 1, ultimate joint of peduncle shorter than penultimate. flagellum 13-jointed. Gnathopod 1 small, 6th joint ovate, palm not defined. Gnathopod 2, 5th joint short, cup-shaped, 6th more than twice as long as broad, widening a little to the palm, which is little oblique, defined by an acute tooth separated by a cavity from 2 or 3 strong tubercles; finger strongly curved, short and thick; left gnathopod 2 smaller than right; larger hand with an acute tooth and 4 or 5 denticles, the smaller without denticles. Peraeopods 1 and 2 short, slender. Peraeopod 3 longer, not very stout. Peraeopods 4 and 5 long and stout. Peraeopods 3—5, 2^{d} joint not much expanded. Uropods 1 and 2 reaching equally far, not to end of peduncle of uropod 3. Uropod 3 very large, peduncle long, rami more than twice as long as peduncle, ovate-lanceolate, with serrate margins. Halves of telson-long and narrow, with a deep terminal notch. L. 12·5 mm.

Port Jackson and Port Stephens [East-Australia].

9. **M. tenella** (Dana) 1852 *Gammarus tenellus*, J. D. Dana in: P. Amer. Ac., *v.* 2 p. 212 | 1853 & 55 *G. t.*, J. D. Dana in: U. S. expl. Exp., *v.* 13 II p. 952 t. 65 f. 7 | 1862 *Moera tenella*, Bate, Cat. Amphip. Brit. Mus., p. 193 t. 35 f. 3 | 1888 *Maera t.*, T. Stebbing in: Rep. Voy. Challenger, *v.* 29 p. 1710 | 1893 *M. grossimana* (part.), A. Della Valle in: F. Fl. Neapel, *v.* 20 p. 727.

Body slender, smooth. Side-plates shallow. Pleon segment 3, posterolateral corners seemingly a little produced but scarcely acute. Eyes small, round. Antenna 1, 1st joint not stout, 2d very long, flagellum little longer than peduncle, accessory flagellum half as long as primary, 7- or 8-jointed. Antenna 2 very slender, short, flagellum as long as ultimate joint of peduncle. Gnathopod 1 small, 6th joint subovate. Gnathopod 2 stout, 5th joint small, narrow, cup-shaped, 6th broad, much longer than broad, widening a little distally; palm nearly straight truncate, not excavate, defined by a small acute tooth. Peraeopods 1 and 2 remarkably slender. Peraeopods 3—5, 2d joint narrowing downward; peraeopod 5 rather shorter than 4th, finger acutely produced behind the nail. Uropod 3 quite long, extending much beyond uropod 2. Telson oblong, apical setae as long as the telson. L. 8 mm.

Tropical Pacific (coral reefs of Fiji Island Viti Levu).

10. **M. lovéni** (Bruz.) 1859 *Gammarus l.*, R. M. Bruzelius in: Svenska Ak. Handl., n. ser. *v.* 3 nr. 1 p. 59 t. 2 f. 9 | 1862 *Moera loveni*, Bate, Cat. Amphip. Brit. Mus., p. 193 t. 35 f. 1 | 1868 *Maera lovéni*, A. M. Norman in: Ann. nat. Hist., ser. 4 *v.* 2 p. 416 t. 21 f. 11, 12 | 1893 *M. loveni*, A. Della Valle in: F. Fl. Neapel, *v.* 20 p. 729 | 1894 *M. lovéni*, G. O. Sars, Crust. Norway, *v.* 1 p. 519 t. 182 f. 2.

Body long. very slender, smooth. Head, lateral lobes narrowly rounded. Side-plates small, 1st the largest, produced forward in a linguiform lobe, 4th twice as broad as deep, scarcely larger than 5th. Pleon segment 3, posterolateral corners almost quadrate, with a little acute point. Eyes small, round, inconspicuous in spirit. Antenna 1 long, slender, 1st joint long, 2d longer, flagellum shorter than peduncle, 17—30-jointed, accessory flagellum 5—7-jointed. Antenna 2 about half as long as antenna 1, ultimate joint of peduncle shorter than penultimate, flagellum not longer than ultimate, 7—12-jointed. Mandible, 3d joint of palp shorter than 2d. Maxilla 1, inner plate with 3 or 4 setae at apex. Gnathopod 1, 5th and 6th joints subequal, setose on both margins, 6th subovate, widest at the convex, ill-defined palm, finger with setae on outer margin and cilia on the inner. Gnathopod 2 rather powerful, 5th joint triangular, 6th large, oblong quadrangular. widest at the serrate, nearly transverse palm, defined by a small tooth, finger armed as in gnathopod 1. Peraeopods 1 and 2 very slender. Peraeopods 3—5, 2d joint little expanded, narrowing downward; peraeopods 4 and 5 elongate. Uropod 3 elongate, reaching much beyond the others, outer ramus about 2½ times as long as peduncle, inner nearly as long as outer. Telson rather shorter than peduncle of uropod 3, cleft nearly to base, lobes conical, rather divergent, with spine between the shorter inner and longer outer tooth. L. ♂ reaching 26 mm.

Arctic Ocean, North-Atlantic, North-Sea, Skagerrak and Kattegat (Disco [West-Greenland], depth 301 m; Spitzbergen, depth 87 m; Sweden, Norway, Denmark, Scotland).

11. **M. othonis** (M.-E.) 1830 *Gammarus o.*, H. Milne Edwards in: Ann. Sci. nat., *v.* 20 p. 368, 373 t. 10 f. 11—13 | 1889 *Maera o.*, A. M. Norman in: Ann. nat. Hist., ser. 6 *v.* 4 p. 125 | 1893 *M. o.*, A. Della Valle in: F. Fl. Neapel, *v.* 20 p. 729 t. 60 f. 8 | 1894 *M. o.*, G. O. Sars, Crust. Norway, *v.* 1 p. 518 t. 182 f. 1 | 1895 *M. o.*, A. O. Walker in: P. Liverp. biol. Soc., *v.* 9 p. 308 | 1847 *Gammarus longimanus*, W. Thompson in: Ann. nat. Hist., *v.* 20 p. 242 | 1857 *G. l.*, Bate in: Ann. nat. Hist., ser. 2 *v.* 19 p. 145 | 1876 *Megamoera longimana*, A. Boeck, Skand. Arkt. Amphip., *v.* 2 p. 382 | 1847 *Gam-*

marus elongatus, (H. Frey &) R. Leuckart, Wirbell. Th., p. 160 | 1859 *G. laevis,* R. M. Bruzelius in: Svenska Ak. Handl., n. ser. *v.* 3 nr. 1 p. 60 t. 2 f. 10 | 1862 *Megamoera longimana* + *M. othonis,* Bate & Westwood, Brit. sess. Crust., *v.* 1 p. 403 f.; p. 405 f. | 1896 *Maera brooki,* T. Scott in: P. phys. Soc. Edinb., *v.* 13 p. 173 t. 5 f. 1—6.

Body moderately compressed, back smooth. Head, lateral lobes angularly produced. Side-plate 1 produced in front, obtusely acuminate, 2^d deeper, 4th quadrate, smaller than 3^d. Pleon segment 3, postero-lateral corners acutely produced and serrate above and below. Eyes oval reniform, dark. Antenna 1 slender, more than $^2/_3$ length of body, 2^d joint longer and narrower than 1st, 3^d short, flagellum longer than peduncle, 24—48-jointed, accessory flagellum 6- or 7-jointed. Antenna 2 about half as long, ultimate and penultimate joints of peduncle subequal, flagellum short, 12—24-jointed. Upper lip, apical margin faintly concave. Mandibular palp slender, 3^d joint as long as or longer than 2^d. Maxilla 1, inner plate narrow, with 3 setae at apex. Gnathopod 1 slender, setose. 6th joint about as long as 5th, oblong oval, palm very oblique, ill-defined, finger small. Gnathopod 2 in ♀ rather stronger than gnathopod 1, 5th joint distally widened and quadrate, 6th rather longer, narrowly oblong, palm oblique, not strongly defined. Gnathopod 2 in ♂ much stronger, 5th joint short, broadly cup-shaped, 6th elongate, oblong oval, palm minutely serrate and imperfectly defined or non-existent; finger half as long as 6th joint or longer than it. Peraeopods 1 and 2 slender and short. Peraeopods 3—5, 2^d joint well expanded, oval, forming a free hind corner, hind margin serrate. Uropod 3 reaching much beyond the others, rami narrowly lanceolate, subequal. Telson rather longer than peduncle of uropod 3, lobes narrowly conical, acute, divergent, with spinule in notch a little way up inner margin. Colour whitish, blotched with rose. L. 11—35 mm.

North-Atlantic, North-Sea and Skagerrak (Bohuslän; Norway; Heligoland; British Isles; West-France); Mediterranean (Marseilles).

12. **M. mastersii** (Hasw.) 1879 *Megamoera m.,* Haswell in: P. Linn. Soc. N. S. Wales, *v.* 4 p. 265 t. 11 f. 1 | 1899 *Maera m.,* T. Stebbing in: Tr. Linn. Soc. London, ser. 2 *v.* 7 p. 426 | 1884 *Megamoera thomsoni,* Miers in: Rep. Voy. Alert, p. 318 t. 34 f. B | 1893 *Ceradocus rubromaculatus* (part.), A. Della Valle in: F. Fl. Neapel, *v.* 20 p. 721.

Body slender, back smooth. Head, lateral lobes rather narrowly produced. Side-plates 1—4 rotundo-quadrate, not very deep but much deeper than the 3 following. Pleon segment 3, postero-lateral corners quadrate with their hind margin denticulate. Eyes narrowly reniform. Antenna 1, 2^d joint of peduncle little longer than 1st, 3^d short, flagellum as long or nearly as long as peduncle, accessory flagellum 4-jointed. Antenna 2, peduncle as long as that of antenna 1 or a little longer, flagellum subequal to ultimate joint of peduncle. Gnathopod 1, 5th and 6th joints slender, equal; palm straight, smooth, oblique, hind margin strongly convex, with 4 small teeth. Gnathopod 2 much larger, 4th joint produced into a small acute tooth, 6th oval, palm oblique, irregularly denticulate, defined by a small tooth. Peraeopods 3—5, 2^d joint well expanded, oval, with free corner, hind margin serrate, peraeopod 5 scarcely so long as 4th. Uropod 3 reaching much beyond the others, peduncle short and broad, rami equal, broadly lanceolate. Telson cleft to base, lobes bluntly conical, a little divergent, with 3 small teeth at the extremity and a notch armed with a single seta near the distal end of the inner border (Haswell), or with a row of spinules along the inner margin to the apex (Miers). Colour (in spirit) light yellowish brown. L. 11 mm.

Port Jackson [East-Australia]; Torres Strait (Albany Island, Prince of Wales' Channel and Thursday Island). Depth 7—17 m.

M. albida (Dana) 1852 *Gammarus albidus*, J. D. Dana in: P. Amer. Ac., *v.* 2 p. 210 | 1853 & 55 *G. a.*, J. D. Dana in: U. S. expl. Exp., *v.* 13 II p. 948 t. 65 f. 4 | 1862 *Megamoera albida*, Bate, `Cat. Amphip. Brit. Mus., p. 231 t. 40 f. 7 (inaccurate copy) | 1893 *Maera? a.*, A. Della Valle in: F. Fl. Neapel, *v.* 20 p. 730.

Uropods 1—3 all long; uropod 1 extends a little beyond the others.

Tropical Pacific (Samoan Island Tongatabu). In shallow waters of the lagoon, among sea weed.

M. aspera (Dana) 1852 *Gammarus asper*, J. D. Dana in: P. Amer. Ac., *v.* 2 p. 209 | 1853 & 55 *G. a.*, J. D. Dana in: U. S. expl. Exp., *v.* 13 II p. 945 t. 65 f. 2 | 1862 *Megamoera aspera*, Bate, Cat. Amphip. Brit. Mus., p. 230 t. 40 f. 5 | 1893 *Elasmopus? asper*, A. Della Valle in: F. Fl. Neapel, *v.* 20 p. 737 | 1899 *Maera aspera*, T. Stebbing in: Tr. Linn. Soc. London, ser. 2 *v.* 7 p. 426.

Pleon segments dorsally denticulate, denticulation irregular and not confined to the posterior margin. Mandibular palp small, slender, 1st joint short, 2d as long as 3d. Peraeopods 3—5, 2d joint oblong, hind margin serrate. L. 12 mm.

Sooloo Sea. Depth 12 m.

M. danae (Stimps.) 1853 *Leptothoe d.*, Stimpson in: Smithson. Contr., *v.* 6 nr. 5 p. 46 t. 3 f. 32 | 1862 *Moera d.*, Bate, Cat. Amphip. Brit. Mus., p. 190 t. 34 f. 6 | 1888 *Maera d.*, T. Stebbing in: Rep. Voy. Challenger, *v.* 29 p. 277 | 1893 *M. d.*, A. Della Valle in: F. Fl. Neapel. *v.* 20 p. 731.

Fundy Bay (Grand Manan). Laminarian zone, on sandy patches among weedy rocks.

M. fusca (Bate) 1864 *Moera f.*, Bate in: P. zool. Soc. London, p. 667 | 1888 *Maera f.*, T. Stebbing in: Rep. Voy. Challenger, *v.* 29 p. 345, 1709 | 1893 *M. f.*, A. Della Valle in: F. Fl. Neapel, *v.* 20 p. 731.

North-Pacific (Esquimalt Harbour [Vancouver Island]). Depth 19 m, from sponge.

M. indica (Dana) 1853 & 55 *Gammarus? indicus*, J. D. Dana in: U. S. expl. Exp., *v.* 13 II p. 961 t. 66 f. 4 | 1893 *G. i.*, A. Della Valle in: F. Fl. Neapel, *v.* 20 p. 454 | 1862 *Megamoera indica*, Bate, Cat. Amphip. Brit. Mus., p. 232 t. 40 f. 9 | 1899 *Maera i.*, T. Stebbing in: Tr. Linn. Soc. London, ser. 2 *v.* 7 p. 426.

Suggestive of a Sunamphitoe (p. 645); but see figure of uropod 3. L. 8 mm.

Balabac Passage [North of Borneo]. Shores of a small coral island.

M. kürgensis (Gerstf.) 1858 *Gammarus k.*, Gerstfeldt in: Mém. prés. Ac. St.-Pétersb., *v.* 8 p. 290 | 1888 *G. k.*, T. Stebbing in: Rep. Voy. Challenger, *v.* 29 p. 309 | 1893 *Amathilla? k.*, A. Della Valle in: F. Fl. Neapel, *v.* 20 p. 766.

L. 4—5 mm.

Arctic Ocean (pond on the Kürga [Siberia]).

M. massavensis (Kossm.) 1880 *Moera m.*, Kossmann, Reise Roth. Meer., *v.* 21 Malacost. p. 133 t. 14 f. 9—11 | 1888 *M. m.*, T. Stebbing in: Rep. Voy. Challenger, *v.* 29 p. 516 | 1893 *Maera truncatipes* (part.), A. Della Valle in: F. Fl. Neapel, *v.* 20-p. 725.

Probably identical with M. tenella (p. 438). L. 4 mm.

Red Sea.

M. pubescens (Dana) 1852 *Amphithoe p.*, J. D. Dana in: P. Amer. Ac., *v.* 2 p. 214 | 1853 & 55 *Gammarus? p.*, J. D. Dana in: U. S. expl. Exp., *v.* 13 II p. 960 t. 66 f. 3 | 1862 *Gammarella p.*, Bate, Cat. Amphip. Brit. Mus., p. 181 t. 33 f. 1 | 1893 *Elasmopus? p.*, A. Della Valle in: F. Fl. Neapel, *v.* 20 p. 737.

Tropical Pacific (Kingsmills). Coral reef.

34. Gen. **Elasmopus** A. Costa

1853 *Elasmopus* (Sp. un.: *E. rapax*), A. Costa in: Rend. Soc. Borbon., n. ser. *v.*2 p. 170, 175 | 1871 *E.*, A. Boeck in: Forh. Selsk. Christian., 1870 p. 212 | 1888 *E.*, T. Stebbing in: Rep. Voy. Challenger, *v.* 29 p. 1018 | 1893 *E.*, A. Della Valle in: F. Fl. Neapel, *v.* 20 p. 732 | 1894 *E.*, G. O. Sars, Crust. Norway, *v.* 1 p. 520 | 1862 *Megamoera* (part.), Bate (& Westwood), Brit. sess. Crust., *v.* 1 p. 400.

Head with lateral lobes rounded. Side-plates 1—4 usually well developed and the 4th excavate behind and deeper than 5th. Antenna 1 longer than antenna 2, 2d joint subequal to 1st, accessory flagellum seldom long. Antenna 2, flagellum short. Mouth-parts in general as in Maera (p. 433), but 3d joint of mandibular palp usually falcate, and pectinate with spinules. Maxilla 1, outer plate sometimes with only 7 spines. Gnathopods 1 and 2 as in Maera. Peraeopods 3—5 robust. Uropod 3 reaching little or not at all beyond the others, rami broad, not long. Telson rather small, deeply cleft.

9 accepted and 4 doubtful species.

Synopsis of accepted species:

1 { One or more of the segments dorsally dentate — **2**.
{ None of the segments dorsally dentate — **4**.

2 { Only one segment dorsally dentate . . . . . . 1. **E. subcarinatus** . p. 441
{ Four segments dorsally dentate — **3**.

3 { Gnathopod 2, palm very oblique . . . . 2. **E. diemenensis** . p. 442
{ Gnathopod 2, palm nearly transverse . . . . . . 3. **E. suensis** . . . . p. 442

4 { Gnathopod 2, 6th joint fringed with very long setae 4. **E. brasiliensis** . . p. 443
{ Gnathopod 2, 6th joint not fringed with very long
{ setae — **5**.

5 { Gnathopod 2, 6th joint with cup-like hollow to
{ receive finger-point . . . . 5. **E. pocillimanus** . p. 443
{ Gnathopod 2, 6th joint without cup-like hollow — **6**.

6 { Uropod 3, rami obtusely lanceolate . . . . . . 6. **E. delaplata** . . . p. 444
{ Uropod 3, rami truncate — **7**.

7 { Antenna 1, accessory flagellum very short . . . 7. **E. rapax** . . . . . p. 444
{ Antenna 1, accessory flagellum well developed — **8**.

8 { Gnathopod 2, palm transverse . . . . 8. **E. viridis** . . . . p. 445
{ Gnathopod 2, palm oblique . . . . 9. **E. boeckii** . . . . p. 445

1. E. subcarinatus (Hasw.) 1879 *Megamoera sub-carinata*, Haswell in: P. Linn. Soc. N. S. Wales, *v.* 4 p. 335 t. 21 f. 4 | 1884 *Moera s.*, Chilton in: N. Zealand J. Sci., *v.* 2 p. 230 | 1888 *Elasmopus s.*, *E. persetosus*, T. Stebbing in: Rep. Voy. Challenger, *v.* 29 p. 1019 t. 98 | 1893 *E. subcarinatus*, A. Della Valle in: F. Fl. Neapel, *v.* 20 p. 733 | 1882 *Moera petriei*, G. M. Thomson in: Tr. N. Zealand Inst.―*v.* 14 p. 236 t. 18 f. 3 | 1883 *M. p.*, Chilton in: Tr. N. Zealand Inst., *v.* 15 p. 82 t. 2 f. 4a.

Pleon segment 4 behind the dorsal depression bicarinate, the 2 apical teeth slightly inclined one toward the other. Head, lateral lobes broadly rounded, a small accessory lobe below the principal. Side-plate 1 somewhat produced forward, 4th the largest, much deeper than 5th. Pleon segment 3, postero-lateral corners acute, a little produced. Eyes large, reniform or oval, dark, advancing on to lateral lobes of head. Antenna 1 elongate, 1st joint about as long as 2d, much stouter, 3d not long nor yet very short, flagellum longer than peduncle, 50—60-jointed, accessory flagellum 6-jointed. Antenna 2 much shorter, ultimate joint of peduncle shorter than penultimate, flagellum shorter than peduncle, 14—17-jointed. Upper lip

442 Elasmopus

faintly emarginate at apex. Lower lip with unusual distal narrowing of inner lobes. Mandible with slender palp, 3^d joint longest, straight, (not falcate, as seems usual in the genus), and with few setae but no long row of spinules. Maxilla 1, inner plate with 3 setae on apex, outer with 7 spines. Maxilla 2, inner plate fringed on distal part of inner margin. Gnathopod 1 setose, 5th joint a little shorter than 6th, 6th oblong, palm slightly oblique, convex, pectinate, well-defined, finger closing against small palmar spines on inner surface of the joint. Gnathopod 2 much larger in $\male$, 5th joint short, broad, cup-shaped, 6th still broader, breadth $^2/_3$ of length, fringed behind with (Chilton: devoid of) very long setae, especially in $\male$ (as in E. brasiliensis), palm forming a broad spinulose process near the finger-hinge, followed by a cavity, a strong tooth, a feeble oblique emargination and a defining denticle; finger finely crenulate, strongly bent, with process touching the central palm-tooth, apex reaching the palmar denticle. In $\female$ gnathopod 2 has 5th joint rather longer than broad, 6th scarcely broader than 5th, nearly twice as long as broad, palm oblique, not dentate, both joints very setose, finger slender, acute. Peraeopods 3—5, 2^d joint rounded oblong, well expanded, especially in peraeopod 5, 4th joint large, 5th and 6th also stout, finger short. Uropods 1—3, rami a little unequal, with obtuse apices. Uropod 3, peduncle short, rami broad, laminar, inner slightly the shorter, with spines chiefly on inner margin, outer with spines chiefly on outer margin, each with spines on the truncate apex. Telson scarcely as long as peduncle of uropod 3, not quite as long as broad, lobes widely divergent, outer tooth of apex produced much beyond the inner, 2 unequal spines intervening. Colour whitish with brown dots, antennae banded with brown. L. about 14 mm.

South-Pacific (Port Jackson, from low water to 66 m; Sydney Harbour; Botany Bay; Port Stephens; off Melbourne; Port Pegasus and Lyttelton Harbour [New Zealand], and at a depth of 2071 m).

2. **E. diemenensis** (HasW.) 1879 *Megamoera d.*, Haswell in: P. Linn. Soc. N. S. Wales, *v.* 4 p. 266 t. 11 f. 3 | 1893 *Elasmopus rapax* (part.)?, A. Della Valle in: F. Fl. Neapel, *v.* 20 p. 736.

Pleon segments 1—4 each with a pair of strong spines [teeth] on its posterior margin near the middle dorsal line. Eyes reniform. Antenna 1, 2^d joint narrower and longer than 1st, flagellum longer than peduncle, accessory flagellum 4-jointed. Antenna 2, peduncle shorter than peduncle of antenna 1, flagellum shorter than peduncle. Gnathopod 1, 5th and 6th joints subequal, setose, 6th irregularly ovate, palm simple, oblique, ill-defined. Gnathopod 2 much larger, 5th joint short, cup-shaped, 6th very large, seemingly piriform, finger closing against its inner surface. Peraeopods 3—5, 2^d joint rounded oblong, well expanded, especially in peraeopod 5, 4th—6th joints very stout, finger small. Uropod 3, rami subequal, twice as long as peduncle, laminar. Telson with the halves laterally compressed, each terminating in 2 acute spines [teeth], and armed with a few short setae. L. 22 mm.

Bass Strait (Tasmania).

3. **E. suensis** (HasW.) 1879 *Megamoera s.*, Haswell in: P. Linn. Soc. N. S. Wales, *v.* 4 p. 335 t. 21 f. 5 | 1884 *M. s.?*, *M. haswelli*, Miers in: Rep. Voy. Alert, p. 317, 318 | 1885 *M. s.* (part.), Haswell in: P. Linn. Soc. N. S. Wales, *v.* 10 p. 103 | 1899 *Elasmopus s.*, T. Stebbing in: Tr. Linn. Soc. London, ser. 2 *v.* 7 p. 426 | 1893 *Ceradocus fasciatus* (part.)?, A. Della Valle in: F. Fl. Neapel, *v.* 20 p. 723, 729.

Body rather robust, peraeon segment 7 and pleon segments 1 and 2, each with 2 small dorsal teeth, segment 3 dorsally emarginate, not dentate, segment 4 dorsally produced into 2 strong teeth. Head with a small lateral tooth behind the lateral lobes. Side-plates 1—4 rather deep, 1st in front rounded or subacute, not much produced. Pleon segment 3, postero-lateral corners truncated and armed with 3—5 teeth. Eyes oval, black. Antenna 1 elongate, 2d joint rather longer than 1st, 3d very short, accessory flagellum with 3 rather long joints. Antenna 2 shorter, ultimate joint of peduncle shorter than penultimate. Gnathopod 1 as in E. subcarinatus (p. 441). Gnathopod 2, 5th joint short, broad, cup-shaped, 6th large, oblong, rounded at base, slightly widened distally, palm nearly transverse, not defined by a tooth, but diversified by a shallow emargination and 3 or 4 very obscure indications of teeth. Peraeopods 1 and 2 very slender, finger long. Peraeopods 3—5 as in E. subcarinatus. Uropod 3, rami subfoliaceous, rather narrow-ovate, and not greatly elongated. Telson, lobes subcylindrical, tipped with a few setae. Colour (in spirit) light brownish pink. L. 8—9 mm.

Torres Strait (Sue Island, Albany Island); Port Denison [East-Australia]. Depth 6—8 m.

4. E. brasiliensis (Dana) 1853 & 55 *Gammarus b.*, J. D. Dana in: U. S. expl. Exp., v.13 ii p.956 t.65 f.10 | 1888 *G. b.*, *Elasmopus* (part.)?, T. Stebbing in: Rep. Voy. Challenger, v.29 p.267, 516 | 1893 *G. b.*, *E.*(part.)?, A. Della Valle in: F. Fl. Neapel, v.20 p.737, 927 | 1880 *Moera b.*, Kossmann, Reise Roth. Meer., v.2i Malacost. p.132 | 1862 *Gammarella b.* + ? *M. pectenicrus*, *M. pectinicrus*, Bate, Cat. Amphip. Brit. Mus., p.180 t.32 f.9; p.192 t.34 f.8.

Body smooth. Eyes small, and (Bate) irregular. Antenna 1, flagellum rather longer than peduncle, many-jointed, accessory flagellum very small and short (Dana), 2-jointed (Bate). Antenna 2 half as long as antenna 1, flagellum short. Mandible, 3d joint of palp (Dana's figure) falcate, with long row of spinules. Gnathopod 1 small, 5th and 6th joints subequal, 6th subovate, hirsute on hind margin. Gnathopod 2 very stout, 5th joint short, cup-shaped, 6th large, piriform, densely furnished behind with long slender setae, palm oblique, not defined (Bate mentions a tubercle near the finger-hinge), finger long, curved. Gnathopod 2 in ♀ much smaller than in ♂, 6th joint subovate, setose. Peraeopods 1—5 apparently much as in E. subcarinatus (p. 441) and other species; (Bate: peraeopod 4, 2d joint abruptly narrow distally). Uropods 1—3 subequal. L. 8—9 mm.

Tropical Atlantic (Rio Janeiro); tropical Pacific (New Guinea)?

5. E. pocillimanus (Bate) 1862 *Moera p.*, Bate, Cat. Amphip. Brit. Mus., p.191 t.34 f.7 | 1888 *Maera p.*, T. Stebbing in: Rep. Voy. Challenger, v.29 p.335 | 1893 *Elasmopus p.*, A. Della Valle in: F. Fl. Neapel, v.20 p.733 t.1 f.4; t.22 f.23—25 | ?1874 *Moera levis*, (S. I. Smith in:) A. E. Verrill in: Rep. U. S. Fish Comm., v.1 p.559.

Body very robust. Side-plate 1 rhomboidal. Pleon segment 3, postero-lateral corners produced, their hind margin crenulate. Eyes large, elliptic, tending to reniform, violet-brown. Antenna 1 nearly as long as body (Della Valle) or half as long (Bate), flagellum as long as peduncle, accessory flagellum rudimentary. Antenna 2 much shorter, flagellum very short. Gnathopod 1. 5th and 6th joints equal, not large, finger small; 6th joint differing on right and left side. Gnathopod 2, 5th joint short, cup-shaped, 6th large, long, a little narrowed distally, palm without processes or defining tooth, but in place of the latter having a bowl-shaped excavation of the inner surface for receiving the tip of the finger, which (Della Valle) is obtuse. Peraeopods 1—5 much as in the preceding species. Uropods 1—3 extending nearly to the same length. Uropod 3, rami scarcely longer than the peduncle, fringed and tipped with short spines. Colour pellucid, lightly banded with crimson. L. 10 mm.

Mediterranean (Italy); North-Atlantic (New Jersey, Long Island Sound, Vineyard Sound)?

6. **E. delaplata** Stebb. 1888 *E. d.*, T. Stebbing in: Rep. Voy. Challenger, *v.* 29 p. 1025 t. 99 | 1893 *E. rapax* (part.)?, A. Della Valle in: F. Fl. Neapel, *v.* 20 p. 736.

Body smooth. Head as in E. subcarinatus (p. 441). Side-plate 1 obtusely much outdrawn, 2^d and 3^d narrowing downward, 4th excavate behind, deeper than 5th. Pleon segment 3, postero-lateral corners subacute, a little outdrawn, their hind margin rather strongly serrate. Eyes small, oval, near margin of lateral lobes of head, white in spirit. Antenna 1 elongate, 1st joint stouter and longer than 2^d, 3^d at least $^2/_3$ as long as 2^d, flagellum longer than peduncle, 35-jointed, accessory flagellum with 3 long joints and a short one. Antenna 2 much shorter, ultimate joint of peduncle rather shorter than penultimate, flagellum 16-jointed. Mandible, 3^d joint of palp elongate, falcate, with long fringe of spinules. Maxilla 1, inner plate with 1 seta and 3 or 4 setules. outer plate with 7 spines. Maxilla 2, inner plate not fringed on inner margin. Gnathopods 1 and 2 nearly as in E. subcarinatus, but in gnathopod 2 palm with an irregular toothed eminence near finger-hinge, the strongly bent finger touching this but leaving a gap between its inner margin and the sinuously sloping remainder of the palm, and closing down between 2 small processes on inner surface of the 6th joint. Peraeopods 1 and 2 slender. Peraeopods 3—5 moderately robust, but less stout than in most species, 2^d joint oblong oval. Uropod 1 reaching beyond uropod 2, uropod 3 a little beyond uropod 1. Uropod 3, rami broad, lanceolate, the ends a little obtuse, not truncate, outer ramus rather longer than inner. Telson rather longer than broad. oval, cleft nearly to base, lobes only apically dehiscent, each with spine and cilium in notch of outer margin a little above the apex. L. 18 mm.

South-Atlantic (Monte Video). Depth 1130 m.

7. **E. rapax** A. Costa 1853 *E. r.*, A. Costa in: Rend. Soc. Borbon., n. ser. *v.* 2 p. 175 | 1889 *E. r.* (part.?), A. M. Norman in: Ann. nat. Hist , ser. 6 *v.* 4 p. 124 t. 11 f. 1—8 (?f. 3) | 1894 *E. r.*, G. O. Sars, Crust. NorWay, *v.* 1 p. 521 t. 183 | 1857 *Gammarus brevicaudatus* (non H. Milne Edwards 1840!), Bate in: Ann. nat. Hist., ser. 2 *v.* 19 p. 145 | 1862 *Megamoera brevicaudata*, Bate (& Westwood), Brit. sess. Crust., *v.* 1 p. 409 f. | 1862 *M. b.*, Bate, Cat. Amphip. Brit. Mus., p. 228 t. 40 f. 2 | ? 1866 *Maera b.*, Cam. Heller in: Denk. Ak. Wien, *v.* 26 II p. 42 t. 3 f. 27, 28 | 1871 *Elasmopus latipes*, A. Boeck in: Forh. Selsk. Christian., 1870 p. 212 | 1876 *E. l.*, A. Boeck, Skand. Arkt. Amphip., *v.* 2 p. 393 t. 24 f. 1 | 1887 *E. l.*, Chevreux in: Bull. Soc. zool. France, *v.* 12 p. 229 f. 3 | 1893 *E. affinis* + *E. rapax*, A. Della Valle in: F. Fl. Neapel, *v.* 20 p. 734 t. 1 f. 9, t. 22 f. 1—15; p. 736 t. 22 f. 16—22.

Body robust, smooth. Head, lateral lobes broadly rounded. obtuse accessory lobe below the principal. Side-plate 1 angularly but very slightly produced in front, 4th excavate behind, much deeper than 5th. Pleon segment 3, postero-lateral corners subquadrate, their hind margin indented. Eyes rounded oval, rather large, very dark (Sars), rosy, pale with white spots (Della Valle). Antenna 1, 1st and 2^d joints subequal in length, 3^d more than half as long as 2^d, flagellum rather shorter than peduncle, many-jointed, accessory flagellum 2-jointed. Antenna 2 much shorter, ultimate and penultimate joints of peduncle subequal, together longer than flagellum. Gnathopod 1, 6th joint longer than 5th, oval quadrangular, palm nearly transverse, well-defined. Gnathopod 2 in ♂, 5th joint short, broad, cup-shaped, 6th powerful, somewhat piriform,

palm with rounded, denticulate lobe near finger-hinge, and 2 widely separated
dentiform projections from within the margin, finger very strong and curved;
(Della Valle: 2 obtuse tubercles on the palm margin of *E. affinis*). In the
♀ the 5ᵗʰ joint is less broad, the 6ᵗʰ more regularly ovate, finger impinging
against spines on inner surface but not into a groove. Peraeopods 1 and 2
of moderate size. Peraeopods 3—5 very robust (Sars: peraeopod 4 the
longest, but?). Uropod 3 reaching somewhat beyond the others, rami stout,
with truncate spinose apices, outer ramus rather the larger. Telson little
longer than broad, cleft nearly to base, lobes oblong, scarcely divergent,
3 or 4 spines in subapical notch of outer margin. Colour greyish or yellowish,
sometimes with dark spots (Della Valle). L. 8—10 mm.

North-Atlantic (Christianiafjord, British Isles, France, Azores); Mediterranean.

8. **E. viridis** (Hasw.) 1879 *Moera v.*, Haswell in: P. Linn. Soc. N. S. Wales,
v. 4 p. 333 t. 21 f. 1 | 1899 *Elasmopus v.*, T. Stebbing in: Tr. Linn. Soc. London, ser. 2
t. 7 p. 426 | 1883 *Moera incerta*, Chilton in: Tr. N. Zealand Inst., *v.* 15 p. 83 t. 3 f. 3 |
1893 *Maera truncatipes* (part.), A. Della Valle in: F. Fl. Neapel, *v.* 20 p. 725.

Body smooth. Side-plates 1—4 very shallow. Eyes round. Antenna 1,
2ᵈ joint slightly narrower and a little longer or shorter than the elongate 1ˢᵗ;
3ᵈ short; flagellum not so long as peduncle, 15—20-jointed; accessory
flagellum half as long as primary, with 5—8 long joints. Antenna 2, peduncle
subequal to peduncle of antenna 1; ultimate joint of peduncle shorter than penultimate; flagellum not quite as long as ultimate joint of peduncle, 7—9-jointed.
Gnathopod 1 small, 5ᵗʰ joint about as long as ovate 6ᵗʰ, palm rather oblique,
moderately well defined. Gnathopod 2 in ♂, 5ᵗʰ joint cup-shaped, 6ᵗʰ very
large, oblong, a little widened distally, palm transverse, crenulate, with small
central gap, defined by one tooth (not 3), finger with central cavity and
pronounced tooth matching hollow of palm; 6ᵗʰ joint in ♀ with the palm
straight (Chilton: palm defined by a short stout tooth, with short stout
setae along the whole.) Peraeopods 1 and 2 rather slender; peraeopods 3—5
stout (Chilton: finger with a produced point behind the nail). Uropod 3
reaching little beyond the others, rami broad, truncate, apically spinose, outer
sometimes the longer, with 2 groups of spines on outer margin. Telson
cleft nearly to base, lobes oblong, scarcely dehiscent, with spines in bidentate
apices. Colour light green. L. 5—6 mm.

South-Pacific (Port Jackson [East-Australia]; Lyttelton Harbour [New Zealand]).

9. **E. boeckii** (Hasw.) 1879 *Megamoera b.*, Haswell in: P. Linn. Soc. N. S. Wales,
v. 4 p. 336 t. 21 f. 6 | 1893 *Maera? b.*, A. Della Valle in: F. Fl. Neapel, *v.* 20 p. 732 |
1899 *Elasmopus b.*, T. Stebbing in: Tr. Linn. Soc. London, ser. 2 *v.* 7 p. 426.

Eyes oblong. Antenna 1, 1ˢᵗ and 2ᵈ joints subequal, 3ᵈ short, flagellum
longer than peduncle, accessory flagellum with 4 long joints. Antenna 2
much shorter, flagellum a little longer than ultimate joint of peduncle.
Gnathopod 1, 6ᵗʰ joint ovate, hirsute, palm undefined. Gnathopod 2 much
larger, 5ᵗʰ joint not very broad, cup-shaped, 6ᵗʰ described as ovate (figured
as oblong), twice as long as broad, palm oblique, slightly excavate, with
4 small teeth, finger rather more than ⅓ as long as 6ᵗʰ joint. Peraeopods 3—5
rather stout, serrated. Uropod 3 short, broad, truncate, spinose. L. 4 mm.

Port Jackson [East-Australia].

E. crassimanus (Miers) 1884 *Moera crassimana*, Miers in: Rep. Voy. Alert,
p. 316 | 1899 *Elasmopus crassimanus*, T. Stebbing in: Tr. Linn. Soc. London, ser. 2
v. 7 p. 426.

Body smooth. Pleon segments 1—3, postero-lateral corners acute, little produced. Gnathopod 1, 5th joint little shorter than 6th. Gnathopod 2, 5th joint broad, 6th longer than broad, narrower distally, with the distal margin very oblique, not acute, but presenting a broad surface, against which the strong arcuate finger closes, and armed with 4 spines or lobes, of which the inner 2 are small. Peraeopods 3—5 rather robust, 2d joint not serrated. Uropod 3, rami subequal, broader and slightly shorter than the others.

Port Jackson [East-Australia].

E. erythraeus (Kossm.) 1880 *Moera erythraea*, Kossmann, Reise Roth. Meer., *v.* 21 Malacost. p. 132 t. 14 f. 1—8 | 1888 *Elasmopus erythraeus*, T. Stebbing in: Rep. Voy. Challenger, *v.* 29 p. 516 | 1893 *E. rapax* (part.), A. Della Valle in: F. Fl. Neapel, *v.* 20 p. 736.

Probably the young of another species of Elasmopus. L. about 6 mm.

Red Sea.

E. miersi (Wrześn.) 1879 *Maera m.*, Wrześniowski in: Zool. Anz., *v.* 2 p. 348 | 1888 *M. m.*, T. Stebbing in: Rep. Voy. Challenger, *v.* 29 p. 502 | 1893 *M. m.*, A. Della Valle in: F. Fl. Neapel, *v.* 20 p. 732 | 1899 *Elasmopus m.*. T. Stebbing in: Tr. Linn. Soc. London, ser. 2 *v.* 7 p. 426.

L. 9 mm.

Chimbote Bay [Peru]. Under stones in tide-pools.

E. peruvianus (Dana) 1852 *Amphithoe peruviana*, J. D. Dana in: P. Amer. Ac., *v.* 2 p. 215 | 1853 & 55 *Gammarus? peruvianus*, J. D. Dana in: U. S. expl. Exp., *v.* 13 II p. 958 t. 66 f. 2 | 1893 *Elasmopus* (part.)?, A. Della Valle in: F. Fl. Neapel, *v.* 20 p. 454 | 1862 *Megamoera peruviensis*, Bate, Cat. Amphip. Brit. Mus., p. 231 t. 40 f. 8.

Near E. brasiliensis (p. 443), compare also E. delaplata (p. 444). Peraeopods 1 and 2 about as long as peraeopod 5, which is longer than peraeopod 4. L. about 12 mm.

Tropical Pacific (Island of San Lorenzo [Peru]). Among sea-weed on the shore.

35. Gen. **Plesiogammarus** Stebb.

1899 *Plesiogammarus* (Sp. typ.: *P. gerstaeckeri*), T. Stebbing in: Tr. Linn. Soc. London, ser. 2 *v.* 7 p. 426.

Near to Gammarus (p. 460), with 6th joint of gnathopod 1 smaller than that of gnathopod 2, but with marginal inflation of many segments, some dorsal setae but no dorsal spines, peduncle of antenna 1 longer than peduncle of antenna 2, 2d joint of peraeopods 3—5 long and narrow, uropod 1 reaching end of the short uropod 3 and telson not cleft to the base.

1 species.

1. **P. gerstaeckeri** (Dyb.) 1874 *Gammarus g.*, B. Dybowsky in: Horae Soc. ent. Ross., *v.* 10 suppl. p. 108 t. 14 f. 5 | 1899 *Plesiogammarus g.*, T. Stebbing in: Tr. Linn. Soc. London, ser. 2 *v.* 7 p. 426 | 1893 *Gammarus locusta* (part.)?, A. Della Valle in: F. Fl. Neapel, *v.* 20 p. 928.

Peraeon segments 1—7 and pleon segments 1—3 with clear marginal swellings, pleon segment 3 on upper surface and segments 4—6 on hind margin carrying a few tolerably long setae. Head with short rostral point, at the sides inflated over the eyes. Side-plates 1—4 small, regularly rounded below. Eyes of varying size and shape, usually reniform, more rarely roundish or punctiform, white. Antenna 1 as long as body or sometimes shorter, 4 times as long as antenna 2, peduncle stouter and somewhat longer than peduncle of antenna 2, 1st joint stout, as long as 2d and 3d combined, flagellum 47-jointed, accessory flagellum 3- or 4-jointed. Antenna 2, ultimate joint of peduncle shorter than penultimate, thickened distally and there carrying a circlet of plumose setae, flagellum 4-jointed. Gnathopods 1 and 2, 5th joint short, 6th slenderly oblong, in gnathopod 2 considerably larger than in gnathopod 1 and slightly widening at the palm, which in both is much shorter than the hind margin. Peraeopods 3—5, 2^d joint narrow, almost rod-like, a little wider above than below, 4—5 times as long as broad, without setae on hind margin, while the 4th and 6th joints of these limbs, the margins of peduncle in uropods 1 and 2 and the lateral surfaces of the pleon segments are beset with numerous long setae. Uropod 1 reaching end of uropod 3 or somewhat beyond. Uropod 2 reaching scarcely $^2/_3$ length of uropod 3. Uropod 3 of $^1/_{12}$—$^1/_{10}$ length of body, rami equal, beset with simple setae. Telson broader than long, divided beyond the middle but not nearly to the base, lobes with numerous apical setae. Colour yellow. L. reaching 16 mm.

Lake Baikal. Depth 20—100 m.

36. Gen. Iphigenella O. Sars

1896 *Iphigenella* (Sp. typ.: *I. acanthopoda*) (*Iphigeneia* O. Grimm in MS.), G. O. Sars in: Bull. Ac. St.-Pétersb., ser. 5 *v.* 4 p. 478.

Body almost smooth, robust, pleon segments 4 and 5 with a few dorsal hairs and a pair of subdorsal spinules, segment 6 with 2 pairs of spinules. Side-plates rather large, not setiferous. Antenna 1 the longer, with accessory flagellum. Mouth-parts nearly as in Gammarus (p. 460), but with inner lobes indicated in lower lip. Gnathopod 1 with short 5th and large 6th joint. Gnathopod 2 with 5th and 6th joints elongate, slender. Peraeopods 1—5 narrowly subchelate. Peraeopods 3—5 rather short and stout, .2^d joint well expanded. Uropod 3 not very large, outer ramus much longer than peduncle, 2-jointed, inner ramus scale-like. Telson narrow, cleft to the base.

1 species.

1. **I. acanthopoda** O. Sars 1896 *I. a.,* G. O. Sars in: Bull. Ac. St.-Pétersb., ser. 5 *v.* 4 p. 478 t. 12 f. 1—17.

Body moderately compressed. Pleon segments 4—6 slightly carinate. Head scarcely rostrate, lateral lobes obliquely truncate. Side-plates 1—4 rather deep, 4th the largest, much deeper than broad, emarginate, 5th with front and hind lobes subequal. Pleon segment 3, postero-lateral corners acutely subquadrate. Eyes small, oval, dark. Antenna 1 more than $^1/_3$ as long as body, peduncle shorter than in antenna 2, 1st joint longer than 2^d and 3^d combined, flagellum twice as long as peduncle, 17-jointed, accessory flagellum 4-jointed. Antenna 2 rather stouter, ultimate and penultimate joints of peduncle subequal, flagellum little shorter than peduncle, 10-jointed. Gnathopod 1, 5th joint short, cup-shaped, 6th powerful, irregularly oval, widening to the rather oblique, sinuous palm, which is about equal to the hind margin and has a strong spine near the middle and 2 at the obtusely rounded palmar angle, finger long and falciform. Gnathopod 2, 5th joint long, a little wider and shorter than the sublinear 6th, both setose, palm transverse, extremely short, finger to match. Peraeopods 1 and 2 alike, rather slender. Peraeopods 3—5 subequal, shorter than peraeopods 1 and 2, 2^d joint in peraeopods 3 and 4 quadrangular oval, widest above, in peraeopod 5 obliquely expanded, produced below in a broadly rounded lobe. In all peraeopods the 6th joint has a wide apex armed with stout spines, forming a kind of palm, against which the short, strongly curved finger impinges. Uropods 1 and 2,

outer ramus shorter than inner, each with several spines at apex. Uropod 3
projecting much beyond the others, outer ramus nearly thrice as long as
peduncle, with 2 sets of spines on outer, 3 on inner margin, 2^d joint spiniform,
almost as large as the little inner ramus. Telson nearly twice as long as broad,
lobes little tapering or divergent, each with 3 spinules on the obtuse apex.
L. 9 mm.

Caspian Sea. Sometimes on Astacus leptodactylus, probably semiparasitic.

37. Gen. Pandorites O. Sars

1895 *Pandorites* (Sp. typ.: *P. podoceroides*) (*Pandora* (non Bruguière 1797,
Mollusca!) O. Grimm in MS.), G. O. Sars in: Bull. Ac. St.-Pétersb., ser. 5 *v.* 3 p. 287.

Body little compressed, smooth, pleon segments 4—6 short and stout,
with a few dorsal hairs and spinules. Side-plate 4 the largest, only slightly
emarginate. Eyes on lateral lobes of head. Antennae 1 and 2 slender,
short, equal, antenna 1 with accessory flagellum. Upper lip rounded. Lower
lip without inner lobes, and mouth-parts in general as in Gammarus (p. 460).
Gnathopods 1 and 2 alike in ♂ and in ♀, very unequal, in gnathopod 1
6th joint small, ovately piriform, in gnathopod 2 very large, expanding to
the long, oblique palm. Peraeopods 1—5 not very long. Peraeopod 5,
2^d joint much expanded, with produced lobe at lower hind corner. Uropod 3
very small, outer ramus short, its 2^d joint minute, inner ramus scale-like.
Telson short, cleft to the base.

1 species.

1. **P. podoceroides** O. Sars 1895 *P. p.*, G. O. Sars in: Bull. Ac. St.-Pétersb.,
ser. 5 *v.* 3 p. 287 t. 19.

In form resembling a species of Ischyrocerus (p. 657), slender,
with broadly rounded back. Head with angular front slightly projecting,
lateral lobes evenly rounded, prominent; post-antennal corners acute,
curving forward. Side-plate 1 small, narrowing distally, more setose than
the rest, 4th almost square, but with convex lower and slightly concave hind
margin. Pleon segments 2 and 3, postero-lateral corners smooth, quadrate.
Eyes close to margin of lateral lobes of head, oval, dark. Antenna 1, 1st joint
longer than 2^d and 3^d combined, 2^d scarcely twice as long as 3^d, flagellum not quite
as long as peduncle, 7-jointed, accessory flagellum small, 4-jointed. Antenna 2,
ultimate joint of peduncle shorter than penultimate, flagellum 5-jointed.
Gnathopod 1 setose, 5th joint short, cup-shaped, 6th rather tumid, palm
scarcely defined except by the palmar spines. Gnathopod 2 less setose,
2^d joint with very convex hind margin, 5th joint short, with short and narrow
setiferous lobe, 6th somewhat flattened, palm longer than hind margin, much
curved, very oblique, its edge sharpened and fringed with setules; of the
palmar spines one very slender and long, finger long and falciform. Peraeopods 1 and 2 very setose, 1st rather the longer. Peraeopods 3—5 rather
strong, not elongate, 2^d joint in peraeopod 3 oblong oval, in peraeopod 4
longer, scarcely broader, narrowing distally, with sinuous hind margin; 2^d joint
in peraeopod 5 greatly expanded, widest below the middle, the obscurely
serrate hind margin unarmed, its lower lobe not greatly produced downward,
front margin densely setose below. Uropods 1 and 2 rather short and stout,
rami subequal, each with apical spines, also 1 lateral spine, except on outer
ramus of uropod 2. Uropod 3 reaching little beyond the others, outer ramus
scarcely longer than peduncle, with 1 lateral and 2 apical spines, the tiny

2^d joint tipped with setules, inner ramus less than $^1/_2$ as long as outer, with 1 apical seta. Telson little longer than peduncle of uropod 3, broader than long, lobes not divergent, a spine and hair on each broad apex. L. ♀ 11, ♂ 13 mm.

Caspian Sea. Depth 13—90 m.

38. Gen. **Pherusa** Leach

1813,14 *Pherusa* (Sp. un.: *P. fucicola*), Leach in: Edinb. Enc., *v.* 7 p. 432 | 1891 *P.*, A. O. Walker in: Ann. nat. Hist., ser. 6 *v.* 7 p. 418; *v.* 8 p. 81 | 1891 *P.*, Pocock in: Ann. nat. Hist., ser. 6 *v.* 7 p. 530 | 1857 *Gammarella* (Sp. un.: *G. orchestiformis*), Bate in: Ann. nat. Hist., ser. 2 *v.* 19 p. 143 | 1893 *G.*, G. O. Sars, Crust. Norway, *v.* 1 p. 446.

Body smooth. Head without acute rostrum. Antenna 1 the longer, 2^d joint of peduncle elongate, accessory flagellum developed. Upper lip smoothly rounded. Lower lip with inner lobes. Mouth-organs normal. Mandibular palp slender, 2^d and 3^d joints elongate. Maxilla 1, inner plate with about 18 setae, outer with 11 spines. Maxilla 2, inner margin of inner plate fringed with setae. Gnathopods 1 and 2 subchelate. Gnathopod 2 differing greatly in the two sexes. Peraeopods 1 and 2 slender. Peraeopods 3—5 with 2^d joint expanded, especially in peraeopod 5. Uropods 1—3 short. Uropod 3, outer ramus little longer than peduncle, inner much shorter. Telson deeply cleft.

1 species.

1. **P. fucicola** Leach 1813/14 *P. f.*, Leach in: Edinb. Enc., *v.* 7 p. 432 | 1816 *P. f.*, Leach in: Enc. Brit., ed. 5 suppl. 1 p. 426 t. 21 f. | 1862 *P. f.* (part.), Bate & Westwood, Brit. sess. Crust., *v.* 1 p. 255 f. z dorsal view | 1893 *Melita f.*, A. Della Valle in: F. Fl. Neapel, *v.* 20 p. 709 t. 1 f. 8; t. 24 f. 1—19 | 1898 *M. f.*, Sowinski in: Mém. Soc. Kiew, *v.* 15 p. 486 t. 11 f. 10—19; t. 12 f. 1—4 | 1830 *Gammarus brevicaudus* + *Amphithoe f.*, H. Milne Edwards in: Ann. Sci. nat., *v.* 20 p. 369; p. 377 | 1840 *G. brevicaudatus*, H. Milne Edwards, Hist. nat. Crust., *v.* 3 p. 53 | 1857 *Amphithoe micrura* + *A. semicarinata* + *G. obtusunguis* (part.: ♂ juv.) + *G. punctimanus*, A. Costa in: Mem. Acc. Napoli, *v.* 1 p. 209 t. 3 f. 2; p. 210 t. 3 f. 3; p. 219 t. 3 f. 8; p. 222 t. 3 f. 6 | 1857 *Gammarella orchestiformis*, Bate in: Ann. nat. Hist., ser. 2 *v.* 19 p. 143 | 1862 *G. normanni* (♀), Bate & Westwood, Brit. sess. Crust., *v.* 1 p. 333 | 1862 *G. brevicaudata* + *Gammarus punctatus* (laps., corr.: *G. punctimanus*), Bate, Cat. Amphip. Brit. Mus., p. 180 t. 32 f. 8; p. 224 | 1874 *Gammarella b.*, T. Stebbing in: Ann. nat. Hist., ser. 4 *v.* 14 p. 13 t. 2 f. 3a—g | 1888 *G. b.*, T. Barrois, Cat. Crust. Açores, p. 47 t. 4 f. 5—12 | 1889 *G. b.*, A. M. Norman in: Ann. nat. Hist., ser. 6 *v.* 4 p. 128 | 1885 *G. longicornis*, R. Koehler in: Bull. Soc. Nancy, *v.* 17 p. 67.

Pleon abruptly depressed and abbreviated behind segment 4, which is more or less obtusely carinate but not at all dentate, segments 5 and 6 with a couple of minute dorsal spinules. Eyes small, round, tinged with red. Antenna 1, 2^d joint nearly as long as 1st, flagellum longer than peduncle, sometimes nearly twice as long, accessory flagellum with 4 tolerably long joints. Antenna 2 much shorter, slender, ultimate joint of peduncle rather shorter than penultimate, flagellum about 20-jointed, subequal to peduncle. Gnathopod 1, 6th joint shorter than 5th, widening somewhat to the palm, which is almost transverse, matching the short finger. Gnathopod 2 in ♀, 6th joint longer than 5th, not widening, palm short, transverse, finger short. Gnathopod 2 in ♂ very unlike that of ♀, 5th joint very short, 6th very large, almond-shaped, narrowest at the finger-hinge, the long oblique palm, except in the young form, scarcely at all distinct from the hind margin, fringed with setae and carrying within the margin a row of spinules, finger very large, curved, sometimes overlapping the whole length of the hand, but shorter in young ♂. Peraeopods 1 and 2, 4th joint scarcely dilated, longer than 5th, 5th than 6th. Peraeopods 3 and 4, 2d joint moderately expanded, more above than below. Peraeopod 5, 2d joint very wide at the middle. Pleopods with many coupling spines on the peduncle. Uropods 1—3 reaching about equally far back. Uropod 3, outer ramus seemingly with minute 2d joint, inner scarcely half as long or half as wide as outer. Telson subquadrate, cleft about $^3/_4$ of length, with spinule in each rounded apex. Colour yellowish brown. L. 7—12(?) mm.

North-Atlantic (South-England, shore at Very low tides; Channel Islands; Scotland, Firth of Clyde; France; Azores); Mediterranean; Black Sea. DoWn to 84 m.

39. Gen. **Niphargoides** O. Sars

1894 *Niphargoides* (Sp. typ.: *N. caspius*), G. O. Sars in: Bull. Ac. St.-Pétersb., ser. 5 *v.* 1 p. 371.

Akin to Pontoporeia (p. 127) and Gammarus (p. 460) rather than Niphargus (p. 405). Body. nearly smooth, robust. Head not rostrate. Side-plates 1—4 not very large, setiferous, 4th the largest, excavate behind. Antennae 1 and 2 short, subequal; antenna 1 with accessory flagellum usually more than 1-jointed. Lower lip with inner lobes. Mandibular palp large. Maxillipeds, 3d joint of palp slender. Mouth-parts otherwise as in Gammarus. Gnathopods 1 and 2 well developed, subchelate, 2d usually the stronger. Peraeopod 5, 2d joint much more expanded than in peraeopods 3 and 4, hind margin, like the peraeopods in general, strongly setiferous. Uropods 1 and 2 with equal rami. Uropod 3 not very large, outer ramus much the larger, 2-jointed, inner ramus scale-like. Telson cleft to the base.

7 species.

Synopsis of species:

1 { Uropod 3, 2d joint of outer ramus rudimentary — **2.**
Uropod 3, 2d joint of outer ramus distinctly developed — **4.**

2 { Gnathopods 1 and 2, 6th joint narroWing toWards the palm . . . . 1. **N. caspius** . . . . p. 450
Gnathopods 1 and 2, 6th joint not narroWing toWards the palm — **3.**

3 { Gnathopods 1 and 2, hind margin longer than palm 2. **N. corpulentus**. . p. 451
Gnathopods 1 and 2, hind margin shorter than palm 3. **N. compactus** . . p. 451

4 { Uropod 3, 2d joint of outer ramus much shorter than inner ramus . . . . 4. **N. grimmi** . . . . p. 452
Uropod 3, 2d joint of outer ramus not much shorter than inner ramus — **5.**

5 { Lobes of telson strongly diVergent . . . . 5. **N. quadrimanus** . p. 452
Lobes of telson not strongly diVergent — **6.**

6 { Gnathopods 1 and 2 subequal . . . . 6. **N. aequimanus** . p. 453
Gnathopod 1 considerably smaller than gnathopod 2 7. **N. borodini** . . . p. 453

1. **N. caspius** O. Sars 1894 *N. c. (Niphargus c.* O. Grimm in MS.), G. O. Sars in: Bull. Ac. St.-Pétersb., ser. 5 *v.* 1 p. 372 t. 16.

Body rather elongate, but robust and uncompressed, with only 2 dorsal spinules on pleon segment 6. Head, lateral lobes rounded, little prominent.

Side-plates 1—4 densely setiferous, 1st not expanded distally. Pleon segments 2 and 3, postero-lateral corners nearly quadrate, with a short oblique row of setules on outer surface, descending to the angle. Eyes not large, oval reniform, dark. Antenna 1 about $^1/_7$ as long as body, 1st joint massive, much longer than 2^d and 3^d combined, 2^d about twice as long as 3^d, flagellum 7-jointed, accessory flagellum 3-jointed. Antenna 2 slightly longer, antepenultimate joint of peduncle stout, with densely setose angular projection below, ultimate joint shorter than penultimate, flagellum 5-jointed. Gnathopods 1 and 2 similar, 5th joint short, cup-shaped, 6th elongate, gradually tapering to the very oblique palm, which is much shorter than hind margin, and chiefly defined by 2 strong spines; gnathopod 2 much the larger. Peraeopods 1—5 densely setiferous, 1st and 2^d with 4th joint large, widening distally, 3^d and 4th with 2^d joint oblong oval, narrowing distally, peraeopod 5, 2^d joint much more expanded, broadly oval, hind margin evenly curved. Uropods 1 and 2, each ramus armed at apex with 4 blunt spines, and with 1 on the lateral margin. Uropod 3, peduncle with fringe of spines, outer ramus not twice as long as peduncle, rather broad, distal half fringed with plumose setae, 2 spines on outer margin, 2^d joint minute; inner ramus shorter than peduncle, tipped with 1 spine. Telson, lobes rather divergent, each with 3 spines on the obtusely truncate apex. L. 11 mm.

Caspian Sea. Depth 66—75 m.

2. **N. corpulentus** O. Sars 1895 *N. c.*, G. O. Sars in: Bull. Ac. St.-Pétersb., ser. 5 *v.* 3 p. 275 t. 17 f. 1—19.

♀ unknown. — ♂. In general like N. caspius, differing as follows. Body more robust and with pleon segments 4—6 slightly raised dorsally, carrying a few small hairs, and with 2 subdorsal spinules on pleon segments 5 and 6; side-plates 1—4 larger and less densely setiferous; pleon segment 2 without the oblique row of setules. Antenna 1, flagellum 8-jointed, accessory flagellum 4-jointed; antenna 2 scarcely longer, flagellum 6-jointed. Gnathopods 1 and 2 less powerful, 6th joint not tapering, but oblong quadrangular, with palm less oblique and defined by a distinct angle, carrying 2 or 3 unequal spines. Peraeopods 1 and 2, 5th joint rather wider; peraeopod 3, 2^d joint more regularly oval; peraeopod 4, 2^d joint narrow, elongate, with the hind margin slightly sinuous; peraeopod 5, 2^d joint greatly expanded, broadly heart-shaped, widest below, the long setae of the strongly curved hind margin springing from small serrations. Uropods 1 and 2, rami less blunt, with the spines more distributed; uropod 3, outer ramus about twice as long as peduncle, inner tipped with 2 spines, and carrying 3 setules on inner margin. Telson with lobes rather less divergent, each broad apex carrying 5 spines. L. 14 mm.

Caspian Sea.

3. **N. compactus** O. Sars 1895 *N. c.*, G. O. Sars in: Bull. Ac. St.-Pétersb., ser. 5 *v.* 3 p. 278 t. 17 f. 14—19.

♀ unknown. — ♂. Agreeing nearly with N. corpulentus. Body extremely robust and compact, peraeon segments 6 and 7 and pleon segments 1—3 each with transverse dorsal furrow, pleon segment 5 carrying 2, segment 6 having 4 subdorsal spinules. Side-plate 1 distally expanded. Antenna 1, 2^d joint more than thrice as long as 3^d, longer than the 9-jointed flagellum, accessory flagellum 4-jointed. Antenna 2, ultimate and penultimate joints of peduncle comparatively slender, flagellum very short, 6-jointed. Gnathopods 1 and 2 similar, powerful, with piriform hand, thus differing in shape from those in other species, the 6th joint being very tumid at the base, the palm very oblique, well defined, much longer than the hind margin, gnathopod 2 much the larger, with the palmar angle more projecting; finger long, curved. Uropod 3, outer ramus nearly thrice as long as peduncle, sublaminar, fringed almost all round, 2^d joint almost evanescent, inner ramus with 7 plumose setae on inner margin and 2 spines at apex. Telson with 4 spines at each apex. L. 17 mm.

North Caspian Sea.

4. N. grimmi O. Sars 1896 N. g., G. O. Sars in: Bull. Ac. St.-Pétersb., ser. 5 v. 4 p. 471 t. 11 f. 1—12.

Resembling N. caspius (p. 450). Appendages much less hirsute. Body robust, pleon segments 5 and 6 each with 2 dorsal spinules. Head, lateral lobes narrowly rounded. Side-plates 1—4 fringed with rather short setae. Pleon segment 2, postero-lateral corners quadrate, segment 3, postero-lateral corners acutely produced, with a few setules on the margin. Antenna 1, flagellum 5-jointed, accessory flagellum 2-jointed, the 2^d joint minute. Antenna 2 not longer, flagellum 3-jointed. Gnathopods 1 and 2, 6th joint rather elongate, oblong, slightly widening to the palm, which is not very oblique, well defined, with 2 palmar spines, one of them very long; gnathopod 2 nearly twice as large as gnathopod 1. Peraeopod 3, 2^d joint piriform. Peraeopod 4, 2^d joint little widened, hind margin sinuous. Peraeopod 5, 2^d joint greatly expanded, little longer than broad, widest at about the middle, fringed behind with hair-like setules. Uropod 3, outer ramus with fringing setae not close set, 2^d joint very small, but well defined and obvious, inner ramus nearly as long as peduncle, tipped with 2 spines. Telson rather longer than broad, lobes little divergent, 4 spines on each apex. L. 8 mm.

Caspian Sea. Depth 47—169 m.

5. N. quadrimanus O. Sars 1895 N. q., N. qvadrimanus, G. O. Sars in: Bull. Ac. St.-Pétersb , ser. 5 v. 3 p. 281; t. 18 f. 1—13.

Body not very robust, somewhat compressed, pleon devoid of subdorsal spinules. Head, lateral lobes broadly rounded. Side-plate 1 scarcely expanded distally, 4th about as wide as deep. Pleon segment 3, postero-lateral corners quadrate, with no oblique row of setules. Eyes rather small, oval reniform, dark. Antenna 1, 1st joint twice as long as 2^d and 3^d combined, 2^d twice as long as 3^d, flagellum about 11-jointed, accessory flagellum 6-jointed. Antenna 2 equal to antenna 1, peduncle geniculate between antepenultimate and penultimate joints, both of which are expanded below to setiferous lobes, ultimate joint nearly as long as penultimate, both spiniferous, flagellum 10-jointed. Gnathopods 1 and 2, 6th joint oblong quadrangular, palm not very oblique, defined by a distinct angle armed with 3 unequal spines; in gnathopod 1 the hand inclines to oval, but in the larger gnathopod 2 its great breadth, encreasing slightly towards the palm, gives it an almost square appearance. Peraeopods 1 and 2, 4th joint large, setiferous chiefly on hind margin, 5th joint not much expanded, 6th, as usual in the genus, linear. Peraeopod 3, 2^d joint regularly oval. Peraeopod 4, 2^d joint much longer, and below narrower, hind margin sinuous. Peraeopod 5, 2^d joint greatly expanded, greatest width near distal end, hind margin fringed with short setae, lower front angle and 4th and 5th joints, as in peraeopod 4, very hirsute. Uropods 1 and 2, rami subequal, each with 5 apical spines and one lateral spine. Uropod 3 projecting far beyond the others, outer ramus with 2 sets of spines on outer margin, fringing setae scattered and simple, 2^d joint not much shorter than inner ramus, which has 2 apical spines. Telson rather small, lobes strongly divergent, each with 1 apical spinule. L. 10 mm.

Caspian Sea. Depth 13—37 m.

6. **N. aequimanus** O. Sars 1895 *N. a.*, *N. aeqvimanus*, G. O. Sars in: Bull. Ac. St.-Pétersb., ser. 5 *v.* 3 p. 285; t. 18 f. 14—23.

Resembling N. quadrimanus. Lateral lobes of head narrowly rounded. Side-plate 1 distally widened. Antenna 1, flagellum 7-jointed, accessory flagellum 5-jointed; antenna 2, flagellum 7-jointed. Gnathopods 1 and 2 nearly equal, as well as similar, 6th joint oblong quadrangular, palm nearly transverse, much shorter than hind margin. Peraeopods 3 and 5, 2^d joint almost as broad as long, the setae on hind margin in peraeopod 5 fewer than in N. quadrimanus. Uropods 1 and 2 without the lateral spine; uropod 3 with more elongate ramus, thrice as long as peduncle, inner ramus with 1 apical spine. Telson with broader lobes, not divergent, 2 unequal spinules on each apex. L. 5 mm.

Caspian Sea. Depth 19 m.

7. **N. borodini** O. Sars 1897 *N. b.*, G. O. Sars in: Annuaire Mus. St -Pétersb.,' *v.* 2 p. 290 t. 15 f. 4—9.

♀. Body somewhat elongated, back broadly rounded, quite smooth. Head, lateral corners produced, narrowly rounded. Side-plates 1—4 comparatively large, 1st not distally widened, 4th deeper than wide. Pleon segment 3, postero-lateral corners subquadrate, with no oblique row of setules. Eyes not very large, oval reniform, dark. Antenna 1, 1st joint very large, twice as long as 2^d and 3^d combined, distally rather narrowed, flagellum not half as long as peduncle, 8-jointed, accessory flagellum 6-jointed. Antenna 2 as long as antenna 1, antepenultimate and penultimate joints of peduncle with setiferous lobes, ultimate joint linear, nearly as long as the 9-jointed flagellum. Gnathopods 1 and 2 strong, 3^d—5th joints short, 6th large, tending to quadrate in form, but widening distally to a somewhat oblique palm, defined by a spiniferous obtuse point; hand, palm, and finger considerably larger in gnathopod 2 than in gnathopod 1. Peraeopods 1—5 and uropods 1 and 2 as in N. quadrimanus. Uropod 3 projecting much beyond the others, outer ramus more than twice as long as peduncle, 1st joint spinose, 2^d longer than the scale-like inner ramus, setose all round, inner ramus with 3 apical spines. Telson with oblong oval lobes, scarcely divergent, each with 3 spinules on blunt apex. L. 13 mm. — ♂ unknown.

Caspian Sea.

40. Gen. **Phreatogammarus** Stebb.

1899 *Phreatogammarus* (Sp. typ.: *P. fragilis*). T. Stebbing in: Tr. Linn. Soc. London, ser. 2 *v.* 7 p. 427.

Near to Gammarus (p. 460). Without eyes. Upper lip broader than deep. Mandible, 1st joint of palp not very short. Gnathopods 1 and 2 equal. Peraeopods 1 and 2 much shorter than peraeopod 3, peraeopod 5 the longest, with its 6th joint longer than any of the other joints. Uropod 3 long, with equal 1-jointed cylindrical rami.

1 species.

1. **P. fragilis** (Chilton) 1882 *Gammarus f.*, Chilton in: N. Zealand J. Sci.'
v. 1 p. 44 | 1882 *G. f.*, Chilton in: Tr. N. Zealand Inst., *v.* 14 p. 179 t. 9 f. 11—18 | 1889
G. f., Moniez in: Rev. biol. Nord France, *v.* 1 p. 253 | 1890 *G. f.*, Wrześniowski in: Z.
Wiss. Zool., *v.* 50 p. 611, 698 | 1894 *G. f.*, Chilton in: Tr. Linn. Soc. London, ser. 2 *v.* 6
p. 227 t. 21 f. 1—25 | 1899 *Phreatogammarus f.*, T. Stebbing in: Tr. Linn. Soc. London,
ser. 2 *v.* 7 p. 427 | 1893 *Gammarus fluviatilis* (part.), A. Della Valle in: F. Fl. Neapel,
v. 20 p. 763.

Body rather slender; pleon segments 4—6 each with 4 or 5 dorsal long
spine-like setae. Head without rostrum, lateral angles broadly rounded, little
produced. Side-plates rather shallow. Pleon segments 2 and 3 with postero-
lateral angles blunt. Antenna 1 subequal to body, 1st joint long, not quite as
long as 2^d and 3^d combined, flagellum many-jointed, more than twice as long as
peduncle, accessory flagellum 6—9-jointed. Antenna 2 more than half as long as
antenna 1, ultimate joint of peduncle slightly longer than penultimate, flagellum
subequal to peduncle, about 20-jointed. Upper lip nearly semicircular, much
broader than deep. Lower lip without inner lobes. Mandible, 1st joint of palp
nearly half as long as 2^d, distally a little widened. Gnathopods 1 and 2 in ♀ rather
large, 6th joint ovate, palm very oblique, finger long, curved. Peraeopods 1
and 2 slender. Peraeopods 3—5 long, 2^d joint not very wide, narrowing
downward, finger small. Peraeopods 4 and 5 with 6th joint much the longest.
Peraeopod 5 nearly as long as body. Uropod 3 extending much beyond the
others, the rami about twice as long as peduncle, scarcely tapering, each with 5
groups of 3 spines and an apical group of 6 or 7 spines. Telson short, the
lobes almost rectangular, each with 4 or 5 spines on the obtusely truncate apex
and 2 or 3 a little above on outer margin. Colour white, semitransparent.
L. reaching 15 mm.

New Zealand. Wells.

41. Gen. **Ommatogammarus** Stebb.

1899 *Ommatogammarus*, T. Stebbing in: Tr. Linn. Soc. London, ser. 2 *v.* 7 p 427.

Near to Gammarus (p. 460). Dorsal spines only on pleon segments
4—6. Eyes irregular, with indented outline (Fig. 86). Antenna 1 the longer,
but with peduncle usually shorter than that of antenna 2; accessory flagellum
of more than 1 joint (Fig. 86). Upper lip narrowed to a rounded apex.
Lower lip, inner lobes rudimentary. Mandible, 3^d joint of palp not very
elongate. Maxilla 1, inner plate fringed with numerous setae, outer carrying
11 spines, 2^d joint of palp with about 10 spine-teeth on one maxilla and
short spines on the other. Maxillipeds, outer plates reaching far along
2^d joint of palp, spine-teeth and setae numerous. Gnathopod 1, 6th joint
not smaller than that of gnathopod 2. Uropod 3, outer ramus about twice
as long as inner, with simple setae on its outer margin, inner ramus with
feathered setae on both margins. Telson cleft to the base.

4 species.

Synopsis of species:

1 {
Dorsal margin of head abruptly bent downward — **2.**
Dorsal margin of head not abruptly bent down-
ward — **3.**
}

2 {
Eyes with hind margin acutely indentured . . . 1. **O. albinus** . . . . p. 455
Eyes with hind margin obtusely lobed . . . . . 2. **O. flavus** . . . . p. 455
}

3 {
Eyes broad . . . . 3. **O. carneolus** . . . p. 456
Eyes narrow . . . . 4. **O. amethystinus** . p. 456
}

1. O. albinus (Dyb.) 1874 *Gammarus a.*, B. Dybowsky in: Horae Soc. ent. Ross., *v.* 10 suppl. p. 71 t. 9 f. 3 | 1899 *Ommatogammarus a.*, T. Stebbing in: Tr. Linn. Soc. London, ser. 2 *v.* 7 p. 427 | 1893 *Gammarus fluviatilis* (part.)?, A. Della Valle in: F. Fl. Neapel, *v.* 20 p. 927.

Dorsal spines on pleon segments 4—6 very delicate, and solitary at the places corresponding to lateral groups. Head abruptly bent downward, the dorsal line being connected by a short curve with a frontal part at right angles to it. Eyes deep, closely adjoining the upright front of head, front margin slightly concave (Fig. 86), hinder cut into very unequal lobes, some of which are acute, clear flesh-coloured or white with a dash of rose-red. Antenna 1 (Fig. 86) about $\frac{1}{3}$ as long as body and twice as long as antenna 2, peduncle rather shorter (or in ♀ sometimes a little longer) than peduncle of antenna 2, 1st joint thick, longer than 2d and 3d combined, flagellum 3 or 4 times as long as the short peduncle, 38-jointed, accessory flagellum 6—8-jointed. Antenna 2 (Fig. 86), ultimate and penultimate joints of peduncle subequal, flagellum 10—13-jointed. Gnathopod 1, 6th joint piriform, very broad at base, hind margin much shorter than the very oblique but well defined palm. Gnathopod 2, 5th and 6th joints longer and more slender than in gnathopod 1, 6th joint oblong, but gradually widening to the oblique palm, which is shorter than hind margin. Peraeopods 3—5, 2d joint with convex hind margin but narrowed near the distal end, where it forms a free angle. Uropods 1 and 2 reaching the second third of uropod 3. Uropod 3, outer ramus about twice as long as inner, 2d joint distinct. Colour more or less yellowish white. L. 28 mm.

Accessory flagellum

Fig. 86. *O. albinus.*
Head, antennae 1 and 2.
[After B. Dybowsky.]

Lake Baikal. Depth 300—1300 m.

2. O. flavus (Dyb.) 1874 *Gammarus f.*, B. Dybowsky in: Horae Soc. ent. Ross., *v.* 10 suppl. p. 72 t. 9 f. 1 | 1899 *Ommatogammarus f.*, T. Stebbing in: Tr. Linn. Soc. London, ser. 2 *v.* 7 p. 427 | 1893 *Gammarus fluviatilis* (part.)?, A. Della Valle in: F. Fl. Neapel, *v.* 20 p. 928.

Dorsal spines very delicate, on pleon segments 4 and 5 in 4 little groups, on segment 6 only in 2, spines in a group varying between 1 and 3. Head bent at right angles as in O. albinus, but the connecting curve larger. Pleon segments 2 and 3, postero-lateral corners acutely produced. Eyes very large, occupying nearly half the surface of head and nearly meeting at top, front margin closely adjoining upright front of head, hind margin divided into small rounded lobes, lower straight or a little concave, black or, in specimens from a great depth, reddish. Antenna 1 about $\frac{1}{2}$ as long as body, in ♂ less, in ♀ more, than twice as long as antenna 2, peduncle shorter than peduncle of antenna 2 (or in ♀ sometimes a little longer), flagellum 3 or 4 times as long as peduncle, 35—43-jointed, accessory flagellum 4- or 5-jointed. Antenna 2, ultimate and penultimate joints of peduncle subequal, flagellum 13—19-jointed. Gnathopod 1, 6th joint described as slenderly piriform, but rather oblong, the hind margin being considerably longer than the oblique, very concave palm, which is armed with a strong spine at the centre. Gnathopod 2, 5th joint rather longer than in gnathopod 1, 6th oblong, considerably shorter but nearly as broad as in gnathopod 1, hind margin slightly indented near the short, concave, nearly transverse palm. Peraeopods 3—5, 2d joint narrower than in O. albinus, the hind margin at the narrowed distal end not forming a free angle. Uropod 1 reaching end of inner ramus of uropod 3. Uropod 3, outer ramus broad, about twice as long as the narrow inner one, 2^d joint indistinct. Colour yellow to clear honey-yellow. L. 30 mm.

Lake Baikal. Depth 100—1300 m.

3. **O. carneolus** (Dyb.) 1874 *Gammarus c.*, B. Dybowsky in: Horae Soc. ent. Ross., *v.* 10 suppl. p. 73 | 1899 *Ommatogammarus c.*, T. Stebbing in: Tr. Linn. Soc. London, ser. 2 *v.* 7 p. 427 | 1893 *Gammarus fluviatilis* (part.)?, A. Della Valle in: F. Fl. Neapel, *v.* 20 p. 928.

Dorsal spines of pleon segments 4—6 arranged as in O. amethystinus. Eyes large, unique in form, like the capital letter B reversed, with 2 great lobes in front, the hinder and lower margins being also broken up into lobes, ruby-red. Antenna 1 more than $^1/_2$ as long as body, nearly twice as long as antenna 2, peduncle shorter than peduncle of antenna 2, flagellum 3 or 4 times as long as peduncle, 31—45-jointed, accessory flagellum 4- or 5-jointed. Antenna 2, flagellum 11—15-jointed, peduncle and flagellum closely beset with simple setae, which are sometimes longer than the flagellum itself. Gnathopod 1, 6th joint slender, piriform; gnathopod 2, 6th joint oblong; in both gnathopods pretty closely covered with setae often exceeding the length of the joint. Uropod 1 not reaching the middle of uropod 3. Uropod 3, outer ramus about twice as long as inner. Colour flesh-red, variable in intensity. L. 18 mm.

Lake Baikal. Depth 300—700 m.

4. **O. amethystinus** (Dyb.) 1874 *Gammarus a.*, B. Dybowsky in: Horae Soc. ent. Ross., *v.* 10 suppl. p. 74 t. 9 f. 6 | 1899 *Ommatogammarus a.*, T. Stebbing in: Tr. Linn. Soc. London, ser. 2 *v.* 7 p. 427 | 1893 *Gammarus fluviatilis* (part.)?, A. Della Valle in: F. Fl. Neapel, *v.* 20 p. 927.

Dorsal spines very delicate, 1 or 2 on the obscure elevations of the hind margins of pleon segments 4 and 5, corresponding to middle and lateral groups, segment 6 having only 2 spinules. Dorsal line of head slightly convex, rostral point depressed. Eyes small, narrow, hind margin broken up into lobes, reddish or bright rose-red to whitish red. Antenna 1 $^2/_3$—$^5/_6$ as long as body, nearly twice as long as antenna 2, peduncle shorter than peduncle of antenna 2, flagellum 3 or 4 times as long as peduncle, 35—62-jointed, accessory flagellum 4- or 5-jointed. Antenna 2, ultimate and penultimate joints of peduncle subequal, flagellum 12—16-jointed. Gnathopod 1, 6th joint slenderly piriform, the concave palm forming a sinuous line with the hind margin. Gnathopod 2, 5th joint decidedly longer than that of gnathopod 1, 6th evenly oblong, shorter than that of gnathopod 1, but scarcely narrower than its widest part. Peraeopods 3—5, 2^d joint with convex hind margin produced below to a rather long obtusely ending process. Uropod 1 reaching middle of uropod 3. Uropod 3, outer ramus about twice as long as inner, its 2^d joint distinct. Colour amethystine to bluish red. L. 24 mm.

Lake Baikal. Depth 500—1300 m.

42. Gen. **Odontogammarus** Stebb.

1899 *Odontogammarus*, T. Stebbing in: Tr. Linn. Soc. London, ser. 2 *v.* 7 p. 427.

In general like Gammarus (p. 460). Lower front angle of side-plate 5 forming a tooth. Peduncle of antenna 1 not shorter than peduncle of antenna 2, 3^d joint as long as 2^d. Gnathopod 1, 6th joint not smaller than that of gnathopod 2. Peraeopods 3—5, 2^d joint produced at lower hind angle into a tooth. Uropod 3 not very long, the 2-jointed outer ramus (as in Gammarus) longer than the inner.

2 species.

Synopsis of species:

In side-plate 5 the tooth bent slightly outWard . . . . 1. O. calcaratus . . . p. 457
In side-plate 5 the tooth directed straight doWnWard . 2. O. margaritaceus . p. 457

1. **O. calcaratus** (Dyb.) 1874 *Gammarus c.*, B. Dybowsky in: Horae Soc. ent. Ross., *v.* 10 suppl. p. 54 t. 7 f. 4 | 1893 *G. c.* (part.), A. Della Valle in: F. Fl. Neapel, *v.* 20 p. 759 | 1899 *Odontogammarus c.*, T. Stebbing in: Tr. Linn. Soc. London, ser. 2 *v.* 7 p. 427.

Pleon segments 4 and 5 each with 3 little groups of dorsal spines on hind margin, segment 6 with 2 little groups, 2 or 3 spines to a group. Side-plate 5 in front half produced far down in a tooth-like angle, bent slightly outward from the body, the hind angle little prominent so that to the rear the lower margin forms almost a straight line; side-plate 6 in front half produced over 2^d joint of peraeopod 4, slightly pointed, not forming a tooth; side-plate 7 rounded in front. Eyes small, reniform, black. Antenna 1 rather more than $^1/_2$ as long as body and twice as long as antenna 2, peduncle as long as or a little longer than peduncle of antenna 2, 1st joint scarcely longer and 2^d perhaps a little shorter than 3^d, flagellum twice as long as peduncle, 45—62-jointed, accessory flagellum 8—10-jointed. Antenna 2, ultimate joint of peduncle longer than penultimate, flagellum shorter than peduncle, 22-jointed, with calceoli in ♂. Characters of gnathopods 1 and 2, 6th joint differing in text and figure. Peraeopods 3 and 4, 2^d joint with front margin very convex, much more so than in peraeopod 5, in all the hind margin is slightly concave, ending below with a long tooth-like angle, reaching far over the 3^d joint (wanting in young from the brood pouch). Uropod 1 reaching nearly to the middle of uropod 3. Uropod 3, 2^d joint of outer ramus about $^1/_7$ of 1st joint, inner ramus $^3/_4$ as long as outer, both with plumose setae on both margins. Colour dorsally greenish yellow, iridescent, on sides dark yellow, here also nacreous. L. 36 mm (including 6 mm for uropod 3).

Lake Baikal. Depth 50—100 m.

2. **O. margaritaceus** (Dyb.) 1874 *Gammarus m.*, B. Dybowsky in: Horae Soc. ent. Ross., *v.* 10 suppl. p. 21, 56 | 1899 *Odontogammarus m.*, T. Stebbing in: Tr. Linn. Soc. London, ser. 2 *v.* 7 p. 427 | 1893 *Gammarus calcaratus* (part.)?, A. Della Valle in: F. Fl. Neapel, *v.* 20 p. 929.

Like O. calcaratus, but head more convex and in front more obtuse; side-plate 5 deeper, with front and hind angles nearly equally long, running out into triangular pointed lappets, of which the front one is somewhat the sharper and longer; side-plate 6, with front corner rounded, hind one produced acutely downward. Eyes irregularly linear-reniform, with broken hind margin, white, scarcely or not at all visible in spirit. Antenna 1, flagellum 53—63-jointed, accessory flagellum 7-jointed; antenna 2, flagellum 19—27-jointed. Peraeopods 3—5, 2^d joint broader, with hind corner more weakly developed. Uropod 3 shorter, but inner ramus relatively longer, outer margin of outer ramus without plumose setae and its 2^d joint diminutive. Colour bright yellowish playing into bluish.

Lake Baikal. Depth 150—1000 m.

43. Gen. **Dikerogammarus** Stebb.

1899 *Dikerogammarus*, T. Stebbing in: Tr. Linn. Soc. London, ser. 2 *v*.7 p.428.

Agreeing in general with Gammarus (p. 460). Pleon segments 4 and 5 each raised dorsally to a spiniferous tubercle. Antenna 1 the longer, accessory flagellum well developed. Gnathopods 1 and 2 larger in ♂ than in ♀, 2ᵈ larger than 1ˢᵗ.

5 species.

Synopsis of species:

1 { Uropod 3, rami Very unequal — **2.**
  { Uropod 3, rami not Very unequal — **4.**

2 { Head arched longitudinally . . . . 1. **D. macrocephalus** . p. 458
  { Head not arched longitudinally — **3.**

3 { Peraeopods 3—5, 2ᵈ joint broadly expanded . . 2. **D. haemobaphes** . . p. 458
  { Peraeopods 3—5, 2ᵈ joint not Very broadly ex-
  { panded . . . . 3. **D. grimmi** . . . . . p. 459

4 { Gnathopod 2 much larger than gnathopod 1 . 4. **D. verreauxii** . . . p. 459
  { Gnathopod 2 not much larger than gnathopod 1 5. **D. fasciatus** . . . . p. 460

1. D. macrocephalus (O. Sars) 1896 *Gammarus m.*, G. O. Sars in: Bull. Ac. St.-Pétersb, ser. 5 *v*.4 p.453 t. 7 f. 1—11 | 1899 *Dikerogammarus m.*, T. Stebbing in: Tr. Linn. Soc. London, ser. 2 *v*. 7 p. 428.

Body robust, back smoothly rounded, except pleon segments 4 and 5, tubercle of pleon segment 4 tipped with 4, of segment 5 with 2 spines; segment 4 without lateral spines, segment 5 left doubtful, segment 6 with 2 on each side at base of telson. Head large. dorsal surface strongly vaulted, with peculiar areolated appearance from points of insertion of strong muscles converging to the mouth area. Side-plates not particularly large. Pleon segment 1, postero-lateral corners rounded, in segments 2 and 3 acutely produced. Eyes very small, distinctly reniform. Antenna 1 slender, not half as long as body. 1ˢᵗ joint as long as 2ᵈ and 3ᵈ combined, flagellum about 27-jointed, accessory flagellum 6-jointed. Antenna 2 stronger, rather setose, flagellum 10-jointed. Gnathopods 1 and 2, 6ᵗʰ joint oblong oval, in gnathopod 2 (at least in ♂) fully twice as large as in gnathopod 1, in both palm oblique, well defined, rather shorter than hind margin. Peraeopods 1—5 . rather slender, not very long. Peraeopods 3 and 4, 2ᵈ joint wider above than below, hind margin sinuous. Peraeopod 5, 2ᵈ joint much larger, broadly oval. hind expansion produced below to an obtusely pointed lobe. Uropods 1 and 2. rami subequal, not long, with apical groups of spines, and here and there a lateral spine. Uropod 3 reaching far beyond the others, peduncle short, outer ramus long, sublinear, fringed densely all round with slender setae, and having 3 groups of spines on each margin, 2ᵈ joint minute; inner ramus little over $\frac{1}{8}$ as long as peduncle. Telson broader than long, each lobe having 3 spinules on narrowly truncate apex L. ♂ 24 mm.

Caspian Sea. Depth 66 m.

2. D. haemobaphes (Eichw.) ?1771 *Oniscus pulex, Cancer p.* (err., non Linné 1758!), Pallas, Reise Ruß., *v*.1 p. 477 | 1841 *Gammarus haemobaphes*, Eichwald in: N. Mém. Soc. Moscou, *v*.7 p.230 t. 37 f. 7 a—c | 1899 *Dikerogammarus h.*, T. Stebbing in: Tr. Linn. Soc. London, ser. 2 *v*.7 p.428 | 1894 *Gammarus h.* + *G. robustoides* (*G. aralo-*

caspius O. Grimm in MS.), G. O. Sars in: Bull. Ac. St.-Pétersb., ser. 5 *v.* 1 p. 215 t. 8; p. 358 t. 12 | 1862 *G. caspius* (non *Oniscus c.* Pallas 1771!), Bate, Cat. Amphip. Brit. Mus., p. 214 t. 38 f. 3 | 1875 *G. aralensis*, Uljanin in: Fedtschenko, Turkestan, Crust. p. 1 t. 5 f. 15.

Very like D. macrocephalus, differing in the following points. Pleon segments 4 and 5, dorsal conical tubercle rather smaller, each with 2 apical spines, segment 4 with a spinule also on each side. Head normal. Pleon segment 3, postero-lateral corners rather quadrate than acutely produced. Eyes well developed. Antenna 1, accessory flagellum 7—9-jointed. Antenna 2, flagellum about 8-jointed. · Peraeopods 3—5 rather stout. Peraeopod 3, 2^d joint broader above than below, but nearly quadrangular, lower hind angle slightly produced. Peraeopod 4, 2^d joint much more expanded above than below. Uropod 3, outer ramus with 1 group of spines on inner margin, the small inner ramus more than half as long as the short peduncle. Telson with only 1 or 2 apical spines on the very narrow apex of each lobe. Colour brownish green, hind margins of segments laterally tinged with pink; eyes dark. L. 9—16 mm.

Black Sea; Caspian Sea (shore to 75 m).

3. **D. grimmi** (O. Sars) 1896 *Gammarus g.*, G. O. Sars in: Bull. Ac. St.-Pétersb., ser. 5 *v.* 4 p. 448 t. 6 f. 1—10 | 1899 *Dikerogammarus g.*, T. Stebbing in: Tr. Linn. Soc. London, ser. 2 *v.* 7 p. 428.

Very like D. haemobaphes. Body rather slender. Pleon segments 4 and 5, tubercle very prominent, narrow, on segment 4 tipped with 4 spines and flanked by 2 unequal spines on each side, on segment 5 tipped with only 2 spines, segment 6 carrying 3 spines on each side. Pleon segment 3, postero-lateral corners acutely produced. Eyes oblong oval, scarcely reniform, usually dark, but variable. Antenna 1, accessory flagellum 7-jointed. Gnathopods 1 and 2 still more unequal in ♂ than in ♀, gnathopod 2 in ♂ very large, 6th joint very tumid at the base. Peraeopods 1—5 rather slender, 2^d joint in peraeopods 3—5 conspicuously narrower than in D. haemobaphes, in peraeopod 5 this joint becoming oblong quadrangular, with nearly straight, much serrate hind margin, produced to a narrowly rounded lobe. Uropod 3, outer ramus very long and narrow, fringed with setae, but without conspicuous spines; inner ramus about half as long as peduncle, almost spine-like in narrowness. Telson with 2 spinules on each obtusely pointed apex. L. ♂ 27 mm.

Caspian Sea. Depth 66—203 m.

4. **D. verreauxii** (Bate) 1862 *Gammarus v.*, *G. verrauxii*, (H. Milne Edwards in MS.?) Bate, Cat. Amphip. Brit. Mus., p. 210 t. 37 f. 5 | 1899 *Dikerogammarus verreauxii*, T. Stebbing in: Tr. Linn. Soc. London, ser. 2 *v.* 7 p. 428 + 1893 *Gammarus locusta* (part.)?, A. Della Valle in: F. Fl. Neapel, *v.* 20 p. 746.

Pleon segments 4 and 5 with dorsal tubercles (without spines in figure). Eyes ovate. Antenna 1 about $^1/_4$ as long as body, flagellum shorter than peduncle. Antenna 2 shorter, peduncle not longer than in antenna 1. Gnathopod 2 twice as long as gnathopod 1, 6th joint long ovate, tapering, palm the entire length of the inferior margin, straight, finger (in figure) very long. Peraeopods 3—5 not very unequal, 2^d joint not much dilated, narrowed below (in figure). Uropod 2 considerably shorter than uropod 1, uropod 3 a little longer (reaching a little further?) than uropod 1, rami equal. L. 37 mm.

New Holland.

5. **D. fasciatus** (Say) 1818 *Gammarus f.*, Say in: J. Ac. Philad., *v.* 1 II p. 374 |
1830 *G. f.*, H. Milne Edwards in: Ann. Sci. nat., *v.* 20 p. 367 | ?1862 *G. f.*, Bate, Cat.
Amphip. Brit. Mus., p. 210 t. 37 f. 6 | 1874 *G. f.*, S. I. Smith in: Rep. U. S. Fish Comm.,
v. 2 p. 653 | 1899 *Dikerogammarus f.*, T. Stebbing in: Tr. Linn. Soc. London, ser. 2
v. 7 p. 428.

Pleon segments 4 and 5 slightly angulated dorsally at hind margin,
each armed with 3 fascicles of spines, the median raised on a distinct protuberance, segment 6 also with 3 fascicles. Antenna 1, 2^d joint surrounded
with long setae, which reach the 5th joint of flagellum, accessory flagellum
5- or 6-jointed. Antenna 2 setose. Gnathopod 1 in $\male$, 6th joint much
narrowed distally, palm very oblique, defined from the spineless hind margin
by small spines, and having a stout spine at the middle, in $\female$ only slightly
narrowed distally, palm less oblique. Gnathopod 2 in $\male$, 5th joint rather
longer than in gnathopod 1, 6th with margins nearly parallel, but widening
a little to the palm, which is a little oblique, slightly concave, with a lamellar
edge, carrying a stout spine near the middle, and defined by groups of small
spines; in $\female$ 5th joint proportionally longer, as wide as the narrow, parallel-
sided 6th, which has a short, nearly transverse palm. Uropod 3, inner ramus
usually with 1 or 2 spines on the inner margin. Telson, each lobe with a
spine and 1 or 2 hairs on outer margin as well as a few spines and hairs
at the apex. L. 10—15 mm.

Northern United States of America. Streams and ponds.

44. Gen. **Gammarus** F.

1758 *Cancer* (part), Linné, Syst. Nat., ed. 10 p. 625 | 1775 *Gammarus* (part.),
J. C. Fabricius, Syst. Ent., p. 418 | 1813/14 *G.*, Leach in: Edinb. Enc., *v.* 7 p. 402, 432 |
1862 *G.*, Bate, Cat. Amphip. Brit. Mus., p. 203 | 1874 *G.* (part.), B. Dybowsky in: Horae
Soc. ent. Ross., *v.* 10 suppl. p. 19 | 1876 *G.*, A. Boeck, Skand. Arkt. Amphip., *v.* 2 p. 364 |
1886 *G.*, Gerstaecker in: Bronn's Kl. Ordn., *v.* 5 II p. 511 | 1888 *G.*, T. Stebbing in:
Rep. Voy. Challenger, *v.* 29 p. 1005, 1673 | 1893 *G.*, A. Della Valle in: F. Fl. Neapel,
v. 20 p. 756 | 1894 *G.*, G. O. Sars, Crust. Norway, *v.* 1 p. 496 | 1793 [Subgen.] *Gammarellus*
(part.), J. F. W. Herbst, Naturg. Krabben Krebse, *v.* 2 p. 106.

Body without carinae, teeth or tubercles; dorsal spinules generally in
median and lateral groups on pleon segments 4—6, but sometimes wanting on 4th
or on both 4th and 5th. Head without conspicuous rostrum. Side-plates 1—4
usually rather deep. Eyes present. Antenna 1 almost always longer than antenna 2,
though with shorter peduncle, accessory flagellum almost always with more
than one joint. Antenna 2, flagellum not or little longer than peduncle.
Upper lip as broad as deep, faintly emarginate at apex. Lower lip without
inner lobes. Mandible normal. Maxilla 1 (Fig. 87 p. 476), inner plate with several
setae, outer with (11?) more or less serrate spines, 2^d joint of palp with
spine-teeth on the apex of one maxilla and slender spines on the other.
Maxillipeds normal. Gnathopods 1 and 2 subchelate, stronger in $\male$ than
in $\female$, gnathopod 2 usually the larger, in the $\male$ often much the larger.
Peraeopods 3—5, 2^d joint variable, much or little expanded. Branchial
vesicles simple, with narrow attachment. Uropod 3, outer ramus usually
long and having a small 2^d joint. Telson cleft, usually to the base.

30 accepted species, 1 doubtful.

Synopsis of accepted species:

$1\begin{cases}\text{Antenna 1, accessory flagellum of only one joint} & \text{1. }\textbf{G. guernei} \ldots \ldots \text{ p. 462}\\ \text{Antenna 1, accessory flagellum of more joints}\\ \quad\text{than one — 2.}\end{cases}$

2 { Antenna 1 shorter than antenna 2 — **3.**
{ Antenna 1 not shorter than antenna 2 — **5.**

3 { Antenna 2, ultimate joint of peduncle shorter
than penultimate · . . . . 2. **G. abbreviatus** . . . p. 462
{ Antenna 2, ultimate joint of peduncle longer
than penultimate — **4.**

4 { Gnathopod 2, 6th joint narrow, subcylindric . 8. **G. annulatus** . . . . p. 463
{ Gnathopod 2, 6th joint not narrow, subrectangular 4. **G. natator** . . . . . p. 463

5 { Pleon segment 4 Without dorsal spinules --- **6.**
{ Pleon segment 4 With dorsal spinules — **16.**

6 { Pleon segment 5 Without dorsal spinules — **7.**
{ Pleon segment 5 With dorsal spinules — **14.**

7 { Uropod 3, outer ramus little longer than peduncle 5. **G. obesus** . . . . . p. 464
{ Uropod 3, outer ramus much longer than
peduncle — **8.**

8 { Uropod 3, inner ramus less than half as long
as outer — **9.**
{ Uropod 3, inner ramus fully half as long as
outer — **13.**

9 { Uropod 3, outer ramus With rather large
2d joint — **10.**
{ Uropod 3, outer ramus With Very small 2d joint — **11.**

10 { Uropod 3, inner ramus scarcely ¹/₄ of outer . 6. **G. macrurus** . . . . p. 464
{ Uropod 3, inner ramus nearly ¹/₂ of outer . . 7. **G. compressus** . . . p. 465

11 { Peraeopods With Very few setae . . . . . . 8. **G. subnudus** . . . . p. 465
{ Peraeopods With numerous setae — **12.**

12 { Pleon segments 4 and 5 dorsally upraised,
gibbous . . . . 9. **G. deminutus** . . . p. 466
{ Pleon segments 4 and 5 not dorsally upraised
or gibbous . . . . 10. **G. similis** . . . . . p. 466

13 { Mandibular palp of normal size . . . . 11. **G. weidemanni** . . . p. 467
{ Mandibular palp abnormally large . . . . . 12. **G. maeoticus** . . . . p. 467

14 { Peraeopod 5, 2d joint produced behind 3d . 13. **G. crassus** . . . . . p. 467
{ Peraeopod 5, 2d joint not produced behind 3d — **15.**

15 { Peraeopod 5, 2d joint broad, abruptly contracted
distally . . . . 14. **G. warpachowskyi** . p. 468
{ Peraeopod 5, 2d joint not broad, not abruptly
contracted . . . . 15. **G. pauxillus** . . . . p. 469

16 { Pleon segments 2 and 3, postero-lateral corners
obliquely truncate . . . . . . · . . . . 16. **G. kietlinskii** . . . p. 469
{ Pleon segments 2 and 3, postero-lateral corners
not truncate — **17.**

17 { Uropod 3, inner ramus less than half as long as
outer — **18.**
{ Uropod 3, inner ramus not less than half
as long as outer — **24.**

18 { Uropod 3, outer ramus remarkably long — **19.**
{ Uropod 3, outer ramus not remarkably long — **21.**

19 { Gnathopod 1, 6th joint much broader than that
of gnathopod 2 . . . . 17. **G. andrussowi** . . · p. 469
{ Gnathopod 1, 6th joint not much broader than
that of gnathopod 2 — **20.**

1. **G. guernei** Chevreux 1889 *G. g.*, Chevreux in: Bull. Soc. zool. France, *v.* 14 p. 294 f. | 1893 *G. pungens* (part.), A. Della Valle in: F. Fl. Neapel, *v.* 20 p. 764.

In form resembling G. pulex (p. 474). Pleon segments 4—6 with small dorsal fascicles of spines, 6 spines on segment 4, 8 on segment 5, 2 on segment 6. Head (in figure) with lateral angles scarcely at all prominent. Side-plates 1—4 rather deep. Pleon segments 1—3, postero-lateral angles acute. Eyes small, reniform. Antenna 1 not quite half as long as body, peduncle not long, 2d joint rather shorter than 1st, and 3d than 2d, flagellum quite smooth, 19-jointed, accessory flagellum 1-jointed. Antenna 2 considerably shorter, flagellum without setae, 9-jointed. Gnathopods 1 and 2 similar, but 5th and 6th joints longer in gnathopod 2, 6th joint oblong, palm well defined, short, a little oblique. Peraeopod 1, 4th joint long and wide, strongly setose on hind margin, 5th joint much narrower and shorter, likewise setose. Peraeopod 2 shorter and slighter, not setose. Peraeopods 3—5 rather long, 2d joint well expanded, oval. Uropod 3 reaching far beyond the others, outer ramus very long, with small 2d joint, inner ramus rudimentary, oval. Telson very short, each broadly oval lobe having a fascicle of spines at middle of outer margin, and another of 5 spines at apex. L. 6 mm.

Azores (Flores). Torrents.

2. **G. abbreviatus** O. Sars 1894 *G. a.*, G. O. Sars in: Bull. Ac. St.-Pétersb., ser. 5 *v.* 1 p. 365.

Body rather short and stout, pleon segment 4 smooth, segments 5 and 6 each with 1 or 2 very small dorsal spinules. Head, lateral lobes slightly prominent, evenly rounded. Side-plates 1—4 distally conspicuously crenulate and setose, 1st obliquely expanded below, 4th very broad. Pleon segments 2 and 3, postero-lateral corners acute. Eyes oval reniform. Antenna 1 very short, not $^1/_4$ as long as body, 1st joint large, longer than 2^d and 3^d combined, flagellum about as long as peduncle, 9-jointed, accessory flagellum 4- or 5-jointed. Antenna 2 rather longer, setose, ultimate joint of peduncle shorter than penultimate, flagellum about 7-jointed. Gnathopods 1 and 2 in ♀ similar, 2^d slightly the larger, 6th joint oblong oval, palm rather oblique, pretty well defined. Gnathopods 1 and 2 in ♂ stronger, much more unequal, 6th joint in gnathopod 1 narrowing distally, palm more oblique, in gnathopod 2 much larger, oblong oval, with the usual spine in middle of palm. Peraeopods 1 and 2 strongly built, densely setose, 4th joint large and expanded. Feraeopods 3—5, 2^d joint expanded, rounded quadrangular in peraeopod 3, expanded above, tapering below in peraeopod 4, much larger and regularly oval in peraeopod 5, produced downward behind in narrowly rounded lobe. Uropods 1 and 2 reaching beyond peduncle of uropod 3. Uropod 3 not very long, outer ramus densely setose, with small but obvious 2^d joint, inner ramus about as long as the short peduncle. Telson about as long as broad, each lobe with 3 apical spines, otherwise smooth. L. ♀ 12, ♂ 13 mm.

Caspian Sea. Among Zostera.

3. **G. annulatus** S. I. Sm. 1873 *G. a.*, (S. I. Smith in:) A. E. Verrill in: Rep. U. S. Fish Comm., *v.* 1 p. 557 | 1893 *G. locusta* (part.), A. Della Valle in: F. Fl. Neapel, *v.* 20 p. 760.

Pleon segment 4 with median fascicle of 2 large and 2 small spines, segments 5 and 6 with lateral fascicles as well as median. Head, lateral lobes truncate, somewhat prominent. Eyes scarcely reniform, lower border remote from lateral lobes of head. Antenna 1 shorter than antenna 2. Antenna 2, ultimate joint of peduncle longer than penultimate, flagellum slender, with rather elongate joints. Gnathopod 1, 6th joint rather elongated, palm very oblique. Gnathopod 2, 6th joint very narrow, elongated, sub-cylindrical, slightly flattened on inner side, palm longitudinal, scarcely distinct from hind margin. Colour greyish white, hind margins of segments annulated with brown, red spots on some of the pleon segments. L. 12—18 mm.

North-Atlantic (New Haven [Connecticut], and Eastport [Maine]). . Under stones, shore.

4. **G. natator** S. I. Sm. 1873 *G. n.*, (S. I. Smith in:) A. E. Verrill in: Rep. U. S. Fish Comm., *v.* 1 p. 558 | 1893 *G. locusta* (part.), A. Della Valle in: F. Fl. Neapel, *v.* 20 p. 760.

Pleon segment 4 with median fascicles of spines, segments 5 and 6 with median and lateral fascicles. Side-plates 1—3, lower margin setose. Pleon segments 2 and 3, postero-lateral angles acutely produced. Eyes large, long, slightly reniform. Antenna 1 shorter than antenna 2, setose, many of the setae plumose. Antenna 2 also setose, ultimate joint of peduncle longer than penultimate, flagellum about $^2/_3$ length of peduncle. Gnathopod 1 in ♂ more slender than gnathopod 2, 6th joint oval, palm continuous with hind margin, defined by 2 small spines, with stout spine in middle of laminar margin, finger strongly curved. Gnathopod 2, 6th joint subrectangular, palm slightly oblique, armed as in gnathopod 1, but with notch for the stout spine. In ♀ gnathopods 1 and 2 have 6th joint smaller and slenderer, somewhat oval, nearly alike in both gnathopods. Uropod 3, rami long, lanceolate, fringed with long plumose setae, outer with slender, almost spiniform 2^d joint, inner as long as 1st joint of outer. Telson, each lobe nearly thrice as long as broad. L. 10—12 mm.

Vineyard Sound [United States of America]. At surface, usually among floating sea-Weed.

5. **G. obesus** O. Sars 1894 *G. o.*, G. O. Sars in: Bull. Ac. St.-Pétersb., ser. 5 *v.* 1 p. 368 t. 15.

Body extremely short and stout, more so than in any other known Gammarus, dorsally smooth, except for a pair of very small spinules on pleon segment 6. Head, lateral lobes little prominent, evenly rounded. Side-plates 1—4 large and deep, distally fringed with setae, 1st rather broadly expanded distally, 4th not as broad as deep. Pleon segments 2 and 3, postero-lateral corners obtusely quadrate. Eyes oval, reniform, not large. Antenna 1 about $^1/_8$ length of body, 1st joint large, longer than 2^d and 3^d combined, flagellum rather longer than peduncle, 15-jointed, accessory flagellum 2-jointed. Antenna 2 rather shorter, flagellum as long as ultimate and penultimate joints of peduncle combined, 7-jointed. Gnathopods 1 and 2, 6th joint oblong oval, palm well defined, not very oblique, more so in ♂ than in ♀, gnathopod 2 a little stronger than gnathopod 1, and both gnathopods rather stronger in ♂ than in ♀. Peraeopods 1—5 rather strong, densely setose, 5th joint short, finger strong. Peraeopods 3 and 4, 2^d joint narrowing distally, so that the lower hind corner is not free. Peraeopod 5, 2^d joint widely expanded, widened distally, with broadly rounded lobe. Uropods 1 and 2 with only apical spines. Uropod 3 unusually short and stout, reaching little beyond the others, outer ramus scarcely longer than peduncle, carrying plumose setae, 2^d joint minute, inner ramus scale-like, with 1 apical spine. Telson much broader than long, each lobe having 1 spine and 2 setules on the narrowly truncated apex. L. ♀ 8, ♂ 9 mm.

Caspian Sea. Shallow Water.

6. **G. macrurus** O. Sars 1894 *G. m.*, G. O. Sars in: Bull. Ac. St.-Pétersb., ser. 5 *v.* 1 p. 350 t. 10 f. 17—27.

♀. Body slender, dorsally smooth except for a pair of very small spinules on pleon segment 6. Head, lateral lobes broadly rounded, prominent. Side-plates 1—4 rather large, fringed with scattered setules; 1st a little expanded, 4th very large, as broad as deep. Pleon segments 2 and 3, postero-lateral corners produced to a rather obtuse point. Eyes not very large, oblong oval, dark. Antenna 1 scarcely more than $^1/_4$ length of body, 1st joint large, nearly twice as long as 2^d and 3^d combined, flagellum about as long as peduncle, 6-jointed, accessory flagellum 3-jointed. Antenna 2 a little shorter, antepenultimate joint of peduncle wide, projecting, ultimate joint shorter than penultimate, flagellum 4-jointed. Gnathopods 1 and 2 small and feeble, 6th joint subequal to 5th in length and scarcely broader, these 2 joints being rather longer, but not wider, in gnathopod 2 than in gnathopod 1. Peraeopods 1 and 2, 4th joint rather expanded, this and the 5th setose. Peraeopod 3, 2^d joint rounded quadrangular. Peraeopod 4, 2^d joint narrower and scarcely longer, narrowed close to distal end. Peraeopod 5, 2^d joint much larger, oval, hinder expansion produced downward in a broadly rounded lobe. Uropods 1 and 2, rami equal, narrow. Uropod 3 long, reaching much beyond the others, outer ramus long, with setae on inner margin, its 2^d joint fully half as long as 1st and as long as the peduncle, inner ramus shorter than peduncle, tipped with 2 spines. Telson much longer than broad, tapering, each narrow lobe having 1 lateral and 1 apical spinule. L. 6 mm. — ♂ not known.

Caspian Sea.

7. **G. compressus** O. Sars 1894 *G. c.*, G. O. Sars in: Bull. Ac. St.-Pétersb., ser. 5 v. 1 p. 353 t. 11 f. 1—10.

♀. Body slender, much compressed, dorsally smooth except for pair of minute spinules on pleon segment 6. Head without rostrum, lateral lobes narrowly rounded, little prominent. Side-plates 1—4 large, setulose, 1st widely expanded forward, 4th as broad as deep. Pleon segments 2 and 3, postero-lateral corners subquadrate. Eyes not large, narrowly oblong, dark. Antenna 1 scarcely more than $^1/_4$ as long as body, 1st joint massive, longer than 2^d and 3^d combined, flagellum subequal to peduncle, 9-jointed, accessory flagellum 3-jointed. Antenna 2 subequal to antenna 1, ultimate joint of peduncle shorter than penultimate, flagellum subequal to peduncle, 6-jointed. Gnathopods 1 and 2 rather small and feeble, palm short and nearly transverse, 5th and 6th joints rather narrower and longer in gnathopod 2 than in gnathopod 1. Peraeopods 1 and 2 strongly built, especially peraeopod 2, 4th joint large, expanded, 5th oval, both densely fringed with setae on hind margin. Peraeopods 3—5 subequal in length, 4th joint rather broad, 2^d joint rounded oval in peraeopod 3, narrower in peraeopod 4, with very sinuous hind margin, much larger and oval in peraeopod 5, ending in a produced, rounded lobe. Uropods 1 and 2, rami subequal, narrow, with spines only at apex. Uropod 3 of moderate size, reaching beyond the others, outer ramus about twice as long as peduncle, 2^d joint about half as long as 1st, inner ramus nearly half as long as outer, tapering, tipped with a spine and 2 or 3 setules. Telson fully as long as broad, lobes gradually diverging, each with 2 spines at the somewhat truncated apex. L. 7 mm. — ♂ unknown.

Caspian Sea.

8. **G. subnudus** O. Sars 1896 *G. s.*, G. O. Sars in: Bull. Ac. St.-Pétersb., ser. 5 v. 4 p. 451 t. 6 f. 11—19.

♀. Back broadly rounded, dorsally smooth except for a few small hairs at end of pleon segments 4—6. Head, lateral lobes broad, obliquely truncated, lower corner the more prominent. Side-plates 1—4 large, not setose, 1st not expanded distally, 4th broad. Pleon segments 2 and 3, postero-lateral corners quadrate. Eyes of medium size, oblong oval, dark. Antenna 1 scarcely more than $^1/_3$ as long as body, 1st joint scarcely as long as 2^d and 3^d combined, flagellum subequal to peduncle, 16-jointed, accessory flagellum 3-jointed. Antenna 2 a little shorter, flagellum shorter than ultimate and penultimate joints of peduncle. Gnathopod 1, 5th joint short, 6th oblong oval, palm rather oblique, defining angle obtuse. Gnathopod 2 similar in shape, but 6th joint more than twice as large as in gnathopod 1. Peraeopods 1—5 not very slender, very scantily furnished with spines and setae, 2^d joint in peraeopod 3 rounded quadrangular, in peraeopod 4 gradually but greatly narrowing distally so as to have no free corner, in peraeopod 5 large, oval, with rounded lobe much produced downward. Uropod 3 of moderate size, but reaching much beyond the others, outer ramus broad, about twice as long as peduncle, fringed with setae, 2^d joint small but distinct, inner ramus very small, scale-like. Telson broader than long, the lobes each narrowing to a blunt apex, tipped with a spine and setule. L. 8 mm. — ♂ unknown.

Caspian Sea' (Bay of Baku). Depth 4—11 m.

9. **G. deminutus** Stebb.*) 1894 *G.* minutus (non *G. pulex* m. P. Gervais 1835!), G. O. Sars in: Bull. Ac. St.-Pétersb., ser. 5 *v.* 1 p. 347 t. 10 f. 1—26.

Body rather short and stout. Pleon segments 4 and 5 dorsally strongly convex, gibbous, 4th partly overlapping 5th dorsally, each with a very small fascicle of hairs, segment 6 with a pair of spinules. Head, lateral lobes broadly rounded, rather prominent. Side-plates 1—4 deep, but less so in ♂ than in ♀, setose, 1st slightly expanded, 4th not nearly as broad·as deep. Pleon segments 2 and 3, postero-lateral corners acutely quadrate. Eyes not large, oblong oval, dark. Antenna 1 little more than ¹/₄ as long as body, 1st joint large, nearly twice as long as 2d and 3d combined. flagellum subequal to peduncle, 8-jointed, accessory flagellum 3-jointed. Antenna 2 a little shorter, antepenultimate joint of peduncle thick, distally projecting, ultimate joint shorter than penultimate, flagellum about half as long as peduncle, 4-jointed. Gnathopods 1 and 2 in ♀ small, subequal, 6th joint oval quadrangular, palm short, almost transverse. Gnathopods 1 and 2 in ♂ stronger, more unequal, 6th joint in gnathopod 1 much narrower and with palm more oblique than in gnathopod 2. Peraeopods 1 and 2, 4th joint rather large, this and 5th strongly setose on hind margin. Peraeopods 3—5 little differing in length, 2d joint in peraeopod 3 subquadrate, lower hind corner nearly rectangular, in peraeopod 4 much narrower, somewhat expanded above. with no free corner below, hind margin sinuous, in peraeopod 5 unusually large, front margin little convex, hind margin extremely so, with broadly rounded lobe produced downward. Uropods 1 and 2, rami with apical spines only. Uropod 3 of moderate size, reaching beyond the others. peduncle short, outer ramus narrow, with 1 fascicle of spines near middle of outer margin, others at apex of 1st joint, 2d joint very small, tipped with 3 setules, inner ramus very small, with 1 apical spinule. Telson with divergent lobes, each having 2 small spinules on the obtuse apex. L. ♀ 4, ♂ 5 mm.

Caspian Sea.

10. **G. similis** O. Sars 1894 *G. s.*, G. O. Sars in: Bull. Ac. St.-Pétersb., ser. 5 *v.* 1 p. 355 t. 11 f. 11—20.

Very like G. compressus (p. 465), less compressed, back broad, smooth, except for 2 minute spinules on pleon segment 6. Head, lateral lobes obtusely rounded, prominent. Side-plates 1—4 setose, 1st scarcely expanded, 4th deeper than broad. Pleon segments 2 and 3, postero-lateral corners subquadrate. Eyes of moderate size. oblong reniform. Antennae 1 and 2 equal, scarcely more than ¹/₄ as long as body. Antenna 1, 1st joint large, nearly twice as long as 2d and 3d combined, flagellum rather longer than peduncle, 11-jointed, accessory flagellum 4- or 5-jointed. Antenna 2, flagellum nearly as long as peduncle, 7-jointed. Gnathopods 1 and 2 in ♀ small, 6th joint oblong, palm little oblique, 5th and 6th joints longer and narrower in gnathopod 2 than in gnathopod 1. Gnathopods 1 and 2 in ♂ more powerful, palm more oblique, 5th and 6th joints little longer but much broader in gnathopod 2 than in gnathopod 1. Peraeopod 5, 2d joint comparatively larger and more expanded than in G. compressus. Uropod 3 extending much beyond the others, outer ramus long, setose on inner margin, 2d joint

---

*) Nom. nov.

small, inner ramus scarcely $^1/_3$ as long as outer, about as long as the short peduncle. Telson rather broader than long, lobes divergent, each with 2 spinules on rounded apex. L. 9 mm.

Caspian Sea. From stones on shore, and among Zostera.

11. **G. weidemanni** O. Sars 1896 *G. w.*, G. O. Sars in: Bull. Ac. St.-Pétersb., ser. 5 *v.* 4 p. 462 t. 9 f. 1—11.

Body rather robust, back broadly rounded, smooth except for 2 pairs of minute spinules on pleon segment 6. Head, lateral lobes obliquely rounded, upper corner the more prominent. Side-plates 1—4 setose, 1st rather expanded, 2^d a little narrowed distally, 4th as broad as deep. Pleon segments 2 and 3, postero-lateral corners acutely quadrate. Eyes of moderate size, oval reniform, dark. Antennae 1 and 2 equal, about $^1/_4$ as long as body. Antenna 1, 1st joint massive, longer than 2^d and 3^d combined, flagellum longer than peduncle, 16-jointed, accessory flagellum 5-jointed. Antenna 2, antepenultimate joint of peduncle projecting, ultimate joint shorter than penultimate, flagellum as long as these two combined, 8-jointed. Mandibular palp about as long as the trunk, setose, 3^d joint curved, narrowly truncate at apex. Gnathopods 1 and 2, 5th joint short, 6th rectangular, palm rather oblique; rather strong in ♀, but stronger in ♂, in both sexes gnathopod 2 larger than gnathopod 1, but similar in shape. Peraeopods 1 and 2, 4th joint large, especially in peraeopod 2, both long and broad, 5th joint short and broad, both densely setose. Peraeopods 3—5 rather stout, spinose, 2^d joint in peraeopod 3 rounded quadrangular, with rounded lobe produced downward, in peraeopod 4 a little longer, upper part wider, but narrowed below, leaving no free corner, in peraeopod 5 much larger, hind margin evenly convex, produced downward in broadly rounded lobe. Uropod 3 reaching much beyond the others, outer ramus about twice as long as peduncle, broad, fringed with setae, 2^d joint minute, inner ramus longer than peduncle, more than half as long as outer, with setae and 2 spines on inner margin. Telson scarcely longer than broad, lobes tapering, divergent, each with 3 spines on blunt apex. L. ♂ 11 mm.

Caspian Sea (Bay Karabugas). On sandy bottom near shore.

12. **G. maeoticus** Sowinski 1894 *G. m.*, Sowinski in: Mém. Soc. Kiew, *v.* 13 (p. 6) t. 1 A, 2 | 1896 *G. m.*, G. O. Sars in: Bull. Ac. St.-Pétersb., ser 5 *v.* 4 p. 465 t. 9 f. 12—20.

In many respects agreeing with G. weidemanni, but some of the differences are striking. Head, lateral lobes narrowly rounded, prominent. Side-plate 1 not at all expanded distally. Pleon segments 2 and 3, postero-lateral corners simply, not acutely, quadrate. Eyes rather small, decidedly reniform. Antennae 1 and 2 subequal, stout, $^1/_4$ as long as body. Antenna 1, flagellum rather shorter than peduncle, 8- or 9-jointed, accessory flagellum 5-jointed. Antepenultimate and penultimate joints of peduncle distally widened, setose, ultimate joint also setose, scarcely shorter than penultimate, flagellum 6-jointed. Mandibular palp more than twice as long as the trunk, 3^d joint as long as 2^d, both broad, setose. Gnathopods 1 and 2 in ♀ weak, subequal, in ♂ much more powerful, unequal. Gnathopod 1 in ♂, 6th joint narrowly oblong. Gnathopod 2 in ♂, 6th joint large and broad, oval quadrangular. L. ♂ 12 mm.

Sea of Azov; Caspian Sea. Shallow water.

13. **G. crassus** O. Sars 1894 *G. c.* (O. Grimm in MS.), G. O. Sars in: Bull. Ac. St.-Pétersb., ser. 5 *v.* 1 p. 362 t. 13.

Body rather short and stout, back broadly rounded. Pleon segment 4
with dorsal fascicle of setules but no spines; segments 5 and 6 each
with 2 dorsal groups containing 1 or 2 spinules a piece. Head, lateral lobes
truncate, a little prominent. Side-plates 1—4 with setules wide apart; side-
plate 1 very little expanded, 4th scarcely as broad as deep. Pleon segments
2 and 3, postero-lateral corners acutely produced. Eyes oval reniform, dark.
Antenna 1 less than $^1/_3$ as long as body, 1st joint large, fully as long as 2^d and 3^d
combined, flagellum slender, longer than peduncle, 16-jointed, accessory
flagellum 4-jointed. Antenna 2 a little shorter, ultimate joint of peduncle
shorter than penultimate, flagellum 8—11-jointed. Gnathopod 1 in ♀, 6th joint
oval, palm rather oblique, in gnathopod 2 rather larger, oblong, palm less oblique.
Gnathopod 1 in ♂, 6th joint narrowly oblong, palm sinuous, oblique, in gnathopod 2 nearly twice as large, broadly oblong, palm less sinuous, both gnathopods
more powerful than in ♀. Peraeopods 1 and 2, 4th joint rather large, 4th and 5th
setose. Peraeopod 3, 2^d joint rounded quadrangular. Peraeopod 4, 2^d joint widely
expanded, except distally, where there is scarcely a free corner. Peraeopod 5,
2^d joint very large, hind margin serrate, evenly convex, the expansion much
produced downward in a broad obtusely truncate lobe. Uropods 1 and 2,
rami nearly equal, with apical spines, the inner also with 1 or 2 lateral
spines. Uropod 3 reaching much beyond the others, outer ramus twice as long
as peduncle, fringed with setae, 2^d joint very small, inner ramus scale-like,
little more than half as long as peduncle. Telson scarcely as long as broad,
each obtuse apex having 2 spines and 2 setules. L. ♀ 11, ♂ 12 mm.

Caspian Sea. Shallow Water to 202 m.

14. **G. warpachowskyi** O. Sars 1894 *G. w.*, G. O. Sars in: Bull. Ac. St.-Pétersb.,
ser. 5 *v.* 1 p. 343 t. 9.

Body rather slender and compressed, pleon segments 4—6 each with
dorsal median fascicle of setules, segment 5 with 2 spinules among the setules,
and oblique lateral rows of 3 spines and 2 or 3 spinules, segment 6 with
lateral groups of 2 spines. Head, lateral lobes angularly prominent. Side-
plates 1—4 with few setules, 1st scarcely expanded distally, 2^d and 3^d
rounded distally, so that the lower edges are not continuous, 4th not so
broad as deep. Pleon segments 2 and 3, postero-lateral corners slightly
produced. Eyes oblong oval, dark, close to margin of head. Antenna 1 scarcely
more than $^1/_4$ as long as body, 3^d joint half as long as 1st, flagellum longer
than peduncle, 9-jointed, accessory flagellum with 2 joints, the last minute.
Antenna 2 shorter, ultimate joint of peduncle little or in ♂ not at all shorter than
penultimate, flagellum half as long as peduncle, with 5 joints in ♀, 7 in ♂, which
has both peduncle and flagellum brush-like. Gnathopods 1 and 2 in ♀ rather
small, 6th joint oval quadrangular, palm nearly transverse, in gnathopod 2
the 5th and 6th joints are rather longer, giving a more slender appearance,
in the ♂ both gnathopods much stronger, not very unequal, though gnathopod 2
is the larger, 6th joint large, oblong, palm slightly oblique, shorter than the
hind margin, finger strong. Peraeopods 1 and 2 rather slender. Peraeopods
3—5 comparatively short and stout, 2^d joint in peraeopod 3 subquadrate,
in peraeopod 4 broadly expanded above, narrowed below, in peraeopod 5
much larger, with broad expansion abruptly contracted near the distal end.
Uropods 1 and 2, rami equal, with apical spines, inner also with spinule at
middle of inner margin. Uropod 3 reaching beyond the others, outer ramus
with fascicles of spines but no plumose setae, 2^d joint small, distinct, inner
ramus small, shorter than the short peduncle. Telson short, broad, lobes

little divergent, each with 3 spinules on convex margin, and 1 spinule with 2 setules on rounded apex. L. ♀ about 6, ♂ 7 mm.

Caspian Sea. Depth 0—5 m.

15. **G. pauxillus** O. Sars 1896 *G. p.* (O. Grimm in MS.), G. O. Sars in: Bull. Ac. St.-Pétersb., ser. 5 *v.* 4 p. 467 t. 10 f. 1—17.

Body slender and compressed, dorsally smooth except for a few small hairs at end of pleon segments 4—6, segments 5 and 6 each with 1 spinule on either side. Head, lateral lobes obliquely truncated, upper corner the more prominent. Side-plates 1—4 not very large, without setae, 1st a little expanded distally. 2d—4th with front corners rounded off. Pleon segments 2 and 3, postero-lateral corners quadrate. Eyes rather large, oblong oval, dark. Antenna 1 slender, more than half as long as body, 2d joint nearly as long as 1st, flagellum twice as long as peduncle, filiform, 16—20-jointed, accessory flagellum 3-jointed. Antenna 2 scarcely more than half as long, flagellum 5—7-jointed. Gnathopods 1 and 2 in ♀ small and feeble, 6th joint oblong, in gnathopod 2 rather longer than in gnathopod 1, slightly widened distally, palm less oblique. Gnathopods 1 and 2 in ♂ powerful, especially gnathopod 2, 6th joint oblong oval, palm much shorter than hind margin. Peraeopods 1 and 2 slender, with few setae. Peraeopods 3—5, 2d joint not greatly expanded, narrowing distally, hind margin not forming a free corner. Uropod 3 extending much beyond the others, outer ramus long, narrow, with fascicles of spines, 2d joint spiniform, tipped with 2 unequal setules, inner ramus very small, shorter than the short peduncle. Telson very short, nearly twice as broad as long, the broadly rounded apex of each lobe smooth, but the outer margin just above carrying 2 setules. L. ♀ about 4, ♂ 6 mm.

Caspian Sea. Depth 75—202 m.

16. **G. kietlinskii** Dyb. 1874 *G. k.*, B. Dybowsky in: Horae Soc. ent. Ross., *v.* 10 suppl. p. 57 t. 1 f. 1 | 1893 *G. fluviatilis* (part.)?, A. Della Valle in: F. Fl. Neapel, *v.* 20 p. 929.

Pleon segments 4—6 each carrying a median and 2 lateral groups of spines on slight elevations of the hind margin, in segment 4 each lateral group having 2 or 3 spines, the median having 2 spines, each on a separate elevation. Head, rostrum short, depressed. Side-plates 1—4 small, low, lower margins not continuous. Pleon segments 2 and 3, postero-lateral corners obliquely truncate from behind forward. Eyes reniform, narrow, somewhat widened below, black. Antenna 1 about half as long as body, flagellum longer than peduncle, 75-jointed, accessory flagellum 14-jointed. Antenna 2 not much shorter, flagellum shorter than peduncle, 80-jointed, with large calceoli in both sexes. Gnathopods 1 and 2 with 6th joint equal in both sexes, more or less piriform. Peraeopods 3—5, 2d joint very slender, expansion not great above and thence tapering. Uropods 1 and 2 reaching end of shorter ramus of uropod 3. Uropod 3 about $^1/_7$ as long as body, rami thickly fringed with fascicles of simple setae and having a few plumose setae on inner margin of outer and both margins of inner ramus, the latter $^1/_2$ or $^3/_5$ as long as outer ramus. Colour red, margins of segments and side-plates, joints and apices of appendages yellow or olive-greenish. L. 81 mm.

Lake Baikal. Depth 50 m.

17. **G. andrussowi** O. Sars 1896 *G. a.*, G. O. Sars in: Bull. Ac. St.-Pétersb., ser. 5 *v.* 4 p. 469 t. 10 f. 18—26.

♀. Body slender and compressed, pleon segments 4 and 5 each with a median fascicle of 2—4 spinules and 2 lateral groups of 3 spinules, segment 6 with a pair of spinules. Head, lateral lobes broad, vertically truncated. Side-plates 1—4 without setae, 1st little expanded distally, but broader than 2^d, 4th rather broad. Pleon segments 2 and 3, postero-lateral corners quadrate. Eyes small, oval reniform. Antenna 1 slender, more than half as long as body, 1st joint not as long as 2^d and 3^d combined, flagellum twice as long as peduncle, 20-jointed, accessory flagellum 4-jointed. Antenna 2 scarcely more than half as long as antenna 1. Gnathopod 1, 5th joint short, 6th large, oblong oval, palm oblique but well defined. Gnathopod 2 slender. 5th and 6th joints subequal, long, sublinear, palm small, almost transverse, the 6th joint being shorter and very much narrower than in gnathopod 1, but 5th and 6th joints together longer than the same two in gnathopod 1. Peraeopods 1 and 2 rather slender. Peraeopods 3—5. 2^d joint rather broad and expanded, with free corner to hind margin in peraeopod 4 as well as 3 and 5. in peraeopod 5 much larger, with the expansion obtusely truncated below. Uropod 3 extending much beyond the others, outer ramus long, sublinear, with fascicles of spines on each margin, 2^d joint small, spiniform, inner ramus scale-like, shorter than the short peduncle. Telson broader than long, lobes rather divergent. each with 1 lateral spine and 3 spinules on the obtuse apex. L. 5 mm. — ♂ unknown.

Caspian Sea.

18. **G. ischnus** Stebb. 1896 *G. tenellus* (non. J. D. Dana 1852!), G. O. Sars in: Bull. Ac. St.-Pétersb., ser. 5 *r*. 4 p. 455 t. 7 f. 12—24 | 1899 *G. ischnus*, T. Stebbing in: Tr. Linn. Soc. London, ser. 2 *v*. 7 p. 428.

Body exceedingly slender and compressed, pleon segments 4—6 each with 1 median and 2 lateral fascicles of slender, upturned spinules, usually 2 to a group, sometimes together with a setule. Head, lateral lobes very obliquely truncated, upper angle acutely prominent. Side-plates 1—4 not large, quite smooth, 1st scarcely expanded, distally rounded, 4th not very broad. Pleon segment 2, postero-lateral corners acutely quadrate, in segment 3 acutely produced. Eyes well developed, but rather small, oblong oval, close to margin of head. Antenna 1 more than half as long as body. peduncle long. 1st joint not as long as 2^d and 3^d combined, flagellum twice as long as peduncle, 20-jointed, accessory flagellum 4-jointed. Antenna 2 considerably shorter, setose, especially in ♂, ultimate and penultimate joints of peduncle equal, flagellum 11-jointed. Gnathopods 1 and 2 rather feeble in ♀, stronger in ♂, not very unequal, but gnathopod 2 the longer, 6th joint oblong, with palm rather oblique in gnathopod 1 and in ♂ somewhat concave, in gnathopod 2 nearly transverse. Peraeopods 1 and 2 rather slender, much smaller than the others. Peraeopods 3—5. 2^d joint narrowly oval, in peraeopod 5 scarcely differing from that in 4th. not forming a free corner in any peraeopod. Uropods 1 and 2, outer ramus rather shorter than inner. Uropod 3 very long, outer ramus sublinear, with 5 fascicles of spines on each side. 2^d joint spiniform, well defined, tipped with setules, inner ramus scale-like, shorter than the short peduncle. Telson small, much broader than long, lobes divergent, with 2 spinules on outer margin of wide upper part. thence rapidly tapering to obtuse apex with 3 spinules. L. ♀ 6 mm, ♂ a little longer.

Caspian Sea (south of Baku). Depth 11 m.

19. **G. placidus** O. Sars 1896 *G. p.* (O. Grimm in MS.), G. O. Sars in: Bull. Ac. St.-Pétersb., ser. 5 *v*. 4 p. 457 t. 8 f. 1—12.

Body exceedingly slender, less compressed than in G. ischnus; pleon segments 4—6 each with 1 median and 2 lateral fascicles, the latter containing each 4 spinules in segment 4, 3 in segment 5, 2 in segment 6. Head, lateral lobes conspicuously acute, the point sometimes slightly deflexed. Side-plates 1—4 not very large, 1st quadrangular, 2d and 3d slightly narrowed distally, 4th at least as broad as deep. Pleon segments 2 and 3, postero-lateral corners acutely quadrate. Eyes dark, very narrow. deep, below a little inflated and bent forward. Antenna 1 slender, nearly $^3/_4$ as long as body, 2d joint fully as long as 1st, flagellum twice as long as peduncle, about 33-jointed, accessory flagellum 7-jointed. Antenna 2 shorter, more densely setose, ultimate joint of peduncle (in figure) longer than penultimate, flagellum about 16-jointed. Gnathopods 1 and 2 in ♀ not powerful, setose, subequal, but 5th and 6th joints in gnathopod 1 rather shorter, 6th oblong oval, palm rather oblique, 6th joint in gnathopod 2 oblong, palm nearly transverse (in young ♂ gnathopod 2 considerably larger than gnathopod 1). Peraeopods 1 and 2 rather slender, 1st the longer. Peraeopods 3—5 stouter, rather long, often reflexed, 2d joint narrowly oblong oval, without a free corner, in peraeopod 5 of same size and shape as in 4th. Uropods 1 and 2, rami spinose. Uropod 3 exceedingly long, outer ramus nearly 4 times as long as peduncle, sublinear, with many fascicles of spines and a few setae on both margins, 2d joint spiniform, inner ramus extremely small, not half as long as peduncle. Telson short, nearly twice as broad as long, lobes not divergent, outer margin almost angularly bent below its 1 spinule, apex obtusely truncated, carrying 1 spinule. L. ♀ 13 mm.

Caspian Sea. Depth 4—75 m.

20. **G. platycheir** O. Sars 1896 G. p., G. O. Sars in: Bull. Ac. St.-Pétersb., ser. 5 v. 4 p. 460 t. 8 f. 14—17.

♂(?). Body rather robust, back broadly vaulted, pleon segments 4—6 dorsally each with 2 adjacent spinules, segment 6 also with a spinule on either side. Head, lateral lobes rather broad, vertically truncated. Side-plates 1—4 rather deep, setose, 1st smallest, 1st and 2d distally narrowed, 3d much larger, oblong, 4th nearly as broad as deep. Pleon segments 2 and 3, postero-lateral corners quadrate, 2d on anterior lateral margin closely fringed with delicate, curved setae. Eyes not large, oval reniform. Antenna 1 very short. not $^1/_4$ as long as body, 1st joint as long as 2d and 3d combined, flagellum shorter than peduncle, 12-jointed, accessory flagellum 4-jointed. Antenna 2 as long or a little longer, flagellum 11-jointed. Gnathopod 1 much the smaller. Gnathopod 2, 5th joint small, 6th very large, flattened, gradually expanding to the long, oblique, evenly curved palm, which is densely fringed with spiniform setules,' and defined by 3 strong spines, among which the tip of the long falciform finger closes down. Peraeopod 2 more setose than peraeopod 1. Peraeopod 3 much shorter than peraeopods 4 and 5, 2d joint in peraeopods 3 and 4 narrowed below, without a free corner, in peraeopod 5 much larger, oblong oval, hind margin closely fringed with setules. Uropod 3 reaching beyond the others, outer ramus subfoliaceous, twice as long as peduncle, fringed with setae, 2d joint very small, inner ramus about $^1/_3$ as long as outer, with 2 spinules at apex. Telson about as long as broad, cleft not quite to the base, the broad lobes distally divergent, with 3 spinules on the blunt apex of each. L. 16 mm.

Caspian Sea (mouth of Volga).

21. **G. pungens** M.-E. 1840 G. p., H. Milne Edwards, Hist. nat. Crust., v. 3 p. 47 | 1893 G. p. (part.). A. Della Valle in: F. Fl. Neapel, v. 20 p. 764 t. 24 f. 35 | 1862

$\dot{G}.\,p.$, *Niphargus*(part.)?, Bate. Cat. Amphip. Brit. Mus., p. 217 | 1888 *G. (N.) p.*, T. Stebbing in: Rep. Voy. Challenger, *v.* 29 p. 253 | 1890 *N. p.*, Wrześniowski in: Z. Wiss. Zool., *v.* 50 p. 603, 673 | 1865 *Gammarus veneris*, Cam. Heller in: Verb. Ges. Wien, *v.* 15 p. 981 | 1894 *G. v.*, Chevreux in: Bull. Soc. zool. France, *v.* 19 p. 171 f. | 1895 *G. v.*, Chevreux in: Rev. biol. Nord France, *v.* 7 p. 159.

Near to G. marinus. Body slender and compressed, pleon segments 4 and 5 each with 1 or 2 spinules at 4 points of back, segment 6 with spines apparently only at 2 points. Head, lateral lobes rather obliquely truncate, post-antennal corners almost acute. Pleon segment 3, postero-lateral corners produced acutely backward. Eyes narrow, oblong reniform, dark in spirit. Antennae 1 and 2 liable to variations in many respects. Antenna 1 nearly half as long as body, 1st joint not longer than 2^d, 3^d about half as long as 2^d, flagellum twice as long as peduncle, 34-jointed, accessory flagellum 4—6-jointed. Antenna 2 considerably shorter, with longer peduncle, penultimate joint of peduncle stouter than ultimate, but scarcely so long, flagellum sub-equal to peduncle, 16-jointed, the calceoli short and broad. Maxillipeds, 3^d joint of palp slightly curved and distally widened. Gnathopods 1 and 2, 6th joint oblong oval, palm rather oblique, slightly sinuous and serrulate, ill-defined except by the palmar spines, finger in gnathopod 1 closing more decidedly against inner surface of hand than in the larger gnathopod 2. Peraeopods 1—5 all slender, 2^d joint in peraeopods 3—5 distally narrowed, free corner quadrate, not very prominent. Uropod 3 reaching much beyond the others, outer ramus more than thrice as long as peduncle, 1st joint with spines at 4 points of outer and 3 of inner margin, both margins fringed with long conspicuous setae, 2^d joint also setose, very small, inner ramus shorter than peduncle, very slender, smooth on inner margin, with spines and setae on outer and apex. Telson cleft to base, little longer than broad, each lobe with 1 spine high up on outer margin and 2 at the narrowed truncate apex. Colour grey or brownish, rarely with rusty spots. L. 6—12 mm.

Italy (hot springs?); Sicily; Cyprus (fresh water, 15—16 m above the sea); Syria.

22. **G. marinus** Leach 1815 *G. m.*, Leach in: Tr. Linn. Soc. London, *v.* 11 p. 359 | 1862 *G. m.*, Bate, Cat. Amphip. Brit. Mus., p. 215 t. 38 f. 4 | 1889 *G. m.*, A. M. Norman in: Ann. nat. Hist., ser. 6 *v.* 4 p. 138 t. 12 f. 12 | 1893 *G. m.*, A. Della Valle in: F. Fl. Neapel, *v.* 20 p. 762 t. 60 f. 28 | 1894 *G. m.*, G. O. Sars, Crust. Norway, *v.* 1 p. 497 t. 175 | 1830 *G. olivii*, H. Milne Edwards in: Ann. Sci. nat., *v.* 20 p. 367, 372 t. 10 f. 9, 10 | 1837 *G. gracilis*, H. Rathke in: Mém. prés. Ac. St.-Pétersb., *v.* 3 p. 374 t. 5 f. 7—10 | 1840 *G. affinis*, H. Milne Edwards, Hist. nat. Crust., *v.* 3 p. 47 | 1843 *G. poecilurus* + *G. kröyerii*, H. Rathke in: N. Acta Ac. Leop., *v.* 20₁ p. 68 t. 4 f. 2; p. 69 t. 4 f. 1 | 1851 *G. locustoides*, F. Brandt in: Middendorff, Reise Sibirien, *v.* 21 p. 139 t. 6 f. 30 a—c.

Body slender and compressed, pleon segments 4—6 each dorsally carrying 2 upward converging rows of spinules, each row having 6—9 spinules in 2 slightly separated groups. Head, lateral lobes vertically truncate. Side-plates 1—4 not very deep, 4th scarcely as broad as deep. Pleon segments 2 and 3, postero-lateral corners acutely quadrate. Eyes narrow, long, oblong reniform, dark. Antenna 1 nearly half as long as body, 1st joint as long as 2^d and 3^d combined, flagellum more than twice as long as peduncle, 33-jointed, accessory flagellum about 7-jointed. Antenna 2 sometimes considerably shorter, flagellum rather longer than peduncle. Gnathopod 1, oblong oval, palm slightly oblique, both 5th and 6th joints rather longer and stronger in ♂ than in ♀. Gnathopod 2 in ♀, 5th joint longer than in gnathopod 1, 6th oblong, with palm nearly transverse, not larger than in gnathopod 1. Gnathopod 2 in ♂ not larger than gnathopod 1, palm nearly transverse.

Peraeopods 3—5 rather short and stout, 2^d joint in peraeopod 3 with lower hind corner rounded, free, in peraeopods 4 and 5 narrowed distally, without a free corner. Uropod 1 reaching beyond uropod 2, uropod 3 beyond uropod 1. Uropod 3, outer ramus long, with spines and setae on both margins, 2^d joint small, spiniform, inner ramus very narrow, scarcely $\frac{1}{8}$ as long as outer. Telson rather small, lobes distally divergent, each with 2 marginal spines near base, and 3 on the truncate apex, 1 or 2 setules usually near outer margin a little above apex. Colour greenish or yellowish brown. L. 15 mm.

North-Atlantic with adjoining seas (Europe, from Trondhjemsfjord round the marine coasts to the Black Sea; North-East-America). Littoral.

23. **G. simoni** Chevreux 1894 *G. s.*, Chevreux in: Bull. Soc. zool. France, *v.* 19 p. 171 f. 2—10.

Pleon segments 1—3 each with 2 or 3 dorsal setules, segments 4—6 with dorsal spinules, segment 4 with 2 near together, segment 6 with 2 wide apart, segment 5 with 2 pairs, each pair having between then 2 setules. Head, lateral lobes obtusely truncate. Side-plate 4 rather less broad than deep. Pleon segments 2 and 3, postero-lateral corners produced, acute. Eyes small, oval. Antenna 1 in $\male$ $\frac{2}{3}$ as long as body, peduncle short, 1st joint as long as 2^d and 3^d combined, flagellum twice as long as peduncle, 30—35-jointed, accessory flagellum (1-) 3—5-jointed. Antenna 2 in $\male$ considerably shorter, peduncle robust, ultimate and penultimate joints tufted with long setae, flagellum 12-jointed, $\frac{2}{3}$ (or less) of it setose like peduncle, apparently without calceoli. Antenna 1 in $\female$ as in $\male$. Antenna 2 in $\female$ shorter, less robust, setae few and short. Gnathopods 1 and 2 in $\female$ feeble, 6th joint oblong, tending to oval, with oblique palm in gnathopod 1, rather longer and narrower, with palm almost transverse in gnathopod 2. Gnathopod 1 in $\male$, side plate with sinuous front margin (in figure), 6th joint large, piriform, with 6 spines on hind margin and palm, finger long, curved, its apex closing against side of 6th joint. Gnathopod 2 in $\male$ scarcely larger, 6th joint oval, with 3 spines on the convex palm, which the finger fits. Peraeopods 1—5 short and robust. Uropod 2 reaching nearly as far as uropod 1. Uropod 3, outer ramus long and broad, fringed with spines and plumose setae, 2^d joint small, spiniform, with apical setae, inner ramus rather broad, about $\frac{1}{4}$ as long as 1st joint of outer, with 2 apical spines. Telson with broad divergent lobes, each having a spine on outer margin near the base, 2 spines and some setules on rounded apex, and in $\male$ a sub-central spinule. L. $\female$ 6, $\male$ reaching 8 mm.

Algeria and Tunis. Fresh Water.

24. **G. duebenii** Lilj. 1851 *G. d.*, W. Liljeborg in: Öfv. Ak. Förb., *v.* 8 p. 22 | 1888 *G. d.*, T. Stebbing in: Rep. Voy. Challenger, *v.* 29 p. 252 | 1894 *G. duebeni*, G. O. Sars, Crust. Norway, *v.* 1 p. 502 t. 177 f. 1 | 1895 *G. duabēni*, T. Scott in: Rep. Fish. Board Scotl., *v.* 13 p. 180, 244 | 1862 *G. locusta* (part.), Bate, Cat. Amphip. Brit. Mus., p. 206 | 1876 *G. l.*, A. Boeck, Skand. Arkt. Amphip., *v.* 2 p. 367 | 1893 *G. l.* (part.), A. Della Valle in: F. Fl. Neapel, *v.* 20 p. 760 | 1874 *G. marinus* (part.), Ritzema Bos, Bijdr. Crust. Hedriophthal., p. 44 | 1889 *G. campylops* (err., non Leach 1815!), A. M. Norman in: Ann. nat. Hist., ser. 6 *v.* 4 p. 139 t. 12 f. 13 | 1889 *G. locusta var. C*, Hoek in: Tijdschr. Nederl. dierk. Ver., ser. 2 *v.* 2 p. 219 t. 10 f. 13, 13'.

In form like G. locusta (p. 476). Pleon segments 4—6 each with 2 spines in median dorsal group and 3 in each lateral group, also clothed with numerous rather long hairs. Head, lateral lobes vertically truncate. Side-plates 1—4 smaller than in G. locusta, more like those in G. marinus. Pleon segments 2 and 3, postero-lateral corners acutely quadrate. Eyes not very large, reniform, dark. Antenna 1 not quite half as long as body, 1st joint scarcely as long as 2^d and 3^d combined, flagellum about twice as long as peduncle, accessory flagellum about 6-jointed. Antenna 2 considerably shorter, flagellum with calceoli in ♂. Gnathopods 1 and 2 in ♀ nearly as in G. locusta, but 6th joint of gnathopod 2 not so narrow. Gnathopods 1 and 2 in ♂ not so strongly developed as in G. locusta, and less unequal. Peraeopods 3—5 not very slender, 2^d joint in peraeopod 3 with lower hind corner rounded. Uropod 3, rami densely setose, inner much more slender than outer, and reaching nearly $^2/_3$ of its length. Telson rather broader than long, each lobe with 4 apical, 1 subapical, and 2—4 lateral spines, the last near the base, the spines accompanied by long diverging setae. Colour generally dark, with the usual pink marks on sides of pleon segments 1—3. L. ♂ 15 mm, ♀ smaller.

South Greenland (Warm springs); Norway (beach, under stones and algae. and in brackish pools above high water mark); Kattegat; Dutch coast; Guernsey; England; Ireland; lochs of Shetland, Outer Hebrides, and Perthshire.

25. **G. pulex** (L.) 1758 Cancer p. (part.), Linné, Syst. Nat., ed. 10 p. 633 | 1763 C. p., Scopoli, Ent. Carniol., p. 412 | 1775 Gammarus p.. J. C. Fabricius, Syst. Ent., p. 418 | 1876 G. p., A. Boeck, Skand. Arkt. Amphip., v. 2 p. 373 t. 24 f. 7 | 1888 G. p., T. Stebbing in: Rep. Voy. Challenger, v. 29 p. 44, 253, 1703 | 1894 G. p., G. O. Sars, Crust. Norway, v. 1 p. 503 t. 177 f. 2 | 1777 Astacus p., Pennant, Brit. Zool., ed. 4 v. 4 p. 17 | 1778 Squilla p., Geer, Mém. Hist. Ins, v. 7 p. 525 | 1793 Cancer (Gammarellus) p. (part.), J. F. W. Herbst, Naturg. Krabben Krebse, v. 2 p. 130 (not figure!) | 1808 C. (Gammarus) p., Montagu in: Tr. Linn. Soc. London, v. 9 p. 93 t. 4 f. 2 | 1815 Gammarus aquaticus, Leach in: Tr. Linn. Soc. London, v. 11 p. 359 | 1830 G. fluviatilis, H. Milne Edwards in: Ann. Sci. nat., v. 20 p. 368 | 1893 G. f. (part), A. Della Valle in: F. Fl. Neapel, v. 20 p. 763 | 1835 G. fossarum, C. L. Koch, C. M. A., v. 5 nr. 2 | 1839 G. stagnalis, Andrzeiowski in: Bull. Soc. Moscou. nr. 1 p. 23 | 1863 G. lacustris, G. O. Sars in: Nyt Mag. Naturv., v. 12 p. 206 | 1867 G. neglectus, G. O. Sars, Crust. d'Eau douce Norvège, p. 46 t. 4, 5; t. 6 f. 1—20 | 1889 G. locusta var. B, Hoek in: Tijdschr. Nederl. dierk. Ver.. ser. 2 v. 2 p. 214 t. 10 f. 12, 12'.

Pleon segments 4—6 each with 2 spines in median dorsal groups and a single spine on each side. Head, lateral lobes vertically truncated, less broad than in G. duebenii (p. 473). Pleon segments 2 and 3, postero-lateral corners simply quadrate. Eyes small, oval, black. Antenna 1 nearly half as long as body, 1st joint not as long as 2^d and 3^d combined, flagellum twice as long as peduncle, 25—28-jointed in ♂, accessory flagellum only 4-jointed. Antenna 2 considerably shorter. flagellum 12-jointed, in ♂ with calceoli. Gnathopods 1 and 2 in ♀ rather small, 6th joint in gnathopod 1 with oblique palm, in gnathopod 2 narrow, oblong, palm nearly transverse. Gnathopods 1 and 2 in ♂ rather stronger, 6th joint in gnathopod 1 piriform, in gnathopod 2 of about the same size, but rather oblong. widening slightly to the palm which is far less oblique. Peraeopods 1—5 slender, 2^d joint in peraeopod 3 with lower hind corner rounded, in peraeopods 4 and 5 without a free corner, but hind margin convex. Uropod 3 long, rami fringed with long setae, some plumose, inner ramus subequal in length to 1st joint of outer (Sars) or $^3/_4$ as long as the whole ramus (Chevreux), with a single spine on inner margin near base. Telson rather small, each lobe with 2 apical spines and 1 on outer margin near the base. Colour dark or light brownish green. L. ♂ reaching 20 mm, ♀ smaller.

Europe. Lakes and streams. generally distributed, found even in lakes 900—1200 m above the sea.

26. **G. delebecquei** Chevreux & Guerne 1892 *G. d.*, Chevreux & Guerne in: Bull. Soc. zool. France, *v.* 17 p. 136 | 1893 *G. fluviatilis* (part.), A. Della Valle in: F. Fl. Neapel, *v.* 20 p. 889.

General form of body, armature of pleon segments 4—6, postero-lateral corners in segments 2 and 3 as in G. pulex. Eyes much larger, faintly reniform. Antenna 1 more than half as long as body, flagellum 32—35-jointed, more than twice as long as peduncle, accessory flagellum 4-jointed. Antenna 2 shorter than 1st, but more slender and elongate than in G. pulex, flagellum shorter than ultimate and penultimate joints of peduncle combined, 12-jointed, with setae of uniform length. Gnathopods 1 and 2 as in G. pulex, but with 6th joint a little narrower, and peraeopods 1—5 rather more slender. Peraeopods 4 and 5, 2d joint with sinuous hind margin, rather abruptly narrowing distally. Uropod 3, rami more sparsely fringed with setae, the outer with several groups of spines on outer margin, the inner slender, not more than half as long as the 1st joint of outer. Telson apparently as in G. pulex. L. ♀ 8, ♂ 12 mm.

France (from a Warm spring at the bottom of the Lac d'Annecy).

27. **G. hyacinthinus** Dyb. 1874 *G. h.*, B. Dybowsky in: Horae Soc. ent. Ross., *v.* 10 suppl. p. 70 | 1893 *G. fluviatilis* (part.)?, A. Della Valle in: F. Fl. Neapel, *v.* 20 p. 929.

Pleon segments 4—6 each with dorsal spinules in 4 groups of 1, 2 or 3 a-piece. Eyes rather large, reniform, black. Antenna 1 not half as long as body, flagellum thrice as long as peduncle, 28-jointed, accessory flagellum 4-jointed. Antenna 2 $^3/_4$ as long as antenna 1, flagellum 15-jointed. Gnathopods 1 and 2 not large, 6th joint in gnathopod 1 piriform, rather longer than the oblong 6th joint in gnathopod 2. Peraeopods 3—5, 2d joint tolerably broad, front and hind margins convex, the hind one ending below in a pointed corner. Uropods 1 and 2 reaching $^2/_3$ length of uropod 3. Uropod 3 not very long, outer ramus with plumose setae only on the inner margin, inner about $^4/_5$ as long as outer, with plumose setae on both margins. Colour variable, greenish, yellowish, or bright reddish. L. 15 mm.

Lake Baikal. Depth 100—300 m.

28. **G. syriacus** Chevreux 1895 *G. s.*, Chevreux in: Rev. biol. Nord France, *v.* 7 p. 160 f.

Body rather robust. Pleon segments 1—3 with some stiff dorsal setae, segments 4—6 each with a transverse elevation of hind margin armed with a variable number of spines, usually 2 median and 1 on each side in segments 4 and 5, and only 1 on each side in segment 6; the spines accompanied by setae. Head, lateral lobes vertically truncate, the sinus below smaller than usual. Side-plates 1—4 deep, 4th much deeper than broad. Pleon segments 1—3, postero-lateral corners acutely and uncinately produced, chiefly in segment 3. Eyes of moderate size, reniform. Antenna 1 in ♂ more than half as long as body, flagellum feebly armed, more than twice as long as peduncle, 40-jointed (21-jointed in ♀), accessory flagellum 4-jointed. Antenna 2 considerably shorter, ultimate joint of peduncle rather longer than penultimate, flagellum about 14-jointed, apparently without calceoli. Gnathopods 1 and 2 as in G. pulex. Peraeopods 1 and 2 with fascicles of long setae on hind margin. Peraeopods 3—5 robust, not elongate, 2d joint in peraeopod 3 with lower hind corner quadrate, in peraeopods 4 and 5 hind corner not free. Uropod 3, outer ramus fringed on outer margin with long plumose setae, inner ramus over $^2/_3$ as long as outer, with 1 or 2 spines and a fringe of short plumose setae on inner margin. Telson much longer than broad, lobes each with 2 long spines and some long setae on the apex, which is obliquely truncate from within; higher up are .2 fascicles of setae. L. ♀ 10, ♂ reaching 19 mm.

Syria. Fresh Water.

29. **G. locusta** (L.) 1758 *Cancer l.*, Linné, Syst. Nat., ed. 10 p. 634 | 1775 *Gammarus l.*, J. C. Fabricius, Syst. Ent., p. 418 | 1862 *G. l.*, Bate, Cat. Amphip. Brit. Mus., p. 206 t. 36 f. 6 | 1889 *G. l.*, A. M. Norman in: Ann. nat. Hist., ser. 6 *v.* 4 p. 137 t. 12 f. 11 | 1889 *G. l.*, Hoek in: Tijdschr. Nederl. dierk. Ver., ser. 2 *v.* 2 p. 206 t. 7 f. 10"; t. 10 f. 10, 10' | 1893 *G. l.* (part.), A. Della Valle in: F. Fl. Neapel, *v.* 20 p. 759 t. 2 f. 1; t. 24 f. 20—34; t. 45 f. 1—11 | 1890 & 94 *G. l.*, G. O. Sars, Crust. Norway, *v.* 1 t. 1; p. 499 t. 176 f. 1 | 1808 *Cancer (G.) l.*, Montagu in: Tr. Linn. Soc. London, *v.* 9 p. 92 t. 4 f. 1 | 1766 *Oniscus pulex* (part.), Pallas, Misc. zool., p. 190 | 1853 *Gammarus p.*, Stimpson in: Smithson. Contr., *v.* 6 nr. 5 p. 55 | 1820 *G. arcticus* (Leach in MS.), Scoresby, Account arct. Regions, *v.* 1 p. 541 t. 16 f. 14 | 1821 & 24 *G. boreus*, E. Sabine in: W. E. Parry, J. Voy., Suppl. p. 51; p. 229 | 1830 *G. ornatus*, H. Milne Edwards in: Ann. Sci. nat., *v.* 20 p. 367, 369 t. 10 f. 1—8 | 1851 *G. sitchensis*, F. Brandt in: Middendorff, Reise Sibirien, *v.* 2r p. 137 t. 6 f. 28 a—c | 1855 *G. mutatus*, W. Liljeborg in: Vetensk. Ak. Handl., 1853 p. 447.

Pleon segments 4—6 each with median dorsal elevation carrying 3—5 spinules, lateral groups well separated, each with 3 or 4 spinules, lateral lobes angularly produced. Side-plates 1—4 larger than in G. marinus (p. 472), especially in ♀ ; side-plate 4 nearly as broad as deep. Pleon segments 2 and 3, postero-lateral corners acutely produced. Eyes rather large, reniform, black with chalky white coating. Antenna 1 in ♀ not quite half as long as body, in ♂ rather longer, 1st joint subequal to 2d and 3d combined, flagellum twice as long as peduncle, accessory flagellum about 8-jointed. Antenna 2 shorter, flagellum about as long as ultimate and penultimate joints of peduncle combined, with calceoli in ♂. Gnathopods 1 and 2 in ♀, 6th joint oblong, in gnathopod 2 longer and narrower. Gnathopod 1 in ♂, 6th joint piriform; gnathopod 2, 6th joint much more powerful, irregularly oblong, palm oblique though less so than in gnathopod 1, in both with strong spine at centre. Peraeopods 3—5 rather slender and elongate, 2d joint in peraeopod 3 forming a free corner, which is produced to an acute point. Uropod 3 elongate, rami fringed with plumose setae and several spines, inner ramus nearly equal to 1st joint of outer. Telson longer than broad, each lobe with 3 apical spines and a seta, a spine and seta near the apex, and near the base 2 or 3 spines and one or two setae. Colour more or less dark brownish green, with pinkish mark on sides of pleon segments 1—3. L. ♂ 20— (arctic specimens) 48 mm, ♀ usually considerably smaller.

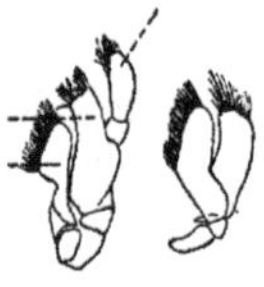

Palp

Outer plate

Inner plate

Fig. 87. Fig. 88.
Fig. 87 & 88. G. locusta.
Maxillae 1 and 2.
[After H. J. Hansen.]

Arctic Ocean, North-Atlantic with adjoining seas (East of United States of America, Europe). From shore to 100 m.

30. **G. camylops** Leach 1813/14 *G. c.*, Leach in: Edinb. Enc., *v.* 7 p. 403 | 1815 *G. campylops*, Leach in: Tr. Linn. Soc. London, *v.* 11 p. 360 | 1862 *G. c.*, Bate & Westwood, Brit. sess. Crust., *v.* 1 p. 375 f. | 1894 *G. c.*, G. O. Sars, Crust. Norway, *v.* 1 p. 500 t. 176 f. 2 | 1819 *G. camptolops*, (Leach in:) Samouelle, Ent. Compend., p. 104 | 1862 *G. c.*, Bate, Cat. Amphip. Brit. Mus., p. 209 t. 37 f. 3 | 1830 *G. camphylops*, H. Milne Edwards in: Ann. Sci. nat., *v.* 20 p. 367.

Pleon segments 4—6 each with 2 median dorsal spinules and 2 on each side. Head, lateral lobes rather obliquely truncate, with very small

:sinus below. Side-plates 1—4 rather small, 4ᵗʰ not nearly so broad as deep. Pleon segments 2 and 3, postero-lateral corners acute, but little produced. Eyes constricted in the middle, slightly sigmoid, very dark. Antenna 1 fully half as long as body, 1ˢᵗ joint rather longer than 2ᵈ and 3ᵈ combined, flagellum nearly thrice as long as peduncle, accessory flagellum about 6-jointed. Antenna 2 shorter, flagellum in ♂ with no apparent calceoli. Gnathopods 1 and 2 as in G. locusta, except that gnathopod 2 has 6ᵗʰ joint in ♀ rather shorter and in ♂ less obliquely truncated. Peraeopods 3—5 rather slender and elongate, 2ᵈ joint in peraeopod 3 with lower hind corner nearly quadrate. Uropod 3, inner ramus not nearly as long as 1ˢᵗ joint of outer, and with fewer spines than in G. locusta. Telson scarcely so long as broad, lobes divergent, each with 3 apical spines, and on outer margin 1 spine· near apex and 1 near base. Colour semipellucid, pale greenish with the usual pink markings faintly indicated. L. ♂ scarcely over 6 mm, ♀ smaller.

North-Atlantic, North-Sea and Skagerrak (South-Norway, in oyster-bed above sea-level;.in shallow Water at Christianiafjord; Isle of Arran; Shetland; Belfast); Kattegat.

G. tunetanus E. Sim. 1886 *G. t.,* E. Simon in: Expl. Tunisie, Crust. p. 6 | 1893 *G. fluviatilis* (part.)?, A. Della Valle in: F. Fl. Neapel, *v.* 20 p. 768.

Said to differ from G. pulex (p. 474) by having integument more sparingly and Very minutely punctate, head a little longer and in front a little attenuate, eyes more elongate reuiform, extending a little over the base of the antennae. accessory flagellum of antenna 1 longer, 6-jointed, reaching 6ᵗʰ joint of primary flagellum, 3ᵈ, 4ᵗʰ and 5ᵗʰ joints a little longer than the rest, equal to one another, flagellum of antenna 2 shorter, 8-jointed, all the joints a little longer than broad. From G. locusta it is said to differ by the much shorter flagellum of antenna 2, With the joints fewer and almost smooth.

Tunis (Kérouan).

45. Gen. **Poekilogammarus** Stebb.

1899 *Poekilogammarus,* T. Stebbing in: Tr. Linn. Soc. London, ser. 2 *v.* 7 p. 428.

In general like Gammarus (p. 460), but usually with dorsal hairs or spinules on all segments of peraeon and pleon, and head rostrate. Antenna 1 with peduncle longer than peduncle of antenna 2, 3ᵈ joint longer than 2ᵈ. Upper lip with wide, almost straight, apical margin; lower lip as in Axelboeckia (p. 391), the principal lobes separated by what may be regarded as rudimentary inner lobes. Maxilla 1, inner plate with about 6 setae. Maxillipeds, outer plates not reaching far along 3ᵈ joint of palp. Gnathopod 1, 6ᵗʰ joint larger than in gnathopod 2. Uropod 3 with equal rami, both carrying plumose setae, outer ramus 1-jointed.

4 species.

Synopsis of species:

1. **P. pictus** (Dyb.) 1874 *Gammarus p.* + *Var. α* + *Var. β*, B. Dybowsky in: Horae Soc. ent. Ross.. *v.* 10 suppl. p. 103 t. 12 f. 3; p. 104 t. 12 f. 2 (juv.) | 1899 *Poekilogammarus p.*, T. Stebbing in: Tr. Linn. Soc. London, ser. 2 *v.* 7 p. 428 | 1893 *Gammarus fluviatilis* (part.)?, A. Della Valle in: F. Fl. Neapel, *v.* 20 p. 930.

Pleon segments 4—6 with delicate dorsal spines. Head in front almost straight, rostral point scarcely indicated. Eyes moderately large, very prominent, as if set on tubercles, oval, black or brown. Antenna 1 longer than $^1/_2$ body (in young specimens longer than body), more than twice as long as antenna 2, 2^d joint only about $^2/_3$ as long as 3^d, flagellum 41—58-jointed, accessory flagellum 6-jointed. Antenna 2, ultimate joint of peduncle rather shorter than penultimate, flagellum 9—11-jointed. Gnathopods 1 and 2 little differing in size or even in shape, 6th joint oblong, with palm more oblique in gnathopod 1, hind margin longer in gnathopod 2, and 'the joint slightly widening to the palm. Peraeopods 3—5, 2^d joint not very wide, narrowing gradually from above downward, hind margin setose. Uropod 1 reaching end of uropod 3. Uropod 3 moderately long, the 1-jointed equal rami fringed with plumose setae. Colour whitish yellow, variegated, with brownish or greenish yellow markings. L. 32 mm.

Lake Baikal. Depth 50—100 m.

2. **P. orchestes** (Dyb.) 1874 *Gammarus o.*, B. Dybowsky in: Horae Soc. ent. Ross., *v.* 10 suppl. p. 104 | 1899 *Poekilogammarus o.*, T. Stebbing in: Tr. Linn. Soc. London, ser. 2 *v.* 7 p. 428 | 1893 *Gammarus pictus* (part.)?, A. Della Valle in: F. Fl. Neapel, *v.* 20 p. 929.

Segments óf peraeon each provided with 2—4 delicate setae; segments of pleon completely covered with them. Head with middle dorsal line convex, rostrum short, projecting with a slightly downward directed curve, its apex obtuse. Eyes rather small, prominent, ovate, a little narrowed above. black. Antenna 1 a little shorter than body, almost twice as long as antenna 2, 2^d joint about $^2/_5$ shorter than 3^d, flagellum not twice as long as peduncle, 21-jointed, accessory flagellum 2-jointed. Antenna 2, ultimate and penultimate joints of peduncle subequal, flagellum shorter than peduncle, 7-jointed. Gnathopod 1, 6th joint piriform. Gnathopod 2, 6th joint oblong. Peraeopods 3—5, 2^d joint slender, in front convex with long pendent setae, behind flatly concave or straight, with 5 or 6 setules. Uropods 1 and 2 scarcely reaching middle of uropod 3. Uropod 3 rather long, $^1/_5$—$^1/_4$ as long as body. Colour yellow, spotted with yellow·brown. L. about 10 mm.

Lake Baikal. Depth 150 m.

3. **P. talitrus** (Dyb.) 1874 *Gammarus t.*, B. Dybowsky in: Horae Soc. ent. Ross., *v.* 10 suppl. p. 105 t. 11 f. 5 | 1899 *Poekilogammarus t.*, T. Stebbing in: Tr. Linn. Soc. London, ser. 2 *v.* 7 p. 428 | 1893 *Gammarus pictus* (part)?, A. Della Valle in: F. Fl. Neapel, *v.* 20 p. 930.

Segments of peraeon each with 2—4 delicate setae; segments of pleon with delicate, irregularly arranged setae on surface and hind margin. Head with a rather long, straight, acute rostrum, downward bent only near the point. Eyes moderately large, prominent, elongate oval, slightly narrowed below, not far apart above, black. Antenna 1 about thrice as long as antenna 2, in ♂ about as long as body, in ♀ rather shorter, 2^d joint about $^1/_8$ shorter than 3^d, flagellum more than twice as long as peduncle, 28—39-jointed, accessory flagellum 3—6-jointed. Antenna 2, ultimate and penultimate joints of peduncle subequal, flagellum shorter than peduncle, 11-jointed. Gnathopod 1, 6th joint piriform (in figure oblong oval). Gnathopod 2, 6th joint narrowly oblong, nearly 3 times as long as 5th joint. Peraeopods 3—5, 2[d] joint moderately broad, in peraeopod 5 with 7 or 8 rather long setae on front and hind margins (in figure rather wider above than below, hind and lower margins nearly straight). Uropod 1 reaching much beyond uropod 2, but scarcely to middle of uropod 3. Uropod 3 $^1/_4$—$^1/_3$ as long as body. Colour bright yellow with bright brown spots. L. about 14 mm.

Lake Baikal. Depth 100—200 m.

4. **P. araneolus** (Dyb.) 1874 *Gammarus a. + G. a. var. quinquefasciatus + G. a. var. ephippiatus*, B. Dybowsky in: Horae Soc. ent. Ross., *v.*10 suppl. p. 106 t. 11 f. 3; p. 107 t. 11 f. 7; p. 107 t. 11 f. 8 | 1899 *Poekilogammarus a.*, T. Stebbing in: Tr. Linn. Soc. London, ser. 2 *v.*7 p. 428 | 1893 *Gammarus fluviatilis* (part.)?, *G. pictus* (part.)?, A. Della Valle in: F. Fl. Neapel, *v.*20 p. 927; p. 928, 930.

All segments of peraeon and pleon carrying short dorsal setae or very thin spinules; on peraeon segments 1—6 the setae generally on the hind margin, on peraeon segment 7 and pleon segments 1—6 over all the dorsal surface (fewer setae in *var. ephippiatus*). Head as in P. orchestes. Eyes large, prominent, ovate, slightly narrowed below, black. Antenna 1 only half as long as body, twice as long as antenna 2, 2[d] joint a little shorter than 3[d], flagellum about twice as long as peduncle, 19—25-jointed, accessory flagellum 3- or 4-jointed. Antenna 2, ultimate and penultimate joints of peduncle subequal, flagellum shorter than peduncle, 5—8-jointed. Gnathopod 1, 6th joint piriform. Gnathopod 2, 5th joint not elongate, 6th oblong, in ♂ widening to the palm. Peraeopods 3—5, 2[d] joint tolerably broad, front and hind margins feebly convex, hind margin with short setae. Uropod 1 reaching end of uropod 3, uropod 2 only end of peduncle of uropod 3. Uropod 3 tolerably long, $^1/_7$ as long as body. Colour bright green or bright greenish yellow with brownish yellow spots; in *var. quinquefasciatus*, transversely banded, head and peraeon segments 1—3 reddish brown, segments 4 and 5 bright yellow, segments 6 and 7 and pleon segments 1 and 2 brown, pleon segments 3—6 and telson bright yellow, uropod 3 brown; in *var. ephippiatus* brownish yellow or dark green, with a saddle-like brightening of peraeon segments 6 and 7. L. 11—14 mm.

Lake Baikal. Depth 10—50 m.

46. Gen. **Echinogammarus** Stebb.

1793 [Subgen.] *Gammarellus* (part.). J. F. W. Herbst, Naturg. Krabben Krebse, *v.*2 p. 106 | 1899 *Echinogammarus*, T. Stebbing in: Tr. Linn. Soc. London, ser. 2 *v.*7 p. 428.

In general like Gammarus (p. 460), but with dorsal spines on segments anterior to pleon segment 4 (Fig. 89 p. 481), antenna 1 the longer, with peduncle shorter than peduncle of antenna 2, and 6th joint of gnathopod 1 almost always larger than that of gnathopod 2.

27 accepted and 1 doubtful species.

Synopsis of accepted species:

1 {
Uropod 3, inner ramus much less than half as long as outer — **2.**
Uropod 3, inner ramus about half as long as outer — **10.**
Uropod 3, inner ramus much more than half as long as outer — **14.**
}

2 { Gnathopod 1, 6th joint smaller than that of gnathopod 2 . . . .   1. **E. berilloni** . . . . p. 481
  { Gnathopod 1, 6th joint larger than that of gnathopod 2 —. **3.**

3 { Pleon closely beset with spines . . . .   2. **E. verrucosus** . . . p. 481
  { Pleon with spines not closely set — **4.**

4 { Uropod 3 short . . . .   3. **E. saphirinus** . . . p. 482
  { Uropod 3 long — **5.**

5 { Uropod 3, outer ramus abnormally long . . .   4. **E. czerskii** . . . . . p. 482
  { Uropod 3, outer ramus not abnormally long — **6.**

6 { Antenna 1 considerably more than half as long as body — **7.**
  { Antenna 1 about half as long as body — **9.**

7 { Uropod 3 with setae short, separate . . . .   5. **E. maackii** . . . . . p. 483
  { Uropod 3 with setae long, in fascicles — **8.**

8 { Pleon segments 1 and 2, dorsal spinules not confined to hind margin . . . .   6. **E. lividus** . . . . . p. 483
  { Pleon segments 1 and 2, dorsal spinules wanting unless on hind margin . . . .   7. **E. viridis** . . . . . p. 484

9 { Pleon segment 3 with long dorsal hairs . . .   8. **E. cyaneus** . . . . . p. 484
  { Pleon segment 3 without dorsal hairs . . .   9. **E. ochotensis** . . . p. 484

10 { Pleon segment 1 without spines . . . . . . 10. **E. testaceus** . . . . p. 485
   { Pleon segment 1 with spines — **11.**

11 { Gnathopod 2, 6th joint piriform . . . . . . 11. **E. sophiae** . . . . . p. 485
   { Gnathopod 2, 6th joint oblong — **12.**

12 { Pleon segment 3 without lateral ridge . . . . 12. **E. fuscus** . . . . . p. 486
   { Pleon segment 3 with lateral ridge — **13.**

13 { Back of pleon warty . . . . 13. **E. murinus** . . . . p. 486
   { Back of pleon not warty . . . . 14. **E. aheneus** . . . . . p. 487

14 { Peraeon segment 7 dorsally spinose — **15.**
   { Peraeon segment 7 not dorsally spinose — **19.**

15 { Eyes shaped like a retort . . . . 15. **E. sarmatus** . . . . p. 487
   { Eyes reniform — **16.**

16 { Peraeon segment 7, spines in a continuous series 16. **E. capreolus** . . . . p. 488
   { Peraeon segment 7, spines in groups — **17.**

17 { Peraeon segment 7, spines in 6 groups . . . 17. **E. ussolzewii** . . . p. 488
   { Peraeon segment 7, spines in 2 groups — **18.**

18 { Eyes narrowly reniform . . . . 18. **E. stenophthalmus** . p. 489
   { Eyes broadly reniform . . . . 19. **E. schamanensis** . . p. 489

19 { Eyes irregularly shaped . . . . 20. **E. leptocerus** . . . p 490
   { Eyes regularly shaped — **20.**

20 { Antenna 1 little longer than antenna 2, about half as long as body — **21.**
   { Antenna 1 much longer than antenna 2. much more than half as long as body — **22.**

21 { No pleon segment with more than one row of dorsal spinules . . . . 21. **E. toxophthalmus** . p. 490
   { Some segments of pleon with more than one row of dorsal spinules . . . . 22. **E. vittatus** . . . . . p. 491

22 { Eyes white; antenna 1 more than twice as long as body . . . . 23. **E. petersii** . . . . . p. 491
   { Eyes black or reddish; antenna 1 less than twice as long as body — **23.**

1. **E. berilloni** (Catta) 1878 *Gammarus b.*, Catta in: Bull. Soc. Borda | 1896
G. b., Chevreux in: Bull. Soc. zool. France, *v.* 21 p. 29 f. | 1899 *Echinogammarus b.*,
T. Stebbing in: Tr. Linn. Soc. London, ser. 2 *v.* 7 p. 429 | 1893 *Gammarus barilloni*,
A. Della Valle in: F. Fl. Neapel, *v.* 20 p. 765.

Form of body nearly as in Gammarus pulex (p. 474). Peraeon segment 7 sometimes with dorsal spines; pleon segments 1—3 with dorsal
spines in great numbers mingled with setae (only with setae in one ♂ specimen (Chevreux)); 30—50 spines on segment 2, 4—8 dorsal spines on
segment 4, 2—4 on segment 5, 2 or 3 on segment 6. Head, post-antennal
angles acute. Pleon segment 3, postero-lateral corners acute. Eyes large,
reniform. Antenna 1 not much longer than antenna 2, 1st joint not much
longer than 2d, flagellum 25—30-jointed, accessory flagellum with 6 very
short joints. Antenna 2, ultimate and penultimate joints of peduncle long,
subequal, flagellum shorter than both combined, without calceoli. Gnathopod 1,
6th joint oval, palm oblique, scarcely distinct from hind margin. Gnathopod 2
in both sexes much larger than gnathopod 1, in ♂ 6th joint of the same shape
as in gnathopod 1, in ♀ more oblong, palm well defined from hind margin, emarginate in the centre. Peraeopod 1 strongly setose on hind margin, setae much
longer than on peraeopod 2. Peraeopods 3—5 shorter and more robust than in
Gammarus pulex, with fascicles of long setae. Uropod 3, outer ramus very
elongate, with long setae and fascicles of spines on the margin, 2d joint
obsolete, inner ramus not $\frac{1}{6}$ length of outer, carrying a few setae and an
apical spinule. Telson broader than long, lobes divergent, each with 2 spines
and many setae at narrowly truncate apex, and usually with 1 spine on outer
margin. Colour greenish brown or reddish brown. L. ♀ 7 mm, ♂ 10—12 mm.

Pyrenees, to a height of 750 m; Jersey. Streams.

2. **E. verrucosus** (Gerstf.) 1858 *Gammarus v.*, Gerstfeldt in: Mém. prés. Ac.
St.-Pétersb., *v.* 8 p. 282 | 1862 *G. v.*, Bate, Cat. Amphip. Brit. Mus., p. 219 t. 39 f. 1 | 1874
G. v., B. Dybowsky in: Horae Soc. ent. Ross., *v.* 10 suppl. p. 67 t. 4 f. 12 |
1888 *G. v.*, T. Stebbing in: Rep. Voy. Challenger, *v.* 29 p. 309 | 1899
Echinogammarus v., T. Stebbing in: Tr. Linn. Soc. London, ser. 2 *v.* 7
p. 429 | 1893 *Gammarus pungens* (part.)?, A. Della Valle in: F. Fl.
Neapel, *v.* 20 p. 768.

Fig. 89.
E. verrucosus.
Dorsal View of
pleon. [After
B. Dybowsky.]

Pleon segments 1—6 (Fig. 89) pectinate with rows of
spinules on rather obliquely set ridge-like or tubercular dorsal
prominences, usually forming 3 transverse series (spines 54) on
segment 1, 5 series (spines 89) on segment 2, 5 series (spines 100)
on segment 3, 2 series on each of segments 4—6 (spines 32,
24, 12). Pleon segment 3, postero-lateral corners approximately
quadrate (produced into long upturned tooth, Bate, but?). Eyes flat,
almost linear reniform, black. Antenna 1 not much longer than
antenna 2, flagellum 35—59-jointed, accessory flagellum 4—10-
jointed. Antenna 2, flagellum 15—26-jointed, short, broad, flattened,
with calceoli in both sexes. Gnathopods 1 and 2, 6th joint tolerably roundish-
triangular (authors' accounts differ). Peraeopods 3—5, 2d joint moderately

expanded, hind margin slightly sinuous, not produced. Uropods 1 and 2 short. Uropod 3 long, outer ramus 6 or 7 times as long as the inner, with spines and fascicles of long setae on both margins. Telson (Fig. 89) as in E. berilloni (p. 481). Colour greenish to yellowish, with a narrow brownish stripe on hind margin of each segment. L. reaching 43 mm.

Lake Baikal, River Angara.

3. E. saphirinus (Dyb.) 1874 *Gammarus s.*, B. Dybowsky in: Horae Soc. ent. Ross., *v.* 10 suppl. p. 98 | 1899 *Echinogammarus s.*, T. Stebbing in: Tr. Linn. Soc. London, ser. 2 *v.* 7 p. 429 | 1893 *Gammarus fluviatilis* (part.), A. Della Valle in: F. Fl. Neapel, *v.* 20 p. 930.

♀ unknown. — ♂. Peraeon segment 7 with 2 little groups of 2 or 3 spines on hind margin; pleon segments 1—6 each with dorsal spines in 2—4 groups adjoining hind margin, spines 18—22 on segment 1, 18—24 on segment 2, 18—20 on segment 3, 6—12 on segment 4, 8—6 on segment 5, 4 on segment 6. Eyes irregularly reniform, long, narrow, upper end a little emarginate, reaching high on top of head, reddish, invisible in spirit. Antenna 1 nearly as long as body, thrice as long as antenna 2, flagellum 87-jointed, accessory flagellum 8-jointed. Antenna 2, flagellum 16-jointed. Gnathopod 1, 6th joint piriform, larger than the oblong 6th joint of gnathopod 2. Peraeopods 3—5, 2^{d} joint broad, hind margin convex, carrying 8—10 short setae, lower corner projecting. Uropods 1 and 2 short, but uropod 1 reaching as far back as the short uropod 3, in which the inner ramus is about $^2/_5$ as long as outer; both carrying a few simple setae. Telson not described. Colour a very delicate sapphire-blue. L. 18 mm.

Lake Baikal. Depth 300 m.

4. E. czerskii (Dyb.) 1874 *Gammarus c.*, B. Dybowsky in: Horae Soc. ent. Ross., *v.* 10 suppl. p. 94 t. 1 f. 2; t. 3 f. 8 | 1899 *Echinogammarus c.*, T. Stebbing in: Tr. Linn. Soc. London, ser. 2 *v.* 7 p. 429 | 1893 *Gammarus pungens* (part.)?, A. Della Valle in: F. Fl. Neapel, *v.* 20 p. 928.

Pleon segments 1—6 with regular little pectinate groups of spines on weak elevations scarcely oblique to the adjacent hind margin, on segment 1 usually 4 groups (2 median, 2 lateral), each with 3 or 4 spines, on segments 2 and 3 a row of 4 groups, preceded by 2 or 3 rows consisting each of 2 groups, on segment 4 a row of 3 groups, preceded by a row of 2 groups, on segments 5 and 6 a row of 3 groups. Each group has usually 3 or 4 spines, sometimes on segment 6 only 1 or 2. Eyes long reniform, widened below, black. Antenna 1 longer than the body, 1st joint rather shorter than 2^{d}, and only half as long as ultimate joint of peduncle of antenna 2, flagellum 90-jointed, accessory flagellum 7—10-jointed. Antenna 2, ultimate joint of peduncle rather longer than penultimate, flagellum 20—28-jointed. Gnathopod 1, 6th joint piriform, much larger than the narrowly oblong 6th joint of gnathopod 2. Peraeopods 3—5, 2^{d} joint narrow, hind margin slightly convex, lower angle very shortly produced. Uropods 1 and 2 long, uropod 1 reaching the first or even the last quarter of uropod 3. Uropod 3 very long, easily detachable, $^2/_5$ as long as body, outer ramus 9 times as long as inner, slightly bent, fringed with long, simple setae. Colour, red above, violet below, appendages violet. L. about 28 mm.

Lake Baikal. Depth 5—8 m.

5. **E. maackii** (Gerstf.) 1858 *Gammarus m.*, Gerstfeldt in: Mém. prés. Ac. St.-Pétersb., v. 8 p. 283 | 1862 *G. m.*, Bate, Cat. Amphip. Brit. Mus., p. 217 t. 38 f. 8 | 1874 *G. m.*, B. Dybowsky in: Horae Soc. ent. Ross., v. 10 suppl. p. 97 | 1888 *G. m.*, T. Stebbing in: Rep. Voy. Challenger, v. 29 p. 309 | 1899 *Echinogammarus m.*, T. Stebbing in: Tr. Linn. Soc. London, ser. 2 v. 7 p. 429 | 1893 *Gammarus pungens* (part.), A. Della Valle in: F. Fl. Neapel, v. 20 p. 764.

Pleon segments 1—6 with spines pectinately arranged on ridges; on segment 1 are 2 oblique lateral ridges, each with 5—8 spines, on segments 2 and 3 the lateral ridges are longer, each with 7 or 8 spines commonly divided into two sets, of which the front is smaller than the hinder, also there are often 2 little median bumps, or on segment 3 two pairs of them, each with 2 or 3 spines; segments 4—6 have the ridges short, each with 3—6 spines, and 2 little median humps, each with 1 or 2 spines. Eyes narrowly reniform, black. Antenna 1 about $^3/_4$ as long as body, 1st joint shorter than ultimate joint of peduncle of antenna 2, flagellum 40—54-jointed, accessory flagellum 4- or 5-jointed. Antenna 2, flagellum 14—17-jointed. Gnathopods 1 and 2, 6th joint piriform, larger in gnathopod 1 than in gnathopod 2 (authors' accounts differ). Peraeopods 3—5, 2^d joint moderately broad, hind margin sinuous, ending in blunt, little projecting corner. Uropod 1 reaching end of peduncle of uropod 3. Uropod 3 long, $^1/_3$ as long as body, distinguished by paucity of setae, which are short and scattered, inner ramus scarcely $^1/_{10}$ as long as outer. Colour, body green, appendages red. L. about 28 mm.

River Angara at Irkutsk; Lake Baikal, depth $^1/_2$—2 m.

6. **E. lividus** (Dyb.) 1874 *Gammarus l.*, B. Dybowsky in: Horae Soc. ent. Ross., v. 10 suppl. p. 68 t. 6 f. 1 | 1899 *Echinogammarus l.*, T. Stebbing in: Tr. Linn. Soc. London, ser. 2 v. 7 p. 429 | 1893 *Gammarus pungens* (part.), A. Della Valle in: F. Fl. Neapel, v. 20 p. 929.

Pleon segments 1—6 with spines, segment 1, hind row in 4 groups, each with 4—6, front in 2 groups, each with 4 spines; segment 2, hind row as in segment 1, preceded by 3 rows, each forming 2 widely separated groups; segment 3 like segment 2, but with occasional addition of some intermediate groups; segment 4, hind row with a middle and 2 lateral groups, front with 2 middle and 2 lateral groups; segments 5 and 6 each with a middle and 2 lateral groups in 1 row. Eyes flat, long reniform, below a little widened and truncate or slightly emarginate, black. Antenna 1 about $^3/_4$ as long as body, peduncle shorter than peduncle of antenna 2, 1st joint rather shorter than 2^d, flagellum with 46—60 joints in ♀, 52—70 in ♂, accessory flagellum 5—9-jointed. Antenna 2, ultimate joint of peduncle slightly longer than penultimate, flagellum shorter than both combined, with 6—20 joints in ♀, 16—26 in ♂, carrying calceoli in both sexes. Gnathopod 1, 5th joint as broad as long, about $^2/_5$ as long as the narrowly piriform 6th. Gnathopod 2, 5th and 6th joints together as long as those of gnathopod 1 but much narrower, 5th joint $^3/_4$ as long as the narrowly oblong 6th, which has a short, well defined, almost truncate palm. Peraeopods 3—5, 2^d joint moderately broad, hind margin slightly convex or sinuous, ending often in a minutely produced angle. Uropod 1 reaching $^1/_3$ of uropod 3, which is rather long, outer ramus 6 or 7 times as long as inner, with spines and fascicles of long simple setae on both margins. Telson (in figure) much longer than broad, each lobe with a spine high up on outer margin and 3 spines on the narrow apex. Colour violet-blue or greenish violet, often with metallic glance, antennae 1 and 2 and uropod 3 red or violet-brown. L. 40 mm.

Lake Baikal. Depth 1—10 m.

7. **E. viridis** (Dyb.) 1874 *Gammarus v.* + *G. v. var. canus* + *G. v. var. olivaceus,*
B. Dybowsky in: Horae Soc ent. Ross., *v.* 10 suppl. p. 95 t. 6 f. 2; p. 95 t. 5 f. 3, t. 4 f. 4;
p. 95 | 1899 *Echinogammarus v.*, T. Stebbing in: Tr. Linn. Soc. London, ser. 2 *v.* 7 p. 429 |
1893 *Gammarus fluviatilis* (part.)?, A. Della Valle in: F. Fl. Neapel, *v.* 20 p. 931.

Pleon segments 1 and 2. spines usually arranged in 2 groups close
to hind margin (wanting on segment 1 in *var. canus,* and on both segments
in *var. olivaceus*); segment 3, spines in 2—4 transverse series, on the hindmost forming oblique lateral groups, but middle groups almost parallel to
the hind margin, in the front rows middle groups sometimes wanting;
segment 4, spines in 1 or 2 series, consisting each of 2 middle and 2 lateral
groups; segments 5 and 6 each with only 1 such row. The ridges or humps
generally feeble, in young specimens scarcely indicated; number of spines
in each group variable, encreasing with age. Eyes reniform, black. Antennae 1
and 2, gnathopods 1 and 2 and peraeopods 3—5 differing little from those
of E. lividus (p. 483), but antenna 1, accessory flagellum 4—6-jointed, antenna 2,
ultimate joint of peduncle more decidedly longer than penultimate, gnathopod 1, 6th joint thrice as long as 5th, the 2 combined longer than 5th and
6th of gnathopod 2, which are comparatively wider than in E. lividus, as is
also 2^d joint of peraeopods 3—5. Antenna 1 comparatively shorter, with
fewer joints to the flagellum in the two varieties. Uropod 1 almost reaches the
end of the inner ramus of uropod 3, which is tolerably long, the outer ramus
about 3—5 times as long as (figure; in text: $^1/_6$—$^1/_4$ longer than) the inner,
carrying marginal spines and simple setae. Telson (in figure) less elongate than
in E. lividus. Colour varying from bright grass-green to dark olive-green,
appendages red, horn-yellow to horn-brown. L. 24—31 mm.

Lake Baikal, shore to 20 m; River Angara.

8. **E. cyaneus** (Dyb.) 1874 *Gammarus c.*, B. Dybowsky in: Horae Soc. ent.
Ross., *v.* 10 suppl. p. 92 | 1899 *Echinogammarus c.*, T. Stebbing in: Tr. Linn. Soc.
London, ser. 2 *v.* 7 p. 429 | 1893 *Gammarus pungens* (part.)?, A. Della Valle in: F. Fl.
Neapel, *v.* 20 p. 928.

Pleon segment 1 with 2—6 delicate spinules on hind margin; segment 2
with 6—8 somewhat more conspicuous, but still setiform spines on hind margin;
segment 3, whole dorsal surface set with long setae. among which the spines,
if present, completely disappear; segments 4—6 less setose, spines stronger,
on segment 4 forming 2 middle and 2 lateral groups, on segments 5 and 6,
1 median and 2 lateral groups, number of spines to a group 2—4. Eyes
slightly reniform, black. Antenna 1 about $^1/_2$ as long as body, flagellum
24—35-jointed, accessory flagellum 4- or 5-jointed. Antenna 2, flagellum
11—13-jointed. Gnathopod 1 piriform, gnathopod 2 oblong. Gnathopod 1
in ♂ longer and broader than gnathopod 2, in ♀ rather broader but not
longer. Peraeopods 3—5, 2^d joint tolerably broad, hind margin slightly
convex, carrying 8—10 short setae, ending in a projecting acute angle. Uropods 1
and 2 about reaching end of inner ramus of uropod 3. Uropod 3 moderately
long, outer ramus 3 or 4 times longer than inner, both with simple marginal
setae. Colour bluish. L. 18 mm.

Lake Baikal. Near the shore. under stones.

9. **E. ochotensis** (F. Brandt) 1851 *Gammarus o.*, F. Brandt in: Middendorff,
Reise Sibirien, *v.* 21 p. 140 t. 6 f. 31a—c | 1862 *G. o.*, Bate, Cat. Amphip. Brit. Mus.,
p. 216 t. 38 f. 5 | 1899 *Echinogammarus o.*, T. Stebbing in: Tr. Linn. Soc. London, ser. 2
v. 7 p. 429 | 1893 *Gammarus marinus* (part.), A. Della Valle in: F. Fl. Neapel, *v.* 20 p. 762.

Pleon segments 1 and 2 more or less clearly furnished in the middle of the hinder dorsal margin with a median forward and 2 lateral hinder groups of spinules on low elevations; segment 3 having these preceded by 2 or 3 pairs of similar elevations carrying 4 or 5 spinules each; segments 4 and 5 each with a pair of slightly curved ridges pectinate with 5 or 6 spines, segment 4 having also 2 smaller lateral elevations with 2 or 3 spines each; segment 6 with a median pair of spines and 3 in a row on each side. Eyes slightly reniform. Antenna 1 about $^1/_2$ as long as body, rather longer than antenna 2, flagellum 30—32-jointed, much longer than peduncle, accessory flagellum about $^1/_5$ length of primary. Antenna 2, ultimate and penultimate joints of peduncle long, subequal, flagellum 20-jointed. Gnathopods 1 and 2 subequal, 6th joint moderately convex, almost rhomboidal, palm set with spines, more or less straight truncate (Brandt). Peraeopods tolerably strong, with short setae and spines. Uropod 1 moderately long and broad, reaching about the middle of uropod 3. Uropod 3, outer ramus about twice as long as telson, thrice as long as inner ramus. Telson conical, each lobe with spines on the blunt apex. L. reaching 25 mm.

Ochotsk Bay [Siberia].

10. **E. testaceus** (Dyb.) 1874 *Gammarus t.*, B. Dybowsky in: Horae Soc. ent. Ross., *v.* 10 suppl. p. 60 | 1899 *Echinogammarus t.*, T. Stebbing in: Tr. Linn. Soc. London, ser. 2 *v.* 7 p. 429 | 1893 *Gammarus fluviatilis* (part)?, A. Della Valle in: F. Fl. Neapel, *v.* 20 p. 931.

Pleon segment 1 without spines; segment 2 with a pair of lateral groups of 5 or 6 spines each; segments 3 and 4 and often 5 with 2 transverse rows, the hinder composed of 2 middle and 2 lateral groups, the front generally only of 2 lateral groups; segments 3—6 or 4 and 5 sometimes having the middle and lateral group on each side coalesced; segment 6 with spines. Eyes narrowly reniform, somewhat widened below, black. Antenna 1 shorter than body, twice as long as antenna 2, flagellum with 32 joints in ♀, to 50 in ♂, accessory flagellum with 4 joints in ♀, to 7 in ♂. Antenna 2, flagellum with 11 joints in ♀, to 23 in ♂, carrying calceoli. Gnathopod 1, 6th joint piriform, somewhat larger than the oblong 6th of gnathopod 2. Peraeopods 3—5, 2d joint broad, hind margin convex, ending below with projecting angle. Uropod 1 reaches middle of uropod 3, inner ramus about $^1/_3$—$^2/_5$ as long as the 2-jointed outer, which has simple setae on its outer margin, but plumose ones also on the inner. Colour dark brownish, passing into greenish, the whole body spotted with yellow or bright greenish. L. ♀ 12—13, ♂ 21 mm.

Lake Baikal (southern shore).

11. **E. sophiae** (Dyb.) 1874 *Gammarus s.*, B. Dybowsky in: Horae Soc. ent. Ross., *v.* 10 suppl. p. 61 | 1899 *Echinogammarus s.*, T. Stebbing in: Tr. Linn. Soc. London, ser. 2 *v.* 7 p. 429 | 1893 *Gammarus fluviatilis* (part.)?, A. Della Valle in: F. Fl. Neapel, *v.* 20 p. 930.

Pleon segments 1—3, spines on hind margin in a pair of lateral groups, 2 or 3 spines to a group; segments 4—6 each with only 3 or 4 delicate spinules, 1 or 2 in the median line, and 1 on each side in the place of lateral groups. Eyes small, oval or reniform, black. Antenna 1 nearly $^2/_3$ as long as body, 1st joint shorter than 2d, flagellum 17—30-jointed, accessory flagellum 3—6-jointed. Antenna 2, flagellum 8—14-jointed, with calceoli. Gnathopods 1 and 2, 6th joint piriform, larger and stronger in gnathopod 1 than in gnathopod 2, and somewhat tumid at the base. Peraeopods 3—5, 2d joint slender, hind margin slightly sinuous, not produced to an angle below. Uropod 1 reaches beyond middle of uropod 3. Uropod 3, outer ramus about twice as long as inner, with plumose setae on the inner margin, 2^d joint very small, hidden among the long terminal spines. Colour yellowish, playing into reddish. L. ♂ 9—15 mm.

Lake Baikal. Depth 200 m.

12. **E. fuscus** (Dyb.) 1874 *Gammarus f.*, B. Dybowsky in: Horae Soc. ent. Ross., *v.* 10 suppl. p. 63 t. 5 f. 2 | 1899 *Echinogammarus f.*, T. Stebbing in: Tr. Linn. Soc. London, ser. 2 *v.* 7 p. 429 | 1893 *Gammarus fluviatilis* (part.)?, A. Della Valle in: F. Fl. Neapel, *v.* 20 p. 928.

Pleon segments 1—6 with dorsal spines in groups on slight elevations of the hind margin; segments 1 and 2 with 2 middle and 2 lateral groups, each with 4 spines; in all the segments the middle and lateral groups sometimes unite on either side; segments 3 and 4 have an additional cross-row of 2 groups, with variable number of spines. Pleon segments 1 and 2, but not segment 3, having the lateral lobe marked off. Eyes rather small, reniform (in figure widened below), black. Antenna 1 about $^2/_3$ as long as body, only a little longer than antenna 2, 1st joint longer than 2^d and 3^d combined, as long as ultimate joint of peduncle of antenna 2 (in figure; shorter in text). flagellum 30-jointed, accessory flagellum 7-jointed. Antenna 2, flagellum 20-jointed, with calceoli. Gnathopods 1 and 2, 6th joint moderately large, in gnathoped 1 piriform, in gnathopod 2 rather smaller, oblong. but with the palm rather oblique. Peraeopods 3—5, 2^d joint rather broad, hind margin convex or somewhat sinuous, ending below in a short, acutely projecting point. Uropods 1 and 2 reaching middle of uropod 3. Uropod 3, inner ramus $^1/_2 - ^2/_3$ as long as outer, both with simple setae on outer margin, but also plumose setae on inner; 2^d joint of outer ramus (in figure) distinct. Telson (in figure) with the lobes narrow, each having 3 spines at apex. Colour violet-brownish, with greenish spots; hind margin of head and front half of peraeon segment 1 brighter coloured; appendages banded. L. ♂ reaching 39 mm.

Lake Baikal. Depth 30—100 m.

13. **E. murinus** (Dyb.) 1874 *Gammarus m.*, B. Dybowsky in: Horae Soc. ent. Ross., *v.* 10 suppl. p. 64 t. 5 f. 1 | 1899 *Echinogammarus m.*, T. Stebbing in: Tr. Linn. Soc. London, ser. 2 *v.* 7 p. 429 | 1893 *Gammarus fluviatilis* (part.)?, A. Della Valle in: F. Fl. Neapel, *v.* 20 p. 929.

Pleon segments 1—6 with dorsal spines on low elevations of the hind margin; segment 1 with 1 cross-row of 2 middle and 2 lateral groups, each containing 4 or 5 spines; segment 2 with 3 cross-rows, the hindmost as in segment 1, but here with 4 spines to a group, preceding row similar or with only 2 middle groups, and the other row generally with only 2 middle groups, spines varying from 1—3 in a group; segment 3 like segment 2, but often with the addition of solitary spinules and simple setae; segment 4 with 2 cross-rows, the spines more delicate. and often giving place to setae; segments 5 and 6 each with 1 cross-row, the lateral groups usually having 4 spines. the median 1 or 2 pairs of setae. Pleon segments 1—3 having the lateral lobes distinctly marked off, black. Antenna 1 nearly $^2/_3$ as long as body, 1st joint as long as 2^d and 3^d combined, and as long as ultimate joint of peduncle of antenna 2, flagellum with 39—53 joints in ♀, 57 in ♂, accessory flagellum with 12 joints in ♀, 10 in ♂. Antenna 2, ultimate joint of peduncle rather shorter than penultimate, flagellum 28-jointed, with calceoli. Gnathopods 1 and 2 as in E. fuscus. Peraeopods 3—5, 2ᵈ joint tolerably broad, but narrower and longer than in E. fuscus. Uropods 1 and 2 reaching middle of uropod 3. Uropod 3, outer ramus strong, with only simple setae on outer margin, plumose setae also on inner; inner ramus narrow, ⁵/₈ as long as outer, with plumose setae on both margins. Telson elongate, (in figure: each lobe with 1 spine high up on outer margin, 1 low down on inner, 3 on narrow apex). Colour bright violet-brownish or dark grey, with bright olive-green spots, appendages banded. L. about 23 mm.

Lake Baikal. Depth 30—100 m.

14. **E. aheneus** (Dyb.) 1874 *Gammarus a.* + *G. a. var. setosus* + *G. a. subvar. miniatus* + *G. a. subvar. succineus*, B. Dybowsky in: Horae Soc. ent. Ross., *v.* 10 suppl. p. 65, 66; t. 7 f. 1, 2; t. 6 f. 3 | 1899 *Echinogammarus a.*, T. Stebbing in: Tr. Linn. Soc. London, ser. 2 *v.* 7 p. 429 | 1893 *Gammarus calcaratus* (part.)?, A. Della Valle in: F. Fl. Neapel, *v.* 20 p. 927.

Pleon segments 1—6 armed with spines, varying in number and arrangement; segment 1 with 1 or 2 cross-rows of spines, segment 2 with 2—4 cross-rows, segment 3 with 4 or 5, segment 5 with 1 or 2, segment 6 with 1 cross-row of spines; segment 4 not specified. Besides the spines occur setae, which sometimes take the place of the spines or conceal them. Pleon segments 1—3 having the lateral lobes distinctly marked off; also on lower margins of side-plates there are usually setae, few or whole fascicles. Eyes reniform or biscuit-shaped, often slightly widened below, black. Antenna 1 nearly or quite as long as body, 1ˢᵗ joint little or not longer than 2ᵈ, 3ᵈ not very short, flagellum 41—116-jointed, accessory flagellum 7—13-jointed. Antenna 2, ultimate joint of peduncle long, longer than penultimate, flagellum 20—33-jointed, with calceoli. Gnathopod 1, 6ᵗʰ joint piriform, larger than the oblong 6ᵗʰ of gnathopod 2. Peraeopods 3—5, 2ᵈ joint moderately broad, hind margin slightly convex or somewhat concave, ending below in a little angle. Uropod 1 not reaching middle of uropod 3. Uropod 3 long and strong, outer ramus 2-jointed, more than twice as long as inner, slightly curved inward, with only simple setae on outer margin; inner ramus with fascicles of simple setae interspersed with plumose ones on its margins. Telson (in figure) as in E. murinus. Colour red or yellow. L. reaching about 37 mm.

Lake Baikal. Depth 50—500 m.

15. **E. sarmatus** (Dyb.) 1874 *Gammarus s.*, B. Dybowsky in: Horae Soc. ent. Ross., *v.* 10 suppl. p. 86 t. 1 f. 3; t. 8 f. 4 | 1899 *Echinogammarus s.*, T. Stebbing in: Tr. Linn. Soc. London, ser. 2 *v.* 7 p. 429 | 1893 *Gammarus fluviatilis* (part.), A. Della Valle in: F. Fl. Neapel, *v.* 20 p. 930.

♀ unknown. — ♂. Peraeon segment 7 and pleon segments 1—6 with dorsal spines, forming a single cross-row on hind margin, formed of 2 lateral groups with 2—5 spines in each, and 2 middle groups with 1—4 spines in each. Eyes retort-shaped, almost pointed above, concave in front, the widened lower part with lower margin slightly concave, white, becoming rosy on exposure. Antenna 1 longer than body, 1ˢᵗ joint long, 2ᵈ and 3ᵈ successively a little shorter, flagellum thrice as long as peduncle, 130-or 131-jointed, accessory flagellum 39-jointed. Antenna 2 very little shorter than antenna 1, ultimate joint of peduncle longer than penultimate, flagellum 110—115-jointed, twice as long as peduncle. All the legs long and delicate. Gnathopods 1 and 2, 6th joint not large, piriform (in figure of gnathopod 2, 6th joint oblong, with a very oblique palm), the 5th joint in both gnathopods rather elongate; gnathopod 2 rather smaller than gnathopod 1. Peraeopods 3—5, 2[d] joint slender, twice or (figure) more than twice as long as broad, the expansion forming a little lobe at top. Uropods 1 and 2 with rami longer than peduncle, almost reaching as far as end of the long uropod 3. Uropod 3, outer ramus with 1st joint twice as long as peduncle, 2[d] joint nearly $^1/_7$ as long as 1st, inner ramus not much shorter· than 1st joint of outer, both with numerous plumose setae on both margins. Colour white or faintly flesh-toned, on exposure becoming more rosy. L. about 42 mm.

Lake Baikal. Depth 1300 m.

16. **E. capreolus** (Dyb.) 1874 *Gammarus c.* + *G. c. var. chloris*, B. Dybowsky in: Horae Soc. ent. Ross, *v.* 10 suppl. p. 87 t. 11 f. 1 | 1899 *Echinogammarus capreolus*, T. Stebbing in: Tr. Linn Soc. London, ser. 2 *v.* 7 p. 429 | 1893 *Gammarus fluviatilis* (part.), A. Della Valle in: F. Fl. Neapel, *v.* 20 p. 928.

Peraeon segment 7 and pleon segments 1—6 with dorsal spines; on peräeon segment 7 and pleon segments 1—3 the spines are very delicate, occupying the hind margin in too close a row to be separated into groups; on segments 4—6 they form 2 lateral groups, each with 2—5 spines, and 2 middle groups, each with 1—3 spines. Eyes rather large, slightly reniform. wider below than above. Antenna 1 longer than the body, 1st joint stout, nearly as long as 2[d] and 3[d] combined, flagellum 9—20 times as long as peduncle, 64—95-jointed, accessory flagellum 5—10-jointed. Antenna 2, ultimate and penultimate joints of peduncle subequal, much more slender than 1st joint of antenna 1, flagellum 29—43-jointed, 2 or 3 times as long as peduncle. Legs delicate, long, brittle. Gnathopod 1, 6th joint slenderly piriform (oblong in figure); gnathopod 2, 5th and 6th joints longer but narrower than in gnathopod 1, 6th narrowly oblong, with short transverse palm. Peraeopod 3 only a little shorter than peraeopod 4. Peraeopods 3—5, 2[d] joint longer than broad, hind margin feebly convex, ending below in a short angle. Uropod 1 longer than uropod 2, reaching end of uropod 3. Uropod 3, outer ramus twice as long as peduncle, its 2[d] joint $^1/_5$ as long as 1st, inner ramus nearly as long as 1st joint of outer, both with plumose setae on margins. Colour horn-yellow. L. 14—19 mm. — *Var. chloris*: eyes larger and broader, legs more short and sturdy, 6th joint of gnathopods 1 and 2 stronger and broader, 2[d] joint of peraeopods 3—5 broader· and shorter, uropods 1—3 shorter, colour greenish, spotted with brown. L. 11—15 mm.

Lake Baikal. Depth 100—200 m.

17. **E. ussolzewii** (Dyb.) 1874 *Gammarus ussolzevii, G. ussolzewii, G. ussolzevi* + *G. ussolzewii var. abyssorum*, B. Dybowsky in: Horae Soc. ent. Ross., *v.* 10 suppl. p 28, 89, 190 t. 9 f. 2 | 1893 *Gammarus fluviatilis* (part.)?, A. Della Valle in: F. Fl. Neapel, *v.* 20 p. 931 | 1899 *Echinogammarus uzzolzewii*, T. Stebbing in: Tr. Linn. Soc. London, ser. 2 *v.* 7 p. 429.

Peraeon segment 7 and pleon segments 1—6 with dorsal spines; on peraeon segment 7 hind margin with 2 middle, 2 lateral and 2 marginal groups of 3 or 4 spines each; pleon segments 1—3. spines of the groups more numerous, so as often to form a continuous row on the hind margin, preceded by another row of 4 groups, 2 lateral and 2 marginal, with 3 or 4 spines each; segments 4—6 with 4 groups, containing each 2—4 spines. Eyes tolerably large, narrowly reniform, slightly widened below, black. Antenna 1 about $^1/_3$ longer than body, 1st joint stout, 3^d not very short, flagellum about 5 times as long as peduncle, joints reaching 200 in number, accessory flagellum 14—17-jointed. Antenna 2 about $^1/_3$ length of antenna 1, ultimate joint of peduncle longer than penultimate, flagellum rather longer than peduncle, 20—42-jointed. Gnathopods 1 and 2 nearly equally long, gnathopod 1 with 6th joint piriform, gnathopod 2, 6th joint narrowly oblong, with short transverse palm in ♀, smaller than 6th joint of gnathopod 1, but in ♂ widening to the oblique palm, and both broader and longer than 6th joint of gnathopod 1. Peraeopods 3—5, 2^d joint moderately broad, hind margin slightly convex or sinuous, or nearly straight, ending below with a projecting acute angle. Uropods 1 and 2 reaching end of shorter ramus of uropod 3. Uropod 3, inner ramus $^2/_3$—$^3/_4$ as long as outer, plumose setae on both margins; outer ramus having them only on its inner margin; its 2^d joint distinct. Colour reddish yellow. L. 34 mm. — *Var. abyssorum* is distinguished by longer extremities, slender 6th joint to gnathopods 1 and 2, longer uropod 3, with relatively shorter inner ramus, colour bright rose-red or reddish yellow, eyes whitish (becoming dark in spirit) or reddish, and plumose setae on outer margin of outer ramus of uropod 3. L. 41 mm.

Lake Baikal. Depth 150—1000 m.

18. **E. stenophthalmus** (Dyb.) 1874 *Gammarus s.*, B. Dybowsky in: Horae Soc. ent. Ross., v. 10 suppl. p. 90 | 1899 *Echinogammarus s.*, T. Stebbing in: Tr. Linn. Soc. London, ser. 2 v. 7 p. 429 | 1893 *Gammarus fluviatilis* (part.)?, A. Della Valle in: F. Fl. Neapel, v. 20 p. 930.

Peraeon segment 7 and pleon segments 1—6 with dorsal spines on slight ridge-like elevations of the hind margin; on peraeon segment 7 only 2 middle rows, on all the pleon segments 2 middle and 2 lateral, on segments 2 and 3 the spines more numerous than on the others, so as to form a continuous row; spines in a group 4—6, rarely 1—3. Eyes narrowly to linear reniform, scarcely widened below, 4 times as deep as broad. Antenna 1 as long as body, thrice as long as antenna 2, 1st joint thick, almost as long as ultimate joint of peduncle of antenna 2, flagellum more than 4 times as long as peduncle, 121-jointed, accessory flagellum 8-jointed. Antenna 2, flagellum shorter than peduncle, 23-jointed. Gnathopod 1, 6th joint slenderly piriform. Gnathopod 2, 6th joint oblong. Peraeopods 3—5, 2^d joint rather broad, with convex front and hind margin, the latter ending below in an acute angle. Uropods 1 and 2 reaching about the middle of uropod 3. Uropod 3, outer ramus without plumose setae on outer margin, inner $^2/_3$ as long, with plumose setae on both margins. L. over 30 mm.

Lake Baikal. Depth 200 m.

19. **E. schamanensis** (Dyb.) 1874 *Gammarus s.*, B. Dybowsky in: Horae Soc. ent. Ross., v. 10 suppl. p. 91 | 1899 *Echinogammarus s.*, T. Stebbing in: Tr. Linn. Soc. London, ser. 2 v. 7 p. 429 | 1893 *Gammarus calcaratus* (part.)?, A. Della Valle in: F. Fl. Neapel, v. 20 p. 930.

Peraeon segment 7 and pleon segments 1—6 with dorsal spines, in regular and usually separate groups in 1 cross-row on each hind margin; peraeon segment 7 with 2 middle groups, all the rest with 2 lateral and 2 middle groups, the latter coalescent on pleon segments 4—6, spines 2—6 in middle groups, 3—6 in lateral groups. Eyes large and broad, reniform, black. Antenna 1 almost as long as body, peduncle thicker and only a little shorter than that of antenna 2, 3^d joint relatively longer than in nearly related species, flagellum twice as long as peduncle, 75—84-jointed, accessory flagellum 7- or 8-jointed. Antenna 2 about $^3/_5$ as long as antenna 1, flagellum shorter than peduncle, 26—35-jointed. Gnathopod 1, 6th joint broad, piriform, gnathopod 2, 6th joint oblong. Peraeopods 3—5, 2^d joint about $^1/_3$ longer than broad, hind margin sinuous, acute lower corner not prolonged into a tooth. Uropods 1 and 2 reaching about $^2/_3$ of uropod 3. Uropod 3, outer ramus without plumose setae on outer margin, inner ramus $^2/_3$ as long as outer. Colour reddish yellow. L. about 23 mm.

Lake Baikal. Depth 200 m.

20. **E. leptocerus** (Dyb.) 1874 *Gammarus l.* + *G. l. var. nematocerus*, B. Dybowsky in: Horae Soc. ent. Ross., *v.* 10 suppl. p. 85 t. 8 f. 2; p. 85 t. 8 f. 3 | 1899 *Echinogammarus l.*, T. Stebbing in: Tr. Linn Soc. London, ser. 2 *v.* 7 p. 429 | 1893 *Gammarus fluviatilis* (part.)?, A. Della Valle in: F. Fl. Neapel, *v.* 20 p. 929.

Pleon segments 1—3 with about 18 dorsal spines in 3 or 4 little groups or in a continuous row on hind margin of each; segments 4—6 with 7—9 spines in 3 groups. Eyes large and broad, irregularly reniform, hind margin uneven and lobed, black. Antenna 1 a little longer than body, nearly thrice as long as antenna 2, 1st joint stout, as long as 2^d and 3^d combined, flagellum about 6 times as long as peduncle, 43—90-jointed. accessory flagellum 4—7-jointed. Antenna 2, ultimate joint of peduncle slender, longer than penultimate, flagellum a little shorter than peduncle, 11—18-jointed. Gnathopods 1 and 2, 6th joint rather large, in gnathopod 1 piriform, in gnathopod 2 oblong, in ♂ widening to the palm. Peraeopods 3—5, 2^d joint slender, lower hind angle unimportant. Uropod 1 reaching middle of uropod 3. Uropod 3, inner ramus only a little shorter than outer, both with plumose setae on their margins. Colour reddish yellow. L. 14—15 mm. — *Var. nematocerus* is distinguished by more slender build of body, longer and more slender legs, flagellum of antenna 2 longer than peduncle, eyes more narrowed above, more numerous setae on uropod 3, and whitish colour. L. 22—23 mm.

·Lake Baikal. Depth 150—670 m.

21. **E. toxophthalmus** (Dyb.) 1874 *Gammarus t.*, B. Dybowsky in: Horae Soc. ent. Ross., *v.* 10 suppl. p 77 | 1899 *Echinogammarus t.*, T. Stebbing in: Tr Linn. Soc. London, ser. 2 *v.* 7 p. 429 | 1893 *Gammarus fluviatilis* (part)?, A. Della Valle in: F. Fl. Neapel, *v.* 20 p. 931.

Pleon segments 1—6 with dorsal spines in a single cross-row of regular groups on low elevations of each segment's hind margin, 1—3 spines in each of the 2 middle groups, 4—8 in the 2 lateral groups, except on segment 6 where the number is reduced to 2. Eyes long, reniform, much curved, widened below, not widely separated above, black. Antenna 1 about $^2/_3$ as long as body, little longer than antenna 2; 1st joint stout, considerably shorter than ultimate joint of peduncle of antenna 2, flagellum about 2$^1/_2$ times as long as peduncle, 45-jointed, accessory flagellum 5-jointed. Antenna 2, flagellum a little shorter than peduncle, 26-jointed. Gnathopods 1 and 2, 6th joint slender, in gnathopod 1 piriform, in gnathopod 2 oblong. Peraeopods 3—5, 2^d joint moderately broad, hind margin almost straight, ending below in a shortly projecting angle. Uropod 1 reaching nearly the end of uropod 3. Uropod 3, inner ramus about $^1/_6$ shorter than outer, both with plumose setae on both margins. Colour clear violet. L. 20 mm.

Lake Baikal. Depth 120 m.

22. **E. vittatus** (Dyb.) 1874 *Gammarus v.*, B. Dybowsky in: Horae Soc. ent. Ross., *v.* 10 suppl. p. 82 | 1899 *Echinogammarus v.*, T. Stebbing in: Tr. Linn. Soc. London, ser. 2 *v.* 7 p. 429 | 1893 *Gammarus fluviatilis* (part.)?, A. Della Valle in: F. Fl. Neapel, *v.* 20 p. 931.

Pleon segment 1 with dorsal spines in a single cross-row formed by 2 quite small lateral groups and 2 middle spines, which seem to stand immediately on the dorsal surface; segment 2 with 3 cross-rows, the front one composed of 2 middle groups, the next and the one on the hind margin of 2 middle and 2 lateral groups, each containing 4 or 5 spines; segment 3 like segment 2, but with addition in front of 2 little groups on the median line; segment 4 with 2 cross-rows, the front one composed of 2 middle groups, the hinder of 2 middle and 2 lateral groups, having 2—5 spines in each; segments 5 and 6 each with 1 cross-row formed by 2 middle and 2 lateral groups. Eyes moderately large, reniform, black. Antenna 1 not $\frac{1}{2}$ as long as body, $\frac{1}{4}$—$\frac{1}{3}$ longer than antenna 2, flagellum not twice as long as peduncle, 33—39-jointed, accessory flagellum 5- or 6-jointed. Antenna 2, flagellum shorter than peduncle. 12—15-jointed. Gnathopod 1, 6th joint slenderly piriform, somewhat larger than oblong 6th joint of gnathopod 2. Peraeopods 3—5, 2d joint moderately broad, hind margin prolonged below in an obtuse angle, the margin convex in peraeopod 3, a little sinuous in peraeopods 4 and 5. Uropod 1 reaching end of shorter ramus of uropod 3. Uropod 3 moderately long, inner ramus $\frac{2}{3}$—$\frac{3}{4}$ as long as outer, both fringed with numerous fascicles of setae, and both having among the simple setae of the fascicles on the inner margin plumose setae. Colour bright yellow green, or bright olive-green, each segment with a narrow stripe of brown on hind margin, antennae brown, with narrow greenish bands. L. 23 mm.

Lake Baikal. Under stones.

23. **E. petersii** (Dyb.) 1874 *Gammarus p.*, B. Dybowsky in: Horae Soc. ent. Ross., *v.* 10 suppl. p. 83 t. 10 f. 1 | 1899 *Echinogammarus p.*, T. Stebbing in: Tr. Linn. Soc. London, ser. 2 *v.* 7 p. 429 | 1893 *Gammarus calcaratus* (part.)?, A. Della Valle in: F. Fl. Neapel, *v.* 20 p. 930.

Peraeon segment 7 very rarely having a couple of delicate spinules on the hind margin; pleon segments 1—6 each with 1 cross-row of dorsal spines, on segments 1—3 forming 6 groups, 2 middle, 2 lateral and 2 marginal, on segments 4—6 only 4 groups, 3—5 spines in a group; exceptionally some delicate isolated dorsal spinules occur on segments 2 and 3. Head with slightly convex upper outline. Eyes moderately large, reniform, a little widened below and narrowed above, white, difficult to make out in spirit, not darkening with exposure to light. Antenna 1 thrice as long as body, 5 times as long as antenna 2, extremely brittle, 1st joint stout, longer than 2d and 3d combined, shorter than ultimate joint of peduncle of antenna 2, flagellum never found complete. 17 times as long as peduncle, at least 350-jointed, accessory flagellum fully as long as peduncle, 16-jointed. Antenna 2, ultimate joint of peduncle slender, much longer than penultimate, flagellum longer than peduncle, 27—57-jointed. Legs delicate. Gnathopod 1, 6th joint piriform. Gnathopod 2, 6th joint oblong. In figures gnathopod 1 in ♀ has the 6th joint shorter, broader and with longer hind margin than in ♂, and gnathopod 2 in ♂ has the 6th widening to the palm much more than in ♀, and consequently with a longer palm and longer finger. Peraeopods 3—4 (5?), 2d joint moderately broad, hind margin little convex, ending below with a weakly projecting angle; 2d joint of peraeopod 5 (in figure) narrow, its breadth greatest at top, equalling only $\frac{1}{5}$ of its length, the hind margin almost straight, ending below in a little obtuse point. Uropods 1 and 2 reaching end of shorter ramus of uropod 3. Uropod 3 long, inner ramus about $^2/_3$ as long as outer, of which the 2^d joint is well developed, $^1/_7$ as long as 1st joint; both rami with plumose setae on both margins. Colour delicate, clear violet or reddish white. L. about 29 mm.

Lake Baikal. Depth 700–1300 m.

24. **E. violaceus** (Dyb.) 1874 *Gammarus v.* + *G. v. var. virescens*, B. Dybowsky in: Horae Soc. ent. Ross., *v.* 10 suppl. p. 75 t. 10 f. 3; p. 76 t. 12 f. 5 | 1899 *Echinogammarus violaceus*, T. Stebbing in: Tr. Linn. Soc. London, ser. 2 *v.* 2 p 429 | 1893 *Gammarus calcaratus* (part)?, A. Della Valle in: F. Fl. Neapel, *v.* 20 p. 931.

Pleon segments 1—6 each with 1 cross-row of dorsal spines on the hind margin pectinately set on ridge-like elevations in 2 middle groups with 3—5 spines each, and 2 lateral groups with 4—7 spines each. Head, rostral point short and rounded. Eyes small, reniform, slightly widened below, black. Antenna 1 in ♂ as long as body, more than twice as long as antenna 2, 1st joint not greatly longer than 2^d, nor 2^d than 3^d, flagellum about 4 times as long as peduncle, 91—148-jointed, accessory flagellum 7- or 8-jointed. Antenna 2, ultimate joint of peduncle longer than penultimate, flagellum shorter than peduncle, 18—25-jointed. Gnathopod 1, 6th joint broad, piriform. Gnathopod 2, 6th joint slender, oblong. In figure gnathopod 1 has 5th joint as broad as long, 6th joint twice as long as 5th, broadly oblong oval; gnathopod 2 has 5th joint long and narrow, not much shorter than the 6th, the 2 combined being much longer than the much broader 5th and 6th of gnathopod 1 combined. Peraeopods 3—5, 2^d joint broad, in peraeopod 5 heart-shaped, about $^1/_3$ longer than broad, hind margin convex, prolonged below into a short angle. Uropods 1 and 2 short, reaching almost $^2/_3$ length of uropod 3. Uropod 3, inner ramus about $^2/_7$—$^1/_3$ shorter than outer, with plumose setae on both margins, while the outer ramus has them only on the inner. Colour dark violet-red or brownish red, appendages brightly banded. L. reaching 37 mm. — *Var. virescens* distinguished by smaller eyes, shorter uropod 3, broader 2^d joint to peraeopods 3—5 and a smaller number of spines.

Lake Baikal. Depth 20—100 m.

25. **E. ibex** (Dyb.) 1874 *Gammarus i.*, B. Dybowsky in: Horae Soc. ent. Ross., *v.* 10 suppl. p. 78 | 1899 *Echinogammarus i.*, T. Stebbing in: Tr. Linn. Soc. London, ser. 2 *v.* 7 p. 429 | 1893 *Gammarus fluviatilis* (part.)?, A. Della Valle in: F. Fl. Neapel, *v.* 20 p. 929.

Pleon segments 1—6 each with 4 groups of dorsal spines on very slight elevations of the hind margin, the 2 middle groups with 1—3 spines each, the lateral groups with 4—8. Head, frontal line almost straight. Eyes large, broad, reniform, slightly widened below, not far apart above, black. Antenna 1 longer than body, twice as long as antenna 2, 1st joint stout, shorter than ultimate joint of peduncle of antenna 2, flagellum 5 times as long as peduncle, 73—80-jointed, accessory flagellum 8-jointed. Antenna 2, flagellum shorter than peduncle, 24—30-jointed. Gnathopods 1 and 2, 6th joint moderately large, in ♂ slightly widening to the palm, in gnathopod 1 piriform, in gnathopod 2 oblong. Legs delicate and thin. Peraeopods 3—5, 2^d joint not broad, front margin carrying long setae, the feebly convex hind margin 10—16 short setae, and ending in a short projecting angle. Uropods 1 and 2 long, but scarcely reaching middle of uropod 3. Uropod 3 in ♂ almost $^1/_3$ as long as body, outer ramus about $^1/_3$ longer than inner, both having plumose setae on both margins. Colour uniform yellow, or with horn-yellow markings. L. 14 mm.

Lake Baikal. Depth 150—200 m.

26. **E. parvexii** (Dyb.) 1874 *Gammarus p.*, B. Dybowsky in: Horae Soc. ent. Ross., *v.* 10 suppl. p. 81 t. 10 f. 2 | 1899 *Echinogammarus p* , T. Stebbing in: Tr. Linn. Soc. London, ser. 2 *v.* 7 p. 429 | 1893 *Gammarus calcaratus* (part.)?, A. Della Valle in: F. Fl. Neapel, *v.* 20 p. 929.

Pleon segments 1—6 each with 4 groups of dorsal spines on elevations of the hind margin, the middle groups each with 1 or 2 spines, the lateral each with 2—4. Eyes tolerably large, biscuit-shaped or reniform, wide apart above, black. Antenna 1 nearly twice as long as body, thrice as long as antenna 2, 1st joint not as long as 2d and 3d combined, flagellum 7—9 times as long as the thin peduncle, 310-jointed, accessory flagellum 17-jointed. Antenna 2, ultimate joint of peduncle longer than penultimate, flagellum shorter than peduncle, 26-jointed. Gnathopod 1, 5th joint as broad as long, 6th very broadly piriform, the basal half tumid, finger strongly curved. Gnathopod 2, 6th joint long and narrow, palm short, almost transverse. Peraeopods 3—5, 2d joint narrow, in figure of peraeopod 5 more than twice as long as broad, hind margin flatly concave, produced to a subacute point. Uropods 1 and 2 long, reaching nearly the middle of uropod 3, rami longer than peduncle. Uropod 3, outer, ramus without plumose setae on outer margin, 2d joint between $\frac{1}{7}$ and $\frac{1}{6}$ as long as 1st, inner ramus $\frac{2}{3}$ as long as 1st joint of outer. Colour orange-yellow. L. about 25 mm.

Lake Baikal. Depth 170 m.

27. **E. polyarthrus** (Dyb.) 1874 *Gammarus longicornis* (non J. C. Fabricius 1779!) + *G. l. var. p.*, B. Dybowsky in: Horae Soc. ent. Ross., *v.* 10 suppl. p. 79; p. 80 t. 10 f. 2 b′, c′ | 1899 *Echinogammarus p.*, T. Stebbing in: Tr. Linn. Soc. London, ser. 2 *v.* 7 p. 429 | 1893 *Gammarus calcaratus* (part.)?, A. Della Valle in: F. Fl. Neapel, *v.* 20 p. 929.

Pleon segments 1—6 each with 4 groups of dorsal spines on ridge-like elevations of the hind margin, the 2 middle groups each with 2—4 spines, the lateral with 2—6, except on segment 6, where there are only 4 spines for the 4 groups; also on segments 4 and 5 the middle groups are weakly developed. Eyes large, little deeper than broad, reniform, slightly widened below, narrowed above, black or reddish. Antenna 1 in ♂ longer than the body, 3—4 times as long as antenna 2, flagellum 5—8 times as long as the stout peduncle, 250—340-jointed, accessory flagellum 16-jointed. Antenna 2, ultimate joint of peduncle longer than 1st joint of antenna 1, flagellum shorter than peduncle, 25—29-jointed. Gnathopod 1, 6th joint piriform base not tumid. Gnathopod 2, 6th joint oblong. Peraeopods 3—5, 2d joint not very broad, hind margin flatly concave, ending in a short, projecting point. Uropods 1 and 2 reaching middle of uropod 3. Uropod 3, inner ramus about $\frac{2}{3}$ length of outer, both with plumose setae on both margins. Colour brighter or darker reddish yellow. L. 32 mm.

Lake Baikal. Depth 170—700 m; the so-called variety, from 300—700 m.

E. mutilus (Abildg.) 1789 *Gammarus m.*, Abildgaard in: O. F. Müller, Zool. Dan., ed. 3 *v.* 3 p. 60 t. 116 f. 1—11 | 1888 *G. m.*, T. Stebbing in: Rep. Voy. Challenger, *v.* 29 p. 56 | 1793 *Cancer (Gammarellus) m.*, J. F. W. Herbst, Naturg. Krabben Krebse, *v.* 2 p. 120 t. 35 f. 7 | 1893 *Gammarus locusta* (part.), A. Della Valle in: F. Fl. Neapel, *v.* 20 p. 759.

Peraeon segment 7 and pleon segments 1—3 furnished dorsally with backward directed spines on hind margin. Antennae 1 and 2 subequal, flagellum long in both. Accessory flagellum of antenna 1 very long, 24-jointed. Peduncle of antenna 2 reaching beyond that of antenna 1. Uropod 3, peduncle short, rami rather long, nearly equal, fringed with long setae.

North-Atlantic (Farö Isles). On the shore.

47. Gen. **Heterogammarus** Stebb.

1899 *Heterogammarus*, T. Stebbing in: Tr. Linn. Soc. London, ser. 2 *v.* 7 p. 429.

In general like Gammarus (p. 460), without dorsal teeth or carinae, or notable processes of head or side-plate 1, with accessory flagellum of antenna 1 more than 1-jointed, uropod 3, outer ramus with 2 joints, and telson cleft; but separated from Gammarus by one or more of the following characters: peduncle of antenna 1 longer than peduncle of antenna 2, 6th joint of gnathopod 1 larger than 6th joint of gnathopod 2, uropod 1 very short.

8 species.

Synopsis of species:

1 { Uropod 3 long, $^1/_7$—$^1/_4$ as long as body — **2.**
{ Uropod 3 not long, $^1/_{17}$—$^1/_{10}$ as long as body — **5.**

2 { Antenna 1 shorter than antenna 2 . . . . 1. **H. stanislavii.** . p. 494
{ Antenna 1 longer than antenna 2 — **3.**

3 { Peraeopod 5, 2^d joint with well developed wing . 2. **H. sophianosii** . p. 494
{ Peraeopod 5, 2^d joint with wing scarcely developed — **4.**

4 { Uropod 3, inner ramus about $^2/_3$ as long as outer 3. **H. capellus.** . . . p. 495
{ Uropod 3, inner ramus scarcely $^1/_4$ as long as outer 4. **H. ignotus** . . . p. 495

5 { Gnathopod 1, 6th joint larger than that of
{   gnathopod 2 — **6.**
{ Gnathopod 1, 6th joint subequal to that of
{   gnathopod 2 — **7.**

6 { Peraeopods 3 and 4, 2^d joint broad . . . . 5. **H. flori** . . . . p. 496
{ Peraeopods 3 and 4, 2^d joint narrow . . . . . . 6. **H. albulus** . . . p. 496

7 { Antenna 1 half as long as body . . . . 7. **H. bifasciatus** . p. 496
{ Antenna 1 $^1/_4$ as long as body . . . . 8. **H. branchialis** . p. 497

1. H. stanislavii (Dyb.) 1874 *Gammarus s.*, B. Dybowsky in: Horae Soc. ent. Ross., *v.* 10 suppl. p. 58 | 1899 *Heterogammarus s.*, T. Stebbing in: Tr. Linn. Soc. London, ser. 2 *v.* 7 p. 429 | 1893 *Gammarus stanislavi, G. pungens* (part.)?, A. Della Valle in: F. Fl. Neapel, *v.* 20 p. 930.

Juv. Only pleon segments 4—6 carrying spines. these situated on 3 slight elevations of the hind margin, the median with 2 or 3, the lateral each with 1 or 2 spines. Head, rostrum obsolete. Side-plates small and low. Pleon segments 2 and 3, postero-lateral corners acute. Eyes broadly reniform, slightly widened below, black. Antenna 1 only about $^3/_4$ as long as antenna 2, peduncle shorter and thinner than peduncle of antenna 2, flagellum 25-jointed, accessory flagellum 5- or 6-jointed. Antenna 2 about half as long as body, flagellum 53-jointed. Gnathopod 1, 6th joint piriform, widened at base, larger than 6th joint of gnathopod 2, which is also piriform, but not specially widened at base. Peraeopods 3—5, 2^d joint rather slender, somewhat expanded proximally. Uropod 1 reaching beyond uropod 2, to the end of inner ramus of uropod 3. Uropod 3 long, $^1/_4$ as long as body, outer ramus 4 or 5 times as long as inner, with plumose setae only on inner margin, inner ramus having them on both margins. Colour bright reddish yellow. L. 18—20 mm.

Lake Baikal. Depth 100 m.

2. H. sophianosii (Dyb.) 1874 *Gammarus s.* + *G. s. var. scirtes*, B. Dybowsky in: Horae Soc. ent. Ross., *v.* 10 suppl. p. 101 t. 10 f. 4; p. 102 t. 11 f. 2 | 1899 *Heterogammarus sophianosii*, T. Stebbing in: Tr. Linn. Soc. London, ser. 2 *v.* 7 p. 429 | 1893 *Gammarus fluviatilis* (part.)?, A. Della Valle in: F. Fl. Neapel, *v.* 20 p. 930.

Pleon segments 4—6 with few, delicate spines on hind margin, other segments without dorsal armature. Head, rostrum obsolete. Eyes not large, little prominent, reniform, black. Antenna 1 half as long as body, twice as long as antenna 2, peduncle not thicker and very little longer than peduncle of antenna 2, flagellum 46—50-jointed, accessory flagellum 7- or 8-jointed. Antenna 2, flagellum not longer than ultimate joint of peduncle, 10- or 11-jointed. Gnathopods 1 and 2 in ♂. 6th joint piriform, in ♀ that of gnathopod 1 piriform, that of gnathopod 2 oblong, apparently the difference of size being slight, and the palm in all cases very oblique. Peraeopods 3—5, 2[d] joint heart-shaped, hind margin of wing carrying a few short setae, and ending below without forming a free corner. Uropods 1 and 2 reaching almost the middle of uropod 3. Uropod 3 $^1/_7 - ^1/_6$ as long as body, inner ramus nearly as long as outer, both fringed with numerous plumose setae. Colour olive-green or brownish, with delicate brown or dark red markings. L. 41—46 mm. — Var. scirtes is distinguished especially by the liveliness of its walking and hopping. L. 12·5 mm.

Lake Baikal. Depth 1—50 m; on sandy coasts.

3. **H. capellus** (Dyb.) 1874 Gammarus c., B. Dybowsky in: Horae Soc. ent. Ross., v. 10 suppl. p. 100 | 1899 Heterogammarus c., T. Stebbing in: Tr. Linn. Soc. London, ser. 2 v. 7 p. 429 | 1893 Gammarus fluviatilis (part.)?, A. Della Valle in: F. Fl. Neapel, v. 20 p. 928.

Pleon segments 4—6 usually with 3 little groups of spines on the hind margin, 1 or 2 spines in each group, other segments without dorsal armature. Head, rostrum a feeble point. Eyes slightly prominent, reniform, black. Antenna 1 rather longer than the body, 4 times as long as antenna 2, peduncle stouter and longer than peduncle of antenna 2, flagellum 46—52-jointed, accessory flagellum 4-jointed. Antenna 2, flagellum 8- or 9-jointed. Gnathopod 1, 6th joint slenderly piriform, equal to the oblong 6th joint of gnathopod 2. Peraeopods 3—5, 2[d] joint narrow, heart-shaped. Uropod 1 reaching middle of uropod 3, uropod 2 scarcely reaching end of peduncle of uropod 3. Uropod 3 about $^1/_6$ as long as body, inner ramus about $^2/_3$ as long as outer, both fringed with plumose setae. Colour reddish yellow with scattered reddish brown spots. L. about 13 mm.

Lake Baikal. Depth 100 m.

4. **H. ignotus** (Dyb.) 1874 Gammarus i., B. Dybowsky in: Horae Soc. ent. Ross., v. 10 suppl. p. 109 t. 4 f. 3 | 1899 Heterogammarus i., T. Stebbing in: Tr. Linn. Soc. London, ser. 2 v. 7 p. 429 | 1893 Gammarus pungens (part.)?, A. Della Valle in: F. Fl. Neapel, v. 20 p. 929.

Surface of body dorsally hairy or carrying scattered, short setae; only on hind margin of .pleon segments 4—6 there are somewhat longer and stouter setae. Eyes punctiform, white, in spirit not visible. Antenna 1 longer than half the body, fully twice as long as antenna 2, peduncle longer than peduncle of antenna 2, flagellum 24-jointed, accessory flagellum 3-jointed. Antenna 2, ultimate and penultimate joints of peduncle subequal, flagellum 9- or 10-jointed. Gnathopod 1, 6th joint piriform, broader but shorter than the narrowly oblong 6th joint of gnathopod 2; 5th joint also shorter in gnathopod 1. Peraeopods 3—5, 2[d] joint narrow, the wing in no part broad, distally almost evanescent. Uropod 1 reaching nearly middle of uropod 3. Uropod 3 $^1/_4$ as long as body, outer ramus with 2[d] joint well developed, inner about $^1/_4$ as long as outer, both with isolated simple setae, but without plumose setae. Colour white. L. about 10 mm.

Lake Baikal. Depth 800 m.

5. **H. flori** (Dyb.) 1874 *Gammarus f., G. florii*, B. Dybowsky in: Horae Soc. ent. Ross., *v.*10 suppl. p. 20, 52; 54, 188 | 1899 *Heterogammarus flori*, T. Stebbing in: Tr. Linn. Soc. London, ser. 2 *v.* 7 p. 429 ╎ 1893 *Gammarus fluviatilis* (part.)?, A. Della Valle in: F. Fl. Neapel, *v.* 20 p. 928.

Pleon segments 3 and 4 with dorsal fascicles of long setae, segments 5 and 6 each with 4 or 5 groups of spines, usually 5 spines in each of the oblique lateral groups. Eyes white, in general roundish, small, in spirit not visible. Antenna 1 $^2/_5$ as long as body, less than twice as long as antenna 2, peduncle subequal to that of antenna 2, its 1st joint with only a couple of short setae on front end, flagellum 22—27-jointed, accessory flagellum 3-jointed. Antenna 2, flagellum 7- or 8-jointed. Side-plates low, with 1 or 2 marginal setae. Gnathopod 1, 6th joint broadly oblong, larger than the narrowly oblong 6th joint of gnathopod 2. Peraeopods 3 and 4, 2^d joint about $^1/_3$ longer than broad, hind margin slightly sinuous, with short setae in peraeopod 3, long, separate setae in peraeopod 4. Peraeopod 5, 2^d joint widened below and produced far downward in a rounded lobe, hind margin with close set setae. Uropods 1 and 2 not reaching end of short uropod 3. Uropod 3 $^1/_{13}$ as long as body, inner ramus about $^2/_3$ as long as outer, with simple setae on inner margin, outer ramus with them on both margins, each ramus with 4 or 5 apical setae, neither with plumose setae. Colour white. L. nearly 16 mm.

Lake Baikal. Depth 50—100 m.

6. **H. albulus** (Dyb.) 1874 *Gammarus flori var. albula*, B. Dybowsky in: Horae Soc. ent. Ross., *v.* 10 suppl. p. 53 ╎ 1899 *Heterogammarus albulus*, T. Stebbing in: Tr. Linn. Soc. London, ser. 2 *v.* 7 p. 429.

Peraeon segments and pleon segments 1—3 each with 3 groups of spines, 2 spines in the centre group, 3 in each of the oblique lateral groups. Side-plates far lower, pleon segments 4—6 shorter, legs with broader joints than in H. flori; eyes as in that species. Antenna 1 longer than half the body, less than twice as long as antenna 2, peduncle stouter and rather shorter than peduncle of antenna 2, its 1st joint with several strong spines on front end, flagellum 16—24-jointed, accessory flagellum 3- or 4-jointed, shorter than in H. flori. Antenna 2, flagellum 6—8-jointed. Gnathopods 1 and 2, 6th joint oblong, in gnathopod 1 larger than in gnathopod 2, and considerably larger than in gnathopod 1 of H. flori, with 5 or 6 long spines along the palm and 3 still longer on the hind margin. Peraeopod 3, 2^d joint twice as long as broad, hind margin feebly convex, without setae. Peraeopod 4, 2^d joint 2$^1/_2$ as long as broad, hind and front margin close set with long setae. Peraeopod 5, 2^d joint much narrower above than below, produced far downward in a rounded lobe. Uropods 1 and 2 short, but reaching beyond uropod 3. Uropod 3 $^1/_{10}$ as long as body, inner ramus about $^3/_5$ as long as outer, carrying simple setae only at apex, outer having such also on its outer margin. Colour white. L. ♀ 8 mm, ♂ 13 mm.

Lake Baikal. Depth 300 m.

7. **H. bifasciatus** (Dyb.) 1874 *Gammarus b.*, B. Dybowsky in: Horae Soc. ent. Ross., *v.* 10 suppl. p. 102 t. 12 f. 6 | 1899 *Heterogammarus b.*, T. Stebbing in: Tr. Linn. Soc London, ser. 2 *v.* 7 p. 429 | 1893 *Gammarus fluviatilis* (part.)?, A. Della Valle in: F. Fl. Neapel, *v.* 20 p. 927.

Pleon segments 4—6 each with 2 groups of spinules. Head, rostrum represented by a projecting curve. Eyes not very large, slightly prominent, reniform, ash-grey. Antenna 1 half as long as body, 2 or 3 times as long as antenna 2, peduncle stouter and a little longer than peduncle of antenna 2, flagellum 24—29-jointed, accessory flagellum 3-jointed. Antenna 2, ultimate joint of peduncle shorter than penultimate, flagellum 5- or 6-jointed. Gnathopods 1 and 2 in $\male$, 6th joint piriform. Gnathopod 1 in $\female$, 6th joint piriform; gnathopod 2, 6th joint oblong. Peraeopods 3—5, 2[d] joint tolerably broad, in peraeopod 5 scarcely $1/_6$ longer than broad, the wing below forming a free, often lobe-like corner. Uropod 1 reaching end of uropod 3. Uropod 3 about $1/_{11}$ as long as body, outer ramus with 2 spines and 2 simple setae on outer margin, plumose setae on inner; inner ramus very little shorter, with plumose setae on both margins. Colour dark brown, with cross-bands of brighter spots on head and mid-peraeon. L. about 12 mm.

Lake Baikal. Depth $1/_2$—2 m, under stones.

8. **H. branchialis** (Dyb.) 1874 *Gammarus b.*, B. Dybowsky in: Horae Soc. ent. Ross., *v.* 10 suppl. p. 110 t. 14 f. 4 | 1899 *Heterogammarus b.*, T. Stebbing in: Tr. Linn. Soc. London, ser. 2 *v.* 7 p. 429 | 1893 *Gammarus fluviatilis* (part.)?, A. Della Valle in: F. Fl. Neapel, *v.* 20 p. 927.

Pleon segments 1—3 with a few fine dorsal setae, segments 4—6 each with 2 regular groups of 2 or 3 spines. Head, rostrum a short rounded point. Side-plates 1—4 setiferous. Eyes rather large, oval, narrowed below, black. Antenna 1 about $1/_4$ as long as body, more than twice as long as antenna 2, peduncle stouter but little longer than peduncle of antenna 2, flagellum 15—17-jointed, accessory flagellum 2-jointed. Antenna 2, ultimate joint of peduncle shorter than penultimate, flagellum 4- or 5-jointed. Gnathopods 1 and 2, 6th joint nearly equal, piriform. Peraeopods 3—5, 2[d] joint tolerably broad, alike in the 3 pairs, hind margin more convex than front, with widely separated setae, but these often in $\male$ wanting on peraeopod 5. Uropods 1 and 2 nearly reaching end of uropod 3. Uropod 3 $1/_{17}$—$1/_{12}$ as long as body, outer ramus with 3—5 fascicles of simple setae on outer margin, 3 pairs of setae on inner; inner ramus about half as long as outer, with 3 pairs of long setae on inner margin. Colour changing with that of the host. L. 9—14 mm.

Lake Baikal. Only taken in marsupium of larger Gammarids, or among the branchiae of the males.

48. Gen. **Parapallasea** Stebb.

1899 *Parapallasea*, T. Stebbing in: Tr. Linn. Soc. London, ser. 2 *v.* 7 p. 429.

Median carina not represented on peraeon or on pleon segments 1—3. Side-plate 4 broader, but not less deep than the preceding, emarginate behind. Antenna 1, flagellum longer than peduncle, accessory flagellum elongate. Peraeopods 3—5 with 2[d] joint little expanded. Telson deeply cleft. Other characters agreeing with Pallasea (p. 374).

3 species.

Synopsis of species:

1. **P. borowskii** (Dyb.) 1874 *Gammarus b.* $+$ *G. b. var. dichrous* $+$ *G. b. subvar. abyssalis*, B. Dybowsky in: Horae Soc. ent. Ross., *v.* 10 suppl. p. 139 t. 2 f. 3 | 1893 *Ceradocus? b.*, A. Della Valle in: F. Fl. Neapel, *v.* 20 p. 927 | 1899 *Parapallasea b.*, T. Stebbing in: Tr. Linn. Soc. London, ser. 2 *v.* 7 p. 429.

Peraeon and pleon segments with weak marginal swellings, peraeon segments 5—7 and pleon segments 1—6 each with a pair of lateral teeth on hind margin, as a rule pointing backward and outward, not unfrequently the first 2 pairs bent uncinately outward or forward; the teeth short, only on pleon segments 1—3 somewhat longer, on pleon segments 4—6 hump-like, each with 3 spinules, and between these humps is a median flat carina with 1 or 2 spinules on the hinder end. Head smooth, upper profile convex, rostrum short. Eyes slightly protuberant, reniform, black. Antenna 1 more than half as long as body, twice as long as antenna 2, 1st joint long, but not as long as 2^d and 3^d combined, flagellum 52—71-jointed, accessory flagellum 10- or 11-jointed. Antenna 2, flagellum 17—20-jointed. Gnatho-pod 1, 6th joint piriform. Gnathopod 2, 6th joint oblong, with convex hind margin. Peraeopods 3—5, 2^d joint narrowly heart-shaped, in peraeopod 5 twice as long as broad. Uropod 3, outer ramus about $^1/_3$ longer than inner, both with long stout setae on the margins. Telson broader than long, divided to the base, apices of the conical lobes wide apart, tipped with spinules. Colour bright flesh-red or orange-red, differing in the varieties. L. reaching 56 mm.

Lake Baikal. Depth 50 to below 1000 m.

2. **P. lagowskii** (Dyb.) 1874 *Gammarus l.*, B. Dybowsky in: Horae Soc. ent. Ross., *v.* 10 suppl. p. 140 t. 2 f. 2 | 1893 *Ceradocus? l.*, A. Della Valle in: F. Fl. Neapel, *v.* 20 p. 929 | 1899 *Parapallasea l.*, T. Stebbing in: Tr. Linn. Soc. London, ser. 2 *v.* 7 p. 429.

Body laterally compressed, peraeon and pleon segments with marginal swellings, which are much more weakly developed on the pleon, lateral processes thick, on peraeon segments 1—5 as outgrowths with their ends directed outward and forward, on peraeon segments 6 and 7 and pleon segments 1—3 as thick, pointed, backward directed teeth, on pleon segments 4—6 as strong humps, each with 4 or 5 spinules. Head feebly convex, rostrum almost flat. Side-plates 1—4 rather deep. Eyes extremely small, almost point-like, white. Antenna 1 more than half as long as body, twice as long as antenna 2, 1st joint long, flagellum 63-jointed, accessory flagellum 10-jointed. Antenna 2, flagellum 22-jointed. Gnathopod 1, 6th joint piriform. Gnathopod 2, 6th joint oblong, with thickened hind margin. Peraeopods 3—5, 2^d joint heart-shaped, in peraeopod 5 nearly thrice as long as broad. Uropod 3, outer ramus about $^1/_6$ longer than inner. Telson cleft to the base. Colour white or bright flesh-red, often with darker spots. L. reaching 64 mm.

Lake Baikal. Depth 800—1300 m.

3. **P. puzyllii** (Dyb.) 1874 *Gammarus p.*, B. Dybowsky in: Horae Soc. ent. Ross., *v.* 10 suppl. p. 141 t. 3 f. 4 | 1893 *Ceradocus? p.*, A. Della Valle in: F. Fl. Neapel, *v.* 20 p. 930 | 1899 *Parapallasea p.*, T. Stebbing in: Tr. Linn. Soc. London, ser. 2 *v.* 7 p. 429.

Peraeon segments 1—6 and pleon segments 1—3 each divided into a dorsal compartment, with concave hind margin, and 2 lateral compartments at right angles to the dorsal, the separation effected by a pair of low, wing-like, outward directed, lateral carinae; on pleon segments 4—6 the lateral carinae are tubercular, surmounted by 4 or 5 spinules. Head with the convex upper surface separated from the sides by obscure lateral carinae, the rostral point small, depressed. Side-plates 1—4 moderately deep. Eyes small, roundish, very prominent, advanced in front of rostrum between antennae 1 and 2, black. Antenna 1 nearly half as long as body, almost twice as long as antenna 2, 1st joint elongate, subequal to 2^d and 3^d combined, flagellum 40—47-jointed, accessory flagellum 5—7-jointed. Antenna 2, flagellum 14—16-jointed. Gnathopod 1, 6th joint piriform. Gnathopod 2, 6th joint oblong. Peraeopods 3—5, 2^d joint elongate, heart-shaped, with rather long remote setae on hind margin. Uropod 3, outer ramus about $^2/_5$ longer than inner. Telson cleft to the base. Colour red. L. reaching 53 mm.

Lake Baikal. Depth 50—500 m.

49. Gen. **Amathillina** O. Sars

1894 *Amathillina* (Sp. typ.: *A. cristata*) (*Amathillinella* O. Grimm in MS.), G. O. Sars in: Bull. Ac. St.-Pétersb., ser. 5 *v.* 1 p. 201.

Body more or less carinate, pleon segments 5 and 6 with subdorsal spinules. Head, rostrum short, lateral lobes short, obtuse, post-antennal corners acute or rectangular. Side-plates 1—4 of moderate size. 4th much the largest, distinctly emarginate. Pleon segments 2 and 3, postero-lateral corners quadrate. Antenna 1 the longer, 3^d joint not very short, accessory flagellum 2—6-jointed. Mouth-parts normal. Gnathopods 1 and 2 subequal. Peraeopod 4 the longest, its 2^d joint moderately expanded, but much less than that of peraeopod 5. Uropod 3 rather short, outer ramus with small narrow 2^d joint, inner ramus small, scale-like. Telson short, broad, cleft to the base.

5 species.

Synopsis of species:

1. A. cristata O. Sars 1894 *A. c.* (*Amathillinella c.* O. Grimm in MS.), G. O. Sars in: Bull. Ac. St.-Pétersb., ser. 5 *v.* 1 p. 202 t. 5; t. 6 f. 1—8.

Body stout, little compressed, carina low on peraeon segment 1, gradually encreasing to broad triangular projections, important in size on peraeon segments 6 and 7 and pleon segments 1 and 2; the carina of pleon segment 3 evenly rounded. Pleon segments 4—6 each dorsally carrying a few simple hairs, segments 5 and 6 each with subdorsal spinules, 2 on each side. Head, rostrum blunt, lateral lobes broadly truncate, post-antennal corners acutely projecting. Eyes rather small, narrowly reniform, dark. Antenna 1, peduncle long, 1st joint not greatly longer than 2^d, nor 2^d than 3^d, flagellum longer than peduncle, 20—25-jointed, accessory flagellum about 5-jointed. Antenna 2 much shorter, flagellum more than half as long as peduncle, 12-jointed. Gnathopods 1 and 2 in ♀ small, subequal, setose;

6th joint in gnathopod 1 rather broader and more expanded distally than in gnathopod 2, where it has a rather narrow oblong oval form, palm in both somewhat oblique, defined by an obtuse angle carrying 2 short spines. Gnathopods 1 and 2 in ♂ much stronger, nearly equal, 6th joint much longer and broader than 5th, somewhat expanded distally, especially in gnathopod 2, palm concave, defined by a nearly rectangular corner armed with 2 spines. Peraeopods 1—5 rather more elongated in ♂ than in ♀, with 2^d joint of peraeopods 3—5 less expanded, especially in peraeopod 5; finger strong, ending in a sharp, curved point. In peraeopods 3 and 4, 2^d joint oval, distally narrowed, but in peraeopod 5 broadly expanded behind, hind margin serrate and ciliate, the broad distal lobe overlapping the 3^d joint. Uropods 1 and 2 spinose, peduncle longer than the subequal rami, uropod 1 much the longer. Uropod 3 short, peduncle much longer than broad, 1st joint of outer ramus not much longer than peduncle, armed with 2 fascicles of spines on each side, 2^d joint narrow, setiferous, $^1/_8$ as long as 1st, inner ramus very small. Telson broader than long, nearly semicircular, lobes not dehiscent, each with a spinule on outer margin, and a spinule and hairs at apex. L. ♀ about 13, ♂ nearly 15 mm.

Caspian Sea. Depth 4—66 m.

2. **A. pusilla** O. Sars 1896 *A. p.*, G. O. Sars in: Bull. Ac. St.-Pétersb., ser. 5 *v.* 4 p. 446 t. 5 f. 15—25.

Body very short and stout, especially in ♀. Peraeon segments 1—5 dorsally quite smooth; segments 6 and 7 and pleon segments 1—3 carinate, each produced dorsally to a distinct laminar projection, the first 4 triangular, the last rounded, gibbous. Pleon segments 4—6 with dorsal hairs. Head larger, side-plates 1—4 smaller than in the other species. Eyes oval reniform. Antenna 1 very slender and elongate, 1st joint nearly as long as 2^d and 3^d combined, flagellum twice as long as peduncle, about 16-jointed, accessory flagellum in ♀ 2-, in ♂ 3-jointed. Antenna 2 much the shorter. Gnathopod 1 in ♀, 6th joint rather broad, oval, palm oblique; gnathopod 2, 6th joint unusually narrow, oblong linear, palm very short, almost transverse. Gnathopods 1 and 2 in ♂ powerful, nearly equal, 6th joint in both large and tumid. Peraeopod 5, 2^d joint well expanded, hind margin serrate rather strongly, the distal lobe more produced in the ♀ than in the ♂, rather narrowly rounded. Uropod 3, peduncle thick, scarcely longer than broad, outer ramus little longer than peduncle, with terminal but no lateral spines, its 2^d joint and the inner ramus very small. Telson broader than long, lobes not dehiscent, each with a minute spinule on the obtusely truncate apex. L. 4 mm.

Caspian Sea. Sublitoral.

3. **A. maximoviczi** O. Sars 1896 *A. m.*, *A. maximovitschi*, G. O. Sars in: Bull. Ac. St.-Pétersb., ser. 5 *v.* 4 p. 444; t. 5 f. 1—14.

Body stout, peraeon not carinate, pleon segments 1—3 each with triangular dorsal crest, low in all, lowest in segment 3, segments 1—6 with dorsal hairs. Side-plates 1—4 rather small; pleon segments 2 and 3, postero-lateral corners minutely produced. Eyes of moderate size, oblong oval, not reniform. Antenna 1, 1st joint much thicker and a little longer than 2^d, 2^d a little longer than 3^d, flagellum about twice as long as peduncle, 20-jointed, accessory flagellum 3-jointed. Antenna 2 in ♀ scarcely half as long as antenna 1, in ♂ rather longer than in ♀. Gnathopods 1 and 2 rather small in ♀, much stronger in ♂, with 6th joint large, somewhat expanded distally, palm oblique, defined by an obtuse corner with several strong spines. Peraeopods 1—5 comparatively short and stout. Peraeopod 3, 2^d joint oval, narrowed distally. Peraeopod 4, 2^d joint, especially in ♀, unusually broad, rounded quadrangular. Peraeopod 5, 2^d joint much expanded, the rounded lower lobe descending below the 3^d joint, a little in ♂, much in ♀. Uropod 3, peduncle rather longer than broad, outer ramus nearly twice as long as peduncle, its 2^d joint and the inner ramus small. Telson rather broader than long, the narrowly rounded apices rather divergent, each lobe with 2 spinules at apex and a hair on outer margin. L. about 6 mm.

Caspian Sea (bay of Karabugas).

4. **A. spinosa** O. Sars 1896 *A. s.* (*A. cristata var. spinata* O. Grimm in MS.), G. O. Sars in: Bull. Ac. St.-Pétersb., ser. 5 *v.* 4 p. 442 t. 4 f. 7—16.

Agreeing with A. cristata (p. 499) in gnathopods 1 and 2 and many other respects. Body moderately slender and compressed, peraeon segments 1—7 and pleon segments 1—3 each raised to a well defined dorsal backward pointing triangular expansion, successively larger, pleon segment 4 ending dorsally in a small but well defined rounded expansion, segments 5 and 6 having only a few dorsal hairs. Eyes reniform, obliquely placed, dark. Antenna 1, accessory flagellum in ♂ 6-jointed. Peraeopods 3—5 differing from those of A. cristata in having the 2^d joint comparatively narrower, and nearly alike in both sexes. Uropod 3 very short and thick, outer ramus scarcely longer than peduncle, only armed with 3 spinules and some setules, 2^d joint very small, inner ramus scarcely longer than broad. Telson short, much broader than long, lobes rather divergent, each with a setule on outer margin and 2 hairs at rounded apex. L. ♂ 25 mm.

Caspian Sea (southern part). Depth 203 m.

5. **A. affinis** O. Sars 1894 *A. a.*, G. O. Sars in: Bull. Ac. St.-Pétersb., ser. 5 *v.* 1 p. 207 t. 6 f. 9—19.

Near to A. cristata (p. 499), and still nearer to A. maximoviczi. Body short and stout, especially in ♀, peraeon segments 1—4 or 1—5 quite smooth, peraeon segments 6 and 7, pleon segments 1—3 each with acutely projecting triangular dorsal carina, pleon segments 4—6 with no projection but fine hairs and small subdorsal spinules. Eyes distinctly reniform. Antenna 1, including 1st joint, slender, flagellum once and a half as long as peduncle, with 15 joints in ♀, to 30 in ♂, accessory flagellum 3-jointed. Gnathopod 1 in ♂, 6th joint rather tumid in the middle; otherwise the gnathopods 1 and 2 are nearly as in A. cristata. Peraeopods 1—5 and uropods 1—3 nearly as in A. maximoviczi. Telson rather longer than broad, lobes a little divergent, with no armature, except 3 fine hairs on each obtusely pointed apex. L. ♀ 6, ♂ 8 mm.

Caspian Sea. Depth 4–6 m.

50. Gen. **Carinogammarus** Stebb.

1862 *Gammaracanthus* (part.), Bate, Cat. Amphip. Brit. Mus., p 201 | 1899 *Carinogammarus*, T. Stebbing in: Tr. Linn. Soc. London, ser. 2 *v.* 7 p. 429.

Distinguished from Gammarus (p. 460) by having carinate segments; carina medio-dorsal only; relative proportions of peduncles of antennae 1 and 2, of gnathopods 1 and 2 and of inner and outer ramus of uropod 3 variable.

9 accepted and 2 doubtful species.

Synopsis of accepted species:

1 { Uropod 3, inner ramus not short — **2.**
  { Uropod 3, inner ramus short — **6.**

2 { Body carinate throughout — **3.**
  { Body only partially carinate — **4.**

3 { Antenna 1, accessory flagellum 3-jointed; gnatho-
  {   pod 2, 6th joint oblong . . . . 1. **C. cinnamomeus** . p. 502
  { Antenna 1, accessory flagellum 12-jointed; gnatho-
  {   pod 2, 6th joint piriform . . . . 2. **C. wagii** . . . . . p. 502

4 { Eyes halfmoon-shaped . . . . 3. **C. pulchellus** . . p. 503
  { Eyes oval — **5.**

5 { Peraeopod 5, 2d joint widened distally . . . . . 4. **C. seidlitzii** . . . p. 503
  { Peraeopod 5, 2d joint narrowed distally . . . . 5. **C. rhodophthalmus** p. 504

6 { Pleon segments 4 and 5 with erect spinose dorsal
  {   tubercle . . . . 6. **C. caspius** . . . . p. 504
  { Pleon segments 4 and 5 without erect dorsal
  {   tubercle — **7.**

7 { At least 2 of the peraeon segments carinate . . 7. **C. atchensis** . . . p. 505
  { At most 1 of the peraeon segments carinate — **8.**

8 { Pleon segments 1—3, carina tipped with spines . 8. **C. subcarinatus** . p. 505
  { Pleon segments 1—3, carina acute, without spines 9. **C. roeselii** . . . . p. 506

1. **C. cinnamomeus** (Dyb.) 1874 *Gammarus c.*, B. Dybowsky in: Horae Soc. ent. Ross., *v.* 10 suppl. p. 114 t. 7 f. 3 | 1899 *Carinogammarus c.*, T. Stebbing in: Tr. Linn. Soc. London, ser. 2 *v.* 7 p. 429 | 1893 *Gammarus fluviatilis* (part.)?, A. Della Valle in: F. Fl. Neapel, *v.* 20 p. 928.

Body laterally compressed, its upper surface rough, with delicate hairs in various parts; all segments dorsally provided with a distinct though weak median carina, occupying the whole length of the segments in peraeon segments 4—7 and pleon segments 1—3; hind margin of pleon segments 4—6 provided with 3 groups of tolerably strong spines, set on hump-like elevations. Head, front rounded, little prominent. Side-plates without setae on lower margin. Eyes rather large, prominent, reniform, black. Antenna 1, peduncle thicker and somewhat longer than that of antenna 2, or rarely shorter; with flagellum nearly $\frac{1}{2}$ as long as body, and almost twice as long as antenna 2; flagellum 21—28-jointed, accessory flagellum 3-jointed. Antenna 2, flagellum 9—12-jointed. Gnathopod 1, 6th joint piriform. Gnathopod 2, 6th joint oblong. Peraeopods 3—5, 2d joint heart-shaped, with very short setae wide apart on hind margin. Uropod 3 rather long, the outer 2-jointed ramus $\frac{1}{3}$—$\frac{1}{2}$ longer than the inner, 2d joint more than $\frac{1}{4}$ as long as 1st, inner ramus longer than peduncle, both rami with plumose setae on inner and simple on outer margin. Colour cinnamon-brown. L. 18 mm.

Lake Baikal. Depth 50—100 m.

2. **C. wagii** (Dyb.) 1874 *Gammarus w.*, B. Dybowsky in: Horae Soc. ent. Ross., *v.* 10 suppl. p. 121 t. 1 f. 4 | 1893 *Ceradocus? w.*, A. Della Valle in: F. Fl. Neapel, *v.* 20 p. 931 | 1899 *Carinogammarus w.*, T. Stebbing in: Tr. Linn. Soc. London, ser. 2 *v.* 7 p. 429.

Median carina developed over the whole body, on the head appearing as an obscure rounded ridge-line, and gradually encreasing in height to pleon segment 3, from peraeon segment 5 running out into an acute, back-ward and upward directed angle; this carina occupies the whole dorsal length of each segment, is tolerably high, strongly compressed laterally, with convex front and concave hind margin; on pleon segments 4—6 becoming hump-like and gradually weaker; marginal swellings are perceptible on the peraeon-segments and pleon segments 1—4, and on either side of the median carina in pleon segments 4—6 are strong humps, carrying 2 or 3 spinules each. Head, rostral point short. Side-plates 1—4 rather deep, 5th not deep. Eyes slightly prominent, reniform, black. Antenna 1 about $^2/_8$ as long as body, more than twice as long as antenna 2, peduncle longer than peduncle of antenna 2, and longer than the 70-jointed flagellum, accessory flagellum 11—13-jointed, longer than the 12-jointed flagellum of antenna 2. Gnathopods 1 and 2, 6th joint rather large, piriform, finger long. Peraeopods 3—5, 2[d] joint heart-shaped, longitudinally carinate (in figure). Uropod 3 reaching back much beyond the others, the rami subequal, long, not foliaceous, both with numerous plumose setae and (in figure) with short stout spines, 2[d] joint of outer ramus rudimentary. Colour bright yellow, with delicate marking. L. 46 mm.

Lake Baikal. Depth 70—150 m.

3. **C. pulchellus** (Dyb.) 1874 *Gammarus p.*, B. Dybowsky in: Horae Soc. ent. Ross., *v.* 10 suppl. p. 118 t. 5 f. 4 | 1899 *Carinogammarus p.*. T. Stebbing in: Tr. Linn. Soc. London, ser. 2 *v.* 7 p. 429 | 1893 *Gammarus fluviatilis* (part.)?, A. Della Valle in: F. Fl. Neapel, *v.* 20 p. 930.

Peraeon segment 7 and segments of pleon with a low median carina, which is obscurely developed on pleon segment 6; surface of body with delicate hairs on the hind part of the segments. Head with a small subacute rostrum. Side-plates 1—4 rather deep, with numerous setules on lower margin, 4th broad, well emarginate behind; pleon segment 3, postero-lateral corners quadrate. Eyes rose-red, so placed on lateral lobes as to show partly in front between antennae 1 and 2, halfmoon-shaped, convex margin in front, hind margin irregular, cut into lobes. Antenna 1 about $^1/_3$ as long as body, nearly twice as long as antenna 2, 1st joint thick, flagellum with 20—29 rather long joints, accessory flagellum with 3 or 4. Antenna 2, flagellum 6—9-jointed. Mandible, 3[d] joint of palp curved, much shorter than 2[d]. Gnathopods 1 and 2 small, slender, 5th joint rather long, especially in gnathopod 2, 6th joint oblong, with palm oblique in gnathopod 1, but transverse in gnathopod 2, the joint being also slightly widened at the palm, finger short. Peraeopod 3, 2[d] joint nearly circular. Peraeopods 4 and 5, 2[d] joint wide, with a distal narrowing, many spinules and setae on the expanded part. Uropod 1 is said to reach as far back as uropod 3, or in ♀ somewhat farther. Uropod 3, inner ramus longer than 1st joint of outer, both beset with spines and plumose setae, 2[d] joint of outer about $^1/_3$ as long as 1st. Telson with a few setules on the obtuse but narrow apices of the tapering lobes. Colour very delicately reddish white. L. reaching 23 mm.

Lake Baikal. Depth 100—700 m.

4. **C. seidlitzii** (Dyb.) 1874 *Gammarus s.*, B. Dybowsky in: Horae Soc. ent. Ross., *v.* 10 suppl. p. 119 t. 5 f. 5 | 1899 *Carinogammarus s.*, T. Stebbing in: Tr. Linn. Soc. London, ser. 2 *v.* 7 p. 429 | 1893 *Gammarus fluviatilis* (part.)?, A. Della Valle in: F. Fl. Neapel, *v.* 20 p. 930.

Peraeon segments 6 and 7 and pleon segments 1—3 with a weak median carina forming a tubercle at about the middle of the segments; the spinules on the 5th and 6th pleon segments are extremely delicate; head, peraeon, and pleon segments 1—3 dorsally beset with delicate hairs, longest on hind margin of peraeon segments. Head narrow, rostrum subacute, not large. Side-plates 1—4 with long setae on lower margin. Eyes moderately large, not very

prominent, oval, a little widened .below, red. Antenna 1 rather less than
$^1/_2$ as long as body, twice as long as antenna 2, 1st joint stout, flagellum
27—34-jointed, accessory flagellum 4- or 5-jointed. Antenna 2, flagellum
9-jointed. Gnathopods 1 and 2 seemingly almost as in C. pulchellus (p. 503),
but 6th joint of gnathopod 1 more piriform, and palm in gnathopod 2
less completely transverse. Peraeopod 3, 2^d joint carried in motion of
the animal at an acute angle to the long axis of the body, the expansion
ending in a lobe below. Peraeopod 5, 2^d joint little longer than broad, its
greatest breadth being at the distal end. Uropods 1 and 2 not long, reaching
nearly to end of uropod 3. Uropod 3, inner ramus longer than 1st joint
of outer. The deeply cleft telson is described as long. Colour white. L. reaching
about 17 mm.

Lake Baikal. Depth 50—100 m.

5. **C. rhodophthalmus** (Dyb.) 1874 *Gammarus r.* + *G. r. var. microphthalmus*,
B. Dybowsky in: Horae Soc. ent. Ross., *v.* 10 suppl. p. 116, 117 t. 14 f. 10 | 1899 *Carino-
gammarus r.*, T. Stebbing in: Tr. Linn. Soc. London, ser. 2 *v.* 7 p. 429 | 1893 *Gammarus
fluviatilis* (part.)?, A. Della Valle in: F. Fl. Neapel, *v.* 20 p. 930.

Peraeon segments 6 and 7 and pleon segments 1—4 or 1—5 with
a weakly indicated median tubercular carina, pleon segment 6 with 2 groups
of delicate spinules; (in *var. microphthalmus* carina extending to pleon segment
5, or only to 3; pleon segment 6 with a pair of setae). Head, front shortly
produced, obtusely triangular. Side-plates 1—4 setose on lower margin,
4th broad, well emarginate; pleon segment 3, postero-lateral corners quadrate.
Eyes very large, prominent, broadly oval, occupying nearly half the lateral
surface of head, the front part turned forward and inward, ruby-red, (in the
variety much smaller, but still tolerably large, black or reddish black).
Antenna 1 half (in the variety more than half) as long as body, twice as long
as antenna 2, 1st joint stout, flagellum 37—44- (in the variety 8—34-)
jointed, accessory flagellum 3—5-jointed. Antenna 2, flagellum 7—11-jointed.
Gnathopods 1 and 2 small, subequal, 5th joint not elongate, 6th piriform in
gnathopod 1 (in figure oblong, with oblique palm; in the variety described as
oval); 6th joint in gnathopod 2 oblong, narrower than in gnathopod 1, palm
much less oblique. Peraeopods 3—5, 2^d joint with the expansion abruptly
narrowing below, though less so in peraeopod 5 than in 3 and 4; in all
the hind margin armed with long setae. Uropod 3, inner ramus about as
long as 1st joint of outer (in variety the rami subequal). Colour white or
bright yellow. L. 14—20 mm.

Lake Baikal. Depth 8—100 m.

6. **C. caspius** (Pall.) 1771 *Oniscus c.*, Pallas, Reise Ruß., *v.* 1 p. 477 ! 1841
Gammarus c., Eichwald in: N. Mém. Soc. Moscou, *v.* 7 p. 230 | 1894 *G. c.*, G O. Sars
in: Bull. Ac. St.-Pétersb., ser. 5 *v.* 1 p. 210 t. 7 | 1899 *Carinogammarus c.*, T. Stebbing
in: Tr. Linn. Soc. London, ser. 2 *v.* 7 p. 430 | 1862 *Gammarus semicarinatus*, Bate,
Cat. Amphip. Brit. Mus., p. 204 t. 36 f. 3 | 1893 *G. fluviatilis* (part.), A. Della Valle in:
F. Fl. Neapel, *v.* 20 p. 763.

Peraeon segment 7 sometimes slightly carinate and produced to a small
tooth; pleon segments 1—3 strongly carinate, each produced into a large
acute backward pointing tooth; segments 4 and 5 each with an elevated
flat-topped dorsal tubercle, capped by 2 pairs of spines, segment 6 with 1
pair of subdorsal spinules. Head, rostrum almost obsolete, lateral lobes broad,
obtusely truncate. Side-plate 4 much the largest. Pleon segments 2 and 3,
postero-lateral corners acute. Eyes oblong reniform, sometimes dark. Antenna 1

slender, nearly half as long as body, 1st joint subequal to 2[d] and 3[d] combined, flagellum longer than peduncle, accessory flagellum 5-jointed. Antenna 2 much shorter, ultimate joint of peduncle shorter than penultimate, flagellum about 10-jointed. Gnathopods 1 and 2, 5th joint short, cup-shaped, 6th oblong oval, palm somewhat oblique, shorter than hind margin, defined by obtuse angle with strong spine; gnathopods stronger in ♂ than in ♀, and gnathopod 2 considerably larger than gnathopod 1. Peraeopod 1 longer than peraeopod 2. Peraeopod 3 shorter than 4 and 5, 2[d] joint little longer than broad, slightly narrowed distally. Peraeopod 4, 2[d] joint longer, distally more narrowed. Peraeopod 5, 2[d] joint longer, oblong quadrangular, hind margin as in peraeopods 3 and 4 serrate. Uropods 1 and 2, rami subequal, with lateral and apical spines. Uropod 3 reaching much beyond the others, short peduncle with 4 spines on apical margin, outer ramus nearly thrice as long as peduncle, fringed round with long plumose setae, 2[d] joint minute; inner ramus scale-like, $1/2$ as long as peduncle, with 1 spinule on inner margin and 1 on apex. Telson scarcely so long as broad, each subconical lobe with 2 spinules and some hairs on the obliquely truncate apex. L. ♀ 13, ♂ 16 mm.

Caspian Sea. Depth 1—37 m.

7. **C. atchensis** (F. Brandt) 1851 *Gammarus a.*, F. Brandt in: Middendorff, Reise Sibirien. *v.* 21 p. 138 t. 6 f. 29 | 1862 *G. a*, Bate, Cat. Amphip. Brit. Mus., p. 217 t. 38 f. 7 | 1899 *Carinogammarus a.*. T. Stebbing in: Tr. Linn. Soc. London, ser. 2 *v.* 7 p. 430 | 1893 *Melita palmata* (part.)?, A. Della Valle in: F. Fl. Neapel, *v.* 20 p. 765.

Peraeon segments 1—3 dorsally rounded, segments 4 and 5 obscurely, 6 and 7 feebly carinate; pleon segments 1—3 pretty strongly carinate, segment 4 slightly carinate in front, behind, like segments 5 and 6, rounded; segments 1—3 with rows of spinules on hind margin, and between these and the carina-tooth a low eminence also beset with rows of spinules; segment 4 with 2 curved eminences on each side, pectinate with 4—7 spinules; segment 5 with 1 or 2 or even 3 such eminences, carrying 6 or 7 or fewer spinules, segment 6 with a front inner very small pair of eminences, and a more spiny outer larger pair behind. Head, rostrum very short. Eyes long elliptic or somewhat reniform. Antenna 1 scarcely half as long as body, flagellum longer than peduncle, 33-jointed, accessory flagellum 7-jointed, about $1/4$ as long as primary. Antenna 2 shorter, flagellum 19-jointed. Gnathopods 1 and 2 subequal, or gnathopod 1 rather the larger, 5th joint short, cup-shaped, 6th rhomboidal, little longer than broad, a little widened to the palm, which is furnished with little teeth or spines and obliquely truncate in gnathopod 1, but almost transverse in gnathopod 2. Peraeopods 1—5 stoutly built, rather short, with fascicles of spines but few setae. Uropods 1 and 2, rami short, broad, strong and spinose. Uropod 3 as long as the preceding or longer, rami spinose, lanceolate, inner not half as long as outer, scale-like (in figure). Telson cleft to the base, as long as short peduncle of uropod 3, each subconical lobe with 3 spinules at apex. L. 25 mm. (According to Brandt; Bate differs in many respects.)

Behring Sea (Isles of Atcha and Unalaschka).

8. **C. subcarinatus** (Bate) 1862 *Gammarus s.* (Stimpson in MS.), Bate, Cat. Amphip. Brit. Mus., p 205 t. 36 f. 5 | 1899 *Carinogammarus s.*, T. Stebbing in: Tr. Linn. Soc. London, ser. 2 *v.* 7 p. 430 | 1893 *Gammarus marinus* (part.)?, A. Della Valle in: F. Fl. Neapel, *v.* 20 p. 765.

Pleon segments 1—3 having the dorsal median line slightly elevated into a carina, with a fascicle of short spines on the blunt apex of the carina (figure), a fascicle of spines on either side, encreasing in importance on successive

segments to end of pleon, segments 4—6 with no median carina. Eyes oval-
Antennae 1 and 2 subequal, $^1/_3$ as long as body. Gnathopod 1 the larger,
5th joint short, cup-shaped, 6th broadly oblong oval, palm slightly concave,
slightly oblique, set with several short spine-teeth, finger with an obtuse
projection near middle of inner margin. Gnathopod 2, 6th joint narrowly
oblong, hind margin serrate and setose, palm transverse, sinuous, armed as
in gnathopod 1; finger as in gnathopod 1. Peraeopods 1—5 subequal,
spinose. Peraeopods 3—5, 2d joint seemingly piriform. Uropod 3, outer
ramus thrice as long as inner. Telson not spinose. L. 25 mm.

Behring Strait.

9. **C. roeselii** (Gerv.) 1755 „*Squilla (Astacus) fluviatilis*“, Rösel, Insecten-
Belustig., *v.* 3 p. 351 t. 62 f. 1—7 | 1835 *Gammarus roëselii*, P. Gervais in: Ann. Sci.
nat., ser. 2 *v.* 4 p. 128 | 1850 *G. fluviatilis* (non H. Milne Edwards 1830!), *G. röselii*,
Hosius in: Arch. Naturg., *v.* 161 p. 234 t. 3. 4 | 1899 *Carinogammarus f.*, T. Stebbing
in: Tr. Linn. Soc. London, ser. 2 *v.* 7 p. 430 | 1841 *Gammarus pulex* (err., non J. C.
Fabricius 1775!), C. L. Koch, C. M. A., *v.* 36 nr. 21.

Peraeon segment 7 produced to a small tooth; pleon segments 1—3
each rounded in front, behind produced to an acute, carinate tooth, devoid
of spines, segments 4 and 5 with slight median elevation, carrying apical
spines, and a raised group of spines on either side, slightly to the rear of
the middle one, segment 6 with a pair of slight subdorsal elevations carrying
spines. Head, rostrum obsolete, lateral lobes obtuse, post-antennal corners
produced acutely forward. Side-plates 1—4 rather deep, 4th the largest.
Pleon segment 3, postero-lateral corners acute. Eyes reniform, black. Antenna 1,
flagellum 34-jointed, accessory flagellum with 4 joints, of which 1st und 4th
are very short. Antenna 2, ultimate joint of peduncle longer than penultimate,
flagellum 15-jointed. Gnathopods 1 and 2 subequal, 5th joint short, 6th piriform
in gnathopod 1, oblong in gnathopod 2. Uropods 1 and 2, outer ramus
rather shorter than inner. Uropod 3 probably as in C. caspius (p. 504).
Telson completely cleft, lobes divergent, tipped with spines. Colour greyish
brown with red spots, conspicuous on tips of dorsal teeth. L. 14 mm.

Europe. Rivers and ponds.

C. macrophthalmus (Stimps.) 1853 *Gammarus m.*, Stimpson in: Smithson.
Contr., *v.* 6 nr. 5 p. 55 | 1862 *Gammaracanthus m.*, Bate, Cat. Amphip. Brit. Mus., p. 203 |
1893 *Ceradocus? m.*, A. Della Valle in: F. Fl. Neapel, *v.* 20 p. 929 | 1894 *Amathilla? m.*,
G. O. Sars, Crust. Norway, *v.* 1 p. 494 | 1899 *Carinogammarus? m.*, T. Stebbing in:
Tr. Linn. Soc. London, ser. 2 *v.* 7 p. 430.

Carinate only at the pleon. Side-plates small. Eyes large, subreniform, approximate. Antenna 1, accessory flagellum scarcely perceptible. Uropods 1 and 2, outer
ramus the shorter. Uropod 3, rami broad lanceolate, shorter than in Gammarellus
homari (p. 387). Colour bright crimson, or red and white mottled. L. 12 mm.

Fundy Bay (Grand Manan). Low-water mark. and Laminarian zone.

C. mucronatus (Say) 1818 *Gammarus m.*, Say in: J. Ac. Philad., *v.* 1 II p. 376 |
1873 *G. m.*, (S. I. Smith in:) A. E. Verrill in: Rep. U. S. Fish Comm., *v.* 1 p. 559 |
1862 *Gammaracanthus m.*, Bate, Cat. Amphip. Brit. Mus., p. 203 | 1894 *Amathilla? m.*,
G. O. Sars, Crust. Norway, *v.* 1 p. 494 | 1899 *Carinogammarus? m.*, T. Stebbing in:
Tr. Linn. Soc. London, ser. 2 *v.* 7 p. 430 | 1893 *Gammarus fluviatilis* (part.)?, A. Della
Valle in: F. Fl. Neapel, *v.* 20 p. 729.

Pleon segments 1—3 mucronate above, but (Smith) not distinctly carinate, segments 3—5 carrying fascicles of spines. Eyes blackish, irregularly reniform, truncate
above. Antennae 1 and 2 subequal. Antenna 1, flagellum 20-jointed, accessory
flagellum reaching end of 4th joint of primary. L. less than 12.5 mm.

North-Atlantic (North Carolina, Cape Cod, usually in brackish water; Florida).

51. Gen. **Gammaracanthus** Bate

1862 *Gammaracanthus* (part.), Bate, Cat. Amphip. Brit. Mus., p. 201 | 1871 *G.*, A. Boeck in: Forh. Selsk. Christian., 1870 p. 214 | 1894 *G.*, G. O. Sars, Crust. Norway, *v.* 1 p. 493 | 1894 *G.*, T. Stebbing in: Bijdr. Dierk., *v.* 17 p. 41.

Body carinate throughout, carina in many segments produced into a tooth, lower margin of peraeon segments carinate, this carina continued along middle of pleon segments 1—3 or 1—4. Head, rostrum long, acute, lateral corners small, rounded, with prominent process below. Side-plates 1—4 ridged from above downward, parallel-sided, 5th nearly as deep as 4th, its front lobe the deeper. Eyes rather prominent, not very large. Antenna 1 with accessory flagellum. Antenna 2 much shorter, slender. Upper lip rounded. Lower lip, inner lobes incompletely separated from outer. Mandible normal, 3[d] joint of palp as long as 2[d]. Maxilla 1, several (6) setae on inner plate, 11 spines on outer. Maxilla 2, inner margin fringed. Maxillipeds, palp elongate. Gnathopods 1 and 2, 5th joint short, cup-shaped, 6th powerful, finger long. Peraeopods 3 and 4 much longer than 1st, 2[d] or 5th. Peraeopods 3—5, 2[d] joint little expanded. Uropod 1, rami shorter than peduncle, equal. Uropod 3, rami longer than peduncle, setose, laminar, ending obtusely. Telson very small, bilobed.

3 species.

Synopsis of species:

1 {	Pleon segments 2 and 3, postero-lateral corners quadrate	1. **G. lacustris** . .	p. 507
	Pleon segments 2 and 3, postero-lateral corners acutely produced — **2.**		
2 {	Rostrum curved, produced beyond 1st joint of antenna 1	2. **G. loricatus** . .	p. 508
	Rostrum almost straight, not produced beyond 1st joint of antenna 1	3. **G. caspius** .	p. 508

1. G. lacustris O. Sars 1861 *Gammarus loricatus* (err., non Sabine 1821!), S. Lovén in: Öfv. Ak. Förh., *v.* 18 p. 287 | 1867 *Gammaracanthus l. var. lacustris*, G. O. Sars, Crust. d'Eau douce Norvège, p. 73 t. 7 f. 1--8 | 1888 *G. lacustris*, T. Stebbing in: Rep. Voy. Challenger, *v.* 29 p. 1699 | 1894 *G. relictus*, G. O. Sars, Crust. Norway, *v.* 1 p. 494 t. 174 | 1896 *G. r.*, G. O. Sars in: Bull. Ac. St.-Pétersb., ser. 5 *v.* 4 p. 439.

Body slender and compressed, dorsal carina rather low in front, projections scarcely beginning till peraeon segment 5, those following to pleon segment 5 very acute. Head, rostrum very slightly curved, not reaching end of 1st joint of antenna 1, lateral corners very small, projection below tuberculiform. Pleon segment 3, postero-lateral corners quadrate. Eyes rather small, rounded, dark. Antenna 1 nearly $\frac{1}{2}$ as long as body, 3[d] joint not very short, flagellum rather longer than peduncle, about 22-jointed, accessory flagellum 4-jointed. Antenna 2 little more than $\frac{1}{2}$ as long as antenna 1, basal joint tumid, ultimate joint of peduncle shorter than penultimate, flagellum 7-jointed. Gnathopod 1 fully as large as gnathopod 2, 5th joint produced to a small narrow setiferous lobe, 6th expanding to beginning of palm, which is convex, scarcely longer than hind margin, defined by an obtuse angle and several spines, one rather long. Gnathopod 2, 6th joint oblong oval, broadest near the base, tapering distally, palm very oblique, much longer than hind margin. Peraeopods 1 and 2 little longer than gnathopods 1 and 2. Peraeopods 3 and 4 extremely slender and elongate, 5th joint much longer than 4th. Peraeopod 5, 2[d] joint a little more expanded proximally than in other peraeopods.

Uropod 3, rami fringed with plumose setae. Telson nearly twice as broad as long, cleft wide, angular, more than half the length. L. ♀ 35 mm.

Sweden, Norway, Russia. Lakes.

2. **G. loricatus** (Sab.) 1821 & 24 *Gammarus l.*, E. Sabine in: W. E. Parry, J. Voy., Suppl. p. 53 t. 1 f. 7; p. 231 t. 1 f. 7 | 1838 *G. l.*, Kröyer in: Danske Selsk. Afh., v. 7 p. 250 t. 1 f. 4 | 1866 *G. l.*, Goës in: Öfv. Ak. Förh., v. 22 p. 531 | 1862 *Gammaracanthus l.*, Bate, Cat. Amphip. Brit. Mus., p. 202 t. 36 f. 2 | 1876 *G. l.*, A. Boeck, Skand. Arkt. Amphip., v. 2 p. 400 | 1894 *G. l.*, T. Stebbing in: Bijdr. Dierk., v. 17 p. 40 | 1893 *Ceradocus l.*, A. Della Valle in: F. Fl. Neapel, v. 20 p. 719.

Near to G. lacustris (p. 507). Body rather stouter. Dorsal carina well developed from rostrum to pleon segment 5 inclusive, variably shaped on the anterior peraeon segments. Head, rostrum curved, projecting beyond 1st joint of antenna 1, lateral corners rounded, not very large, projection below boldly prominent. Pleon segments 2 and 3, lateral corners acutely produced. Eyes rather small, oval reniform, dark., Antenna 1, flagellum 30-jointed, accessory flagellum 4-jointed. Antenna 2, flagellum 12—16-jointed. Gnathopods 1 and 2 similar to those of G. lacustris, but more robust and gnathopod 2 decidedly larger than gnathopod 1. Peraeopods 3—5 a little stronger than in G. lacustris. Colour yellowish white. L. reaching 43—58 mm.

Arctic Ocean (Greenland, Spitzbergen, Nova Zembla, Siberia). Depth 79 m.

3. **G. caspius** O. Sars 1880 *G. c.* (nom. nud.), O. Grimm in: Arch. Naturg., v. 46 I p. 119 | 1896 *G. c.*, G. O Sars in: Bull. Ac. St.-Pétersb., ser. 5 v. 4 p. 439 t. 4 f. 1—6.

Very like G. loricatus, but rostrum nearly straight, only reaching apex of 1st joint of antenna 1, projections below lateral corners of head less prominent; pleon segments 2 and 3, postero-lateral corners less produced. Antenna 2, flagellum shorter than ultimate joint of peduncle, 8-jointed. L. 36 mm.

Caspian Sea. Depth 203 m.

52. Gen. Acanthogammarus Stebb.

1899 *Acanthogammarus*, T. Stebbing in: Tr. Linn. Soc. London, ser. 2 v. 7 p. 430.

Body with median more or less dentate carina, and also lateral or marginal carinae more or less developed. Head, rostrum very short. Side-plate 5 much shallower than 4th. Antenna 1 the longer, accessory flagellum usually much developed, always with more than 1 joint. Mouth-parts normal. Peraeopods 3 and 4, 2^d joint narrowed below. Uropod 3, rami subequal, not foliaceous. Telson deeply cleft.

6 species.

Synopsis of species:

1. **A. cabanisii** (Dyb.) 1874 *Gammarus c*, B. Dybowsky in: Horae Soc. ent.
Ross., *v.* 10 suppl. p. 122 t. 13 f. 5 | 1899 *Acanthogammarus c.*, T. Stebbing in: Tr. Linn.
Soc. London, ser. 2 *v.* 7 p. 430 | 1893 *Pallasea cancellus* (part.)?, A. Della Valle in: F.
Fl. Neapel, *v.* 20 p. 928.

Median carina on all segments, on peraeon segments 1—5 in the form
of low, narrow, tooth-like processes, looking in side-view like little triangles,
on peraeon segments 6 and 7 and pleon segments 1—3 in form of acute,
backward directed divided spines or of narrow, tolerably acute, and high
2-jointed teeth; all segments have also a flat, distinct marginal swelling,
and the pleon segments also on the hind margin a bump, carrying 1 or 2
spines. Head smooth above, not inflated in the ocular region, rostrum
acute, slightly directed upward, separated by a deep cavity from the lateral
lobes. Side-plates 1—4 distally acute, the 1st almost spine-like, greatly
advanced forward over the cheeks. Eyes small, point-like, white. Antenna 1
longer than the body, 4—6 times as long as antenna 2, 1st joint prismatic,
tapering, much longer than 2d and 3d combined, twice as long as ultimate
joint of peduncle in antenna 2, flagellum very long, 84—108-jointed, accessory
flagellum 8-jointed. Antenna 2, ultimate joint of peduncle serrato-dentate
on inner margin, flagellum tapering, with 17 joints set obliquely, of serrate
appearance and furnished (?) with calceoli. Gnathopods 1 and 2, 6th joint
is described as piriform, with broad base (in figure rather oblong oval, with
no palm distinguishable). Peraeopods 3—5, 2d joint nearly thrice as long
as broad, almost linear, with the wing slightly developed above. Peraeopods
4 and 5 shorter than the body. Uropod 1 reaching much beyond uropod 3.
Uropod 3, outer ramus about $\frac{1}{7}$ shorter than the inner, both having numerous
plumose setae. Telson cleft to the middle, apices acute (in figure blunt).
Colour white. L. 55 mm.

Lake Baikal. Depth 200—700 m.

2. **A. zieńkowiczii** (Dyb.) 1874 *Gammarus z.*, B. Dybowsky in: Horae Soc.
ent. Ross., *v.* 10 suppl. p. 124 t. 3 f. 5 | 1899 *Acanthogammarus z.*, T. Stebbing in: Tr.
Linn. Soc. London, ser. 2 *v.* 7 p. 430 | 1893 *Ceradocus? zienkoviczii*, A. Della Valle in:
F. Fl. Neapel, *v.* 20 p. 931.

Median carina more or less strongly developed throughout peraeon and
pleon, on peraeon segment 1 as a bump with 3 teeth, 2 side by side bending
forward in front, a 3d behind bending back; on peraeon segments 2—6
commonly as a ridge with blunt hump in front and acute, forward bent tooth
behind; on pleon segments 1—3 the carina is higher, and has a 3d, acute,
straight tooth; on segment 4 there is a hump and a straight tooth, on segments
5 and 6 usually no tooth; on peraeon segments 1—5 the margins are swollen;
on all the others, on the hind margin over the marginal swellings, there
is a hump which runs out into an acute, outward and backward directed
tooth. Head smooth, rostrum short, acute. Side-plates 1—4 shallow,
1st—3d distally narrowed (not in figure) and rounded; 4th broad, the hind
emargination (in figure) extending the whole depth of the plate. Eyes small,
point-like, violet. Antenna 1 nearly twice as long as body and seven times
as long as antenna 2, 1st joint as in A. cabanisii, flagellum of great length,
108—119-jointed, accessory flagellum 5-jointed. Antenna 2, ultimate joint
of peduncle thickened, projecting over base of flagellum and armed with
long stout plumose setae, flagellum 7-jointed. Gnathopod 1, 6th joint piriform.
Gnathopod 2 somewhat larger, 6th joint oblong, hind margin and palm convex.
Peraeopods 3—5 very long and fragile, 2d joint almost linear. Peraeopods
4 and 5 longer than the body. Uropod 3, outer ramus about $\frac{1}{10}$ shorter

than inner, both with numerous plumose setae. Telson divided. Colour rose-red or violet, with yellowish shimmer above. L. 32 mm.

Lake Baikal. Depth 300—700 m.

3. **A. godlewskii** (Dyb.) 1874 *Gammarus g.* + *G. g. var. victorii*, B. Dybowsky in: Horae Soc. ent. Ross., *v.* 10 suppl. p. 143, 144 t. 1 f. 6 | 1893 *Ceradocus? g.*, A. Della Valle in: F. Fl. Neapel, *v.* 20 p. 928 | 1899 *Acanthogammarus g.*, T. Stebbing in: Tr. Linn. Soc. London, ser. 2 *v.* 7 p. 430.

Median carina usually on peraeon segments 1—4 forming short, blunt teeth, on peraeon segments 5—7 and pleon segments 1—3 long, acute teeth, those on peraeon segment 7 and pleon segment 1 the longest, on pleon segment 4 a short, blunt tooth; on pleon segments 5 and 6 the carina is weak, not dentate; the coalesced marginal and lateral carinae on peraeon segments 1—3 represented by short teeth standing out horizontally from the body and directed somewhat backward; on segment 4 the tooth is of unique size, (especially in *var. victorii*); on segments 5—7 the teeth are short; on pleon segments 1—4 giving place to flat swellings, and on segments 5 and 6 unrepresented. Head rough above, with a depression at the front, rostrum short, acute. Side-plates 1—4 with concave lower margin, 1st produced acutely forward, 2^{d} and 3^{d} rather narrowed distally; 4th pentagonal, much deeper than 5th, with lateral projecting spine-like tooth near lower margin. Eyes prominent, reniform, black. Antenna 1 longer than half the body, more than twice as long as antenna 2, peduncle thicker and longer than that of antenna 2, or in variety more slender and shorter, 3^{d} joint in ♂ as long as 1st, or rather longer, flagellum 50—60-jointed, accessory flagellum 10-jointed. Antenna 2, antepenultimate joint of peduncle in ♂ cylindric, in ♀ flattened and about 5 times as long as broad, flagellum 11-jointed. Gnathopods 1 and 2, 6th joint piriform, in gnathopod 2 hind margin more convex than in gnathopod 1. Peraeopods 3—5, 2^{d} joint narrow, longitudinally carinate; in peraeopods 3 and 4 the hind margin slightly sinuous, in 5th slightly convex, in all fringed with short, simple setae. Uropod 1 reaching beyond uropod 3. Uropod 3, inner ramus little shorter than outer, both with numerous plumose setae. Telson divided to the centre, or (in the variety) more deeply. Colour reddish or brownish yellow. L. reaching 76 mm.

Lake Baikal. Depth 10—150 m.

4. **A. radoszkowskii** (Dyb.) 1874 *Gammarus r.*, B. Dybowsky in: Horae Soc. ent. Ross., *v.* 10 suppl. p. 149 t. 13 f. 3 | 1893 *Ceradocus? r.*, A. Della Valle in: F. Fl. Neapel, *v.* 20 p. 930 | 1899 *Acanthogammarus r.*, T. Stebbing in: Tr. Linn. Soc. London, ser. 2 *v.* 7 p. 430.

Median carina on peraeon segments 1—7 and pleon segments 1—3 tubercular, on pleon segments 4 and 5 forming a strong, broad tooth, with humps on the front surface, on pleon segment 6 reduced to a feeble elevation; lateral carinae on peraeon segments 1—7 and pleon segments 1—3 consisting of little humps, successively smaller; marginal carinae represented by feeble swellings, those on pleon segments 4 and 5 each carrying 2 or 3 spine-points. Head, upper surface level, rostrum very short, much less advanced than the subquadrate lateral lobes. Side-plate 1 slightly expanded below and directed forward, 4th broad, convex in front and below, behind produced outward into a large semicylindrical blunt tooth. Pleon segments 2 and 3 (in figure), postero-lateral corners quadrate. Eyes small, often point-like, irregular in shape, white. Antenna 1 scarcely $^{1}/_{3}$ as long as body, about twice as long as antenna 2, peduncle not thicker but shorter than peduncle of antenna 2, flagellum

34-jointed, accessory flagellum 4-jointed. Antenna 2, flagellum 10-jointed. Gnathopods 1 and 2, 5th joint short, 6th tending to oblong, but widening somewhat to the palm, especially in gnathopod 1, which has the palm more oblique than that of gnathopod 2. Peraeopods 3 and 4, 2^d joint heart-shaped, with long setae on hind margin. Peraeopod 5, 2^d joint well expanded, oblong, but with the wing not quite reaching the top of the 3^d joint. Uropod 1 not reaching beyond uropod 3. Uropod 3, inner ramus very little shorter than outer, both with plumose setae on inner, and simple setae on outer margin. Telson divided. Colour brownish. L. 45 mm.

Lake Baikal. Depth 100—200 m.

5. **A. armatus** (Dyb.) 1874 *Gammarus a.*, B. Dybowsky in: Horae Soc. ent. Ross., *v.* 10 suppl. p. 146 t. 12 f. 1 | 1893 *Ceradocus? a.*, A. Della Valle in: F. Fl. Neapel, *v.* 20 p. 927 | 1899 *Acanthogammarus a.*, T. Stebbing in: Tr. Linn. Soc. London, ser. 2 *v.* 7 p. 430.

Median carina on peraeon segments 1—7 composed of lateral compressed humps, successively larger, on pleon segments 1—3 represented by 2 groups of spines, arranged in sets of 3—5 on low elevations, on pleon segments 4—6 represented only by 4 spines on the hind margin of each segment; lateral carinae of peraeon, placed high up on the back, composed of strong ridge or tooth-like prominences, strongly compressed from before backward, directed obliquely to the long axis of the animal and bent backward, the upper rounded margin beset with 2—7 spines. Head smooth, rostrum short. Side-plates carry only 2 short, spine-like setae. Pleon segment 3, postero-lateral corners rounded. Eyes rather large, oval reniform, slightly prominent, black. Antenna 1 about $^3/_4$ as long as body, 4—5 times as long as antenna 2, with peduncle stouter and twice as long, 1st joint little longer than either 2^d or 3^d, which are subequal, flagellum 30—47-jointed, accessory flagellum 3—5-jointed. Antenna 2, flagellum 5—8-jointed. Gnathopod 1, 6th joint piriform. Gnathopod 2, 6th joint oblong. Peraeopods 3—5, 2^d joint heart-shaped, lower hind corner quadrate. Uropod 1 not reaching beyond uropod 3. Uropod 3, outer ramus $^1/_5$—$^1/_4$ longer than inner, both with plumose setae on inner margin. Telson divided, the division (in figure) reaching to the base, the lobes conical, not so long as peduncle of uropod 3. Colour bright horn-brown with darker marbling and a lighter band across peraeon. L. reaching over 24 mm.

Lake Baikal. Depth 10 m.

6. **A. parasiticus** (Dyb.) 1874 *Gammarus p.*, B. Dybowsky in: Horae Soc. ent. Ross., *v.* 10 suppl. p. 147 t. 3 f. 3 | 1893 *Ceradocus? p.*, A. Della Valle in: F. Fl. Neapel, *v.* 20 p. 929 | 1899 *Acanthogammarus p.*, T. Stebbing in: Tr. Linn. Soc. London, ser. 2 *v.* 7 p. 430.

Median carina on peraeon segments 1—7 forming tooth-like elevations, on pleon segments 1—3 keel-like, all beset with a tolerably constant number of spines; the divided tooth of peraeon segment 1 with 3 pairs of spines, segment 2 with 4 or 5, segment 3 with 4—6, the rest with 2 or 3; pleon segments 1—3 each with 2 spines at the base and several others; pleon segments 4—6 without spines; marginal carinae immediately over the side-plates, on peraeon segments 1—7 tooth-like elevations, each with 3—5 spines, on pleon segments 1—3 represented by very weakly developed tubercular humps, each with 1 spine. Head with short rostrum, and 16—18 denticles distributed over the upper surface and sides. Side-plates 1—4 with only isolated short setae. Eyes roundish, very prominent, black.

.Antenna 1 as long as body, almost thrice as long as antenna 2, with peduncle
.stouter and almost twice as long, flagellum 27-jointed, accessory flagellum
2-jointed. Antenna 2, flagellum 8-jointed. Gnathopod 1, 6th joint piriform.
.Gnathopod 2, 6th joint oblong, widened towards the palm. Peraeopods 3—5,
2^{d} joint not widely expanded, hind margin acute-angled at top, then convex,
concave and spinose below, outer surface longitudinally carinate, the carina
in peraeopods 3 and 4 armed with 2 or 3 spines and produced below. Uropod 3,
outer ramus scarcely longer than the inner, with plumose setae only on
inner margin, the inner ramus having them on both margins. Telson divided.
Colour greenish to yellowish, with golden yellow tips to carinae and teeth.
L. 12 mm.

Lake Baikal. On or near Spongia baicalensis.

Gammari nominatim, reapse incertae sedis.

Gammarus caudisetus Viv.   1805 *G. c.*, Viviani, Phosphor. Maris, p. 7 t. 1
f. 3, 4 | 1888 *G. c.*, *?Hyperia medusarum*, T. Stebbing in: Rep. Voy. Challenger, *v.* 29
p. 76, 78 | 1893 *G.? c.*, A. Della Valle in: F. Fl. Neapel, *v.* 20 p. 766 | 1872 *Hyperia c.*,
A. Boeck, Skand. Arkt. Amphip., *v.* 1 p. 39.

Genoa harbour.

G. chilensis Nic.   1849 *G. c.*, H. Nicolet in: Gay, Hist. Chile, *v.* 3 p. 239.

Body smooth, without spines or tubercles. Eyes round, very small. Antenna 1
very slender, much longer than antenna 2, 1st joint stout, cylindric. Antenna 2 considerably stouter, peduncle very long, hirsute, flagellum short. Gnathopods 1 and 2
subequal, 6th joint little dilated, elongate, suboval. Peraeopods 4 and 5 very long,
spinulose. L. 8 mm.

Sea of Chili.

G. circinnatus Viv.   1805 *G. c.*, Viviani, Phosphor. Maris, p. 9 t. 2 f. 9, 10 | 1888
G. c., *Amphithoe? c.*, T. Stebbing in: Rep. Voy. Challenger, *v.* 29 p. 76, 78 | 1893 *Elas-
mopus? c.*, A. Della Valle in: F. Fl. Neapel, *v.* 20 p. 766.

Body subcylindrical. Pleon segments 1—3, postero-lateral corners angular.
Antenna 1 short, much longer than antenna 2, 2^{d} joint in figure much shorter than 1st.
Limbs all slender. Uropod 3 slightly suggestive of Ampithoe (p. 631). Colour yellowish
to faint reddish.

Genoa harbour.

G. crassimanus Viv.   1805 *G. c.*, Viviani, Phosphor. Maris, p. 10 t. 2 f. 7, 8 | 1872
G.? c., A. Boeck, Skand. Arkt. Amphip., *v.* 1 p. 39 | 1888 *G. c.*, T. Stebbing in: Rep.
Voy. Challenger, *v.* 29 p. 77 | 1893 *G. c.*, A. Della Valle in: F. Fl. Neapel, *v.* 20 p. 726 |
1862 *Moera truncatipes* (part.)?, Bate, Cat. Amphip. Brit. Mus., p. 189.

Antenna 1, 2^{d} joint longer and thinner than 1st. Antenna 2 only half as long.
Eyes rather broad, black. Front feet with granular chela.

Genoa harbour.

G. dubius Johnst.   1827 *G. d.*, G. Johnston in: Zool. J., *v.* 3 p. 178 | 1888 *G. d.*,
T. Stebbing in: Rep. Voy. Challenger, *v.* 29 p. 131 | 1893 *G. d.*, *Amphithoe* (part.)?,
A. Della Valle in: F. Fl. Neapel, *v.* 20 p. 766.

Eyes roundish, black. Antennae 1 and 2 slender, nearly equal, about half as
long as body. Antenna 1, 1st joint longer than 2^{d}. Gnathopods 1 and 2 like those
of Ampithoe rubricata (p. 639), but rather smaller, 6th joint oblong, not much dilated.
Anterior legs short, hinder long. Uropods 1—3 long. Telson divided. L. 4—6 mm.

North-Sea (Berwick). Rock pools among Confervae.

G. flabellifer Stimps. 1855 *G. f.*, Stimpson in: P. Ac. Philad., *v.* 7 p 382 | 1862 *G. f.*, Bate, Cat. Amphip. Brit. Mus., p. 222 | 1893 *G. locusta* (part.)?, A. Della Valle in: F. Fl. Neapel, *v.* 20 p. 766.

Slender, smooth posteriorly. Eyes small, round, black. Antenna 1 half as long as body, flagellum 20-jointed, accessory flagellum 5-jointed. Gnathopods 1 and 2, 6th joint oblong elliptic, With a fusiform area beloW surrounded by short setae. Uropod 3, rami long, lamelliform, elliptical, equal, spreading like a fan. L. 12 mm.

North-Pacific (Loo Choo Islands).

G. fontinalis A. Costa 1883 *G. f.*, A. Costa in: Atti Acc. Napoli, ser. 2 *v.* 1 nr. 2 p. 82, 106 | 1884 *G. f.*, A. Costa in: Bull. Soc. ent. Ital., *v.* 15 p. 340 | 1893 *G. f.*, A. Della Valle in: F. Fl. Neapel, *v.* 20 p. 766.

Very near to Carinogammarus roeselii (p. 506); differs by Want of the dorsal spines of the pleon segments 4—6, these only on the posterior margin presenting a few backWard directed spinules.

Sardinia. In a spring.

G. minus Say 1818 *G. m.*, Say in: J. Ac. Philad., *v.* 1 II p. 375 | 1840 *G. fasciatus* (part.)?, *G. minimus* (laps.), H. Milne EdWards, Hist. nat. Crust., *v.* 3 p. 46.

United States of America. In brooks under stones.

G. peloponnesius Guér. 1832 & 35 *G. p.*, *G. peloponnesiacus*, Guérin(-Méne-Ville) in: Exp. Morée, *v.* 3 I sect. 2 p. 45; Atlas p. 3 Zool. t. 27 f. 5 | 1888 *G. peloponnesius*, T. Stebbing in: Rep. Voy. Challenger, *v.* 29 p. 147 | 1893 *G. locusta* (part.), A. Della Valle in: F. Fl. Neapel, *v.* 20 p. 760.

Pleon segment 5 with stiff dorsal spinules. Eyes reniform. Antenna 1 shorter than antenna 2, the latter With calceoli. Gnathopods 1 and 2 subequal, subchelate. Peraeopods 1—5 rather long, equal. L. 13—16 mm.

Greece.

G. podurus Abildg. 1789 *G. p.*, Abildgaard in: O. F. Müller, Zool. Dan., ed. 3 *v.* 3 p. 59 t. 116 f. 1—6 | 1872 *G. p.*, A. Boeck, Skand. Arkt. Amphip., *v.* 1 p. 38 | 1793 *Cancer (Gammarellus) p.*, J. F. W. Herbst, Naturg. Krabben Krebse, *v.* 2 p. 119 t. 35 f. 6 | 1830 *Amphithoe podura*, H..Milne EdWards in: Ann. Sci. nat., *v.* 20 p. 376 | 1862 *Pherusa p.*, Bate, Cat. Amphip. Brit. Mus., p. 145 | 1893 *Gammarus locusta* (part.), A. Della Valle in: F. Fl. Neapel, *v.* 20 p. 759.

Pleon segments 4 and 5 With dorsal spinules. Antenna 1 apparently shorter than antenna 2, With no accessory flagellum (in figure). Gnathopods 1 and 2, 6th joint oval, not Very large.

Oeresund.

G. savii M.-E. 1830 *G. s.*, H. Milne EdWards in: Ann. Sci. nat., *v.* 20 p. 369 | 1862 *Moera s.*, Bate, Cat. Amphip. Brit. Mus., p. 191 | 1893 *Maera? s.*, A. Della Valle in: F. Fl. Neapel, *v.* 20 p. 731.

Not a species of Gammarus. Pleon segment 4 dorsally produced to an acute tooth, the other pleon segments perfectly smooth. Pedüncle of antenna 2 reaching much beyond that of antenna 1. Gnathopod 1 Very small. Gnathopod 2 much larger, but still small, much as in G. locusta (p. 476). Uropods 1—3 reaching equally far back. Uropod 3 Very short, rami Very small, stiliform. Telson a little horizontal plate.

North-Atlantic (coasts of la Vendée [France]).

G. sp., Viv. 1805 *G. longicornis* (non J. C. Fabricius 1779!), ViViani, Phosphor. Maris, p. 8 t. 2 f. 3, 4 | 1888 *G. l.*, T. Stebbing in: Rep. Voy. Challenger, *v.* 29 p. 76 f. | 1893 *Dexamine gibbosa* (part.)?, A. Della Valle in: F. Fl. Neapel, *v.* 20 p. 578.

Antennae 1 and 2 Very long and slender, each With 2 long joints in peduncle. Antenna 1 the longer. Uropods 1—3, rami laminar, elliptic. Colour yellowish.

Genoa. In Weedy bays.

G. sp., Risso 1826 *G. marinus* (non Leach 1815!), A. Risso, Hist. nat. Eur. mérid., *v.* 5 p. 96 | 1888 *G. m.*, T. Stebbing in: Rep. Voy. Challenger, *v.* 29 p. 128.

Colour intense grey, with deep grey dots, antennae and legs paler.

Mediterranean.

---

31. Fam. **Dexaminidae**

1813/14 *Dexameridae*, Leach in: Edinb. Enc., *v.* 7 p. 432 | 1876 Subfam. *Dexaminae*, A. Boeck, Skand. Arkt. Amphip., *v.* 2 p. 310 | 1888 *Dexaminidae*, T. Stebbing in: Rep. Voy. Challenger, *v.* 29 p. 573, 900 | 1893 *Dexaminidi* (part.), A. Della Valle in: F. Fl. Neapel, *v.* 20 p. 556.

Pleon segments 5 and 6 coalesced. Antenna 1 with long 2^d joint, without accessory flagellum. Upper lip rounded. Lower lip varying. Mandible (Fig. 90 p. 520) without palp. Maxilla 1 varying, inner plate with only 1 or 2 setae or setules. Maxilla 2, inner plate the smaller, not fringed on inner margin. Maxillipeds (Fig. 92 p. 522), outer plates very long, palp rather short, finger small or wanting. Gnathopods 1 and 2 feeble, subchelate. Peraeopods 1—5, fingers of all commonly pointing backward. Uropod 3, rami subequal, extending beyond uropod 2. Telson elongate, deeply cleft.

Marine.

5 genera, 10 accepted species and 5 doubtful.

Synopsis of genera:

1 { Maxillipeds, palp 3-jointed; lower lip with inner lobes rudimentary — **2.**
Maxillipeds (Fig. 92), palp 4-jointed; lower lip with inner lobes well developed — **3.**

2 { Peraeopods 1—5, 4th joint shorter than 5th and 6th combined . . . . 1. Gen. **Dexamine** . . . p. 514
Peraeopods 1—5, 4th joint longer than 5th and 6th combined . . . . 2. Gen. **Tritaeta** . . . . . p. 517

3 { Maxilla 1, palp 1-jointed . . . . 3. Gen. **Paradexamine** . p. 518
Maxilla 1, palp 2-jointed — **4.**

4 { Maxilla 1, 2^d joint of palp large; maxillipeds, inner plates well developed . . . . . . 4. Gen. **Polycheria** . . p. 519
Maxilla 1, 2^d joint of palp small; maxillipeds (Fig. 92), inner plates rudimentary . . . . 5. Gen. **Guernea**. . . . p. 521

1. Gen. **Dexamine** Leach

1813/14 *Dexamine* (Sp. un.: *D. spinosa*), Leach in: Edinb. Enc., *v.* 7 p. 432 | 1888 *D.*, T. Stebbing in: Rep. Voy. Challenger, *v.* 29 p. 945 | 1893 *D.* (part.), A. Della Valle in: F. Fl. Neapel, *v.* 20 p. 572 | 1894 *D.*, G. O. Sars, Crust. Norway, *v.* 1 p. 473 | 1846 *Dexamene*, L. Agassiz, Nomencl. zool., Index p. 121 | 1851 *Amphithonotus* (part.), (A. Costa in:) F. W. Hope, Cat. Crost. Ital., p. 45 | 1861 *Amphitonotus*, E. Grube, Ausfl. Triest, p. 136.

Body rather stout in ♀; some segments with dorsal projections. Head, rostral projection very small, post-antennal corners obsolete. Side-plates of moderate size. Antenna 1 with long peduncle, in ♀ longer than antenna 2.

Lower lip with rudimentary inner lobes. Mandible, spine-row very small. Maxilla 1, inner plate with 1 seta, outer with 11 spines, palp's single joint rather large, apically smooth on one maxilla, denticulate on the other. Maxilla 2 rather small, armed only at the apices. Maxillipeds, inner plates rather small but well developed, outer plates almost covering the slender 3-jointed palp, fringed with spine-teeth on inner margin. Gnathopod 2 rather longer than 1ˢᵗ. Peraeopods 3—5, 2ᵈ joint successively less expanded. Uropod 3, rami narrowly lanceolate.

4 species accepted, 4 doubtful.

Synopsis of accepted species:

1 { Pleon segment 3 dorsally unidentate — **2.**
{ Pleon segment 3 dorsally tridentate — **3.**

2 { Peraeopod 5, 2ᵈ joint laminar . . . . 1. **D. spinosa** . . . . . p. 515
{ Peraeopod 5, 2ᵈ joint sublinear. . . . . . . 2. **D. thea** . . . . p. 516

3 { Pleon segment 2 tricarinate . . . . 3. **D. spiniventris** . . . p. 516
{ Pleon segment 2 not tricarinate . . . . . . 4. **D. blossevilliana** . . p. 517

1. **D. spinosa** (Mont.) 1813 *Cancer (Gammarus) spinosus*, Montagu in: Tr. Linn. Soc. London, *v.* 11 p. 3 t. 2 f. 1 | 1818 *G. s.*, Lamarck, Hist. An. s. Vert., *v.* 5 p. 181 | 1813/14 *Dexamine spinosa*, Leach in: Edinb. Enc., *v.* 7 p. 432 | 1862 *D. s.*, Bate & Westwood, Brit. sess. Crust., *v.* 1 p. 237 f. | 1862 *D. s.*, Bate, Cat. Amphip. Brit. Mus., p. 130 t. 24 f. 1 | 1876 *D. s.*, A. Boeck, Skand. Arkt. Amphip., *v.* 2 p. 312 t. 11 f. 5 | 1893 *D. s.* (part.), A. Della Valle in: F. Fl. Neapel, *v.* 20 p. 573 t. 5, 12; t. 18 f. 1—17, 19 | 1893 & 94 *D. s.*, G. O. Sars, Crust. Norway, *v.* 1 t. 166 f. 2; t. 167; p. 475 | 1840 *Acanthonotus? s.*, H. Milne Edwards, Hist. nat. Crust., *v.* 3 p. 25 | 1855 *Amphithoe s.*, Gosse, Man. mar. Zool., *v.* 1 p. 141 f. 266 | ?1826 *Atylus corallinus*, A. Risso, Hist. nat. Eur. mérid., *v.* 5 p. 99 | 1830 *Amphithoe marionis*, H. Milne Edwards in: Ann. Sci. nat., *v.* 20 p. 375 | 1857 *Amphithonotus m.*, A. Costa in: Mem. Acc. Napoli, *v.* 1 p. 195 | 1843 *Amphithoë tenuicornis*, H. Rathke in: N. Acta Ac. Leop., *v.* 201 p. 77 t. 4 f. 3 | 1859 *Gammarus speciosus* (laps., corr.: *G. spinosus*) + *Dexamine t.*, R. M. Bruzelius in: Svenska Ak. Handl., n. ser. *v.* 3 nr. 1 p. 79 | 1851 *Amphithonotus acanthophthalmus*, (A. Costa in:) F. W. Hope, Cat. Crost. Ital., p. 45 | 1868 *Dexamine spiniventris var. pontica*, Czerniavski in: Syezda Russ. Est., Syezda 1 Zool. p. 111 t. 8 f. 16 | 1898 *D. spinosa var.?*, Sowinski in: Mém. Soc. Kiew, *v.* 15 p. 490 t. 12 f. 5, 6.

Peraeon robust, pleon carinate, segments 1—4 each produced to a strong dorsal tooth, hinder part of segments 5 and 6 with 3 small dorsal ridges each ending in a small tooth. Head, rostrum short, blunt at apex, lateral corners triangular, acute. Side-plate 1 a little concave in front, 4ᵗʰ deeper than the rest. Pleon segment 3, postero-lateral corners sharply produced. Eyes oblong reniform, larger in ♂ than in ♀, dark brown with whitish coating. Antenna 1, 1ˢᵗ joint produced to an obtuse tooth (or acute, Sowinski), 2ᵈ much narrower and over once and a half as long, 3ᵈ very short (or not very short, Sowinski), flagellum about twice as long as peduncle, 30—45-jointed. Antenna 2 in ♀ shorter, in ♂ longer than antenna 1, ultimate joint of peduncle longer than penultimate, flagellum filiform. Gnathopod 1, 2ᵈ joint curved, 5ᵗʰ shorter than 6ᵗʰ, widened distally, 6ᵗʰ expanding to the palm which is oblique, as long as hind margin, well defined by the palmar spines at the obtuse angled junction. Gnathopod 2, 2ᵈ joint longer, curved, 5ᵗʰ little widened distally, as long as 6ᵗʰ, which is similar to that of gnathopod 1 but longer. Peraeopods 1 and 2 alike, 6ᵗʰ joint longer than 5ᵗʰ, finger rather strong. Peraeopods 3—5, 6ᵗʰ joint shorter than 5ᵗʰ. Peraeopod 3, 2ᵈ joint oblong oval, lower hind corner produced in a rounded lobe, 4th joint subfusiform, much broader and longer than 5th. Peraeopod 4, 2^d joint piriform, abruptly narrowed below, 4th as long as 5th. Peraeopod 5 rather shorter, 2^d joint oval, narrowest below, 4th shorter than 5th. Branchial vesicles pleated. Uropod 3, rami about twice as long as peduncle. Telson thrice as long as broad, cleft to $^3/_4$ length, each half with 3 lateral spines and 1 on obtuse apex. Colour a mixture of chestnut-brown, pink, light yellow and pure white (Sars), sometimes brilliant with mingled green and red. L. ♀ reaching 14 mm, ♂ rather smaller.

Arctic Ocean, North-Atlantic with adjoining seas (Europe from Vadsö to the Black Sea; Azores). Within tide-marks to 75 m.

2. **D. thea** Boeck 1861 *D. t.*, A. Boeck in: Forb. Skand. Naturf., Møde 8 p. 658 | 1876 *D. t.*, A. Boeck, Skand. Arkt. Amphip., *v.* 2 p. 315 t. 12 f. 1 | 1885 *D. t.*, J. S. Schneider in: Norske Selsk. Skr., 1884 p. 20 t. 2 | 1893 *D. t.*, *D. spinosa* (part.)?, A. Della Valle in: F.. Fl. Neapel, *v.* 20 p. 579 | 1893 & 94 *D. t.*, G. O. Sars, Crust. Norway, *v.* 1 t. 168 f. 1; p. 477 | 1862 *D. tenuicornis* (err., non *Amphithoë* t. H. Rathke 1843!), Bate & Westwood, Brit. sess. Crust., *v.* 1 p. 240 f. | 1871 *D. heibergi*, A. Boeck in: Forh. Selsk. Christian., 1870 p. 187 | 1876 *D. h.*, A. Boeck, Skand. Arkt. Amphip., p. 316 t. 12 f. 3.

Peraeon, pleon and head nearly as in D. spinosa (p. 515), but eyes smaller, oval. Antenna 1, 1st joint not apically produced, 3^d joint not extremely short, flagellum 16-jointed. Antenna 2 in ♀ much shorter, ultimate joint of peduncle subequal to penultimate (Sars) or shorter than it (Schneider), flagellum little longer than peduncle, 11-jointed. Gnathopods 1 and 2 and peraeopods 1 and 2 nearly as in D. spinosa. Peraeopod 3, 2^d joint with hinder margin and lower lobe subangular, 4th much longer and broader than 5th, 5th slightly shorter than 6th. Peraeopod 4, 2^d joint little longer than broad, very convex behind, and finely serrate, 4th longer than 5th, 5th than 6th. Peraeopod 5, 2^d joint almost linear, 5th longer than 4th or 6th, finger strong in this and the other peraeopods. Branchial vesicles simple. Uropod 3, rami twice as long as peduncle, rather narrow. Telson not quite thrice as long as broad, cleft nearly to the base, spines as in D. spinosa, with addition of 2 subdorsal pairs. Colour yellowish, semipellucid, mottled with brown and pink. L. 4—6 mm.

Arctic Ocean, North-Atlantic, North-Sea and Skagerrak (from Vadsö to Christiania-fjord; north of Great Britain; France); Kattegat. Depth to 113 m.

3. **D. spiniventris** (A. Costa) 1853 *Amphithonotus s.*, A. Costa in: Rend. Soc. Borbon., n. ser. *v.* 2 p. 173 | 1857 *A. s.*, A. Costa in: Mem. Acc. Napoli, *v.* 1 p. 186 t. 2 f. 1 | 1864 *Dexamine s.*, E. Grube in: Arch. Naturg., *v.* 30 ɪ p. 195 | 1866 *D. s.*, Cam. Heller in: Denk. Ak. Wien, *v.* 26 ɪɪ p. 30 | 1893 *D. spinosa* (part.), A. Della Valle in: F. Fl. Neapel, *v.* 20 p. 573 t. 5 f. 9; t. 18 f. 18.

Pleon moderately robust, pleon segments 1—4 and 6 carinate, each produced to a dorsal tooth, in segment 4 and sometimes in 3 preceded by a smaller tooth, segments 1—3 and 6 having a pair of latero-dorsal ridges ending in small teeth, sometimes obsolete on segment 1, the part of the coalesced segment corresponding to segment 5 separated as usual from segment 6 by a depression. Head, rostrum very short, lateral corners angular. Pleon segment 3, postero-lateral corners sharply produced. Eyes reniform, not very large. Antenna 1, 1st joint produced to a sharp tooth, 2^d less than double as long as 1st (Della Valle; Heller: thrice), flagellum 40—50-jointed. Antenna 2 shorter (at least in ♀), ultimate joint of peduncle longer than penultimate, flagellum about 40-jointed. Gnathopods 1 and 2 and peraeopods 1—5, uropods and telson much as in D. spinosa (p. 515). Colour dull yellow. L. 7—8 mm.

Mediterranean.

4. D. blossevilliana Bate 1862 *D. b.*, Bate, Cat. Amphip. Brit. Mus., p. 131
t. 24 f. 2 | 1893 *D. spinosa* (part.), A. Della Valle in: F. Fl. Neapel, *v.* 20 p. 573.

Very near to D. spinosa (p. 515), but distinguished as follows. Pleon
segment 3 with a strong dorsal tooth on each side of the central one;
segment 4 (in figure) with a small dorsal tooth in front of its apical one.
Eyes very large, quadrate. Telson very long, cleft nearly to the base, each
external margin carrying 4 fasciculi of hairs, each apex serrated and carrying
a subapical spine. L. 10 mm.

Habitat unknown.

D. anisopus (Grube) 1861 *Amphithöe (Amphitonotus) a.*, E. Grube, Ausfl. Triest,
p. 136 | 1864 *Dexamine a.*, E. Grube in: Arch. Naturg., *v.* 30 ı p. 197 | 1893 *D. spinosa*
(part.), A. Della Valle in: F. Fl. Neapel, *v.* 20 p. 573.

Probably identical with D. spiniventris; but peraeopods 1 and 2 not symmetrical.
L. over 6 mm.

Adriatic.

D. leptonyx (Grube) 1861 *Amphithöe (Amphitonotus) l.*, E. Grube, Ausfl. Triest,
p. 136 | 1864 *Dexamine l.*, E. Grube in: Arch. Naturg., *v.* 30 ı p. 198 | 1893 *Atylus? l.*,
A. Della Valle in: F. Fl. Neapel, *v.* 20 p. 703.

Probably identical with D. spinosa (p. 515); but 1st joint of antenna 1 is said
to be unarmed. L. 8 mm.

Adriatic.

D. miersii Hasw. 1885 *D. m.*, Haswell in: P. Linn. Soc. N. S. Wales, *v.* 10
p. 102 t. 13 f. 8—12 | 1893 *D. m.*, A. Della Valle in: F. Fl. Neapel, *v.* 20 p. 578.

Torres Strait (Thursday Island).

D. scitulus Harford 1877 *D. s.*, Harford in: P. Calif. Ac., *v.* 7 p. 116 | 1893
D. s., A. Della Valle in: F. Fl. Neapel, *v.* 20 p. 579.

Magdalena Bay [Lower California]. Depth 11 m.

2. Gen. **Tritaeta** Boeck

1871 *Lampra* (Sp. un.: *L. gibbosa*) (non Jac. Hübner 1816, Lepidoptera!), A.
Boeck in: Forh. Selsk. Christian., 1870 p. 188 | 1876 *Tritaeta*, A. Boeck, Skand. Arkt.
Amphip., *v.* 2 p. 317 | 1888 *T.* (part.), T. Stebbing in: Rep. Voy. Challenger, *v.* 29 p. 941 |
1894 & 95 *T.*, G. O. Sars, Crust. Norway, *v.* 1 p. 478, 698.

Body rather stout in ♀, only pleon segment 4 dorsally produced.
Head, rostrum almost obsolete. Side-plates shallow, some of irregular angular
form. Antennae 1 and 2 subequal; antenna 1 with long peduncle. Mouth-
parts and gnathopods 1 and 2 nearly as in Dexamine (p. 514). Peraeopods
1—5 strong, subequal, 4th joint very long, 5th and 6th short, the spiniferous
widening of the 5th joint giving all a subchelate character. Peraeopods 3—5,
2d joint not much expanded, least in peraeopod 5. Uropods as in Dexamine.

1 species.

1. T. gibbosa (Bate) ?1861 *Amphithöe brevitarsis*, E. Grube, Ausfl. Triest,
p. 135 | 1864 *Dexamine b.*, E. Grube in: Arch. Naturg., *v.* 30 ı p. 196 | 1862 *Atylus
gibbosus*, Bate (& Westwood), Brit. sess. Crust., *v.* 1 p. 248 f. | 1862 *A. g.*, Bate, Cat.
Amphip. Brit. Mus., p. 137 t. 26 f. 3 | 1871 *Lampra gibbosa*, A. Boeck in: Forb. Selsk.
Christian., 1870 p. 188 | 1876 *Tritaeta g.*, A. Boeck, Skand. Arkt. Amphip., *v.* 2 p. 318
t. 12 f. 2 | 1890 *T. g.*, A. O. Walker in: P. Liverp. biol. Soc., *v.* 4 p. 249 t. 16 f. 4, 6 |

1893, 94 & 95 *T. g.*, G. O. Sars, Crust. Norway, *v.* 1 t. 168 f. 2; p. 479; p. 698 t. VIII f. 1 | 1895 *T. g.*, A. O. Walker in: P. Liverp. biol. Soc., *v.* 9 p. 306 | 1881 *Dexamine dolichonyx*, Nebeski in: Arb. Inst. Wien, *v.* 3 p. 145 t. 13 f. 40 | 1888 *Tritaeta d.*, T. Stebbing in: Rep. Voy. Challenger, *v.* 29 p. 520, 941, 945.

Pleon segments 1—3 in ♂ slightly raised dorsally. Side-plates 1 and 2 subquadrate, 3^d and 4th acutely produced at the lower corners, 5th with front lobe the deeper, rounded. Pleon segment 3, postero-lateral corners acute, not greatly produced. Eyes large, rounded oval, larger in ♂, reddish brown with whitish coating. Antenna 1 long, 1st joint much shorter than 2^d, flagellum twice as long as peduncle, about 18-jointed. Antenna 2 in ♀ scarcely so long, ultimate joint of peduncle shorter than penultimate, flagellum about 18-jointed; in ♂ longer than antenna 1, penultimate and antepenultimate joints of peduncle densely fringed above, flagellum about 24-jointed. Gnathopod 1, 5th joint distally widened, rather shorter than 6th, which is a short oval, widening to the palm; palm slightly oblique, fairly well defined. Gnathopod 2 longer, 5th joint slender, little widened distally, quite as long as 6th, which is longer and more slender than in gnathopod 1. Gnathopod 1 often has a deep sinus in the front margin of the 6th joint in ♂ and at least occasionally also in ♀, perhaps a copulatory feature. Peraeopods 1—5, 4th joint longer than 5th and 6th combined, 5th armed with 5 strong spines on expanded extremity; 6th rather longer in peraeopods 1 and 2, equal or shorter in 3^d—5th, in all facing backward, not expanded distally; finger strong and curved. Peraeopod 3, 2^d joint narrowly oblong oval, in peraeopod 4 narrower, with angular projection proximally, in peraeopod 5 sublinear, slightly widened proximally. Branchial vesicles simple in ♀, pleated in ♂. Uropod 3, rami scarcely twice as long as peduncle, narrowly lanceolate, spinose. Telson rather over twice as long as broad, cleft nearly to base, in ♀ with 3 marginal and 3 submarginal spines on each side, and 1 spine on each obtuse apex, in ♂ with an additional marginal spine (Sars). Colour dark brownish with opaque white patches. L. 6 mm.

North-Atlantic and North-Sea (Norway, France, British Isles, Azores); Mediterranean.

3. Gen. **Paradexamine** Stebb.

1899 *Paradexamine* (Sp. typ.: *Dexamine pacifica*), T. Stebbing in: Ann. nat. Hist., ser. 7 *v.* 4 p. 210.

In general character like Dexamine (p. 514), but distinguished as follows. Lower lip with inner lobes well developed and mandibular processes upturned. Maxilla 1 with the 1-jointed palp uniform in left and right maxilla. Maxillipeds with small distinct finger to the palp.

1 species.

1. **P. pacifica** (G. M. Thoms.) 1879 *Dexamine p.*, G. M. Thomson in: Tr. N. Zealand Inst., *v.* 11 p. 238 t. 10B f. 4 | 1886 *D. p.*, G. M. Thomson & Chilton in: Tr. N. Zealand Inst., *v.* 18 p. 149 | 1899 *D. p.*, *Paradexamine sp. typ.*, T. Stebbing in: Ann. nat. Hist., ser. 7 *v.* 4 p. 210 | 1893 *D. spinosa* (part.), A. Della Valle in: F. Fl. Neapel, *v.* 20 p. 574.

Peraeon and pleon segment 1 dorsally rounded, pleon segments 2—4 carinate, segment 1 sometimes produced to a dorsal tooth or 3 teeth, segments 2 and 3 produced to 3 large dorsal teeth, largest in segment 3, segment 4 produced into a large tooth slightly upturned at the acute apex, a dorsal depression on 3[d], 4th and the coalesced segments, the latter carrying 1 or 2 pairs of spines, and having the dorsal and lower apices acute on each side of base of telson. Head, rostrum small, acute, slightly depressed, lateral corners produced into a small acute point. Side-plates of moderate size. Pleon segments 2 and 3, postero-lateral corners acutely produced. Eyes rectangular oval, remaining dark in spirit. Antenna 1 long, 1st joint stout (Thomson: with a tooth at the end), 2[d] varying from little longer to nearly twice as long; flagellum with 40—50 joints, 1st sometimes longer than 3[d] of peduncle. Antenna 2 about $^2/_3$ length of antenna 1, slender, ultimate joint of peduncle about half as long as penultimate, flagellum with 25—30 joints, several rather long. Upper lip with 2 hairy lobes, which fold one on the other (at least when the lip is detached). Lower lip with principal lobes broad, the mandibular processes not as usual produced backward, but represented only by a small acute process directed forward. Mandible, cutting edge with a straight piece ending in 3 teeth, accessory plate divided into many denticles, spine-row of 2 or 3 very small spines, molar with slender spines fringing one border of the right mandible, palp entirely wanting. Maxilla 1, inner plate with 1 or 2 setules, outer plate with 11 spines, palp with about a dozen flexible spines. Maxilla 2, inner plate shorter and narrower than outer, each with slender spines on oblique apex. Maxillipeds, inner plates of moderate size, with many slender spines but seemingly without the 3 stout spine-teeth commonly found on the apex, outer plates long, about reaching middle of palp's 3[d] joint, with spine-teeth on inner margin and slender spines on outer. Gnathopods 1 and 2, 5th joint longer than 6th, 6th with row of spines along inner surface, widening a little to palm, which is oblique, microscopically pectinate, well defined by palmar spines, finger with subapical tooth. Gnathopod 2 rather the longer. Peraeopods 1 and 2 slender, 5th joint shorter than 4th. Peraeopod 3, 2[d] joint broad oval, hind margin smooth, 4th joint longer and 5th shorter than 6th. Peraeopod 4, 2[d] joint much narrowed below, 4th longer than 5th, 5th longer than 6th. Peraeopod 5, 2[d] joint much smaller than in preceding peraeopods, narrowed below, 4th and 5th joints subequal; finger curved, rather strong in all the peraeopods. Uropod 1 rather long, outer ramus little shorter than inner; uropod 2, outer ramus shorter than inner; uropod 3, rami short, subequal. Telson reaching beyond uropod 3, deeply cleft. L. reaching 8 mm.

South-Pacific (New Zealand; Jervis Bay [East-Australia]).

4. Gen. **Polycheria** Hasw.

1879 *Polycheria*, Haswell in: P. Linn. Soc. N. S. Wales, *v.* 4 p. 345 | 1893 *P.*, A. Della Valle in: F. Fl. Neapel, *v.* 20 p. 579 | 1881 *Polychiria*, E. v. Martens in: Zool. Rec., *v.* 16 Crust. p. 31 | 1882 *Polycheria, Polychelia*, G. M. Thomson in: Tr. N. Zealand Inst., *v.* 14 p. 233 | 1898 *Polycheria, Polycharia*, Calman in: Ann. N. York Ac., *v.* 11 p. 261, 268, 288.

Agreeing with Tritaeta (p. 517), but distinguished as follows. Lower lip with well formed inner lobes. Maxilla 1 with 2-jointed palp. Maxillipeds having a small distinct finger to the palp. Peraeopods 1—5 (Fig. 91 p. 520) having the 5th joint linear and the 6th expanded distally, to form a distinct concave palm. Telson with acute apices.

2 species.

Synopsis of species:

1. **P. antarctica** (Stebb.) 1875 *Dexamine a.*, T. Stebbing in: Ann. nat. Hist., ser. 4 *v.* 15 p. 184 t. 15$_A$ f. 1 | 1878 *Atylus antarcticus*, T. Stebbing in: Ann. nat. Hist., ser. 5 *v.* 2 p. 370 | 1893 *Polycheria antarctica*, A. Della Valle in: F. Fl. Neapel, *v.* 20 p. 580 t. 58 f. 83, 84 | 1888 *Tritaeta a.* + *T. kergueleni*, T. Stebbing in: Rep. Voy. Challenger, *v.* 29 p. 941; p. 941 t. 83.

Pleon segments carinate except front part of segment 1, 4th produced to a dorsal tooth, 6th raised on each side of base of telson. Head, rostrum obsolete. Side-plates 1 and 2 rather produced and acute at lower front corner, 3d and 4th with subacute lobe in front. Pleon segment 3, postero-lateral corner forming a short acute tooth, which is sometimes obsolete. Eyes large, rounded oval, and (Haswell) red. Antenna 1, 1st joint scarcely half as long as 2d, 3d very short, flagellum rather longer than peduncle, with 20 joints, most with long sensory filaments. Antenna 2 in ♀ equal to antenna 1, ultimate joint of peduncle rather shorter than penultimate, flagellum shorter than peduncle, 11-jointed. Upper lip broadly and smoothly rounded. Mandible (Fig. 90), spine-row with 3 spines on left, 2 on right mandible. Maxilla 1, inner plate with 2 setae, outer with 10 spines, palp with slender spines on apex, 1st joint very short, 2d very long. Maxilla 2, a few setae on inner margin of inner plate. Maxillipeds, outer plates very large, reaching beyond middle of 3d joint of palp, inner margin fringed with about 20 small spine-teeth. Gnathopod 1, 5th joint subfusiform, longer than the 6th, 6th short oval, palm not very oblique, finely pectinate, well defined, 4th—6th joints

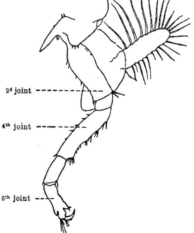

Fig. 90.
P. antarctica.
'Left mandible.

Fig. 91. **P. antarctica.**
Peraeopod 1.

2d joint
4th joint
6th joint

with long spines on hind margin, 6th with them also on front. Gnathopod 2 similar, but 2d, 4th, 5th and 6th joints longer. Peraeopods 1—5 nearly alike; 2d joint in peraeopods 1 (Fig. 91) and 2 broader, not longer than 4th, in 3d to 5th a little widened proximally, and a little longer than 4th; in all peraeopods 4th as long as 5th and 6th combined, 6th in peraeopods 1 and 2 rather longer than 5th, in all peraeopods 6th with inner margin produced into a tooth tipped with 2 spines, against which the much curved finger impinges. Branchial vesicles simple in ♀, pleated in ♂. Uropod 3, outer ramus shorter than inner (Stebbing), or rami equal (Haswell). Telson elongate, cleft nearly to the base, with several marginal spines, apices tipped each with a spine. Colour yellowish. L. 3—8 mm.

Antarctic Ocean (lat. 77° 30′ S., depth 548 m, in sponge); South-Pacific (Port Jackson [East-Australia], depth 4 m; Paterson Inlet [New Zealand], depth 19 m); southern Indian Ocean (Kerguelen, depth 51 and 232 m).

2. **P. tenuipes** Hasw. 1879 *P. t.* + *?P. brevicornis*, Haswell in: P. Linn. Soc. N. S. Wales, *v.* 4 p. 345 t. 22 f. 8; p. 346 | 1882 *P. obtusa*, G. M. Thomson in: Tr. N. Zealand Inst., *v.* 14 p. 233 t. 17 f. 3 | 1898 *P. osborni*, *Polycharia o.*, Calman in: Ann. N. York Ac., *v.* 11 p. 268 t. 32 f. 2; p. 288.

Closely resembles P. antarctica, but differs in the following respects. Dorsal processes of pleon segments 4 and 6 much less prominent. Side-plate 4 with anterior process reduced to a short blunt lobe. Maxilla 1 with 1st joint very short, 2^{d} very long (1-jointed: Calman). Maxillipeds with outer plates longer, nearly equalling palp, and armed with only about 11 spine-teeth on inner margin. Gnathopod 1 with palmar edge very short, not more than $^1/_3$ as long as finger; gnathopod 2 with 6th joint twice as long as broad, palmar edge about $^1/_2$ as long as finger. Peraeopod 3 with front and hind margins of 6th joint nearly parallel, and thumb-like process much less prominent. L. ♀ 7 mm.

Puget Sound.

5. Gen. **Guernea** Chevreux

1868 *Helleria* (Sp. uu.: *H. coalita*) (non Ebner 1868, Isopoda!), A. M. Norman in: Anu. nat. Hist., ser. 4 v. 2 p. 418 | 1887 *Guernea*, Chevreux in: Bull. Soc. zool. France, v. 12 p. 302 | 1890 *G.*, T. Stebbing in: Ann. nat. Hist., ser. 6 v. 5 p. 192 | 1893 *G.*, A. Della Valle in: F. Fl. Neapel, v. 20 p. 570 | 1887 *Prinassus* (Sp. un.: *P. nordenskiöldii*), H. J. Hansen in: Vid. Meddel., ser. 4 v. 9 p. 82.

Pleon segments 5 and 6 coalesced, abruptly truncated behind. Head, rostrum obsolete. Side-plate 5 the largest. Antenna 1 in ♂ much shorter than antenna 2. Upper lip rounded. Lower lip with inner lobes well developed. Mandible, cutting edge obscurely dentate, with accessory plate but no apparent spine-row, molar represented by a laminar fold. Maxilla 1, inner plate with 1 setule, outer with 7 spines, palp with 1st joint slightly longer than 2^{d}. Maxilla 2, inner plate with 2 (1?) setae, outer plate longer, with a few setae. Maxillipeds, inner plates rudimentary, outer plates reaching middle of 3^{d} joint of palp, fringed with spine-teeth on distal part of inner margin, finger of palp small. Gnathopods 1 and 2, 6th joint subquadrate. Peraeo-pods 1 and 2, 4th joint much longer than 5th; 2^{d} joint smaller in peraeo-pod 4 than in 3 or 5; peraeopod 5 with 4th and 5th joints wide and plumose. Uropod 3, rami lanceolate, not much longer than peduncle. Telson rather longer than broad.

2 species accepted, 1 doubtful.

Synopsis of accepted species:

Gnathopod 1, 6th joint decidedly longer than broad;
    peraeopod 5, 2d joint much narrowed distally. . . 1. G. coalita . . . . . p. 521
Gnathopod 1, 6th joint scarcely longer than broad;
    peraeopod 5, 2d joint little narrowed distally . . . 2. G. nordenskiöldii . p. 522

1. **G. coalita** (Norm.) 1868 *Helleria c.*, A. M. Norman in: Ann. nat. Hist., ser. 4 v. 2 p. 418 t. 22 f. 8; t. 23 f. 1—6 | 1888 *Guernea c.*, T. Stebbing in: Rep. Voy. Challenger, v. 29 p. 595, 596, 386 | 1893 *G. c.* (part.), Chevreux & E. L. Bouvier in: Ann. Sci. nat., ser. 7 v. 15 p. 123 | 1893 *G. c.* (part.), A. Della Valle in: F. Fl. Neapel, v. 20 p. 570 t. 31 f. 20—33; t. 58 f. 80 | 1887 *G. c.* + *G. laevis*, Chevreux in: Bull. Soc. zool. France, v. 12 p. 303 f. 1, 2; p. 328.

Peraeon dorsally rounded, pleon somewhat compressed, with strongly marked dorsal division between segments 3 and 4, segments 5 and 6 completely fused, but with a dorsal depression between two humps which, like segment 4, are serrate; telson attached to the lower margin of the steeply truncate back of the compound segment. Head, lateral corners obtusely produced. Side-plates 1—4 not very large, successively larger, lower margin rounded, narrow, subserrate, 5th large, the hind lobe deeper than side-plate 4; side-plate 7 larger than 6th. Pleon segment 3, postero-lateral corners rounded. Eyes round. Antenna 1 in ♀, 1st joint stout, not much longer than 2[d], 3[d] short, flagellum shorter than the short peduncle, 7-jointed, in ♂ 1st joint of peduncle shorter than 2[d], flagellum rather longer than peduncle, about 9-jointed. Antenna 2 in ♀ equal to antenna 1, ultimate joint of peduncle much narrower and shorter than penultimate, flagellum 8- or 9-jointed, in ♂ ultimate and penultimate joints of peduncle subequal, flagellum slender. Upper lip flatly rounded, smooth. Lower lip, principal lobes with an acute apex, inner lobes prominent, mandibular processes obsolete. Maxilla 2, inner plate with well rounded apex. Gnathopod 1, 2[d] joint very narrow and bent at the base, 5th joint a little shorter than 6th, 6th decidedly longer than broad, widening a little distally, palm scarcely oblique, well defined, a little convex, finger shutting closely upon it. Gnathopod 2 similar, but base of 2[d] joint little bent, less narrow, 5th rather longer than 6th, 6th longer and less widened than in gnathopod 1. Peraeopods 1 and 2, 6th joint narrower but longer than 5th. Peraeopod 3, 2[d] joint greatly expanded below, bowed out in front, 4th rather dilated, following joints slender. Peraeopod 4, 2[d] joint much smaller, piriform, widest above, other joints nearly as in peraeopod 3. Peraeopod 5, 2[d] joint very large, straight in front, very convex behind, narrowed below, 4th and 5th expanded, with setae on both margins, 6th very slender. The hexagonal markings of the integument are very conspicuous on the expansion of 2[d] joint in peraeopods 3—5. Uropods 1 and 2 rather small, rami shorter than peduncle, inner ramus shorter than outer, each tipped with a spine. Uropod 3, rami equal, not long, but longer than peduncle, smooth in ♀, with plumose setae in ♂. Telson ovate, cleft almost to the base. Colour yellow with red and greenish dots and bands. L. 1·25—2 mm.

North-Atlantic and North-Sea (Shetland; Moray Firth; Firth of Clyde; Irish sea; Croisic; Saint-Jean-de-Luz); Mediterranean (Cannes; Naples).

2. G. nordenskiöldii (H.-J. Hansen) 1887 *Prinassus n.*, H. J. Hansen in: Vid. Meddel., ser. 4 v. 9 p. 82 t. 2 f. 7; t. 3 f. 1 | 1890 *Guernea coalita* (part.)?, T. Stebbing in: Ann. nat. Hist., ser. 6 v. 5 p. 193 | 1893 *G. c.* (part.), Chevreux & E. L. Bouvier in: Ann. Sci. nat., ser. 7 v. 15 p. 123 | 1893 *G. c.* (part.), A. Della Valle in: F. Fl. Neapel, v. 20 p. 572.

Like G. coalita (p. 521), but distinguished as follows. Pleon segments 4—6 more strongly carinate, the compound segment not dorsally depressed or serrate. Antenna 1, flagellum 6-jointed; antenna 2, flagellum 3-jointed. Gnathopod 1, 6th joint scarcely longer than broad, much widened to the rather oblique palm, 6th joint in gnathopod 2 rather longer but otherwise nearly as in gnathopod 1. Peraeopod 5 with 2[d] joint narrower below than above, but still broadly rounded below. L. 3·3 mm.

Palp

Outer plate

Davis Strait (Sukkertoppen, depth 113 m, and Christianshaab, depth 28—56 m [Greenland]).

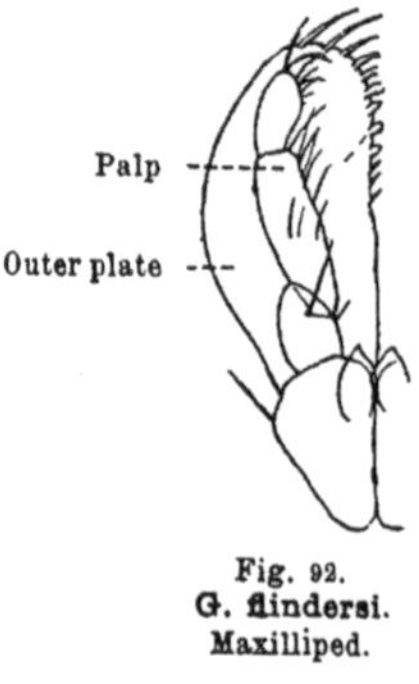

G. flindersi (Stebb.) 1888 *Dexamine f.*, T. Stebbing in: Rep. Voy. Challenger, v. 29 p. 946 t. 187 c | 1893, A. Della Valle in: F. Fl. Neapel, v. 20 p. 578.

Fig. 92.
G. flindersi.
Maxilliped.

Perhaps to be referred to an independent genus. Maxillipeds (Fig. 92), inner plates rudimentary, outer very large, completely covering the 4-jointed palp, seemingly fused in the basal half, then armed each with 3 or 4 interlocking spine-teeth and finally serrate and armed with 5 long curved spines; finger of palp very small. L. very small.

Flinders Passage [East-Australia]. Depth 15 m.

32. Fam. Talitridae

1813/14 *Orchestidae*, Leach in: Edinb. Enc., *v.* 7 p. 432 | 1853 *O.*, J. D. Dana in: U. S. expl. Exp., *v.* 13ıı p. 826, 846 | 1857 Subfam. *Talitrini*, A. Costa in: Mem. Acc. Napoli, *v.* 1 p. 173 | 1862 *Orchestidae*, Bate, Cat. Amphip. Brit. Mus., p. 4 | 1871 *O.*, A. Boeck in: Forb. Selsk. Christian., 1870 p. 91 | 1888 *O.*, T. Stebbing in: Rep. Voy. Challenger, *v.* 29 p. 602 etc. | 1890 *Orchestiidae*, G. O. Sars, Crust. Norway, *v.* 1 p. 21 | 1893 *Orchestidi*, A. Della Valle in: F. Fl. Neapel, *v.* 20 p. 489.

Head without pronounced rostrum, mouth-parts strongly projecting below. Side-plates 2—4 rather large, 5th bilobed (side-plates 1—4 often with a small backward projection near the top of hind margin). Antenna 1 usually much shorter than antenna 2, without accessory flagellum. Antenna 2 having the basal joint coalesced with the head. Upper lip large, distally rounded, median line usually prominent. Lower lip without inner lobes. Mandible without palp, otherwise normal. Maxilla 1, inner plate slender, tipped with 2 plumose setae, outer plate with 9 apical spines, palp 1-jointed, small, from rudimentary to a length reaching base of spines of outer plate. Maxilla 2, both plates with apical fringe of slender spines, which on the inner plate is bounded by a plumose seta on the inner margin. Gnathopod 2 in the ♂ either feebly chelate or (more generally) strongly subchelate (Fig. 96, 97, 99 p. 565, 580, 583), in the ♀ either feebly chelate or subchelate. Uropod 3 usually with a single ramus. Telson (Fig. 98, 100 p. 580, 583) usually short.

Represented on almost all coasts, chiefly between tide-marks or not far above or below them, but sometimes reaching considerable heights inland; also found at sea, perhaps generally in connection with floating objects; of the fresh water forms some found down to considerable depths in lakes, others in streamlets up to very great heights.

13 genera, 101 accepted species and 45 doubtful.

Synopsis of genera:

1 { Pleopod 3 degraded . . . . 2. Gen. **Talitroides** . . p. 527
{ Pleopod 3 not degraded — **2.**

2 { Maxillipeds, 4th joint of palp wanting or quite
{   rudimentary — **3.**
{ Maxillipeds, 4th joint of palp distinct — **6.**

3 { Gnathopod 1 simple in the ♂ — **4.**
{ Gnathopod 1 subchelate in the ♂ — **5.**

4 { Gnathopod 2 feebly chelate in the ♂ . . . . 1. Gen. **Talitrus** . . . p. 524
{ Gnathopod 2 strongly subchelate in the ♂ . . 3. Gen. **Orchestoidea** . p. 527

5 { Gnathopod 1 (Fig. 93) subchelate in the ♀ . 4. Gen. **Orchestia** . . p. 530
{ Gnathopod 1 (Fig. 94) simple in the ♀ . . . . 5. Gen. **Talorchestia** . p. 543

6 { Uropod 3, 1-jointed — **7.**
{ Uropod 3 (Fig. 98), not 1-jointed — **8.**

7 { Telson partially cleft . . . . 6. Gen. **Ceina** . . . . p. 554
{ Telson entire . . . . 7. Gen. **Chiltonia** . . p. 555

8 { Uropod 3 with two rami — **9.**
{ Uropod 3 (Fig. 98) with one ramus — **10.**

9 { Telson divided . . . . 8. Gen. **Parhyale** . . . p. 556
{ Telson entire . . . . 9. Gen. **Neobule** . . . p. 556

10 { Maxillipeds, 4th joint of palp not unguiform . 10. Gen. **Parorchestia** . p. 557
{ Maxillipeds, 4th joint of palp unguiform — **11.**

11 {
Gnathopod 2 having the 5th joint masked by
the 4th in the ♂ . . . . 11. Gen. **Hyale** . . . . p. 559
Gnathopod 2 (Fig. 97) having the 5th joint
produced between the 4th joint and 6th in
the ♂ — 12.

12 {
Telson (Fig. 98) undivided . . . . 12. Gen. **Hyalella** . . . p. 574
Telson (Fig. 100) more or less divided . . . . 13. Gen. **Allorchestes** . p. 581

1. Gen. **Talitrus** Latr.

1793 [Subgen.] *Gammarellus* (part.), J. F. W. Herbst, Naturg. Krabben Krebse,
v. 2 p. 106 | 1802 *Talitrus* (part.), (Latreille in:) Bosc, Crust., v. 1 p. 78; v. 2 p. 148 |
1802 & 03 *T.* (part.), Latreille, Hist. Crust. Ins., v. 3 p. 38; v. 6 p. 294 | 1888 *T.*, T. Stebbing
in: Rep. Voy. Challenger, v. 29 p. 1682 | 1890 *T.*, G. O. Sars, Crust. Norway, v. 1 p. 22 |
1891 *T.*, T. Stebbing in: Ann. nat. Hist., ser. 6 v. 8 p. 324 | 1893 *T.*, A. Della Valle in:
F. Fl. Neapel, v. 20 p. 492 | 1835 *Thalitrus*, Guérin(-Méneville) in: Exp. Morée, Atlas
p. 3 | 1851 Subgen. *Talitrorchestia* (Sp. un.: *Talitrus cloquetii*), F. Brandt in: Bull.
phys.-math. Ac. St.-Pétersb., v. 9 p. 137.

Peraeon dorsally broad, pleon compressed. Side-plate 1 narrow, 5th
broad and deep. Antenna 1 much shorter than peduncle of antenna 2.
Antenna 2, basal joint or joints soldered to head, with no gland-cone,
ultimate joint of peduncle the longest joint. Epistome forming an obtuse
angle with upper lip. Upper lip distally rounded. Lower lip with tuft of
setules at inner corner of principal lobes. Maxilla 1, palp minute, 2-jointed.
Maxillipeds, palp rather short and broad, 4th joint wanting. Gnathopod 1
stronger than gnathopod 2, simple, 5th joint strong but linear. Gnathopod 2
feeble and similar in both sexes, 5th joint expanded proximally, 6th produced
beyond a minute chela-forming finger. Peraeopod 2 with short, notched
finger. Peraeopods 4 and 5 with expansion only of 2d joint. Branchial
vesicles twisted or bent. Marsupial plates small, lanceolate. Telson simple.

4 species accepted, 2 doubtful.

Synopsis of accepted species:

1 {
Peraeopods 3 and 4, 2d joint narrowing distally . . 1. **T. sylvaticus** . p. 524
Peraeopods 3 and 4, 2d joint regularly expanded — 2.

2 {
Antenna 1 reaching much beyond the penultimate
joint of peduncle of antenna 2 . . . . 2. **T. alluaudi** . . p. 525
Antenna 1 not reaching much beyond penultimate
joint of peduncle of antenna 2 — 3.

3 {
Antenna 2, peduncle stout . . . . 3. **T. saltator** . . p. 525
Antenna 2, peduncle slender . . . . 4. **T. gulliveri** . . p. 526

1. **T. sylvaticus** HasW. 1879 *T. s.*, Haswell in: P. Linn. Soc. N. S. Wales, v. 4
p. 246 t. 7 f. 1 | 1885 *T. s.* (*T. affinis*, laps. corr.: *T. assimilis*); Haswell in: P. Linn.
Soc. N. S. Wales, v. 10 p. 95 t. 10 f. 1 | 1893 *T. s.*, G. M. Thomson in: P. R. Soc. Tasmania,
1892 p. 15 t. 4 f. 1—10 | 1893 *T. s.*, A. Della Valle in: F. Fl. Neapel, v. 20 p. 512 | 1880
T. assimilis, Haswell in: P. Linn. Soc. N. S. Wales, v. 5 p. 97 t. 5 f. 1 | 1898 *Orchestia
sylvicola* (err., non J. D. Dana 1852!), G. M. Thomson in: Tr. N. Zealand Inst., v. 31 p. 203.

Side-plate 5 much the broadest. Eyes small, round. Antenna 1, 1st joint
compressed, 2d the longest, 3d very short. flagellum rather shorter than peduncle.
Antenna 2 rather long, ultimate joint of peduncle nearly twice as long as
penultimate, flagellum longer than peduncle. Maxillipeds, palp with rudiment
of a blunt terminal joint (Thomson). Gnathopod 1, 6th joint slender, tapering,
finger (in figure) very small. Gnathopod 2, 4th joint obtusely prominent,
more so in ♂ than in ♀, 5th and 6th joints considerably longer in ♀ than

in ♂, apex of 6th rather narrowly and shortly produced beyond the minute finger in both sexes. Peraeopods 1 and 2 subequal (Haswell; but in figure peraeopod 1 is considerably the larger). Peraeopod 3 much shorter than peraeopods 4 and 5; branchial vesicles much twisted. Peraeopods 3 and 4, 2[d] joint piriform, with scarcely any free corner. Peraeopod 5, 2[d] joint nearly as broad as long. Pleopods 1 and 2 feebly developed, 3[d] wanting (?). Uropod 3 very short, conical ramus ending in a single spine. Telson spinulose, very slightly emarginate at apex. Colour usually dark slate, occasionally dull yellow. L. 11 mm.

New South Wales (on moist ground in wood and scrubs; at Rootyhill, over 50 km from coast); Tasmania (on Mount Kosciusko and to a height of 760 m on Mount Wellington).

2. T. alluaudi Chevreux 1896 *T. a.*, Chevreux in: Feuille Natural., v. 26 p. 7 f. 1—4.

Body little compressed. Side-plate 5 large, as deep as the preceding. Eyes rather large, round. Antenna 1 very long, reaching beyond middle of ultimate joint of peduncle of antenna 2, flagellum as long as peduncle, 6-jointed. Antenna 2, antepenultimate joint of peduncle rather long, penultimate joint ²/₃ as long as ultimate, flagellum shorter than peduncle, 9—11-jointed. Gnathopod 1 robust, 5th joint distally widened, 6th spinose, stout and short. Gnathopod 2, 5th joint scarcely longer than 6th, which has the apex broadly rounded and considerably produced beyond the minute finger. Peraeopods 3 and 4, 2[d] joint narrowly oval. Peraeopods 4 and 5 short, robust. Peraeopod 5, 2[d] joint almost circular. Uropods 1 and 2 normal. Uropod 3 extremely small, scarcely half as long as telson, peduncle with long apical spine, ramus minute, conical, about half as long as peduncle. Telson large, quadrate, as broad as long, not emarginate, with 2 apical spines and 4 on each side. L. ♀ 5—6, ♂ 7 mm.

Seychelles (in rotten trunks of cocoanut-trees and in the humus of forests); Paris (hot-houses of the Jardin des Plantes).

3. T. saltator (Mont.) 1766 *Oniscus locusta* (err., non *Cancer l.* Linné 1758!), Pallas, Misc. zool., p. 191 t. 14 f. 15 | ? 1777 *Cancer l.*, Pennant, Brit. Zool., ed. 4 v. 4 p. 21 | 1793 *C. (Gammarellus) l.*, J. F. W. Herbst, Naturg. Krabben Krebse, v. 2 p. 127 t. 36 f. 1 | 1802 *Gammarus l.*, *Talitrus* (part.), Latreille, Hist. Crust. Ins., v. 3 p. 39 | 1888 *T. l.*, T. Stebbing in: Rep. Voy. Challenger, v. 29 p. 1722 | 1890 *T. l.*, G. O. Sars, Crust. Norway, v. 1 p. 23 t. 9 | ˙1893 *T. l.*, A. Della Valle in: F. Fl. Neapel, v. 20 p. 492, 947 t. 57 f. 52, 53 | 1808 *Cancer (Gammarus) saltator*, Montagu in: Tr. Linn. Soc. London, v. 9 p. 94 t. 4 f. 3 | 1830 *Talitrus s.*, H. Milne Edwards in: Ann. Sci. nat., v. 20 p. 364 | 1813/14 *T. littoralis*, Leach in: Edinb. Enc., v. 7 p. 402 | 1832 & 35 *T. platycheles*, *T. platichelis*, Guérin (-Méneville) in: Exp. Morée, v. 31 sect. 2 p. 44; Atlas p. 3 Zool. t. 27 ̃ ˙——˙ | 1893 *T. locusta forma mediterranea*, Chevreux in: Bull. Soc. zool. France, v. 18 p. 124.

Body robust, in ♂ rather elongate. Head, rostral point minute. Side-plate 1 triangular, 2[d]—4th quadrate, 5th nearly as deep as 4th. Pleon segment 3, postero-lateral corners quadrate. Eyes with dark pigment, irregularly rounded, often wider apart than their diameter. Antenna 1 about reaching end of penultimate joint of peduncle of antenna 2, 2[d] joint the longest, flagellum much shorter than peduncle, 7- or 8-jointed. Antenna 2 much longer in ♂ than in ♀, ultimate joint of peduncle much longer than penultimate, narrowest at base, flagellum subequal to peduncle, with 25 joints in ♀, 35 or more in ♂, short and stout, with short, stout, apical spines. Gnathopod 1 spinose, 2[d] joint channelled in front, 3[d] notched at top of front margin, 5th long, stout, free part with margins parallel, 6th much shorter, slightly tapering, finger short. Gnathopod 2, 2^d joint membranaceous, narrowly oval, 3^d longer than 4th, 5th widest in proximal half of free part, then from an angle of the hind margin narrowing to apex, 6th shorter than 5th, front margin straight, apex narrowly rounded, produced beyond the minute, longitudinally placed finger. Peraeopod 1 considerably longer than peraeopods 2 or 3. Peraeopod 2, finger short, conspicuously notched. Peraeopod 3 short, stout. Peraeopods 4 and 5 rather elongate. Peraeopods 3—5, 2^d joint narrowest above, front margin very convex, hind expansion strongly developed below, the joint much smaller than the side-plate in peraeopod 3, but much larger in peraeopods 4 and 5. Pleopods 1—3 slender, rami about as long as peduncle. Uropods 1 and 2, both rami spinose on margin and apex. Uropod 3, ramus subequal to peduncle, with spinules on margin and slender spine at apex. Telson broader than long, apically rounded, with a few spinules. Colour like sand or a deal board, or different shades of horn in section. L. ♀ 15, ♂ 16—25 mm.

North-Atlantic and adjoining seas (European coast, at least from the south of Norway to Naples; Azores).

4. **T. gulliveri** Miers 1876 *T. g.*, Miers in: Ann. nat. Hist., ser. 4 *v.* 17 p. 406 | 1879 *T. g.*, Miers in: Phil. Tr., *v.* 168 p. 495 | 1893 *T.? g.*, A. Della Valle in: F. Fl. Neapel, *v.* 20 p. 511.

Body slender, smooth. Head small, anterior margin straight. Pleon segments 2 and 3, postero-lateral angle somewhat prominent and acute, its hind margin straight. Eyes round, black. Antenna 1 about reaching end of penultimate joint of peduncle of antenna 2, 1st joint shorter than 2^d or 3^d, flagellum as long as peduncle, with 4 joints, last minute. Antenna 2, not as long as body, much shorter and more slender than in T. saltator (p. 525), penultimate joint of peduncle more than half as long as ultimate, flagellum rather longer than peduncle, 18-jointed. Gnathopods 1 and 2 small and weak. Gnathopod 1, 5th, 6th and 7th joints short, subequal, not dilated. Gnathopod 2, palm slightly dilated and the finger quite rudimentary as in T. saltator. Peraeopods 4 and 5 considerably longer than the preceding peraeopods, 2^d joint moderately developed. Uropods 1 and 2, rami subequal. Uropod 3 quite rudimentary. L. about 10 mm.

None of the specimens had largely developed subcheliform gnathopod 2, but, if all were ♀, that would leave the genus indefinite.

Island of Rodriguez. Under stones in damp places.

T. cloquetii (Aud.) 1826 *Orchestia c.*, Audouin in: Descr. Égypte, *v.* 1 iv p. 93; Crust. t. 11 f. 9 | 1862 *O. c.*, *O. cloquettii*, Bate, Cat. Amphip. Brit. Mus., p. 22 t. 4 f. 1 (a very inaccurate copy) | 1893 *O.? cloquetii*, A. Della Valle in: F. Fl. Neapel, *v.* 20 p. 509 | 1830 *Talitrus c.*, H. Milne Edwards in: Ann. Sci. nat., *v.* 20 p. 364 | 1835 *Thalitrus c.*, Guérin (-Méneville) in: Exp. Morée, Atlas p. 3 Zool. t. 27 f. 4e | 1851 *Talitrus c.*, *Talitrorchestia*, F. Brandt in: Bull. phys.-math. Ac. St.-Pétersb., *v.* 9 p. 135, 137, 138.

Savigny's figure merely represents a misapprehension of T. saltator (p. 525) in regard to the gnathopod.

Red Sea? (Egypt).

T. fissispinosus (Kossm.) 1880 *Orchestia fissispinosa*, Kossmann, Reise Roth. Meer., *v.* 21 Malacost. p. 129 t. 13 f. 1—5 | 1893 *O.? f.*, A. Della Valle in: F. Fl. Neapel, *v.* 20 p. 509 | 1888 *Talitrus fissispinosus*, T. Stebbing in: Rep. Voy. Challenger, *v.* 29 p. 515.

Probably young ♀. L. 5 mm.

Red Sea.

2. Gen. **Talitroides** Bonnier

1898 *Talitroïdes*, (J. Bonnier in:) Willem in: Ann. Soc. ent. Belgique, *v.* 42 p. 208.

Near to Talitrus (p. 524). Palp of maxilla 1 much reduced. Gnathopod 1 simple. Gnathopod 2 alike in both sexes. Pleopods 1 and 2 with inner ramus rudimentary, reduced to a simple tubercle. Pleopod 3 consisting of a small process representing the peduncle without rami. Eggs very few and comparatively large, contained in a marsupium formed by very short lamellae attached to the peraeopods 1—3.

1 species.

1. **T. bonnieri** Stebb.[1]

With the characters of the genus.

Origin unknown (found in a conservatory at Ghent).

3. Gen. **Orchestoidea** Nic.

1849 *Orchestoidea* (Sp. un.: *O. tuberculata*), H. Nicolet in: Gay, Hist. Chile, *v.* 3 p. 229 | 1888 *O.*, T. Stebbing in: Rep. Voy. Challenger, *v.* 29 p. 231 | 1891 *O.*, T. Stebbing in: Ann. nat. Hist., ser. 6 *v.* 8 p. 328 | 1898 *O.*, Calman in: Ann. N. York Ac., *v.* 11 p. 265 | 1850 *Talitronus*, J. D. Dana in: Amer. J. Sci.. ser. 2 *v.* 9 p. 295 | 1852 *T.* (Sp. un.: *T. insculptus*), J. D. Dana in: P. Amer. Ac., *v.* 2 p. 202 | 1851 *Megalorchestes*, *Megalorchestia*. F. Brandt in: Bull. phys.-math. Ac. St.-Pétersb., *v.* 9 p. 142, 310.

Like Talitrus (p. 524), except that gnathopod 2 in ♂ is powerfully subchelate instead of feebly chelate.

5 species.

Synopsis of species:

1 { Body with teeth or tubercles — **2.**
 Body smooth — **3.**

2 { Body tuberculate . . . . 1. **O. tuberculata** . . p. 527
 Body with dorsal teeth on pleon . . . . 2. **O. fischerii** . . . p. 528

3 { Eyes well separated — **4.**
 Eyes approximate . . . . 5. **O. brasiliensis** . . p. 529

4 { Gnathopod 2 strongly subchelate; only ♂ known 3. **O. californiana** . . p. 528
 Gnathopod 2 feebly chelate; only ♀ known . . . 4. **O. pugettensis** . . p. 529

1. **O. tuberculata** Nic. 1849 *Talitrus chilensis* (♀) + *O. t.*, H. Nicolet in: Gay, Hist. Chile, *v.* 3 p. 229; p. 231 Crust. t. 2 f. 4 | 1852 *Talitrus ornatus* (♀) + *Talitronus insculptus*, J. D. Dana in: P. Amer. Ac., *v.* 2 p. 201; p. 202 | 1853 & 55 *Orchestia (Talitrus) insculpta* + *O. tuberculata*, J. D. Dana in: U. S. expl. Exp., *v.* 13 п p. 855 t. 57 f. 1a—m (♂), n—r (♀); p. 1595 | 1893 *O. t.*, A. Della Valle in: F. Fl. Neapel, *v.* 20 p. 496 t. 57 f. 55.

In ♂ peraeon segments, side-plates, 2d joint of peraeopods 3—5 and sometimes pleon segments 1—3 sculptured or marked with transverse ridges and series of tubercles; in ♀ the sculpturing is almost evanescent, chiefly observable as a few faint sulcations or rugosities on side-plates and 2d joint of peraeopods 3—5. Side-plate 1 narrow, curved forward, 2d—5th broad, 5th decidedly less deep than 4th. Eyes subrotund, large. Antenna 1 very

---

[1] Spec. nov. After Jules Bonnier. — Willem published Bonnier's account and name of the genus, without describing or naming the species, for which the genus is created.

small, not reaching end of penultimate joint of antenna 2; joints of peduncle successively shorter, flagellum 5—7-jointed. Antenna 2 in ♂ longer than half the body, powerful, penultimate joint of peduncle long and stout, ultimate still longer, variable; flagellum shorter than peduncle, many-jointed. Antenna 2 in ♀ much smaller than in ♂, flagellum subequal to peduncle, 12—19-jointed. Gnathopod 1 in ♂, 5th joint much longer than 6th, narrowing a little to distal end, 6th linear, subcylindrical, without palm, finger small. Gnathopod 2 in ♂, 5th joint very small, masked behind by 4th, 6th massive, palm spinulose, rather oblique, nearly straight, with an excavation near the middle, corresponding with a slight projection in the margin of the long curved finger. Gnathopod 1 in ♀, 5th joint narrowly oval, 6th narrowing to the palm-less apex, finger small. Gnathopod 2 in ♀ feeble, 6th joint elliptical, the chela-forming finger longitudinal, nearly reaching apex of 6th joint. Peraeopod 1 decidedly longer than peraeopod 2, the latter with notched finger. Peraeopod 3 short, peraeopod 5 a little shorter than peraeopod 4, moderately stout. Uropods 1—3 very spinose; uropod 1, both rami armed with spines. Colour yellowish white, peduncle of antenna 2 orange. L: 18—20 mm.

South-Pacific (near Valparaiso). Burrowing in the sand; ♀ under sea-weed thrown up by the tides.

2. O. fischerii (M.-E.) ?1828 *Orchestia f.*, H. Milne Edwards in: Mém. Soc. Hist. nat. Paris, *v.* 5 t. 25 f. 14 | 1830 *O. f.*, H. Milne Edwards in: Ann. Sci. nat., *v.* 20 p. 361, 362 | 1840 *O. f.*, H. Milne Edwards, Hist. nat. Crust., *v.* 3 p. 19 t. 29 f. 4 | 1893 *O. f.*, A. Della Valle in: F. Fl. Neapel, *v.* 20 p. 497 t. 57 f. 56 | 1862 *Orchestoidea f.*, Bate, Cat. Amphip. Brit. Mus., p. 11 t. 2 f. 1.

Body stout, pleon segments 2 and 3 each with 3 large vertical dorsal teeth, placed near together, one on each side of the median line, the 3d being by implication central, segments 4 and 5 scabrous above. Eyes large, round. Antenna 1 very short, stout, tapering. Antenna 2 nearly as long as peraeon. Maxillipeds, 1st and 2d joints of palp expanded. Gnathopod 1 in ♂ small, not cheliferous, and ending in a small very narrow hand. Gnathopod 2 in ♂, 6th joint powerful, with long oblique palm, having near the centre a prominent process, finger very large. Peraeopods 1—3 small. Peraeopod 4 very large, 2d joint greatly developed, raised behind above its point of attachment and overlapping much of peraeopod 5 and the pleon, the other joints long but of the ordinary form. Peraeopod 5 much shorter, 2d joint as broad as long. (According to Bate the eyes are small, ultimate joint of peduncle of antenna 2 thrice as long as the short penultimate, pleon segments 1—3 each with 2 dorsal teeth, but he does not say that he has examined a specimen.) L. about 13 mm.

Morea, bay of Calamati (Petalidi [Greece]).

3. O. californiana (F. Brandt) 1851 *Megalorchestes californianus* (nom. nud.), *Megalorchestia californiana*, F. Brandt in: Bull. phys.-math. Ac. St.-Pétersb., *v.* 9 p. 142, 311 t. | 1898 *Orchestoidea californiana*, Calman in: Ann. N. York Ac., *v.* 11 p. 265 t. 31 f. 1 | 1853 & 55 *Orchestia (Talitrus?) scabripes*, J. D. Dana in: U. S. expl. Exp., *v.* 13 II ˙p. 860; t. 57 f. 4 | 1857 *Megalorchestia s. + M. californiana*, Stimpson in: Boston J. nat. Hist., *v.* 6 p. 516 | 1862 *Orchestoidea s. + O. c.*, Bate, Cat. Amphip. Brit. Mus., p. 11 t. 1 f. 3; p. 14 | 1877 *Megalorchestia franciscana*, Lockington in: P. Calif. Ac., *v.* 7 p. 47 | 1893 *Orchestia gammarellus* (part.), A. Della Valle in: F. Fl. Neapel, *v.* 20 p. 500, 508.

Not improbably the ♂ of O. pugettensis. Body smooth. Side-plates rather large, 5th hardly shorter than 4th (Dana; in Brandt's figure side-plate 5 is extremely wide (from front to back), whereas the basal joint of peraeopod 3, which is said to be extremely dilated, is in the figure not half as wide as the side-plate). Eyes large, a little reniform (Dana); of moderate size, suboval, slightly narrowed above, wide apart, black (Brandt). Antenna 1 about reaching middle of penultimate joint of peduncle of antenna 2, flagellum shorter than peduncle, 7- or 8-jointed. Antenna 2 much longer than body; ultimate joint of peduncle sometimes more than twice as long as penultimate, flagellum longer than peduncle, 35—37 (Dana: 20—22)-jointed. Gnathopod 1, joints scabrous (Lockington), 5th with narrow triangular sub-apical process of hind margin, 6th much shorter than 5th, narrow, without palm, finger short. Gnathopod 2, 5th joint small, masked by 4th; 6th large, suboval; palm scabrous, moderately oblique (Lockington: almost transverse), the centre of it forming a low oblong tooth or process; finger long and much curved, its apex touching the usual groove in the hand. Peraeopods 4 and 5 subequal, longer than the rest. Peraeopods 3—5, 2^d joint well expanded; surface of peraeopods scattered over with minute spinules (not mentioned by Brandt). Uropod 1, outer ramus without marginal spines (Dana), equally spinulose with the inner one (Stimpson). Telson simple, cordiform, dorsally longitudinally impressed. L. 22—29 mm.

North-Pacific (North California). Among detritus at high-tide level.

4. O. pugettensis (Dana) 1853 & 55 *Orchestia (Talitrus) p.*, J. D. Dana in: U. S. expl. Exp., v. 13 n p. 859; t. 57 f. 3 a—d | 1857 *O. p.*, Stimpson in: Boston J. nat. Hist., v. 6 p. 516 | 1862 *Orchestoidea p.*, Bate. Cat. Amphip. Brit. Mus., p. 8 t. 2 f. 3 | 1893 *Orchestia gammarellus* (part.), A. Della Valle in: F. Fl. Neapel, v. 20 p. 499.

Probably the ♀ of O. californiana (p. 528). Back broad. Head almost completely truncate in front. Side-plates 2—4 square, much larger than 1st, 5th broad, front lobe about as deep as preceding side-plates, 6th strongly produced backward, lower margin ridged. Pleon segment 3, postero-lateral corners rather sharply quadrate. Eyes rudely subrotund, as far apart as their longest diameter. Antenna 1 very short, flagellum shorter than peduncle. Antenna 2, ultimate joint of peduncle nearly twice as long as penultimate, flagellum rather longer than peduncle, of 30 joints, broader than long, with short setules. Gnathopod 1 spinose, 2^d joint slightly curved, channelled in front, 5th long, widening to a subapical pellucid process, 6th much shorter and narrower than 5th, but moderately long, slightly curved, without palm but with a slight longitudinal pellucid swelling of hind margin, finger straight. Gnathopod 2, 2^d joint very large, a long and broad oval, the membranous expansion fringed with spinules, 4th joint having at hinder apex a little triangular (seemingly 2-jointed, Dana) process, 5th a little longer than 6th, with membranous bulging of free part, finger small, longitudinal, not reaching produced apex of 6th joint; branchial vesicles small; marsupial plates short, narrow, with a few apical setae. Peraeopods 1 and 2, marsupial plates long, narrow, fringed with long setae. Peraeopod 2, finger abruptly narrowed midway. Peraeopods 3—5, 2^d joint broadly expanded. Peraeopods 4 and 5 long, not specially expanded below 2^d joint. Uropod 1 long, peduncle and both rami very spinose (Dana: outer ramus naked). Uropod 3, ramus narrow, longer than peduncle. Telson about as long as broad. L. 16 mm.

North-Pacific (North California).

5. O. brasiliensis (Dana) 1853 & 55 *Orchestia (Talitrus) b.*, J. D. Dana in: U. S. expl. Exp., v. 13 n p. 857; t. 57 f. 2 a—h | 1862 *Orchestoidea b.*, Bate. Cat. Amphip. Brit. Mus., p. 13 t. 2 f. 4 | 1893 *Orchestia gammarellus* (part.), A. Della Valle in: F. Fl. Neapel, v. 20 p. 499.

Body smooth. Side-plate 5 much shallower than 4th. Pleon segment 3, postero-lateral corners quadrate. Eyes rather large, approximate in front. Antenna 1 very short, flagellum 3-jointed. Antenna 2 in ♀ not half as long as body, flagellum scarcely as long as peduncle, about 16-jointed. Gnathopod 1 in ♂ as in ♀. Gnathopod 2 in ♂, 6th joint large, subovate, palm slightly convex, spinulose, very oblique, defined by a well marked angle but not toothed, finger elongate. Gnathopod 1 in ♀ longer than gnathopod 2, finger thick, curved, not much shorter than the palmless 6th joint, which in figure appears longer than 5th. Gnathopod 2 in ♀, 5th joint with very bulging hind margin, 6th with straight front and bulging hind margin, distally narrowly produced much beyond the minute longitudinal finger. Peraeopods 4 and 5 much longer than the preceding peraeopods, 2[d] joint broadly expanded, the rest slender, spinules numerous. Uropod 1 rather long. Uropod 3 short, ramus as long as peduncle. L. 12 mm.

Rio Janeiro harbour.

4. Gen. **Orchestia** Leach

1793 [Subgen.] *Gammarellus* (part.), J. F. W. Herbst, Naturg. Krabben Krebse, v. 2 p. 106 | 1802 *Talitrus* (part.), (Latreille in:) Bosc, Crust., v. 1 p. 78 | 1802 *T.* (part.), Latreille, Hist. Crust. Ins., v. 3 p. 38 | 1813/14 *Orchestia* (Sp. un.: *O. littorea*), Leach in: Edinb. Enc., v. 7 p. 402 | 1888 *O.*, T. Stebbing in: Rep. Voy. Challenger, v. 29 p. 602, 1678 | 1890 *O.*, G. O. Sars, Crust. Norway, v. 1 p. 24 | 1893 *O.*, A. Della Valle in: F. Fl. Neapel, v. 20 p. 494 | 1847 *Scamballa* (part.), (Leach in MS.) A. White, Crust. Brit. Mus., p. 86 | 1876 *Orchestes* (non Illiger 1798, Coleoptera!), R. T. Maitland in: Tijdschr. Nederl. dierk. Ver., v. 2 p. 11.

Fig. 93.
O. selkirki, ♀.
Gnathopod 1.

Like Talitrus (p. 524) except that gnathopod 1 (Fig. 93) in ♂ and in ♀ is less strongly developed and is subchelate instead of simple, while gnathopod 2 in ♂ is powerfully subchelate instead of feebly chelate. Antenna 2 in ♂ perhaps never attains so strong a development as is met with in Talitrus and Talorchestia (p. 543); the maxillipeds seem sometimes to have an obscure rudiment of the 4th joint of the palp, and the front lobe of side-plate 5 may be as deep as side-plate 4. The notching of the finger in peraeopod 2 is seldom so sharp and decided as in Talorchestia.

Generally distributed in temperate and tropical seas between tidemarks among sea-weed and stones, as distinguished from the sand-dwelling Talitrus and Talorchestia, also found inland at various heights up to 1375 m above sea-level.

23 accepted and 6 doubtful species.

Synopsis of accepted species:[*]

$\left\{\begin{array}{l}\text{Peraeopod 5 in adult } \mathring{\jmath}, \text{ 4}^{th} \text{ and 5}^{th} \text{ joints strongly}\\ \quad \text{incrassated — 2.}\\ \text{Peraeopod 5 in adult } \mathring{\jmath}, \text{4}^{th} \text{ and 5}^{th} \text{joints not strongly}\\ \quad \text{incrassated — 6.}\end{array}\right.$

1

$\left\{\begin{array}{l}\text{Gnathopod 2 in } \mathring{\jmath}, \text{ palm extremely oblique . . .} \quad \text{1. } \textbf{O. mediterranea}. \text{ p. 531}\\ \text{Gnathopod 2 in } \mathring{\jmath}, \text{ palm not extremely oblique — 3.}\end{array}\right.$

2

---

[*] Since this table is based upon structures of ♂, the species 22 and 23, of which ♂ specimens are unknown, are omitted; they differ by the palm of gnathopod 1: feebly developed in 22, O. parvispinosa p. 541, well developed in 23, O. montana p. 542.

3 { Gnathopod 2 in ♂, finger Without a strong tooth — **4.**
  { Gnathopod 2 in ♂, finger With a strong tooth — **5.**

4 { Peraeopod 5 in adult ♂, 5th joint as broad as long,
  {   6th straight . . . .   2. **O. gammarellus** . p. 532
  { Peraeopod 5 in adult ♂, 5th joint longer than broad,
  {   6th curVed . . . .   3. **O. chevreuxi** . . p. 533

5 { Gnathopod 2 in ♂, tooth of finger central . . .   4. **O. montagui** . . p. 533
  { Gnathopod 2 in ♂, tooth of finger close to hinge   5. **O. tucurauna** . . p. 534

6 { Gnathopod 1 in ♂, 4th joint with apical process — **7.**
  { Gnathopod 1 in ♂, 4th joint Without apical
  {   process — **8.**

7 { Gnathopod 2 in ♂, palm With two excaVations .   6. **O. bottae** . . . . p. 534
  { Gnathopod 2 in ♂, palm smoothly conVex . . .   7. **O. traskiana** . . p. 534

8 { Body segments acquiring transVerse dorsal ridges — **9.**
  { Body segments not acquiring dorsal ridges — **10.**

9 { Gnathopod 2 in ♂, palm slightly oblique . . . .   8. **O. serrulata** . . p. 535
  { Gnathopod 2 in ♂, palm Very oblique . . . . .   9. **O. guernei** . . . p. 536

10 { Peraeopod 5, 2d joint scarcely expanded . . . .   10. **O. marmorata** . p. 536
   { Peraeopod 5, 2d joint regularly expanded — **11.**

11 { Gnathopod 2 in ♂, palm deeply concaVe . . . .   11. **O. capensis** . . . p. 537
   { Gnathopod 2 in ♂, palm not deeply concaVe — **12.**

12 { Gnathopod 2 in ♂, palm with strong tooth near
   {   finger-hinge — **13.**
   { Gnathopod 2 in ♂, palm Without tooth near finger-
   {   hinge — **14.**

13 { Gnathopod 2 in ♂, palm With one strong tooth .   12. **O. chiliensis** . . p. 537
   { Gnathopod 2 in ♂, palm with two teeth . . . .   13. **O. dentata** . . . p. 537

14 { Gnathopod 2 in ♂, 6th joint narroWing to the palm   14. **O. selkirki** . . . p. 538
   { Gnathopod 2 in ♂, 6th joint not narroWing to the
   {   palm — **15.**

15 { Gnathopod 1 in ♂, 5th joint much longer than 6th — **16.**
   { Gnathopod 1 in ♂, 5th joint not much longer
   {   than 6th — **17.**

16 { Side-pate 5 deep . . . .   15. **O. pickeringii** . . p. 538
   { Side-plate 5 not deep . . . .   16. **O. nitida** . . . . p. 539

17 { Gnathopod 2 in ♂, palm Very oblique — **18.**
   { Gnathopod 2 in ♂, palm not Very oblique — **19.**

18 { Gnathopod 1 in ♂, finger coVering Whole apex of
   {   6th joint . . . .   17. **O. humicola** . . . p. 539
   { Gnathopod 1 in ♂, finger not coVering Whole apex
   {   of 6th joint . . . .   18. **O. floresiana** . . p. 539

19 { Antenna 2 slender, flagellum cylindric . . . . .   19. **O. grillus** . . . . p. 540
   { Antenna 2 stout, flagellum flattened — **20.**

20 { Gnathopod 1 in ♂, 6th joint With prominent process   20. **O. platensis** . . p. 540
   { Gnathopod 1 in ♂, 6th joint Without prominent
   {   process . . . .   21. **O. sulensoni** . . p. 541

1. **O. mediterranea** A. Costa 1857 *O. m.*, A. Costa in: Mem. Acc. Napoli, *v.*1 p. 181 | 1866 *O. m.*, Cam. Heller in: Denk. Ak. Wien, *v.*26II p. 4 t. 1 f. 7 | 1881 *O. m.*, Nebeski in: Arb. Inst. Wien, *v.*3 p. 143 | 1895 *O. m.*, Chevreux in: Rev. biol. Nord France, *v.*7 p. 158 | 1899 *O. m.*, T. Scott in: Rep. Fish. Board Scotl., *v.*17 p. 264 t. 13 f. 9—11 | 1857 *O. laevis*, Bate in: Ann. nat. Hist., ser. 2 *v.*19 p. 136 | 1862 *O. trigonocheirus*, Bate, Cat. Amphip. Brit. Mus., p. 26 t. 4 f. 6 | 1893 *O. chilensis* (part.), *O. spinimana*, A. Della Valle in: F. Fl. Neapel, *v.*20 p. 498 t. 2 f. 8; t. 15 f. 31—38; p. 129, 248, 941.

Body rather compressed. Side-plate 1 much smaller than 2[d], 5th nearly as deep as 4th. Pleon segment 3, postero-lateral corners quadrate. Eyes subrotund or transversely oval, dark, usually nearer together than their width. Antenna 1 reaching end of penultimate joint of peduncle of antenna 2, joints of peduncle subequal in length, flagellum shorter than peduncle, 6- or 7-jointed. Antenna 2 $^1/_3$ as long as body or less, penultimate joint of peduncle $^3/_4$ as long as ultimate, flagellum subequal to peduncle, 20-jointed. Gnathopod 1 in ♂, 5th joint with the subapical pellucid process scabrous, broadly rounded, 6th much shorter, hind margin sinuous, distal pellucid process scabrous, point of finger not reaching its extremity, though extending beyond the genuine palm-margin. Gnathopod 2 in ♂, 4th and 5th joints very small, 6th very large, piriform, narrow at the finger-hinge, near to which the long, extremely oblique, sinuous, spinulose palm at least usually forms a small tooth, finger very long and sinuous, its apex overlapping the palm at commencement of short hind margin, although the palm itself ends with the short channelling adapted to receive the finger-point. Gnathopod 1 in ♀, 5th joint without pellucid process, 6th shorter than 5th, spinose, oblong, with a small pellucid widening at apex of hind margin, which is considerably overlapped by the rather stout, acute finger. Gnathopod 2 in ♀, 2[d] joint expanded, laminar, narrowing distally, 4th joint with a distinct, though not much produced, rounded process at apex of hind margin, 5th produced along much of free part of hind margin, process longitudinally truncate, 6th widening to a broadly rounded process which does not extend far beyond the palm proper and apex of finger. Peraeopod 3, 2[d] joint broadly oval, 4th and 5th rather stout. Peraeopod 4, 2[d], 4th and 5th joints much longer, 2[d] oblong oval. Peraeopod 5, 2[d] joint much wider than in peraeopod 4, hind margin with lower corner broadly rounded, 4th joint in ♂ triangular, distally much widened, with apex of hind margin truncate, 5th joint in ♂ not longer than broad, immediately from narrow point of attachment widened on both sides beyond the broad distal end of 4th joint, then narrowing to apex which is still much wider than the linear 6th joint; in ♀ 4th and 5th joints not at all wider than those of peraeopod 4. Uropod 1, rami considerably shorter than peduncle. Uropod 2, rami subequal to peduncle, outer ramus shorter than inner. Uropod 3, ramus rather shorter than the stout peduncle. Telson triangular, fringed with spinules. Colour greenish brown. L. 15 mm.

Mediterranean; North-Atlantic (France, England, Wales, Ireland). Sometimes considerably beyond high-water mark.

2. **O. gammarellus** (Pall.) 1766 *Oniscus g.*, Pallas, Misc. zool., p. 191 t. 14 f. 25 | 1791 *Gammarus g.*, A. G. Olivier in: Enc. méth., v. 6 p. 187 | 1793 *Cancer g.*, J. F. W. Herbst, Naturg. Krabben Krebse. v. 2 p. 129 t. 36 f. 2, 3 | 1802 & 03 *Oniscus g.*, *Talitrus* (part.), Latreille, Hist. Crust. Ins., v. 3 p. 39; v. 6 p. 300 t. 56 f. 5, 6 | 1872 *Orchestia g.*, A. Boeck, Skand. Arkt. Amphip., v. 1 p. 102 | 1888 *O. g.*, T. Stebbing in: Rep. Voy. Challenger, v. 29 p. 1714 | 1893 *O. g.* (part.), A. Della Valle in: F. Fl. Neapel, v. 20 p. 499 t. 2 f. 11; t. 15 f. 1—12, 39—43 | 1808 *Cancer (Gammarus) littoreus*, Montagu in: Tr. Linn. Soc. London, v. 9 p. 96 t. 4 f. 4 | 1812 *Astacus l.*, Pennant, Brit. Zool., ed. 5 v. 4 p. 32 | 1813/14 *Orchestia littorea*, Leach in: Edinb. Enc., v. 7 p. 402 t. 221 f. 6 | 1890 *O. l.*, G. O. Sars, Crust. Norway, v. 1 p. 24 t. 10 | 1876 *Orchestes l.*, R. T. Maitland in: Tijdschr. Nederl. dierk. Ver., v. 2 p. 11 | 1840 *Orchestia littoralis*, H. Lucas in: Hist. An. artic., Crust. Arach. Myr., p. 225 | 1848 *O. euchore*, Friedr. Müller in: Arch. Naturg., v. 14ı p. 53 t. 4 | ?1852 *O. dispar*, J. D. Dana in: P. Amer. Ac., v. 2 p. 204 | ?1853 & 55 *O. d.*, J. D. Dana in: U. S. expl. Exp., v. 13 ıı p. 878; t. 59 f. 6 a—m | 1868 *O. brevidigitata*, Bate & Westwood, Brit. sess. Crust., v. 2 p. 497 f. | 1886 *O. chilensis* (err., non *O. chiliensis* Milne Edwards 1840!), G. M. Thomson & Chilton in: Tr. N. Zealand Inst., v. 18 p. 145.

Body rather compressed. Side-plate 1 small, 2^d—4th quadrate, the small process of hind margin distinct, 5th broad, in front nearly as deep as 4th. Pleon segment 3, postero-lateral corners quadrate. Eyes irregularly rounded, black, less than the diameter apart. Antenna 1 seldom reaching beyond penultimate joint of peduncle of antenna 2, 1st joint shorter than 2^d or 3^d, flagellum 6—8-jointed. Antenna 2 about $^1/_3$ as long as body, ultimate joint of peduncle sometimes considerably longer than penultimate, flagellum about 20-jointed. Gnathopod 1 in $\male$, 6th joint shorter than 5th, both distally widened, 6th usually resting on produced part of 5th. finger matching transverse palm, but not covering the produced part of 6th joint. Gnathopod 2 in $\male$, 2^d joint channelled in front, not lobed, 5th very small, 6th large, widening to the palm, which is almost transversely arcuate between a depression adjoining finger-hinge and an obtuse defining projection. Gnathopod 1 in $\female$, 5th joint distally widened, 6th narrowly oblong, the short transverse palm overlapped by finger. Gnathopod 2 in $\female$, 2^d joint membranaceous, broader above than below, 5th with free hinder part bulging, 6th produced, but not very greatly. beyond the small, rather obliquely placed finger; marsupial plates rather narrow, long, fringed with long setae. Peraeopod 2 shorter than peraeopod 1, finger with sinuous inner margin. Peraeopods 3—5 successively longer, 2^d joint well expanded. Peraeopod 5 in $\male$, 4th joint greatly widening distally, 5th greatly expanded, widest near base, both these joints having flattened edges before and behind, and in adolescent $\male$ showing numerous gradations to the linear form found in $\female$ and young; 6th joint straight. Uropod 1 rather long, rami shorter than peduncle, both with lateral spines. Uropod 3, ramus much shorter than peduncle. Telson soft, thick, about as broad as long, apex slightly emarginate. Colour $\male$ brownish, $\female$ greenish. L. about 17 mm.

North-Atlantic With adjoining seas (Europe, from the Baltic to the Black Sea; Algiers; Madeira; Azores); ?South-Pacific (Illawarra [New South Wales], sea shores).

3. **O. chevreuxi** Guerne 1887 *O. c.* ($\female$), Guerne in: Naturaliste, *v.*9 (p. 6) | 1888 *O. c.* ($\female$), Guerne in: Bull. Soc. zool. France, *v.*13 p 59 f. 1, 3, 5, 7 | 1888 *O. c.* ($\male$), Chevreux in: Bull. Soc. zool. France, *v.*13 p. 92 f. 1, 3, 5 | 1888 *O. littorea* (part.)?, T. Barrois in: Bull. Soc. zool. France, *v.*13 p. 19 | 1893 *O. gammarellus* (part.), A. Della Valle in: F. Fl. Neapel, *v.*20 p. 499.

Distinguished from O. gammarellus in $\male$ by: peduncle of antenna 2 longer and more robust, flagellum also more elongate (the 18—20 joints being mostly longer than broad), gnathopod 2, 6th joint more broadly oval, palm much more convex, with deep groove adjoining the defining angle, finger shorter and more robust, peraeopod 5 more elongate, with 4th and 5th joints robust but much longer than wide, 6th joint long, curved; in $\female$ by: antennae 1 and 2 longer, gnathopod 2, 2^d joint much less widened, 4th joint carrying 2 spines on apex, 5th and 6th joints more elongate, peraeopod 5 longer, with very long, slightly curved 6th joint, uropods 1 and 2 longer, uropod 3 with much stronger spines on peduncle, telson with stronger spinules, and its narrower apex much more sharply emarginate. Colour violet, almost rose, deeper above. L. about 15 mm.

Azores (Fayal, in the caldiera or crater); Teneriffe (in the forest of Las Mercedes, under detritus on the banks of streamlets. not less than 500 m above sea-level).

4. **O. montagui** Aud. 1826 *O. m.*, Audouin in: Descr. Égypte, *v.*1 IV p. 93 Crust. t. 11 f. 7 | 1866 *O. m.*, Cam. Heller in: Denk. Ak. Wien, *v.*26 II p. 2 t. 1 f. 3, 4 | 1866 *O. m.*, E. Grube in: Arch. Naturg., *v.*32 I p 380 t. 9 f. 1, 2 | 1868 *O. m.*, Czerniavski in:

Syezda Russ. Est., Syezda 1 Zool. p. 118 t. 8 f. 34—39 | 1893 *O. m.*, Chevreux in: Bull. Soc. zool. France, *v.* 18 p. 128 f. 2 | 1895 *O. m.*, Chevreux in: Rev. biol. Nord France, *v.* 7 p. 158 | 1837 *O. littorea* (err., non Leach 1813/14!), H. Rathke in: Mém. prés Ac. St.-Pétersb., *v.* 3 p. 371 t. 5 f. 1—6 | 1862 *O. l.* (part.), Bate, Cat. Amphip. Brit. Mus., p. 27 | 1893 *O. gammarellus* (part.), A. Della Valle in: F. Fl. Neapel, *v.* 20 p. 499.

Extremely variable, by gradual transitions completely united with O. bottae (Czerniavski); still nearer to O. gammarellus (p. 532) by the expanded 4th and 5th joints in peraeopod 5 of adult ♂. Finger of gnathopod 2 in ♂ powerfully arched at the hinge, with a projection in middle of inner margin, which becomes very prominently developed, forming a blunt tooth; in Savigny's figure (in: Descr. Égypte) the 6th joint has the palm defined by an obvious tooth, which other authors do not represent. L. 15 mm (Rathke).

Mediterranean and Black Sea.

5. **O. tucurauna** Fr. Müll. 1864 *O. t.*, *O. tucuratinga*, Fritz Müller, Für DarWin, p. 54 f. 50, 51 | 1893 *O. gammarellus* (part.), A. Della Valle in: F. Fl. Neapel, *v.* 20 p. 499, 510.

♂ (young, but mature). Antenna 2 slender, joints of flagellum distinct; gnathopod 2 with palm uniformly convex, peraeopod 5 slender and like peraeopod 4; subsequently 2, 3 or 4 proximal joints of flagellum of antenna 2 are fused together, gnathopod 2 acquires a deep emargination in palm close to the finger-hinge and a process of the finger fitting this cavity, while the 4th and 5th joints of peraeopod 5 are swelled to a considerable incrassation. The figure of gnathopod 2 in ♂ shows a very broad 6th joint, with spinulose palm, slightly oblique, not defined by any tooth. The figure of gnathopod 2 in ♀ shows a well expanded 2^{d} joint, 5th bulging, not much longer than 6th, which is broadly rounded at apex and little produced beyond the minute longitudinal finger. L.?

Brazil?

6. **O. bottae** M.-E. 1840 *O. b.*, H. Milne Edwards, Hist. nat. Crust., *v.* 3 p. 17 | 1851 *O. b.*, F. Brandt in: Bull. phys.-math. Ac. St.-Pétersb., *v.* 9 p. 141 | 1868 *O. b.* + *?O. b. var. feminaeformis* (juv.), Czerniavski in: Syezda Russ. Est., Syezda 1 Zool. p. 117 t. 8 f. 28, (29?), 30—32 | 1865 *O. cavimana*, Cam. Heller in: Verh. Ges. Wien, *v.* 15 p. 979 t. 17 | 1879 *O. c.*, Hoek in: Tijdschr. Nederl. dierk. Ver., *v.* 4 p. 131 t. 9 f. 8—10 | 1881 *O. c.*, Nebeski in: Arb. Inst. Wien, *v.* 3 p. 142 t. 2 f. 10 etc.; t. 3 f. 21 etc. | 1895 *O. bottae* + *O. c.*, Chevreux in: Rev. biol. Nord France, *v.* 7 p. 156 f. 1—4; p. 158.

Closely related to O. grillus (p. 540). Eyes roundish. Antenna 1, joints of peduncle successively longer, flagellum 6-jointed. Antenna 2 less than ⅟₂ as long as body, flagellum 14—22-jointed. Gnathopod 1 in ♂, 5th and 6th joints strongly produced at hinder apex; also the 4th joint has at least sometimes a small prominence at hinder apex; finger overlapping palm, but not reaching end of apical process. Gnathopod 2 in ♂, 6th joint large, broadly suboval, palm oblique, divided into 3 protuberances, the broadest central, separated (in adults) by a deep excavation from that next finger-hinge, and a shallow one from the small defining tubercle; finger long, with margin bulging at a point corresponding to excavation in palm. Peraeopod 5, 4th and 5th joints not specially widened at any stage. Colour very dark, almost black (Heller). L. 12—21 mm.

Black Sea; Palestine; Cyprus (Mt. Olympus, 1255 m aboVe sea-leVel); Holland (in garden remote from sea); France.

7. **O. traskiana** Stimps. 1857 *O. t.*, Stimpson in: P. Calif. Ac., *v.* 1 p. 90 | 1857 *O. t.*, Stimpson in: Boston J. nat. Hist., *v.* 6 p. 517 | 1862 *O. t.*, Bate, Cat. Amphip. Brit. Mus., p. 19 t. 3 f. 4 | 1893 *O. t.*, A. Della Valle in: F. Fl. Neapel, *v.* 20 p. 510.

Side-plates 2—4 quadrate, much larger and deeper than 1st or 5th. Eyes squarely rounded, black, less than the diameter apart. Antenna 1 reaching end of penultimate joint of peduncle of antenna 2, flagellum shorter than peduncle, 5-jointed. Antenna 2, about $^1/_3$ as long as body or a little more, ultimate joint of peduncle not much longer than penultimate, flagellum subequal to peduncle, 16-jointed in $\male$, 12-jointed in $\female$. Gnathopod 1 in $\male$, 5th joint not very much longer than 6th, 4th, 5th and 6th joints produced behind into pellucid bosses, successively larger, narrow on joints 4 and 5, broad on joint 6, which has a transverse palm, finger small, only reaching end of true palm, not overlapping the boss. Gnathopod 2 in $\male$, 2^d joint with front margin very slightly convex, not lobed, 3^d joint lobed, 5th very small, 6th large, ovate, widening slightly to the oblique, convex, spinulose palm, which ends in a slight notch at the defining angle, finger large, curved. Gnathopod 1 in $\female$, 4th and 5th joints not produced into processes, 5th subapically widened, 6th narrowly oblong, with a scarcely perceptible scabrous pellucid apical widening, which as well as the little transverse palm is overlapped by the finger. Gnathopod 2 in $\female$, expansion of 2^d joint wider above than below, 4th joint squared at distal hind corner, 5th with very convex free part of hind margin, 6th not much shorter, well produced beyond the small longitudinal finger. The young from the marsupium have gnathopods 1 and 2 almost exactly as in the mother. Peraeopod 2, finger with sinuous inner margin. Peraeopods 3—5, 2^d joint broadly oval, successively larger, hind margin spinulose, 4th joint moderately wide. Uropods 1 and 2 not very spinose, rami not much shorter than peduncle, both rami with marginal spines. Uropod 3, ramus slender, nearly as long as peduncle. Telson spinulose about apex. Colour light grey, sometimes greenish or brownish, always very pale. L. $\male$ 15 mm, $\female$ smaller.

North-Pacific (San Francisco). Among the rejectamenta along high-water mark.

8. O. serrulata Dana 1852 *O. s.*, J. D. Dana in: P. Amer. Ac., *v.* 2 p. 204 | 1853 & 55 *O. s.*, J. D. Dana in: U. S. expl. Exp., *v.* 13 II p. 870; t. 58 f. 7 a—l ($\male$), m—o ($\female$?) | 1893 *O. s.*, A. Della Valle in: F. Fl. Neapel, *v.* 20 p. 498 t. 57 f. 61, 62 | 1862 *O. aucklandiae* + *O. s.*, Bate, Cat. Amphip. Brit. Mus., p. 17 t. 1 a f. 3; p. 31 t. 5 f. 4 | 1886 *O. a.*, G. M. Thomson & Chilton in: Tr. N. Zealand Inst., *v.* 18 p. 145 | 1893 *O. a.*, A. Della Valle in: F. Fl. Neapel, *v.* 20 p. 505 t. 57 f. 65 | 1885 *O. ornata*, Filhol in: Recu. Passage Vénus, *v.* 3 II Zool. p. 463 t. 53 f. 2.

In $\male$ peraeon segment 1 encircled by a raised ridge or corrugation, segments 2—7 (Thomson & Chilton: 2—5) similarly corrugated, except that the front ridge is withdrawn from the front of the segment, and sometimes a little broken; this character (not mentioned by Dana and Bate) is wanting in $\female$ or barely indicated. Side-plate 1 small, front margin straight, the others or at least 2^d—5th in $\male$ with encircling ridge, 2^d—4th having a conspicuous ridge down the centre. Pleon segment 3, postero-lateral corners quadrate, in $\female$ with a minutely produced point. Eyes subrotund, dark, in both sexes wider apart than their diameter. Antenna 1 in $\male$ reaching half-way along penultimate joint of peduncle of antenna 2, in $\female$ to the end of it; 3 joints of peduncle subequal, flagellum 6- or 7-jointed. Antenna 2 smooth, much larger in $\male$ than in $\female$, peduncle angular in section, ultimate joint not greatly longer than penultimate, both long in $\male$; flagellum 17- or 18-jointed, shorter than peduncle in $\male$, longer in $\female$. Gnathopod 1 in $\male$, 5th joint strongly produced into subapical tubercle, 6th shorter, but with apical process broadly produced, so that the finger does not reach the end of it. Gnathopod 2 in ♂, 6th joint massive, widening to the palm, which is spinulose, slightly oblique, in shape variable, sometimes deeply excavate near finger-hinge, then running with straight slope to an acute defining tooth, sometimes more gently sinuous between hinge and tooth (in Dana's figure ending in a broad excavation with untoothed defining angle); finger long to match the palm, its point passing the tooth. Gnathopod 1 in ♀, 6th joint shorter than 5th, spinose, slightly narrower at apex than base, but still affording a small palm to the overlapping finger. Gnathopod 2 in ♀, 2^{d} joint evenly and nowhere widely expanded, 3^{d} rather long, 5th not much longer than 6th, 6th roundly produced a little way beyond the finger. Peraeopods 3 and 4, 2^{d} joint oval. Peraeopod 5, 2^{d} joint notably serrulate on hind margin, which, instead of being rounded below as in peraeopods 3 and 4, is squared or angularly produced. In peraeopods 4 and 5, the 4th and 5th joints are moderately stout, somewhat angular in section. Uropods 1—3, all the rami with marginal spines, in uropod 1 rami shorter than peduncle, in uropods 2 and 3 subequal to it. Telson with the usual little apical notch, and a few spinules. L. ♂ sometimes exceeding 25 mm, ♀ very much smaller.

South-Pacific (New Zealand). Among sea-weed and under stones between tide-marks.

9. **O. guernei** Chevreux 1889 *O. g., Talorchestia?*, Chevreux in: Bull. Soc. zool. France, *v.* 14 p. 332 f.

♀ unknown. — ♂. Body tumid, all segments dorsally elevated, peraeon segments 6 and 7 and pleon segment 1 each with 2 transverse dorsal ridges. Side-plates not very deep. Pleon segment 3, postero-lateral corners simply quadrate (in figure). Eyes ovate. Antenna 1 reaching end of penultimate joint of peduncle of antenna 2. Antenna 2 half as long as body, ultimate joint of peduncle much longer than penultimate, flagellum 18-jointed. Gnathopod 1 rather robust, 5th and 6th joints strongly produced apically. Gnathopod 2, 6th joint large, very broad, palm much longer than the very short hind margin, finger robust, strongly curved, matching the long palm. Peraeopod 4 very strong, 4th and 5th joints much longer than broad. Peraeopod 5 not known. Uropods 1—3 and telson nearly as in O. gammarellus (p. 532). L. 9 mm.

Bay of Horta [Azores (Fayal)]. In the sand.

10. **O. marmorata** (HasW.) 1880 *Talorchestia? m.*, HasWell in: P. Linn. Soc. N. S. Wales, *v.* 5 p. 99 t. 5 f. 3 | 1893 *T. m., Orchestia chilensis* (part.)?, A. Della Valle in: F. Fl. Neapel, *v.* 20 p. 512.

♀ unknown. — ♂. Side-plates with spinuliferous lower margin, 1st very small, sometimes quite concealed by 2^{d}. Eyes irregularly oval, wider than diameter apart, dark in spirit. Antenna 1 slender, flagellum 5-jointed. Antenna 2 more than thrice as long as antenna 1, about $^{1}/_{3}$ as long as body, peduncle stout, flagellum as long as peduncle, flattened, with 17 joints, which are short and most of them broad. Gnathopod 1, 5th joint subequal to 6th, distally produced into a narrow, rounded process, 6th much widened distally, palm transverse, finger well developed, overlapping the true palm but not the broad pellucid process of hind margin of 6th joint. Gnathopod 2, 6th joint large, oval, palm oblique, defined by a minute acute tooth. Peraeopods very thick, spinose, finger of peraeopod 2 with irregular inner margin, 2^{d} joint of peraeopod 5 not dilated behind except at the top. Uropod 3 small, ramus shorter than peduncle. Telson spinulose, with slightly notched apex. The whole of the integument very hard. Colour, marbled red and white. L. 16 mm.

Tasmania.

11. O. capensis Dana 1853 & 55 *O. c.*, J. D. Dana in: U. S. expl. Exp., *v.*13 II p. 866; t. 58 f. 3a, b | 1862 *O. c.*, Bate, Cat. Amphip. Brit. Mus., p. 23 t. 4 f. 2 | 1893 *O. c.*, A. Della Valle in: F. Fl. Neapel, *v.* 20 p 506 t. 57 f. 69.

♀ unknown. — ♂. Side-plates 2—6 very large, 5th nearly as deep as 4th. Eyes subrotund. Antenna 1 reaching end of penultimate joint of peduncle of antenna 2, flagellum 7—10-jointed. Antenna 2 rather more than half as long as body, ultimate joint of peduncle twice as long as penultimate or more, flagellum rather longer than peduncle, 14—16-jointed, setules minute. Gnathopod 1, 5th joint without apical process (Bate), 6th narrow, scarcely widened at apex, which is excavato-truncate, finger hardly longer than apex (Dana) (Bate: longer than apex). Gnathopod 2, 6th joint large, robust, widening to the palm, which is deeply, almost semicircularly excavate between a small tooth adjoining finger-hinge and the acute defining tooth. Peraeopods 1—5 all robust, with groups of spines rather crowded, 2d joint in peraeopods 3—5 broadly expanded; peraeopods 4 and 5 longer than the rest, but not elongate. L. 16—18 mm.

Cape of Good Hope.

12. O. chiliensis M.-E. 1840 *O. c.*, H. Milne Edwards, Hist. nat. Crust., *v.* 3 p. 18 | 1888 *O. c.*, T. Stebbing in: Rep. Voy. Challenger, *v.* 29 p. 185 | 1898 *O. c.*, G. M. Thomson in: Tr. N. Zealand Inst., *v.* 31 p. 199 | 1849 *O. chilensis*, Nicolet in: Gay, Hist. Chile, *v.* 3 p. 233 | 1853 & 55 *O. c.?*, J. D. Dana in: U. S. expl. Exp., *v.* 13 II p. 867; t. 58 f. 4 | 1862 *O. c.*, Bate, Cat. Amphip. Brit. Mus., p. 30 t. 1a f. 8 | 1893 *O. c.* (part.), A. Della Valle in: F. Fl. Neapel, *v.* 20 p. 498.

In general like O. mediterranea (p. 531). Antenna 1, 2d and 3d joints of peduncle each longer than 1st, flagellum shorter than peduncle, 6- or 7-jointed. Antenna 2, ultimate and penultimate joints of peduncle very stout, and the latter little shorter than the former; flagellum shorter than peduncle, 17-jointed. Gnathopod 2 in ♂, 6th joint very large, the palm having a large obtuse tooth near finger-hinge, and thence running rather sinuously and obliquely but not extremely so to meet the hind margin which is longer and better defined than in O. mediterranea; finger long, rather sinuous. Peraeopod 5, 2d joint tending to quadrate at lower hind corner, 4th and 5th joints stout, but little more so than in peraeopod 4, both much longer than broad. Uropods 1 and 2, rami shorter than peduncle. Uropod 3, ramus much shorter than peduncle. L. 16—18 mm.

Pacific (Chili; New Zealand). Under stones and in little pools between tide-marks.

13. O. dentata Filh. 1885 *O. d.*, Filhol in: Recu. Passage Vénus, *v.* 3 II Zool. p. 462 t. 53 f. 1 | 1898 *O. telluris?*, G. M. Thomson in: Tr. N. Zealand Inst., *v.* 31 p. 200.

Antenna 1 reaching end of penultimate joint of peduncle of antenna 2. Antenna 2 rather more than half as long as body, penultimate joint of peduncle inflated in the middle, flagellum less than half as long as peduncle. Gnathopod 1, 5th joint distally widened, the lateral faces compressed, 6th joint more reduced than 5th. Gnathopod 2, 6th joint robust, almost globular, palm with a tolerably strong tooth at centre and a small one near the point to which the finger tip is applied; finger with prominence which closes down immediately behind central tooth of palm. Peraeopods 1—3 successively larger, peraeopod 5 a little longer than peraeopod 4. L. 13 mm.

Cook Strait (Isle of Kapiti).

14. **O. selkirki** Stebb. 1888 *O. s.*, T. Stebbing in: Rep. Voy. Challenger, *v.* 29 p. 603 t. 1, 2 | 1893 *O. gammarellus* (part.), A. Della Valle in: F. Fl. Neapel, *v.* 20 p. 499.

In general closely resembling O. gammarellus (p. 532). Antenna 1 in ♂, flagellum 8- or 9-jointed, in ♀ 6-jointed. Antenna 2 in ♂, flagellum reaching 28 joints. Maxillipeds, a tubercle representing 4th joint of palp. Gnathopod 1 in ♀, see Fig. 93 (p. 530). Gnathopod 2 in ♂, 6th joint very large, widest near base, narrowing slightly to the very sinuous, spiniferous palm, the strong finger having inner margin divided into 2 concave spaces, the first closing over the little concavity and great convexity of the palm, the second leaving a narrow space between itself and the remainder of the palm, which ends in a groove for receiving the finger tip. Gnathopod 2 in ♀, 2d joint not greatly widened, narrowing downward, 6th joint broadly produced beyond the minute, obliquely placed finger. Peraeopod 5, 4th and 5th joints in both sexes long and narrow. Uropod 3, ramus about as long as peduncle. Telson as broad as long, spinulose, the narrow apex a little emarginate. L. 13—17 mm.

South-Pacific (Juan Fernandez). Shore.

15. **O. pickeringii** Dana 1853 & 55 *O. p.*, *O. pickeringi*, J. D. Dana in: U. S. expl. Exp., *v.* 13 II p. 882, 1595; t. 59 f. 9 (♂) | 1862 *O. pickeringii*, Bate, Cat. Amphip. Brit. Mus., p. 32 t. 5 f. 6 | 1879 *O. macleayana*, Haswell in: P. Linn. Soc. N. S. Wales, *v.* 4 p. 250 t. 7 f. 2 | ?1885 *Allorchestes crassicornis var. coogeensis*, Chilton in: P. Linn. Soc. N. S. Wales, *v.* 9 p. 1035 t. 46 f. 1 a, b | 1893 *Orchestia gammarellus* (part.), A. Della Valle in: F. Fl. Neapel, *v.* 20 p. 499.

Body not very stout. Side-plates not deep, 5th nearly or quite as deep as 4th. Pleon segment 3, postero-lateral angles quadrate, with minutely produced point. Eyes variable in size, shape and distance apart, dark. Antenna 1 reaching end of penultimate joint of peduncle of antenna 2, 3 joints of peduncle subequal in length, flagellum 4—6-jointed. Antenna 2 scarcely half as long as body, ultimate joint of peduncle not greatly longer than penultimate, both becoming broad in ♂, flagellum longer(?) than peduncle (Dana), with 12-joints, 1st in ♀ representing 3 or 4 coalesced. Maxilla 1, palp very minute. Gnathopod 1 in ♂, 5th joint much longer than 6th, with narrowly rounded pellucid process, 6th with broadly rounded process, not overlapped by the finger. Gnathopod 2 in ♂, 6th joint large, ovate, palm with 2 (sometimes coalesced) convex spinulose processes, the one near the finger-hinge much the broader (not in Dana's figure); finger stout, evenly curved (in Dana rather sinuous), reaching recess formed by advance of hind margin beyond 2d process of palm. Gnathopod 1 in ♀, 6th joint narrowly oblong, with small apical pellucid process, overlapped by the finger. Gnathopod 2 in ♀, 2d joint well expanded above, 4th scarcely lobed, 6th nearly as broad and long as 5th, the produced apex well rounded. Peraeopod 5, 2d joint as usual considerably broader than in peraeopods 3 and 4; 4th and 5th joints rather stout, cylindrical in peraeopods 4 and 5. Pleopods very slender. Uropod 1, outer ramus with spines only at apex. Uropod 3 small, ramus shorter than peduncle. L. about 12 mm.

Pacific (Hawaiian Islands; California; New South Wales). Among decaying sea-weed.

16. **O. nitida** Dana 1852 *O. n.*, J. D. Dana in: P. Amer. Ac., *v.*2 p.204 | 1853 & 55 *O. n.*, J. D. Dana in: U. S. expl. Exp., *v.*13π p. 868; t. 58 f. 5a—f (f. 6a—d?) | 1862 *O. n.* + *O. fuegensis*, Bate, Cat. Amphip. Brit. Mus., p. 29 t. 5 f. 1; p. 17 t. 1a f. 2 | 1893 *O. gammarellus* (part.) + *O. chilensis* (part.), A. Della Valle in: F. Fl. Neapel, *v.* 20 p. 499, 498.

Body compressed. Side-plates of moderate size, 5th decidedly less deep than 4th. Pleon segment 3, postero-lateral corners slightly recurved. Eyes round. Antenna 1 reaching end of penultimate joint of peduncle of antenna 2 (Dana) (Bate: beyond it), flagellum 5-jointed. Antenna 2 shorter than half the body (Dana); peduncle short, ultimate and penultimate joints subequal, flagellum 14- or 15-jointed, flattened. Gnathopod 1 in ♂, 5th joint considerably longer than 6th, with subapical process, 6th apically widened, obliquely truncate (Dana), the obliquity slight in the figure; finger not quite reaching extremity of apical process. Gnathopod 2 in ♂, 6th joint large, subovate, palm straight, nearly longitudinal, with terminal groove for apex of long finger (Dana), oblique (Bate). Gnathopod 1 in ♀, 6th joint narrowly oblong, finger overlapping the short palm (Bate; not observed by Dana). Gnathopod 2 in ♀, 4th joint with scabrous apical process, 5th with free hind margin, 6th narrowly produced beyond the longitudinal finger. Peraeopods 4 and 5 with numerous short spines on the slender 5th and 6th joints. Uropod 1, outer ramus with only apical spines. Colour green. L. about 8—12 mm.

Tierra del Fuego (among floating Fucus, near the shores); Port Famine.

17. **O. humicola** Marts. 1868 *O. h.*, E. v. Martens in: Arch. Naturg., *v.* 341 p. 56 | 1888 *O. h.*, T. Stebbing in: Rep. Voy. Challenger, *v.* 29 p. 384 | 1892 *O. h.*, M. Weber, Reise Niederl. O.-Ind., *v.* 2 p. 569 | 1893 *O. h.*, A. Della Valle in: F. Fl. Neapel, *v.* 20 p. 509.

Side-plates 1—4 comparatively large, rounded, the 5th somewhat less deep and much narrower than the 4th. Pleon segment 3, postero-lateral corners quadrate, margin above irregularly serrulate. Eyes long-oval, rather more than the smaller diameter apart. Antenna 1, flagellum shorter than peduncle, 3—5-jointed. Antenna 2 half as long as body, flagellum as long as peduncle, 12—15-jointed. Gnathopod 1 in ♂, 5th joint with inconsiderable distal convexity, 6th distally widened, palm sinuous, finger strong, reaching end of apex of 6th joint. Gnathopod 2 in ♂, 6th joint strong, long-oval, palm long, gently convex, spinose, oblique, finger long and strong, ending with abruptly narrowed cylindrical apex, overlapping the palm. Gnathopod 1 in ♀, 5th joint not very long, 6th not distally widened, palm very short, concave, finger large, hooked, overlapping palm. Gnathopod 2 in ♀, 4th joint distally lobed, 5th with free part of hind margin bulging, 6th not far nor yet broadly produced beyond the little oblique palm. Peraeopod 5, 2[d] joint rather strongly serrate or crenulate on upper part of hind margin. Uropod 1, rami considerably shorter than peduncle, outer ramus with only apical spines. Uropod 3, conical ramus as long as peduncle, with an apical spine. L. nearly 8 mm.

Japan. Among damp fallen leaves.

18. **O. floresiana** M. Weber 1892 *O. f.*, M. Weber, Reise Niederl. O.-Ind., *v.* 2 p. 562 f. 9—12.

Pleon segment 3, postero-lateral corners produced to a point, margin above convex, smooth. Eyes round, somewhat less than the diameter apart.

Antenna 1, joints of peduncle successively shorter, flagellum shorter than
peduncle, 4-jointed. Antenna 2 about $^1/_3$ as long as body, ultimate and
penultimate joints of peduncle elongate, flagellum 12—16-jointed. Gnathopod 1 in ♂, 5th and 6th joints strongly produced apically, 6th shorter than
5th, finger overlapping palm, but not reaching end of apical process. Gnathopod 2 in ♂, 6th joint large, oval, palm not very convex, ending in long
groove for apex of the long finger. Gnathopod 1 in ♀, 5th joint longer
than 6th, not greatly widened distally, 6th narrowly oblong, finger a little
longer than palm. Gnathopod 2 in ♀, 5th joint not longer than 6th, hind
margin bulging, 6th broadly produced beyond the little concave palm and
longitudinal finger. Peraeopod 5, 2^d joint with spines on front and many
spinules on hind margin. Uropod 1, rami shorter than peduncle, outer ramus
with only apical spines. Uropod 3, ramus rather shorter than peduncle.
L. 8 mm.

Flores.

19. O. grillus (Bosc) 1802 *Talitrus g.*, Bosc, Crust., v.2 p.152 t.15 f.1,2 |
1803 *T.g.*, Latreille, Hist. Crust. Ins., v.6 p.300 | 1840 *T.gryllus, Orchestia* (part.)?, H. Milne
Edwards, Hist. nat. Crust., v.3 p.17 | 1844 *O. g.*, De Kay, Zool. N.-York, v.6 p.36 t.7
f.19 | 1862 *O. g.*, Bate, Cat. Amphip. Brit. Mus., p.18 t.3 f.2 | 1847 *O. g. (Scamballa
sayana* Leach in MS.), A. White, Crust. Brit. Mus., p.86 | 1873 *O. palustris*, (S. I. Smith
in:) A. E. Verrill in: Rep. U. S. Fish Comm., v.1 p.555 | 1893 *O. gammarellus* (part.)?,
A. Della Valle in: F. Fl. Neapel, v.20 p.511.

Much resembling O. gammarellus (p. 532). Pleon segment 3, postero-
lateral corners acute, a little produced, margin above almost smooth. Eyes
subrotund, rather large, dark, nearer together than their width. Antenna 1
about reaching end of penultimate joint of peduncle of antenna 2. Antenna 2
rather slender, flagellum 17—25-jointed. Gnathopod 1 in ♂, 4th joint with
concave distal margin, 5th joint with subapical process prominent, narrow,
6th with apical widening strongly produced outward and a little downward,
finger matching apical margin. Gnathopod 2 in ♂, 2^d joint not distally
widened, 6th becoming very large, oval, palm spinulose, subequal to hind
margin, oblique, convex between 2 slight depressions, finger large, the point
not very acute. Gnathopod 1 in ♀, 5th and 6th joints spinose, 6th narrow,
parallel-sided, finger matching or overlapping the small, slightly concave
palm. Gnathopod 2 in ♀, 2^d joint broadly laminar, hind margin nearly
straight, front strongly convex except just at distal end, 5th rather broad,
6th distally produced beyond the slightly oblique finger. Peraeopod 2, inner
margin of finger scarcely irregular. Peraeopods 4 and 5 with 4th—6th joints
long and slender, even in the largest specimens of both sexes. Uropod 1,
rami shorter than peduncle, both with marginal spines. Uropod 3, ramus
shorter than peduncle. Telson spinulose. L. reaching in ♀ 18, in ♂ 22 mm.

Atlantic coasts of North America. Under vegetable detritus and sand, also in salt-
marshes and up to fresh water; at times above high-water mark in almost dry places.

20. O. platensis Krøyer 1845 *O. p.*, Krøyer in: Naturh. Tidsskr., ser.2 v.1
p.304 t.2 f.2 | 1862 *O. p.*, Bate, Cat. Amphip. Brit. Mus., p.19 t.3 f.3 | 1867 *O. crassi-
cornis*, A. Costa in: Annuario Mus. Napoli, v.4 p.38 | 1895 *O. c.*, Chevreux in: Rev.
biol. Nord France, v.7 p.154 | 1873 *O. agilis*, (S. I. Smith in:) A. E. Verrill in: Rep.
U. S. Fish Comm., v.1 p.555 t.4 f.14 | 1893 *O. gammarellus* (part.), A. Della Valle in:
F. Fl. Neapel, v.20 p.499 | 1883 *O. tiberiadis*, L. Lortet in: Arch. Mus. Lyon, v.3 p.190 f.
1888 *O. incisimana*, Chevreux in: C.-R. Ass. Franç., Sess. 17 v.2 p.346 t.6 f.1,2.

Pleon segment 3, postero-lateral corners quadrate, margin above serrulate. Eyes elliptic, black, nearer together than the longer diameter. Antenna 1 scarcely reaching end of penultimate joint of peduncle of antenna 2. Antenna 2 less than half as long as body, peduncle stout, ultimate joint longer than penultimate, flagellum shorter than peduncle, tapering rather conspicuously, 12—15-jointed, strikingly flattened. Gnathopod 1 in $\male$, 5th and 6th joints both strongly produced distally on hind margin, the 6th rather the more stoutly and a little downward, finger matching apical margin. Gnathopod 2 in $\male$, 2d joint not widened distally, 6th rather large, broad, palm more or less oblique, smoothly convex or having near the finger-hinge a broad spinulose convexity, followed by a much smaller lobe between 2 little notches, the long curved finger closing down with a narrow apical piece into the pocket adjoining the 2d notch. Gnathopod 1 in $\female$, 5th joint a little subapically widened, 6th slightly narrowed distally, with a feebly developed palm, which is overlapped by the finger. Gnathopod 2 in $\female$, 2d joint laminar, the expansion narrowing distally, 5th with free part of hind margin evenly convex, 6th not much shorter than 5th, the minute finger slightly oblique. Peraeopod 2, finger abruptly narrowed at the nail. Peraeopods 3—5, 4th joint thick, 5th joint in peraeopods 4 and 5 also thick, but without any great widening. Branchial vesicles of gnathopod 2 extremely sinuous. Uropod 1, apical spines on both rami, marginal only on one. Uropod 2 stout. Uropod 3, ramus rather shorter than peduncle. Telson spinulose. Colour dull greyish brown (Krøyer). L. 10—15 mm.

Banks of Rio de la Plata (N. W. of MonteVideo); Atlantic coasts of North-America (from Bay of Fundy to New Jersey); Bermudas; Mediterranean; Lake of Tiberias [Palestine].

21. **O. sulensoni** Stebb. 1899 *O. s.,* T. Stebbing in: Tr. Linn. Soc. London, ser. 2 *v.* 7 p. 400 t. 30 C.

$\female$ unknown. — $\male$. Integument (in spirit) membranaceous, iridescent. Side-plates not deep. Pleon segment 3, postero-lateral corners quadrate, point scarcely produced. Eyes not very large. Antenna 1, 1st joint at least as broad as long, 2d and 3d each slightly longer, flagellum with 4 joints, together about as long as 3d joint of peduncle. Antenna 2, peduncle stout, ultimate joint rather longer than penultimate, flagellum shorter than peduncle, 18—21-jointed. Gnathopod 1, 4th joint without apical process, 5th with prominent, narrowly rounded. distal pellucid process, 6th oblong, widening very slightly to the palm, which has no conspicuous process and is overlapped by the finger. Gnathopod 2, 2d joint comparatively slender, 5th very small, completely masked by 4th, 6th very large, fringed with spinules on the hind margin, widening to the palm, which is moderately oblique, spinulose, smoothly convex between a blunt defining tooth and a deep depression near the finger-hinge, the depression corresponding with a rounded process of the finger's inner margin. Peraeopod 1 conspicuously longer than peraeopod 2, both slender. Peraeopod 2, finger short, with 2 notches. Peraeopods 3 and 4, 2d joint oval. Peraeopod 5 having its 2d joint considerably larger, and the oval modified by the straightness of its hind margin, the remaining joints slender. Uropod 1, upper ramus with lateral spines, rather shorter than lower ramus, which has no lateral spines. Uropod 2, rami equal, both with lateral spines. Uropod 3, ramus not half as long as peduncle. L. 10 mm.

Madeira?

22. **O. parvispinosa** M. Weber 1892 *O. p.,* M. Weber, Reise Niederl. O.-Ind., *v.* 2 p. 566 f. 17—19.

♀. Pleon segment 3 as in O. floresiana (p. 539), with postero-lateral corners less produced. Eyes round, as far apart as the diameter. Antenna 1, flagellum shorter than peduncle, 3-jointed. Antenna 2 about $^1/_3$ as long as body, flagellum 11—13-jointed. Gnathopod 1 said to be nearly as in O. floresiana, but figure gives 6th joint more narrowed apically, with longer finger. Gnathopod 2, 4th joint distally lobed, 5th narrow at base, distally widened into a lobe, 6th rather narrowly produced beyond the very oblique palm. Peraeopod 5, 2d joint broadly expanded, with few spinules on hind margin. Uropod 1, rami shorter than peduncle, both with lateral as well as apical spines. Uropod 3, conical ramus only half as long as peduncle, with long apical spine. L. 8 mm. — ♂ unknown.

Java (Mount Salak, at 1575 m height). Under stones and decaying timber.

23. **O. montana** M. Weber 1892 *O. m.*, M. Weber, Reise Niederl. O.-Ind., *v.* 2 p. 567 f. 20—22.

♀. Pleon segment 3, postero-lateral corners obtusely produced. Eyes roundish, less than the smaller diameter apart. Antenna 1, flagellum shorter than peduncle, 4-jointed. Antenna 2 $^1/_3$ as long as body, flagellum as long as peduncle, 12—15-jointed. Gnathopod 1, 5th joint longer than 6th, free part of hind margin convex, not lobed, 6th oblong, comparatively broad, slightly widened distally, finger matching the transverse palm. Gnathopod 2, 4th and the long 5th joint distally lobed; 6th also long, rather narrow, produced much beyond the small, oblique palm. Peraeopod 5, 2d joint broadly expanded, serrate, spinulose. Uropod 1, outer ramus with only apical spines. Uropod 3, ramus conical, smaller than peduncle, with only apical spines. L. 12 mm. — ♂ unknown.

South Celebes. Under stones and fallen leaves, at 1150 m height.

O. beaucoudraii (M.-E.) 1830 *Talitrus b.*, H. Milne Edwards in: Ann. Sci. nat., *v.* 20 p. 364 | 1862 *Orchestia b.*, *O. littorea* (part.: ♀)?, Bate, Cat. Amphip. Brit. Mus., p. 369.

Bay of St. Malo (Chausay Islands).

O. constricta A. Costa 1857 *O. c.*, A. Costa in: Mem. Acc. Napoli, *v.* 1 p. 183 | 1862 *O. mediterranea* (part.) *(O. constructa)*, Bate, Cat. Amphip. Brit. Mus., p. 24 | 1866 *O. montagui* (err., non Audouin 1826?), Cam. Heller in: Denk. Ak. Wien, *v.* 26 ii p. 3 | 1868 *O. bottae* (err., non Milne Edwards 1840?), Czerniavski in: Syezda Russ. Est., Syezda 1 Zool. p. 117 | 1893 *O. chilensis* (part.)?, A. Della Valle in: F. Fl. Neapel, *v.* 20 p. 509.

♂. Peraeopod 5, 4th and 5th joints not dilated. L. 22 m.

Adriatic (Terra d'Otranto).

O. inaequalis Heller 1861 *O. i.*, Cam. Heller in: SB. Ak. Wien, *v.* 44 i p. 289 | 1888 *O. i.*, T. Stebbing in: Rep. Voy. Challenger, *v.* 29 p. 330 | 1893 *O. i.*, A. Della Valle in: F. Fl. Neapel, *v.* 20 p. 509.

Anomalous O. gammarellus (p. 532) or O. montagui (p. 533). L. 16 mm.

Red Sea.

O. novaezealandiae Bate 1862 *O. n.-z.*, Bate, Cat. Amphip. Brit. Mus., p. 20 t. 3 f. 5 | 1881 *O. sylvicola* (err., non J. D. Dana 1852!), G. M. Thomson in: Tr. N. Zealand Inst., *v.* 13 p. 208 etc. | 1893 *O. gammarellus* (part.)?, A. Della Valle in: F. Fl. Neapel, *v.* 20 p. 510.

♀. Gnathopod 1, 6th joint well developed, longer than 5th. Peraeopod 3 as long as peraeopod 4 or 5, 2d joint long. L. about 6 mm.

New Zealand.

O. ochotensis F. Brandt 1851 *O. o.*, F. Brandt in: Bull. phys.-math. Ac. St.-Pétersb., *v.* 9 p. 140 | 1851 *O. o.*, F. Brandt in: Middendorff, Reise Sibirien, *v.* 21 p. 74 t. 6 f. 18—26 | 1862 *O. o.*, Bate, Cat. Amphip. Brit Mus., p. 369 t. 1a f. 9 | 1893 *O. gammarellus* (part.), A. Della Valle in: F. Fl. Neapel, *v.* 20 p. 499.

L. 12 mm.

Sea of Ochotsk.

O. rectimana Dana 1852 *O. r.*, J. D. Dana in: P. Amer. Ac., *v.* 2 p. 203 | 1853 & 55 *O. tahitensis*, J. D. Dana in: U. S. expl. Exp., *v.* 13 II p. 877; t. 59 f. 5a—g | 1862 *O. t.*, Bate, Cat. Amphip. Brit. Mus., p. 33 t. 5 f. 8.

Perhaps a species af Talorchestia. L. ♀ 6—8 mm.

Tahiti. In damp places at 457 m above sea-level.

5. Gen. **Talorchestia** Dana

1847 *Scamballa* (part.), (Leach in MS.) A. White, Crust. Brit. Mus.. p. 86 | 1852 Subgen. *Talorchestia*, J. D. Dana in: Amer. J. Sci., ser. 2 *v.* 14 p. 310 | 1853 Subgen. *T.*, J. D. Dana in: U. S. expl. Exp., *v.* 13 II p. 851 | 1857 *T.*, Bate in: Ann. nat. Hist., ser. 2 *v.* 19 p. 135 | 1888 *T.*, T. Stebbing in: Rep. Voy. Challenger, *v.* 29 p. 257, 262 | 1891 *T.*, T. Stebbing in: Ann. nat. Hist., ser. 6 *v.* 8 p. 328 | 1893 *T.*, Chevreux in: Bull. Soc. zool. France, *v.* 18 p. 127 | 1885 *Talorchestes, Thalorchestia*, Filhol in: Recu. Passage Vénus, *v.* 3 II Zool. p. 459, 461; Planches p. 28.

Fig. 94.
T. novaehollandiae, ♀.
Gnathopod 1.

Like Orchestia (p. 530), except that gnathopod 1 in ♀ (Fig. 94) is simple, instead of subchelate. Peraeopod 2 usually has the inner margin of the finger more sharply constricted than in allied genera.

Sea-shores, widely distributed, generally in sand.

19 accepted and 2 doubtful species.

Synopsis of accepted species:

1 { Gnathopod 2 in ♂, palm defined by tooth or process — **2.**
Gnathopod 2 in ♂, palm without defining tooth or process — **12.**

2 { Peraeopod 5, 2ᵈ joint exceptionally expanded 1. **T. scutigerula** . . . . p. 544
Peraeopod 5, 2ᵈ joint not exceptionally expanded — **3.**

3 { Gnathopod 2 in ♂ having a very deep palmar cavity — **4.**
Gnathopod 2 in ♂ not having a very deep palmar cavity — **5.**

4 { Gnathopod 2 in ♂, defining tooth of palm immensely long . . . . 2. **T. darwinii** . . . . . p. 545
Gnathopod 2 in ♂, defining tooth of palm not immensely long . . . . 3. **T. deshayesii** . . . . p. 545

5 { Gnathopod 2 in ♂, palm defined by 2 teeth — **6.**
Gnathopod 2 in ♂, palm defined by a single tooth — **7.**

6 { Gnathopod 2 in ♂, palm with a tooth in addition to the defining teeth . . . . . 4. **T. tridentata** . . . . p. 546
Gnathopod 2 in ♂, palm without tooth additional to the defining teeth . . . . . . 5. **T. pravidactyla** . . . p. 546

1. T. scutigerula (Dana) 1853 & 55 *Orchestia s.*, J. D. Dana in: U. S. expl. Exp., *v.* 13 ɪɪ p. 863; t. 58 f. 2 | 1862 *O. s.*, Bate, Cat. Amphip. Brit. Mus., p. 26 t. 4 f. 7 | 1893 *O. s.*, A. Della Valle in: F. Fl. Neapel, *v.* 20 p. 497 t. 57—60.

Body rather compressed. Side-plate 1 in ♂ more quadrate than usual, and not much overlapped by side-plate 2, 4th rather large, much deeper than 5th. Pleon segment 3, postero-lateral corners acutely quadrate. Eyes small, round, dark, farther apart than their width. Antenna 1 reaching end of penultimate joint of peduncle of antenna 2 or beyond it, joints of peduncle very short, flagellum 5-jointed. Antenna 2 from $\frac{1}{3}$—$\frac{1}{2}$ as long as body, ultimate joint of peduncle not nearly twice as long as penultimate, flagellum rather longer than peduncle, about 15—18-jointed. Gnathopod 1 in ♂, 5th joint not greatly longer than 6th, its subapical process well developed, 6th with process so well developed that the finger only reaches the end without overlapping. Gnathopod 2 in ♂, 5th joint extremely small, 6th massive, palm well defined, almost transverse, with a triangular tooth very near the finger-hinge, or in the form described by Dana rather near the centre, finger well arched, inner margin slightly projecting where it meets the palmar process. Gnathopod 1 in ♀, 6th joint too much narrowed distally to make the limb subchelate. Gnathopod 2 in ♀, 6th joint nearly as long as 5th, produced beyond the small finger. Peraeopod 2 with the usual notch in the finger. Peraeopod 3, 2[d] joint rather squarely rounded. Peraeopod 4, 2[d] joint similar to that of peraeopod 3 but larger. Peraeopod 5 longer and stouter than peraeopod 4, 2[d] joint with elliptic shield-like hinder expansion produced above the side-plates and also downward, but much narrower in ♀ than in ♂. Telson not longer than broad, D-shaped. Colour greenish brown. L. 18—21 mm.

Nassau Bay (Tierra del Fuego), among cast up sea-weed; South-Atlantic (Hermit Island; Falkland Islands).

2. **T. darwinii** (Fr. Müll.) 1864 *Orchestia d.*, Fritz Müller, Für DarWin, p. 16 f. 7—9 | 1893 *O. deshayesii* (part.), A. Della Valle in: F. Fl. Neapel, *v.* 20 p. 507.

♀ unknown. — ♂. Side-plate 1 is very small, side-plate 2˙ much larger than any of the others. Eyes rounded, black. Antenna 1 very small, not reaching middle of penultimate joint of peduncle of antenna 2; antenna 2, ultimate and penultimate joints of peduncle equal, flagellum subequal to peduncle, 13-jointed. Gnathopod 1, 5th and 6th joints apically dilated, finger shorter than apex of 6th joint; gnathopod 2, 3[d] joint lobed in front, 5th very small, 6th very large, in younger ♂ having the palm almost chelately produced with bilobed ending, the blunt finger with strong median lobe, the apex slightly overlapping palm, in elder ♂ the hind margin produced into a long tooth slightly beyond finger-hinge, from which it is separated by a deep and wide triangular cavity, the finger broad and straight with a blunt sort of hook overlapping end of above-mentioned tooth. Peraeopods 3 and 4, 2[d] joint narrow, peraeopod 5 longer than the preceding peraeopods, with much broader 2[d] joint, other joints slender. L.? (Description based on Müller's drawings.)

Brazil. Near the shore, under leaves, dung, and the loose earth thrown up by burrowing crabs.

3. **T. deshayesii** (Aud.) 1826 *Orchestia d.*, Audouin in: Descr. Égypte, *v.* 1 iv p. 93 Crust. t. 11 f. 8 | 1847 *O. d.* (*Scamballa kuhliana* Leach in MS.), A. White, Crust. Brit. Mus., p. 86 | 1866 *O. d.*, Cam. Heller in: Denk. Ak. Wien, *v.* 26 II p. 3 t. 1 f. 5, 6 | 1868 *O. d.*, Czerniavski in: Syezda Russ. Est., Syezda 1 Zool. p. 131 t. 8 f. 52, 53 | 1887 *O. d.*, T. Barrois, Note Orchesties, p. 6 f. | 1893 *O. d.* (part.), A. Della Valle in: F. Fl. Neapel, *v.* 20 p. 507 t. 2 f. 5; t. 15 f. 15—30; t. 57 f. 70—73 | 1893 *Talorchestia d.*, Chevreux in: Bull. Soc. zool. France, *v.* 18 p. 127 f. | 1899 *T. d.*, T. Stebbing in: Tr. Linn. Soc. London, ser. 2 *v.* 7 p. 400 t. 30 A | 1848 *Orchestia gryphus*, Friedr. Müller in: Arch. Naturg., *v.* 14 I p. 57, 62 t. 4 f. 18, 28 | 1888 *O. g.*, T. Stebbing in: Rep. Voy. Challenger, *v.* 29 p. 226, 421.

Back moderately broad, peraeon segment 1 longer than 2[d]. Side-plates shallow, 1st less deep and narrower than the following 4; 5th in front about as deep as 4th. Pleon segment 3, postero-lateral corners sharply quadrate. Eyes rounded, dark, often wider apart than their width. Antenna 1 sometimes not reaching end of penultimate joint of peduncle of antenna 2, 2[d] joint subequal to 1st, flagellum shorter than peduncle, 5- or 6-jointed. Antenna 2 very variable, ultimate joint of peduncle sometimes $2\frac{1}{2}$ times as long as penultimate, flagellum shorter than peduncle, with about 20 joints, most of them carrying calceoli (Barrois). Maxillipeds, palp with a rudiment of 4th joint (Della Valle). Gnathopod 1 in ♂, 5th joint at apex of hind margin with conspicuous pellucid process, 6th joint shorter, with similar but stouter process, finger with sinuous inner margin, a little overlapping palm. Gnathopod 2 in ♂, 4th and 5th joints very short, 6th very long, hind margin produced almost from the base into a curved tooth, within which the long finger closes, leaving a cavity between its concave inner margin and the long oblique, almost straight palm; the main part of the hand has a conical shape, the palm and front margin being comparatively little divergent. Young ♂ (Barrois, Heller, Czerniavski?) has the hand of gnathopod 2 with a palm chelately produced, or excavate at right angles to the hind margin, or more or less oblique. Gnathopod 1 in ♀ without processes on 5th and 6th joints, 6th at apex but little stouter than base of finger. Gnathopod 2 in ♀, 2^{d} joint wider above than below, 5th wider proximally than distally, 6th shorter than 5th. Peraeopods 1—5 very spinose, in general nearly as in T. quoyana. Uropod 3, ramus shorter than peduncle. Telson obtusely triangular, slightly emarginate, spinulose. Colour light brownish yellow, with rows of darker spots; sometimes dorsum carmine red. L. reaching 15 mm.

European coasts from Baltic and England (in Devon only in sand) to the Black Sea; Egypt.

4. T. tridentata Stebb. 1899 *T. t.*, T. Stebbing in: Tr. Linn. Soc. London, ser. 2 *v.* 7 p. 398 t. 30ʙ.

Body not very broad. Side-plates 1—4 scabrous, quadrate, 2^{d}—4th deeper than 1st. Eyes roughly oval, rather more than their longer diameter apart. Antenna 1 very small, not reaching middle of penultimate joint of peduncle of antenna 2. Antenna 2 about $^1/_8$ as long as body, ultimate joint of peduncle twice as long as penultimate, flagellum scarcely as long as peduncle, flattened, with about 24 short transverse joints. Gnathopod 1 in ♂ spinose, 5th joint long, the subapical pellucid process narrow, very prominent, 6th joint rather narrow, the apical pellucid process prominent, finger with sinuous inner margin, extending beyond the process of 6th joint. Gnathopod 2 in ♂, 2^{d} joint channelled, apices of front scarcely lobed, 5th joint diminutive, 6th very large, hind margin fringed with spinules, palm oblique, having near the finger-hinge a large triangular spinulose tooth, followed by a sinuous slope, and defined from hind margin by 2 teeth side by side, finger very large, with swelling near the hinge, and beyond this the concave margin fringed with small spinules. Peraeopod 2 much shorter than peraeopod 1, finger with strong prominence near base of nail. Peraeopod 3 very short, 2^{d} joint nearly as broad as long. Peraeopods 4 and 5 not very elongate, but much longer than peraeopod 3. All peraeopods spinose, with rather small branchial vesicles. Uropod 1 long, rami much shorter than peduncle, both with marginal spines. Uropod 2, rami not shorter than peduncle, stoutly spined. Uropod 3, ramus as long as peduncle or longer. Telson short, spinulose. L. 11 mm.

North-Pacific (California).

5. T. pravidactyla Hasw. 1880 *T. p.*, Haswell in: P.˙Linn. Soc. N. S. Wales, *v.* 5 p. 100 t. 5 f. 5.

Body stout. Eyes small, round, black, wide apart. Antenna 1 reaching beyond penultimate joint of peduncle of antenna 2, flagellum in ♀ 7-, in ♂ 10-jointed. Antenna 2 in ♂ more than thrice as long as antenna 1, ultimate joint of peduncle more than twice as long as penultimate, flagellum as long as ultimate joint of peduncle, 21-jointed. Antenna 2 in ♀ much smaller than in ♂, 15-jointed. Gnathopod 1 in ♂, 6th joint rather long, widening a little distally, hind margin produced into a short narrow process, palm transverse, concave, finger rather longer than palm. Gnathopod 2 in ♂, 6th joint large, palm oblique, defined by a blunt tooth, with a second tooth close to it on the distal side, and a rounded elevation about the middle; finger geniculate, its apex lying between the two palmar teeth when the hand is closed. Whether the palmar teeth are side by side or successive is left indefinite by the figure, which in fact shows neither of them. Gnathopod 1 in ♀, 5th joint much longer and broader than the narrow 6th, which has a pellucid process too small to give the limb a subchelate character. Gnathopod 2 in ♀, 6th joint thrice as long as broad, not much produced beyond the finger. Peraeopods 1—5 spiny and robust, peraeopods 3—5, 2^d joint broadly rounded, in peraeopod 5 very broad and somewhat produced downward. Uropods 1—3 and telson plentifully furnished with marginal and apical spines and spinules. L. 15—17 mm.

South-Pacific (Tasmania).

6. **T. limicola** Hasw. 1880 *T. l.*, Haswell in: P. Linn. Soc. N. S. Wales, *v.*5 p. 98 t. 5 f. 2 | 1893 *Orchestia l.*, A. Della Valle in: F. Fl. Neapel, *v.*20 p. 505 t. 57 f. 64.

Antenna 1 very short, flagellum 5-jointed. Antenna 2 four times as long as antenna 1, flagellum as long as peduncle, 12-jointed. Gnathopod 1 in ♂, 5th joint with large subapical process, 6th oblong with distal widening, palm transverse, slightly overlapped by finger. Gnathopod 2 in ♂, 6th joint large, broadly oblong, palm transverse, defined by a broad, apically notched tooth, followed by an excavation to match the great central process of the broad but pointed finger. Gnathopod 1 in ♀, 5th joint little longer than 6th, which narrows to an apex not broader than base of finger. Gnathopod 2 in ♀, 5th and 6th joints subequal, 6th produced considerably beyond the longitudinal finger. L. 9 mm.

Queensland (Bowen). In Mangrove swamps, under decaying wood, etc.

7. **T. quoyana** (M.-E.) 1840 *Talitrus brevicornis* + *Orchestia q.*, H. Milne Edwards, Hist. nat. Crust., *v.*3 p. 15, 19 | 1840 *O. q.*, H. Milne Edwards in: G. Cuvier, Règne an., ed. 3 Crust. t. 59 f. 4 | 1893 *O. q.*, A. Della Valle in: F. Fl. Neapel, *v.*20 p. 506 t. 57 f. 68 | 1886 *Talorchestia q.*, G. M. Thomson & Chilton in: Tr. N. Zealand Inst., *v.*18 p. 146 | 1852 *Talitrus novi-zealandiae*, J. D. Dana in: P. Amer. Ac., *v.*2 p. 201 | 1853 & 55 *Orchestia (Talitrus?) n.* + *T. brevicornis* + *O. (Talorchestia?) quoyana*, J. D. Dana in: U. S. expl. Exp., *v.*13$_{II}$ p. 852 t. 56 f. 5; p. 854 t. 56 f. 6; p. 863 t. 58 f. 1 | 1862 *Talitrus b.*, *Orchestia b.* + *Orchestoidea? n.* + *Talorchestia q.*, Bate, Cat. Amphip. Brit. Mus., p. 9 t. 1a f. 6; p. 10 t. 1 f. 2; p. 16 t. 2 f. 7 | 1885 *Talorchestia armata*, *Thalorchestia a.*, Filhol in: Recu. Passage Vénus, *v.*3$_{II}$ Zool. p. 460 t. 53 f. 3; Planches p. 28.

Back broad. Side-plate 1 produced forward with rounded angle and straight front margin, side-plates 2—4 quadrate, successively broader, 5th less deep than 4th. Pleon segment 3, postero-lateral corners quadrate, with the hind margin finely serrulate. Eyes dark, irregularly rounded, scarcely their own width apart. Antenna 1 reaching little beyond penultimate joint of peduncle of antenna 2, 1st and 2^d joints subequal, each longer than 3^d, flagellum shorter than peduncle, 8-jointed. Antenna 2, ultimate joint of peduncle in ♂ twice as long as penultimate, less in ♀, flagellum subequal to peduncle, with about 30 joints, 27 in ♀, 29 in ♂. Upper lip distally rather narrowly rounded, inner plate with flat distal margin. Maxilla 1 with minute 2-jointed palp. Maxillipeds, 2^d joint of palp very broad. Gnathopod 1 in ♂,

35*

2^d joint channelled in front, 5th much longer than 6th, not lobed, 6th very spinose, distally lobed, palm straight, not oblique, finger reaching beyond it. Gnathopod 2 in ♂, 2^d and 3^d joints channelled in front, 4th short, square, 5th very short, cup-shaped, 6th massive, palm oblique, spinulose, with broad rectangular tooth near finger-hinge, and prominent acute defining tooth, within which the finger closes. Gnathopod 1 in ♀, 6th joint narrowing to the finger, distally scarcely broader than base of finger, densely spinose. Gnathopod 2 in ♀, 2^d joint elongate oval, 3^d and 4th subequal, 5th longer than 6th, the part free from 4th narrowly oval, 6th produced with rounded end beyond the small chela-forming finger. Peraeopod 2 shorter than peraeopod 1, finger notched, and that in both sexes. Peraeopod 3 very short, 2^d joint almost circular, 4th and 5th short but rather broad. Feraeopod 4 long, 2^d joint broadly oval. Peraeopod 5 rather longer, 2^d joint as long as in peraeopod 4 but much wider, the broad crenulate expansion giving it a circular look; 4th—7th joints in peraeopods 4 and 5 long, not especially widened. Pleopods with small rami, shorter than peduncle. 4- or 5-jointed. Uropods 1 and 2 spinose, rami ending obtusely, outer rather the shorter. Uropod 3, the small ramus a little longer than the peduncle. Telson thick, broader than long, faintly emarginate, surrounded with spinules, and looking as if its distal angles were upturned. L. reaching 23 mm.

South-Pacific (New Zealand).

8. **T. quadrimana** (Dana) 1852 *Orchestia q.*, J. D. Dana in: P. Amer. Ac., v. 2 p. 204 | 1853 & 55 *O. q.*, J. D. Dana in: U. S. expl. Exp., v. 13 II p. 879; t. 59 f. 7 | 1893 *O. q.*, A. Della Valle in: F. Fl. Neapel, v. 20 p. 504 t. 57 f. 63 | 1879 *Talorchestia q.*, Haswell in: P. Linn. Soc. N. S. Wales, v. 4 p. 248 t. 7 f. 3 | 1880 *T. q. var.?*, Haswell in: P. Linn. Soc. N. S. Wales, v. 5 p. 100 t. 6 f. 1.

Side-plates 2—4 broad. Eyes large, round. Antenna 1 reaching end of penultimate joint of peduncle of antenna 2. Antenna 2, ultimate joint of peduncle not much longer than penultimate, flagellum subequal to peduncle, joints not longer than broad. Gnathopod 1 in ♂, 6th joint little shorter than 5th, subtriangular, the expanded distal margin a little excavate, not overlapped by the finger. Gnathopod 2 in ♂, 6th joint stout, subquadrate, palm transverse, slightly excavate. Peraeopods 1 and 2 slight in structure. Peraeopods 3—5 successively longer, only the 2^d joint expanded and this in peraeopod 5 (Haswell) much broader than that of the others. Uropod 1, outer ramus unarmed. Uropod 3, ramus slender. Colour white, with irregular light-red spots. L. 12—14 mm.

South-Pacific (New South Wales, Queensland, Port Denison). On sandy beaches, under cast up weed above the reach of ordinary tides.

9. **T. diemenensis** Hasw. 1879 *T. d.*, Haswell in: P. Linn. Soc. N. S. Wales, v. 4 p. 248 t. 7 f. 6 | 1893 *T. d.*, G. M. Thomson in: P. R. Soc. Tasmania, 1892 p. 17 t. 5 f. 6—8 | 1893 *T. d.*, A. Della Valle in: F. Fl. Neapel, v. 20 p. 512 | 1891 *Orchestia d.*, T. Stebbing in: Ann. nat. Hist., ser. 6 v. 8 p. 325 | 1898 *O. gammarellus*, G. M. Thomson in: Tr. N. Zealand Inst., v. 31 p. 198.

Antenna 1 very short. Antenna 2, flagellum as long as peduncle. Gnathopod 1 in ♂, 5th joint distally broad, 6th oblong, twice as long as broad, palm transverse, slightly sinuous, finger short, in figure not nearly as long as apex of 6th joint. Gnathopod 2 in ♂, 6th joint (in figure) nearly as broad as long, palm convex, moderately oblique, defined by a rounded tooth. Gnathopod 1 in ♀, 5th and 6th joints narrower than in ♂, the palmar border with a deep median notch, the finger well developed (Haswell); gnathopod 1 in ♀ simply unguiculate (Thomson). Gnathopod 2 in ♀, 5th and 6th joints subequal, 6th produced some way beyond the minute, obliquely placed finger. Telson triangular, blunt. L. 6--13 mm.

South-Pacific (Tasmania). Under stones in estuaries.

10. **T. megalophthalma** (Bate) ? 1844 *Talitrus quadrifidus*, De Kay, Zool. N.-York, *v.* 6 p. 36 t. 14 f. 27 | 1847 *Orchestia megalophthalmus* (*Scamballa m.* Leach in MS.) (nom. nud.), A. White, Crust. Brit. Mus., p. 86 | 1851 *O. megalophthalmos* (nom. nud.), F. Brandt in: Bull. phys.-math. Ac. St.-Pétersb., *v.* 9 p. 142 | 1862 *O. megalophthalma*, Bate, Cat. Amphip. Brit. Mus., p. 22 t. 3 f. 8 | 1893 *O. m.*, A. Della Valle in: F. Fl. Neapel, *v.* 20 p. 496 t. 57 f. 54 | 1873 *Talorchestia m.*, (S. I. Smith in:) A. E. Verrill in: Rep. U. S. Fish Comm., *v.* 1 p. 556.

Very near to T. longicornis, but back still broader, eyes far larger, these being of exceptional size and only separated by half their width or less, antenna 2 shorter (?), and 6th joint of gnathopod 2 in the ♂ with long palm, scarcely oblique, gnathopod 2 in the ♀ with 4th joint strongly produced at lower hind angle; uropod 3, ramus less strongly spined. L. 19 mm.

North-Atlantic (from Cape Cod to New Jersey [North America]).

11. **T. longicornis** (Say) 1818 *Talitrus l.*, Say in: J. Ac. Philad., *v.* 1 II p. 384 | 1830 *Orchestia l.*, H. Milne Edwards in: Ann. Sci. nat., *v.* 20 p. 361 | 1844 *O. l.*, De Kay, Zool. N.-York, *v.* 6 p. 35 t. 9 f. 28, 28 a | 1847 *O. l.* (*Scamballa l.* Leach in MS.), A. White, Crust. Brit. Mus., p. 86 | 1862 *O. l.*, Bate, Cat. Amphip. Brit. Mus., p. 18 t. 3 f. 1 | 1893 *O. l.*, A. Della Valle in: F. Fl. Neapel, *v.* 20 p. 505 t. 57 f. 66, 67 | 1851 *O. l.*, *Megalorchestes?*, *Megalorchestia? l.*, F. Brandt in: Bull. phys.-math. Ac. St -Pétersb., *v.* 9 p. 142, 312 | 1873 *Talorchestia l.*, (S. I. Smith in:) A. E. Verrill in: Rep. U. S. Fish Comm., *v.* 1 p. 556.

Back broad. Side-plate 1 little produced, narrow, as deep as the 2^{d}, which is slightly emarginate behind, 5th not so deep as 4th but broader. Pleon segment 3, postero-lateral corners quadrate. Eyes oval, narrowed above, or irregularly rounded, about as far apart as their smaller diameter, dark. Antenna 1, peduncle about reaching end of penultimate joint of peduncle of antenna 2. 2^{d} joint of peduncle rather longer than 1st or 3^{d}, flagellum short, 6-jointed. Antenna 2, ultimate joint of peduncle more than twice as long as penultimate, flagellum in a ♀ specimen nearly as long as peduncle, 22-jointed. Maxillipeds with no unguiculate terminal joint. Gnathopod 1 in ♂, 5th joint with subapical pellucid process outward, 6th much shorter and narrower, oblong, with slight distal widening into a pellucid process, finger with smooth inner margin, a little overlapping palm. Gnathopod 2 in ♂, 2^{d} joint abruptly widened from narrow base, 3^{d} larger than 4th or 5th, 6th massive, widening to the palm, which is only a little oblique, defined by a very blunt tooth, between which and the hinge is a convex swelling of the palm, or a considerably concavity (in ♂ specimen 25 m͞m long, Bate) or a large rounded tooth between 2 cavities (Milne-Edwards). Gnathopod 1 in ♀, 5th and 6th joints without the pellucid processes, 6th narrowing to the finger-hinge, being there little wider than base of finger. Gnathopod 2 in ♀, 2^{d} joint broadly and irregularly oval, membranaceous, 4th quadrate, not produced, 5th a little longer than 6th, free part of hind margin very convex, 6th elongate, produced much beyond the minute chela-forming finger. Peraeopod 2, finger rather slender, deeply notched. Other peraeopods much as in T. quoyana (p. 547). Uropod 1, rami equal. Uropod 2, outer ramus the shorter. Uropod 3, ramus longer than peduncle. Telson obtusely triangular, slightly emarginate, with spinules at the 2 points thus formed; a transverse

plate only partially free and not reaching the apex is truncate with small
median slit. Colour pale greyish, or sand-coloured. L. reaching 25 mm
and over.

North-Atlantic (from Cape Cod to New Jersey [North America]). Burrowing in sand.

12. **T. pollicifera** (Stimps.) 1855 *Orchestia p.*, Stimpson in: P. Ac. Philad.,
v. 7 p. 383 | 1857 *O. p.*, Stimpson in: Boston J. nat. Hist., v. 6 p. 517 | 1862 *Talorchestia
p.*, Bate, Cat. Amphip. Brit. Mus., p. 16 | 1893 *Orchestia gammarellus* (part.)?, A. Della
Valle in: F. Fl. Neapel, v. 20 p. 510.

Eyes rather small, round, black. Antenna 1 in ♀ reaching end of
penultimate joint of peduncle of antenna 2. Antenna 2 in ♂ stout, peduncle
twice as long as flagellum. Antenna 2 in ♀ slender, flagellum 12-jointed.
Gnathopod 1 in ♂, pellucid process in joint 6 rather unusually produced.
Gnathopod 2 in ♂, 6th joint ovate, of moderate size. Gnathopod 1 in ♀ simple.
Gnathopod 2 in ♀ with small hands, having a minute lateral finger. Caudal
stylets (uropod 3?) short, rami subconical. Colour pale brownish. L. 15 mm.

North-Pacific (Loo Choo).

13. **T. tumida** G. M. Thoms. 1885 *T. t.*, G. M. Thomson in: N. Zealand J. Sci.,
v. 2 p. 577 | 1887 *T. t.*, T. Stebbing in: Tr. zool. Soc. London, v. 12 vi p. 202 t. 39 f. A |
1888 *T. t.*, G. M. Thomson in: Tr. N. Zealand Inst., v. 21 p. 260 t. 13 f. 4—8 | 1898
Orchestia t. (part.), G. M. Thomson in: Tr. N. Zealand Inst., v. 31 p. 203 | 1893 *O. gam-
marellus* (part.), A. Della Valle in: F. Fl. Neapel, v. 20 p. 501.

Back broad. Side-plate 1 almost concealed by the broader side-plate 2,
4th the widest, irregularly quadrate, deeper than 5th. Pleon segment 3,
postero-lateral corners quadrate. Eyes large, round, less than their own
width apart, turquoise-blue. Antenna 1 a little overlapping penultimate joint
of peduncle of antenna 2, all 3 joints of peduncle very short, flagellum
shorter than peduncle, 7- or 8-jointed in ♂, 5-jointed in ♀. Antenna 2
less than ⅛ as long as body, ultimate joint of peduncle about twice as
long as the short penultimate, flagellum a little shorter than penultimate,
12—15-jointed in ♂, 12-jointed in ♀. Upper lip distally broadly rounded.
Maxilla 1, palp barely perceptible. Gnathopod 1 in ♂, 5th joint considerably
longer than 6th, hind margin subdistally forming a small linear lobe, 6th
joint oblong, with squarish distal lobe, little produced outward, and overlapped by the finger. Gnathopod 2 in ♂, 3d joint longer than the quadrate
4th or the small cup-shaped 5th, 6th massive, palm oblique, very spinulose,
with 2 deep excavations between which a stout process meets a strong
projection of finger's inner margin; in young ♂ the palm and finger are
simpler. Gnathopod 1 in ♀, 6th joint oval, very narrow at junction with
the finger which equals it in length. Gnathopod 2 in ♀, 6th joint flattened,
narrowly oval, distally produced beyond the minute chela-forming finger.
Peraeopod 2 shorter than peraeopod 1, finger notched. Peraeopod 3, 2d joint
subcircular. Peraeopod 4, 2d joint broadly oval, 4th expanding to the bluntly
produced distal end, 5th abruptly widened from a narrow neck and then
narrowing, but still at apex much wider than 6th. Peraeopod 5 shorter than
4th, 2d joint very broad, subcircular, crenulate, other joints not specially
widened. Pleopods, rami 7-jointed, subequal to peduncle. Uropods 1
and 2, rami spinose, obtuse. Uropod 3, ramus longer than peduncle. Telson
as broad as long, with spinules round apical margin. Colour ivory-white.
L. 13 mm.

South-Pacific (Dunedin [New Zealand]). Sand-banks, among roots of littoral
plants, above high-water mark.

14. **T. telluris** (Bate) 1862 *Orchestia t.*, Bate, Cat. Amphip. Brit. Mus., p. 20
t. 3 f. 6; t. 4 f. 4 (including f. 2h on right) | 1886 *O. t.*, G. M. Thomson & Chilton in:
Tr. N. Zealand Inst., *v.* 18 p. 145 | 1893 *O. gammarellus* (part.), A. Della Valle in: F.
Fl. Neapel, *v.* 20 p. 499.

Body rather compressed. Side-plates not very large. Pleon segment 3,
postero-lateral corners quadrate, with little spinules in the shallow serrations
of their hind margin. Eyes round, rather large, dark, less far apart than
their width. Antenna 1 in ♂ reaching rather beyond penultimate joint of
peduncle of antenna 2, joints of peduncle short, 3^d slightly shortest, flagellum
shorter than peduncle, 5-jointed. Antenna 2 short, ultimate joint of peduncle not
twice as long as penultimate, flagellum subequal to peduncle, with about 15 joints
in ♂, rather fewer in ♀. Gnathopod 1 in ♂, 5th joint rather longer than
6th, with a small pellucid process at middle of free hind margin, 6th distally
produced into a much larger process, which is a little overlapped by the
finger. Gnathopod 2 in ♂, 2^d joint rather slender, the next 3 as usual
small, 6th very large, widening to the palm, which forms a triangular tooth
between an excavation near the finger-hinge and an oblique convexity passing
insensibly into the hind margin; the long finger has a corresponding cavity
near the hinge, than a broad protuberance and finally a concavity matching
the convexity of the palm. Gnathopod 1 in ♀, 5th joint with no process, 6th
spinose, narrowing to the finger-hinge and forming no palm, finger $^2/_3$ as long as
6th joint. Gnathopod 2 in ♀, 2^d joint little widened, 4th not produced at hinder
angle, 5th with the free portion uniformly widened, 6th nearly as long as 5th,
subapically widened, produced much beyond the finger. Peraeopod 2 with the
finger as usual notched. Peraeopod 3, 2^d joint small, rounded oval. Feraeopod 4, 2^d joint larger, oval, and the whole limb much longer than peraeopod 3.
Peraeopod 5, 2^d joint larger in both sexes, but especially in ♂, nearly circular
but broader below than above, 4th joint in ♂ triangular, widening to the
distal end, 5th joint in ♂ expanded behind to a process comparable to a
dish-cover and at least sometimes assuming monstrous proportions, its oval
being much larger than the 2^d joint of its own limb or than the great hand
of gnathopod 2; 6th and 7th joints slender as in peraeopod 4. Uropod 1,
rami equal. Uropod 2, rami equal or subequal. Uropod 3, ramus rather
longer than peduncle. Telson triangular. L. ♀ 9, ♂ 12 mm.

South-Pacific (New Zealand, from Bay of Islands to Stewart Island). Sandy
shores, just above tide-marks.

15. **T. gracilis** (Dana) 1852 *Talitrus g.*, J. D. Dana in: P. Amer. Ac., *v.* 2
p. 201 | 1853 & 55 *Orchestia (Talorchestia) g.*, J. D. Dana in: U. S. expl. Exp., *v.* 13 ıı
p. 861 t. 57 f. 5 | 1862 *T. g.*, Bate, Cat. Amphip. Brit. Mus., p. 15 t. 2 f. 5 | 1893 *Orchestia
gammarellus* (part.), *Talitrus* (part.)?, A. Della Valle in: F. Fl. Neapel, *v.* 20 p. 499, 511, 947.

Side-plate 5 nearly as deep as 4th. Eyes in figure rather small, round.
Antenna 1 short, not reaching end of penultimate joint of peduncle of antenna 2, flagellum shorter than peduncle. Antenna 2 in ♂ longer than the body,
penultimate joint of peduncle rather long, but less than half as long as ultimate,
flagellum longer than peduncle. Antenna 2 in ♀ longer than half the body,
penultimate joint of peduncle $^2/_3$ as long as ultimate, flagellum longer than
peduncle. Gnathopod 1 in ♂, 5th joint long, showing no pellucid process, 6th considerably shorter, distally a little widened, apex of hind margin narrow, finger
scarcely overlapping palm. Gnathopod 2 in ♂, 6th joint large, ovate, palm
spinulose, oblique, convex. almost continuous with hind margin, finger long,
adjusted to palm. Gnathopod 1 in ♀ rather stout, 3^d—6th joints described

as subequal, finger quite small. Gnathopod 2 in φ, 2^d joint rather expanded, 5th little longer than 6th, with hind margin bulging at the centre, 6th produced beyond the minute finger to a rounded lobe, which bends forward, so that the front line of the hand is slightly concave. Peraeopod 2 much shorter than peraeopod 1; the finger is gibbous below, or has a prominent angle and is stout. Peraeopods 3—5, 2^d joint narrowly oval. Peraeopod 5 the longest. Uropod 1, outer ramus with spines only at apex. Uropod 3, peduncle rather longer than telson, which the ramus nearly equals. Nearly colourless. L. about 12 mm.

Coral island in the Balabac Passage.

16. **T. spinipalma** (Dana) 1852 *Orchestia s.*, J. D. Dana in: P. Amer. Ac., *v.* 2 p. 203 | 1853 & 55 *O. s.*, J. D. Dana in: U. S. expl. Exp., *v.* 13 II p. 875; t. 59 f. 4 a—e | 1862 *O. s.*, Bate, Cat. Amphip. Brit. Mus., p. 28 t. 4 f. 9 | 1880 *Talorchestia terrae-reginae*, Haswell in: P. Linn. Soc. N. S. Wales, *v.* 5 p. 98 t. 5 f. 4 | 1893 *Orchestia chilensis* (part.), A. Della Valle in: F. Fl. Neapel, *v.* 20 p. 498 | 1898 *O. chiliensis*, G. M. Thomson in: Tr. N. Zealand Inst., *v.* 31 p. 199.

Side plates not very deep, 5th nearly as deep as 4th. Antenna 1 very short, flagellum 3—5-jointed (Dana) (Haswell: 6—8-jointed). Antenna 2, peduncle 4 times as long as antenna 1, flagellum as long as peduncle or shorter, 20-jointed; ultimate joint of peduncle much longer than penultimate. Gnathopod 1 in $\male$, 5th joint longer than 6th, with small apical process, 6th narrowly oblong, very slightly dilated at apex, palm transverse, finger overlapping it. Gnathopod 2 in $\male$, 6th joint large, subovate, palm spinulose, very oblique, ill-defined, having near the hinge a process, which fits into a hollow of the elongate finger. Gnathopod 1 in φ, 5th joint longer than 6th, apex of which is scarcely broader than base of short finger. Gnathopod 2 in φ. 5th joint widened at middle of hind margin, 6th joint short, the rounded apex little produced beyond the small longitudinal finger. Peraeopods 1—5 spinulose, peraeopods 3—5 with 2^d joint well expanded, hind margin of it in peraeopod 5 almost straight with well rounded lower corner; peraeopods 4 and 5 long and slender. Uropod 1, outer ramus with only apical spines. L. 10—12 mm.

Tropical Pacific (Tongatabu, under sea-weed; Port Denison [Queensland], on sandy beach).

17. **T. brito** Stebb. 1891 *T. b.*, T. Stebbing in: Ann. nat. Hist., ser. 6 *v.* 8 p. 324 t. 15 | 1895 *T. b.*, Chevreux in: Rev. biol. Nord France, *v.* 7 p. 158 | 1893 *Orchestia chilensis* (part.), A. Della Valle in: F. Fl. Neapel, *v.* 20 p. 498.

Back moderately broad. Side-plate 1 directed somewhat forward, narrower than the subequal, subquadrate 2^d—4th. Pleon segment 3, posterolateral corners quadrate. Eyes large, irregularly rounded, separated by less than their width, in life white, with dark pigment showing through. Antenna 1 not quite reaching end of penultimate joint of peduncle of antenna 2, 3 joints of peduncle subequal, or sometimes in φ 2^d the longest, flagellum less than half as long as peduncle, 7-jointed in $\male$, 5-jointed in φ. Antenna 2, penultimate joint of peduncle more than half as long as the long and stout ultimate, flagellum with about 30 short joints in $\male$, 22 in φ. Maxilla 1, minute palp 2-jointed. Gnathopod 1 in $\male$, 5th joint long, with pellucid bubble-like process near distal end of hind margin, 6th much shorter, more spinose, with similar process, finger short, projecting beyond palm. Gnathopod 2 in $\male$, 4th joint short, square, 5th small, cup-shaped, 6th massive, palm spinulose, scarcely crenulate, convex, very oblique, closed finger leaving a small gap near hinge, otherwise fitting palm except at overlapping apex. Gnathopod 1 in ♀ nearly as in ♂, but without the pellucid processes of 5th and 6th joints, the latter narrowing to base of finger. Gnathopod 2 in ♀ almost membranaceous, 2d joint rather narrowly oval, 5th subdistally widened, 6th with rounded end produced beyond the minute chela-forming finger. Peraeopods 2 and 3 shorter than the rest and with much smaller finger, deeply notched in peraeopod 2, slightly in peraeopod 3. In peraeopod 3, 2d joint subcircular, in peraeopod 4 broadly oval, in peraeopod 5 as long as in peraeopod 4 but rather wider. In peraeopods 4 and 5 joints 4—6 elongate, spinulose, none expanded. Pleopods, peduncle long, membranaceous, rami with 15 or 16 joints. Uropods 1 and 2, peduncle rather longer than the spinose rami. · Uropod 1 much the longest. Uropod 3, ramus about as long as peduncle. Telson narrowest at truncate distal end, surrounded with groups of spinules. Colour yellowish white, with bars of yellow, side-plates bordered with purple, and sometimes pleon banded with the same. L. ♂ 20 mm, ♀ smaller.

North-Atlantic (North-Devon. burrowing in sand; Verdon [Gironde]).

18. **T. novaehollandiae** Stebb. 1899 *T. n.*, T. Stebbing in: Tr. Linn. Soc. London, ser. 2 *v.* 7 p. 399 t. 31 A.

Body stout. Pleon segment 2, postero-lateral corners with an acute point, segment 3 quadrate. Eyes round, dark, about their diameter apart. Antenna 1 reaching beyond penultimate joint of peduncle of antenna 2, joints of peduncle not elongate, flagellum shorter than peduncle, 6-jointed. Antenna 2 verticillately spinulose. about $^1/_3$ as long as body, ultimate joint of peduncle longer than penultimate, flagellum rather shorter than peduncle, 19-jointed. Gnathopod 1 in ♂, 5th joint longer than 6th, with narrow apical process, 6th short, much widened distally, finger overlapping palm but not process of hind margin. Gnathopod 2 in ♂, 2d joint narrow, 6th very large, slightly widening to the almost transverse palm, which is defined by a small pocket, a broad convexity leading thence to a spinulose concave space near the finger-hinge, over which space the finger arches, the convexity of its sinuous margin touching the convexity of the palm and its apex passing into the defining pocket. Gnathopod 1 in ♀ see Fig. 94 (p. 543), 6th joint short, spinulose, narrowing very gradually to the short finger. Gnathopod 2 in ♀, 2d joint membranaceous, well expanded, 6th joint rather narrow. Peraeopods 1—5 spinulose, 2d joint in peraeopods 3—5 well expanded, largest and broadest in peraeopod 5, with subquadrate ending of hind margin, peraeopod 4 much longer than peraeopod 3, peraeopod 5 considerably longer than peraeopod 4, its 4th and 5th joints rather long and stout. Uropods 1—3, on peduncle and all. rami, have marginal spines. Uropod 3, ramus slender, shorter than peduncle. Telson much longer than broad, composed of separate halves, which appear to fold closely together, each with 2 apical spinules and 2 marginal spines. L. about 10 mm.

South-Pacific (Manly Beach [East-Australia]).

19. **T. martensii** (M. Weber) 1892 *Orchestia m.*, M. Weber, Reise Niederl. O.-Ind., *v.* 2 p. 564 f. 13—16.

Pleon segment 3, postero-lateral corners with produced point. Eyes oval, as far apart as the shorter diameter. Antenna 1, 2d joint of peduncle the longest, flagellum shorter than peduncle, 5-jointed. Antenna 2, ultimate and penultimate joints of peduncle long, flagellum almost as long as peduncle,

21—25-jointed. Gnathopod 1 in ♂, 5th joint apically furnished with narrowly rounded process, 6th distally little widened, with slightly concave palm, overlapped by the finger. Gnathopod 2 in ♂, 6th joint short but broadly ovate, palm slightly convex and oblique, joining the almost straight hind margin in a gentle curve, the long finger closing between 2 rows of spines which line the palm. Gnathopod 1 in ♀, 5th joint without process, 6th narrowing to the apex, which is hardly wider than base of finger. Gnathopod 2 in ♀, 5th joint broad, front margin very convex, hind produced into a central lobe, 6th almost as long as 5th, produced into a broadly rounded lobe bent forward so as to project beyond the minute, longitudinally placed finger. Peraeopod.5, 2^d joint with 13—15 spines on the hind margin, which is straight with rounded corners. Uropod 3, ramus rather shorter than peduncle. L. ♀ 11, ♂ 8 mm.

Flores. Under stones in and at the margin of the rivulet Lella.

T. africana Bate 1862 *T.? a.*, Bate, Cat. Amphip. Brit. Mus., p. 15 t. 2 f. 6.

♀. Side-plate 5 as deep as the preceding and very broad, nearly as broad as length of 3 segments of peraeon. Eyes small. Antenna 1 as long as peduncle of antenna 2. L. 14 mm.

Port Natal [South Africa].

T. cookii Filh. 1885 *T. c., Thalorchestia c.*, Filhol in: Recu. Passage Vénus, *v.* 3 ɪɪ Zool. p. 459 t. 53 f. 4; Planches p. 28 | 1893 *Orchestia? cooki*, A. Della Valle in: F. Fl. Neapel, *v.* 20 p. 512 | 1898 *O. tumida* (part.), G. M. Thomson in: Tr. N. Zealand Inst., *v.* 31 p. 203.

Probably identical with T. tumida (p. 550). Only ♂ described. L. 17 mm.

South-Pacific (New Zealand).

6. Gen. **Ceina** Della Valle

1893 *Ceina* (Sp. un.: *C. egregia*), A. Della Valle in: F. Fl. Neapel, *v.* 20 p. 530 | 1899 *C.*, T. Stebbing in: Tr. Linn. Soc. London, ser. 2 *v.* 7 p. 397.

Antenna 1 longer than peduncle of antenna 2. Maxillipeds, finger of palp broad, subtriangular. Gnathopod 1 in ♂ and ♀, and gnathopod 2 in ♀ subchelate, small. Gnathopod 2 in ♂ much larger, subchelate or (in maturity) chelate. Uropod 3 tubercular, without rami. Telson partially cleft.

1 species.

1. C. egregia (Chilton) 1883 *Nicea e.*, Chilton in: Tr. N. Zealand Inst., *v.* 15 p. 77 t. 2 f. 2a—1 | 1886 *N. e.*, G. M. Thomson & Chilton in: Tr. N. Zealand Inst., *v.* 18 p. 144 | 1888 *N.? e.*, T. Stebbing in: Rep. Voy. Challenger, *v.* 29 p. 1712 | 1893 *Ceina e.*, A. Della Valle in: F. Fl. Neapel. *v.* 20 p. 530 t. 58 f. 14—21.

Body much compressed, segments imbricated and subcarinate, more strongly in ♀ than in ♂, peraeon segment 1 produced somewhat over the head, especially in ♀. Head produced slightly upward at base of antenna 1. Side-plates 1—4 successively deeper, 5th—7th much shallower. Pleon segment 3, postero-lateral corners rounded. Eyes of moderate size, round. Antenna 1, 2^d joint slightly shorter than 1st, 3^d than 2^d, flagellum subequal to peduncle, with sensory filaments. Antenna 2 about $^2/_5$ as long as body, ultimate and penultimate joints of peduncle equal, flagellum rather longer than peduncle. Lips, mandible and maxillae 1 and 2 undescribed. Mandible not showing (in figure) spine-row or molar. Maxillipeds with 2 rounded apical teeth on inner plate, 1st joint of palp produced on outer side beyond the 2^d, 3^d distally widened, not longer than 4th. Gnathopods 1 and 2 in ♀ equal and similar, 5th joint long, subtriangular, 6th rather longer, not wider, oblong, palm slightly oblique, defined by palmar spine, finger matching palm. Gnathopod 1 in ♂ as in ♀. Gnathopod 2 in ♂, 2^d joint narrow, 5th apparently coalesced with the large oblong oval 6th, which in smaller specimens has a short transverse palm, but, the hind margin being gradually produced, at length a complete though small chela is formed. Peraeopods 1—5 subequal, rather stout, almost unarmed, 4th joint rather expanded distally, finger strong, curved. Peraeopods 3—5, 2^d joint rounded oval. Uropods 1 and 2 short, rami subequal to peduncle, spinose. Uropod 3, each member consisting of a single, rounded, unarmed joint; the pleon segment 6 has the appearance of forming a peduncle to the pair. Telson subrectangular, about as broad as long, apically rounded, cleft about to centre. Colour red, sometimes blue; integument thick, opaque. L. 6—7 mm.

Lyttelton Harbour [New Zealand]. On sea-weed (Macrocystis).

7. Gen. **Chiltonia** Stebb.

1899 *Chiltonia* (Sp. un.: *C. mihiwaka*), T. Stebbing in: Tr. Linn. Soc. London, ser. 2 *v.* 7 p. 408.

Side-plates 1—4 deep. Antennae 1 and 2 equal in length. Maxilla 1 without palp, notched at palp's normal position. Maxillipeds, 4th joint of palp short, conical. Mouth-parts otherwise normal as in the family. Gnathopods 1 and 2 subchelate, gnathopod 2 very unlike in ♀ and in ♂. Uropod 3 1-jointed. Telson simple.

1 species.

1. **C. mihiwaka** (Chilton) 1898 *Hyalella m.*, Chilton in: Ann. nat. Hist., ser. 7 *v.* 1 p. 423 t. 18 | 1899 *Chiltonia m.*, T. Stebbing in: Tr. Linn. Soc. London, ser. 2 *v.* 7 p. 408.

Body stout, back of peraeon broad. Side-plate 4 much wider than the preceding, much deeper than the following peraeopods. Pleon segment 3, postero-lateral corners quadrate. Eyes small, round. Antenna 1, peduncle rather long, 1st joint as long as 2^d and 3^d combined, flagellum subequal to peduncle, in ♂ 10-, in ♀ 8-jointed. Antenna 2, gland-cone prominent, ultimate joint of peduncle a little longer than penultimate, flagellum shorter than peduncle, in ♂ 8-, in ♀ 6-jointed. Maxillipeds, inner plates with outermost of the 3 apical spine-teeth large, outer plates smaller than inner, armed with fine setae, 4th joint of palp with a long apical seta. Gnathopod 1, 5th joint as long as 6th, hind margin fringed with about 15 long setae, 6th subrectangular, widening distally, palm nearly transverse, well defined by a projection of hind margin. Gnathopod 2 in ♂, 3^d, 4th and 5th joints small, subequal, 5th not lobed, squared, 6th very large, oblong, slightly widened distally, palm slightly convex, very little oblique, defined by a process of hind margin, finger stout, inner margin slightly projecting near base. Gnathopod 1 in ♀ as in ♂. Gnathopod 2 in ♀ like gnathopod 1, but 5th joint shorter, subtriangular, with only 5 or 6 setae. Peraeopods 3—5, 2^d joint very broad, hind margin very convex, minutely serrate. Uropods 1 and 2 well developed. Uropod 3 very minute, with no distinct peduncle, the single ramus pear-shaped and bearing a few minute setae, the ramus here spoken of perhaps including the peduncle. Telson subrectangular, angles rounded. Colour greyish or nearly white. L. reaching 5 mm.

New Zealand. Mountain streams up to about 457 m; and at 731 m.

8. Gen. **Parhyale** Stebb.

1897 *Parhyale* (Sp. un.: *P. fasciger*), T. Stebbing in: Tr. Linn. Soc. London, ser. 2
v. 7 p. 26.

Distinguished from Hyale (p. 559) only by the uropod 3, which has
a minute inner ramus.

1 species.

1. **P. fascigera** Stebb. 1897 *P. fasciger,* T. Stebbing in: Tr. Linn. Soc. London,
ser. 2 *v.* 7 p. 26 t. 6.

Side-plates 1—4 well developed, 1st widened below, 4th slightly
bi-emarginate behind. Pleon segment 3, postero-lateral corners quadrate.
Eyes large, oval, dark, obliquely approximate. Antenna 1, 1st joint fully as
long as 2d and 3d combined, flagellum longer than peduncle, 10- or 11-jointed.
Antenna 2, ultimate and penultimate joints of peduncle subequal, flagellum
not greatly longer than peduncle, in ♂ 20-, in ♀ 14—16-jointed. Maxillipeds,
2d and 3d joints of palp distally widened, 3d in ♂ having a dense fascicle
of setae at outer apex. Gnathopod 1 in ♂, 5th joint with broad, fringed
distal lobe, 6th as broad and long as 5th, oblong, with row of spines on distal
half of hind margin, palm rather oblique, well defined, finger stout, inner
margin with small setules, the tip acute, passing 2 stout palmar spines.
Gnathopod 1 in ♀, 6th joint less broad than 5th, finger not so stout as in ♂.
Gnathopod 2 in ♂, 6th joint massive, oblong, hind margin with little sub-
distal group of spinules, palm very oblique, spinulose, convex, ending in a
small pocket with palmar spines, receiving apex of long and broad, well
curved finger. Gnathopod 2 in ♀, 6th joint less massive, palm still more oblique,
leaving a very short hind margin, which is fringed with spines; marsupial
plates long, distally acute and closely fringed with rather short setae.
Peraeopods 1—5 not elongate, the spines not large, finger short, with strong
inner setule. Peraeopods 3—5, 2d joint well expanded, hind margin not
strongly serrate. Uropods 1 and 2, outer ramus without marginal spines.
Uropod 3 short, peduncle not as long as telson, outer ramus a little shorter
than peduncle, inner ramus conical, tipped with a very short seta. Telson
divided to the base, the lobes oblong or subtriangular, almost vertically
placed. L. about 7 mm.

Caribbean Sea (St. Thomas, harbour; Antigua).

9. Gen. **Neobule** Hasw.

1879 *Neobule* (Sp. un.: *N. algicola*), Haswell in: P. Linn. Soc. N. S. Wales, *v.* 4
p. 255 | 1880 *Neobula*, J. V. Carus in: Zool. Anz., *v.* 3 p. 291.

Maxillipeds, inner plates developed, but not outer. Gnathopod 2 larger
than gnathopod 1, both subchelate. Uropod 3 biramous. Telson squamiform.

1 accepted and 2 doubtful species.

1. **N. gaimardii** (M.-E.) 1840 *Amphitoe g.,* H. Milne Edwards, Hist. nat.
Crust., *v.* 3 p. 37 | 1879 *Neobule algicola,* Haswell in: P. Linn. Soc. N. S. Wales, *v.* 4
p. 255 t. 8 f. 4 | 1885 *N. (Hyale?) a.,* Haswell in: P. Linn. Soc. N. S. Wales, *v.* 10 p. 96
t. 11 f. 4—6 | 1893 *N. (H.?) a.,* A. Della Valle in: F. Fl. Neapel, *v.* 20 p. 897.

Side-plates large, especially side-plate 4. Eyes round. Antennae 1
and 2 equal. Antenna 1, 1st joint longer and stouter than the others,

3^d very short, flagellum rather longer than peduncle, about 11-jointed. Antenna 2, ultimate joint of peduncle longer than penultimate, flagellum subequal to peduncle, 11-jointed. Gnathopod 1, 5th joint triangular, subequal to the oblong 6th, palm transverse, slightly concave, finger matching palm. Gnathopod 2, 5th joint distally expanded into a broad, somewhat produced lobe, 6th broadly oblong, palm straight, transverse. Peraeopod 5 longer than peraeopod 4, 2^d joint broader than in preceding peraeopods. Uropod 3. rami extremely short (not figured by Haswell). L. about 6 mm.

South-Pacific (New Holland; Kiama [New South Wales]). Among sea-weed between tide-marks.

N. armorica (M.-E.) 1830 *Amphithoe a.*, H. Milne Edwards in: Ann. Sci. nat., v. 20 p. 378 | 1862 *Amphithoë a.*, *Nicea* (part.)?, Bate, Cat. Amphip. Brit. Mus., p. 243 | 1893 *Amphithoe a.*, A. Della Valle in: F. Fl. Neapel, v. 20 p. 424.

Uropod 3 short, ending in 2 conical appendages much shorter than peduncle.

Brittany.

N. reynaudii (M.-E.) 1830 *Amphithoe r.*, H. Milne Edwards in: Ann. Sci. nat., v. 20 p. 378 | 1862 *Amphithoë r.*, Bate, Cat. Amphip. Brit. Mus., p. 243 | 1893 *Amphithoe rubricata* (part.), A. Della Valle in: F. Fl. Neapel, v. 20 p. 459.

Cape of Good Hope.

10. Gen. **Parorchestia** Stebb.

1899 *Parorchestia*, T. Stebbing in: Tr. Linn. Soc. London, ser. 2 v. 7 p. 402.

Like Orchestia (p. 530), but maxillipeds with 4th joint of palp distinct, though very small, conical and having a spine on the truncate apex.

3 species.

Synopsis of species:

1 $\begin{cases} \text{Antenna 1, } 3^d \text{ joint shorter than } 2^d \ldots \ldots \ldots \text{ 1. } \textbf{P. tenuis} \ldots \text{ p. 557} \\ \text{Antenna 1, } 3^d \text{ joint longer than } 2^d - \textbf{2.} \end{cases}$

2 $\begin{cases} \text{Antenna 1, joints of flagellum unusually elongate . 2. } \textbf{P. hawaiensis} \text{ . p. 558} \\ \text{Antenna 1. joints of flagellum not unusually elongate 3. } \textbf{P. sylvicola} \text{ . . p. 558} \end{cases}$

1. **P. tenuis** (Dana) 1852 *Orchestia t.*, J. D. Dana in: P. Amer. Ac., v. 2 p. 202 | 1853 & 55 *O. t.*, J. D. Dana in: U. S. expl. Exp., v. 13 п p. 872; t. 59 f. 1 | 1862 *O. t.*, Bate, Cat. Amphip. Brit. Mus., p. 29 t. 4 f. 10 | 1899 *O. t.*, *Parorchestia* (part.), T. Stebbing in: Tr. Linn. Soc. London, ser. 2 v. 7 p. 402 | 1881 *O. sylvicola* ´(err., non J. D. Dana 1852!), G. M. Thomson in: Tr. N. Zealand Inst., v. 13 p. 212 t. 7 f. 4 | 1884 *Allorchestes recens*, G. M. Thomson in: Tr. N. Zealand Inst., v. 16 p. 235 t. 13 f. 2—5 | 1886 *A. r.*, G. M. Thomson & Chilton in: Tr. N. Zealand Inst., v. 18 p. 145 | 1888 *A. r.*, T. Stebbing in: Rep. Voy. Challenger, v. 29 p. 1639 | 1893 *Orchestia gammarellus* (part.), A. Della Valle in: F. Fl. Neapel, v. 20 p. 501.

Body compressed. Side-plate 1 less broad than any of side-plates 2—5; side-plates 2—4 excavate behind, 5th shallower than 4th. Pleon segment 3, postero-lateral corners quadrate, a little produced acutely. Eyes small, roundish, dark, scarcely their width apart. Antenna 1, 3^d joint of peduncle not so long as 2^d, flagellum subequal to peduncle, 5- or 6-jointed. Antenna 2 half as long as body (Dana) or less; ultimate joint of peduncle a little longer than penultimate, flagellum rather longer than peduncle, 8—14-jointed. Upper lip apically broadly rounded. Maxilla 1, palp minute. Gnathopod 1 in ♂ small, compact, 4th joint having on hind margin an apical scabrous boss, 5th with free part of hind margin broadly produced, scabrous, 6th shorter

than 5th, widening distally into a scabrous boss, so that the finger does not
reach the end of the apical border but only that of the true palm, which is
transverse and setulose. Gnathopod 2 in ♂, 2^d joint slightly widened distally,
5th small, triangular, masked by 4th; 6th large, with the closed finger forming
a broad oval, palm rather oblique, nearly straight, with slight sinuosities,
fringed with spines, well defined but not by a tooth. Gnathopod 1 in ♀,
expansions of 4th and 5th joints smaller than in ♂, 6th narrow at base,
then rather wide, almost evenly oblong and smoothly margined to the palm,
which is nearly straight and transverse, with a finger reaching exactly to
the end of it and of the apical margin. Gnathopod 2 in ♀, 2^d joint not
expanded, 4th and 5th resembling those of gnathopod 1 in ♂, though a little
less robust, 6th nearly as long as 5th, hind margin straight, produced in a
narrowly rounded lobe a little beyond the short, stout, acute finger and the
small, excavate, oblique palm. Peraeopods 1—5 rather short and compact,
finger small. Peraeopods 3—5, 2^d joint well expanded, as broad below as
above. Branchial vesicles oval. Pleopods 1—3, rami as long as peduncle.
Uropods 1 and 2 not elongate, outer ramus without marginal spines. Uropod 3
small, ramus slender, rather shorter than peduncle. Telson rather longer
than broad, proximally squared, then narrowing to entire apex with groups
of spinules. L. about 12 mm.

New Zealand. Among roots of grasses and in a small stream.

2. **P. hawaiensis** (Dana) 1853 & 55 *Orchestia h.*; J. D. Dana in: U. S. expl.
Exp., *v.* 13 II p. 880; t. 59 f. 8 | 1862 *O. h.*, Bate, Cat. Amphip. Brit. Mus., p. 32 t. 5
f. 7 | 1893 *O.? h.*, A. Della Valle in: F. Fl. Neapel, *v.* 20 p. 509 | 1899 *O. h.*, *Parorchestia*
(part.), T. Stebbing in: Tr. Linn. Soc. London, ser. 2 *v.* 7 p. 402.

Side-plates rather large, 5th almost as deep as 4th. Pleon segments 1—3
with no hairs, setae, or notches on the lateral margin. Eyes nearly round.
Antenna 1 reaching nearly end of ultimate joint of peduncle of antenna 2,
3^d joint as long as 1st and 2^d combined, flagellum with 7 joints which are
full 4 times as long as broad. Antenna 2 rather more than half as long as
body, peduncle rather long, flagellum longer than peduncle, with 17 or 18 joints,
which are more than twice as long as broad. Gnathopod 1 in ♀, 6th joint
oblong, but narrower at apex, and not properly truncate, finger a little longer
than the width of the joint. Gnathopod 2 in ♀, 4th joint gibbous and finely
scabrous below, 6th not much shorter than 5th, with distal expansion not very
broad, the minute finger rather obliquely placed. Peraeopods 1—5 slender,
1st and 2^d comparatively long, 3^d not shorter than 2^d. Uropod 1, outer ramus
with only apical spines, one very long. L. 16—18 mm.

Tropical Pacific (Hawaiian Islands).

3. **P. sylvicola** (Dana) 1852 *Orchestia s.*, J. D. Dana in: P. Amer. Ac., *v.* 2
p. 202 | 1853 & 55 *O. s.*, J. D. Dana in: U. S. expl. Exp., *v.* 13 II p. 873; t. 59 f. 2 ♀, f.
3 ♂ | 1862 *O. s.*, Bate, Cat. Amphip. Brit. Mus., p. 21 t. 3 f. 7 | 1881 *O. s.*, G. M. Thom-
son in: Tr. N. Zealand Inst., *v.* 13 p. 208 t. 7 f. 4 | 1886 *O. s.*, G. M. Thomson & Chilton
in: Tr. N. Zealand Inst., *v.* 18 p. 145 | 1893 *O. s.*, A. Della Valle in: F. Fl. Neapel,
v. 20 p. 510 | 1899 *O. s.*, *Parorchestia* (part.), T. Stebbing in: Tr. Linn. Soc. London,
ser. 2 *v.* 7 p. 402.

Body rather compressed. Side-plates not very large, 5th nearly as deep
as 4th. Pleon segment 3, postero-lateral corners quadrate. Eyes round,
rather small, dark, usually as far apart as their width. Antenna 1 reaching
end of penultimate joint of peduncle of antenna 2, but sometimes much
farther, 3^d joint longer or much longer than 2^d or 1st, flagellum not shorter

than peduncle, 8-jointed. Antenna 2 slender, penultimate joint of peduncle more than half as long as ultimate, flagellum with about 20 joints, which are longer than broad. Gnathopod 1 in ♂ slender, 4th joint with obtuse pellucid process of hind margin, 5th with free part of hind margin broadly produced, 6th joint not very long, with $^2/_3$ length of hind margin slightly bulging, and the small finger reaching end of palm, but not of apical border. Gnathopod 2 in ♂, 6th joint large, subovate, palm spinulose, very oblique, slightly convex, not defined by any tooth, with a very small excavation near the hinge, finger with a corresponding thickening, the rest of its inner margin slightly sinuous. Gnathopod 1 in ♀, 4th joint longer than 3^d, 5th much longer than 6th, 6th narrowly oblong, finger a little overlapping the palm. Gnathopod 2 in ♀, 3^d joint longer than 4th, 5th rather longer than 6th, free part of hind margin somewhat expanded, 6th elongate, produced to a narrowly rounded apex, much beyond the small finger, which is placed rather obliquely. Peraeopod 3 considerably shorter than peraeopod 4, 2^d joint in both piriform, with scarcely any free corner. Peraeopod 5 rather longer than peraeopod 4, 2^d joint much wider, as broad as long, a little narrowed below, free corner rounded, distinct. Uropods 1 and 2, outer ramus with spines only at apex, in both sexes (Dana: only in ♂). Uropod 3, peduncle short, scarcely longer than broad, ramus triangular, much narrower than peduncle and rather shorter, with 3 spines. Telson slightly notched at apex. L. 12—16 mm.

New Zealand. From moist soil in the bottom of the extinct Volcano of Taiamai far from the sea, and perhaps in other parts.

11. Gen. **Hyale** H. Rathke

1837 *Hyale* (Sp. un.: *H. pontica*), H. Rathke in: Mém. prés. Ac. St.-Pétersb., *v.*3 p.377 | 1876 *H.*, T. Stebbing in: Ann. nat. Hist., ser. 4 *v.*17 p.337 | 1879 *H.*, Subgen. *H.*, Wrześniowski in: Zool. Anz., *v.*2 p.201 | 1888 *H.*, T. Stebbing in: Rep. Voy. Challenger, *v.*29 p.171 etc. | 1890 *H.*, G. O. Sars, Crust. Norway, *v.*1 p.26 | 1893 *H.*, A. Della Valle in: F. Fl. Neapel, *v.*20 p.517 | 1849 *Nicea* (Sp. un.: *N. lucasii*), H. Nicolet in: Gay, Hist. Chile, *v.*3 p.237 | 1849 *Allorchestes* (part.), J. D. Dana in: Amer. J. Sci., ser.2 *v.*8 p.136 | 1852 *A.* (part.), J. D. Dana in: P. Amer. Ac., *v.*2 p.205 | 1856 *Galanthis* (Sp. un.: *G. lubbockiana*) (non Gistl 1848, Mollusca!), Bate in: Rep. Brit. Ass., Meet. 25 p.57 | 1857 *G.*, Bate in: Ann. nat. Hist., ser. 2 *v.*19 p.136.

Side-plate 4 much deeper than 5th. Antenna 1 longer than peduncle of antenna 2. Maxilla 1, palp 1-jointed, reaching to base of apical spines of outer plate. Maxillipeds, palp 4-jointed. Gnathopod 2 in ♂ (Fig. 96 p. 565), 5th joint small, masked behind by 4th. Gnathopod 2 in ♀, 5th joint produced behind between 4th and 6th. Uropod 3 uniramous. Telson divided.

Chiefly on coasts, also in the Sargasso Sea and on floating objects of various kinds, sometimes apparently in the open ocean.

22 accepted and 20 doubtful species.

Synopsis of accepted species:

1 { Body carinate — **2.**
{ Body not carinate — **3.**

2 { Body without dorsal teeth . . . . 1. **H. carinata** . . . . . p. 561
{ Body with dorsal teeth . . . . 2. **H. ochotensis** . . . . p. 561

3 { Antenna 2 longer than the body . . . . . 3. **H. campbellica** . . . p. 562
{ Antenna 2 not longer than the body — **4.**

4 { Gnathopod 2 in ♂. hind margin of 6th joint extremely short — **5.**
Gnathopod 2 in ♂, hind margin of 6th joint not extremely short — **11.**

5 { Gnathopod 1 in ♂, finger strongly furcate .   4. **H. diplodactyla** . . . p. 562
Gnathopod 1 in ♂, finger not strongly furcate — **6.**

6 { Gnathopod 1 in ♂, 6th joint with hump on front margin . . . .   5. **H. galateae** . . . . . p. 563
Gnathopod 1 in ♂, 6th joint with smooth front margin — **7.**

7 { Gnathopod 1 in ♂, 6th joint with serrulate distal margin . . . .   6. **H. maroubrae** . . . . p. 563
Gnathopod 1 in ♂, 6th joint without serrulate distal margin — **8.**

8 { Gnathopod 2 in ♂, palm densely fringed with setules — **9.**
Gnathopod 2 in ♂, palm not densely fringed with setules — **11.**

9 { Eyes reniform . . . .   7. **H. graminea** . . . . p. 564
Eyes not reniform — **10.**

10 { Gnathopod 2 in ♂, finger not nearly as long as 6th joint . . . .   8. **H. hirtipalma** . . . . p. 564
Gnathopod 2 in ♂, finger nearly as long as 6th joint . . . .   9. **H. macrodactyla** . . . . p. 564

11 { Gnathopod 1 in ♂, finger with subapical constriction . . . .   10. **H. aquilina** . . . . . p. 565
Gnathopod 1 in ♂, finger without subapical constriction — **12.**

12 { Peraeopods 4 and 5, 6th joint with spine and setae on mid hind margin . . . .   11. **H. prevostii** . . . . . p. 565
Peraeopods 4 and 5, 6th joint with smooth hind margin — **13.**

13 { Antenna 2 stout — **14.**
Antenna 2 not stout — **17.**

14 { Peraeopods 1—5, finger distinctly though microscopically pectinate — **15.**
Peraeopods 1—5, finger not distinctly pectinate — **16.**

15 { Uropod 1, marginal spines on only one of the rami . . . .   12. **H. grandicornis** . . . p. 566
Uropod 1, marginal spines on both rami . .   13. **H. novaezealandiae** . p. 567

16 { Gnathopod 1 in ♂, 6th joint distally narrowed   14. **H. grimaldii** . . . . . p. 567
Gnathopod 1 in ♂, 6th joint distally widened   15. **H. crassicornis** . . . p. 568

17 { Peraeopods 3—5, 6th joint with a very conspicuous serrate spine — **18.**
Peraeopods 3—5, 6th joint with no exceptionally conspicuous serrate spine — **19.**

18 { Gnathopod 1 in ♂, 5th joint produced along 6th   16. **H. pontica** . . . . . p. 568
Gnathopod 1 in ♂, 5th joint not produced along 6th . . . .   17. **H. media** . . . . . . p. 569

19 { Peraeopods 1—5, setule on inner margin of finger strong . . . .   18. **H. perieri** . . . . . . p. 570
Peraeopods 1—5, setule on inner margin of finger weak — **20.**

1. **H. carinata** (Bate) 1862 *Allorchestes carinatus*, Bate, Cat. Amphip. Brit.
Mus., p. 87 t. 6 f. 2 (♂ juv.?) | 1866 *Nicea longicornis* (non *Orchestia l.* Krøyer 1845!),
E. Grube in: Arch. Naturg., v. 321 p. 388 | 1866 *N. crassipes*, Cam. Heller in: Denk.
Ak. Wien, v. 26 ii p. 12 t. 1 f. 34, 35 (♂ juv.?) | 1893 *Hyale pontica* (part.: juv.)?, A. Della
Valle in: F. Fl. Neapel, v. 20 p. 528.

Peraeon segments 1—6 dorsally rounded, not broadly, peraeon segment 7
and pleon segments 1—3 slightly carinate. Side-plates 1—4 rather large,
1st widening, 2d narrowing distally. Pleon segment 3, postero-lateral corners
quadrate, with slight projecting point. Eyes small, subrotund, wider than
their diameter apart. Antenna 1 stouter than antenna 2 and not shorter,
1st joint stout, as long as 2d and 3d combined, flagellum 11-jointed. Antenna 2,
ultimate and penultimate joints of peduncle subequal in length, neither of
them as long or nearly as stout as 1st of antenna 1, flagellum 12-jointed. In
the young taken from mother's pouch antenna 1 decidedly longer than antenna 2,
flagellum 3-jointed, antenna 2, flagellum 2-jointed. Lower lip, principal lobes
distally narrowed, divergent. Maxillipeds, apical spine of palp longer than the
conical 4th joint. Gnathopod 1 in ♂, 6th joint small, elongate quadrangular, hind
margin almost straight, with a setose indent near middle, palm nearly transverse, the slightly curved finger matching it (Heller). Gnathopod 2 in ♂,
5th joint not produced between 4th and 6th, 6th broad, somewhat oval, palm
oblique, about equal to the nearly straight hind margin, defined from it by a blunt
angle, finger slightly curved (Heller). Gnathopod 1 in ♀, 2d joint narrow,
reaching end of side-plate, 5th not lobed, scarcely as long as the narrow
oblong 6th, of which the hind margin is fringed with spines, palm convex,
not at all oblique; finger longer than palm. Gnathopod 2 in ♀ with narrow
branchial vesicle, and long broad marsupial plate fringed with long
setae, the limb almost exactly like that of gnathopod 1, 2d joint slightly
larger, 5th rather broader, 6th rather longer (Grube). Peraeopods 1 and 2
rather slender. Peraeopods 3—5 rather robust, 2d joint with hind margin
faintly crenulate, in peraeopods 3 and 4 this joint oblong, a little wider above
than below, 4th joint expanded to twice width of 5th; in peraeopod 5 2d joint
with convex front margin, hinder expansion giving it a circular appearance,
the expansion produced below the 3d joint, 4th less expanded than in the 2
preceding peraeopods; finger in all peraeopods stout and curved. Uropods 1—3
short. Telson short, broad, each half nearly as broad at base as the length,
apex also broad. L. ♀ reaching 6, ♂ reaching 8 mm.

Mediterranean (Italy, Messina); Adriatic.

2. **H. ochotensis** (F. Brandt) 1851 *Allorchestes o.*, F. Brandt in: Middendorff,
Reise Sibirien, v. 21 p. 143 t. 6 f. 27 a—f | 1893 *A. o.*, A. Della Valle in: F. Fl. Neapel,
v. 20 p. 528 | 1888 *Hyale o.*, T. Stebbing in: Rep. Voy. Challenger, v. 29 p. 247.

Body at centre moderately, behind very strongly compressed, in the last ²/₃ having a carina gradually encreasing in strength, peraeon segments 6 and 7 each with small dorsal tooth, pleon segments 1—4 each with a strong one. Side-plates 1—5 rather large. Pleon segment 3, postero-lateral corners seemingly obtusely quadrate. Eyes rather small, almost reniform, black. Antenna 1 very little shorter than antenna 2, which are about half as long as body, both antennae sparingly armed, with flagellum longer than peduncle. Gnathopod 1, 5th joint short, distally wider than 6th, 6th elongate oval, palm oblique, ill-defined, finger bent. Gnathopod 2, 5th joint distally narrower than 6th, 6th somewhat similar in shape to that of gnathopod 1 but much larger, elongate-rhomboidal. Peraeopod 3,. 2^{d}. joint. rather small, almost circular. Peraeopod 4, 2^{d} joint rounded quadrate. Peraeopod 5, 2^{d} joint with expansion large, rounded behind, below produced into a triangular lobe. Uropod 3, ramus narrow, spinose, as long as peduncle. Telson with a spinule on each of the short, rounded rhomboidal, apically thickened lobes. L. reaching 29 mm.

Sea of Ochotsk.

3. H. campbellica (Filh.) 1885 *Allorchestes c.*, Filhol in: Recu. Passage Vénus, *v.* 3 n Zool. p. 466.

Eyes small, round. Antenna 1 reaching base of flagellum of antenna 2, which is a little longer than the body. Gnathopod 1, 5th and 6th joints equal in length. Gnathopod 2, 6th joint robust, oval, a little compressed laterally, finger hooked, the part of the hand to which it is applied showing on the inner face a series of minute denticulations; 5th joint without produced process („spine"). Peraeopods 3—5 have very fine setae at the apices of their joints; hind margin of 4th joint denticulate. L. 8 mm.

South-Pacific (shore of Perseverance Bay [Campbell Island]).

4. H. diplodactyla Stebb. 1899 *H. diplodactylus*, T. Stebbing in: Tr. Linn. Soc. London, ser. 2 *v.* 7 p. 403 t. 31 C.

Side-plates 1 and 2 not very deep, 1st little widened below. Eyes rounded, light-coloured in spirit. Antenna 1 much longer than peduncle of antenna 2, peduncle short, its joints successively much shorter, flagellum in ♂ 14-, in ♀ 9-jointed. Antenna 2, ultimate joint of peduncle longer than penultimate, flagellum in ♂ 26-, in ♀ 17-jointed. Gnathopod 1 in ♂ (Fig. 95), 2^{d} joint short, distal part wide, 4th distally squared, supporting fringed hind lobe of 5th, this lobe projecting beyond a short straight piece of the hind margin, 6th joint widening greatly to the palm, hind margin sinuous, much shorter than smoothly curved front, palm long, not very oblique, excavate, ending in a wide pocket, which receives the deeply furcate end of finger, which is thus wider distally than at its base. Gnathopod 2 in ♂ scarcely differing from that of H. galateae, except that the expanded front margin of the 2^{d} joint is closely fringed with setules, and the 6th is widest near to the base instead of at junction of hind margin with the long very oblique palm. In other respects ♀ and ♂ appear to agree with H. galateae. Telson rather markedly upturned, with a slight twist to the rounded apex of each triangular lobe. L. about 5 mm.

4th joint

6th joint

Fig. 95. **H. diplodactyla**, ♂. Gnathopod 1.

Caribbean Sea (St. Croix).

5. **H. galateae** Stebb. 1899 *H. g.*, T. Stebbing in: Tr. Linn. Soc. London, ser. 2 *v.* 7 p. 402 t. 31 B.

Body and to some extent the appendages scabrous with little hairs or scales like an inverted T (in Pacific, but not in Atlantic specimens). Side-plate 1 widened distally, 2^d and 3^d not very deep. Pleon segment 3, postero-lateral corners quadrate. Eyes large, oval, nearly meeting at the top of head, black. Antenna 1 much longer than peduncle of antenna 2, joints of peduncle small, successively shorter, flagellum in ♂ with 9 or 10, in ♀ with 7 distally widened joints. Antenna 2 about $^1/_3$ as long as body, ultimate joint of peduncle considerably longer than penultimate, flagellum in ♂ 12-, in ♀ 9-jointed. Maxilla 1, palp with a small constriction. Maxillipeds, 4th joint of palp slender, curved. Gnathopod 1 in ♂, 2^d joint short and broad, 4th apically squared, 5th nearly as broad as long, forming a rounded lobe behind, with spines along the somewhat flattened hind margin, 6th rhomboidal, the long front margin having a slight hump at the centre, the hind margin much shorter, making an angle at the widest part of the joint before joining the oblique, spinuliferous palm, from which it is defined by a palmar spine, which is overlapped by the curved finger. Gnathopod 2 in ♂, 2^d and 3^d joints distally lobed in front, 4th with broadly rounded apex, 5th very short, with the little horny-looking process from the hind margin on either side more conspicuous than usual, 6th large, widest where the short, smoothly curved hind margin meets the long, very oblique, nearly straight, spinulose palm, narrowest at hinge of the long, curved finger, which has a strong bulb at base of inner margin. Gnathopods 1 and 2 in ♀ small, 5th joint short, 6th oblong, slightly widened at rather oblique, spinulose palm, gnathopod 2 rather the larger. Peraeopods 1—5 moderately robust, finger with minute inner setule; the oval 2^d joint in peraeopods 3—5 nearly smooth behind, its wing in peraeopod 5 broadly rounded and somewhat produced downward. Uropods 1 and 2 with lateral spines on both rami, in uropod 2 rami rather stout, unequal. Uropod 3, ramus rather shorter than peduncle. Telson with distal half of each lobe triangular. *L.* 4 mm.

Pacific (lat. 38° N., long. 180° E.; lat. 5° N., 137° E.); North-Atlantic (Sargasso Sea; lat. 26° N., long. 59° W.).

6. **H. maroubrae** Stebb. 1899 *H. m.*, T. Stebbing in: Tr. Linn. Soc. London, ser. 2 *v.* 7 p. 405 t. 32 C.

Body rather compressed. Pleon segment 3 with postero-lateral corners quadrate. Eyes rounded, about the diameter apart, darkish. Antenna 1, peduncle short, 1st joint as long as 2^d and 3^d combined, flagellum of 9 slender joints. Antenna 2 about half as long as body, flagellum longer than peduncle, slender, 19-jointed. Gnathopod 1 in ♂, 4th, 5th and 6th joints subequal in length, 6th oblong, scarcely longer than wide, hind margin from near base fringed with spinules, which distally meet a transverse row of smaller spinules, across which the short finger closes, as if along a palm, but half the distal margin of the 6th joint extends beyond these in a microscopically denticulate lobe at right angles to the hind margin though the junction is rounded off. Gnathopod 2 in ♂, 5th joint very small, triangular, 6th very large, broadest proximally, hind margin very short, spine-fringed palm very oblique and long, well defined, the long finger nearly reaching the 4th joint, its apex passing on inner side of palmar spine into a pocket on surface of the joint. Branchial vesicles very small. Peraeopods 1—5 having on 6th joint a distal, blunt-headed, partially serrate spine, and between this and the finger a spine upbent towards the other, of a peculiar fusiform shape. The setule on inner margin of finger very small. Uropod 3, ramus moderately slender, as long as peduncle, each with apical spines only. Telson with lobes distally somewhat acutely triangular. L. 5 mm.

Maroubra Bay near Sydney [East-Australia].

7. **H. graminea** (Dana) 1852 *Allorchestes? g.*, J. D. Dana in: P. Amer. Ac., *v.* 2 p. 208 | 1853 & 55 *A.? g.*, J. D. Dana in: U. S. expl. Exp., *v.* 13 II p. 897; t. 61 f. 3 a—b | 1862 *A. gramineus*, Bate, Cat. Amphip. Brit. Mus., p. 46 t. 7 f. 8 | 1893 *Hyale prevostii* (part.), A. Della Valle in: F. Fl. Neapel, *v.* 20 p. 519.

Eyes reniform. Antenna 1 $^2/_3$ as long as antenna 2, flagellum slender, longer than peduncle, about 14-jointed. Antenna 2 less than half as long as body, flagellum stout, much longer than peduncle, joints hardly oblong. Gnathopod 1, 6th joint narrow, slightly broadest at middle, finger long, stout, folding against hind margin. Gnathopod 2, 6th joint large, narrowly ovate, narrow at finger-hinge, palm setulose, very oblique, not at all convex, scarcely defined from the very short hind margin. Peraeopods 1—5 feebly armed. Colour tints of green and yellow, hind legs partly carmine. L. 12—14 mm.

Tropical Atlantic (Rio Janeiro).

8. **H. hirtipalma** (Dana) 1852 *Allorchestes h.*, J. D. Dana in: P. Amer. Ac., *v.* 2 p. 205 | 1853 & 55 *A. h.*, J. D. Dana in: U. S. expl. Exp., *v.* 13 II p. 888; t. 60 f. 4 | 1879 *Hyale (A.) h.*, Wrześniowski in: Zool. Anz., *v.* 2 p. 200 | 1862 *A. inca*, Bate, Cat. Amphip. Brit. Mus., p. 40 t. 6 f. 7 | 1879 *Nicea fimbriata*, G. M. Thomson in: Tr. N. Zealand Inst., *v.* 11 p. 236 t. 10 B f. 2 | 1886 *N. f.*, G. M. Thomson & Chilton in: Tr. N. Zealand Inst., *v.* 18 p. 144 | 1888 *Hyale f.* + *H. hirtipalma*, T. Stebbing in: Rep. Voy. Challenger, *v.* 29 p. 1705 | 1895 *H. f.*, G. M. Thomson in: Tr. N. Zealand Inst., *v.* 27 p. 211 | 1893 *H. prevostii* (part.), A. Della Valle in: F. Fl. Neapel, *v.* 20 p. 519.

Eyes dark, rather small, subrotund. Antenna 1 half or more than half as long as antenna 2, proportionately longer in small specimens, joints of peduncle subequal, flagellum much longer than peduncle, 13—15-jointed. Antenna 2 less than half as long as body, ultimate joint of peduncle a little longer than penultimate, with a brush of setules on lower margin, flagellum longer than peduncle, with 17—22 joints, each with brush of setules. Maxillipeds, 4th joint of palp short, slender, acute. Gnathopod 1, 5th joint rather short and broad, 6th widening to the palm, which is longer and more oblique in a large specimen than in a small, setulose, with palmar spines (stout teeth, Thomson), finger fringed within with tiny setules. Gnathopod 2, 5th joint small, 6th very large, subovate, narrowing to finger-hinge, palm very oblique, densely fringed with setules, finger not nearly as long as 6th joint. Peraeopods 1—5 all rather long, 5th and 6th joints without spines on convex margin, finger rather short and stout, curved, acute. Peraeopods 3—5, 2^d joint well expanded. Uropods 1—3 not very long. Uropods 1 and 2, outer ramus rather the shorter. Uropod 3, ramus subequal to peduncle. Telson cleft almost to the base. Colour pale yellow. L. reaching 20 mm.

Pacific (Valparaiso; Island of San Lorenzo [Peru]; Dunedin [New Zealand]; Macquarie Island).

9. **H. macrodactyla** Stebb. 1899 *H. macrodactylus*, T. Stebbing in: Tr. Linn. Soc. London, ser. 2 *v.* 7 p. 404 t. 31 D.

Side-plates not deep, 1st distally widened, 4th wide, with deep hind emargination. Pleon segment 3, postero-lateral corners quadrate, scarcely produced. Eyes not large, rounded, wider than diameter apart. Antenna 1, peduncle short, 1st joint subequal to 2d and 3d combined, flagellum with 13 joints, of which the proximal are short. Antenna 2 more than half as long as body, ultimate joint of peduncle longer than the short penultimate, flagellum 25-jointed. Gnathopod 1 in ♂, 2d joint short, broad except at base, 3d with small front lobe, 5th with small hind lobe, 6th oblong oval, palm oblique, spinulose, separated from hind margin by rounded angle carrying a palmar spine, against the inner side of which the apex of finger closes. Gnathopod 2 in ♂ (Fig. 96), 2d joint lobed at distal end of front margin, 3d lobed in front, rounded apex of 4th touching the base of 6th, 6th elongate, widest at base, front margin smoothly curved, palm closely fringed with slender spinules, extending almost whole length of joint, nearly straight, but with an emargination between two slight swellings, one of which adjoins the finger-hinge, the long, blunt, slightly sinuous finger capable of touching apex of 4th joint. In a specimen, which appears to be a young ♂, the oblique palm is straight, finger matching it. Gnathopods 1 and 2 in ♀, 6th joint narrowly oblong. Peraeopods 1—5, 6th joint at inner apex carrying a strong blunt spine, with a similar much shorter one below it, finger curved, inner setule minute. Peraeopods 3 and 5, 2d joint somewhat orbicular. Peraeopod 4, 2d joint oblong oval, rather wider above than below. Uropods 1 and 2, both rami with lateral spines. Uropod 3, ramus as long as peduncle. Telson cleft to base, lobes bluntly triangular. L. about 4 mm.

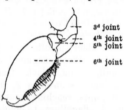

Fig. 96. H. macrodactyla, ♂.
Gnathopod 2.

Tropical Atlantic (St. Thomas' harbour; Rio Janeiro).

10. **H. aquilina** (A. Costa) 1857 *Amphithoe a.*, A. Costa in: Mem. Acc. Napoli, *v.*1 p.202 t.2 f.7 | 1893 *Hyale a.*, A. Della Valle in: F. Fl. Neapel, *v.*20 p.523 t.16 f.43—47 | 1862 *Allorchestes pereiri* (part.), Bate, Cat. Amphip. Brit. Mus., p.42 | 1866 *Nicea fasciculata*, Cam. Heller in: Denk. Ak. Wien, *v.*26 II p.6 t.1 f.10—12 | 1879 *Hyale f.*, Wrześniowski in: Zool. Anz., *v.*2 p.202 | 1888 *H. gazella* (part.), T. Stebbing in: Rep. Voy. Challenger, *v.*29 p.297.

Side-plate 1 broader than deep. Antenna 1 half as long as antenna 2 or a little more, flagellum twice as long as peduncle, 15- or 16-jointed. Antenna 2 about ⅔ as long as body or a little less, ultimate joint of peduncle rather longer than penultimate, flagellum fully twice as long as peduncle, 30—32-jointed. Maxillipeds, 3d joint of palp setose, 4th acute. Gnathopod 1 in ♂ small, 5th and 6th joints subequal or 5th joint a little longer, 6th subrotund, about as broad as long, palm a little sinuous, spinose, almost transverse, finger compressed, laminar, dilated from the base towards the apex, which is constricted, ending in a curved point, producing resemblance to a hawk's bill. Gnathopod 2 in ♂ robust, oblong oval, palm oblique, not strongly defined (Heller: with a small defining tubercle). Gnathopods 1 and 2 in ♀ similar, but 6th joint rather stouter in gnathopod 2; both gnathopods having 5th joint triangular, shorter than oblong 6th, finger slender, curved. Uropod 3, ramus shorter than peduncle. Telson small, lobes conical. L. 8 mm.

Mediterranean; Adriatic.

11. **H. prevostii** (M.-E.) 1830 *Amphithoe p.*, H. Milne Edwards in: Ann. Sci. nat., *v.*20 p.378 | 1862 *Nicea p.*, Bate, Cat. Amphip. Brit. Mus., p.53 | 1866 *Allorchestes prevosti*, E. Grube in: Arch. Naturg., *v.*321 p.386 | 1888 *Hyale prevostii*, T. Stebbing in: Rep. Voy. Challenger, *v.*29 p.144 | 1893 *H. p.* (part.), A. Della Valle in: F. Fl.

Neapel, *v.* 20 p. 519 t. 2 f. 6; t. 16 f. 23—42 | 1843 *Amphithoë p.*, *A. nilssonii*, H. Rathke in: N. Acta Ac. Leop., *v.* 20ɪ p. 81 t. 4 f. 5ᴀ—ᴇ; p. 264 | 1861 *Allorchestes n.*, Bate & WestWood, Brit. sess. Crust., *v.* 1 p. 40 f. | 1876 *Hyale nilssoni*, T. Stebbing in: Ann. nat. Hist., ser. 4 *v.* 17 p. 337 t. 18 f. 1a—h | 1890 *H. n.*, G. O. Sars, Crust. NorWay, *v.* 1 p. 26 t. 11 f. 1 | 1887 *H. nilsoni* (part.) + *H. n. var. minor*, Chevreux in: Bull. Soc. zool. France, *v.* 12 p. 293, 294 | 1888 *H. n.* + *H. n. var. major*, Chevreux in: Bull. Soc. zool. France, *v.* 13 p. 32 | 1893 *H. n.*, Chevreux & E. L. Bouvier in: Ann. Sci. nat., ser. 7 *v.* 15 p. 111 | 1845 *Orchestia nidrosiensis*, Krøyer in: Naturh. Tidsskr., ser. 2 *v.* 1 p. 299 | 1857 *Allorchestes danai*, Bate in: Ann. nat. Hist., ser. 2 *v.* 19 p. 186.

Back smooth. Side-plates with lower margin smooth, side-plate 1 widened below. Pleon segment 3, postero-lateral corners obtusely quadrate. Eyes small, round, black; (large, oval in *var. major*). Antenna 1 in ♀ nearly as long as antenna 2, flagellum 6-jointed, in ♂ flagellum sometimes with 10 joints, reaching considerably beyond peduncle of antenna 2. Antenna 2 in ♂, flagellum 15-jointed. Gnathopod 1 in ♂, 5th joint distally some-what widened, not produced along hind margin of 6th, 6th oblong, hind margin with spinules from the base to a submedian spine, palm slightly oblique, rounded off. Gnathopod 2 in ♂, 7th joint robust, broadest proximally, front margin very convex, hind margin with various spinules but no conspicuous notch, palm rather oblique, somewhat sinuous, well defined, finger closing between two rows of spines to the terminal cavity of the palm, which is armed with minute spines. Gnathopod 1 in ♀ nearly as in ♂. Gnathopod 2 in ♀ like gnathopod 1, but rather larger, 5th joint more expanded distally. Peraeopods 1—5 with numerous spines and setae, but none of exceptional size, the 6th joint having besides others an apical and a subapical one minutely booked at tip and opposable to the strong curved finger. Peraeopods 3—5, 2^{d} joint expanded with smooth hind margin. Peraeopods 4 and 5, 6th joint with group of setae and spine at middle of hind margin. Uropods 1 and 2, rami short, with marginal and strong apical spines. Uropod 3 small, ramus much shorter than peduncle. Telson divided into 2 obtuse lobes with their surfaces adjacent. Colour green with metallic lustre. L. 6—8 mm (almost double in *var. major*).

North-Atlantic and adjoining seas (Europe; Azores). In sea-Weed betWeen the high-Water marks respectiVely of neap- and spring-tides; *var. major* at low-tide.

12. **H. grandicornis** (Krøyer) 1845 *Orchestia g.*, *O. longicornis*, Krøyer in: Naturh. Tidsskr., ser. 2 *v.* 1 p. 292; t. 1 f. 2a—n | 1851 *Allorchestes g.*, F. Brandt in: Bull. phys.-math. Ac. St.-Pétersb., *v.* 9 p. 142 | 1888 *Hyale g.*, T. Stebbing in: Rep. Voy. Challenger, *v.* 29 p. 210 | 1849 *Nicea lucasii*, H. Nicolet in: Gay, Hist. Chile, *v.* 3 p. 238; Crust. t. 2 f. 7 | 1852 *Allorchestes verticillata* + *A. peruviana*, J. D. Dana in: P. Amer. Ao., *v.* 2 p. 205, 206 | 1862 *A. verticillatus*, *A. verticellatus*, Bate, Cat. Amphip. Brit. Mus., p. 43 t. 7 f. 1 | 1893 *Hyale prevostii* (part.), A. Della Valle in: F. Fl. Neapel, *v.* 20 p. 519.

Side-plate 1 somewhat produced forward, front margin faintly concave, angle broadly rounded (in Nicolet's figure subacute). Pleon segment 3, postero-lateral corners quadrate, but not acutely so. Eyes rather large, dark, subrotund, sometimes approximate at top of the head. Antenna 1 about $^{1}/_{5}$ as long as body, more than half as long as antenna 2, joints of peduncle successively shorter, flagellum much longer than peduncle, 12-jointed. An-tenna 2 about $^{1}/_{3}$ as long as body, with small but discernible gland-cone, ultimate joint of peduncle a little longer than penultimate, flagellum longer than peduncle, longer and stouter than flagellum of antenna 1, but some-times with only 12 joints. Upper lip, distal margin almost straight truncate, with rounded corners. Maxillipeds, 4th joint of palp a slender pointed setuliferous finger. Gnathopod 1 in ♂, 2^{d} joint narrow at base, thence to the middle greatly widening, 5th joint sending out a lobe behind, fringed with spinules and not reaching the apex,. but resting in the concave distal margin of the 4th joint; 6th narrowly oblong, with spinuliferous indent near apex of hind margin, palm convex, slightly oblique, a little overlapped by finger. Gnathopod 2 in ♂, 2^d joint not broad, 4th rather produced behind, and helping the small obscure 5th to form a cup for heart-shaped base of the large oval 6th joint, of which the spinose palm is oblique, nearly straight, ending in a pocket, where it meets the short hind margin, which ends in a sort of double tubercle carrying 2 palmar spines; the finger powerful, its apex closing into the pocket. Gnathopods 1 and 2 in ♀ resembling one another and gnathopod 1 in ♂; gnathopod 2 rather the larger. Peraeopod 3 stouter but not longer than peraeopods 1 and 2; peraeopod 4 rather stronger than peraeopod 5; 2^d joint in peraeopods 3—5 oblong oval, more rounded in peraeopod 3, widest in peraeopod 5 with somewhat angular hind margin, in peraeopod 4 hind margin closely fringed with small spinules; 4th joint having tufts of spines on hind margin in all 3 peraeopods. In peraeopods 1—5 finger strong, curved, concave margin microscopically pectinate, teeth larger towards the nail. Pleopod 1, rami longer than peduncle, 15- or 16-jointed. Uropod 1, peduncle rather longer than rami, with a strong apical spine, outer ramus with only apical spines. Uropod 2, peduncle rather shorter than rami. Uropod 3, ramus apically truncate, shorter than peduncle. Telson cleft to the base, each lobe triangular from centre to apex, inner margin straight. Colour green to greyish or brownish. L. 8—12 mm.

South-Pacific (Valparaiso).

13. **H. novaezealandiae** (G. M. Thoms.) 1879 *Nicea n.*, G. M. Thomson in: Tr. N. Zealand Inst., *v.* 11 p. 235 t. 10 B f. 1 a—f | 1888 *Hyale n.*, T. Stebbing in: Rep. Voy. Challenger, *v.* 29 p. 500 | 1895 *H. n.*, G. M. Thomson in: Tr. N. Zealand Inst., *v.* 27 p. 211 | 1886 *Nicea neo-zelanica*, G. M. Thomson & Chilton in: Tr. N. Zealand Inst., *v.* 18 p. 144 | 1893 *Hyale prevostii* (part.), A. Della Valle in: F. Fl. Neapel, *v.* 20 p. 520.

Very near to H. grandicornis, but distinct in the following respects. Gnathopod 1 in ♂ larger, hind lobe of 5th joint larger, more produced, apical, 6th joint broader. Gnathopod 2 in ♂, 2^d joint distally a little widened and lobed. Peraeopods 3—5, 4th joint having the spines along hind margin solitary, not grouped. Peraeopod 4, 2^d joint nearer to the rounded form in 2^d joint of peraeopods 3 and 5, its hind margin also almost smooth and not spinulose. Peraeopods 1—5, microscopical pectination of concave margin of finger more faint, the customary setule stronger, than in H. grandicornis. Uropod 1, both rami with strong marginal spines. Colour yellowish, marbled with red. L. 12—14 mm.

South-Pacific (Otago Harbour [New Zealand], rock pools; Macquarie Island).

14. **H. grimaldii** Chevreux 1891 *H. g.*, Chevreux in: Bull. Soc. zool. France, *v.* 16 p. 257 f. 1—5 | 1893 *H. prevostii* (part.), A. Della Valle in: F. Fl. Neapel, *v.* 20 p. 889.

Side-plates 1—4 deep. Pleon segment 3, postero-lateral corners quadrate. Eyes of moderate size, subrotund, black. Antenna 1, peduncle short, flagellum 9- or 10-jointed in ♂, 7-jointed in ♀. Antenna 2 $^1/_3$ as long as body, very stout, ultimate joint of peduncle rather longer than penultimate, flagellum 17—23-jointed in ♂, 12-jointed in ♀; the distal joints in all the flagella somewhat elongate. Gnathopod 1 in ♂, 5th joint with spine-fringed lobe, 6th oblong, but a little widened at base, palm almost transverse, finger overlapping its rounded angle. Gnathopod 2 in ♂ large,

2^d joint with front distal end forming a very broad rounded lobe, 3^d with similar lobe, 4th bluntly pointed, completely masking the small 5th, 6th very broad, oblong, widest proximally, palm well defined, not very oblique, with rounded tooth near finger-hinge, finger strong. Gnathopod 1 in ♀, 2^d joint a little widened distally, 5th with spine-fringed lobe, distally wider than 6th, which is rather narrowly oblong, slightly widened at palm, finger very little longer than palm. Gnathopod 2 in ♀ like gnathopod 1, marsupial plates very long, rather broad in the middle, fringed with short setae. Peraeopods 1—5, 6th joint with spines at 5 points of inner margin, finger short, curved, inner setule minute. Peraeopods 1 and 2, 4th joint rather expanded. Peraeopods 3—5, 4th joint strongly expanded, stoutly spined on both margins, 5th expanded distally with group of apical spines on hind margin. Peraeopods 3 and 5, 2^d joint broadly oval. Peraeopod 4, 2^d joint oblong, with rounded angles, hind margin almost straight. Uropod 2, outer ramus shorter and narrower than inner. Uropod 3, ramus shorter than peduncle. Telson cleft to the base, the lobes of moderate size, apically narrowed, or (Chevreux) remarkably large, apically inflated. Colour orange brown, antennae and legs violet-rose. L. about 6 mm.

North-Atlantic (lat. 42° N., long. 24° W.; lat. 38° N., long. 64° W.); West-Mediterranean. On floating objects and Thalassochelys.

15. **H. crassicornis** (HasW.) 1879 *Allorchestes c.*, Haswell in: P. Linn. Soc. N. S. Wales, *v.* 4 p. 252 t. 7 f. 5 | 1885 *A. c.*, Haswell in: P. Linn. Soc. N. S. Wales, *v.* 10 p. 95 t. 10 f. 2—5 | 1893 *Hyale pontica* (part.)?, A. Della Valle in: F. Fl. Neapel, *v.* 20 p. 528.

Body slender. Pleon segment 3, postero-lateral corners quadrate. Eyes oval, dark, very near together above. Antenna 1 slender, reaching beyond peduncle of antenna 2, 2^d joint rather shorter than 1st, 3^d scarcely shorter than 2^d, peduncle $^2/_3$ as long as the 11-jointed flagellum. Antenna 2 about $^1/_3$ as long as body, very stout, ultimate joint of peduncle much longer than penultimate, flagellum tapering, rather longer than peduncle, with 13 joints, of which the proximal are very thick, peduncle and flagellum plumose. Gnathopod 1, 5th joint triangular, with spinules on rounded apex, 6th oblong, slightly widened distally, spinuliferous on distal half of hind margin, palm transverse, finger strong, matching palm. Gnathopod 2, 6th joint large, oblong, but with rather oblique, well defined, spinulose palm, over which the finger folds closely, not quite reaching the end. Peraeopod 3 short, 4th and 5th subequal, spines not strong. Uropod 3 short, ramus as long as peduncle. Lobes of telson blunt. — L. about 10 mm.

South-Pacific (Kiama [New South Wales]). Between tide-marks.

16. **H. pontica** H. Rathke 1837 *H. p.*, H. Rathke in: Mém. prés. Ac. St.-Pétersb., *v.* 3 p. 378 t. 5 f. 20—28 | 1862 *H. p.*, Bate, Cat. Amphip. Brit. Mus., p. 87 t. 14 a f. 1 | 1888 *H. p.*, T. Stebbing in: Rep. Voy. Challenger, *v.* 29 p. 163, 499 | 1840 *Amphitoe p.*, H. Milne Edwards, Hist. nat. Crust., *v.* 3 p. 37 | 1875 *Nicea p.*, Catta in: Rev. Sci. nat., *v.* 4 p. 166 | 1856 *Galanthis lubbockiana*, Bate in: Rep. Brit. Ass., Meet. 25 p. 57 t. 17 f. 7 | 1857 *Allorchestes imbricatus* + *G. l.*, Bate in: Ann. nat. Hist., ser. 2 *v.* 19 p. 136 | 1861 *Nicea l.*, Bate & Westwood, Brit. sess. Crust., *v.* 1 p. 47 f. | 1876 *Hyale l.*, T. Stebbing in: Ann. nat. Hist., ser. 4 *v.* 17 p. 337 t. 18 f. 2, a—d | 1890 *H. l.*, G. O. Sars, Crust. Norway, *v.* 1 p. 27 t. 11 f. 2 | 1895 *H. l.*, A. M. Norman in: Ann. nat. Hist., ser. 6 *v.* 15 p. 479 | 1871 *H. nilsoni* (part.), A. Boeck in: Forh. Selsk. Christian., 1870 p. 94 | 1872 & 76 *H. n.*, *H. nilssonii*, A. Boeck, Skand. Arkt. Amphip., *v.* 1 p. 109 t. 3 f. 3; *v.* 2 p. 712 | 1879 *H. imbricata*, Wrześniowski in: Zool. Anz., *v.* 2 p. 202.

Segments of body imbricated, integument firm. Side-plates with lower margin slightly crenulate. Pleon segment 3, postero-lateral corners quadrate, but not acutely. Eyes of moderate size. Antennae 1 and 2 in ♀ subequal. In ♂ antenna 1 considerably shorter than antenna 2, but reaching much beyond peduncle of latter, flagellum longer than peduncle, with 11 joints, most of them distally widened. Antenna 2 in ♂, ultimate and penultimate joints of peduncle subequal, flagellum 15-jointed. Gnathopod 1 in ♂, 5th joint in adult strongly produced along hind margin of 6th, 6th oblong, surface not far from base armed with a broad spine, distal part of hind margin fringed with spinules, palm transverse, rounded off but armed with a stout spine, finger matching palm. Gnathopod 2 in ♂, 6th joint robust, broadest proximally, front very convex, hind margin nearly straight, with small spinulose notch, palm slightly oblique, with small cavity not far from hinge, well defined, bordered with spinules and setules and having 2 palmar spines in a terminal cavity, the finger closing down on the inner side of these spines. Gnathopod 1 in ♀, 5th joint distally wider than 6th, 6th oblong, with spinules at notch of hind margin and also at apex, palm slightly oblique. Gnathopod 2 in ♀ similar to gnathopod 1, but larger, 5th joint more produced. Peraeopods 1—5 robust, 6th joint carrying a large, straight, subapical spine, serrate, tipped with a little hook, and opposable to the strongly curved finger; nearer the base the 6th joint has a much smaller spine and at the apex one resembling the subapical, on a much smaller scale. Peraeopods 3—5, 2d joint broadly expanded, with crenulate hind margin, 6th joint with hind margin unarmed. Uropod 1, outer ramus rather shorter than inner, both with marginal spines. Uropod 3, ramus a little shorter than peduncle, with apical spinules. Telson divided into 2 obtuse-ended lobes, which have their surfaces adjacent. Colour bluish grey or almost black, or yellowish, or darkish green. L. 7 mm.

North-Atlantic and North-Sea (England, France, West-Norway); Mediterranean (Algeria).

17. **H. media** (Dana) 1853 & 55 *Allorchestes m.*, J. D. Dana in: U. S. expl. Exp., v.13 II p.898; t.61 f.4 a—g, l—n | 1862 *A. medius*, Bate, Cat. Amphip. Brit. Mus., p.46 t.7 f.9 | 1879 *Hyale (A.) media*, Wrześniowski in: Zool. Anz., v.2 p.200 | 1893 *H. pontica* (part.), A. Della Valle in: F. Fl. Neapel, v.20 p.524.

Side-plates 1—5 wide, not deep. Pleon segment 3, postero-lateral corners quadrate, scarcely produced. Eyes apparently variable in shape and size. Antenna 1, peduncle short, flagellum 10—14-jointed. Antenna 2, ultimate joint of peduncle rather longer than penultimate, both rather short, flagellum 16-jointed. Gnathopod 1 in ♂, 2d joint strongly widening from very narrow base, 5th with rounded fringed lobe, not produced along the 6th, 6th oblong, margins parallel, hinder with median spinules, finger a little overlapping slightly oblique palm. Gnathopod 2 in ♂, 2d joint distally widened, 3d lobed in front, 6th large, subovate, widest before reaching the oblique well defined palm. Gnathopods 1 and 2 in ♀ about as in H. rubra (p. 572). Peraeopods 1—5 robust, finger curved, inner setule minute. Peraeopods 1 and 2, 6th joint having on inner apex a stout, smooth spine, which is widest at its apex. Peraeopods 3—5 with a very large submedian serrate spine, like that which is a distinguishing feature in H. pontica, but higher up on the margin; at the apex a small spine curves towards the smooth finger. Peraeopods 3 and 5, 2d joint rotund, in peraeopod 5 very broad; peraeopod 4, 2d joint tending to oblong; in all 3 the hind margin nearly smooth, except for one pronounced indent. Pleopods with 2 short broad coupling hooks and

3 cleft spines. Uropods 1 and 2 rather short. Uropod 3, ramus as long as peduncle. Telson cleft to base, lobes triangular. L. 6—12 mm.

Atlantic (Rio Janeiro, harbour; Virgin Island St. Thomas; Cape Verde Islands; Tierra del Fuego?).

18. **H. perieri** (H. Luc.) 1846 *Orchestia p.*, H. Lucas in: Expl. Algérie, p. 52 t. 5 f. 1 | ?1866 *Allorchestes p.*, E. Grube in: Arch. Naturg., v. 321 p. 382 t. 9 f. 2 | 1862 *A. pereiri* (part.), Bate, Cat. Amphip. Brit. Mus., p. 42 t. 6 f. 10 | 1866 *Nicea macronyx*, Cam. Heller in: Denk. Ak. Wien, v. 26ɪɪ p.9 t. 1 f. 21—24 | 1879 *Hyale m.*, Wrześniowski in: Zool. Anz., v. 2 p. 202 | 1868 *Nicea perieri* + *N. p. var. pontica* (non *Hyale pontica* H. Rathke 1837!) + *N. perieri var. brevicornis*, Czerniavski in: Syezda Russ. Est., Syezda 1 Zool. p. 116 t. 8 f. 26, 27 | 1875 *N. prevostii* (part.), Catta in: Rev. Sci. nat., v. 4 p. 166 | 1893 *Hyale p.* (part.), A. Della Valle in: F. Fl. Neapel, v. 20 p. 519.

Eyes oval, or reniform, black, nearly meeting at top. Antenna 1 reaching a little beyond peduncle of antenna 2, flagellum with 10 or 11 joints, all cylindric, longer than broad. Antenna 2 half as long as body, almost completely unarmed, flagellum considerably longer than peduncle, with 16 or 17 cylindric joints, the proximal short, the following longer. Maxillipeds, 3ᵈ joint of palp feebly setose over base of finger, more strongly on inner side. Gnathopod 1, 5ᵗʰ joint rather shorter than 6ᵗʰ, 6ᵗʰ oblong, hind margin with setose submedian notch, palm almost transverse, finger slightly curved, slender, acute. Gnathopod 2 in ♂, 4ᵗʰ joint rather acutely produced to touch the 6ᵗʰ; 6ᵗʰ large, almost piriform, the oblique, setulose palm passing gradually without interruption into the shorter hind margin (Heller's description, but in Heller's figure palm well defined by a cavity and 2 palmar spines; also well defined in Lucas' figure); finger well curved, its point closing against the inner surface. Gnathopod 2 in ♀, 5ᵗʰ joint forming a long and thin process between 4ᵗʰ and 6ᵗʰ joints, 6ᵗʰ more of an oval, with shorter palm. Peraeopods 3—5 slender, 6ᵗʰ joint rather long, almost straight, finger strong, apically bent, the inner setule conspicuous. Colour dull yellow. L. ♀ 5—8, ♂ 6—15 mm.

Mediterranean, Adriatic. Black Sea, tropical Atlantic (Virgin Island St. Thomas).

19. **H. camptonyx** (Heller) 1866 *Nicea c.*, Cam. Heller in: Denk. Ak. Wien, v. 26ɪɪ p. 10 t. 1 f. 25—30 | 1879 *Hyale c.*, Wrześniowski in: Zool. Anz., v. 2 p. 202 | 1893 *H. prevostii* (part.), A. Della Valle in: F. Fl. Neapel, v. 20 p. 519.

Back rounded. Side-plates 1—4 moderately large, 1ˢᵗ slightly widened distally, 2ᵈ square. Pleon segment 3, postero-lateral corners quadrate with small produced point. Eyes small, subrotund, wider than the diameter apart. Antenna 1 reaching beyond middle of flagellum of antenna 2, peduncle short, 1ˢᵗ joint as long as 2ᵈ and 3ᵈ combined, flagellum long, 12—16-jointed. Antenna 2 half as long as body, ultimate joint of peduncle longer than penultimate, flagellum with 25—26 joints, the earlier short, the distal longer. In young from mother's pouch antenna 1 with flagellum of 3 joints, much shorter than antenna 2 with 5-jointed flagellum. Maxillipeds, 4ᵗʰ joint of palp as long as 3ᵈ, with apical seta in ♂ longer than the joint. Gnathopod 1 in ♂, 2ᵈ joint rather broad except proximally, 5ᵗʰ shorter than 6ᵗʰ, distal lobe fringed with spines, 6ᵗʰ oblong, slightly wider at base, front margin convex, hind straight, with group of spinules at centre, finger overlapping convex, oblique, ill-defined palm. Gnathopod 2 in ♂, 2ᵈ joint with distal half of front expanded, 3ᵈ forming a small lobe, 4ᵗʰ and 5ᵗʰ very short, 6ᵗʰ large, oblong oval, widest near base, palm spinose, somewhat oblique, defined by a blunt tooth or an abruptly transverse space from the straight hind margin, finger entering pocket at end of palm. Gnathopod 1 in ♀, 2^d joint narrower than in ♂, 6th not wider at base than distally. Gnathopod 2 in ♀ like gnathopod 1, but 2^d joint rather more expanded distally; marsupial plates long, broad, fringed with long setae. Peraeopods 1—5 rather robust, 6th joint slightly curved, concave margin spinose, convex unarmed, finger strong, curved, inner setule very small. Peraeopod 3, 2^d joint subrotund; peraeopod 4, 2^d joint oval; hind margin in these peraeopods nearly smooth. Peraeopod 5, 2^d joint large, subrotund; hind margin produced below 3^d joint, rather strongly crenulate. Uropods 1—3 short, ramus of uropod 3 shorter than peduncle. Telson small, each half rather longer than breadth at base, apex blunt, narrow. L. about 6 mm.

Mediterranean; Adriatic; North-Atlantic (Portugal; Azores). In sea-weed and on a floating log.

20. **H. schmidtii** (Heller) ? 1862 *Allorchestes microphthalmus*, Bate, Cat. Amphip. Brit. Mus., p. 39 t. 6 f. 6 | 1866 *Nicea schmidtii*, Cam. Heller in: Denk. Ak. Wien, $v.26$ II p. 11 t. 1 f. 31, 32 | 1879 *Hyale s.*, Wrześniowski in: Zool. Anz., $v.2$ p. 202 | 1893 *H. prevostii* (part.), A. Della Valle in: F. Fl. Neapel, $v.20$ p. 520.

Near to H. camptonyx. Antenna 1, flagellum 14- or 15-jointed. Antenna 2 $^2/_3$ as long as body, flagellum 32—40-jointed. Maxillipeds, 4th joint of palp not quite as long as 3^d, apical seta short. Gnathopod 1 in ♂, 6th joint suboval, being broader than in the other species, and the spinules on hind margin less close together, finger long. Gnathopod 2 in ♂, 2^d joint with front margin straight, produced below into a small lobe, 3^d joint with well marked front lobe, 6th joint with very oblique palm, making an angle with hind margin, but not defined by tooth or transverse space. Peraeopods 3—5, 6th joint strong, curved, finger long, acute, and without(?) inner setule. Peraeopod 5, 2^d joint with hind margin almost smooth. Uropod 3, ramus not very much shorter than peduncle. L. 7 mm.

Mediterranean; Adriatic; North-Atlantic (Portugal).

21. **H. nigra** (HasW.) 1879 *Allorchestes niger*, Haswell in: P. Linn. Soc. N. S. Wales, $v.4$ p. 319 | 1885 *A. n.*, Haswell in: P. Linn. Soc. N. S. Wales, $v.10$ p. 96 t. 11 f. 1—3 | 1893 *Hyale* (part.)?, A. Della Valle in: F. Fl. Neapel, $v.20$ p. 528.

Very similar to H. camptonyx. Pleon segment 3, postero-lateral corners acutely quadrate, slightly produced. Eyes round, sometimes tending to oval. Antenna 1 reaching much beyond peduncle of antenna 2, peduncle short, 2^d joint much shorter than 1st, 3^d than 2^d, flagellum much longer than peduncle, 9—11-jointed. Antenna 2 about half as long as body, slender, flagellum very much longer than peduncle, joints in ♂ reaching 25 in number. Gnathopod 1, 5th joint with fringed rounded process not clasping base of 6th, 6th oblong, rather narrow, with median group of spinules on hind margin, palm short, slightly oblique. Gnathopod 2 in ♂, 3^d joint with well developed lobe in front, 4th with apex rather acutely produced, 6th large, broadly oblong oval, palm spinulose, oblique but shorter than hind margin, well defined by a small tooth, finger fringed with setules on inner margin and having a strongly projecting bulb at its base. Gnathopods 1 and 2 in ♀ small, very like one another and the gnathopod 1 in ♂. Peraeopods 1—5 moderately robust, not strongly armed, the broad 2^d joint of peraeopod 5 with crenulate hind margin. Uropods 1 and 2 rather stout. Uropod 3 very small, the ramus very short, much shorter than the peduncle. Telson small, cleft almost to base, the lobes not very broad apically. Colour blackish purple or brown. L. 5 mm.

Port Jackson [East-Australia]. Among sea-weed.

22. **H. rubra** (G. M. Thoms.) 1879 *Nicea r.*, G. M. Thomson in: Tr. N. Zealand Inst., *v* 11 p. 236 t. 10в f. 3 | 1886 *N. r.*, G. M. Thomson & Chilton in: Tr. N. Zealand Inst., *v.* 18 p. 144 | 1888 *Hyale r.*, T. Stebbing in: Rep. Voy. Challenger, *v.* 29 p. 500 | 1893 *H. prevostii* (part.), A. Della Valle in: F. Fl. Neapel, *v.* 20 p. 520.

Side-plates 1—4 rather deep. Pleon segment 3, postero-lateral corners a little outdrawn, acute. Eyes irregularly rounded. Antenna 1 half as long as antenna 2, peduncle very short, flagellum with 22 joints in ♂, fewer in ♀, with sensory filaments in both. Antenna 2 half as long as body or more, ultimate joint of peduncle rather longer than penultimate, flagellum thrice as long as peduncle, with 40—60 joints or more in ♂, with 36 in ♀. Gnathopod 1 in ♂, 5th joint with well rounded lobe of hind margin, 6th oblong with slightly convex margins, palm little oblique, finger slightly overlapping. Gnathopod 2 in ♂, 6th joint very large, broadly oval, hind margin well developed, palm oblique, finger thick at base (Thomson: with 2 tubercles at the joint), closing tightly between 2 rows of spinules. Gnathopods 1 and 2 in ♀ alike, 6th joint rather longer in gnathopod 2, oblong, front margin slightly convex, hind margin with 1 small notch, palm a little oblique, scarcely or little overlapped by finger. Peraeopods 1—5, spines on inner margin of 6th joint apically serrulate, encreasing in size to the antepenultimate, thence diminishing, setule on finger small. Peraeopods 4 and 5 decidedly longer than the preceding peraeopods, 2d joint in peraeopod 5 broadly expanded, the rounded hind margin crenulate. Uropods 1—3 and telson of the usual character. Uropods 1 and 2 with lateral spines on both rami. Colour pink. L. 8—9 mm.

South-Pacific (New Zealand).

H. babirussa (A. Costa) 1857 *Amphithoe b.*, A. Costa in: Mem. Acc. Napoli, *v.* 1 p. 201 t. 2 f. 5 | 1862 *Allorchestes babicus* (laps., corr.: *babirussa*), Bate, Cat. Amphip. Brit. Mus., p. 50 | 1893 *Hyale prevostii* (part.), A. Della Valle in: F. Fl. Neapel, *v.* 20 p. 519.

Bay of Naples.

H. bucchichi (Heller) 1866 *Nicea b.*, Cam. Heller in: Denk. Ak. Wien, *v.* 26 ıı p. 7 t. 1 f. 13—15 | 1888 *Hyale b.*, T. Stebbing in: Rep. Voy. Challenger, *v.* 29 p. 366, 1705 | 1893 *H. prevostii* (part.), A. Della Valle in: F. Fl. Neapel, *v.* 20 p. 519.

Maxillipeds, 4th joint of palp very short, thickened at the base. L. ♀ 7, ♂ 8 mm.

Adriatic (Lesina).

H. gazella (A. Costa) 1857 *Amphithoe g.*, A. Costa in: Mem. Acc. Napoli, *v.* 1 p. 202 t. 2 f. 6 a—e | 1862 *Allorchestes g.*, Bate, Cat. Amphip. Brit. Mus., p. 50 | 1893 *Hyale prevostii* (part.), A. Della Valle in: F. Fl. Neapel, *v.* 20 p. 519.

Bay of Naples.

H. georgiana (Pfeff.) 1888 *Allorchestes georgianus*, Pfeffer in: Jahrb. Hamburg. Anst., *v.* 5 p. 77 t. 1 f. 1 a—n, 4 | 1893 *Hyale prevostii* (part.), A. Della Valle in: F. Fl. Neapel, *v.* 20 p. 520.

Probably identical with H. hirtipalma (p. 564). L. about 16 mm.

South-Atlantic (South Georgia). At low-tide, under stones.

H. gracilis (Dana) 1852 *Allorchestes g.*, J. D. Dana in: P. Amer. Ac., *v.* 2 p. 205 | 1853 & 55 *A. g.*, J. D. Dana in: U. S. expl. Exp., *v.* 13 ıı p. 889; t. 60 f. 5 a—d | 1893 *Hyale prevostii* (part.), A. Della Valle in: F. Fl. Neapel, *v.* 20 p. 519.

Gnathopod 2, 4th joint very acutely produced behind the small 5th. L. 12—16 mm.

Tropical Pacific (Tongatabu). Shallow water among sea-weeds.

H. hawaiensis (Dana) 1853 & 55 *Allorchestes h.*, J. D. Dana in: U. S. expl. Exp., *v.* 13 ɪɪ p. 900; t. 61 f. 5 a—h | 1893 *Hyale prevostii* (part.), A. Della Valle in: F. Fl. Neapel, *v.* 20 p. 519.

L. about 9 mm.

Tropical Pacific (Hawaiian Island of Maui).

H. helleri (Grube) ? 1864 *Allorchestes imbricatus* (juv.) (err., non Bate 1857!), E. Grube, Lussin, p. 72 | 1866 *A. helleri*, E. Grube in: Arch. Naturg., *v.* 321 p. 384 t. 9 f. 3 | 1893 *Hyale prevostii* (part.), A. Della Valle in: F. Fl. Neapel, *v.* 20 p. 520.

Approximate to the larger H. nudicornis. L. 3—5 mm.

Adriatic.

H. nudicornis (Heller) 1866 *Nicea n.*, Cam. Heller in: Denk. Ak. Wien, *v.* 26 ɪɪ p. 8 t. 1 f. 16—19 | 1879 *Hyale n.*, Wrześniowski in: Zool. Anz., *v.* 2 p. 202 | 1893 *H. prevostii* (part.), A. Della Valle in: F. Fl. Neapel, *v.* 20 p. 519.

Possibly identical with Parhyale fascigera (p. 556), but uropod 3 not described. Gnathopod 1 in ♂, finger stout and strong, swollen in the middle, the point overlapping palm.

Adriatic.

H. orientalis (Dana) 1853 & 55 *Allorchestes o.*, J. D. Dana in: U. S. expl. Exp., *v.* 13 p. 896; t. 61 f. 2 a—h | 1893 *Hyale prevostii* (part.), A. Della Valle in: F. Fl. Neapel, *v.* 20 p. 519.

Antenna 1, flagellum moniliform. L. about 6 mm.

Sooloo Sea (off the harbour of Soung).

H. piedmontensis (Bate) 1862 *Allorchestes p.*, Bate, Cat. Amphip. Brit. Mus., p. 35 t. 1 a f. 5 | 1879 *Hyale (A.) p.*, Wrześniowski in: Zool. Anz., *v.* 2 p. 200 | 1893 *H.? p.*, A. Della Valle in: F. Fl. Neapel, *v.* 20 p. 529.

Perhaps young of H. prevostii (p. 565). Gnathopod 1 rather larger than gnathopod 2. L. under 4 mm.

Coast of „Piedmont".

H. pugettensis (Dana) 1853 & 55 *Allorchestes p.*, J. D. Dana in: U. S. expl. Exp., *v.* 13 ɪɪ p. 901; t. 61 f. 6 a—d | 1857 *A. p.*, Stimpson in: Boston J. nat. Hist., *v.* 6 p. 518 | 1893 *Hyale prevostii* (part.), A. Della Valle in: F. Fl. Neapel, *v.* 20 p. 519.

Puget's Sound [North West America].

H. rudis (Heller) 1866 *Nicea r.*, Cam. Heller in: Denk. Ak. Wien, *v.* 26 ɪɪ p. 12 t. 1 f. 33 | 1879 *Hyale r.*, Wrześniowski in: Zool. Anz., *v.* 2 p. 202 | 1893 *H. prevostii* (part.), A. Della Valle in: F. Fl. Neapel, *v.* 20 p. 520.

Peraeopods 3—5, 6th joint straight, slender, finger short, with small setule on inner margin.

Adriatic (Lesina).

H. rupicola (Hasw.) 1879 *Allorchestes r.*, Haswell in: P. Linn. Soc. N. S. Wales, *v.* 4 p. 250 t. 8 f. 1 | 1885 *A. r.*, Haswell in: P. Linn. Soc. N. S. Wales, *v.* 10 p. 95 t. 10 f. 9—12 | 1893 *Hyale r.*, G. M. Thomson in: P. R. Soc. Tasmania, 1892 (p. 18) | 1893, A. Della Valle in: F. Fl. Neapel, *v.* 20 p. 529.

Very closely allied to H. perieri (p. 570). Peraeopods 1—5 with very conspicuous setule on inner margin of finger. L. 9 mm.

South-Pacific (Port Jackson and Botany Bay [East-Australia]; Tasmania?).

H. seminuda (Stimps.) 1857 *Allorchestes s.*, Stimpson in: P. Calif. Ac., *v.*1 p. 90 | 1857 *A. s.*, Stimpson in: Boston J. nat. Hist., *v.* 6 p. 518 | 1893 *Hyale prevostii* (part.)?, A. Della Valle in: F. Fl. Neapel, *v.* 20 p. 529.

Bay of San Francisco. Littoral zone.

H. stebbingi Chevreux 1888 *H. s.*, Chevreux in: Bull. Soc. zool. France, *v.*13 p. 32 | 1888 *H. s.*, T. Barrois, Cat. Crust. Açores, p. 33.

Probably a small variety of H. prevostii (p. 565). L. ♂ 5 mm.

North-Atlantic (Azores).

H. stewarti (Filh.) 1885 *Allorchestes s.*, Filhol in: Bull. Soc. philom., ser. 7 *v.* 9 p. 54 | 1885 *A. s.*, Filhol in: Recu. Passage Vénus, *v.* 3 II Zool. p. 465 t. 53 f. 5.

L. ♂ 17 mm.

Paterson Inlet (Stewart Island by New Zealand).

H. stolzmani Wrześn. 1879 *H. s.*, Wrześniowski in: Zool. Anz., *v.* 2 p. 201 | 1893 *H. pontica* (part.)?, A. Della Valle in: F. Fl. Neapel, *v.* 20 p. 529.

L. ♀ 7, ♂ 9 mm.

Bay of Chimbote [Peru]. Under stones in small tidal pools.

H. tenella (A. Costa) 1857 *Amphithoe t.*, A. Costa in: Mem. Acc. Napoli, *v.* 1 p. 204 t. 2 f. 8 | 1893 *Hyale pontica* (part.)?, A. Della Valle in: F. Fl. Neapel, *v.* 20 p. 526.

L. 5 mm.

Bay of Naples.

H. villosa S. I. Sm. 1876 *H. v.*, (S. I. Smith in:) Kidder in: Bull. U. S. Mus., *v.* 3 p. 58.

Closely allied to H. hirtipalma (p. 564), but gnathopod 1 with 6th joint not widening to the palm. L. nearly 10 mm.

Southern Indian Ocean (Kerguelen Island). Rocky beaches.

H. sp., (Hasw.) 1879 *Allorchestes longicornis* (non *Orchestia l.* Krøyer 1845!), Haswell in: P. Linn. Soc. N. S. Wales, *v.* 4 p. 251 t. 7 f. 4 | 1885 *A. l.*, Haswell in: P. Linn. Soc. N. S. Wales, *v.* 10 p. 95 t. 10 f. 6—8 | 1893, A. Della Valle in: F. Fl. Neapel, *v.* 20 p. 528.

Closely agreeing with H. camptonyx (p. 570). L. about 10 mm.

South-Pacific (Kiama [New South Wales]). Between tide-marks.

12. Gen. **Hyalella** S. I. Sm.

1874 *Hyalella* (Sp. un.: *H. dentata*), S. I. Smith in: Rep. U. S. Fish Comm., *v.* 2 p. 645 | 1888 *H.*, T. Stebbing in: Rep. Voy. Challenger, *v.* 29 p. 172, 433 | 1893 *H.*, A. Della Valle in: F. Fl. Neapel, *v.* 20 p. 512 | ?1877 *Lockingtonia* (Sp. un.: *L. fluvialis*), Harford in: P. Calif. Ac., *v.* 7 p. 53.

Like Hyale (p. 559), except that maxilla 1 has a smaller palp, not reaching base of apical spines of outer plate, gnathopod 2 in ♂ (Fig. 97 p. 580) has the 5th joint produced behind between the 4th and 6th, and the telson (Fig. 98 p. 580) is entire.

Fresh water; in depths of lakes down to 120 m, and at heights above sea-level extending to 4053 m.

15 species.

Synopsis of the species:

1 { With dorsal teeth — **2.**
  { Without dorsal teeth — **7.**

2	{ Dorsal teeth only on the pleon . . . . { Dorsal teeth on both peraeon and pleon — 3.	1. H. azteca	p. 575
3	{ Side-plates 1—4 apically acute — 4. { Side-plates 1—4 not apically acute — 5.		
4	{ Peraeon segments with lateral teeth . . . . . . { Peraeon segments without lateral teeth	2. H. echinus . . . 3. H. longipes . .	p. 576 p. 576
5	{ Dorsal teeth large, present on all peraeon segments { Dorsal teeth small, not present on peraeon seg- ments 1—4 — 6.	4. H. lucifugax . .	p. 576
6	{ Gnathopod 2 in ♂, 6th joint broader than long, palm defined by a tooth . . . . { Gnathopod 2 in ♂, 6th joint longer than broad, palm undefined	5. H. latimana . . 6. H. longipalma .	p. 577 p. 577
7	{ Side-plates 1—4 produced into spine-like processes { Side-plates 1—4 not produced into spine-like pro- cesses — 8.	7. H. armata . . .	p. 577
8	{ Without accessory branchiae (so far as known) — 9. { With accessory branchiae — 11.		
9	{ Uropod 3 nearly reaching end of uropod 2 . . . { Uropod 3 not nearly reaching end of uropod 2 — 10.	8. H. longistila . .	p. 577
10	{ Stout, with coppery lustre . . . . { Not stout, without coppery lustre	9. H. cuprea . . . 10. H. inermis . . .	p. 578 p. 578
11	{ Gnathopod 2 in ♂, finger closing strongly on inner surface of 6th joint . . . . { Gnathopod 2 in ♂, finger with apex only on inner surface of 6th joint — 12.	11. H. jelskii	p. 578
12	{ Peraeopod 1 shorter than peraeopod 2 — 13. { Peraeopod 1 not shorter than peraeopod 2 — 14.		
13	{ Maxilla 1, palp comparatively long . . . . { Maxilla 1, palp very short	12. H. dybowskii . 13. H. lubomirskii .	p. 579 p. 579
14	{ Gnathopod 2 in ♂, hind margin of 6th joint apically bulging, in ♀ 6th joint not apically widened . { Gnathopod 2 in ♂, hind margin of 6th joint not apically bulging, in ♀ 6th joint apically widened	14. H. meinerti . . 15. H. warmingi . .	p. 579 p. 580

1. H. azteca (Sauss.) ?1818 *Ampithoe dentata*, Say in: J. Ac. Philad., *v.* 1 ɪɪ p. 383 | 1874 *Hyalella d.*, S. I. Smith in: Rep. U. S. Fish Comm., *v.* 2 p. 645 t. 2 f. 8—10 | 1876 *Allorchestes dentatus*, Faxon in: Bull. Mus. Harvard, *v.* 3 p. 373 | 1858 *Amphitoe aztecus*, Saussure in: Mém. Soc. Genève, *v.* 14 ɪɪ p. 474 t. 5 f. 33 | 1888 *Hyalella azteca*, T. Stebbing in: Rep. Voy. Challenger, *v.* 29 p. 311 | 1862 *?Allorchestes knickerbockeri* + *Amphithoë a.*, Bate, Cat. Amphip. Brit. Mus., p. 36 t. 6 f. 1; p. 250 | ?1877 *Lockingtonia fluvialis*, Harford in: P. Calif. Ac., *v.* 7 p. 54.

Body slightly compressed, pleon segments 1 and 2 each produced into a dorsal tooth, usually well marked, but apparently sometimes obsolescent. Side-plates 1—4 almost rectangular. Pleon segment 3, postero-lateral corners subquadrate. Eyes nearly round. Antenna 1, flagellum rather longer than peduncle, 7—9-jointed. Antenna 2 a little longer, ultimate and penultimate joints of peduncle long, subequal, flagellum little longer than peduncle, 8—12-jointed. Gnathopod 1 small and slender, palm transverse, nearly straight, defined by a small tooth, behind which apex of finger closes. Gnathopod 2 in ♂, 6th joint very stout, a little longer than broad, palm oblique, slightly convex, with abrupt notch near middle and 2 slight emarginations near defining angle; finger stout, curved. Gnathopod 2 in ♀ weak and slender, 5th and 6th joints long, narrow, 6th more than twice as long as broad, hind margin produced a little beyond front, though not enough to form a proper chela between the small palm and small finger. Peraeopods 3—5, 2^d joint broad, serrate, but smooth in Saussure's figure. Peraeopod 5 little longer than peraeopod 4. Uropod 3, ramus nearly as long as peduncle, slender, tapering, with setae at apex. Telson stout, as long as broad; apical margin rounded, with slender seta on each side. L. about 4—6 mm.

Mexico; North America (Florida, Connecticut etc.). Streams, lakes, stagnant water.

2. **H. echinus** (Faxon) 1876 *Allorchestes e.*, Faxon in: Bull. Mus. HarVard, *v.* 3 p. 367 f. 19—21 | 1888 *Hyalella e.*, T. Stebbing in: Rep. Voy. Challenger, *v.* 29 p. 455 | 1893 *H. e.*, A. Della Valle in: F. Fl. Neapel, *v.* 20 p. 517 t. 58 f. 12, 13.

Body short, very stout, peraeon segments 1—7 and pleon segments 1—4 each having on hind margin 2 medio-dorsally elevated spine-like processes, and peraeon segments 1—7 with pleon segment 1 each having a small marginal process on either side. Head with small deflexed rostral point, a tubercle on each cheek. Side-plates 1—4 large, triangular, with ridge from centre of base to apex. Side-plate 4 with tubercle on hind margin, side-plate 5 with tubercle on each lobe. Eyes round, large, protuberant. Antenna 1, flagellum 6—8-jointed. Antenna 2, flagellum 9-jointed. Gnathopod 2 in ♂, 6th joint large, somewhat widened to the rather oblique, slightly concave palm, which receives point of finger in a terminal notch. Gnathopod 2 in ♀, 5th and 6th joints subequal. Peraeopods 3—5, 2^d joint moderately expanded; peraeopod 3 only $^2/_3$ as long as the elongate, subequal 4th and 5th peraeopods. L. 5—7 mm.

Lake Titicaca. Depth 18—110 m.

3. **H. longipes** (Faxon) 1876 *Allorchestes l.*, Faxon in: Bull. Mus. HarVard, *v.* 3 p. 368 f. 22—25 | 1888 *Hyalella l.*, T. Stebbing in: Rep. Voy. Challenger, *v.* 29 p. 455 | 1893 *H. l.*, A. Della Valle in: F. Fl. Neapel, *v.* 20 p. 515 t. 58 f. 6, 7.

Peraeon segment 1 with front margin forming a short medio-dorsal tooth; peraeon segments 1—7 and pleon segments 1—3 each with hind margin produced to a backward-pointing, medio-dorsal tooth or spine-like process, small on peraeon segments 1—4 (the first 2 teeth sometimes longer, curved forward), large on the others, largest on pleon segments 1 and 2; pleon segment 4 with hind margin forming a median projection either subacute or convex. Side-plates 1—4 long, apically pointed. Pleon segments 1—3, postero-lateral corners slightly produced. Eyes round, protuberant. Antenna 1, flagellum 13-jointed. Antenna 2 somewhat longer, flagellum 14-jointed. Gnathopod 2 in ♂, 6th joint not very large, widening to palm, which is not very oblique, defined by a small projecting tooth. Peraeopods 4 and 5 very long. Uropod 3 very short, with a very slender seta at tip. L. about 10 mm.

Lake Titicaca. Depth 20—72 m.

4. **H. lucifugax** (Faxon) 1876 *Allorchestes l.*, Faxon in: Bull. Mus. HarVard, *v.* 3 p. 369 f. 26 | 1888 *Hyalella l.*, T. Stebbing in: Rep. Voy. Challenger, *v.* 29 p. 455 | 1893 *H. l.*, A. Della Valle in: F. Fl. Neapel, *v.* 20 p. 515 t. 58 f. 4, 5.

♀ unknown. — ♂. Very like H. longipes, with the same number of dorsal processes, but 1st dorsal tooth overlies the head, the next 6 are also large, curved forward. Side-plates 1—4 with rounded apices. Antennae 1 and 2 subequal. L. 11 mm.

Lake Titicaca. Depth 72—110 m.

5. **H. latimana** (Faxon) 1876 *Allorchestes latimanus*, Faxon in: Bull. Mus. Harvard, v. 3 p. 370 f. 27, 28 | 1888 *Hyalella l.*, T. Stebbing in: Rep. Voy. Challenger, v. 29 p. 455 | 1893 *H. latimana*, A. Della Valle in: F. Fl. Neapel, v. 20 p. 515 t. 58 f. 8.·

Body thick, peraeon segments 6 and 7 and pleon segments 1—3 each produced to a small medio-dorsal spiniform tooth. Side-plates 1—4 quadrate, their lower angles rounded, side-plate 4 emarginate behind. Pleon segments 1—3, postero-lateral corners prolonged backward. Eyes nearly round. Antenna 1 $^2/_3$ as long as antenna 2, peduncle reaching middle of ultimate joint of peduncle of antenna 2. Antenna 2 half as long as body, number of joints in flagellum not recorded. Gnathopod 2 in ♂, 6th joint large, broader than long, widening much to the palm, which is moderately oblique, straight, defined by a prominent tooth. Gnathopod 2 in ♀ much smaller. Peraeopod 5 of moderate length. Uropod 3 reaching a little beyond telson. Telson broad. L. 7—12 mm.

Lake Titicaca. Depth 18—36 m.

6. **H. longipalma** (Faxon) 1876 *Allorchestes longipalmus*, Faxon in: Bull. Mus. Harvard, v. 3 p. 371 f. 29—31 | 1888 *Hyalella l.*, T. Stebbing in: Rep. Voy. Challenger, v. 29 p. 455 | 1893 *H. longipalma*, A. Della Valle in: F. Fl. Neapel, v. 20 p. 516 t. 58 f. 9.

Very near to H. latimana, but peraeon segment 5 in addition having a dorsal tooth. Eyes round, antennae 1 and 2 each with flagellum of 15 joints. Gnathopod 2 in ♂, 6th joint large and swollen, palm sinuous and very long, setose, so oblique as to leave a very short hind margin, from which it is defined chiefly by impact of finger. Telson with seta on each side of hind margin. L. 9—13 mm.

Lake Titicaca.

7. **H. armata** (Faxon) 1876 *Allorchestes armatus*, Faxon in: Bull. Mus. Harvard, v. 3 p. 364 f. 1—18 | 1888 *Hyalella armata*, T. Stebbing in: Rep. Voy. Challenger, v. 29 p. 455 | 1893 *H. a.*, A. Della Valle in: F. Fl. Neapel, v. 20 p. 514 t. 58 f. 2, 3.

Body stout, hind margin of segments raised in conspicuous transverse ridge. Side-plates 1—4 produced into spine-like processes (longer in specimens from greater depth), 1st—3^d directed downward and forward, 4th about twice as long as 3^d, at right angles to length and height of body. Eyes round. Antenna 1, peduncle reaching middle of ultimate joint of peduncle of antenna 2, flagellum 12-jointed. Antenna 2 much longer than antenna 1, gland-cone prominent, flagellum 13-jointed. Maxilla 1, palp (in figure) reaching more than half the distance from its base to apex of outer plate. Gnathopod 1, 5th joint as broad (in figure broader than and as long) as 6th, palm transverse, slightly concave, finger curved. Gnathopod 2 in ♂ very large, 6th joint widening much to the palm, which is oblique, straight, with small setae, finger slender, curved. Gnathopod 2 in ♀ smaller, 5th and 6th joints subequal, palm nearly transverse. Peraeopods 3 and 4 apparently with narrow 2^d joint. Peraeopods 4 and 5 elongate. Branchial vesicles simple. Marsupial plates long and rather broad, fringed with short setae, apex pointed. Uropod 3 very small, curved upward, projecting little beyond the broad telson. Integument with rows of microscopic hairs. L. 7—10 mm.

Lake Titicaca. Depth 20—120 m.

8. **H. longistila** (Faxon) 1876 *Allorchestes longistilus*, Faxon in: Bull. Mus. Harvard, v. 3 p. 375 f. 37 | 1893 *Hyalella dentata var.?*, A. Della Valle in: F. Fl. Neapel, v. 20 p. 517.

Body smooth, long, slender. Side-plates 1—4 quadrate. Pleon segments 1—3, postero-lateral corners acute. Eyes nearly round, dark, approximating above. Antenna 1 nearly as long as antenna 2, flagellum 13-jointed, both antennae shorter in ♀ than in ♂. Gnathopod 2 in ♂, 6th joint large, widening to the oblique, rather convex palm, which is defined by a projection; 6th joint in ♀ long and narrow. Peraeopods 3—5 said to be subequal, but in figure showing the usual successive elongation. Uropod 3 very long, extending far beyond the tip of the telson, almost to the end of uropod 2. Telson with 2 long setae on hind margin. L. 3—6 mm.

Brazil (near Campos).

9. **H. cuprea** (Faxon) 1876 *Allorchestes cupreus*, Faxon in: Bull. Mus. Harvard, v. 3 p. 372 f. 32—34 | 1893 *Hyalella cuprea*, A. Della Valle in: F. Fl. Neapel, v. 20 p. 514 t. 58 f. 1.

Body stout, smooth. Side-plates 1—4 rather deep. Pleon segment 3, postero-lateral corners quadrate. Antenna 1, flagellum about 10-jointed. Antenna 2 considerably longer, about $^1/_8$ as long as body. Gnathopod 2 in ♂, 6th joint swollen, palm convex (or, in figure, sinuous), setiferous, oblique, defined by a prominence, finger strong, curved. Gnathopod 2 in ♀, 6th joint long and narrow. Peraeopods 3—5, 2^d joint large, hind margin slightly serrate. Peraeopod 3 much shorter than peraeopods 4 or 5. Colour in many parts a coppery lustre. L. 9—11 mm.

Lake Titicaca.

10. **H. inermis** S. I. Sm. ?1860 *Amphithoe andina*, A. Philippi, Reise Atacama, p. 170 | 1888 *Hyalella a.*, T. Stebbing in: Rep. Voy. Challenger, v. 29 p. 326 | 1874 *H. inermis*, S. I. Smith in: Rep. U. S. geol. Surv. Terr. 1873, p. 609 t. 1 f. 1, 2 | 1891 *H. i.*, T. Stebbing in: E. Whymper, Trav. Great Andes, p. 361 f.; suppl. p. 125 | 1876 *Allorchestes dentatus var. inermis* + *A. d. var. gracilicornis*, Faxon in: Bull. Mus. Harvard, v. 3 p. 373 f.; p. 374 | 1893 *Hyalella dentata* (part.), A. Della Valle in: F. Fl. Neapel, v. 20 p. 516.

Body entirely devoid of dorsal teeth. Appendages as in H. azteca (p. 575), except: Antenna 1 in ♂ reaching middle of flagellum of antenna 2; 3^d joint of peduncle fully as long as 2^d; flagellum longer than peduncle, 8—10-jointed. Antenna 2, flagellum much longer than peduncle, 12-jointed (in *var. gracilicornis* ♀ antenna 2 half as long as body, twice as long as antenna 1; flagellum 13-jointed). Gnathopod 1, 5th and 6th joints stout, 6th widening to palm, which is defined by a blunt tooth, 6th joint bent down on broad apex of 5th; gnathopod 2 has the palm nearly straight, without any abrupt notch near the middle; finger slightly curved, terminating in an acute horny tip. L. 4—6 mm.

Colorado; Ecuador; Peru; Argentina; Chili. In pools, lakes (Titicaca), canals etc., also in saline Water; to a height of 4053 m.

11. **H. jelskii** (Wrześn.) 1879 *Hyale (Allorchestes) j.*, Wrześniowski in: Zool. Anz., v. 2 p. 176, 490 | 1888 *H. j.*, T. Stebbing in: Rep. Voy. Challenger, v. 29 p. 1705 | 1893 *Hyalella j.*, A. Della Valle in: F. Fl. Neapel, v. 20 p. 514.

Head longer than 1st peraeon segment. Eyes small, circular. Antenna 1 more than half as long as antenna 2, flagellum twice as long as peduncle, with 11 joints in ♂ and 9 in ♀. Antenna 2 in ♂ $^2/_3$ as long as body, flagellum twice as long as peduncle, with 18 joints; in ♀ more than $^1/_2$ as long as body, flagellum twice as long as peduncle. 14-jointed. Maxilla 1, palp less than $^1/_8$ distance between its base and apex of inner plate.

Gnathopod 1 as in H. dybowskii, except that 6th joint is only $^2/_3$ as long as 5th. Gnathopod 2 in ♂, 5th joint $^1/_3$ as long as 6th, palm rather oblique, spinose, finger with $^1/_3$ of its length on inner surface of hand when closed. Gnathopod 2 in ♀, 6th joint 2$^1/_3$ times as long as broad. Peraeopods 1 and 2 subequal. Peraeopod 4 slightly longer than peraeopod 5. Accessory branchiae on gnathopod 2 and peraeopods 1—3 successively longer, on peraeopod 4 longest and double, on peraeopod 5 wanting. L. ♀ 4, ♂ 5 mm.

East slope of Cordilleras (Pumamarca). Height 2511 m.

12. **H. dybowskii** (Wrześn.) 1879 *Hyale (Allorchestes) d.*, Wrześniowski in: Zool. Anz., *v.* 2 p. 199 | 1888 *H. d.*, T. Stebbing in: Rep. Voy. Challenger, *v.* 29 p. 1705 | 1893 *Hyalella d.*, A. Della Valle in: F. Fl. Neapel, *v.* 20 p. 514.

Head as long as peraeon segment 1. Eyes oval, widened below. Antenna 1 little more than half as long as antenna 2, peduncle $^2/_3$ as long as flagellum, which has in ♂ 13, in ♀ 10 joints, most of them with a couple of sensory filaments. Antenna 2 half as long as body, peduncle $^2/_3$ as long as flagellum, which has in ♂ 15, in ♀ 14 joints. Maxilla 1, palp about $^2/_3$ as long as distance between its base and the apex of outer plate. Gnathopod 1, 5th joint as long as 6th, 6th narrow at base, front margin convex, hind concave, palm transverse, with 2 not very prominent rounded lobes, finger shorter than palm. Gnathopod 2 in ♂, 5th joint as long as 4th, half as long as 6th, 6th triangular, palm $^2/_3$ as long as the joint, rather oblique, forming a single curved lobe, finger passing its apex only between 2 palmar spines on to surface of joint. Gnathopod 2 in ♀ like gnathopod 1, but more slender, 6th joint small, not twice as long as broad. Peraeopod 1 considerably shorter than peraeopod 2. Peraeopods 4 and 5 equal. Simple tubular accessory branchiae on peraeopods 1—5.

West slope of the Cordilleras (Pancal, Montana de Nancho). Height 2196 m.

13. **H. lubomirskii** (Wrześn.) 1879 *Hyale (Allorchestes) l.*, Wrześniowski in: Zool. Anz., *v.* 2 p. 177 | 1888 *H. l.*, T. Stebbing in: Rep. Voy. Challenger, *v.* 29 p. 1705 | 1893 *Hyalella l.*, A. Della Valle in: F. Fl. Neapel, *v.* 20 p. 514.

Head as long as peraeon segment 1. Eyes rather large, irregularly oval. Antenna 1 half as long as antenna 2, flagellum not twice as long as peduncle, 10-jointed. Antenna 2 little over half as long as body, flagellum more than twice as long as peduncle, 18-jointed. Maxilla 1, palp less than $^1/_3$ distance between its base and apex of inner plate. Gnathopod 1 as in A. jelskii, but 6th joint almost as long as 5th. Gnathopod 2 in ♂, 5th joint less than $^1/_3$ as long as 6th, palm rather oblique, spinose, forming 2 lobes, finger with apex only on inner surface of hand. Gnathopod 2 in ♀, 6th joint not quite twice as long as broad. Peraeopod 1 a little shorter than peraeopod 2. Peraeopod 4 longer than peraeopod 5. Accessory branchiae simple, quite rudimentary on gnathopod 2, very small on peraeopods 1 and 2, longer than the principal branchiae on peraeopods 3—5. L. 5—6 mm.

West slope of Cordilleras (Pacasmayo). Height 2511 m.

14. **H. meinerti** Stebb. 1899 *H. m.*, T. Stebbing in: Tr. Linn. Soc. London, ser. 2 *v.* 7 p. 407 t. 32ʙ.

Side-plates 1—3 deeper than broad. Pleon segment 3, postero-lateral corners acutely quadrate. Eyes black, usually wider apart than the diameter. Antenna 1 slender, more than half as long as antenna 2, 3^d joint nearly as long as 2^d, but much resembling joints of flagellum, flagellum with 9 or 10

elongate joints. Antenna 2 slender. more than half as long as body, ultimate
joint of peduncle long, flagellum with 13—15 elongate joints. Gnathopod 1
in ♂ as in H. inermis (p. 578) but less robust, the 6th joint scarcely widening
to the transverse palm. Gnathopod 2 in ♂ (Fig. 97), 2d joint narrow, 4th
with rounded apex, 5th with the usual cup-forming fringed lobe, 6th much
longer than broad, basal part narrow, rather abruptly widening at the boss
which defines the oblique, slightly sinuous
palm. Gnathopods 1 and 2 in ♀ small,
4th joint with rounded apex, 6th in gnathopod 1 shorter than 5th, but in gnathopod 2 about as long, in both narrow,
oblong, narrowest at base, the short
palm transverse or slightly tending to
form an acute angle with the hind margin.
Peraeopods 3—5 successively · a little
longer, 2d joint oval in peraeopods 3
and 4, much wider in peraeopod 5.
Eggs and marsupial plates large. Acces-
sory branchiae on peraeopods 1—5, ordinary branchiae not perceived on
peraeopod 5. Uropods 1—3 unusually slender; uropods 1 and 2 with lateral
spines on both rami. Uropod 3 (Fig. 98) comparatively long, the tapering
ramus rather longer than peduncle, extending considerably beyond the telson.
Telson (Fig. 98) oblong oval, with pair of spinules on rounded apex. L. 5—6 mm.

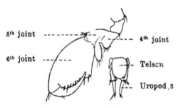

5th joint — — — —
6th joint — — —
4th joint
Telscn
Uropod.3

Fig. 97.   Fig. 98.
Gnathopod 2, ♂.   Uropod 3 and telson.
Fig. 97 & 98. H. meinerti.

Brazil (Laguna di Espino).

15. **H. warmingi** Stebb. 1899 *H. w.*, T. Stebbing in: Tr. Linn. Soc. London,
ser. 2 *v.* 7 p. 406 t. 32 A.

Body rather robust. Side-plate 6 deeply lobed behind. Pleon segment 3, postero-lateral corners a little produced backward, acute. Eyes small.
dark, rounded, more than the diameter apart. Antenna 1, 2d joint little shorter
than 1st, 3d a little shorter than 2d, flagellum with 13 joints in ♂, 10 in ♀.
Antenna 2 more than half as long as body, gland-cone prominent, ultimate
joint of peduncle longer than penultimate, flagellum with 19 joints in ♂,
15 in ♀. Maxilla 1, palp minute. Maxillipeds, 4th joint of palp with curved
spine on the blunt end. Gnathopod 1 in ♂, 5th joint with bulging hind
margin fringed, 6th shorter, widening with sinuous hind margin to the transverse palm, which is defined. by a process within which the finger closes,
an oblique row of spinules on surface. Gnathopod 2 in ♂, 2d joint not
lobed below, 5th short, the fringed cup-forming process rather long, 6th joint
large, oval, palm very oblique, sinuous, forming 3 lobes, of which the centre
one is broadest, the finger closing at the 3d into a pocket, which meets the
hind margin at a well defined angle. Gnathopods 1 and 2 in ♀ similar to
gnathopod 1 in ♂, but smaller, and 6th joint in gnathopod 2 in ♀ rather
longer and more slender than in gnathopod 1. Peraeopods 1—5 tolerably
robust and spinose. Peraeopod 1 not shorter than peraeopod 2. Peraeopods 3—5, 2d joint broadly oval, subequal in peraeopods 3 and 4, considerably
larger in peraeopod 5. Simple accessory branchiae on peraeopods 1—4 not
so long as principal branchiae. Uropods 1 and 2 with lateral spines on both
rami. Uropod 3, ramus as long as peduncle. Telson squared at base, then
broadly rounded with 2 distant setules on broad apical margin. Colour
dusky in spirit. L. 5—6 mm.

Brazil (Lagoa Santa, Watercourse near Rio de Janeiro). Height 1200 m.

13. Gen. **Allorchestes** Dana

1849 *Allorchestes* (part.), J. D. Dana in. Amer. J. Sci., ser. 2 *v.* 8 p. 136 | 1852 *A.* (part.), J. D. Dana in: P. Amer. Ac., *v.* 2 p. 205 | 1888 *A.* (part.), T. Stebbing in: Rep. Voy. Challenger, *v.* 29 p. 1686 | 1880 *Aspidophoreia* (Sp. un.: *A. diemenensis*), Haswell in: P. Linn. Soc. N. S. Wales, *v.* 5 p. 101.

Distinguished from Hyale (p. 559) by gnathopod 2 in ♂ (Fig. 99 p. 583), which has the 5th joint produced behind between the 4th and 6th joints, and from Hyalella (p. 574) by having the telson (Fig. 100 p. 583) more or less divided; sometimes the palp of maxilla 1 agrees with that of Hyale, sometimes with that of Hyalella.

5 accepted and 13 doubtful species.

Synopsis of accepted species:

1 { Maxilla 1. palp not reaching apex of outer
　　 plate — **2.**
　 { Maxilla 1, palp reaching apex of outer plate — **4.**

2 { Gnathopod 1 in ♂, finger much longer than palm　1. **A. novizealandiae** . . p. 581
　 { Gnathopod 1 in ♂, finger matching palm — **3.**

3 { Telson divided to the base . . . . 2. **A. compressus** . . . p. 581
　 { Telson not divided beyond the middle . . . 3. **A. malleolus** . . . . p. 582

4 { Apices of telson acute . . . . 4. **A. plumicornis** . . . p. 583
　 { Apices of telson truncate . . . . 5. **A. humilis** . . . . . p. 584

1. **A. novizealandiae** Dana 1852 *A. n.-z.* (♀) + *A. intrepida* (♂), J. D. Dana in: P. Amer. Ac., *v.* 2 p. 207 | 1853 & 55 *A. n.*, J. D. Dana in: U. S. expl. Exp., *v.* 13 II p. 894; t. 61 f. 1 a—f (♂), g—v (♀) | 1862 *A. novae-zelandiae*, Bate, Cat. Amphip. Brit. Mus., p. 37 t. 6 f. 3 | 1886 *A. neo-zelanica*, G. M. Thomson & Chilton in: Tr. N. Zealand Inst., *v.* 18 p. 144 | 1893 *Hyale prevostii* (part.), A. Della Valle in: F. Fl. Neapel, *v.* 20 p. 519.

Side-plate 4 very large. Antenna 1, flagellum more than twice as long as peduncle, 16-jointed. Antenna 2 in ♀ little longer than antenna 1, more considerably longer in ♂, in ♂ scarcely ¹/₂ as long as body, flagellum longer than peduncle, with 14 very slender joints. Maxillipeds, 3^{d} joint of palp broad. Gnathopod 1 in ♂ quite small, 5th joint fully as long as 6th, distally much broader, with narrow, subacute prolongation of hind margin, 6th oblong, slightly wider at base, palm transverse, excavate, defined by an acute tooth, finger more than twice as long as palm. Gnathopod 2 in ♂, 5th joint narrowly produced between 4th and 6th, 6th large, ovate, palm nearly straight, oblique, longer than hind margin, spinulose, finger long, yet not reaching rounded end of palm. Gnathopod 1 in ♀, 5th joint distally wide, but without acute process, 6th widening a little to truncate palm, finger not longer than palm. Gnathopod 2 in ♀, 4th joint somewhat produced, 5th strongly produced along hind margin of 6th, process with a rounded apex, 6th as in gnathopod 1 but rather larger. Peraeopods 3—5 successively longer, setae minute, 4th joint expanded, especially in peraeopod 3. Uropod 1, one of the rami (in figure) with only apical spines. L. 10 mm.

Pacific (New Zealand, in holes in Wood bored by Teredos; Valparaiso?).

2. **A. compressus** Dana 1852 *A. compressa* + *A. australis*, J. D. Dana in: P. Amer. Ac., *v.* 2 p. 205, 206 | 1853 & 55 *A. a.* + *A. gaimardii?*, J. D. Dana in: U. S. expl. Exp., *v.* 13 II p. 892 t. 60 f. 7 a—o; p. 884 t. 60 f. 1 a—i | 1862 *A. a.* + *A. g.*, Bate, Cat. Amphip. Brit. Mus., p. 45 t. 7 f. 6; p. 41 t. 6 f. 9 | 1880 *Aspidophoreia diemenensis*, Haswell in: P. Linn. Soc. N. S. Wales, *v.* 5 p. 101 t. 6 f. 2 | 1893 *Hyale prevostii* (part.) + *H. pontica* (part.), A. Della Valle in: F. Fl. Neapel, *v.* 20 p. 519, 523, 528, 530.

Body compressed, especially at pleon. Side-plates 1—4 deep, 4th also wide, 5th shallow, 2^d—4th quadrate. Pleon segment 3, postero-lateral corners quadrate, with minutely produced point. Eyes oval, wider apart than the longer diameter. Antennae 1 and 2 of rather variable proportions (in young from mother's pouch subequal). Antenna 1 usually rather longer than peduncle of antenna 2, flagellum 10—20-jointed. Antenna 2, flagellum shorter or not much longer than peduncle, 10—20-jointed, upper lip broader than deep. Maxilla 1, palp minute. Maxillipeds, 2^d and 3^d joints of palp broad. Gnathopod 1 in ♂, 5th joint slightly longer than 6th, widest subapically, with spinules at the projection of front and hind margins, 6th joint oblong, a little widened at almost transverse convex palm, front margin convex, hind rather sinuous, finger fitting palm. Gnathopod 2 in ♂ robust. 2^d joint with small downward produced lobe, 3^d also with front margin lobed, 4th apically produced behind, 5th produced backward in a rather slender and not strongly spined lappet, 6th large, the palm spinulose, very oblique, defined from the slightly bulging hind margin by palmar spines and the small hollow which receives the apex of the strong finger. Gnathopod 1 in ♀ as in ♂, except that 6th joint is more elongate, as long as 5th. Gnathopod 2 in ♀ rather larger than gnathopod 1, 2^d joint not produced downward, 3^d without conspicuous lobe, 4th produced as in ♂, 5th with lappet stretching along part of straight hind margin of 6th, 6th broader than in gnathopod 1, slightly widening to the transverse palm, finger matching palm. Branchial vesicles large, oval, with narrow neck. Marsupial plates broad, oblong, produced at one corner, setae short. Peraeopods 1 and 2 subequal, slender. Peraeopod 3, 2^d joint oblong oval, front margin with spines, nearly straight, hind margin nearly smooth, 4th joint widened, spinose on both margins. Peraeopod 4 considerably longer, 2^d and 4th joints not quite so wide; branchial vesicles in peraeopods 3 and 4 with accessory lobe. Peraeopod 5 shorter than peraeopod 4, especially in ♂, but 2^d joint much larger and more rounded behind. widest subapically and broadly produced behind 3^d joint, 4th not much widened, finger as in all the peraeopods short, curved. Uropod 1, rami decidedly shorter than peduncle. Uropod 2, rami a little shorter than peduncle. Uropod 3, ramus small, conical, shorter than stout peduncle, tipped with a minute spinule. Telson broad, the 2 quadrate lobes, separated by a linear fissure, set at an angle to one another, gable-like. Surface ornamented with coloured spots and white·dots (Haswell). L. 11—20 mm.

Southern Indian Ocean and South-Pacific (south- and West-coasts of Australia, Tasmania).

3. **A. malleolus** Stebb. 1899 *A. m..* T. Stebbing in: Tr. Linn. Soc. London, ser. 2 *v.* 7 p. 409 t. 33 A.

Body moderately compressed. Side-plates 1—4 rather deep, with no projecting point of hind margin. Pleon segment 3, postero-lateral corners bluntly produced. Eyes not large, rounded, at least their diameter apart, dark in spirit. Antenna 1 about ³/₄ as long as antenna 2, joints of peduncle successively shorter, flagellum longer than peduncle, 10—12-jointed. Antenna 2 not more than ¹/₃ as long as body, peduncle stout, flagellum shorter than peduncle, 10—12-jointed. In young from the marsupium antenna 1 not shorter than antenna 2, flagellum of each 2- or 3-jointed. Maxilla 1, palp minute, on well defined interruption of hind margin of outer plate. Maxilla 2, principal seta on inner margin of inner plate not very elongate. Gnathopod 1 in ♂, 2^d joint widening rapidly to the middle, 4th not longer than 3^d, 5th little longer than 6th, widest subapically, with spines on both margins at widest part, 6th widening to a palm-like angle. a part of the sinuous hind margin being adapted to rest on hind process of 5th joint, the margin then abruptly turning to join the short spinulose palm, which is exactly fitted by the stout 2-pointed finger. Gnathopod 2 in ♂ (Fig. 99), 2d joint with no conspicuous distal lobe, 4th produced but not acute, 5th produced into a shallow cup-forming fringed process, 6th oval, finger closing over an oblique, almost straight palm into the usual pocket, armed with 2 palmar spines, the hind margin not at all bulging, carrying spinules at 2 points. Gnathopod 1 in ♀, 6th joint oblong, slightly widening to transverse palm, hind margin sinuous, finger acute, closely fitting palm. Gnathopod 2 in ♀, 4th joint subacutely produced, 5th with broad fringed process produced partly along hind margin of 6th joint, which is oblong, hind margin straight, finger acute, scarcely reaching end of transverse palm. In young from marsupium gnathopods 1 and 2 have a general resemblance

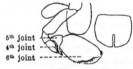

5th joint
4th joint
6th joint

Fig. 99. Fig. 100.
Gnathopod 2, ♂. Telson.
Fig. 99 & 100. A. malleolus.

to gnathopod 1 in ♀. Marsupial plates of gnathopod 2 and peraeopod 1 have one distal corner subacutely produced, in peraeopod 2 they end squarely; in all the fringing setae are short. Peraeopods 1—5 not strongly spined, finger curved. Peraeopods 3—5, 2d joint oblong oval, the front margin nearly straight, hind margin produced downward in rounded lobe; in peraeopod 3 at least as broad and about as long as in the longer peraeopod 4; in peraeopod 4 more oblong than oval, widest proximally; in peraeopod 5 much broader, widest distally. Peraeopods 3—5 in ♀ shorter and stouter than in ♂. 4th joint widened distally. Pleopods with 2 or 3 simple coupling hooks, and on inner margin of 1st joint of inner ramus 4 or 5 spines not cleft but at apex a little dilated and hooked. Uropod 3 small, ramus shorter than peduncle. Telson (Fig. 100) nearly square, with slightly convex sides, median cleft not reaching beyond middle, its sides not divergent. L. about 7 mm.

Yellow Sea, Sea of Japan. In sea-weed.

4. A. plumicornis (Heller) 1866 *Nicea p.*, Cam. Heller in: Denk. Ak. Wien, *v.* 26 II p. 5 t. 1 f. 8, 9 | 1899 *Allorchestes p.*, T. Stebbing in: Tr. Linn. Soc. London, ser. 2 *v.* 7 p. 412 t. 33 C | 1893 *Hyale prevostii* (part.), A. Della Valle in: F. Fl. Neapel, *v.* 20 p. 519.

Body compressed. Eyes an elongate round. Antenna 1 reaching nearly to the middle of flagellum of antenna 2, 1st joint longer than 2d or 3d, flagellum with 18 or 19 successively lengthening joints. Antenna 2, flagellum longer than peduncle, with 22 joints, terminal joints of peduncle and first half of flagellum beset on underside with long fascicles of setae. Maxilla 1, palp reaching base of spines of outer plate. Maxillipeds, penultimate joint of palp setose, ultimate long, acute. Gnathopod 1 in ♂, 4th joint with blunt produced point, 5th with broad distal fringed lobe, 6th oblong oval, rather longer than 5th, palm oblique, spinulose, finger somewhat thickened, its outer margin abruptly curving to an acute apex. Gnathopod 2 in ♂ much larger, 4th joint bluntly produced, 5th very short, but wide, embracing base of 6th, which is oval, with small group of spinules on hind margin, palm oblique, well defined by angle and palmar spines, finger strong, acute, much curved. Peraeopod 5 like peraeopod 4, but rather longer, 6th joint slender, straight, finger acute, little curved, setule of inner margin prominent as in all the peraeopods. Uropods 1—3, all the rami with marginal spines. Uropod 3,

peduncle rather shorter than telson. Telson divided to the base, the 2
triangular lobes inclined one to the other. L. 9—12 mm.

Mediterranean, Adriatic.

5. **A. humilis** Dana 1852 *A. h.*, J. D. Dana in: P. Amer. Ac., *v.* 2 p. 206 |
1853 & 55 *A. h.*, J. D. Dana in: U. S. expl. Exp., *v.* 13 II p. 890; t. 60 f. 6a—e | 1899
A. h., T. Stebbing in: Tr. Linn. Soc. London, ser. 2 *v.* 7 p. 413 t. 33 D | 1893 *Hyale·
prevostii* (part.)?, A. Della Valle in: F. Fl. Neapel, *v.* 20 p. 528.

Eyes a little oblong or round, light-coloured. Antenna 1 a little shorter than
antenna 2, peduncle $^2/_3$ as long as peduncle of antenna 2, flagellum with 6—8 joints,
which are very distinct, somewhat moniliform (distally widened). Antenna 2
about $^1/_3$ as long as body, ultimate and penultimate joints of peduncle subequal,
peduncle with 9 or 10 joints, which are neither very long nor distally widened,
setae all short. Maxilla 1, palp reaching base of spines of outer plate as in
Hyale (p. 559). Maxillipeds, 3^d joint of palp narrow. Gnathopod 1, 5th joint
short, 6th much longer, oblong, widening to the oblique spinulose palm, with
spinules at points of hind margin. Gnathopod 2 similar, 6th joint considerably
larger, though (Dana) not twice as long, palm somewhat more oblique, with
defining spine, spinules as in gnathopod 1, 5th joint with small lobe produced
between 4th and 6th, finger curved and short, shutting close against the palm.
Peraeopods 3—5, 2^d joint nearly orbicular, hind margin slightly crenulate.
Peraeopods 1—5 moderately stout, finger curved, with inner setules, as usual.
Branchial vesicles narrow proximally, then becoming inflated. Uropod 3 very
short, ramus as long as peduncle, each with apical spines. Telson divided beyond
the middle, the apices divergent, truncate, tipped with spinules. L. 5—8 mm.

Pacific (Port Jackson [East-Australia], shallow pools along shores; Saghalien).

A. angustus Dana 1854 *A. a.*, J. D. Dana in: P. Ac. Philad., *v.* 7 p. 177 | 1857
A. a., Stimpson in: Boston J. nat. Hist., *v.* 6 p. 520 | 1893 *A. a.*, *Hyale* (part.)?, A. Della
Valle in: F. Fl. Neapel, *v.* 20 p. 527.

Side-plates 1—4 Very deep, 5th shallow. L. 9 mm.

North-Pacific (California).

A. brevicornis Dana 1852 *A. b.*, J. D. Dana in: P. Amer. Ac., *v.* 2 p. 206 |
1853 & 55 *A. b.*, J. D. Dana in: U. S. expl. Exp., *v.* 13 II p. 893; t. 60 f. 8a—h | 1893
Hyale prevostii ♀ (part.)?, A. Della Valle in: F. Fl. Neapel, *v.* 20 p. 528.

Perhaps young ♂ of a Hyale (p. 559). L. 10 mm.

South-Pacific (New Zealand). Along shores of the Bay of Islands.

A. istricus (Grube) 1861 *Amphithöe (Hyale) istrica*, E. Grube, Ausfl. Triest,
p. 135 | 1864 *Nicea i.*, E. Grube in: Arch. Naturg., *v.* 30 I p. 200 | 1866 *N. i.*, E. Grube
in: Arch. Naturg., *v.* 32 I p. 387 t. 9 f. 5 | 1868 *N. i.*, Czerniavski in: Syezda Russ. Est.,
Syezda 1 Zool. p. 114 t. 8 f. 24, 25 | 1893 *Hyale prevostii* (part.), A. Della Valle in:
F. Fl. Neapel, *v.* 20 p. 519.

L. 6—9 mm.

Adriatic (Trieste), Black Sea.

A. japonicus Stimps. 1855 *A. japonica*, Stimpson in: P. Ac. Philad., *v.* 7 p. 383 |
1893 *Hyale pontica* (part.)?, A. Della Valle in: F. Fl. Neapel, *v.* 20 p. 528.

Eyes large, black, Very broad oval, closely approaching each other above.
L. 12 mm.

North-Pacific (Japan).

A. littoralis Stimps. 1853 *A. l.*, Stimpson in: Smithson. Contr., *v.* 6 nr. 5 p. 49 t. 3 f. 36 | 1873 *Hyale l.*, (S. I. Smith in:) A. E. Verrill in: Rep. U. S. Fish Comm., *v.* 1 p. 556 | 1893 *H. prevostii* (part.), A. Della Valle in: F. Fl. Neapel, *v.* 20 p. 519.

L. 8—10 mm.

North-Atlantic (America from Massachusetts Bay to Grand Manan). Among fucus, etc. and in tide-pools.

A. patagonicus R. O. Cunningh. 1871 *A. p.*, R. O. Cunningham in: Tr. Linn. Soc. London, *v.* 27 p. 498 t. 59 f. 14 | 1888 *Hyalella sp.?*, T. Stebbing in: Rep. Voy. Challenger, *v.* 29 p. 404 | 1893 *H. sp.?*, A. Della Valle in: F. Fl. Neapel, *v.* 20 p. 514.

No description; perhaps a species of Hyalella (p. 574).

Strait of Magellan, Sandy Point. In a freshwater stream.

A. paulensis Heller 1865 *A. p.*, Cam. Heller in: Reise NoVara, *v.* 2 III Crust. p. 128 t. 11 f. 4 | 1888 *Hyale sp.?*, T. Stebbing in: Rep. Voy. Challenger, *v.* 29 p. 383 | 1893 *H. prevostii* (part.)?, A. Della Valle in: F. Fl. Neapel, *v.* 20 p. 529.

L. 12 mm.

Southern Indian Ocean? (St. Paul).

A. penicillatus Stimps. 1855 *A. penicillata*, Stimpson in: P. Ac. Philad., *v.* 7 p. 383 | 1893 *Hyale prevostii* (part.), A. Della Valle in: F. Fl. Neapel, *v.* 20 p. 288, 519.

Antenna 2 with fascicles of long setae. L. 6 mm.

Sea of Japan.

A. plumulosus Stimps. 1857 *A. p.*, Stimpson in: Boston J. nat. Hist., *v.* 6 p. 519 | 1893 *Hyale prevostii* (part.), A. Della Valle in: F. Fl. Neapel, *v.* 20 p. 519.

Antenna 2 with fascicles of plumose setae. L. 10 mm.

San Francisco Bay. Gravelly shores in the littoral zone.

A. rubricornis Stimps. 1855 *A. r.*, Stimpson in: P. Ac. Philad., *v.* 7 p. 383 | 1879 *Hyale (A.) r.*, Wrześniowski in: Zool. Anz., *v.* 2 p. 200 | 1893 *Hyale sp.?*, A. Della Valle in: F. Fl. Neapel, *v.* 20 p. 529.

L. 16 mm.

Sea of Japan.

A. sayi Bate 1862 *A. s.*, Bate, Cat. Amphip. Brit. Mus., p. 39 t. 6 f. 5 | 1893 *Podocerus falcatus* (part.)?, A. Della Valle in: F. Fl. Neapel, *v.* 20 p. 453.

L. 9 mm.

North-America.

A. stylifer Grube 1866 *A. s.*, E. Grube in: Arch. Naturg., *v.* 321 p. 386 t. 9 f. 4 | 1893 *Hyale prevostii* (part.), A. Della Valle in: F. Fl. Neapel, *v.* 20 p. 520.

Gnathopod 2 in ♂, 5th joint Very short, distally produced to a curVed stiliform process betWeen 4th and 6th. L. 5 mm.

Adriatic. Under stones on beach.

A. sp., Heller 1866 *Nicea nilsoni* (err., non *Amphithoë nilssonii* H. Rathke 1843!), Cam. Heller in: Denk. Ak. Wien, *v.* 26 II p. 4.

L. ♂ 5—6 mm.

Adriatic.

---

33. Fam. **Aoridae**

1899 *Aoridae*, T. Stebbing in: Ann. nat. Hist., ser. 7 *v.* 4 p. 211.

Head, lateral lobes little produced. Side-plates of very moderate depth, 4th with hind margin not excavate. Antenna 1 the longer, with 3d joint short (except in Paradryope, p. 602), with accessory flagellum (except in Aoroides). The mouth-organs appear to agree in having border of upper lip round or very faintly emarginate, lower lip with acutely produced mandibular processes. Mandible with 3^d joint of palp longer than 2^d. Maxilla 1 with a single seta on inner plate, 10 spines on outer, 2^d joint of palp elongate; maxilla 2 with inner plate fringed. Maxillipeds well developed. Gnathopods 1 (Fig. 101) and 2 not simple, 1st the larger, and its shape usually differing much in ♂ and ♀. Peraeopods 1 and 2 glandular. Peraeopod 5 the longest. Branchial vesicles simple. Uropods 1—3 biramous, 3^d not elongate. Telson simple.

The position of Paradryope (p. 602) in this family cannot, however, be regarded as absolutely secure, and Della Valle's suggestion that I may have transposed the gnathopods 1 and 2, if it could be substantiated, would give a welcome relief, by permitting the transfer of the genus to the Photidae (p. 603).

Marine. .

7 genera, 25 accepted species and 3 doubtful.

Synopsis of genera:

1 { Antenna 1 without accessory flagellum -. . . . 1. Gen. **Aoroides** . . . . p. 586
{ Antenna 1 with accessory flagellum — **2**.

2 { Gnathopod 1 in ♂ (Fig. 101), 4th joint immensely
{ produced . . . . 2. Gen. **Aora** . . . . . . p. 587
{ Gnathopod 1 in ♂, 4th joint not immensely
{ produced — **3**.

3 { Gnathopod 1 in ♂, 5th joint produced to a
{ strong tooth . . . . 3. Gen. **Microdeutopus** . p. 588
{ Gnathopod 1 in ♂, 5th joint not produced to
{ a strong tooth — **4**.

4 { Uropod 3, rami not minute — **5**.
{ Uropod 3, rami minute — **6**.

5 { Gnathopod 1 in ♂, 5th and 6th joints subequal
{ in width . . . . 4. Gen. **Lembos** . . . . p. 594
{ Gnathopod 1 in ♂, 5th joint much wider than 6th 5. Gen. **Lemboides** . . . p. 600

6 { Pleon segment 6 dorsally evanescent . . . . 6. Gen. **Dryopoides** . . p. 601
{ Pleon segment 6 not dorsally evanescent . . 7. Gen. **Paradryope** . . p. 602

1. Gen. **Aoroides** A. Walker

1898 *Aoroides* (Sp. un.: *A. columbiae*), A. O. Walker in: P. Liverp. biol. Soc., v. 12 p. 284.

♀. As in Aora and Microdeutopus (p. 588), except that antenna 1 is entirely devoid of accessory flagellum, and mandibular palp is very slight and very sparingly furnished with setae on the 3^d joint.

1 species.

1. **A. columbiae** A. Walker 1898 *A. c.*, A. O. Walker in: P. Liverp. biol. Soc., v. 12 p. 285 t. 16 f. 7—10.

♂ unknown. — ♀. Like ♀ of Microdeutopus anomalus (p. 591) except for the generic differences, and side-plate 1 rounded below, peraeopods 3—5 with wider 2^d joint, uropod 3 with peduncle as long as the rami, which are equal, inner ramus with 1 spine about the middle of the inner margin, outer with apical spines only. L. 5 mm.

Puget Sound.

2. Gen. **Aora** Krøyer

1845 *Aora* (Sp. un.: *A. typica*), Krøyer in: Naturh. Tidsskr., ser. 2 *v.* 1 p. 328, 335 | 1888 *A.*, T. Stebbing in: Rep. Voy. Challenger, *v.* 29 p. 1072 (synonymy to date) | 1893 *A.*, A. Della Valle in: F. Fl. Neapel, *v.* 20 p. 406 | 1894 *A.*, G. O. Sars, Crust. Norway, *v.* 1 p. 544 | 1849 *Lalaria* (Sp. un.: *L. longitarsis*), H. Nicolet in: Gay, Hist. Chile, *v.* 3 p. 240 | 1856 *Lonchomerus* (nom. nud.), Bate in: Rep. Brit. Ass., Meet. 25 p. 58 | 1857 *L.* (Sp. un.: *L. gracilis*), Bate in: Ann. nat. Hist., ser. 2 *v.* 19 p. 143 | 1857 *Lalasia*, Bate in: Ann. nat. Hist., ser. 2 *v.* 20 p. 525 | 1859 *Autonoe* (part.), R. M. Bruzelius in: Svenska Ak. Handl., n. ser. *v.* 3 nr. 1 p. 23.

Body slender. Head, lateral lobes obtuse. Side-plates of medium depth, in ♂ 1st somewhat produced forward, 2d rather larger than 3d, 5th less deep than 4th. Antenna 1 much the longer, flagellum long, accessory flagellum well developed. Lower lip with mandibular processes strongly produced. Mandible, both cutting plates dentate, spine-row of 3 or 4 spines, molar strong. Maxilla 1, inner plate small, with 1 apical seta, 2d joint of palp with several spine-teeth. Maxilla 2, inner plate the smaller, its inner margin fringed. Maxillipeds, inner and outer plates well developed, well armed, finger of palp not stout. Gnathopod 1 in ♂ (Fig. 101) elongate, slender, 4th joint produced into a very long spine-like process behind the 5th, which is a little broader and longer than the narrowly oblong 6th, finger much overlapping palm. Gnathopod 1 in ♀, 4th joint not produced, 5th much shorter than 6th. Gnathopod 2 in ♂ and ♀, 5th and 6th joints subequal. Peraeopods 3—5, 2d joint not very widely expanded, 4th peraeopod longer than 3d, 5th than 4th. Marsupial plates large, broad. Uropod 3, rami equal, much longer than peduncle.

1 species.

1. **A. typica** Krøyer 1845 *A. t.*, Krøyer in: Naturh. Tidsskr., ser. 2 *v.* 1 p. 328 t. 3 f. 3a—l | 1849 *Lalaria longitarsis*, H. Nicolet in: Gay, Hist. Chile, *v.* 3 p. 243; Crust. t. 2 f. 8, 8a—f | 1857 *Lonchomerus gracilis*, Bate in: Ann. nat. Hist., ser. 2 *v.* 19 p. 143 | 1862 *Aora g.* + *A. typica*, Bate, Cat. Amphip. Brit. Mus., p. 160 t. 29 f. 7; p. 161 t. 29 f. 8 | 1885 *A. t.*, Chilton in: Ann. nat. Hist., ser. 5 *v.* 16 p. 370 | 1893 *A. g.* + *A. t.*, A. Della Valle in: F. Fl. Neapel, *v.* 20 p. 407 t. 2 f. 9, t. 12 f. 25—39, t. 56 f. 37; p. 409 t. 56 f. 38—40 | 1894 *A. g.*, G. O. Sars, Crust. Norway, *v.* 1 p. 545 t. 193 | 1859 *Autonoe punctata*, R. M. Bruzelius in: Svenska Ak. Handl., n. ser. *v.* 3 nr. 1 p. 24 t. 1 f. 3a—g | 1879 *Microdeutopus maculatus* (♀), G. M. Thomson in: Ann. nat. Hist., ser. 5 *v.* 4 p. 331 t. 16 f. 5—8 | 1881 *Microdentopus m.*, G. M. Thomson in: Tr. N. Zealand Inst., *v.* 13 p. 217 t. 8 f. 7A—C | 1882 *M. m.* (♂), Chilton in: Tr. N. Zealand Inst., *v.* 14 p. 173 t. 8 f. 3a, b | 1879 *Microdeuteropus tenuipes* (♀) + *M. mortoni* (♂), Haswell in: P. Linn. Soc. N. S. Wales, *v.* 4 p. 339 t. 22 f. 1, 2 | 1882 *Microdeutopus t.* + *M. m.*, Haswell, Cat. Austral. Crust., p. 264 | 1888 *Aora kergueleni* + *A. trichobostrychus*, T. Stebbing in: Rep. Voy. Challenger, *v.* 29 p. 1073 t. 109, f. A ♂, f. D ♀; p. 1078 t. 109 f. B ♂, f. C ♀.

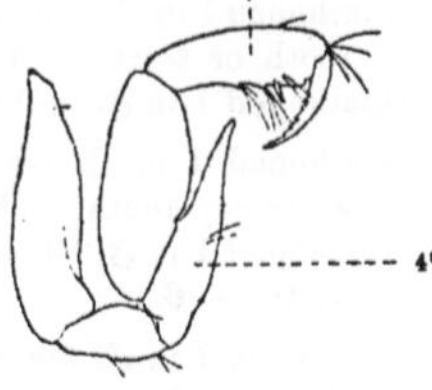

Fig. 101. **A. typica**, ♂. Gnathopod 1.

Pleon segment 3, postero-lateral corners obtusely quadrate. Eyes small, oval, dark. Antenna 1 2/3 as long as body, 2d joint the longest, 3—4 times as long as 3d, flagellum with 20—30 joints or more, accessory flagellum 5-jointed. Antenna 2 much shorter, ultimate joint of peduncle rather longer than penultimate or than flagellum, flagellum 7-jointed. The gnathopod 1 in ♂ (Fig. 101) affords the principal generic character, but the 2d joint in adult ♂ *(typica)* is said to have a triangular process on the front margin, not observed in any of the other forms; in *A. trichobostrychus* the hind margin of the 2d joint has a dense brush of setae, whereas generally it is smooth; the prolongation of the 4th joint, the relative lengths of 5th, 6th and finger are very variable, but how far the proportions are constant is as yet indeterminate. In *A. gracilis* ♂ the 2^{d} joint of the 5th peraeopod is widened near the distal end, while in the ♀; and perhaps in both sexes of some forms, it is oblong oval without such widening. The telson is sometimes longer than broad *(A. trichobostrychus)*, nearly as broad as long *(A. gracilis, A. kergueleni)*, usually with the apex obtusely angular, but sometimes *(A. gracilis* Della Valle, not Sars*)* emarginate. Colour whitish or greyish, with speckling. L. 8 mm *(A. gracilis)*.

North-Atlantic with adjoining seas (Europe), Pacific (South-America, Australia, New Zealand), southern Indian Ocean (Kerguelen Island).

3. Gen. **Microdeutopus** A. Costa

1853 *Microdeutopus* (Sp. un.: *M. gryllotalpa*), A. Costa in: Rend. Soc. Borbon., n. ser. *v.* 2 p. 171 | 1876 *M.*, A. Boeck, Skand. Arkt. Amphip., *v.* 2 p. 564 | 1888 *M.*, T. Stebbing in: Rep. Voy. Challenger, *v.* 29 p. 299, 1711 | 1893 *M.*, A. Della Valle in: F. Fl. Neapel, *v.* 20 p. 410 | 1894 *M.*, G. O. Sars, Crust. Norway, *v.* 1 p. 539 | 1856 *Lembos* (part.), Bate in: Rep. Brit. Ass., Meet. 25 p. 58 | 1857 *L.* (part.), Bate in: Ann. nat. Hist., ser. 2 *v.* 19 p. 142 | 1859 *Autonoe* (part.), R. M. Bruzelius in: Svenska Ak. Handl., n. ser. *v.* 3 nr. 1 p. 23 | 1862 *Stimpsonia* (Sp. un.: *S. chelifera*) (non C. Girard 1853, Nemertini!) + *Microdeutopus* (part.) *(Microdeuteropus)* *(Microdentopus* Bate), Bate & Westwood, Brit. sess. Crust., *v.* 1 p. 284, 287 | 1869 *Microdeuteropus* (part.), A. M. Norman in: Rep. Brit. Ass., Meet. 38 p. 281 | 1874 *M.*, T. Stebbing in: Ann. nat. Hist., ser. 4 *v.* 14 p. 12 | 1893 *Stimpsonella*, A. Della Valle in: F. Fl. Neapel, *v.* 20 p. 421.

Body slender. Head with lateral lobes usually obtuse. Side-plate 1 in ♂ often produced forward, 2^{d} or 3^{d} the deepest, remainder successively shallower. Antenna 1, 2^{d} joint the longest, accessory flagellum distinct. General character as in Aora, but gnathopod 1 in ♂ with 4th joint not greatly produced, 5th very bulky, its hind margin produced to a tooth, which with help of 6th joint and finger makes the limb complexly subchelate.

10 species accepted, 1 doubtful.

Synopsis of accepted species:

1 { Gnathopod 2 in ♂, palm defined by a tooth — **2**.
{ Gnathopod 2 in ♂, palm not defined by a tooth — **3**.

2 { Gnathopod 2 in ♂, 6th joint much shorter than 5th 1. **M. chelifer** . . . . p. 589
{ Gnathopod 2 in ♂, 6th joint not shorter than 5th 2. **M. armatus** . . . . p. 589

3 { Gnathopod 1 in ♂, tooth of 5th joint with accessory
    tooth or teeth — **4**.
{ Gnathopod 1 in ♂, tooth of 5th joint simple — **8**.

4 { Gnathopod 1 in ♂, 5th joint with more than 1
    accessory tooth — **5**.
{ Gnathopod 1 in ♂, 5th joint with only 1 accessory
    tooth — **6**.

5 { Gnathopod 1 in ♂, 5th joint with innermost tooth
    the largest . . . .   3. **M. gryllotalpa** . . p. 590
{ Gnathopod 1 in ♂, 5th joint with middle tooth
    the largest . . . .   4. **M. stationis** . . . p. 590

6 { Gnathopod 1 in ♂, 5th joint with inner tooth the
    smaller . . . .   5. **M. haswelli** . . . p. 591
{ Gnathopod 1 in ♂, 5th joint with outer tooth
    the smaller — **7**.

7 {
  Gnathopod 1 in ♂, 5th joint with accessory tooth at base of principal . . . . 6. **M. anomalus** . . . p. 591
  Gnathopod 1 in ♂, 5th joint with accessory tooth apart from principal . . . . · . . 7. **M. propinquus** . . p. 592

8 {
  Antenna 1, flagellum shorter than peduncle . . 8. **M. megnae** . . . . p. 592
  Antenna 1, flagellum longer than peduncle — **9.**

9 {
  Gnathopod 2 in ♂ very slender and very setose 9. **M. versiculatus** . p. 593
  Gnathopod 2 in ♂ not very slender and not very setose . . . . 10. **M. damnoniensis** . p. 593

1. **M. chelifer** (Bate) 1862 *Stimpsonia chelifera*, Bate (& Westwood), Brit. sess. Crust., *v.* 1 p. 285 f. | 1862 *S. c.*, Bate, Cat. Amphip. Brit. Mus., p. 162 t. 29 f. 9 | 1878 *S. c.*, T. Stebbing in: Ann. nat. Hist., ser. 5 *v.* 1 p. 34 t. 5 f. 2, 3 | 1888 *S. c.*, *Microdeutopus* (part.), T. Stebbing in: Rep. Voy. Challenger, *v.* 29 p. 334 | 1893 *Stimpsonella c.*, A. Della Valle in: F. Fl. Neapel, *v.* 20 p. 424 t. 56 f. 42—45.

Side-plate 1 with an acute front corner in ♂. Antennae 1 and 2 subequal (Bate: antenna 1 the shorter). Antenna 1, 2d joint longer than 1st, 3d very short, flagellum longer than peduncle, accessory flagellum 2-jointed. Antenna 2, antepenultimate joint of peduncle very thick, penultimate very long, at the base winged below so as to equal depth of preceding joint, ultimate about as long as penultimate or flagellum. Gnathopod 1 in ♂ very large, the size chiefly due to the 5th joint, the distal tooth of which varies greatly in length, a deep cavity separating it from a little setose tooth adjacent to base of the somewhat oblong 6th joint; 6th joint gently convex in front with undulating hind margin and no conspicuous palm. Gnathopod 2 in ♂, 5th joint much longer than 6th, 6th as broad as long, widening to the variable palm, which is defined by a tooth sometimes large but capable of vanishing altogether, finger matching palm. Gnathopod 1 in ♀, 6th joint slightly larger than 5th, about twice as long as broad, palm defined by a palmar spine, overlapped by serrate finger. Gnathopod 2 in ♀ similar to gnathopod 1, but much smaller. Peraeopods 1 and 2, 4th and 5th joints much broader than 6th. Peraeopods 3—5, 2d joint narrow; peraeopod 5 much the longest; in all peraeopods the 6th joint has an apical group of setae. Uropod 3, peduncle short, stout, rami equal, little longer than peduncle. Telson with a setule at the lateral notch on each side of the rounded, unproduced apex. L. 8—9 mm.

English Channel (South-Devon).

2. **M. armatus** Chevreux 1886 & 87 *M. a.*, Chevreux in: Bull. Soc. zool. France, *v.* 11 p. XLI: *v.* 12 p. 312 t. 5 f. 11, 12; textf. 6, 7 p. 296 | 1893 *Stimpsonella armata*, A. Della Valle in: F. Fl. Neapel, *v.* 20 p. 422 t. 4 f. 8; t. 11 f. 13—24.

Very slender; ventral surface of peraeon segments 2—4 (sometimes others?) armed each with a spiniform tooth. Head without rostrum, lateral lobes little prominent. Side-plate 1 in ♂ little produced, in ♀ rhomboidal. Eyes small, round, black. Antenna 1 rather shorter than the body, 2d joint longer than 1st, 3d over 1/8 as long as 2d, flagellum longer than peduncle, 11-jointed, accessory flagellum very short, with 2 joints, 2d minute. Antenna 2 much the shorter, ultimate joint of peduncle as long as penultimate, slightly longer than the 4-jointed flagellum. Gnathopod 1 in ♂, 2d joint very broad except at the extremities, 5th massive, hind margin finely crenulate distally and produced to a strong tooth, 6th much shorter and narrower, somewhat oblong, with very irregularly crenate hind margin, finger long, slender. Gnathopod 2 in ♂, 2d joint broad, front margin somewhat indentured and distally produced, 5th almost oblong, 6th rather longer, oblong, palm only slightly oblique, defined by a small tooth, and sometimes carrying a submedian one, finger matching palm. Gnathopod 1 in $\mathcal{Q}$, 2^d joint not expanded, 5th shorter than 6th, 6th oblong but with rather oblique palm defined by a palmar spine and overlapped by finger. Gnathopod 2 in $\mathcal{Q}$ similar to gnathopod 1 but smaller, and the finger not overlapping the palm. Peraeopods 1—5 slender, little armed; 2^d joint in peraeopods 3—5 narrowly oval. Uropods 1 and 2, rami slender, with few spines. Uropod 3, rami equal, a little shorter than peduncle. Telson nearly as broad as long, tapering to a truncate apex, the lateral processes little marked. Colour yellow, with large blotches of violet-brown on peraeon (Chevreux), front part of peraeon wine-red or grey or yellowish, eggs emerald green (Della Valle). L. 3—4 mm.

North-Atlantic (South-East of Brittany); Bay of Naples. Depth 10—20 m.

3. **M. gryllotalpa** A. Costa 1853 *M. g.,* A. Costa in: Rend. Soc. Borbon., n. ser. *v.* 2 p. 178 | 1876 *M. g.,* A. Boeck, Skand. Arkt. Amphip., *v.* 2 p. 565 t. 29 f. 6 | 1884 *M. g.,* H. Blanc in: N. Acta Ac. Leop., *v.* 47 p. 75 t. 9 f. 82—90 | 1893 *M. g.,* A. Della Valle in: F. Fl. Neapel, *v.* 20 p. 411 t. 1 f. 12; t. 11 f. 25—43 | 1894 *M. g.,* G. O. Sars, Crust. Norway, *v.* 1 p. 543 t. 192 f. 2 | 1859 *Autonoe grandimana,* R. M. Bruzelius in: Svenska Ak. Handl., n. ser. *v.* 3 nr. 1 p. 26 t. 1 f. 5 | 1862 *Microdeutopus grandimanus,* Bate, Cat. Amphip. Brit. Mus., p. 378 | 1873 *M. minax,* (S. I. Smith in:) A. E. Verrill in: Rep. U. S. Fish Comm., *v.* 1 p. 562 | ?1885 *Amphithoe salenskii,* J. V. Carus, Prodr. F. Medit., *v.* 1 p. 396.

Body moderately robust. Head without rostrum, lateral lobes blunt, scarcely projecting. Side-plates shallow, 1st moderately produced in $\mathcal{O}$. Pleon segment 3, postero-lateral corners bluntly quadrate. Eyes small, round, dark (Sars), or whitish grey (Della Valle). Antenna 1 considerably more (Sars; Della Valle: less) than half as long as body, 2^d joint rather longer than 1st, 3^d about $^1/_3$ as long as 2^d, flagellum longer than peduncle, 10—22-jointed, accessory flagellum 1- or 2-jointed. Antenna 2 much shorter, ultimate joint of peduncle as long as penultimate, rather longer than 6- or 7-jointed flagellum. Gnathopod 1 in $\mathcal{O}$ massive, 2^d joint widening from narrow neck, 5th extremely large, broad oval, armed with 4 teeth successively larger inwards toward base of 6th joint, which is about $^1/_4$ the size of 5th, somewhat oblong, but narrowing distally, with sinuous hind margin forming 2 or 3 irregular lobes, finger strong, serrate; teeth of 5th joint variable: 2—5. Gnathopod 2 in $\mathcal{O}$, 2^d joint strongly dilated, except at the extremities, 5th and 6th joints long and narrow, setose, palm and finger very small. Gnathopod 1 in $\mathcal{Q}$, 2^d joint not expanded, 5th rather robust, shorter than oblong 6th, palm rather oblique, ill-defined, overlapped by finger. Gnathopod 2 in $\mathcal{Q}$, 2^d joint not expanded, otherwise nearly as in $\mathcal{O}$, but smaller. Peraeopods 1—5 slender, 2^d joint in peraeopods 3—5 not greatly expanded, oblong in 3^d, long oval in 5th. Uropod 3, rami subequal to peduncle, inner rather the shorter. Telson rather longer than broad, apex convex (Sars in figure), or broader than long, apex concave (Della Valle), with 2 or 3 setules on each side of apex. Colour, densely variegated with dark brown. L. 4—8 mm.

North-Atlantic with adjoining seas (Europe, from Norway to Venice (not all localities to be trusted); East of United States of America).

4. **M. stationis** Della Valle 1881 *M. gryllotalpa* (err., non A. Costa 1853!), Nebeski in: Arb. Inst. Wien, *v.* 3 p. 155 t. 13 f. 41 | 1880 *M. g.,* Sowinski in: Mém. Soc. Kiew, *v.* 6 p. 128 t. 5 f. 17a—d | 1893 *M. stationis,* A. Della Valle in: F. Fl. Neapel, *v.* 20 p. 415 t. 5 f. 2, t. 10 f. 31—41 | 1895 *M. s.,* Sowinski in: Mém. Soc. Kiew, *v.* 14 p. 237 t. 4 f. 1—6 | 1898 *M. s.,* Sowinski in: Mém. Soc. Kiew, *v.* 15 p. 480.

General aspect that of an Ampithoe (p. 631). Mouth-parts, peraeopods and uropods as in M. gryllotalpa; uropods more slender. Head rather large. Side-plate 1 in ♂ very much and acutely produced, in ♀ rhomboidal. Eyes small, round. Antenna 1 rather slender, 2^d joint longer than 1st, more than thrice as long as 3^d, flagellum much longer than peduncle, 24-jointed, accessory flagellum longer than 3^d joint of peduncle, 4-jointed. Antenna 2 moderately robust, ultimate joint of peduncle as long as penultimate, much longer than the 8-jointed flagellum (Della Valle), or rather shorter than the 9-jointed flagellum (Sowinski). Gnathopod 1 in ♂, 2^d joint greatly expanded, 5th massive but much longer than broad, produced into a large tooth, flanked by a smaller one (Sowinski: wanting) within the palm, and a minute one at its base on the hind margin; 6th joint rather slender, with hind margin extremely sinuous, not crenate; finger serrate, rather small. Gnathopod 2 in ♂, 2^d joint little expanded, 5th and 6th joints subequal, oblong, palm oblique, ill-defined, finger small. Gnathopod 1 in ♀, 6th joint rather stout, a little longer than 5th, oblong, palm transverse, the rounded angle overlapped by the finger. Gnathopod 2 in ♀, 2^d joint slender, 6th stouter than in ♂ and much longer than 5th, finger matching palm. Telson longer than broad, subelliptical with the apex feebly concave, without processes or setae. Colour blackish grey, except antennae and distal parts of limbs. L. 10—12 mm.

Mediterranean (Naples, in fine sand, depth 10—20 m; Trieste); Black Sea, depth 23—105 m.

5. **M. haswelli** Stebb. 1879 *Microdeuteropus chelifer* (non *Stimpsonia chelifera* Bate 1862!), Haswell in: P. Linn. Soc. N. S. Wales, v. 4 p. 340 t. 22 f. 3 | 1882 *Microdeutopus c.*, Haswell, Cat. Austral. Crust., p. 265 | 1899 *M. haswelli*, T. Stebbing in: Ann. nat. Hist., ser. 7 v. 3 p. 350.

Eyes small, round. Antenna 1 fully half as long as body, flagellum longer than peduncle, accessory flagellum 3-jointed. Antenna 2 subpediform, nearly as long as antenna 1, peduncle stout, flagellum shorter than ultimate joint of peduncle, obscurely multiarticulate. Gnathopod 1 in ♂ very large, 2^d joint apparently not expanded, 5th massive, much longer than broad, widening distally, hind margin produced into a strong tooth, with a shorter and blunter one between it and the base of the 6th joint; 6th joint very much smaller than 5th, front margin convex, hind strongly concave, with small proximal tubercle; finger short, serrate. Gnathopod 2 small, 5th joint ovate, 6th a little longer, palm undefined. Peraeopods 1 and 2 stout. Peraeopods 3—5, 2^d joint long ovate, peraeopods 4 and 5 much longer than 3^d. Uropod 3 very small, rami short, subfoliaceous. Telson conical, armed above with a few slender hairs. L. 5 mm.

Port Jackson [East-Australia].

6. **M. anomalus** (H. Rathke) 1843 *Gammarus a.*, H. Rathke in: N. Acta Ac. Leop., v. 20 ı p. 63 t. 4 f. 7 | 1855 *G. (Gammaropsis) a.*, W. Liljeborg in: Vetensk. Ak. Handl., 1853 p. 457 | 1859 *Autonoe anomala*, R. M. Bruzelius in: Svenska Ak. Handl., n. ser. v. 3 nr. 1 p. 25 t. 1 f. 4 | 1862 *Microdeutopus anomalus*, Bate, Cat. Amphip. Brit. Mus., p. 379 | 1876 *M. a.*, A. Boeck, Skand. Arkt. Amphip., v. 2 p. 567 t. 25 f. 5 | 1893 *M. a.* (part.), A. Della Valle in: F. Fl. Neapel, v. 20 p. 417 t. 56 f. 41 | 1894 *M. a.*, G. O. Sars, Crust. Norway, v. 1 p. 540 t. 191 | 1898 *M. a.*, Sowinski in: Mém. Soc. Kiew, v. 15 p. 480 t. 10 f. 20—24 | 1869 *Microdeuteropus a.*, A. M. Norman in: Rep. Brit. Ass., Meet. 38 p. 281 | 1856 *Lembos cambriensis* (nom. nud.), Bate in: Rep. Brit. Ass., Meet. 25 p. 58 | 1857 *L. c.*, Bate in: Ann. nat. Hist., ser. 2 v. 19 p. 142.

Body very slender. Head, lateral lobes a little produced, rounded. Side-plate 1 sharply and much produced in ♂, quadrate in ♀, with front angle not obtuse. Side-plate 2 deepest in ♂, 3^d in ♀. Pleon segment 3, postero-lateral corners subquadrate, not acutely. Eyes small, round, black with whitish coating. Antenna 1 $^2/_3$ as long as body; 2^d joint considerably longer than 1st, about thrice as long as 3^d; flagellum much longer than peduncle, about 22-jointed (Bruzelius: 20—28), accessory flagellum 3—5-jointed. Antenna 2 much shorter, ultimate joint of peduncle subequal to penultimate, scarcely longer than the 8-jointed flagellum. Gnathopod 1 in ♂, 2^d joint not widened, having (in Sars' figure) a small denticle on upper part of front margin, 5th very large, but much longer than broad, hind margin produced to a strong tooth with secondary tooth at its base outside, this being accompanied by 1 or several small teeth (but only 1 in figures), 6th joint narrow, with strongly concave hind margin, angular near finger-hinge (Sars), or very sinuous, with distal half very convex, not angular (Sowinski); finger long, serrate. Gnathopod 2 in ♂, 6th joint as long as 5th, scarcely so wide, narrowly oblong, palm and finger very short. Gnathopod 1 in ♀, 2^d joint with smooth margins, 5th nearly as long as 6th, which widens a little to the slightly oblique palm, finger overlapping palm. Gnathopod 2 in ♀ nearly as in ♂, but smaller. Peraeopods 1 and 2 slightly armed. Peraeopods 3—5, 2^d joint narrowly oblong, peraeopod 5 elongate, with the usual fascicle of setae at hind apex of 6th joint. Uropod 3, inner ramus rather the shorter, as long as peduncle. Telson a little longer than broad, apex slightly angular, with 3 slender spines on each side (Sars), or convex between distinctly outdrawn subacute points (Sowinski). Colour whitish, sometimes with reddish tinge, mottled with dark brown spots in transverse bands. L. 6—9 mm.

North-Atlantic (South-West-Norway, at least to Trondhjemsfjord, depth 11—56 m; Shetlands, depth 128—164 m); Black Sea, depth 54—64 m.

7. **M. propinquus** O. Sars 1894 *M. danmoniensis* (err., non *Lembos dammoniensis* Bate 1856!), *M. propinqvus*, G. O. Sars, Crust. Norway, *v.* 1 p. 542; t. 192 f. 1.

Closely allied to M. anomalus (p. 591), but differing at follows. Eyes rather larger, rounded oval. Antenna 1, flagellum little longer than peduncle, accessory flagellum of 2 equal joints. Gnathopod 1 in ♂, tooth of 5th joint less elongate, the accessory tooth a little removed from it, 6th joint with hind margin less concave; gnathopod 1 in ♀ less slender, 6th joint considerably longer than 5th. Peraeopod 5 less elongate. Colour whitish, mottled with dark brown. L. ♂ 5 mm.

North-Atlantic (Norway). Very shallow water.

8. **M. megnae** Giles 1888 *M. m.*, G. M. Giles in: J. Asiat. Soc. Bengal, *v.* 57 p. 231 t. 7 f. 1—4.

Head without rostrum; lateral lobes obtuse, little prominent. Side-plates shallow, 1st not strongly produced, 3^d deeper than 2^d or 4th. Pleon segment 3, postero-lateral corners rounded. Eyes small, rounded, black. Antenna 1 about half as long as body, 2^d joint longer than 1st, 3^d about $^1/_2$ as long as 1st, flagellum much shorter than peduncle, 10—14-jointed, accessory flagellum very small, 1-jointed. Antenna 2 stouter than antenna 1, subequal in length; ultimate joint of peduncle as long as penultimate, longer than flagellum of 10—12 very short joints. Mandible, 2^d and 3^d joints of palp subequal, neither much longer than the 1st. Gnathopod 1 in ♂, 2^d joint much expanded, 5th massive, longer than broad, hind margin produced into a slender tooth, 6th joint very short, nearly quadrangular, hirsute, finger strong and a little varicose, but otherwise unarmed. Gnathopod 2 in ♂ short, slender, imperfectly subchelate, 6th joint shorter than 5th, widening (in figure) to an oblique palm; finger reaching the obtuse defining angle. Gnathopod 1 in ♀ much smaller than in ♂, 6th joint larger than 5th, palm feebly defined. Gnathopod 2 in ♀ smaller and more slender than the 1st. Peraeopod 1 shorter than peraeopod 2; finger in both rather long and falciform. Peraeopods 3—5 as usual in the genus. Uropods 1—3 reaching about the same level; peduncle of uropod 3 extremely short. Telson short, armed above with a pair of peculiar conical protuberances bearing each a single strong bristle. Colour dirty white, somewhat pellucid. L. 4—5 mm.

Bay of Bengal (Megna Shoals). Taken in the surface net, depth about 11 m.

9. M. versiculatus (Bate) 1856 *Lembos v.* (nom. nud.), Bate in: Rep. Brit. Ass., Meet. 25 p. 58 | 1857 *L. v.*, Bate in: Ann. nat. Hist., ser. 2 *v.* 19 p. 142 | 1862 *Microdentopus v.*, Bate, Cat. Amphip. Brit. Mus., p. 165 t. 30 f. 5 | 1895 *Microdeutopus v.*, A. O. Walker in: Ann. nat. Hist., ser. 6 *v.* 15 p. 469 | 1869 *Microdeuteropus v.*, A. M. Norman in: Rep. Brit. Ass., Meet. 38 p. 282 | 1874 *M. v.*, T. Stebbing in: Ann. nat. Hist., ser. 4 *v.* 14 p. 12 t. 1 f. 2, 2a—f | 1876 *Autonoë longipes* (part.)?, A. Boeck, Skand. Arkt. Amphip., *v.* 2 p. 574 | 1893 *Microdeutopus anomalus* (part.)?, A. Della Valle in: F. Fl. Neapel, *v.* 20 p. 418.

Head without rostrum, lateral lobes obtuse, little produced. Side-plate 1 acutely produced in ♂. Pleon segment 3, postero-lateral corners obtusely quadrate. Eyes rather large, dark. Antenna 1, 2[d] joint longer than 1st, about twice as long as 3[d]; flagellum 15-jointed, accessory flagellum 1-jointed. Antenna 2, ultimate joint of peduncle subequal to penultimate, rather longer than the 6-jointed flagellum. Gnathopod 1 in ♂ massive, the bulk depending chiefly on the very broad and long 5th joint, of which the hind margin is produced into a strong, slightly outward curving tooth; the small 6th joint, short and stout, with a median tubercle on the hind margin, closely approximates the tooth of the 5th, which is crossed by the short serrate finger. Gnathopod 2 in ♂ very slender, hirsute on 4th—6th joints with long plumose setae, 4th acute at apex, 5th longer than 6th, together forming a geniculation, palm and finger very small. Gnathopod 1 in ♀, 5th joint rather longer than 6th, but about equal in breadth, palm rather oblique, feebly defined, overlapped by finger. Gnathopod 2 in ♀ as in ♂. Peraeopods 1 and 2, 4th and 5th joints a little widened, finger nearly as long as 6th joint. Peraeopods 3—5, 2[d] joint pretty well expanded; peraeopod 5 the longest. Uropod 3, rami little longer than peduncle. L. 4—5 mm.

North-Atlantic and English Channel (Plymouth; Salcombe Harbour; Shetland, depth 128—146 m).

10. M. damnoniensis (Bate) 1856 *Lembos d.* (nom. nud.), Bate in: Rep. Brit. Ass., Meet. 25 p. 58 t. 17 f. 9 | 1857 *L. d.*, Bate in: Ann. nat. Hist., ser. 2 *v.* 19 p. 142 | 1857 *L. danmoniensis*, A. White, Hist. Brit. Crust., p. 180 | 1862 *Miorodentopus gryllotalpa* (err., non *Microdeutopus g.* A. Costa 1853!), Bate, Cat. Amphip. Brit. Mus., p. 163 t. 30 f. 1 | 1876 *Microdeutopus g.*, A. Boeck, Skand. Arkt. Amphip., *v.* 2 p. 565 | 1893 *M. anomalus* (part.) + *M. algicola*, A. Della Valle in: F. Fl. Neapel, *v.* 20 p. 417; p. 418 t. 1 f. 3; t. 11 f. 1—12.

Distinguished from M. propinquus by the gnathopod 1 in ♂, which has no accessory tooth on the 5th joint, and the 6th joint of almost uniform breadth. Antenna 1, accessory flagellum with 1 (Bate) or 2 (Della Valle) joints. Lateral processes of telson rather prominently developed, with several setules on its surface. Colour greenish grey, with some black spots. L. 4·5—6 mm.

English Channel (South England); Bay of Naples, among algae.

M. titii Heller 1866 *M. t.*, Cam. Heller in: Denk. Ak. Wien, *v.* 26 II p. 48 t. 4 f. 8 | 1893 *M. t.*, A. Della Valle in: F. Fl. Neapel, *v.* 20 p. 420.

Perhaps identical with M. damnoniensis (p. 593); but antenna 2 considerably longer than antenna 1. L. 5 mm.

Adriatic (Pirano).

4. Gen. **Lembos** Bate

1855 [Subgen.] *Gammaropsis* (part.), W. Liljeborg in: Vetensk. Ak. Handl., 1853 p. 455 | 1856 *Lembos* (nom. nud.) (part.), Bate in: Rep. Brit. Ass., Meet. 25 p. 58 | 1857 *L.* (part.), Bate in: Ann. nat. Hist., ser. 2 *v.* 19 p. 142 | 1895 *L.*, T. Stebbing in: Ann. nat. Hist., ser. 6 *v.* 16 p. 207 | 1859 *Autonoe* (part.), R. M. Bruzelius in: Svenska Ak. Handl., n. ser. *v.* 3 nr. 1 p. 23 | 1888 *A.*, T. Stebbing in: Rep. Voy. Challenger, *v.* 29 p. 1081 | 1893 *A.*, A. Della Valle in: F. Fl. Neapel. *v.* 20 p. 398 | 1894 *Autonoë*, G. O. Sars, Crust. Norway, *v.* 1 p. 546.

In general agreement with Aora (p. 587), also with well developed flagella of antenna 1, but in gnathopod 1 of the ♂ the 4th joint is not elongate, the 5th and 6th joints are stout, subequal in width, strongly setose, and the palm of the 6th has tooth-like projections. In L. arcticus the maxillipeds have the outer plates exceptionally small.

9 species accepted, 2 doubtful.

Synopsis of accepted species:

1 { Without trace of eyes . . . . 1. **L. longidigitans** . p. 594
   { Not without trace of eyes — **2.**

2 { Maxillipeds, outer plates very short; peraeopods 3—5, 2d joint very narrow . . . . 2. **L. arcticus** . . . . p. 595
   { Maxillipeds, outer plates not very short; peraeopods 3—5, 2d joint moderately narrow — **3.**

8 { Ventral surface of peraeon armed with many spines 3. **L. spiniventris** . . p. 595
   { Ventral surface of peraeon not armed with spines — **4.**

4 { Peraeopods 1 and 2 in ♂ both having 4th joint strongly setose . . . . 4. **L. hirsutipes** . . . p. 596
   { Peraeopod 1, but not peraeopod 2, in ♂ having 4th joint strongly setose . . . . 5. **L. megacheir** . . . p. 596
   { Peraeopods 1 and 2 in ♂ neither having 4th joint strongly setose — **5.**

5 { Gnathopod 1 in ♂, 2d joint densely setose at lower hind corner . . . . 6. **L. longipes** . . . p. 597
   { Gnathopod 1 in ♂, 2d joint not densely setose at lower hind corner — **6.**

6 { Lower lip, hind processes very long; mandible, spines of spine-row very numerous . . . . . 7. **L. philacanthus** . p. 598
   { Lower lip, hind processes of moderate length; mandible, spines of spine-row in moderate number — **7.**

7 { Gnathopod 2, 2d joint strongly dilated . . . . . 8. **L. kerguieleni** . . p. 598
   { Gnathopod 2, 2d joint not dilated . . . . 9. **L. websterii** . . . p. 599

1. **L. longidigitans** (Bonnier) 1896 *Autonoe l.*, J. Bonnier in: Ann. Univ. Lyon, *v.* 26 p. 659 t. 40 f. 1 | 1899 *Lembos l.*, T. Stebbing in: Ann. nat. Hist., ser. 7 *v.* 3 p. 350.

Head without rostrum, lateral lobes not at all prominent. Side-plate 1 in ♂ quadrate. Pleon segment 3, postero-lateral corners much rounded.

Eyes and eye-pigment entirely wanting. Antennae 1 and 2 broken. Mandible, 3^d joint of palp longer than 2^d, distal margin concave. Gnathopod 1 in $\male$, 2^d joint stout except proximally, 5th and 6th nearly equal in width and length, palm towards the middle emarginate and raised into a strong tooth, finger shorter than 6th joint but much overlapping palm. Gnathopod 2 in $\male$ much smaller, 2^d joint little dilated, 6th joint rather larger than 5th, finger a little overlapping the palm. Gnathopod 1 in $\female$, 2^d joint not dilated, 6th wider than 5th, palm very oblique and ill-defined. Gnathopod 2 in $\female$ nearly as in $\male$, but shorter, palm more transverse. Peraeopods 1 and 2 slender, little setose, 2^d—5th joints glandular, finger long, as long as 6th joint. Peraeopods 3—5, 2^d joint oval; peraeopod 5 elongate. Uropod 3, rami nearly equal, little longer than peduncle. Telson rounded oval, with setules at each lateral process. L. about 6 mm.

Bay of Biscay. Depth 950 m.

2. **L. arcticus** (H. J. Hansen) 1887 *Microdeutopus a.*, H. J. Hansen in: Dijmphna Udb., p. 231 t. 22 f. 3 | 1894 *M. a.*, T. Stebbing in: Bijdr. Dierk., v. 17 p. 43 | 1893 *Autonoe arctica*, A. Della Valle in: F. Fl. Neapel, v. 20 p. 406 t. 56 f. 35, 36 | 1895 *Lembos arcticus*, T. Stebbing in: Ann. nat. Hist., ser. 6 v. 16 p. 207.

Side-plate 1 rather smaller than 2^d, its apex acute, not aculeate, 2^d—4th rotundo-quadrate, 5th nearly as deep as 4th, with shallow straight hind lobe. Pleon segment 3, postero-lateral corners quadrate, in 4th produced to a small tooth. Eyes very small. Antenna 1 not reaching end of 3^d pleon segment, 2^d joint longer than 1st, 3^d short, flagellum much longer than peduncle, 35-jointed; accessory flagellum 7-jointed. Antenna 2 about $^1/_2$ as long as antenna 1, ultimate joint of peduncle shorter than penultimate, but at least as long as the 7-jointed flagellum. Mandible, 3^d joint of palp considerably longer than 2^d. Maxillipeds, outer plates much smaller than is usual in the genus. Gnathopod 1, 5th joint shorter than 6th, rather broad except at base, hind margin setose, 6th with groups of setae on distal half of front, and along hind margin, widening to and at the convex transverse palm, this being sharply defined by a small acute tooth, which is crossed by the apex of the finger. Gnathopod 2, 5th joint setose, wider but rather shorter than the narrowly oblong and very setose 6th, which has a short transverse palm, overlapped by the denticulate, spinulose finger. Peraeopods 1 and 2 rather slender, 4th joint rather longer than 5th, 6th longer than 5th and much more slender, finger slender, acute. Peraeopods 3—5, 2^d joint very narrow; peraeopod 4 much longer than 3^d, 5th than 4th. Uropods 1 and 2, rami equal in length, in uropod 2 one stouter than the other. Uropod 3, peduncle short, pentagonal; inner ramus a little longer than the outer. Telson, upper half semicircular, lower triangular, with strong prominence on each side of the base of the triangle, carrying 3 unequal setules. Colour whitish, with 4 longitudinal rows of brown grey blotches. L. 29 mm.

Kara Sea (lat. 70° N.). Depth 37—96 m.

3. **L. spiniventris** (Della Valle) 1893 *Autonoe s.*, A. Della Valle in: F. Fl. Neapel, v. 20 p. 400 t. 5 f. 7; t. 56 f. 17—34 | 1895 *Lembos s.*, T. Stebbing in: Ann. nat. Hist., ser. 6 v. 16 p. 207.

Body moderately robust, peraeon segments 1 and 2 narrow, ventral surface of peraeon armed with many large spines (only in $\male$?). Head short with straight, very acute rostrum. Side-plate 1 in $\male$, front angle very acutely prolonged, in $\female$ subacute, in side-plate 2 in $\male$ acute, not much prolonged, in $\female$ blunt; side-plate 5 notably less deep than 4th. Eyes brown.

Antenna 1, 1st joint $^2/_3$ as long as 2^d, 3^d less than half as long as 1st, flagellum longer than peduncle, 20-jointed, accessory flagellum 8-jointed. Antenna 2 shorter, ultimate joint of peduncle as long as penultimate, longer than the 8-jointed flagellum. Upper lip semicircular. Mandible, 3^d joint of palp longer than 2^d. Maxilla 1, inner plate rudimentary, without setules. Maxillipeds, outer plates rather large, strongly armed. Gnathopod 1 in ♂, 5th joint a little shorter than 6th; 6th oval, palm slightly oblique, well defined, but without a tooth, a small cavity occupying about $^1/_3$ of its length adjacent to the defining angle, which is slightly overlapped by the serrate finger. Gnathopod 2 in ♂ rather smaller, similar, except that the palm is more oblique, simply and slightly convex. Gnathopod 1 in ♀, palm straight, oblique, otherwise nearly as in ♂. Gnathopod 2 in ♀ narrower than gnathopod 1, palm less oblique. Peraeopods 1 and 2 as in L. arcticus (p. 595), or finger rather longer. Peraeopods 3—5 as in L. arcticus, except that the 2^d joint is rather more expanded, especially in peraeopod 5. Pleopods 1—3, peduncle rather large, outer ramus the shorter. Uropod 3, peduncle short and stout, rami twice as long as peduncle, nearly equal. Telson longer than broad, a setule in the indent on each side of rounded apex. Colour, bands of rose alternating with sulphur-yellow or white. L. 5—7 mm.

Bay of Naples. Depth 10—20 m.

4. **L. hirsutipes** Stebb. 1895 *L. h.*, T. Stebbing in: Ann. nat. Hist., ser. 6 *v.* 16 p. 207 t. 8, t. 9 B.

Side-plate 1 in ♂ subacutely produced, in ♀ subrhomboidal. Pleon segment 3, postero-lateral corners obtusely quadrate. Eyes small. Antenna 1, 1st joint rather long, rest unknown. Antenna 2, antepenultimate joint of peduncle much thicker than penultimate, ultimate rather shorter than penultimate and longer than the 4- or 5-jointed flagellum. Gnathopod 1 in ♂, 2^d joint with brush of long setae at hind distal end; 5th joint nearly as broad as 6th but considerably shorter; 6th broad, oblong, palm transverse with deep narrow cleft between a submedian tooth and the rather long and sinuous defining tooth, beyond which the apex of the denticulate finger projects. Gnathopod 2 in ♂, 2^d joint somewhat expanded, oval; 5th joint rather longer and broader than the oblong 6th, both setose on both margins, the 5th having a group of very long setae at the hind apex, and the 6th many such along the front; palm short, not sharply defined, overlapped by the finger. Gnathopod 1 in ♀, 2^d joint slender, unarmed, 6th joint much longer than 5th and slightly broader, palm with a little triangular indentation. Gnathopod 2 in ♀, 2^d joint slender, other joints nearly as in ♂, but 5th and 6th much more sparsely furnished in front though similarly on the hind margin. Peraeopods 1 and 2 in ♂, but not in ♀, having the large 4th joint on both margins densely fringed with long simple setae. Peraeopod 3, 2^d joint moderately expanded, 6th longer than 4th or 5th, slightly curved, reverted; finger short. Peraeopod 4 longer, 6th joint straight, not reverted, finger short. Peraeopod 5? Pleopods 1—3, outer ramus much shorter than inner, both with 10 joints. Uropod 3, rami equal. Telson as long as broad, with 2 setae at each lateral process, apex rounded. L. 4 mm.

Off Cape of Good Hope.

5. **L. megacheir** (O. Sars) 1879 *Autonoë m.*, G. O. Sars in: Arch. Naturv. Kristian., *v.* 4 p. 458 | 1894 *A. m.*, G. O. Sars, Crust. Norway, *v.* 1 p. 550 t. 195 f. 2 | 1895 *Lembos m.*, T. Stebbing in: Ann. nat. Hist., ser. 6 *v.* 16 p. 207 | 1893 *Autonoe longipes* (part.)?, A. Della Valle in: F. Fl. Neapel, *v.* 20 p. 403.

Body rather slender, near to L. longipes. Side-plate 1 as in ♂ and ♀ of L. spiniventris (p. 595), 2^d—4th rotundo-quadrate, 5th nearly as deep as 4th. Pleon segment 3, postero-lateral corners subquadrate. Eyes without visual elements, represented by patch of whitish pigment on each side. Antenna 1 slender, nearly as long as body, 2^d joint much longer than 1st, thrice as long as 3^d, flagellum nearly twice as long as peduncle, about 20-jointed, accessory flagellum 4-jointed. Antenna 2 little over half as long as antenna 1, ultimate joint of peduncle as long as penultimate, longer than flagellum. Gnathopod 1 in ♂, 2^d joint with tuft of short setae at lower hind corner; 3^d—5th densely setose only along hind margin; 5th a little shorter than 6th; 6th broad, oblong oval, but with nearly transverse palm, defined by a sharp tooth separated from a submedian angle by an excavation; apex of finger crossing the defining tooth. Gnathopod 2 in ♂ slender, the rather long and narrow 5th and 6th joints densely setose on both margins, especially the front, palm short, slightly oblique, matching finger. Gnathopod 1 in ♀, palm oblique, not strongly defined. Gnathopod 2 in ♀ much smaller than gnathopod 1, shaped as in the ♂, but slightly furnished on front margin. Peraeopod 1 in ♂, but not peraeopod 2, having 4th joint densely setose on both margins, but especially on the front. Peraeopods 3—5, 2^d joint little expanded; peraeopod 5 very elongate in ♂ and ♀. Uropod 3, inner ramus somewhat larger than the outer. Telson with a setule on either side of the apex. Colour uniformly yellowish. L. ♂ 8 mm.

Arctic Ocean. North-Atlantic and North-Sea (Norway from Stavangerfjord up to Finmark). Depth 94—564 m.

6. **L. longipes** (Lilj.) 1852 *Gammarus l.*, W. Liljeborg in: Öfv. Ak. Förh., *v.* 9 p. 10 | 1855 *G. (Gammaropsis) l.*, W. Liljeborg in: Vetensk. Ak. Handl., 1853 p. 457 | 1859 *Autonoe l.*, R. M. Bruzelius in: Svenska Ak. Handl., n. ser. *v.* 3 nr. 1 p. 28 | 1893 *A. l.* (part.), A. Della Valle in: F. Fl. Neapel, *v.* 20 p. 403 | 1894 *Autonoë l.*, G. O. Sars, Crust. Norway, *v.* 1 p. 549 t. 195 f. 1 | 1895 *A. l.*, A. M. Norman in: Ann. nat. Hist., ser. 6 *v.* 15 p. 490 | 1862 *Microdentopus l.*, Bate, Cat. Amphip. Brit. Mus., p. 166 | 1895 *Lembos l.*, T. Stebbing in: Ann. nat. Hist., ser. 6 *v.* 16 p. 207 | 1871 *Autonoe plumosa*, A. Boeck in: Forh. Selsk. Christian., 1870 p. 239 | 1876 *Autonoë longipes* (part.) + *A. p.*, A. Boeck, Skand. Arkt. Amphip., *v.* 2 p. 572 t. 25 f. 2; p. 574 t. 25 f. 3.

. Body more slender and side-plate 1 in ♂ less produced than in L. websterii (p. 599). Pleon segment 3, postero-lateral corners rotundo-quadrate. Eyes small, rounded, black. Antenna 1 about ⅔ as long as body, 2^d joint much longer than 1st, thrice as long as 3^d; flagellum longer than peduncle, about 18-jointed; accessory flagellum 3- or 4-jointed. Antenna 2 much shorter, ultimate joint of peduncle as long as penultimate, rather longer than the 6-jointed flagellum. Gnathopod 1 in ♂, 2^d joint broad, with dense brush of long setae at lower hind corner, 4th and 5th with fascicles of setae on hind margin, 5th fringed only on distal part of front margin, 6th all along it; 6th longer than 5th, oblong oval; palm bidentate but oblique, so that the tip of the larger defining tooth does not reach the level of the smaller inner tooth. Gnathopod 2 in ♂ nearly as long as gnathopod 1 but much more slender, 5th joint longer than 6th, both long and setose, especially on the front margin; palm short, rather oblique. Gnathopod 1 in ♀ not robust, 6th joint rather longer than 5th, widening a little to the rather oblique palm, defined by an obtuse angle and slender spine. Gnathopod 2 in ♀ rather more slender, 6th joint subequal to 5th, narrow, palm almost transverse. Peraeopods 1—5 slightly armed. Peraeopods 3—5, 2^d joint oblong oval. Peraeopod 5 much the longest. Uropod 3, rami subequal, nearly twice as long as peduncle. Telson rounded oval, somewhat tapering, with 2 or 3 setae on each side of apex. Colour whitish, with light reddish transverse bands, but no specks. L. reaching nearly 12 mm.

Kara Sea; Kattegat; North-Atlantic, North-Sea and Skagerrak (South- and West-Norway at least to Trondhjemsfjord). Depth 19—274 m.

7. L. philacanthus (Stebb.) 1888 *Autonoe philacantha*, T. Stebbing in: Rep. Voy. Challenger, *v.* 29 p. 1082 t. 110 | 1895 *Lembos p.*, T. Stebbing in: Ann. nat. Hist., ser. 6 *v.* 16 p. 207 | 1893 *Autonoe longipes* (part.)?, A. Della Valle in: F. Fl. Neapel, *v.* 20 p. 405.

Head, rostrum scarcely perceptible; lateral lobes small, acute. Side-plate 1 small, not produced forward. Pleon segment 3, postero-lateral corners. rounded. Eyes small, narrow, reniform, set obliquely on lateral lobes of head. Antenna 1, 1st joint longer than head, 2^d much longer than 1st, 3^d scarcely $^1/_2$ as long as 1st; flagellum longer than peduncle, more than 18-jointed; accessory flagellum 7-jointed. Antenna 2 shorter, ultimate joint of peduncle about as long as penultimate, scarcely so long as the 9-jointed flagellum. Upper lip, distal margin almost straight. Lower lip (Fig. 102), mandibular process thin and unusually long. Mandible, spine-row of 12 spines, molar with accessory process; 3^d joint of palp rather shorter than 2^d, with dense group of pectinate spines and cilia near the

 Mandibular process

middle. Maxilla 1, inner plate with 1 plumose seta. Maxillipeds, inner and outer plates broad, well armed. Gnathopod 1, 2^d joint devoid of long setae; 5th and 6th joints massive, spinose, 5th rather longer than broad, rather shorter than 6th, which has the hind margin much

Fig. 102. **L. philacanthus.**
Lower lip.

shorter than the convex front; palm oblique, sinuous, finely but irregularly denticulate, defined by a tooth serrate on the inner side and there carrying a long palmar spine, the sharp apex of the curved serrate finger closing against this on to the surface of the joint. Gnathopod 2 smaller, 6th joint as long as 5th, almost oblong; palm nearly transverse, slightly sinuous, finely pectinate, well defined, the specimen having a palmar tooth in only one of the 2^d gnathopods; finger serrate, matching palm. Peraeopods 1 and 2, 2^d and 4th joints glandular, 4th much longer than 5th or 6th, feebly armed; finger short. Peraeopods 3—5, 2^d joint little expanded. Uropods 1—3 spinose, uropod 2 stout, 3^d with short peduncle, inner ramus the longer. Telson scarcely longer than broad, with 5 setiform spines at each subapical corner. L. ♂ (?) about 11 mm.

Bass Strait (East Moncoeur Island). Depth 71 m.

8. L. kergueleni (Stebb.) 1888 *Autonoe k.*, T. Stebbing in: Rep. Voy. Challenger, *v.* 29 p. 1087 t. 111 | 1895 *Lembos k.*, T. Stebbing in: Ann. nat. Hist., ser. 6 *v.* 16 p. 207 | 1893 *Autonoe longipes* (part.)?, A. Della Valle in: F. Fl. Neapel, *v.* 20 p. 405.

Head with small rostrum; lateral lobes little produced, acute. Side-plate 1 obtusely produced. Pleon segment 3, postero-lateral corners minutely notched, border above bulging. Eyes small. Antenna 1, 1st joint longer than head; rest unknown. Antenna 2, antepenultimate joint of peduncle broad, ultimate joint of peduncle as long as penultimate, rather longer than the 7-jointed flagellum. Lower lip, mandibular process rather long, acute. Mandible, spines of spine-row 5 or 6; 3^d joint of palp fully as long as 2^d. Other mouth-parts nearly as in L. philacanthus. Gnathopod 1 not very setose, such fascicles as there are being chiefly on hind margin of 3^d—6th joints; 5th joint stout, much shorter than 6th, which widens a little to the transverse bidentate palm, the defining tooth separated by a cavity from a smaller submedian tooth, between which and the finger-hinge the palm is sinuously denticulate; apex of almost smooth finger closing against inner margin of the defining tooth. Gnathopod 2 narrower, except the broadly expanded, oval 2^d joint; 5th broader than 6th, a little shorter, with long setae fringing distal $^2/_3$ of front; 6th oblong, the rather convex front margin strongly fringed; palm pectinate, almost transverse, defined by obtuse angle and palmar spines, a little overlapped by the short, stout, serrate finger. Peraeopods 1 and 2 slightly armed. Peraeopods 3—5, 2^d joint not much expanded; peraeopod 4 much longer than 3^d, 5th than 4th; finger not long. Uropod 3, rami short, subequal, longer than peduncle. Telson oval, narrowing distally, with a seta at the notch on each side of the rounded apex. L. about 5 mm.

Cumberland Bay [Kerguelen Island]. Depth 239 m.

9. **L. websterii** Bate 1856 *L. w.* (nom. nud.), Bate in: Rep. Brit. Ass., Meet. 25 p. 58 | 1857 *L. w.*, Bate in: Ann. nat. Hist., ser. 2 *v.* 19 p. 142 | 1895 *L. w.*, T. Stebbing in: Ann. nat. Hist., ser. 6 *v.* 16 p. 207 | 1862 *Microdeutopus w.*, Bate & Westwood, Brit. sess. Crust., *v.* 1 p. 291 f. | 1862 *Microdentopus w.*, Bate, Cat. Amphip. Brit. Mus., p. 164 t. 30 f. 2 | 1869 *Microdeuteropus websteri*, A. M. Norman in: Rep. Brit. Ass., Meet. 38 p. 282 | 1887 *Microdeutopus w.*, Chevreux in: Bull. Soc. zool. France, *v.* 12 p. 312 | 1894 *Autonoe w.*, G. O. Sars, Crust. Norway, *v.* 1 p. 547 t. 194 | 1876 *Microdeuteropus bidentatus*, T. Stebbing in: Ann. nat. Hist., ser. 4 *v.* 17 p. 73 t. 4 f. 1 a; t. 5 f. 1. 1 b | 1876 *Autonoë longipes* (part.), A. Boeck, Skand. Arkt. Amphip., *v.* 2 p. 572 | 1893 *Autonoe l.* (part.), A. Della Valle in: F. Fl. Neapel, *v.* 20 p. 403 t. 3 f. 13; t. 10 f. 20—30.

Body rather tumid in ♀. Side-plate 1 in ♂ acutely produced, in ♀ subrhomboidal. Pleon segment 3, postero-lateral corners rotundo-quadrate. Eyes very small, rounded oval, dark. Antenna 1 more than $^1/_2$ as long as body, 2^d joint longer than 1st, not thrice as long as 3^d, flagellum longer than peduncle, with about 15 joints, accessory flagellum with 4 or 5 joints, last minute. Antenna 2 much shorter, ultimate joint of peduncle as long as penultimate, longer than the 4- or 5-jointed flagellum. Gnathopod 1 in ♂, 2^d joint without brush of long setae at lower hind corner; 4th and 5th with long setae on hind margin; 5th and 6th densely fringed on convex front margin, both stout, subequal; palm transverse, defined by a strong tooth, separated by a cavity from a smaller inner one, while on the outer side a palmar spine springs from a small prominence (Della Valle: a little tooth); the serrate finger reaching the tip of the spine. Gnathopod 2 in ♂, 2^d joint with lower front corner a little produced outward; 5th and 6th joints long and narrow, densely fringed on both margins, especially the front, palm short, nearly transverse, a little overlapped by the short finger. Gnathopod 1 in ♀, 6th joint much longer than 5th, oblong oval, palm ill-defined, finger serrate. Gnathopod 2 in ♀ much smaller, similar to gnathopod 2 in ♂, but smaller, corner of 2^d joint not produced, front of 5th and 6th not densely fringed. Peraeopods 1 and 2 slightly armed. Peraeopods 3—5, 2^d joint oblong oval, not very broad; peraeopod 5 much the longest. Uropod 3, rami nearly equal, not much longer than peduncle. Telson rounded oval, with 3 spines on each side of the somewhat angular tip. Colour whitish, with narrow transverse bands of dark brown specks. L. 5—6 mm.

North-Atlantic with adjoining seas (south and west of Norway, north and south of Great Britain, France); Mediterranean (Naples). Generally in comparatively shallow water.

L. fuegiensis (Dana) 1853 & 55 *Gammarus f.*, J. D. Dana in: U. S. expl. Exp.,
v. 13 II p. 954; t. 65 f. 8 a—h | 1862 *Moera f.*, *M. fuegeensis*, Bate, Cat. Amphip. Brit.
Mus., p. 194 t. 35 f. 4 | 1893 *Microdeutopus* (part.)?, A. Della Valle in: F. Fl. Neapel,
v. 20 p. 425.

L. 7 mm.

South-Pacific? (Feejee Islands).

L. tenuis (Dana) 1852 *Gammarus t.*, J. D. Dana in: P. Amer. Ac., *v.* 2 p. 211 |
1853 & 55 *G. t.*, J. D. Dana in: U. S. expl. Exp., *v.* 13 II p. 950; t. 65 f. 5 a—c | 1862
Microdentopus t., Bate, Cat. Amphip. Brit. Mus., p. 165 t. 30 f. 4 | 1893 *Microdeutopus t.*,
A. Della Valle in: F. Fl. Neapel, *v.* 20 p. 420 | 1894 *Autonoë? t.*, G. O. Sars, Crust. Norway,
v. 1 p. 547 | 1895 *Lembos? t.*, T. Stebbing in: Ann. nat. Hist., ser. 6 *v.* 16 p. 207.

L. 6 mm.

Sooloo Sea. Depth 12 m.

5. Gen. **Lemboides** Stebb.

1895 *Lemboides* (Sp. un.: *L. afer*), T. Stebbing in: Ann. nat. Hist., ser. 6 *v.* 16 p. 209.

Like Lembos (p. 594), but accessory flagellum of antenna 1 shorter than
3^d joint of peduncle, and in gnathopod 1 of the ♂ the 5th joint is much
broader and longer than the 6th, though not dentate or opposable to the finger.

2 species.

Synopsis of species:

Antenna 1, 2^d joint not longer than 1st . . . . 1. **L. afer** . . . p. 600
Antenna 1, 2^d joint much longer than 1st . . . . 2. **L. australis** . p. 601

1. **L. afer** Stebb. 1895 *L. a.*, T. Stebbing in: Ann. nat. Hist., ser. 6 *v.* 16 p. 209
t. 9 A, 10.

Side-plate 1 in ♂ broader but rather less deep than the following, not
acuminate; no side-plate very large. Pleon segment 3 large, postero-lateral
corners rounded. Eyes small, dark. Antenna 1 about $\frac{1}{3}$ as long as body,
1st joint slightly longer than 2^d; 3^d about $\frac{1}{3}$ as long as 2^d; flagellum
rather longer than peduncle, about 15-jointed; accessory flagellum not as
long as 1st joint of primary, with 2 joints, 2^d minute. Antenna 2 shorter,
antepenultimate joint of peduncle as long as broad, with expansion of lower
edge, ultimate joint of peduncle shorter and much narrower than penultimate,
longer than the 3- or 4-jointed flagellum. Gnathopod 1 in ♂, 4th joint
acute, 5th much broader than 2^d and as long, distal margin straight, not
covered by base of 6th joint, of which the palm is rather oblique, having a
broad denticulate cavity between a strong tooth near the finger-hinge and
a smaller one near to a still smaller defining tooth, against which the serrate
finger impinges. Gnathopod 2 in ♂ more slender, otherwise rather like
gnathopod 1, but 5th joint not broader than 2^d; 6th shorter and rather
narrower than 5th; both plumose; the palm forming a small cavity between
a tooth near the finger-hinge and the defining point, which the rather short
finger just reaches. Gnathopod 1 in ♀, 5th joint densely setose on hind-
margin, only a little longer and broader than the 6th, which is setose on
both margins and has a short palm, overlapped by the serrate finger. Gnathopod 2 in ♀ with long plumose setae along the front, 5th joint distally widened,
6th longer, narrow, the small finger fitting the short convex palm. Peraeopod 1, 4th joint plumose on front margin, more so than in peraeopod 2.
Uropods 1—3, outer ramus shorter than inner; inner ramus of uropod 2

stoutest of all. Telson as long as broad, lateral processes of shallowly rounded apex each with 5 spinules. L. about 6 mm.

Off Cape of Good Hope.

2. **L. australis** (Hasw.) 1879 *Microdeuteropus a.*, HasWell in: P. Linn. Soc. N. S. Wales, *v.* 4 p. 271 t. 11 f. 5 | 1882 *Microdeutopus a.*, HasWell, Cat. Austral. Crust., p. 263 | 1888 *Autonoe a.*, T. Stebbing in: Rep. Voy. Challenger, *v.* 29 p. 1087 | 1899 *Lemboides a.*, T. Stebbing in: Ann. nat. Hist., ser. 7 *v.* 3 p. 350 | 1893 *Autonoe longipes* (part.), A. Della Valle in: F. Fl. Neapel, *v.* 20 p. 403.

Antenna 1 longer than head and peraeon; 2^d joint twice as long as 1st, 3^d very short; flagellum longer than peduncle, accessory flagellum small (in figure). Antenna 2 nearly $^2/_3$ as long as antenna 1, ultimate joint of peduncle shorter than penultimate, longer than flagellum, which is armed with hooked setae. Gnathopod 1 large, subchelate; 4th joint small, narrow; 5th large, armed with a few scattered hairs; 6th smaller (in figure much smaller), irregularly quadrate, rather longer than broad, palm scarcely oblique, deeply excavate, denticulate, defined by a triangular tooth; finger stout, denticulate, apex (in figure) reaching rather beyond the short palm. Gnathopod 2 smaller; 5th and 6th joints subequal, with fascicles of setae along hind margin; 6th joint ovate (in figure oblong), twice as long as broad; palm not defined, nearly transverse; finger stout, short, denticulate. Peraeopod 2 longer than peraeopod 1; finger in both long, slender. Uropod 3, rami shorter than in uropods 1 and 2, lanceolate. Telson large, armed with a few short hairs. L. 7 mm.

Port Jackson [East-Australia].

6. Gen. **Dryopoides** Stebb.

1888 *Dryopoides* (Sp. un.: *D. westwoodi*), T. Stebbing in: Rep. Voy. Challenger, *v.* 29 p. 1145 | 1890 *Dryapoides*, Warburton in: Zool. Rec., *v.* 25 Crust. p. 19.

Like Lembos (p. 594), except that the accessory flagellum of antenna 1 is minute, and that the 2 rami of uropod 3 are rudimentary, pleon segment 6 dorsally evanescent.

1 species.

1. **D. westwoodi** Stebb. 1888 *D. w.*, T. Stebbing in: Rep. Voy. Challenger, *v.* 29 p. 1146 t. 122 | 1889 *D. w.*, J. Bonnier in: Bull. sci. France Belgique, *v.* 20 p. 391 | 1893 *D. w.*, A. Della Valle in: F. Fl. Neapel, *v.* 20 p. 425 | 1890 *Dryapoides westwoodii*, Warburton in: Zool. Rec., *v.* 25 Crust. p. 19.

Back not much arched, pleon segment 4 as long as any preceding segment. Head, lateral lobes somewhat produced, narrowly rounded. Side-plates 1—5 not very deep; 1st produced forward, 2^d deeper than broad, 5th much broader than deep, nearly as deep as 4th. Pleon segment 3, postero-lateral corners rounded. Eyes round, rather large, a little removed from lateral lobes of head. Antenna 1 as long as body, slender, 1st joint longer than head, 2^d longer than 1st, 3^d about $^1/_4$ as long as 2^d; flagellum much longer than peduncle, about 30-jointed; accessory flagellum with 2 joints, 1st small, 2^d minute. Antenna 2 shorter, but with much longer peduncle, ultimate joint of peduncle longer than penultimate or than the 8-jointed flagellum. Gnathopod 1, 5th joint rather shorter than the oval 6th, which has the palm in ♀ scarcely distinct from hind margin, but in ♂ apparently somewhat excavate; finger matching palm, with many decurrent teeth on its inner margin. Gnathopod 2 smaller, 5th joint as long as 6th, and distally a little wider, 6th narrowly oblong, palm transverse, very short, convex, just over-lapped by the short finger. Peraeopods 1 and 2 very glandular; 2^d and 4th joints rather robust; finger more than half as long as 6th joint. Peraeo-pods 3—5, 2^d joint oblong, not much expanded. Peraeopod 4 longer than peraeopod 3, and peraeopod 5 than peraeopod 4; finger in each with inner margin furred, produced into a blunt process carrying a plumose seta, which overlaps the apex. Uropods 1 and 2, peduncle longer than the rami, which in each are subequal, spinose, blunt. Uropod 3, peduncle short, broad, just reaching beyond the telson; the diminutive rami equal, narrowly oval, inner armed with 3 plumose spinules, outer with a longer apical spine and a spinule above it. Telson broader than long, its distal arch ending in a slightly produced blunt point; on each side not far from lateral margins there is a group of plumose setae. L. about 8 mm.

Off Melbourne. Depth 62 m.

7. Gen. **Paradryope** Stebb.

1888 *Paradryope* (Sp. un.: *P. orguion*), T. Stebbing in: Rep. Voy. Challenger, *v.* 29 p. 1151 | 1898 *Ischyrocerus* (part.)?, A. Della Valle in: F. Fl. Neapel, *v.* 20 p. 451.

Pleon segment 6 dorsally well developed. Side-plates shallow. Antennae 1 and 2 with peduncle elongate. Antenna 1, 3^d joint of peduncle longer than 2^d or 1st, accessory flagellum small. Mandibular palp very elongate. Gnatho-pod 1 larger than gnathopod 2. Peraeopods 1—5, 2^d joint little expanded; peraeopod 5 the longest. Uropods 1 and 2, outer ramus considerably shorter than inner. Uropod 3, rami almost rudimentary, outer a little longer than inner. Telson simple.

1 species.

1. P. orguion Stebb. 1888 *P. o.*, T. Stebbing in: Rep. Voy. Challenger, *v.* 29 p. 1151 t. 123.

Back rather broadly rounded. Head, rostrum short, acute; lateral lobes acute, a little produced. Side-plates 1—4 little deeper than the following; 5th and 6th apparently with an acute hind lobe. Pleon segment 3, postero-lateral corners rounded. Eyes small, round, near the lateral lobes of head. Antennae 1 and 2 elongate. Antenna 1, 1st joint rather longer than head; 2^d considerably longer than 1st, 3^d a little longer than 2^d; flagellum shorter than peduncle, 8-jointed; accessory flagellum 1 slender joint. Antenna 2 rather longer, ultimate joint of peduncle a little longer than penultimate; flagellum 9-jointed. Gnathopod 1, 2^d joint shorter and very much narrower than 6th; 5th rather longer than broad, much shorter and narrower than the very large, oval 6th, of which the oblique palm is strongly sculptured, with cavities separating a small submedian process from the finger-hinge and from a large tooth, denticulate on the edge which unites with the hind margin of the joint, the dentate finger closing over this edge against strong palmar spines. Gnathopod 2 smaller, but with longer 5th joint, which is as long as the 6th; 6th oval, rather stout, palm not very oblique, convex, finely pectinate, finger slightly denticulate, closely fitting palm. Peraeopods 1—5 not greatly differing in length; peraeopod 3 scarcely as long as 1st or 2^d, none of the joints much widened. Uropods 1 and 2,

·outer ramus shorter than inner. Uropod 3, peduncle broad, reaching well beyond the telson, rami narrow, almost acute. Telson a little longer than broad, apex obtuse, a little produced; a spinule near centre of each lateral margin. L. about 5 mm.

North-Pacific (lat. 36° N., long. 158° E.). Depth of 4200 m.

34. Fam. **Photidae**

1872 & 76 *Photidae* (part.), A. Boeck. Skand. Arkt. Amphip., *v.*1 p.74; *v.*2 p.546 | 1882 *P.*, G. O. Sars in: Forh. Selsk. Christian., nr. 18 p.29 | 1888 *P.* (part.), T. Stebbing in: Rep. Voy. Challenger, *v.*29 p.1061 | 1894 *P.* (part.), G. O. Sars, Crust. Norway, *v.*1 p.538.

Head, lateral lobes often slightly produced. Side-plates variable in depth and relative size, 2^d not unfrequently the largest, 4th with hind margin not excavate. Antenna 1 often subequal to antenna 2, sometimes longer; accessory flagellum varying from obsolete to long. Mouth-parts normal and in general as in the Aoridae (p.585), but: mandibular processes of lower lip not acutely produced, 3^d joint of mandibular palp usually not longer than 2^d, inner plate of maxilla 1 with a variable number of setae. Gnathopods 1 and 2 (Fig. 104, 105, 107; p.610, 614) either subchelate or simple, but gnathopod 1 not the larger, and sexual difference chiefly affecting gnathopod 2. Peraeopods 1 and 2 (Fig. 106 p.613) glandular. Peraeopods 4 and 5 longer than the rest. Uropods 1 and 2 biramous. Uropod 3 biramous except in Microprotopus (p.604). Telson simple (Fig. 103 p.607).

Marine.

10 genera, 38 accepted species and 5 doubtful.

Synopsis of genera:

1 { Uropod 3 with only 1 ramus . . . . 1. Gen. **Microprotopus** . p.604
   { Uropod 3 with 2 rami — **2.**

2 { Uropod 3 (Fig. 103) with one ramus much
   {   smaller than the other — **3.**
   { Uropod 3 with the rami not very unequal — **4.**

3 { Gnathopods 1 and 2 subchelate . . . . . . 2. Gen. **Photis** p.605
   { Gnathopods 1 and 2 (Fig. 104, 105) simple . 3. Gen. **Haplocheira** . . p.609

4 { Antenna 1, 3^d joint longer than 1st or subequal
   {   to it — **5.**
   { Antenna 1, 3^d joint shorter than 1st — **7.**

5 { Antenna 1, accessory flagellum well developed 4. Gen. **Eurystheus** . . p.610
   { Antenna 1, accessory flagellum rudimentary or
   {   obsolete — **6.**

6 { Gnathopod 2, 5th joint short . . . . 5. Gen. **Podoceropsis** . . p.618
   { Gnathopod 2, 5th joint long . . . . 6. Gen. **Megamphopus** . p.621

7 { Antenna 1, accessory flagellum obsolete . . . 7. Gen. **Goësia** p.622
   { Antenna 1, accessory flagellum developed — **8.**

$8 \left\{ \begin{array}{l} \text{Gnathopod 2 subchelate} \ldots \ldots \ldots \ldots \text{ 8. Gen. } \textbf{Protomedeia} \ . \ . \text{ p. 623} \\ \text{Gnathopod 2 simple — 9.} \end{array} \right.$

$9 \left\{ \begin{array}{l} \text{Gnathopod 2, 5}^{\text{th}} \text{ joint broadly expanded to} \\ \quad \text{the front} \ldots \ldots \ldots \ldots \ldots \ldots \text{ 9. Gen. } \textbf{Xenocheira} \ . \ . \text{ p. 624} \\ \text{Gnathopod 2, 5}^{\text{th}} \text{ joint not broadly expanded . 10. Gen. } \textbf{Leptocheirus} . \ . \text{ p. 625} \end{array} \right.$

1. Gen. **Microprotopus** Norm.

1852 *Dercothoe* (part.), J. D. Dana in: Amer. J. Sci., ser. 2 *v.* 14 p. 313 | 1853 *D.* (part.), J. D. Dana in: U. S. expl. Exp., *v.* 13 п p. 911, 968 | 1867 *Microprotopus* (Sp. un.: *M. maculatus*), A. M. Norman in.: Rep. Brit. Ass., Meet. 36 p. 197, 203 | 1893 *M.*, A. Della Valle in: F. Fl. Neapel, *v.* 20 p. 391 | 1894 *M.*, G. O. Sars, Crust. Norway, *v.* 1 p. 566 | 1879 *Orthopalame* (Sp. un.: *O. terschellingi*), Hoek in: Tijdschr. Nederl. dierk. Ver., *v.* 4 p. 123.

Head, lateral lobes moderately produced, post-antennal corners well marked. Side-plates rather large, 5th with deep front lobe. Antennae 1 and 2 not very elongate, nor very unequal. Antenna 1 with accessory flagellum. Gnathopod 2 especially large in the ♂. Peraeopods 1 and 2, 2^{d} joint a little expanded. Peraeopods 3—5, 2^{d} joint broadly oval. Uropod 3 with a single ramus tipped with spines. Telson small.

2 species accepted, 3 obscure.

Synopsis of accepted species:

Antenna 2, flagellum more than 3-jointed . . . . 1. **M. maculatus** . . p. 604
Antenna 2, flagellum not more than 3-jointed . . . . . 2. **M. longimanus** . p. 605

1. **M. maculatus** Norm. 1867 *M. m.*, A. M. Norman in: Rep. Brit. Ass., Meet. 36 p. 203 |· 1868 *M. m.*, A. M. Norman in: Ann. nat. Hist., ser. 4 *v.* 2 p. 419 t. 23 f. 7—11 | 1874 *M. m.*, T. Stebbing in: Ann. nat. Hist., ser. 4 *v.* 14 p. 13 t. 2 f. 5 | 1876 *M. m.*, A. Boeck, Skand. Arkt. Amphip., *v.* 2 p. 559 t. 26 f. 3 | 1889 *M. m.*, Hoek in: Tijdschr. Nederl. dierk. Ver., ser. 2 *v.* 2 p. 224 | 1890 *M. m.*, Chevreux in: Bull. Soc. zool. France, *v.* 15 p. 148 f. 2, 4, 6, 7 | 1893 *M. m.*, A. Della Valle in: F. Fl. Neapel, *v.* 20 p. 393 t. 56 f. 13—16 | 1894 *M. m.*, G. O. Sars, Crust. Norway, *v.* 1 p. 567 t. 201 | 1898 *M. m.*, Sowiuski in: Mém. Soc. Kiew, *v.* 15 p. 470 | 1879 *Orthopalame terschellingi*, Hoek in: Tijdschr. Nederl. dierk. Ver., *v.* 4 p. 123 t. 9 f. 4—7.

Body, dorsum broadly rounded. Side-plates 1—4 strongly setiferous, 1st and 2^{d} wider in ♂ than in ♀. Pleon segment 3, postero-lateral corners broadly rounded. Eyes round, small, dark. Antenna 1 about $^1/_3$ as long as body, 1st joint subequal to 2^{d}, nearly twice as long as 3^{d}; flagellum little longer than peduncle, 8—10-jointed; accessory flagellum with 2 joints, 2^{d} minute. Antenna 2 subequal to antenna 1, ultimate and penultimate joints of peduncle subequal, flagellum longer than ultimate joint of peduncle, 5—7-jointed. Gnathopod 1 rather small, 5th joint rather long and narrow, setose on hind margin, 6th joint about as long, narrowly oval, more widened distally in ♂ than in ♀, palm defined by an obtuse angle. Gnathopod 2 in ♂, 3^{d} joint longer than 4th; 5th short, broad, cup-shaped; 6th very large and long, front convex, hind or palmar margin straight between a projecting basal and a distal tooth; between the latter and the finger-hinge is a cavity; finger very long, slightly sinuous, reaching basal tooth of 6th joint. Gnathopod 2 in ♀ stouter than gnathopod 1; 4th joint longer than 5th, the 4th and the cup-shaped 5th joint being each produced into a strongly setose lobe; 6th broader than in gnathopod 1 and palm more defined. Uropod 3, peduncle rather stout, ramus subequal to it in length, narrow, tipped with setules and 2 spines. Telson rather broader than long, distally truncate, with a little tooth at each corner carrying 2 unequal spines. Colour sometimes blackish with crowded dark spots. L. 3 mm.

North-Atlantic with adjoining seas (Europe from Bergen in Norway to the Adriatic; Azores). Depth 4—20 m, on sandy bottom.

2. **M. longimanus** Chevreux 1886 & 87 *M. l.*, Chevreux in: Bull. Soc. zool. France, *v.* 11 p. XLI; *v.* 12 p. 311 f. 5; p. 295 t. 5 f. 5—10 | 1890 *M. l.*, Chevreux in: Bull. Soc. zool. France, *v.* 15 p. 148 f. 1, 3, 5 | 1893 *M. l.*, A. Della Valle in: F. Fl. Neapel, *v.* 20 p. 392 t. 56 f. 7—12 | 1894 *M, l.*, G. O. Sars, Crust. Norway, *v.* 1 p. 566 | 1890 *M. maculatus* (part.), J. Bonnier in: Bull. sci. France Belgique, *v.* 22 p. 173 t. 8, 9.

Side-plates 1—4 not strongly setiferous, 2ᵈ in ♂ rotundo-quadrate. Pleon segment 3, postero-lateral corners rotundo-quadrate. Eyes round, not very small, red. Antennae 1 and 2 equal, nearly as in M. maculatus, but flagellum of antenna 1 only 5-jointed, flagellum of antenna 2 only 3-jointed, scarcely as long as ultimate joint of peduncle. Gnathopod 1 nearly as in M. maculatus. Gnathopod 2 in ♂, 3ᵈ joint shorter than 4ᵗʰ, 4ᵗʰ with simple setae on the produced rounded apex, 5ᵗʰ cup-shaped, with long plumose setae on the hind lobe, 6ᵗʰ very large, subrectangular, front margin nearly straight, the opposite one, of which the chief part is palmar, armed according to age with 1—3 strong teeth, finger reaching the tooth nearest the base of the 6ᵗʰ joint. Gnathopod 2 in ♀, 4ᵗʰ and 5ᵗʰ joints broader apically than in ♂, 6ᵗʰ much narrower than 5ᵗʰ, long, tapering, smooth, carrying in front some long plumose setae; finger curved, much shorter than 6ᵗʰ joint. Peraeopods and uropods differing little from those of M. maculatus. Telson apically rounded. Colour yellowish with transverse brown bands. L. 2 mm.

North-Atlantic (Croisic [West-France], Pas-de-Calais). On algae (Rhodomela pinastroides Ag.) on rocks at low-tide, or rocky bottom.

M. emissitius (Dana) 1852 *Gammarus e.*, J. D. Dana in: P. Amer. Ac., *v.* 2 p. 211 | 1853 & 55 *Dercothoe e.* (part.), J. D. Dana in: U. S. expl. Exp., *v.* 13 ɪɪ p. 969; t. 66 f. 9 a—e | 1862 *Dercothoë (Cerapus) e.*, *Dercothoe emistuis*, Bate, Cat. Amphip. Brit. Mus., p. 259 t. 44 f. 7 | 1893 *Protomedeia maculata* (part.), A. Della Valle in: F. Fl. Neapel, *v.* 20 p. 387, 436.

L. 8 mm.

Sooloo Sea. Depth 12 m.

M. hirsuticornis (Dana) 1852 *Gammarus h.*, J. D. Dana in: P. Amer. Ac., *v.* 2 p. 210 | 1853 & 55 *Dercothoe? h.*, J. D. Dana in: U. S. expl. Exp., *v.* 13 ɪɪ p. 972; t. 67 f. 2 | 1862 *Dercothoë (Cerapus) h.*, Bate, Cat. Amphip. Brit. Mus., p. 260 t. 44 f. 9 | 1893 *Protomedeia maculata* (part.), A. Della Valle in: F. Fl. Neapel, *v.* 20 p. 436.

Bay of Rio Janeiro.

M. minutus Sowinski 1894 *M. m.*, Sowinski in: Mém. Soc. Kiew, *v.* 13 p. 329 t. 4 f. 1—15 | 1898 *M. m.*, Sowinski in: Mém. Soc. Kiew, *v.* 15 p. 470.

The description of this species could not be obtained.

Sea of Azov.

2. Gen. **Photis** Krøyer

1842 *Photis* (Sp. un.: *P. reinhardi*), Krøyer in: Naturh. Tidsskr., *v.* 4 p. 155 | 1876 *P.*, A. Boeck, Skand. Arkt. Amphip., *v.* 2 p. 553 | 1888 *P.*, T. Stebbing in: Rep. Voy. Challenger, *v.* 29 p. 1063 | 1893 *P.*, A. Della Valle in: F. Fl. Neapel, *v.* 20 p. 394 | 1894

P., G. O. Sars, Crust. Norway, *v.* 1 p. 568 | 1862 *Eiscladus* (Sp. un.: *E. longicaudatus*),
Bate & Westwood, Brit. sess. Crust., *v.* 1 p. 411 | 1869 *Heiscladus*, A. M. Norman in:
Rep. Brit. Ass., Meet. 38 p. 255, 259, 284 | 1874 *Heiscladius*, M'Intosh in: Ann. nat.
Hist., ser. 4 *v.* 14 p. 269.

Body smooth. Head, lateral lobes somewhat produced. Side-plates 1—5
rather large, 5th scarcely less deep than the preceding pairs. Eyes on lateral
lobes of head. Antennae 1 and 2 subequal, peduncle elongate. Antenna 1
with 3d joint subequal to 1st, a rudiment of accessory flagellum sometimes
present. Mouth-parts as in Eurystheus (p. 610), except that 3d joint of
mandibular palp is less elongate. Gnathopods 1 and 2 stronger in ♂ than
in ♀, 2d stronger than 1st, with short 5th joint. Peraeopod 3 short, upturned,
2d joint broad, finger very short, clasped against apex of 6th joint, and having
a denticle on its outer margin. Uropod 3 (Fig. 103), rami very unequal,
the outer having 2 joints, 2d minute; the inner ramus very small. Telson
(Fig. 103) small, broader than long.

5 species accepted, 1 obscure.

Synopsis of accepted species:

$\Bigg\{$
1 $\Bigg\{$ Lateral lobes of head little produced; uropod 3,
  outer ramus not longer than peduncle — **2.**
  Lateral lobes of head much produced; uropod 3,
  outer ramus longer than peduncle — **4.**

2 $\Big\{$ Peraeopod 3, 2d joint widened distally . . . . . 1. **P. brevicaudata** . p. 606
  Peraeopod 3, 2d joint narrowed distally — **3.**

3 $\Big\{$ Gnathopod 1, 6th joint longer than 5th . . . . . 2. **P. reinhardi** . . . p. 607
  Gnathopod 1, 6th joint not longer than 5th . . . 3. **P. macrocarpa** . p. 607

4 $\Big\{$ Body slender, side-plates not very deep . . . . . 4. **P. longicaudata** . p. 608
  Body stout, side-plates very deep . . . . 5. **P. tenuicornis** . . p. 608

1. P. brevicaudata Stebb. 1888 *P. b.*, T. Stebbing in: Rep. Voy. Challenger,
v. 29 p. 1068 t. 108 | 1893 *P. reinhardi* (part.), A. Della Valle in: F. Fl. Neapel, *v.* 20 p. 395.

♀. Head, lateral lobes acute, little produced. Side-plates 1—5 very
deep, 5th scarcely less deep than 4th and much broader. Pleon segment 3,
postero-lateral corners obtusely quadrate. Eyes small, round, dark, lenses
numerous. Antenna 1, 3d joint intermediate in length between 1st and
longer 2d, flagellum shorter than peduncle, 7- or 8-jointed. Antenna 2,
ultimate and penultimate joints of peduncle equal, flagellum 6-jointed.
Gnathopod 1, 5th joint stout, rather shorter than the oval 6th; palm finely
pectinate, continuous with hind margin, only defined by the palmar spines;
finger with 4 decurrent teeth. Gnathopod 2, 5th joint as usual short, cup-
shaped; 6th broad, oblong oval, palm obliquely excavate, defined by a palmar
spine at the well marked angle; finger dentate, matching palm. Peraeopods 1
and 2 rather stout and setose; 6th joint not very slender, not $1\frac{1}{2}$ as long
as 5th. Peraeopod 3, 2d joint as broad as long, widened distally; 5th joint
rather longer than 4th. Peraeopod 4, 2d joint as long, but not quite as
broad as in peraeopod 3, remaining joints longer than in peraeopod 3.
Peraeopod 5 little longer than peraeopod 4, 2d joint considerably narrower.
Uropod 3, outer ramus rather shorter than peduncle, its 2d joint tipped with
a long spine. Telson very short, much broader than long, apex rounded.
L. less than 4 mm.

Off Melbourne. Depth 60 m.

2. **P. reinhardi** Krøyer 1842 *P. r.*, Krøyer in: Naturh. Tidsskr., *v.*4 p. 155 |
1876 *P. r.*, A. Boeck, Skand. Arkt. Amphip., *v.*2 p. 554 t. 26 f. 1 | 1893 *P. r.* (part.),
A. Della Valle in: F. Fl. Neapel, *v.*20 p. 395 | 1894 *P. r.,* G. O. Sars, Crust. Norway,
*v.*1 p. 569 t. 202 | 1866 *Amphithoe r.,* Goës in: Öfv. Ak. Förh., *v.*22 p. 532 | 1852 *Amphi-
thoë pygmaea,* W. Liljeborg in: Öfv. Ak. Förh., *v.*9 p. 9 | ?1895 *Photis pollex,* A. O.
Walker in: P. Liverp. biol. Soc., *v.*9 p. 312 t. 19 f. 16—19.

Body rather stout, with broadly vaulted back. Head, lateral lobes
acute, not much produced, rounded below. Side-plates 1—5 setose on lower
margin; 1st a little narrowed distally; 5th very large, distally obliquely rounded.
Pleon segment 3, postero-lateral corners rounded. Eyes small, rounded, not
close to margin of head. Antennae 1 and 2 rather strong and setose.
Antenna 1 about half as long as body; 3d joint longer than 1st, shorter
than 2d; flagellum shorter than peduncle, about 9-jointed. Antenna 2 a
little shorter, ultimate and penultimate joints of peduncle subequal, flagellum
shorter than peduncle, about 9-jointed. Gnathopod 1 in ♂ robust, 5th joint
distally widened, a little shorter than the broadly oblong oval 6th; palm
nearly transverse, a little excavate, defined by an obtuse angle, finger matching
palm, denticulate. Gnathopod 2 in ♂, 2d joint little produced at lower
front corner, 5th broadly cup-shaped; 6th large, palm transverse, defined by
a strong tooth, followed by a cavity and than by 2 tubercles near the finger-
hinge; finger strong, matching palm. Gnathopod 1 in ♀ as in ♂, but rather
smaller, palm not excavate. Gnathopod 2 in ♀ as in ♂, but smaller, with
sinuous palm defined by a projecting angle with palmar spine. Peraeopods 1—5
comparatively stout. Peraeopods 1 and 2 rather setose. Peraeopod 3, 2d joint
as broad as long, narrowest distally. Peraeopods 4 and 5 nearly equal in
length; 2d joint oblong oval, narrowing a little distally. Uropod 3, outer
ramus scarcely so long as peduncle, its 2d joint tipped with a long spine.
Telson rather broader than long, subtriangular, with small process on each
side of apex. Colour greyish white, with bands of light brown. L. 5 mm.

Arctic Ocean, North-Atlantic, North-Sea and Skagerrak (Greenland; Iceland;
Norway, depth 37—94 m; ?Liverpool Bay, depth 4—19 m); Kattegat.

3. **P. macrocarpa** Stebb. 1888 *P. macrocarpus,* T. Stebbing in: Rep. Voy.
Challenger, *v.*29 p. 1064 t. 107 | 1893 *P. reinhardi* (part.), A. Della Valle in: F. Fl.
Neapel, *v.*20 p. 395.

Head, lateral lobes as in P. reinhardi. Side-plates 1—5 not specially
setose, 1st distally widened, 5th with front lobe large but less deep than
4th. Pleon segment 3, postero-lateral corners obtusely quadrate. Eyes small,
round, with very few lenses. Antennae 1 and 2 about $\frac{2}{8}$
as long as body. Antenna 1, 1st joint rather long, subequal
to 3d, shorter than 2d; flagellum subequal to peduncle,
14-jointed. Antenna 2, ultimate joint of peduncle shorter
than penultimate; flagellum subequal to peduncle, 12-jointed.
Gnathopod 1, 5th joint rather longer than the oval 6th,
palm minutely pectinate, only defined by palmar spines,
finger rather long and broad, with 7 decurrent teeth.
Gnathopod 2 in ♂, palm excavate, defined by a tooth
with palmar spine. Gnathopod 2 in ♀, 5th joint stout,
but longer than broad, 6th broad, oblong oval; palm
rather oblique, shorter than hind margin, a little overlapped
by the dentate finger. Peraeopods 1 and 2, 4th joint long, broad, setose
in front, 6th about 1 1/2 as long as 5th; finger over half as long as 6th joint.
Peraeopod 3, 2d joint rather longer than broad, much narrowed distally,

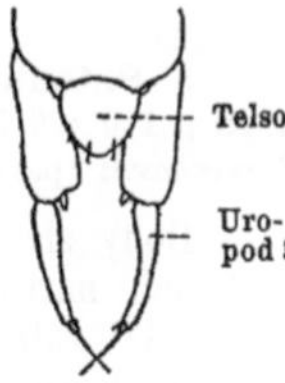

Telson

Uro-
pod 3

Fig. 103.
P. macrocarpa.
Uropod 3 and telson.

5th scarcely so long as 4th or 6th. Peraeopod 4 longer than peraeopod 3. Peraeopod 5 much longer than peraeopod 4; 2^d joint rather narrowly oblong. Uropod 3 (Fig. 103), outer ramus about as long as peduncle, 2^d joint tipped with a long straight spine. Telson (Fig. 103) about as broad as long, with setules on either side of the triangular apex. L. less than 4 mm.

Southern Indian Ocean (Kerguelen Island).

4. P. longicaudata (Bate & Westw.) 1862 *Eiscladus longicaudatus*, Bate & Westwood, Brit. sess. Crust., *v.* 1 p. 412 f. | 1869 *Heiscladus l.*, A. M. Norman in: Rep. Brit. Ass., Meet. 38 p. 284 | 1874 *Heiscladius l.*, M'Intosh in: Ann. nat. Hist., ser. 4 *v.* 14 p. 269 | 1877 *Photis longicaudata*, Meinert in: Naturh. Tidsskr., ser. 3 *v.* 11 p. 142 | 1887 *P. l.*, Chevreux in: Bull. Soc. zool. France, *v.* 12 p. 311 | 1894 *P. l.*, G. O. Sars, Crust. Norway, *v.* 1 p. 571 t. 203 f. 1 | 1895 *P. longicaudatus*, A. O. Walker in: Ann. nat. Hist., ser. 6 *v.* 15 p. 471 | 1871 *P. lütkeni*, A. Boeck in: Forh. Selsk. Christian., 1870 p. 233 | 1876 *P. l.*, A. Boeck, Skand. Arkt. Amphip., *v.* 2 p. 556 t. 26 f. 2 | 1893 *P. reinhardi* (part.), A. Della Valle in: F. Fl. Neapel, *v.* 20 p. 395 t. 3 f. 3, t. 10 f. 1—19.

Body more slender than in P. reinhardi (p. 607) and side-plates less deep. Head, lateral lobes greatly projecting, narrowly rounded. Side-plate 5 distally narrowly rounded. Pleon segment 3, postero-lateral corners rounded. Eyes very small, round, close to margin of lateral lobes of head. Antennae 1 and 2 slender, not densely setose, subequal, more than half as long as body, flagellum 8- or 9-jointed. Antenna 2, antepenultimate joint of peduncle unusually narrow and elongate. Gnathopod 1, 5th joint not greatly widened distally, about as long as 6th, which widens to the oblique, obtuse-angled palm. Gnathopod 2 in ϑ, 2^d joint produced into a rounded decurrent lobe at distal front corner; 5th joint cup-shaped; 6th widening to the palm, which is defined by a projecting angle, being also somewhat excavated in the middle, and exhibiting, on either side of the excavation, a slight angular projection (Sars), deeply excavated and slightly ciliated (Bate). Peraeopods 1—5 much more slender than in P. reinhardi. Peraeopods 1 and 2, 6th joint very narrow, nearly twice as long as 5th. Peraeopod 5 considerably longer than the other peraeopods. Uropod 3, outer ramus longer than peduncle, 2^d joint tipped only with 2 slender setae. Telson very small, shaped as in P. reinhardi. Colour whitish with light brown bands, flagella of antennae crimson. L. ϑ scarcely over 4 mm (Sars), reaching 12 mm (Bate, Walker).

North-Atlantic with adjoining seas (Norway, Denmark, Great Britain, France); Mediterranean (Naples). Depth 10—56 m.

5. P. tenuicornis O. Sars 1882 *P. t.*, G. O. Sars in: Forh. Selsk. Christian., nr. 18 p. 110 t. 6 f. 4 | 1894 *P. t.*, G. O. Sars, Crust. Norway, *v.* 1 p. 572 t. 203 f. 2 | 1893 *P. reinhardi* (part.)?, A. Della Valle in: F. Fl. Neapel, *v.* 20 p. 397.

Body short, stout, back broadly rounded. Head, lateral corners well produced, narrowly rounded. Side-plates 1—5 large and deep, especially in ♀; 1st rather narrowed distally; 5th large, distally expanded. Pleon segment 3, postero-lateral corners subquadrate, slightly produced. Eyes very small, close to margin of lateral lobes of head. Antennae 1 and 2 equal, very slender, sparsely setose, about $^1/_3$ as long as body. Antenna 1, 1st joint shorter than 2^d, a little longer than 3^d; flagellum nearly as long as peduncle, 5-jointed. Antenna 2, flagellum as long as ultimate and penultimate joints of peduncle, 5-jointed. Gnathopod 1 in ϑ, 5th joint as long as 6th, distally widened; 6th oval, but with the oblique palm deeply excavate, defined by a projecting angle, to which the apex of finger reaches. Gnathopod 2 in ♂, 2^d joint with rounded lobe at lower front corner, 5th joint broadly cup-shaped, 6th very large, palm oblique, much longer than hind margin, defined by a strongly projecting triangular lobe, minutely crenulate and bisinuate, with 2 angular projections, fiuger strong, impinging within the sinus adjacent to defining process. Gnathopod 1 in ♀, 5th joint slender, longer than the oval 6th. Gnathopod 2 in ♀ stouter, 5th joint short, cup-shaped, 6th stout, palm oblique, defined by an obtuse angle. Peraeopods 1—5 nearly as in P. longicaudata, but 5th less elongated. Uropod 3, outer ramus longer than peduncle, 2^d joint tipped with slender setae. Telson extremely small, nearly twice as broad as long, with projection on each side of apex. Colour whitish grey, with slightly darker hue on side-plates. L. scarcely 4 mm.

Arctic Ocean and North-Atlantic (Greenland; Norway, depth 56—75 m).

P. producta (Stimps.) 1855 *Dercothoe? productus*, Stimpson in: P. Ac. Philad., *v.* 7 p. 382 | 1893 *Photis reinhardi* (part.)?, A. Della Valle in: F. Fl. Neapel, *v.* 20 p. 397.

Gnathopods 1 and 2 subequal, 6th joint oblong. L. 16 mm.

North-Pacific (Tanegasima).

3. Gen. **Haplocheira** Hasw.

1879 *Haplocheira* (Sp. un.: *H. typica*), Haswell in: P. Linn. Soc. N. S. Wales, *v.* 4 p. 273 | 1885 *H.*, Haswell in: P. Linn. Soc. N. S. Wales, *v.* 10 p. 106 | 1888 *H.*, T. Stebbing in: Rep. Voy. Challenger, *v.* 29 p. 1171 | 1881 *Haplochira*, E. v. Martens in: Zool. Rec., *v.* 16 Crust. p. 32.

Head, lateral lobes not greatly produced. Side-plates of medium depth, 5th less deep than 4th, 1st—4th with setae on lower margin. Antennae 1 and 2 short, subequal, peduncle longer than flagellum. Antenna 1 with accessory flagellum. Upper lip faintly emarginate. Mandible, 3^d joint of palp shorter than 2^d. Maxilla 1, inner plate with fringe of many setae, outer with 9 spines; 2^d joint of palp long. Maxilla 2, inner plate fringed along inner margin. Maxillipeds, inner plates rather broad, outer narrow and rather short, palp elongate. Gnathopod 1 (Fig. 104 p. 610) simple or scarcely subchelate. Gnathopod 2 (Fig. 105 p. 610) simple, 5th and 6th joints long and slender, fringed with very long setae. Peraeopods 3—5 robust, not elongate, 2^d joint well expanded. Uropods 1—3 rather stout, spinose, peduncle produced to a long spine-like process. Uropod 3 short, peduncle stout, inner ramus minute, much shorter than outer. Telson with hook at each distal angle.

1 species.

1. H. barbimana (G. M. Thoms.) 1879 *Gammarus barbimanus*, G. M. Thomson in: Tr. N. Zealand Inst., *v.* 11 p. 241 t. 10D f. 1 | 1886 *Corophium barbimanum*, G. M. Thomson & Chilton in: Tr. N. Zealand Inst., *v.* 18 p. 143 | 1893 *Leptocheirus barbimanus*, A. Della Valle in: F. Fl. Neapel, *v.* 20 p. 433 t. 57 f. 4, 5 | 1879 *Haplocheira typica*, Haswell in: P. Linn. Soc. N. S. Wales, *v.* 4 p. 273 t. 11 f. 2 | 1885 *H. t.*, Haswell in: P. Linn. Soc. N. S. Wales, *v.* 10 p. 106 t. 16 f. 4—8 | 1884 *Corophium lendenfeldi*, Chilton in: Tr. N. Zealand Inst., *v.* 16 p. 262 t. 20 f. 1a—e | 1888 *Haplocheira plumosa* + *H. barbimanus*, T. Stebbing in: Rep. Voy. Challenger, *v.* 29 p. 1172 t. 126; p. 1177.

Body not much compressed laterally. Head, lateral lobes rounded or pointed. Pleon segment 3, postero-lateral corners obtusely quadrate. Eyes oval. Antenna 1, 1st and 2^d joints subequal, 3^d rather over half as long as 2^d; flagellum 9—12-jointed; accessory flagellum 2—5-jointed Antenna 2 stouter, ultimate joint of peduncle rather shorter than penultimate, flagellum not longer than ultimate, with 4—6 joints, last tipped with curved spines. Gnathopod 1 (Fig. 104), 4th joint very short, with setae on hind margin; 5th stouter than 6th, about $^2/_3$ as long, densely setose on hind margin; 6th slender, tapering, with long setae on both margins, palm practically wanting; finger with a subapical tooth.   Gnathopod 2 (Fig. 105) much like gnatho-

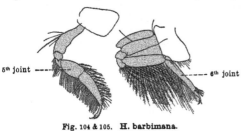

pod 1, but 5th joint as long as the narrow tapering 6th, both fringed on hind margin with double row of very elongate plumose setae; finger short, with subapical tooth. Peraeopod 3, 2^d joint with a bulge at the proximal hind corner. Peraeopod 5, 2^d joint widened distally. Uropods 1 and 2, rami not very unequal, those of uropod 2 the

5th joint

6th joint

Fig. 104 & 105. H. barbimana.
Gnathopods 1 and 2.

stouter; in all the spines are strong. Uropod 3, the outer ramus scarcely as long as peduncle, inner very small. Telson with sides more or less converging to the rather broad, almost transverse apex, with 2 spinules and a setule on the surface near each apical tooth. Colour greyish.   L. 5—6 mm.

South-Pacific (Lyttelton Harbour [New Zealand]; Port Jackson [East-Australia], under stones at low-water mark); southern Indian Ocean (Kerguelen Island, depth 222 m).

4. Gen. **Eurystheus** Bate

1855 [Subgen.] *Gammaropsis* (part.), W. Liljeborg in: Vetensk. Ak. Handl., 1853 p. 455 | 1861 *G.*, A. Boeck in: Forh. Skand. Naturf., Møde 8 p. 659 | 1876 *G.*, A. Boeck, Skand. Arkt. Amphip., v. 2 p. 580 | 1888 *G.*, T. Stebbing in: Rep. Voy. Challenger, v. 29 p. 1092 | 1894 *G.*, G. O. Sars, Crust. Norway, v. 1 p. 557 | 1856 *Eurystheus* (nom. nud.), Bate in: Rep. Brit. Ass., Meet. 25 p. 58 | 1857 *E.* (Sp. un.: *E. tridentatus*), Bate in: Ann. nat. Hist., ser. 2 v. 19 p. 143 | 1859 *Autonoe* (part.), R. M. Bruzelius in: Svenska Ak. Handl., n. ser. v. 3 nr. 1 p. 23 | 1873 *Eurytheus*, A. Marschall, Nomencl. zool., p. 409 | 1884 *Paranaenia*, Chilton in: Tr. N. Zealand Inst., v. 16 p. 258 | 1898 *Maeroides* (Sp. un.: *M. thompsoni*), A. O. Walker in: P. Liverp. biol. Soc., v. 12 p. 282.

Body slender. Head, lateral lobes projecting. Side-plates of moderate size, 2^d—4th varying in relative depth, 4th not emarginate behind, 5th with deep front lobe. Eyes, when present, well developed, often rather large. Antennae 1 and 2 slender, nearly equal, with slender setae. Antenna 1, 3^d joint elongate, accessory flagellum always distinct, usually of several joints. Upper lip more or less produced in front, distal margin a little insinuate or rounded. Mandibular palp very large, 3^d joint lamellar, strongly setose. Maxilla 1, inner plate distinct, its setae varying from 1 to 11, outer plate with 10 spines; 2^d joint of palp elongate. Maxilla 2, inner plate with inner margin fringed. Gnathopods 1 and 2 subchelate (Fig. 107 p. 614), 2^d stronger in ♂ than in ♀, and usually with some variation in shape. Peraeopods 1 and 2 (Fig. 106 p. 613), 2^d and 4th joints sometimes a little widened. Peraeopod 4 longer than peraeopod 3, peraeopod 5 longer than peraeopod 4; 2^d joint in all somewhat expanded. Uropod 2 shorter than uropod 1, uropod 3 shorter than uropod 2. Telson not elongate.

13 species.

Synopsis of species:

1 { Eyes lageniform or oblong — **2.**
    Eyes not lageniform or oblong — **3.**

2 { Eyes lageniform . . . .   1. **E. atlanticus** . . p. 611
    Eyes oblong . . . .   2. **E. afer** . . . . . p. 612

3 { Pleon segment 4 dorsally dentate — **4.**
    Pleon segment 4 not dorsally dentate — **7.**

4 { Telson emarginate . . . .   3. **E. thompsoni** . . p. 612
    Telson not emarginate — **5.**

5 { Pleon segment 4 with 1 medio-dorsal tooth — **6.**
    Pleon segment 4 with 3 medio-dorsal teeth . .   4. **E. crassipes** . . p. 612

6 { Gnathopod 2 in ♀, palm without defining tooth .   5. **E. thomsoni** . . p. 613
    Gnathopod 2 in ♀, palm with defining teeth . .   6. **E. ostroumowi** . p. 614

7 { Gnathopod 2 in ♂, 2ᵈ joint greatly expanded . .   7. **E. exsertipes** . . p. 614
    Gnathopod 2 in ♂, 2ᵈ joint not greatly expanded — **8.**

8 { Limbs of peraeon thickly coated with setae and
      fine hairs . . . .   8. **E. hirsutus** . . . p. 615
    Limbs of peraeon not thickly coated with setae
      and fine hairs — **9.**

9 { Side-plate 3 in ♂ subacutely produced forward .   9. **E. dentifer** . . . p. 615
    Side-plate 3 in ♂ not subacutely produced for-
      ward — **10.**

10 { Gnathopod 2 in ♂, finger closing on to the sur-
        face of 6ᵗʰ joint — **11.**
     Gnathopod 2 in ♂, finger closing on to the margin
        of 6ᵗʰ joint — **12.**

11 { Gnathopod 2 in ♂, 6ᵗʰ joint narrow, finger short . 10. **E. longimanus** . p. 616
     Gnathopod 2 in ♂, 6ᵗʰ joint broad, finger long .   11. **E. palmatus** . . . p. 616

12 { Telson apically emarginate . . . .   12. **E. chiltoni** . . . p. 617
     Telson not emarginate . . . ' . . . .   13. **E. maculatus** . . p. 617

1. **E. atlanticus** (Stebb.) 1888 *Gammaropsis atlantica*, T. Stebbing in: Rep.
Voy. Challenger, *v.* 29 p. 1101 t. 114 | 1893 *Protomedeia? a.,* A. Della Valle in: F. Fl.
Neapel. *v.* 20 p. 441.

♀. Head, lateral lobes narrow, acute, strongly produced. Side-plate 1
rather produced forward, but obtusely; 2ᵈ broader than deep. Pleon segment 3,
postero-lateral corners rounded. Eyes lageniform, close to margin of head,
the narrow neck uppermost. Antenna 1, 3ᵈ joint subequal to 1ˢᵗ, shorter than
2ᵈ; flagellum 17-jointed, accessory flagellum 6-jointed. Antenna 2, antepenultimate joint of peduncle a little concave above as if to receive lateral
lobe of head; ultimate and penultimate joints subequal, flagellum 10-jointed.
Gnathopod 1, 5ᵗʰ joint rather shorter and narrower than 6ᵗʰ; 6ᵗʰ oblong
oval, palm oblique, slightly defined, longer than hind margin, slightly overlapped by the finger. Gnathopod 2 much larger; 5ᵗʰ joint much shorter than
6ᵗʰ, cup-shaped; 6ᵗʰ oblong, palm slightly oblique, irregularly convex, crenate,
defined by a tooth, which is overlapped by apex of serrulate finger. Peraeopods 3—5, 2ᵈ joint well expanded, narrowing distally; finger short. Uropod 3,
peduncle a little longer than the rami. Telson a little longer than broad,
with triangular apex, a plumose setule at the angles of its base and a spine
on the surface near each angle. Colour, dark stellate markings over much
of the surface, including the mouth-parts. L. 7 mm. — ♂ unknown.

Tropical Atlantic (St. Vincent [Cape Verde Islands]).

2. **E. afer** (Stebb.) 1888 *Gammaropsis afra*, T. Stebbing in: Rep. Voy. Challenger, *v.* 29 p. 1097 t. 113 | 1893 *Protomedeia? a.* (part), A. Della Valle in: F. Fl. Neapel, *v.* 20 p. 440.

♀. Head, lateral lobes narrow, acute, moderately produced. Side-plate 1 not produced forward. Pleon segment 3, postero-lateral corners slightly rounded. Eyes oblong, vertical, close to margin of head. Antenna 1, 2^d joint much longer than 1st or 3^d; accessory flagellum 6-jointed. Antenna 2, ultimate joint of peduncle rather longer than penultimate; flagellum 13-jointed. Upper lip smoothly rounded. Gnathopod 1, 2^d joint with a distal widening; 5th joint nearly as long and distally nearly as broad as 6th; 6th oblong oval, palm slightly oblique, shorter than hind margin, finger serrulate, matching palm. Gnathopod 2, 5th joint short, cup-shaped, 6th rather broadly oval, with palm oblique, very sinuous, defined by a simple angle, finger closing against a palmar tooth on the surface. Peraeopods 1 and 2, 4th joint much longer than 5th. Peraeopod 3, 2^d joint above almost as broad as the length, narrowing distally. Peraeopod 5, 2^d joint not much expanded, but broader above than below; finger short. Uropod 3, peduncle rather longer than the short rami, inner a little shorter than the outer. Telson scarcely longer than broad, with some plumose setules on lateral margins and a spine at each angle of base of triangular apex. L. 7 mm. — ♂ unknown.

Southern Indian Ocean (Cape Agulhas [South-Africa]). Depth 270 m.

3. **E. thompsoni** (A. Walker) 1898 *Maeroides t.*, A. O. Walker in: P. Liverp. biol. Soc., *v.* 12 p. 283 t. 16 f. 3—6 | 1899 *Gammaropsis t.*, T. Stebbing in: Ann. nat. Hist., ser. 7 *v.* 3 p. 350.

♀ unknown. — ♂. Pleon segments 4 and 5 each with 2 dorsal teeth on hind margin, a setule at base of each tooth. Head, lateral lobes acute. Side-plates of average depth. Pleon segment 3, postero-lateral corners produced to a small tooth, margin above bulging. Eyes large, long oval, dark, entering lateral lobes of head. Antennae 1 and 2 about half as long as body. Antenna 1, 1st joint little more than half as long as 2^d, as long as 3^d; flagellum subequal to peduncle; accessory flagellum 7-jointed. Antenna 2, ultimate joint of peduncle longer than penultimate; flagellum subequal to peduncle. Gnathopod 1, 5th joint as long as 2^d, longer than the oval, setose 6th. Gnathopod 2 very powerful; 5th joint as broad as long, cup-shaped, 6th large, oblong, palm scarcely oblique, defined by an angular prominence, and having a submedian double tooth and another larger one near the hinge of the finger, across which it projects a pointed lobe on the outer surface; 5th and 6th joints both setose on hind margin; finger (in figure) strong, a little overlapping the palm. Peraeopods strong, 4th and 5th described as equal; 2^d joint broad at base, distally narrowing, hind margin slightly serrate. Uropods 1—3 extending back the same distance. Uropod 3, peduncle nearly as long as the equal spinose rami. Telson widely but not deeply cleft, a spine and a seta at the end of each division, figured only in lateral view. Colour yellowish with grey dorsal freckles and darker spots on 2^d joint of peraeopods. L. 10 mm.

Puget Sound.

4. **E. crassipes** (Hasw.) 1880 *Moera c.*, Haswell in: P. Linn. Soc. N. S. Wales, *v.* 5 p. 103 t. 7 f. 2 | 1899 *Gammaropsis c.*, T. Stebbing in: Ann. nat. Hist., ser. 7 *v.* 3 p. 350 | 1893 *Ceradocus fasciatus* (part.)?, A. Della Valle in: F. Fl. Neapel, *v.* 20 p. 723.

Pleon segment 4 with 3 small distant teeth on hind dorsal margin. Head, lateral corners not acute. Side-plates all small, 6th like 5th, scarcely smaller. Pleon segment 3, postero-lateral corners quadrate, not acute. Eyes blackish or brownish in spirit, not large. Antennae 1 and 2 subequal, fringed with long slender setae. Antenna 1, 3^d joint longer than 1st, nearly equal to 2^d, flagellum shorter than peduncle, about 20-jointed; accessory flagellum 7—10-jointed. Antenna 2, ultimate and penultimate joints of peduncle subequal; flagellum shorter than peduncle, 18—24-jointed. Upper lip acutely produced in front. Mandible, spines of spine-row numerous. Maxilla 1, inner plate fringed with setae. Gnathopod 1 in ♂, 5th joint longer but rather narrower than 6th; 6th widest where palm joins hind margin with strong, scarcely interrupted convexity; finger slightly overlapping palm. Gnathopod 1 in ♀, 6th joint not wider than 5th. Gnathopod 2 in ♂, 4th joint small, quadrate, 5th small, cup-shaped, very short, 6th very large, widening to the nearly transverse palm, defined by a strong tooth, near to which is another conspicuous tooth of variable size, followed by a 2^d cavity and a squarish prominence; finger massive, closing down between the 2 teeth, and having on its inner margin a prominence not nearly large enough to fill the 2^d cavity of palm, and a bulge near the hinge. Sometimes one of the gnathopods 2 is much smaller than the other, and has the 5th joint relatively larger, the 6th with less prominently sculptured palm as in the ♀. Peraeopods 1—3 short. Peraeopods 3—5, finger short, stout, much curved. Peraeopod 3, 2^d joint oval, narrowed a little distally. Peraeopod 4 much longer than the preceding peraeopods, stout; 4th—6th joints widening with age, 2^d relatively narrow, little longer than 4th or 6th, oblong, but with hind margin convex above and below the concave middle part. Peraeopod 5 subequal to 4th in length, less broad, 2^d joint oblong, sinuous hind margin serrate, produced subacutely downward. Branchial vesicles narrow at base, broadly rounded distally. Uropod 3 short, the subequal rami nearly as long as peduncle. Telson small, as broad as long, apical border with an acute central projection, a spine on each of the pair of subapical, sublateral elevations. L. 8 mm.

Port Jackson and Jervis Bay [East-Australia].

5. E. thomsoni (Stebb.) 1888 *Gammaropsis t.*, T. Stebbing in: Rep. Voy. Challenger. *v.*29 p.1103 t.115 | 1893 *Protomedeia? afra* (part.), A. Della Valle in: F. Fl. Neapel, *v.*20 p.440.

♀. Pleon segments 4 and 5 with medio-dorsal emargination, 4th with a small tooth in centre of emargination. Head, lateral lobes narrow, subangular. Side-plate 1 a little produced forward, obtusely, 2^d much larger, about as broad as deep. Pleon segment 3, postero-lateral corners forming a slightly upturned tooth, hind margin bulging above. Eyes rather large, close to margin of head. Antennae 1 and 2 broken. Upper lip slightly and unsymmetrically bilobed. Gnathopod 1, 5th joint longer than 6th, spinose behind; 6th spinose on both sides, oval, more convex behind than in front; palm finely pectinate, slightly defined; finger long, serrate, closely fitting the palm. Gnathopod 2, 2^d joint not expanded, 5th triangular, cup-shaped; 6th large, widening to the palm, which is long, oblique, denticulate, 2 of the teeth prominent, the defining angle obtuse, carrying palmar spines, finger almost smooth, matching the palm. Peraeopods 1 and 2 (Fig. 106) glandular, of the usual pattern. Peraeopods 3—5, 2^d joint

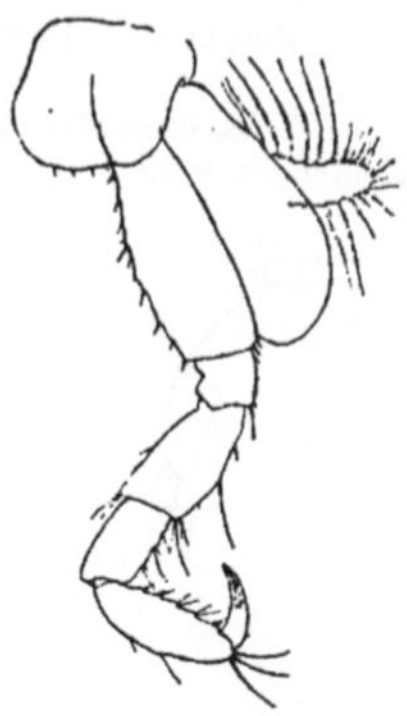

Fig. 106. **E. thomsoni**, ♀.
Peraeopod 2.

a sort of oblong oval, narrowed distally, hind margin rather sinuous in peraeopods 4 and 5; finger short. Uropod 3, peduncle as long as the equal rami.
Telson little longer than broad, almost round, with a spine at each apical
angle. L. 6 mm.

South-Pacific (east of New Zealand). Depth 2000 m?

6. E. ostroumowi (Sowinski) 1898 *Protomedeia o.*, Sowinski in: Mém. Soc. Kiew,
v. 15 p. 475 t. 10 f. 1—19.

Pleon segments 4 and 5 each with 1 dorsal tooth, preceded by a
spinule or setule (hind margin of segment 5 in figure emarginate between
2 teeth?). Head, lateral lobes outdrawn to a small acute point. Side-
plate 1 with 2 little teeth on lower margin, 2^d—4th with denticle at lower
hind corner. Pleon segment 3, postero-lateral corner acute, slightly upturned,
with small sinus above. Eyes large, reniform, almost filling the lateral lobes of
head. Antenna 1, 1st and 3^d joints equal in length, shorter than 2^d; flagellum
subequal to 2^d and 3^d joints combined, 9-jointed; accessory flagellum 4-jointed.
Antenna 2 about as long as antenna 1, ultimate joint of peduncle a little shorter
than penultimate; flagellum a little longer than penultimate joint, 9-jointed.
Mandible, 2^d and 3^d joints of palp equal. Maxilla 1, inner plate fringed with
numerous setae. Maxilla 2 and maxillipeds normal. Gnathopod 1 as in E. thomsoni (p. 613). Gnathopod 2 in ♀ nearly as in E. thomsoni, but 6th joint
scarcely widening to palm, which is defined by a small tooth. Gnathopod 2
in ♂, 6th joint very large, widening to palm, scarcely longer than broad,
palm defined by a strong tooth, and having a median sinus between two
denticulate slopes, finger strong, matching palm, which is subject to some
variations. Peraeopods 1 and 2, none of the joints specially robust. Feraeo-
opod 3, 2^d joint oblong oval. Peraeopod 4, 2^d joint much larger than in
peraeopod 3 but of the same shape, 4th (in ♂ only) monstrously expanded
behind so as to resemble a 2^d joint rather than a 4th, nearly as broad as
long, narrowed distally, almost as large as the 2^d joint, but differently
shaped. Uropod 3, peduncle as long as outer ramus, which is rather shorter
than inner. Telson as in E. thomsoni. L. a little under 11 mm.

Bosphorus. Down to 85 m.

7. E. exsertipes (Stebb.) 1888 *Gammaropsis e.*, T. Stebbing in: Rep. Voy.
Challenger, *v.* 29 p. 1093 t. 112 | 1893 *Protomedeia e.*, A. Della Valle in: F. Fl. Neapel,
v. 20 p. 440 t. 57 f. 12.

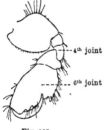

Fig. 107.
E. exsertipes, ♂.
Gnathopod 2.

Head, lateral lobes narrow, acute. Side-plate 1
rounded in front, produced a little forward; 2^d in ♂
broader than deep, 5th nearly as deep as 4th. Pleon
segment 3, postero-lateral corners rounded. Eyes small,
subrotund. Antenna 1 as long as the body, 1st joint
as long as the head, 2^d much longer, 3^d intermediate;
flagellum rather shorter than peduncle, 17-jointed;
accessory flagellum with 4 slender joints. Antenna 2
rather shorter, ultimate joint of peduncle a little shorter
than the elongate penultimate, flagellum about half
as long as peduncle, 12-jointed. Upper lip faintly
emarginate. Gnathopod 1, 5th joint a little shorter
than 2^d, a little longer than 6th, 6th narrowly oval,
front with spines in rows, hind margin and palm a continuous convexity,
palm finely pectinate, finger serrate, in ♂ even longer than the 6th joint.

Gnathopod 2 in ♂ (Fig. 107) little longer but much broader than gnathopod 1; 2ᵈ joint attached close to lower margin of side-plate, widely expanded, somewhat narrowed distally; 5ᵗʰ short, cup-shaped; 6ᵗʰ large, longer than broad, widening to the oblique palm, which is defined by a pretty strong tooth, whence it has a straight slope till near the finger-hinge; finger serrate, strongly curved, closing on to the surface without reaching defining tooth of palm. Gnathopod 2 in ♀, 2ᵈ joint broad, but not abnormally so, palm defined by a small acute tooth, the oblique margin finely pectinate, not irregular. Peraeopods 1 and 2 not very stout. Peraeopods 3—5, 2ᵈ joint not greatly expanded, narrowed distally; peraeopod 5 the longest. Uropod 1, peduncle longer than rami. Uropod 2, rami very unequal, peduncle longer than the outer, scarcely so long as the inner. Uropod 3, peduncle longer than the narrow subequal rami. Telson shorter than peduncle of uropod 3, longer than broad, subtriangular, with a spine near centre of each side, the margin below on each side furry with scale-like spinules. L. 10 mm.

Southern Indian Ocean (Kerguelen Island).

8. **E. hirsutus** Giles 1887 *E. h.*, G. M. Giles in: J. Asiat. Soc. Bengal, *v.* 56 p. 227 t. 8 | 1888 *Gammaropsis h.*, T. Stebbing in: Rep. Voy. Challenger, *v.* 29 p. 1643 | 1893 *Protomedeia maculata* (part.)?, A. Della Valle in: F. Fl. Neapel, *v.* 20 p. 439.

Pleon segment 3 dorsally longer than any other segment. Head, lateral lobes blunt. Side-plates all (in figure) exceedingly shallow, 5ᵗʰ—7ᵗʰ almost transversely linear (not mentioned in text). Eyes rather small, red brown. Antenna 1 rather more than ¹/₉ as long as body; 3ᵈ joint (in figure) scarcely so long as 1ˢᵗ, 2ᵈ longer than 1ˢᵗ; flagellum shorter than peduncle, 11-jointed; accessory flagellum with 4 short joints. Antenna 2 a little shorter, ultimate joint of peduncle slightly shorter than penultimate, slightly longer than the 7-jointed flagellum. Gnathopod 1, 6ᵗʰ joint oval, nearly as long as 5ᵗʰ; finger weakly serrate. Gnathopod 2, 2ᵈ joint not expanded, 5ᵗʰ cup-shaped, 6ᵗʰ rather broadly oval, palm oblique, defined by a little tooth and having a small median one; finger feebly serrate, curved. Peraeopods 1 and 2, 2ᵈ joint a little expanded, narrowly oval; peraeopod 1 rather the longer and stouter. Peraeopods 3—5, 2ᵈ joint greatly expanded; peraeopod 5 the longest, 4ᵗʰ much longer than 3ᵈ. Uropod 3 very short, rami not much longer than peduncle. Telson a short compressed lamina armed with a number of short tooth-like spines like those on the uropods. Colour, nearly transparent with a few patches of reddish brown. L. 4 mm.

Bay of Bengal. Surface.

9. **E. dentifer** (HasW.) 1879 *Moera dentifera*, Haswell in: P. Linn. Soc. N. S. Wales, *v.* 4 p. 332 t. 20 f. 4 | 1884 *Paranaenia typica* + *P. d.*, Chilton in: Tr. N. Zealand Inst., *v.* 16 p. 259 t. 19 f. 1 (♂ juv.?); p. 260 t. 21 f. 2 | 1893 *P. d.* (part.), *Protomedeia* (part.)?, A. Della Valle in: F. Fl. Neapel, *v.* 20 p. 441 | 1899 *Gammaropsis d.*, T. Stebbing in: Ann. nat. Hist., ser. 7 *v.* 3 p. 350.

Head, lateral corners acute. Side-plate 3 in ♂ produced forward rather acutely under the 2ᵈ, the long lower margin having a series of elliptical markings or pellucid spaces. Pleon segment 3, postero-lateral corners quadrate with minutely produced point. Antenna 1, 3ᵈ joint equal to 1ˢᵗ, shorter than 2ᵈ; flagellum shorter than peduncle, reaching 13 joints; accessory flagellum with 5 or 6 long joints. Antenna 2, ultimate and penultimate joints of peduncle subequal; flagellum 13-jointed. Gnathopod 1 small, 5ᵗʰ joint rather longer than 6th; 6th widest at base, which is straight, palm slightly distinguished from serrate hind margin; finger overlapping palm. Gnathopod 2 in ♂, 2^d joint channelled in front, each front margin ending in a rounded lobe; 4th subacutely produced behind the short cup-shaped 5th; 6th very large, widening to the palm, which is very oblique, nearly straight, bordered by several fascicles of long setae, and defined by a long tooth; finger long, strongly curved, dilated at base. Gnathopod 2 in ♀ with 5th joint longer than in ♂, palm of 6th ill-defined (Chilton, in figure). Peraeopods 1—5 of the usual relative lengths, none robust; 2^d joint in peraeopods 3—5 but little dilated. Uropod 3, rami longer than peduncle, inner slightly the longer. Telson very small, truncate, each of the subapical projections carrying a spine. Colour light olive with minute black dots. L. 5 mm.

South-Pacific (Port Jackson [East-Australia]; Lyttelton Harbour [New Zealand]).

10. E. longimanus (Chilton) 1884 *Paranaenia l.*, Chilton in: Tr. N. Zealand Inst., *v.* 16 p. 261 t. 20 f. 2a—c | 1899 *Gammaropsis l.*, T. Stebbing in: Ann. nat. Hist., ser. 7 *v.* 3 p. 350 | 1893 *Paranaenia dentifera* (part.), *Protomedeia* (part.)?, A. Della Valle in: F. Fl. Neapel, *v.* 20 p. 441.

Side-plates normal, 3^d not produced forward as in ♂ of E. dentifer (p. 615). Gnathopod 1 in ♂, 4th joint ending distally in an acute point; 5th considerably longer than 6th, thickly fringed with setae, chiefly in short transverse rows, 6th not broader than 5th, tufts of setae on both sides, palm slightly concave, defined by a short stout spine; finger much longer than palm, slightly curved, acute. Gnathopod 2 in ♂, 4th joint distally acutely produced, 5th triangular, more than half as long as 6th; 6th not wider than 5th, rectangular, twice as long as broad, with 3 rows of setae in tufts, palm transverse, rounded off; finger short, curved, impinging on lateral surface of 6th joint. Gnathopod 1 in ♀, palm slightly convex, not defined. Gnathopod 2 in ♀ smaller than that of ♂, palm slightly concave. Uropod 3, rami only slightly longer than peduncle. L. 5 mm.

Lyttelton Harbour [New Zealand].

11. E. palmatus (Stebb. & D. Roberts.) 1891 *Podoceropsis p.*, T. Stebbing & D. Robertson in: Tr. zool. Soc. London, *v.* 13 I p. 36 t. 6 | 1899 *Gammaropsis palmata*, T. Stebbing in: Ann. nat. Hist., ser. 7 *v.* 3 p. 350 | 1893 *Podoceropsis megacheir* (part.), A. Della Valle in: F. Fl. Neapel, *v.* 20 p. 453 t. 57 f. 23, 24 | 1898 *P. m.*, SoWinski in: Mém. Soc. Kiew, *v.* 15 p. 466 t. 9 f. 1—8 | 1894 *Gammaropsis nana, P. palmata?*, G. O. Sars, Crust. Norway, *v.* 1 p. 561 t. 199 f. 2; p. 562 | 1895 *G. n.*, A. O. Walker in: P. Liverp. biol. Soc., *v.* 9 p. 311.

Body moderately slender. Head, lateral lobes acutely produced. Side-plates 1—5 rather deep; 5th with front lobe nearly as deep as 4th. Pleon segment 3, postero-lateral corners obtusely quadrate. Eyes dark brown, rounded (Sars), oval (Walker). Antenna 1 less than half as long as body, 3^d joint a little longer than 1st, 2^d longer than 3^d: flagellum shorter than peduncle, with 5 or 6 joints; accessory flagellum with 2 or 3 joints, the last minute. Antenna 2 a little shorter, ultimate and penultimate joints of peduncle subequal; flagellum about half as long as peduncle, 5-jointed (Sars, in figure). Upper lip but slightly produced in front. Gnathopod 1, 5th and 6th joints subequal, 5th fully as long as 6th or somewhat shorter, 6th with palm more or less deeply excavate, triangular oval, finger with minute setules and a denticle or perfectly smooth. Gnathopod 2 in ♂, 2^d joint not expanded, 5th cup-shaped, 6th broadly oblong, palm nearly transverse, but strongly rounded at the angle or bisinuate; finger strongly curved, much longer than the palm, closing against inner surface of 6th joint. Gnathopod 2 in ♀ little stronger than gnathopod 1, 6th joint longer than 5th, oblong oval, not broad, palm defined by a distinct angle, its margin minutely but irregularly crenulate, finger a little overlapping the defining angle. Peraeopods 1 and 2, 2^d joint a little widened, narrowly oval. Peraeopod 3, 2^d joint wide proximally, narrowing distally. Peraeopod 4 longer, 2^d joint longer but narrower. Peraeopod 5 similar to peraeopod 4, but of larger dimensions. Uropod 3, peduncle longer than the equal, acute rami. Telson small, as broad as long, apical margin slightly concave with a couple of spinules on surface ridge at each side. Colour whitish with light brown transverse bands (Sars). L. ♀ scarcely exceeding 3, ♂ 4 mm.

Firth of Clyde; Christianiafjord, depth 38 m; Bosphorus, depth 36 m.

12. **E. chiltoni** (G. M. Thoms.) 1897 *Maera c.*, G. M. Thomson in: Ann. nat. Hist., ser. 6 v. 20 p. 447 t. 10 f. 1—5 | 1899 *Gammaropsis c.*, T. Stebbing in: Ann. nat. Hist., ser. 7 v. 3 p. 350.

Body slender, rather compressed. Head, lateral lobes produced into an obtusely pointed process. Side-plate 1 elongated, widening and rounded below, produced forward at their infero-anterior angle. Eyes subreniform, broader below than above, close to margin of lateral lobes of head. Antenna 1, 1st joint stout, 2^d about twice as long as 1st, 3^d about $^2/_3$ as long as 2^d; flagellum unknown; accessory flagellum nearly as long as ultimate joint of peduncle. Antenna 2, peduncle subequal to peduncle of antenna 1, ultimate and penultimate joints of peduncle subequal; flagellum slightly longer than ultimate joint. Mandibular palp, 2^d and 3^d joints widening distally, 2^d longer than 3^d, 3^d with long setae on broad apex. Maxilla 1, inner plate short, acute, without setae, outer with 10 spines, palp apparently 3-jointed, the middle joint short(?). Maxilla 2, inner plate fringed with short setae, chiefly round the inner margin. Maxillipeds normal, 4th joint of palp short· and rounded, with stout apical setae, but without a claw. Gnathopod 1 small, 4th joint produced into an acute tip (in figure rotundo-quadrate at apex); 5th long, widening a little distally, 6th shorter, narrow ovate, palm oblique, dentate (no teeth in figure); finger curved, acute, about $^2/_3$ as long as 6th joint (in figure subequal), with a few minute denticulations along inner edge. Gnathopod 2 very large, 4th joint produced forward and channelled; 5th large, triangular, cup-shaped, as broad as long; 6th very large, hind margin straight, front very convex, producing a great bulge a little way from the base, ending in a small tooth at the finger-hinge; palm nearly transverse, with large triangular median tooth, followed by a cavity and ending with a slightly concave margin, finger strong, folding down a little beyond the palm. Peraeopods 1 and 2 slender, 3^d stout, 4th and 5th successively longer, rather slender. Uropod 1 the longest, peduncle with large apical spine, rami subequal, shorter than peduncle. Uropod 2, inner ramus subequal to peduncle, rather longer than outer. Uropod 3 shortest, rami subequal, rather shorter than peduncle. Telson subquadrate, sides converging, a rounded emargination extending $^1/_3$ of the length, between 2 obtuse apices, each with a spinule. L. 4—5 mm.

South-Pacific (New Zealand). Depth 15 m.

13. **E. maculatus** (Johnst.) 1827 *Gammarus m.*, G. Johnston in: Zool. J., v. 3 p. 176 | 1888 *Gammaropsis m.*, T. Stebbing in: Rep. Voy. Challenger, v. 29 p. 131, 286 | 1893 *Protomedeia maculata* (part.), A. Della Valle in: F. Fl. Neapel, v. 20

p. 436 t. 14 f. 28—40; t. 57 f. 8—11 | 1855 *Gammarus (Gammaropsis) erythrophthalmus*, W. Liljeborg in: Vetensk. Ak. Handl., 1853 p. 455 | 1871 *Gammaropsis e.* (part.?), A. Boeck in: Forh. Selsk. Christian.; 1870 p. 241 | 1876 *G. e.* (part.?), A. Boeck, Skand. Arkt. Amphip., *v.* 2 p. 581 t. 25 f. 6 | 1889 *G. e.*, Hoek in: Tijdschr. Nederl. dierk. Ver., ser. 2 *v.* 2 p. 226 t. 9 f. 4 k | 1894 *G. erythrophthalma*, G. O. Sars, Crust. Norway, *v.* 1 p. 558 t. 198 | 1895 *G. e.*, A. O. Walker in: Ann. nat. Hist., ser. 6 *v.* 15 p. 470 | 1859 *Autonoe e.*, R. M. Bruzelius in: Svenska Ak. Handl., n. ser. *v.* 3 nr. 1 p. 27 | ?1866 *Maera e.*, Cam. Heller in: Denk. Ak. Wien, *v.* 26 II p. 42 | 1856 *Eurystheus tridentatus* (nom. nud.), Bate in: Rep. Brit. Ass., Meet. 25 p. 58 | 1857 *E. t.*, Bate in: Ann. nat. Hist., ser. 2 *v.* 19 p. 143 | 1862 *E. erythrophthalma* + *E. bispinimanus* + *Gammarus maculatus*, Bate, Cat. Amphip. Brit. Mus., p. 196 t. 35 f. 7; p. 197 t. 35 f. 8; p. 223 | 1882 *Gammaropsis melanops*, G. O. Sars in: Forh. Selsk. Christian., nr. 18 p. 111 t. 6 f. 5 | 1894 *G. m.*, G. O. Sars, Crust. Norway, *v.* 1 p. 560 t. 199 f. 1.

Body rather slender. 4th pleon segment (Boeck) armed with 2 minute medio-dorsal teeth. Head, lateral lobes acute. Side-plate 2 largest in ♂. Pleon segment 3, postero-lateral corners quadrate with small tooth, margin above more or less bulging. Eyes rather large, oval reniform, red or very dark. Antenna 1, 1st joint subequal to 3^d, or shorter; flagellum 10—15-jointed; accessory flagellum 4—6-jointed. Antenna 2 rather shorter. Upper lip produced acutely or obtusely in front. Gnathopod 1 slender; 5th and 6th joints subequal; palm ill-defined; finger denticulate (Bruzelius). Gnathopod 2 much larger; 2^d and 5th joints robust in ♂, 6th with palm oblique, having 3 (2?) teeth or tubercles in the ♂, 2 in the ♀. Peraeopods 3—5, 2^d joint well expanded or narrowly oval. Uropod 3, rami about as long as peduncle. Telson as broad as long, with a spine and usually setules on each side of apex. Colour pale yellow with dusky bands. L. 6—12 mm.

Arctic Ocean and North-Atlantic (Greenland, depth 38—112 m; from Norway to France; Azores).

5. Gen. **Podoceropsis** Boeck

1861 *Podoceropsis* (Sp. un.: *P. sophia*), A. Boeck in: Forh. Skand. Naturf., Møde 8 p. 666 | 1876 *P.*, A. Boeck, Skand. Arkt. Amphip., *v.* 2 p. 583 | 1888 *P.*, T. Stebbing in: Rep. Voy. Challenger, *v.* 29 p. 1108 | 1893 *P.* (part.), A. Della Valle in: F. Fl. Neapel, *v.* 20 p. 451 | 1894 *P.*, G. O. Sars, Crust. Norway, *v.* 1 p. 574 | 1862 *Noenia*, Bate (& Westwood), Brit. sess. Crust., *v.* 1 p. 471 | 1862 *Naenia* (non J. F. Stephens 1829, Lepidoptera!), Bate, Cat. Amphip. Brit. Mus., p. 271 | 1871 *Xenoclea* (Sp. un.: *X. batei*), A. Boeck in: Forh. Selsk. Christian, 1870 p. 234.

Like Eurystheus (p. 610), except that antenna 1 has no accessory flagellum or only a rudiment of it, and that gnathopod 2 in the ♂ has the 5th joint always short and in the ♀ has the 6th joint conspicuously wider than that of gnathopod 1.

5 species.

Synopsis of species:

1 { Eyes wanting — 2.
{ Eyes present — 3.

2 { Gnathopod 2, palm oblique . . . . 1. **P. abyssi** . . . . . . p. 619
{ Gnathopod 2, palm transverse . . . . 2. **P. kermadeci** . . . . p. 619

3 { Peraeopods 1 and 2, 6th joint very long . . 3. **P. lindahlii** . . . . . p. 619
{ Peraeopods 1 and 2, 6th joint not very long — 4.

4 { Side-plate 5 not as deep as side-plate 4 . . . 4. **P. sophia** . . . . . . p. 620
{ Side-plate 5 fully as deep as side-plate 4 . . 5. **P. nitida** . . . . . . p. 620

1. **P. abyssi** Chevreux 1887 *P. a.*, Chevreux in: Bull. Soc. zool. France, *v.* 12 p. 577 | 1893 *P. a.*, A. Della Valle in: F. Fl. Neapel, *v.* 20 p. 452 | 1896 *Gammaropsis abyssorum*, J. Bonnier in: Ann. Univ. Lyon, *v.* 26 p. 661 t. 40 f. 2.

Body narrow, elongate. Head with inter-antennal process produced into a point. Side-plates small, 1st—4th rectangular, 2d the largest. Pleon segment 3, postero-lateral corners rounded. Eyes wanting. Antenna 1 $^2/_3$ as long as body; 1st joint as long as head, 2d twice as long as 1st, 3d as long as 2d; flagellum with 7 very long joints; accessory flagellum consisting of 1 extremely short joint. Antenna 2 nearly as long as antenna 1; ultimate and penultimate joints of peduncle equal; flagellum 5-jointed. Maxillipeds, 4th joint of palp long, unguiform. Gnathopod 1 rather large, 5th joint triangular, $^2/_3$ as long as 6th, 6th broadly oval; palm carrying 3 large, obtuse, finely crenulate teeth and ending in a smooth acute tooth; finger nearly as long as 6th joint. Gnathopod 1 in ♀ weaker than in ♂, gnathopod 2, 5th joint short, 6th similar to that of gnathopod 1 but larger, palm carrying a sharp tooth, followed by several non-crenulate tubercles, finger relatively much shorter. Gnathopod 2 in ♂ stronger on the right side than on the left. Peraeopods 3—5, 2d joint long and narrow. Uropods 1—3 with peduncle rather short; rami of uropods 1 and 2 with a few spines; rami of uropod 3 bare, equal, shorter than peduncle (Bonnier, in figure). Telson small, triangular, carrying a large stiff hair towards the middle of its upper part. L. 3—4 mm.

Bay of Biscay. Depth 510—950 m.

2. **P. kermadeci** Stebb. 1888 *P. k.*, T. Stebbing in: Rep. Voy. Challenger, *v.* 29 p. 1109 t. 116 | 1893 *P. sophiae* (part:), A. Della Valle in: F. Fl. Neapel, *v.* 20 p. 452.

Head, lateral lobes acute, not very prominent. Side-plates 1—4 not very deep, 2d the largest, broader than deep. No eyes perceived. Antenna 1, 1st joint longer than the head, a little shorter than 2d, subequal to 3d; flagellum with 6 joints, the 1st longest; a rudimentary accessory flagellum, tipped with 2 setules. Antenna 2 rather shorter, like antenna 1 carrying long setae; ultimate joint of peduncle slightly shorter than penultimate; flagellum with 5 joints, 1st longer than the rest combined. Upper lip faintly emarginate, obtusely produced in front. Maxilla 1, inner plate with 3 plumose setae on inner margin, 2 apical setules. Maxillipeds, outer plates rather short, not strongly armed; 2d joint of palp long, finger short, blunt, tipped with spines. Gnathopod 1 rather feeble, 5th joint fully as long and broad as 6th; 6th oblong oval, palm finely pectinate, almost transverse; defining angle obtuse, much overlapped by the curved finger. Gnathopod 2, 2d joint with concave front and convex hind margin, 5th short, cup-shaped, much narrower than the 6th, which is of great size, nearly as broad as long, widest at the transverse palm, this being much sculptured, with blunt defining tooth, to which an acute one is adjacent, separated by a well marked cavity from the serrate and dentate portion which meets the hinge of the finger; finger curved, almost smooth, and matching the palm. Peraeopods 1 and 2 rather long and slender, 6th joint longer than 4th, nearly as long as 2d; finger half as long as 6th joint. Rest wanting. Existing portion 5 mm long.

South-Pacific (north of Kermadec Islands). Depth 1152 m.

3. **P. lindahlii** H. J. Hansen 1887 *P. l.*, H. J. Hansen in: Vid. Meddel., ser. 4 *v.* 9 p. 157 t. 6 f. 2, 2a | 1893 *P. l.*, A. Della Valle in: F. Fl. Neapel, *v.* 20 p. 453 | 1894 *P. lindahli*, G. O. Sars, Crust. Norway, *v.* 1 p. 574.

Head, lateral lobes somewhat produced, rounded. Side-plate 1 (in figure) very small, 2^d—4th subequal, 5th deeper. Pleon segment 3, postero-lateral corners broadly rounded. Eyes large, ovate, black. Antennae 1 and 2 subequal, less than half as long as body. Antenna 1 with long setae, 1st joint $^2/_3$ as long as 2^d, 3^d intermediate in length; flagellum about 8-jointed, with no accessory flagellum. Antenna 2, flagellum about 8-jointed. Gnathopod 1 rather feeble, 6th joint a little shorter and not broader than 5th, finger much longer than the somewhat oblique palm. Gnathopod 2 rather robust, 5th joint short, cup-shaped, 6th $^2/_3$ as broad as long; palm somewhat oblique, evenly rounded. Peraeopods 1 and 2 elongate, almost longer than peraeopod 5; 6th joint unarmed, very elongate, nearly as long as 4th and 5th combined. Peraeopods 4 and 5, 2^d joint rather narrow, narrowing distally, 4th and 6th joints subequal. Uropod 3, peduncle rather longer than the subequal rami. Telson furnished with 2 setae. Sex and age doubtful. L. 5 mm.

Davis Strait (West-Greenland). Depth 91 m.

4. P. sophia Boeck 1861 *P. s.*, A. Boeck in: Forh. Skand. Naturf., Møde 8 p. 666 | 1888 *P. s.*, T. Stebbing in: Rep. Voy. Challenger, *v.* 29 p. 322 | 1871 *P. sophiae*, A. Boeck in: Forh. Selsk. Christian., 1870 p. 242 | 1876 *P. s.*, A. Boeck. Skand. Arkt. Amphip., *v.* 2 p. 584 t. 25 f. 7 | 1893 *P. s.* (part.), A. Della Valle in: F. Fl. Ncapel, *v.* 20 p. 452 t. 57 f. 21, 22 | 1894 *P. s.*, G. O. Sars, Crust. Norway, *v.* 1 p. 574 t. 204 | 1895 *P. s.*, A. O. Walker in: Ann. nat. Hist., ser. 6 *v.* 15 p. 473 | 1862 *Noenia tuberculosa* + *N. undata*, Bate (& Westwood), Brit. sess. Crust., *v.* 1 p. 472, 477 f. | 1862 *Naenia t.* + *N. u.*, Bate, Cat. Amphip. Brit. Mus., p. 271 t. 46 f. 2; p. 272 t. 46 f. 5.

Head, lateral lobes sharply quadrate. Side-plate 1 small, rounded in front, 5th not so deep as 4th. Pleon segment 3, postero-lateral corners with minutely produced obtuse point. Eyes large, broadly oval, bright red. Antennae 1 and 2 equal, nearly $^2/_3$ as long as body. Antenna 1, 1st joint shorter than 3^d, 2^d much longer; flagellum shorter than peduncle, about 10-jointed. Antenna 2, ultimate and penultimate joints of peduncle elongate, equal; flagellum 7- or 8-jointed. Upper lip slightly emarginate, very obtusely produced in front. Mandibular palp rather slender. Gnathopod 1 slender, 5th joint as long as 6th; 6th oval, hind margin with ill-defined palm much more convex than front; finger rather long, denticulate. Gnathopod 2 in ϑ, 5th joint short, cup-shaped, 6th large, oblong oval, twice as long as broad; palm oblique, longer than hind margin, not strongly defined, with 2 tooth-like projections not far from finger-hinge; finger strong, curved. Gnathopod 2 in φ, 6th joint very large, oval; palm longer than hind margin, minutely serrulate, nearly straight, armed with 3 (Bate: 2) successive spines (Sars); finger long, serrate. Peraeopods 1 and 2, 4th joint nearly as long as 5th and 6th combined. Peraeopods 3—5, 2^d joint oblong oval. Uropod 3, rami a little longer than peduncle, with acute, unarmed apex (Sars), tipped with spines (Bate, in figure). Telson ovoid, a little longer than broad, with 2 spinules on each side of apex. Colour, reddish and yellow bands and patches along the sides; ova in pouch bright red. L. 6 mm.

North-Atlantic, North-Sea, Skagerrak and Kattegat (Norway, Denmark, Holland, Great Britain, France). Depth 56—282 m.

5. P. nitida (Stimps.) 1853 *Podocerus nitidus*, Stimpson in: Smithson. Contr., *v.* 6 nr. 5 p. 45 | 1862 *Noenia rimapalmata* (*N. rimapalma* Bate) + *N. excavata* (Bate), Bate & Westwood, Brit. sess. Crust., *v.* 1 p. 474 f.; p. 476 f. | 1862 *Naenia rimapalma* + *N. e.*, Bate, Cat. Amphip. Brit. Mus., p. 272 t. 46 f. 3, 4 | 1877 *Podoceropsis rimapalmata* + *N. e.*, Meinert in: Naturh. Tidsskr., ser. 3 *v.* 11 p. 152 | 1879 *N. r.*, T. Stebbing in: Rep.

Devonsh. Ass., *v.* 11 p. 520 | 1894 *Podoceropsis excavata*, G. O. Sars, Crust. Norway, *v.* 1 p. 576 t. 205 | ?1867 *Noenia caudadentata* (nom. nud.), A. M. Norman in: Nat. Hist. Tr. Northumb., *v.* 1 p. 16 | 1871 *Xenoclea batei*, A. Boeck in: Forh. Selsk. Christian., 1870 p. 235 | 1876 *X. b.*, A. Boeck, Skand. Arkt. Amphip., *v.* 2 p. 561 t. 25 f. 8 | 1890 *Podoceropsis b.*, Meinert in: Udb. Hauchs, *v.* 3 p. 178 | 1876 *Xenoclea megachir*, S. I. Smith & Harger in: Tr. Connect. Ac., *v.* 3 p. 32 t. 8 f. 1—4 | 1893 *Podoceropsis sophiae* (part.), A. DellaValle in: F. Fl. Neapel, *v.* 20 p. 452.

Body rather stouter than in P. sophia; head with lateral lobes more acutely produced with less convex lower margin. Side-plate 1 not very small, 2^d in $\male$ broader than deep, 5th fully as deep as 4th. Pleon segment 3, postero-lateral corners with obtuse point scarcely produced. Eyes of moderate size, rounded oval, dark. Antennae 1 and 2 equal, more than $^1/_2$ as long as body; peduncles as in P. sophia; flagellum of antenna 1 12—16-jointed, of antenna 2 13-jointed. Upper lip feebly emarginate, produced in front to a sharp point. Mandibular palp not very slender. Gnathopod 1 very slender, 6th joint scarcely as long as 5th, subfusiform, finger very long. Gnathopod 2 in $\male$, 5th joint short, cup-shaped, 6th very large, oblong oval; palm minutely tuberculate, with a median sinus between 2 lobes, that near the finger-hinge truncate and a little emarginate, the other dentiform not far from the defining angle, which is scarcely reached by the broad curved finger. Gnathopod 2 in $\female$ as in $\male$, except that the cavity in the palm is wider, not bounded by a separate lobe near finger-hinge, and the tooth on the other side is closer to the defining angle. Peraeopods 1—2, 4th joint not nearly as long as 5th and 6th combined. Peraeopod 3, 2^d joint nearly as broad as long. Peraeopods 4 and 5, 2^d joint oblong oval. Uropod 3, rami rather unequal, a little shorter than peduncle, the inner the larger, tipped with 2 setules. Telson as broad as long, with 3 slender spines on either side of apex. Colour, variegated with dark brown, especially along the sides. L. 7—8 mm.

North-Atlantic, North-Sea, Skagerrak and Kattegat (Norway, Denmark, Holland, Great Britain, Hake Bay [Grand Manan]). Depth 56—75 m.

6. Gen. **Megamphopus** Norm.

1869 *Megamphopus* (Sp. typ.: *M. cornutus*), A. M. Norman in: Rep. Brit. Ass., Meet. 38 p. 282 | 1894 *M.*, G. O. Sars, Crust. Norway, *v.* 1 p. 563.

Body slender. Head, lateral lobes projecting. Side-plate 2 in $\male$ the largest, 3^d and 4th successively smaller. Eyes well developed. Antennae 1 and 2 very slender. Antenna 1 the shorter, with very small accessory flagellum. Mouth-parts as in Eurystheus (p. 610), except that 3^d joint of mandibular palp is relatively shorter. Gnathopods 1 and 2 with 5th and 6th joints elongate, strongly developed though weakly subchelate in $\male$, feebly developed, almost simple, and subequal in $\female$. In other respects scarcely differing from Eurystheus.

1 species.

1. **M. cornutus** Norm. 1869 *M. c.*, A. M. Norman in: Rep. Brit. Ass., Meet. 38 p. 282 | 1894 *M. c.*, G. O. Sars, Crust. Norway, *v.* 1 p. 564 t. 200 | 1895 *M. c.*, A. O. Walker in: P. Liverp. biol. Soc., *v.* 9 p. 311 | 1871 *Protomedeia longimana*, A. Boeck in: Forh. Selsk. Christian., 1870 p. 240 | 1876 *P. l.*, A. Boeck, Skand. Arkt. Amphip., *v.* 2 p. 578 t. 25 f. 4; t. 29 f. 5 | 1892 *P. l.*, D. Robertson in: P. nat. Hist. Soc. Glasgow, n. ser. *v.* 3 p. 220 | 1878 *P. intermedia*, T. Stebbing in: Ann. nat. Hist., ser. 5 *v.* 2 p. 367 t. 15 f. 3 | 1888 *Podoceropsis* (part.)?, T. Stebbing in: Rep. Voy. Challenger, *v.* 29 p. 1108 | 1893 *P.* (part.)?, A. Della Valle in: F. Fl. Neapel, *v.* 20 p. 451.

Head, lateral lobes acutely produced. Side-plate 1 in ♀ oblong, in ♂ with an acutely produced and deflexed apex; 2^d—4th subequal, with 5th a little less deep than 4th in ♀; but in ♂ 2^d much larger than the rest, 4th smaller than 3^d, not deeper than 5th. Pleon segment 3, postero-lateral corners broadly rounded. Eyes rather large, oval, dark brown. Antenna 1 less than $^1/_2$ as long as body, 1st joint short and thick, as long as 3^d, 2^d much longer; flagellum 8—10-jointed; accessory flagellum consisting of 1 small joint. Antenna 2 rather longer, antepenultimate joint of peduncle not very short, ultimate rather longer than penultimate; flagellum 10-jointed. Upper lip rounded, middle of margin truncate, bluntly produced in front. Maxilla 1, inner plate with 1 apical seta. Gnathopod 1 in ♂, 2^d joint slender, 5th nearly as long, stouter, hind margin ending in a small obtuse tooth; 6th shorter, oblong oval, palm scarcely defined from short hind margin, finely crenulate, set with spines; finger very long, compressed, its breadth nearly uniform except at the ends. Gnathopod 2 in ♂ considerably stronger, 2^d joint abruptly curved at base, 5th stout, long, 6th somewhat stouter, longer, front margin convex, hind straight, setose; the short palm with a median tubercle; the finger similar to that of gnathopod 1, folding on to surface of 6th joint parallel and close to hind margin. Gnathopod 1 in ♀, 6th joint nearly as long as 5th, subfusiform; palm but slightly defined; finger slender. Gnathopod 2 in ♀ not much stronger, 5th joint as long as the tapering 6th, in which the palm is scarcely at all defined. Peraeopods 1 and 2 rather compact. Peraeopods 3—5 more slender, 2^d joint oblong oval. Uropod 3, rami longer than peduncle, outer shorter than inner. Telson subquadrate, rather broader than long, with spine at each angle of base of the triangular apex. Colour whitish, pellucid, darkened by red and brown specks. L. 5—8 mm.

Arctic Ocean, North-Atlantic and adjoining seas (from Lofoten Isles to the Bristol Channel). Depth 11—56 m.

7. Gen. **Goësia** Boeck

1871 *Goësia* (Sp. un.: *G. depressa*), A. Boeck in: Forh. Selsk. Christian., 1870 p. 231 | 1876 *G.*, A. Boeck, Skand. Arkt. Amphip., *v.* 2 p. 550.

Head, lateral lobes little produced. Side-plates of medium depth, 5th not so deep as 4th, the lower margin fringed with setae. Antenna 1 the longer, peduncle shorter than flagellum, accessory flagellum obsolete. Maxillipeds (Goës, in figure), palp 3-jointed. Gnathopod 1 as in Leptocheirus (p. 625). Gnathopod 2 subchelate, but not strongly, setose, the 6th joint as broad as the 5th, which is not very elongate. Otherwise as in Leptocheirus.

1 species.

1. **G. depressa** (Goës) 1866 *Autonoë d.*, Goës in: Öfv. Ak. Förb., *v.* 22 p. 532 t. 41 f. 32 | 1871 *Goësia d.*, A. Boeck in: Forh. Selsk. Christian., 1870 p. 231 | 1876 *G. d.*, A. Boeck, Skand. Arkt. Amphip., *v.* 2 p. 550.

Side-plate 1 narrowed below, 2^d and 3^d deeper than 4th. Pleon segment 3, postero-lateral corners acutely produced. Eyes very small, oval. Antenna 1 more than $^2/_3$ as long as body, 2^d joint rather longer than 1st, 3^d small; flagellum 20—25-jointed, accessory flagellum not apparent in figure. Antenna 2 much shorter; ultimate and penultimate joints of peduncle subequal; flagellum 10—15-jointed. In the figure 3^d joint of mandibular palp is longer than 2^d, maxilla 1 has a long inner plate devoid of setae, maxilla 2 has the inner plate not fringed on inner margin, and the maxilliped's

palp has no finger, but these points are doubtful without description.
Gnathopod 1, 3^d joint prominent, 3^d—6th strongly setose on hind margin,
5th scarcely so long or broad as 6th; palm of 6th short, transverse, overlapped
by the small pectinate finger. Gnathopod 2, 2^d and 6th joints with very
long plumose setae fringing front margin; 5th joint a little longer, not
broader, than 6th, widening distally; 6th narrowing to small oblique palm,
defined by a strong palmar spine; finger rather strong, curved, smooth.
Peraeopods 1 and 2 rather long. Peraeopod 3 short, peraeopod 4 a little,
and peraeopod 5 much longer; 2^d joint of peraeopods 3—5 narrowly oblong
oval. Uropods 1 and 2 rather stout; rami a little unequal. Uropod 3 very
short; rami equal, elliptic or ovate, scarcely as long as the short, stout
peduncle. Telson very short, twice as broad as long; sides converging to
the slightly sinuous apex, which has a pair of plumose setules at each angle.
L. 12 mm.

Arctic Ocean (Spitzbergen). Depth 9 m.

8. Gen. Protomedeia Krøyer

1842 *Protomedeia* (Sp. un.: *P. fasciata*), Krøyer in: Naturh. Tidsskr., *v.* 4 p. 154 |.
1855 [Subgen.] *Gammaropsis* (part.), W. Liljeborg in: Vetensk. Ak. Handl., 1853 p.455 |
1859 *Autonoe* (part.), R. M. Bruzelius in: Svenska Ak. Handl., n. ser. *v.* 3 nr. 1 p. 23.

Side-plates rather small, 2^d in ♂ larger than 3^d or 4th, 5th deeper
than 4th, especially in ♂. Antenna 1 longer than antenna 2; accessory
flagellum well developed. Upper lip, margin rounded. Mandible, molar
strong, 3^d joint of palp subequal to 2^d, very setose. Maxilla 1, inner plate
small, with a few hairs on the margin, outer plate with 10 spines; 2^d joint
of palp long, with truncate apex. Maxilla 2, inner plate not fringed with
spines or setae along the inner margin, though there are some scattered
hairs; outer plate the longer, widening to the triangular closely fringed
apex. Maxillipeds well armed, 4th joint of palp very short. Gnathopod 1
slender, with 5th and 6th joints subequal. Gnathopod 2 in ♂ much larger
than gnathopod 1, 5th joint larger than 6th; in ♀ similar to gnathopod 1
and not much larger. Peraeopods 1 and 2 with slender 6th joint and finger;
peraeopod 1, as the gnathopods 1 and 2, very setose. Peraeopods 3—5,
2^d joint moderately expanded; peraeopod 4 longer than 3^d, 5th longer than 4th.
Uropod 3 the shortest, rami unequal, spinose. Telson rounded, with hook
at each side of the broad apex.

1 species.

1. P. fasciata Krøyer 1842 *P. f.*, Krøyer in: Naturh. Tidsskr., *v.* 4 p. 154 |
1876 *P. f.*, A. Boeck, Skand. Arkt. Amphip., *v.* 2 p. 576 t. 25 f. 1 | 1884 *P. f.*, J. S.
Schneider in: Tromsø Mus. Aarsh., *v.* 7 p. 124 | 1893 *P. f.*,–A. Della Valle in: F. Fl.
Neapel, *v.* 20 p. 435 t. 57 f. 6, 7 | 1894 *P. f.*, G. O. Sars, Crust. Norway, *v.* 1 p. 552 t. 196 |
1855 *Gammarus (Gammaropsis) macronyx*, W. Liljeborg in: Vetensk. Ak. Handl.,
1853 p. 458 | 1859 *Autonoe m.*, R. M. Bruzelius in: Svenska Ak. Handl., n. ser. *v.* 3 nr. 1
p. 29 t. 1 f. 6 | 1866 *Autonoë m.*, Goës in: Öfv. Ak. Förb., *v.* 22 p. 531 t. 40 f. 31 | 1862
Microdentopus m., *Microdeutopus m.* + *Protomedeia fasciata*, Bate, Cat. Amphip. Brit.
Mus., p. 167, 379; p. 172.

Body rather slender, subdepressed. Peraeon segment 2 much longer
in ♂ than in ♀. Head, lateral lobes little produced. Side-plate 1 subrhomboidal in ♀, front corner a little more produced in ♂, lower margin
setose; 2^d—4th nearly alike in ♀; 2^d in ♂ both broader and more rounded; 5th in
♀ little deeper than 4th, in ♂ much deeper and with front lobe more squared.

Pleon segment 3, postero-lateral corners rotundo-quadrate. Eyes very small, rounded oval, black. Antenna 1 scarcely over half as long as body; 3^d joint about half as long as 1st or 2^d; flagellum much longer than peduncle, about 14-jointed (Sars) (Schneider: 22-jointed); accessory flagellum 5—7-jointed. Antenna 2, ultimate joint of peduncle subequal to penultimate or flagellum, the latter 6- or 7-jointed. Upper lip with apex entire (indented in Sars, figure). Gnathopod 1 in $\male$, 2^d joint distally widened, 5th rather longer than 6th, which is oblong oval, with short rounded palm overlapped by small finger. Gnathopod 2 in $\male$, 5th joint very broad, longer than broad, distally truncate; 6th narrower and shorter, yet broadly oval, with strongly serrate hind margin and palm; finger rather strong. Gnathopod 1 in $\female$, 2^d joint not expanded distally; 5th joint scarcely longer than 6th. Gnathopod 2 in $\female$, 2^d joint long, slender, fringed on both margins with long plumose setae, 6th joint rather longer than 5th, slightly tapering, palm very short, finger small. Peraeopods 1 and 2, 4th joint about as long as 5th and 6th combined and much wider, in peraeopod 1 this and the 2^d joint densely setose; finger as long as 6th joint or longer. Peraeopods 3—5, 2^d joint oblong oval, narrowing a little downward. Uropod 3, inner ramus the shorter, both spinose, outer also tipped with a slender seta. Telson about as broad as long, having on either side of the broadly rounded apical margin a little spine-like hook and 2 or 3 slender spines. Colour whitish with straight brown bands crossing the segments and corresponding blotches on side-plates. L. 8 mm.

Arctic Ocean, North-Atlantic, North-Sea, Skagerrak and Kattegat (Greenland, Spitzbergen, Iceland, Finmark [Norway], Sweden, Denmark). Depth 11—73 m.

9. Gen. **Xenocheira** Hasw.

1879 *Xenocheira* (Sp. un.: *X. fasciata*), Haswell in: P. Linn. Soc. N. S. Wales, *v.* 4 p. 272 | 1893, A. Della Valle in: F. Fl. Neapel, *v.* 20 p. 433, 948 | 1881 *Xenochira*, E. v. Martens in: Zool. Rec., *v.* 16 Crust. p. 31.

Side-plates all shallow. Antenna 1 much the longer, 3^d joint not elongate; with accessory flagellum. Maxillipeds described as non-unguiculate (palp 3-jointed). Gnathopod 1 subchelate. Gnathopod 2 simple, the 5th joint broadly expanded in front of the neighbouring joints, and described as articulating at its base with the 3^d joint. Gnathopods 1 and 2 both setose.

1 species.

1. **X. fasciata** Hasw. 1879 *X. f.*, Haswell in: P. Linn. Soc. N. S. Wales, *v.* 4 p. 272 t. 11 f. 6 | 1885 *X. f.*, Haswell in: P. Linn. Soc. N. S. Wales, *v.* 10 p. 105 t. 16 f. 1—3.

Body slender, perhaps subdepressed; pleon segment 3 the longest. Head, lateral lobes scarcely prominent. Side-plates 1—4 small, 5th (in figure) nearly as deep, broader. Eyes round. Antenna 1 more than half as long as body, 1st joint rather longer than head, 2^d a little longer than 1st, 3^d about $1/_3$ as long as 2^d; flagellum longer than peduncle. Antenna 2 with longer peduncle; ultimate joint of peduncle shorter than penultimate, longer than the 9-jointed flagellum. Maxillipeds, outer plates and palp with a close fringe of long hairs. Gnathopod 1, 5th joint short, triangular, masked behind by the triangular 4th; 6th twice as long as 5th, narrow, convex in front, with straight hind margin; finger (in figure, 1885) smooth, small, but overlapping the short, almost transverse palm. Gnathopod 2, 3^d joint subtriangular, articulating with both 4th and 5th, 5th with a broad free distal margin, fringed with setae which reach to end of the long narrow setose 6th; the narrowly rounded apex of the 6th joint is occupied by a very small, bent finger. Peraeopod 2 (in figure) slender. Peraeopod 3 (in figure) shorter, with piriform 2^d joint. Uropods 1 and 2 armed with a few acute spines. Uropod 3, rami narrow, with marginal spinules and apical setae. Telson scale-like, very short. L. 7 mm.

Port-Jackson [East-Australia].

10. Gen. **Leptocheirus** Zadd.

1844 *Leptocheirus* (Sp. un.: *L. pilosus*), Zaddach, Syn. Crust. Pruss., p. 7 | 1876 *L.*, A. Boeck, Skand. Arkt. Amphip., *v.* 2 p. 547 | 1888 *L.*, T. Stebbing in: Rep. Voy. Challenger, *v.* 29 p. 1707 | 1893 *L.*, A. Della Valle in: F. Fl. Neapel, *v.* 20 p. 426 | 1894 *L.*, G. O. Sars, Crust. Norway, *v.* 1 p. 554 | 1853 *Leptochirus* (non Germar 1824, Coleoptera!), J. D. Dana in: U. S. expl. Exp., *v.* 13 II p. 910 | 1853 *Ptilocheirus* (Sp. un.: *P. pinguis*), Stimpson in: Smithson. Contr., *v.* 6 nr. 5 p. 55 | 1871 *Boeckia* (Sp. un.: *B. typica*), A. W. Malm in: Öfv. Ak. Förh., *v.* 27 p. 543.

Body not strongly compressed. Pleon segment 3 the longest. Head without conspicuous rostrum. Side-plate 2 the largest, sometimes entirely concealing the 1st; side-plate 5 with front lobe much deeper than the hind one. Antenna 1 the longer, 3^d joint of peduncle not exceptionally long, accessory flagellum usually present. Upper lip subquadrate, faintly emarginate. Lower lip said to have the inner lobes sometimes partially coalesced. Mandible with 6—16 spines in spine-row, 1st joint of palp not much shorter, 3^d rather longer than 2^d. Maxilla 1, inner plate rather large, usually only with 1 seta at tip, outer with 11 spines (L. pinguis), but perhaps more commonly 10 (said to have 14 in Boeckia typica = L. hirsutimanus). Maxilla 2, inner plate the shorter, narrowed distally, fringed on inner margin. Maxillipeds, inner and outer plates well developed, but rather narrow, finger of palp short, obtuse. Gnathopod 1, 2^d joint fringed on front, 3^d—6th on hind margin; 3^d bulging behind; 6th with palm well or ill-defined but short, finger short. Gnathopod 2 simple, 2^d, 5th and 6th joints with long fringes in front, 5th longer than the tapering 6th; finger weak. Peraeopods 3—5, 2^d joint expanded; peraeopod 5 much the longest. Uropods 1 and 2, peduncle produced into a long spine-like apical process, rami robust, spinose. Uropod 3, rami short, longer than peduncle. Telson short.

8 species accepted, 1 obscure.

Synopsis of accepted species:

1 {	Gnathopod 1 longer and much stouter than gnathopod 2	1. **L. cornuaurei** . . . . p. 626
	Gnathopod 1 not longer and not much stouter than gnathopod 2 — **2**.	
2 {	Antenna 1 without accessory flagellum . . .	2. **L. aberrans** . . . . . p. 626
	Antenna 1 with accessory flagellum — **3**.	
3 {	Side-plates 1—4 with spiniferous serrations of hind margin	3. **L. pinguis** . . . . . p. 627
	Side-plates 1—4 without spiniferous serrations of hind margin — **4**.	
4 {	Pleon segment 4 not dentate	4. **L. hirsutimanus** . . p. 627
	Pleon segment 4 dentate — **5**.	
5 {	Pleon segment 4 with 2 dorsal teeth — **6**.	
	Pleon segment 4 with 3 dorsal teeth — **7**.	

1. **L. cornuaurei** Sowinski 1898 *L. c.*, Sowinski in: Mém. Soc. Kiew, *v.* 15 p. 470 t. 9 f. 9—22.

♀ unknown. — ♂. Side-plate 1 oblong, with rounded margin, 2^d large, distally much widened and outdrawn. Antenna 1, 2^d joint much longer and narrower than 1st, 3^d more than $^1/_3$ as long as 2^d; flagellum and accessory flagellum missing. Antenna 2, flagellum $^1/_3$ longer than ultimate joint of peduncle, of 12 joints, the last rudimentary. Mandible with spine-row of 6 spines. Maxilla 1 with a single seta on inner plate. Gnathopod 1 thicker and longer than gnathopod 2; 2^d joint comparatively short and strong, narrowed at base, (in figure without setae on front margin), 4th much narrower than 3^d, 5th stout, not twice as long as broad, a little shorter and narrower than the oblong 6th, which also is not twice as long as broad; palm rather oblique, not strongly defined, matching the stout finger; joints 3—5 densely setose on hind margin, but 6th joint not densely. Gnathopod 2 of much the same structure as in L. guttatus (p. 629); 4th joint subequal in length to the tapering 6th, which is only slightly shorter than the 5th; all with the usual armature; finger small. Peraeopods 1 and 2, 6th joint a little shorter and much narrower than 4th, finger half as long as 6th. Peraeopods 3—5, 2^d joint well expanded, in peraeopod 3 narrowed distally; 6th joint longer than 4th, finger short. Uropod 2, rami with few spines; in uropod 3 both rami ending with a group of setae. Telson broader than long, the apical margin concave, crenulate, with 2 pairs of setules. L. 5 mm.

Bosphorus (Golden Horn [Constantinople]).

2. **L. aberrans** (Ohlin) 1895 *Protomedeia a.*, Ohlin in: Acta Univ. Lund., *v.* 31 nr. 6 p. 53 f. 7—14.

Body slender, but back tolerably broad, rounded. Head, lateral corners produced, subacute. Side-plate 1 as deep as 2^d but only half as broad, 2^d larger than 3^d or 4th, 5th with front lobe little deeper than hind one. Pleon segment 3, postero-lateral corners produced into a small acute tooth. Eyes apparently absent. Antenna 1 longer than 2^d, longer than half the body; 1st joint a little shorter than 2^d, 3^d about $^1/_3$ length of 2^d; flagellum rather longer than peduncle, 11-jointed; accessory flagellum absent. Antenna 2 $^1/_4$—$^1/_3$ shorter than antenna 1; antepenultimate joint of peduncle broad, ultimate and penultimate joints equal and together as long as the 5-jointed flagellum. Gnathopod 1, 2^d joint shorter than 5th and 6th combined, fringed on front margin, 3^d bulging behind, with very long subapical setae, 4th not longer than 3^d, 4th—6th fringed on hind margin, 6th as long as 5th, a little wider, narrowly oblong; the short palm nearly transverse, a little overlapped by the finger. Gnathopod 2, 2^d joint fringed on front with very long setae, 5th considerably longer than 6th, 6th with palm very oblique, ill-defined except by a palmar spine; finger smooth, rather stout, as long as palm. Peraeopods 1 and 2, finger long, acute, as long as 6th joint. Peraeopod 5 the longest, nearly as long as body, 2^d, 4th and 5th joints equal, 6th nearly as long as 4th and 5th combined. Uropods 1—3, peduncle in all reaching a little beyond the telson. Uropod 3, rami short, not longer than distal breadth of peduncle, outer very little the longer, each with 4 apical setae. Telson short, broader than long, faintly emarginate. L. 5 mm.

Baffins Bay (lat. 73⁰ N.). Depth 22—32 m.

3. **L. pinguis** (Stimps.) 1853 *Ptilocheirus p.*, Stimpson in: Smithson. Contr., *v.*6 nr.5 p.56 | 1888 *Leptocheirus p.*, T. Stebbing in: Rep. Voy. Challenger. *v.*29 p.279 | 1893 *L. p.*, A. Della Valle in: F. Fl. Neapel, *v.*20 p.432 t.57 f.1—3 | 1894 *L. p.*, G. O. Sars, Crust. Norway, *v.*1 p.555 | 1862 *Protomedeia fimbriata* (*Gammarus fimbriatus* Stimpson in MS.) + *P. p.*, Bate, Cat. Amphip. Brit. Mus., p.169 t.31 f.1; p.170 t.31 f.2.

Body very broad anteriorly, narrowing behind, peraeon not much compressed, with back only slightly vaulted; peraeon segment 1 in ♂ as long as 2[d] and 3[d] combined, in ♀ only as long as 2[d]; pleon segment 4 produced on each side to an obtuse dorso-lateral tooth or angle, 5th and 6th with similar projections, more acute, accompanied by several spinules. Head broader than long, with minute rostrum; lateral lobes produced, not large, rounded. In ♀ side-plate 1 linguiform, strongly produced forward, obtusely pointed; 1st—4th with hind margin serrate, spiniferous; 2[d] overlapping only a small part of the 1st, deeper and much broader than 1st and deeper than broad; 3[d] and 4th as deep as 2[d] but less broad; 5th with front lobe narrow, distally narrowed, less deep than 4th; 6th and 7th very small. In ♂ (Stimpson) side-plate 1 large, subrhomboidal, 2[d] much the largest, projecting downward, furrowed along the middle. Pleon segment 3, postero-lateral corners a little obtusely quadrate. Eyes obliquely oval, or reniform, dark. Antenna 1 scarcely half as long as body; 1st and 2[d] joints rather long, equal, 3[d] about ¹/₈ as long as 2[d]; flagellum longer than peduncle, more than 23-jointed, accessory flagellum 8-jointed. Antenna 2 a little shorter, ultimate joint of peduncle shorter than penultimate, scarcely as long as the 14-jointed flagellum. Upper lip with pointed process at each upper angle. Mandible with 15 or 16 spines in the spine-row. Maxilla 1, outer plate with 11 spines. Maxillipeds, outer plates with inner margin densely fringed, the spines slender. Gnathopod 1, 6th joint as long as 5th, gently widening to the transverse, convex, setulose palm, over which the denticulate finger is closely fitted without quite reaching the angle. Gnathopod 2, besides the usual fringes the 4th joint has long setae on lower front margin; 5th and 6th are densely fringed on hind margin; 6th much more than half as long as 5th, only a little tapering but without palm, about twice as long as the slender setulose finger. Peraeopods 1 and 2, 4th joint longer than 5th or 6th. Peraeopods 3—5, 2[d] joint broadly oblong oval. Pleopods, peduncle short, coupling spines 2, with dentate margins, rami very long, inner with 30 joints, outer considerably shorter, with 28. Uropods 1 and 2, rami a little unequal, not especially robust, rather longer than peduncle, its process not being included. Uropod 3, process of peduncle short. Telson very short, broader than long, centre of apex convex; each angle raised, carrying an oblique surface row of spinules, while higher up on each side, the surface has a row of 5 spines. Colour, transverse bands of dark grey on a light ground, antennae and legs white, except expansion of 2[d] joint in peraeopods 3—5. L. reaching 16 mm.

North-Atlantic (New England and north to Labrador). Low-Water mark to at least ?7½ m.

4. **L. hirsutimanus** (Bate) 1862 *Protomedeia hirsutimana*, Bate (& Westwood), Brit. sess. Crust., *v.*1 p.298 f. | 1862 *P. hirsutimanus*, Bate, Cat. Amphip. Brit. Mus., p.168 t.30 f.6 | ?1866 *P. h.*, Cam. Heller in: Denk. Ak. Wien, *v.*26ɪɪ p.34 | ?1866

P. h., E. Grube in: Arch. Naturg., *v.* 321 p. 402 | 1887 *Ptilocheirus h.*, Chevreux in:
Bull. Soc. zool. France, *v.* 12 p. 309 | 1888 *Leptocheirus h.*, T. Stebbing in: Rep. Voy.
Challenger, *v.* 29 p. 561 | 1895 *L. h.*, A. O. Walker in: P. Liverp. biol. Soc., *v.* 9 p. 310 |
1895 *L. h.*, A. O. Walker in: Ann. nat. Hist., ser. 6 *v.* 15 p. 469 | 1871 *L. pilosus* (non
Zaddach 1844!), A. Boeck in: Forh. Selsk. Christian., 1870 p. 230 | 1894 *L. p.*, G. O.
Sars, Crust. Norway, *v.* 1 p. 555 t. 197 | 1871 *Boeckia typica*, A. W. Malm in: Öfv. Ak.
Förh., *v.* 27 p. 544 t. 5 f. 1, 1 a—g.

Body rather stout, somewhat compressed, but back broadly rounded;
pleon without teeth. Head, lateral lobes rounded, little prominent. Side-
plate 1 very small, acutely quadrate, covered by the very large, greatly
produced, in front broadly rounded 2^d side-plate; 3^d and 4th narrowly oblong,
deep; narrow front lobe of 5th not quite so deep, as 4th. Pleon segment 3,
postero-lateral corners rounded, slightly crenulate above. Eyes small, rounded,
dark. Antenna 1 not nearly half as long as body; 1st joint slightly longer
than 2^d or equal to it, 3^d half as long as 2^d; flagellum 16—20-jointed; accessory
flagellum 4—6 jointed. Antenna 2 rather shorter; ultimate and penultimate
joints of peduncle subequal; flagellum 9-jointed. Gnathopod 1, 5th joint
not twice as long as 6th, which is broader distally, with short palm and
small finger (Boeck), or: 6th (in ♀) about as long as 5th and scarcely
broader, palm somewhat oblique, finger (in figure) matching or scarcely
overlapping it (Sars). Gnathopod 2, 5th joint more than twice as long
as 6th, long oval, junction with 6th a little oblique, 6th tapering; finger
short, nearly straight, setuliferous. Peraeopods 1 and 2, 2^d and 4th joints
moderately wide, 6th long and narrow, finger short, slender, scarcely more
than $^1/_8$ as long as 6th joint. Peraeopod 3, 2^d joint narrowed distally, front
very convex, hind margin very sinuous; 5th joint longer than 4th or 6th;
finger very small. Peraeopods 4 and 5 successively longer; 2^d joint oblong
oval and as in peraeopod 3 fringed behind with plumose setae; finger short,
bidentate. Uropods 1 and 2 unusually strong, especially the 2^d, which have
short broad peduncle, rami very stout, each with double row of stout spines.
Uropod 3 very small, rami a little longer than peduncle, tapering, with
lateral spines and apical setae. Telson rather broader than long, with a
spine at each lateral angle of the apex. Colour dark grey, with lighter
flecks of various sizes. L. 8 mm.

North-Atlantic with adjoining seas (from Norway to France).

5. **L. dellavallei** Stebb. 1864 *Protomedeia fasciata* (non Krøyer 1842!),
A. Costa in: Annuario Mus. Napoli, *v.* 2 p. 155 t. 2 f. 8 | 1893 *Leptocheirus pilosus* (part.,
non Zaddach 1844!), A. Della Valle in: F. Fl. Neapel, *v.* 20 p. 427 t. 4 f. 10; t. 12 f. 1—14 |
1899 *L. dellavallei*, T. Stebbing in: Ann. nat. Hist., ser. 7 *v.* 3 p. 350.

Body rather slender, pleon segment 4 with 2 dorso-lateral teeth and
perhaps a median one, all accompanied by spinules. Side-plate 1 with lower
front angle produced into a long acute process, length and acuteness a little
variable. Side-plate 2 very large, broader than deep, rounded in front.
Eyes round. Antenna 1, 1st joint stout, 2^d a little longer, 3^d rather over
$^1/_3$ as long as 2^d; flagellum about 12-jointed; accessory flagellum 2-jointed,
as long as 1st joint of primary. Antenna 2 much shorter, ultimate and
penultimate joints of peduncle equal. Gnathopod 1, 5th joint longer than
6th, which is oblong, a little widened near the short truncate palm; finger
curved, reaching beyond the palm, with a decurrent spine near the apex.
Gnathopod 2 nearly as in L. hirsutimanus (p. 627), but 5th and 6th joints
rather narrower, 6th more than half as long as 5th. Peraeopods 1 and 2, 6th
joint long and slender, finger nearly as long or (Della Valle) even longer.

Peraeopod 3, 2^d joint broadly oval, both margins convex (in figure); 4th joint longer than 5th or 6th; finger small. Peraeopods 4 and 5, 2^d joint broadly oval; peraeopod 5 much the longest. Uropods 1—3 reaching nearly the same level, the rami long, unequal, the spines long and slender. Uropod 1, rami slender, longer than peduncle. Uropod 2, peduncle rather short, as in uropod 1 with long apical spine, rami longer than peduncle, the longer also much the stouter. Uropod 3, peduncle with tooth at inner angle, rami longer than peduncle. Telson broad, but short, lateral angles of apex very prominent, each with a spine, the apex triangular, produced to an obtuse point. Colour lemon-yellow, sprinkled with many red-brown spots. L. 6—7 mm.

Bay of Naples. Depth 10—20 m.

6. **L. pectinatus** (Norm.) 1869 *Protomedeia pectinata*, A. M. Norman in: Rep. Brit. Ass., Meet. 38 p. 283 | 1887 *Ptilocheirus pectinatus*, Chevreux in: Bull. Soc. zool. France, v. 12 p. 309 | 1888 *Leptocheirus p.*, T. Stebbing in: Rep. Voy. Challenger, v. 29 p. 1707 | 1894 *L. p.*, G. O. Sars, Crust. Norway, v. 1 p. 555 | 1893 *L. pilosus* (part.), A. Della Valle in: F. Fl. Neapel, v. 20 p. 427 | 1895 *L. p.*, A. O. Walker in: P. Liverp. biol. Soc., v. 9 p. 310.

Agreeing with L. dellavallei (p. 628) except in the following particulars. Side-plate 1 quadrate, lower front corner scarcely produced, slightly obtuse. Antenna 1, 3^d joint $^2/_3$ (sometimes less) as long as 2^d; flagellum about 10-jointed. Antenna 2 relatively longer, the ultimate and penultimate joints of peduncle being more elongate. Peraeopods 1 and 2 with 6th joint and finger not quite so long and slender as in L. dellavallei. Peraeopods 3—5 probably in agreement with that species. Uropods 1—3, spines seemingly less elongate. Uropod 3, tooth at inner angle of peduncle more produced. Telson, apical border between the prominent angles not triangular but gently convex. L. about 5 mm.

North-Atlantic (Shetland Islands).

7. **L. guttatus** (Grube) 1864 *Protomedeia guttata*, E. Grube in: Jahresber. Schles. Ges., v. 41 p. 63 | 1866 *P. g.*, E. Grube in: Arch. Naturg., v. 32 I p. 408 t. 10 f. 3 | 1888 *Leptocheirus guttatus*, T. Stebbing in: Rep. Voy. Challenger, v. 29 p. 366 | 1894 *L. g.*, G. O. Sars, Crust. Norway, v. 1 p. 555.

♀. Pleon segment 4 with 3 medio-dorsal small teeth, the 2 lateral teeth produced further back than the middle one. Side-plates 1—4 not so deep as in L. hirsutimanus (p. 627). Side-plate 1 (figure) quite clear of side-plate 2, which is subequal to 3^d or 4th; 5th produced a little below 4th. Eyes suborbicular. Antenna 1 almost $^1/_3$ as long as body; 1st and 2^d joints equal; flagellum 7-jointed, accessory flagellum not perceived. Antenna 2 not much shorter, flagellum 3-jointed. Gnathopod 2, 5th and 6th joints less elongate than in L. hirsutimanus; finger as slender. and acute as, though less long than, that of peraeopod 1. Peraeopods 1 and 2, 4th joint subquadrate. Peraeopod 3 short, 2^d joint oval. Peraeopod 5 much longer than preceding peraeopods, but not extremely elongate. Colour tawny, with round spots, few but large and conspicuous, of bright reddish brown in tranverse bands on the head and segments (1st of peraeon and last 4 of pleon excepted), including side-plates and 2^d joint of peraeopods 3—5. In other respects the species is said to resemble L. hirsutimanus. L. 5 mm.

Adriatic.

8. **L. tricristatus** (Chevreux) 1886 & 87 *Ptilocheirus t.*, Chevreux in: Bull. Soc. zool. France, v. 11 p. XL; v. 12 p. 310 t. 5 f. 3, 4 | 1888 *Leptocheirus t.*, T. Stebbing in: Rep. Voy. Challenger, v. 29 p. 1708 | 1894 *L. t.*, G. O. Sars, Crust. Norway, v. 1 p. 555.

Body more elongate and side-plates smaller than in L. hirsutimanus, pleon segment 4 with 3 medio-dorsal, sharp and strong teeth. Pleon segment 3, postero-lateral corners strongly rounded, margin slightly crenulate. Eyes small, oval, black. Antenna 1, 2^d joint longer than 1st, 3^d scarcely half as long as 2^d; accessory flagellum 3-jointed. Antenna 2, ultimate joint of peduncle shorter than penultimate. Gnathopod 1, 6th joint as long as 5th, widening to the finger, which is as long as the palm (in figure: 6th joint oval with scarcely defined palm). Gnathopod 2, 2^d joint almost as long as 3^d—6th combined, setae still longer; 4th, 5th and 6th in length equal one to the other (in figure: 5th much longer than 4th or 6th); finger slightly curved. Uropods 1 and 2, peduncle with strong apical spine. Uropod 2, inner ramus the longer, with 5 large teeth at the apex. Uropod 3 very short, rami laminar, outer with apical setae, inner shorter, with apical tooth. Telson prismatic, strongly concavé above. Colour simply yellow or with dorsal brown bands. L. 7 mm.

France (South-West of Brittany).

L. pilosus Zadd. 1844 L. p., Zaddach, Syn. Crust. Pruss., p. 8.

Probably identical with one or other of the species here described. L. 4 mm. Baltic.

---

35. Fam. Isaeidae

1853 Subfam. *Isaeinae*, J. D. Dana in: U. S. expl. Exp., v. 13 II p. 913.

Side-plate 5 the largest. Antenna 1, accessory flagellum well developed. Gnathopod 2 the larger, subchelate. Peraeopods 1 and 2 not glandular. Peraeopods 1—5 subchelate, 3^d—5th equal. Uropod 3 biramous. Telson simple.

Marine.

1 genus with 1 species.

1. Gen. Isaea M.-E.

1830 *Isaea* (Sp. un.: *I. montagui*), H. Milne Edwards in: Ann. Sci. nat., v. 20 p. 380 | 1893 *I.*, A. Della Valle in: F. Fl. Neapel, v. 20 p. 679 | 1862 *I., Iscoea*, Bate, Cat. Amphip. Brit. Mus., p. 122; t. 22 f. 1.

Body robust. Side-plate 5 as deep as the 4th. Antenna 1 the longer, with well developed accessory flagellum. Maxilla 1, outer plate with 11 spines (Della Valle, in figure). Gnathopods 1 and 2 subchelate, 2^d stronger. Peraeopods 1—5, 6th joint distally dilated. Uropod 3 not elongate, rami equal, narrow.

1 species.

1. **I. montagui** M.-E. 1830 *I. m.*, H. Milne Edwards in: Ann. Sci. nat., v. 20 p. 380 | 1840 *I. m.*, H. Milne Edwards, Hist. nat. Crust., v. 3 p. 26 t. 29 f. 11 | 1862 *I. m.*, *Iscoea montagua*, Bate, Cat. Amphip. Brit. Mus., p. 122; t. 22 f. 1 | 1887 *Isaea montagui*, Chevreux in: Bull. Soc. zool. France, v. 12 p. 301 | 1893 *I. m.*, A. Della Valle in: F. Fl. Neapel, v. 20 p. 679 t. 6 f. 7; t. 13 f. 30—42.

Head, lateral lobes not strongly produced. Side-plates 1—5 rather deep. Eyes crimson. Antenna 1, 1st joint a little shorter than 2^d, 3^d about $^2/_3$ as long as 2^d; flagellum subequal to peduncle, 16-jointed; accessory flagellum 6-jointed. Antenna 2, ultimate joint of peduncle rather shorter than penultimate and than the 8—12-jointed flagellum. Upper lip rounded. Mandible,

3^d joint of palp as long as 2^d. Maxilla 1, inner plate with pointed apex, and no conspicuous seta. Maxilla 2, inner plate fringed. Maxillipeds well armed, finger as long as 3^d joint. Gnathopod 1 small, 5th joint as long as the oval 6th; finger matching the oblique convex palm. Gnathopod 2, 5th joint cup-shaped, 6th robust, palmar margin rather oblique, divided into 4 teeth; finger robust. Peraeopods 1 and 2, 2^d joint narrow. Peraeopods 3—5, 2^d joint well expanded, especially in peraeopod 3. In all peraeopods the dilated 6th joint has a sort of oblique, serrate palm, set with strong spines, whereby these limbs acquire a subchelate character. Uropods 1—3, rami subequal, spinose. Uropod 3 the shortest, the slender pointed rami about as long as peduncle. Telson shorter than peduncle of uropod 3, its apex rather abruptly pointed. Colour reddish yellow, more or less in transverse bands. L. 5 mm.

North-Atlantic and Mediterranean (from Great Britain to the Adriatic). On Mamaia squinado (Herbst).

---

36. Fam. **Ampithoidae**

1899 *Amphithoidae*, T. Stebbing in: Ann. nat. Hist., ser. 7 *v.* 4 p. 211.

Head, lateral lobes not very prominent. Side-plates regular; 4th with hind margin not excavate; 5th with broad front lobe, as deep as 4th (Fig. 109 p. 642). Antenna 1 with 3^d joint short; accessory flagellum wanting or small. Lower lip (Fig. 108 p. 636) with front lobes deeply notched. Mandibular palp stout, slender, or wanting; molar usually well developed. Mouth-parts otherwise as in Aoridae (p. 585). Gnathopods 1 and 2 not simple, 2^d usually the larger, usually larger in $\male$ than in $\female$ and more or less differently shaped. Peraeopods 1 and 2 glandular. Peraeopod 3 reverted. Peraeopod 5 the longest. Uropods 1—3 biramous. Uropod 3 with short rami, the outer uncinate. Telson simple (Fig. 110 p. 643).

Marine.

6 genera, 22 accepted species and 19 doubtful.

Synopsis of genera:

1 { Mandible with palp — **2.**
    { Mandible without palp —- **5.**

2 { Antenna 1 (Fig. 109) Without accessory fla-
    { gellum — **3.**
    { Antenna 1 With accessory flagellum — **4.**

3 { Peraeopods 3—5, 6th joint not strongly Widened
    { at apex . . . . 1. Gen. **Ampithoe** . . . p. 631
    { Peraeopods 3—5 (Fig. 109), 6th joint strongly
    { Widened at apex . . . . 2. Gen. **Pleonexes** . . . p. 642

4 { Uropod 3, outer ramus With 2 books . . . . 3. Gen. **Grubia** . . . . p. 644
    { Uropod 3, outer ramus With 1 hook . . . . 4. Gen. **Amphithoides** . p. 645

5 { Mandible, molar Well developed . . . . 5. Gen. **Sunamphitoe** . p. 645
    { Mandible, molar eVanescent . . . . 6. Gen. **Biancolina** . . p. 646

1. Gen. **Ampithoe** Leach

1813/14 *Ampithöe* (Sp. un.: *A. rubricata*), Leach in: Edinb. Enc., *v.* 7 p. 403, 432 | 1829 *Ampithoe*, Latreille in: G. CuVier, Règne an., n. ed. *v.* 4 p. 121 | 1816 *Cymadusa* (Sp. un.: *C. filosa*), SaVigny, Mém. An. s. Vert., *v.* 1 p. 109 | 1816 *Amphithoë*, Latreille in: Nouv. Dict., ed. 2 *v.* 1 p. 470 | 1825 *Amphithoe*, A. G. Desmarest, Consid. gén. Crust.,

p. 268 | 1849 *Amphithöe*, J. D. Dana in: Amer. J. Sci., ser. 2 v. 8 p. 137 | 1862 *Amphithoë* (part.), Bate, Cat. Amphip. Brit. Mus., p. 233 | 1888 *A.*, T. Stebbing in: Rep. Voy. Challenger, v. 29 p. 1113 | 1893 *Amphithoe* (part.), A. Della Valle in: F. Fl. Neapel, v. 20 p. 454 | 1894 *Amphithoë*, G. O. Sars, Crust. Norway, v. 1 p. 578 | 1836 *Amphitoe*, Guérin-Méneville, Iconogr. Règne an., v. 3 Crust. p. 23 | 1840 *A.* (part.), H. Milne Edwards, Hist. nat. Crust., v. 3 p. 28 | 1845 *Amphitöe*, H. Goodsir in: Ann. nat. Hist., v. 15 p. 75 | 1852 *Amphitoë*, J. D. Dana in: P. Amer. Ac., v. 2 p. 213.

Head without rostrum. Side-plates 1—5 well developed, 5th as wide as 4th, with a very small hind lobe. Antenna 1 without accessory flagellum, usually longer than antenna 2, though with shorter peduncle. Mouth-parts prominent below the head. Upper lip distally rounded. Lower lip (Fig. 108 p. 636), inner lobes well developed; outer lobes bifid; mandibular processes prominent. Mandible normal, principal and secondary plate multidentate, spines in spine-row numerous, molar of moderate size; 3d joint of palp sometimes widened distally and crowded with setae, at others not widened and slightly armed. Maxilla 1, inner plate very small, usually with 1—3 setae, outer plate with 10 spines; 2d joint of palp with several apical spines. Maxilla 2, outer plate the larger, inner distally narrowed, inner margin fringed. Maxillipeds, outer plates large, well fringed with spine-teeth; palp not very elongate. Gnathopod 1 subchelate, usually the smaller. Gnathopod 2 usually subchelate, stronger in ♂ than in ♀ and generally of a different shape. Peraeopods 1 and 2, 2d joint expanded, sometimes greatly, for the cement-glands, the secretion from which issues through the apex of the finger to supply fibres for constructive purposes. Peraeopods 4 and 5 longer than the others. Uropod 3, outer ramus carrying 2 reverted spines. Telson short, usually or always having the angles of the apex minutely hooked.

Among algae; occasionally also on floating algae.

17 accepted species, 14 obscure.

Synopsis of accepted species:

1 {	Gnathopod 1 in ♂ larger than gnathopod 2 .	1. A. megaloprotopus . p. 633
	Gnathopod 1 in ♂ smaller than gnathopod 2—2.	
2 {	Gnathopod 2 in ♂ chelate	2. A. lacertosa . . . . p. 633
	Gnathopod 2 in ♂ subchelate — 3.	
3 {	Peraeopods 1 and 2, character of 2d joint not knoWn — 4.	
	Peraeopods 1 and 2, 2d joint broadly expanded — 8.	
	Peraeopods 1 and 2, 2d joint not broadly expanded — 11.	
4 {	Gnathopod 2 in ♂, palm oblique — 5.	
	Gnathopod 2 in ♂, palm transVerse — 6.	
5 {	Gnathopod 2 in ♀, palm with tubercle near finger-hinge	3. A. cinerea . . . . . p. 634
	Gnathopod 2 in ♀, palm with prominence at defining angle	4. A. longimana . . . p. 634
6 {	Gnathopod 2 in ♂, palm simply concaVe . .	5. A. quadrimana . . . p. 635
	Gnathopod 2 in ♂, palm margin diversified — 7.	
7 {	Gnathopod 2 in ♂, palm with broad, low, median tooth	6. A. valida . . . . . p. 635
	Gnathopod 2 in ♂, palm With narrow, bifid, median tooth	7. A. mitsukurii . . . p. 635

1. A. megaloprotopus (Stebb.) 1895 *Amphithoe m.*, T. Stebbing in: Ann. nat. Hist., ser. 6 *v.* 15 p. 397 t. 14, 15 B.

Side-plate 1 very large, oblong with rounded angles, produced forward so as completely to cover the mouth-parts; side-plate 2 not half as large. Eyes rather small, rounded, black (in spirit). Antenna 1, 1ˢᵗ joint strongly setose below, 2ᵈ much thinner, a little shorter, 3ᵈ small; flagellum longer than peduncle, with more than 23 joints. Antenna 2, penultimate joint of peduncle long. Mandible, 5 or 6 spines in the spine-row, 3ᵈ joint of palp nearly as long as 2ᵈ and 1ˢᵗ combined. Maxilla 1, inner plate with 3 small setae on inner margin. Maxillipeds, base very stout. Gnathopod 1, 2ᵈ joint narrow, channelled in front, with prominent distal lobe, 5ᵗʰ cup-shaped, short, 6ᵗʰ massive, as long as 2ᵈ, oblong, with rather convex front; palm nearly transverse, a little concave, sharply defined, a little overlapped by the finger. Gnathopod 2 like gnathopod 1, but 6ᵗʰ joint rather shorter and narrower, palm more sinuous, the short, curved finger closing down within its defining point. Peraeopods 1 and 2, 2ᵈ joint narrowly oval. Peraeopods 3 and 4, 2ᵈ joint broad. Peraeopod 5, 2ᵈ joint narrower in proportion to length than in peraeopods 3 and 4, but only constricted close to distal end. Branchial vesicles broadly flask-shaped. Pleopods with 2 coupling-spines, 6 cleft spines, rami 17—19-jointed. Uropods 1 and 2 spinose. Uropod 3, rami little more than half as long as peduncle, with the usual 2 upturned spines on outer ramus. Telson slightly broader than long, with lateral setules, apex bluntly triangular between 2 small upturned hooks, with a backward pointing seta adjoining each hook on inner side. L. about 11 mm.

Tropical Atlantic (Antigua). From sea-weed on rocks.

2. A. lacertosa (Bate) 1858 *Amphithoe l.*, Bate in: Ann. nat. Hist., ser. 3 *v.* 1 p. 362 | 1862 *A. l.*, Bate, Cat. Amphip. Brit. Mus., p. 236 t. 41 f. 5 | 1893 *A. l.*, A. Della Valle in: F. Fl. Neapel, *v.* 20 p. 461 t. 57 f. 37.

Side-plate 1 very large, extending forward to front of head, 2^d much smaller. Eyes ovate. Antenna 1 about half as long as body; flagellum a little longer than peduncle. Antenna 2 rather shorter, peduncle not longer than peduncle of antenna 1. Gnathopod 1, 5th joint slightly dilated, longer than 6th, 6th not broader than 5th, ovate, palm oblique, not defined; finger short. Gnathopod 2 much larger, 5th joint short, cup-shaped, 6th large, quadrate, hind margin produced so as to form a kind of thumb, though not equalling the finger in length. Peraeopods 1—5, all having the dilated base [2^d joint] tapering to the distal extremity. Uropod 3 reaching beyond the other 2 uropods; peduncle (in figure very much thicker than peduncles of uropods 1 and 2 and longer than their rami) with 2 short blunt spines on upper margin, and 5 or 6 on upper part of apex; outer ramus with 2 strong recurved hooks. Telson acute. L. 21 mm.

Arctic Seas.

3. **A. cinerea** (Hasw.) 1879 *Amphithoë c.* (♀) + *A. grandimanus* (non *A. grandimana* A. Boeck 1861!) (♂), Haswell in: P. Linn. Soc. N.S.Wales, *v.* 4 p. 269 t. 11 f. 4; p. 270 | 1893 *A. rubricata* (part.)?, A. Della Valle in: F. Fl. Neapel, *v.* 20 p. 459.

Eyes round, projecting, almost colourless. Antenna 1 more than half as long as body; 2^d joint of peduncle longer than 1st, 3^d very short; flagellum much longer than peduncle. Antenna 2 shorter; flagellum shorter than ultimate and penultimate joints of peduncle. Gnathopod 1, 6th joint long ovate, setose, palm oblique, undefined. Gnathopod 2 in ♂ much larger than gnathopod 1; 6th joint broad, irregularly ovoid; palm oblique, deeply excavated, its border waved, defined posteriorly by a strong tooth. Gnathopod 2 in ♀, 6th joint broader but rather shorter than in gnathopod 1, palm oblique, convex, devoid of teeth, but with a tubercle near finger-hinge. Peraeopods 1 and 2 subequal, stoutish. Peraeopod 3, 2^d joint subcircular. Peraeopods 4 and 5, 2^d joint oval. Uropod 3 not extending so far as uropod 2; outer ramus short, with 2 hooks, inner slightly longer, broader, with a few short setae. Telson subtriangular, blunt. Colour ashy grey. L. 16 mm.

Port Jackson [East-Australia].

4. **A. longimana** (S. I. Sm.) 1873 *Amphithoë l.,* (S. I. Smith in:) A. E. Verrill in: Rep. U. S. Fish Comm., *v.* 1 p. 563 | 1893 *A. rubricata* (part.), A. Della Valle in: F. Fl. Neapel, *v.* 20 p. 456.

Eyes round, black (in spirit). Antenna 1 in ♂ slender, as long as body, 2^d joint rather longer than 1st, 3^d about half as long as 2^d; flagellum about twice as long as peduncle. Antenna 2 stouter, rather shorter; ultimate joint of peduncle considerably longer than penultimate, subequal to flagellum. Antenna 2 in ♀ shorter and more slender. Gnathopods 1 and 2 in ♂ stout and long. Gnathopod 1 in ♂, 5th joint as long as 6th, both setose, 6th much more than twice as long as broad, oblong; palm very short, transverse, much overlapped by stout finger. Gnathopod 2 in ♂, 5th joint short, cup-shaped, 6th as long as in gnathopod 1, much broader; palm oblique, with deep sinus close to projecting defining angle. Gnathopod 1 in ♀, 5th and 6th joints shorter, proportionally broader; palm more oblique. Gnathopod 2 in ♀, 6th joint short, somewhat oval; palm with slight defining prominence. Peraeopods 4 and 5, 2^d joint with hind margin unarmed. L. 6—9 mm.

North-Atlantic (New Jersey, Long Island, Vineyard Sound). ,

5. **A. quadrimana** (HasW.) 1879 *Amphithoë quadrimanus*, HasWell in: P. Linn. Soc. N.S.Wales, v. 4 p. 337 t. 21 f. 7 | 1882 *A. q.*, HasWell, Cat. Austral. Crust., p. 266 | 1893 *A. rubricata* (part.), A. Della Valle in: F. Fl. Neapel, v. 20 p. 456.

Side-plates 1—7 all very shallow, 1st—4th deeper than 5th—7th, 5th not larger than 6th (in figure, but in contradiction to the generic definition by Haswell 1882). Eyes small, round. Antenna 1 more than half as long as body; flagellum thrice as long as peduncle. Antenna 2 half as long as antenna 1, antepenultimate joint of peduncle stout, distally produced below to a rounded protuberance; this and ultimate and penultimate joints of peduncle fringed below with plumose setae, flagellum as long as peduncle. Mouth-parts projecting downward prominently (in figure). Gnathopod 1, 6th joint subquadrate, a little widening distally; palm nearly transverse, not defined. Gnathopod 2, 4th and 5th joints narrow at apex of hind margin, 6th large, subquadrate, twice as long as broad; palm nearly transverse, concave. Uropod 3 reaching beyond uropod 2; outer ramus armed with 3 hooks, inner laminar, with slender spines. Telson armed with about 6 slender spines. L. 6 mm.

Port Jackson [East-Australia].

6. **A. valida** (S. I. Sm.) 1873 *Amphithoë v.*, (S. I. Smith in:) A. E. Verrill in: Rep. U. S. Fish Comm., v. 1 p. 563 | 1893 *A. rubricata* (part.?), A. Della Valle in: F. Fl. Neapel, v. 20 p. 459.

Eyes round, black (in spirit). Antenna 1 in ♂, 2d joint little longer than 1st, 3d short, slender. Antenna 2 subequal to antenna 1; ultimate and penultimate joints of peduncle subequal. Gnathopod 1 in ♂ short, compressed; 5th joint as broad as 6th, 6th broad, oval, palm and hind margin together nearly a semicircle; finger fitting palm. Gnathopod 2 in ♂ very large; 5th joint small, 6th oblong, distally widened, very large and thickened, outer surface convex, inner flattened; palm transverse, with a broad low median tooth, and rounded prominence at the defining angle, within which the stout curved finger closes. Gnathopod 1 in ♀ slightly more elongated. Gnathopod 2 in ♀ smaller than in ♂; palm slightly oblique. Colour bright green. L. 10—13 mm.

North-Atlantic (New Jersey and Long Island Sound).

7. **A. mitsukurii** (Della Valle) 1893 *Amphithoe m.*, A. Della Valle in: F. Fl. Neapel, v. 20 p. 460 t. 57 f. 30—32.

Gnathopod 2 in ♂, 6th joint very large and long, widened distally; palm transverse, defined by a somewhat produced blunt tooth, and having a little tubercle near the finger-hinge, between which and the defining tooth is a bilobed tubercle; finger curved, its apex touching defining tooth of palm. Gnathopod 1 (in figure) with produced side-plates; 5th joint very broad, longer than the oval 6th. Peraeopods 3—5, 6th joint little dilated distally. Telson broader than long, narrowing distally, with almost straight apical margin between the uncinate angles. L. 17 mm.

Tokio [Japan].

8. **A. flindersi** (Stebb.) 1888 *Amphithoë f.*, T. Stebbing in: Rep. Voy. Challenger, v. 29 p. 1120 t. 118 | 1893 *A. rubricata* (part.), A. Della Valle in: F. Fl. Neapel, v. 20 p. 456.

(Juv.?) Side-plate 1 larger than the 2d, below obtusely produced forward. Pleon segment 3, postero-lateral corners obtusely quadrate, with shallow indent just above the angle. Eyes small, oval in horizontal direction. Antennae 1 and 2 defective. Lower lip (Fig. 108 p. 636), outer branch of bifid lobes advanced in front of the inner. Mandible, 3^d joint of palp as long as 2^d, not expanded, with 5 setae (or spines) on apex. Maxilla 1, inner plate with 3 setae on inner margin; 2^d joint of palp with 5 spines on the dentate apex. Gnathopod 1, 5th joint rather shorter than 6th, distally lobed, 6th oblong, a little widened distally, palm oblique, defined by a palmar spine, the angle rounded off; finger with fine pectination followed by 5 successively larger teeth. Gnathopod 2, 5th joint cup-shaped, considerably shorter than 6th, which is stouter but not longer than in gnathopod 1; palm minutely pectinate; finger as in gnathopod 1. Peraeopods 1 and 2, 2^d joint broad proximally, narrower distally, 4th joint widening distally,

Fig. 108.
A. flindersi.
Lower lip.

but without rounded lobe. Pleopods with 2 cleft spines, with 8 or 9 joints in the rami. Uropod 1, peduncle little longer than inner ramus, outer the shorter. Uropod 3 reaching beyond the others; peduncle longer than the rami; outer ramus with 2 short hooked spines, inner oval, slightly longer than the outer, edged with 4 or 5 spines or spinules. Telson short, almost circular, with an upturned corner on either side of the broadly rounded apex. L. about 4 mm.

Flinders Passage (lat. 10^0 S., long. 142^0 E.). Depth 15 m.

9. **A. humeralis** (Stimps.) 1864 *Amphithoe h.*, Stimpson in: P. Ac. Philad., p. 156 | 1888 *A. h.*, T. Stebbing in: Rep. Voy. Challenger, v. 29 p. 351 | 1898 *A. h.*, Calman in: Ann. N. York. Ac., v. 11 p. 271 t. 33.

Body robust, rather compressed. Head, lateral lobes rounded. Side-plate 5 large, subquadrate. Eyes small, rounded, dark. Antenna 1 nearly as long as body; flagellum more than twice as long as peduncle (Stimpson). Antenna 1 more than half as long as body; 1st and 2^d joints subequal, 3^d very small; flagellum $2^1/_2$ times as long as peduncle (Calman). Antenna 2 half as long as body; flagellum not longer than antepenultimate joint (penultimate?) of peduncle (Stimpson). Antenna 2 stout; ultimate joint of peduncle a little shorter than penultimate; flagellum rather more than $1^1/_2$ length of peduncle (Calman). Mouth-parts (Calman) nearly as in A. femorata, but mandibular palp has 3^d joint longer than 2^d (not so in figure), not expanded. Lower lip with bifid front lobes. Gnathopods 1 and 2 similar in the two sexes, rather slender, and densely setose. Gnathopod 1, 5th joint longer than 6th, 6th oblong, about $2^1/_2$ times as long as broad, hind margin with shallow distal concavity near angle of very short transverse palm; finger serrate, overlapping palm. Gnathopod 2, 5th joint slightly longer than 6th, lobed behind, 6th hardly more than twice as long as broad, palm short, transverse, overlapped by finger. Peraeopods 1 and 2, 2^d joint very large, broadly oval, 4th widened distally and lobed. Peraeopod 3 very short, 2^d joint broad. Peraeopod 5 much longer than peraeopod 4. Uropod 3 not reaching beyond uropod 2; peduncle thrice as long as rami; outer ramus with 2 hooks, inner lamellate, truncate, bearing setae. Telson small, obtusely triangular, with a few setae on each side. L. 26—30 mm.

Puget Sound. About low-water mark.

10. **A. femorata** (Krøyer) ?1840 *Amphitoe gaudichaudii*, H. Milne Edwards, Hist. nat. Crust., v. 3 p. 31 | 1845 *Amphithoe femorata*, Krøyer in: Naturh. Tidsskr., ser. 2 v. 1 p. 335 t. 3 f. 4 a—i | 1893 *A. rubricata* (part.) + *Grubia crassicornis* (part.)?, A. Della Valle in: F. Fl. Neapel, v. 20 p. 456; p. 466.

In many respects resembling Sunamphitoe pelagica (p. 645). Body robust, dorsally rounded, smooth. Side-plates (or some of them) with setae on lower margin. Eyes small, subrotund, whitish. Antenna 1 more than half as long as body; 1st joint thick, 2^d very little shorter, 3^d small; flagellum 30—40-jointed. Antenna 2 shorter, stout, almost pediform; ultimate and penultimate joints of peduncle equal; flagellum scarcely half as long as peduncle, tapering, 12- or 13-jointed. Lips not described. Mandibular palp long, thin; 2^d joint the longest, 3^d with apical serrulate setae. Maxilla 1, inner plate with 1 seta, outer with 10 spines; palp slender, with 5 or 6 serrate spines. Maxilla 2 and maxillipeds normal. Gnathopod 1, 5th joint shorter than 6th, lobed and setose on hind margin, 6th narrowly oblong; palm short, transverse, overlapped by finger. Gnathopod 2 in ♂ considerably larger; 6th joint pretty strongly tapering. Gnathopod 2 in ♀ like gnathopod 1, but with shorter 5th and broader 6th joint. Peraeopods 1 and 2, 2^d joint broad, oval, laminar, 4th widened distally. Peraeopod 3 stout, 2^d joint broader than long, the following joints stout, not widened, 6th spinose on both margins; finger short, strong, much curved. Peraeopod 4 longer and more slender, 2^d joint long oval, finger fitted for grasping. Peraeopod 5 slightly longer than peraeopod 4. Branchial vesicles oval. Marsupial plates narrow, long, with long setae. Pleopods, rami long, 30-jointed. Uropods 1 and 2, outer ramus shorter than inner, broader, and more spinose, inner of uropod 2 at least as long as peduncle, the rest shorter. Uropod 3 shorter but stouter than uropod 2; peduncle twice as long as rami; outer ramus with the usual 2 hooks, and several rows of microscopic spinules, inner rather shorter but broader, nearly circular, armed with some setae and 2 little stout spines. Telson small, triangular, with 2 setae on the margin. Colour in life a very dark olive-green. L. reaching 19 mm.

South-Pacific (Valparaiso near the shore); Atlantic (Brazil)?

11. **A. brevipes** (Dana) 1852 *Amphithoe b.*, J. D. Dana in: P. Amer. Ac., v.2 p.216 | 1853 & 55 *A. b.* + ?*A. peregrina* (juv.), J. D. Dana in: U. S. expl. Exp., v.13ıı p.936, 941; t. 64 f. 5 a—n; p. 940 t. 64 f. 4 a, 4 b. | 1862 *A. falklandi* (juv.?) + *A. p.* + *A. b.*, Bate, Cat. Amphip. Brit. Mus., p. 237 t. 41 f. 6; p. 247 t. 43 f. 1; p. 248 t. 43 f. 2, 2 i, 2 u, 1 u | 1893 *A. rubricata* (part.)?, A. Della Valle in: F. Fl. Neapel, v. 20 p. 456, 459.

Apparently very near to A. femorata. Head, lateral lobes very little produced. Side-plate 1 scarcely produced forward, 5th the largest. Eyes round. Antenna 1 about half as long as body; 1st joint longest; flagellum more than twice as long as peduncle. Antenna 2 much shorter, ultimate and penultimate joints of peduncle equal; flagellum 15-jointed. Mandibular palp rather slender; 3^d joint with 5 apical setae. Gnathopod 1, 5th joint shorter than 6th, 6th oblong; palm short, transverse; finger longer than palm. Gnathopod 2 in ♂, 5th joint small, cup-shaped, 6th large, sub-ovate, narrowest at finger-hinge, near to which there is a minute acute tooth; the palm in general being undistinguished from the hind margin, finger long. Gnathopod 2 in ♀ scarcely distinguishable from gnathopod 1. Peraeopods 1 and 2, 2^d joint dilated, 4th joint distally lobed. Peraeopods 4 and 5 considerably longer than peraeopod 3. Uropod 3 normal, reaching beyond uropod 2. Telson not described. L. 18 mm.

South-Atlantic (Tierra del Fuego, depth 9 m; Falkland Islands); South-Pacific (Valparaiso)?

12. **A. brasiliensis** (Dana) 1853 & 55 *Amphithoe b.*, J. D. Dana in: U. S. expl. Exp., v. 13ıı p. 943; t. 64 f. 6 a—n | 1893 *A. rubricata* (part.), A. Della Valle in: F. Fl. Neapel, v. 20 p. 456.

Body compressed. Side-plates large. Antenna 1 much longer than
half body; flagellum very long, slender. Antenna 2 a little shorter, hirsute;
peduncle hardly shorter than flagellum. Gnathopod 1, 5th joint (in figure)
shorter and much narrower than 6th, 6th with front margin slightly convex, hind
margin and palm (described as oblique-transverse) together making (in figure) a
great bulge without any defining mark; finger half as long as 6th joint. Gnathopod 2 not much larger, more hirsute, 5th joint distally cup-shaped, 6th oblique-
transverse at apex; palm hardly excavate, long hirsute, acute at lower limit
(in figure: widening to a slightly oblique, well defined palm, which is very
different in appearance from that of gnathopod 1). The mouth-parts and
uropod 3 as figured are fairly appropriate to Ampithoe. Maxilla 1 shows
no inner plate, but a tuft of 5 setae at the place proper to the inner plate.
L. 16 mm.

Tropical Atlantic (Rio Janeiro).

13. **A. kergueleni** (Stebb.) 1888 *Amphithoë k.*, T. Stebbing in: Rep. Voy.
Challenger, *v.* 29 p. 1116 t. 117 | 1893 *A. rubricata* (part.)?, A. Della Valle in: F. Fl.
Neapel, *v.* 20 p. 463.

♀. Side-plate 1 much widened below. Pleon segment 3, postero-
lateral corners obtusely quadrate. General structure in close agreement with
A. rubricata. Eyes rounded oval. Antenna 1, 2^d joint rather longer than
1st, 3^d more than $^1/_3$ as long as 2^d; flagellum much longer than peduncle,
33-jointed. Antenna 2 shorter, ultimate joint of peduncle rather shorter
than penultimate; flagellum 20-jointed. Lower lip, front lobes bifid but less
sharply than in the European species. Maxilla 1, inner plate with many
plumose setae on inner margin; 2^d joint of palp with 9 spine-teeth round
the apex. Gnathopod 1, 5th joint elongate, nearly as long as 6th, 6th oblong,
widening slightly towards the rather oblique, finely pectinate palm, which
has a palmar spine at the rounded defining angle; finger overlapping palm,
its inner margin cut into 14 teeth. Gnathopod 2, 5th joint much shorter
than 6th, distally cup-shaped, 6th rather broadly oblong; palm oblique, concave,
well defined; finger when touching the defining angle not closing the space
between its dentate margin and the palm. Other appendages nearly as in
A. rubricata. Telson broader than long, narrowing to apex, which forms an
almost straight border between the acute corners. L. about 6 mm. — ♂
unknown.

Southern Indian Ocean (Kerguelen Island).

14. **A. japonica** (Stebb.) 1888 *Amphithoë j.*, T. Stebbing in: Rep. Voy. Challenger,
v. 29 p. 1124 t. 138 A | 1893 *A. rubricata* (part.), A. Della Valle in: F. Fl. Neapel, *v.* 20 p. 456.

♀. Eyes small, irregularly round. Antenna 2, 2^d joint longer than 1st,
3^d about $^1/_8$ as long as 2^d, carrying a little setuliferous (seemingly jointed)
tubercle; flagellum much longer than peduncle, attaining to 46 joints. Antenna 2,
ultimate joint of peduncle rather shorter than penultimate; flagellum longer
than both combined, 25-jointed. Mouth-parts as in A. rubricata. Gnathopod 1, 5th joint twice as long as broad, rather longer than 6th, 6th oblong
oval, palm oblique, defined by a palmar spine at the rounded defining angle;
finger serrate, a little overlapping the palm. Gnathopod 2, 5th joint cup-
shaped, shorter but distally rather wider than the 6th, 6th oblong, with a
convex palm, very slightly oblique, nearly matched by the finger, the apex
of which scarcely overlaps it. Peraeopods 1 and 2, 2^d joint long, not greatly
widened. Peraeopod 3, 2^d joint as broad as long, 6th spinose, its apex on

the inner side forming 2 small laminar projections. Peraeopod 5, 2^d joint piriform. Pleopod 1 having on peduncle about a dozen coupling spines, on 1st joint of inner ramus 9 cleft spines; outer ramus with 22 joints, inner with 23. Uropods 1 and 2 spinose, peduncle longer than rami. Uropod 3, peduncle much longer than rami, spinose; inner ramus with several spines on margins and some small stout surface spines. Telson scarcely longer than broad, with denticle at each angle of the convex distal margin. L. 14 mm. — ♂ unknown.

Bay of Kobé [Japan]. Depth 15 m.

15. **A. vaillantii** (H. Luc.) 1846 *Amphithoe v.*, H. Lucas in: Expl. Algérie, An. artic. *v.* 1 p. 54 Crust. t. 5 f. 3 | 1857 *A. penicillata*, A. Costa in: Mem. Acc. Napoli, *v.* 1 p. 207 t. 2 f. 9 | 1866 *A. p.*, Cam. Heller in: Denk. Ak. Wien, *v.* 26 II p. 43 t. 3 f. 29—34 | 1876 *A. p.*, Catta in: Ann. Sci. nat., ser. 6 *v.* 3 nr. 1 p. 27 t. 2 f. 2 i | 1881 *A. p.*, Nebeski in: Arb. Inst. Wien, *v.* 3 p. 149 t. 13 f. 42 d | 1862 *A. desmarestii*, Bate. Cat. Amphip. Brit. Mus., p. 238 t. 41 f. 8 | ?1868 *A. vaillantii var. pontica*, Czerniavski in: Syezda Russ. Est., Syezda 1 Zool. p. 102 t. 7 f. 19—27 | ?1880 *A. erythraea*, Kossmann, Reise Roth. Meer., *v.* 21 Malacost. p. 134 t. 14 f. 12, 13 | 1893 *A. rubricata* (part.), A. Della Valle in: F. Fl. Neapel, *v.* 20 p. 456.

Antennae 1 and 2 equal, setose (Lucas). Antenna 1 in ♂ longer than antenna 2; 1st joint scarcely longer than 2^d, 3^d a third as long as 1st; flagellum twice as long as peduncle, 20—30-jointed. Antenna 2, peduncle long, stout; flagellum short, 18-jointed. Eyes small, rounded, deep brown. Gnathopod 1 elongate, 2^d joint distally lobed, 5th joint distally quadrate, 6th longer, setose, oblong; palm straight (Lucas), very short, slightly excavate (Czerniavski), finger long. Gnathopod 2 very long, 2^d joint strongly lobed distally; 6th joint with palm deeply emarginate, defined by a conspicuous tooth, the front much longer than the hind margin and (Lucas' figure) bluntly produced beyond base of the comparatively short, strongly curved, serrate finger (Czerniavski figures a much shorter palm of gnathopod 2 in ♂, and adds, that in ♀ gnathopod 1 has the 5th joint short, palm of 6th not convex, and gnathopod 2 has a broader 5th joint, the 6th not elongate, nor apically produced). Peraeopods 1—5 and uropods 1—3 in agreement with A. rubricata. Telson obtusely triangular. Colour yellow, punctate with green or dark spots. L. 12—17 (Lucas), 8—10 mm (Czerniavski).

Mediterranean (Algeria, under cast-up algae); Black Sea?, Red Sea?

16. **A. rubricata** (Mont.) 1808 *Cancer (Gammarus) rubricatus*, Montagu in Tr. Linn. Soc. London, *v.* 9 p. 99 t. 5 f. 1 | 1812 *Astacus r.*, Pennant, Brit. Zool., ed. 5 *v.* 4 p. 33 | 1813/14 *Amphithöe rubricata*, Leach in: Edinb. Enc., *v.* 7 p. 403, 432 | 1874 *Amphithöe r.*, T. Stebbing in: Ann. nat. Hist., ser. 4 *v.* 14 p. 113 t. 11 f. 2, 2a | 1893 *A. r.* (part.), A. Della Valle in: F. Fl. Neapel, *v.* 20 p. 456 t. 2 f. 2; t. 13 f. 1—17; t. 57 f. 25, 26 | 1894 *A. r.*, T. Stebbing in: Bijdr. Dierk., *v.* 17 p. 44 | 1894 *A. r.*, G. O. Sars, Crust. Norway, *v.* 1 p. 579 t. 206 | 1827 & 28 *Gammarus punctatus* (non *Amphithoe punctata* Say 1818!), G. Johnston in: Zool. J., *v.* 3 p. 177, 490 | ?1837 *Amphithoë picta*, H. Rathke in: Mém. prés. Ac. St.-Pétersb., *v.* 3 p. 379 t. 5 f. 15—19 | 1840 „*Amphitoé rouge*", H. Milne Edwards, Hist. nat. Crust., *v.* 3 p. 33 | 1843 *Amphithoe podoceroides*, H. Rathke in: N. Acta Ac. Leop., *v.* 201 p. 79 t. 4 f. 4 | 1876 *A. p.*, A. Boeck. Skand. Arkt. Amphip., *v.* 2 p. 588 t. 26 f. 5; t. 27 f. 3 | 1846 *A. albomaculata*, Krøyer in: Naturh. Tidsskr., ser. 2 *v.* 2 p. 67 | 1846 *A. a.*, Krøyer in: Voy. Nord, Crust. t. 11B f. 1a—u | 1857 *Amphitoë littorina*, Bate in: Ann. nat. Hist., ser. 2 *v.* 19 p. 147 | 1862 *Amphithoë rubricata + A. l. + Sunamphithoë podoceroides*, Bate, Cat. Amphip. Brit. Mus., p. 233 t. 41 f. 1; p. 234 t. 41 f. 2; p. 251 t. 43 f. 7.

Body long, compressed, but not always slender. Head, lateral lobes blunt. Side-plate 1 bluntly produced somewhat forward, 2^d—5th rotundo-quadrate below. Pleon segment 3, postero-lateral corners obtusely quadrate. Eyes small, rounded oval, dark red. Antenna 1, 1st and 2^d joints subequal, 3^d less than half as long as 2^d; flagellum slender in ♀, sometimes twice as long as peduncle, but in ♂ sometimes not longer than peduncle, 23 — 35-jointed. Antenna 2 stout, in ♀ decidedly shorter than antenna 1, but less so in ♂; ultimate joint of peduncle as long as penultimate; flagellum about half as long as peduncle in ♀, less than that in ♂, 9—15-jointed. Lower lip, outer part of bifid lobe distally rounded. Mandible, 3^d joint of palp longer than 2^d, widening distally and densely beset with setae on apex and distal part of inner margin. Gnathopod 1, 2^d joint lobed at apex of channelled front, 5th a little shorter than 6th, 6th oblong oval, palm oblique, finger rather longer than palm, serrulate. Gnathopod 2 in ♂, 5th joint much shorter than the large 6th, which has the palm more or less excavate, defined by a small but distinct tooth, within which the curved finger impinges. Gnathopod 2 in ♀ like gnathopod 1, but with wider 5th and 6th joints. Peraeopods 1 and 2, 2^d joint narrowly oval, 4th distally widened but not apically lobed. Peraeopod 3, 2^d joint broader than long. Peraeopods 4 and 5, 2^d joint oblong oval, not much expanded. Peraeopods 3—5, 6th joint not narrowing distally, with tuft of setae at back of apex; finger up-curved. Uropods 1 and 2 spinose, in uropod 1 peduncle longer than rami. Uropod 3, peduncle twice as long as rami, with transverse row of spines on apical margin; outer ramus with 2 strong hooks, inner rather narrower with 2 lateral spines, 1 apical, and several setae. Telson as broad as long, apex rounded, between 2 little tubercles, within which are 3 setules on either side. Colour rather variable, from red to green, with dark stellate markings, often with white spots along middle of back. It forms a dwelling by knitting together various fragments, the cement-fibres apparently supplied from the glandcells of peraeopods 1 and 2. L. reaching 20 mm.

North-Atlantic with adjoining seas (Europe). Among algae between tide-marks, and at small depths.

17. **A. inda** (M.-E.) 1830 *Amphithoe i.*, H. Milne Edwards in: Ann. Sci. nat., *v.* 20 p. 376 | 1840 *Amphitoe indica*, H. Milne Edwards, Hist. nat. Crust., *v.* 3 p. 31 | 1888 *Amphithoe i.*, G. M. Giles in: J. Asiat. Soc. Bengal, *v.* 57 p. 240 t. 10 f. 1—7 | ?1852 *A. rubella*, J. D. Dana in: P. Amer. Ac., *v.* 2 p. 215 | 1853 & 55 *A. r.*, J. D. Dana in: U. S. expl. Exp., *v.* 13 II p. 936; t. 64 f. 1 a—d | 1862 *A. indica* + *A. r.*, Bate, Cat. Amphip. Brit. Mus., p. 240 t. 42 f. 3; p. 246 t. 42 f. 8 | 1893 *A. rubricata* (part.)?, *A. inda*, A. Della Valle in: F. Fl. Neapel, *v.* 20 p. 459, 463.

Body rather stout. Side-plates 1—5 broad. Pleon segment 3, postero-lateral corners subquadrate. Antennae 1 and 2 equal, or antenna 1 rather the longer. Antenna 1 more than $^1/_3$ as long as body; 2^d joint nearly as long as 1st, 3^d very small; flagellum with 13 or 14 short joints. Antenna 2, ultimate joint of peduncle a little longer than penultimate, flagellum with 9—11 short joints. Mandibular palp small. Gnathopod 1 small, 5th joint shorter than 6th, 6th oblong, narrow, narrower at apex; finger longer than palm. Gnathopod 2, 6th joint stout, broad, subrectangular; palm transverse, little excavate and unevenly so, defining angle prominent and acute but not produced; finger moderately large and somewhat serrate. Peraeopods 1 and 2, 2^d joint not much widened. Peraeopod 3, 2^d joint broad. Peraeopods 4 and 5 much longer, with 2^d joint somewhat less broad. Peraeopods 3—5, distal end of 6th joint with 2 blunt spines, including between them a rounded depression, and suited to subserve the guiding of a thread (Giles). Uropods 1—3 normal, reaching back about the same distance; inner ramus of uropod 3 with 1 spine and a few setules. Telson small, laminar, somewhat upturned, and of a roundedly conical outline (Giles), triangular (Milne Edwards). Colour purple, with patches of golden-yellow. L. 5—6 mm.

Indian Ocean; Middle of Bay of Bengal, on drift; ?Sooloo Sea, depth 12 m.

A. australiensis (Bate) 1862 *Amphithoë a.*, Bate, Cat. Amphip. Brit. Mus., p. 237 t. 41 f. 7 | 1893 *A. rubricata* (part.), A. Della Valle in: F. Fl. Neapel, *v.* 20 p. 456. L. 12 mm.

Southern Indian Ocean (South-Australia).

A. brusinae (Heller) 1866 *Amphithoe b.*, Cam. Heller in: Denk. Ak. Wien, *v.* 26 II p. 44 t. 4 f. 2, 3 | 1893 *A. rubricata* (part.)?, A. Della Valle in: F. Fl. Neapel, *v.* 20 p. 459. L. 5 mm.

Adriatic (Lissa).

A. chilensis (Nic.) 1849 *Amphitoe c.*, H. Nicolet in: Gay, Hist. Chile, *v.* 3 p. 235 Crust. t. 2 f. 5 a—d | 1893 *Amphithoë c.*, A. Della Valle in: F. Fl. Neapel, *v.* 20 p. 463. L. 16 mm.

Pacific (Chili).

A. filicornis (Dana) 1853 & 55 *Amphithoe f.*, J. D. Dana in: U. S. expl. Exp., *v.* 13 II p. 944; t. 65 f. 1 a—g | 1893 *A. rubricata* (part.), A. Della Valle in: F. Fl. Neapel, *v.* 20 p. 456.

Tropical Atlantic (Rio Janeiro).

A. filigera (Stimps.) 1855 *Amphithoe f.*, Stimpson in: P. Ac. Philad., *v.* 7 p. 382 | 1893 *Grubia crassicornis* (part.)?, A. Della Valle in: F. Fl. Neapel, *v.* 20 p. 467. L. 12 mm.

North-Pacific (Loo Choo).

A. filosa (Sav.) 1816 *Cymadusa f.*, Savigny, Mém. An. s. Vert., *v.* 1 p. 51, 109; t. 4 f. 1 a, b, e, i, o, u | 1826 *Amphithoë f.*, Audouin in: Descr. Égypte, *v.* 1 IV p. 93 Crust. t. 11 f. 4, ?5 | 1893 *Grubia crassicornis* (part.)?, A. Della Valle in: F. Fl. Neapel, *v.* 20 p. 466.

Only figured; no description.

Mediterranean or Red Sea? (Egypt).

A. maculata (Stimps.) 1853 *Amphithoe m.*, Stimpson in: Smithson. Contr., *v.* 6 nr. 5 p. 53 | 1893 *A. rubricata* (part.)? A. Della Valle in: F. Fl. Neapel, *v.* 20 p. 459. L. 16 mm.

Fundy Bay (Grand Manan). On rocky bottoms in laminarian zone and at low water.

A. orientalis (Dana) 1853 & 55 *Amphithoe o.*, J. D. Dana in: U. S. expl. Exp., *v.* 13 II p. 937; t. 64 f. 2 a—f (3 f on plate). L. 5—6 mm.

Bay of Manila [Philippine Islands]. From floating kelp.

A. pausilipae (M.-E.) 1830 *Amphithoe p.*, H. Milne Edwards in: Ann. Sci. nat., *v.* 20 p. 376 | 1840 *Amphitoe pausilipii*, H. Milne Edwards, Hist. nat. Crust., *v.* 3 p. 30 | 1851 *Amphithoe gracilis*, (A. Costa in:) F. W. Hope, Cat. Crost. Ital., p. 45 | 1857 *A. g.* + *A. pausylipi*, *A. pausilippii*, A. Costa in: Mem. Acc. Napoli, *v.* 1 p. 208 t. 3 f. 4; p. 206 | 1893 *Grubia crassicornis* (part.)?, A. Della Valle in: F. Fl. Neapel, *v.* 20 p. 466, 467.

Bay of Naples.

A. punctata Say 1818 *A. p.*, Say in: J. Ac. Philad., *v.* 1 ɪɪ p. 383 | 1840 *Amphitoe p.*, H. Milne Edwards, Hist. nat. Crust., *v.* 3 p. 35 | 1893 *Amphithoe? p.*, A. Della Valle in: F. Fl. Neapel, *v.* 20 p. 464.

Egg Harbour [United States of America.]

A. ramondi (Aud.) 1826 *Amphithoë r.*, Audouin in: Descr. Égypte, *v.* 1 ɪv p. 93 Crust. t. 11 f. 6 | 1893 *Grubia crassicornis* (part.)?, A. Della Valle in: F. Fl. Neapel, *v.* 20 p. 466.

Mediterranean or Red Sea? (Egypt).

A. stimpsoni (Boeck) 1872 *Amphithoe s.*, A. Boeck in: Forh. Selsk. Christian., 1871 p. 43, 49 t. 1 f. 5 | 1893 *A. rubricata* (part.)?, A. Della Valle in: F. Fl. Neapel, *v.* 20 p. 459.

Resembling A. japonica (p. 638), but antenna 1, 2ᵈ joint shorter than 1ˢᵗ, and gnathopod 1 with 5ᵗʰ joint shorter than 6ᵗʰ. L. 13 mm.

San Francisco.

A. tongensis (Dana) 1852 *Amphitoë t.*, J. D. Dana in: P. Amer. Ac., *v.* 2 p. 216 | 1853 & 55 *Amphithoe t.*, J. D. Dana in: U. S. expl. Exp., *v.* 13 ɪɪ p. 939; t. 64 f. 3 a—c | 1893 *Grubia crassicornis* (part.)?, A. Della Valle in: F. Fl. Neapel, *v.* 20 p. 467.

L. 12 mm.

Tropical Pacific (Tongatabu). Along the shores of coral islets, in shallow water, among sea-weed.

A. sp., (Bate & Westw.) 1862 *Amphithoë albomaculata* (err., non Krøyer 1846!), Bate & Westwood, Brit. sess. Crust., *v.* 1 p. 426 f.

North-Atlantic (East of the Shetland Islands). Depth 128—164 m.

2. Gen. **Pleonexes** Bate

1836 *Anisopus* (Sp. un.: *A. dubius*) (non Meigen 1803, Diptera!), R. Templeton in: Tr. ent. Soc. London, *v.* 1 p. 185 | 1856 *Pleonexes* (nom. nud.), Bate in: Rep. Brit. Ass., Meet. 25 p. 59 | 1857 *P.* (Sp. un.: *P. gammaroides*), Bate in: Ann. nat. Hist., ser. 2 *v.* 19 p. 147 | 1894 *P.*, G. O. Sars, Crust. Norway, *v.* 1 p. 581.

Like Ampithoe (p. 631), except that in peraeopods 3—5 the 6ᵗʰ joint is subchelately widened at the apex (Fig. 109). Front lobes of lower lip bifid. Mandibular palp narrow, 3ᵈ joint setose only at apex. Peraeopods 1 and 2 have the 2ᵈ joint rather broadly oval.

1 accepted species, 3 doubtful.

1. **P. gamma-roides** Bate 1856 *P. g.* (nom. nud.), Bate in: Rep. Brit. Ass., Meet. 25 p. 59 | 1857 *P. g.*, Bate in: Ann. nat. Hist., ser. 2 *v.* 19 p. 147 |

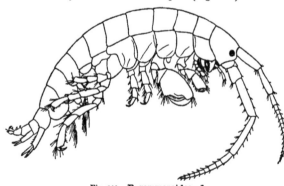

Fig. 109. **P. gammaroides**, ♂.
Lateral View. [After G. O. Sars.]

1894 *P. g.*, G. O. Sars, Crust. Norway, *v.* 1 p. 582 t. 207 | 1862 *Amphithoë g.*, A. *gammeroides*, Bate, Cat. Amphip. Brit. Mus., p. 235 t. 41 f. 4 | 1874 *Sunamphithoë gammaroides*, T. Stebbing in: Ann. nat. Hist., ser. 4 *v.* 14 p. 114 t. 11, 12 f. 3, 3 a—f | 1866 *Amphithoë bicuspis*

(err., non Kröyer 1838!), Cam. Heller in: Denk. Ak. Wien, v. 26 ɪɪ p. 44 t. 4 f. 1 | 1881
A. b., Nebeski in: Arb. Inst. Wien, v. 3 p. 149 t. 13 f. 42 f | 1893 A. gammaroides + A. b.,
A. Della Valle in: F. Fl. Neapel, v. 20 p. 462 t. 57 f. 36; p. 461 t. 57 f. 33—35 | 1871
Sunamphithoë hamulus (err., non Bate 1887!) + S. longicornis, A. Boeck in: Forh.
Selsk. Christian., 1870 p. 245 | 1876 S. h. + S. l., A. Boeck, Skand. Arkt. Amphip., v. 2
p. 594 t. 27 f. 1; p. 596 t. 27 f. 2.

Body (Fig. 109) smooth, compressed. Head tending to bend a little
downward, lateral lobes rounded. Side-plates not very deep, 1st a little
produced forward, 5th the largest. Pleon segment 3, postero-lateral corners
obtusely quadrate. Eyes small, round, red. Antenna 1 variable, longer or
shorter than antenna 2; 1st joint larger than 2ᵈ, 3ᵈ less than half as long as 2ᵈ;
flagellum 15-jointed. Antenna 2, ultimate and penultimate joints of peduncle
subequal; flagellum about 10-jointed. Lower lip with outer branch of bifid
lobes produced beyond the inner (Sars: inner lobe obsolete). Mandible,
spines in spine-row 8 or 9; 3ᵈ joint of palp slightly longer and not broader
than 2ᵈ. Gnathopod 1, 2ᵈ joint with distal lobe, 5th not

Uropod 3

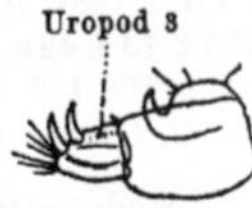

Fig. 110.
P. gammaroides.
Uropod 3 and telson.
[After G. O. Sars.]

longer than broad, 6th oblong oval; palm slightly oblique,
finger matching it. Gnathopod 2 in ♂, 5th joint short,
broad, cup-shaped, 6th large, quadrate; palm oblique, strongly
sinuous diversely on 2 edges; finger serrulate, strongly
arched, touching defining point of palm (Sars describes
palm as somewhat flexuous and defined by a distinct angle).
Gnathopod 2 in ♀ differing from gnathopod 1 by having
the 6th joint much larger, oval quadrate, palm oblique,
almost straight. Peraeopods 1 and 2, 2ᵈ joint rather broadly oval, 4th
longer than broad. Peraeopod 3, 2ᵈ joint scarcely so broad as long. Feraeopods 4 and 5, 2ᵈ joint long oval. Peraeopods 3—5, 6th joint distally expanded,
fitting into cavity of upturned finger towards which spines project planted within
the rounded apical lobe of the 6th joint; one of the spines curved. Uropods 1—3,
peduncle longer than rami. Uropod 3 (Fig. 110), outer ramus with 2 upturned
spines; inner laminar with several setae. Telson (Fig. 110) not longer than
broad, with a small subapical hook on either side of a convex apex (Sars:
terminating in 2 juxtaposed, very strong, hooked projections; in figure longer
than broad, with a very narrow, slightly incised apex). Colour bright yellowish
green, with scattered black dots or stellate markings. L. 6—7 mm.

North-Atlantic with adjoining seas (Norway, Great Britain, France, Azores).

P. dubius (R. Templ.) 1836 Anisopus d., R. Templeton in: Tr. ent. Soc. London,
v. 1 p. 185 t. 20 f. 1 | 1862 Amphithoë dubia, Bate, Cat. Amphip. Brit. Mus., p. 245 t. 42
f. 7 | 1893 A. d., A. Della Valle in: F. Fl. Neapel, v. 20 p. 464 | 1888 Sunamphithoë d.,
T. Stebbing in: Rep. Voy. Challenger, v. 29 p. 167.

L. about 4 mm.

Tropical Indian Ocean (Mauritius)?

P. validus (Czern.) 1868 Sunamphithoë valida, Czerniavski in: Syezda Russ.
Est., Syezda 1 Zool. p. 101 t. 6 f. 36 | 1893 Amphithoe? v., A. Della Valle in: F. Fl.
Neapel, v. 20 p. 464.

Perhaps identical with P. gammaroides. L. about 6 mm.

Black Sea.

P. virescens (Stimps.) 1853 Amphithoe v., Stimpson in: Smithson. Contr., v. 6
nr. 5 p. 53 | 1862 Amphithoë punctata (part.), Bate, Cat. Amphip. Brit. Mus., p. 241 | 1893
Grubia crassicornis (part.)?, A. Della Valle in: F. Fl. Neapel, v. 20 p. 467.

L. 11 mm.

Fundy Bay (Grand Manan).

41*

3. Gen. **Grubia** Czern.

1868 *Grubia* (Sp. un.: *G. taurica*), Czerniavski in: Syezda Russ. Est., Syezda 1 Zool. p. 103 | 1888 *G.*, T. Stebbing in: Rep. Voy. Challenger, *v.* 29 p. 377 | 1893 *G.*, A. Della Valle in: F. Fl. Neapel, *v.* 20 p. 464.

Like Ampithoe (p. 631), except as follows. Antenna 1 has a 1-jointed accessory flagellum, antenna 2 an elongate flagellum. Mandible, the 3^d joint of the slender palp not distally expanded. Maxilla 1, the inner plate with 3 small lateral setae. Peraeopod 3 with the 2^d joint longer than broad.

1 accepted species, 1 imperfectly described.

1. **G. crassicornis** (A. Costa)

1857 *Amphithoe c.* + *A. elongata*, A. Costa in: Mem. Acc. Napoli, *v.* 1 p. 206 t. 3 f. 1a—d; p. 209 t. 3 f. 5 | 1893 *Grubia c.*, A. Della Valle in: F. Fl. Neapel, *v.* 20 p. 464, 918; t. 2 f. 12; t. 13 f. 18, 29 | 1866 *Podocerus largimanus* + *P. longicornis*, Cam. Heller in: Denk. Ak. Wien, *v.* 26 II p. 46 t. 4 f. 6; p. 47 t. 4 f. 7 | 1881 *Amphithoë largimana* + *A. longicornis*, Nebeski in: Arb. Inst. Wien, *v.* 3 p. 150 t. 13 f. 42c, 42b | 1868 *Grubia taurica*, Czerniavski in: Syezda Russ. Est., Syezda 1 Zool. p. 103 t. 8 f. 1—10 | 1888 *G. t.*, T. Stebbing in: Rep. Voy. Challenger, *v.* 29 p. 377 | 1875 *G. taurica var. massiliensis*, Catta in: Rev. Sci. nat., *v.* 4 p. 165.

Body rather robust. Head, lateral lobes little prominent. Side-plate 1 little produced forward, 2^d—5th somewhat rounded below. Pleon segment 3, postero-lateral corners subquadrate, with small point distinct from margin above. Eyes small, rounded, red with scattered points of white. Antenna 1 as long as body; 1st and 2^d joints long, equal, 3^d about $^1/_3$ of 2^d; flagellum 29—50-jointed; accessory flagellum scarcely half as long as 1st joint of primary, but slender, distinct. Antenna 2 slender, $^2/_3$ as long as antenna 1 or more; ultimate and penultimate joints of peduncle equal; flagellum with 23—30 joints, sometimes as long as peduncle. Mandible, 3^d joint of palp set with pectinate spines, but not densely. Maxilla 1, 2^d joint of palp with 8 small spines round apex and part of inner margin; outer plate with 10 apical spines. Gnathopod 1, 5th joint nearly as long as 6th, distally squared, 6th oval, palm only defined by a palmar spine, which is crossed by tip of faintly serrate finger. Gnathopod 2 in ♂, 5th joint short, cup-shaped. 6th oblong oval, the palm strongly sinuous, scarcely (to strongly) defined from the straight hind margin; the finger overlapping the palmar margin and closing on to the surface of the joint. Gnathopod 2 in ♀ similar to gnathopod 1, but with shorter 5th and enlarged 6th joint. Peraeopods 1 and 2, 2^d joint narrowly oval, and others as in Ampithoe rubricata (p. 639). Peraeopods 3—5, 2^d joint narrowing distally, successively longer and narrower, 4th—6th joints rather long and narrow. Pleopods with 2 coupling spines. Uropods 1—3 nearly as in Ampithoe rubricata. Uropod 3, outer ramus the shorter, hinder reverted spine the shorter; the oval inner ramus beset with slender spines and 1 rather stout one. Telson nearly as broad as long; sides converging to the 2 reverted points; apex projecting a little between these, obtuse angled. Colour yellowish green with red spots. L. 6—12 mm.

Mediterranean, Black Sea.

G. **setosa** (Hasw.)

1879 *Amphithoë s.*, Haswell in: P. Linn. Soc. N.S.Wales, *v.* 4 p. 270 | 1885 *Amphithoe s.*, Chilton in: P. Linn. Soc. N.S.Wales, *v.* 9 p. 1040 | 1893 *A. rubricata* (part.), A. Della Valle in: F. Fl. Neapel, *v.* 20 p. 456.

Antenna 1 with a very short accessory flagellum.

South-Pacific (Botany Bay, rock-pools; Sydney Harbour).

4. Gen. **Amphithoides** Kossm.

1880 *Amphithoïdes* (Sp. un.: *A. longicornis*), Kossmann, Reise Roth. Meer., *v.* 21 Malacost. p. 135 | 1888 *A.*, T. Stebbing in: Rep. Voy. Challenger, *v.* 29 p. 516, 517.

Like Ampithoe (p. 631), except that antenna 1 has an accessory flagellum, uropod 3 has only 1 well developed hook on the outer ramus, and the telson is flat, unarmed.

1 species accepted, 1 doubtful.

1. A. longicornis Kossm. 1880 *A. l.*, Kossmann, Reise Roth. Meer, *v.* 21 Malacost. p. 135 | 1893 *Grubia crassicornis* (part.), A. Della Valle in: F. Fl. Neapel, *v.* 20 p. 464.

♀. Side-plates as in Ampithoe (p. 631). Antenna 1 as long as body; 2^d joint rather longer than 1st, 3^d much shorter; flagellum 23-jointed; accessory flagellum 2-jointed, shorter than 1st joint of primary. Antenna 2, 2^d and 3^d joints of peduncle [? ultimate and penultimate] very long; flagellum 17-jointed. Gnathopods 1 and 2 almost exactly alike in shape and size, but 5th joint rather longer and more slender in gnathopod 1, and the palm defined by a palmar spine only in gnathopod 2; 6th joint in both widening to the convex palm, finger faintly serrate. Marsupial plates broad. Uropod 3, 2^d hook of outer ramus indicated by a scarcely visible blunt tubercle. L. 4 mm.

Red Sea.

A. comptus (S. l. Sm.) 1873 *Amphithoë compta*, (S. I. Smith in:) A. E. Verrill in: Rep. U. S. Fish Comm., *v.* 1 p. 564 | 1893 *A.? c.*, A. Della Valle in: F. Fl. Neapel, *v.* 20 p. 463.

Mouth-parts, uropod 3, and telson undescribed. L. reaching 13 mm.

North-Atlantic (from North Carolina to Cape Cod, among eel-grass; Vineyard Sound, at surface).

5. Gen. **Sunamphitoe** Bate

1856 *Sunamphitoë* (nom. nud.), Bate in: Rep. Brit. Ass., Meet. ·25 p. 59 | 1857 *S.*, Bate in: Ann. nat. Hist., ser. 2 *v.* 19 p. 147 | 1888 *S.*, T. Stebbing in: Rep. Voy. Challenger, *v.* 29 p. 1722 | 1857 *Synamphithoe*, A. White, Hist. Brit. Crust., p. 201 | 1862 *Sunamphithoë*, Bate (& Westwood), Brit. sess. Crust., *v.* 1 p. 429 | 1894 *S.*, G. O. Sars, Crust. Norway, *v.* 1 p. 584.

Like Ampithoe (p. 631), but decisively distinguished by absence of mandibular palp; in the bifid front lobes of lower lip the outer lobe less prominent; gnathopod 2 in ♂ much larger than gnathopod 2 in ♀ and differently shaped; hooks of telson small.

1 species.

1. S. pelagica (M.-E.) 1830 *Amphithoe p.*, H. Milne Edwards in: Ann. Sci. nat., *v.* 20 p. 378 | 1840 *Amphitoe p.*, H. Milne Edwards, Hist. nat. Crust., *v.* 3 p. 36 | 1856 *Sunamphitoë hamulus* (nom. nud.) + *S. conformatus* (nom. nud.), Bate in: Rep. Brit. Ass., Meet. 25 p. 59 | 1857 *S. h.* (♀?) + *S. conformata* (♂), Bate in: Ann. nat. Hist., ser. 2 *v.* 19 p. 147 | 1857 *Synamphithoe h.* + *S. c.*, A. White, Hist. Brit. Crust., p. 202 | 1862 *Sunamphithoë h.* + *S. c.*, Bate & Westwood, Brit. sess. Crust., *v.* 1 p. 430 f.; p. 432 f. | 1862 *S. h.* + *S. c.*, Bate, Cat. Amphip. Brit. Mus., p. 250 t. 43 f. 5; p. 251 t. 43 f. 6 | 1874 *S. c.*, T. Stebbing in: Ann. nat. Hist., ser. 4 *v.* 14 p. 116 t. 12 f. 4, 4a—d | 1894 *S. c.*, G. O. Sars, Crust. Norway, *v.* 1 p. 585 t. 208 | ? 1861 *Amphithoë grandimana*, A. Boeck in: Forh. Skand. Naturf., Møde 8 p. 668 | 1876 *A. g.*, A. Boeck, Skand. Arkt. Amphip., *v.* 2 p. 591 t. 26 f. 4 | 1893 *Amphithoe hamulus* + *Grubia crassicornis* (part.), A. Della Valle in: F. Fl. Neapel, *v.* 20 p. 463 t. 57 f. 28, 29; p. 464.

Side-plates 1—4 with setules at lower hind corner; side-plate 1 scarcely produced forward, 5th the largest. Body smooth, compressed. Head, lateral lobes broadly rounded. Pleon segment 3, postero-lateral corners obtusely quadrate. Eyes small, rounded, red. Antenna 1 longer than antenna 2; 1st joint the largest, 3d small; flagellum elongate, in ♂ reaching more than 40 joints. Antenna 2, ultimate joint of peduncle shorter than penultimate, flagellum shorter than peduncle, in ♂ reaching 17 setose joints. Mandible just as in Ampithoe rubricata (p. 639), except for absence of palp. Gnathopod 1, 2d joint apically a little widened, 5th distally widened, 6th rather longer than 5th, oblong; palm transverse, short; finger projecting beyond palm, faintly serrate. Gnathopod 2 in ♂, 5th joint cup-shaped, short, its hind lobe narrow, 6th large, oblong oval, a setose lobe or tubercle projecting near hinge of finger, and from this the microscopically denticulate palm joining or forming the hind margin without defining point; the long, slightly sinuous finger about 3/4 as long as 6th joint. Gnathopod 2 in ♀ like gnathopod 1, but with 5th joint shorter, and 6th rather broader. Peraeopods 1 and 2, 2d joint broadly oval, 4th about as broad as long; finger blunt at tip. Peraeopod 3, 2d joint broader than long, 4th and 5th shorter than 6th. Peraeopods 4 and 5, 2d joint oblong oval, narrowing a little distally, 4th—6th joints moderately long. In peraeopods 3—5 the finger is curved, its inner margin marked as if for serration, of the apical spines on the 6th joint one follows the curve of the finger (Bate in 1857 describes these limbs as scarcely prehensile in S. hamulus). Pleopods 1—3 with 3 coupling spines. Uropods 1 and 2, rami not very long. Uropod 3, outer ramus microscopically denticulate on upper margin, of the upturned spines the apical the smaller, inner ramus with several setules and 2 spinules. Telson broader than long, with a little tubercle or upturned point on either side of triangular apex. Colour greenish yellow with stellate markings. L. 8—17 mm.

North-Atlantic, North-Sea and English Channel (Norway; France; North and South Devon, between tide-marks; Azores).

6. Gen. **Biancolina** Della Valle

1893 *Biancolina* (Sp. un.: *B. algicola*), A. Della Valle in: F. Fl. Neapel, v. 20 p. 562.

Antenna 1 longer than antenna 2, without accessory flagellum, flagellum longer than peduncle. Upper lip rounded as in Ampithoe (p. 631). Lower lip, outer lobes indented on outer not on inner margin, inner lobes distinct to the base, little divergent apically. Mandible short, cutting edge with about 10 unequal teeth; accessory plate with 4 or 5 teeth, smaller on right than on left mandible, spine-row of 3 or 4 minute spines; molar evanescent; palp wholly wanting. Maxilla 1 powerful, inner plate small, with 1 seta, outer long, with 9 spines so crowded as to make the counting rather uncertain. Maxilla 2, both plates slender, feebly armed, outer a little the longer and broader. Maxillipeds, inner plates rather long and narrow on elongate base, outer reaching beyond 2d joint of palp, armed with few but strong spine-teeth; palp's joints 1—3 short, stout; finger small, conical, with well formed nail. Gnathopod 1 forming a small imperfect chela. Gnathopod 2 massive, subchelate. Peraeopods 1—5, 2d joint somewhat expanded, finger curved. Peraeopod 3, 6th joint reverted. Uropod 3, peduncle stout, rami short, lamellar, outer with 2 hooks. Telson small.

1 species.

1. **B. cuniculus** (Stebb.) 1874 *Amphithoë c.*, T. Stebbing in: Ann. nat. Hist., ser. 4 *v.* 14 p. 112 t. 11 f. 1, 1 a—e | 1893 *A. c.* + *Biancolina algicola* (♀ juv.), A. Della Valle in: F. Fl. Neapel, *v.* 20 p. 460 t. 57 f. 38; p. 562 t. 3 f. 11, t. 32 f. 38—53 | 1899 *B. c.*, T. Stebbing in: Ann. nat. Hist., ser. 7 *v.* 3 p. 350.

Integument brittle. Head with bulging cheeks, rabbit-like in profile, attachment of antennae prominent. Side-plates not deep, but shaped as in Ampithoe (p. 631). Pleon segment 3, postero-lateral corners rounded. Eyes small, round, red. Antenna 1, 1st joint a little longer than 2^d, which is twice as long as 3^d; flagellum sometimes twice as long as peduncle, with 18 unequal joints, some with sensory filaments. Antenna 2, ultimate joint of peduncle rather shorter than penultimate, flagellum shorter than peduncle, 5-jointed. Gnathopod 1, 2^d joint expanded distally, 3^d and 4th equal, 5th widening a little distally, scarcely longer than 6th, which has a small slightly produced palm, much overlapped by the small and stout but acute finger. Gnathopod 2 massive (at least in ♂); 2^d joint expanded, 5th short, cup-like, 6th very large, broad and long; palm excavate, defined by a tooth; finger large, arcuate, obliquely truncate, leaving cavity between its apex and its minute nail. Peraeopods 1 and 2, 2^d joint expanded, oval, 4th short and wide, 6th longer than 5th. Peraeopod 3 short, 2^d joint rounded, expanded proximally, $3^d—5^{th}$ short. Peraeopods 4 and 5 rather long, 2^d joint somewhat expanded, narrowed distally, 4th and 5th moderately long, 6th longer. Uropod 1, peduncle long, setose, with an oval process at apex, the slightly unequal rami rather shorter than peduncle. Uropod 2 shorter; some setae on peduncle, which is as long as the rami. Uropod 3, peduncle short, much longer than the small broad rami, inner with 4 slight spinules. Colour bright yellow. L. 4·5 mm.

Bay of Naples, depth 1—2 m; Torbay, within tide-marks.

---

37. Fam. Jassidae

1899 *Ischyroceridae*, T. Stebbing in: Ann. nat. Hist., ser. 7 *v.* 4 p. 211.

Head, lateral lobes often somewhat prominent. Side-plates variable in relative proportions, 4th with hind margin usually not excavate (Fig. 111 p. 650). Antennae 1 and 2 variable in relative proportions; accessory flagellum of antenna 1 distinct or indistinct, but never large. Upper lip with pointed epistome. Lower lip and mouth-parts in general as in Aoridae (p. 585), except that the 3^d joint of mandibular palp is shorter than 2^d, laminar, and that the maxillipeds in Wyvillea (p. 648) have inner and outer plates rudimentary and palp 3-jointed. Gnathopods 1 and 2 subchelate, 2^d the larger, often greatly modified in ♂. Peraeopods 1 and 2 glandular. Peraeopod 3 reverted. Peraeopod 5 the longest. Uropods 1—3 biramous. Uropod 3, rami very short, the outer uncinate and usually surmounted by denticles. Telson simple.

Marine.

5 genera, 21 accepted species and 5 doubtful.

Synopsis of genera:

1 { Maxillipeds, palp 3-jointed . . . . 1. Gen. **Wyvillea** . . . p. 648
{ Maxillipeds, palp 4-jointed — **2**.

2 { Antenna 1, accessory flagellum indistinct . . 2. Gen. **Parajassa** . . . p. 649
{ Antenna 1, accessory flagellum distinct — **3**.

8 { Side-plate 5 not much deeper than side-plate 6.   3. Gen. **Microjassa**. . . p. 651
    { Side-plate 5 much deeper than side-plate 6 — **4.**

4 {
Gnathopod 2 in ♂, hind margin of 6th joint
    produced into a tooth; gnathopod 2 in ♀
    much larger than gnathopod 1 . . . . .   4. Gen. **Jassa** . . . . . p. 652
Gnathopod 2 in ♂, hind margin of 6th joint
    not produced into a tooth; gnathopod 2 in
    ♀ not much larger than gnathopod 1 . .   5. Gen. **Ischyrocerus** . p. 657

1. Gen. **Wyvillea** Hasw.

1879 *Wyvillea* (Sp. un.: *W. longimanus*), Haswell in: P. Linn. Soc. N.S. Wales,
v. 4 p. 336 | 1888 *W.*, T. Stebbing in: Rep. Voy. Challenger, *v.* 29 p. 513 | 1880 *Macleayia*,
Haswell in: Ann. nat. Hist., ser. 5 *v.* 5 p. 32.

Side-plates scarcely so deep as their segments. Antenna 1 shorter
than antenna 2, with accessory flagellum. Mandible with palp. Maxillipeds
exunguiculate, inner and outer plates rudimentary; palp 3-jointed. Gnathopod 2
very large. Uropod 3 apparently as in Jassa (p. 652). Telson entire or emarginate.

2 species.

Synopsis of species:

Telson with blunt apex . . . .   1. **W. longimana** . . p. 648
Telson with apex emarginate . . . .   2. **W. haswelli** . . . p. 648

1. W. longimana Hasw.   1879 *W. longimanus*, Haswell in: P. Linn. Soc. N.S.
Wales, *v.* 4 p. 337 t. 22 f. 7 | 1884 *Podocerus l.*, Chilton in: Tr. N. Zealand Inst., *v.* 16
p. 255 t. 17 f. 2 a—e | 1879 *P. cylindricus* (err., non Say 1818?), T. W. Kirk in: Tr. N.
Zealand Inst., *v.* 11 p. 402.

Eyes round. Antenna 1 scarcely $^1/_8$ as long as body; 2^d joint twice
as long as 1st, a little longer than 3^d; flagellum as long as 2^d joint of
peduncle, 7-jointed; accessory flagellum nearly $^1/_4$ as long as primary, 2-jointed.
Antenna 2 about half as long as body, stout, subpediform; peduncle as long
as antenna 1; flagellum as long as ultimate joint of peduncle, armed distally
with curved spines. Gnathopod 1 small, 5th joint about half as long as 6th,
6th ovoid, distally narrowed; palm very oblique, defined only by 2 palmar
spines; finger serrate (Chilton). Gnathopod 2 in ♂ very large; 5th joint
small, triangular, 6th cylindrical, curved, margins parallel, 4 times as long
as broad, with a blunt tooth at each end of the concave, rather hirsute
margin, which should be considered as the palm; proximal tooth not decurrent.
Gnathopod 2 in ♀ (Chilton) not larger than gnathopod 1, and similar thereto,
but with 5th joint shorter, and 6th rather more narrowed distally; finger in
both gnathopods roughened rather than serrate. Peraeopods 1—5 all short
and broad, 1st and 2^d shorter than the rest. Uropod 3, peduncle long,
narrowing slightly distally; rami very short, inner stiliform, outer ending in
3—6 upturned teeth (Chilton). Telson conical, blunt. Colour pale yellow,
with many black dots and markings (Chilton). L. 3—14 mm.

South-Pacific (Port Jackson [East-Australia]; Worser Bay and Lyttelton Harbour
[New Zealand]).

2. W. haswelli (G. M. Thoms.)   1897 *Maera h.*, G. M. Thomson in: Ann. nat.
Hist., ser. 6 *v.* 20 p. 449 t. 10 f. 6—10 | 1899 *Wyvillea h.*, T. Stebbing in: Ann. nat. Hist.,
ser. 7 *v.* 3 p. 350.

Body slender and compressed. Head long, without rostrum; lateral
lobes small, acutely produced. Eyes produced well forward on lateral lobes

of head; lenses numerous. Antenna 1, 1st joint thicker but rather shorter than 2d or 3d, 3d a little shorter than 2d; flagellum slender, broken; accessory flagellum 4-jointed, about as long as 3d joint of peduncle. Antenna 2 broken. Mandible normal; palp with very short 1st joint, 3d rather shorter than 2d, broad ended, with setae on apex. Gnathopod 1 small, 5th joint distally a little expanded, 6th rather shorter, oval, palm oblique, ill-defined, finger a little over half as long as 6th joint. Gnathopod 2 very large, 4th joint small, triangular, hind margin produced acutely, 5th joint very short, similarly produced very acutely, 6th very long, narrow at base, widening gradually to a transverse, slightly denticulate palm, defined by a small tooth; the straight hind margin having also a small tooth or projection at its base; the great finger projects far over the palm, its falcate end closing between the acute ends of the 4th and 5th joints. Peraeopod 4 rather slender. Uropods 1 and 2 reaching nearly as far as uropod 3. Telson (in figure) broader than long, rounded, with an excavation little more than $1/_3$ the length, between 2 rounded apices, each with a spinule. L. 4 mm.

South-Pacific (New-Zealand). Depth 15 m.

2. Gen. Parajassa Stebb.

1813/14 *Jassa* (part.), Leach in: Edinb. Enc., *v.*7 p.433 | 1859 *J.*, R. M. Bruzelius in: Svenska Ak. Handl., n. ser. *v.*3 nr. 1 p.18 | 1871 *Janassa* (Sp. un.: *J. variegata*) (non G. Münster 1839, Pisces!), A. Boeck in: Forh. Selsk. Christian., 1870 p.249 | 1876 *J.*, A. Boeck, Skand. Arkt. Amphip., *v.*2 p. 608 | 1894 *J.*, G. O. Sars, Crust. Norway, *v.*1 p.598 | 1895 *J.*, A. O. Walker in: Ann. nat. Hist., ser. 6 *v.*15 p.472 | 1899 *Parajassa*, T. Stebbing in: Ann. nat. Hist., ser. 7 *v.*3 p.240.

Like Jassa (p. 652), except that the accessory flagellum of antenna 1 is nearly obsolete, none of the side-plates are particularly deep, and the 5th side-plate is almost as deep as the 4th.

2 species.

Synopsis of species:

1. P. pelagica (Leach) 1813/14 *Jassa p.*, Leach in: Edinb. Enc., *v.* 7 p.433 | 1899 *Parajassa p.*, T. Stebbing in: Ann. nat. Hist., ser. 7 *v.*3 p. 240 | 1843 *Podocerus capillatus*, H. Rathke in: N. Acta Ac. Leop., *v.*20 r p.89 t. 4 f.8 | 1859 *Jassa capillata*, R. M. Bruzelius in: Svenska Ak. Handl., n. ser. *v.*3 nr. 1 p.19 | 1894 *Janassa c.*, G. O. Sars, Crust. Norway, *v.*1 p.599 t.214 | 1895 *J. c.*, A. O. Walker in: P. Liverp. biol. Soc., *v.*9 p.316 | 1871 *J. variegata* (err., non *Podocerus variegatus* Leach 1813/14!), A. Boeck in: Forh. Selsk. Christian., 1870 p.250 | 1876 *J. v.*, A. Boeck, Skand. Arkt. Amphip., *v.*2 p.608 t.28 f.1; t.29 f.2 and? 3 | 1893 *Podocerus falcatus* (part.), A. Della Valle in: F. Fl. Neapel, *v.*20 p.445.

Body (Fig. 111 p. 650) rather tumid, especially in ♀. Head rather small, front lobes narrowly rounded. 1st and 3d side-plates produced below the 2d. Pleon segment 3, postero-lateral corners rounded. Eyes small, rounded, dark. Antennae 1 and 2 in ♂ und ♀ densely setose with setae separate and in fascicles. Antenna 1 more than $1/_3$ as long as body; 3d joint longer than 1st, as long as 2d, rather longer than flagellum, of which 1st joint is very large, 2d and 3d minute; accessory flagellum a mere tubercle. Antenna 2 longer and much stronger; ultimate joint of peduncle slightly longer than penultimate, as long as flagellum, which is similar to that of antenna 1, but larger. Gnathopod 1, 5ᵗʰ joint triangular, 6ᵗʰ oval, widest in the middle where the hind margin joins the straight oblique palm, defined by an obtuse angle and palmar spines. Gnathopod 2 nearly alike in ♂ and ♀; 6ᵗʰ joint broad, in ♀ the internal edge having a lunar notch (Leach), the finger closing within the well marked defining angle, while in the ♂ this angle is produced into an almost thumb-like tooth, over-lapped by the strong finger.

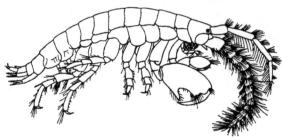

Fig. 111. **P. pelagica.** ♂.
Lateral View. [After G. O. Sars.]

Peraeopods 1 and 2 short and stout, 2ᵈ joint a little widened. Peraeopods 3—5 slightly encreasing successively in length; 2ᵈ joint oblong oval; in peraeopod 5 the lower hind corner slightly produced. Uropod 3 reaching rather beyond the others, outer ramus uncinate, without marginal serrations, inner with a lateral spinule. Telson about as long as broad, with small projection and adjacent setule at each side of the apical triangle. Colour grey or dark, with transverse bars of brown. L. ♀ 7, ♂ 9 mm.

Arctic Ocean and North-Atlantic (Norway up to Finmark, Shetland Islands); Kattegat; German Sea; Liverpool Bay.

2. **P. tristanensis** (Stebb.) 1888 *Podocerus t.*, T. Stebbing in: Rep. Voy. Challenger, *v.* 29 p. 1141 t. 121 | 1893 *Ischyrocerus t.*, A. Della Valle in: F. Fl. Neapel, *v.* 20 p. 450 t. 57 f. 20 | 1899 *Parajassa t.*, T. Stebbing in: Ann. nat. Hist., ser. 7 *v.* 3 p. 240.

Body not robust. Head, lateral lobes not very prominent, with acute point. Side-plates 1—5 about equal in depth. Pleon segment 3, postero-lateral corners obtuse. Eyes rounded oval, dark in spirit. Antennae 1 and 2 subequal, slender, with long setae, not· densely grouped, some plumose. Antenna 1, 3ᵈ joint much longer than 1ˢᵗ, subequal to 2ᵈ, shorter than flagellum, of which the 1ˢᵗ joint is about half as long as the other 3 combined; accessory flagellum seemingly absent. Antenna 2, ultimate joint of peduncle as long as penultimate, both together rather longer than the 4-jointed flagellum. Gnathopod 1, 2ᵈ joint not longer than 5ᵗʰ, which is nearly or quite as long as the rather narrowly oblong oval 6ᵗʰ; palm ill-defined, shorter than the hind margin; finger overlapping palm, finely pectinate, with a denticle near the acute apex. Gnathopod 2 in ♂, 2ᵈ joint short, with greatly expanded distal lobe, 5ᵗʰ short, broadly cup-shaped, 6ᵗʰ large, not twice as long as broad; palm with broad tooth near finger-hinge, then shallowly excavate, parallel to front margin, and subequal to hind margin, which ends in a short tooth, against which the apex of the stout, much curved, smooth finger impinges. Gnathopod 2 in ♀ differing from gnathopod 1 by the short, cup-shaped 5ᵗʰ joint, the 6ᵗʰ somewhat larger with a better defined palm, and the pectinate finger armed with a longer denticle. Peraeopods 1 and 2 rather short and compact, the 2ᵈ and 4ᵗʰ joints a little expanded. Peraeopod 5 rather long; 2ᵈ joint oblong, not much expanded, shorter than the 6ᵗʰ.

Pleopods, rami with 5 joints. Uropod 3, peduncle broad, not twice as long as rami; outer ramus the shorter, with 2 minute spines at its slightly bent tip; inner ramus with a rather larger apical spine. Telson as broad as long, rounded triangular, with a small projection and adjacent setule at each side of the apical triangle. L. about 3 mm.

South-Atlantic (Tristan da Cunha). Depth 201 m.

3. Gen. Microjassa Stebb.

1899 *Microjassa* (Sp. un.: *Podocerus cumbrensis*), T. Stebbing in: Ann. nat. Hist., ser. 7 *v.* 3 p. 240.

Like Jassa (p. 652), except in regard to side-plates, antenna 2 and maxillipeds. Side-plates 2—4 much deeper than the rest, 4th large, emarginate behind for the small 5th. Antenna 2 but little stronger than antenna 1. Maxillipeds, outer plates scantily armed. As in Ischyrocerus (p. 657) gnathopods 1 and 2 of ♀ little unequal, but gnathopod 2 of ♂ much larger than gnathopod 1, and differing in shape as well as size from gnathopod 2 of ♀.

1 species.

1. M. cumbrensis (Stebb. & D. Roberts.) 1891 *Podocerus c.*, T. Stebbing & D. Robertson in: Tr. zool. Soc. London, *v.* 13 ɪ p. 38 t. 6 ʙ | 1895 *P. c.*, A. O. Walker in: P. Liverp. biol. Soc., *v.* 9 p. 316 | 1899 *P. c.*, *Microjassa*, T. Stebbing in: Ann. nat. Hist., ser. 7 *v.* 3 p. 240 | 1893 *P. falcatus* (part.), *P. anguipes* (part.)?, A. Della Valle in: F. Fl. Neapel, *v.* 20 p. 445, 944.

Head, rostrum small, blunt; lateral lobes obtusely pointed. Side-plate 1 very small, almost concealed by the much larger and deeper 2d; 3d and 4th as deep as 2d and much broader; 4th of equal breadth and depth, with shallow emargination behind; 5th—7th small, shallow. Pleon segment 3, postero-lateral angles bluntly produced. Eyes round. Antenna 1, 1st joint short and stout, 2d long, a little longer than 3d; flagellum $\frac{1}{2}$ as long as peduncle, with 3 slender joints; accessory flagellum 1-jointed, about $\frac{1}{3}$ as long as 1st joint of primary. Antenna 2 stouter and a little longer than antenna 1, otherwise very similar; ultimate and penultimate joints of peduncle subequal in length; flagellum as long as ultimate joint of peduncle, with 3 slender joints. Mandible, cutting plate quadridentate, accessory plate similar on left, serrate on right, molar with small denticulate plate in front; 2d and 3d joints of palp broad, 3d the shorter. Maxillipeds, outer plates with 3 spine-teeth on inner and 4 on distal margin, 4th joint of palp short, blunt, tipped with long spines. Gnathopod 1, 2d joint a little bent, 5th cup-shaped, scarcely longer than 4th, 6th oval, about twice as long as 5th; palm defined by 3 palmar spines, slightly overlapped by apex of finger. Gnathopod 2 in ♂, 2d joint channelled in front, shorter than 6th; 5th joint short, coalesced with 6th; 6th joint very large, oblong, attaining a length $2\frac{1}{2}$ times the width; hind margin produced into a long tooth, forming a deep cavity between itself and the straight palmar margin, which is produced to a smaller tooth near the finger-hinge; over the 2 cavities arches an immensely long, curved, sinuous finger, with a small prominence on inner margin. Gnathopod 2 in ♀ is like gnathopod 1, but a little larger. Peraeopods 1 and 2, 2d joint long oval, 4th rather broad, widening distally, 5th broader but much shorter than 6th. Peraeopod 3, 2d joint longer than broad, widest proximally. Peraeopod 4 like the 3d, but 2d joint with hind margin more strongly serrate.

Peraeopod 5 longer, 2^d joint widest at the middle. Branchial vesicles very small. Marsupial plates rather large. Pleopods 1—3 with 2 coupling spines, inner ramus with 5—7 joints, outer with 6—8. Uropod 1, peduncle longer than rami. Uropod 2, peduncle longer than outer, shorter than inner ramus. Uropod 3, peduncle stout, longer than peduncle of uropod 2; outer ramus having on distal half about 9 minute denticles, encreasing towards apex; inner ramus rather longer, straight, tipped with spine. Telson triangular, as broad as long, with a setule on either side of rounded apex. Colour brown. L. 3 mm.

Firth of Clyde, depth 38 mm; Irish Sea (North Wales).

4. Gen. Jassa Leach

1813/14 *Jassa* (part.), Leach in: Edinb. Enc., *v.* 7 p. 433 | 1899 *J.,* T. Stebbing in: Ann. nat. Hist., ser. 7 *v.* 3 p. 239 | 1852 *Cratophium,* J. D. Dana in: Amer. J. Sci., ser. 2 *v.* 14 p. 309 | 1853 *C.,* J. D. Dana in: U. S. expl. Exp., *v.* 13 II p. 832, 840.

Body more slender in ♂ than in ♀. Head with small rostral point and lateral lobes somewhat produced. Side-plate 2 broader than deep in ♂; 3^d and 4th in both sexes deeper than 2^d or 5th; 4th not perceptibly emarginate behind. Antennae 1 and 2 setose. peduncle long, flagellum much shorter. Antenna 1, 1st joint shorter than 2^d or 3^d; accessory flagellum very small. Antenna 2 usually considerably longer and stouter than antenna 1, especially in ♂. Upper lip rounded, epistome (so far as known) with pointed process. Lower lip normal. Mandible normal, palp very large, 3^d joint shorter than 2^d, distally widened, strongly setose. Maxilla 1, inner plate very small with inconspicuous marginal setae, outer plate with 9 apical spines; 2^d joint of palp long, with apical spine-teeth. Maxilla 2, inner plate the shorter, with lateral fringe. Maxillipeds normal, compact. Gnathopod 2 in both sexes much larger than gnathopod 1, stronger in ♂ than in ♀; the hind margin of 6th joint in ♂ produced into a strong tooth. Peraeopods 1 and 2, 2^d joint slightly expanded. Peraeopods 3—5 successively longer, with moderately expanded 2^d joint. Uropods 1 and 2, outer ramus shorter than inner. Uropod 3, peduncle stout, much longer than the rami, outer of which ends in a hook or hooked spine and usually has little teeth on the upper margin, inner ramus straight. Telson triangular.

7 species accepted, 4 obscure.

Synopsis of accepted species:

$1\begin{cases} \text{Gnathopod 2 in ♂, finger with prominent tooth} \\ \quad \text{on inner margin — 2.} \\ \text{Gnathopod 2 in ♂, finger without prominent tooth} \\ \quad \text{on inner margin — 3.} \end{cases}$

$2\begin{cases} \text{Gnathopod 2 in ♂, } 6^{th} \text{ joint with subbasal thick-} \\ \quad \text{ended tooth} \dots \dots \dots \dots \dots \dots \text{1. J. dentex} \dots \dots \text{p. 653} \\ \text{Gnathopod 2 in ♂, } 6^{th} \text{ joint with basal thin-} \\ \quad \text{ended tooth} \dots \dots \dots \dots \dots \dots \text{2. J. falcatiformis} \dots \text{p. 653} \end{cases}$

$3\begin{cases} \text{Gnathopod 2 in ♂, } 6^{th} \text{ joint with thumb-like} \\ \quad \text{process near the base — 4.} \\ \text{Gnathopod 2 in ♂, } 6^{th} \text{ joint with thumb-like process} \\ \quad \text{remote from base — 6.} \end{cases}$

4 {	Antenna 2, ultimate joint of peduncle much longer than penultimate	3. J. ingens	. . . . p. 653
	Antenna 2, ultimate joint of peduncle not much longer than penultimate — 5.		
5 {	Gnathopod 2 in ♂, thumb-like process with simple apex	4. J. pulchella	. . . p. 654
	Gnathopod 2 in ♂, thumb-like process with bifid apex	5. J. pusilla	. . . . p. 655
6 {	Peraeopod 4 in ♂ with joints 4—6 narrow . . .	6. J. ocius	. . . . . p. 655
	Peraeopod 4 in ♂ with joints 4—6 broad . . .	7. J. frequens	. . . p. 656

1. J. dentex (Czern.) 1868 *Podocerus d.*, Czerniavski in: Syezda Russ. Est., Syezda 1 Zool. p. 100 t. 6 f. 35 | 1880 *P. d.*, Sowinski in: Mém. Soc. Kiew, *v.* 6 p. 110 t. 4 f. 8 | 1899 *Jassa d.*, T. Stebbing in: Ann. nat. Hist., ser. 7 *v.* 3 p. 350 | 1893 *Podocerus herdmani*, A. O. Walker in: P. Liverp. biol. Soc., *v.* 7 p. 37 f. 13 | 1895 *P. h.*, *P. falcatus var. ?*, A. O. Walker in: Ann. nat. Hist., ser. 6 *v.* 15 p. 473 | 1895 *P. h.*, *P. f. var. ?*, A. O. Walker in: P. Liverp. biol. Soc., *v.* 9 p. 314 | 1893 *P. ocius* (part.), A. Della Valle in: F. Fl. Neapel, *v.* 20 p. 448 | 1898 *P. o.*, Sowinski in: Mém. Soc. Kiew., *v.* 15 p. 460 | 1894 *P. odontonyx*, G. O. Sars, Crust. Norway, *v.* 1 p. 597 t. 213 f. 2.

♀. In general like J. pusilla (p. 655). — ♂. Head, lateral lobes rather broad, at apex minutely emarginate. Side-plate 1 rather produced forward, 2^d twice as broad as deep. Eyes rather large, rounded, brownish or black. Antenna 1 about half as long as body; 2^d joint as long as 3^d, nearly twice as long as 1st; flagellum 3—5-jointed, accessory flagellum extremely minute. Antenna 2 a little longer, more robust; flagellum 3- or 4-jointed. Gnathopod 1 nearly as in J. pulchella (p. 654) and J. pusilla. Gnathopod 2, 6th joint oblong oval, with a stout tooth close up to the finger-hinge, and another prolonged from the hind margin, with a rather deep cavity between; both teeth sometimes bifid; finger strong and curved, with a variable tooth at middle of inner margin. Peraeopods, uropods and telson seemingly like those in J. pusilla. Colour, variegated with dark brown. L. 3—5 mm.

Arctic Ocean, North-Atlantic and Irish Sea (Trondhjemsfjord and Hammerfest; Colwyn Bay; Liverpool Bay); Black Sea. Depth 1—282 m.

2. J. falcatiformis (Sowinski) 1898 *Podocerus f.* + *Ischyrocerus constantino-politanus* (juv.), Sowinski in: Mém. Soc. Kiew., *v.* 15 p. 461 t. 8 f. 9—16; p. 463 t. 8 f. 17—25.

Adult ♀ unknown. — ♂. Antenna 1, 2^d joint more than twice as long as 1st, a little longer than 3^d, a little shorter than the 3-jointed flagellum. Gnathopod 2, 6th joint long and rather slender, narrowing to the finger-hinge, front margin very convex, a slender tooth, fully half the length of the joint, rising from the base behind, alongside the straight palm, which is interrupted a little beyond the basal tooth by a small projecting tooth, apposed to a similar tooth on the inner margin of the finger; finger as long as the 6th joint, with 5 setules on its distal concavity. Uropod 3, peduncle long and thick, outer ramus uncinate, distally crenulate with 9 small close-set blunt teeth. L. 4 mm.

Bosphorus. Depth 32—36 m.

3. J. ingens (Pfeff.) 1888 *Podocerus i.*, Pfeffer in: Jahrb. Hamburg. Anst., *v.* 5 p. 131 t. 3 f. 1 | 1893 *P. falcatus* (part.), A. Della Valle in: F. Fl. Neapel, *v.* 20 p. 445.

Head, lateral lobes bluntly triangular. Side-plate 1 produced a little forward, 2^d broader than deep, 3^d and 4th successively deeper; front lobe of 5th nearly as deep as 4th. Pleon segment 3, postero-lateral corners obtuse. Eyes longitudinally oval. Antenna 1, 1st joint $^3/_4$ as long as 2^d, 3^d as long as 2^d; flagellum a little longer than 3^d joint of peduncle, with 7 joints, 1st much the longest. Antenna 2 much longer, about $^8/_4$ as long as body; ultimate joint of peduncle very long, 1$^1/_2$ times as long as penultimate, not setose; flagellum shorter than ultimate joint of peduncle, 6-jointed. Gnathopod 1, 6th joint piriform, the narrow part apical; finger finely serrate. Gnathopod 2 in ♂ very large and powerful; the 6th joint long, slightly curved, with tooth of moderate length springing from near the base and having in a notch on the outer side a tuft of spines or setae; a triangular tooth is near the finger-hinge, and against this a prominence of the long finger works, while its apex closes down on to the outer notch of the other tooth. Gnathopod 2 in ♀ differs from ♂ in that the proximal tooth is represented by a rounded, truncate prominence of the hind or palmar margin. Peraeopods 1 and 2 comparatively short and weak; 4th joint distally widened. Peraeopods 3—5 successively longer, 2^d joint well expanded, its hind margin in peraeopod 3 weakly concave, in 4th straight, in 5th weakly convex. Telson small, rounded triangular, rather broader than long, with 1 spine on each side of the apex. Colour grey and whitish, variegated with brown. L. 12—26 mm.

South-Atlantic (South Georgia). At lowest ebb of spring-tides.

4. J. pulchella Leach 1813/14 *J. p.*, Leach in: Edinb. Enc., *v.*7 p.433 | 1899 *J. p.*, T. Stebbing in: Ann. nat. Hist., ser. 7 *v.*3 p.239 | 1830 *Podocerus pulchellus*, H. Milne Edwards in: Ann. Sci. nat.. *v.*20 p.384 | 1843 *P. calcaratus*, H. Rathke in: N. Acta Ac. Leop., *v.*201 p.91 t.4 f.9 | 1851 *Ischyrocerus (P.) c.*, *I. anguipes* (err., non *I.angvipes* Kröyer 1838!), W. Liljeborg in: Öfv. Ak. Förh., *v.*8 p.23 | 1853 & 55 *Cratophium validum*, J. D. Dana in: U. S. expl. Exp., *v.*13ii p.841; t.56 f.2 | 1886 *Podocerus validus*, G. M. Thomson & Chilton in: Tr. N. Zealand Inst., *v.*18 p.143 | 1857 *P. falcatus* (err., non *Cancer (Gammarus) f.* Montagu 1808!), Bate in: Ann. nat. Hist., ser.2 *v.*19 p.148 | 1876 *P. f.* (part.), A. Boeck, Skand. Arkt. Amphip., *v.*2 p.603 t.27 f.4,7 | 1879 *P. f.*, Hoek in: Tijdschr. Nederl. dierk. Ver., *v.*4 p.120 t.8 f.13—15; t.9 f.1—3 | 1881 *P. f.*, Nebeski in: Arb. Inst. Wien, *v.*3 p.151 t.13 f.44 | 1887 *P. f.*, J. Bonnier in: Bull. sci. Nord, *v.*18 p.340 | 1893 *P. f.* (part.), A. Della Valle in: F. Fl. Neapel, *v.*20 p.445 t.14 f.1—10; t.57 f.13—16 | 1894 *P. f.*, G. O. Sars, Crust. Norway, *v.*1 p.594 t.212 | 1895 *P. f.*, A. O. Walker in: Ann. nat. Hist., ser. 6 *v.*15 p.472 | 1862 *P. pulchellus + P. validus + P. f. + P. pelagicus* (err., non *Jassa pelagica* Leach 1813/14!), Bate, Cat. Amphip. Brit. Mus., p.253 t.43 f.8; p.253 t.43 f.9; p.255 t.44 f.1; p.255 t.44 f.2 | 1879 *P. australis*, Haswell in: P. Linn. Soc. N.S.Wales, *v.*4 p.338 t.21 f.8 | 1888 *P. falcatus + P. validus*, T. Stebbing in: Rep. Voy. Challenger, *v.*29 p.1132 t.119; p.1135 t.138B.

Body more slender and compressed in ♂ than in ♀. Head, lateral lobes small, rounded, prominent. Side-plate 1 somewhat angularly produced forward, 2^d in ♂ broader than deep, 3^d and 4th in ♂ considerably deeper than 2^d or 5th. Pleon segment 3, postero-lateral corners quadrate, with minute projecting point. Eyes small, rounded, dark. Antenna 1 not half as long as body, 3^d joint considerably longer than 1st, shorter than 2^d or flagellum; flagellum 5—9-jointed. Antenna 2 much longer and stouter, especially in ♂; ultimate joint of peduncle usually longer than penultimate; flagellum with 3—6 joints, 1st much the longest; flagellum and at least ultimate joint of peduncle often carrying fascicles of plumose setae. Gnathopod 1, 6th joint oval or somewhat piriform; palm straight, oblique, defined slightly by an obtuse angle and palmar spines. Gnathopod 2 in ♂, 5th joint masked by 4th, cup-shaped, very small, obscurely separated from 6th, 6th very large, narrow, nearly straight, palm nearly parallel to front margin, with distal tooth near finger-hinge; the short hind margin ending in fully adult ♂ in a long tooth (in younger ♂ a denticle nearer the distal than the proximal end of the joint), ornamented on the outer side with tufts of setae, its apex overlapped by that of the long slightly sinuous finger; inner margin of finger sometimes on proximal half minutely tuberculate. Gnathopod 2 in ♀ much larger than gnathopod 1, irregular oval, or long piriform, hind margin very short, the long palm having a distal tooth near the finger-hinge. thence passing with a long excavation to 2 prominences, of which the 2^d is beset with 3 palmar spines, among which the apex of the long serrate finger impinges. Peraeopods 1 and 2 rather short and stout, 2^d joint a little expanded, 4th distally widened. Peraeopods 3—5 successively longer, 2^d joint oval. Uropod 3, outer ramus uncinate, with 2 teeth on upper distal margin. Telson very small, triangular, with 2 setae on each side of the acute apex. Colour, brown or red patches, of various shapes and sizes. L. 5—10 mm.

Atlantic with adjoining seas (Europe from Trondhjemsfjord to Naples; Azores; Rio Janeiro); Southern Indian Ocean (Kerguelen Island); Pacific (Philippines; Port Jackson [East-Australia]; lat. 43° S., long 82° W.).

5. **J. pusilla** (O. Sars) 1876 *Podocerus falcatus* (part.), A. Boeck, Skand. Arkt. Amphip., v. 2 p. 605 t. 28 f. 2 | 1893 *P. f.* (part.), A. Della Valle in: F. Fl. Neapel, v. 20 p. 445 | 1895 *P. f. var.?*, A. O. Walker in: Ann. nat. Hist., ser. 6 v. 15 p. 473 | 1882 *P. minutus* (non *Ischyrocerus m.* W. Lilljeborg 1855!), G. O. Sars in: Forb. Selsk. Christian., nr. 18 p. 112 t. 6 f. 6, 6a | 1894 *P. pusillus*, G. O. Sars, Crust. Norway, v. 1 p. 596 t. 212 f. 1.

In general like J. pulchella, but shorter and stouter. Head, lateral lobes rather broad and angular in front. Side-plate 1 in ♂ rather sharply produced, 5th nearly as deep as 4th. Pleon segment 3, postero-lateral corners simply quadrate. Eyes rather large, rounded, dark brown. Antenna 1, 3^d joint about twice as long as 1st, as long as 2^d, rather shorter than the 5-jointed flagellum. Antenna 2 rather longer, more robust, but not very stout; flagellum slender, 4-jointed, fascicles of setae not plumose. Gnathopod 2 in adult ♂ differing from that of J. pulchella in that the basal tooth of 6th joint is much shorter and broader and has a distinctly bifid apex. Gnathopod 2 in ♀, 6th joint more regularly oval, palm less strongly sinuous, defined by a very slight angle, the 2 prominences much smaller and less widely separated than in J. pulchella. Peraeopods 1—5 more slender, less densely setiferous. Uropod 3, outer ramus less coarsely hooked. Telson small, acutely produced at apex. Colour, variegated with irregular patches of brown. L. about 5 mm.

Arctic Ocean, North-Atlantic, North-Sea, Irish Sea and Skagerrak (South- and West-Norway, depth 37—188 m, on hydroids; British Isles; France).

6. **J. ocius** (Bate) 1862 *Podocerus o.*, Bate (& Westwood), Brit. sess. Crust., v. 1 p. 450 f. | 1862 *P. o.*, Bate, Cat. Amphip. Brit. Mus., p. 257 t. 44 f. 5 | 1866 *P. o.*, Cam. Heller in: Denk. Ak. Wien, v. 26 ɪɪ p. 45 | 1868 *P. o.*, Czerniavski in: Syezda Russ. Est., Syezda 1 Zool. p. 99 | 1881 *P. o.*, Nebeski in: Arb. Inst. Wien, v. 3 p. 152—154 t. 13 f. 43 | 1893 *P. o.* (part.), A. Della Valle in: F. Fl. Neapel, v. 20 p. 448 t. 14 f. 11—27 | 1895 *P. o.*, A. O. Walker in: Ann. nat. Hist., ser. 6 v. 15 p. 473 | 1895 *P. o.*, A. O. Walker in: P. Liverp. biol. Soc., v. 9 p. 316.

In general like J. pulchella, but distinguished by the gnathopod 2 in ♂ and ♀, and uropod 3. Antennae 1 and 2 rather slender. Antenna 1, flagellum with 4 joints, 1st the longest, thrice as long as the little 1-jointed accessory flagellum. Antenna 2, flagellum with 3 joints, 1st about twice as long as 2^d and 3^d combined, 2^d more than twice as long as 3^d. Gnathopod 1, 6th joint piriform or oval, palm not defined except by the palmar spines; finger slightly denticulate near the nail. Gnathopod 2 in ♂, 6th joint widening distally, hind margin long, ending in a considerable tooth, the slightly oblique palm having another rather smaller tooth adjacent to this, and after a cavity a broader tooth near the finger-hinge. Gnathopod 2 in ♀ smaller than in ♂ but very similar, only that the tooth of the hind margin is less advanced than the adjacent tooth of the palm. Peraeopods 3—5 rather more slender than in J. pulchella. Uropod 3, outer ramus with uncinate apex but without teeth on upper edge. Telson with a seta on each side of acute apex. Colour pale yellow. L. 3—5 mm.

- Bristol Channel (North Devon); Irish Sea (Isle of Man); Mediterranean; Black Sea.

7. **J. frequens** (Chilton) 1883 *Podocerus f.*, Chilton in: Tr. N. Zealand Inst., v. 15 p. 85 t. 3 f. 2 | 1884 *P. latipes* (♂), Chilton in: Tr. N. Zealand Inst., v. 16 p. 258 t. 19 f. 2 a—d | 1893 *P. frequens, P. l.*, A. Della Valle in: F. Fl. Neapel, v. 20 p. 447, 448.

Eyes moderately large, round. Antennae 1 and 2 equal, setose, slender. Antenna 1, 3^d joint longer than 1st, as long as 2^d, shorter than flagellum; flagellum 8-jointed, accessory flagellum 2- or 3-jointed. Antenna 2, ultimate joint of peduncle as long as penultimate, shorter than the 8-jointed flagellum; spines at apex of flagellum little curved. Gnathopod 1, 5th joint longer than 6th, 6th narrowly oval; proximal half of finger serrate. Gnathopod 2 in ♂, 5th joint short, cup-shaped, 6th stout, oblong, the hind margin ending in a square slightly emarginate process, a little overlapped by the broad proximally serrate finger; the not very oblique palm having a deep indent adjacent to process of hind margin. Gnathopod 2 in ♀ and probably young ♂ having 6th joint without the square process, but the palm a little concave. Peraeopods 1—5 rather stout, especially peraeopod 4, which in the adult ♂ has all the joints expanded, the 2^d—5th as broad as long. Uropod 3, peduncle stout, narrowing at apex, rami small, slender, nearly naked. Telson with 2 curved spines, and anterior to these 1 or 2 simple setae. L. about 3 mm.

Lyttelton Harbour [New Zealand].

J. californica (Boeck) 1872 *Podocerus californicus*, A. Boeck in: Forh. Selsk. Christian., 1871 p. 41 t. 1 f. 6 | 1893 *P. falcatus* (part.), A. Della Valle in: F. Fl. Neapel, v. 20 p. 445.

L. 7 mm.

North-Pacific (California).

J. falcata (Mont.) 1808 *Cancer (Gammarus) falcatus*, Montagu in: Tr. Linn. Soc. London, v. 9 p. 100 t. 5 f. 2 | 1812 *Astacus f.*, Pennant, Brit. Zool., ed. 5 v. 4 p. 34 | 1813/14 *Cancer (Gammarus) f.,* Jassa (part.)?, Leach in: Edinb. Enc., v. 7 p. 433 | 1899 *J. f.*, A. O. Walker in: Ann. nat. Hist., ser. 7 v. 3 p. 395 | 1829—43 *J. pelagica* (err., non Leach 1813/14!), Guérin-Méneville, Iconogr. Règne an., v. 2 Crust. t. 27 f. 3; v. 3 Crust. p. 23.

Possibly identical with J. dentex (p. 653). The elongate curved finger of gnathopod 2 has (in figure) a prominent tooth on the inner margin not far from the finger-hinge.

English Channel (South Devon).

J. orientalis (Dana) 1852 *Gammarus o.*, J. D. Dana in: P. Amer. Ac., v. 2 p. 212 | 1853 & 55 *Cratophium orientale*, J. D. Dana in: U. S. expl. Exp., v. 13 II p. 843; t. 56 f. 3a—b | 1862 *Podocerus orientalis*, Bate, Cat. Amphip. Brit. Mus., p. 258 t. 44 f. 6 | 1893 *P. falcatus* (part.), A. Della Valle in: F. Fl. Neapel, v. 20 p. 445.

Gnathopod 2, 6th joint subovate, palm nearly longitudinal, tridentate. L. 6 mm.

Eastern entrance of the Strait of Sunda.

J. ornata (Miers) 1875 *Podocerus ornatus*, Miers in: Ann. nat. Hist., ser. 4 v. 16 p. 75 | 1879 *P. o.*, Miers in: Phil. Tr., v. 168 p. 210 t. 11 f. 6 | 1893 *P. ? o.*, A. Della Valle in: F. Fl. Neapel, v. 20 p. 454.

L. 13 mm.

Swain's Bay [Kerguelen Island].

5. Gen. **Ischyrocerus** Kröyer

1838 *Ischyrocerus* (Sp. un.: *I. angvipes*), Kröyer in: Danske Selsk. Afh., v. 7 p. 283, 287 | 1894 *I.*, G. O. Sars, Crust. Norway, v. 1 p. 587.

Like Jassa (p. 652), except as follows. Side-plate 2 in ♂ not differently shaped from side-plate 3. Gnathopod 2 in the ♀ little larger than gnathopod 1. Gnathopod 2 in the ♂ with hind margin of the 6th joint not produced into a prominent decurrent tooth.

9 species accepted, 1 obscure.

Synopsis of accepted species:

1 { Eyes present — 2.
  { Eyes Wanting — 7.

2 { Uropod 3, rami more than half as long as peduncle   1. **I. nanoides** . . p. 657
  { Uropod 3, rami less than half as long as peduncle — 3.

3 { Gnathopod 2 in ♂, 6th joint more than twice as long
  {   as broad . . . .   2. **I. anguipes** . . p. 658
  { Gnathopod 2 in ♂, 6th joint not more than twice as
  {   long as broad — 4.

4 { Peraeopods 3—5 rather slender — 5.
  { Peraeopods 3—5 rather robust — 6.

5 { Pleon segment 3, postero-lateral corners square . .   3. **I. megacheir** . p. 659
  { Pleon segment 3, postero-lateral corners round . .   4. **I. assimilis** . . p. 659

6 { Eyes of moderate size . . . .   5. **I. latipes** . . . p. 660
  { Eyes unusually large . . . .   6. **I. megalops** . . p. 660

7 { Antenna 1, 1st joint as long as the head . . . . .   7. **I. tenuicornis** . p. 660
  { Antenna 1, 1st joint not as long as the head — 8.

8 { Antennae 1 and 2 short, robust; accessory flagellum
  {   nearly as long as 1st joint of primary . . . . .   8. **I. brevicornis** . p. 661
  { Antennae 1 and 2 long, slender; accessory flagellum
  {   not nearly as long as 1st joint of primary . . .   9. **I. tuberculatus** p. 661

1. **I. nanoides** (H. J. Hansen) 1887 *Podocerus n.*, H. J. Hansen in: Vid. Meddel., ser. 4 v. 9 p. 162 t. 6 f. 4—4b | 1893 *Protomedeia maculata* (part.), A. Della Valle in: F. Fl. Neapel, v. 20 p. 436.

♀. Body elongate, narrow. Head, lateral lobes acute, little produced. Side-plates rather small, in figure front lobe of 5th nearly as deep as 4th. Pleon segment 3, postero-lateral corners broadly rounded. Eyes rather small, reddish. Antennae 1 and 2 long, slender, with long setae. Antenna 1, 2d joint very little longer than 3^d, flagellum 1$^1/_2$ times as long as 3^d joint of peduncle, 7- or 8-jointed; accessory flagellum as long as 1st joint of primary, 1-jointed. Antenna 2 longer, $^2/_3$ as long as body; flagellum about 6-jointed. Gnathopods 1 and 2 nearly alike; 6th joint not large, oblong, palm very oblique, evenly convex, the defining angle not very conspicuous, marked by 2 palmar spines. Peraeopod 4 much longer than peraeopod 3 and peraeopod 5 than peraeopod 4; all rather slender. Uropod 3, peduncle rather long, longer than peduncle of uropod 2, stout, with long rami, more than half as long as peduncle. L. 4—5 mm. — ♂ unknown.

Baffin Bay (lat. 71° N., long. 59° W.).

2. I. anguipes Kröyer 1838 *I. angvipes*, Kröyer in: Danske Selsk. Afh., *v.* 7 p. 283 t. 3 f. 14 a--m | 1840 *I. anguipes*, H. Milne Edwards, Hist. nat. Crust., *v.* 3 p. 56 | 1859 *Podocerus a.*, R. M. Bruzelius in: Svenska Ak. Handl., n. ser. *v.* 3 nr. 1 p. 21 | 1874 *P. a.*, Buchholz in: Zweite D. Nordpolarf., *v.* 2 p. 378 Crust. t. 13 f. 2; t. 14 | 1876 *P. a.*, A. Boeck, Skand. Arkt. Amphip., *v.* 2 p. 603 t. 27 f. 5, 6 | 1893 *P. a.*, A. Della Valle in: F. Fl. Neapel, *v.* 20 p. 444 t. 57 f. 18 | 1894 *P. a.*, T. Stebbing in: Bijdr. Dierk., *v.* 17 p. 44 | ?1843 *Gammarus zebra*, H. Rathke in: N. Acta Ac. Leop., *v.* 201 p. 74 t. 3 f. 4 | 1851 *Ischyrocerus z.*, W. Liljeborg in: Öfv. Ak. Förh., *v.* 8 p. 23 | 1855 *Podocerus z.*, W. Liljeborg in: Vetensk. Ak. Handl., 1853 p. 446 | ?1853 *Cerapus fucicola*, Stimpson in: Smithson. Contr., *v.* 6 nr. 5 p. 48 t. 3 f. 34 | 1855 *Ischyrocerus minutus*, W. Liljeborg in: Öfv. Ak. Förh., *v.* 12 p. 128 | 1862 *Podocerus cylindricus* (err., non Say 1818!), Bate, Cat. Amphip. Brit. Mus., p. 256 t. 44 f. 4 | 1884 *P. falcatus* (err., non *Cancer (Gammarus) f.* Montagu 1808!), H. Blanc in: N. Acta Ac. Leop., *v.* 47 p. 79 t. 9 f. 96—101 | ?1889 *P. isopus*, A. O. Walker in: P. Liverp. biol. Soc., *v.* 3 p. 209 t. 11 f. 11—13 | ?1890 *P. i.*, A. O. Walker in: P. Liverp. biol. Soc., *v.* 4 p. 250 t. 16 f. 7 | ?1894 *P. megacheir*, T. Stebbing in: Bijdr. Dierk., *v.* 17 p. 44 | 1894 *Ischyrocerus anguipes, I. angvipes + I. minutus*, G. O. Sars, Crust. Norway, *v.* 1 p. 588 t. 209; p. 589 t. 210 f. 1 | 1895 *I. anguipes + I. m.*, Ohlin in: Acta Univ. Lund., *v.* 31 nr. 6 p. 56.

Body rather slender. Head, lateral lobes somewhat obtuse but with a minutely projecting point. Side-plates encreasing in depth from 1st—4th in ♀, but in ♂ 2^d—4th equal in depth; 5th with front lobe rather less deep than 4th. Pleon segment 3, postero-lateral corners rather obtusely quadrate. Eyes small, rounded, dark brown. Antenna 1 about $^1/_3$ as long as body; 3^d joint of peduncle much longer than 1st, little shorter than 2^d; flagellum about half as long as peduncle, attaining 9 joints, but often with fewer; accessory flagellum shorter than 1st joint of primary, with 2 joints, 2^d minute. Antenna 2 longer, especially in ♂; flagellum about as long as ultimate joint of peduncle, 5—7-jointed. Gnathopod 1 rather feeble, 6th joint oval, distally narrow; the oblique palm defined by palmar spines, finger serrate. Gnathopod 2 in ♂ very large; 5th joint very small, cup-shaped, 6th elongate, curved; the concave, hirsute hind margin or palm parallel to the convex front, and forming near the finger-hinge a truncate, finely denticulate process; the finger bulging a little after passing this process; palm margin occasionally straight or even a little convex. Gnathopod 2 in ♀ similar to gnathopod 1, but more robust. Peraeopods 1—5 rather strongly built. Peraeopods 3—5, 2^d joint oblong oval, hind margin smooth or crenulated; finger microscopically pectinate for some distance. Uropod 3, peduncle long, rami short, outer tapering, tipped with small spines, inner (not outer) rather broader, apically bent, having on upper margin 3 or 4 denticles. Telson little longer than broad, obtusely triangular, with dorsal transverse row of 1—4 slender spines on each side. Colour very variable. L. 4—15 mm.

Arctic Ocean (widely distributed); North-Atlantic, North-Sea and Skagerrak (Norway and Bohuslän; Grand Manan; Liverpool Bay?); Kattegat; West Baltic, depth 18 m.

3. **I. megacheir** (Boeck) 1871 *Podocerus m.*, A. Boeck in: Forh. Selsk. Christian., 1870 p. 247 | 1876 *P. m.*, A. Boeck, Skand. Arkt. Amphip., *v.* 2 p. 602 t. 29 f. 4 | 1893 *Podoceropsis m.* (part.), A. Della Valle in: F. Fl. Neapel, *v.* 20 p. 453 | 1894 *Ischyrocerus m.*, G. O. Sars, Crust. Norway, *v.* 1 p. 592 t. 211 | ? 1895 *I. m.*, A. M. Norman in: Ann. nat. Hist., ser. 6 *v.* 15 p. 492.

Body rather long and slender. Head, lateral lobes acute. Side-plates not very deep, 1st quadrate in front, 4th scarcely larger than 3^d, 5th nearly as deep as 4th. Pleon segment 3, postero-lateral corners quadrate. Eyes rather large, rounded oval, light brown. Antennae 1 and 2 about $^2/_3$ as long as body, with long setae, antenna 2 rather the longer. Antenna 1, 1st joint shorter than head, 3^d elongate, as long as 2^d or a little longer; flagellum longer than 3^d joint of peduncle, with 9—10 joints in ♂ and 4—8 in ♀; accessory flagellum as long as 1st joint of primary, with minute 2^d joint. Antenna 2, ultimate joint of peduncle longer than penultimate, shorter than flagellum, which has 8—11 joints. Gnathopod 1, 6th joint oval, with palm ill-defined, larger in ♂ than in ♀. Gnathopod 2 in ♂, 6th joint large and tumid, about thrice (or less) as long as broad; palm slightly flexuous and a little crenulate, defined by an obtuse angle with small palmar spines; finger strong, apex impinging within border of palm. Gnathopod 2 in ♀ similar to gnathopod 1, but with shorter 5th and more oblong and stouter 6th joint. Peraeopods 1—5 slender; 2^d joint in peraeopods 3—5 oblong oval, narrowing downward. Uropod 3, peduncle elongate, outer ramus but slightly hooked at the tip, with 1 denticle on upper edge. Telson as long as broad, with 2 setae on each side of sub-acute apex. Colour whitish, pellucid, with light transverse orange bands L. 7—12 mm.

Arctic Ocean and North-Atlantic (Norway, Iceland, Bear Island, Spitzbergen). Depth 100—1444 m.

4. **I. assimilis** (O. Sars) 1879 *Podocerus a.*, G. O. Sars in: Arch. Naturv. Kristian., *v.* 4 p. 450 | 1885 *P. a.*, G. O. Sars in: Norske Nordhavs-Exp., *v.* 6 Crust. I p. 205 t. 17 f. 1 a—c | 1893 *Podoceropsis megacheir* (part.), A. Della Valle in: F. Fl. Neapel, *v.* 20 p. 453.

Body rather elongate. Head, lateral lobes acutely produced. Side-plates of moderate depth, 1st rounded in front, 5th notably not so deep as 4th. Pleon segment 3, postero-lateral corners obtusely rounded. Eyes rather small, longitudinally oval, dark brown. Antennae 1 and 2 powerful, subequal, more than half as long as body. Antenna 1, 1st joint as long as head, 2^d and 3^d much longer, equal; flagellum longer than 3^d joint of peduncle, 9-jointed; accessory flagellum 1-jointed, very small, scarcely more than $^1/_3$ as long as 1st joint of primary. Antenna 2, ultimate joint of peduncle rather longer than penultimate, both elongate, flagellum as long as ultimate, 8-jointed. Gnathopod 1, 6th joint oval, palm quite undefined, unless by extent of finger. Gnathopod 2, especially in ♂, powerful, 6th joint very large and tumid, as long as 4 preceding joints together, oblong oval; the long nearly straight hind or palmar margin fringed with partly plumose setae; finger strong, falciform, closing upon inner surface of hand. Peraeopods 3—5, 2^d joint oblong; peraeopod 4 much longer than 3^d, 5th than 4th. Uropod 3, peduncle rather elongate, rami remarkably small. Telson short, but thick, tubular. Colour whitish, with brown spots or shadings. L. reaching 8 mm.

North-Atlantic and Arctic Ocean (west of Helgeland, depth 850 m; south of Bear Island, depth 66 m).

5. **I. latipes** Krøyer 1842 *I. l.*, Krøyer in: Naturh. Tidsskr., *v.* 4 p. 162 | 1893
I. l., A. Della Valle in: F. Fl. Neapel, *v.* 20 p. 450 t. 57 f. 19 | 1894 *I. l.*, G. O. Sars,
Crust. NorWay, *v.* 1 p. 591 | 1862 *Podocerus l.*, Bate, Cat. Amphip. Brit. Mus., p. 257 |
1876 *P. l.*, A. Boeck, Skand. Arkt. Amphip., *v.* 2 p. 600 t. 29 f. 1 | 1887 *P. l.*, H. J.
Hansen in: Vid. Meddel., ser. 4 *v.* 9 p. 161 t. 6 f. 3—3 b.

Body elongate, depressed, back broad. Head, lateral lobes rounded.
Side-plates 3 and 4 much larger than 2^d, 5th much less deep than the 4th.
Pleon segment 3, postero-lateral corners quadrate. Eyes obliquely oval, not
large. Antenna 1 more than half as long as body, setose, 1st joint about
as long as head, 2^d much longer, 3^d as long as 2^d; flagellum scarcely
longer than 3^d joint of peduncle, 9-jointed; accessory flagellum 1-jointed,
about half as long as 1st joint of primary. Antenna 2 longer, ultimate joint
of peduncle little longer than penultimate, flagellum little longer than
ultimate, 7-jointed. Maxillipeds, palp (Boeck) elongate, the 3^d joint (in figure)
more than twice as long as broad. Gnathopod 1, 5th joint distally widened
and setose, 6th oval, nearly triangular, $^1/_3$ longer than broad, palm convex;
finger long, curved, serrate. Gnathopod 2 in ♂ much larger, 5th joint very
short, cup-shaped, 6th not twice as long as broad, oval, plump; hind and
palmar margin evenly convex, fringed with setae; finger long and curved.
Gnathopod 2 in ♀ (Hansen), 6th joint very short and broad, palm oblique,
long, with an emargination, defining angle rounded. Peraeopods 3—5, 2^d
joint broad, in peraeoped 3 longer than broad. Uropod 3, peduncle very
long, rami very short, the outer (Boeck) having 2 blunt teeth on the end,
inner without teeth. Telson longer than broad, oval, apically pointed, with
a spine on each side of the apex. L. 14—15 mm.

Arctic Ocean (Greenland). Depth 15—188 m.

6. **I. megalops** O. Sars 1894 *I. m.*, G. O. Sars, Crust. Norway, *v.* 1 p. 591 t. 210 f. 2.

♂. Body not very slender but somewhat compressed. Head, lateral
lobes angular. Side-plates rather large, 5th much less deep than 4th. Pleon
segment 3, postero-lateral corners obtusely quadrate. Eyes unusually large,
rounded oval, dark. Antennae 1 and 2 rather strongly built. Antenna 1
about half as·long as body; 1st joint shorter than head, 2^d and 3^d not
very long, but each longer than 1st; flagellum longer than 3^d joint of
peduncle, 8-jointed; accessory flagellum extremely minute, not nearly half
as long as 1st joint of primary. Antenna 2 scarcely longer than antenna 1;
ultimate joint of peduncle a little longer .than penultimate, flagellum subequal to penultimate, 6-jointed. Gnathopods 1 and 2 rather stout and similar,
but gnathopod 2 much the larger; 6th joint broad, oval; palm oblique, simple,
defined by a slightly marked angle, with several palmar spines. Feraeopods 1—5 rather stout; 2^d joint in peraeopod 3 very broad proximally.
Uropod 3 nearly as in I. anguipes (p. 658). Telson triangular, with a few
simple dorsal setae, apex obtusely pointed. Colour whitish, with indistinct
brownish bands. L. 7 mm.

Arctic Ocean (Hammerfest [Norway]). On hydroids; depth about 75 m.

7. **I. tenuicornis** (O. Sars) 1879 *Podocerus longicornis* (err., non Cam. Heller
1866!), G. O. Sars in: Arch. Naturv. Kristian., *v.* 4 p. 461 | 1885 *P. tenuicornis*, G. O.
Sars in: Norske Nordhavs-Exp., *v.* 6 Crust. I p. 209 t. 17 f. 3 | 1893 *Podoceropsis sophiae*
(part.), A. Della Valle in: F. Fl. Neapel, *v.* 20 p. 452.

Body rather compressed. Head, lateral corners greatly produced, acute.
Side-plates well developed, front lobe of 5th notably less deep than 4th.

Pleon segment 3, postero-lateral corners acute. Eyes wanting. Antennae
1 and 2 nearly as long as body, with long setae. Antenna 1, 1st joint
longer than head, 2^d much longer than 1st, 3^d nearly as long as 2^d;
flagellum with 5 elongate joints; accessory flagellum 1-jointed, half as long as
1st joint of primary. Antenna 2, ultimate joint of peduncle long, slightly
shorter than penultimate; flagellum as in antenna 1. Gnathopods 1 and 2
rather feeble, 6th joint oval, without spines, merely furnished with delicate
bristles; 2^d gnathopod not much larger than 1st. Peraeopods 3—5 almost
equal in length; 2^d joint oblong oval. Uropods 1—3 slender. Colour uniform
white. L. 3 mm.

Arctic Ocean (north-West of Finmark). Cold area; depth 2090 m.

8. **I. brevicornis** (O. Sars) 1879 *Podocerus b.*, G. O. Sars in: Arch. Naturv.
Kristian., *v.* 4 p. 460 | 1882 *P. b.*, Hoek in: Niederl. Arch. Zool., suppl. 1 nr. 7 p. 63 |
1885 *P. b.*, G. O. Sars in: Norske Nordhavs-Exp., *v.* 6 Crust. I p. 207 t. 17 f. 2 a—c | 1893
Podoceropsis sophiae (part.), A. Della Valle in: F. Fl. Neapel, *v.* 20 p. 452.

Body rather stout. Head, lateral lobes acute. Side-plates pretty well
developed, 5th notably less deep than 4th. Pleon segment 3, postero-lateral
corners quadrate. Eyes wanting. Antennae 1 and 2 short, robust, not half
as long as body, with short setae. Antenna 1, 1st joint shorter than head,
3^d longer than 1st, shorter than 2^d; flagellum 7- or 8-jointed; accessory
flagellum 1-jointed, nearly as long as 1st joint of primary. Antenna 2,
flagellum shorter than ultimate joint of peduncle, 6- or 7-jointed. Gnathopods 1 and 2 robust, similar, but 2^d, especially in ♂, considerably larger;
6th joint broad, compressed, the oblique palm well defined from the hind
margin, with powerful palmar spines in gnathopod 1, and a distinct angular
projection in gnathopod 2. . Peraeopods 3—5, 2^d joint rather broad, especially
in peraeopod 5. Uropods 1—3 not very elongate, in particular uropod 3.
Colour whitish. L. 6—9 mm.

North-Atlantic and Arctic Ocean (Storeggen Bank and in open sea; lat. 63° N.,
long. 5° E.; round Bear Island and Spitzbergen, depth 275—1444 m; lat. 74° N., long.
45° E., depth 301 m).

9. **I. tuberculatus** (Hoek) 1882 *Podocerus t.*, Hoek in: Niederl. Arch. Zool.,
suppl. 1 nr. 7 p. 64 t. 3 f. 32 | 1893 *P. t.*, A. Della Valle in: F. Fl. Neapel, *v.* 20 p. 443
t. 57 f. 17 | 1894 *P. t.*, T. Stebbing in: Bijdr. Dierk., *v.* 17 p. 44 | 1888 *P. hoeki*, T. Stebbing
in: Rep. Voy. Challenger, *v.* 29 p. 1136 t. 120.

Head, lateral lobes obtuse with minute point in front. Side-plate 4
the largest. Pleon segment 3, postero-lateral corners obtusely quadrate.
Eyes wanting. Antenna 1, 3^d joint a little longer than 1st, decidedly
shorter than 2^d; flagellum 6-jointed, longer than 3^d joint of peduncle;
accessory flagellum half or less than half as long as 1st joint of primary,
1-jointed. Antenna 2 longer, ultimate joint of peduncle slightly longer
than penultimate or than the 5-jointed flagellum. Gnathopod 1, 5th joint cup-
shaped, 6th broad, proximally oblong, but narrowing distally; the palm longer
than hind margin, very oblique, straight, finely serrate, defined by palmar
spines, among which the curved serrate finger closes. Gnathopod 2 in ♀ like
gnathopod 1, but with the 6th joint larger, and in the ♂ of the same general
character, but with 5th joint comparatively smaller, 6th large, more oblong, with
the palm obliquely sinuous, irregularly tuberculate, defined by a small tooth,
within which the broad, somewhat denticulate finger closes. Peraeopods 1
and 2 rather stout. Peraeopods 3—5, 2^d joint oblong, in peraeopod 4 narrowing
distally. Uropod 3 rather short, peduncle about 2$^1/_2$ times as long as rami,

of which one is apically bent, but apparently without any row of denticles on the upper edge. Telson rounded triangular. L. 5 mm.

Barents Sea (lat. 71°—77° N., long. 50° E., depth 126—320 m); South-Pacific (lat. 40° S., long. 178° E., depth 2071 m).

I. monodon (Heller) 1866 *Podocerus m.*, Cam. Heller in: Denk. Ak. Wien, *v.* 26 ii p. 45 t. 4 f. 4, 5 | 1893 *P. falcatus* (part.), A. Della Valle in: F. Fl. Neapel, *v.* 20 p. 445.

Finger of gnathopod 2 not denticulate. L. 5 mm.

Adriatic (Lesina).

---

38. Fam. **Corophiidae**

1849 *Corophidae*, J. D. Dana in: Amer. J. Sci., ser. 2 *v.* 8 p. 139 | 1876 *C.*, A. Boeck, Skand. Arkt. Amphip., *v.* 2 p. 619 | 1888 *Corophiidae*, T. Stebbing in: Rep. Voy. Challenger, *v.* 29 p. 1154 | 1894 *C.*, G. O. Sars, Crust. Norway, *v.* 1 p. 606 | 1893 *Corofidi* (part.), A. Della Valle in: F. Fl. Neapel, *v.* 20 p. 351.

Body usually more or less depressed. Pleon small. Side-plates usually small and often not in continuity. Antennae 1 and 2 of variable proportions, with or without accessory flagellum. Mouth-parts generally normal, except that mandibular palp is not always 3-jointed, and the inner plate of maxilla 1 is sometimes evanescent. Gnathopods 1 and 2 variable in character and relative proportions. Peraeopods 1 and 2 usually glandular. Peraeopod 5 the longest. Pleopods often with peduncle internally expanded. Uropod 1 biramous. Uropod 2 biramous or uniramous. Uropod 3 small, weakly biramous, uniramous, or even without rami. Telson simple (Fig. 115, 118 p. 686, 691), sometimes lobate.

Marine, but extending into brackish or even almost fresh Water.

11 genera, 44 accepted species and 11 doubtful.

Synopsis of genera:

$$10 \begin{cases} \text{Mandibular palp 1-jointed} \dots \dots \dots & \text{10. Gen. } \textbf{Siphonoecetes} \quad \text{p. 681} \\ \text{Mandibular palp 2-jointed} \dots \dots \dots & \text{11. Gen. } \textbf{Corophium} \; . \; . \; \text{p. 685} \end{cases}$$

1. Gen. **Concholestes** Giles

1888 *Concholestes* (Sp. un.: *C. dentalii*), G. M. Giles in: J. Asiat. Soc. Bengal, *v.* 57 p. 237 | 1890 *C.*, G. M. Giles in: J. Asiat. Soc. Bengal, *v.* 59 p. 63 | 1893 *C., Siphonoecetes* (part.), A. Della Valle in: F. Fl. Neapel, *v.* 20 p. 895, 924.

Mouth-parts imperfectly known; otherwise agreeing with Siphonoecetes (p. 681), except that uropod 2 is uniramous, and that uropod 3 has no ramus.

1 species.

1. **C. dentalii** Giles 1888 *C. d.*, G. M. Giles in: J. Asiat. Soc. Bengal, *v.* 57 p. 238 t. 7 f. 7—11 | 1893 *C. d., Siphonoecetes* (part.)?, A. Della Valle in: F. Fl. Neapel, *v.* 20 p. 895, 924 | 1890 *C. dentallii*, G. M. Giles in: J. Asiat. Soc. Bengal, *v.* 59 p. 63.

Body slender, head and peraeon thrice as long as pleon; peraeon segment 1 dorsally produced in a small setose lobe over base of head. Head almost truncate in front, but with 2 little median points. Side-plates as in Siphonoecetes (p. 681). Eyes very small, on front angles of head. Antenna 1 stout, less than half as long as body, setose, flagellum as long as 2^d joint of peduncle, 5-jointed. Antenna 2 very stout, nearly as long as peraeon, setose; ultimate and penultimate joints of peduncle subequal; flagellum short and stout, 2-jointed, ending in 2 curved spines. Mandible of simple form and palpate, maxillipeds small and unguiculate. Gnathopod 1 feebly subchelate; 5th joint slightly longer and broader than 6th, which is described as without palm, the grasping power lying between the stiffish hairs of its hind margin and the serrate finger. Gnathopod 2 little longer but much stouter; 5th joint short, broad, cup-shaped, 6th swollen at the base, hind margin very short, palm oblique, strongly armed with 3 formidable teeth; finger stout, strongly serrate, while a powerful secondary tooth projects obliquely on either side, nearly as strong as the central tooth, and giving the finger a trifid appearance. Peraeopods 1 and 2 short, 2^d joint expanded, finger long and straight. Peraeopods 3 and 4, 2^d joint narrow, 5th stout, cylindric, armed at apex with 1 short stout spine, the rounded apex being also densely clothed with short recurved hooks; the retroverted 6th joint attached to front margin of 5th; finger forming a small but strong hook. Peraeopod 5 longer, normal, 2^d joint not much expanded, distally tapering. Marsupial plates narrow, fringed with long setae. Pleopods 1—3 small, peduncle much broader than long. Uropod 1, peduncle stout, armed with a few stout spines, rami half as long as peduncle, stout, spinose. Uropod 2 shorter, peduncle very short, broad; ramus rounded, with apical recurved hooks. Uropod 3 short, blunt, apically spiniferous. L. about 8 mm.

Bay of Bengal (off the Seven Pagodas, near Madras). Depth 13 m, on a sandy bottom. Makes its home in the shell of Dentalium lacteum, lining it with cemented sandy particles.

2. Gen. **Paracorophium** Stebb.

1899 *Paracorophium* (Sp. un.: *Corophium excavatum*), T. Stebbing in: Ann. nat. Hist., ser. 7 *v.* 3 p. 350.

Body compressed. Head with produced lateral lobes. Side-plates continuous, 1st not produced forward. Eyes small, on lateral lobes of head. Antenna 1 slender, without accessory flagellum; flagellum with several joints.

Antenna 2 robust; flagellum slight, of more than 3 joints. Mandible with 3-jointed palp. Gnathopod 1 as in Corophium (p. 685). Gnathopod 2 nearly as in Corophium, but the long process of 4th joint fringed on its front or inner margin, while the 5th is fringed on its hind margin, the 2 joints therefore, though fitting together, having no look of coalescence; 6th joint with small palm. Peraeopod 3 the shortest, setose; 6th joint with strong spines. Peraeopods 4 and 5 successively much longer. Peraeopods 3—5, 2^d joint widely expanded. Uropod 1, and still more uropod 2, stout, with strong spines, biramous. Uropod 3 small, outer ramus nearly as long as peduncle, inner oval, minute. Telson entire, short.

1 species.

1. **P. excavatum** (G. M. Thoms.) 1884 *Corophium e.*, G. M. Thomson in: Tr. N. Zealand Inst., *v.* 16 p. 236 t. 12 f. 1—8 | 1899 *C. e.*, *Paracorophium*, T. Stebbing in: Ann. nat. Hist., ser. 7 *v.* 3 p. 241, 350.

Pleon segments 4 and 5 are dorsally coalesced (in figure). Head with small rostral point, lateral lobes narrowly rounded. Eyes small, round, on lateral lobes. Antennae 1 and 2 subequal, more than $^1/_2$ as long as body. Antenna 1, 1st and 2^d joints long, 1st much the longer (figure), 3^d very short; flagellum as long as peduncle, 10-jointed. Antenna 2, joints of peduncle short and thick; flagellum short, 6-jointed. Mandibular palp longer than trunk, 1st joint rather long. Gnathopod 1 rather small, 3^d joint short, with tuft of long setae, 4th very short, 5th long, fringed with long setae on hind margin, 6th rather shorter, slightly widened distally; palm transverse, finger scarcely overlapping it. Gnathopod 2 rather longer, slender; 4th joint produced into a scoop-like process, fringed on each margin with long setae, and into which the carpus [5th joint] is fitted closely when the limb is folded; 5th joint slightly widened distally, densely fringed along hind margin, 6th subequal to 5th, margins nearly parallel, hind produced into a small tooth; palm transverse, sinuous, much overlapped by the finger. Peraeopods 1 and 2 rather short, simple, nearly destitute of spines or setae. Peraeopod 3, 2^d joint piriform, 4th longer than 5th or 6th, 6th with a row of strong spines along the reverted front margin; finger short. Peraeopod 4 twice as long as 3^d; peraeopod 5 still longer, finger as in 3^d and 4th reverted. Uropods 1 and 2 each with strong spines on peduncle and rami. Uropod 3 with peduncle as broad as long, and carrying a few setae but no spines; inner ramus not half as long as outer. Telson broader than long, apically rounded. Colour dirty grey. L. 4 mm.

New Zealand (Brighton Creek (salt water) near Dunedin).

3. Gen. **Camacho** Stebb.

1888 *Camacho* (Sp. un.: *C. bathyplous*), T. Stebbing in: Rep. Voy. Challenger, *v.* 29 p. 1178.

Head, mouth-parts, peraeon with its side-plates and gnathopods 1 and 2 (♀) nearly as in Xenodice (p. 699), but differing as follows. Mandible with spines in spine-row numerous. Maxilla 1, inner plate with a single apical seta. Maxillipeds with finger of palp as long as the 3^d joint. Antenna 1 with elongate 1st joint (the rest unknown). Pleon segment 4 not especially elongate. Pleopods, peduncle distally widened. Uropods 1—3 biramous. Uropod 3 with short broad peduncle and small rami, the outer longer than the peduncle, the inner minute. Telson simple.

1 species.

1. **C. bathyplous** Stebb. 1888 *C. b.*, T. Stebbing in: Rep. Voy. Challenger, *v.* 29 p. 1179 t. 127.

♀. Body elongate, head and peraeon subdepressed, pleon compressed; peraeon segments 3—7 laterally dimpled. Head, lateral lobes very small, subacute. Side-plates all shallow, for the most part not contiguous, 1st—4th with front corner directed forward. Pleon segment 3, postero-lateral corners obtusely quadrate. Eyes wanting. Antenna 1, 1st joint considerably longer than the head. Antenna 2, antepenultimate joint of peduncle at least half as long as 1st joint of antenna 1. Upper lip broad. Mandible, 10 spines in spine-row; spines of 3^d joint of palp very elongate. Maxilla 1, outer plate with 11 spines, 2^d joint of palp with 8 apical spine-teeth. Gnathopods 1 and 2, 3^d and 4th joints short, 5th but little shorter than 2^d, rather longer and narrower than the oval 6th, both strongly armed, with slender pectinate spines; finger matching the convex ill-defined palm; 2^d and 5th joints longer in gnathopod 2 than in gnathopod 1. Branchial vesicles and marsupial plates narrow. Pleopods 1—3, peduncle short, the members of each pair meeting distally; slender rami wide apart; coupling spines much dentate, cleft spines long; 1st joint of inner ramus dilated proximally. Uropod 1, peduncle longer than outer ramus. Uropod 2 shorter, peduncle rather longer than the subequal, spinose rami. Uropod 3, peduncle almost broader than long, expanded beyond the rami, the pair meeting under the telson. Telson rather broader than long, nearly circular, but with a produced angle on each side of the convex apex. L. 16 mm.

South-Pacific (lat. 41° S., long. 178° E.). Depth 2011 m.

4. Gen. **Cerapus** Say

1817 *Cerapus* (Sp. un.: *C. tubularis*), Say in: J. Ac. Philad., *v.* 1ɪ p. 49 | 1888 *C.*, T. Stebbing in: Rep. Voy. Challenger, *v.* 29 p. 100, 1157 | 1893 *C.*, A. Della Valle in: F. Fl. Neapel, *v.* 20 p. 876 | 1894 *C.*, G. O. Sars, Crust. Norway, *v.* 1 p. 606 | 1840 *Cerapodina* (Sp. un.: *C. abdita*), H. Milne Edwards, Hist. nat. Crust., *v.* 3 p. 62.

Body slender; depressed pleon small, its after part strongly flexed. Head with distinct rostrum; eyes at lateral corners. Side-plates 1—4 very small, 5th and 6th larger, 5th—7th bilobed, front lobe the deeper. Antenna 1 without accessory flagellum. Antennae 1 and 2, flagellum short. Upper lip not bilobed. Mouth-parts normal. Mandible, 3^d joint of palp fully as long as 2^d. Maxilla 1, inner plate not or little setose, outer with 9 or 10 apical spines. Maxilla 2, inner plate not fringed on inner margin. Gnathopod 1 subchelate. Gnathopod 2 (Fig. 112 p. 667) complexly subchelate and powerful in ♂, in ♀ feeble and simple. Peraeopods 1 and 2 (Fig. 113 p. 667) short, 2^d joint broad and long, glandular, front margin convex, base of joint expanded in peraeopod 1. Peraeopods 3—5 short, recurved, finger very short, bidentate. Peraeopod 3 (Fig. 114 p. 667), 2^d joint rather expanded, short, 4th distally expanded on both sides of 5th. Peraeopods 4 and 5, 2^d joint variable as to expansion. Branchial vesicles narrow, attached only to peraeopods 1—3. Marsupial plates 1—3 narrow, 4th large. Pleopod 1 large, 2^d and 3^d successively smaller, with 1 ramus dwindled. Uropod 1 normal. Uropods 2 and 3 uniramous, the ramus short, uncinate. Telson short, broad, bilobed, densely spinulose above. It is doubtful whether all the above characters are certainly applicable throughout the genus.

6 species.

Synopsis of species:

1 { Antenna 1, 1st joint distally widened — **2.**
    { Antenna 1, 1st joint not distally widened — **3.**

2 { Antennae 1 and 2, flagella 2-jointed . . . . . . 1. **C. crassicornis** . . p. 666
    { Antennae 1 and 2, flagella 4- or 5-jointed . . . 2. **C. sismithi** . . . p. 666

3 { Antennae 1 and 2, flagella 3-jointed : . . . . . 3. **C. tubularis** . . . p. 667
    { Antennae 1 and 2, flagella 4—6-jointed — **4.**

4 { Antenna 1, 3d joint of peduncle much longer
    {   than 2d . . . . 4. **C. flindersi** . . . p. 668
    { Antenna 1, 3d joint of peduncle not much longer
    {   than 2d — **5.**

5 { Gnathopod 2 in ♂, inner tooth of 5th joint
    {   prominent, acute . . . . 5. **C. abditus** . . . . p. 668
    { Gnathopod 2 in ♂, inner tooth of 5th joint not
    {   prominent, obtuse . . . . 6. **C. calamicola** . . p. 669

1. **C. crassicornis** (Bate) 1857 *Siphonocetus c.,* Bate in: Ann. nat. Hist., ser. 2
v. 19 p. 149 | 1862 *Siphonoecetes c.,* Bate & Westwood, Brit. sess. Crust., *v.* 1 p. 469 f. |
1882 *Cerapus c.,* G. O. Sars in: Forh. Selsk. Christian., nr. 18 p. 113 t. 6 f. 8 | 1893 *C. c.,*
A. Della Valle in: F. Fl. Neapel, *v.* 20 p. 378 t. 55 f. 52 | 1894 *C. c.,* G. O. Sars, Crust.
Norway, *v.* 1 p. 607 t. 217.

Pèraeon segments 1 and 2 in ♀ shorter than any of the following 5;
but in ♂ segment 1 is not shorter and segment 2 much longer. Head,
rostrum rather large, acute; lateral lobes short, obtuse. Side-plate 5 much
the largest, its front lobe broad and deep. Pleon segment 3, postero-lateral
corners rounded. Eyes small, round, dark. Antennae 1 and 2 very setose,
short, subequal. Antenna 1 the stouter, slightly the longer, nearly $^1/_3$ as
long as body; 1st joint about as long as 2d and 3d combined, much broader,
widened distally and produced to a triangular lobe over 2d joint, which is
as long as 3d; flagellum scarcely longer than 3d joint of peduncle, with 2 joints,
last minute. Antenna 2, ultimate and penultimate joints of peduncle subequal; flagellum as in antenna 1. Gnathopod 1, 5th joint nearly as long as
6th, distally wide with setose lobe, 6th wide at base; palm oblique, scarcely
defined; finger long. Gnathopod 2 in ♂, 2d joint expanded to a broad
oval, 5th large, scarcely widening distally, front margin sinuous, partially
serrate, hind margin produced near centre to a strong tooth, distally forming
a transverse palm, defined by a tooth with smaller one within; 6th joint
shorter, much narrower, projecting finger-wise over palm of 5th, while the
long finger impinges on middle tooth of 5th. Gnathopod 2 in ♀ feeble, 2d joint
slightly widened; 5th joint triangular, broader than the narrowly oval or fusiform
6th, which is without distinct palm. Peraeopods 4 and 5 with 2d joint little
expanded, especially that of peraeopod 5. Uropod 1, outer ramus much
larger than inner, outer margin spinulose. Uropod 2 only about half as
long as 1st, ramus oblong oval. Uropod 3, peduncle rather thick, ramus
extremely minute. Telson with 2 dorsal rows of sharp, upturned spinules.
Colour pale yellow. L. ♀ 4, ♂ 5 mm.

Inhabits a free membranous tube, cylindrical, open at both ends, 1st joint of
antenna 1 acting as an operculum.

North-Sea (Northumberland; Jaederen [Norway], depth 75—94 m).

2. **C. sismithi** Stebb. 1888 *C. s.,* T. Stebbing in: Rep. Voy. Challenger, *v.* 29
p. 1158 t. 124 | 1893 *C. s.,* A. Della Valle in: F. Fl. Neapel, *v.* 20 p. 379 t. 55 f. 53—57 |
1894 *C. s.,* G. O. Sars, Crust. Norway, *v.* 1 p. 607.

♀ unknown. — ♂. Peraeon segment 2 not longer than any of the succeeding peraeon segments. Head, rostrum acute, slightly depressed, reaching beyond the slightly rounded lateral lobes. Side-plate 5 very broad, front lobe rather deeper than in the other side-plates. Eyes small, round. Antenna 1 less than half as long as body; 1st joint shorter than 2d and 3d combined, distally widened; rounded process above produced over 2d joint, process below acute; 3d joint shorter than 2d; flagellum rather longer than 1st joint

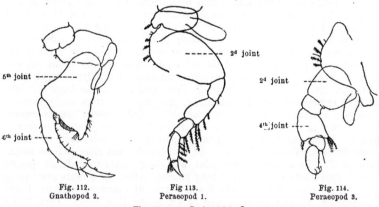

Fig. 112.
Gnathopod 2.

Fig 113.
Peraeopod 1.

Fig. 114.
Peraeopod 3.

Fig. 112—114. C. sismithi, ♂.

of peduncle, 5-jointed. Antenna 2 a little longer, ultimate and penultimate joints of peduncle subequal; flagellum longer than ultimate joint of peduncle, 4-jointed. Gnathopod 1, 5th joint distally rather wider than 6th, which is widest near base; palm serrate, finely pectinate; finger occupying apex of 6th joint, pectinate and with decurrent subapical tooth. Gnathopod 2 (Fig. 112). 2d joint widest near base, subrectangular, 5th much broader and longer than 2d, wide at base, much wider distally; front margin sinuous, hind rather longer, ending in a tooth, a large convex denticulate process occupying its palmar or distal margin between the tooth and base of strap-shaped, much curved 6th joint, which is nearly as long as 5th, 4—5 times as long as broad, with tooth near base and adpressed teeth or spines near apex of inner margin; 'finger less than ¹/₂ as long as 6th joint. Peraeopod 1 (Fig. 113), the large 2d joint widest above, proximal front angle broadly rounded, 4th joint much longer than broad. Peraeopod 2, 2d joint broadly oval, 4th and 5th joints longer than in peraeopod 1. Peraeopods 4 and 5, 2d joint well expanded, though successively rather narrower and longer than 2d joint of peraeopod 3. Peraeopod 3 see Fig. 114. Uropods 1—3 and telson as in C. crassicornis. L. 3 mm.

In cylindrical tubes of sand.

Cumberland Bay [Kerguelen Island]. Depth 226 m.

3. **C. tubularis** Say 1817 *C. t.*, Say in: J. Ac. Philad., *v.* 1ɪ p. 50, 96 t. 4 f. 7—11 | 1880 *C. t.*, S. I. Smith in: Tr. Connect. Ac., *v.* 4 p. 277 t. 2a | 1888 *C. t.*, T. Stebbing in: Rep. Voy. Challenger, *v.* 29 p. 101, 522 | 1893 *C. t.*, A. Della Valle in: F. Fl. Neapel, *v.* 20 p. 378 t. 55 f. 46—51 | 1894 *C. t.*, G. O. Sars, Crust. Norway, *v.* 1 p. 607.

Body broad, depressed, shallow, in ♂ tapering slightly and continuously from head to telson, in ♀ 4th and 5th peraeon segments each about twice as long as 1st or 7th. Head with slight dorsal carina in front, rostrum small. Side-plate 5 the largest. Eyes small, black. Antennae 1 and 2 subequal, shorter in $\female$ than in $\male$. Antenna 1 in $\male$ rather more than half as long as body; 1st joint stout, laterally compressed, lower margin with carina prominent near base, 2^{d} and 3^{d} subequal, each rather longer than 1st and subequal to 3-jointed flagellum; 1st joint of flagellum rather longer than 2^{d} and 3^{d} combined. Antenna 2 slightly shorter, scarcely stouter; ultimate joint of peduncle a little longer than penultimate, flagellum scarcely as long as ultimate, 3-jointed, 1st joint considerably longer than 2^{d} and 3^{d} combined. Gnathopod 1 about as in C· sismithi (p. 666). Gnathopod 2 in $\male$ in general shape as in C. sismithi, but front margin of 5th joint convex, tooth of hind margin more produced, its palmar margin shorter, occupied by a cavity and a small tooth near the base of the 6th joint, which is only about twice as long as broad, hind margin irregular, minutely denticulate, finger stout, serrulate, rather shorter than 6th joint. Peraeopod 1 in $\male$, the large glandular 2^{d} joint rectangular, widest proximally, the projecting front angle narrowly rounded; 4th joint broader than long, 5th a little shorter, nearly square. Peraeopod 2 in $\male$, 2^{d} joint broader in the middle, other joints as in peraeopod 1. Peraeopod 1 in $\female$, 2^{d} joint proportionally broader than in $\male$, and angle different. Peraeopod 2 in $\female$, 2^{d} joint broader and more oval in outline than in $\male$. Peraeopod 3, 4th joint with long, narrow, spatulate process behind, tipped with 1 short and 3 long plumose setae; the process overlapping the 5th joint, which is shaped like it and is apically squamose, carrying 1 seta. Peraeopod 4, 2^{d} joint about as long as 4th, longer than broad. Peraeopod 5 a little more slender. Rest like C. crassicornis (p. 666). Colour almost black from crowded specks of dark purplish pigment, extremities of appendages colourless, semi-translucent. L. 4—5 mm.

Tube free, reaching 7 mm, black, cylindrical, slightly enlarged at one or both ends.

North-Atlantic (Egg Harbor [New Jersey]; Vineyard Sound, depth 15—19 m; Noank Harbor [Connecticut]).

4. **C. flindersi** Stebb. 1888 C. f., T. Stebbing in: Rep. Voy. Challenger, v. 29 p. 1163 t. 125 | 1893 C. f., A. Della Valle in: F. Fl. Neapel, v. 20 p. 380 | 1894 C. f., G. O. Sars, Crust. Norway, v. 1 p. 607.

$\female$. Nearly allied to C. tubularis (p. 667). Antenna 1, 2^{d} joint much shorter than 1st or 3^{d} joint; flagellum as long as 3^{d} joint, 4-jointed. Antenna 2, penultimate joint of peduncle proximally wider than antepenultimate, considerably shorter than ultimate; flagellum rather longer than ultimate joint of peduncle, with 4 joints, 1st the longest. Maxilla 1, outer plate with 10 spines on apex, as compared with 9 in C. sismithi (p. 666). Peraeopod 1 as in C. tubularis with 4th joint broader than long; but peraeopod 2 with 4th joint much longer than broad, fully as long as 5th and 6th combined. Peraeopod 3 with the lobe-like 5th joint less produced than in C. tubularis; probably to peraeopod 3 belong the broad marsupial plates, as in C. crassicornis (p. 666). L. 5—6 mm.

Flinders Passage [North-Australia].

5. **C. abditus** R. Templ. 1836 C. a., R. Templeton in: Tr. ent. Soc. London, v. 1 p. 188 t. 20 f. 5 a—k | 1888 C. a., T. Stebbing in: Rep. Voy. Challenger, v. 29 p. 168 | 1893 C. a., A. Della Valle in: F. Fl. Neapel, v. 20 p. 379 | 1840 *Cerapodina abdita*, H. Milne Edwards, Hist. nat. Crust., v. 3 p. 62.

Body elongate. Head large, laterally subtriangular, dorsally quadrate, most dilated anteriorly, a minute rostrum projecting forwards; front in lateral view almost vertical. Eyes a little prominent, small, round, black. Antenna 1 tapering, about $^2/_3$ as long as body; 3 joints of peduncle subequal or 1st rather the shortest, 1st in figure with sub-basal widening; flagellum 5-jointed. Antenna 2, ultimate and penultimate joints of peduncle longer and stouter than those in antenna 1; flagellum 5-jointed. Antennae 1 and 2, setae of peduncle in double rows, longer towards apices of joints, joints of flagella with spines. Gnathopod 1, 5th joint scarcely as long as 6th, distally a little widened and (in figure) produced to a short tooth, 6th narrowly oblong, at the apex transversely truncate but very narrow; the much curved finger, though short, reaching much beyond it. Gnathopod 2, 5th joint extremely large, subrectangular, hind margin produced to a strong tooth, a smaller one adjacent, filling up the space between it and base of the slightly curved 6th joint, which is widest at base, has a concave smooth hind margin and its apex occupied by the strong, much curved, but not very long finger. Peraeopods 1 and 5 in harmony with genus. L. about 3 mm.

Tube, nearly 5 mm long, papyraceous, cylindrical, head of animal protrusible alternately from either end.

At Mauritius or on the way thither from England.

6. **C. calamicola** (Giles) 1885 *Cyrtophium c.*, G. M. Giles in: J. Asiat. Soc. Bengal, *v.* 54 p. 54 t. 1 | 1888 *Cerapus c.*, T. Stebbing in: Rep. Voy. Challenger, *v.* 29 p. 563 | 1893 *C. c.*, A. Della Valle in: F. Fl. Neapel, *v.* 20 p. 380.

Body elongate, peraeon segments 4—6 the longest. Head subquadrate, rather long, rostrum small. Side-plates 5 and 6 (in figure) much the largest. Eyes (in figure) small, dark. Antenna 1 more than half as long as body; 3 joints of peduncle subequal, 1st the stoutest, slightly widened near base; flagellum 3—6-jointed. Antenna 2 rather longer, ultimate joint of peduncle rather longer than penultimate; flagellum with 3—6 equal joints (in figure 1st much the longest). Gnathopod 1, 6th joint long, ovate; finger as long as 6th joint, inner margin very finely serrate. Gnathopod 2 very large, 5th joint triangular, hind margin produced to a strong incurved tooth, separated by a rather broad cavity from a smaller, blunter tooth, near base of the long, ovate 6th joint (in figure rather rectangular); finger as long as 6th joint, curved, acute, inner margin with a peculiar serrature of square, chisel-edged teeth. Peraeopod 3, 2^{d} joint (in figure) much longer than broad; finger having hind margin provided with two curious short finger-like processes. Peraeopods 4 and 5, 2^{d} joint broadly oval, much more expanded than in peraeopod 3; finger rounded and provided with a large tuft of hairs. Pleopod 3 the smallest. Uropod 1, inner ramus 2-jointed (?1-jointed, tipped with a spine). Uropod 2, inner ramus rudimentary (wanting?). Colour golden brown with deep chocolate blotches. L. 3—5 mm.

Tube 5—10 mm long, brown banded with light yellow and black.

North-West of Bay of Bengal (Orissa coast).

5. Gen. **Chevreuxius** Bonnier

1896 *Chevreuxius* (Sp. un.: *C. grandimanus*), J. Bonnier in: Ann. Univ. Lyon, *v.* 26 p. 663.

♀ unknown. — ♂. Body elongate, depressed. Side-plates very small, not in contact. Eyes wanting. Antennae 1 and 2 with the peduncle long

and slender, longer than flagellum. Accessory flagellum of antenna 1 very
short. Lower lip, inner lobes well developed. Mandible, 3^d joint of palp
the longest. Maxilla 1, inner plate small, with 1 seta. Gnathopod 1 complexly subchelate, the finger impinging against either the small 6th or the
very large 5th joint. Gnathopod 2 much smaller, subchelate. Uropods 2
and 3 short, one-branched. Telson entire.

    1 species.

    1. **C. grandimanus** Bonnier 1896 *C. g.*, J. Bonnier in: Ann. Univ. Lyon, *v.*26
p. 663 t. 40 f. 3.

    Head large, rostrum and inter-antennal processes little prominent. Pleon
segments 1—3, postero-lateral corners rounded. Antenna 1, 1st joint stoutest,
little shorter than 2^d, with 3 spines on under margin; 3^d joint rather long,
though little more than half as long as 2^d; flagellum 8-jointed; accessory flagellum
with 2 joints, 2^d minute. Antenna 2 a little longer, not stouter; ultimate
joint of peduncle longer than the long penultimate; flagellum not long,
with 4 joints, with 2 strong spines at end of 2^d. Upper lip distally rounded
(slightly notched in figure). Maxilla 1, with 10 spines on apex of inner
plate. Maxillipeds, outer plates much larger than inner; 4th joint of
palp short. Gnathopod 1, 2^d joint channelled in front. 3^d and 4th very
short, 5th enormous, hind margin produced to a tooth, base of 6th not
nearly occupying truncate distal margin of 5th, its hind margin forming a
blunt tooth, which defines the excavate, very oblique palm, completely overlapped by the finger. Gnathopod 2 very slender; the narrow 5th joint longer
than the equally narrow 6th, which form a little excavate palm, defined by
a tooth and scarcely overlapped by the short finger. Peraeopods 1 and 2,
4th joint rather long. Peraeopods 3—5, 2^d joint narrowly oval. Peraeopod 3
the shortest. Uropod 1, peduncle long and stout, longer than the rami, all
armed with spines; inner ramus the shorter. Uropod 2, peduncle shorter
but stout, the ramus short and narrow. Uropod 3 very short, peduncle
stout, ramus narrow, ending in 2 setules. Telson broader than long, with
upturned denticle at each angle of the truncate apex. L. nearly 4 mm.

    Bay of Biscay. Depth 950 m.

6. Gen. **Ericthonius** M.-E.

    1830 *Ericthonius* (Sp. un.: *E. difformis*), H. Milne Edwards in: Ann. Sci. nat.,
v. 20 p. 382 | 1888 *E*, T. Stebbing in: Rep. Voy. Challenger, *v.* 29 p. 142, 285, 1672 | 1837
Erichthonius, Burmeister, Handb. Naturg.. p. 569 | 1893 *E.*, A. Della Valle in: F. Fl.
Neapel. *v.* 20 p. 381 | 1894 *E*, G. O. Sars, Crust. Norway, *v.* 1 p. 601 | 1840 *Erichtonius*,
H. Lucas in: Hist. An. artic., Crust. Arach. Myr., p. 231 | 1852 *Pyctilus*, J. D. Dana in:
P. Amer. Ac., *v.* 2 p. 218 | 1853 *P.*, J. D. Dana in: U. S. expl. Exp., *v.* 13 II p. 911, 973.

    Side-plates small, with a tendency of 2^d and 5th to exceed the others
in size. Head rather elongate; lateral lobes produced. Pleon segments 1—3
not wide or deep; postero-lateral corners rounded; segment 6 longer than
segment 5. Eyes on lateral lobes. Antennae 1 and 2 slender, subequal,
setose; peduncle long; flagellum of several joints. Antenna 1 without accessory
flagellum. Antenna 2 attached much behind antenna 1; antepenultimate
joint of peduncle long. Upper lip with rounded entire margin, and acute
process on surface. Lower lip with inner lobes. Mandibular palp long,
3^d joint lamellar, densely setose. Maxilla 1, inner plate with a few setae,
outer with 9 apical spines; 2^d joint of palp long. Maxilla 2, inner plate

with fringed inner margin. Maxillipeds, palp rather narrow. Gnathopod 1 alike in ♂ and ♀, subchelate, 5th joint not shorter than 6th. Gnathopod 2 larger, in ♂ complexly subchelate, 5th joint very large, produced into a tooth; in ♀ normal, 5th joint much smaller than 6th, produced into a narrow lobe. Peraeopods 1 and 2, 2^d joint expanded. Peraeopods 3—5, 2^d joint not greatly expanded, the external expansion oblong, the inner oval; in peraeopod 3 the finger short, reverted, with denticle on convex margin. Branchial vesicles small, absent from gnathopod 2. Marsupial plates broad. Pleopods 1—3 normal. Uropods 1 and 2 biramous. Uropod 3, ramus single, small, tipped with 2 upturned spinules. Telson short, broad, with the lateral lobes densely spinulose on surface.

6 species accepted, 3 doubtful.

Synopsis of accepted species:

1 { Gnathopod 2 in ♂, 5th joint bidentate — **2.**
   { Gnathopod 2 in ♂, 5th joint unidentate — **3.**

2 { Peraeopod 3, 2^d joint not produced doWnWard . .1. **E. brasiliensis** . . p. 671
   { Peraeopod 3, 2^d joint acutely produced doWnWard 2. **E. pugnax** . . . . p. 672

3 { Gnathopod 2 in ♂, palmar margin of 5th joint
   {   deeply incised . . . . 3. **E. macrodactylus** p. 672
   { Gnathopod 2 in ♂, palmar margin of 5th joint not
   {   deeply incised — **4.**

4 { Eyes small . . . . 4. **E. difformis** . . . p. 672
   { Eyes large — **5.**

5 { Gnathopod 2 in ♂, 6th joint with broadly lobed
   {   hind margin . . . . 5. **E. hunteri** . . . . p. 673
   { Gnathopod 2 in ♂, 6th joint With narrowly tubercled
   {   hind margin . . . . 6. **E. megalops** . . . p. 673

1. E. brasiliensis (Dana) 1853 & 55 *Pyctilus b.*, J. D. Dana in: U. S. expl. Exp., *v.* 13 II p. 976; t. 67 f. 5a—h | 1853 *Ericthonius bidens*, A. Costa in: Rend. Soc. Borbon., n. ser, *v.* 2 p. 177 | 1857 *Erichthonius rapax*, Stimpson in: Boston. J. nat. Hist., *v.* 6 p. 515 | 1872 *Cerapus r.*, A. Boeck in: Forh. Selsk. Christian., 1871 p. 40 t. 1 f. 2 | 1857 *Ericthonius difformis*, Bate in: Ann. nat. Hist., ser. 2 *v.* 19 p. 148 | 1893 *Erichthonius d.* (part.), A. Della Valle in: F. Fl. Neapel, *v.* 20 p. 381 t. 1 f. 10; t. 9 f. 1—20 | 1862 *Cerapus abditus* (err., non Templeton 1836!), Bate & Westwood, Brit. sess. Crust., *v.* 1 p. 455 f. | 1876 *C. a.*, A. Boeck, Skand. Arkt. Amphip., *v.* 2 p. 613 t. 28 f. 4 | 1894 *Erichthonius a.*, G. O. Sars, Crust. NorWay, *v.* 1 p. 602 t. 215 | 1898 *E. a.?*, SoWinski in: Mém. Soc. Kiew, *v.* 15 p. 458 | 1862 *Cerapus a.* + *C. brasiliensis*, Bate, Cat. Amphip. Brit. Mus., p. 263 t. 45 f. 2; p. 267 t. 45 f. 8 | 1864 *C. latimanus*, E. Grube in: Jahresber. Schles. Ges., *v.* 41 p. 63 | 1873 *C. minax*, (S. I. Smith in:) A. E. Verrill in: Rep. U. S. Fish Comm., *v.* 1 p. 565.

Body moderately slender, with broadly vaulted back. Head, lateral lobes broad with a small apical point. Side-plates contiguous, 2^d and 5th the largest, 2^d with close-set short linear markings round part of margin. Eyes rather large, rounded, prominent, bright red. Antenna 1 more than half as long as body, 2^d joint longer than 3^d, 3^d than 1st, flagellum nearly as long as peduncle, about 12-jointed. Antenna 2 subequal to antenna 1, ultimate joint of peduncle a little longer than penultimate; flagellum about 9-jointed. Gnathopod 1 in ♂, 5th joint broad, rather longer than 6th, 6th gently convex in front, behind quadrately rounded, with palm subequal to hind margin; finger matching palm. Gnathopod 2 in ♂, 2^d joint widening from a narrow neck, 5th very large, the palmar part produced into 2 teeth, the inner the smaller; 6th joint much narrower but not much shorter than basal part of 5th, hind margin slightly concave in the middle; finger very large, falciform; but both 6th joint and finger variable in size and shape. Gnathopod 2 in ♀, 5th joint produced along hind margin of 6th in a narrow lobe distally armed with setae and recurved spines, 6th joint large, palm rather oblique, longer than hind margin, finger matching. Peraeopods 1 and 2, 2^d joint broadly oval, finger smooth. Peraeopods 3—5 successively longer; 2^d joint successively narrower in proportion to length, oblong (Dana: nearly orbicular in peraeopod 3, narrow in peraeopod 5); in all finger with denticle on hinder convex margin. Pleopods 1—3 normal, with coupling spines and cleft spines. Uropods 1 and· 2, peduncle longer than rami, margins more or less pectinate, spines small except the apical; in uropod 2 the rami laminar. Uropod 3, peduncle broad at base, much longer than the ramus, which ends in 2 upturned spines. Telson twice as broad as long, triangularly pointed between 2 rounded lobes, densely spinulose on the surface. Colour brownish or orange, with dots. L. 4—12 mm.

Occupies tubes affixed to hydroids and algae.

Atlantic with adjoining seas (Europe from South- and West-Norway (depth 19—75 m) to Adriatic and ?Bosphorus; Rio Janeiro; Vineyard Sound); North-Pacific (San Francisco, depth 4 m).

2. **E. pugnax** (Dana) 1852 *Erichthonius p.*, *Pyctilus p.*, J. D. Dana in: P. Amer. Ac., v. 2 p. 218 | 1853 & 55 *P. p.*, J. D. Dana in: U. S. expl. Exp., v. 13ıı p. 975; t. 67 f. 4a—d | 1862 *Cerapus p.*, Bate, Cat. Amphip. Brit. Mus., p. 267 t. 45 f. 7 | 1893 *Erichthonius difformis* (part.), A. Della Valle in: F. Fl. Neapel, v. 20 p. 381.

Closely related to E. brasiliensis (p. 671), but the 6th joint of gnathopod 2 in ♂ sparingly erose within, hind margin (in figure) composed of 3 smoothly rounded tubercles; peraeopod 3 is said to have the 2^d joint of the same form as in E. macrodactylus, with a narrow acute prolongation of the hind margin. L. ?

Sooloo Sea.

3. **E. macrodactylus** (Dana) 1852 *Erichthonius m.*, *Pyctilus m.*, J. D. Dana in: P. Amer. Ac., v. 2 p. 218 | 1853 & 55 *P. m.*, J. D. Dana in: U. S. expl. Exp., v. 13ıı p. 974; t. 67 f. 3a—c | 1862 *Cerapus m*, Bate, Cat. Amphip. Brit. Mus., p. 266 t. 45 f. 6 | 1893 *Erichthonius difformis* (part.), A. Della Valle in: F. Fl. Neape,l v. 20 p. 381.

Distinguished from E. difformis especially by the long tooth of the 5th joint in gnathopod 2 being separated from the base of the 6th joint by a very deep cavity, and by the 2^d joint of peraeopod 3 having a narrow acute prolongation of the hind margin. Side-plate 5 smaller than either the 4th or 6th (in figure, but?). L. ?

Sooloo Sea.

4. **E. difformis** M.-E. 1830 *E. d.*, H. Milne Edwards in: Ann. Sci. nat., v. 20 p. 382 | 1888 *E. d.*, T. Stebbing in: Rep. Voy. Challenger, v. 29 p. 595 | 1859 *Erichtonius d.*, R. M. Bruzelius in: Svenska Ak. Handl., n. ser. v. 3 ur. 1 p. 17 | 1879 *Cerapus d,.* Hoek in: Tijdschr. Nederl. dierk. Ver., v. 4 p. 119 t. 5 f. 14, 15; t. 6 f. 3; t. 8 f. 11, 12 | 1889 *C. d.*, Hoek in: Tijdschr. Nederl. dierk. Ver., ser. 2 v. 2 p. 229 t. 10 f. 15 | 1894 *Erichthonius d.*, G. O. Sars, Crust. Norway, v. 1 p. 604 t. 216 f. 1 | 1842 *Podocerus leachii*, Krøyer in: Naturh. Tidsskr., v. 4 p. 163 | 1853 *Cerapus whitei*, Gosse, Rambles Devonsh., p. 382 | 1857 *Podocerus punctatus*, Bate in: Ann. nat. Hist., ser. 2 v. 19 p. 148 | 1862 *Cerapus difformis* + *Dercothoë p.*, C. p., Bate & Westwood, Brit. sess. Crust., v. 1

p. 457 f.; p. 461 f. | 1862 *D. p.*, *C. p.* + *C. d.* + *C. leachii*, Bate, Cat. Amphip. Brit. Mus., p. 260 t. 44 f. 6; p. 265 t. 45 f. 5; p. 268 | 1868 *C. macrodactylus var. pontica*, Czerniavski in: Syezda Russ. Est., Syezda 1 Zool. p. 97 | 1871 *C. longimanus*, A. Boeck in: Forh. Selsk. Christian., 1870 p. 252.

Possibly not distinct from E. brasiliensis (p. 671). Body more slender than in E. brasiliensis, especially in ♂ (Sars). Head, lateral lobes with the apical point sometimes well marked (not noted by Sars). Side-plates in ♂ differing from those in ♀, the 2^d standing notably apart, and considerably larger than the others. Eyes rather small, rounded, dark red to black. Antennae 1 and 2 elongate, 2^d and 3^d joints in antenna 1 subequal. The chief distinction from E. brasiliensis rests with gnathopod 2 in ♂, which is of great length, the neck or proximal part of joint 2 elongated, 5th joint thrice as long as broad, slightly constricted at the middle, the tooth or thumb elongate, nearly straight, separated from the 6th joint by a narrow palmar cavity, the tooth sometimes having a subapical inner tooth; 6th joint rather narrow, hind margin sinuous in the middle, with a projection at base; finger with setae on both margins and a group of very long ones near the tip. Colour greyish, mottled with brown spots. L. 4—12·5 mm.

Tubes attached.

North-Atlantic with adjoining seas (Europe from Trondhjemsfjord to the Black Sea; Azores; United States of America, depth 0—200 m).

5. **E. hunteri** (Bate) ? 1853 *Cerapus rubricornis*, Stimpson in: Smithson. Contr., v. 6 nr. 5 p. 46 t. 3 f. 33 | ? 1867 *C. r.*, *C. rubiformis*, Packard in: Mem. Boston Soc., v. 1 p. 297 | 1862 *C· hunteri*, Bate, Cat. Amphip. Brit. Mus., p. 264 t. 45 f. 3 | 1876 *C. h.*, A. Boeck, Skand. Arkt. Amphip., v. 2 p. 618 t. 28 f. 5 | 1894 *Ericthonius h.*, T. Stebbing in: Bijdr. Dierk., v. 17 p. 45 | 1894 *Erichthonius h.*, G. O. Sars, Crust. Norway, v. 1 p. 605 t. 216 f. 2 | 1880 *Ericthonius difformis* (part.), S. I. Smith in: Tr. Connect. Ac., v. 4 p. 279 | 1893 *Erichthonius d.* (part.), A. Della Valle in: F. Fl. Neapel, v. 20 p. 382.

Closely related to E. difformis, but differing as follows. Body in ♂ much broader than high (Stimpson). Side-plate 5 larger than side-plate 2, with front lobe very large. Eyes rather large. Antennae 1 and 2 perhaps more setose, and very hairy (Stimpson); each with flagellum of 12 joints; 3^d joint of antenna 1 usually shorter than 2^d. Gnathopod 1, 6th joint rather shorter than 5th, variably expanded. Gnathopod 2 in ♂ more robust; 5th joint not constricted, stout; the tooth not elongate (not always so short as represented by Sars); 6th joint broad, approximate to tooth of 5th, its hind margin lamellar, divided in the middle by a small incision, the two lobes thus formed, however, being rather variable in their relative proportions, finger not very large or rather large, without any notable setae. Peraeopods 1 and 2, 2^d joint expanded, but less broad in proportion to length than in E. brasiliensis (p. 671). Peraeopods 1—3, 2^d joint oblong, successively narrower. Peraeopods 4 and 5, finger with denticle on hind margin scarcely perceptible. Uropod 3 with a rather longer and more slender ramus. L. 5—15 mm.

Arctic Ocean, North-Atlantic and North-Sea (Norway; North-East-England; North-America; lat. 73° N., long. 34° E.); Kattegat.

6. **E. megalops** (O. Sars) 1879 *Cerapus m.*, G. O. Sars in: Arch. Naturv. Kristian., v. 4 p. 461 | 1885 *Erichthonius m.*, G. O. Sars in: Norske Nordhavs-Exp., v. 6 Crust. I p. 210 t. 17 f. 4 a—b | 1893 *E. difformis* (part.), A. Della Valle in: F. Fl. Neapel, v. 20 p. 383.

Apparently very near to E. hunteri (p. 673), but in gnathopod 2 in ♂ the tooth of 5th joint reaches nearly the extremity of the 6th and the hind margin of the 6th joint is crenulate or furnished with several tuberculiform projections, as described by Stimpson in E. brasiliensis (p. 671). Telson small, tubular, armed at the apex with a number of small spines. Colour whitish, variegated with brown. L. about 7 mm.

Arctic Ocean (North-West of Finmark, depth 1217 m; South of Jan Mayen, depth 179 m).

E. fasciatus (Stimps.) 1853 *Cerapus f.*, Stimpson in: Smithson. Contr., v. 6 nr. 5 p. 49 t. 3 f. 35 | 1893 *Erichthonius difformis* (part.)?, A. Della Valle in: F. Fl. Neapel, v. 20 p. 387.

L. nearly 8 mm.

Hake Bay [Grand Manan]. Depth 66 m, on a gravelly bottom

E. peculans (Dana) 1852 *Amphitoe p.*, J. D. Dana in: P. Amer. Ac., v. 2 p. 213 | 1853 & 1855 *Dercothoe speculans*, J. D. Dana in: U. S. expl. Exp., v. 13 II p. 971; t. 67 f. 1a—h | 1862 *D. s., Cerapus s.*, Bate, Cat. Amphip. Brit. Mus., p. 260 t. 44 f. 8 | 1893 *Erichthonius difformis* (part.), A. Della Valle in: F. Fl. Neapel, v. 20 p. 381.

Perhaps the ♀ of E. pugnax or E. macrodactylus (p. 672.)

Sooloo Sea. Depth 12 m.

E. sp., Dana 1853 & 55 *Dercothoe emissitius* (part.)? (err., non *Gammarus e.* J. D. Dana 1852!), J. D. Dana in: U. S. expl. Exp., v. 13 II p. 970; t. 66 f. 10a—e.

Sooloo Sea.

7. Gen. **Cerapopsis** Della Valle

1893 *Cerapopsis* (Sp. un.: *C. longipes*), A. Della Valle in: F. Fl. Neapel, v. 20 p. 356, 388.

Side-plates 2—5 large. Antenna 1 without accessory flagellum. Antennae 1 and 2 with flagellum few-jointed. Mouth-parts as in Ericthonius (p. 670), except as follows. Mandibular palp has the 3d joint rather long, apically rounded. Maxilla 1 has the inner plate longer, tipped with 1 large and 1 small seta, and outer plates of maxillipeds are more prolonged. Gnathopod 1 with palm undefined. Gnathopod 2 becoming chelate in ♂, remaining subchelate, with palm undefined, in ♀. Gnathopods 1 and 2 much larger in ♂ than in ♀. Peraeopods 1 and 2, 2d joint narrow. Peraeopods 3 and 4, 2d joint broad, finger reverted. Peraeopod 5, 2d joint not very broad. Uropods 1 and 2 biramous. Uropod 3, peduncle not expanded, ramus single. Telson entire.

1 species.

1. **C. longipes** Della Valle 1893 *C. l.*, A. Della Valle in: F. Fl. Neapel, v. 20 p. 388 t. 3 f. 10; t. 9 f. 20—40; t. 56 f. 1.

Body compressed, rather robust. Side-plate 1, lower front angle more produced in ♂ than in ♀, 2d covering 2d joint of limb in ♀ but not in ♂, 3d trapezoidal, 4th rectangular, 5th with front lobe as deep as 4th. Antenna 1, 1st joint stout, ²/₃ as long as 2d, 3d as long as 1st; flagellum less than half as long as peduncle, with 3 rather long joints. Antenna 2 a little longer than antenna 1, antepenultimate joint of peduncle rather long, penultimate shorter than ultimate, flagellum little longer than ultimate, 3-jointed. Gnathopod 1 in ♂, 2d joint slender, 4th short, 5th more than 4 times as long as broad, 6th also narrow and a little shorter; finger slender, a little shorter than the 6th joint. Gnathopod 2 in adult ♂ much stouter and rather longer; 2d joint comparatively short, robust, 3^d—5th short, 6th enormously developed, hind margin produced to a bifurcate process, the very large curved finger forming a very elongate cone or an actual cylinder, and combining with the variable process of the 6th joint to form a true didactyle chela. Gnathopod 1 in $\female$, 5th joint as broad as 6th, and a little longer, 6th almond-shaped, finger large but shorter than 6th joint, which in young $\male$ it outstrips in length. Gnathopod 2 in $\female$ a little larger than gnathopod 1; 3^d and 4th joints short, 5th also short, broader than long, 6th almond-shaped, finger robust but short. Peraeopods 1 and 2 slender, finger small and slight. Peraeopod 3, 2^d joint almost circular, its margins entire; the other joints short, compact, 6th with apical seta, finger short, stout. Peraeopod 4 rather longer; 2^d joint less broad, 6th with apical spine simulating a second finger. Peraeopod 5 rather longer, 2^d joint scarcely half as long as broad, finger arched. Pleopods 1—3, peduncle rather stout, inner ramus narrower but rather longer than outer, inner margin of 1st joint smooth; outer ramus proximally very broad, Uropods 1 and 2 not very spinose. Uropod 3, peduncle longer than ramus, which is slightly incurved and tapering, with apical spine. Telson triangular, with 2 setules on each side of upper surface, apex rounded. Colour in $\male$, head and segments yellow in front, brown behind, side-plates 1—4 each with a brown blotch; in $\female$ back light brown, yellow blotches on sides; eggs (large, very few) reddish yellow. L. 3 mm.

Bay of Naples. In fine sand, depth 10—20 m.

8. Gen. **Neohela** S. I. Sm.

1861 *Hela* (Sp. un.: *H. monstrosa*) (non Münster 1840, Decapoda!), A. Boeck in: Forh. Skand. Naturf., Møde 8 p. 668 | 1876 *H.*, A. Boeck, Skand. Arkt. Amphip., *v.* 2 p. 643 | 1886 *H.*, Gerstaecker in: Bronn's Kl. Ordn., *v.* 5 ₁₁ p. 495 | 1881 *Neohela*, S. I. Smith in: P. U. S. Mus., *v.* 3 p. 448 | 1888 *N.*, T. Stebbing in: Rep. Voy. Challenger, *v.* 29 p. 322, 325, 1215 | 1893 *N.*, A. Della Valle in: F. Fl. Neapel, *v.* 20 p. 342 | 1894 *N.*, G. O. Sars, Crust. NorWay, *v.* 1 p. 623 | 1882 *Helella*, G. O. Sars in: Forh. Selsk. Christian., nr. 18 p. 31.

Body very slender, tapering backward; pleon slight, subcylindric, abruptly narrower than peraeon. Head quadrate. Side-plates small and shallow, not contiguous. Antennae 1 and 2 very long and slender, 1st with well developed accessory flagellum. Upper lip bilobed. Maxilla 2 not fringed on side of inner plate, otherwise mouth-parts in general as in Unciola (p. 676). Gnathopods 1 and 2 subchelate; 2^d joint narrow, 5th not short, finger elongate. Peraeopods 1—5 long and slender, 2^d joint linear. Pleopods 1—3 very slender and feeble. Uropods 1 and 2 biramous, slender, the rami spinose; uropod 1 much the longer. Uropod 3 small, peduncle very small, the single ramus much longer. Telson imperfectly defined from pleon segment 6, smooth, distally tapering.

2 species.

Synopsis of species:

Eyes imperfectly developed, represented by whitish pigment 1. **N. monstrosa** . p. 675
Eyes prominent, salmon-coloured . . . . 2. **N. phasma** . . p. 676

1. **N. monstrosa** (Boeck) 1861 *Hela m.*, A. Boeck in: Forh. Skand. Naturf., Møde 8 p. 668 | 1876 *H. m.*, A. Boeck, Skand. Arkt. Amphip., *v.* 2 p. 643 t. 32 f. 1 | 1881 *Neohela m.*, S. I. Smith in: P. U. S. Mus., *v.* 3 p. 448 | 1887 *N. m.*, H. J. Hansen in: Vid. Meddel., ser. 4 *v.* 9 p. 168 | 1893 *N. m.* (part.), A. Della Valle in: F. Fl. Neapel. *v.* 20 p. 343 t. 55 f. 19—24 | 1894 *N. m.*, G. O. Sars, Crust. NorWay, *v.* 1 p. 624 t. 224 | 1882 *Helella m.*, G. O. Sars in: Forh. Selsk. Christian., nr. 18 p. 31.

43*

Peraeon segments 1—4 laterally rather expanded and angularly produced in front. Head, rostrum obsolete; front corners acute, little produced, sides projecting outward in a sharp tooth. Side-plates 1—3 acute in front. Pleon segment 3 without postero-lateral angles. Eyes represented on each side by a small patch of opaque whitish pigment. Antenna 1 longer than body; 2^d joint 2—3 times as long as 1st, 3^d rather longer than 1st; flagellum longer than peduncle, many-jointed, accessory flagellum not as long as 3^d joint of peduncle, 6—10-jointed. Antenna 2 longer; ultimate and penultimate joints of peduncle very long, spinulose all round like peduncle of antenna 1; flagellum longer than peduncle, many-jointed. Gnathopod 1, 5th joint densely setose, about as long as 2^d or 6th, 6th also densely setose, widening gradually to the transverse palm, which has 2 sharp teeth in the middle, and is defined by a 3^d, the prolongation of the hind margin; finger projecting much beyond palm, its outer margin densely setose. Gnathopod 1 in ♀ like that of ♂, but less strong. Gnathopod 2 smaller; 5th joint shorter than 2^d, much narrower than. 6th, which widens to the oblique straight palm, defined by an obtuse angle; finger slender, setose on outer margin, reaching much beyond palmar spines. Peraeopods 1 and 2, 5th and 6th joints subequal; finger slight. Peraeopod 3, 5th joint much shorter than 6th, finger slight. Peraeopods 4 and 5 much longer than preceding; 5th joint short, finger falciform, strong. Uropod 1, outer ramus in ♂, not in ♀, widened. Uropod 3, ramus nearly thrice as long as peduncle. Telson triangular, with obtuse apex. Colour yellowish, semi-pellucid. L. ♀ 25, ♂ 28—30 mm.

Arctic Ocean, North-Atlantic, North-Sea and Skagerrak (West- and North-Norway, depth 188—514 m; Spitzbergen; Farö Isles, down to 2288 m; Baffin Bay); Kattegat.

2. **N. phasma** S. I. Sm. 1881 *N. p.*, S. I. Smith in: P. U. S. Mus., *v.* 3 p. 448 | 1894 *N. p.*, G. O. Sars, Crust. Norway, *v.* 1 p. 624, 625 | 1893 *N. monstrosa* (part.), A. Della Valle in: F. Fl. Neapel, *v.* 20 p. 343.

♂. Head, rostrum slightly prominent, obtusely angular. Eyes large and prominently convex, salmon-coloured (in spirit). Antenna 1, 2^d joint more than thrice as long as 1st, 3^d considerably longer than 1st; accessory flagellum as long as 3^d joint of peduncle, 9-jointed. L. 26 mm.

North-Atlantic (South of New England). Depth 680 m.

9. Gen. **Unciola** Say

1818 *Unciola* (Sp. un.: *U. irrorata*), Say in: J. Ac. Philad., *v.* 1 II p. 388 | 1888 *U.*, T. Stebbing in: Rep. Voy. Challenger, *v.* 29 p. 1168 | 1893 *U.*, A. Della Valle in: F. Fl. Neapel, *v.* 20 p. 336 | 1894 *U.*, G. O. Sars, Crust. Norway, *v.* 1 p. 619 | 1845 *Glauconome* (Sp. un.: *G. leucopis*) (non Goldfuss 1826, Bryozoa!), Krøyer in: Naturh. Tidsskr., ser. 2 *v.* 1 p. 491, 501 | 1876 *G.*, A. Boeck, Skand. Arkt. Amphip., *v.* 2 p. 636 | 1862 *Dryope, Driope* (non *Dryope* Robineau-Desvoidy 1830, Diptera!), Bate, Cat. Amphip. Brit. Mus., p. 276; t. 47.

Body slender, rather depressed; peraeon long, pleon segment 6 very short. Head square, with somewhat projecting front corners. Side-plates all shallow. Pleon segment 3, postero-lateral corners acute, usually with sinus above. Eyes, when present, small, on front corners of head. Antenna 1 the longer, with small accessory flagellum. Antenna 2 the stouter, and stouter in ♂ than in ♀. Upper lip unsymmetrically bilobed. Lower lip normal. Mandible normal, 3^d joint of palp narrow, shorter than 2^d. Maxilla 1, inner plate small, with few setae, outer with 9 apical spines; 2^d joint of

palp long. Maxilla 2, inner plate with lateral fringe. Maxillipeds normal;
outer plates broad, with rather strong spine-teeth. Gnathopod 1 much the
stronger, differing to some extent in the 2 sexes, subchelate; the finger
setose on outer margin. Gnathopod 2 feebly or sometimes scarcely subchelate. Peraeopods 1—5 rather slender; 2^d joint little expanded, finger
with setules on inner margin. Pleopods with strongly serrate coupling spines
on short, rather stout peduncle. Uropods 1 and 2 biramous. stout, with
strong spines. Uropod 3 very short, with small outer ramus; the inner never
articulated, but sometimes marked off from peduncle, at other times represented
only by backward prolongation of peduncle. Telson lamellar, rounded.

8 species.

Synopsis of species:

1. **U. laticornis** H. J. Hansen 1887 *U. l.*, H. J. Hansen in: Vid. Meddel., ser. 4
v. 9 p. 166 t. 6 f. 7—7b | 1889 *U. l.*, J. Bonnier in: Bull. sci. France Belgique, *v.* 20
p. 388, 396 | 1894 *U. l.*. G. O. Sars, Crust. Norway, *v.* 1 p. 619 | 1893 *U. irrorata* (part.),
A. Della Valle in: F. Fl. Neapel, *v.* 20 p. 338.

♀ unknown. — ♂. Head, rostrum rather short, lateral corners somewhat produced, truncate. Pleon segment 3, postero-lateral angles acute, with
sinus above. Eyes distinct, pale. Antenna 2, antepenultimate joint of peduncle
much dilated, rather longer than broad, infero-distally broadly rounded, not
much produced; penultimate joint much dilated, scarcely twice as long as
broad, proximally thick, distally much compressed. Gnathopod 1, 2^d joint
short, not twice as long as broad, 6th much longer than 2^d, not remarkably
broad at base, the basal process rather small, as broad as long, subacute,
unarmed, a nearly straight margin between this and the finger-hinge being
interrupted by a small tooth; finger reaching the basal process. Gnathopod 2,
5th joint scarcely longer than 6th, which is rectangular, not twice as long
as broad; palm nearly transverse. Peraeopods long and slender. Uropod 1,
peduncle about 1$^1/_2$ times as long as outer ramus; outer ramus rather longer than

inner, each armed with a few stout spines. Uropod 3, peduncle forming a rather broad, oblique plate; its hinder angle reaching beyond apex of telson, armed with a large spine and some setae; inner ramus in no way marked off from peduncle, outer attached at middle of outer side of peduncle, rather large, armed with a minute spine and some long setae. Telson as broad as long; apical margin much more curved than basal. L. 5·7 mm.

Davis Strait (lat. 69° N., long. 58° W.). Depth 339 m.

2. U. leucopis (Krøyer) 1845 *Glauconome l*, Krøyer in: Naturh. Tidsskr., ser. 2 *v.* 1 p. 491 t. 7 f. 2a—e | 1846 *G. l.*, Krøyer in: Voy. Nord., Crust. t. 19 f. 1a—u | 1894 *Unciola l.*, G. O. Sars, Crust. Norway, *v.* 1 p. 620 t. 222 | 1896 *U. l.*, J. Bonnier in: Ann. Univ. Lyon, *v.* 26 p. 666 | 1862 *U. l.*, *U. leucopes*, Bate, Cat. Amphip. Brit. Mus., p. 279 t. 47 f. 3 | 1880 *U. irrorata* (part.), S. I. Smith in: Tr. Connect. Ac., *v.* 4 p. 280 | 1882 *U. i.*, G. O. Sars in: Forh. Selsk. Christian., nr. 18 p. 114 | 1887 *U. i.*, H. J. Hansen in: Dijmphna Udb., p. 232 | 1887 *U. i.*, H. J. Hansen in: Vid. Meddel., ser. 4 *v.* 9 p. 164 t. 6 f. 5, 5a | 1889 *U. i.* (part.), J. Bonnier in: Bull. sci. France Belgique, *v.* 20 p. 393 | 1893 *U. i.* (part.), A. Della Valle in: F. Fl. Neapel, *v.* 20 p. 338.

Body rather strongly constructed, peraeon segments transversely furrowed, pleon segments 1—3 each with rounded prominence on each side. Head, rostrum acute, rather long; front corners obtuse, rather prominent. Side-plates 1—4 subacute in front, 5th deeper than the rest. Pleon segment 3, postero-lateral corners acute with sinus above. Eyes represented by a small patch of opaque whitish pigment on each side. Antenna 1 nearly ²/₃ as long as body; 2d joint rather longer than 1st, not twice as long as 3d; flagellum rather shorter than peduncle, about 16-jointed; accessory flagellum 5-jointed. Antenna 2, antepenultimate joint of peduncle nearly as in U. petalocera (p. 681), but penultimate joint not so widely expanded at base; ultimate joint nearly as long as penultimate; flagellum nearly half as long as peduncle, about 10-jointed. Gnathopod 1 nearly as in U. irrorata, except that in ♂ the palm is angularly prominent between 2 cavities. Gnathopod 2 rather feeble, densely setose; 6th joint about as long as 5th, narrowly oblong; palm very short, transverse. Peraeopods 3—5, 2d joint slightly expanded, densely setose. Uropods 1 and 2 rather stout; rami in each subequal, with strong apical spine. Uropod 3 very small; process of peduncle broad, narrowing gradually to apex tipped with 4 setae; inner ramus not marked off, outer very small, not nearly reaching end of process of peduncle, tipped with 4 setae. Telson round, with setule on each side. Colour yellowish grey. L. 13 mm.

Arctic Ocean (widely distributed, depth to 300 m; Varangerfjord [Norway], depth 170—226 m).

3. U. irrorata Say 1818 *U. i.*, Say in: J. Ac. Philad., *v.* 1 II p. 389 | 1840 *U. i.*, H. Milne Edwards, Hist. nat. Crust., *v.* 3 p. 69 | 1853 *U. i.*, Stimpson in: Smithson. Contr., *v.* 6 nr. 5 p. 45 | 1862 *U. i.*, Bate, Cat. Amphip. Brit. Mus., p. 279 | 1873 *U. i.*, (S. I. Smith in:) A. E. Verrill in: Rep. U. S. Fish Comm., *v.* 1 p. 340, 567 t. 4 f. 19 | 1880 *U. i.*, S. I. Smith in: Tr. Connect. Ac., *v.* 4 p. 280 | ?1888 *U. i.*, T. Stebbing in: Rep. Voy. Challenger, *v.* 29 p. 1169 t. 138c | 1889 *U. i.* (part.), J. Bonnier in: Bull. sci. France Belgique, *v.* 20 p. 393 | 1893 *U. i.* (part.), A. Della Valle in: F. Fl. Neapel, *v.* 20 p. 338 t. 55 f. 37—41.

Body dorsally broad, flattened. Head, rostrum distinct, acute, lateral corners blunt, moderately prominent. Side-plates very shallow, 1st—4th with front corner acute. Pleon segment 3, postero-lateral corners acute, with sinus above. Eyes small, rounded. Antenna 1 elongate, 2^d joint longer than 1st, nearly thrice as long as 3^d; flagellum nearly as long as peduncle, in $\male$ attaining 23 joints; accessory flagellum with 5 joints, the last minute. Antenna 2 in $\male$ and $\female$ not distinguishable from those in U. petalocera (p. 681); flagellum in $\male$ attaining 16 joints. Gnathopod 1, 2^d joint very broad except at the base, 5th joint short, broad, produced to a prominent tooth with irregular apex, 6th very broad. especially in $\male$, with basal process projecting beyond process of 5th joint, its blunt apex carrying a stout spine; the long oblique palm minutely crenulate throughout, with a very slight cavity at each end; finger reaching basal process of 6th joint, teeth of inner margin not very conspicuous, setae of outer numerous and strong. Gnathopod 2, 5th joint decidedly longer than 6th, both very setose, 6th narrowly oval or almost tapering, palm almost obsolete; finger very small. Feraeopods 1—5 slender; 2^d joint very little expanded, rather more so in peraeopods 1 and 2 than in 3—5, setose in the latter. Uropods 1 and 2 not elongate, spines few but stout, especially the apical ones; rami nearly equal to one another, more than half as long as peduncle. Uropod 3, inner process of peduncle not very broad, with an oblique line distinctly defining it as representing the inner ramus, tipped with 2 or 3 setae; outer ramus longer than inner and produced beyond it, fringed with 7 setae. Telson rounded. Colour red or brown, mottled with white; antennae 1 and 2 annulated and gnathopod 1 marked with bright red (Smith). L. reaching 15 mm.

North-Atlantic (North America from Labrador to New Jersey). Depth 0—800 m.

4. **U. crassipes** H. J. Hansen 1887 *U. c.*, H. J. Hansen in: Vid. Meddel., ser. 4 *v.* 9 p. 165 t. 6 f. 6, 6a | 1896 *U. c.*, J. Bonnier in: Ann. Univ. Lyon, *v.* 26 p. 667 | 1893 *U. irrorata* (part.). A. Della Valle in: F. Fl. Neapel, *v.* 20 p. 338.

$\female$ unknown. — $\male$. Head, rostrum rather long; front corners truncate, rather prominent. Side-plate 1 acute in front. Pleon segment 3, posterolateral corners with a rather small acute but not produced point, without conspicuous sinus between it and the bulging hind margin. Eyes perhaps wanting. Antenna 1, 2^d joint thrice as long as 3^d; accessory flagellum (in figure) 4- or 5-jointed. Antenna 2, antepenultimate and penultimate joints of peduncle scarcely so stout as in U. leucopis, somewhat compressed. Gnathopods 1 and 2 seemingly as in U. leucopis. Uropod 1, peduncle about $^2/_3$ longer than the rami, which have stout spines. Uropod 3, inner process of peduncle, though not separated from the base by an articulating membrane, indicating a small inner ramus, tipped with 2 spinules; outer ramus broader and much longer than inner and reaching beyond it, with a spinule and some setae. Telson shortly oval. L. nearly 9 mm.

Baffin Bay (lat. 71° N., long. 59° W.). Depth 376 m.

5. **U. planipes** Norm. 1867 *U. p.*, A. M. Norman in: Nat. Hist. Tr. Northumb., *v.* 1 p. 14 t. 7 f. 9—13 | 1893 *U. p.*, A. Della Valle in: F. Fl. Neapel, *v.* 20 p. 341 t. 55 f. 42—45 | 1894 *U. p.*, G. O. Sars, Crust. Norway. *v.* 1. p. 621 t. 223 | 1868 *U. leucopes* (err., non *Glauconome leucopis* Krøyer 1845!), Bate & Westwood, Brit. sess. Crust., *v.* 2 p. 518 | 1871 *Glauconome kröyeri + G. steenstrupi*, A. Boeck in: Forb. Selsk. Christian., 1870 p. 259, 260 | 1876 *G. k. + G. s.*, A. Boeck, Skand. Arkt. Amphip., *v.* 2 p. 639 t. 30 f. 1; p. 640.

Body slender, dorsally smooth; peraeon segment 2 in $\male$ with long ventral spine. Head, rostrum distinct, rather flattened. lateral corners angular, little prominent. Side-plate 1 produced acutely forward, 2^d deeper than the rest, subacute in front in ♂, not in ♀. Pleon segment 3, postero-lateral corners acute, with sinus above. Eyes represented only by an irregular patch of whitish pigment on each side. Antenna 1 long and slender, 2^d joint longer than 1st, nearly thrice as long as 3^d, flagellum subequal to peduncle, 14-jointed, accessory flagellum with 2 joints, 2^d minute. Antenna 2 in ♂, antepenultimate joint of peduncle thick, lower distal end round, slightly produced, penultimate also thick, ultimate not very slender; in ♀ these joints are much more slender; flagellum as long as ultimate joint of peduncle in ♀, shorter in ♂, 6-jointed. Gnathopod 1 in ♂, 5th joint as broad as long, 6th moderately stout, hind margin not extremely short; palm defined by an obtuse angle against which the apex of closed finger impinges, leaving 2 palmar cavities. Gnathopod 1 in ♀, 2^d, 5th and 6th joints less broad than in ♂, with the palm oblique but nearly straight. Gnathopod 2 slender, stouter in ♂ than in ♀; 5th joint large, much longer than 6th, densely setose on hind margin, 6th tapering distally, without palm; finger very small. Peraeopods 1—5 very slender. Peraeopods 3—5, 2^d joint narrow, not setose. Branchial vesicles very small. Uropods 1 and 2 not very robust, spinose. Uropod 3, laminar inner process of peduncle rather narrow, armed with a single spine; inner ramus not indicated, outer produced beyond process of peduncle, sublinear, with 5 long apical setae and 1 seta on outer margin. Telson subrotund, with 2 spinules on each side. Colour pale yellow, banded with light orange; whitish patches along the sides. L. ♀. 5, ♂ 6 mm.

Arctic Ocean, North-Atlantic, North-Sea and Skagerrak (Norway from Christiania-fjord to Vadsö, depth 94— 564 m; Greenland, depth 90 m; British Isles); Kattegat.

6. **U. crenatipalma** (Bate) 1855 *U. irrorata* (err., non Say 1818!), Gosse in: Ann. nat. Hist., ser. 2 v. 16 p. 307 | 1862 *Dryope i., Driope i. + Dryope crenatipalma, Driope c.*, Bate, Cat. Amphip. Brit. Mus., p. 276 t. 47 f. 1; p. 277 t. 47 f. 2 | 1863 *Dryope i. + D. crenatipalmata*, Bate & Westwood, Brit. sess. Crust., v. 1 p. 488 f.; p. 490 f. | 1874 *D. crenatipalma*, T. Stebbing in: Rep. Devonsh. Ass., v. 6 p. 770 | 1889 *Unciola crenatipalmata*, J. Bonnier in: Bull. sci. France Belgique, v. 20 p. 392 t. 12, 13 | 1893 *U. c.*, Chevreux & E. L. Bouvier in: Ann. Sci. nat., ser. 7 v. 15 p. 138 t. 2 f. 12 | 1893 *U. c.*, A. Della Valle in: F. Fl. Neapel, v. 20 p. 340 t. 55 f. 32—36 | 1896 *U. c.*, J. Bonnier in: Ann. Univ. Lyon, v. 26 p. 666.

Body rather broad, depressed. Head, rostrum short, acute; lateral corners with inner point acute. Side-plates 1 and 2 with front corner acute. Pleon segment 3, postero-lateral corners acute, with sinus above. Eyes small, rounded, remaining dark in spirit. Antenna 1, 2^d joint rather longer than 1st, less or not more than twice as long as 3^d; flagellum shorter than peduncle, 11-jointed; accessory flagellum with 2 joints, 2^d minute. Antenna 2, ante-penultimate and penultimate joints of peduncle in ♂ stout but not laminar, ultimate nearly as long as penultimate, flagellum as long as ultimate, 9-jointed. Gnathopod 1, 2^d joint broad except at base, 5th short, broad, produced into a blunt setose process, 6th broad, produced into a blunt basal process reaching beyond process of 5th joint; the long oblique palm in ♂ forming 3 cavities, in ♀ convex between 2 very shallow cavities, in both sexes minutely crenulate; finger meeting basal process, its inner margin denticulate, outer with several setae. Gnathopod 2 rather feeble; 5th joint rather shorter than 6th, which is rectangular, having in ♂ an almost transverse, slightly convex palm, over which the small finger closes tightly, while in ♀ the palm is deeply excavate, the closed finger leaving a conspicuous gap. Peraeopods 1—5 nearly as in U. irrorata (p. 678). Uropods 1 and 2, rami somewhat unequal. Uropod 3 very small, inner process of peduncle tipped with a spine and seta and indicated as the inner ramus by a constriction on each side, but without line of demarcation; outer ramus tipped with setae, broader than long, reaching apex of inner; or (Chevreux & Bouvier) about as broad as long, reaching apex of the inner. Telson round. L. about 8 mm.

North-Atlantic with adjoining seas (England, France, Spain).

7. U. petalocera (O. Sars) 1876 *Glauconome planipes?* (err., non *Unciola p.*, A. M. Norman 1867!), G. O. Sars in: Arch. Naturv. Kristian., *v.* 2 p. 360 | 1879 *G. petalocera*, G. O. Sars in: Arch. Naturv. Kristian., *v.* 4 p. 462 | 1885 *Unciola p.*, G. O. Sars in: Norske Nordhavs-Exp., *v.* 6 Crust. I p. 212 t. 17 f. 5a—1 | 1889 *U. p.*, J. Bonnier in: Bull. sci. France Belgique, *v.* 20 p. 396 | 1894 *U. p.*, G. O. Sars, Crust. Norway, *v.* 1 p. 619 | 1893 *U. irrorata* (part.), A. Della Valle in: F. Fl. Neapel, *v.* 20 p. 338.

Body slender and in general much resembling U. planipes (p. 679). Head truncate in front, without distinct rostrum, lateral corners acute. Side-plates 1 — 4, front corner acute, and in side-plate 1 prolonged. Pleon segment 3, postero-lateral corners acute, rather strongly produced. Eyes not visible. Antenna 1 slender, elongate, 2^d joint longer than 1st, and fully thrice as long as 3^d, flagellum about as long as peduncle, 20-jointed. accessory flagellum 2-jointed. Antenna 2 in $\male$, antepenultimate joint of peduncle greatly expanded (in $\female$ not), lower distal end rounded and rather strongly produced, penultimate strongly expanded, lamellate, narrowing distally; flagellum a little longer than ultimate joint of peduncle, with 9 joints (or in figure of $\male$ about 16). Gnathopod 1 in $\male$, 6th joint very broad and strongly compressed, with 2 deep cavities in the palm, which is defined by an obtuse angle; finger remarkably long, falciform, its apex nearly reaching the 4th joint. Gnathopod 1 in $\female$, 6th joint piriform; palm faintly incurved, very oblique, indistinctly defined from the short hind margin; finger closing between 2 sets of palmar spines. Gnathopod 2 rather feeble, setose, 5th joint a little shorter than 6th, which is subrectangular, palm nearly transverse; finger small. Peraeopods 1—5 slender, 1st—3^d subequal, 4th and 5th considerably longer. Peraeopods 3—5, 2^d joint linear, not setose. Uropods 1 and 2 nearly as in U. planipes. Uropod 3, peduncle produced within to a pointed lobe, carrying a single spine; inner ramus not indicated, outer narrowly oval, produced beyond inner process of peduncle, fringed with setules. Telson semi-elliptical, with 2 short apical bristles. Colour whitish. L. 10 mm.

North-Atlantic (lat. 63—75° N.). Depth 658—1237 m.

8. U. incerta Bonnier 1896 *U. i.*, J. Bonnier in: Ann. Univ. Lyon, *v.* 26 p. 666 t. 40 f. 4.

$\female$. Resembling in general U. crenatipalma. Eyes wanting. Antenna 1, 1st joint of peduncle armed below with double row of spinules, 2^d not quite twice as long as 3^d; accessory flagellum with 3 joints, 1st and 2^d equal, 3^d nodiform. Uropod 3, peduncle broadly produced on inner side, armed at apex with a spine and 2 setae; inner ramus not indicated, outer small, not reaching end of process of peduncle, apex truncate, armed with 2 spines and 8 setae. L. about 5 mm. — $\male$ unknown.

Bay of Biscay. Depth 180 m.

10. Gen. **Siphonoecetes** Krøyer

1845 *Siphonoecetes* (Sp. un.: *S. typicus*), Krøyer in: Naturh. Tidsskr., ser. 2 *v.* 1 p 481, 491 | 1876 *S.*, A. Boeck, Skand. Arkt. Amphip., *v.* 2 p. 630 | 1888 *S.*, T. Stebbing in: Rep. Voy. Challenger, *v.* 29 p. 212 | 1893 *S.*, A. Della Valle in: F. Fl. Neapel, *v.* 20 p. 357 | 1894 *S.*, G. O. Sars, Crust. Norway, *v.* 1 p. 609 | 1857 *Siphonocetus*, Bate in:

Ann. nat. Hist., ser. 2 v. 19 p. 149 | 1862 *Siphonoecetes (Siphonaecetus)*, Bate & Westwood, Brit. sess. Crust., v. 1 p. 463, 467 | 1862 *Siphonoecetus* (part.), Bate, Cat. Amphip. Brit. Mus., p. 268 | 1871 *S.*, A. Boeck in: Forh. Selsk. Christian., 1870 p. 257 | 1873 *Siphonocoetus*, A. Marschall, Nomencl. zool:, p. 420 | 1887 *Siphonaecetes*, Chevreux in: Bull. Soc. zool. France, v. 12 p. 290, 317.

Body slender. peraeon much longer than pleon. Head subquadrate. Side-plates very shallow, setulose, 1ˢᵗ subacute in front, 5ᵗʰ, 6ᵗʰ much produced backward. Eyes, when present, placed on front corners of head. Antennae 1 and 2 setose. Antenna 1, peduncle much longer than flagellum, its 3 joints subequal, without accessory flagellum. Antenna 2 much longer, stouter; flagellum short, of 1 long and 2 short joints, with unguiform spines at apex. Upper lip rounded or faintly bilobed. Lower lip with the mandibular processes narrow. Mandible with palp of 1 setose joint, molar with accessory plate, otherwise normal. Maxilla 1, inner plate obsolete, outer with 7 apical spines; palp long, with 6 spines on apex of 2ᵈ joint and some setae on outer margin. Maxilla 2, inner plate fringed on inner margin. Maxillipeds normal; last joint of palp very small, tipped with spines. Gnathopod 1 without distinct palm. Gnathopod 2 stronger, distinctly subchelate. Peraeopods 1 and 2, 2ᵈ joint strongly expanded, glandular, 4ᵗʰ joint broad, cordiform, 5ᵗʰ minute; finger long, straight. Peraeopods 3 and 4, 2ᵈ and 4ᵗʰ joints not much expanded, 5ᵗʰ short, broad, distally scabrous; finger reverted, bidentate. Peraeopod 5 normal; 2ᵈ joint not broad, densely setose, finger as in peraeopods 3 and 4. Branchial vesicles and marsupial plates narrow. Pleopods 1—3, peduncle very broadly expanded; the 2 coupling spines very slender, with 3 or 4 teeth on each side; 1ˢᵗ joint of inner ramus with smoothly concave inner margin. Uropod 1, peduncle longer than outer ramus, each spinulose on outer margin; inner ramus shorter than outer, spinulose on inner margin. Uropod 2 similar to uropod 1, but smaller. Uropod 3 short, peduncle broadly produced on inner side; ramus very small, tipped with setae. Telson broader than long, apically rounded, 2 small patches of microscopic spinules flanking the apex. Occupying tubes, constructed or adapted.

5 accepted species, 1 obscure.

Synopsis of accepted species:

1. **S. typicus** Krøyer 1845 *S. t.*, Krøyer in: Naturh. Tidsskr., ser. 2 v. 1 p. 481 t. 7 f. 4 a—f | 1846 *S. t*, Krøyer in: Voy. Nord, Crust. t. 20 f. 1 a—v | 1893 *S. t.* (part.), A. Della Valle in: F. Fl. Neapel, v. 20 p. 358.

Body cylindric but ventrally flattened; peraeon about ³/₅ of total length. Head, rostrum and front angles not very prominent. Side-plate 5 about thrice as broad as deep, outdrawn behind into a very long point. Eyes not visible. Antenna 1 about half as long as body; joints of peduncle successively a little shorter, together more than thrice as long as flagellum, which is 4—6-jointed. Antenna 2 nearly as long as body, nearly twice as long as antenna 1, ultimate joint of peduncle shorter than penultimate, a little longer than antepenultimate; flagellum shorter than ultimate joint of peduncle, with 1 long and 2 very short joints, last broader than long, tipped with 2 divergent spines. Antennae 1 and 2 very setose. Gnathopod 1, 5th joint a little longer and distally much broader than the conical 6th, which has strong spines on hind margin; finger about half as long as 6th joint, with 7 or 8 serrations on inner margin, and with a tubercle near the apex. Gnathopod 2 slightly shorter but far stouter than gnathopod 1, 2^d joint $^2/_3$ as broad as long, 4th produced with blunt apex, 5th cup-shaped, its narrow lobe tipped with a strong spine looking like a continuation of the lobe, 6th oval, hind margin much more convex than front, with 7 large spines on it successively larger to the palm, which is shorter and has 2 sharp tubercles; finger curved, acute, with about 10 teeth, and near the apex a tubercle and 2 long setae. Peraeopods 1 and 2 not differing from those of S. colletti. Uropod 3, peduncle broadly produced on inner side, ramus scarcely produced beyond it, oval or somewhat conical, with 12 strong setae. Telson much broader than long, bare. Colour pale yellow, marbled with brown. L. 6 mm.

Arctic Ocean (Greenland). Depth 30—35 m.

2. **S. pallidus** O. Sars 1882 S. p., G. O. Sars in: Forh. Selsk. Christian., nr. 18 p. 113 t. 6 f. 7 a—d, 7 x | 1894 S. p., G. O. Sars, Crust. Norway, v. 1 p. 611 t. 218 f. 2 | 1893 S. typicus (part.), A. Della Valle in: F. Fl. Neapel, v. 20 p. 358.

Head, rostrum small; lateral corners extremely narrow, not widened apically. Eyes replaced on each side by a small patch of an opaque whitish pigment. Antenna 1 not reaching end of penultimate joint of peduncle of antenna 2; joints of peduncle successively shorter; flagellum about half as long as peduncle, with 7 joints, last minute. Antenna 2 nearly as long as body, ultimate joint of peduncle shorter than penultimate. Gnathopod 1, 6th joint much narrower than 5th and scarcely tapering distally. Gnathopod 2, 6th joint armed with 5 spines on hind margin. Uropod 3, inner expansion of peduncle comparatively narrow, ramus very small, not reaching beyond expansion of peduncle. Colour pale yellowish. L. about 4 mm.

Arctic Ocean and North-Atlantic (West-Norway up to Hasvig [Finmark]). Depth 94—282 m. Usually in shells of Dentalium.

3. **S. colletti** Boeck 1862 S. typicus (err., non Krøyer 1845!), Bate & Westwood, Brit. sess. Crust., v. 1 p. 465 f. | 1893 S. t. (part.), A. Della Valle in: F. Fl. Neapel, v. 20 p. 358 | 1871 S. colletti, A. Boeck in: Forh. Selsk. Christian., 1870 p. 258 | 1876 S. c., A. Boeck, Skand. Arkt. Amphip., v. 2 p. 633 t. 28 f. 9 | 1894 S. c. (S. mucronatus Aug. Metzger in MS), G. O. Sars, Crust. Norway, v. 1 p. 610 t. 218 f. 1. | 1871 S. cuspidatus, Aug. Metzger in: Jahresber. Ges. Hannover, v. 21 p. 30 | 1873 S. c. (sp. nov.!), (S. I. Smith in:) A. E. Verrill in: Rep. U. S. Fish Comm., v. 1 p. 501, 566.

Head, rostrum acute (Sars: short, triangular; Smith: long, slender); lateral lobes much produced, apically swollen, carrying the well developed, rounded, dark eyes. Antenna 1 considerably exceeding half length of body, reaching beyond penultimate joint of peduncle of antenna 2 (Sars), about to middle of that joint (Smith); 3 joints of peduncle subequal; flagellum decidedly (Sars) or scarcely (Smith) longer than a joint of peduncle, 5—7-jointed.

Antenna 2 about as long as body, ultimate and penultimate joints of peduncle subequal or ultimate the shorter; flagellum shorter than ultimate joint of peduncle, with 3 joints, the last 2 very small. Gnathopod 1, 6th joint about as long· as 5th, conically tapering, hind margin armed with 3 strong spines. Gnathopod 2 rather strong; 2[d] joint rather short and thick, 4th produced to obtuse setose prominence, 5th having on its narrow lobe a stout spine and several setae, 6th large, oblong oval, hind margin armed with 6 spines, larger as they approach the short, not very oblique palm. Gnathopods 1 and 2 of the very same structure in the two sexes (Sars), gnathopod 2 in ♀ more slender than in the ♂ (Smith). Peraeopods 1 and 2, 4th joint very large and expanded, nearly cordiform in outline. Uropod 1 (in figure), peduncle spinose on outer, not on distal margin, outer ramus $^3/_4$ as long as peduncle, with 11 spines on straight outer margin, inner ramus much shorter. Uropod 2 with a spine-fringed, distal lobe, outer ramus not much shorter than peduncle, fringed with spines on outer margin. Uropod 3, peduncle expanded inside to a broad setiferous lobe; ramus more than half as long as peduncle, obliquely truncate apex with 1 spine and 7 setae. Colour greyish white, varied with yellowish brown, antenna 2 with yellow and brown bands. L. 4—8 mm.

North-Atlantic, North-Sea and Skagerrak (South- and West-Norway, depth 11—37 m; Shetland Isles, depth 73—164 m; Firth of Forth; Vineyard Sound and Buzzard Bay); Kattegat.

4. **S. dellavallei** Stebb. 1893 *S. typicus* (part.), A. Della Valle in: F. Fl. Neapel, *v.* 20 p. 358 t. 4 f. 11—13, t. 7 f. 23—28 | 1899 *S. dellavallei*, T. Stebbing in: Ann. nat. Hist., ser. 7 *v.* 3 p. 241, 350.

Differs from S. pallidus (p. 683) in having a well developed acute rostrum and the eyes well developed, dark in spirit, but conspicuous from their abundant white pigment. Antenna 1 scarcely reaching ultimate joint of peduncle of antenna 2; flagellum less than half as long as peduncle, with 5 joints, the last minute. Antenna 2. ultimate joint of peduncle shorter than penultimate. Side-plate 1 rather blunt in front. Appendages in general little different from those of S. pallidus. Colour yellowish white with patches of brown, antennae annulate with white and brown. L. 5—6 mm.

Constructing free tubes.

Bay of Naples. Depth 10—20 m, on fine sand.

5. **S. sabatieri** Rouv. 1894 *S. s.*, Rouville in: C.-R. Ass. France, Sess. 23 *v.* 1 p. 173.

In general agreement with S. colletti (p. 683), except as to uropods 1—3. Uropod 1. peduncle much longer than rami, not obviously spinose on outer margin, but closely fringed with spines on distal border; outer ramus not much longer than inner, with 6 or 7 spines along convex outer margin. Uropod 2, peduncle as in uropod 1; rami slender, outer rather the longer, about half as long as peduncle; both rami smooth except for 2 apical spines. Uropod 3, peduncle expanded on inner side to a narrow setiferous lobe; ramus not half as long as peduncle, little more than a cylindrical tubercle, tipped with a couple of setae. In peraeopods 1 and 2 the 2[d] joint is very largely expanded, and in peraeopods 3 and 4 the bulbous 5th joint appears to be scabrous all over. L. slightly under 4 mm.

Gulf of Lion (Étang de Thau). In small univalve shells.

S. kröyeranus (Bate) 1857 *Siphonocetus k.*, Bate in: Ann. nat. Hist., ser. 2 *v.* 19 p. 149 | 1862 *Siphonoecetes whitei* (err., non *Cerapus w.* Gosse 1853!), Bate & Westwood, Brit. sess. Crust., *v.* 1 p. 467 f. | 1862 *Siphonoecetus w.*, Bate, Cat. Amphip. Brit. Mus., p. 270 t. 45 f. 10.

English Channel (Weymouth).

11. Gen. **Corophium** Latr.

?1793 *Cymothoa* (part.), J. C. Fabricius, Ent. syst., *v.*2 p.503 | 1806 *Corophium* (Sp. un.: *C. longicorne*), Latreille, Gen. Crust. Ins., *v.*1 p.58 | 1888 *C.*, T. Stebbing in: Rep. Voy. Callenger, *v.*29 p.79, 1670 | 1893 *C.*, A. Della Valle in: F. Fl. Neapel, *v.*20 p.362 | 1894 *C.*, G. O. Sars, Crust. Norway, *v.*1 p.612 | 1895 *C.*, G. O. Sars in: Bull. Ac. St.-Pétersb., ser. 5 *v.*3 p.291 | 1813/14 *C.*, *Corophrium*, Leach in: Edinb. Enc., *v.*7 p.432 | 1830 *Corophia*, H. Milne Edwards in: Ann. Sci. nat., *v.*20 p.384 | 1851 *Audouinia* (nom. nud.), (A. Costa in:) F. W. Hope, Cat. Crost. Ital.. p.24.

Body depressed throughout. Head with narrow lateral lobes. Side-plates small, discontinuous, 1st conically produced, tipped with setae. Eyes small or imperfectly developed, on lateral lobes of head. Antenna 1 without accessory flagellum; flagellum slender, with several joints. Antenna 2 (Fig. 116, 117 p. 691) strong, pediform, usually much longer in ♂ than in ♀; flagellum short, 3-jointed, with apical hooked spines. Upper lip broad. Lower lip normal. Mandible with 2-jointed palp, slender, each joint carrying a strong plumose seta; other parts normal. Maxilla 1, inner plate nearly obsolete, outer with 7 spines on apical margin, 2^d joint of palp long. Maxilla 2, inner plate fringed on inner margin. Maxillipeds, inner plates narrow, without apical spine-teeth, outer long, narrow, with slender spines on inner margin; finger of palp small, with apical spine. Gnathopod 1 slender. the short projecting 3^d joint and long 5th densely fringed with long setae; 6th joint narrow, with short palm. Gnathopod 2 rather larger, 4th joint closely attached to hind margin of 5th; its own convex hind margin fringed with very long plumose setae in 2 rows; 6th joint sublinear, without palm. Feraeopods 1 and 2, 2^d and 4th joints somewhat expanded, 5th very short; finger slender. Peraeopods 3 and 4, 2^d joint moderately expanded, 4th produced in front of the short 5th, which carries 2 oblique rows of spines, 6th slender, not long; finger short, reverted. Peraeopod 5 long and slender, 2^d joint fringed on both margins with long setae. No branchial vesicles on gnathopod 2. Marsupial plates narrow. Pleopods 1—3, peduncle greatly expanded on inner side; the 2 coupling spines with several teeth; inner ramus the longer, without cleft spines on 1st joint. Uropods 1 and 2, rami rather short, with strong spines on outer margin. Uropod 3, peduncle short, ramus single, laminar, with some fringing setae. Telson entire, small, distinct (Fig. 115, 118, p. 686, 691).

12 species accepted, 7 obscure.

Synopsis of accepted species:

1 { Pleon segments 4—6 distinct — **2.**
  { Pleon segments 4—6 coalesced — **11.**

2 { Uropod 3, ramus oval — **3.**
  { Uropod 3, ramus parallel-sided — **6.**

3 { Antenna 2 not chelate — **4.**
  { Antenna 2 chelate — **5.**

4 { Telson obtusely pointed. . . . . 1. **C. volutator.** . . . p. 686
  { Telson transversely truncate. . . . . 2. **C. nobile** . . . . p. 687

5 { Pleon segments 4—6 smooth . . . . 3. **C. chelicorne** . . p. 687
  { Pleon segments 4—6 spinulose . . . . 4. **C. spinulosum** . p. 688

6 { Uropod 3, ramus linear — **7.**
  { Uropod 3, ramus not linear — **8.**

7 { Antenna 1 in ♂, 1st joint without process . . '. 5. **C. affine** . . . . p. 688
  { Antenna 1 in ♂, 1st joint with hooked process . 6. **C. runcicorne** . . p. 689

1. C. volutator (Pall.) 1710 „*Pulex marinus cornutus*", Jo. Ray, Hist. Ins., p. 43 | ?1761 *Oniscus bicaudatus*, Linné, Fauna Svec., ed. 2 p. 500 | ?1793 *Cymothoa bicaudata*, J. C. Fabricius, Ent. syst., v. 2 p. 507 | 1893 *Corophium bicaudatum*, A. Della Valle in: F. Fl. Neapel, v. 20 p. 372 t. 56 f. 2—6 | 1766 *Oniscus volutator*, Pallas, Misc. zool., p. 192 t. 14 f. 20 | 1888 *Corophium v.*, T. Stebbing in: Rep. Voy. Challenger, v. 29 p. 21, 29, 34 | 1893 *C. v.*, Chevreux & E. L. Bouvier in: Ann. Sci. nat., ser. 7 v. 15 p. 140 | 1767 *Cancer grossipes*, Linné, Syst. Nat., ed. 12 v. 2 p. 1055 | 1777 *Gammarus g.*, J. C. Fabricius, Gen. Ins., p. 248 | 1836 *Corophium g.*, R. Templeton in: Mag. nat. Hist., v. 9 p. 12 | 1876 *C. g.*, A. Boeck, Skand. Arkt. Amphip., v. 2 p. 623 t. 28 f. 6 | 1894 *C. g.*, G. O. Sars, Crust. Norway, v. 1 p. 614 t. 219 | 1896 *C. g.*, Sowinski in: Mém. Soc. Kiew, v. 15 p. 373 (distribution) | 1777 *Astacus linearis*, Pennant, Brit. Zool., ed. 4 v. 4 p. 17 t. 16 f. 31 | 1779 *Gammarus longicornis*, J. C. Fabricius. Reise Norweg., p. 258 | 1793 *G. l.*, J. C. Fabricius, Ent. syst., v. 2 p. 515 | 1806 *Corophium longicorne*, Latreille. Gen. Crust. Ins., v. 1 p. 59 | ?1874 *C. bonelli*, Ritzema Bos, Bijdr. Crust. Hedriophthal., p. 54.

Body moderately slender, pleon segments 4—6 distinct. Head, rostrum small; lateral lobes narrowly rounded, not very prominent. Side-plate 1 with about 5 setae on produced apex. Eyes very small, dark. Antenna 1 in ♂ nearly half, in ♀ little more than $^1/_3$ as long as body; 1st joint longer than 2d and 3d combined, in ♂ serrulate below, in ♀ with 2 stout spines; flagellum nearly as long as peduncle, 12—14-jointed. Antenna 2 in ♂ as long as body; penultimate joint of peduncle stout and long, lower-margin produced to a short tooth bounding an apical sinus; ultimate joint much narrower, nearly as long, without any tooth, much longer than the 3-jointed flagellum. Antenna 2 in ♀ about half as long as body, except for shortening of ultimate and penultimate joints of peduncle, constructed as in ♂. Gnatho-pod 1, 5th joint rather longer than 6th; 6th widening a little distally; palm nearly transverse and matching finger. Gnathopod 2, 5th joint rather shorter than 6th; finger rather long, perfectly smooth. Peraeopods 1 and 2, finger scarcely longer than 6th joint. Peraeo-

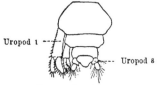

Uropod 1

Uropod 8

Fig. 115. C. volutator.
Uropods 1—3 and telson.
[After G. O. Sars.]

pods 3 and 4, 4th joint setose on both margins. Peraeopod 5, 2d joint oval, well expanded. Uropod 3, peduncle broader than the rounded oval, setose, ramus (Fig. 115). Telson almost an equilateral triangle, with obtusely pointed apex. Colour whitish, marbled with brown on back and antenna 2. L. 6 mm.

North-Atlantic with adjoining seas (Europe from West-Norway to the Adriatic). Forming tubular galleries in the mud of tidal swamps.

2. **C. nobile** O. Sars 1895 *C. n.*, G. O. Sars in: Bull. Ac. St.-Pétersb., ser. 5 *v.* 3 p. 29.2 t. 20, 21 | 1896 *C. n.*, Sowinski in: Mém. Soc. Kiew, *v.* 15 p. 375.

Segments of pleon distinct. Head with short broad triangular rostrum. Side-plate 1 the largest, produced forward to a narrowly rounded lobe, carrying many setae, some of them plumose. Eyes small, rounded, dark; visual elements seeming to be imperfectly developed. Antenna 1 very slender, in ♂ more than half as long as body, rather longer than in ♀; peduncle setose, densely in ♂; 1st joint not quite so long as 2d and 3d combined, with 2 teeth on lower margin; 3d joint scarcely more than half as long as 2d; flagellum with about 20 joints, filiform, in ♀ subequal to peduncle, in ♂ longer. Antenna 2 subpediform, in ♂ more than $^3/_4$ as long as body, much smaller in ♀; first 2 joints coalesced, as long as 3d; penultimate joint of peduncle large, tumid in ♂, in both sexes produced into 2 strong teeth, the outer the longer; ultimate joint sublinear, much shorter than penultimate, a somewhat recurved tooth of its lower margin adapted to pass between the 2 apical teeth of the penultimate; flagellum not quite as long as ultimate joint of peduncle, setose, with 1 long and 2 very short joints, the last tipped with 2 curved spines. Upper lip somewhat quadrate with dentiform projection from surface. Gnathopods 1 and 2 as in C. volutator. Peraeopods 1 and 2 comparatively slender; 2d joint slightly expanded, 4th subequal to 5th and 6th combined, its front margin in ♂ fringed with long diverging setae, 6th very narrow, conically tapering, finger subequal to it, slender, acute. Peraeopod 3, 2d joint narrowly oval, hind margin straight, almost smooth, 4th with front apex produced, 5th with 2 oblique rows of curved spines on the outer side, the lower row of 6 ending at the produced hind corner; 6th joint longer, linear; finger short, curved. Peraeopod 4 similar, but with longer joints and hind margin of 2d curved, fringed with plumose setae. Peraeopod 5 much longer, more than half as long as body; 2d joint broadly oval, narrowest distally, densely setose, 4th—6th linear, 6th the longest; finger short, curved, acute. Uropods 1—3 as in C. volutator. Telson transversely truncate. Colour, dark brown in bands or mottled; antenna 2 obliquely banded. L. ♀ 10, ♂ 11 mm.

Caspian Sea. Depth to about 75 m.

3. **C. chelicorne** O. Sars 1895 *C. c.*, G. O. Sars in: Bull. Ac. St.-Pétersb., ser. 5 *v.* 3 p. 299 t. 22.

Pleon segments distinct, division between 5th and 6th rather obscure. Head, front convex, without rostrum. Side-plate 1 with only 3 slender setae at apex. Eyes very small, rounded, dark. Antenna 1 rather short, in ♀ only about $^1/_3$ as long as body, sparingly setose; 1st joint as long as 2d and 3d combined, with about 7 spinules on lower margin in ♀, not in ♂; 2d joint more than twice as long as 3d in ♂, not in ♀; flagellum shorter than peduncle, with 15 joints in ♂, 10 in ♀. Antenna 2 strong, in ♂ $^2/_3$ as long as body; the large penultimate joint of peduncle with small inner tooth at apex, and an outer tooth produced to the end of ultimate joint, which is only about half as long as penultimate and has a short stout recurved prominence below middle of lower margin and a strong apical tooth, capable of crossing the chela-forming tooth of penultimate; flagellum 3-jointed, subequal to ultimate joint of peduncle. Gnathopods 1 and 2 as in C. volutator. Peraeopods 1 and 2, 4th joint rather longer than 5th and 6th combined, 5th very short, 6th not quite linear, finger rather shorter. Peraeopods 3 and 4 as in C. nobile. Peraeopod 5, 6th joint nearly twice as long as 5th. Uropods as in C. nobile (p. 687), except that rami of uropods 1 and 2 are without spines on inner margin. Telson and colour as in C. nobile. L. ♀ 7, ♂ 8 mm.

Caspian Sea. Depth 10—80 m.

4. **C. spinulosum** O. Sars 1896 *C. s.*, G. O. Sars in: Bull. Ac. St.-Pétersb., ser. 5 *v.* 4 p. 481 t. 12 f. 18—25.

Segments of pleon distinct, 4th—6th with spinules along hind margin, 4th and 5th with lateral spinules. Head with frontal margin only slightly convex. Side-plate 1 with 4 apical setae and several setules. Eyes very small, rounded. Antenna 1 slender, sparsely setose, in ♀ nearly half as long as body; 1st joint about as long as 2^{d}, with 2 spines on lower margin; flagellum rather longer than peduncle, about 14-jointed. Antenna 2 in ♂, penultimate joint of peduncle with large process, broad at base, with small denticle on its inner side, its acute apex chelately crossing the small apical tooth of the ultimate; in ♀ process of penultimate joint much smaller, reaching about to middle of ultimate, which is almost simply cylindric; flagellum rather shorter than ultimate joint of peduncle. Gnathopod 2, finger smooth, with only a few slender hairs. Peraeopods 1 and 2, 4th joint slightly expanded, not very setose; finger very long and slender, rather longer than 6th joint. Peraeopod 5, 2^{d} joint densely setose, the slender terminal joints having spinules as well as setae. Uropods 1 and 2 with numerous spines on both margins of peduncle and rami. Uropod 3, the ramus armed with spines as well as setae. Telson ending obtusely (in figure), distinguished by a rather conspicuous erect spine on each side of its base. L. ♀ 9·5 mm.

Caspian Sea. Depth 45 m.

5. **C. affine** Bruz. 1859 *C. a.*, R. M. Bruzelius in: Svenska Ak. Handl., n. ser. *v.* 3 nr. 1 p. 16 | 1876 *C. a.*, A. Boeck, Skand. Arkt. Amphip., *v.* 2 p. 629 t. 28 f. 7 | 1893 *C. a.*, A. Della Valle in: F. Fl. Neapel, *v.* 20 p. 371 t. 55 f. 60 | 1894 *C. a.*, G. O. Sars, Crust. Norway, *v.* 1 p. 618 t. 221 f. 2 | ? 1898 *C. a.*, Sowinski in: Mém. Soc. Kiew, *v.* 15 p. 457 | 1869 *C. tenuicorne* (♀), A. M. Norman in: Rep. Brit. Ass., Meet. 38 p. 286.

Body rather slender, pleon segments 4—6 distinct. Head without distinct rostrum; lateral lobes prominent, acute. Eyes represented by a small patch of opaque white pigment on each side. Antenna 1 rather long and slender; 1st joint narrowing distally, with 6 recurved spines at base, and 4 slender spines below, 2^{d} very slender, scarcely shorter than 1st; flagellum less than half as long as peduncle, 6-jointed. Antenna 2 in ♂ about $^{3}/_{4}$ as long as body; penultimate joint of peduncle very large, distal tooth quite small, inner tooth a small tubercle; ultimate joint not nearly half as long, narrow, with sinuous lower margin, produced distally in a tooth-like process; 1st joint of flagellum with distal nodiform projection. Antenna 2 in ♀ rather feeble, not longer than antenna 1; glandular-process at base very long; antepenultimate joint of peduncle with single spine, penultimate simply cylindrical with 3 slender spines, ultimate more than half as long, with 4 spines distributed. Gnathopod 1, 6th joint very narrow, not at all widened distally; finger denticulate, with prominent subapical tooth. Gnathopod 2 as in C. crassicorne (p. 690); finger with denticle near base. Peraeopods 1 and 2, 4th joint little expanded; finger a little longer than 5th and 6th joints combined. Peraeopod 5 very long and slender, 2^{d} joint very little expanded, its hind margin slightly concave. Uropods 1 and 2, rami slender, apical spines unusually long. Uropod 3, peduncle longer than broad, ramus linear, setae almost confined to the apex. Telson nearly as long as broad; apex narrowly truncate. Colour uniformly pale yellowish. L. 4 mm.

Arctic Ocean, North-Atlantic, North-Sea and Skagerrak (Norway southward from Lofoten Islands; Bohuslän; East Frisian Coast, depth 19—56 m; Shetland Isles); Kattegat; Bosphorus?

6. **C. runcicorne** Della Valle 1893 *C. r., C. runcinatum*, A. Della Valle in: F. Fl. Neapel, *v.* 20 p. 13, 369 t. 4 f. 7, t. 8 f. 1—16, 19 | 1898 *C. runcicorne*, Sowinski in: Mém. Soc. Kiew, *v.* 15 p. 456.

In close agreement with C. affine, but with a few distinguishing points. Antenna 1 in ♂, 1st joint with strong curved tooth directed forward near the base; flagellum 8—10-jointed. Antenna 2 in ♂, ultimate joint of peduncle apparently without the distal tooth; 1st joint of flagellum much longer than 2d and 3d combined, and with 4 teeth or nodules on its lower margin. Gnathopod 1, finger much longer than apex of 6th joint, and with subapical tooth very large. Gnathopod 2, finger without denticle, but broad at base, and abruptly narrowed near apex. Peraeopods 1 and 2, finger much longer than 5th and 6th joints combined. Telson apically rounded. Colour pale yellowish, slightly brownish on the back. L. 4—5 mm.

Bay of Naples, depth 10—20 m; Bosphorus, depth 44—85 m.

7. **C. robustum** O. Sars 1895 *C. r., C. bidentatum*, G. O. Sars in: Bull. Ac. St.-Pétersb., ser. 5 *v.* 3 p. 304; t. 23 f. 10—16 | 1896 *C. r.*, Sowinski in: Mém. Soc. Kiew, *v.* 15 p. 376.

Back broad, flattened; pleon segments distinct, but 5th from 6th not very sharply. Head with obtuse-angled front. Side-plate 1 as in C. chelicorne (p. 687). Eyes small, dark. Antenna 1 in ♀ not ⅓ as long as body, in ♂ with densely setose peduncle; 2d joint shorter in ♀ than in ♂, in both considerably shorter than 1st; flagellum shorter than peduncle, about 12-jointed. Antenna 2, even in ♀ very robust, in ♂ more than ⅔ as long as body; penultimate joint of peduncle nearly as long as ultimate and flagellum combined, apically produced to a moderately long, slightly curved, spiniform projection, with small bilobed expansion within at its base; ultimate joint with recurved tooth above the middle and spiniform apical tooth; flagellum scarcely as long as ultimate joint of peduncle. Peraeopods 1 and 2, 2d joint laminar, with long setae on front margin. 4th rather broad, about as long as 5th and 6th combined, in ♂ densely setose in front. Peraeopod 5 considerably more than half as long as body. Uropods 1 and 2, rami without spines on inner margin. Uropod 3, ramus parallel-sided. Telson truncately triangular. Colour dark. L. ♀ 7, ♂ 8 mm.

Caspian Sea. Depth to 75 m.

8. **C. mucronatum** O. Sars 1895 *C. m.*, G. O. Sars in: Bull. Ac. St.-Pétersb., ser. 5 *v.* 3 p. 307 t. 24 f. 1—7 | 1896 *C. m.*, Sowinski in: Mém. Soc. Kiew, *v.* 15 p. 375.

Segments of pleon distinct, 5th from 6th not very sharply. Head with sharp rostral point advanced beyond the narrow front corners. Eyes small, rounded, dark. Antenna 1 more than ⅛ as long as body, little setose; 1st joint longer than 2d and 3d combined, with 3 distant spinules; flagellum subequal to peduncle, 12-jointed. Antenna 2 little more than half as long as body, strong; penultimate joint of peduncle inflated, with mucronate projection reaching beyond middle of ultimate, and having a small tooth at its base within, near to which the cylindric ultimate joint of peduncle puts forth a small recurved tooth; flagellum rather shorter than ultimate joint of peduncle. Peraeopods 1 and 2, 2^d joint decidedly laminar, 4th distally wide, with slender setae on front, finger subequal to the stoutish 6th joint. Peraeopod 5 about half as long as body; 4th—6th joints rather broad and compressed, bordered with fascicles of slender setae. L. ♂ 6 mm.

Caspian Sea.

9. **C. curvispinum** O. Sars 1895 *C. c.*, G. O. Sars in: Bull. Ac. St.-Pétersb., ser. 5 *v.* 3 p. 302 t. 23 f. 1—9 | 1896 *C. c.*, SoWinski in: Mém. Soc. KieW, *v.* 15 p. 375.

Segments of pleon distinct, 5th from 6th not very sharply. Head slightly angular at centre; front corners narrow, rather prominent. Eyes small. Antenna 1 in ♂ very setose; 2^d joint of peduncle fully as long as 1st; flagellum scarcely longer than 2^d and 3^d joints of peduncle combined, 12-jointed. Antenna 1 in ♀ scarcely $^1/_3$ as long as body, sparingly setose; 1st joint of peduncle as long as 2^d and 3^d combined, with 4 or 5 spinules below; flagellum shorter than peduncle, 9-jointed. Antenna 2 in ♂ nearly as long as body, penultimate joint of peduncle widening distally, produced to a strongly incurved spiniform tooth, with slightly cleft tooth at its base within; ultimate joint of peduncle nearly as long as penultimate, cylindric, a recurved tooth near the base meeting the incurved point of the penultimate; flagellum shorter than ultimate joint of peduncle. Antenna 2 in ♀ much less strongly developed. Gnathopods 1 and 2 of the usual character. Peraeopods 1 and 2 short and stout, 2^d joint laminar, 4th nearly as broad as long, setose on both margins; finger as long as 6th joint. Peraeopod 5 slender, about half as long as body. Uropod 2 short and stout, spines on the rami not numerous. Uropod 3 not very stout. L. ♀ 6, ♂ 7 mm.

Caspian Sea.

10. **C. monodon** O. Sars 1895 *C. m.*, G. O. Sars in: Bull. Ac. St.-Pétersb., ser. 5 *v.* 3 p. 309 t. 24 f. 8—16 | 1896 *C. m.*, SoWinski in: Mém. Soc. KieW, *v.* 15 p. 374.

Body rather slender, especially in ♂. Pleon segments all distinct. Head, rostral projection almost a right angle, front corners narrow, little prominent. Eyes large, rounded, dark. Antenna 1 less than half as long as body; peduncle in ♂ setose below, in ♀ sparingly setose; 1st joint longer than 2^d and 3^d combined, with 1 spinule at end of the joint; flagellum in ♂ shorter than peduncle, 12-jointed, in ♀ as long as peduncle, 10-jointed. Antenna 2 in ♂ about $^2/_3$ as long as body; penultimate joint of peduncle not greatly inflated, almost cylindric, produced to a simple narrowly mucronate tooth, not nearly reaching middle of ultimate and without inner tooth; ultimate joint with rudiment only of a tooth near base; flagellum scarcely half as long as ultimate joint of peduncle. Antenna 2 in ♀ not longer than antenna 1, about $^1/_3$ as long as body, without even rudiment of a tooth on ultimate joint of peduncle. Peraeopods 1 and 2 slender; peraeopod 5 about half as long as body. Uropod 2 very small, compared with uropod 1. Uropod 3, ramus much narrower than peduncle, with 1 spine among a few apical setae. Peraeon segment 7 and the rest of hind part of body almost devoid of pigment. L. ♀ 4, ♂ 5 mm.

Caspian Sea. Depth to 75 m.

11. **C. crassicorne** Bruz. 1859 *C. c.*, R. M. Bruzelius in: Svenska Ak. Handl., n. ser. *v.* 3 nr. 1 p. 15 t. 1 f. 2 | 1876 *C. c.*, A. Boeck, Skand. Arkt. Amphip., *v.* 2 p. 626 t. 28 f. 8 | 1879 *C. c.* (part.), Hoek in: Tijdschr. Nederl. dierk. Ver., *v.* 4 p. 115 t. 5 f. 16,

t. 8 f. 4—10 | 1889 *C. c.*, Hoek in: Tijdschr. Nederl. dierk. Ver., ser. 2 *v.* 2 p. 230 | 1893 *C. c.*, A. Della Valle in: F. Fl. Neapel, *v.* 20 p. 367 t. 55 f. 58, 59 | 1894 *C. c.*, G. O. Sars, Crust. Norway, *v.* 1 p. 615 t. 220 | 1896 & 98 *C. c.*, Sowinski in: Mém. Soc. Kiew, *v.* 15 p. 373, 455 | 1862 *C. spinicorne* (♀) (non Stimpson 1857!), Bate, Cat. Amphip. Brit. Mus., p. 281 t. 47 f. 5 | 1863 *C. bonellii* (part.: ♀), Bate & Westwood, Brit. sess. Crust., *v.* 1 p. 497 f. | ? 1868 *C. b.*, Czerniavski in: Syezda Russ. Est., Syezda 1 Zool. p. 96.

Body broad, pleon segments 4—6 coalesced. Head, rostrum short, distinct, acute; lateral lobes prominent, acute. Side-plate 1 with 3 setae on the produced apex. Eyes small, rounded, dark brown with whitish coating. Antenna 1 scarcely $^1/_3$ as long as body in ♀ and less than $^1/_2$ in ♂; 1st joint nearly as long as 2^d and 3^d combined, thick above, with 4 recurved spines at base, compressed below with row of about 6 spines; 3^d joint much shorter than 2^d; flagellum shorter than peduncle, 6-jointed. Antenna 2 in ♂ (Fig. 117)

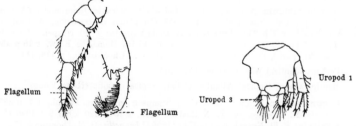

Flagellum

Flagellum

Uropod 1

Uropod 3

Fig. 116.   Fig. 117.
Antenna 2, ♀.   Antenna 2, ♂.
Fig. 118.
Uropods 1—3 and telson.
Fig. 116—118. *C. crassicorne.* [After G. O. Sars.]

stout, scarcely as long as body; penultimate joint of peduncle large, with 2 apical teeth, the outer reaching the sub-basal tooth of the ultimate, which is much shorter and narrower than the penultimate, and has a small apical tooth, a little separated from line of lower margin; 1st joint of flagellum with long setae. Antenna 2 in ♀ (Fig. 116) little longer than antenna 1, stout; antepenultimate joint of peduncle with 3 or 4 stout spines, penultimate with 8—12 such spines along the sharp edge of its broad laminar expansion; ultimate joint of peduncle little more than half as long, with a strong median spine. Gnathopods 1 and 2 as in C. volutator (p. 686), except that the slender finger of gnathopod 2 has an inner denticle. Peraeopods 1—5 as in C. volutator, except that in peraeopods 1 and 2 the finger is relatively longer, in peraeopods 3 and 4 the 4th joint less setose, and that peraeopod 5 is somewhat more slender. Uropod 3, ramus oblong oval, nearly twice as large as peduncle (Fig. 118). Telson rather more obtusely rounded than in C. volutator. Colour whitish mottled with light brown. L. about 5 mm.

Arctic Ocean, North-Atlantic with adjoining seas (Jan Mayen; Europe from Vadsö [Norway] to West-France, depth 11—37 m); Bosphorus. Found in Norfolk in almost fresh water.

12. **C. bonellii** (M.-E.) 1830 *Corophia b.*, H. Milne Edwards in: Ann. Sci. nat., *v.* 20 p. 385 | 1840 *Corophium b.*, H. Milne Edwards, Hist. nat. Crust., *v.* 3 p. 67 | 1862 *C. b.*, Bate, Cat. Amphip. Brit. Mus., p. 282 | 1893 *C. bonelli*, A. Della Valle in: F. Fl. Neapel, *v.* 20 p. 368 | 1894 *C. b.*, G. O. Sars, Crust. Norway, *v.* 1 p. 616 t. 221 f. 1.

♀. Pleon segments 4—6 coalesced. Head without proper rostrum, front produced in a broad triangle; lateral lobes short, apically rounded. Eyes large, very dark. Antenna 1 about $^1/_3$ as long as body; 1st joint slightly shorter than 2^d and 3^d combined, with small hooked spine near base and

3 spines along lower margin; flagellum more than half as long as peduncle, 6-jointed. Antenna 2 little longer than antenna 1, not very stout; penultimate joint of peduncle nearly cylindric, with 3 strong spines on margin, ultimate rather shorter, with a spine beyond centre of lower margin. Gnathopod 1, 6th joint widening a little distally, palm rather oblique, finger extending very little beyond it. Gnathopod 2, finger with subapical denticle. Peraeopods 1 and 2, 4th joint rather broad, 5th very short, finger subequal to 5th and 6th combined. Peraeopod 5, 2[d] joint oval, the others very slender. Uropod 3, peduncle very short, ramus broadly oval, fringed with 7—9 setae. Telson twice as broad as long, transversely truncate at apex. Colour whitish, dorsally banded with dark brown. L. 6 mm. — ♂ unknown.

North-Atlantic, North-Sea. Skagerrak and English Channel (South- und West-Norway up to the Trondhjemsfjord; South-England; West-France). Depth 11—19 m.

C. acherusicum A. Costa ?1851 *Audouinia acherusica* (nom. nud.), (A. Costa in:) F. W. Hope, Cat. Crost. Ital., p. 24 | 1857 *Corophium acherusicum*, A. Costa in: Mem. Acc. Napoli, *v.* 1 p. 232 | 1893 *C. a.*, A. Della Valle in: F. Fl. Neapel, *v.* 20 p. 364 t. 1 f. 11; t. 8 f. 17, 18, 20—41 | ?1898 *C. a.*, Sowinski in: Mém. Soc. Kiew, *v.* 15 p. 455 | 1897 *C. bonellii* (part.), (T. Stebbing in:) H. J. Hansen, Choniostom., p. 113, 114.

Perhaps identical with C. bonellii (p. 691).

Mediterranean; tropical Atlantic (Cuba); tropical Pacific (Hong Kong).

C. contractum Stimps. 1855 *C. c.*, Stimpson in: P. Ac. Philad., *v.* 7 p. 383 | ?1881 *C. c.*, G. M. Thomson in: Tr. N. Zealand Inst., *v.* 13 p. 220 t. 8 f. 9 | 1893 *C. c.*, A. Della Valle in: F. Fl. Neapel, *v.* 20 p. 374 | ?1886 *C. bonellii*, G. M. Thomson & Chilton in: Tr. N. Zealand Inst., *v.* 18 p. 142.

Possibly identical with C. bonellii (p. 691). L. 3—4 mm.

Pacific (Japan, New Zealand).

C. cylindricum (Say) 1818 *Podocerus cylindricus*, Say in: J. Ac. Philad., *v.* 1 II p. 387 | 1873 *Corophium cylindricum*, (S. I. Smith in:) A. E. Verrill in: Rep. U. S. Fish Comm., *v.* 1 p. 370, 566 | 1888 *C. c.*, T. Stebbing in: Rep. Voy. Challenger, *v.* 29 p. 104 | 1893 *C. c.*, A. Della Valle in: F. Fl. Napel, *v.* 20 p. 376.

North-Atlantic (from New Jersey to Vineyard Sound).

C. dentatum Fr. Müll. 1864 *C. d.*, Fritz Müller, Für Darwin, p. 51 | 1893 *C. d.*, A. Della Valle in: F. Fl. Neapel, *v.* 20 p. 375.

Brazil?

C. quadriceps Dana 1852 *C. q.*, J. D. Dana in: P. Amer. Ac., *v.* 2 p. 219 | 1853 & 55 *C. ?q.*, J. D. Dana in: U. S. expl. Exp., *v.* 13 II p. 836; t. 55 f. 8 | 1888 *C. ?q.*, T. Stebbing in: Rep. Voy. Challenger, *v.* 29 p. 255 | 1893 *C. q.*, A. Della Valle in: F. Fl. Neapel, *v.* 20 p. 374.

Probably a young form. L. about 2 mm.

Harbour of Rio Janeiro.

C. salmonis Stimps. 1857 *C. s.*, Stimpson in: Boston J. nat. Hist., *v.* 6 p. 514 | 1888 *C. s.*, T. Stebbing in: Rep. Voy. Challenger, *v.* 29 p. 303 | 1893 *C. s.*, A. Della Valle in: F. Fl. Neapel, *v.* 20 p. 375.

Puget Sound.

C. spinicorne Stimps. 1857 *C. s.*, Stimpson in: Boston J. nat. Hist., *v.* 6 p. 514 | 1857 *C. s.*, Stimpson in: P. Calif. Ac., *v.* 1 p. 89 | 1893 *C. s.*, A. Della Valle in: F. Fl. Neapel, *v.* 20 p. 375.

L. 10 mm.

San Francisco Bay.

39. Fam. **Cheluridae**

1847 *Cheluridae*, G. J. Allman in: Ann. nat. Hist., *v.* 19 p. 361 | 1876 *C.*, A. Boeck, Skand. Arkt. Amphip., *v.* 2 p. 645 | 1888 *C.*, T. Stebbing in: Rep. Voy. Challenger, *v.* 29 p. 218 | 1894 *C.*, G. O. Sars, Crust. Norway, *v.* 1 p. 626 | 1893 *Cheluridi*, A. Della Valle in: F. Fl. Neapel. *v.* 20 p. 345.

Pleon segments 4—6 coalesced. Side-plates small. Antenna 2 with blade-like flagellum. Mouth-parts on the whole normal. Pleopods with peduncle produced on the inner side. Uropods 2 and 3 abnormal (Fig. 119 p. 694).

Marine.

1 genus with 1 species.

1. Gen. **Chelura** Phil.

1839 *Chelura* (Sp. un.: *C. terebrans*), A. Philippi in: Arch. Naturg., *v.* 51 p. 120 | 1876 *C.*, A. Boeck, Skand. Arkt. Amphip., *v.* 2 p. 646 | 1888 *C.*, T. Stebbing in: Rep. Voy. Challenger, *v.* 29 p. 181, 217, 1670 | 1893 *C.*, A. Della Valle in: F. Fl. Neapel, *v.* 20 p. 346 | 1894 *C.*, G. O. Sars, Crust. Norway, *v.* 1 p. 626 | 1847 *Nemertes* (nom. nud.), A. White, Crust. Brit. Mus., p. 90.

Body (Fig. 119 p. 694) broad, subdepressed, pleon segments 1 and 2 very short. Head without rostrum. Antenna 1 short, with accessory flagellum. Antenna 2 longer and much stouter. Mouth-parts prominent. Upper lip with the margin entire. Lower lip with inner lobes thin, adpressed. Mandible, molar strong; palp rather short, 3^d joint about as long as 2^d. Maxilla 1, inner plate narrow, tipped with 3 short setae, outer with 9 spines, palp long. Maxilla 2, inner plate having 2 setae on inner margin, of which the more distal is the longer. Maxillipeds, inner plates rather long, its apical spines slender, seemingly unaccompanied by spine-teeth. Gnathopod 1 the stronger. Peraeopods 3—5, 2^d joint little expanded. Uropod 2 subdorsal in position, with greatly expanded peduncle and very short rami. Uropod 3, peduncle very short, inner ramus rudimentary, outer large and laminar. Telson entire. Only one pair of hepato-pancreatic coeca.

1 species.

1. **C. terebrans** Phil.

1839 *C. t.*, A. Philippi in: Arch. Naturg., *v.* 51 p. 120 t. 3 f. 5 | 1847 *C. t.*, G. J. Allman in: Ann. nat. Hist., *v.* 19 p. 361 t. 13, 14 | 1880 *C. t.*, S. I. Smith in: P. U. S. Mus., *v.* 2 p. 233 f. | 1888 *C. t.*, T. Stebbing in: Rep. Voy. Challenger, *v.* 29 p. 1695 | 1893 *C. t.*, A. Della Valle in: F. Fl. Neapel, *v.* 20 p. 347 t. 6 f. 3; t. 7 f. 1—22 | 1894 *C. t.*, G. O. Sars, Crust. Norway, *v.* 1 p. 627 t. 225 | 1847 *Nemertes nesaeoides* (nom. nud.), A. White, Crust. Brit. Mus., p. 90 | 1863 *Chelura terebans*, Bate & Westwood, Brit. sess. Crust., *v.* 1 p. 503 f. | 1868 *C. pontica*, Czerniavski in: Syezda Russ. Est., Syezda 1 Zool. p. 95 t. 7 f. 1—18 | 1868 *Limnoria xylophaga*, E. Hesse in: Ann. Sci. nat., ser. 5 *v.* 10 p. 101.

Back broadly vaulted, with setules rising from hind margin of the segments. Pleon segment 3 dorsally produced to a conical curved process much longer in ♂ than in ♀, with a short process on each side of it; the following segment (4th—6th in coalescence) is dorsally flat, rectangular, in ♂ nearly half, in ♀ about $^1/_3$ as long as the rest of the body. Head, lateral lobes rounded, post-antennal corners well defined. Side-plates successively less deep, 1st rounded quadrate, 5th and 6th bilobed. Pleon segment 3, postero-lateral corners quadrate. Eyes on lateral lobes of head, small, rounded, dark. Antenna 1 about $^1/_5$ as long as body; flagellum rather setose, as long as 2^d and 3^d joints of peduncle combined, 6-jointed; accessory flagellum with 2 joints, 2^d minute. Antenna 2 curving; ultimate joint of peduncle little longer than penultimate or antepenultimate, like the flagellum densely setose; flagellum about $2^1/_2$ as long as broad, consisting of 1 massive joint tipped with 1 or 2 minute ones almost hidden in the setose clothing. Gnathopod 1 chelate, small, 5th joint shorter than 6th, which is oblong, a little curved, the short thumb not strongly produced, stouter than the small finger, which inclines towards it, giving almost a subchelate appearance. Gnathopod 2 longer but more slender, 5th joint nearly as long as the narrow, subfusiform 6th, which ends in a chela smaller but more definite than that of gnathopod 1. Peraeopods 1—5 not elongate; 2d joint slightly expanded, 4th distally

widened, 5th short, 6th distally narrowed; finger short, curved; in peraeopods 3—5 the 2d and 4th joints fringed with long setae. Branchial vesicles rather large. Marsupial plates narrow. Pleopods 1—3 have rather elongate coupling spines on expanded inner part of peduncle; 1st joint of inner

Fig. 119. C. terebrans, ♂. Lateral view.
[After G. O. Sars.]

ramus with inner margin smooth. Uropod 1, peduncle more than twice as long as rami; inner ramus the broader, with 3 apical spinules. Uropod 2 attached close to uropod 1; peduncle enormously expanded, the expansion in ♀ long as well as broad and densely hirsute, in ♂ broader than long, with serrate margin; in both sexes rami very short, laminar, apically serrate. Uropod 3, attached far from the others; peduncle short, inner ramus oval, minute, not visible from above; outer very large, in ♂ very long, broad, spinulose, narrowing distally, in ♀ much shorter, but still large, broadly oval, with margin serrate and spinulose. Telson subcarinate dorsally, about as broad as long, irregularly tapering to an acute point, furnished with a few setules. Colour light brown. L. ♀ 5, ♂ 6 mm.

North-Atlantic with adjoining seas (Europe from Norway to the Black Sea; North America). In submerged or partially submerged or waterlogged timber.

40. Fam. **Podoceridae**

1849 *Dulichidae*, J. D. Dana in: Amer. J. Sci., ser. 2 *v.* 8 p. 135, 140 | 1857 *Dyopedidae*, Bate in: Ann. nat. Hist., ser. 2 *v.* 19 p. 150 | 1876 *Dulichidae*, A. Boeck, Skand. Arkt. Amphip., *v.* 2 p. 649 | 1888 *Dulichiidae*, T. Stebbing in: Rep. Voy. Challenger, *v.* 29 p. 1182 | 1893 *Dulichidi*, A. Della Valle in: F. Fl. Neapel, *v.* 20 p. 314 | 1894 *Dulichiidae*, G. O. Sars, Crust. Norway, *v.* 1 p. 628.

Pleon segment 4 elongate, 5th and 6th very short, one of them some-times missing. Side-plates small, shallow (Fig. 122, 126 p. 705, 712). Antennae 1 and 2 elongate, setose; the relative length varying; flagellum shorter than peduncle; accessory flagellum present or absent. Upper lip slightly bilobed. Lower lip with inner lobes. Mouth-parts normal. Gnathopod 2 usually much larger than gnathopod 1 and much larger in ♂ than in ♀. Peraeo-pods 3—5 usually elongate, 2d joint not or little expanded. Pleopods 1—3 (Fig. 120 p. 696), peduncle not expanded. Uropod 1 normal, of uropods 2 and 3 one or other missing or rudimentary or abnormal. Telson entire (Fig. 121, 127, p. 696, 714).

Marine.

8 genera, 30 accepted species and 1 obscure.

Synopsis of genera:

1 {
Antenna 1 Without accessory flagellum — **2.**
Antenna 1 (Fig. 122 p. 705) With accessory flagellum — **4.**

2 {
Pleon With only 5 distinct segments preceding the telson . . . . 1. Gen. **Laetmatophilus** p. 695
Pleon With 6 distinct segments preceding the telson — **3.**

3 {
Pleon segment 5 carrying uropods . . . . . 2. Gen. **Cyrtophium** . . p. 697
Pleon segment 5 not carrying uropods . . . 3. Gen. **Leipsuropus** . . p. 698

4 {
3 pairs of uropods (Fig. 125 p. 707) present — **5.**
Only 2 pairs of uropods (Fig. 127 p. 714) present — **7.**

5 {
Antenna 1 longer than antenna 2 . . . . . 4. Gen. **Xenodice** . . . p. 699
Antenna 1 shorter than antenna 2 (Fig. 122, 123 p. 705, 707) — **6.**

6 {
Gnathopods 1 and 2 subchelate . . . . . . 5. Gen. **Podocerus** . . . p. 700
Gnathopods 1 and 2 (Fig. 124 p. 707) simple 6. Gen. **Icilius** . . . . . p. 706

7 {
Last pair of uropods normal . . . . 7. Gen. **Dulichia** . . . . p. 708
Last pair of uropods (Fig. 127) rudimentary 8. Gen. **Paradulichia** . . p. 713

1. Gen. Laetmatophilus Bruz.

1859 *Laetmatophilus* (Sp. un.: *L. tuberculatus*), R. M. Bruzelius in: Svenska Ak. Handl., n. ser. *v.* 3 nr. 1 p. 10 | 1886 *L.*, Gerstaecker in: Bronn's Kl. Ordn., *v.* 5 ɪɪ p. 493 | 1888 *L.*, T. Stebbing in: Rep. Voy. Challenger, *v.* 29 p. 1197 | 1893 *L.*, A. Della Valle in: F. Fl. Neapel, *v.* 20 p. 316 | 1894 *L.*, G. O. Sars, Crust. Norway, *v.* 1 p. 629 | 1873 *Laetmophilus*, A. Marschall, Nomencl. zool., p. 411 | 1885 *Laematophilus*, HasWell in: P. Linn. Soc. N. S. Wales, *v.* 10 p. 107, 110.

Peraeon with marked constriction between segments 1 and 2; segments behind constriction much broader in ♀ than in ♂. Pleon consisting of 5 segments, each carrying a single pair of legs; last 3 segments reflexed. Head with frontal process carrying antenna 1. Side-plates small, 2^d largest in ♂, 3^d in ♀. Antennae 1 and 2 strong, setose, with flagellum few-jointed; no accessory flagellum to antenna 1; antenna 2 the longer. Upper lip bilobed. Mandible with large palp; 3^d joint setose, broad ended. Maxilla 1. inner plate small; palp 2-jointed. Maxillipeds, outer plates reaching much beyond the smaller inner plates; 4th joint of palp bluntly conical. Gnathopod 1 much smaller than gnathopod 2, 5th and 6th joints subequal. Gnathopod 2, 5th joint small, 6th broader in ♀, longer in ♂; finger large. Peraeopods 1—5 subequal; 2^d joint not expanded, 6th joint the longest; finger strong, curved. Branchial vesicles very small. Marsupial plates large and broad. Pleopods 1—3 (Fig. 120 p. 696) feeble. Uropod 1 (Fig. 121 p. 696), inner ramus much longer than outer. Following uropod (probably uropod 3) consisting of a small simple plate. Telson simple, rounded.

4 species.

Synopsis of species:

1 {
Without dorsal tubercles . . . . 1. **L. purus** . . . p. 696
With dorsal tubercles — **2.**

2 {
Dorsal tubercles obtuse . . . . *ı* 2. **L. tuberculatus** p. 696
Dorsal tubercles spiniform — **3.**

3 {
Tubercles not more than 2 in transVerse row . . . 3. **L. armatus** . . p. 697
Tubercles as many as 4 in transVerse row . . . . 4. **L. hystrix** . . . p. 697

1. **L. purus** Stebb. 1888 *L. p.*, T. Stebbing in: Rep. Voy. Challenger, *v.* 29
p. 1198 t. 132 | 1894 *L. p.*, G. O. Sars, Crust. Norway, *v.* 1 p. 630 | 1893 *L. tuberculatus*
(part.), A. Della Valle in: F. Fl. Neapel, *v.* 20 p. 317.

Back appearing transversely corrugated; head and peraeon segments 1—5
each having a dorsal depression; peraeon segment 3 with small ventral
process; pleon segment 4 cylindrical, longest of all segments. Eyes round, dark,
on prominent lobes. Antenna 1, 1st joint of peduncle $^1/_2$ as long as 3^d, 2^d longer
than 3^d, flagellum as long as 3^d, with 3 joints, 1st nearly thrice as long as
2^d and 3^d combined. Antenna 2 stouter, ultimate joint of peduncle longer than
penultimate, subequal to 1st and 2^d of peduncle of antenna 1 combined; flagellum
of 1 long joint with 1 or 2 microscopic apical joints. Gnathopod 1, 4th joint
short, setose, 5th ovoid, strongly fringed with plumose setae or spines on

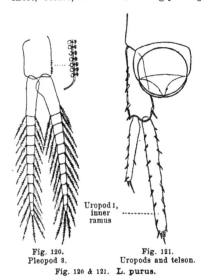

hind margin, 6th widening almost
abruptly at base, with very short hind
margin, and long, oblique, slightly
convex, crenate palm, fringed with
plumose spines, and having a row of
10 close-set serrate spines leading to
the defining angle; finger matching
palm, distal half of inner margin with
about 12 decurrent spine-teeth. Gnathopod 2 in ♂, 2^d joint short, broad,
the channelled front with projecting
apices, 3^d, 4th and 5th joints very
short, 4th joint with acute apex, 6th
very long, a sort of oblong oval,
having near the finger-hinge a broad
lobe followed by a narrow blunt tooth;
the long strong finger closing over
these, and leaving a narrow gap between its smooth inner margin and
that of the hand, on the surface of
which it applies its apex near the
base, the meeting of hind margin
and palm being indefinite. Peraeo-

Uropod 1,
inner
ramus

Fig. 120.   Fig. 121.
Pleopod 3.   Uropods and telson.
Fig. 120 & 121. *L. purus.*

pods 3—5, 2^d joint a little longer than the 4th, shorter than the 5th.
Pleopods 1—3 (Fig. 120), peduncle much shorter than rami, with 6 or 7 pairs
of coupling spines; joints of rami 11—13, no cleft spines observed. Uropod 1
(Fig. 121), peduncle as long as outer ramus, which is less than $^2/_3$ as long
as the broader inner ramus; both with spines on both margins. Terminal
uropod narrowly oval, when turned obliquely inward covered by the almost
circular telson (Fig. 121). L. about 6 mm.

Off Cape of Good Hope.

2. **L. tuberculatus** Bruz. 1859 *L. t.*, R. M. Bruzelius in: Svenska Ak. Handl.,
n. ser. *v.* 3 nr. 1 p. 11 t. 1 f. 1 | 1876 *L. t.*, A. Boeck, Skand. Arkt. Amphip. *v.* 2 p. 663
t. 29 f. 7 | 1877 *L. t.*, Meinert in: Naturh. Tidsskr., ser. 3 *v.* 11 p. 166 | 1893 *L. t.* (part.),
A. Della Valle in: F. Fl. Neapel, *v.* 20 p. 317 t. 55 f. 1—3 | 1894 *L. t.* (part.), G. O. Sars, Crust.
Norway, *v.* 1 p. 630 t. 226 | ? 1898 *L. t.*, Sowinski in: Mém. Soc. Kiew, *v.* 15 p. 451 t. 8 f. 1—8.

Head with small dorsal tubercle. Segments of peraeon transversely
furrowed, 1st with 2 dorsal tubercles, one behind other, 2^d with small
tubercle in front of furrow; the remainder rugged in outline, scarcely tubercular.

Eyes rounded, slightly protuberant, with light yellowish pigment. Antennae 1 and 2 about as in L. purus. Antenna 1 about $^3/_4$ as long as body. Gnathopod 1 as in L. purus, except 6th joint gradually widening to the palm, which is not longer than hind margin, somewhat oblique, setose, defined by an obtuse angle, slightly overlapped by the apparently not denticulate finger. Gnathopod 2 in ♂ as in L. purus, except that the tooth on the 6th joint is more acute. Gnathopod 2 in ♀, apex of 4th joint more produced than in ♂; 6th joint broad, rounded oval; palm convex, defined by a short tooth; finger strong, curved, matching palm. Peraeopods 1—5 subequal, slender; 2d joint much longer than 4th, subequal to 6th, which is longer than 5th; finger strong, curved. Pleopods (in Sars' figure) with 9-jointed rami. Colour yellowish. L. 4—5 mm.

Arctic Ocean, North-Atlantic, North-Sea, Skagerrak and Kattegat (Scandinavian coasts). Depth 45—600 m.

3. **L. armatus** (Norm.) 1869 *Cyrtophium armatum*, A. M. Norman in: Rep. Brit. Ass., Meet. 38 p. 285 | 1888 *Laetmatophilus a.*, Chevreux & Guerne in: C.-R. Ac. Sci., v. 106 p. 626 | 1888 *L. a.*, T. Stebbing in: Rep. Voy. Challenger, v. 29 p. 1628 | 1894 *L. a.*, G. O. Sars, Crust. Norway, v. 1 p. 632 t. 227 f. 1 | 1895 *L. a.*, A. M. Norman in: Ann. nat. Hist., ser. 6 v. 15 p. 493 | 1871 *L. spinosissimus*, A. Boeck in: Forh. Selsk. Christian., 1870 p. 266 | 1893 *L. tuberculatus* (part.), A. Della Valle in: F. Fl. Neapel, v. 20 p. 317.

Head with dorsal tubercle acutely produced. Peraeon segment 1 with 2 spiniform tubercles, one behind other, segment 2 with 1 median, followed by 2 side by side, segments 3—7 and pleon segments 1 and 2 each with a transverse pair. Appendages closely agreeing with those of L. tuberculatus. Colour light yellowish, transversely banded with vivid orange. L. scarcely 4 mm.

Arctic Ocean, North-Atlantic and North-Sea (West-Norway to Lofoten Islands, depth 94—565 m; Shetland Isles, depth 180—200 m).

4. **L. hystrix** (Hasw.) 1880 *Cyrtophium? h.*, Haswell in: P. Linn. Soc. N. S. Wales, v. 5 p. 104 t. 7 f. 3 | 1885 *Laematophilus h.*, Haswell in: P. Linn. Soc. N. S. Wales, v. 10 p. 110 | 1893 *Laetmatophilus tuberculatus* (part.), A. Della Valle in: F. Fl. Neapel, v. 20 p 317.

Head, peraeon segments 1—7 and pleon segments 1 and 2 armed with spiniform tubercles. Of these there is a median row, flanked on the peraeon and perhaps also on the head by 2 lateral rows on each side; lateral borders of peraeon segments 2—6 produced outward and upward, acuminate. Side-plates with an acute point on lower margin. Appendages appear in close agreement with L. purus, but in gnathopod 1 in ♂ the 6th joint is without the long row of spines at defining angle of palm. Gnathopod 2 in ♂ not observed. Gnathopod 2 in ♀ as in L. tuberculatus. Rami of pleopods with 6—8 joints. Uropod 1, outer ramus $^2/_3$ as long as inner. L. about 4 mm.

Port Jackson and Port Stephens [East-Australia].

2. Gen. **Cyrtophium** Dana

1852 *Cyrtophium*, J. D. Dana in: Amer. J. Sci., ser. 2 v. 14 p. 309 | 1853 *C.* (Sp. un.: *C. orientale*), J. D. Dana in: U. S. expl. Exp., v. 13 II p. 831, 839.

Agreeing with Podocerus (p. 700) except that antenna 1 has no accessory flagellum, and the longer ramus in uropods 1 and 2 is broad comparatively.

2 accepted species, 1 doubtful.

Synopsis of accepted species:

Gnathopod 2 in ♂, palm and hind margin smoothly continuous   1. **C. orientale** . . p. 698

Gnathopod 2 in ♂, palm well defined, forming three strong
teeth . . . .   2. **C. minutum**   . p. 698

1. **C. orientale** Dana 1853 & 55 *C. o.*, J. D. Dana in: U. S. expl. Exp., *v.* 13 II
p. 840; t. 56 f. 1 a—d | 1893 *Platophium o.* (part.), A. Della Valle in: F. Fl. Neapel,
v. 20 p. 332.

Body elliptical, dorsally corrugated, not carinate. Antennae 1 and 2
subpediform, subequal. Antenna 1, flagellum with 3 subequal joints, together
subequal to 3^d joint of peduncle. Antenna 2, flagellum with 3 joints, the 1st long,
stiliform, longer than ultimate joint of peduncle, the other 2 minute. Gnathopod 2, 6th joint stout, subelliptical, nearly straight below and hirsute; finger
nearly as long as 6th joint. Peraeopod 5 (in figure) almost completely
unarmed; 2^d joint quite narrow. Uropods 1 and 2, apical spines of great
length, numerous spines on inner margin of the broad longer ramus, the
other subterete. Telson (in figure) bluntly triangular. L. about 4 mm.

Strait of Singapore.

2. **C. minutum** HasW. 1879 *C. m.*, Haswell in: P. Linn. Soc. N. S. Wales, *v.* 4
p. 343 t. 22 f. 6 | 1885 *C. m.*, Haswell in: P. Linn. Soc. N. S. Wales, *v.* 10 p. 109 t. 18
f. 1—5, 9 | 1888 *C. m.*, T. Stebbing in: Rep. Voy. Challenger, *v.* 29 p. 1193 | 1893 *Plato-
phium m.*, A. Della Valle in: F. Fl. Neapel, *v.* 20 p. 334.

Body corrugated, most strongly on pleon segments 1—3. Eyes prominent.
Antennae 1 and 2 subequal. Antenna 1, flagellum with 3 joints, the 1st longer
than 2^d and 3^d combined, the 3^d very small. Antenna 2, flagellum much
shorter than ultimate joint of peduncle, with 2 joints, 1st long, stout, 2^d minute,
tipped with curved spine. Mandible, 3^d joint of palp less dilated than in
Podocerus (p. 700). Gnathopod 1 as in Podocerus, except that the palm is
less strongly distinguished from hind margin of 6th joint. Gnathopod 2
in ♂, 4th joint distally produced to an acute point and very hirsute,
5th rudely squared, not projecting, 6th with front margin smoothly curved,
not notched for spines; palm defined from short hind margin by long
tooth, which the apex of finger reaches, arching over 2 others. Peraeopods 1—5 all nearly alike; 2^d joint short and narrow, with a little projection
at upper corner, 5th longer than 4th, 6th than 5th; the spinules on the
6th joint very unimportant in appearance. Pleopods 1—3 each with 4—6
coupling spines. Uropods 1—3 and telson about as in C. orientale. Colour
pellucid with brown dots, and across the head a brown band. L. 2—3 mm.

Port Jackson [East-Australia].

C. laeve Heller 1866 *C. l.*, Cam. Heller in: Denk. Ak. Wien, *v.* 26 II p. 49 t. 4
f. 9—12 | 1893 *Platophium brasiliense* (part.), A. Della Valle in: F. Fl. Neapel, *v.* 20 p. 329.

Adriatic (Lesina).

3. Gen. **Leipsuropus** Stebb.

1899 *Leipsuropus* (Sp. un.: *Cyrtophium parasiticum*), T. Stebbing in: Ann. nat
Hist., ser. 7 *v.* 3 p. 241.

Like Cyrtophium (p. 697), but pleon segment 5, though present, devoid
of legs.

1 species.

1. **L. parasiticus** (HasW.) 1879 *Cyrtophium parasiticum*, Haswell in: P. Linn. Soc. N. S. Wales, v.4 p. 274 t.12 f.1 | 1885 *C. p.*, Haswell in: P. Linn. Soc. N. S. Wales, v. 10 p. 107 t. 17 f. 1—7 | 1893 *Platophium p.* (part.). A. Della Valle in: F. Fl. Neapel, v. 20 p. 335 | 1899 *Cyrtophium p.*, *Leipsuropus*, T. Stebbing in: Ann. nat. Hist., ser. 7 v. 3 p. 241.

Body without medio-dorsal processes, laterally tuberculate, peraeon dorsally ovate, widest at 4th segment, division between segments 6 and 7 dorsally obscure, pleon segment 5 very short. Side-plates squared, with rounded corner, 4th widest, slightly emarginate below. Eyes round, rather prominent. Antenna 1 elongate, slender, 1st joint scarcely half as long as 2d, 3d a little shorter than 2d; flagellum as long as 2d joint of peduncle, with 4 joints, 1st rather longer than the other 3, 3d longer than 2d. Antenna 2 longer and much stouter; ultimate joint of peduncle rather longer than penultimate; flagellum as long as penultimate joint of peduncle, 1 long stout joint with a minute terminal (Haswell: flagellum with traces of division into 9 joints). Gnathopod 1, 2d joint shorter than 5th, 3d—6th setose on hind margin, 5th broad except at base, 6th rather shorter and narrower than 5th, oval, hind margin extremely short, forming an angle with the long, convex, oblique palm; finger long and strong, not quite reaching end of palm. Gnathopod 2 in ♂, 2d joint much shorter than 6th, 3d—5th very small, 6th powerful, elongate, narrowest at base, front margin convex, hind nearly straight but with a shallow emargination, variable in position, much or little or not at all overlapped by the strong and long, smooth finger; one of the points of the emargination much or little projecting, and perhaps to be regarded as defining the palm from the straight hind margin. Gnathopod 2 in ♀ (in figure), 6th joint widening greatly to a nearly transverse, slightly sinuous, well defined palm, matched by the finger. Peraeopods 1—5 of uniform structure, very slightly armed; 2d joint short, not expanded, 4th joint very short, 5th stout, longer than the 2d, rather shorter than the 6th; finger curved, acute. Pleopods 1—3, peduncle of pleopod 1 with rectangular inner corners, peduncles of 2d and 3d successively shorter, inner corners rounded, so that the inner rami are not contiguous to one another; coupling spines 4 pairs; rami 7- or 8-jointed. Uropod 1 slender; peduncle slightly overlapping telson; outer ramus slightly longer than peduncle, nearly ²/₃ as long as inner, with spinules on outer margin, inner with spinules on both margins. Uropod 3, the narrowly oval plates not reaching end of the obtusely pointed ovate telson. L. 4—5 mm.

Port Jackson [East-Australia].

4. Gen. **Xenodice** Boeck

1871 *Xenodice* (Sp. un.: *X. frauenfeldti*), A. Boeck in: Forb. Selsk. Christian., 1870 p. 266 | 1876 *X.*, A. Boeck, Skand. Arkt. Amphip., v.2 p 665 | 1888 *X.*, T. Stebbing in: Rep. Voy. Challenger, v. 29 p. 402 | 1893 *X.*, A. Della Valle in: F. Fl. Neapel, v. 20 p. 318 | 1894 *X.*, G. O. Sars, Crust. Norway, v.1 p. 632 | 1895 *Zenodice*, A. M. Norman in: Ann. nat. Hist., ser. 6 v. 15 p. 493.

Body slender. Pleon segment 4 much longer than 5th and 6th. Head produced over base of antenna 2. Side-plates small, subequal. Antennae 1 and 2 elongate; flagellum many-jointed; antenna 1 the longer, with accessory flagellum. Upper lip faintly bilobed. Mandibular palp large; 3d joint setose, broad-ended. Maxilla 1, inner plate fringed with 7 or 8 setae. Maxilla 2, inner plate with long row of setae fringing inner margin. Maxillipeds, outer plates reaching much beyond small inner plates. Gnathopods 1 and 2 not very unequal, not much stronger in ♂ than in ♀. Peraeopods 1—5 very slender and long; 2^d joint the longest, linear. Peraeopods 3—5 successively longer. Branchial vesicles very small. Pleopods well developed. Uropods 1 and 2 normal. Uropod 3 rudimentary, not reaching end of squamiform telson.

1 species.

1. **X. frauenfeldti** Boeck 1871 X. f., A. Boeck in: Forb. Selsk. Christian., 1870 p. 267 | 1876 X. f., A. Boeck, Skand. Arkt. Amphip., v. 2 p. 666 | 1893 X. f., A. Della Valle in: F. Fl. Neapel, v. 20 p. 319 | 1894 X. f., G. O. Sars. Crust. Norway, v. 1 p. 633 t. 227 f. 2 | 1895 Zenodice f., A. M. Norman in: Ann. nat. Hist., ser. 6 v. 15 p. 493.

Body long, nearly cylindric, smooth; pleon segment 4 produced latero-dorsally into 2 microscopic teeth. Head long, front truncate. Side-plates sub-quadrate. Pleon segment 3, postero-lateral corners quadrate. Eyes represented by 2 patches of whitish pigment. Antenna 1 nearly as long as body, 1st joint of peduncle about half as long as 2^d or 3^d, flagellum rather shorter than 2^d and 3^d combined, about 12-jointed; accessory flagellum 4- or 5-jointed. Antenna 2 rather shorter, ultimate and penultimate joints of peduncle subequal; flagellum more than half as long as peduncle, 9-jointed. Gnathopods 1 and 2 setose. Gnathopod 1 in ♂, 5th joint longer and wider than the oval, distally narrowed 6th; palm oblique, defined by a palmar spine, slightly overlapped by the finger. Gnathopod 2 in ♂, 5th joint as long as 6th, which is widest where the straight, very oblique palm is defined by a strong tooth. Gnathopod 1 in ♀ more slender than in ♂, but similar. Gnathopod 2 in ♀ like gnatho-pod 1, but rather stouter; marsupial plates long and narrow. Peraeopods 1 and 2 nearly twice as long as the gnathopods; peraeopods 3—5 still longer. Uropod 1 longer than uropod 2, the rami slender in both, outer rather the shorter. Uropod 3, the small single joint scarcely reaching end of the obtusely pointed telson. Colour greyish white, transversely banded with faint orange. L. 8—14 mm.

Arctic Ocean, North-Atlantic and North-Sea (Trondhjemsfjord, Lofoten Islands, Rödberg, Hardangerfjord). Depth 150—565 m.

5. Gen. **Podocerus** Leach

1813/14 Podocerus (Sp. un.: P. variegatus), Leach in: Edinb. Enc., v. 7 p. 433 | 1899 P., T. Stebbing in: Ann. nat. Hist., ser. 7 v. 3 p. 237 | 1852 Platophium, J. D. Dana in: Amer. J. Sci., ser. 2 v. 14 p. 309 | 1853 P. (Sp. un.: P. brasiliense), J. D. Dana in: U. S. expl. Exp., v. 13 II p. 831, 837 | 1888 P., T. Stebbing in: Rep. Voy. Challenger, v. 29 p. 257, 1184 | 1893 P. (part.), A. Della Valle in: F. Fl. Neapel, v. 20 p. 327 | 1894 P. (part), G. O. Sars, Crust. Norway, v. 1 p. 629, 630 | 1885 Dexiocerella, Haswell in: P. Linn. Soc. N. S. Wales, v. 10 p. 107.

Pleon narrow, ventrally flexed; its 4th segment elongate, 5th and 6th very short. Head quadrate. Side-plates 1—7 small, shallow. Pleon segment 3, postero-lateral corners distally rounded. Eyes prominent, placed at front corners of the head (Fig. 122 p. 705). Antenna 1 shorter than antenna 2, fringed below with long setae; flagellum much shorter than peduncle; accessory flagellum usually 1-jointed, always small. Antenna 2 longer in ♂ than in ♀; flagellum much shorter than peduncle, few-jointed. Upper lip slightly bilobed. Lower lip, inner lobes well developed. Mandible normal, with spine-row of 2 or 3 spines; palp much longer than trunk, 3^d joint short, distally widened and fringed with many spines. Maxilla 1, inner plate obsolete, outer with 9 spines on the distal margin, palp with a few spine-teeth on apex of the long 2^d joint. Maxilla 2, inner plate the shorter, spines almost confined to distal margin. Maxillipeds normal, 4th joint of palp short, not unguiform, tipped with spines. Gnathopods 1 and 2 subchelate; gnathopod 1 much the smaller. Gnathopod 2, 5th joint small, in ♂ sometimes coalescing with the large 6th joint, which is usually more elongate than in the ♀, in both sexes having the front margin notched for spines. Peraeopods 1—5, 2d joint never very large; finger strong and curved. Branchial vesicles tending to develop extra lobes or become twisted. Marsupial plates of great size. Uropods 1 and 2, peduncle and rami spinose; one ramus decidedly shorter than the other; 1 spine in the apical group very long. Uropod 3, a small hollow plate facing the telson's lateral margin. Telson entire, with a process on the upper surface carrying spinules.

10 species.

Synopsis of species:

1 { Body not carinate — **2.**
{ Body more or less carinate — **9.**

2 { Peraeopods 1 and 2, 2d joint expanded . . .   1. **P. cheloniae** . . . . p. 701
{ Peraeopods 1 and 2, 2d joint not conspicuously
{   expanded — **3.**

3 { Antenna 1, accessory flagellum 4(?)-jointed .   2. **P. andamanensis** . . p. 702
{ Antenna 1, accessory flagellum 1(?)-jointed — **4.**

4 { Gnathopod 1 in ♀, palm shorter than hind
{   margin of 6th joint . . . .   3. **P. inconspicuus** . . p. 702
{ Gnathopod 1 in ♀, palm longer than hind
{   margin of 6th joint — **5.**

5 { Gnathopod 2 in ♂, palm with strong teeth — **6.**
{ Gnathopod 2 in ♂, palm without strong teeth — **8.**

6 { Gnathopod 2 in ♂, palm defined by a strong
{   tooth . . . .   4. **P. lobatus** . . . . . p. 703
{ Gnathopod 2 in ♂, palm not defined by a
{   strong tooth — **7.**

7 { Antenna 2 in ♂ not very elongate . . . . .   5. **P. chelonophilus** . p. 703
{ Antenna 2 in ♂ very elongate . . . .   6. **P. variegatus** . . . p. 703

8 { Peraeopods 1—5, 6th joint with stout spines
{   on lower half of inner margin . . . . . .   7. **P. brasiliensis** . . . p. 704
{ Peraeopods 1—5, 6th joint with stout spines
{   on upper half of inner margin . . . . . .   8. **P. laevis** . . . . . . p. 704

9 { Carinate processes from head to pleon segment 3   9. **P. danae** . . . . . . p. 705
{ Carinate processes from peraeon segment 6 to
{   pleon segment 2 . . . .   10. **P. cristatus** . . . . p. 706

1. P. cheloniae (Stebb.) 1888 *Platophium c.*, T. Stebbing in: Rep. Voy. Challenger, *v.* 29 p. 1190 t. 130 | 1893 *P. chelonophilum* (part.), Chevreux & Guerne in: Bull. Soc. ent. France, p. 118 | 1899 *Podocerus chelonophilus* (part.), T. Stebbing in: Ann. nat. Hist., ser. 7 *v.* 3 p. 239 | 1893 *Platophium brasiliense* (part.), A. Della Valle in: F. Fl. Neapel, *v.* 20 p. 329.

Body elliptical, smooth. Eyes round. Antenna 1 short; flagellum as long as 2d, longer than 3d joint of peduncle, 4-jointed; accessory flagellum 1-jointed. Antenna 2 rather longer, robust; flagellum ³/₄ as long as ultimate joint of peduncle, 3-jointed. Gnathopod 1, 5th joint about as long and broad as the oval 6th; palm and hind margin scarcely distinguished, both spinose; finger with 2 teeth on inner margin. Gnathopod 2, 4th joint bluntly produced, masking the small 5th, 6th large, broadly oval, palm convex, slightly distinguished from the hind margin by a minute tooth or notch, with 2 palmar spines; finger matching palm. Peraeopods 1 and 2, 2^d joint with convex front margin forming a winged expansion. Peraeopods 3—5, 2^d joint expanded behind in peraeopod 3 just as in front of the 1st and 2^d, in 4th and 5th with more flattened hind margin. Pleopods 1—3, coupling spines reaching 9. Telson rather broader than long, distally rounded; projection from upper surface broad, carrying 2 spines. Colour, dark stellate marking or round spots, scattered or crowded. L. 6—7 mm.

Atlantic. On Chelonia imbricata (L.).

2. **P. andamanensis** (Giles) 1890 Cyrtophium andamanense, G. M. Giles in: J. Asiat. Soc. Bengal, v. 59 p. 72 t. 2 f. 7 | 1899 Podocerus andamanensis, T. Stebbing in: Ann. nat. Hist., ser. 7 v. 3 p. 239 | 1893 Platophium orientale (part.), A. Della Valle in: F. Fl. Neapel, v. 20 p. 895.

Body slightly corrugated, not carinate; peraeon segment 5 much shorter than the 1st, 2^d or 7th. Eyes small. Antenna 1, 2^d joint considerably longer than 1st or 3^d, or than the 4-jointed flagellum; accessory flagellum minute, with 4 short joints (?). Antenna 2 much longer and stouter; ultimate joint of peduncle subequal to penultimate (in figure longer), flagellum very short, with 2 stout long joints, armed with strong hooked spines. Gnathopod 1 very small; the articulation between 3^d and 4th joints very oblique; finger probably fused with 6th joint, the subchela being formed between these and the dilated 5th joint. Gnathopod 2 very much larger; 4th joint not produced (figure), 5th coalesced with the long oval or fusiform 6th, which is devoid of long setae; finger strongly curved, little more than half, as long as 6th joint. Peraeopod 3 as long as peraeopods 1 and 2, but stouter; peraeopods 4 and 5 successively longer. Pleopods 1—3 exceptionally small. Uropod 1, peduncle as long as the longer ramus; the shorter ramus scarcely more than half as long. Uropod 2 shorter but stouter, its rami in like manner unequal and spinose. Uropod 3 reduced to a rudimentary tubercle. Telson small and laminar, armed with a few short, stiff hairs. Colour dirty white, sparely sprinkled with minute dark brown spots. L. 3 mm.

Bay of Bengal (Andaman Islands). Taken in the surface net.

3. **P. inconspicuus** (Stebb.) 1888 Platophium inconspicuum, T. Stebbing in: Rep. Voy. Challenger, v. 29 p. 1194 t. 131 | 1899 Podocerus inconspicuus, T. Stebbing in: Ann. nat. Hist., ser. 7 v. 3 p. 239 | 1893 Platophium parasiticum (part.), A. Della Valle in: F. Fl. Neapel, v. 20 p. 335.

♀. Body elliptical, slightly imbricated. Eyes broadly oval, large. Antennae 1 and 2 imperfect. Gnathopod 1. 5th joint elongate, narrowing distally, rather longer than 6th, 6th widening to the palm, which is shorter than the hind margin and forms an obtuse angle with it, carrying a palmar spine; the finger not overlapping palm, with a row of setules near the apex. Gnathopod 2, 5th joint small, 6th large, abruptly wider than 5th, hind margin very short, palm very long, oblique, defined by a small apical tooth, serrulate near finger-hinge, and then fringed with stout spines; finger stout, matching palm. Marsupial plates of great size. Peraeopods 1—5 unknown. Pleopods 1—3 each with 2 coupling spines. Uropod 3 not reaching end of telson. Telson distally rounded, with 2 spines rising from the surface at about the centre. Colour, dark pigment flakes, remaining dark in spirit. L. about 3 (not 6) mm. — ♂ unknown.

Port Jackson [East-Australia]. Depth 3—18 m.

4. P. lobatus (Hasw.) 1885 *Dexiocerella lobata*, Haswell in: P. Linn. Soc. N. S. Wales, *v.* 10 p. 110 t. 18 f. 6—8 | 1888 *Cyrtophium lobatum*, Chevreux & Guerne in: C.-R. Ac. Sci., *v.* 106 p. 627 | 1888 *Platophium l.*, T. Stebbing in: Rep. Voy. Challenger, *v.* 29 p. 1184 | 1899 *Podocerus lobatus*, T. Stebbing in: Ann. nat. Hist., ser. 7 *v.* 3 p. 239 | 1893 *Platophium orientale* (part.), A. Della Valle in: F. Fl. Neapel, *v.* 20 p. 333.

Segments 1—4 elevated in the medio-dorsal line. Antenna 1, 3^d joint of peduncle rather shorter (in figure rather longer) than 2^d; flagellum a little longer than 3^d joint of peduncle, 5 (in figure 6)-jointed; accessory flagellum 1-jointed. Antenna 2 as long as the body, stout; penultimate joint of peduncle narrow at base, then broad, ultimate considerably longer, flagellum about half as long as ultimate joint of peduncle, with 3 joints, 1st much longer than 2^d and 3^d combined. Gnathopod 1 as in Cyrtophium minutum (p. 698). Gnathopod 2 very large, 4th joint with apex of hind margin acutely produced, 6th large, irregularly ovoid, not twice as long as the greatest breadth, hind margin very short, palm defined by a strong tooth, and having near the finger-hinge a denticulated lobe, followed by a conical tooth, as in various other species. Peraeopods 1—5 unknown. Uropods 1—3 and telson as in P. cristatus (p. 706). L. about 4 mm.

South-Pacific (Broughton Islands near Port Stephens [East-Australia]).

5. P. chelonophilus (Chevreux & Guerne) 1888 *Cyrtophium chelonophilum*, Chevreux & Guerne in: C.-R. Ac. Sci., *v.* 106 p. 626 | 1893 *Platophium c.* (part.), Chevreux & Guerne in: Bull. Soc. ent. France, p. 118 | 1899 *Podocerus chelonophilus* (part.), T. Stebbing in: Ann. nat. Hist., ser. 7 *v.* 3 p. 239 | 1893 *Platophium brasiliense* (part.), A. Della Valle in: F. Fl. Neapel, *v.* 20 p. 329.

Body elliptical, dorsally smooth, not at all imbricated. Head rectangular, much broader than long. Eyes round, prominent. Antenna 1 very short, $^2/_8$ as long as antenna 2; flagellum not longer than 3^d joint of peduncle, 6-jointed; accessory flagellum 1-jointed. Antenna 2 very robust; flagellum scarcely half as long as ultimate joint of peduncle, 3-jointed. Gnathopod 1 in $\male$, 6th joint triangular, hind margin almost at right angles with palm. Gnathopod 2 in $\male$, 6th joint very large, oval, having on the hind margin (which is undistinguished from the palm) a sharp tooth followed by 2 great denticulate tubercles; finger stout, strongly curved, shorter than hind margin. Gnathopod 1 in $\female$ as in $\male$, but smaller. Gnathopod 2 in $\female$, 6th joint short, rounded, as broad as long, hind margin smooth; finger slender, regularly curved. Peraeopods 3—5 large, robust, successively longer; 2^d joint short, oval, scarcely expanded behind. Telson squamiform, a little broader than long. L. $\female$ 7, $\male$ 9 mm.

North-Atlantic (Azores). On Thalassochelys caretta (L.).

6. P. variegatus Leach 1813/14 *P. v.*, Leach in: Edinb. Enc., *v.* 7 p 433 | ?1836—40 *P. v.*, H. Milne Edwards in: G. Cuvier, Règne an., ed. 3 Crust. p. 179 t. 61 f. 4 | 1899 *P. v.*, T. Stebbing in: Ann. nat. Hist., ser. 7 *v.* 3 p. 237, 350 | 1857 *Cyrtophium darwinii*, Bate in: Ann. nat. Hist., ser. 2 *v.* 19 p. 148 | 1862 *C. d.*, Bate, Cat. Amphip. Brit. Mus., p. 274 t. 46 f. 8 | 1874 *C. d.*, T. Stebbing in: Rep. Devonsh. Ass., *v.* 6 p. 770 | 1888 *Platophium d.*, T. Stebbing in: Rep. Voy. Challenger, *v.* 29 p. 292 | 1893 *P. brasiliense* (part.), A. Della Valle in: F. Fl. Neapel, *v.* 20 p. 329 t. 2 f. 7; t. 7 f. 39—58 | 1894 *Laetmatophilus tuberculatus* (part.), G. O. Sars, Crust. Norway, *v.* 1 p. 630.

Near to P. brasiliensis (p. 704), but body more strongly corrugated or imbricated, and more broadly elliptical in $\female$. Antenna 1, 2^d joint very little longer than 3^d; flagellum with 6 joints carrying numerous hyaline filaments in $\male$, with 4 joints in $\female$; accessory flagellum 1-jointed. Antenna 2 large in $\male$;

ultimate joint of peduncle very elongate; flagellum with 4 joints, the 1st much the longest. Gnathopod 1 as in P. brasiliensis, but in both sexes less robust. Gnathopod 2 in ♂, 6th joint with the front margin more convex than in P. brasiliensis, and the opposite margin armed near the finger-hinge with a broad denticulate process, followed by a strong tooth; the plumose setae as in the species compared; finger reaching nearly to the base of the 6th joint. Gnathopod 2 in ♀, 4th joint with a broader apex, 6th rather less broad, finger longer than in P. brasiliensis. Pleopods with 5 coupling spines. Colour dark red, not rarely with a brilliant patch of purple or lilac on the back. L. 3—4 mm.

North-Atlantic (South-West-England, West-France); Mediterranean (Naples, Adriatic).

7. **P. brasiliensis** (Dana) 1853 & 55 *Platophium brasiliense,* J. D. Dana in: U. S. expl. Exp., *v.* 13$_{II}$ p. 838; t. 55 f. 9a—l | 1888 *P. b.,* T. Stebbing in: Rep. Voy. Challenger, *v.* 29 p. 265 | 1893 *P. b.* (part.), A. Della Valle in: F. Fl. Neapel, *v.* 20 p. 329 | 1899 *Podocerus brasiliensis,* T. Stebbing in: Ann. nat. Hist., ser. 7 *v.* 3 p. 239.

Body not carinate, in ♂ narrowly, in ♀ more broadly, elliptical. Antenna 1, flagellum with 4 (Dana: 3—5) joints, together about as long as 2^d joint of peduncle. Antenna 2, ultimate joint of peduncle very long; flagellum with 4 joints, each with curved spine at apex, 4th joint very small, its curved spine the strongest. Gnathopod 1 stronger in ♂ than in ♀; 5th joint shorter than 6th in ♂, not in ♀, in each forming a broad lobe at middle of hind margin, 6th with very short hind margin, abruptly widening to the long sloping palm; 4th—6th joints armed with numerous spines on and about the hind margin, 6th with several rows on surface adjoining front; finger matching palm, curved, serrate. Gnathopod 2 in ♂, 4th joint scarcely at all produced at distal hind corner, 5th small, scarcely distinct from 6th, 6th very long (not short), narrowly oblong oval, fringed along the straight hinder edge, which is palm and hind margin all in one, with very long plumose hairs; the crenulation near the finger-hinge almost obsolete; finger more than half as long as 6th joint. Gnathopod 2 in ♀, 4th joint considerably produced at the hinder apex, 5th small. triangular, but quite distinct, 6th not very long, broadly oval, the palm defined from the short hind margin by 2 slender prominent spines, the joint armed with numerous seta-like spines but no plumose setae. Peraeopods 1—5, 6th joint spinose on both margins, on the front the strong spines being on the lower half; finger curved, strong. Peraeopods 3—5, 2^d joint narrow, and narrowing distally. Pleopods ·1—3 with only 2 coupling spines. Uropod 3, ramus very small. Telson with 8 spines radiating round the projecting distal margin. L. 6 mm.

Tropical Atlantic (Rio Janeiro, Antigua).

8. **P. laevis** (Hasw.) 1885 *Dexiocerella l.,* Haswell in: P. Linn. Soc. N. S. Wales, *v.* 10 p. 111 t. 18 f. 10—12 | 1888 *Platophium laeve,* T. Stebbing in: Rep. Voy. Challenger, *v.* 29 p. 1184 | 1899 *Podocerus laevis,* T. Stebbing in: Ann. nat. Hist., ser. 7 *v.* 3 p. 239 | 1888 *Cyrtophium haswelli,* Chevreux & Guerne in: C.-R. Ac. Sci., *v.* 106 p. 627 | 1893 *Platophium parasiticum* (part.), A. Della Valle in: F. Fl. Neapel, *v.* 20 p. 335.

Body smooth, elliptical in ♀. Antenna 1, 3^d joint of peduncle very little shorter than 2^d; flagellum 5- or 6-jointed; accessory flagellum very small, 1-jointed. Antenna 2 nearly as long as body in ♂, in ♀ much shorter; peduncle very stout, ultimate joint as long as 2 preceding joints together (in ♂, not so long in ♀); flagellum $\frac{1}{2}$ as long as ultimate joint of peduncle, with 3 indistinct joints (Haswell), in ♀ more than half as long as ultimate joint of peduncle, with a minute 4th joint. Mandible, 3d joint of palp very broad-ended. Gnathopod 1, 5th joint with the hind lobe very broad, 6th in ♂, but not in ♀, rather longer than 5th, hind margin very short, at a decided angle with palm, which is not defined by a tooth; finger as usual denticulate; front margin of 6th joint more strongly spined in ♀ than in ♂. Gnathopod 2 in ♂, 4th joint produced at apex of hind margin into a short tooth; 5th joint obscurely separated from the large oval 6th, of which the palm is serrulate towards finger-hinge and indistinctly defined from the short hind margin by a small notch and spine, which is scarcely reached by apex of finger, the fringing setae not very long. Gnathopod 2 in ♀, 4th joint with tip truncate, 6th more broadly oval than in ♂; palm more convex, not serrulate, defined by a little tooth and strong group of spines. Peraeopods 1—5 stout, with strong, curved finger; inner margin of 6th joint having the strong spines on the upper half. Peraeopods 1 and 2, 2d joint with very inconspicuous dilatation. Peraeopods 3—5, 2d joint well expanded. Pleopods 1—3 each with 2 coupling spines. Uropods 1 and 2, rami rather less elongate than usual. Uropod 3, the plate not reaching end of telson. Telson, the upper process conical, rather strongly produced, tipped with 2 spinules. L. 8 mm, or less.

Port Molle, among sea-weed, and Maroubra Bay [East-Australia].

9. **P. danae** (Stebb.) 1888 *Platophium d.*, T. Stebbing in: Rep. Voy. Challenger. *v.* 29 p. 1185 t. 128, 129 | 1899 *Podocerus d.*, T. Stebbing in: Ann. nat. Hist., ser. 7 *v.* 3 p. 239 | 1893 *Platophium orientale* (part.), A. Della Valle in: F. Fl. Neapel, *v.* 20 p. 332 t. 55 f. 17, 18.

In the medio-dorsal line from head to 3d pleon segment extends a series of carinate teeth or processes, 2 small ones on peraeon segment 1,

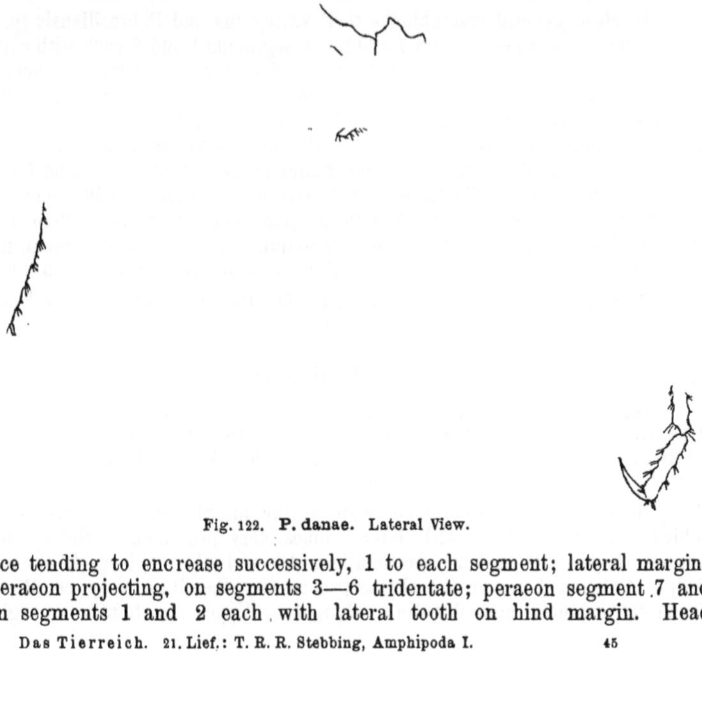

Fig. 122. **P. danae.** Lateral View.

thence tending to encrease successively, 1 to each segment; lateral margins of peraeon projecting, on segments 3—6 tridentate; peraeon segment 7 and pleon segments 1 and 2 each with lateral tooth on hind margin. Head

truncate in front (Fig. 122 p. 705). Eyes very prominent. Antenna 1, flagellum
with 8—10 joints carrying many hyaline filaments, together longer than 2^d joint
of peduncle; accessory flagellum 1-jointed. Antenna 2, ultimate joint of
peduncle very long; flagellum slender, with 4 joints, the 1st much longer
than the other 3 united, all 4 together subequal to penultimate joint of peduncle.
Gnathopod 1 slight, 5th joint fully as long as 6th, general structure as in
P. brasiliensis (p. 704). Gnathopod 2 in ♂, 4th joint slightly but bluntly
produced, 5th distinct, triangular, short, 6th thrice as long as 5th, or as its
own breadth, with no distinction between hind margin and palm, the two
closely fringed with rather long plumose setae, and broken into 2 or 3 teeth
near the hinge of the setuliferous finger, which is about half as long as
6th joint, curved and stout. Gnathopod 2 in ♀, 4th joint little and not
acutely produced, 6th broadly oval, palm defined from the short hind margin
by an acute tooth and strong palmar spine, against which the broad curved
finger impinges. Peraeopods 1—5 nearly as in P. brasiliensis, but the finger
fringed with a few setules on the convex margin. Pleopods 1—3 each with
2 coupling spines. Uropod 3, the concave plate bordered with 6 or 7 spinules.
Telson with bluntly conical projection from its upper surface not quite reaching
its distal margin; apex of the cone tipped with 2 spines. L. about 14 mm.

Southern Indian Ocean (Kerguelen Island). Depth 232 m.

10. **P. cristatus** (G. M. Thoms.) 1879 *Cyrtophium cristatum*, G. M. Thomson
in: Ann. nat. Hist., ser. 5 $v. 4$ p. 331 t. 16 f. 9—15 | 1881 *C. c.*, G. M. Thomson in: Tr.
N. Zealand Inst., $v. 13$ p. 219 t. 8 f. 8 | 1888 *Platophium c.*, T. Stebbing in: Rep. Voy.
Challenger, $v. 29$ p. 500 | 1899 *Podocerus cristatus*, T. Stebbing in: Ann. nat. Hist., ser. 7
$v. 3$ p. 239 | 1879 *Cyrtophium dentatum*, Haswell in: P. Linn. Soc. N. S. Wales, $v. 4$ p. 342
t. 22 f. 5 | 1885 *Dexiocerella dentata*, Haswell in: P. Linn. Soc. N. S. Wales, $v. 10$ p. 109 t. 17
f. 8—12 | 1893 *Platophium orientale* (part.), A. Della Valle in: F. Fl. Neapel, $v. 20$ p. 332

In close general resemblance to P. variegatus and P. brasiliensis (p. 703,
704). Peraeon segments 6 and 7 and pleon segments 1 and 2 each with carinate
process in medio-dorsal line. Antenna 1, flagellum 6—8-jointed, accessory
flagellum 1-jointed (Haswell 1879: biarticulate, and 1885: consisting of several
coalescent joints). Antenna 2, flagellum with 3 joints, the last tipped with
a strong spine. Gnathopod 2 in ♂, 4th joint with apex of hind margin
acutely produced, 6th having near the finger-hinge a denticulate lobe followed
by a conical tooth, which like the lobe is of variable size; finger nearly reaching base
of 6th joint. Peraeopods 3—5 with 2^d joint slightly broader than in the
other species. Pleopods 1—3 each with 2 coupling spines. Telson conical, tipped
at the apex with 2 spinules in juxtaposition. Colour grey or red. L. 5—6 mm.

Southern Indian Ocean and South-Pacific (Dunedin Harbour [New Zealand],
depth 7—9 m; eastern and southern Australia).

6. Gen. Icilius Dana

1849 *Icilius*, J. D. Dana in: Amer. J. Sci., ser. 2 $v. 8$ p. 140 | 1852 *I.* (Sp. un.:
I. ovalis), J. D. Dana in: P. Amer. Ac., $v. 2$ p. 220 | 1886 *I.*, Gerstaecker in: Bronn's
Kl. Ordn., $v. 5$ II p. 497 | 1888 *I.*, T. Stebbing in: Rep. Voy. Challenger, $v. 29$ p. 1202 |
1893 *I.*, A. Della Valle in: F. Fl. Neapel, $v. 20$ p. 327, 344.

Body (Fig. 123) depressed, wide at the middle; pleon segments 3—6
folded ventrally. Head with lateral lobes very prominent. Side-plates of
various shapes, 1st and 2^d the smallest, 4th and 5th the largest, none very
deep. Antenna 1 much the shorter; accessory flagellum very small. An-
tenna 2, peduncle and flagellum elongate. Upper lip faintly emarginate.

Lower lip, mandibular processes short, obtuse. Mandible normal; 3^d joint of palp not longer than 2^d. Maxilla 1, inner plate with 2—4 setae on narrow apex, outer with 11 spines; 2^d joint of palp with numerous spines and spine-teeth. Maxilla 2, inner plate the shorter, inner margin only partially fringed; both plates broad, rather short. Maxillipeds, inner plates broad, outer not very large, fringing spines slender; 2^d joint of palp broad, finger slender. Gnathopods 1 and 2 (Fig. 124) simple, slender; 5th and 6th joints elongate, fringed with long setae. Peraeopods 1—5, 2^d joint not widely expanded; peraeopod 5 longest. Branchial vesicles simple. Marsupial plates long and broad. Uropod 3, inner ramus longer than peduncle (Fig. 125), outer not longer, very small. Telson simple.

1 species.

1. **I. ovalis** Dana 1852 *I. o.*, J. D. Dana in: P. Amer. Ac., *v.*2 p.220 | 1888 *I. o.*, T. Stebbing in: Rep. Voy. Challenger, *v.*29 p.255, 265, 1706 | 1893 *I. o.*, A. Della Valle in: F. Fl. Neapel, *v.*20 p.345 t. 55 f. 25—31 | 1853 & 55 *I. ellipticus*, J. D. Dana in: U. S. expl. Exp., *v.*13 ɪɪ p.844; t. 56 f.4a—g | 1879 *I. australis* + *I.punctatus*, HasWell in: P. Linn. Soc. N. S.Wales, *v.*4 p.274 t. 12 f. 2; p.343 t. 23 f. 1 | 1882 *I. a.,* HasWell, Cat. Austral. Crust., p.275 t. 4 f. 4, a, b | 1888 *I. danae* + *I. ellipticus* + *I. a.*, T. Stebbing in: Rep. Voy. Challenger, *v.*29 p. 1203 t. 133; p. 1208.

Back broadly oval (Fig. 123), peraeon segments 1 and 2 very short, pleon segments 1—3 medio-dorsally acutely produced (Dana); peraeon segment 7 and pleon-segments 1 and 2 with 2 sets of submedian spinules (sometimes producing the effect of median processes). Head broader than long, with small rostrum and little lobe above the rounded ocular lobe, beneath which there is a spiniform process. Side-plates 1 and 2 very small, hind margin pointed, 3^d and 4th with bluntly pointed spinulose hind apex, 5th bilobed. Pleon segments 1 and 2, postero-lateral corners sharply pointed, the margin above produced into a 2^d acute process. Pleon segment 3 smaller, postero-lateral corners incurved, acute. Eyes prominent, hemispherical, red. Antenna 1, 1st and 2^d joints subequal, 3^d much shorter; flagellum rather longer than peduncle; accessory flagellum a small oblong joint, tipped with 2 setae.

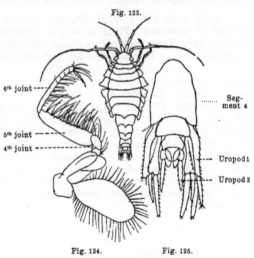

Fig. 123.

6th joint

5th joint
4th joint

Seg-
ment 4

Uropod 1

Uropod 3

Fig. 124.   Fig. 125.

Fig. 123—125. I. ovalis.

Fig. 123. Dorsal View of ♀. — Fig. 124. Gnathopod 2. — Fig. 125. Dorsal View of pleon.

Antenna 2 about twice as long as 1st, as long as body; ultimate joint of peduncle longer than penultimate; flagellum much longer than elongate peduncle. Gnathopods 1 and 2 (Fig. 124), 3^d and 4th joints very short, 5th very long, longer than 2^d, slightly tapering, more in gnathopod 2

45*

than in 1st; 6th joint longer than 2d, slightly shorter than 5th, of uniform narrowness; finger half as long as 5th joint, inner margin setuliferous. Peraeopods 1 and 2 robust, more so in ♂ than ♀, spinose; 4th joint very short, 5th subequal to the short 2d, 6th longer, in ♂ decidedly, in ♀ less decidedly, subchelate; the finger closing against a little spinose palm or widened apex of 6th joint. Peraeopods 3 and 4 equal, similar to preceding peraeopods but without palm; 2d joint channelled behind, outer margin cut distally into 2 teeth. Peraeopod 5 much longer, 2d joint somewhat expanded behind with a proximal lobe, distally as in peraeopods 3 and 4. Pleopods 1—3, peduncle rather short and broad. Uropods 1 and 2, outer ramus the shorter, both spinose, finely pectinate; peduncle longer than rami in uropod 1, subequal to longer ramus in uropod 2. Uropod 3, peduncle short, produced obtusely on inner side; inner ramus rather long, outer short, narrow (not shown in Fig. 125 p. 707). Telson subacute, rather longer than broad. Colour, covered with minute red or grey spots. L. (distended) about 9 mm.

Balabac Passage (north of Borneo), depth 56 m; Port Jackson [East-Australia], on calcareous sponges; Port Phillip (Melbourne), depth 60 m.

7. Gen. **Dulichia** Krøyer

1845 *Dulichia* (Sp. un.: *D. spinosissima*), Krøyer in: Naturh. Tidsskr., ser. 2 *v.* 1 p. 512, 521 | 1888 *D.,* T. Stebbing in: Rep. Voy. Challenger, *v.* 29 p. 213 | 1893 *D.,* A. Della Valle in: F. Fl. Neapel, *v.* 20 p. 320 | 1894 *D.,* G. O. Sars, Crust. Norway, *v.* 1 p. 634 | 1857 *Dyopedos, Dulichia,* Bate in: Ann. nat. Hist., ser. 2 *v.* 19 p. 150, 271.

Body (Fig. 126 p. 712) slender; peraeon segment 1 the shortest, 6th and 7th coalesced. Pleon of only 5 segments besides the telson, the 4th elongate. Head produced in front. Side-plates small, not contiguous. Antennae 1 and 2 long and slender, peduncle much longer than flagellum. Antenna 1 the longer; accessory flagellum very small. Upper lip slightly bilobed. Mandibular palp not very large, 3d joint narrow. Maxilla 1, inner plate small. Maxilla 2, inner plate fringed on inner margin. Gnathopod 1 not subchelate, 5th joint larger than 6th. Gnathopod 2 in ♂ subchelate, 6th joint powerful. Gnathopod 2 in ♀ similar to gnathopod 1, but with shorter 5th joint. Peraeopods 1 and 2 small, 2d rather longer than 1st. Peraeopods 3—5 stronger, rather or very long, peraeopod 5 the longest; 2d joint linear, 4th elongate. Branchial vesicles narrow, usually on gnathopod 2 and peraeopods 1—3. Marsupial plates very broad, especially the 2 middle ones. Pleopods large; peduncle long and strong. Uropods 1 and 2 with narrow, linear, unequal rami. Uropod 3 entirely wanting. Telson oval.

10 species.

Synopsis of species:

3 { Peraeopods 1 and 2, 2d joint not produced into a tooth. . . . . 2. **D. monacantha** . p. 710
Peraeopods 1 and 2, 2d joint produced into a tooth 3. **D. arctica** . . . . p. 710

4 { Eyes Without visual elements — 5.
Eyes With Visual elements — 6.

5 { Gnathopod 2 in ♂, defining tooth of palm at right angles to long axis of the 6th joint . . . . . 4. **D. macera** . . . p. 710
Gnathopod 2 in ♂, defining tooth of palm not at right angles to long axis of the. 6th joint . . 5. **D. hirticornis** . . p. 711

6 { Eyes Very small . . . . 6. **D. normani** . . . p. 711
Eyes not Very small — 7.

7 { Gnathopod 2 in ♂, 6th joint Without thumb-like process near base . . . . 7. **D. nordlandica** . p. 711
Gnathopod 2 in ♂, 6th joint With thumb-like process near base — 8.

8 { Gnathopod 2 in ♂, 6th joint With thumb-like process contorted . . . . 8. **D. falcata** . . . . p. 712
Gnathopod 2 in ♂, 6th joint With thumb-like process not contorted — 9.

9 { Gnathopod 2 in ♂, finger Very long . . . . . . 9. **D. porrecta** . . . p. 712
Gnathopod 2 in ♂, finger not Very long . . . . 10. **D. tuberculata** . p. 713

1. **D. spinosissima** Krøyer 1845 *D. s.*, Krøyer in: Naturh. Tidsskr., ser. 2 v. 1 p. 512 t. 6 f. 1a—h | 1846 *D. s.*, Krøyer in: Voy. Nord, Crust. t. 22 f. 1a—n | 1876 *D. s.*, A. Boeck, Skand. Arkt. Amphip., v. 2 p. 651 | 1893 *D. s.*, A. Della Valle in: F. Fl. Neapel, v. 20 p. 324 t. 55 f. 15, 16 | 1894 *D. s.*, G. O. Sars, Crust. Norway, v. 1 p. 635 t. 228.

Peraeon segments produced to acute lateral points; last peraeon segment and pleon segments 1 and 2 each with a pair of tubercular medio-dorsal prominences; pleon segment 3 produced backward in a long slightly curved tooth. Head very large; carina produced to a lamellar acute-ending rostrum; post-antennal corners acutely produced. Side-plates projecting acutely forward, 2d acutely bilobed, 3d and 4th acute below. Eyes large, prominent, acute at centre, dark red. Antenna 1 densely setose (setae not very long), nearly as long as the body (Krøyer: about 1 1/2 of body); 2d joint more than twice as long as 1st, obtusely produced at apex, 3d rather longer, as long as the 5-jointed flagellum; accessory flagellum 3-jointed. Antenna 2 shorter and narrower, ultimate and penultimate joints of peduncle each obtusely produced at apex, flagellum shorter than ultimate, 3-jointed. Gnathopod 1, 5th joint very setose on hind margin, narrowing distally, longer and rather broader than the setose 6th, which slightly widens distally; finger long, denticulate on inner margin. Gnathopod 2 in ♂, 2d joint produced in front to a triangular distal lobe, 5th cup-shaped, setose on hind margin, 6th broad, oval quadrangular, hind margin pretty well developed, palm defined by a strong tooth, and having a smaller one near the finger-hinge; finger strong, densely setose, overlapping palm. Gnathopod 2 in ♀ less powerful; 6th joint rather broadly oval, without teeth. Peraeopods 1 and 2 small and feeble; 2d joint very little dilated. Peraeopods 3—5 long, strong; 4th joint subequal to 5th and 6th combined, which are densely spinose; finger strong. Uropods 1 and 2, rami narrow, acute, spinose. Telson rather longer than broad, with 2 setules on each side, apex narrowly rounded. Colour whitish, pellucid, with irregular brown and light yellow markings. L. ♀ 12—31 mm.

Arctic Ocean (Greenland, Spitzbergen, Finmark, Tromsö). Depth 20—300 m.

2. **D. monacantha** Metzg. 1875 *D. m.*, Aug. Metzger in: Jahresber. Comm.
D. Meere, *v.* 2/3 p. 296 t. 6 f. 8 a, b | 1888 *D. m.*, T. Stebbing in: Rep. Voy. Challenger.
v. 29 p. 445 | 1894 *D. m.*, G. O. Sars, Crust. Norway. *v.* 1 p. 638 t. 230 f. 1 | 1893 *D. tuber-culata* (part.), A. Della Valle in: F. Fl. Neapel, *v.* 20 p. 321.

Body nearly as in D. porrecta (p. 712), especially in ♀, perhaps rather
more robust in ♂. Head very slightly produced in front. Side-plate 1
in ♂ produced forward in long horn-like process, in ♀ small, subquadrate.
Side-plate 2 much larger in ♂ than in ♀, broadly quadrate with blunt front
corner. Eyes perhaps a little smaller than in D. porrecta, and antennae 1
and 2 rather less long, but in general as in that species. Gnathopod 1 as
in the species compared, but gnathopod 2 in ♂ not nearly so slender and
elongate; 5th joint rather massive, 6th only a little longer than 5th, palm
rather short, defined by a long tooth, the cavity between that and the tooth
near finger-hinge being deep but not wide; finger of moderate length, overlapping palm, with a tubercle on inner margin very near the hinge. Peraeopods 1—5 as in D. porrecta. Uropod 1, peduncle armed on outer margin
only with 4 distant spines. Colour light yellowish, variegated with dark
brown. L. 5—7 mm.

Arctic Ocean and North-Atlantic (Finmark, depth 38—95 m; West-Norway);
Skagerrak, depth 217 m; Kattegat (Great Belt and Aarhus Bay, depth 12—21 m).

3. **D. arctica** J. Murdoch 1885 *D. a.*, J. Murdoch in: P. U. S. Mus., *v.* 7 p. 521 |
1893 *D. tuberculata* (part.). A. Della Valle in: F. Fl. Neapel, *v.* 20 p. 321.

Body smooth. Head slightly produced, forming an obtuse angle. Side-
plate 1 produced into a sharp spine projecting forward, the rest unarmed.
Eyes small, round, black. Gnathopod 2, 2^{d} joint dilated and armed with
2 teeth, 6th large, subtriangular, and armed on the edge with 2 long stout
teeth. Peraeopods 3—5 not specially long; 4th joint as long as 5th and
6th combined. Uropod 2, outer ramus nearly twice as long as peduncle,
inner a little longer. Colour greyish. L. ?.

Arctic Ocean (Alaska). Depth 9 m.

4. **D. macera** O. Sars 1879 *D. m.*, G. O. Sars in: Arch. Naturv. Kristian., *v.* 4
p. 464 | 1885 *D. m.*, G. O. Sars in: Norske Nordhavs-Exp., *v.* 6 Crust. I p. 220 t. 18
f. 2, 2 a | 1893 *D. m.*, A. Della Valle in: F. Fl. Neapel, *v.* 20 p. 322 t. 55 f. 11.

Body very slim, unarmed. Peraeon segment 1 very short, the rest longer,
subequal. Head rather small, obtusely conical. Side-plate 2 the largest, but
all small, wide apart. Eyes rudimentary, represented only by an irregular
aggregation of white pigment on each side of head. Antenna 1 probably
much longer than body; flagellum 6-jointed; accessory flagellum 6-jointed.
Antenna 2 rather longer than the body, densely setose with short setae;
flagellum longer than ultimate joint of peduncle, 5-jointed. Gnathopod 1
slight, 5th joint longer and stouter than the slenderly oval 6th. Gnathopod
2 in ♂ almost completely unarmed; 2^{d} joint proximally very narrow, distally
rather dilated, 5th longer than broad, 6th oblong oval, with process standing out
at right angles to its long axis close to base, at distal end having a sharp tooth,
finger strong, falciform, not reaching the basal process. Peraeopods 1 and 2,
2^{d} joint linear. Peraeopods 3 and 4 (and 5 ?) very long and slender, 4th joint
much longer than 5th and 6th combined; finger apparently weak. Colour
whitish, translucent. L. 10·5 mm.

Arctic Ocean (north of Vesteraalen Islands; between Bear Island and Finmark,
depth 842—1638 m). In the cold area.

5. **D. hirticornis** O. Sars 1876 *D. h.*, G. O. Sars in: Arch. Naturv. Kristian., *v.* 2 p. 261 | 1885 *D.'h.*, G. O. Sars in: Norske Nordhavs-Exp., *v.* 6 Crust. I p. 218 t. 18 f. 1, 1 a | 1893 *D. h.*, A. Della Valle in: F. Fl. Neapel, *v.* 20 p. 523 t. 55 f. 13 | 1895 *D. hirsuticornis*, G. O. Sars, Crust. Norway, *v.* 1 p. 700.

Body robust, somewhat depressed, smooth. Peraeon segments 1 and 2 short; the rest longer, subequal. Head not strongly produced, obtusely conical. Side-plates all very small, squamiform, subequal. Eyes very small, not prominent, rounded oval, oblique, with a very light whitish yellow pigment. Antenna 1 robust, as long as body, densely setose with long setae; 3^d joint of peduncle longer than 2^d, or than the 4-jointed flagellum; accessory flagellum very small, 3-jointed. Antenna 2 shorter and more slender, but armed like the 1st; ultimate and penultimate joints of peduncle subequal; flagellum shorter than in antenna 1. Gnathopod 1 very setose; 6th joint about as long as 5th, narrowly oval. Gnathopod 2 in ♂, 6th joint very setose, hind margin running out into a long palm-defining tooth, separated by a deep and rather wide interval from the smaller tooth near hinge of the strong, somewhat sinuous finger. Gnathopod 2 in ♀ like gnathopod 1. Peraeopods 1 and 2, 2^d joint somewhat dilated. Peraeopods 3—5 subequal, robust, not very long, finger strong, falciform. Colour nearly translucent, with faint yellowish pigmentation. L. ♀ 11 mm.

Arctic Ocean and North-Atlantic (north-west of Stat, far from coast; near Storeggen Bank; north-west of Finmark). Depth 775—1165 m.

6. **D. normani** O. Sars 1895 *D. n.*, G. O. Sars, Crust. Norway, *v.* 1 p. 699 t. VIII f. 2 | 1895 *D. n.*, A. M. Norman in: Ann. nat. Hist., ser. 6 *v.* 15 p. 493.

♀. Body rather slender, quite smooth. Peraeon segments successively longer to the 4th. Head angularly produced in front. Side-plates all very small. Eyes extremely minute, though well developed and with dark pigment. Antenna 1 rather strong, very long, densely fringed with long setae; 3^d joint of peduncle much longer than 2^d, longer than the 5-jointed flagellum; accessory flagellum minute, 3-jointed. Antenna 2 armed like antenna 1, somewhat shorter; ultimate and penultimate joints of peduncle subequal. Gnathopods 1 and 2 apparently as in D. porrecta (p. 712). Peraeopods 1 and 2, 2^d joint expanded, between oval and fusiform. Peraeopods 3—5 not very long nor very spinose. Colour in spirit simply greyish white. L. 5 mm.

Trondhjemsfjord.

7. **D. nordlandica** Boeck 1871 *D. n.*, A. Boeck in: Forh. Selsk. Christian., 1870 p. 263 | 1876 *D. n.*, A. Boeck, Skand. Arkt. Amphip., *v.* 2 p. 653 t. 29 f. 11 | 1893 *D. n.*, A. Della Valle in: F. Fl. Neapel, *v.* 20 p. 324 t. 55 f. 14 | 1894 *D. n.*, G. O. Sars, Crust. Norway, *v.* 1 p. 641 t. 231 f. 2; t. 232 f. 1.

Body extremely slender, long, smooth. Head not very large, not much produced. Side-plates all very small. Eyes without visual elements, represented by patch of white pigment on each side. Antennae 1 and 2 extremely slender and long, setose with short setae. Antenna 1 longer than the body; flagellum in ♂ with 5 joints, in ♀ with 4. Gnathopod 1, 4th joint longer than usual, 5th longer than the fusiform 6th. Gnathopod 2 in ♂, 2^d joint dilated only in distal half, 5th joint short, 6th with front margin very convex, hind slightly concave, distally produced to a sharp tooth with smaller one between it and hinge; the finger far overlapping both, with tubercle at its base, its apex reaching the 5th joint or nearly. Gnathopod 2 in ♀ like gnathopod 1, but with 5th joint not longer than 6th. Peraeopods 1 and 2,

2^d joint narrow. Peraeopods 3—5 very slender and long; peraeopod 5 scarcely longer than the 2 preceding, its 6th joint much longer than the 5th, slightly expanded at base and armed with a row of strong spines. Uropod 2, peduncle more than half as long as peduncle of uropod 1. Colour pellucid, pale yellowish. L. 5 mm.

Arctic Ocean, North-Atlantic, North-Sea and Skagerrak (Lofoten Islands; South- and West-Norway). Depth 188—565 m.

8. **D. falcata** (Bate) 1857 *Dyopedos falcatus*, Bate in: Ann. nat. Hist., ser. 2 *v.* 19 p. 151 | 1862 *Dulichia falcata*, Bate, Cat. Amphip. Brit. Mus., p. 348 t. 54 f. 10 | 1876 *D. f.*, A. Boeck, Skand. Arkt. Amphip., *v.* 2 p. 652 t. 29 f. 10 | 1894 *D. f.*, G. O. Sars, Crust. Norway, *v.* 1 p. 640 t. 231 f. 1 | 1893 *D. porrecta* (part.), A. Della Valle in: F. Fl. Neapel, *v.* 20 p. 322.

Body (Fig. 126) slender, long, smooth. Head somewhat produced in front. Side-plate 1 very small, quad-

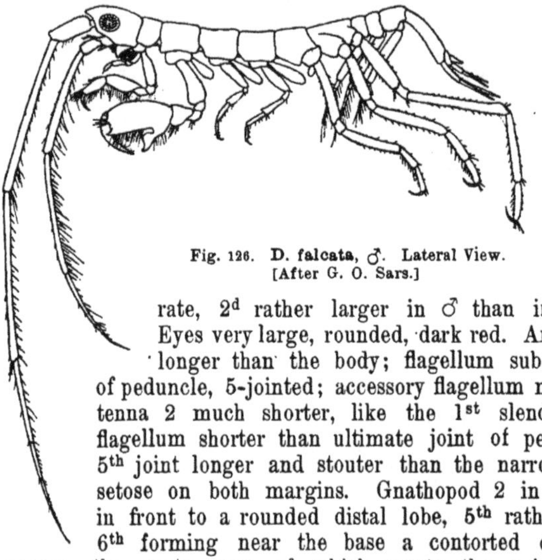

Fig. 126. **D. falcata**, ♂. Lateral View.
[After G. O. Sars.]

rate, 2^d rather larger in ♂ than in ♀, slightly bilobed. Eyes very large, rounded, dark red. Antenna 1 slender, much longer than the body; flagellum subequal to ultimate joint of peduncle, 5-jointed; accessory flagellum minute, 3-jointed. Antenna 2 much shorter, like the 1st slender and feebly setose; flagellum shorter than ultimate joint of peduncle. Gnathopod 1, 5th joint longer and stouter than the narrowly oval 6th, which is setose on both margins. Gnathopod 2 in ♂, 2^d joint expanded in front to a rounded distal lobe, 5th rather longer than broad, 6th forming near the base a contorted outstanding thumb-like process, the acute apex of which meets the point of the finger; the slightly concave remainder of the long palm running parallel with the convex front margin and ending in a sharp tooth near the tubercle of the finger. Gnathopod 2 in ♀, 5th joint rather shorter than 6th, dilated at middle of hind margin, otherwise resembling gnathopod 1. Peraeopods 1 and 2, 2^d joint scarcely expanded. Peraeopod 5 notably longer than peraeopods 3 and 4; its 5th joint much longer than the 6th. Uropod 1, peduncle nearly smooth. Uropod 2, peduncle fully half as long as peduncle of uropod 1. Colour whitish, pellucid, with dark claret-red markings. L. 6—8 mm.

Arctic Ocean, North-Atlantic, North-Sea and Skagerrak (Moray Firth; from Christianiafjörd to Vadsö [Norway]). Depth 37—94 m.

9. **D. porrecta** (Bate) 1857 *Dyopedos porrectus*, Bate in: Ann. nat. Hist., ser. 2 *v.* 19 p. 151 | 1862 *Dulichia porrecta*, Bate, Cat. Amphip. Brit. Mus., p. 348 t. 54 f. 9 | 1876 *D. p.*, A. Boeck, Skand. Arkt. Amphip., *v.* 2 p. 658 t. 30 f. 2, 3 | 1893 *D. p.* (part.), A. Della Valle in: F. Fl. Neapel, *v.* 20 p. 322 t. 55 f. 12 | 1894 *D. p.*, G. O. Sars, Crust. Norway, *v.* 1 p. 637 t. 229.

Body smooth, much more slender in ♂ than in ♀. Peraeon segment 1 much shorter than the rest. Head somewhat produced. Side-plate 1 small, quadrate, 2^d larger in ♂ than in ♀, the rest very small. Eyes rather large, a little prominent, rounded, dark red. Antenna 1 about as long as the body; 3^d joint rather longer than 2^d, subequal to 5-jointed flagellum; accessory flagellum very small, 3-jointed. Antenna 2 rather shorter; flagellum subequal to ultimate joint of peduncle. Both antennae setose with moderately long setae. Gnathopod 1, 5th joint expanded somewhat at the base, longer than the narrowly oval 6th. Gnathopod 2 in ♂ very long; 2^d joint long and slender, 5th small, longer than broad, 6th elongate, densely setose along palm, which is defined from the short hind margin by a very long tooth, widely diverging from the small one adjacent to the tubercle on inner and very setose margin of the very elongate finger. Gnathopod 2 in ♀ as long as gnathopod 1; 5th joint only about half as long as 6th, which is somewhat tapering; finger rather short. Peraeopods 1 and 2, 2^d joint considerably dilated in the middle. Peraeopods 3—5 not very elongate, sparingly spinose. Uropod 1, peduncle minutely spinulose on outer margin; inner ramus considerably longer than outer, inner margin minutely spinulose between the spines. Colour whitish, pellucid, with narrow transverse brownish stripes. L. ♀ 5, ♂ 6 mm.

Arctic Ocean, North-Atlantic, North-Sea, Skagerrak and Kattegat (West-Greenland; Iceland; Lofoten Islands; South- and West-NorWay; Danish Waters; Shetlands; Scotland). Depth 7—113 m.

10. **D. tuberculata** Boeck 1871 *D. t.* + *D. curticauda*, A. Boeck in: Forh. Selsk. Christian., 1870 p.263, 264 | 1876 *D. t.* + *D. c.*, A. Boeck, Skand. Arkt. Amphip., v.2 p.655 t.30 f.4; p.657 t.29 f.9 | 1893 *D. t.* + *D. c.*, A. Della Valle in: F. Fl. Neapel, v.20 p.321 t.55 f.6—10; p.325 | 1885 *D. t.*, G. O. Sars in: Norske Nordhavs-Exp., v.6 Crust. I p.215 t.17 f.6, 6x | 1894 *D. curticauda*, G. O. Sars, Crust. NorWay, v.1 p.639 t.230 f.2 | 1879 *D. septentrionalis*, G. O. Sars in: Arch. Naturv. Kristian., v.4 p.463.

Body smooth. Head somewhat produced. Side-plate 1 small, quadrate, 2^d little longer in ♂ than in ♀. Eyes large, rounded, convex, dark red. Antenna 1 about as long as body; flagellum longer than ultimate joint of peduncle; accessory flagellum minute, 3-jointed, middle joint much the longest. Antenna 2 much shorter. Both antennae sparsely setose with short setae. Gnathopod 1 as in D. porrecta. Gnathopod 2 in ♂; 2^d joint widened except at base, 5th triangular, small, 6th very large, produced not far from base to a narrow, acute, outward bent tooth; the palm thence forward concave, parallel to convex front margin, and ending in a sharp tooth near the tubercle on inner margin of the short finger. Gnathopod 2 in ♀ like gnathopod 1, except that the 5th joint is not longer than the 6th, and bulges at middle. Peraeopods 1 and 2, 2^d joint very little expanded. Feraeopods 3—5 elongate; 5th joint of peraeopod 5 much longer than 6th. Uropod 1, peduncle with about 8 spines on outer margin. Uropod 2, peduncle scarcely half as long as peduncle of uropod 1. Colour whitish, with some reddish brown markings. L. ♂ 5 mm.

Arctic Ocean, North-Atlantic. North-Sea and Skagerrak (from Christianiafjord to Vadsö [NorWay], in moderate depths; Spitzbergen; West-Greenland). Depth 19—28 m.

8. Gen. **Paradulichia** Boeck

1871 *Paradulichia* (Sp. un.: *P. typica*), A. Boeck in: Forh. Selsk. Christian., 1870 p.265 | 1876 *P.*, A. Boeck, Skand. Arkt. Amphip., v.2 p.660 | 1888 *P.*, T. Stebbing in: Rep. Voy. Challenger, v.29 p.402 | 1893 *P.*, A. Della Valle in: F. Fl. Neapel, v.20 p.319 | 1894 *P.*, G. O. Sars, Crust. Norway, v.1 p.642.

♀. In close agreement with Dulichia (p. 708), except that uropod 2 is quite rudimentary, peduncle and rami being represented by a single conically tapering joint (Fig. 127). Antennae not so elongate as in Dulichia. Maxilla 2 has the inner plate not fringed with setae on the inner margin. — ♂ unknown.

1 species.

1. **P. typica** Boeck 1871 *P. t.,* A. Boeck in: Forh. Selsk. Christian.. 1870 p.265 | 1876 *P. t.,* A. Boeck, Skand. Arkt. Amphip., *v.*2 p. 660 t. 29 f. 8 | 1893 *P. t.,* A. Della Valle in: F. Fl. Neapel, *v.*20 p. 319 | 1894 *P. t.,* G. O. Sars, Crust. Norway, *v.*1 p. 642 t. 232 f. 2.

♀. Body smooth; peraeon rather tumid, segments 1 and 2 short. Head slightly produced in front. Side-plates all very small. Eyes large, prominent, rounded, dark red. Antenna 1, 3^d joint of peduncle the longest, subequal to 5-jointed flagellum; accessory flagellum with 3 joints, 1st subequal to 2^d and 3^d combined. Antenna 2 much shorter and thinner; flagellum shorter than ultimate joint of peduncle. Both antennae densely setose, setae not very long. Gnathopod 2, 6th joint rather more broadly oval than in Dulichia (p. 708). Peraeopods 1 and 2, 2^d joint very slightly dilated. Peraeopods 3—5 not greatly elongated; in peraeopod 5 the 4th joint as long as 5th and 6th combined, 5th little longer than 6th. Uropod 1, peduncle strong, not elongate, outer margin fringed with about 8 spines; rami slender, densely fringed with small spines, inner ramus rather the longer, twice as long as peduncle. Uropod 2 (Fig. 127) reaching just beyond the oval telson, and armed with 1 apical and 2 lateral spines. Colour pale yellow, with brownish patches. L. 5 mm.

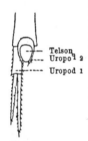

Fig. 127.
P. typica, ♀.
Uropods and telson.
[After G. O. Sars.]

Telson
Uropod 2
Uropod 1

North-Atlantic and North-Sea (Hardangerfjord, depth 56 m; Aalesund, depth 94—188 m).

---

41. Fam. **Hyperiopsidae**

1886 *Hyperiopsidae,* Bovallius in: N. Acta Soc. Upsal., ser. 3 *v.*13 nr. 9 p. 3, 31.

Head bulbous. Side-plates rather shallow. Eyes imperfectly developed, but extensive. Antenna 1 with peduncle short; 1st joint of flagellum elongate, setose; accessory flagellum well developed. Mouth-parts normal, except that maxilla 1 is without inner plate (Chevreux in MS.). Gnathopods 1 and 2 simple. Peraeopods 1 and 2, 4th joint widened and elongate. Peraeopods 3—5, 2^d joint narrow. Peraeopod 5 elongate, almost filiform. Uropods 1—3 biramous, rami equal in uropod 1 and uropod 3, unequal in uropod 2. Telson very small, with a short apical incision.

Marine.

1 genus, 1 species.

1. Gen. **Hyperiopsis** O. Sars

1885 *Hyperiopsis* (Sp. un.: *H. vøringii*), G. O. Sars in: Norske Nordhavs-Exp., *v.*6 Crust. I p.231 | 1886 *H.,* Bovallius in: N. Acta Soc. Upsal., ser. 3 *v.*13 nr. 9 p. 31 | 1888 *H.,* T. Stebbing in: Rep. Voy. Challenger, *v.*29 p. 572.

With the characters of the family.

1 species.

1. **H. vøringii** O. Sars 1885 *H. v.*, G. O. Sars in: Norske Nordhavs-Exp., *v.* 6 Crust. I p. 231 t. 20 f. 21 | 1886 *H. voeringii*, Bovallius in: N. Acta Soc. Upsal., ser. 3 *v.* 13 nr. 9 p. 32 t. 2 f. 40 | 1899 *H. vöringi*, Chevreux in: Bull. Soc. zool. France, *v.* 24 p. 148.

Body broad, smooth. Pleon segments 5 and 6 elongate. Head with slight projection over base of antenna 1. Side-plates 1—4 only about half as deep as their segments. Pleon segment 3, postero-lateral angles subquadrate. Eyes without refractive elements or distinctly developed pigment. Antenna 1 not quite half as long as body, 1st joint short and thick, 2^d and 3^d very small; flagellum with 1 large joint carrying sensory filaments, followed by 12 short joints; accessory flagellum of 4 joints, 1st the largest. Antenna 2 shorter and more slender. Mandible, molar strong, palp very elongate. Maxilla 1, palp large, 1st joint more than half as long as 2^d. Maxilla 2, inner plate not fringed on inner margin. Maxillipeds, inner and outer plates well developed, but outer scarcely reaching middle of 2^d joint of the elongate palp, 4th joint of which is unguiform. Gnathopods 1 and 2 feeble, similar, not subchelate; 6th joint slender, scarcely longer than 5th. Gnathopod 2 rather the longer. Peraeopods 1 and 2, 4th joint tending to fusiform, nearly translucent, more than twice as long as 5th and 6th combined, 6th conical, carrying a mobile finger. Peraeopod 5 almost devoid of setae and with indistinct articulation of its long slender joints. Uropods 1 and 2, peduncle longer than the rami. Uropod 3, peduncle subequal to the rami. Telson as broad as long; apical lobes rounded. Whole organism fragile. L. about 11 mm.

Arctic Ocean and North-Atlantic (Norway, depth 1130 m; from stomach of Rodichthys regina Collett, depth 2410 m; Lofoten Isles, depth 1095 m).

Gammarideorum genera dubia et species dubiae.

Lepleurus Raf. 1820 *L.* (Sp. un.: *L. rivularis*), Rafinesque, Ann. Nat., p. 6.

L. rivularis Raf. 1820 *L. r.*, Rafinesque, Ann. Nat., p. 6 | 1825 *L. r.*, A. G. Desmarest, Consid. gén. Crust., p. 274.

Pennsylvania and at Shannon run. In brooks.

Lusyta Nardo 1847 *L.* (Sp. un.: *Cancer algensis*), Nardo, Prosp. Fauna Venet., p. 20 | 1847 *L.*, Nardo, Sinon. Spec. Chiereghini, p. 10.

L. algensis Nardo 1847 *Cancer a.* (Chiereghin in MS.), *Lusyta*, Nardo, Prosp. Fauna Venet., p. 20 | 1847 *L. a.*, Nardo, Sinon. Spec. Chiereghini, p. 10 | 1869 *L. a.*, *Lysita a.*, Nardo in: Mem. Ist. Veneto, *v.* 14 p. 331 t. 15 f. 7a—c; p. 283 | 1893 *Erichthonius difformis* (part.)?, A. Della Valle in: F Fl. Neapel, *v.* 20 p. 386.

Head feebly rostrate. Eyes laterally prominent. Antenna 1 the longer. Gnathopods 1 and 2 rather large. Peraeopod 3 with finger reverted. Peraeopod 5 the longest.

Adriatic (Venice). Tubicolous on Zostera marina.

Pephredo Raf. 1814 *P.* (Sp. un.: *P. heteroclitus*) (nom. nud.), Rafinesque, Précis Découv. somiol., p. 26 | 1815 *P.* (nom. nud.), Rafinesque, Anal. Nat., p. 101 | 1817 *P.* (Sp. un.: *P. potamogeti*), Rafinesque in: Amer. monthly Mag., *v.* 2 p. 41.

P. potamogeti Raf. 1817 *P. p.*, Rafinesque in: Amer. monthly Mag., *v.* 2 p. 41. Hudson River and Fishkill.

Psammylla Raf. 1817 *P.* (Sp. un.: *P. littoralis*), Rafinesque in: Amer. monthly Mag., *v.* 2 p. 41.

P. littoralis Raf. 1817 *P. l.*, Rafinesque in: Amer. monthly Mag., *v.* 2 p. 41. North-Atlantic (Long-Island, New-York) and Hudson River.

Sperchius Raf. 1820 *S.* (Sp. un.: *S. lucidus*), Rafinesque, Ann. Nat., p. 6.

S. lucidus Raf. 1820 *S. l.*, Rafinesque, Ann. Nat., p. 6 | 1825 *S. l.*, A. G. Desmarest, Consid. gén. Crust., p. 273.

Lexington [Kentucky]. In springs and brooks.

Addenda et Corrigenda.

Pag. 12. — Add to the literature of the 2. Gen. **Trischizostoma** Boeck:
1905 *Guerinella*, Chevreux in: Bull. Mus. Monaco, nr. 85 p. 7.

Pag. 13.—Add to the literature of species 1.**Trischizostoma nicaeense** (A. Costa):
1905 *Guerinella nicaeensis*, Chevreux in: Bull. Mus. Monaco, nr. 85 p. 7.

Pag. 22. — Add to the literature of species 3. **Sophrosyne hispana** (Chevreux):
1900 *S. h.*, Chevreux in: Résult. Camp. Monaco, *v.* 16 p. 13 t. 3 f. 1.

Pag. 23. — Put after the 12. Gen. **Valettia** Stebb.:

12*. Gen. **Vijaya** A. Walker

1904 *Vijaya* (Sp. un.: *V. tenuipes*), A. O. Walker in: Herdman, Rep. Ceylon Pearl Fish., *v.*2 p. 231. 241.

Cfr. Amaryllis (p. 23).

1. **V. tenuipes** A. Walker 1904 *V. t.*, A. O. Walker in: Herdman, Rep. Ceylon Pearl Fish., *v.* 2 p. 241 t. 1 f. 5.

Cfr. Glycerina affinis (p. 61).

Pag. 29 and 30. — Add to the species of the 15. Gen. **Cyphocaris** Boeck:
3. **C. richardi** Chevreux 1905 *C. r.*, Chevreux in: Bull. Mus. Monaco, nr. 24 p. 1 f. 1, 2 a—g.

4. **C. alicei** Chevreux 1905 *C. a.*, Chevreux in: Bull. Mus. Monaco, nr. 27 p. 1 f. 1, 2 a—m.

Cfr. C. challengeri (p. 29).

Pag. 30. — Put after the 15. Gen. **Cyphocaris** Boeck:

15*. Gen. **Paracyphocaris** Chevreux

1905 *Paracyphocaris* (Sp. typ.: *P. praedator*), Chevreux in: Bull. Mus. Monaco, nr. 32 p. 1.

1. **P. praedator** Chevreux 1905 *P. p.*, Chevreux in: Bull. Mus. Monaco, nr. 32 p. 1 f. 1, 2 a—g, 3 a—g.

Pag. 31. — Add to the literature of species 2. **Cyclocaris guilelmi** Chevreux:
1900 *C. g.*, G. O. Sars in: Nansen, Norweg. North Polar Exp., *v.* 1 nr. 5 p. 20 t. 2, 3 | 1900 *C. faroensis*, A. M. Norman in: Ann. nat. Hist., ser. 7 *v.* 5 p. 197 t. 6 f. 5—15.

Pag. 31. — Put after the 16. Gen. **Cyclocaris** Stebb.:

16*. Gen. **Lysianopsis** S. J. Holmes

1903 *Lysianopsis* (Sp. un.: *L. alba*), S. J. Holmes in: Amer. Natural., *v.* 37 p. 276 | 1905 *L.*, S. J. Holmes in: Bull. U. S. Bureau Fish., *v.* 24 p. 475. . Near Lysianella (p. 31).

1. **L. alba** S. J. Holmes 1903 *L. a.*, S. J. Holmes in: Amer. Natural., *v.* 37 p. 276 | 1905 *L. a.*, S. J. Holmes in: Bull. U. S. Bureau Fish., *v.* 24 p. 475 t. 5 f. 1, f. in text.

Pag. 33 and 34. — Add to the species of the 19. Gen. **Pseudalibrotus** Della Valle:

4. **P. nanseni** O. Sars 1900 *P. n.*, G O. Sars in: Nansen, Nòrweg. North Polar Exp., *v.* 1 nr. 5 p. 26 t. 4, 5.

5. **P. glacialis** O. Sars 1900 *P. g.* + *P. g. var. leucopis,* G. O. Sars in: Nansen, Norweg. North Polar Exp., *v.* 1 nr. 5 p. 31 t. 6.

Pag. 34. — Put after the 19. Gen. **Pseudalibrotus** Della Valle:

19*. Gen. **Parambasia** A. Walker & A. Scott

1903 *Parambasia* (Sp. un.: *P. forbesii*), A. O. Walker & A. Scott in: H. O. Forbes, Nat. Hist. Sokotra, p. 217, 221.

1. **P. forbesii** A. Walker & A. Scott 1903 *P. f.*, A. O. Walker & A. Scott in: H. O. Forbes, Nat. Hist. Sokotra, p. 221 t. 14 A ·f. 5—·5 m.

Pag. 39. — Add to the literature of species 5. **Lysianassa cinghalensis** (Stebb.):

1904 *Lysianax c.*, A. O. Walker in: Herdman, Rep. Ceylon Pearl Fish., *v.* 2 p. 242 t. 1 f. 6 | 1903 *L. urodus,* A. O. Walker & A. Scott in: H. O. Forbes, Nat. Hist. Sokotra, p. 220 t. 14 A f. 4—4 g.

Pag. 37 and 40. — Add to the species of 21. Gen. **Lysianassa** M.-E.:

8. **L. coelochir** (A. Walker) 1904 *Lysianax c.*, A. O. Walker in: Herdman, Rep. Ceylon Pearl Fish., *v.* 2 p. 243 t. 1 f. 7.

Pag. 42. — Add to the species of 23. Gen. **Normanion** Bonnier:

3. **N. abyssi** Chevreux 1903 *N. a.*, Chevreux in: Bull. Soc. zool. France, *v.* 28 p. 82 f. 1 a—f.

Pag. 48. — Put after the 28. Gen. **Menigrates** Boeck:

28*. Gen. **Charcotia** Chevreux

1906 *Charcotia* (Sp. un.: *C. obesa*), Chevreux in: Bull. Soc. zool. France, *v.* 30 p. 163.

1. **C. obesa** Chevreux 1906 *C. o.*, Chevreux in: Bull. Soc. zool. France, *v.* 30 p. 163 f. 3.

Pag. 49 and 51. — Add to the species of 29. Gen. **Aristias** Boeck:

6. **A. topsenti** Chevreux 1900 *A. t.*, Chevreux in: Résult. Camp. Monaco, *v.* 16 p. 18 t. 3 f. 2.

Pag. 51 and 52. — Add to the species of 30. Gen. **Ambasia** Boeck:

2*. **A. murmanica** Brüggen 1905 *A. m.*, E. Brüggen in: Trav. Soc. St.-Pétersb., *v.* 36 p. 3, 8; t. f. 1.

Pag. 51. — The name of the 1ˢᵗ species of 30. Gen. **Ambasia** Boeck is to be changed into: **A. atlantica** (M.-E.). Add to the literature of this species:

1830 *Gammarus atlanticus*, H. Milne Edwards in: Ann. Sci. nat., *v.* 20 p. 368 | 1840 *Lysianassa atlantica*, H. Milne Edwards, Hist. nat. Crust., *v.* 3 p. 22 | 1862 *L. a.*, Bate, Cat. Amphip. Brit. Mus., p. 68 t. 10 f. 10 | 1857 *L. marina*, Bate in: Ann. nat. Hist., ser. 2 *v.* 19 p. 138 | 1857 *L. m.* + *Opis typica* (err., non Krøyer 1846!), A. White, Hist. Brit. Crust., p. 168; p. 165 | 1898 *Ambasia danielsseni*, A. O. Walker in: P. Liverp. biol. Soc., *v.* 12 p. 171 | 1900 & 05 *A. d.*, *Lysianassa atlantica*, A. M. Norman in: Ann. nat. Hist., ser. 7 *v.* 5 p. 144; *v.* 16 p. 79.

Pag. 52. — Put after the 30. Gen. **Ambasia** Boeck:

30*. Gen. **Schisturella** Norm. ǀ

1900 *Schisturella* (Sp. un.: *S. pulchra*), A. M. Norman in: Ann. nat. Hist., ser. 7 *v.* 5 p. 208.

Created for **Ambasia pulchra** (H. J. Hansen) (p. 52):

1. **S. pulchra** (H. J. Hansen)

Add to the literature of this species:

1900 *Schisturella pulchra*, A. M. Norman in: Ann. nat. Hist., ser. 7 *v.* 5 p. 208.

Pag. 54. — Add to the literature of species 1. **Anonyx nugax** (Phipps):

1904 *A. n.*, S. J. Holmes in: P. Calif. Ac., ser. 3 *v.* 3 p. 513 t. 35 f. 17—19; t. 36 f. 20 | 1877 *Lysianassa fisheri*, Lockington in: P. Calif. Ac., *v.* 7 p. 48.

Pag. 56. — Put after the 32. Gen. **Anonyx** Kröyer:

32*. Gen. **Socarnella** A. Walker

1904 *Socarnella* (Sp. un.: *S. bonnieri*). A. O. Walker in: Herdman, Rep. Ceylon Pearl Fish., *v.* 2 p. 231, 239.

Cfr. Amaryllis (p. 23), Socarnes (p. 56).

1. **S. bonnieri** A. Walker 1904 *S. b.*, A. O. Walker in: Herdman, Rep. Ceylon Pearl Fish., *v.* 2 p. 239 t. 1 f. 4.

Pag. 58 and 60. — Add to the species of the 34. Gen. **Hippomedon** Boeck:

7. **H. bidentatus** Chevreux 1903 *H. b.*, Chevreux in: Bull. Soc. zool. France, *v.* 28 p. 87 f. 4 a—f.

8. **H. serratus** S. J. Holmes 1903 *H. s.*, S. J. Holmes in: Amer. Natural., *v.* 37 p. 278 | 1905 *H. s.*, S. J. Holmes in: Bull. U. S. Bureau Fish., *v.* 24 p. 473 t. 4 f. 2 and f. in text.

Pag. 61. — Put after the 35. Gen. **Glycerina** Hasw.:

35*. Gen. **Paracallisoma** Chevreux

1903 *Paracallisoma* (Sp. un.: *P. alberti*), Chevreux in: Bull. Soc. zool. France, *v.* 28 p. 84.

Cfr. Scopelocheirus (p. 61).

1. **P. alberti** Chevreux 1903 *P. a.*, Chevreux in: Bull. Soc. zool. France, *v.* 28 p. 84 f. 2, 3 a—d.

Pag. 65 and 66. — Add to the species of the 39. Gen. **Centromedon** O. Sars:

    5. **C. crenulatus** Chevreux 1900 *C. crenulatum*, Chevreux in: Résult. Camp. Monaco, *v.* 16 p. 26 t. 5 f. 3.

Pag. 67. — Add to the species of the 40. Gen. **Cheirimedon** Stebb.:

    3. **C. fougneri** A. Walker 1903 *C. f.*, A. O. Walker in: J. Linn. Soc., *v.* 29 p. 41 t. 7 f. 1—6.

    4. **C. hansoni** A. Walker 1903 *C. h.*, A. O. Walker in: J. Linn. Soc., *v.* 29 p. 42 t. 7 f. 7—12.

    5. **C. dentimanus** Chevreux 1906 *C. d.*, Chevreux in: Bull. Soc. zool. France, *v.* 30 p. 159 f. 1.

Pag. 68. — Add to the species of the 41. Gen. **Tryphosella** Bonnier:

    2. **T. abyssi** Norm. 1900 *T. a.*, A. M. Norman in: Ann. nat. Hist., ser. 7 *v.* 5 p. 205 t. 6 f. 16—20.

Pag. 69. — Add to the literature of species 2. **Tryphosa kergueleni** (Miers):

    1903 *Hoplonyx k.*, A. O. Walker in: J. Linn. Soc., *v.* 29 p. 51.

Pag. 69 and 72. — Add to the species of the 42. Gen. **Tryphosa** Boeck:

    10. **T. adarei** A. Walker 1903 *T. a.*, A. O. Walker in: J. Linn. Soc., *v.* 29 p. 49 t. 8 f. 38—44.

    11. **T. murrayi** A. Walker 1903 *T. m.*, A. O. Walker in: J. Linn. Soc., *v.* 29 p. 50 t. 9 f. 45—51.

    12. **T. nugax** S. J. Holmes 1904 *T. n.*, S. J. Holmes in: Harriman Alaska Exp., p. 234 f. 119, 120.

    13. **T. cucullata** A. Walker 1904 *T. c.*, A. O. Walker in: Herdman, Rep. Ceylon Pearl Fish., *v.* 2 p. 244 t. 4 f. 8.

Pag. 73. — Add to the literature of the 45. Gen. **Tmetonyx** Stebb.:

    1900 *Haplonyx* (laps.?, corr.: *Hoplonyx*), A. M. Norman in: Ann. nat. Hist., ser. 7 *v.* 5 p. 209, 211, 212.

Pag. 76. — Add to the literature of species 7. **Tmetonyx leucophthalmus** (O. Sars)

    1903 *Hoplonyx leucophtalmus*, Chevreux in: Bull. Soc. zool. France, *v.* 28 p. 97.

Pag. 74 and 77. — Add to the species of the 45. Gen. **Tmetonyx** Stebb.:

    9. **T. exiguus** (Chevreux) 1902 *Hoplonyx e.*, Chevreux in: C.-R. Ass. Franç., Sess. 30 *v.* 2 p. 696 t. 5 f. 1 a—k.

    10. **T. stebbingi** (A. Walker) 1903 *Hoplonyx s.*, A. O. Walker in: J. Linn. Soc., *v.* 29 p. 52 t. 9 f. 52—57*.

Pag. 79. — Add to the literature of species 3. **Lepidepecreum clypeatum** Chevreux:

    1900 *L. c.*, Chevreux in: Résult. Camp. Monaco, *v.* 16 p. 28 t. 4 f. 2.

Pag. 81 and 83. — Add to the species of the 48. Gen. **Orchomenella** O. Sars:

    6. **O. dilatata** Chevreux 1903 *O. d.*, Chevreux in: Bull. Soc. zool. France, *v.* 28 p. 90 f. 5 a—f:

    Near O. laevis (p. 81).

7. **O. pinguides** A. Walker 1903 *O. p.*, A. O. Walker in: J. Linn. Soc., *v.* 29 p. 46 t. 8 f. 24—30.

Very near O. pinguis (p. 82).

8. **O. franklini** A. Walker 1903 *O. f.*, A. O. Walker in: J. Linn. Soc., *v.* 29 p. 47 t. 8 f. 31—36.

Near O. minuta (p. 82).

9. **O. macronyx** Chevreux 1906 *O. m.*, Chevreux in: Bull. Soc. zool. France, *v.* 30 p. 161 f. 2.

Pag. 84 and 85. — Add to the species of the 49. Gen. **Orchomenopsis** O. Sars:

5. **O. proxima** Chevreux 1903 *O. p.*, Chevreux in: Bull. Soc. zool. France, *v.* 28 p. 93 f. 6 a—c.

Very near O. obtusa (p. 85).

6. **O. chevreuxi** Stebb.[1]) 1903 *O. excavata* (non *Orchomene excavatus* O. Sars 1891, laps. pro: *cavimanus*), Chevreux in: Bull. Soc. zool. France, *v.* 28 p. 94 f. 7 a—g.

Near O. zschauii (p. 85).

7. **O. nodimana** A. Walker 1903 *O. nodimanus*, A. O. Walker in: J. Linn. Soc., *v.* 29 p. 44 t. 7 f. 13—17.

8. **O. rossi** A. Walker 1903 *O. r.*, A. O. Walker in: J. Linn. Soc., *v.* 29 p. 45 t. 7 f. 18—23.

Nearly related to O. obtusa (p. 85).

Pag. 85. — Put after the 49. Gen. **Orchomenopsis** O. Sars:

50. Gen. **Katius** Chevreux

1905 *Katius* (Sp. un.: *K. obesus*), Chevreux in: Bull. Mus. Monaco, nr. 35 p. 1.

Near Orchomenopsis (p. 83).

1. **K. obesus** Chevreux 1905 *K. o.*, Chevreux in: Bull. Mus. Monaco, nr. 35 p. 1 f. 1, 2 a—h, 3 a—h.

Pag. 87. — Add to the literature of **Anonyx schmardae** Heller:

1902 *Socarnes schmardai*, Chevreux in: C.-R. Ass. Franç., Sess. 30 *v.* 2 p. 693.

Pag. 87. — **Lysianassa fisheri** Lockington enters into the synonymy of **Anonyx nugax** (Phipps), see above p. 719.

Pag. 87. — **Lysianassa marina** Bate enters into the synonymy of **Ambasia atlantica** (M.-E.), see above p. 719.

Pag. 100. — Add to the literature of species 1. **Ampelisca eschrichtii** Krøyer:

1905 *A. eschrichti*, S. J. Holmes in: Bull. U. S. Bureau Fish., *v.* 24 p. 525 | 1853 *Pseudophthalmus pelagicus*, Stimpson in: Smithson. Contr., *v.* 6 nr. 5 p. 57 | 1862 *Ampelisca pelagica*, Bate, Cat. Amphip. Brit. Mus., p. 94.

Pag. 103. — Add to the literature of species 8. **Ampelisca uncinata** Chevreux:

1900 *A. u.*, Chevreux in: Résult. Camp. Monaco, *v.* 16 p. 42 t. 6 f. 3.

---

[1]) Nom. nov. After E. Chevreux.

Pag. 104 and 107. — The species 17. **Ampelisca serraticaudata** Chevreux enters into the synonymy of species 11. **A. rubella** A. Costa [fide: Chevreux in: Résult. Camp. Monaco, v. 16 p. 44].

Pag. 105. — Add to the literature of species 12. **Ampelisca amblyops** O. Sars:

1887 *A. anomala* (err., non O. Sars 1882!), Chevreux in: Bull. Soc. zool. France, *v.* 12 p. 567 [fide: Chevreux in: Résult. Camp. Monaco, *v.* 16 p. 44].

Pag. 109. — Add to the literature of species 20. **Ampelisca spinimana** Chevreux:

1900 *A. s.*, Chevreux in: Résult. Camp. Monaco, *v.* 16 p. 39 t. 6 f. 2.

Pag. 99 and 111. — Add to the species of the 1. Gen. **Ampelisca** Krøyer:

26. **A. compressa** S. J. Holmes 1903 *A. c.*, S. J. Holmes in: Amer. Natural., *v.* 37 p. 273 | 1905 *A. c.*, S. J. Holmes in: Bull. U. S. Bureau Fish., *v.* 24 p. 480 t. 6 f. 1 and f. in text.

27. **A. tridens** A. Walker 1904 *A. t.*, A. O. Walker in: Herdman, Rep. Ceylon Pearl Fish., *v.* 2 p. 249 t. 2 f. 11, t. 4 f. 11.

28. **A. scabripes** A. Walker 1904 *A. s.*, A. O. Walker in: Herdman, Rep. Ceylon Pearl Fish., *v.* 2 p. 250 t. 2 f. 12.

29. **A. brachyceras** A. Walker 1904 *A. b.*, A. O. Walker in: Herdman, Rep. Ceylon Pearl Fish., *v.* 2 p. 251 t. 2 f. 13.

30. **A. cyclops** A. Walker 1904 *A. c.*, A. O. Walker in: Herdman, Rep. Ceylon Pearl Fish., *v.* 2 p. 253 t. 2 f. 14.

31. **A. chevreuxi** A. Walker 1904 *A. c.*, A. O. Walker in: Herdman, Rep. Ceylon Pearl Fish., *v.* 2 p. 254 t. 3 f. 15.

Pag. 100 and 111. — **Ampelisca pelagica** (Stimps.) enters into the synonymy of species 1. **A. eschrichtii** Krøyer.

Pag. 114. — Add to the literature of species 7. **Byblis guernei** Chevreux:

1900 *B. g.*, Chevreux in: Résult. Camp. Monaco, *v.* 16 p. 45 t. 7 f. 1.

Pag. 116. — The name of the 1st species of 3. Gen. **Haploops** Lilj. is to be changed into: **H. dellavallei** Chevreux. Add to the literature of this species:

1900 *H. dellavallei*, Chevreux in: Résult. Camp. Monaco, *v.* 16 p. 47.

Pag. 118 and 273. — A. O. Walker 1904 (in: Herdman, Rep. Ceylon Pearl Fish., v. 2 p. 246) creates the family **Argissidae** for the genera **Platyischnopus** Stebb. (p. 122) and **Argissa** Boeck (p. 276).

Pag. 123 and 124. — Add to the species of the 2. Gen. **Platyischnopus** Stebb.:

3. **P. herdmani** A. Walker 1904 *P. h.*, A. O. Walker in: Herdman, Rep. Ceylon Pearl Fish., *v.* 2 p. 247 t. 2 f. 10.

Pag. 131. — Add to the literature of species 5. **Urothoe poucheti** Chevreux:

1900 *U. p.*, Chevreux in: Résult. Camp. Monaco, *v.* 16 p. 31 t. 5 f. 4.

Pag. 129 and 132. — Add to the species of the 7. Gen. **Urothoe** Dana:

8. **U. spinidigitus** A. Walker 1904 *U. s.*, A. O. Walker in: Herdman, Rep. Ceylon Pearl Fish., *v.* 2 p. 245 t. 1 f. 9.

9. **U. poseidonis** Reib. 1905 *U. p.*, J. Reibisch in: Wiss. Meeresunters., *v.* 8 Abt. Kiel p. 163 t. 4 f. 17—21.

Pag. 138. — **Paraphoxus maculatus** (Chevreux) is an accepted species; add to the literature of this species:

> 1900 *P. m.*, Chevreux in: Résult. Camp. Monaco, *v.* 16 p. 34 t. 5 f. 5.

Pag. 137 and 138. — Add to the species of the 3. Gen. **Paraphoxus** O. Sars:

> 3. **P. spinosus** S. J. Holmes   1903 *P. s.*, S. J. Holmes in: Amer. Natural., *v.* 37 p. 276 | 1905 *P. s.*, S. J. Holmes in: Bull. U. S. Bureau Fish., *v.* 24 p. 477 f.

Pag. 142. — Add to the literature of species 5. **Harpinia excavata** Chevreux:

> 1900 *H. e.*, Chevreux in: Résult. Camp. Monaco, *v.* 16 p. 37 t. 6 f. 1.

Pag. 140 and 145. — Add to the species of the 5. Gen. **Harpinia** Boeck:

> 13. **H. latipes** Norm.   1900 *H. l.*, A. M. Norman in: Ann. nat. Hist., ser. 7 *v.* 5 p. 338 f.

Pag. 147. — Add to the literature of species **Pontharpinia uncirostrata** (Giles):

> 1904 *Leptophoxus uncirostratus*, A. O. Walker in: Herdman, Rep. Ceylon Pearl Fish., *v.* 2 p. 249.

Pag. 149, 151 and 161. — **Probolium spence-batei** Stebb. (p. 161) is to be accepted among the species of 1. Gen. **Amphilochus** Bate: 6. **A. spence-batei** (Stebb.). Add to the literature of this species:

> 1900 *Amphilochus anomalus*, Chevreux in: Résult. Camp. Monaco, *v.* 16 p. 48 t. 7 f. 2.

Pag. 150. — Synonyma of species 2. **Amphilochus neapolitanus** Della Valle are: **A. melanops** A. Walker (p. 152), fide A. M. Norman in: Ann. nat. Hist., ser. 7 v. 6 p. 34 t. 3 f. 1, and: **A. melanops** A. Walker (p. 152), **A. brunneus** Della Valle (p. 151) and ? **A. marionis** Stebb. (p. 151), fide A. O. Walker in: J. Linn. Soc., v. 28 p. 300.

Pag. 151—153. — **Amphilochus longimanus** Chevreux belongs to the 2. Gen. **Amphilochoides** O. Sars:

> 4. **A. longimanus** (Chevreux)   1888 *Amphilochus l.*, Chevreux in: Bull. Soc. zool. France, *v.* 13 p. 41 | 1900 *Amphilochoides l.*, Chevreux in: Résult. Camp. Monaco, *v.* 16 p. 50 t. 7 f. 3.

Pag. 152 and 161. — The 1st species of 2. Gen. **Amphilochoides** O. Sars changes its name into: 1. **A. serratipes** (Norm.), fide A. M. Norman in: Ann. nat. Hist., ser. 7 v. 6 p. 35. Add to its literature from p. 161:

> 1869 *Probolium serratipes*, A. M. Norman in: Rep. Brit. Ass., Meet. 38 p. 273.

Pag. 158. — Add to the species of 6. Gen. **Cyproidea** Hasw.:

> 2. **C. otakensis** (Chilton)   1900 *Cyproidia o.*, Chilton in: Ann. nat. Hist., ser. 7 *v.* 5 p. 243 t. 5.

Pag. 160. — Add to the literature of **Peltocoxa brevirostris** (T. & A. Scott):

> 1900 *Stegoplax b.*, A. M. Norman in: Ann. nat. Hist., ser. 7 *v.* 6 p. 38.

Pag. 161. — Put after the 9. Gen. **Paracyproidea** Stebb.:

10. Gen. **Gallea** A. Walker

> 1904 *Gallea* (Sp. un.: *G. tecticauda*), A. O. Walker in: Herdman, Rep. Ceylon Pearl Fish., *v.* 2 p. 232, 256.

> 1. **G. tecticauda** A. Walker   1904 *G. t.*, A. O. Walker in: Herdman, Rep. Ceylon Pearl Fish., *v.* 2 p. 256 t. 3 f. 16, t. 8 f. 16.

Pag. 162 and 163. — Add to the species of 1. Gen. **Seba** Bate:

    3. **S. armata** Chevreux 1900 *S. a.*, Chevreux in: Résult. Camp. Monaco, *v.* 16 p. 111 t. 13 f. 1.

    Add to the literature of this species from species 2. **S. saundersii** Stebb. (p. 163):

    1889 *Grimaldia armata*, Chevreux in: Bull. Soc. zool. France, *v.* 14 p. 284 f. | 1899 *Seba a.*, Chevreux in: C.-R. Ass. Franç., Sess. 27 *v.* 2 p. 483.

Pag. 164 and 168. — Add to the species of 2. Gen. **Leucothoe** Leach:

    12. **L. euryonyx** A. Walker 1901 *L. e.*, A. O. Walker in: J. Linn. Soc., *v.* 28 p. 302 t. 27 f. 24—26.

    13. **L. hornelli** A. Walker 1904 *L. h.*, A. O. Walker in: Herdman, Rep. Ceylon Pearl Fish., *v.* 2 p. 258 t. 3 f. 17.

    14. **L. stegoceras** A. Walker 1904 *L. s.*, A. O. Walker in: Herdman, Rep. Ceylon Pearl Fish., *v.* 2 p. 259 t. 3 f. 17 a.

Pag. 167. — Synonyma of species 11. **Leucothoe lilljeborgii** Boeck are: **L. furina** Chevreux (non Savigny) (p. 165), **L. incisa** D. Roberts. (p. 167) and **L. serratipalma** (laps. pro: *serraticarpa* Della Valle) fide A. M. Norman in: Ann. nat. Hist., ser. 7 v. 6 p. 47, and: **L. incisa** D. Roberts. fide J. Reibisch in: Wiss. Meeresunters., v. 8 Abt. Kiel p. 179 t. 5 f. 5.

Pag. 170 and 171. — Add to the species of the 1. Gen. **Anamixis** Stebb.:

    2. **A. stebbingi** A. Walker 1904 *A. s.*, A. O. Walker in: Herdman, Rep. Ceylon Pearl Fish., *v.* 2 p. 259 t. 3 f. 18.

Pag. 172. — Add to the literature of 1. Gen. **Metopa** Boeck:

    1900 *Metopina* (Sp. typ.: *Metopa palmata*) (non Macquart 1835, Diptera!), A. M. Norman in: Ann. nat. Hist., ser. 7 v. 6 p. 45 | 1902 *Sthenometopa*, A. M. Norman in: Ann. nat. Hist., ser. 7 *v.* 10 p. 480.

Pag. 173. — Add to the literature of species 2. **Metopa robusta** O. Sars:

    1900 *Metopina r.*, A. M. Norman in: Ann. nat. Hist., ser. 7 *v.* 6 p. 45 | 1902 *Sthenometopa r.*, A. M. Norman in: Ann. nat. Hist., ser. 7 *v.* 10 p. 480.

Pag. 174. — Add to the literature of species 5. **Metopa palmata** O. Sars:

    1900 *Metopina p.*, A. M. Norman in: Ann. nat. Hist., ser. 7 *v.* 6 p. 45 | 1902 *Sthenometopa p.*, A. M. Norman in: Ann. nat. Hist., ser. 7 *v.* 10 p. 481.

Pag. 175. — Add to the literature of species 6. **Metopa clypeata** (Krøyer):

    1900 *Metopina c.*, A. M. Norman in: Ann. nat. Hist., ser. 7 *v.* 6 p. 45 | 1902 *Sthenometopa c.*, A. M. Norman in: Ann. nat. Hist., ser. 7 *v.* 10 p. 481.

Pag. 172 and 182. — Add to the species of the 1. Gen. **Metopa** Boeck:

    22. **M. abscisa** Norm. 1868 *Montagua clypeata* (err., non *Leucothoe c.* Krøyer 1842!), Bate & Westwood, Brit. sess. Crust., *v.* 2 p. 499 | 1900 *Metopa abscisa*, A. M. Norman in: Ann. nat. Hist., ser. 7 *v.* 6 p. 42 t. 3 f. 6—10.

    23. **M. quadrangula** Reib. 1905 *M. q.*, J. Reibisch in: Wiss. Meeresunters., *v.* 8 Abt. Kiel p. 178 t. 5 f. 1—4.

Pag. 187, 190, 195. — The 7ᵗʰ species of 4. Gen. **Proboloides** Della Valle is synonym to the 6ᵗʰ species of 1. Gen. **Stenothoe** Dana, and changes its name into **P. clypeatus** (Stimps.) with the following literature:

7. **P. clypeatus** (Stimps.) 1853 *Stenothoe clypeata*, Stimpson in: Smithson. Contr., *v.* 6 nr. 5 p. 51 | ?1862 *S. clypeatus*, Bate, Cat. Amphip. Brit. Mus., p. 61 t. 9 f. 7 | 1887 *Metopa groenlandica*, H. J. Hansen in: Vid. Meddel., ser. 4 *v.* 9 p. 94 t. 3 f. 7—7e | 1905 *M. g.*, S. J. Holmes in: Bull. U. S. Bureau Fish., *v.* 24 p. 483 t. 6 f. 3, f. in text. | 1893 *Stenothoe clypeata* + *M. g.*, A. Della Valle in: F. Fl. Neapel, *v.* 20 p. 569; p. 640 t. 59 f. 55.

Pag. 192 and 200. — Add to the genera of 11. Fam. **Stenothoidae:**

2. Gen. **Parametopa** Chevreux

1901 *Parametopa* (Sp. un.: *P. kervillei*), Chevreux in: Bull. Soc. Rouen, *v.* 36 p. 233.

1. **P. kervillei** Chevreux 1901 *P. k.*, Chevreux in: Bull. Soc. Rouen, *v.* 36 p. 233 t. 3.

3. Gen. **Stenothoides** Chevreux

1900 *Stenothoides* (Sp. typ.: *S. perrieri*). Chevreux in: Résult. Camp. Monaco, *v.* 16 p. 55.

1. **S. perrieri** Chevreux 1900 *S. p.*, Chevreux in: Résult. Camp. Monaco, *v.* 16 p. 55 t. 8 f. 2.

Pag. 193 and 199. — Add to the species of 1. Gen. **Stenothoe** Dana:

16. **S. setosa** Norm. 1900 *S. s.*, A. M. Norman in: Ann. nat. Hist., ser. 7 *v.* 6 p. 39 t. 3 f. 2—4.

Near S. dollfusi (p. 196).

17. **S. cypris** S. J. Holmes 1903 *S. c.*, S. J. Holmes in: Amer. Natural., *v.* 37 p. 278 | 1905 *S. c.*, S. J. Holmes in: Bull. U. S. Bureau Fish., *v.* 24 p. 484 f.

18. **S. minuta** S. J. Holmes 1903 *S. m.*, S. J. Holmes in: Amer. Natural., *v.* 37 p. 278 | 1905 *S. m.*, S. J. Holmes in: Bull. U. S. Bureau Fish., *v.* 24 p. 485 f.

19. **S. alaskensis** S. J. Holmes 1904 *S. a.*, S. J. Holmes in: Harriman Alaska Exp., *v.* 10 p. 236 f. 121, 122.

20. **S. gallensis** A. Walker 1904 *S. g.*, A. O. Walker in: Herdman, Rep. Ceylon Pearl Fish., *v.* 2 p. 261 t. 3 f. 19.

Very near S. valida (p. 194).

Pag. 196. — Add to the literature of species 8. **Stenothoe dollfusi** Chevreux:

1900 *S. d.*, Chevreux in: Résult. Camp. Monaco, *v.* 16 p. 53 t. 8 f. 1.

Pag. 198. — Add to the literature of species 12. **Stenothoe marina** (Bate):

1904 *S. m. var. sinhalensis*, A. O. Walker in: Herdman, Rep. Ceylon Pearl Fish., *v.* 2 p. 261.

Pag. 200. — Put after the 11. Fam. **Stenothoidae:**

11*. Fam. **Ingolfiellidae**

1903 *Ingolfiellidae*, H. J. Hansen in: J. Linn. Soc., *v.* 29 p. 117.

1. Gen. **Ingolfiella** H. J. Hansen

1903 *Ingolfiella*, H. J. Hansen in: J. Linn. Soc., *v.* 29 p. 118, 128.

1. I. abyssi H. J. Hansen 1903 *I. a.*, H. J. Hansen in: J. Linn. Soc., *v.* 29 p. 118 t. 14 f. 1—18, t. 15 f. 19—21.

2. I. littoralis H. J. Hansen 1903 *I. l.*, H. J. Hansen in: J. Linn. Soc., *v.* 29 p. 124 t. 15 f. 22—33.

Pag. 201. — Add to the literature of species 1. **Pereionotus testudo** (Mont.):

1903 *P. t.*, A. O. Walker & A. Scott in: H. O. Forbes, Nat. Hist. Sokotra, p. 227 t. 14 B f. 4 a, b.

Pag. 204. — Add to the literature of species 1. **Iphinotus typicus** (G. M. Thoms.):

1902 *I. typica*, G. M. Thomson in: Ann. nat. Hist., ser. 7 *v.* 10 p. 464.

Pag. 205. — Put at the end of 12. Fam. **Phliantidae:**

Gen. **Kuria** A. Walker & A. Scott

1903 *Kuria* (Sp. un.: *K. longimanus*), A. O. Walker & A. Scott in: H. O. Forbes, Nat. Hist. Sokotra, p. 217, 228.

Cfr. Bircenna (p. 205).

K. longimana A. Walker & A. Scott 1903 *K. longimanus*, A. O. Walker & A. Scott in: H. O. Forbes, Nat. Hist. Sokotra, p. 228 t. 14 B f. 5—5 n.

Pag. 234. — Add to the literature of species **Liljeborgia pugettensis** (Dana):

1904 *Gammarus p.*, S. J. Holmes in: Harriman Alaska Exp., *v.* 10 p. 239.

Pag. 237, 238. — Add to the species of 1. Gen. **Perioculodes** O. Sars:

2. P. serra A. Walker 1904 *P. s.*, A. O. Walker in: Herdman, Rep. Ceylon Pearl Fish., *v.* 2 p. 262 t. 4 f. 20.

Near P. longimanus (p. 237).

Pag. 244. — Add to the species of 6. Gen. **Oediceros** Krøyer:

3. O. newnesi A. Walker 1903 *O. n.*, A. O. Walker in: J. Linn. Soc., *v.* 29 p. 53 t. 9 f. 62—66, t. 10 f. 67, 68.

Very near O. saginatus (p. 244).

Pag. 248. — Add to the literature of 9. Gen. **Arrhis** Stebb.:

? 1880 *Aceropsis* (nom. nud.), Stuxberg in: Bih. Svenska Ak., *v.* 5 nr. 22 p. 63.

Pag. 254. — The name of the 2^d species of 13. Gen. **Acanthostepheia** Boeck is to be changed into: 2. **A. behringiensis** (Lockington), of which **A. pulchra** Miers is a synonym. Add to the literature of species 2. **A. behringiensis** (Lockington):

1904 *A. behringanus*, S. J. Holmes in: P. Calif. Ac., ser. 3 *v.* 3 p. 315 t. 36 f. 25—28.

Pag. 259. — Add to the literature of species 1. **Monoculodes gibbosus** Chevreux:

1900 *M. g.*, Chevreux in: Résult. Camp. Monaco, *v.* 16 p. 59 t. 8 f. 3.

Pag. 259 and 267. — Add to the species of 17. Gen. **Monoculodes** Stimps.:

19. **M. edwardsi** S. J. Holmes 1905 *M. e.*, S. J. Holmes in: Bull. U. S. Bureau Fish., *v.* 24 p. 487 f.

Pag. 276. — Add to the species of 2. Gen. **Tiron** Lillj.:

2. **T. thompsoni** A. Walker 1904 *T. t.*, A. O. Walker in: Herdman, Rep. Ceylon Pearl Fish., *v.* 2 p. 263 t. 4 f. 21.

Pag. 285—287, 308. — Add to the genera of 22. Fam. **Calliopiidae**:

1*. Gen. **Bouvierella** Chevreux

1900 *Bouvierella* (Sp. typ.: *B. carcinophila*), Chevreux in: Résult. Camp. Monaco, *v.* 16 p. 70.

Created for **Laothoes carcinophilus** (Chevreux) (p. 287):

1. **B. carcinophila** (Chevreux).

Add to the literature of this species:

1900 *B. c.*, Chevreux in: Résult. Camp. Monaco, *v.* 16 p. 71 t. 9 f. 3.

Pag. 285 and 294. — Put after the 6. Gen. **Leptamphopus** O. Sars:

6*. Gen. **Oradarea** A. Walker

1903 *Oradarea* (Sp. un.: *O. longimana*), A. O. Walker in: J. Linn. Soc., *v.* 29 p. 40, 56.

Near Amphithopsis (p. 289), but perhaps nearer Leptamphopus (p. 293).

1. **O. longimana** A. Walker 1903 *O. l.*, A. O. Walker in: J. Linn. Soc., *v.* 29 p. 56 t. 10 f. 77—89.

Strangely like Leptamphopus novaezealandiae (p. 294).

Pag. 299 and 300. — Add to the species of 11. Gen. **Atylopsis** Stebb.:

3. **A. latipalpus** A. Walker & A. Scott 1903 *A. l.*, A. O. Walker & A. Scott in: H. O. Forbes, Nat. Hist. Sokotra, p. 222 t. 11A f. 7—71.

Pag. 304. — Add to the literature of 15. Gen. **Apherusa** A. Walker:

1858 *Phaedra* (Sp. un.: *P. antiqua*), Bate in: Quart. J. geol. Soc., *v.* 15 p. 138, 140.

Pag. 304 and 308. — Add to the species of 15. Gen. **Apherusa** A. Walker:

8. **A. gracilis** S. J. Holmes 1903 *A. g.*, S. J. Holmes in: Amer. Natural., *v.* 37 p. 287 | 1905 *A. g.*, S. J. Holmes in: Bull. U. S. Bureau Fish., *v.* 24 p. 495 f.

9. **A. clevei** O. Sars 1904 *A. c.*, G. O. Sars in: Publ. Expl. Mer, nr. 10 p. 3 t. 1.

10. **A. ovalipes** Norm. 1906 *A. o.*, A. M. Norman (& T. Scott), Crust. Devon Cornwall, p. 75 t. 8 f. 1—8.

A. antiqua (Bate) 1858 *Phaedra a.*, Bate in: Quart. J. geol. Soc., *v.* 15 p. 138 t. 6 f. 8 | 1905 *P. a.*, *Apherusa?*, A. M. Norman in: Ann. nat. Hist., ser. 7 *v.* 16 p. 81.

Pag. 312 and 315. — Add to the species of 2. Gen. **Neopleustes** Stebb.:

    8. **N. pacificus** (A. Walker) 1898 *Paramphithoe pacifica*, A. O. Walker in: P. Liverp. biol. Soc., *v.* 12 p. 281.

    N. pugettensis (Dana) 1853 & 55 *Iphimedia p.*, J. D. Dana in: U. S. expl. Exp., *v.* 13 II p. 932; t. 63 f. 6 a—g | 1898 *Paramphithoe p.*, A. O. Walker in: P. Liverp. biol. Soc., *v.* 12 p. 281.

Pag. 320. — Put after the 5. Gen. **Sympleustes** Stebb.:

6. Gen. **Dautzénbergia** Chevreux

    1900 *Dautzenbergia* (Sp. un.: *D. grandimana*), Chevreux in: Résult. Camp. Monaco, *v.* 16 p. 73.

Created for Sympleustes grandimanus (Chevreux) (p. 318):

    1. **D. grandimana** (Chevreux)

Add to the literature of this species:

    1900 *D. g.*, Chevreux in: Résult. Camp. Monaco, *v.* 16 p. 73 t. 10 f. 1.

Pag. 321 and 324. — Add to the species of 1. Gen. **Epimeria** A. Costa:

    5. **E. inermis** A. Walker 1903 *E. i.*, A. O. Walker in: J. Linn. Soc., *v.* 29 p. 54 t. 10 f. 69.

Pag. 328 and 329. — Add to the species of 1. Gen. **Atylus** Leach:

    2. **A. walkeri** Stebb.[1]) 1903 *A. antarcticus* (non T. Stebbing 1878!), A. O. Walker in: J. Linn. Soc., *v.* 29 p. 58 t. 11 f. 91—97.

Pag. 329 and 334. — Add to the species of 2. Gen. **Nototropis** A. Costa:

    8. **N. minikoi** (A. Walker) 1905 *Paratylus m.*, A. O. Walker in: Gardiner, Fauna Mald. Laccad., *v.* 2 p. 925 f. 141 I—v.

    9. **N. granulosus** (A. Walker) 1904 *Paratylus g.*, A. O. Walker in: Herdman, Rep. Ceylon Pearl Fish., *v.* 2 p. 265.

Pag. 334 and 338. — Add to the genera of 26. Fam. **Melphidippidae:**

3. Gen. **Hornellia** A. Walker

    1904 *Hornellia* (Sp. un.: *H. incerta*), A. O. Walker in: Herdman, Rep. Ceylon Pearl Fish., *v.* 2 p. 233, 268.

    1. **H. incerta** A. Walker 1904 *H. i.*, A. O. Walker in: Herdman, Rep. Ceylon Pearl Fish., *v.* 2 p. 269 t. 4 f. 27.

Pag. 338 and 355. — Add to the genera of 27. Fam. **Eusiridae**

7. Gen. **Eusirogenes** Stebb.

    1904 *Eusirogenes* (Sp. un.: *E. dolichocarpus*), T. Stebbing in: Tr. Linn. Soc. London, ser. 2 *v.* 10 II p. 13, 15.

    1. **E. dolichocarpus** Stebb. 1904 *E. d.*, T. Stebbing in: Tr. Linn. Soc. London, ser. 2 *v.* 10 II p. 15 t. 2 A.

---

[1]) Nom. nov. After Alfred O. Walker.

Pag. 339 and 343. — Add to the species of 1. Gen. **Eusirus** Krøyer:

> 9. **E. laevis** A. Walker 1903 *E. l.*, A. O. Walker in: J. Linn. Soc., *v.* 29 p. 55 t. 10 f. 70—76.
>
> Cfr. E. antarcticus (p. 340).

Pag. 345 and 346. — Add to the species of 3. Gen. **Eusiroides** Stebb.:

> 4. **E. sarsi** Chevreux 1900 *E. s.*, Chevreux in: Résult. Camp. Monaco, *v.* 16 p. 65 t. 9 f. 2.
>
> 5. **E. orchomenipes** A. Walker 1904 *E. o.*, A. O. Walker in: Herdman, Rep. Ceylon Pearl Fish., *v.* 2 p. 264 t. 4 f. 23.

Pag. 350. — Add to the literature of species 3. **Rhachotropis grimaldii** (Chevreux):

> 1900 *Rachotropis g.*, Chevreux in: Résult. Camp. Monaco, *v.* 16 p. 68 t. 9 f. 1.

Pag. 355 and 356. — Add to the species of 1. Gen. **Batea** Fr. Müll.:

> 2. **B. secunda** S. J. Holmes 1903 *B. s.*, S. J. Holmes in: Amer. Natural., *v.* 37 p. 284 | 1905 *B. s.*, S. J. Holmes in: Bull. U. S. Bureau Fish., *v.* 24 p. 499 f.

Pag. 362. — Add to the species of 6. Gen. **Atyloides** Stebb.:

> 2. **A. gabrieli** Sayce 1901 *A. g.*, Sayce in: P. R. Soc. Victoria, *v.* 13 p. 230 t. 37, 38.
>
> 3. **A. fontanus** Sayce 1902 *A. fontana*, Sayce in: P. R. Soc. Victoria, *v.* 15 p. 49 t. 5.

Pag. 362. — Add to the literature of species 1. **Atyloides serraticauda** Stebb.:

> 1903 *A. s.*, A. O. Walker in: J. Linn. Soc., *v.* 29 p. 58 t. 11 f. 90.

Pag. 364 and 411. — Add to the genera of 30. Fam. **Gammaridae:**

21*. Gen. **Bathyonyx** Vejd.

> 1905 *Bathyonyx* (Sp. un.: *B. devismesi*), Vejdovský in: SB. Böhm. Ges., nr. 28 p. 2.
>
> 1. **B. devismesi** Vejd. 1905 *B. d.*, Vejdovský in: SB. Böhm. Ges., nr. 28 p. 2 t. 1 f. 1—15, t. 2 f. 16—19.

21**. Gen. **Pseudoniphargus** Chevreux

> 1901 *Pseudoniphargus* (Sp. un.: *P. africanus*), Chevreux in: Bull. Soc. zool. France, *v.* 26 p. 211.
>
> 1. **P. africanus** Chevreux 1901 *P. a.*, Chevreux in: Bull. Soc. zool. France, *v.* 26 p. 211 f. 1, 2 a—h, k.

Pag. 370. — Add to the literature of 5. Gen. **Crangonyx** Bate:

> 1902 *Stygonectes* (Sp. typ.: *Crangonyx flagellatus*), W. P. Hay in: P. U. S. Mus., *v.* 25 p. 430.

Pag. 371 and 373. — Add to the species of 5. Gen. **Crangonyx** Bate:

> 7. **C. bowersii** C. J. Ulr. 1902 *C. b.*, C. J. Ulrich in: Tr. Amer. micr. Soc., *v.* 23 p. 85 t. 14.

Pag. 371. — Add to the literature of species 2. **Crangonyx flagellatus** Benedict:

1902 *C. f., Stygonectes sp. typ.*, W. P. Hay in: P. U. S. Mus., *v.* 25 p. 430.

Pag. 373. — Add to the literature of species 5. **Crangonyx bifurcus** O. P. Hay:

1902 *Niphargus b.*, W. P. Hay in: P. U. S. Mus., *v.* 25 p. 429.

Pag. 374. — Add to the literature of 7. Gen. **Pallasea** Bate:

1901 *Pleuracanthus* (non Gray 1832, Coleoptera!) (part.) + *Dybowskia* (non Dall 1876. Mollusca!) (part.), Garjajeff in: Trudui Kazan. Univ., *v.* 35 p. 16.

Pag. 376. — Add to the literature of species 3. **Pallasea cancellus** (Pall.):

1901 *Dybowskia c.*, Garjajeff in: Trudui Kazan. Univ., *v.* 35 p. 40.

Pag. 377. — Add to the literature of species 4. **Pallasea gerstfeldtii** (Dyb.):

1901 *Dybowskia cancellus var. g.*, Garjajeff in: Trudui Kazan. Univ., *v.* 35 p. 40.

Pag. 377. — Add to the literature of species 5. **Pallasea quadrispinosa** O. Sars:

1901 *Dybowskia kesslerii var. europeus,* Garjajeff in: Trudui Kazan. Univ., *v.* 35 p. 39.

Pag. 378. — Add to the literature of species 6. **Pallasea kesslerii** (Dyb.):

1901 *Dybowskia k.*, Garjajeff in: Trudui Kazan. Univ., *v.* 35 p. 39.

Pag. 378. — Add to the literature of species 7. **Pallasea baikali** Stebb.:

1901 *Pleuracanthus lovenii,* Garjajeff in: Trudui Kazan. Univ., *v.* 35 p. 42.

Pag. 379. — Add to the literature of species 8. **Pallasea brandtii** (Dyb.):

1901 *Dybowskia b.*, Garjajeff in: Trudui Kazan. Univ., *v.* 35 p. 40.

Pag. 379. — Add to the literature of species 9. **Pallasea grubii** (Dyb.):

1901 *Dybowskia g.*, Garjajeff in: Trudui Kazan. Univ., *v.* 35 p. 39.

Pag. 380. — Add to the literature of species 10. **Pallasea cancelloides** (Gerstf.):

1901 *Dybowskia c.*, Garjajeff in: Trudui Kazan. Univ., *v.* 35 p. 40.

Pag. 375 and 380. — Add to the species of 7. Gen. **Pallasea** Bate:

P. dryshenskii (Garjajeff) 1901 *Dybowskia d.*, Garjajeff in: Trudui Kazan. Univ., *v.* 35 p. 33 t. 2 f. 19, t. 3 f. 68—72.

P. meyerii (Garjajeff) 1901 *Dybowskia m.*, Garjajeff in: Trudui Kazan. Univ., *v.* 35 p. 36 t. 2 f. 17, t. 3 f. 58—62.

P. nigra (Garjajeff) 1901 *Pleuracanthus niger,* Garjajeff in: Trudui Kazan. Univ., *v.* 35 p. 40 t. 2 f. 21, t. 3 f. 79—83.

P. viridis (Garjajeff) 1901 *Dybowskia v.*, Garjajeff in: Trudui Kazan. Univ., *v.* 35 p. 32 t. 2 f. 18, t. 3 f. 63—67.

Pag. 381. — Add to the literature of species 1. **Weyprechtia heuglini** (Buchh.):

1900 *Wayprechtia h.*, A. Birula in: Annuaire Mus. St.-Pétersb., *v.* 4 p. 442 f. 3—9.

Pag. 388. — Add to the literature of 13. Gen. **Eucrangonyx** Stebb.:

1902 *Bactrurus* (Sp. un.: *Crangonyx mucronatus*), W. P. Hay in: P. U. S. Mus., *v.* 25 p. 430.

Pag. 388. — Add to the literature of species 1. **Eucrangonyx mucronatus** (S. A. Forb.):

1902 *Crangonyx m., Bactrurus*, W. P. Hay in: P. U. S. Mus., *v.* 25 p. 430.

Pag. 390. — Add to the literature of species 5. **Eucrangonyx antennatus** (Pack.):

1902 *Niphargus a.*, W. P. Hay in: P. U. S. Mus., *v.* 25 p. 430 f. 6 a—m.

Pag. 391. — Add to the literature of 14. Gen. **Axelboeckia** Stebb.:

1901 *Ctenacanthus* (non Agassiz 1837, Pisces!) (part.), Garjajeff in: Trudui Kazan. Univ., *v.* 35 p. 15.

Pag. 392. — Add to the literature of species 2. **Axelboeckia carpenterii** (Dyb.):

1901 *Ctenacanthus c.*, Garjajeff in: Trudui Kazan. Univ., *v.* 35 p. 21.

Pag. 404. — Add to the literature of 20. Gen. **Neoniphargus** Stebb.:

1901 *Unimelita*, Sayce in: P. R. Soc. Victoria, *v.* 13 p. 237.

Pag. 404 and 405. — Add to the species of 20. Gen. **Neoniphargus** Stebb.:

2. **N. spenceri** (Sayce) 1901 & 02 *Unimelita s.*, Sayce in: P. R. Soc. Victoria, *v.* 13 p. 238 t. 40; *v.* 15 p. 57.

3. **N. fultoni** Sayce 1902 *N. f.*, Sayce in: P. R. Soc. Victoria, *v.* 15 p. 57 t. 7.

Pag. 404. — Add to the literature of species 1. **Neoniphargus thomsoni** Stebb.:

1901 *Niphargus montanus, Unimelita* (part.), Sayce in: P. R. Soc. Victoria, *v.* 13 p. 237 | 1901 *N. m.*, Sayce in: Ann. nat. Hist., ser. 7 *v.* 8 p. 562.

Pag. 405 and 409. — Add to the species of 21. Gen. **Niphargus** Schiödte:

11. **N. mortoni** G. M. Thoms. 1893 *N. m.*, G. M. Thomson in: P. R. Soc. Tasmania, 1892 p. 68 t. 4 f. 11, 12; t. 5 f. 1—5.

Showing approximation to Iphigenella (p. 447).

12. **N. pulchellus** Sayce 1900 *N. p.*, Sayce in: P. R. Soc. Victoria, *v.* 12 p. 152 t. 15, 16.

13. **N. plateaui** Chevreux 1901 *N. p.* + *N. p. elongatus* + *N. p. robustus*, Chevreux in: Bull. Soc. zool. France, *v.* 26 p. 168 f. 1, 2; p. 173 f. 3; p. 234 f. 1.

14. **N. ladmiraulti** Chevreux 1901 *N. l.*, Chevreux in: Bull. Soc. zool. France, *v.* 26 p. 174 f. 1—4.

Pag. 408. — Add to the literature of species 6. **Niphargus viréi** Chevreux:

: 1901 *N. v.*, Chevreux in: Bull. Soc. zool. France, *v.* 26 p. 197 f. 1 a, 2 b—g.

Pag. 408. — Add to the literature of species 7. **Niphargus fontanus** Bate:

1901 *N. f.*, Chevreux in: Bull. Soc. zool. France. *v.* 26 p. 201 f. 1, 2 a—d.

Pag. 408. — Add to the literature of species 8. **Niphargus kochianus** Bate:

1904 *N. k.*, Kane in: Ann. nat. Hist., ser. 7 *v.* 14 p. 274 t. 8 f. 1—3.

Pag. 409. — Add to the literature of species **Niphargus caspary** (Pratz):

1905 *N. c., N. casparyi*, Vejdovský in: SB. Böhm. Ges., nr. 28 p. 18.

Pag. 410. — Add to the literature of species **Niphargus rhipidiophorus** (Catta):

> 1901 *Gammarus r. ?*, Chevreux in: Bull. Soc. zool. France, *v.* 26 p. 216
> f. 1, 2 a—f.

Pag. 410. — Add to the literature of species **Niphargus subterraneus** (Leach):

> 1904 *N. s.*, *N. fontanus* Bate ?, Kane in: Ann. nat. Hist., ser. 7 *v.* 14 p. 280
> t. 8 f. 4, 5.

Pag. 411. — Add to the species of 22. Gen. **Eriopisa** Stebb.:

> 2. **E. sechellensis** Chevreux 1901 *E. s.*, Chevreux in: Mém. Soc. zool.
> France, *v.* 14 p. 403 f. 19—23.

Pag. 417. — Add to the species of 27. Gen. **Parelasmopus** Stebb.:

> 2. **P. setiger** Chevreux 1901 *P. s.*, Chevreux in: Mém. Soc. zool. France,
> *v.* 14 p. 412 f. 32—39.

Pag. 417. — Add to the literature of species 1. **Parelasmopus suluensis** (Dana):

> 1904 *P. s.*, A. O. Walker in: Herdman, Rep. Ceylon Pearl Fish., *v.* 2
> p. 278 t. 6 f. 38.

Pag. 421 and 428. — Add to the species of 30. Gen. **Melita** Leach:

> 13. **M. richardi** Chevreux 1900 *M. r.*, Chevreux in: Résult. Camp.
> Monaco, *v.* 16 p. 81 t. 10 f. 3.

> 14. **M. parvimana** S. J. Holmes 1903 *M. p.*, S. J. Holmes in: Amer.
> Natural., *v.* 37 p. 279 | 1905 *M. p.*, S. J. Holmes in: Bull. U. S. Bureau Fish.,
> *v.* 24 p. 506 f.

> 15. **M. zeylanica** Stebb. 1904 *M. z.*, T. Stebbing in: Spolia-Zeyl., *v.* 2
> p. 22 t. 5.

Pag. 429. — Add to the literature of species **Melita inaequistylis** (Dana):

> 1904 *Maera tenuicornis*, A. O. Walker in: Herdman, Rep. Ceylon Pearl
> Fish., *v.* 2 p. 273 t. 5 f. 33.

Pag. 433 and 439. — Add to the species of 33. Gen. **Maera** Leach:

> 13. **M. dubia** Calm. 1898 *M. d.*, Calman in: Ann. N. York Ac., *v.* 11
> p. 269 t. 32 f. 3.
>
> Near to M. furcicornis (p. 437) and M. fusca (p. 440).

> 14. **M. hirondellei** Chevreux 1900 *M. h.*, Chevreux in: Résult. Camp.
> Monaco, *v.* 16 p. 84 t. 11 f. 1.

> 15. **M. othonides** A. Walker 1904 *M. o.*, A. O. Walker in: Herdman,
> Rep. Ceylon Pearl Fish., *v.* 2 p. 271 t. 5 f. 29.
>
> Very near M. othonis (p. 438).

Pag. 441 and 445. — Add to the species of 34. Gen. **Elasmopus** A. Costa:

> 10. **E. insignis** Chevreux 1901 *E. i.*, Chevreux in: Mém. Soc. zool.
> France, *v.* 14 p. 406 f. 24—31.

> 11. **E. sokotrae** A. Walker & A. Scott 1903 *E. s.*, A. O. Walker & A.
> Scott in: H. O. Forbes, Nat. Hist. Sokotra, p. 223 t. 14 B f. 1—1 i.
>
> Near E. insignis Chevreux.

> 12. **E. dubius** A. Walker 1904 *E. d.*, A. O. Walker in: Herdman, Rep.
> Ceylon Pearl Fish., *v.* 2 p. 276 t. 5 f. 35.
>
> Compared with E. rapax (p. 444) and *Moera festiva* (p. 431).

13. **E. spinimanus** A. Walker 1904 *E. s.*, A. O. Walker in: Herdman, Rep. Ceylon Pearl Fish., *v.* 2 p. 277 t. 5 f. 36.

Compared with E. rapax and E. affinis (p. 444).

14. **E. serrula** A. Walker 1904 *E. s.*, A. O. Walker in: Herdman, Rep. Ceylon Pearl Fish., *v.* 2 p. 277 t. 8 f. 37.

Compared with E. rapax (p. 444).

15. **E. latibrachium** A. Walker 1905 *E. l.*, A. O. Walker in: Gardiner, Fauna Mald. Laccad., *v.* 2 p. 928 t. 88 f. 6—10.

Pag. 460 and 477. — Add to the species of 44. Gen. **Gammarus** F.:

31. **G. limnaeus** S. I. Sm. 1871 *G. lacustris* (non G. O. Sars 1863!), S. I. Smith in: Amer. J. Sci., ser. 3 *v.* 2 p. 453 | 1874 *G. limnaeus*, S. I. Smith in: Rep. U. S. Fish. Comm., *v.* 2 p. 651 t. 2 f. 6, 7.

Near G. pulex (p. 474).

32. **G. sarsii** Sowinski 1898 *G. s.*, Sowinski in: Bull. Ac. St.-Pétersb., *v.* 8 p. 369 t. 2 f. 7—16, t. 3 f. 13.

33. **G. australis** Sayce 1901 *G. a.*, Sayce in: P. R. Soc. Victoria, *v.* 13 p. 233 t. 39.

34. **G. tetrachantus** Garbini 1902 *G. t.*, Garbini in: Zool. Anz., *v.* 25 p. 153 f. 1.

35. **G. haasei** Sayce 1902 *G. h.*, Sayce in: P. R. Soc. Victoria, *v.* 15 p. 53 t. 6.

Compared with G. australis Sayce (p. 733) and Niphargus mortoni G. M. Thoms. (p. 731).

36. **G. propinquus** W. P. Hay 1902 *G. p.*, W. P. Hay in: P. U. S. Mus., *v.* 25 p. 224.

Compared with Dikerogammarus fasciatus (p. 460).

37. **G. purpurascens** W. P. Hay 1902 *G. p.*, W. P. Hay in: P. U. S. Mus., *v.* 25 p. 433 f. 7 a—n.

Compared with G. limnaeus S. I. Sm. (p. 733).

Pag. 462. — Add to the literature of species 1. **Gammarus guernei** Chevreux:

1900 *G. g.*, Chevreux in: Résult. Camp. Monaco, *v.* 16 p. 76 t. 10 f. 2.
?Near Niphargus rhipidiophorus (Catta) (p. 410).

Pag. 473. — Add to the literature of species 24. **Gammarus duebenii** Lilj.:

1897 *G. d. var. wilkitskii*, Birula in: Annuaire Mus. St.-Pétersb., p. 108.

Pag. 497. — Add to the literature of 48. Gen. **Parapallasea** Stebb.:

1901 *Pleuracanthus* (non Gray 1832, Coleoptera!) (part.), Garjajeff in: Trudui Kazan. Univ., *v.* 35 p. 16.

Pag. 498. — Add to the literature of species 1. **Parapallasea borowskii** (Dyb.):

1901 *Pleuracanthus b.* + *P. b. var. abyssalis* + *P. b. var. dichraas* (*Gammarus borawkii var. dichrou*, laps. pro: *G. borowskii var. dichrous*), Garjajeff in: Trudui Kazan. Univ., *v.* 35 p. 42, 43.

Pag. 498. — Add to the literature of species 2. **Parapallasea lagowskii** (Dyb.):

1901 *Pleuracanthus l.*, Garjajeff in: Trudui Kazan. Univ., *v.* 35 p. 43.

Pag. 498. — Add to the literature of species 3. **Parapallasea puzyllii** (Dyb.):

1901 *Pleuracanthus p.*, Garjajeff in: Trudui Kazan. Univ., *v.* 35 p. 43.

Pag. 501. — Add to the literature of 50. Gen. **Carinogammarus** Stebb.:

1901 *Ctenacanthus* (non Agassiz 1837, Pisces!) (part.), Garjajeff in: Trudui Kazan. Univ., *v.* 35 p. 15.

Pag. 502. — Add to the literature of species 2. **Carinogammarus wagii** (Dyb.):

1901 *Ctenacanthus w.*, Garjajeff in: Trudui Kazan. Univ., *v.* 35 p. 21.

Pag. 506. — Add to the literature of species **Carinogammarus mucronatus** (Say):

1905 *C. m.*, S. J. Holmes in: Bull. U. S. Bureau Fish., *v.* 24 p. 503 f. | 1905 *C. m.*, Paulmier in: Bull. N.-York Mus., Bull. 91 Zool. 12 p. 161 f. 30.

Pag. 501 and 506. — Add to the species of 50. Gen. **Carinogammarus** Stebb.:

C. roseus (Garjajeff) 1901 *Ctenacanthus r.*, Garjajeff in: Trudui Kazan. Univ., *v.* 35 p. 19 t. 2 f. 11, t. 3 f. 27—31.

C. ruber (Garjajeff) 1901 *Ctenacanthus r.*, Garjajeff in: Trudui Kazan. Univ., *v.* 35 p. 17 t. 2 f. 9, 10, t. 3 f. 22—26.

Pag. 508. — Add to the literature of 52. Gen. **Acanthogammarus** Stebb.:

1901 *Polyacanthus* (non Cuvier & Valenciennes 1831, Pisces!) + *Dybowskia* (non Dall 1876, Mollusca!) (part.) + *Ctenacanthus* (non Agassiz 1837, Pisces!) (part.), Garjajeff in: Trudui Kazan. Univ., *v.* 35 p. 15, 16.

Pag. 508 and 512. — Add to the species of 52. Gen. **Acanthogammarus** Stebb.:

7. **A. labbei** Chevreux 1903 *A. l.*, Chevreux in: Bull. Mus. Paris, *v.* 9 p. 224.

A. albus (Garjajeff) 1901 *Polyacanthus a.*, Garjajeff in: Trudui Kazan. Univ., *v.* 35 p. 29 t. 2 f. 14; t. 3 f. 41—45.

A. balkirii (Garjajeff) 1901 *Polyacanthus b.*, Garjajeff in: Trudui Kazan. Univ., *v.* 35 p. 21 t. 1 f. 1; t. 2 f. 15; t. 3 f. 46—51 | 1903 *Acanthogammarus b.*, Chevreux in: Bull. Mus. Paris, *v.* 9 p. 224.

A. flavus (Garjajeff) 1901 *Polyacanthus f.*, Garjajeff in: Trudui Kazan. Univ., *v.* 35 p. 25 t. 2 f. 12; t. 3 f. 32—36.

A. korotneffii (Garjajeff) 1901 *Polyacanthus k.*, Garjajeff in: Trudui Kazan. Univ., *v.* 35 p. 27 t. 2 f. 16; t. 3 f. 53—57.

A. maximus (Garjajeff) 1901 *Polyacanthus m.*, Garjajeff in: Trudui Kazan. Univ., *v.* 35 p. 23 t. 2 f. 13; t. 3 f. 37—40.

Pag. 509. — Add to the literature of species 1. **Acanthogammarus cabanisii** (Dyb.):

1901 *Ctenacanthus c.*, Garjajeff in: Trudui Kazan. Univ., *v.* 35 p. 21.

Pag. 509. — Add to the literature of species 2. **Acanthogammarus zieńkowiczii** (Dyb.):

1901 *Ctenacanthus z.*, Garjajeff in: Trudui Kazan. Univ., *v.* 35 p. 21.

Pag. 510. — Add to the literature of species 3. **Acanthogammarus godlewskii** (Dyb.):

1901 *Polyacanthus g.* + *P. g. var. victorii*, Garjajeff in: Trudui Kazan. Univ., *v.* 35 p. 31.

Pag. 510. — Add to the literature of species 4. **Acanthogammarus radosz-kowskii** (Dyb.):

1901 *Ctenacanthus r.*, Garjajeff in: Trudui Kazan. Univ., *v.* 35 p. 21.

Pag. 511.—Add to the literature of species 5. **Acanthogammarus armatus** (Dyb.):

1901 *Dybowskia armata* + *D. a. var. ongureni.* Garjajeff in: Trudui Kazan. Univ., *v.* 35. p. 38, 39 t. 2 f. 20; t. 3 f. 73—78.

Pag. 511. — Add to the literature of species 6. **Acanthogammarus parasiticus** (Dyb.):

1901 *Polyacanthus p.*, Garjajeff in: Trudui Kazan. Univ., *v.* 35 p. 31.

Pag. 515 and 517. — Add to the species of 1. Gen. **Dexamine** Leach:

5. **D. serraticrus** A. Walker 1904 *D. s.*, A. O. Walker in: Herdman, Rep. Ceylon Pearl Fish., *v.* 2 p. 265 t. 4 f. 24.

Near D. spinosa (p. 515).

Pag. 519 and 521. — Add to the species of 4. Gen. **Polycheria** Hasw.:

3. **P. atolli** A. Walker 1905 *P. a.*, A. O. Walker in: Gardiner, Fauna Mald. Laccad., *v.* 2 p. 926 t. 88 f. 1—5.

Pag. 521. — Add to the literature of species 1. **Guernea coalita** (Norm.):

?1904 *G. laevis*, A. O. Walker in: Herdman. Rep. Ceylon Pearl Fish., *v.* 2 p. 267 t. 4 f. 26.

Pag. 525. — Add to the literature of species 2. **Talitrus alluaudi** Chevreux:

1901 *T. a.*, Chevreux in: Mém. Soc. zool. France, *v.* 14 p. 389 f. 1—7.

Pag. 530 and 542. — Add to the species of 4. Gen. **Orchestia** Leach:

24. **O. anomala** Chevreux 1901 *O. a.*, Chevreux in: Mém. Soc. zool. France, *v.* 14 p. 393 f. 8—12.

Seems Very near Parorchestia hawaiensis (Dana), see p. 558 and 735.

25. **O. excavata** Chevreux 1902 *O. e.*, Chevreux in: Bull. Mus. Paris, *v.* 8 p. 521.

Pag. 545. — Add to the literature of species 3. **Talorchestia deshayesii** (Aud.):

1900 *Orchestoidea d.*, A. M. Norman in: Ann. nat. Hist., ser. 7 *v.* 5 p. 139.

Pag. 552. — Add to the literature of species 17. **Talorchestia brito** Stebb.:

1900 *Orchestoidea b.*, A. M. Norman in: Ann. nat. Hist., ser. 7 *v.* 5 p. 140.

Pag. 555. — Add to the species of 7. Gen. **Chiltonia** Stebb.:

2. **C. australis** (Sayce) 1901 *Hyalella a.*, Sayce in: P. R. Soc. Victoria, *v.* 13 p. 226 t. 36 | 1902 *Chiltonia a.*, Sayce in: P. R. Soc. Victoria, *v.* 15 p. 47.

3. **C. subtenuis** Sayce 1902 *C. s.*, Sayce in: P. R. Soc. Victoria, *v.* 15 p. 48 t. 4.

Pag. 558. — Add to the literature of species 2. **Parorchestia hawaiensis** (Dana):

1900 *Parorchestia h.*, T. Stebbing in: Fauna Haw., *v.* 2 p. 529 t. 21 c.

Pag. 559 and 572. — Add to the species of 11. Gen. **Hyale** H. Rathke:

23. **H. chiltoni** G. M. Thoms. 1899 *H. c.*, G. M. Thomson in: Tr. N. Zealand Inst., *v.* 31 p. 206.

24. **H. brevipes** Chevreux 1901 *H. b.*, Chevreux in: Mém. Soc. zool France, *v.* 14 p. 400 f. 15—18.

Pag. 564. — Add to the literature of species 9. **Hyale macrodactyla** Stebb.:

1901 *H. macrodactylus,* Chevreux in: Mém. Soc. zool. France, *v.* 14 p. 397 f. 13, 14.

Pag. 565. — Add to the literature of species 11. **Hyale prevostii** (M.-E.):

1900 *H. prevosti,* Chevreux in: Résult. Camp. Monaco, *v.* 16 p. 7 t. 1 f. 3 | 1903 *H. nilssoni,* A. O. Walker & A. Scott in: H. O. Forbes, Nat. Hist. Sokotra, p. 219 t. 14 A f. 3 a--3 e | 1904 *H. n. var.? kuriensis,* A. O. Walker in: Herdman, Rep. Ceylon Pearl Fish., *v.* 2 p. 238.

Pag. 567. — Add to the literature of species 14. **Hyale grimaldii** Chevreux:

1900 *H. g.,* Chevreux in: Résult. Camp. Monaco, *v.* 16 p. 10 t. 2 f. 2.

Pag. 570. — Add to the literature of species 19. **Hyale camptonyx** (Heller):

1900 *H. c.,* Chevreux in: Résult. Camp. Monaco, *v.* 16 p. 12 t. 2 f. 3.

Pag. 574. — Add to the literature of species **Hyale stebbingi** Chevreux:

1900 *H. s.,* Chevreux in: Résult. Camp. Monaco, *v.* 16 p. 8 t. 2 f. 1.

Pag. 574 and 580. — Add to the species of 12. Gen. **Hyalella** S. I. Sm.:

16. **H. richardi** Chevreux 1902 *H. r.,* Chevreux in: Bull. Soc. zool. France, *v.* 27 p. 223 f. 1, 2 a—e.

A littoral species from Isle of Alboran, between Spain and Morocco.

17. **H. neveu-lemairei** Chevreux 1904 *H. n.-l.,* Chevreux in: Bull. Soc. zool. France, *v.* 29 p. 131 f. 1, 2 a—d.

From Lake Titicaca.

18. **H. pernix** (Moreira) 1903 *Allorchestes p.,* C. Moreira in: Arch. Mus. Rio Jan., *v.* 12 p. 187 t. 1, 2.

Rio de Janeiro, Brazil, 2240 m above sea-level.

Pag. 585. — Add to the literature of species **Allorchestes littoralis** Stimps.:

1905 *A. l., Hyale l.,* S. J. Holmes in: Bull. U. S. Bureau Fish., *v.* 24 p. 472 f.; t. 3 f. 2.

Pag. 586 and 603. — Add to the genera of 33. Fam. **Aoridae:**

8. Gen. **Coremapus** Norm.

1905 *Coremapus* (Sp. typ.: *C. versiculatus*), A. M. Norman in: Ann. nat. Hist., ser. 7 *v.* 16 p. 78.

1. **C. versiculatus** (Bate) 1905 *C. v.,* A. M. Norman in: Ann. nat. Hist., ser. 7 *v.* 16 p. 78.

Add to this species the literature of **Microdeutopus versiculatus** (Bate) from p. 593.

Pag. 593. — Add to the literature of species 10. **Microdeutopus damnoniensis** (Bate):

1905 *M. danmonensis,* S. J. Holmes in: Bull. U. S. Bureau Fish., *v.* 24 p. 515 f.

Pag. 594 and 599. — Add to the species of 4. Gen. **Lembos** Bate:

10. **L. smithi** (S. J. Holmes) 1903 *Autonoe s.*, S. J. Holmes in: Amer. Natural., *v.* 37 p. 290 | 1905 *A. s.*, S. J. Holmes in: Bull. U. S. Bureau Fish., *v.* 24 p. 516 f.

11. **L. podoceroides** A. Walker 1904 *L. p.*, A. O. Walker in: Herdman, Rep. Ceylon Pearl Fish., *v.* 2 p. 279 t. 6 f. 39.

Near *L.* websterii (p. 599).

12. **L. chelatus** A. Walker 1904 *L. c.*, A. O. Walker in: Herdman, Rep. Ceylon Pearl Fish., *v.* 2 p. 280 t. 6 f. 40.

Pag. 603 and 630. — Add to the genera of 34. Fam. **Photidae** the following 4 genera:

11. Gen. **Cheiriphotis** A. Walker

1904 *Cheiriphotis* (Sp. un.: *C. megacheles*), A. O. Walker in: Herdman, Rep. Ceylon Pearl Fish., *v.* 2 p. 234, 283.

Near Microprotopus Norm. (p. 604).

1. **C. megacheles** (Giles) 1885 *Melita m.* (♂), G. M. Giles in: J. Asiat. Soc. Bengal, *v.* 54 p. 70 t. 3 | 1887 *Eurystheus hirsutus* (♀), G. M. Giles in: J. Asiat. Soc. Bengal, *v.* 56 p. 227 t. 8 | 1904' *Cheiriphotis megacheles*, A. O. Walker in: Herdman, Rep. Ceylon Pearl Fish., *v.* 2 p. 284 t. 6 f. 42.

Add to this species the literature of **Eurystheus hirsutus** Giles from p. 615.

12. Gen. **Audulla** Chevreux

1901 *Audulla* (Sp. un.: *A. chelifera*), Chevreux in: Mém. Soc. zoöl. France, *v.* 14 p. 431.

1. **A. chelifera** Chevreux 1901 *A. c.*, Chevreux in: Mém. Soc. zool. France, *v.* 14 p. 432 f. 56—65 | 1903 *A. c.*, A. O. Walker & A. Scott in: H. O. Forbes, Nat. Hist. Sokotra, p. 225 t. 14 B f. 2 a, b.

13. Gen. **Bonnierella** Chevreux

1900 *Bonnierella* (Sp. un.: *B. abyssi*), Chevreux in: Résult. Camp. Monaco, *v.* 16 p. 97.

1. **B. abyssi** (Chevreux) 1900 *B. a.*, Chevreux in: Résult. Camp. Monaco, *v.* 16 p. 97 t. 11 f. 3.

Add to the literature of this species that of **Podoceropsis abyssi** Chevreux from p. 619, except *Gammaropsis abyssorum* J. Bonnier.

2. **B. abyssorum** (Bonnier) 1900 *B. a.*, Chevreux in: Résult. Camp. Monaco, *v.* 16 p. 99.

14. Gen. **Chevalia** A. Walker

1904 *Chevalia* (Sp. un.: *C. aviculae*), A. O. Walker in: Herdman, Rep. Ceylon Pearl Fish., *v.* 2 p. 234, 288.

Intermediate between Fam. Photidae and Fam. Corophiidae.

1. **C. aviculae** A. Walker 1904 *C. a.*, A. O. Walker in: Herdman, Rep. Ceylon Pearl Fish., *v.* 2 p. 288 t. 7 f. 50; t. 8 f. 50.

Pag. 606 and 609. — Add to the species of 2. Gen. **Photis** Krøyer:

    6 P. longimana A. Walker 1904 *P. longimanus*, A. O. Walker in: Herdman, Rep. Ceylon Pearl Fish., *v.* 2 p. 286 t. 7 f. 44.

    P. nana A. Walker 1904 *P. n.*, A. O. Walker in: Herdman, Rep. Ceylon Pearl Fish., *v.* 2 p. 287 t. 7 f. 45.

    Generic position doubtful.

Pag. 611 and 618. — Add to the species of 4. Gen. **Eurystheus** Bate:

    14. E. dentatus (Chevreux) 1900 *Gammaropsis dentata*, Chevreux in: Résult. Camp. Monaco, *v.* 16 p. 93 t. 12 f. 1.

    15. E. tenuicornis (S. J. Holmes) 1904 *Gammaropsis t.*, S. J. Holmes in: Harriman Alaska Exp., *v.* 10 p. 239 f. 124.

    Compared with Megamphopus (*Megamophus*, laps.!) cornutus Norm. (p. 621).

    16. E. zeylanicus (A. Walker) 1904 *Gammaropsis z.*, A. O. Walker in: Herdman, Rep. Ceylon Pearl Fish.. *v.* 2 p. 282 t. 6 f. 41.

    17. E. gardineri (A. Walker) 1905 *Gammaropsis g.*, A. O. Walker in: Gardiner, Fauna Mald. Laccad., *v.* 2 p. 929 t. 88 f. 11—14, 16, 17.

Pag. 623 and 624. — Add to the species of 8. Gen. **Protomedeia** Krøyer:

    2. P. grandimana Brüggen 1905 *P. g.*, Brüggen in: Trav. Soc. St.-Pétersb., *v.* 36 nr. 1 p. 6, 9; t. f. 5.

Pag. 627. — Add to the literature of species 3. **Leptocheirus pinguis** (Stimps.):

    1905 *Ptilocheirus p.*, S. J. Holmes in: Bull. U. S. Bureau Fish., *v.* 24 p. 522 f.; t. 12 f. 3.

Pag. 632 and 641. — Add to the species of 1. Gen. **Ampithoe** Leach:

    18. A. alluaudi (Chevreux) 1901 *Amphithoe a.*, Chevreux in: Mém. Soc. zool. France, *v.* 14 p. 418 f. 40—45.

    19. A. intermedia (A. Walker) 1904 *Amphithoë i.*, A. O. Walker in: Herdman, Rep. Ceylon Pearl Fish., *v.* 2 p. 290 t. 7 f. 46.

    Between A. rubricata (Mont.) and A. vaillantii (H. Luc.) (p. 639).

Pag. 634. — Add to the literature of species 4. **Ampithoe longimana** (S. I. Sm.):

    1905 *Amphithoë l.*, S. J. Holmes in: Bull. U. S. Bureau Fish., p. 509 f.; t. 13 f. 2.

Pag. 635. — Add to the literature of species 6. **Ampithoe valida** (S. I. Sm.):

    1905 *Amphithoe v.*, Paulmier in: Bull. N. York Mus., Bull. 91 Zool. 12, p. 164 f.

Pag. 642 and 643. — Add to the species of 2. Gen. **Pleonexes** Bate:

    2. P. ferox Chevreux 1902 *P. f.*, Chevreux in: C.-R. Ass. Franç., Sess. 30 *v.* 2 p. 697 t. 5 f. 2a—2i.

Pag. 644. — Add to the species of 3. Gen. **Grubia** Czern.:

    2. G. hirsuta Chevreux 1900 *G. h.*, Chevreux in: Bull. Soc. zool. France, *v.* 25 p. 95 f. 1—5.

    3. G. microphthalma Chevreux 1901 *G. m.*, Chevreux in: Mém. Soc. zool. France, *v.* 14 p. 422 f. 46—49 | 1905 *G. m.*, A. O. Walker in: Gardiner, Fauna Mald. Laccad., *v.* 2 p. 930 f. 142.

    4. G. longicornis A. Walker & A. Scott 1903 *G. l.*, A. O. Walker & A. Scott in: H. O. Forbes, Nat. Hist. Sokotra, p. 226 t. 14 B f. 3a—3d.

    Perhaps a synonym of Amphithoides longicornis Kossm. (p. 645).

Pag. 644 and 645. — **Amphithoides comptus** (S. I. Sm.) is to be accepted among the species of the 3. Gen. **Grubia** Czern.: 5. **G. compta** (S. I. Sm.). Add to the literature of this species:

> 1905 *G. c.*, S. J. Holmes in: Bull. U. S. Bureau Fish., *v.* 24 p. 510 f.

Pag. 645. — Put after the 4. Gen. **Amphithoides** Kossm.:

4*. Gen. Paragrubia Chevreux

> 1901 *Paragrubia* (Sp. un.: *P. vorax*), Chevreux in: Mém. Soc. zool. France, *v.* 14 p. 426.

> 1. **P. vorax** Chevreux 1901 *P. v.*, Chevreux in: Mém. Soc. zool. France, *v.* 14 p. 427 f. 50—55.

Pag. 645. — Add to the literature of species 1. **Sunamphitoe pelagica** (M.-E.):

> 1900 *Sunamphithoe p.*, Chevreux in: Résult. Camp. Monaco, *v.* 16 p. 102 t. 11 f. 4.

Pag. 652. — Add to the literature of 4. Gen. **Jassa** Leach:

> 1905 *Bruzeliella* (Sp. typ.: *B. falcata*), A. M. Norman in: Ann. nat. Hist., ser. 7 *v.* 16 p. 83.

Pag. 654. — Add to the literature of species 4. **Jassa pulchella** Leach:

> 1816 *Jassa p.*, Leach in: Enc. Brit., ed. 5 suppl. p. 426 (with no other species of *Jassa*) | 1829 *J. p.*, Latreille in: G. Cuvier. Règne an., n. ed. *v.* 4 p. 122 (with no other species of *Jassa*) | 1905 *Bruzeliella falcata*, A. M. Norman in: Ann. nat. Hist., ser. 7 *v.* 16 p. 83, 92.

Pag. 655. — Add to the literature of species 5. **Jassa pusilla** (O. Sars):

> 1905 *Bruzeliella p.*, A. M. Norman in: Ann. nat. Hist., ser. 7 *v.* 16 p. 84, 92.

Pag. 655. — Add to the literature of species 6. **Jassa ocius** (Bate):

> 1905 *Bruzeliella ocia*, A. M. Norman in: Ann. nat. Hist., ser. 7 *v.* 16 p. 84, 92.

Pag. 652 and 656. — Add to the species of 4. Gen. **Jassa** Leach:

> 8. **J. goniamera** A. Walker 1903 *J. g.*, A. O. Walker in: J. Linn. Soc., *v.* 29 p. 61 t. 11 f. 98—106 a.

> 9. **J. marmorata** S. J. Holmes 1903 *J. m.*, S. J. Holmes in: Amer. Natural., *v.* 37 p. 289 | 1905 *J. m.*, S. J. Holmes in: Bull. U. S. Bureau Fish., *v.* 24 p. 511 f.

Pag. 657 and 662. — Add to the species of 5. Gen. **Ischyrocerus** Kröyer:

> 10. **I. commensalis** Chevreux 1900 *I. c.*, Chevreux in: Résult. Camp. Monaco, *v.* 16 p. 104 t. 12 f. 2.

Pag. 662 and 693. — Add to the genera of 38. Fam. **Corophiidae**:

12. Gen. Grandidierella Coutière

> 1904 *Grandidierella* (Sp. un.: *G. mahafalensis*), Coutière in: Bull. Soc. philom., ser. 9 *v.* 6 p. 166.

> 1. **G. mahafalensis** Coutière 1904 *G. m.*, Coutière in: Bull. Soc. philom., ser. 9 *v.* 6 p. 166 f. 1—19.

> Compared with Camacho bathyplous Stebb. (p. 665), and Chevreuxius grandimanus Bonnier (p. 670).

Pag. 667. — Add to the literature of species 3. **Cerapus tubularis** Say:

1905 *C. t*, S. J. Holmes in: Bull. U. S. Bureau Fish., *v.* 24 p. 517 f.

Pag. 668 and 669. — Add to the literature of species 6. **Cerapus calamicola** (Giles) the literature of species 4. **C. flindersi** Stebb. (♀), which is to be dropped, and:

1892 *Cerapus flindersi* (♂), Chilton in: Rec. Austral. Mus., *v.* 2 p. 1 t. 1 | 1904 *C. calamicola*, A. O. Walker in: Herdman, Rep. Ceylon Pearl Fish., *v.* 2 p. 293.

Pag. 671 and 674. — Add to the species of 6. Gen. **Ericthonius** M.-E.:

7. **E. rubricornis** (Stimps.)   1905 *Erichthonius r.*, S. J. Holmes in: Bull. U. S. Bureau Fish., *v.* 24 p. 518 f.

Add to the literature of this species from species 5. **Ericthoniųs hunteri** (Bate) (p. 673):

1853 *Cerapus rubricornis*, Stimpson in: Smithson. Contr., *v.* 6 nr. 5 p. 46 f. 33 | 1867 *C. r., C. rubiformis*, Packard in: Mem. Boston Soc., *v.* 1 p. 297.

Pag. 671. — Add to the literature of species 1. **Ericthonius brasiliensis** (Dana):

1905 *Erichthonius minax*, S. J. Holmes in: Bull. U. S. Bureau Fish., *v.* 24 p. 519 f.

Pag. 672. — Add to the literature of species 3. **Ericthonius macrodactylus** (Dana):

1904 *Erichthonius m.*, A. O. Walker in: Herdman, Rep. Ceylon Pearl Fish., *v.* 2 p. 292 t. 7 f. 48.

Pag. 682 and 684. — Add to the species of 10. Gen. **Siphonoecetes** Krøyer:

6. **S. orientalis** A. Walker 1904 *S. o.*, A. O. Walker in: Herdman, Rep. Ceylon Pearl Fish., *v.* 2 p. 294 t. 7 f. 49.

7. **S. smithianus** Rathbun 1873 *S. cuspidatus* (non Aug. Metzger 1871!), (S. I. Smith in:) A. E. Verrill in: Rep. U. S. Fish Comm., *v.* 1 p. 501, 566 | 1905 *S. smithianus*, M. J. Rathbun in: Pap. Boston Soc., *v.* 7 p. 74.

Pag. 683. — Subtract from the literature of species 3. **Siphonoecetes colletti** Boeck:

1873 *S. cuspidatus* (non Aug. Metzger 1871!), (S. I. Smith in:) A. E. Verrill in: Rep. U. S. Fish Comm., *v.* 1 p. 501, 566.

Pag. 685 and 692. — Add to the species of 11. Gen. **Corophium** Latr.:

13. **C. maeoticum** Sowinski 1898 *C. m.*, Sowinski in: Bull. Ac. St. Pétersb., *v.* 8 p. 362 t. 1 f. 1—5.

14. **C. triaenonyx** Stebb. 1904 *C. t.*, T. Stebbing in: Spolia Zeyl., *v.* 2 p. 25 t. 6 A.

Pag. 692. — Add to the literature of species **Corophium acherusicum** A. Costa:

1900 *C. a.*, Chevreux in: Résult. Camp. Monaco, *v.* 16 p. 109.

Pag. 692. — Add to the literature of species **Corophium cylindricum** (Say):

1905 *C. c.*, S. J. Holmes in: Bull. U. S. Bureau Fish., *v.* 24 p. 521 f. | 1905 *C. c.*, Paulmier in: Bull. N.-York Mus., Bull. 91 Zool. 12 p. 167 f. 37.

Pag. 701 and 706. — Add to the species of 5. Gen. **Podocerus** Leach:

    11. **P. synaptochir** (A. Walker) 1904 *Platophium s.*, A. O. Walker in: Herdman, Rep. Ceylon Pearl Fish., *v.* 2 p. 296 t. 8 f. 52.

    12. **P. zeylanicus** (A. Walker) 1904 *Platophium zeylanicum*, A. O. Walker in: Herdman, Rep. Ceylon Pearl Fish., *v.* 2 p. 297 t. 8 f. 53.

Pag. 703. — Add to the literature of species 5. **Podocerus chelonophilus** (Chevreux & Guerne):

    1900 *Platophium chelonophilum, Podocerus chelonophilus,* Chevreux in: Résult. Camp. Monaco, *v.* 16 p. 115 t. 13 f. 2, t. 14 f. 7; p. 166.

Pag. 704. — Add to the literature of species 8. **Podocerus laevis** (Hasw.):

    1904 *Platophium laeve*, A. O. Walker in: Herdman, Rep. Ceylon Pearl Fish., *v.* 2 p. 295 t. 7 f. 51.

Pag. 714 and 715. — Add to the species of 1. Gen. **Hyperiopsis** O. Sars:

    2. **H. australis** A. Walker 1906 *H. a.*, A. O. Walker in: Ann. nat. Hist., ser. 7 *v.* 17 p. 454.

Liljeborgia aequabilis Stebb. 1888 *L. a.*, T. Stebbing in: Rep. Voy. Challenger, *v.* 29 p. 988.

Moera approximans Hasw. 1879 *M. a.*, Haswell in: P. Linn. Soc. N. S. Wales, *v.* 4 p. 334 t. 21 f. 3.

Cerapus bidens Czern. 1868 *C. b.*, Czerniavski in: Syezda Russ. Est., Syezda 1 Zool. p. 98.

Amphithoe boecki Della Valle 1893 *A. b.*, A. Della Valle in: F. Fl. Neapel, *v.* 20 p. 462, 918 t. 57 f. 36.

Maera bruzelii Stebb. 1888 *M. b.*, T. Stebbing in: Rep. Voy. Challenger, *v.* 29 p. 1014 t. 97.

Orchestia californiensis Dana 1854 *O. c.*, J. D. Dana in: P. Ac. Philad., *v.* 7 p. 177.

Lada chalubinskii Wrześn. 1879 *L. chalubińskii*, Wrześniowski in: Zool. Anz., *v.* 2 p. 322.

Atylus comes Giles 1888 *A. c.*, G. M. Giles in: J. Asiat. Soc. Bengal, *v.* 57 p. 243 t. 10 f. 8—10.

Amphithoe costata M.-E. 1830 *A. c.*, H. Milne Edwards in: Ann. Sci. nat., *v.* 20 p. 379 t. 10 f. 14—16.

Calliope didactyla G. M. Thoms. 1879 *C. d.*, G. M. Thomson in: Tr. N. Zealand Inst., *v.* 11 p. 240 t. 10 c f. 3 a—e.

Maera diversimanus Miers 1884 *M. d.*, Miers in: Rep. Voy. Alert, p. 567 t. 52 f. D.

Amphitopsis dubia Vosseler 1889 *A. d.*, Vosseler in: Arch. Naturg., *v.* 551 p. 156 t. 8 f. 32—36.

Enone Risso 1826 *E.* (Sp. un.: *E. punctata*), A. Risso. Hist. nat. Eur. mérid., *v.* 5 p. 96.

Tryphosa erosa Meinert 1890 *T. e.*, Meinert in: Udb. Hauchs, *v.* 3 p. 155 t. 1 f. 25—29.

Maera fasciculata (G. M. Thoms.) 1880 *Megamaera f.*, G. M. Thomson in: Ann. nat. Hist., ser. 5 *v.* 6 p. 5 t. 1 f. 5 | 1893 *Maera f.*, G. M. Thomson in: P. R. Soc. Tasmania, 1892 p. 28.

Orchestia gayi Nic. 1849 *O. g.*, H. Nicolet in: Gay, Hist. Chile, *v.* 3 p. 234.

Iphimediopsis geniculata Della Valle 1893 *I. g., I. eblanae*, A. Della Valle in: F. Fl. Neapel, *v.* 20 p. 19, 933.

Podoprionides incerta A. Walker 1906 *P. i.*, A. O. Walker in: Ann. nat. Hist., ser. 7 *v.* 17 p. 457.

Melita insatiabilis Fr. Müll. 1864 *M. i.*, Fritz Müller, Für Darwin, p. 18 f. 11.

Lada Wrześn. 1879 *L.* (Sp. un.: *L. chalubińskii*), Wrześniowski in: Zool. Anz., *v.* 2 p. 322.

Melita leonis J. Murdoch 1885 *M. l.*, J. Murdoch in: P. U. S. Mus., *v.* 7 p. 521.

Amphithonotus levis G. M. Thoms. 1879 *A. l.*, G. M. Thomson in: Ann. nat. Hist., ser. 5 *v.* 4 p. 330 t. 16 f. 1—4.

Melita messalina Fr. Müll. 1864 *M. m.*, Fritz Müller, Für Darwin, p. 18 f. 10.
Gammarus multifasciatus Bate 1862 *G. m.*, (Stimpson in MS.) Bate, Cat. Amphip.
     Brit. Mus., p. 211 t. 37 f. 7.
Gammarus nolens Johnst. 1827 *G. n.*, G. Johnston in: Zool. J., *v.* 3 p. 179 | 1857
     . . *Anonyx? n.*, A. White, Hist. Brit. Crust., p. 169.
Protomedeia nordmannii (M.-E.) 1862 *P. n.*, Bate, Cat. Amphip. Brit. Mus., p. 171.
Ichthyomyzocus ornatus E. Hesse 1873 *I. o.*, E. Hesse in: Ann. Sci. nat., ser. 5 *v.* 17
     nr. 7 p. 5 t. 4 f. 1.
Concholestes pallidus (O. Sars) 1890 *C. p.*, G. M. Giles in: J. Asiat. Soc. Bengal,
     v. 59 p. 64.
Pleustes parvus Boeck 1876 *P. p.*, A. Boeck, Skand. Arkt. Amphip., *v.* 2 p. 305 t. 23
     f. 5 (*? P. panoplus* juv.).
Cerapus pelagicus (Leach) 1840 *C. p.*, H. Milne Edwards, Hist. nat. Crust., *v.* 3 p. 61.
Gammarus plumicornis A. Costa 1853 *G. p.*, A. Costa in: Rend. Soc. Borbon., n. ser.
     v. 2 p. 176.
Melita podager (M.-E.) 1862 *M. p.*, Bate, Cat. Amphip. Brit. Mus., p. 184 t. 33 f. 5.
Podoprionides A. Walker 1906 *P.* (Sp. un.: *P. incerta*), A. O. Walker in: Ann.
     nat. Hist., ser. 7 *v.* 17 p. 457.
Grayia pugettensis (Dana) 1862 *G. p.*, Bate, Cat. Amphip. Brit. Mus., p. 101 t. 14 f. 4.
Oedicerus puliciformis Giles 1888 *O. p.*, G. M. Giles in: J. Asiat. Soc. Bengal,
     v. 57 p. 248 t. 7 f. 5, 6.
Enone punctata (Risso) 1826 *E. p.*, A. Risso, Hist. nat. Eur. mérid., *v.* 5 p. 96.
Gammarus redmanni Bate 1862 *G. r.*, (Leach in MS.) Bate, Cat. Amphip. Brit.
     Mus., p. 212 t. 37 f. 9.
Gammarus robustus S. I. Sm. 1874 *G. r.*, S. I. Smith in: Rep. U. S. geol. Surv.
     Terr., 1873 p. 610 t. 2 f. 7—12.
Talitrus rubropunctatus Risso 1816 *T. r.*, Risso, Crust. Nice, p. 127.
Gammarus spetsbergensis Vosseler 1889 *G. s.*, Vosseler in: Arch. Naturg., *v.* 55ı
     p. 158 t. 8 f. 25—31.
Lysianassa spinifera Stimps. 1853 *L. s.*, Stimpson in: Smithson. Contr., *v.*. 6 nr. 5 p. 49.
Gammarus spinipes Johnst. 1829 *G. s.*, G. Johnston in: Zool. J., *v.* 4 p. 417.
Ichthyomyzocus squatinae E. Hesse 1873 *I. s.*, E. Hesse in: Ann. Sci. nat., ser. 5 *v.* 17
     nr. 7 p. 12 t. 4 f. 19—27.
Oniscus stroemianus O. Fabr. 1780 *O. s.*, O. Fabricius, Fauna Groenl., p. 261.
Gammarus subtener Stimps. 1864 *G. s.*, Stimpson in: P. Ac. Philad., p. 157.
Talitrus tripudians Krøyer 1845 *T. t.*, Krøyer in: Naturh. Tidsskr., ser. 2 *v.* 1 p. 311 t. 3
     f. 2 a—e.
Gammarus truncatus Viv. 1805 *G. t.*, Viviani, Phosphor. Maris, p. 8 t. 2 f. 5, 6.
Gammarus unguiserratus A. Costa 1853 *G. u.*, A. Costa in: Rend. Soc. Borbon.,
     n. ser. *v.* 2 p. 176.
Iphimedia vulgaris Stimps. 1853 *I. v.*, Stimpson in: Smithson. Contr., *v.* 6 nr. 5 p. 53.

Alphabetical Index.

Nomenclator generum et subgenerum.

Acanthechinus Thomas R. R. Stebbing in: Rep. Voy. Challenger, v. 29 p. 883. 1888. Sp.: *A. tricarinatus.* „ἄκανθα, a spine, and ἐχῖνος, a hedge-hog or sea-urchin".

Acanthogammarus Thomas R. R. Stebbing in: Tr. Linn. Soc. London, ser. 2 v. 7 p. 430. 1899 V. Sp.: *A. cabanisii, A. zieńkowiczii, A. godlewskii, A. rodoszkowskii, A. armatus, A. parasiticus.* „alludes to the dentate carinae".

Acanthonotosoma [pro: *Acanthonotozoma'* A. Boeck 1876]. W. S. M. D'Urban in: Ann. nat. Hist., ser. 5 v. 6 p. 255. 1880.

Acanthonotozoma Axel Boeck, Skand. Arkt. Amphip., v. 2 p. 237. 1876. Sp. typ.: *Acanthonotus cristatus.* „ἄκανθα (Torn), νῶτος (Ryg), σῶμα (Legeme)".

Acanthonotus ([Richard] Owen in MS.) James Clark Ross in: John Ross, App. sec. Voy., nat. Hist. p. 90. 1835. Sp.: *A. cristatus.*

Acanthosoma ([Richard] Owen in MS.) James Clark Ross in: John Ross, App. sec. Voy., nat. Hist. p. 91. 1835. Sp.: *A. hystrix.*

Acanthostepheia Axel Boeck in: Forh. Selsk. Christian., 1870 p. 163. 1871. Sp.: *A. malmgreni.*

Acanthostephia [pro: *Acanthostepheia* A. Boeck 1871]. Eduard von Martens in: Zool. Rec., v. 8 p. 190. 1873.

Acanthozoma pro: *Acanthonotozoma* A. Boeck 1876. Axel Boeck, Skand. Arkt. Amphip., v. 2 p. 229, 712. 1876.

Acanthozone Axel Boeck in: Forh. Selsk. Christian., 1870 p. 184. 1871. Sp.: *A. cuspidata.*

Aceroides G. O. Sars, Crust. Norway, v. 1 p. 340 t. 120. 1892. Sp.: *Aceropsis latipes.*

Aceropsis Anton Stuxberg in: Bih. Svenska Ak., v. 5 nr. 22 p. 63. 1880. [*nom. nud.*]

Aceropsis pro: *Aceroides* O. Sars 1892. G. O. Sars, Crust. Norway, v. 1 t. 120. 1892.

Aceros Axel Boeck in: Forh. Skand. Naturf., Møde 8 p. 651. 1861. Sp. typ.: *Oediceros obtusus.*

Acidostoma William Lilljeborg in: N. Acta Soc. Upsal., ser. 3 v. 6 nr. 1 p. 18 (tabell.), 34. 1865. Sp.: *A. obesum.* „From ἀκίς a point and στόμα mouth".

Acidostomum pro: *Acidostoma* W. Lilljeborg 1865. Antonio Della Valle in: F. Fl. Neapel, v. 20 p. 19, 916. 1893.

Acontiostoma Thomas R. R. Stebbing in: Rep. Voy. Challenger, v. 29 p. 709. 1888. Sp. typ.: *A. marionis.* „from ἀκόντιον, a dart, στόμα, a mouth".

Actinacanthus pro: *Acanthechinus* T. Stebbing 1888. T. R. R. Stebbing in: Tierreich, v. 21 p. 326. 1906 IX. „ἀκτίς, ray, ἄκανθα, spine".

Aedicerus [pro: *Oediceros* Krøyer 1842]. William A. Haswell, Cat. Austral. Crust., p. 238, 315. 1882.

Aglaura C. S. Rafinesque, Anal. Nat., p. 101. 1815. [*nom. nud.*]

Alibrotus [Henri] Milne Edwards, Hist. nat. Crust., v. 3 p. 23. 1840. Sp.: *A. chauseicus.*

Alicella Ed. Chevreux in: Bull. Soc. zool. France, v. 24 p. 154. 1899. Sp.: *A. gigantea.*

Allorchestes Jacobus D. Dana in: Amer.
J. Sci., ser. 2 *v.* 8 p. 136. 1849 VII. —
Jacobus D. Dana in: P. Amer. Ac.,
v. 2 p. 205. 1852. Sp.: *A. compressa,
A. verticillata, A. hirtipalma, A. gracilis,
A. peruviana, A. humilis, A. australis,
A. brevicornis, A. novi-zealandiae, A.
intrepida, A. orientalis, ? A. graminea.*

Allorchestina *Subgen.* J. F. Brandt in:
Bull. phys.-math. Ac. St.-Pétersb., *v.* 9
p. 141. 1851 I 9. Sp.: *Orchestia nidro-
siensis, ? O. perieri.*

Amanonyx C. Spence Bate in: Rep. Brit.
Ass., Meet. 25 p. 58. 1856. Sp.: *A.
guerinianus.* [*nom. nud.*]

Amaryllis William A. Haswell in: P.
Linn. Soc. N. S. Wales, *v.* 4 p. 253. 1879.
Sp.: *A. macrophthalmus, A. brevicornis.*

Amathia Heinrich Rathke in: Mém.
prés. Ac. St.-Pétersb., *v.* 3 p. 375. 1837.
Sp.: *A. carinata.* „nach einer Meer-
nympfe".

Amathilla pro: *Amathia* H. Rathke 1837.
C. Spence Bate and J. O. Westwood,
Brit. sess. Crust., *v.* 1 p. 359. 1862 VII 1.

Amathillina G. O. Sars in: Bull. Ac.
St.-Pétersb., ser. 5 *v.* 1 p. 201, 203. 1894.
Sp. typ.: *A. cristata.*

Amathillinella pro: *Amathillina* G. O.
Sars 1894. ([O.] Grimm in MS.) G. O.
Sars in: Bull. Ac. St.-Pétersb., ser. 5
v. 1 p. 201. 1894.

Amathillopsis Camil Heller in: Denk.
Ak. Wien, *v.* 35 p. 35. 1875. Sp.: *A.
spinigera.*

Ambasia Axel Boeck in: Forh. Selsk.
Christian., 1870 p. 97. 1871. Sp.: *A.
danielssenii.*

Ampelisca Henrik Krøyer in: Naturh.
Tidsskr., *v.* 4 p. 154. 1842. Sp.: *A.
eschrichtii.* „Nomen mulieris apud
Plautum in Rudente".

Ampelisia [pro: *Ampelisca* Krøyer 1842].
H. Krøyer in: Voy. Nord, Crust. t. 23
f. 1. 1846.

Amphilochoides G. O. Sars, Crust.
Norway, *v.* 1 p. 220. 1892. Sp. typ.:
Amphilochus odontonyx.

Amphilochus C. Spence Bate & J. O.
Westwood, Brit. sess. Crust., *v.* 1
p. 179. [1862 1.] Sp.: *A. manudens.*

Amphithoë [pro: *Ampithöe* Leach 1813/14].
[Pierre André] Latreille in: Nouv.
Dict., ed. 2 *v.* 1 p. 470. 1816.

Amphithoïdes Robby Kossmann, Reise
Roth. Meer., *v.* 21 Malacost. p. 135. 1880.
Sp.: *A. longicornis.*

Amphithonotus (A. Costa in:) F. G.
Hope, Cat. Crost. Ital., p. 45. 1851.
Sp.: *A. marionis, A. panopla, A. carinatus, A. acanthophthalmus, Acanthonotus guttatus.*

Amphithopsis Axel Boeck in: Forh.
Skand. Naturf., Møde 8 p. 661. 1861. Sp.:
*A. bicuspis, A. elegans, A. laeviuscula,
A. tridentata, A. glaber, A. longicaudata.*

Amphitoe [pro: *Ampithöe* Leach 1813/14].
F. E. Guérin-Méneville, Iconogr.
Règne an., *v.* 3 Crust. p. 23. 1836.

Amphitonotus [pro: *Amphithonotus* A.
Costa 1851]. Adolph Eduard Grube,
Ausfl. Triest, p. 136. 1861.

Amphitopsis [pro: *Amphithopsis* A. Boeck
1861]. Julius Vosseler in: Arch.
Naturg., *v.* 551 p. 156. 1889.

Ampithöe [William Elford Leach in:]
Edinb. Enc., *v.* 7 p. 403, 432. [1813/14.]
Sp.: *A. rubricata.*

Amplisca [pro: *Ampelisca* Krøyer 1842].
Édouard Chevreux in: Bull. Soc.
zool. France, *v.* 12 p. 574. 1887.

Anamixis Thomas R. R. Stebbing in:
Tr. Linn. Soc. London, ser. 2 *v.* 7 p. 35.
1897 V. Sp.: *A. hanseni.*

Andania Axel Boeck in: Forh. Selsk.
Christian., 1870 p. 128. 1871. Sp.: *A.
abyssi, A. nordlandica.*

Andaniella G. O. Sars, Crust. Norway, *v.* 1
p. 210. 1891. Sp.: *A. pectinata.*

Andaniexis pro: *Andania* A. Boeck 1871.
T. R. R. Stebbing in: Tierreich. *v.* 21
p. 94. 1906 IX.

Andaniodes [pro: *Andaniotes* T. Stebbing
1897]. J. Victor Carus in: Zool. Anz.,
Bibliogr. *v.* 2 p. 622. 1897 XI 29.

Andaniopsis G. O. Sars, Crust. Norway,
v. 1 p. 208. 1891. Sp.: *A. nordlandica.*

Andaniotes Thomas R. R. Stebbing in:
Tr. Linn. Soc. London, ser. 2 *v.* 7 p. 30.
1897 V. Sp.: *A. corpulentus.*

Anisopus Robert Templeton in: Tr. ent.
Soc. London, *v.* 1 p. 185. 1836. Sp.:
A. dubius.

50*

Anonyx Henrik Kröyer in: Danske Selsk. Afh., *v.* 7 p. 242. 1838. Sp.: *A. vahlii, A. lagena, A. appendiculosus.* „Af a priv. og ονυξ, Negl".

Aora Henrik Krøyer in: Naturh. Tidsskr., ser. 2 *v.* 1 p. 328, 335. 1845. Sp.: *A. typica.* „Navnet paa en Nymfe".

Aoroides Alfred O. Walker in: Tr. Liverp. biol. Soc., *v.* 12 p. 284. 1898. Sp.: *A. columbiae.*

Apherusa Alfred O. Walker in: Ann. nat. Hist., ser. 6 *v.* 8 p. 83. 1891. [Sp.: *Pherusa jurinii.*]

Apocrangonyx Thomas R. R. Stebbing in: Tr. Linn. Soc. London, ser. 2 *v.* 7 p. 422. 1899 V. Sp. typ.: *A. lucifugus.*

Araneops Achille Costa in: Rend. Soc. Borbon., n. ser. *v.* 2 p. 169, 171. 1853. Sp.: *A. diadema, A. brevicornis.*

Argissa Axel Boeck in: Forh. Selsk. Christian., 1870 p. 125. 1871. Sp.: *A. typica.*

Aristias Axel Boeck in: Forh. Selsk. Christian., 1870 p. 106. 1871. Sp.: *A. tumidus.*

Arrhis pro: *Aceros* A. Boeck 1861. T. R. R. Stebbing in: Tierreich, *v.* 21 p. 248. 1906 IX. „ἄρρις, without nose".

Asope C. S. Rafinesque, Anal. Nat., p. 101. 1815. [*nom. nud.*]

Aspidophoreia William A. Haswell in: P. Linn. Soc. N. S. Wales, *v.* 5 p. 101. 1880. Sp.: *A. diemenensis.*

Aspidopleurus G. O. Sars, Crust. Norway, *v.* 1 p. 203. 1891. Sp.: *A. gibbosus, ? Stegocephalus kessleri.*

Astyra Axel Boeck in: Forh. Selsk. Christian., 1870 p. 133. 1871. Sp.: *A. abyssi.*

Atyloides Thomas R. R. Stebbing in: Rep. Voy. Challenger, *v.* 29 p. 913. 1888. Sp.: *A. australis, A. assimilis, A. serraticauda.*

Atylopsis Thomas R. R. Stebbing in: Rep. Voy. Challenger, *v.* 29 p. 924. 1888. Sp.: *A. magellanicus, A. dentatus, A. emarginatus.*

Atylus William Elford Leach, Zool. Misc., *v.* 2 p. 21. 1815. Sp.: *A. carinatus.*

Audouinia (A. Costa in:) F. G. Hope, Cat. Crost. Ital., p. 24. 1851. Sp.: *A. acherusica.* [*nom. nud.*]

Audulla Ed. Chevreux in: Mém. Soc. zool. France, *v.* 14 p. 431. 1901. Sp.: *A. chelifera.* „Anagramme de Alluaud".

Autonoe Ragnar M. Bruzelius in: Svenska Ak. Handl., n. ser. *v.* 3 nr. 1 p. 23. 1859. Sp.: *A. punctata, A. anomala, A. grandimana, A. erythrophthalma, A. longipes, A. macronyx.* „En dotter af Nereus och Doris".

Axelboeckia pro: *Boeckia* .G. O. Sars 1894. Thomas R. R. Stebbing in: Tr. Linn. Soc. London, ser. 2 *v.* 7 p. 423. 1899 V. „in honour of the late Axel Boeck".

Bactrurus William Perry Hay in: P. U. S. Mus., *v.* 25 p. 430. 1902 IX 23. Sp.: *Crangonyx mucronatus.*

Baikalogammarus Thomas R. R. Stebbing in: Tr. Linn. Soc. London, ser. 2 *v.* 7 p. 425. 1899 V. Sp. typ.: *B. pullus.* „alludes to Lake Baikal".

Barentsia Thomas R. R. Stebbing in: Bijdr. Dierk., *v.* 17/18 p. 25. 1894. Sp.: *B. hoeki.* „Willem Barents".

Batea Fritz Müller in: Ann. nat. Hist., ser. 3 *v.* 15 p. 276. 1865. Sp.: *B. catharinensis.*

Bathymedon G. O. Sars, Crust. Norway, *v.* 1 p. 332. 1892. Sp.: *B. longimanus, B. saussurei, B. obtusifrons.*

Bathyonyx Fr. Vejdovský in: SB. Böhm. Ges., 1905 nr. 28 p. 2. 1905. Sp.: *B. devismesi.*

Bathyporea [pro: *Bathyporeia* Lindström 1856]. C. Spence Bate in: Ann. nat. Hist., ser. 2 *v.* 19 p. 271. 1857.

Bathyporeia G. Lindström in: Öfv. Ak. Förh., *v.* 12 p. 59. 1855. Sp.: *B. pilosa.*

Bellia C. Spence Bate in: Ann. nat. Hist., ser. 2 *v.* 7 p. 318. 1851. Sp.: *B. arenaria.*

Biancolina Antonio Della Valle in: F. Fl. Neapel, *v.* 20 p. 562. 1893. Sp.: *B. algicola.* „in omaggio al sig. S. Lobianco".

Bircenna Charles Chilton in: Tr. N. Zealand Inst., *v.* 16 p. 264, 265. 1884 V. Sp.: *B. fulvus (B. fulva).*

Boeckia A. W. Malm in: Öfv. Ak. Förh., *v.* 27 p. 543. 1871. Sp.: *B. typica.*

Boeckia ([O.] Grimm in MS.) G. O. Sars in: Bull. Ac. St.-Pétersb., ser. 5 *v.* 1 p. 182. 1894. Sp.: *B. spinosa.* „the name of ... Boeck".

Bonnierella Ed. Chevreux in: Résult. Camp. Monaco, *v.* 16 p. 97. 1900. Sp.: *B. abyssi.* „Jules Bonnier".

Boruta L. Wrześniowski in: Pam. Fizyjogr., *v.* 8 p. 264. 1888. Sp.: *B. tenebrarum.*

Boscia [pro: *Melita* Leach 1813/14]. [William Elford Leach in:] Edinb. Enc., *v.* 7 p. 435. [1813/14.]

Bouvierella Ed. Chevreux in: Résult. Camp. Monaco, *v.* 16 p. 70. 1900. Sp. typ.: *B. carcinophila.* „E. L. Bouvier".

Bovallia Georg Pfeffer in: Jahrb. Hamburg. Anst., *v.* 5 p. 95. 1888. Sp.: *B. gigantea.*

Brachyuropus Thomas R. R. Stebbing in: Tr. Linn. Soc. London, ser. 2 *v.* 7 p. 424. 1899 V. Sp.: *B. grewingkii, B. reichertii.* „alludes to the shortness of the third uropods".

Brandtia C. Spence Bate, Cat. Amphip. Brit. Mus., p. 129. 1862. Sp.: *B. latissima.*

Bruzelia Axel Boeck in: Forb. Selsk. Christian., 1870 p. 149. 1871. Sp.: *B. typica.*

Bruzeliella A. M. Norman in: Ann. nat. Hist., ser. 7 *v.* 16 p. 83. 1905 VII. Sp. typ.: *B. falcata.*

Byblis Axel Boeck in: Forb. Selsk. Christian., 1870 p. 228. 1871. Sp.: *B. gaimardi.*

Callimerus T. R. R. Stebbing in: Ann. nat. Hist., ser. 4 *v.* 18 p. 445. 1876. Sp.: *C. acudigitata.*

Calliope (Leach in MS.) C. Spence Bate in: Rep. Brit. Ass., Meet. 25 p. 58. 1856. Sp.: *C. leachii.*

Calliopius pro: *Calliope* Bate 1856. William Lilljeborg in: N. Acta Soc. Upsal., ser. 3 *v.* 6 nr. 1 p. 18 (tabella), 19. 1865.

Callisoma Oronzio-Gabriele Costa, Fauna Reg. Napoli, Crost., Cat. p. 5. [1840.] Sp.: *C. punctata.* [*nom. nud.*]

Callisoma Achille Costa, Fauna Reg. Napoli, fasc. Marz. 1851 p. 1. 1851. Sp.: *C. punctatum, C. hopei.* „καλος *pulcher* bello, e σωμα *corpus* corpo".

Camacho Thomas R. R. Stebbing in: Rep. Voy. Challenger, *v.* 29 p. 1178. 1888. Sp.: *C. bathyplous.* „from a personage mentioned in Don Quixote".

Cancer Carolus Linnaeus, Syst. Nat., ed. 10 p. 344, 625. 1758. Sp.: *C. cursor, C. raninus, C. mutus, C. minutus, C. ruricola, C. vocans, C. craniolaris, C. philargius, C. rhomboides, C. maculatus, C. pelagicus, C. nucleus, C. lactatus, C. maenas, C. depurator, C. feriatus, C. granulatus, C. pagurus, C. chabrus, C. araneus, C. cuphaeus, C. muscosus, C. personatus, C. maja, C. longimanus, C. horridus, C. cristatus, C. superciliosus, C. cornutus, C. longipes, C. spinifer, C. cruentatus, C. hepaticus. C. calappa, C. grapsus, C. aeneus, C. punctatus, C. dorsipes, C. symmysta, C. bernhardus, C. diogenes, C. gammarus, C. astacus, C. carcinus, C. pennaceus, C. squilla, C. crangon, C. carabus, C. cancharus, C. pilosus, C. norvegicus, C. homarus, C. arctus, C. mantis, C. scyllarus, C. pulex, C. locusta, C. salinus, C. stagnalis.*

Cardenio Thomas R. R. Stebbing in: Rep. Voy. Challenger, *v.* 29 p. 806. 1888. Sp.: *C. paurodactylus.* „The generic name is taken from a character in Don Quixote".

Cardenis [pro: *Cardenio* T. Stebbing 1888]. Cecil Warburton in: Zool. Rec., *v.* 25 Crust. p. 18. 1890.

Cardiophilus G. O. Sars in: Bull. Ac. St.-Pétersb., ser. 5 *v.* 4 p. 474. 1896. Sp.: *C. baeri.*

Caridophilus [pro: *Cardiophilus* O. Sars 1896]. J. V. Carus in: Zool. Anz., Regist. 16—20 p. 89. 1899.

Carinogammarus Thomas R. R. Stebbing in: Tr. Linn. Soc. London, ser. 2 *v.* 7 p. 429. 1899 V. Sp.: *C. cinnamomeus, C. wagii, C. pulchellus, C. seidlitzii. C. rhodophthalmus, C. caspius, C. atchensis, C. subcarinatus, C. fluviatilis, ? C. macrophthalmus, ? C. mucronatus.*

Carolobatea Thomas R. R. Stebbing in: Ann. nat. Hist., ser. 7 *v.* 4 p. 208. 1899. Sp. typ.: *Halimedon schneideri.* „in recollection of the late Charles Spence Bate".

Ceina Antonio Della Valle in: F. Fl. Neapel, *v.* 20 p. 530. 1893. Sp.: *C. egregia.* „derivato con una leggiera trasposizione di lettere dal primitivo *Nicea*".

Centromedon G. O. Sars, Crust. Norway, *v.* 1 p. 99. 1891. Sp. typ.: *C. pumilus*

Ceradocus Achille Costa in: Rend. Soc. Borbon., n. ser. *v.* 2 p. 170, 177. 1853. Sp.: *C. orchestiipes.*

Cerapodina [Henri] Milne Edwards, Hist. nat. Crust., *v.* 3 p. 62. 1840. Sp.: *C. abdita.*

Cerapopsis Antonio Della Valle in: F. Fl. Neapel, *v.* 20 p. 356, 388. 1893. Sp.: *C. longipes.*

Cerapus Thomas Say in: J. Ac. Philad., *v.* 1ɪ p. 49. 1817 VIII. Sp.: *C. tubularis.* „κέρας, a horn, and πούς, a foot".

Charcotia Ed. Chevreux in: Bull. Soc. zool. France, *v.* 30 p. 163. 1906 I 31. Sp.: *C. obesa.* „Dr. Charcot".

Cheirimedon Thomas R. R. Stebbing in: Rep. Voy. Challenger, *v.* 29 p. 638. 1888. Sp.: *C. crenatipalmatus.* „χείρ, the hand, and μέδων, a lord".

Cheiriphotis Alfred O. Walker in: Herdman, Rep. Ceylon Pearl Fish., *v.* 2 p. 234, 283. 1904. Sp.: *C. megacheles.*

Cheirocratus Alfred Merle Norman in: Nat. Hist. Tr. Northumb., *v.* 1 p. 12. 1867. Sp.: *C. mantis.* „χείρ and κρατέω, strong in the hand".

Chelura A. Philippi in: Arch. Naturg., *v.* 5ɪ p. 120. 1839. Sp.: *C. terebrans.*

Chevalia Alfred O. Walker in: Herdman, Rep. Ceylon Pearl Fish., *v.* 2 p. 234, 288. 1904. Sp.: *C. aviculae.*

Chevreuxius Jules Bonnier in: Ann. Univ. Lyon, *v.* 26 p. 663. 1896. Sp.: *C. grandimanus.*

Chiltonia Thomas R. R. Stebbing in: Tr. Linn. Soc. London, ser. 2 *v.* 7 p. 397, 408. 1899 V. Sp.: *C. mihiwaka.* „in compliment to Dr. Charles Chilton".

Chimaeropsis Fr. Meinert in: Udb. Hauchs, *v.* 3 p. 167. 1890. Sp.: *C. danica.*

Chironesimus G. O. Sars, Crust. Norway, *v.* 1 p. 108. 1891. Sp.: *C. debruynii.*

Chloris William A. Haswell in: Ann. nat. Hist., ser. 5 *v.* 5 p. 33. 1880.

Chosroës Thomas R. R. Stebbing in: Rep. Voy. Challenger, *v.* 29 p. 1208. 1888. Sp.: *C. incisus.* „An Armenian King".

Cleïppides Axel Boeck in: Forh. Selsk. Christian., 1870 p. 201. 1871. Sp.: *C. tricuspis.*

Cleonardo Thomas R. R. Stebbing in: Rep. Voy. Challenger, *v.* 29 p. 959. 1888. Sp.: *C. longipes, Tritropis appendiculata.* „from a personal name in Don Quixote".

Colomastix Adolph Eduard Grube, Ausfl. Triest, p. 125, 137. 1861. Sp.: *C. pusilla.*

Concholestes G. M. Giles in: J. Asiat. Soc. Bengal, *v.* 57 p. 237. 1888 X 10. Sp.: *C. dentalii.*

Constantia B. N. Dybowsky in: Horae Soc. ent. Ross., *v.* 10 suppl. p. 186. 1874. Sp.: *C. branickii, C. b. var. alexandri.*

Coremapus A. M. Norman in: Ann. nat. Hist., ser. 7 *v.* 16 p. 78. 1905 VII. Sp. typ.: *C. versiculatus.* „κόρημα, a brush, and πούς, a foot".

Corophia [pro: *Corophium* Latreille 1806]. H. Milne-Edwards in: Ann. Sci. nat., *v.* 20 p. 384. 1830.

Corophium P. A. Latreille, Gen. Crust. Ins., *v.* 1 p. 58. 1806. Sp.: *C. longicorne.*

Corophrium [pro: *Corophium* Latreille 1806]. [William Elford Leach in:] Edinb. Enc., *v.* 7 p. 432. [1813/14.]

Costantia [pro: *Constantia* B. Dybowsky 1874]. B. N. Dybowsky in: Horae Soc. ent. Ross., *v.* 10 suppl. p. 50. 1874.

Crangonyx C. Spence Bate in: P. Dublin Univ. zool. bot. Ass., *v.* 1 p. 237. 1859. Sp.: *C. subterraneus.* „κραγγών and νύξ".

Cratippus C. Spence Bate, Cat. Amphip. Brit. Mus., p. 275. 1862. Sp.: *C. tenuipes.*

Cratophium James D. Dana in: Amer. J. Sci., ser. 2 *v.* 14 p. 309. 1852 XI. — James D. Dana in: U. S. expl. Exp., *v.* 13 ɪɪ p. 832, 840. [1853.] Sp.: *C. validum, C. orientale.*

Cressa Axel Boeck in: Forh. Selsk. Christian., 1870 p. 145. 1871. Sp.: *C. schiødtei, C. minuta.*

Ctenacanthus W. Garjajeff in: Trudui Kazan. Univ., *v.* 35 nr. 6 p. 15. 1901. Sp.: *C. ruber, C. roseus, C. carpenterii, C. wagii, C. cabanisii, C. zieńkowiczii, C. radoszkowskii.*

Cuvieria [pro: *Leucothöe* Leach 1813/14]. [William Elford Leach in:] Edinb. Enc., *v.* 7 p. 435. [1813/14.]

Cychreus C. S. R afi nesque, Anal. Nat., p. 101. 1815. [*nom. nud.*]

Cyclocaris Thomas R. R. Stebbing in: Rep. Voy. Challenger, *v.*29 p. 664. 1888. Sp.: *C. tahitensis.* „from κύκλος, a circle, and κάρα, head".

Cymadusa [Subgen.?] Jules-César Savigny, Mém. An. s. Vert., *v.*1 p. 109. 1816 I. Sp.: *C. filosa.*

Cymothoa Joh. Christ. Fabricius, Ent. syst., *v.*2 p. 503. 1793. Sp.: *C. paradoxa, C. imbricata, C. falcata, C. asilus, C. guadeloupensis, C. oestrum, C. entomon, C. aquatica, C. marinus, C. linearis, C. chelipes, C. bicaudata, C. scopulorum, C. americana, C. psora, C. physodes, C. spinosa, C. acuminata, C. emarginata, C. albicornis, C. ceti, C. oceanica, C. serrata, C. assimilis.*

Cyphocaris (Lütken in MS.) Axel Boeck in: Forh. Selsk. Christian., 1870 p. 103. 1871. Sp.: *C. anonyx.*

Cypridoidea pro: *Cyproidea* HasWell [1880]. Edward Caldwell Rye in: Zool. Rec., *v.*16 Index p. 4. 1881.

Cyproidea [pro: *Cyproidia* HasWell 1879]. William A. Haswell in: Ann. nat. Hist., ser. 5 *v.*5 p. 31. 1880.

Cyproidia William A. Haswell in: P. Linn. Soc. N. S. Wales, *v.*4 p. 320. 1879. Sp.: *C. ornata, C. lineata.*

Cyrtophium James D. Dana in: Amer. J. Sci, ser. 2 *v.*14 p. 309. 1852 XI. — James D. Dana in: U. S. expl. Exp., *v.*13 II p. 831, 839. [1853]. Sp.: *C. orientale.*

Danaia C. Spence Bate in: Ann. nat. Hist., ser. 2 *v.*19 p. 137. 1857. Sp.: *D. dubia.*

Darwinea C. Spence Bate in: Rep. Brit. Ass., Meet. 25 p. 58. 1856. Sp.: *D. compressus.* [*nom. nud.*]

Darwinia [pro: *Darwinea* Bate 1856]. C. Spence Bate in: Ann. nat. Hist., ser. 2 *v.*19 p. 141. 1857. Sp.: *D. compressa.*

Dautzenbergia Ed. Chevreux in: Résult. Camp. Monaco, *v.*16 p. 73. 1900. Sp.: *D. grandimana.* „Dautzenberg".

Dercothoe James D. Dana in: Amer. J. Sci., ser. 2 *v.*14 p. 313. 1852 XI. — James D. Dana in: U. S. expl. Exp.,

*v.*13 II p. 911, 968. [1853.] Sp.: *D. emissitius, D. speculans, ? D. hirsuticornis.* „δερκω, to look".

Dermophilus Édouard van Beneden & Émile Bessels in: Mém. cour. Ac. Belgique, *v.*34 nr. 4 p. 26. 1870. Sp.: *D. lophii.*

Desmophilus [pro: *Dermophilus* E. Beneden & Bessels 1870]. Thomas H. Huxley, Man. Anat. Invert., p. 367. 1877.

Dexamene pro: *Dexamine* Leach 1813/14. L. Agassiz, Nomencl. zool., Index p. 121. 1846.

Dexamine [William Elford] Leach in: Edinb. Enc., *v.*7 p. 432. [1813/14.] Sp.: *D. spinosa.*

Dexiocerella William A. Haswell in: P. Linn. Soc. N. S. Wales, *v.*10 p. 107. 1885 VI. Sp.: *D. dentata, D. lobata, D. laevis.*

Dikerogammarus Thomas R. R. Stebbing in: Tr. Linn. Soc. London, ser. 2 *v.*7 p. 428. 1899 V. Sp.: *D. macrocephalus, D. haemobaphes, D. grimmi, D. verreauxii, D. fasciatus.* „alludes to the two horn-like eleVations on the pleon".

Dinoa C. S. Rafinesque, Anal. Nat., p. 101. 1815. [*nom. nud.*]

Driope [pro: *Dryope* Bate 1862]. C. Spence Bate, Cat. Amphip. Brit. Mus., t. 47. 1862.

Dryapoides [pro: *Dryopoides* T. Stebbing 1888]. Cecil Warburton in: Zool. Rec., *v.*25 Crust. p. 19. 1890.

Dryope C. Spence Bate, Cat. Amphip. Brit. Mus., p. 276. 1862. Sp.: *D. irrorata, D. crenatipalma.*

Dryopoides Thomas R. R. Stebbing in: Rep. Voy. Challenger, *v.* 29 p. 1145. 1888. Sp.: *D. westwoodi.* „Dryope ... and εἶδος, likeness".

Dulichia Henrik Krøyer in: Naturh. Tidsskr., ser. 2 *v.*1 p. 512, 521. 1845. Sp.: *D. spinosissima.* „δολιχος (lang)".

Dybowskia W. Garjajeff in: Trudui Kazan. Univ., *v.*35 nr. 6 p. 16. 1901. Sp. & Subsp.: *D. viridis, D. dryshenkii, D. meyerii, D. armata, D. a. var. ongureni, D. grubii, D. kesslerii, D. k. var. europeus, D. brandtii, D. cancellus, D. c. var. gerstfeldtii, D. cancelloides.*

Dyopedos C. Spence Bate in: Ann. nat. Hist., ser. 2 *v.* 19 p. 150. 1857. Sp.: *D. porrectus, D. falcatus.*

Echinogammarus Thomas R. R. Stebbing in: Tr. Linn. Soc. London, ser. 2 *v.* 7 p. 428. 1899 V. Sp.: *E. berilloni, E. verrucosus, E. maackii, E. ochotensis, E. saphirinus, E. czerskii, E. lividus, E. viridis, E. cyaneus, E. testaceus, E sophiae, E. fuscus, E. murinus, E. aheneus, E. sarmatus, E. capreolus, E. uzzolzewii, E. stenophthalmus, E. sehamanensis, E. leptocerus, E. toxophthalmus, E. vittatus, E. petersii, E. violaceus, E. ibex, E. parvexii, E. polyarthrus, ? Gammarus mutilus.* „alludes to the numerous spines on the body".

Egidia Achille Costa in: Rend. Soc. Borbon., n. ser. *v.* 2 p. 170, 172. 1853. Sp.: *E. pulchella.*

Eiscladus C. Spence Bate & J. O. Westwood, Brit. sess. Crust., *v.* 1 p. 411. [1862 XI 1.] Sp.: *E. longicaudatus.* „Εἰς one, κλαδος branch".

Elasmopus Achille Costa in: Rend. Soc. Borbon., n. ser. *v.* 2 p. 170, 175. 1853. Sp.: *E. rapax.*

Enone A. Risso, Hist. nat. Eur. mérid., *v.* 5 p. 96. 1826. Sp.: *E. punctata.*

Ephipphora [pro: *Ephippiphora* White 1847]. J. D. Dana in: Amer. J. Sci., ser. 2 *v.* 8 p. 428. 1849 XI.

Ephippiphora Adam White in: P. zool. Soc. London, *v.* 15 p. 124. 1847 [XI 10]. Sp.: *E. kroyeri.*

Epidesura Axel Boeck in: Forh. Skand. Naturf., Møde 8 p. 659. 1861. Sp. typ.: *Amphithoë compressa.*

Epimeria (A. Costa in:) F. G. Hope, Cat. Crost. Ital., p. 46. 1851. Sp.: *E. tricristata.*

Eratea C. S. Rafinesque, Anal. Nat., p. 101. 1815. [*nom. nud.*]

Erichthonius [pro: *Ericthonius* H. Milne Edwards 1830]. Hermann Burmeister, Handb. Naturg., p. 569. 1837.

Erichtonius [pro: *Ericthonius* H. Milne Edwards 1830]. [Hippolyte] Lucas in: Hist. An. artic., Crust. Arach. Myr. p. 231. 1840.

Ericthonius H. Milne Edwards in: Ann. Sci. nat., *v.* 20 p. 382. 1830. Sp.: *E. difformis.*

Eriopis Ragnar M. Bruzelius in: Svenska Ak. Handl., n. ser. *v.* 3 nr. 1 p. 37, 64. 1859. Sp.: *E. elongata.* „En dotter af Jason och Medea".

Eriopisa pro: *Eriopis* R. M. Bruzelius 1859. Thomas R. R. Stebbing in: Ann. nat. Hist., ser. 6 *v.* 5 p. 193. 1890.

Eriopsis pro: *Eriopis* R. M. Bruzelius 1859. August Wrześniowski in: Z. wiss. Zool., *v.* 50 p. 632. 1890 X 10.

Euandania Thomas R. R. Stebbing in: Ann. nat. Hist., ser. 7 *v.* 4 p. 206. 1899. Sp. typ.: *Andania gigantea.*

Eucrangonyx Thomas R. R. Stebbing in: Tr. Linn. Soc. London, ser. 2 *v.* 7 p. 423. 1899 V. Sp.: *E. mucronatus, E. vejdovskyi, E. packardii, E. gracilis, E. antennatus.*

Euonyx Alfred Merle Norman in: Rep. Brit. Ass., Meet. 36 p. 197, 202. 1867. Sp.: *E. chelatus.*

Eurymera Georg Pfeffer in: Jahrb. Hamburg. Anst., *v.* 5 p. 102. 1888. Sp.: *E. monticulosa.*

Euryporeia pro: *Eurytenes* W. Lilljeborg 1865. G. O. Sars, Crust. Norway, *v.* 1 p. 85. 1891.

Eurystheus C. Spence Bate in: Rep. Brit. Ass., Meet. 25 p. 58. 1856. Sp.: *E. tridentatus.* [*nom. nud.*]

Eurystheus C. Spence Bate in: Ann. nat. Hist., ser. 2 *v.* 19 p. 143. 1857. Sp.: *E. tridentatus.*

Eurytenes William Lilljeborg in: N. Acta Soc. Upsal., ser. 3 *v.* 6 nr. 1 p. 11. 1865. Sp.: *E. magellanicus.* „εὐρυτενής, which signifies widely stretched".

Eurythenes [pro: *Eurytenes* W. Lilljeborg 1865]. (S. I. Smith in:) Samuel H. Scudder, Nomencl. zool., suppl. L. p. 135. 1882.

Eurytheus [pro: *Eurystheus* Bate 1857]. Augustus de Marschall, Nomencl. zool., p. 409. 1873.

Eusinus [pro: *Eusirus* Krøyer 1845]. Augustus de Marschall, Nomencl. zool., p. 409. 1873.

Eusirogenes T. R. R. Stebbing in: Tr. Linn. Soc. London, ser. 2 *v.* 10 II p. 13, 15. 1904 XI. Sp.: *E. dolichocarpus.*

Eusiroides Thomas R. R. Stebbing in: Rep. Voy. Challenger, *v.* 29 p. 969. 1888. Sp.: *E. caesaris, E. pompeii, E. crassi, ?Atylus monoculoides, ?A. lippus.*

Eusiropsis Thomas R. R. Stebbing in: Tr. Linn. Soc. London, ser. 2 *v.* 7 p. 39. 1897 V. Sp.: *E. riisei.*

Eusirus Henrik Krøyer in: Naturh. Tidsskr., ser. 2 *v.* 1 p. 501, 511. 1845. Sp.: *E. cuspidatus.* „Ευσειρος, en Søn af Poseidon og Okeaniden Idothea".

Exoediceros Thomas R. R. Stebbing in: Ann. nat. Hist., ser. 7 *v.* 4*p. 208. 1899. Sp. typ.: *Oediceros fossor.*

Exunguia (Alfred Merle Norman in:) George SteWardson Brady & DaVid Robertson in: Ann. nat. Hist., ser. 4 *v.* 3 p. 359. 1869. Sp.: *E. stilipes.* „*ex* and *unguis*, Without a nail".

Galanthis C. Spence Bate in: Rep. Brit. Ass., Meet. 25 p. 57. 1856. Sp.: *G. lubbockiana.*

Gallea Alfred O. Walker in: Herdman, Rep. Ceylon Pearl Fish., *v.* 2 p. 232, 256. 1904. Sp.: *G. tecticauda.* „From the port of Galle".

Gammaracanthus C. Spence Bate, Cat. Amphip. Brit. Mus., p. 201. 1862. Sp.: *G. loricatus, G. mucronatus, G. macrophthalmus.*

Gammarella C. Spence Bate in: Ann. nat. Hist., ser. 2 *v.* 19 p. 143. 1857. Sp.: *G. orchestiformis.*

Gammarellus [*Subgen.*] Johann Friedrich Wilhelm Herbst, Naturg. Krabben Krebse, *v.* 2 p. 106. 1793. Sp.: *Cancer (G.) setiferus, C. (G.) chinensis, C. (G.) pedatus, C. (G.) armiger, C. (G.) oculatus, C. (G.) bipes, C. (G.) trixapus, C. (G.) homari, C. (G.) harangum, C. flexuosus, C. (G.) ampulla, C. (G.) nugax, C. (G.) paludosus, C. (G.) podurus, C. (G.) mutilus, C. (G.) stagnalis, C. (G.) grossipes, C. (G.) cancellus, C. (G.) locusta, C. gammarellus, C. (G.) pulex, C. (G.) arenarius, C. (G.) crassicornis, C. (G.) strömianus, C. (G.) spinicarpus, C. (G.) sedentarius, C. (G.) cicada, C. (G.) serratus, C. (G.) medusarum, C. (G.) corniger, C. (G.) abyssinus, C. (G.) linearis, C. (G.) ventricosus.*

Gammaropsis [*Subgen.*] V. [Wilhelm] Liljeborg in: Vetensk. Ak. Handl., 1853 p. 455. 1855. Sp.: *Gammarus erythrophthalmus, G. anomalus, G. longipes, G. macronyx.*

Gammarus Io. Christ. Fabricius. Syst. Ent., p. 418. 1775. Sp.: *G. locusta, G. pulex, G. linearis, G. salinus, G. stagnalis.*

Gitana Axel Boeck in: Forb. Selsk. Christian., 1870 p. 132. 1871. Sp.: *G. sarsi, G. rostrata.*

Gitanopsis G. O. Sars, Crust. Norway, *v.* 1 p. 223. 1892. Sp.: *G. bispinosa, G. inermis, G. arctica.*

Glauconome Henrik Krøyer in: Naturh. Tidsskr., ser. 2 *v.* 1 p. 491, 501. 1845. Sp.: *G. leucopis.* „En af Nereiderne".

Glycera William A. Haswell in: P. Linn. Soc. N. S.Wales, *v.* 4 p. 256, 322. 1879. Sp.: *G. tenuicornis.*

Glycerina William A. Haswell, Cat. Austral. Crust., p. 233. 1882. Sp.: *G. tenuicornis.* „Altered from *Glycera*".

Gmelina ([O.] Grimm in MS.) G. O. Sars in: Bull. Ac. St.-Pétersb., ser. 5 *v.* 1 p. 191. 1894. Sp.: *G. costata, G. kusnezowi.*

Gmelinopsis G. O. Sars in: Bull. Ac. St.-Pétersb., ser. 5 *v.* 4 p. 434. 1896. Sp.: *G. tuberculata, G. aurita.*

Goësia Axel Boeck in: Forh. Selsk. Christian., 1870 p. 231. 1871. Sp.: *G. depressa.*

Goplana [August] Wrześniowski in: Zool. Anz., *v.* 2 p. 299. 1879 VI 9. Sp.: *G. polonica, G. ambulans.* „In polnischer Sprache bezeichnet *Goplana* eine Wasser-Nymphe".

Gossea C. Spence Bate (& J. O. Westwood), Brit. sess. Crust., *v.* 1 p. 276. [1862 IV 1.] Sp.: *G. microdentopa.* „in compliment to Mr. Gosse".

Grandidierella H. Coutière in: Bull. Soc. philom., ser. 9 *v.* 6 p. 166. 1904. Sp.: *G. mahafalensis.*

Graya [pro: *Grayia* Bate 1862]. C. Spence Bate, Cat. Amphip. Brit. Mus., t. 14 a, 16. 1862.

Grayia C. Spence Bate (& J. O. Westwood), Brit. sess. Crust., *v.* 1 p. 151. [1862 I 1.] Sp.: *G. imbricata.* „in compliment to Dr. J. E. Gray".

Grimaldia Ed. Chevreux in: Bull. Soc. zool. France, *v.* 14 p. 283. 1889. Sp.: *G. armata.* „Grimaldi".

Grubia Voldemarus Czerniavski in: Syezda Russ. Est., Syezda 1 Zool. p. 103. 1868. Sp.: *G. taurica.* „Dedicata cel. A. E. Grubio".

Guerina pro: *Guerinia* A. Costa 1853. Antonio Della Valle in: F. Fl. Neapel, *v.* 20 p. 775. 1893.

Guerinella pro: *Guerinia* A. Costa 1853. Ed. Chevreux in: Bull. Mus. Monaco, nr. 35 p. [7]. 1905 V 5.

Guerinia (Hope in MS.) Achille Costa, Descr. 3 Crost. dal Hope, p. 3. 1853. Sp.: *G. nicaeensis.*

Guernea pro: *Helleria* A. M. Norman 1868. Edouard Chevreux in: Bull. Soc. zool. France, *v.* 12 p. 302. 1887. „Jules de Guerne".

Gulbarentsia pro: *Barentsia* T. Stebbing 1894. Thomas R. R. Stebbing in: Bijdr. Dierk., *v.* 17/18 p. 2. 1894.

Hakonboeckia Thomas R. R. Stebbing in: Tr. Linn. Soc. London, ser. 2 *v.* 7 p. 425. 1899 V. Sp. typ.: *H. strauchii.* „in compliment to Hakon Boeck".

Halibrotus pro: *Alibrotus* H. Milne Edwards 1840. L. Agassiz, Nomencl. zool., Index p. 14, 171. 1846.

Halice Axel Boeck in: Forh. Selsk. Christian., 1870 p. 152. 1871. Sp.: *H. abyssi, H. grandicornis.*

Halicoides Alfred O. Walker in: Ann. nat. Hist., ser. 6 *v.* 17 p. 344. 1896. Sp.: *H. anomala.*

Halicreion Axel Boeck in: Forh. Selsk. Christian., 1870 p. 173. 1871. Sp.: *H. longicaudatus.*

Halicrion [pro: *Halicreion* A. Boeck 1871]. Eduard von Martens in: Zool. Rec., *v.* 8 p. 190. 1873.

Halimedon Axel Boeck in: Forh. Selsk. Christian., 1870 p. 169. 1871. Sp.: *H. mølleri, H. saussurei, H. longimanus, H. brevicalcar.*

Halirages Axel Boeck in: Forb. Selsk. Christian., 1870 p. 194. 1871. Sp.: *H. bispinosus, H. borealis, H. tridentatus, H. fulvocinctus.*

Haliragoides G. O. Sars, Crust. Norway, *v.* 1 p. 432. 1893. Sp.: *H. inermis.*

Halirhages [pro: *Halirages* A. Boeck 1871]. Anton Stuxberg in: Bih. Svenska Ak., *v.* 5 nr. 22 p. 64. 1880.

Haplocheira William A. Haswell in: P. Linn. Soc. N.S. Wales, *v.* 4 p. 273. 1879. Sp.: *H. typica.*

Haplochira pro: *Haplocheira* Haswell 1879. Eduard von Martens in: Zool. Rec., *v.* 16 Crust. p. 32. 1881.

Haplonyx [pro: *Hoplonyx* O. Sars 1891]. [Alfred Merle] Norman in: Ann. nat. Hist., ser. 7 *v.* 5 p. 209, 211, 212. 1900 II.

Haploops [William] Liljeborg in: Öfv. Ak. Förb., *v.* 12 p. 135. 1855. Sp.: *H. tubicola, H. carinata.*

Harmomia [pro: *Harmonia* Haswell 1879]. William A. Haswell in: P. Linn. Soc. N.S. Wales, *v.* 4 p. 330. 1879.

Harmonia William A. Haswell in: P. Linn. Soc. N. S. Wales, *v.* 4 p. 330, 349. 1879. Sp.: *H. crassipes (Harmonia c.).*

Harpina Axel Boeck in: Forb. Selsk. Christian., 1870 p. 135. 1871. Sp.: *H. plumosa, H. crenulata.*

Harpinia pro: *Harpina* A. Boeck 1871. Axel Boeck, Skand. Arkt. Amphip., *v.* 2 p. 218. 1876. „Ἅρπυια (et graesk Kvindenavn)".

Harpinioides Thomas R. R. Stebbing in: Rep. Voy. Challenger, *v.* 29 p. 936. 1888. Sp.: *H. drepanocheir.* „Harpinia, Boeck, and εἶδος, likeness".

Harpinoides [pro: *Harpinioides* T. Stehbing 1888]. Cecil Warburton in: Zool. Rec., *v.* 25 Crust. p. 19. 1890.

Haustorius P. L. St. Müller in: Slabber, Phys. Belustig., p. 48. 1775. Sp.: *H. arenarius.*

Heiscladius [pro: *Eiscladus* Bate & Westwood 1862]. W. C. M'Intosh in: Ann. nat. Hist., ser. 4 *v.* 14 p. 269. 1874.

Heiscladus [pro: *Eiscladus* Bate & Westwood 1862]. Alfred Merle Norman in Rep. Brit. Ass., Meet. 38 p. 255, 259, 284. 1869.

Hela Axel Boeck in: Forh. Skand. Naturf., Møde 8 p. 668. 1861. Sp.: *H. monstrosa.*

Helella [pro: *Hela* A. Boeck 1861]. (Smith in MS.) G. O. Sars in: Forh. Selsk. Christian., 1882 nr. 18 p. 31. 1882.

Helleria Alfred Merle Norman in: Ann. nat. Hist., ser. 4 *v.* 2 p. 418. 1868. Sp.: *H. coalita.* „dedicated to Prof. Heller".

Heterogammarus Thomas R. R. Stebbing in: Tr. Linn. Soc. London, ser. 2 *v.* 7 p. 429. 1899 V. Sp.: *H. stanislavii, H. sophianosii, H. capellus, H. ignotus, H. flori, H. bifasciatus, H. branchialis, H. albulus.* „alludes to the character of the genus as a second self to *Gammarus*".

Hippias C. S. Rafinesque, Anal. Nat., p. 101. 1815. [*nom. nud.*]

Hippomedon Axel Boeck in: Forh. Selsk. Christian., 1870 p. 102. 1871. Sp.: *H. holbølli, H. abyssi.*

Hirondella [pro: *Hirondellea* Chevreux 1889]. J. V. Carus in: Zool. Anz., Regist. 11—15 p. 142. 1893.

Hirondellea Ed. Chevreux in: Bull. Soc. zool. France, *v.* 14 p. 285. 1889. Sp.: *H. trioculata.* „Hirondelle".

Hoplonyx G. O. Sars, Crust. Norway, *v.* 1 p. 91. 1891. Sp.: *Anonyx cicadoides, Hoplonyx cicada, H. similis, H. acutus, H. albidus, H. leucophthalmus, H. caeculus.*

Hornellia Alfred O. Walker in: Herdman, Rep. Ceylon Pearl Fish., *v.* 2 p. 233, 268. 1904. Sp.: *H. incerta.* „after Mr. Jas. Hornell".

Hyale Heinrich Rathke in: Mém. prés. Ac. St.-Pétersb., *v.* 3 p. 377. 1837. Sp.: *H. pontica.* „nach einer Nymphe aus dem Gefolge der Diana".

Hyalella Sidney I. Smith in: Rep. U. S. Fish Comm., *v.* 2 p. 645. 1874. Sp.: *H. dentata.*

Hyalellopsis Thomas R. R. Stebbing in: Tr. Linn. Soc. London, ser. 2 *v.* 7 p. 422. 1899 V. Sp. typ.: *H. czyrniańskii.*

Hyperiopsis G. O. Sars in: Norske Nordhavs-Exp., *v.* 6 Crust. I p. 231. 1885. Sp.: *H. vøringii.*

Ichnopus Achille Costa in: Rend. Soc. Borbon., n. ser. *v.* 2 p. 169, 172. 1853. Sp.: *I. taurus.*

Ichthyomyzocus [C. E.] Hesse in: Ann. Sci. nat., ser. 5 *v.* 17 nr. 7 p. 1, 5. 1873. Sp.: *I. ornatus, I. morrhuae, I. lophii, I. squatinae.* „De ἰχθύϩ, poisson; μύζω, je suce".

Icilius James D. Dana in: Amer. J. Sci., ser. 2 *v.* 8 p. 140. 1849 VII. — Jacobus D. Dana in: P. Amer. Ac., *v.* 2 p. 220. 1852. Sp.: *I. ovalis.*

Icridium [Adolph Eduard] Grube in: Jahresber. Schles. Ges., *v.* 41 p. 58. 1864. Sp.: *I. fuscum.*

Iduna Axel Boeck in: Forh. Skand. Naturf., Møde 8 p. 656. 1861. Sp.: *Gammarus brevicornis, G. fissicornis (Iduna f.).*

Idunella G. O. Sars, Crust. Norway, *v.* 1 p. 536. 1894. Sp.: *I. aeqvicornis.*

Idurella [pro: *Idunella* O. Sars 1894]. R. I. Pocock in: Zool. Rec., *v.* 32 Crust. p. 41. 1896.

Ingolfiella H. J. Hansen in: J. Linn. Soc., *v.* 29 p. 118, 128. 1903 X 31. Sp.: *I. abyssi, I. littoralis.*

Iphigeneia pro: *Iphigenella* G. O. Sars 1896. ([O.] Grimm in MS.) G. O. Sars in: Bull. Ac. St.-Pétersb., ser. 5 *v.* 4 p. 478. 1896 V.

Iphigenella G. O. Sars in: Bull. Ac. St.-Pétersb., ser. 5 *v.* 4 p. 478. 1896 V. Sp. typ.: *I. acanthopoda.*

Iphigenia George M. Thomson in: Tr. N. Zealand Inst., *v.* 14 p. 237. 1882 V. Sp.: *I. typica.*

Iphimedia Heinrich Rathke in: N. Acta Ac. Leop., *v.* 201 p. 85. 1843. Sp.: *I. obesa.* „Nach einer Geliebten Neptuns".

Iphimediopsis Antonio Della Valle in: F. Fl. Neapel, *v.* 20 p. 585, 933. 1893. Sp.: *I. eblanae (I. geniculata).*

Iphinotus Thomas R. R. Stebbing in: Tr. Linn. Soc. London, ser. 2 *v.* 7 p. 414, 419. 1899 V. Sp.: *I. chiltoni.* „from the prefix ἰφι-, and νῶτος, back".

Iphiplateia Thomas R. R. Stebbing in: Tr. Linn. Soc. London, ser. 2 *v.* 7 p. 414. 1899 V. Sp.: *I. whiteleggei.* „from the prefix ἰφι-, signifying strength, and πλατεῖα, broad".

Isaea H. Milne Edwards in: Ann. Sci. nat., *v.* 20 p. 380. 1830. Sp.: *I. montagui.*

Ischyroceras [pro: *Ischyrocerus* Kröyer 1838]. James D. D a n a in: Amer. J. Sci., ser. 2 *v.* 8 p. 138. 1849 VII.

Ischyrocerus Henrik K r ö y e r in: Danske Selsk. Afh., *v.* 7 p. 283, 287. 1838. Sp.: *I. angvipes.* „Af ισχυρος, staerk, og κερας, Horn".

Iscoea [pro: *Isaea* H. Milne Edwards 1830]. C. Spence B a t e, Cat. Amphip. Brit. Mus., t. 22 f. 1. 1862.

Isoea [pro: *Isaea* H. Milne Edwards 1830]. [Hippolyte] L u c a s in: Hist. An. artic., Crust. Arach. Myr. p. 230. 1840.

Isolus C. S. R a f i n e s q u e, Anal. Nat., p. 101. 1815. [*nom. nud.*]

Janassa Axel B o e c k in: Forh. Selsk. Christian., 1870 p. 249. 1871. Sp.: *J. variegata.*

Jassa [William Elford] L e a c h in: Edinb. Enc., *v.* 7 p. 433. [1813/14]. Sp.: *J. pulchella, J. pelagica,* ? *Cancer (Gammarus) falcatus.*

Katius Ed. C h e v r e u x in: Bull. Mus. Monaco, nr. 35 p. 1. 1905 V 5. Sp.: *K. obesus.* „mot arabe *Kat,* chat".

Kerguelenia Thomas R. R. S t e b b i n g in: Rep. Voy. Challenger, *v.* 29 p. 1219. 1888. Sp.: *K. compacta.*

Kroyea [pro: *Kröyera* Bate 1857]. C. Spence B a t e, Cat. Amphip. Brit. Mus., t. 17. 1862.

Kröyera C. Spence B a t e in: Ann. nat. Hist., ser. 2 *v.* 19 p. 140. 1857. Sp.: *K. carinata.*

Kroyeria [pro: *Kröyera* Bate 1857]. Adolph Eduard G r u b e, Lussin, p. 72. 1864.

Kuria A. O. W a l k e r & Andrew S c o t t in: H. O. Forbes, Nat. Hist. Sokotra, p. 217, 228. 1903. Sp.: *K. longimanus.*

Lada [August] Wrześniowski in: Zool. Anz., *v.* 2 p. 322. 1879 VI 23. Sp.: *L. chalubińskii.* „in der slavischen Mythologie die Liebesgöttin".

Laematophilus [pro: *Laetmatophilus* R. M. Bruzelius 1859]. William A. H a s w e l l in: P. Linn. Soc. N. S. Wales, *v.* 10 p. 107, 110. 1885.

Laetmatophilus Ragnar M. B r u z e l i u s in: Svenska Ak. Handl., n. ser. *v.* 3 nr. 1 p. 10. [1859.] Sp.: *L. tuberculatus.* „λαῖτμα och φίλος".

Laetmophilus [pro: *Laetmatophilus* R. M. Bruzelius 1859]. Augustus de M a r s c h a l l, Nomencl. zool., p. 411. 1873.

Lafystius Henrik K r ø y e r in: Naturh. Tidsskr., *v.* 4 p. 156. 1842. Sp.: *L. sturionis.* „Λαφυσιος, gulosus".

Lalaria [Hercule N i c o l e t in:] Gay, Hist. Chile, *v.* 3 p. 240. 1849. Sp.: *L. longitarsis.*

Lalasia [pro: *Lalaria* H. Nicolet 1849]. C. Spence B a t e in: Ann. nat. Hist., ser. 2 *v.* 20 p. 525. 1857.

Lampra Axel B o e c k in: Forb. Selsk. Christian., 1870 p. 188. 1871. Sp.: *L. gibbosa.*

Laothoë pro: *Laothoës* A. Boeck 1871. G. O. S a r s, Crust. Norway, *v.* 1 p. 453. 1893.

Laothoës Axel B o e c k in: Forh. Selsk. Christian., 1870 p. 202. 1871. Sp.: *L. meinerti.*

Laphystiopsis G. O. S a r s, Crust. Norway, *v.* 1 p. 386. 1893. Sp.: *L. planifrons.*

Laphystius pro: *Lafystius* Krøyer 1842. L. A g a s s i z, Nomencl. zool., Index p. 200, 202. 1846.

Leipsuropus Thomas R. R. S t e b b i n g in: Ann. nat. Hist., ser. 7 *v.* 3 p. 241. 1899. Sp.: *Cyrtophium parasiticum.* „signifying an omission of a uropod".

Lemboides Thomas R. R. S t e b b i n g in: Ann. nat. Hist., ser. 6 *v.* 16 p. 209. 1895. Sp.: *L. afer.*

Lembos C. Spence B a t e in: Rep. Brit. Ass., Meet. 25 p. 58. 1856. Sp.: *L. cambriensis, L. damnoniensis, L. versiculatus, L. websterii.* [*nom. nud.?*]

Lembos C. Spence B a t e in: Ann. nat. Hist., ser. 2 *v.* 19 p. 142. 1857. Sp.: *L. cambriensis, L. versiculatus, L. websterii, L. damnoniensis.*

Lepidactylis Thomas S a y in: J. Ac. Philad., *v.* 1 II p. 379. 1818. Sp.: *L. dytiscus.*

Lepidactylus [pro: *Lepidactylis* Say 1818]. Fr. M e i n e r t in: Udb. Hauchs, *v.* 3 p. 160. 1890.

Lepidepecreum C. Spence B a t e & J. O. W e s t w o o d, Brit. sess. Crust., *v.* 2 p. 509. 1868 [XII 31]. Sp.: *L. carinatum, Anonyx longicornis.*

Lepleurus C. S. Rafinesque, Ann. Nat., p. 6. 1820. Sp.: *L. rivularis.*

Leptamphopus G. O. Sars, Crust. Norway, v. 1 p. 458. 1893. Sp.: *L. longimanus.*

Leptocheirus Ernestus Gustavus Zaddach, Syn. Crust. Pruss., p. 7. 1844. Sp.: *L. pilosus.*

Leptochela Axel Boeck, Skand. Arkt. Amphip., v. 2 p. 190. 1876. Sp.: *Opis leptochela.*

Leptochirus [pro: *Leptocheirus* Zaddach 1844]. James D. Dana in: Amer. J. Sci., ser. 2 v. 8 p. 137. 1849 VII.

Leptophoxus G. O. Sars, Crust. Norway, v. 1 p. 146. 1891. Sp.: *L. falcatus.*

Leptothoe William Stimpson in: Smithson. Contr., v. 6 nr. 5 p. 46. 1853. Sp.: *L. danae.*

Leucothöe [William Elford Leach in:] Edinb. Enc., v. 7 p. 386, 403, 432, 435. [1813/14.] Sp.: *L. articulosa (Cuviera a.).*

Liljeborgia C. Spence Bate (& J. O. Westwood), Brit. sess. Crust., v. 1 p. 202. [1862 II 1.] Sp.: *L. pallida, L. shetlandica.* „in compliment to . . . . Professor Liljeborg".

Lilljeborgia [pro: *Liljeborgia* Bate 1862]. Axel Boeck in: Forh. Selsk. Christian., 1870 p. 154. 1871.

Lockingtonia W. G. W. Harford in: P. Calif. Ac., v. 7 p. 53. 1877. Sp.: *L. fluvialis.* „dedicate . . . to Mr. W. N. Lockington".

Lonchomerus C. Spence Bate in: Rep. Brit. Ass., Meet. 25 p. 58. 1856. Sp.: *L. gracilis.* [*nom. nud.*]

Lonchomerus C. Spence Bate in: Ann. nat. Hist., ser. 2 v. 19 p. 143. 1857. Sp.: *L. gracilis.*

Lusyta Gio. Domenico Nardo, Prosp. Fauna Venet., p. 20. 1847. Sp.: *Cancer algensis.*

Lycesta Jules-César Savigny, Mém. An. s. Vert., v. 1 p. 109. 1816 I. Sp.: *L. furina.*

Lycianassa [pro: *Lysianassa* H. Milne Edwards 1830]. Thomas Bell in: Belcher, Last arct. Voy., v. 2 p. 406. 1855.

Lysianassa H. Milne Edwards in: Ann. Sci. nat., v. 20 p. 364. 1830. Sp.: *L. costae, L. chauseica.*

Lysianassina *Subgen.* Achille Costa in: Annuario Mus. Napoli, v. 4 p. 43. 1867. Sp.: *Lysianassa filicornis, L. longicornis.*

Lysianax pro: *Lysianassa* H. Milne Edwards 1830. Thomas R. R. Stebbing in: Rep. Voy. Challenger, v. 29 p. 681, 1676. 1888.

Lysianella G. O. Sars in: Porh. Selsk. Christian., 1882 nr. 18 p. 78. 1882. Sp.: *L. petalocera.*

Lysianopsis S. J. Holmes in: Amer. Natural., v. 37 p. 276. 1903 IV. Sp.: *L. alba.*

Lysita pro: *Lusyta* Nardo 1847. Gio. Domenico Nardo in: Mem. Ist. Veneto, v. 14 p. 283, 340. [1869].

Macleayia William A. Haswell in: Ann. nat. Hist., ser. 5 v. 5 p. 32. 1880.

Macrohectopus pro: *Constantia* Dybowsky 1874. T. R. R. Stebbing in: Tierreich, v. 21 p. 394. 1906 IX. „Μακρός, long, ἔκτος, sixth, πούς, foot".

Maera [William Elford Leach in:] Edinb. Enc, v. 7 p. 386, 403, 432, 436. [1813/14.] Sp.: *M. grossimana (Mülleria g.).*

Maeroides Alfred O. Walker in: P. Liverp. biol. Soc., v. 12 p. 282. 1898. Sp.: *M. thompsoni.*

Maerza [pro: *Maera* Leach 1813/14]. [Pierre André] Latreille in: G. Cuvier, Règne an., v. 3 p. 47. 1817.

Megalonoura ([Alfred Merle] Norman in MS.) (A. O. Walker in:) W. A. Herdman in: P. Liverp. biol. Soc., v. 3 p. 39. 1889. Sp.: *M. agilis.*

Megalorchestes J. F. Brandt in: Bull. phys.-math. Ac. St.-Pétersb., v. 9 p. 142. 1851 I 9. Sp.: *M. californianus, ? Orchestia longicornis.*

Megalorchestia [pro: *Megalorchestes* F. Brandt 1851]. J. F. Brandt in: Bull. phys.-math. Ac. St.-Pétersb., v. 9 p. 310. 1851 V 31. Sp.: *M. californiana, ? M. longicornis.*

Megaloura pro: *Megaluropus* [Hoek 1889]. (A. Merle Norman in MS.) P. P. C. Hoek in: Tijdschr. Nederl. dierk. Ver., ser. 2 v. 2 p. 198. 1889 III.

Megaluropus (A. Merle Norman in MS.) P. P. C. Hoek in: Tijdschr. Nederl. dierk. Ver., ser. 2 v. 2 p. 197. 1889 III. Sp.: *M. agilis.*

Megamaera [pro: *Megamoera* Bate 1862].
C. Spence Bate, Cat. Amphip. Brit.
Mus., t. 39. 1862.

Megamoera C. Spence Bate (& J. O.
Westwood), Brit. sess. Crust., *v.* 1
p. 400. [1862 XI 1.] Sp.: *M. semiserrata, M. longimana, M. othonis, ? M.
alderi, M. brevicaudata.*

Megamphopus Alfred Merle Norman in:
Rep. Brit. Ass., Meet. 38 p. 282. 1869.
Sp. typ.: *M. cornutus.*

Melita [William Elford Leach in:] Edinb.
Enc., *v.* 7 p. 386, 403, 432, 435. [1813/14.]
Sp.: *M. palmata (M. palmeta) (Boscia
nigricans).*

Melite [pro: *Melita* Leach 1813/14]. [Pi-
erre André] Latreille in: G. Cuvier,
Règne an., *v.* 3 p. 47. 1817.

Melphidippa Axel Boeck in: Forb. Selsk.
Christian., 1870 p. 218. 1871. Sp.: *M.
spinosa, M. longipes, M. borealis.*

Melphidippella G. O. Sars, Crust. Norway, *v.* 1 p. 487. 1894. Sp.: *M. macera.*

Menigrates Axel Boeck in: Forh. Selsk.
Christian., 1870 p. 113. 1871. Sp.: *M.
obtusifrons.*

Mesopleustes Thomas R. R. Stebbing in:
Ann. nat. Hist., ser. 7 *v.* 4 p. 209. 1899.
Sp. typ.: *Pleustes abyssorum.*

Metaphoxus Jules Bonnier in: Ann.
Univ. Lyon, *v.* 26 p. 630. 1896. Sp.:
M. typicus.

Metopa Axel Boeck in: Forh. Selsk.
Christian., 1870 p. 140. 1871. Sp.: *M.
clypeata, M. glacialis, M. alderii, M.
bruzelii, M. affinis, M. longicornis, M.
megacheir, M. longimana, M. nasuta.*

Metopella G. O. Sars, Crust. Norway,
v. 1 p. 274. 1892. Sp.: *Metopa longimana, M. neglecta, M. nasuta.*

Metopina [Alfred Merle] Norman in:
Ann. nat. Hist., ser. 7 *v.* 6 p. 45. 1900 VII.
Sp. typ.: *Metopa palmata.*

Metopoides Antonio Della Valle in:
F. Fl. Neapel, *v.* 20 p. 907. 1893. Sp.:
*Metopa magellanica, M. parallelocheir,
M. ovata, M. compacta.*

Microcheles Henrik Krøyer in: Naturh.
Tidsskr., ser. 2 *v.* 2 p. 58, 66. 1846. Sp.:
M. armata.

Microdentopus pro: *Microdeutopus* A.
Costa 1853. C. Spence Bate (& J. O.
Westwood), Brit. sess. Crust., *v.* 1
p. 287. [1862 IV 1.]

Microdeuteropus pro: *Microdeutopus* A.
Costa 1853. C. Spence Bate & J. O.
Westwood, Brit. sess. Crust., *v.* 1
p. 287. [1862 IV 1.]

Microdeutopus Achille Costa in: Rend.
Soc. Borbon., n. ser. *v.* 2 p. 171, 178.
1853. Sp.: *M. gryllotalpa.*

Microjassa Thomas R. R. Stebbing in:
Ann. nat. Hist., ser. 7 *v.* 3 p. 240. 1899.
Sp.: *Podocerus cumbrensis.*

Microplax pro: *Iduna* A. Boeck 1861.
William Lilljeborg in: N. Acta Soc.
Upsal., ser. 3 *v.* 6 nr. 1 p. 18 (tabella),
19. 1865.

Microprotopus Alfred Merle Norman
in: Rep. Brit. Ass., Meet. 36 p. 197, 203.
1867. Sp.: *M. maculatus.*

Micruropus Thomas R. R. Stebbing in:
Tr. Linn. Soc. London, ser. 2 *v.* 7 p. 424.
1899 V. Sp.: *M. puella, M. inflatus,
M. vortex, M. talitroides, M. littoralis,
M. glaber, M. rugosus, M. wahlii, M.
fixsenii, M. perla, M. klukii, M. pachytus.* „refers to the smallness of the
third uropods".

Moera [pro: *Maera* Leach 1813/14].
[Henri] Milne Edwards, Hist. nat.
Crust., *v.* 3 p. 54. 1840.

Monoculodes William Stimpson in:
Smithson. Contr., *v.* 6 nr. 5 p. 54. 1853.
Sp.: *M. demissus.*

Monoculopsis G. O. Sars, Crust. Norway,
v. 1 p. 310. 1892. Sp.: *M. longicornis.*

Montagua C. Spence Bate in: Rep. Brit.
Ass., Meet. 25 p. 57. 1856. Sp.: *M.
monoculoides, M. marinus, M. pol-
lexianus, M. dubius.*

Montaguana pro: *Montagua* Bate 1856.
Charles Chilton in: Tr. N. Zealand
Inst., *v.* 15 p. 78. 1883.

Mülleria [pro: *Maera* Leach 1813/14].
[William Elford Leach in:] Edinb.
Enc., *v.* 7 p. 436. [1813/14.]

Naenia [pro: *Noenia* Bate 1862]. C.
Spence Bate, Cat. Amphip. Brit. Mus.,
p. 271. 1862.

Nannonyx G. O. Sars, Crust. Norway,
v. 1 t. 24. 1890. Sp.: *N. goësii.*

Nemertes ([William Elford] Leach in MS.) [Adam White], Crust. Brit. Mus., p. 90. 1847. Sp.: *N. nesaeoides*. [*nom. nud.*]

Neobula [pro: *Neobule* Haswell 1879]. J. Victor Carus in: Zool. Anz., *v.* 3 p. 291. 1880 VI 21.

Neobule William A. Haswell in: P. Linn. Soc. N. S. Wales, *v.* 4 p. 255. 1879. Sp.: *N. algicola*.

Neohela pro: *Hela* A. Boeck 1861. S. I. Smith in: P. U. S. Mus., *v.* 3 p. 448. 1881.

Neoniphargus Thomas R. R. Stebbing in: Tr. Linn. Soc. London, ser. 2. *v.* 7 p. 424. 1899 V. Sp. typ.: *N. thomsoni*.

Neopleustes T. R. R. Stebbing in: Tierreich, *v.* 21 p. 311. 1906 IX. Sp. typ.: *Amphitoe pulchella*. „νέος, new, and *Pleustes*".

Nicea [Hercule Nicolet in:] Gay, Hist. Chile, *v.* 3 p. 237. 1849. Sp.: *N. lucasii*.

Nicippe Ragnar M. Bruzelius in: Svenska Ak. Handl., n. ser. *v.* 3 nr. 1 p. 37, 99. [1859]. Sp.: *N. tumida*. „Namn på en dotter af Pelops".

Niphargoides G. O. Sars in: Bull. Ac. St.-Pétersb., ser. 5 *v.* 1 p. 371. 1894 XII. Sp. typ.: *N. caspius*.

Niphargus J. C. Schiödte in: Danske Selsk. Skr., ser. 5 *v.* 2 p. 26. [1849.] Sp.: *N. stygius*. „Νίφαργος".

Noenia C. Spence Bate (& J. O. Westwood), Brit. sess. Crust., *v.* 1 p. 471. [1862 XII 1.] Sp.: *N. tuberculosa, N. rimapalmata, N. excavata, N. undata, ? Gammarus spinipes*.

Normania Axel Boeck in: Forh. Selsk. Christian., 1870 p. 119. 1871. Sp.: *N. qvadrimana*.

Normanion pro: *Normania* A. Boeck 1871. Jules Bonnier in: Bull. sci. France Belgique, *v.* 24 p. 167. 1893 V 5.

Nototropis Achille Costa in: Rend. Soc. Borbon., n. ser. *v.* 2 p. 170, 173. 1853. Sp.: *Notrotopis spinulicauda*.

Notrotopis [pro: *Nototropis* A. Costa 1853]. Achille Costa in: Rend. Soc. Borbon., n. ser. *v.* 2 p. 173. 1853.

Odius pro: *Otus* Bate & Westwood 1862. William Lilljeborg in: N. Acta Soc. Upsal., ser. 3 *v.* 6 nr. 1 p. 18 (tabella), 19. 1865.

Odontogammarus Thomas R. R. Stebbing in: Tr. Linn. Soc. London, ser. 2 *v.* 7 p. 427. 1899 V. Sp.: *O. calcaratus, O. margaritaceus*. „alludes to the tooth on the fifth sideplates".

Oediceroides Thomas R. R. Stebbing in: Rep. Voy. Challenger, *v.* 29 p. 843. 1888. Sp.: *O. rostrata (O. conspicua), O. cinderella, O. ornata*.

Oediceropsis William Lilljeborg in: N. Acta Soc. Upsal., ser. 3 *v.* 6 nr. 1 p. 18 (tabella), 19. 1865. Sp.: *O. brevicornis*.

Oediceros Henrik Krøyer in: Naturh. Tidsskr., *v.* 4 p. 155. 1842. Sp.: *O. saginatus*. „Οἰδέω, tumeo et κέρας, cornu".

Oedicerus [pro: *Oediceros* Krøyer 1842]. James D. Dana in: Amer. J. Sci., ser. 2 *v.* 8 p. 138. 1849 VII.

Ommatogammarus Thomas R. R. Stebbing in: Tr. Linn. Soc. London, ser. 2 *v.* 7 p. 427. 1899 V. Sp.: *O. albinus, O. flavus, O. carneolus, O. amethystinus*. „alludes to curious character of the eyes".

Onesimoides Thomas R. R. Stebbing in: Rep. Voy. Challenger, *v.* 29 p. 647. 1888. Sp.: *O. carinatus*. „to call attention to the relationship beTween this genus and *Onesimus*, Boeck".

Onesimus [pro: *Onisimus* A. Boeck 1871]. Axel Boeck, Skand. Arkt. Amphip., *v.* 1 t. 4, 5, 6. 1872.

Onisimus Axel Boeck in: Forh. Selsk. Christian., 1870 p. 111. 1871. Sp.: *O. litoralis, O. plautus, O. edwardsii*.

Opis Henrik Krøyer in: Naturh. Tidsskr., *v.* 4 p. 149. 1842. Sp.: *O. eschrichtii*. „Nomen virginis Hyperboreae".

Opisa Axel Boeck, Skand. Arkt. Amphip., *v.* 2 p. 190. 1876. Sp. typ.: *Opis typica (Opisa eschrichti)*. „Ὧπισα (en Pige hos Herodot)".

Oradarea Alfred O. Walker in: J. Linn. Soc., *v.* 29 p. 40, 56. 1903 VII 30. Sp.: *O. longimana*. „*Ora* = beach, and *Adare*".

Orchestes [pro: *Orchestia* Leach 1813/14]. [William Elford Leach in:] Edinb. Enc., *v.* 7 p. 402. [1813/14.]

Orchestia [William Elford Leach in:] Edinb. Enc., *v.* 7 p. 402, 482. [1813/14.] Sp.: *O. littorea*.

Orchestoidea [Hercule Nicolet in:] Gay, Hist. Chile, *v.*3 p.229. 1849. Sp.: *O. tuberculata.*

Orchomena [pro: *Orchomene* A. Boeck 1871]. Eduard von Martens in: Zool. Rec., *v.*8 p.188. 1873.

Orchomene Axel Boeck in: Forb. Selsk. Christian., 1870 p.114. 1871. Sp.: *O. pingvis, O. serratus, O. minutus, O. goësii, O. umbo.*

Orchomenella G.O.Sars, Crust. Norway, *v.*1 p.66. 1890. Sp.: *O. minuta, O. pingvis, O. ciliata, O. groenlandica, ?Tryphosa barbatipes.*

Orchomenopsis G. O. Sars. Crust. Norway, *v.*1 p.73. 1891. Sp.: *Orchomene musculosus, O. abyssorum, ?O. excavatus, Orchomenopsis obtusa.*

Orthopalame P. P. C. Hoek in: Tijdschr. Nederl. dierk. Ver., *v.*4 p.123. 1879. Sp.: *O. terschellingi.* „Von ὀρθός (recht, gerade) und Παλάμη (palma, Palme)".

Otus C. Spence Bate & J. O. Westwood, Brit. sess. Crust., *v.*1 p.223. [1862II1.] Sp.: *O. carinatus.* „son of Iphimedia".

Palinnotus Thomas R. R. Stebbing in: Ann. nat. Hist., ser.7 *v.*5 p.16. 1900. Sp. typ.: *P. thomsoni.*

Pallasea C. Spence Bate, Cat. Amphip. Brit. Mus., p.200, 380. 1862. Sp.: *P. cancellus, P. cancelloides.*

Pallasia [pro: *Pallasea* Bate 1862]. Axel Boeck in: Forh. Selsk. Christian., 1870 p.206. 1871.

Pallasiella [pro: *Pallasea* Bate 1862]. G. O. Sars, Crust. Norway, *v.*1 p.505. 1894.

Pandora pro: *Pandorites* G. O. Sars 1895. ([O.] Grimm in MS.) G. O. Sars in: Bull. Ac. St.-Pétersb., ser.5 *v.*3 p.287. 1895X.

Pandorites G. O. Sars in: Bull. Ac. St.-Pétersb., ser.5 *v.*3 p.287. 1895X. Sp. typ.: *P. podoceroides.*

Panoplaea [pro: *Panoploea* G. M. Thomson 1880]. G. M. Thomson in: Tr. N. Zealand Inst., *v.*13 p.212. 1881.

Panoploea George M. Thomson in: Ann. nat. Hist., ser.5 *v.*6 p.2. 1880. Sp.: *P. spinosa, P. debilis.*

Paracalliope Thomas R. R. Stebbing in: Ann. nat. Hist., ser.7 *v.*4 p.210. 1899. Sp. typ.: *Calliope fluviatilis.*

Paracallisoma Ed. Chevreux in: Bull. Soc. zool. France, *v.*28 p.84. 1903II. Sp.: *P. alberti.*

Paraceradocus Thomas R. R. Stebbing in: Tr. Linn. Soc. London, ser.2 *v.*7 p.426. 1899V. Sp. typ.: *P. miersii.*

Paracorophium Thomas R. R. Stebbing in: Ann. nat. Hist., ser.7 *v.*3 p.350. 1899. Sp.: *Corophium excavatum.*

Paracrangonyx Thomas R. R. Stebbing in: Tr. Linn. Soc. London, ser.2 *v.*7 p.422. 1899V. Sp. typ.: *P. compactus.*

Paracyphocaris Ed. Chevreux in: Bull. Mus. Monaco, nr.32 p.1. 1905IV15. Sp. typ.: *P. praedator.*

Paracyproidea Thomas R. R. Stebbing in: Ann. nat. Hist., ser.7 *v.*4 p.207. 1899. Sp. typ.: *Cyproidea lineata.*

Paradexamine Thomas R. R. Stebbing in: Ann. nat. Hist., ser.7 *v.*4 p.210. 1899. Sp. typ.: *Dexamine pacifica.*

Paradryope Thomas R. R. Stebbing in: Rep. Voy. Challenger, *v.*29 p.1151. 1888. Sp.: *P. orguion.*

Paradulichia Axel Boeck in: Forh. Selsk. Christian., 1870 p.265. 1871. Sp.: *P. typica.*

Paragrubia Ed. Chevreux in: Mém. Soc. zool. France, *v.*14 p.426. 1901. Sp.: *P. vorax.*

Parajassa pro: *Janassa* A. Boeck 1871. Thomas R. R. Stebbing in: Ann. nat. Hist., ser.7 *v.*3 p.240. 1899.

Paraleptamphopus Thomas R. R. Stebbing in: Ann. nat. Hist., ser.7 *v.*4 p.209. 1899. Sp.: *Calliope subterranea, Pherusa caerulea.*

Paraleucothoe Thomas R. R. Stebbing in: Ann. nat. Hist., ser.7 *v.*4 p.208. 1899. Sp. typ.: *Leucothoe novae-hollandiae.*

Parambasia A. O. Walker & Andrew Scott in: H. O. Forbes, Nat. Hist. Sokotra, p.217,221. 1903. Sp.: *P.forbesi.*

Paramera [pro: *Paramoera* Miers 1875]. (Edward J. Miers in:) Samuel H. Scudder, Nomencl. zool., suppl. L. p.247. 1882.

Parametopa Ed. Chevreux in: Bull. Soc. Rouen, *v.*36 p.233. 1901. Sp.: *P. kervillei.*

Paramicruropus Thomas R. R. Stebbing in: Tr. Linn. Soc. London, ser. 2 *v.* 7 p. 423. 1899 V. Sp.: *P. solskii, P. taczanowskii.*

Paramoera Edward J. Miers in: Ann. nat. Hist., ser. 4 *v.* 16 p. 75. 1875. Sp.: *Melita fresnelii, M. tenuicornis, Paramoera australis.*

Paramphithoe Ragnar M. Bruzelius in: Svenska Ak. Handl., n. ser. *v.* 3 nr. 1 p. 37, 68. 1859. Sp.: *P. panopla, P. pulchella, P. hystrix, P. compressa, P. bicuspis, P. tridentata, P. elegans, P. laeviuscula, P. norvegica.*

Paranaenia Charles Chilton in: Tr. N. Zealand Inst., *v.* 16 p. 258. 1884 V. Sp.: *P. typica, P. dentifera, P. longimanus.*

Parandania Thomas R. R. Stebbing in: Ann. nat. Hist., ser. 7 *v.* 4 p. 206. 1899. Sp. typ.: *Andania boecki.*

Parapallasea Thomas R. R. Stebbing in: Tr. Linn. Soc. London, ser. 2 *v.* 7 p. 429. 1899 V. Sp.: *P. borowskii, P. lagowskii, P. puzyllii.*

Parapherusa pro: *Harmonia* Haswell 1879. T. R. R. Stebbing in: Tierreich, *v.* 21 p. 383. 1906 IX. „Παρδ, beside, *Pherusa*".

Paraphoxus G. O. Sars. Crust. Norway, *v.* 1 p. 148. 1891. Sp.: *P. oculatus.*

Parapleustes R. Buchholz in: Zweite D. Nordpolarf., *v.* 2 p. 299, 337, 398. 1874. Sp.: *P. glacilis (P. gracilis).*

Pararistias David Robertson in: P. nat. Hist. Soc. Glasgow, n. ser. *v.* 3 p. 201. 1892. Sp.: *P. audouinianus.*

Paratryphosites Thomas R. R. Stebbing in: Ann. nat. Hist., ser. 7 *v.* 4 p. 206. 1899. Sp. typ.: *Lysianassa abyssi.*

Paratylus G. O. Sars. Crust. Norway, *v.* 1 p. 462. 1893. Sp.: *P. swammerdami, P. falcatus, P. vedlomensis, P. smitti, P. nordlandicus, et aliae.*

Pardalisca Henrik Krøyer in: Naturh. Tidsskr., *v.* 4 p. 153. 1842. Sp.: *P. cuspidata.* „Nomen ancillae apud Plautum in Casina".

Pardaliscella G. O. Sars, Crust. Norway, *v.* 1 p. 407. 1893. Sp.: *P. boeckii.*

Pardaliscoides Thomas R. R. Stebbing in: Rep. Voy. Challenger, *v.* 29 p. 1725. 1888. Sp.: *P. tenellus.*

Parelasmopus Thomas R. R. Stebbing in: Rep. Voy. Challenger, *v.* 29 p. 1029. 1888. Sp.: *P. suluensis.*

Parharpinia Thomas R. R. Stebbing in: Ann. nat. Hist., ser. 7 *v.* 4 p. 207. 1899. Sp. typ.: *Phoxus villosus.*

Parhyale Thomas R. R. Stebbing in: Tr. Linn. Soc. London, ser. 2 *v.* 7 p. 26. 1897 V. Sp.: *P. fasciger.*

Paroediceros G. O. Sars, Crust. Norway, *v.* 1 p. 291. 1892. Sp.: *Oediceros macrocheir, O. curvirostris, O. sp., Paroediceros lynceus, P. propinqvus.*

Paronesimus Thomas R. R. Stebbing in: Bijdr. Dierk., *v.* 17/18 p. 2, 14. 1894. Sp.: *P. barentsi.* „near to Onisimus".

Paropisa Thomas R. R. Stebbing in: Ann. nat. Hist., ser. 7 *v.* 4 p. 206. 1899. Sp. typ.: *Opisa hispana.*

Parorchestia Thomas R. R. Stebbing in: Tr. Linn. Soc. London, ser. 2 *v.* 7 p. 397, 402. 1899 V. Sp.: *Orchestia tenuis, O. hawaiensis, O. sylvicola.* „παρά, near, and *Orchestia*".

Peltocoxa J. D. Catta in: Rev. Sci. nat., *v.* 4 p. 161. 1875. Sp.: *P. marioni.*

Pephredo C. S. Rafinesque-Schmaltz, Précis Découv. somiol., p. 26. 1814. Sp.: *P. heteroclitus.* [*nom. nud.*]

Pephredo C. S. Rafinesque in: Amer. monthly Mag., *v.* 2 p. 41. 1817 XI. Sp.: *P. potamogeti.*

Pereionotus C. Spence Bate & J. O. Westwood, Brit. sess. Crust., *v.* 1 p. 226. 1862 II 1. Sp.: *P. testudo.*

Perioculodes G. O. Sars, Crust. Norway, *v.* 1 p. 312. 1892. Sp.: *P. longimanus.*

Perrierella Ed. Chevreux & E.-L. Bouvier in: Bull. Soc. zool. France, *v.* 17 p. 50. 1892. Sp.: *P. crassipes.*

Phaedra C. Spence Bate in: Quart. J. geol. Soc., *v.* 15 p. 138, 140. [1858.] Sp.: *P. antiqua.*

Pherusa [William Elford] Leach in: Edinb. Enc., *v.* 7 p. 432. [1813/14.] Sp.: *P. fucicola.*

Phippsia pro: *Aspidopleurus* O. Sars 1891. T. R. R. Stebbing in: Tierreich, *v.* 21 p. 89. 1906 IX. „After Constantine John Phipps".

Phlias E. Guérin[-Méneville] in: Mag. Zool., Cl. 7 t. 19. 1836. Sp.: *P. serratus.* „Phlias, l'un des Argonautes".

Photis Henrik Krøyer in: Naturh. Tidsskr., *v.* 4 p. 155. 1842. Sp.: *P. reinhardi.* „Nomen ancillae apud Apuleium in Asino aureo".

Phoxocephalus pro: *Phoxus* Krøyer 1842. Thomas R. R. Stebbing in: Rep. Voy. Challenger, *v.* 29 p. 810. 1888.

Phoxus Henrik Krøyer in: Naturh. Tidsskr., *v.* 4 p. 150. 1842. Sp.: *P. holbölli*, *P. plumosus.* „Φοξος, capite acuto".

Phreatogammarus Thomas R. R. Stebbing in: Tr. Linn. Soc. London, ser. 2 *v.* 7 p. 427. 1899 V. Sp. typ.: *P. fragilis.* „means a well-*Gammarus*".

Platamon Thomas R. R. Stebbing in: Rep. Voy. Challenger, *v.* 29 p. 642. 1888. Sp.: *P. longimanus.* „from the Greek word πλαταμών, a broad space".

Platophium James D. Dana in: Amer. J. Sci., ser. 2 *v.* 14 p. 309. 1852 XI. — James D. Dana in: U. S. expl. Exp., *v.* 13 π p. 831, 837. [1853.] Sp.: *P. brasiliense.*

Platyischnopus Thomas R. R. Stebbing in: Rep. Voy. Challenger, *v.* 29 p. 830. 1888. Sp.: *P. mirabilis.* „πλατύς, broad, ἰσχνός, narrow, πούς, a foot".

Platyschnopus [pro: *Platyischnopus* T. Stebbing 1888]. Antonio Della Valle in: F. Fl. Neapel, *v.* 20 p. 784. 1893.

Pleonexes C. Spence Bate in: Rep. Brit. Ass., Meet. 25 p. 59. 1856. Sp.: *P. gammaroides.* [*nom. nud.*]

Pleonexes C. Spence Bate in: Ann. nat. Hist., ser. 2 *v.* 19 p. 147. 1857. Sp.: *P. gammaroides.*

Plesiogammarus Thomas R. R. Stebbing in: Tr. Linn. Soc. London, ser. 2 *v.* 7 p. 426. 1899 V. Sp. typ.: *P. gerstaeckeri.*

Pleuracanthus W. Garjajeff in: Trudui Kazan. Univ., *v.* 35 nr. 6 p. 16. 1901. Sp. & Subsp.: *P. niger*, *P. lovenii*, *P. borowskii*, *P. b. var. abyssalis*, *P. b. var. dichraas*, *P. puzyllii*, *P. lagowskii.*

Pleustes C. Spence Bate in: Ann. nat. Hist., ser. 3 *v.* 1 p. 362. 1858. Sp.: *P. tuberculata.*

Plexaura C. S. Rafinesque, Anal. Nat., p. 101. 1815. [*nom. nud.*]

Podoceropsis Axel Boeck in: Forh. Skand. Naturf., Møde 8 p. 666. 1861. Sp.: *P. sophia.*

Podoceros [pro: *Podocerus* Leach 1813/14]. A. Goës in: Öfv. Ak. Förh., *v.* 22 p. 532. 1866.

Podocerus [William Elford Leach in:] Edinb. Enc., *v.* 7 p. 433. [1813/14.] Sp.: *P. variegatus.*

Podoprion Edouard Chevreux in: Mém. Soc. zool. France, *v.* 4 p. 6. 1891. Sp.: *P. bolivari.*

Podoprionella G. O. Sars, Crust. Norway, *v.* 1 p. 687. 1895. Sp.: *P. norvegica.*

Podoprionides Alfred O. Walker in: Ann. nat. Hist., ser. 7 *v.* 17 p. 457. 1906 V. Sp.: *P. incerta.*

Poekilogammarus Thomas R. R. Stebbing in: Tr. Linn. Soc. London, ser. 2 *v.* 7 p. 428. 1899 V. Sp.: *P. pictus*, *P. orchestes*, *P. talitrus*, *P. araneolus.* „alludes to the Variegated colouring of the several species".

Polyacanthus W. Garjajeff in: Trudui Kazan. Univ., *v.* 35 nr. 6 p. 16. 1901. Sp. & Subsp.: *P. belkinii*, *P. maximus*, *P. flavus*, *P. korotneffii*, *P. albus*, *P. godlewskii*, *P. g. var. victorii*, *P. parasiticus.*

Polycharia [pro: *Polycheria* Haswell 1879]. W. T. Calman in: Ann. N. York Ac., *v.* 11 p. 288. 1898 VIII 31.

Polychelia pro: *Polycheria* Haswell 1879. George M. Thomson in: Tr. N. Zealand Inst., *v.* 14 p. 233. 1882 V. „from its many claws".

Polycheria William A. Haswell in: P. Linn. Soc. N. S. Wales, *v.* 4 p. 345. 1879. Sp.: *P. tenuipes*, *P. brevicornis.*

Polychiria pro: *Polycheria* Haswell 1879. Eduard von Martens in: Zool. Rec., *v.* 16 Crust. p. 31. 1881.

Pontharpinia Thomas R. R. Stebbing in: Tr. Linn. Soc. London, ser. 2 *v.* 7 p. 32. 1897 V. Sp.: *P. pinguis.*

Pontiporeia [pro: *Pontoporeia* Krøyer 1842]. James D. Dana in: U. S. expl. Exp., *v.* 13 π p. 912. [1853].

Pontocrates Axel Boeck in: Forh. Selsk. Christian., 1870 p. 171. 1871. Sp.: *P. norvegicus, P. haplocheles.*

Pontogeneia Axel Boeck in: Forh. Selsk. Christian., 1870 p. 193. 1871. Sp.: *P. inermis.*

Pontogenia [pro: *Pontogeneia* A. Boeck 1871]. Eduard von Martens in: Zool. Rec., *v.* 8 p. 190. 1873.

Pontoporeia Henrik Krøyer in: Naturh. Tidsskr., *v.* 4 p. 152. 1842. Sp.: *P. femorata.* „Ποντοπόρεια (pontivaga), nomen Nereidis apud Hesiodum".

Pontoporia pro: *Pontoporeia* Krøyer 1842. L. Agassiz, Nomencl. zool., Index p. 305. 1846.

Prinassus H.J. Hansen in: Vid. Meddel., ser 4 *v.* 9 p. 82. 1887. Sp.: *P. nordenskiöldii.* „Πρινασσός, graesk Bynavn".

Priscilla Axel Boeck in: Porh. Selsk. Christian., 1870 p. 124. 1871. Sp.: *P. armata.*

Priscillina pro: *Priscilla* A. Boeck 1871. Thomas R. R. Stebbing in: Rep. Voy. Challenger, *v.* 29 p. 1680. 1888.

Probolium Achille Costa in: Rend. Soc. Borbon., n. ser. *v.* 2 p. 170, 173. 1853. Sp.: *P. polyprion.*

Proboloides Antonio Della Valle in: F. Fl. Neapel, *v.* 20 p. 907. 1893. Sp.: *Metopa gregaria, M. calcarata, M. crenatipalmata, M. nasutigenes.*

Protomedeia Henrik Krøyer in: Naturh. Tidsskr., *v.* 4 p. 154. 1842. Sp.: *P. fasciata.* „Πρωτομέδεια, filia Nerei et Doridis".

Psammylla C. S. Rafinesque in: Amer. monthly Mag., *v.* 2 p. 41. 1817 XI. Sp.: *P. littoralis.* „abbreViated from *Psammopsylla*".

Pseudalibrotus Antonio Della Valle in: F. Fl. Neapel, *v.* 20 p. 798. 1893. Sp.: *P. littoralis.*

Pseudoniphargus Ed. Chevreux in: Bull. Soc. zool. France, *v.* 26 p. 211. 1901 XII. Sp.: *P. africanus.*

Pseudophthalmus William Stimpson in: Smithson. Contr., *v.* 6 nr. 5 p. 57. 1853 III. Sp.: *P. pelagicus, P. limicola.*

Pseudotiron Ed. Chevreux in: Bull. Soc. zool. France, *v.* 20 p. 166. 1895. Sp.: *P. bouvieri.*

Pseudotryphosa G.O. Sars, Crust.Norway, *v.* 1 p. 83. 1891. Sp.: *P. umbonata, Tryphosa antennipotens.*

Pterygocera [pro: *Pterygocerus* Latreille 1825]. [Pierre André] Latreille in: G. CuVier, Règne an., n. ed. *v.* 4 p. 124. 1829.

Pterygocerus [Pierre André] Latreille in: Enc. méth., *v.* 10 p. 236. 1825. Sp.: *Oniscus arenarius.*

·Ptilocheirus William Stimpson in: Smithson. Contr., *v.* 6 nr. 5 p. 55. 1853. Sp.: *P. pinguis.*

Pyctilus Jacobus D. Dana in: P. Amer. Ac., *v.* 2 p. 218. 1852. Sp.: *Erichthonius macrodactylus (P. m.), E. pugnax (P. p.)*: „πύκτης, pugil".

Rachotropis [pro: *Rhachotropis* S. I. Smith 1883]. Jules Bonnier in: Ann. Univ. Lyon, *v.* 26 p. 653. 1896.

Rhachotropis pro: *Tritropis* A. Boeck 1871. Sidney I. Smith in: P. U. S. Mus., *v.* 6 p. 222. 1883 X 5. „'Ράχις et τρόπις".

Rozinante Thomas R. R. Stebbing in: Bijdr. Dierk., *v.* 17/18 p. 2, 38. 1894. Sp. typ.: *Paramphithoë fragilis.* „Don Quixote's famous charger".

Sancho Thomas R. R. Stebbing in: Tr. Linn. Soc. London, ser. 2 *v.* 7 p. 42. 1897 V. Sp.: *S. platynotus.* „from a character famous in fiction".

Scamballa ([William Elford] Leach in MS.) [Adam White], Crust. Brit. Mus., p. 86. 1847. Sp.: *S. longicornis, S. kuhliana, S. sayana, S. tristensis, S. megalophthalmus.*

Schisturella [Alfred Merle] Norman in: Ann. nat. Hist., ser. 7 *v.* 5 p. 208. 1900 II. Sp.: *S. pulchra.* σχιστός, divided, οὐρά, tail".

Schraderia Georg Pfeffer in: Jahrb. Hamburg. Anst., *v.* 5 p. 141 t. 2 f. 5. 1888. Sp.: *S. gracilis.*

Scopelocheirus C. Spence Bate in: Rep. Brit. Ass., Meet. 25 p. 58. 1856. Sp.: *S. breviatus.*

Scopelocheirus C. Spence Bate in: Ann. nat. Hist., ser. 2 *v.* 19 p. 138. 1857. Sp.: *S. crenatus.*

Seba (A. Costa in MS.) C. Spence Bate, Cat. Amphip. Brit. Mus., p. 159. 1862. Sp.: *S. innominata.*

Siphonaecetes [pro: *Siphonoecetes* Krøyer 1845]. Édouard Chevreux in: Bull. Soc. zool. France, *v.* 12 p. 290, 317. 1887.

Siphonaecetus [pro: *Siphonoecetes* Krøyer 1845]. C. Spence Bate & J. O. Westwood, Brit. sess. Crust., *v.* 1 p. 463, 467. 1862 XII 1.

Siphonocetus [pro: *Siphonoecetes* Krøyer 1845]. C. Spence Bate in: Ann. nat. Hist., ser. 2 *v.* 19 p. 149. 1857.

Siphonocoetus [pro: *Siphonoecetes* Krøyer 1845]. Augustus de Marschall, Nomencl. zool., p. 420. 1873.

Siphonoecetes Henrik Krøyer in: Naturh. Tidsskr., ser. 2 *v.* 1 p. 481, 491. 1845. Sp.: *S. typicus.* „Σιφων, tubus, et οικετης, incola".

Siphonoecetus [pro: *Siphonoecetes* Krøyer 1845]. C. Spence Bate, Cat. Amphip. Brit. Mus., p. 268. 1862.

Socarnella Alfred O. Walker in: Herdman, Rep. Ceylon Pearl Fish., *v.* 2 p. 231, 239. 1904. Sp.: *S. bonnieri.*

Socarnes Axel Boeck in: Forh. Selsk. Christian., 1870 p. 99. 1871. Sp.: *S. vahli.*

Socarnioides [pro: *Socarnoides* T. Stebbing 1888]. Frank E. Beddard in: Zool. Rec., *v.* 25 Index p. 15. 1890.

Socarnoides Thomas R. R. Stebbing in: Rep. Voy. Challenger, *v.* 29 p. 690. 1888. Sp.: *S. kergueleni.*

Sophrosyne Thomas R. R. Stebbing in: Rep. Voy. Challenger, *v.* 29 p. 652. 1888. Sp.: *S. murrayi.* „from σωφροσύνη, temperance".

Sperchius C. S. Rafinesque, Ann. Nat., p. 6. 1820. Sp.: *S. lucidus.*

Spinifer ([Carl] Holbøll in MS.) Henrik Krøyer in: Naturh. Tidsskr., *v.* 4 p. 151. 1842. Sp.: *S. spinosissimus, S. flagelliformis, Phoxus plumosus.*

Stebbingia Georg Pfeffer in: Jahrb. Hamburg. Anst., *v.* 5 p. 110. 1888. Sp.: *S. gregaria.*

Stegocephaloides G. O. Sars, Crust. Norway, *v.* 1 p. 201. 1891. Sp.: *S. christianiensis, S. auratus.*

Stegocephalus Henrik Krøyer in: Naturh. Tidsskr., *v.* 4 p. 150. 1842. Sp.: *S. inflatus.* „Στεγω, tego et κεφαλη, Caput".

Stegoplax G. O. Sars in: Forh. Selsk. Christian., 1882 nr. 18 p. 88. 1882. Sp.: *S. longirostris.*

Stenia James D. Dana in: Amer. J. Sci., ser. 2 *v.* 8 p. 136. 1849 VII. — Jacobus D. Dana in: P. Amer. Ac., *v.* 2 p. 209. 1852. Sp.: *S. magellanica.*

Stenopleura Thomas R. R. Stebbing in: Rep. Voy. Challenger, *v.* 29 p. 949. 1888. Sp.: *S. atlantica.* „from στενός, narrow, and πλευρά, side".

Stenopleustes G. O. Sars, Crust. Norway, *v.* 1 p. 354. 1893. Sp.: *S. malmgreni, S. nodifer.*

Stenothoe James D. Dana in: Amer. J. Sci., ser. 2 *v.* 14 p. 311. 1852 XI. — James D. Dana in: U. S. expl. Exp., *v.* 13 II p. 923. [1853.] Sp.: *S. validus.*

Stenothoides Ed. Chevreux in: Résult. Camp. Monaco, *v.* 16 p. 55. 1900. Sp. typ.: *S. perrieri.*

Stenyolus C. S. Rafinesque, Anal. Nat., p. 101. 1815. [*nom. nud.*]

Sthenometopa pro: *Metopina* A. M. Norman 1900. A. M. Norman in: Ann. nat. Hist., ser. 7 *v.* 10 p. 480. 1902 XII.

Stimpsonella pro: *Stimpsonia* Bate 1862. Antonio Della Valle in: F. Fl. Neapel, *v.* 20 p. 421. 1893.

Stimpsonia C. Spence Bate (& J. O. Westwood), Brit. sess. Crust., *v.* 1 p. 284. [1862 IV 1.] Sp.: *S. chelifera.* „in compliment to the distinguished naturalist of the United States' Exploring Expedition to the North Pacific".

Stomacontion Thomas R. R. Stebbing in: Ann. nat. Hist., ser. 7 *v.* 4 p. 205. 1899. Sp. typ.: *Acontiostoma pepinii.*

Stygobromus E. D. Cope in: Amer. Natural., *v.* 6 p. 409, 413, 422. 1872. Sp.: *S. vitreus.*

Stygodromus [pro: *Stygobromus* E. D. Cope 1872]. Eduard von Martens in: Zool. Rec., *v.* 10 p. 189. 1875.

Stygonectes William Perry Hay in: P. U. S. Mus., *v.* 25 p. 430. 1902 IX 23 Sp. typ.: *Crangonyx flagellatus.*

Sulcator pro: *Bellia* Bate 1851. C. Spence Bate in: Ann. nat. Hist., ser. 2 *v.* 13 p. 504. 1854.

Sunamphithoë [pro: *Sunamphitoë* Bate 1857]. C. Spence Bate (& J. O. Westwood), Brit. sess. Crust., *v.* 1 p. 429. 1862 XI 1.

Sunamphitoë C. Spence Bate in: Rep. Brit. Ass., Meet. 25 p. 59. 1856. Sp.: *S. hamulus, S. conformatus.* [*nom. nud.*]

Sunamphitoë C. Spence Bate in: Ann. nat. Hist., ser. 2 *v.* 19 p. 147. 1857. Sp.: *S. hamulus, S. conformata.*

Sympleustes Thomas R. R. Stebbing in: Ann. nat. Hist., ser. 7 *v.* 4 p. 209. 1899. Sp.: *Amphithoe latipes, Amphithopsis glaber, A. pulchella, A. olrikii, A. grandimana.*

Synamphithoe [pro: *Sunamphitoë* Bate 1857]. Adam White, Hist. Brit. Crust., p. 201. 1857.

Synchelidium G. O. Sars, Crust. Norway, *v.* 1 p. 317. 1892. Sp.: *S. brevicarpum, S. haplocheles, S. intermedium.*

Synopia James D. Dana in: Amer. J. Sci., ser. 2 *v.* 14 p. 315. 1852. — James D. Dana in: U. S. expl. Exp., *v.* 13 II p. 994. [1853.] Sp.: *S. ultramarina, S. gracilis, S. angustifrons.*

Synopioides Thomas R. R. Stebbing in: Rep. Voy. Challenger, *v.* 29 p. 999, 1224. 1888. Sp.: *S. macronyx, S. secundus.*

Synurella (A. W. Wrzesniowski in:) Hoyer in: Z. wiss. Zool., *v.* 28 p. 403. 1877 III 8. Sp.: *S. polonica.*

Syrrhoë A. Goës in: Öfv. Ak. Förh., *v.* 22 p. 527. 1866. Sp.: *S. crenulata, S. bicuspis [?].*

Syrrhoites G. O. Sars, Crust. Norway, *v.* 1 p. 391. 1893. Sp.: *S. serrata.*

Talitroïdes (J. Bonnier in:) Victor Willem in: Ann. Soc. ent. Belgique, *v.* 42 p. 208. 1898. Sp.: —.

Talitronus J. D. Dana in: Amer. J. Sci., ser. 2 *v.* 9 p. 295. 1850 V. — Jacobus D. Dana in: P. Amer. Ac., *v.* 2 p. 202. 1852. Sp.: *T. insculptus.*

Talitrorchestia *Subgen.* J. F. Brandt in: Bull. phys.-math. Ac. St.-Pétersb., *v.* 9 p. 137. 1851 I 9. Sp.: *Talitrus cloquetii.*

Talitrus ([Pierre André] Latreille in:) L. A. G. Bosc, Crust., *v.* 1 p. 78 [& *v.* 2 p. 148, 152]. X [1802]. Sp.: *Gammarus locusta, Oniscus gammarellus; [Talitrus locusta, T. grillus].*

Talorchestes [pro: *Talorchestia* J. D. Dana 1852]. H. Filhol in: Recu. Passage Vénus, *v.* 3 II Zool. p. 459. 1885.

Talorchestia *Subgen.* James D. Dana in: Amer. J. Sci., ser. 2 *v.* 14 p. 310. 1852 XI. — James D. Dana in: U. S. expl. Exp.. *v.* 13 II p. 851, 861. [1853.] Sp.: *Orchestia (T.) gracilis, O. (T.?) quoyana.*

Teraticum Charles Chilton in: Tr. N. Zealand Inst., *v.* 16 p. 257. 1884 V. Sp.: *T. typicum.*

Tessarops Alfred Merle Norman in: Ann. nat. Hist., ser. 4 *v.* 2 p. 412. 1868. Sp.: *T. hastata.*

Tetradeion Thomas R. R. Stebbing in: Ann. nat. Hist., ser. 7 *v.* 4 p. 207. 1899. Sp. typ.: *Cyproidia crassa.* „from the Greek τετραδεῖον, a set of four".

Tetromatus C. Spence Bate in: Rep. Brit. Ass., Meet. 25 p. 58. 1856. Sp.: *T. typicus, T. bellianus.* — C. Spence Bate in: Ann. nat. Hist., ser. 2 *v.* 19 p. 139. 1857. Sp.: *T. typicus, T. bellianus.*

Thalitrus [pro: *Talitrus* Latreille 1802]. [Félix Edouard] Guérin [-Méneville] in: Exp. Morée, Atlas p. 3. 1835.

Thalorchestia [pro: *Talorchestia* J. D. Dana 1852]. H. Filhol in: Recu. Passage Vénus, *v.* 3 II Zool. p. 461, Planches p. 28. 1885.

Thersites C. Spence Bate in: Rep. Brit. Ass., Meet. 25 p. 59. 1856. Sp.: *T. guilliamsonia, T. pelagica.* [*nom. nud.*]

Thersites C. Spence Bate in: Ann. nat. Hist., ser. 2 *v.* 19 p. 146. 1857. Sp.: *T. guilliamsoniana, T. pelagica.*

Thiella C. S. Rafinesque, Anal. Nat., p. 101. 1815. [*nom. nud.*]

Thoelaos pro: *Laothoës* A. Boeck 1871. Antonio Della Valle in: F. Fl. Neapel, *v.* 20 p. 592. 1893.

Tiron William Lilljeborg in: N. Acta Soc. Upsal., ser. 3 *v.* 6 nr. 1 p. 18 (tabella), 19. 1865. Sp.: *T. acanthurus.* „Τείρων Proper name".

Tmetonyx pro: *Hoplonyx* O. Sars 1891. T. R. R. Stebbing in: Tierreich, *v.* 21 p. 73, 720 1906 IX. „τμητός, shaped by cutting, ὄνυξ, nail".

Trischizostoma Axel Boeck in: Forb. Skand. Naturf., Møde 8 p. 637. 1861. Sp.: *T. raschii*.

Tritaeta pro: *Lampra* A. Boeck 1871. Axel Boeck, Skand. Arkt. Amphip., *v.* 2 p. 317. 1876. „Τριταία, et Egennavn".

Tritropis Axel Boeck in: Forh. Selsk. Christian., 1870 p. 158. 1871. Sp.: *T. aculeata, T. helleri, T. fragilis.*

Tryphosa Axel Boeck in: Forh. Selsk. Christian.. 1870 p. 117. 1871. Sp.: *T. nanus, T. høringii, T. nanoides, T. longipes.*

Tryphosella Jules Bonnier in: Bull. sci. France Belgique, *v.* 24 p. 170, 174. 1893 V 5. Sp.: *T. sarsi, T. compressa, T. hörringii, T. angulata, T. nanoïdes, T. antennipotens, T. barbatipes.*

Tryphosites G. O. Sars, Crust. Norway, *v.* 1 p. 81. 1891. Sp. typ.: *Anonyx longipes.*

Typhosa [pro: *Tryphosa* A. Boeck 1871]. Jules Bonnier in: Bull. sci. France Belgique, *v.* 24 p. 170. 1893 V 5.

Unciola Thomas Say in: J. Ac. Philad., *v.* 1 II p. 388. 1818. Sp.: *U. irrorata.*

Unimelita O. A. Sayce in: P. R. Soc. Victoria, *v.* 13 p. 237. 1901 III. Sp.: *U. spenceri, Niphargus montanus.*

Uristes James D. Dana in: Amer. J. Sci., ser. 2 *v.* 8 p. 136. 1849 VII. — Jacobus D. Dana in: P. Amer. Ac., *v.* 2 p. 209. 1852. Sp.: *U. gigas.*

Urothoe James D. Dana in: Amer. J. Sci., ser. 2 *v.* 14 p. 311. 1852 XI. — James D. Dana in: U. S. expl. Exp., *v.* 13 II p. 920. [1853.] Sp.: *U. rostratus, U. irrostratus.*

Urothoides Thomas R. R. Stebbing in: Tr. zool. Soc. London, *v.* 13 I p. 26. 1891. Sp.: *U. lachneëssa.*

Valettia Thomas R. R. Stebbing in: Rep. Voy. Challenger, *v.* 29 p. 723. 1888. Sp.: *V. coheres.* „in compliment to the Baron Adolphe de la Valette".

Vertumnus (Leach in MS.) [Adam White], Crust. Brit. Mus., p. 89. 1847. Sp.: *V. cranchii.* [*nom. nud.*]

Vijaya Alfred O. Walker in: Herdman, Rep. Ceylon Pearl Fish., *v.* 2 p. 231, 241. 1904. Sp.: *V. tenuipes.* „*Vijáya*, an ancient king in Ceylon".

Wayprechtia [pro: *Weyprechtia* Stuxberg 1880]. A. Birula in: Annuaire Mus. St.-Pétersb., *v.* 4 p. 426, 442. 1900.

Westwoodea C. Spence Bate in: Rep. Brit. Ass., Meet. 25 p. 58. 1856. Sp.: *W. caeculus, W. carinatus.* [*nom. nud.*]

Westwoodia [pro: *Westwoodea* Bate 1856]. C. Spence Bate in: Ann. nat. Hist., ser. 2 *v.* 19 p. 139. 1857. Sp.: *W. caecula.*

Westwoodilla pro: *Westwoodia* Bate 1857. C. Spence Bate & J. O. Westwood, Brit. sess. Crust., *v.* 1 p. 154. [1862 I 1.]

Weyprechtia Anton Stuxberg in: Bih. Svenska Ak., *v.* 5 nr. 22 p. 27. 1880. Sp.: *W. mirabilis.*

Wyvillea William A. Haswell in: P. Linn. Soc. N. S. Wales, *v.* 4 p. 336. 1879. Sp.: *W. longimanus.* „in honour of Prof. Sir C. Wyville Thomson".

Xenocheira William A. Haswell in: P. Linn. Soc. N. S. Wales, *v.* 4 p. 272. 1879. Sp.: *X. fasciata.*

Xenochira pro: *Xenocheira* Haswell 1879. Eduard von Martens in: Zool. Rec., *v.* 16 Crust. p. 31. 1881.

Xenoclea Axel Boeck in: Forb. Selsk. Christian., 1870 p. 234. 1871. Sp.: *X. batei.*

Xenodice Axel Boeck in: Forh. Selsk. Christian.. 1870 p. 266. 1871. Sp.: *X. frauenfeldti.*

Zacoreus C. S. Rafinesque, Anal. Nat., p. 101. 1815. [*nom. nud.*]

Zaramella [pro: *Zaramilla* T. Stebbing 1888]. Frank E. Beddard in: Zool. Rec., *v.* 25 Index p. 17. 1890.

Zaramilla Thomas R. R. Stebbing in: Rep. Voy. Challenger, *v.* 29 p. 866. 1888. Sp.: *Z. kergueleni.* „from an imaginary personage in Don Quixote".

Zenodice [pro: *Xenodice* A. Boeck 1871]. [Alfred Merle] Norman in: Ann. nat. Hist., ser. 6 *v.* 15 p. 493. 1895.

Lightning Source UK Ltd.
Milton Keynes UK
UKHW02f0723160818
327336UK00008B/313/P